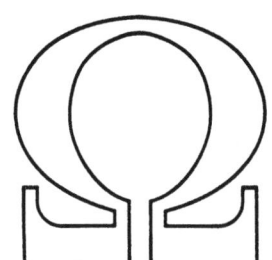 Perspectives in Mathematical Logic

Ω-Group:

R. O. Gandy H. Hermes A. Levy G. H. Müller
G. E. Sacks D. S. Scott

Ω-Bibliography of Mathematical Logic

Edited by Gert H. Müller

In Collaboration with Wolfgang Lenski

Volume I

Classical Logic

Wolfgang Rautenberg (Editor)

Springer-Verlag
Berlin Heidelberg GmbH

Gert H. Müller
Wolfgang Lenski
Mathematisches Institut, Universität Heidelberg
Im Neuenheimer Feld 288, D-6900 Heidelberg

Wolfgang Rautenberg
Mathematisches Institut, Freie Universität
Arnimallee 2, 1000 Berlin 33

The series *Perspectives in Mathematical Logic* is edited by the Ω-Group of the Heidelberger Akademie der Wissenschaften. The Group initially received a generous grant (1970-1973) from the Stiftung Volkswagenwerk and since 1974 its work has been incorporated into the general scientific program of the Heidelberger Akademie der Wissenschaften (Math. Naturwiss. Klasse).

ISBN 978-3-662-22256-0   ISBN 978-3-662-22254-6 (eBook)
DOI 10.1007/978-3-662-22254-6

Library of Congress Cataloging in Publication Data
[Omega]-bibliography of mathematical logic.
(Perspectives in mathematical logic)
Includes indexes.
Contents: v. 1. Classical logic / Wolfgang Rautenberg, ed. - v. 2. Non-classical logics / Wolfgang Rautenberg, ed. - v. 3. Model theory / Heinz-Dieter Ebbinghaus, ed. [etc.]
1. Logic, Symbolic and mathematical-Bibliography. I. Müller, G. H. (Gert Heinz), 1923 - II. Lenski, Wolfgang, 1952 -. III. Title: Bibliography of mathematical logic. IV. Series.
Z6654.M26047 1987 [QA9] 016.5113 86-31426

This work is subject to copyright. All rights are reserved, whether the whole or part of the material is concerned, specifically those of translation, reprinting, re-use of illustrations, broadcasting, reproduction by photocopying machine or similar means, and storage in data banks. Under § 54 of the German Copyright Law where copies are made for other than private use, a fee is payable to "Verwertungsgesellschaft Wort", Munich.

© Springer-Verlag Berlin Heidelberg 1987
Ursprünglich erschienen bei Springer-Verlag Berlin Heidelberg New York 1987
Softcover reprint of the hardcover 1st edition 1987

2141/3140-543210

*Dedicated
to
Alonzo Church*

whose bibliographic work for the
Journal of Symbolic Logic
was a milestone in the
development of modern logic.

Table of Contents

Preface IX

Introduction XV

User's Guide XIX

Ω-Classification Scheme XXVII

Subject Index 1

    Syntax of logical languages B03 3
    Classical propositional logic and boolean functions B05 5
    Classical first-order logic B10 25
    Higher-order logic and type theory B15 38
    Fragments of classical logic B20 48
    Abstract deductive systems B22 62
    Classical foundations of number systems B28 68
    Logical foundations of other classical theories; axiomatics B30 76
    Mechanization of proofs and logical operations B35 85
    Infinitesimal analysis in pure mathematics H05 103
    Other applications of infinitesimal analysis H10 113
    Other applications of logic B80 117
    Proceedings . . . . Part of B97 121
    Textbooks, surveys B98 124

Author Index 135

Source Index 389

    Journals 391
    Series 416
    Proceedings 421
    Collection volumes 440
    Publishers 449

Miscellaneous Indexes 459

    External classifications 461
    Alphabetization and alternative spellings of author names 481
    International vehicle codes 483
    Transliteration scheme for Cyrillic 485

Preface

Gert H. Müller

The growth of the number of publications in almost all scientific areas, as in the area of (mathematical) logic, is taken as a sign of our scientifically minded culture, but it also has a terrifying aspect. In addition, given the rapidly growing sophistication, specialization and hence subdivision of logic, researchers, students and teachers may have a hard time getting an overview of the existing literature, particularly if they do not have an extensive library available in their neighbourhood: they simply do not even know what to ask for! More specifically, if someone vaguely knows that something vaguely connected with his interests exists somewhere in the literature, he may not be able to find it even by searching through the publications scattered in the review journals. Answering this challenge was and is the central motivation for compiling this Bibliography.

The Bibliography comprises (presently) the following six volumes (listed with the corresponding Editors):

I.	Classical Logic	W. Rautenberg
II.	Non-classical Logics	W. Rautenberg
III.	Model Theory	H.-D. Ebbinghaus
IV.	Recursion Theory	P. G. Hinman
V.	Set Theory	A. R. Blass
VI.	Proof Theory; Constructive Mathematics	
	J. E. Kister; D. van Dalen & A. S. Troelstra.	

Each volume is divided into four main parts:

1) The *Subject Index* is arranged in sections by topics, usually corresponding to sections in the classification scheme; each section is ordered chronologically by year, and within a given year the items are listed alphabetically by author with the titles of the publications and their full classifications added.

2) The *Author Index* is ordered alphabetically by author, and contains the full bibliographical data of each publication together with its review numbers in Mathematical Reviews (MR), Zentralblatt für Mathematik und ihre Grenzgebiete (Zbl), Journal of Symbolic Logic (JSL), and Jahrbuch über die Fortschritte der Mathematik (FdM). We much regret that we were not able to include reviews from Referativnyj Zhurnal Matematika in this edition.

3) The *Source Index* gives the full bibliographical data of each source (journals and books) for which only abbreviated forms are used in the Author Index.

4) The *Miscellaneous Indexes* contain various further indexes and tables to aid the reader in using the Bibliography.

For a more detailed technical description of the Bibliography see the *Table of Contents* and the *User's Guide*.

The uniform classification of all entries is a central feature of the Bibliography. The basic framework is the 03 section of the (1985 version of the) 1980 classification scheme of Mathematical Reviews and Zentralblatt für Mathematik und ihre Grenzgebiete. However, this has been modified in a number of ways. Indeed, the 1980 scheme was designed for the classification of works written after 1980, whereas the majority of entries in the Bibliography come before this date. In some areas this has made the classification of older works difficult, and we have tried to cope with this by adding a few new sections and altering slightly the interpretation of others. We have not designated the classifications assigned to a work as primary and secondary, because of the difficulty in doing so in many cases. Each volume contains the full annotated classification scheme together with a description of its general features. In their *introductions* the Editors discuss specifically their interpretations of the classification sections falling in their respective volumes.

The Subject Index is another central feature of the Bibliography. Reading through this Index gives a *historical perspective* for each classification section and provides a rather quick overview of the literature in it. By browsing through the entries of a specific area the reader may be rewarded by finding things (literature, subjects, questions) he was not aware of or had forgotten.

An obvious question now is the extent to which one can rely on the *completeness* and *correctness* of the Bibliography and on the *accuracy of the classifications*. We comment on each of these aspects separately.

In an effort to be as complete as possible, we consulted all sources available to us and decided in favour of inclusion in doubtful cases (so that certainly some papers with little bearing on mathematical logic are listed here and there). As the historical starting point for the Bibliography we chose the appearance of Frege's *Begriffsschrift* (1879). A certain restriction on scope stems from our decision to concentrate on mathematical logic and in particular on those areas defined by the titles of the six volumes. A major source of material was provided by the review journals mentioned above; we used them both to identify publications in the less known journals and to find review numbers and other bibliographical data of items found in other sources. We also made use of various lists of literature contained in books, survey articles, mimeographed notes, etc. Some especially valuable newer sources were:

W. Hodges: A Thousand Papers in Model Theory and Algebra

M. A. McRobbie, A. Barcan and P. B. Thistlewaite: Interpolation Theorems: A Bibliography

D. S. Scott and J. M. B. Moss †: A Bibliography of Books on Symbolic Logic, Foundations of Mathematics and Related Subjects

C. A. B. Peacocke and D. S. Scott: A Selective Bibliography of Philosophical Logic.

Various strategies and crosschecks were used to ensure the completeness of the bibliographical data and in particular of the reviews mentioned above. For each item listed in the Bibliography we tried to include any translations, reprintings in alternative sources and errata, and to give cross references for a work appearing in several parts.

On the whole this Bibliography was compiled and organized for use by the practising mathematician; there is no claim that the most rigorous standards of librarianship are met.

It is hard to say how successful our striving for completeness was. This is especially true for the most recent literature. No 1986 items were included. We checked all the main journals in logic, the reviews in MR, Zbl and JSL and Current Mathematical Publications for literature published up to the end of 1985, but undoubtedly some gaps remain.

As for correctness, in any ordinary book we can tolerate a number of printing errors because of our knowledge of the language and the context, but, when one organizes data connected by (abstract) pointers in a computer program, almost every typing error has far-reaching consequences. Various consistency tests were used to check the program and the input data. There are, however, many other sources for mistakes and errors.

For some items our references contained incomplete or ambiguous information. Although we tried to complete the bibliographical data, this was often difficult, particularly in cases where, for example, the source was obscure or the publisher was given only by location. Another source of errors lies in the identification of author names. An author may publish using abbreviations of his first, his second or both of his given names. This is generally not a problem for authors with uncommon surnames, but if the surname is, e.g., Smith or Brown the possibility of misidentification arises. We may have identified two different authors or failed to identify two or more different forms of an author name.

It is unavoidable in a project of this scope that there will be errors, particularly in the classification, so perhaps it is worthwhile to explain briefly the process by which the classification was done. Items entered before 1981 were originally classified according to a scheme unrelated to the current one. To begin the conversion to the 1980 scheme we used the computer to change old categories to their new versions wherever there was a well-defined correspondence. Then every entry was checked and if necessary reclassified by hand. From 1981 each new entry was classified shortly after being entered in the database. For the most part this was done on the basis of titles, reviews, and other information, but without consulting the works themselves. This was necessary to preserve the finiteness of the enterprise, but it has inevitably led to errors, certainly in some cases egregious ones. These were constantly being corrected during the final editing process, but many will remain.

Although the Editors have to some extent used different strategies in classifying the entries falling into their respective volumes, finally a reasonable degree of uniformity has arisen. The user is referred to the Editors' introductions for further details on the classifying procedure.

A special apology goes to the native speakers of languages with diacritical marks. Our central difficulty was to get the right spelling of names used in different forms in such a variety of sources. In addition, entering diacritical marks in a computer introduces yet another source of errors; so they have almost all been ignored (the User's Guide and the Miscellaneous Indexes contain details of those that have been transliterated). We appreciate that, although the absence of, for example, accents in the text of a French title may not create undue problems, the lack of diacritical marks in author names is particularly unfortunate. We hope that this omission will not be too misleading.

The future

By its nature a bibliography has lasting value to the extent it succeeds in "completing the past". But it should also serve for some years as an aid to current research. We have various plans to extend the scope of the present Bibliography by including new areas such as universal algebra, sheaves, philosophical logic (subdividing the present volumes I and II appropriately), and philosophy of mathematics. The present six volumes cover only approximately 80% of the data on our computer files.

The possibility of extending the classification scheme by developing a so-called thesaurus system was discussed on several occasions. Certainly this would be desirable; to some extent *Alonzo Church* tried to create such a system in connection with his bibliography in the Journal of Symbolic Logic. However there are difficult scientific problems connected with the creation of such a system and their solution requires much time and expertise.

Another way to extend the Bibliography which would perhaps better serve the purpose of providing an overview of certain special areas would be to commission a series of survey papers to appear from time to time as, say, an additional issue of the Journal of Symbolic Logic; each paper would include an annotated listing of the literature taken from the Bibliography.

There are plans to establish a bibliographical centre for Mathematical Logic and adjacent areas. A central function of such a centre would be to collect information on all new publications (including mimeographed notes, theses, etc.) as well as to correct errors and omissions in the current data. It is hoped that all logicians would provide information concerning their own publications as they appear. A continuation of the Bibliography together with supplements (to appear periodically) would be prepared at the centre. We also hope to make available an on-line system. From these activities and the flow of information from the individual logician to the centre and vice versa a "living Bibliography" would emerge. This would provide a way to determine the main trends in the progress (or decline) of specified directions of work. So a centre would exist at which it would be possible to gain some oversight of the rapidly developing field of mathematical logic.

Acknowledgements

Work on the Bibliography started at the same time as the Ω-Group came into being, early in 1969. To begin with, index cards were used for storing bibliographical data; it was *Horst Zeitler* and *Diana Schmidt* who convinced me that we are living in the 20th century and that the data should be computerized. They, together with *Ann Singleterry-Ferebee*, first brought the Bibliography to a workable computerized form at the end of the seventies. In this period I also had the help of *Ulrich Felgner* and *Klaus Gloede*, in particular in classifying the literature. At about this time, others contributed in many useful ways. In particular, important problems of principle were highlighted by a long list of intriguing questions from *Dana Scott*: "How do you classify this or that item...?" *Robert Harrison* worked faithfully collecting data for the Source Index.

The second period, beginning in the early eighties, was characterized by the programming necessary to manage the data. This was carried out by *Ulrich Burkhardt* (†) and *Werner Wolf* and finally by the *outstanding work* of *Rolf Bogus*. In this period we also changed the classification system and for this I had the continuous and intensive help of *Andreas Blass* and *Peter Hinman*. In addition, both of them, together with *Heinz-Dieter Ebbinghaus*, gave me much advice about organization and technical arrangements. Over the last four years the work of large groups of students has been essential for collecting reviews, entering corrections and new items into the computer, etc. Again and again I have been overwhelmed by their idealism and energy. Among them I wish to mention particularly the continuous help of *Elisabeth Wette* and *Ulrike Wieland*.

The Bibliography would not have reached publishable form without the work of my collaborator *Wolfgang Lenski* (in the second period). It would have been unthinkable for me to interfere anywhere in the process of the growth of the Bibliography without discussing the matter with him beforehand. He has accumulated a detailed knowledge of every aspect of the project and has devoted his talents for many years to the common enterprise.

My secretary, *Elfriede Ihrig*, has willingly assisted in the work of the Ω-Group and the Bibliography from the beginning, over many years, filled with ups and downs and with all kinds of tasks. She has always maintained her warmhearted balance.

To all I express my personal warm thanks!

The *Journal of Symbolic Logic* sent information concerning papers for which reviews were never published. We also acknowledge permission to use computer tapes with lists of literature covering certain periods of time from *Mathematical Reviews* and from *Zentralblatt für Mathematik und ihre Grenzgebiete*.

Yuzuru Kakuda and *Tosiyuki Tugue* collected and prepared the Japanese literature for us. *Petr Hajek* and *Gerd Wechsung* helped us with updating the bibliographical references of so many sources not available to us. *Mo Shaokui* corrected data on the Chinese literature and added items of which we were not aware.

The Editors filled many gaps, corrected all mistakes which came to their attention and undertook the burden of checking - and changing if necessary - the classification of the entries in their special areas. Here again I would like to mention *Rolf Bogus* and *Wolfgang Lenski* who organized the enormous exchange service for the transfer of literature among the Editors and the inputting of the many changes and corrections. *Andreas Blass* and *Peter Hinman* were also instrumental in this exchange; their preliminary classification of each item added to the Bibliography during the final editing process meant that the Editors had mainly to look at items inside their own areas. *Jane Kister* read through the whole Source Index correcting mistakes and suggesting valuable changes in it.

In collecting and organizing the data for the Bibliography we have received much help from various sources, and especially in letters from colleagues all over the world, containing information and suggestions. I apologize for being unable to answer them all individually, but all were read carefully.

We thank all those concerned.

As everybody can guess, the whole enterprise was indeed expensive. *Financial support was provided by the Heidelberger Akademie der Wissenschaften in the framework of the Ω-Group project.*

Special thanks go to the firm APPL, who transformed our computer tapes to the present printed form, and to the editorial and production staff of SPRINGER-Verlag for their continuous help, notably in the traditionally fine realization of the six volumes.

Finally, through working so many years on this project I have come to understand and appreciate more and more the immense work of Alonzo Church in building his Bibliography of Logic and its adjacent areas, together with a detailed classification, that is contained in so many volumes of the Journal of Symbolic Logic. Understanding comes from doing.

Introduction

Wolfgang Rautenberg

What distinguishes mathematical logic from traditional logic is not merely the use of symbols and some calculation. These features occur to some extent already in ancient logic. The essential difference is the distinction between object-language and meta-language and the investigation of the object-language (its syntax and semantics) with mathematical tools. The logical operations concern the (formalized) object-language, while the logical means of the meta-language are often unspecified and sometimes allowed to be different from those of the object-language. From this point of view it seems appropriate that this Bibliography starts with Frege's Begriffsschrift of 1879.

As is well known, the formalization of classical (two-valued) logic is essentially a result of the endeavor of Frege, Russell and others to create a logistic foundation for mathematics. Therefore much of the literature of this volume is related to the foundations of mathematics.

There are many good reasons to argue that "logic" should properly mean first-order classical logic. This is not merely a convenient convention for mathematicians and (mathematical) logicians. Linguists, philosophers, computer scientists, physicists and others need to agree on a common logical base (a metalogic) if they want to communicate with each other. This is usually the logic of common sense, essentially two-valued logic.

The present volume comprises those parts of Section B (general logic) of the Omega classification scheme which deal with two-valued logic. An exception is B22 which concerns the common features of all kinds of logic. The volume also contains the subsection B35 on mechanized deduction, which is closely related to various formalization techniques of first-order logic. B25 has been removed to Volume III and B40 to Volume VI. Subsections B65 and B70 are not included in the present edition of the Bibliography.

Comments on the Classification Scheme

The general principles followed in selecting and classifying the data in this volume included the following. Short contributions in dictionaries or encyclopedias were generally not included. Nor were papers, for example in pedagogical journals, whose only goal was popularization. Contributions to philosophy, the theory of science, etc. were included only if they were (or appeared to be) relevant to mathematical logic. Many of the entries in, for example, the model theory volume explicitly deal with first-order logic, but this did not automatically give them the classification B10 unless one of the author's goals was in the area designated by B10. A similar principle was followed in the case of multiple classifications within Volumes I and II.

Because of the large volume of material, the possibility of classification errors cannot be excluded, particularly in the case of multiple classifications. A particular problem was the boundary between the literature of mathematical logic and philosophically oriented, critical, or descriptive literature. Whether a particular paper near this boundary was included depended not only on decisions by the editor but also on what was included in the database and on the accessibility of the paper.

The information available for classifying entries varied considerably from one entry to another. Many entries were classified on the basis of reviews. Others were classified on the basis of previously assigned classifications or other information, and some were classified merely on the basis of their titles when neither the article itself nor a review was available. In such situations, some errors are unavoidable.

It was originally planned to publish Section B as a single volume (like each of C through F). This turned out to be impossible in a volume of reasonable size, so this section has been split into two volumes. We hope that this splitting, which occurred near the end of the preparation of the bibliography, has not led to new unforeseen errors.

In spite of all difficulties, we hope to have rather completely covered the literature from 1879 to 1985 (with understandably decreasing completeness for 1984 and 1985). The reader is invited to send criticisms or proposals for the next edition to the editor of this volume.

Syntax of logical languages   B03

This section is intended to contain those papers that are concerned almost exclusively with the (usually simple) syntactic structure of logical languages. It does not include works about programming languages, since such data were not collected for the Bibliography.

Classical propositional logic and boolean functions   B05

This area includes the theory of Boolean functions and has numerous applications (see B70 and B80). It also includes some entries dealing with border areas like the complexity of classes of Boolean functions, but this literature was not systematically collected, so the selection may appear haphazard.

Classical first-order logic   B10

First-order logic is the central core of mathematical logic, and the Gödel completeness theorem is a foundation for most of the subdisciplines of mathematical logic. Topics going beyond completeness (Löwenheim-Skolem theorems, interpolation, omitting types, etc.) are in the model theory volume (Volume III). Some modifications of first-order logic (e.g. allowing the empty domain) are found under B60.

Higher-order logic and type theory   B15

This field, which goes back to the Principia Mathematica of Whitehead and Russell, was originally conceived as a way to avoid the logical and set-theoretic antinomies. The possibility of a set-theoretic foundation for mathematics has reduced the significance of type theory for classical logic, but type theory continues to play an important role in, e.g., proof theory. The interested reader should therefore also consult section F35. Fragments of type theory (e.g., theories in second-order logic) are also treated in sections C85 and E70.

Fragments of classical logic   B20

By fragments we mean logical systems that do not have the full expressive power of the corresponding classical system (negationless logic, Aristotelian logic, etc.). They should not be confused with systems, like intuitionistic logic, that have the full expressive power of classical logic but reduced deductive power. B20 contains the numerous papers and books that deal with fragments of two-valued propositional and predicate logic. Works on fragments of intuitionistic logic are in B55, and works dealing with the reduction problem and decision problem for classes of formulas with restricted quantifier prefixes, etc., are in B25, C40 (Volume III) and D35 (Volume IV).

Abstract deductive systems   B22

This area was founded primarily by the work of A. Tarski. Almost all logical systems are determined by certain axioms and rules of a structural nature (possibly given by schemata). This section contains papers analyzing such systems at a high level of generality and abstraction. Questions about adequate semantics play a role here, and there are close connections with algebraic logic and many-valued logic. Most of the applications of the general results in B22 are to areas covered in Volume II.

Classical foundations of number systems   B28
Logical foundations of other classical theories; axiomatics   B30

The classical and axiomatic foundations of number systems are of special importance from the point of view of mathematical logic and were therefore split from B30 and given a section of their own, B28. Among other mathematical theories, there is a clear disparity in foundational work, with geometry having a distinct advantage, largely for historical reasons. We have made some effort to reduce this disparity. We have not included foundational work that is concerned with a specific field, like geometry or probability theory, but not with mathematical logic or its applications. For example, a reformulation of Hilbert's axioms about the betweenness relation is a routine mathematical matter and has in this sense nothing to do with mathematical logic. The same is true of innumerable contributions modifying axioms in particular fields, contributions that should properly be considered as belonging to those fields. For the axiomatic foundations of set theory, the reader is referred to E30.

The classification B30 has also been used for contributions to the foundations of mathematics, the axiomatization of metamathematics, etc.; most of these were written before the 1980 classification and many are only of historical interest for contemporary mathematical logic.

After discussions with the other editors, it was decided to exclude contributions to the foundations of physics, the methodology of science and other disciplines if they have no immediate connection with mathematical logic. Boundaries here are sometimes difficult to discern, and we ask the reader's indulgence for our all too subjective judgments.

Mechanization of proofs and logical operations   B35

As the logical consequences of a concrete axiom system are recursively enumerable but not in general decidable, there arose a need, at first within mathematical logic, for practical computational methods. More recently, the field of automatic theorem proving has become an essential foundation of artificial intelligence. It also has many other applications in computer science. The part of the literature contained in this Bibliography includes much application-oriented work, but we do not claim completeness in this direction.

Infinitesimal analysis in pure mathematics   H05
Other applications of infinitesimal analysis   H10

The characteristic feature of the infinitary analysis (non-standard analysis) introduced by A. Robinson is the method of proving theorems in classical analysis by means of arguments in non-standard models and a transfer principle. This feature distinguishes non-standard analysis from older analysis using infinitesimals (Leibniz, Euler) and the various possible generalizations of analysis. For the user equipped with a basic knowledge of logic, infinitary analysis is a highly efficient tool; it can also help to clarify physical phenomena whose treatment with traditional analytic methods is very difficult.

Other applications of logic   **B80**

This section contains primarily applications outside mathematics and mathematical logic. For applications within mathematics, see for example Volume III. Applications to computer science are only partially included in this Bibliography; see the section on algorithmic and dynamic logic in Volume II.

Proceedings   **B97**
Textbooks, surveys   **B98**

In general, proceedings are listed ih this volume only if they are primarily concerned with one of the subsections listed above. In contrast, we have included in B98 those monographs and textbooks that have to do with mathematical logic or could be of interest to logicians. The B98 subsections in Volumes I and II are identical.

Acknowledgments

First of all I want to express my thanks to Gert Müller for his patience in the course of preparing this Bibliography. I also thank the other editors for valuable hints and suggestions and my student Manfred Borzechowski for technical assistance. Finally, I want to express my particular thanks to Andreas Blass who translated this introduction into English.

User's Guide

Wolfgang Lenski & Gert H. Müller

§1. Introduction

After some opening remarks, the organization of this Guide follows the main division of the volume: *Subject Index, Author Index, Source Index, Miscellaneous Indexes*. For each part we give first a general explanation followed by a more detailed description of typical entries in the index in question. The reader will probably find the User's Guide most helpful when he comes across an unclear entry in the Bibliography: he can then turn directly to the corresponding section in this Guide for an explanation of the abbreviations and conventions used.

§2. General remarks

The main languages of the Bibliography are English, French, German, Italian, and Spanish. For other languages translations (of titles, names of sources, etc.) are used – with some few exceptions in cases for which we had no translation. These translations were taken from various available sources or made by the Editors.

For practical reasons, all entries are in the Roman alphabet and diacritical marks have not been used. Thus, for languages other than English certain conventions have been adopted.

The *transliteration of Cyrillic* names and titles, the treatment of *diacritical marks* and the *alphabetization and alternative spelling* of author names are explained in detail in the Miscellaneous Indexes.

The *abbreviations* of *sources* were either taken from one of the various reviewing journals or invented by us. Although we had to abbreviate long titles, we hope that in most cases the abbreviation will suggest the full title in a sufficiently understandable way. How successful we were is left to the user to decide.

The *review numbers* given with the entries in the Author Index are from *Mathematical Reviews* (MR), *Zentralblatt für Mathematik und ihre Grenzgebiete* (Zbl), *Journal of Symbolic Logic* (JSL), and *Jahrbuch über die Fortschritte der Mathematik* (FdM). We made a serious attempt to include all reviews of any given item but we have doubts concerning our success. We also tried to avoid listing two reviews for a given item in those cases in which the second "review" simply points to the original review and does not give any additional information.

In case of *multipart publications* pointers are given to the other parts, as far as they are known, in the *Remarks* to the publication in question. It is not always the case that the different parts of a publication all have the same classifications. Thus it may happen that, for example, part I has a classification in this volume and part II does not. In this case the Remarks for part I indicate the author(s) and year of publication of part II. The user will need to consult the other volumes for further bibliographic information on part II.

The general way to search through the Bibliography is to use certain *pointers*: From the Subject Index to the Author Index the pointer is [Author, Year, Title]; from the Author Index to the Source Index the pointers are 5-digit codes; e.g. (J 1234) is a code which sends the user to the J-section of the Source Index.

A *word* of *caution*: In order to use the Bibliography for quotations in future publications it is necessary to use both the Author Index and the Source Index; it is not generally sufficient to quote just from the Author Index. For example, for the paper "AANDERAA, S.O. [1974] *On k-tape versus (k−1)-tape real time computation*" a quotation of the source of this paper as given in the Author Index listing of this item, "Complexity of Computation; 1973 New York 75-96", would with high probability be misleading: one might try to find a book of this title published in New York in 1973 whereas in fact "1973 New York" denotes the date and place of the conference in the proceedings of which Aanderaa's paper appears. The volume was actually published in 1974. The source code (**P** 0761) for "Complexity of Computation; 1973 New York" should be used to find the full details of the source in the Source Index. The abbreviations of the sources may themselves be misleading without the corresponding additional details (e.g. country codes) given in the Source Index. For example, many abbreviations for conference proceedings do not include an abbreviation for "Proceedings of ...". Thus "Proceedings of the Third Brazilian Conference on Mathematical Logic" is abbreviated by "Brazil Conf Math Logic (3); 1979 Recife"; a reader without the Bibliography at hand might search in vain for the volume under "Brazil" in his library whereas in fact it might be found alphabetically under "Proceedings".

The Source Index includes, as far as they are known to us, *International Standard Serial Numbers* (ISSN) or *Book Numbers* (ISBN) and *Library of Congress* (LC) numbers. They may help in finding the source in question in libraries or bookstores.

To facilitate searches for works spanning two or more of the major subfields of logic, the first of the Miscellaneous Indexes lists the entries in the present volume that also occur in other volumes of the Bibliography.

Accidental occurrences of features not explained in the User's Guide are left as exercises to the user. HINT: Write to us (in any case), please.

§3. Subject Index

This is a listing of publication items ordered

> first by the (special) *classification sections,*
> then by the *year of appearance,*
> and finally *alphabetically* by *author,*

showing the author, title and the codes of all classification sections which apply to the given publication.

- The *titles* are given in the main languages of the Bibliography; if the original title is in another language, this is indicated in parentheses, e.g. (Russian), but only a translation of the title is given. Information on summaries in languages other than the original is included.
- If a publication is by *multiple authors,* it occurs only *once,* under the alphabetically first name. (But see also the Author Index.)
- In order to get the full bibliographical data of a publication, use the author, year and title to find the item in the Author Index.
- The classification sections listed in each volume have been selected by the individual editors. Sections B96–F96 have been systematically omitted; for the collected works of an author refer to the Author Index.

§4. Author Index

This is a listing of publication items ordered

> alphabetically by *author*, and for a given author chronologically by the *year of appearance*, and therein alphabetically by *title* of the item.

- The titles are given in the main languages as in the Subject Index.
- The *names* of the *authors* are written in the Roman alphabet using the Transliteration Table (see Miscellaneous Indexes) if necessary. There may be many versions of the name in use for a given author; (e.g. different combinations of the given name(s) or initials; different names used before and after marriage; different transliterations). The Miscellaneous Indexes include a table of different versions (known to us) and the corresponding form used in this Bibliography.
- Here publications with *multiple authors* are listed under *each* author but in the alphabetically later cases only the year is given and there is a pointer to the full entry given under the first author.
- The last entries for an author may contain a *reference to other name(s)* under which he/she also has publications in the Bibliography or *to other volumes* of the Bibliography where he/she has publications not mentioned in the present volume. A complete list of the author's papers contained in the six volumes is obtained by consulting the other volumes.
- In the following we explain the *individual entries* in more detail by giving an *idealized* example using fictitious names and sources showing all features that might occur; in a given case some features may not appear either because they do not apply or because our information is incomplete. The typefaces of the example and the order of its fields are as in the Author Index but, for *expository reasons only, here* we list all features on separate lines numbered by (1), (2), ...; we list explicitly those fields that begin a new line in the Author Index itself. (The foregoing description applies not only to the explanation of the Author Index treated here but also to the explanation of the Source Index later on.)

Example

(1) AUTHOR, K.J. & COMPANION, CECIL X. [1972]
(2) *On coding and decoding (Russian) (English and French summaries)*
(3) (**J** 9999) or (**S** 9998) or (**P** 9997) or (**C** 9996) or (**X** 9995)
(4) J Math 1*1-10
  or Math Logic Series 1
  or Logic Conf; 1999 London 3-10
  or Math Publ xxv+200pp
(5) • ERR/ADD ibid 2*3-4 or (**J** 8888) Arch of Logic 2*3-4
(A new line begins here.)
(6) • LAST ED [1983] (**X** 9900) Logic Publ xx+100pp
(7) • REPR [1981] (**J** 9901) Math Logic J 2*3-8
(8) • TRANSL [1979] (**J** 9902) Math Transl 1*4-8
(A new line begins here.)
(9) ⋄ B05 B20 C12 ⋄
(10) • REV MR 99a:03001 Zbl 999#03001 JSL 99.321 FdM 99.123
(11) • REM This is an illustrative example
(12) • ID 12345

Explanations

(1) lists the authors followed by the year (in brackets) of publication of the item. Exceptionally a full given name (e.g. CECIL) is used to distinguish several authors with the same surname and initials.

(2) gives the title of the item followed, if the original is not in an official language of the bibliography, by the original language in parentheses and an indication of summaries in languages other than the original.

(3) is a pointer (or "source code") to the *Source Index;* there are five types: *Journal* (**J**), *Series* (**S**), *Proceedings Volume* (**P**), *Collection Volume* (**C**), and *Publisher* (**X**); one such code appears in (3). In order to find the full bibliographical data of the source use the pointer to locate the source in the Source Index; e.g. (**J** 9999) is given in the J-section of the Source Index.

Note: For a small number of items the source code is 0000, 1111, 2222 or 3333 (*not* preceded by J,S,P,C, or X). The code 0000, respectively 1111, indicates that the item is a thesis, respectively technical report. The code 2222, respectively 3333, is used for those cases in which the source, respectively publisher, is unknown. In each such case any further source information available is given in the Remarks (see line (11)).

(4) contains the abbreviation of the source indicated by the code in (3) followed by the paging as appropriate. Certain uniform features of the form of abbreviation used for proceedings and collection volumes should help the reader to recognise the volume. Abbreviations for proceedings (**P**) volumes end with an indication of the year and place of the corresponding conference, e.g. 1973 New York. Likewise, a name in parentheses, e.g. (Goedel), in an abbreviation of a collection (**C**) volume indicates the honorand to whom the volume is dedicated. A name followed by a colon, e.g. "Wang:", at the beginning of a collection volume abbreviation, indicates the author of all papers in the collection. The paging takes one of the following forms:

1*1-10   : Volume 1, pages 1-10 (for journals or series)
1/2*1-10 : Volume 1, Issue 2, pages 1-10 (for journals)
3-10   : pages 3-10 (for proceedings or collection volumes)
xx+200pp : initial paging + paging of a book (following a publisher or series)

(5) The • here and later is intended to make the entries easier to read. It is used to separate different types of information. After the • is the bibliographical information for published errata or addenda to the item. The two ways ERR/ADD can be given correspond to the cases in which its source is the same as in (4) (indicated by "ibid") and that in which it is in a different source; in the latter case the entry is of the same form as in (3) and (4).

The remaining information is not strictly part of the bibliographical data but contains useful additions.

(6), (7), (8) list the most recent edition, reprintings and translations, respectively, given by source as in (3) and (4); note that (7) and/or (8) may contain several entries for one publication.

(9) The classification codes enclosed in ◇ always begin a new line. Note that the codes are given in alphabetical/numerical order; no distinction of primary and secondary classification is made. (The classifications often differ from those assigned to the item in MR or Zbl.)

(10) lists the reviews. Sometimes two reviews are given from one reviewing journal. This may happen, e.g., when an item and its erratum/addendum are reviewed separately or when two different editions of a book have independent reviews.

(11) contains additional information not appropriate for coding in one of the standard fields.

(12) Each entry ends with its *identification number*. It is not used elsewhere in the main body of this volume except occasionally in the Introduction and the Remarks of another item where it may be used to pinpoint an item not uniquely identified by author(s) and year. The identification number is used (together with author(s) and year) as a pointer in the External Classification Code Index. We ask that the identification number be used in any correspondence with the Editors concerning this publication, as the bibliographical data base is indexed by these numbers.

§5. Source Index

This index contains the bibliographical data of the sources of the publications listed in this volume. It is subdivided into the following parts.

J (*Journals*), **S** (*Series*), **C** (*Collection volumes*), **P** (*Proceedings*), **X** (*Publishers*).

• Each part is ordered by the 4-digit source code numbers. (There is no significance to the particular 4-digit number assigned to a given source other than as a way to find the entry in the source index. Numbers were assigned as the sources were entered into the data base and so the numbering does not correspond to alphabetical order or order of publication.) Each 4-digit number is used *only once* as a source code so that, e.g., 0007 is a source code for a journal and the number 0007 is not used as a code for a series, proceedings, collection volume or publisher.

• Titles are given in the original language, using the transliteration system (see Miscellaneous Indexes) where necessary, followed, if necessary, by a translation into one of the main languages in parentheses. Sometimes if the original title is unknown to us, we give only a translated title in parentheses. Sometimes a source, e.g., a journal, has more than one title (English, French, German); in this case all titles are given, separated by *. These measures were taken to ease the search in libraries. In order to explain the entries in the Source Index we again use idealized examples and apply the conventions described in §4 above.

Journals

Example of a journal entry:

    (1)   **J** 8888 Math Div • F
(A new line begins here)
    (2)   *Mathematica Diversa * Mathematiques Diverses*
    (3)   [1900ff] or [1905-1935] ISSN 0007-0882
(A new line begins here.)
    (4)   • CONT OF (**J** 8885) J Math Ser A
    (5)   • CONT AS (**J** 8887) J Math Ser C
    (6)   • TRANSL IN (**J** 9904) Math Transl
    (7)   • TRANSL OF (**J** 9905) Matemat
((4) - (7) may contain more than one entry)
(A new line begins here.)
    (8)   • REL PUBL (**J** 9903) Mathematica (Subseria)
    (9)   • REM This journal is a fiction

Explanations

(1) Source code and abbreviation of the journal as used in the Author Index followed by the *international vehicle code* of the country in which the journal is published. A list of these codes is included in the Miscellaneous Indexes.

(2) The form of title(s) (and translations) are explained above.

(3) [1900ff] indicates that this journal has appeared continuously since 1900; [1905-1935] indicates that the journal appeared from 1905 to 1935. The International Standard Serial Number (ISSN) is given whenever possible.

(4), (5) give the predecessors (continuation of) and successors (continued as) of the journal in (2). In some cases in (4) or (5) the source code may be missing; this means that there are no entries in the Author Index which refer to the continued source. (It is mentioned, however, for the convenience of the user.)

(6) lists the translation journals of (2) and (7) gives the journal of which (2) is a translation; the source code is shown only if the translation in question is used as a source in this Bibliography. (6) and (7) do not both occur in a single journal entry.

(8) lists further entries in the Bibliography related to this journal, e.g. a subseries of the journal.

(9) is intended for additional information of various kinds.

Series

It is often hard to determine what should and what should not be characterised as a series. Some serials that we have chosen to treat as series may elsewhere be considered to be journals. In other cases, in particular certain publication series of university mathematics departments, the series includes all publications of its publisher and so might reasonably be identified with the publisher. Despite these considerations, we have chosen to list series separately to accord with the form of quotation often used in the modern literature.

Example of a series entry:

(1) S 8999 Notae Log • NL
(A new line begins here)
(2) *Notae Logicae* * *Notas Logicas*
(3) [1900ff] or [1905-1935]
(4) • ED: EDITOR, A.A. & COEDITOR, B.B.
(5) • SER (S 8998) Notes in Phil
(6) • PUBL (X 9950) Logic Publ Co: Heidelberg
(7) • ALT PUBL (X 9951) Math Publ Inc: London
(A new line begins here.)
(8) • CONT OF (S 9975) Notes in Logic A
(9) • CONT AS (S 9901) Notes in Logic B
(10) • TRANSL IN (S 9902) Notes de Logique
(11) • TRANSL OF (S 9903) Logical Notes
(A new line begins here.)
(12) • ISSN 0011-11122 (or ISBN 0011-11123) LC-No 73-10000
(13) • REL PUBL (S 9900) Notae Logicae (Subseria)
(14) • REM The origins of this series are somewhat obscure

Explanations

Entries (1), (2), (3), (8)-(11), (13), and (14) correspond to (1), (2), (3), (4)-(7), (8), (9), respectively, of the *journal entry* described above.

(4) lists the editors of the series (given in the same form as in line (1) of the Author Index example).

(5) Occasionally a series is itself a subseries of another series or journal. This is indicated in (5) (with an S or J as appropriate).

(6) gives the publisher of (2). For those publishers not listed in the publisher section of the Source Index, an abbreviation is sometimes used if either the abbreviation is readily understandable or the full name is not known.

(7) Some sources are published by two or more publishers; ALT PUBL lists the alternative publisher(s).

(12) lists the ISSN (or ISBN) and the Library of Congress number.

Proceedings and Collection Volumes

Example of a proceedings or collection volume:

    (1)   P 9920 Atti Congr Mat; 1971 London, ON • CDN
            or
            C 9921 Atti Congr Mat • D

(A new line begins here.)

    (2)   [1972]
    (3)   *Atti del Congresso di Matematica* * *Actes du Congres de Mathematique*
    (4)   • ED: EDITOR, A.A. & COEDITOR, B.B.
    (5)   • SER (S 8999) Notes in Logic
    (6)   • PUBL (X 9950) Logic Publ Co: Heidelberg
    (7)   • ALT PUBL (X 9951) Math Publ Co: London

(A new line begins here.)

    (8)   • DAT&PL 1971 Aug;London, ON, CDN
    (9)   • ISBN 0-012-34567-X, LC-No 84-98765
    (10)   • REL PUBL (P 9947) Atti Congr Mat Vol Spez

(A new line begins here.)

    (11)   • TRANSL IN [1973] Conf de Logique Math (3); London, ON, CDN
            • PUBL (X 9949) Livres: Paris
    (12)   • TRANSL OF [1971] Konf Math Logik (3); London, ON, CDN
            • PUBL (X 9948) Buchverlag: Stuttgart

(A new line begins here.)

    (13)   • REM Not all the articles appear in the translation

Explanations

(1), (3), (4) - (7), (9), (11), (12), (13) correspond to (1), (2), (4)-(7), (12), (10), (11) and (14), respectively, of a *series entry*. In (11), (12) PUBL denotes the publisher of the translation or original, respectively.

(2) denotes the year of publication of the volume (and not, in the case of a proceedings, the year of the conference).

(8) is used for proceedings volumes to indicate the date (year and month) and place of the conference, given by the city, the state (for the USA and elsewhere) and the country using its code as defined above. Note in case of *Proceedings* (**P**) volumes in (1) the country code of the place of the conference is repeated for conformity reasons, whereas for *Collection* (**C**) volumes the country code in (1) refers to the location of the publisher as in the case of *Journals* and *Series*.

(10) lists further entries in the Bibliography related to this volume, e.g. another proceedings volume of the same conference or a journal of which the volume is a special issue.

Publisher

Example of a publisher entry:

(1) X 9950 *Logic Publishing Company* (Heidelberg, D & London, GB) ISBN 0-01
(2) • REL PUBL (X 9930) Editions Logiques: Paris, F
(3) • REM In London called Logic Publishing Corporation

Explanations

(1) lists the source code and full name of the publisher followed, in parenthesis, by the cities from which the publisher publishes and the ISBN. As in (8) of a **P** or **C** entry, codes are used for countries (see Miscellaneous Indexes).

(2) lists those publishers who have connections with the publisher listed in (1).

§6. Miscellaneous Indexes

This part contains the following indexes:

1. External classifications
2. Alphabetization and alternative spellings of author names
3. International vehicle codes
4. Transliteration scheme for Cyrillic

In each case a description of the contents and use are given in the corresponding introductory texts.

Ω-Classification Scheme

Andreas R. Blass
Peter G. Hinman

The classification scheme used for the Ω-Bibliography is a modified version of the section "03: Mathematical Logic and Foundations" of the 1985 Mathematics Subject Classification of *Mathematical Reviews* and *Zentralblatt für Mathematik und ihre Grenzgebiete*. For the sake of uniformity we have labeled all sections with a letter followed by a two-digit number; the prefix 03 is superfluous and therefore omitted. This decision has led to the creation of new sections to replace 03-01 through 03-06 (cf. X96-X98 and A10) and several sections with prefix other than 03 which have substantial logical content. Examples of the latter sort are B70 (to replace 94C10) and B75 (to replace the "logical part" of 68B10) (68Q55 and 68Q60 since 1985).

An important category of differences between the two schemes arises from the fact that whereas the MR/Zbl system is intended to classify works written after 1980, the majority of entries in the Ω-Bibliography were written before 1980. The subject matter of Mathematical Logic has, of course, changed immensely over the years, and today's categories are not always sufficient to distinguish properly important lines of earlier research. To deal with this problem we have added a few new sections (e.g. B22, B28, B65, C07, E07, and E47), renamed others (e.g. B35, C35, and E10), and altered slightly the interpretation of others (e.g. B25 and D65). To aid the reader in learning our conventions we have added descriptors to the section names. Topics preceded by a + (−) sign are specifically included (excluded) from a section. When this is in conflict with current MR/Zbl practice, this fact is also noted.

A

A05 **Philosophical and critical**

A10 **History, Biography, Bibliography**
MR uses 03-03 and 01A for history and biography
MR puts bibliography under specific fields.

B   GENERAL LOGIC

B03 **Syntax of logical languages**

B05 **Classical propositional logic and boolean functions**
  + Axiomatizations of classical propositional logic
  + Boolean functions (machine manipulation is also in B35); MR puts these in G05 and in 06E30 and 94C10.
  − Fragments of propositional logic: see B20
  − Switching circuits: see B70; MR also uses 94C10

B10 **Classical first-order logic**
  + Many-sorted logic
  + Syntax and semantics up to the Completeness Theorem
  − Model theory: see Cn, particularly C07
  − Proof theory: see Fn

B15 Higher-order logic and type theory
+ Higher-order algebraic and other theories
− Higher-order model theory: see C85
− Set theory with classes: see E30 and E70
− Intuitionistic theory of types: see F35

B20 Fragments of classical logic
+ Fragments of propositional and of first-order logic
+ Fragments used in model theory, set theory, etc.
+ Syllogistic
− Classical propositional logic: see B05
− Weak axiomatizations without restrictions on formulas: see B55, B60, F50 ("Fragment" refers to reduced expressive power, not reduced deductive power; MR heading "Subsystems of classical logic" includes both)

B22 Abstract deductive systems
+ Consequence relations
MR uses B99

B25 Decidability of theories and sets of sentences
+ Decidability of satisfiability
+ Decidable Diophantine problems
− Decidable word problems: see D40
− Other decidability results: see subject of problem, e.g. D05, or D80; MR includes these results here.
− Undecidability results: see D35, D40, D80, etc.

B28 Classical foundations of number systems
+ Natural numbers, real numbers, ordinal numbers
+ Axiomatic foundations and set-theoretic foundations
MR uses B30

B30 Logical foundations of other classical theories; axiomatics
+ Axiomatic method
+ Geometry, probability, physics, etc.
+ Models for non-mathematical theories
− Foundations of parts of logic: see that part.
MR heading: "Foundations and axiomatics of classical theories" includes also B28

B35 Mechanization of proofs and logical operations
+ Theorem proving, proof checking by machine
+ Minimization algorithms for Boolean functions
+ Optimization of logical operations
MR sometimes uses 03-04 or 68G15 (68T15 since 1985)

B40 Combinatory logic and lambda-calculus
+ Models of lambda-calculus

B45 Modal and tense logic
+ Intensional logic; see also A05
+ Normative and deontic logic
+ Other non-truth-functional systems

B46 Relevance and entailment
+ Fragments
− Primarily modal logic
MR uses B45

B48 Probability and inductive logic
　　See also A05 and C90
　　+ Confirmation theory
　　− Foundations of probability: see B30; MR uses B48

B50 Many-valued logic
　　+ Matrix interpretations of propositional connectives unless used only as a tool for investigating classical propositional logic.
　　− Boolean valued set theory: see E40
　　− Probability logic: see B48 or C90

B51 Quantum logic
　　− Algebraic study of Quantum logic: see G12
　　MR uses only G12

B52 Fuzzy logic
　　+ Vagueness logic
　　− Papers demonstrating the fuzziness of the author's thought processes

B53 Paraconsistent logic
　　+ Discussive and dialectical logic
　　MR uses B60

B55 Intermediate and related logics
　　+ (Fragments of) propositional and predicate logics between intuitionistic or minimal and classical

B60 Other logics
　　− Intuitionistic logic: see F50 (MR uses B20)

B65 Logic of natural languages
　　− Computer languages: see B75
　　− Formal grammars unless applied to natural languages: see D05
　　− Natural language as a tool for the study of thought, reality, etc.: see A05
　　MR uses B65, B99, and 68Fn (68Sn since 1985)

B70 Logic in computer design; switching circuits
　　+ Hardware related to logic
　　MR uses 94Cn

B75 Logic of algorithmic and programming languages
　　+ Algorithmic and dynamic logic; MR uses B70 (formerly B45)
　　+ Logical analysis of programs
　　+ Logical aspects of database query languages and information retrieval
　　+ Semantics of programming languages related to logic
　　+ Software related to logic
　　− Specific algorithms: see subject of algorithm
　　MR uses B60, B70, 68Bn, 68Fn, and 68H05 (68Pn, 68Qn, and 68Tn since 1985)

B80 Other applications of logic
　　MR uses B99

B96 Collected works
　　+ Selected works
　　− Collections (almost) entirely in one subfield: see that subfield
　　MR uses 01A75, 03-03, and 03-06

B97 Proceedings
+ Collections of papers by various authors, even if they do not derive from any actual conference
- Proceedings (almost) entirely in one subfield: see that subfield
- Proceedings not concentrated in this field: see Source Index

MR uses 03-06

B98 Textbooks, surveys

MR uses 03-01 and 03-02

B99 None of the above or uncertain, but in this section

C   MODEL THEORY

C05 Equational classes, universal algebra
+ Quasi-varieties, if the emphasis is algebraic
- Word problems: see D40

C07 Basic properties of first-order languages and structures
+ Completeness, compactness, Löwenheim-Skolem, and omitting types theorems for ordinary first-order logic; MR uses C50 for omitting types
+ General properties of first-order theories
+ Homomorphisms, automorphisms, and isomorphisms of first-order structures
- Analogues of these for stronger languages: see C55, C70, C75, etc.

C10 Quantifier elimination and related topics

C13 Finite structures
+ The spectrum problem
+ Probabilities of sentences being true in finite structures

C15 Denumerable structures

C20 Ultraproducts and related constructions
+ Applications of ultraproducts
+ Reduced products, limit ultrapowers, etc.
- General products: see C30

C25 Model-theoretic forcing
+ Existentially closed structures, model companions, etc.
- Model complete theories: see C35
- Set-theoretic forcing: see E35, E40

C30 Other model constructions
+ Contructions involving indiscernibles
+ Products, diagrams

C35 Categoricity and completeness of theories
+ Model completeness
- Gödel's completeness theorem: see C07
- Completeness of axiomatizations of other logics: see those logics, e.g., B45

C40 Interpolation, preservation, definability
+ Definability in classes of structures
- Definability in recursion theory: see appropriate Dn.
- Definability in set theory: see E15, E45, and E47

C45 Stability and related concepts
+ Rank, total transcendence (even before stability was defined)

C50 Models with special properties
+ Saturated, rigid, etc.

C52 Properties of classes of models

C55 Set-theoretic model theory
+ Cardinality and ordering of models
+ Generalized Löwenheim-Skolem results
− Applications of set theory to some part of model theory: see that part
− Models of set theory: see C62
− Original Löwenheim-Skolem theorem: see C07

C57 Recursion-theoretic model theory
+ Model theory of recursive, arithmetical, etc. structures, types, etc.
− Recursion theory without substantial model-theoretic content: see D45
MR uses D45

C60 Model-theoretic algebra
+ Applications of model theory to specific algebraic theories
− Applications of set theory to algebra: see E75
− Decidability questions for algebraic theories: see B25, D35, and D40
− Model theory of orderings: see C65
− Universal algebra: see C05

C62 Models of arithmetic and set theory
+ Admissible sets as models: see also C70 and D60
+ Nonstandard models of arithmetic, when model theory is emphasized
+ Omega-models of higher-order arithmetic
− Models introduced only for consistency results: see F25 and E35
− Nonstandard models of arithmetic, when non-standardness is emphasized: see H15 or H20
MR uses C62, C65, F30, or H15

C65 Models of other mathematical theories
+ Other applications of model theory outside logic
+ Theories of orderings
− Uses of models for purely foundational studies: see B30

C70 Logic on admissible sets
+ All sorts of "effective" infinitary logic

C75 Other infinitary logic
+ Infinitary logic even if not model theory, e.g., infinite terms in proof theory and infinitary definability in set theory

C80 Logic with extra quantifiers and operators
− Hilbert epsilon-theorems: see B10
− Modal or many-valued operators: see B45 or B50

C85 Second- and higher-order model theory
+ Weak second-order theories (quantification over finite sets)

C90 Nonclassical models
+ Boolean-valued models
+ Sheaf models
+ Kripke models (also in B45 or F50)
+ Probability models (often also in B48)
+ Topological models (unless the topological structure is condensed into a quantifier: see C80); MR uses C85
− Models of lambda calculus: see B40

C95 Abstract model theory
+ Lindström's theorem, delta-logics, etc.

C96 Collected works
+ Selected works
− Collections (almost) entirely in one subfield: see that subfield
MR uses 01A75, 03-03, and 03-06

C97 Proceedings
+ Collections of papers by various authors, even if they do not derive from any actual conference
− Proceedings (almost) entirely in one subfield: see that subfield
− Proceedings not concentrated in this field: see Source Index
MR uses 03-06

C98 Textbooks, surveys
MR uses 03-01 and 03-02

C99 None of the above or uncertain, but in this section

D RECURSION THEORY

D03 Thue and Post systems, etc.
+ Markov's normal algorithms

D05 Automata and formal grammars in connection with logical questions
+ Cellular automata
+ Finite automata
+ Generalized automata
+ Regular events
− Grammar of natural languages: see B65
MR uses 68 for most of these topics

D10 Turing machines and related notions
+ Potentially infinite automata
+ Probabilistic Turing machines

D15 Complexity of computation
+ Chaitin-Kolmogorov-Solomonoff complexity
+ Finer classification of decidable problems
+ Generalized complexity
+ Resource-bounded computability and reducibility
+ Speed-up theorems
− Complexity of derivations and proofs: see F20
− Complexity of specific non-logical problems (excluded from the Ω-Bibliography)
− Syntactic complexity, complexity of Boolean functions, etc.
MR uses also 68Q15

D20 Recursive functions and relations, subrecursive hierarchies
+ Computable functions of real numbers; MR uses D65 and F60
+ General theory of algorithms
+ Partial recursive functions
+ Primitive recursion

D25 Recursively enumerable sets and degrees
+ Finer classification of undecidable r.e. problems
+ Many-one, truth table, etc., degrees of r.e. sets
+ Sets whose theory is closely related to that of r.e. sets, e.g., productive sets: see also D50
− Generalizations of recursive enumerability: see D60 and D65
− Partial functions with r.e. graphs: see D20

D30 Other degrees; reducibilities
+ Degrees in generalized recursion and constructibility: see also D55, D60, D65, and E45
+ Jump operators
− Subrecursive reducibilities: see D15 and D20

D35 Undecidability and degrees of sets of sentences
+ Hilbert's tenth problem and extensions
+ Reduction classes of the predicate calculus (also in B20)
− Decidability results: see B25
− Halting problems, word problems, etc.: see D03, D05, D10, D30, D40, or D80

D40 Word problems, etc.
+ Conjugacy, isomorphism, and other algorithmic problems in algebra
+ Decidability and undecidability
+ Other algorithmic questions in classical algebra
− Problems concerning production systems or formal grammar: see D03 and D05
− Recursive functions on words: see D20

D45 Theory of numerations, effectively presented structures
+ Numberings of (partial) recursive functions
+ Numerations in the sense of Ershov
+ Recursive algebra, except when it is about recursive equivalence types: see D50
+ Recursive order types
− Classical recursive analysis: see F60
− Model theory of recursive structures: see C57
− Recursive arithmetic: see F30

D50 Recursive equivalence types of sets and structures, isols
+ Concepts traditionally associated with isols, e.g., regressiveness and immuneness

D55 Hierarchies
+ Arithmetical, Borel, analytical, projective, etc. hierarchies
− Descriptive Set Theory in which hierarchical questions are not central: see E15
− Hierarchies of definability in set theory: see E47
− Incidental use of hierarchies outside recursion theory
− Subrecursive hierarchies: see D15 and D20

D60 Recursion theory on ordinals, admissible sets, etc.
+ Beta-recursion on inadmissible ordinals
− Classification of ordinary recursive functions using ordinals: see D20
− Ordinal notations: see D45 and F15
− Other aspects of admissibility: see C62, C70, or E45

D65 Higher-type and set recursion
+ Primitive recursive set functions
− Functionals in Proof Theory: see F10
− Recursion on the hereditarily finite sets: see D20
− Recursion with all arguments and parameters of type ≤ 1: see D20; MR includes this in D65 as long as there are type 1 arguments

D70 Inductive definability
+ Constructions equivalent to inductive definitions, e.g. set derivatives, game sentences, etc.
+ Recursion theory of inductive definitions and their duals
− Inductive definitions in proof theory: see F35 and F50
− Mechanics of inductive definitions: see B28, E20, or E30

D75 Abstract and axiomatic recursion theory
+ Algebras of (partial) recursive functions; MR uses D20
+ Recursion over general structures

D80 Applications
+ Decidability or undecidability results in areas outside logic and algebra
+ Effective versions of problems outside logic and algebra

D96 Collected works
+ Selected works
− Collections (almost) entirely in one subfield: see that subfield
MR uses 01A75, 03-03, and 03-06

D97 Proceedings
+ Collections of papers by various authors, even if they do not derive from any actual conference
− Proceedings (almost) entirely in one subfield: see that subfield
− Proceedings not concentrated in this field: see Source Index
MR uses 03-06

D98 Textbooks, surveys
MR uses 03-01 and 03-02

D99 None of the above or uncertain, but in this section

E   SET THEORY

E05 Combinatorial set theory
+ Partition relations, ideals, ultrafilters, trees named after people; MR uses also 04A20
− Finite combinatorics (excluded from the Ω-Bibliography); MR uses 05Xn

E07 Relations and orderings
+ Relation algebras: see also G15; MR uses G15
− Theories about ordering: see C65
MR uses E20, 04A05, 04A20, or 06An

E10 Ordinal and cardinal numbers
+ Cardinal algebras, ordinal algebras
+ Dedekind finite cardinals
− Cardinal exponentiation and the (generalized) continuum hypothesis: see E50; MR sometimes uses 04A10
− Combinatorial aspects of cardinals and ordinals: see E05
− Large cardinals: see E55

E15 Descriptive set theory
+ Definability properties of sets (in the real line or similar spaces)
+ Effective descriptive set theory
− General topology, measure theory, etc.: see E75
MR sometimes uses 04A15
See also D55

E20 Other classical set theory
+ Set algebra

E25 Axiom of choice and related propositions
+ Weak axioms of choice and their negations
MR sometimes uses 04A25

E30 Axiomatics of classical set theory and its fragments
+ Zermelo-Fraenkel set theory and minor variants
+ Gödel-Bernays set theory (also in E70)
− Morse-Kelley set theory (a second order theory: see E70)
− New Foundations, etc.: see E70

E35 Consistency and independence results
+ Forcing used to prove consistency

E40 Other aspects of forcing and Boolean-valued models
+ Forcing in generalized recursion theory: see also D60 and D65
− Model theoretic forcing: see C25

E45 Constructibility, ordinal definability and related notions
+ Other inner models, e.g. the core model

E47 Other notions of set-theoretic definability
+ Lévy hierarchy, indescribability
− Formalization of branches of mathematics within set theory

E50 Continuum hypothesis and Martin's axiom
+ Cardinal exponentiation
+ Variants of Martin's axiom
MR sometimes uses 04A30

E55 Large cardinals
+ Effective (denumerable) analogues of large cardinals
+ Weakly inaccessible and larger cardinals
− Axioms of infinity provable in ZFC
− Large proof-theoretic ordinals: see F15

E60 Determinacy and related principles which contradict the axiom of choice
+ Infinite exponent partition relations
+ Projective determinacy, definable determinacy
+ Other uses of infinite games in set theory and logic
− Applications of games outside set theory and logic
− Weak axioms that merely contradict choice

E65 Other hypotheses and axioms
+ Reflection principles
+ Combinatorial principles

E70 Nonclassical and second-order set theories
+ Leśniewski's Ontology and Mereology; MR uses B60
+ Nonstandard theories, e.g. New Foundations, Ackermann
+ Set theories formulated in non-classical logic
+ Theory of real classes (Morse-Kelley, and Gödel-Bernays set theory); MR uses E30

E72 Fuzzy sets

E75 Applications
+ Independence from set theory of mathematical propositions (also in E35)
+ Results in other branches of mathematics obtained by set theoretic methods
− Set-theoretical foundations of mathematics: see B28 and B30

E96 Collected works
+ Selected works
− Collections (almost) entirely in one subfield: see that subfield

MR uses 01A75, 03-03, and 03-06

E97 Proceedings
+ Collections of papers by various authors, even if they do not derive from any actual conference
− Proceedings (almost) entirely in one subfield: see that subfield
− Proceedings not concentrated in this field: see Source Index

MR uses 03-06

E98 Textbooks, surveys
MR uses 03-01 and 03-02

E99 None of the above or uncertain, but in this section

F PROOF THEORY AND CONSTRUCTIVE MATHEMATICS

F05 Cut elimination and normal form theorems
+ Hilbert's epsilon symbol
− Cut elimination and normal form theorems for modal systems: see B45

F07 Structure of proofs
− Proof schemas used rather than studied: see B10, C07, etc.

F10 Functionals in proof theory
− Typed lambda-calculus: see B40

F15 Recursive ordinals and ordinal notations
+ Ordinal notations even if not proof theory
+ Transfinite progressions of theories (Turing, Feferman; also in F30)

F20 Complexity of proofs
− Complexity of non-proof-theoretic procedures: see D15
− Purely qualitative (rather than quantitative) properties of proofs: see F07

F25 Relative consistency and interpretations
- Consistency of systems of arithmetic: see F30 and F35
- Set theoretic consistency results: see E35

F30 First-order arithmetic and fragments
+ Gödel incompleteness theorems
+ Metamathematics of intuitionistic arithmetic
+ Provability logic; MR uses also B45 and F40
+ Provably recursive functions; MR uses also D20
+ Recursive arithmetic
- Model theory of arithmetic: see C62 and H15

F35 Second- and higher-order arithmetic and fragments
+ Metamathematics of intuitionistic analysis
+ Proof theory of systems of type theory
+ Proof theory of generalized inductive definitions
- Model theory : see C62

F40 Gödel numberings in proof theory
+ Any use of Gödel numbering of syntax
- Gödel numberings in recursion theory: see D20 and D45

F50 Metamathematics of constructive systems
+ Intuitionistic logic and subsystems; MR uses also B20
+ Model theoretic methods applied to constructive systems
+ Rèalizability
- Metamathematics of predicative systems: see F65

F55 Constructive and intuitionistic mathematics
+ Bishop school of constructivism
- Metamathematics: see F50

F60 Constructive recursive analysis
+ Classical recursive analysis
+ Soviet school of constructivism
- Metamathematics: see F50

F65 Other constructive mathematics
+ Constructive trends not covered by F55 or F60
+ Predicative mathematics
+ Metamathematics of predicative systems
- Other metamathematics: see F50

F96 Collected works
+ Selected works
- Collections (almost) entirely in one subfield: see that subfield
MR uses 01A75, 03-03, and 03-06

F97 Proceedings
+ Collections of papers by various authors, even if they do not derive from any actual conference
- Proceedings (almost) entirely in one subfield: see that subfield
- Proceedings not concentrated in this field: see Source Index
MR uses 03-06

F98 Textbooks, surveys
MR uses 03-01 and 03-02

F99 None of the above or uncertain, but in this section

G ALGEBRAIC LOGIC

G05 Boolean algebras
+ Boolean rings, etc.
- Boolean functions : see B05; MR puts Boolean functions in G05, 06E30, and sometimes 94C10
- Pseudo-Boolean algebras : see G10

G10 Lattices and related structures
+ Heyting algebras; MR uses also 06D20
+ Semilattices, continuous lattices; MR uses 06B35
- Studies of "The lattice of..." where the lattice structure is not the main point

G12 Quantum logic
See also B51

G15 Cylindric and polyadic algebras, relation algebras

G20 Łukasiewicz and Post algebras
+ Lattices (or weaker structures) corresponding to many-valued logic

G25 Other algebras related to logic
+ Boolean algebras with provability and other operators
+ Implicative algebras, BCK algebras, etc.

G30 Categorical logic, topoi
+ Almost any connection between categories and logic, e.g. categories of models, logical foundations of category theory
- Pure category theory (Excluded from the Ω-Bibliography); MR uses 18Xn

G96 Collected works
+ Selected works
- Collections (almost) entirely in one subfield: see that subfield
MR uses 01A75, 03-03, and 03-06

G97 Proceedings
+ Collections of papers by various authors, even if they do not derive from any actual conference
- Proceedings (almost) entirely in one subfield: see that subfield
- Proceedings not concentrated in this field: see Source Index
MR uses 03-06

G98 Textbooks, surveys
MR uses 03-01 and 03-02

G99 None of the above or uncertain, but in this section

H NONSTANDARD MODELS

H05 Infinitesimal analysis in pure mathematics

H10 Other applications of infinitesimal analysis
+ Economics, physics, etc.

H15 Nonstandard models of arithmetic
+ Work emphasizing nonstandard methods
- Work emphasizing model theory : see C62

H20 Other nonstandard models

H96 Collected works
+ Selected works
− Collections (almost) entirely in one subfield: see that subfield
MR uses 01A75, 03-03, and 03-06

H97 Proceedings
+ Collections of papers by various authors, even if they do not derive from any actual conference
− Proceedings (almost) entirely in one subfield: see that subfield
− Proceedings not concentrated in this field: see Source Index
MR uses 03-06

H98 Textbooks, surveys
MR uses 03-01 and 03-02

H99 None of the above or uncertain, but in this section

Subject Index

B03 Syntax of logical languages

1879
FREGE, F.L.G. *Begriffsschrift, eine der arithmetischen nachgebildete Formelsprache des reinen Denkens* ⋄ A05 B03 B05 B10 ⋄

1889
PEANO, G. *The principles of arithmetic, presented by a new method (Latin)* ⋄ B03 B28 E75 F30 ⋄

1894
PEANO, G. *Notations de la logique mathematique. Introduction au formulaire de mathematique* ⋄ A10 B03 ⋄

1896
FREGE, F.L.G. *Ueber die Begriffsschrift des Herrn Peano und meine eigene* ⋄ A05 A10 B03 ⋄

1900
PEANO, G. *Formules de logique mathematique* ⋄ B03 B10 ⋄

1931
MENGER, K. *Eine elementare Bemerkung ueber die Struktur logischer Formeln* ⋄ B03 ⋄

1936
QUINE, W.V.O. *Definition of substitution* ⋄ B03 ⋄

1937
CURRY, H.B. *On the use of dots as brackets in logical expressions* ⋄ B03 ⋄
HETPER, W. *Grundlagen der Semantik (Polish)* ⋄ B03 ⋄

1938
HERMES, H. *Semiotik. Eine Theorie der Zeichengestalten als Grundlage fuer Untersuchungen von formalisierten Sprachen* ⋄ B03 ⋄

1939
DUCASSE, C.J. *Symbols, signs and signals* ⋄ B03 ⋄

1941
SCHROETER, K. *Ein allgemeiner Kalkuelbegriff* ⋄ B03 ⋄

1942
THOMPSON, S.M. *Syllogistic logic in linear notation* ⋄ A10 B03 ⋄
TURING, A.M. *The use of dots as brackets in Church's system* ⋄ B03 ⋄

1949
DANTZIG VAN, D. *Signifies and its relation to semiotics* ⋄ B03 ⋄
FEYS, R. *A simple notation for relations* ⋄ B03 E07 ⋄

1950
KALICKI, J. *On the structure of bracket-free formulae* ⋄ B03 ⋄

1952
CURRY, H.B. *On the definition of substitution, replacement and allied notions in an abstract formal system* ⋄ B03 ⋄

1953
BAR-HILLEL, Y. *A quasi-arithmetical notation for syntactic description* ⋄ B03 ⋄
LOEB, M.H. *Concatenation as basis for a complete system arithmetic* ⋄ B03 ⋄

1954
BAR-HILLEL, Y. *Can translation be mechanized?* ⋄ B03 B65 ⋄
BURKS, A.W. & WARREN, D.W. & WRIGHT, J.B. *An analysis of a logical machine using parenthesis-free notation* ⋄ B03 D05 ⋄
JOHANSSON, I. *Symboles logiques dans l'enseignement des theories deductives* ⋄ B03 ⋄

1956
HAERTIG, K. *Explizite Definitionen einiger Eigenschaften von Zeichenreihen* ⋄ B03 ⋄

1957
CHOMSKY, N. *Syntactic structures* ⋄ B03 ⋄
HAERTIG, K. *Ein Spezialfall der Substitution als Grundbeziehung der elementaren Semiotik* ⋄ B03 ⋄
KOTARBINSKA, J. *The concept of sign (Polish) (Russian and English summaries)* ⋄ A05 B03 ⋄
REVZIN, I.I. *Some questions of the formalization of syntax I,II (Russian)* ⋄ B03 ⋄

1958
BACKUS, J.W. *The syntax and semantics of the proposed international algebraic language of the Zurich ACM-GAMM conference* ⋄ B03 ⋄

1960
PATTERSON, G.W. *What is a code?* ⋄ B03 ⋄
PAWLAK, Z. *New method of parenthesis-free notations of formulae* ⋄ B03 ⋄

1962
HAMBLIN, C.L. *Translation to and from Polish notation* ⋄ B03 ⋄

1964
MAKANIN, G.S. *A new solvable case of the decision problem for first-order predicate calculus (Russian)* ⋄ B03 B25 ⋄

1966

BLIKLE, A.J. *Formalisation of parenthesis-free languages* ◊ B03 ◊

MORGAN DE, A. *Some suggestions in logical phraseology* ◊ B03 ◊

1967

EHRENFEUCHT, A. & PAWLAK, Z. *Some remarks on the bracket free notation* ◊ B03 ◊

ONO, H. *Normalizing procedure of sequences over ordinary logical constants* ◊ B03 ◊

WEISS, M. *Axiomatische Untersuchungen zur elementaren Theorie der freien Halbgruppen mit Substitution als undefiniertem Grundbegriff* ◊ B03 D05 ◊

1968

CARNES, R.D. & WILCOX, W.C. *An infixed, punctuation-free notation* ◊ B03 G05 ◊

OHAMA, S. *On a formalism which makes any sequence of symbols well-formed* ◊ B03 ◊

SCHREIBER, P. *Lexikographische Ordnung als Grundbegriff der Semiotik* ◊ B03 ◊

1970

GEACH, P.T. *A program for syntax* ◊ B03 ◊

MAYOH, B.H. *The relation between an object and its name: notation systems and their fixed point theorems* ◊ A05 B03 D20 D35 D45 F30 F60 ◊

1972

BERG, J. *Bolzano's theory of an ideal language* ◊ A10 B03 ◊

BRUSENTSOV, N.P. *An improved parenthesis-free notation for formulas (Russian)* ◊ B03 ◊

RICKEY, V.F. *Axiomatic inscriptional syntax. I. general syntax* ◊ B03 ◊

THIEL, C. *Auf dem Weg zur Vereinheitlichung der logischen Symbolik* ◊ B03 ◊

1973

APRILE, G. *Sulla sintassi della notazione polacca* ◊ B03 ◊

RICKEY, V.F. *Axiomatic inscriptional syntax; part II: The syntax of protothetic* ◊ B03 ◊

SOTIROV, V. *Is only one bracket sufficient? (Bulgarian)* ◊ B03 ◊

1974

BEHRMANN, E. & CRAEMER, D. *An algorithm for well-formed formulas in infix-notation* ◊ B03 ◊

CORCORAN, J. & FRANK, W. & MALONEY, M.J. *String theory* ◊ A10 B03 ◊

GANDY, R.O. *Set-theoretic functions for elementary syntax* ◊ B03 D20 D65 E47 ◊

SOBEL, J.H. *Principia mathematica description theory: the classical and an alternative notation* ◊ B03 ◊

TAMARI, D. *Formulae for well formed formulae and their enumeration* ◊ B03 ◊

1975

CRESSWELL, M.J. *Note on the use of sequences in "Logics and languages"* ◊ B03 E75 ◊

FAL'K, V.N. & MOROKHOVEK, YU.E. *A generalized non-parenthesis-free language (Russian)* ◊ B03 ◊

STANFORD, P.H. *Polish circles* ◊ B03 ◊

1976

EVENDEN, J. *A Begriffsschrift for sentential logic* ◊ B03 ◊

SOBEL, J.H. *Alternative notations for principia mathematica description theory: possible modifications* ◊ B03 ◊

1977

BERG, J. *Bolzano's contribution to logic and philosophy of mathematics* ◊ A05 A10 B03 ◊

MAKANIN, G.S. *The problem of the solvability of equations in a free semigroup (Russian)* ◊ B03 B25 D40 ◊

MAKANIN, G.S. *The problem of solvability of equations in a free semigroup (Russian)* ◊ B03 B25 D40 ◊

1978

BERGMAN, GEORGE M. *Terms and cyclic permutations* ◊ B03 D40 ◊

CALUDE, C. *On the category of recursive languages* ◊ B03 D45 F40 G30 ◊

GREENSTEIN, C.H. *Dictionary of logical terms and symbols* ◊ B03 ◊

KRIZEK, P. *On rewriting sentences into formulas* ◊ A05 B03 ◊

POTTS, T.C. *Fregean grammar: A formal outline* ◊ A10 B03 ◊

VITYAEV, E.E. *Regularities in the language of empirical systems (Russian)* ◊ B03 ◊

1979

FARAT, V.M. & KOZHEVNIKOVA, G.P. *Algorithms for translation of algebraic expressions into an improved parenthesis-free notation and an analysis of their effectiveness(Russian)* ◊ B03 D15 ◊

FOMINA, N.I. & KALINICHENKO, L.A. *A method of syntactic analysis of constructions of a predicate logic language in data base control systems (Russian)* ◊ B03 ◊

SOBEL, J.H. *Sentential notations: Unique decomposition* ◊ B03 ◊

1980

MAKANIN, G.S. *Equations in a free semigroup (Russian)* ◊ B03 B25 D40 ◊

SHABANOV-KUSHNARENKO, YU.P. *Analytic methods for implicit determination of finite alphabetical operators (Russian)* ◊ B03 ◊

1981

CALUDE, C. & PAUN, G. *Global syntax and semantics for recursively enumerable languages* ◊ B03 D25 ◊

1982

PKHAKADZE, SH.S. *Some problems of the theory of abbreviating symbols (Russian) (English and Georgian summaries)* ◊ B03 ◊

1983

SATO, M. *Theory of symbolic expressions I* ◊ B03 ◊

B05 Classical propositional logic and boolean functions

1879
CLIFFORD, W.K. *Lectures and essays. 2 Vols* ◊ B05 ◊
FREGE, F.L.G. *Begriffsschrift, eine der arithmetischen nachgebildete Formelsprache des reinen Denkens* ◊ A05 B03 B05 B10 ◊

1904
HUNTINGTON, E.V. *Sets of independent postulates for the algebra of logic* ◊ B05 B30 G05 ◊

1906
MACCOLL, H. *Symbolic logic and its applications* ◊ B05 B98 ◊

1910
WHITEHEAD, A.N. & RUSSELL, B. *Principia mathematica. Vol.I* ◊ B05 B10 B15 B28 E30 E98 ◊

1913
SHEFFER, H.M. *A set of five independent postulates for boolean algebras, with application to logical constants* ◊ B05 G05 ◊
VASIL'EV, N.A. *Logic and metalogic (Russian)* ◊ A05 B05 ◊

1916
HORN VAN, C.E. *An axiom in symbolic logic* ◊ B05 ◊
NICOD, J. *A reduction in the number of primitive propositions of logic* ◊ B05 ◊

1920
TAYLOR, J.S. *Sheffer's set of five postulates for boolean algebras in terms of the operation "rejection" made completely independent* ◊ B05 G05 ◊

1921
LUKASIEWICZ, J. *Two-valued logic (Polish)* ◊ A05 B05 B50 ◊
POST, E.L. *Introduction to a general theory of elementary propositions* ◊ B05 B20 B50 ◊

1923
TAJTELBAUM, A. *Sur le terme primitif de la logistique* ◊ A05 B05 ◊

1924
BERNSTEIN, B.A. *Complete sets of representations of two-elements algebras* ◊ B05 ◊
TAJTELBAUM, A. *Sur les truth-functions au sens de MM. Russell et Whitehead* ◊ A05 B05 ◊

1925
ZYLINSKI, E. *Some remarks concerning the theory of deduction* ◊ B05 ◊

1926
BERNAYS, P. *Axiomatische Untersuchungen des Aussagenkalkuels der "Principia Mathematica"* ◊ A05 B05 ◊
BERNSTEIN, B.A. *Sets of postulates for the logic of propositions* ◊ B05 ◊

1927
BELL, E.T. *Arithmetic of logic* ◊ B05 G05 ◊
BERNSTEIN, B.A. *The dual of a logical expression* ◊ B05 ◊
ZHEGALKIN, I.I. *On the technique of proposition calculus in symbolic logic (Russian) (French summary)* ◊ B05 ◊

1928
CHURCH, A. *On the law of excluded middle* ◊ A05 B05 F50 ◊
ZHEGALKIN, I.I. *Arithmetization of symbolic logic (Russian) (French summary)* ◊ B05 ◊

1929
BERNSTEIN, B.A. *Irredundant sets of postulates for the logic of propositions* ◊ B05 ◊
ZHEGALKIN, I.I. *Arithmetization of symbolic logic II (Russian) (French summary)* ◊ B05 ◊

1930
HAERLEN, H. *Ueber Axiomensysteme als Satzfunktionen* ◊ B05 ◊
LUKASIEWICZ, J. & TARSKI, A. *Untersuchungen ueber den Aussagenkalkuel* ◊ A05 B05 ◊
TARSKI, A. *Untersuchungen ueber den Aussagenkalkuel* ◊ A05 B05 ◊

1931
BERNSTEIN, B.A. *Application of boolean algebras to proving consistency and independence of postulates* ◊ B05 G05 ◊
BERNSTEIN, B.A. *Whitehead and Russell's theory of deduction as a mathematical science* ◊ A05 B05 ◊
GOEDEL, K. *Eine Eigenschaft der Realisierung des Aussagenkalkuels* ◊ B05 ◊
GOEDEL, K. *Ueber Unabhaengigkeitsbeweise im Aussagenkalkuel* ◊ B05 ◊
SKOLEM, T.A. *Ueber die symmetrisch allgemeinen Loesungen im identischen Kalkuel* ◊ B05 ◊

1932
BERNSTEIN, B.A. *Note on the condition that a boolean equation has a unique solution* ◊ B05 ◊
BERNSTEIN, B.A. *On proposition *4.78 of "Principia Mathematica"* ◊ B05 E20 ◊
BERNSTEIN, B.A. *On unit-zero boolean representations of operations and relations* ◊ B05 G05 ◊

BERNSTEIN, B.A. *On Nicod's reduction in the number of primitives of logic* ⋄ A05 B05 ⋄
LUKASIEWICZ, J. *Ein Vollstaendigkeitsbeweis des zweiwertigen Aussagenkalkuels* ⋄ A05 B05 ⋄
QUINE, W.V.O. *A note on Nicod's postulate* ⋄ A10 B05 ⋄
WAJSBERG, M. *Ein neues Axiom des Aussagenkalkuels in der Symbolik von Sheffer* ⋄ B05 ⋄

1933
BERNSTEIN, B.A. *On section A of "Principia Mathematica"* ⋄ A05 A10 B05 ⋄
BERNSTEIN, B.A. *Remarks on propositions *1.1 and *3.35 of "Principia Mathematica"* ⋄ A05 B05 ⋄
PICH, W. *Ueber Unabhaengigkeitsbeweise im Aussagenkalkuel* ⋄ B05 ⋄
SCHUETTE, K. *Ueber einen Teilbereich des Aussagenkalkuels* ⋄ B05 ⋄

1934
KALMAR, L. *Ueber die Axiomatisierbarkeit des Aussagenkalkuels* ⋄ B05 ⋄
QUINE, W.V.O. *Ontological remarks on the propositional calculus* ⋄ A05 B05 ⋄

1935
BAYLIS, C.A. & BENNETT, A.A. *A calculus for propositional concepts* ⋄ B05 ⋄
FLOYD, W.F. & WOODGER, J.-H. *A simple method of testing truth-functions* ⋄ B05 ⋄
NOTCUTT, B. *A set of independent postulates for propositional functions of one variable* ⋄ A05 B05 ⋄
TARSKI, A. *Ueber die Erweiterungen der unvollstaendigen Systeme des Aussagenkalkuels* ⋄ B05 B22 ⋄
ZERMELO, E. *Grundlagen einer allgemeinen Theorie der mathematischen Satzsysteme (erste Mitteilung)* ⋄ B05 C75 E20 ⋄

1936
HERMES, H. & SCHOLZ, H. *Ein neuer Vollstaendigkeitsbeweis fuer das reduzierte Fregesche Axiomensystem des Aussagenkalkuels* ⋄ B05 ⋄
MCKINSEY, J.C.C. *On boolean functions of many variables* ⋄ B05 G05 ⋄
MCKINSEY, J.C.C. *On the generation of the functions Cpq and Np of Lukasiewicz and Tarski by means of a single binary operation* ⋄ B05 ⋄
MCKINSEY, J.C.C. *On the independence of Hilbert and Ackermann's postulates for the calculus of propositional functions* ⋄ B05 ⋄
MCKINSEY, J.C.C. *Reducible boolean functions* ⋄ B05 ⋄
PICH, W. *Ueber Unabhaengigkeitsbeweise im Aussagenkalkuel* ⋄ B05 ⋄
SMITH, HENRY BRADFORD *The algebra of propositions* ⋄ B05 B45 ⋄

1937
HETPER, W. *Problem of completeness of the system of elementary semantics (Polish)* ⋄ B05 E70 ⋄
KOBRZYNSKI, Z. *La theorie des determinants logiques* ⋄ A05 B05 ⋄
MCKINSEY, J.C.C. *A condition that a first boolean function vanish wherever a second does not* ⋄ B05 G05 ⋄

1938
DUTHIE, W. *Boolean functions of bounded variation* ⋄ B05 ⋄
HETPER, W. *Le calcul des proposition etabli sans axiomes (Polish)* ⋄ B05 ⋄
MIHAILESCU, E.G. *Recherches sur les formes normales* ⋄ B05 ⋄
MIHAILESCU, E.G. *Recherches sur l'equivalence et la reciprocite dans le calcul des propositions* ⋄ B05 ⋄
MIHAILESCU, E.G. *Sur le calcul des propositions* ⋄ B05 ⋄
QUINE, W.V.O. *Completeness of the propositional calculus* ⋄ B05 ⋄
TARSKI, A. *Der Aussagenkalkuel und die Topologie* ⋄ B05 F50 G05 G10 ⋄

1939
HUNTINGTON, E.V. *Note on a recent set of postulates for the calculus of propositions* ⋄ B05 ⋄
MCGILL, V.J. *Concerning the laws of contradiction and excluded middle* ⋄ B05 ⋄
NOVIKOV, P.S. *On some existence theorems (Russian)* ⋄ B05 C75 ⋄
SOBOCINSKI, B. *Axiomatisierung des "konjunktiv-negativen" Systems des Aussagenkalkuels (Polish)* ⋄ B05 ⋄
USHENKO, A.P. *The calculus of propositions and self-contradiction* ⋄ B05 ⋄
WERNICK, W. *An enumeration of logical functions* ⋄ B05 ⋄

1940
BOCHVAR, D.A. *Ueber einen Aussagenkalkuel mit abzaehlbaren logischen Summen und Produkten (Russian summary)* ⋄ B05 C75 ⋄
MUELLER, H.R. *Algebraischer Aussagenkalkuel* ⋄ B05 ⋄
POLYA, G. *Sur les types des propositions composees* ⋄ B05 ⋄
WERNICK, W. *Functional dependence in the calculus of propositions* ⋄ B05 ⋄

1941
PANKAJAM, S. *On the formal structure of the propositional calculus I* ⋄ B05 ⋄
POST, E.L. *The two-valued iterative systems of mathematical logic* ⋄ B05 B20 ⋄

1942
BRONSTEIN, D.J. *A correction to the sentential calculus of Tarski's introduction to logic* ⋄ B05 ⋄
PANKAJAM, S. *On the formal structure of the propositional calculus II* ⋄ B05 ⋄
WERNICK, W. *Complete sets of logical functions* ⋄ B05 ⋄

1943
DEXTER, G.E. *The calculus of non-contradiction* ⋄ A05 B05 ⋄
FUENTES MIRAS, J.R. *Das Entscheidungsproblem im Aussagenkalkuel (Spanisch)* ⋄ B05 ⋄
HOFF-HANSEN, E. *A mathematical Interpretation of the classical propositional calculus (Norwegian)* ⋄ B05 ⋄

NOVIKOV, P.S. *On the consistency of certain logical calculus (Russian summary)* ◊ B05 B28 C75 ◊
SCHROETER, K. *Axiomatisierung der Fregeschen Aussagenkalkuele* ◊ B05 ◊
SKOLEM, T.A. *Some remarks on the preceding article of E.Hoff-Hansen (Norwegian)* ◊ B05 C05 G25 ◊

1945
DESTOUCHES-FEVRIER, P. *Rapport entre le calcul des problemes et le calcul des propositions* ◊ A05 B05 F50 ◊

1946
GOETLIND, E. *On some equivalence propositions in two valued logic (Norwegian)* ◊ B05 ◊

1947
GOETLIND, E. *An axiom system for the calculus of propositions (Swedish)* ◊ B05 ◊

1948
CHURCH, A. *Conditioned disjunction as a primitive connective for the propositional calculus* ◊ B05 ◊
JASKOWSKI, S. *Sur les variables propositionelles dependantes* ◊ B05 ◊
JASKOWSKI, S. *Trois contributions au calcul des propositions bivalent* ◊ B05 ◊
RIDDER, J. *Logic of propositions* ◊ B05 ◊
WENDELIN, H. *Ein Kriterium fuer die Erweiterbarkeit einer Implikation zu einer Aequivalenz* ◊ B05 ◊
WRIGHT VON, G.H. *On the idea of logical truth I* ◊ A05 B05 B20 B25 ◊

1949
BAUER, F.L. *Zur Algebraik des Logikkalkuels* ◊ A05 B05 ◊
LEVIN, N.P. *Computational logic* ◊ B05 G05 ◊
RASIOWA, H. *Sur un certain systeme d'axiomes du calcul des propositions* ◊ B05 ◊
ROSE, A. *A reduction in the number of the axioms of the propositional calculus* ◊ B05 ◊

1950
GRENIEWSKI, H. *Certain notions of the theory of numbers as applied to the propositional calculus* ◊ B05 ◊
GRENIEWSKI, H. *Functors of the propositional calculus* ◊ B05 ◊
GRENIEWSKI, H. *Groups and fields definable in the propositional calculus* ◊ B05 C60 ◊
KALICKI, J. *Note on truth tables* ◊ B05 ◊
ROSE, A. *Completeness of Lukasiewicz-Tarski propositional calculi* ◊ B05 ◊
SAMPEI, Y. *On the orthogonal expansion of the boolean polynomial and its applications I* ◊ B05 ◊
SHUKLA, R. & SINGH, R.K.P. *A note on Goetlind's axiom system for the calculus of propositions* ◊ B05 ◊
WANG, HAO *A proof of independence* ◊ B05 ◊
WRIGHT VON, G.H. *On the idea of logical truth II* ◊ A05 B05 B20 B25 ◊

1951
ALVES, M.T. *A theorem of metamathematics* ◊ B05 ◊
ANGSTL, H. *Ueber Gleichungen im Aussagenkalkuel* ◊ B05 ◊
GRENIEWSKI, H. *Arithmetics of natural numbers as part of the bi-valued propositional calculus* ◊ B05 F30 ◊
LOS, J. *An algebraic proof of completeness for the two-valued propositional calculus* ◊ B05 G05 ◊
LYNDON, R.C. *Identities in two-valued calculi* ◊ B05 B20 C05 ◊
ROSE, A. *A formalization of the c-o-propositional calculus* ◊ B05 ◊
ROYCE, J. *An extension of the algebra of logic* ◊ B05 ◊
ROYSE, JOSIAH *The relation of the principles of logic to the foundations of geometry* ◊ B05 ◊

1952
ASHENHURST, R.L. *The application of counting techniques* ◊ B05 ◊
BERKELEY, E.C. *A summary of symbolic logic and its practical applications* ◊ B05 ◊
KALICKI, J. *A test for the equality of truth-tables* ◊ B05 ◊
PIAGET, J. *Essai sur les transformations des operations logiques. Les 256 operations ternaires de la logique bivalente des propositions* ◊ B05 ◊
QUINE, W.V.O. *The problem of simplifying truth functions* ◊ B05 ◊
SKOLEM, T.A. *On the proofs of independence of the axioms of the classical sentential calculus* ◊ B05 ◊
VEITCH, E.W. *A chart method for simplifying truth functions* ◊ B05 ◊
YABLONSKIJ, S.V. *On complete systems of functions of the algebra of logic (Russian)* ◊ B05 ◊
YABLONSKIJ, S.V. *On the superpositions of functions of the algebra of logic (Russian)* ◊ B05 ◊

1953
ELLIS, D. *Remarks on boolean functions* ◊ B05 ◊
HINTIKKA, K.J.J. *A new approach to sentential logic* ◊ B05 ◊
LUKASIEWICZ, J. *Comment on K.J.Cohen's remark* ◊ B05 F50 ◊
MEREDITH, C.A. *Single axioms for the system (C,N), (C,O) and (A,N) of the two-valued propositional calculus* ◊ B05 ◊
QUINE, W.V.O. *Two theorems about truth-functions* ◊ B05 B45 ◊
REICHBACH, J. *Ueber den auf Alternative und Negation aufgebauten Aussagenkalkuel (Polish and Russian summaries)* ◊ B05 ◊
ROSE, A. *Conditioned disjunction as a primitive connective for the erweiterter Aussagenkalkuel* ◊ B05 ◊
SAMPEI, Y. *On the orthogonal expansion of the boolean polynomial and its application II* ◊ B05 ◊
SLEPLAN, D. *On the number of symmetry types of boolean functions of n variables* ◊ B05 ◊
SLUPECKI, J. *Ueber die Regeln des Aussagenkalkuels (Polish and Russian summaries)* ◊ B05 ◊
WANG, SHIQIANG *An axiom system for the proposition calculus (Chinese)* ◊ B05 ◊

1954
CURRY, H.B. *Generalization of the deduction theorem* ◊ B05 ◊
GUITEL, G. *Sur une representation symbolique du processus logique d'une demonstration* ◊ B05 ◊

KALICKI, J. *An undecidable problem in the algebra of truth-tables* ⋄ B05 D35 ⋄
LORENZEN, P. *Zur Begruendung der zweiwertigen Aussagenlogik* ⋄ A05 B05 ⋄
PAP, A. *Propositions, sentences, and the semantic definition of truth* ⋄ A05 B05 B10 ⋄
PARRY, W.T. *A new symbolism for the propositional calculus* ⋄ B05 ⋄
POVAROV, G.N. *On functional separability of Boolean functions (Russian)* ⋄ B05 ⋄
ROSE, A. *A formalisation of the 2-valued propositional calculus with self-dual primitives* ⋄ B05 ⋄
STANDLEY, G.B. *Ideographic computation in the propositional calculus* ⋄ B05 ⋄
YABLONSKIJ, S.V. *Realization of a linear function in the class of Π-schemes (Russian)* ⋄ B05 ⋄

1955

BERNSTEIN, B.A. & PARKER, W.L. *On uniquely solvable boolean equations* ⋄ B05 G05 ⋄
BROUWER, L.E.J. *The effect of intuitionism on classical algebra of logic* ⋄ B05 F50 ⋄
BURKS, A.W. *Propositional statements* ⋄ B05 ⋄
GRIZE, J.-B. *L'implication et la negation vues au travers des methodes de Gentzen et de Fitch* ⋄ B05 ⋄
HARARY, F. *Note on an enumeration theorem of Davis and Slepian* ⋄ B05 ⋄
HENKIN, L. *Boolean representation through propositional calculus* ⋄ B05 ⋄
IZUMI, Y. & WADA, T. *Sur la notion de la perfection* ⋄ A05 B05 ⋄
KANGER, S. *A note on partial postulate sets for propositional logic* ⋄ B05 ⋄
MO, SHAOKUI *Some axiom systems for propositional calculus (Chinese) (English summary)* ⋄ B05 ⋄
MOISIL, G.C. *Sur le fonctionnement des schemas a boutons reels (Romanian) (Russian and French summaries)* ⋄ B05 B70 ⋄
NELSON, RAYMOND J. *Simplest normal truth functions* ⋄ B05 ⋄
NELSON, RAYMOND J. *Weak simplest normal truth functions* ⋄ B05 ⋄
NINOMIYA, I. *On the number of types of symmetric boolean output matrices* ⋄ B05 B70 ⋄
POSTLEY, J.A. *A method for the evaluation of a system of boolean algebraic equations* ⋄ B05 ⋄
POVAROV, G.N. *To the study of symmetric Boolean functions from the point of view of the theory of relay-contact circuits (Russian)* ⋄ B05 B70 ⋄
QUINE, W.V.O. *A way to simplify truth functions* ⋄ B05 ⋄
SOBOCINSKI, B. *Note on a problem of Paul Bernays* ⋄ B05 ⋄

1956

ASSER, G. *Einfuehrung in die mathematische Logik. Teil I: Aussagenkalkuel* ⋄ B05 ⋄
BERGMANN, G. *Propositional functions* ⋄ A05 B05 ⋄
BING, K. *On simplifying truth-functional formulas* ⋄ B05 ⋄
ELLIS, D. *Remarks on boolean functions II* ⋄ B05 ⋄
ITO, MAKOTO *General solution of boolean equation with m-variables* ⋄ B05 G05 ⋄
KLIMOVSKY, G. *Tres enunciados equivalentes al teorema de Zorn* ⋄ B05 E25 G05 ⋄
KOROBKOV, V.K. *Realization of symmetric functions in the class of π-circuits (Russian)* ⋄ B05 ⋄
MIHAILESCU, E.G. *Formes normales dans le calcul des propositions bivalentes (Romanian) (Russian and French summaries)* ⋄ B05 ⋄
MIHAILESCU, E.G. *Formes normales dans l'ensemble S(C) (Romanian) (Russian and French summaries)* ⋄ B05 ⋄
MIHAILESCU, E.G. *Formes normales dans l'ensemble S(D) du calcul bivalent des propositions (Romanian) (Russian and French summaries)* ⋄ B05 ⋄
MUELLER, R.K. *A topological method for the determination of the minimal forms of a boolean function* ⋄ B05 ⋄
STAHL, G. *La suficiencia de la logica bivalente para la fisca de los cuantos* ⋄ B05 G12 ⋄
WANG, XIANJUN *The truth functions in mathematical logic is the logical abstraction of composite propositions (Chinese)* ⋄ B05 ⋄

1957

ARANGO, H. & SANTOS, J. *La operacion puente en las algebras de Boole* ⋄ B05 G05 ⋄
BLANCHE, R. *Sur la structuration du tableau de connectifs interpropositionnels binaires* ⋄ B05 ⋄
BORKOWSKI, L. *Recent investigations on the propositional calculus (Polish)* ⋄ B05 ⋄
CHISHOLM, R.M. & SYMONDS, B.K. *Inference by complementary elimination* ⋄ B05 ⋄
HIZ, H. *Inferential equivalence and natural deduction* ⋄ A05 B05 ⋄
LOS, J. *Remarks on Henkin's paper: boolean representation through propositional calculus* ⋄ B05 ⋄
MARKOV, A.A. *On the inversion complexity of a system of functions (Russian)* ⋄ B05 ⋄
MCNAUGHTON, R. *On the measure of normal formulas* ⋄ B05 B70 ⋄
YABLONSKIJ, S.V. *On classes of functions of the algebra of logic admitting a simple schematic realization (Russian)* ⋄ B05 B70 ⋄

1958

ABHYANKAR, S. *Minimal "sum of products of sums" expressions of boolean functions* ⋄ B05 ⋄
BEHMANN, H. *Ein logischer Abakus* ⋄ A05 B05 ⋄
BETH, E.W. *On the completeness of the classical sentential logic* ⋄ B05 ⋄
FREUDENTHAL, H. *Logique mathematique appliquee* ⋄ A05 B05 B45 B70 ⋄
GOODSTEIN, R.L. *Models of propositional calculi in recursive arithmetic* ⋄ B05 B50 F30 F50 ⋄
HIRSCHHORN, E. *Simplification of a class of boolean functions* ⋄ B05 ⋄
KELLNER, W.G. & MULLIN, A.A. *A residue test for boolean functions* ⋄ B05 ⋄
KURMIT, A.A. *Independence of a system of axioms for the propositional calculus (Russian) (Latvian summary)* ⋄ B05 ⋄

LEJEWSKI, C. *On implicational definitions (Polish and Russian summaries)* ◇ B05 B20 ◇

MLEZIVA, M. *Theory of propositions (Czech)* ◇ B05 ◇

PORTE, J. *Deux systemes simples pour le calcul des propositions* ◇ B05 ◇

PORTE, J. *Schemas pour le calcul des propositions fonde sur la conjonction et la negation* ◇ B05 ◇

PRIOR, A.N. *Pierce's axioms for propositional calculus* ◇ B05 ◇

SCHUETTE, K. *Aussagenlogische Grundeigenschaften formaler Systeme* ◇ B05 F05 F07 ◇

STAHL, G. *An opposite and an expanded System* ◇ B05 ◇

STAHL, G. *Enfoque moderno de la logica clasica* ◇ B05 B10 ◇

VOJSHVILLO, E.K. *Method of simplification of the forms of truth functional expressions (Russian)* ◇ B05 ◇

1959

ABHYANKAR, S. *Absolute minimal expressions of boolean functions* ◇ B05 ◇

ANDERSON, A.R. & BELNAP JR., N.D. *A simple treatment of truth functions* ◇ B05 ◇

ANDREOLI, G. *Proprieta delle funzioni simmetriche elementari nelle algebre di Boole e nelle algebre dei livelli* ◇ B05 G05 ◇

DEVIDE, V. *Elementare Aufzaehlung der minimalen erzeugenden Operationssysteme der Aussagenlogik* ◇ B05 ◇

DUNHAM, B. & FRIDSHAL, R. *The problem of simplifying logical expressions* ◇ B05 ◇

HUZINO, S. *On the existence of Sheffer stroke class in the sequential machines* ◇ B05 D05 ◇

ITO, MAKOTO *On boolean equation with many unknown elements and generalized Poretzky's formula* ◇ B05 ◇

MIHAILESCU, E.G. *Formules normales des foncteurs de la logique classique (Romanian) (Russian and French summaries)* ◇ B05 ◇

MIHAILESCU, E.G. *Les formes normales dans le calcul bivalent des propositions* ◇ B05 ◇

MIHAILESCU, E.G. *Sur quelques theorems de la logique classique* ◇ B05 ◇

MLEZIVA, M. *Die Unabhaengigkeit des Axiomensystems des Aussagenkalkuels von Hermes und Scholz* ◇ B05 ◇

POVAROV, G.N. *On symmetries of boolean functions (Russian)* ◇ B05 ◇

QUINE, W.V.O. *On cores and prime implicants of truth functions* ◇ B05 ◇

SLUPECKI, J. *The Lukasiewicz function (Polish)* ◇ B05 B20 ◇

STYAZHKIN, N.I. *Vereinfachung gewisser Algorithmen des klassischen Aussagenkalkuels durch P.S. Poretskij (Russian)* ◇ B05 ◇

WANG, HAO *Circuit synthesis by solving sequential boolean equations* ◇ B05 B70 ◇

1960

ADAM, ANDRAS *Zur Theorie der Wahrheitsfunktionen* ◇ B05 ◇

ANGELL, R.B. *Note on a less restricted type of rule of inference* ◇ B05 ◇

ANGELL, R.B. *The sentential calculus using rule of inference R_e* ◇ B05 ◇

ASSER, G. & RAUTENBERG, W. *Ein Verfahren zur Axiomatisierung der Kontradiktionen gewisser zweiwertiger Aussagenkalkuele* ◇ B05 ◇

BETH, E.W. & LEBLANC, H. *A note on the intuitionist and the classical propositional calculus* ◇ B05 F50 ◇

HAERTIG, K. *Zur Axiomatisierung der Nicht-Identitaeten des Aussagenkalkuels* ◇ B05 ◇

LEBLANC, H. *On requirements for conditional probability functions* ◇ B05 B30 B48 ◇

LEBLANC, L. *Dualite pour les egalites Booleennes* ◇ B05 ◇

LUPANOV, O.B. *On the complexity of the realization by formulas of the function of logical algebra (Russian)* ◇ B05 ◇

LYNGHOLM, C. & YOURGRAU, W. *A double-iteration property of boolean functions* ◇ B05 ◇

LYUBCHENKO, G.G. *Representing boolean functions by formulae (Ukrainian) (Russian and English summaries)* ◇ B05 ◇

MIHAILESCU, E.G. *Sur quelques theoremes dans les calculus des propositions bivalentes* ◇ B05 ◇

MONTEIRO, A. *Matrices de Morgan characteristiques pour le calcul propositionnel classique* ◇ B05 G05 ◇

PORTE, J. *Un systeme pour le calcul des propositions classiques ou la regle de detachement n'est pas valable* ◇ B05 ◇

SCHMIDT, H.A. *Mathematische Gesetze der Logik. I. Vorlesungen ueber Aussagenlogik* ◇ B05 B98 F50 F98 ◇

STUERMANN, W.E. *Plotting boolean functions* ◇ B05 ◇

TUROWICZ, A.B. *Sur une methode algebrique de verification des theoremes de la logique des enonces (Polish and Russian summaries)* ◇ B05 ◇

ZELEZNIKAR, A. *Solvability problems of propositional equations (Serbo-Croatian summary)* ◇ B05 ◇

1961

AKERS JR., S.B. *A truth table method for the synthesis of combinational logic* ◇ B05 ◇

BASU, M.S. & CHOUDHURY, A.K. *On detection of group invariance or total symmetry of a boolean function* ◇ B05 ◇

BEHMANN, H. *Das Vereinfachungsproblem fuer aussagenlogische Normalformen* ◇ B05 ◇

CHU, J.T. *A generalization of a theorem of Quine for simplifying truth functions* ◇ B05 ◇

GREBENSHCHIKOV, V.N. *The coalition of systems of equations of Boolean algebra and their solution (Russian)* ◇ B05 ◇

HENRY, DESMOND PAUL *The truncation of truth-functional calculation* ◇ B05 ◇

KAZAKOV, V.D. *On higher level minimal expressions for boolean functions* ◇ B05 ◇

LEBLANC, H. *The algebra of logic and the theory of deduction* ◇ B05 ◇

LUKASIEWICZ, J. *Remarks on Nicod's axiom and on "generalizing deduction" (Polish)* ◇ B05 ◇

LUPANOV, O.B. *Implementing the algebra of logic functions in terms of bounded depth formulas in the basis of &, ∨, ¬ (Russian)* ◇ B05 ◇

LUPANOV, O.B. *On the realization of functions of logical algebra by formulae of finite classes (formulae of limited depth) in the basis &, ∨, ¬ (Russian)* ⋄ B05 ⋄

LYUBCHENKO, G.G. *On finding formulas of minimum length for functions of algebraic logic (Ukrainian) (Russian summary)* ⋄ B05 ⋄

MCNAUGHTON, R. *Unate truth functions* ⋄ B05 B70 ⋄

PRICE, ROBERT *The stroke function in natural deduction* ⋄ B05 F07 ⋄

PRIOR, A.N. *Some axiom-pairs for material and strict implication* ⋄ B05 B45 ⋄

SADOWSKI, W. *A proof of completeness of the two-valued propositional calculus (Polish) (Russian and English summaries)* ⋄ B05 ⋄

SANCHIS, L.E. *Nueva demonstracion de la completicidad functional del calculo proposicional bivalente* ⋄ B05 ⋄

SCHUETTE, K. *Ein formales System der klassischen Aussagenlogik mit einer einzigen Grundverknuepfung* ⋄ B05 ⋄

SHESTOPAL, G.A. *On the number of simple bases of Boolean functions (Russian)* ⋄ B05 ⋄

SIEMENS JR., D.F. *An extension of "Fitch's rules"* ⋄ B05 ⋄

SOBOCINSKI, B. *On the single axioms of protothetic II,III* ⋄ A05 B05 B15 ⋄

WALIGORSKI, S. *Calculation of the Quine's table for truth functions* ⋄ B05 ⋄

WHEELER, R.F. *Complete propositional connectives* ⋄ B05 ⋄

1962

ADAM, ANDRAS *Ueber die monotone Superposition der Wahrheitsfunktionen* ⋄ B05 ⋄

ALLEN, L.E. *Wff'n proof: the game of modern logic* ⋄ B05 ⋄

ARNOLD, B.H. *Logic and boolean algebra* ⋄ B05 B98 G05 ⋄

CAMION, P. *Traitement de l'information par l'algebre de Boole* ⋄ B05 G05 ⋄

EVENDEN, J. *A lattice-diagram for the propositional calculus* ⋄ B05 ⋄

FARIS, J.A. *Truth-functional logic* ⋄ B05 ⋄

GINDIKIN, S.G. & MUCHNIK, A.A. *The completeness of a system made up of nonreliable elements realizing a function of algebraic logic (Russian)* ⋄ B05 ⋄

HOCKNEY, R. *An intersection algorithm giving all irredundant normal forms from a prime implicant list* ⋄ B05 B35 ⋄

KOROBKOV, V.K. & REZNIK, T.L. *Certain algorithms for the computation of monotonic functions in the algebra of logic (Russian)* ⋄ B05 ⋄

LANTSCHOOT VAN, E. *New concepts in mathematical logic* ⋄ B05 ⋄

LEBLANC, H. *Boolean algebra and the propositional calculus* ⋄ B05 G05 ⋄

LEBLANC, L. *Duality for boolean equalities* ⋄ B05 ⋄

MENGER, K. *Function algebra and propositional calculus* ⋄ B05 G05 ⋄

MIHAILESCU, E.G. *The properties of the logical difference operator with respect to recirpocity and to disjunction (Romanian) (Russian and French summaries)* ⋄ B05 ⋄

NIDDITCH, P.H. *Propositional calculus* ⋄ B05 ⋄

PETER, R. *Ueber die "kuerzeste" Form von Booleschen Funktionen (Russian summary)* ⋄ B05 ⋄

POGORZELSKI, W.A. & SLUPECKI, J. *A proof of the completeness of the classical propositional calculus on the ground of an axiomatic methodology (Polish)* ⋄ B05 G05 ⋄

PORTE, J. *Quelques extensions du theoreme de deduction* ⋄ B05 ⋄

PORTE, J. *Un systeme logistique tres faible pour le calcul propositionnel classique* ⋄ B05 ⋄

SCHARLE, T.W. *A diagram of the functors of the two-valued propositional calculus* ⋄ B05 ⋄

TURQUETTE, A.R. *A general theory of k-place stroke functions in 2-valued logic* ⋄ B05 ⋄

VASIL'EV, YU.L. *Irreducible disjunctive normal forms for certain classes of truth functions (Russian)* ⋄ B05 ⋄

WALIGORSKI, S. *On normal equivalents of truth functions* ⋄ B05 G05 ⋄

WHEELER, R.F. *An asymptotic formula for the number of complete propositional connectives* ⋄ B05 ⋄

1963

ALLEN, L.E. *Wff: the beginner's game of modern logic* ⋄ B05 ⋄

ARNOLD, R.F. & HARRISON, M.A. *Algebraic properties of symmetric and partially symmetric boolean functions* ⋄ B05 ⋄

ATAKA, H. *Some theorems on weighted-majority functions* ⋄ B05 ⋄

BAZILEVSKIJ, YU.YA. *The theory of sequential logical functions* ⋄ B05 ⋄

CANTY, J.T. *Completeness of Copi's method of deduction* ⋄ B05 C07 ⋄

CARRUCCIO, E. *Equazioni logiche nel calcolo delle proposzioni* ⋄ B05 ⋄

GAVRILOV, M.A. *The present state of the theory of relay circuits (Russian)* ⋄ B05 B70 ⋄

GOTO, E. & TAKAHASI, H. *Some theorems useful in threshold logic for enumerating boolean functions* ⋄ B05 ⋄

HARRISON, M.A. *The number of transitivity sets of Boolean functions* ⋄ B05 ⋄

IBUKI, K. & NAEMURA, K. & NOZAKI, A. *General theory of complete sets of logical functions* ⋄ B05 B70 ⋄

JASKOWSKI, S. *Ueber Tautologien, in welchen keine Variable mehr als zweimal vorkommt* ⋄ B05 ⋄

KOCHKAREV, B.S. *The improvement of E. N. Hilbert's estimates for the number of monotone functions of the algebra of logic (Russian)* ⋄ B05 ⋄

KOROBKOV, V.K. *Estimation of the number of monotonic functions of the algebra of logic and of the complexity of the algorithm for finding the resolvent set for an arbitrary monotonic function of the algebra of logic (Russian)* ⋄ B05 G10 ⋄

MARKOV, A.A. *On the inversion complexity of a system of boolean functions (Russian)* ⋄ B05 D55 ⋄

MENGER, K. & SCHULTZ, MARTIN H. *Postulates for the substitutive algebra of the 2-place functors in the 2-valued calculus of propositions* ⋄ B05 G25 ⋄

MEREDITH, C.A. & PRIOR, A.N. *Notes on the axiomatics of the propositional calculus* ⋄ B05 ⋄
MIHAILESCU, E.G. *Generalization of some normal forms* ⋄ B05 ⋄
PINTER, C. *Formes minimales des fonctions booleennes* ⋄ B05 ⋄
REZNIKOFF, I. *Chaines de formules* ⋄ B05 C40 C90 F50 ⋄
ROHLEDER, H. *Eine Halbordnung im Aussagekalkuel* ⋄ B05 ⋄
SALOMAA, A. *On essential variables of functions, especially in the algebra of logic* ⋄ B05 E20 G15 ⋄
SCOGNAMIGLIO, G. *Un metodo di calcolo dei prodotti delle matrici booleane elementari* ⋄ B05 D03 ⋄
URBANO, R.H. *Boolean matrices and the stability of neural nets* ⋄ B05 D05 ⋄
VASIL'EV, YU.L. *Comparison of the complexity of terminal and minimal disjunctive normal forms (Russian)* ⋄ B05 ⋄
WELLS, M.B. *Application of a finite set covering theorem to the simplification of boolean function expressions* ⋄ B05 ⋄
ZHURAVLEV, YU.I. *On inessential variables of functions of a boolean algebra which are not defined everywhere (Russian)* ⋄ B05 ⋄

1964
CHAVCHANIDZE, V.V. *Fundamental relations of the analytic theory of propositional algebra (Russian)* ⋄ B05 ⋄
CRESSWELL, M.J. *Propositional arithmetic* ⋄ B05 D03 ⋄
DEVIDE, V. *Mathematical logic. Part I (The classical logic of propositions) (Croatian) (English summary)* ⋄ B05 B98 ⋄
GOE, G. *Three axiom negation-alternation formulation of the truth-functional calculus* ⋄ B05 ⋄
LORENS, C.S. *Invertible boolean functions* ⋄ B05 ⋄
MARKOV, A.A. *Normal algorithms which compute Boolean functions (Russian)* ⋄ B05 D03 ⋄
NECHIPORUK, EH.I. *On the synthesis of logical nets in incomplete and degenerate basis (Russian)* ⋄ B05 ⋄
PARKER, F.D. *Boolean matrices and logic* ⋄ B05 ⋄
PETERS, R. *Two remarks concerning Menger's and Schultz's postulates for the substitutive algebra of the 2-place functors in the 2-valued calculus of propositions* ⋄ B05 ⋄
PETROSYAN, A.V. *Einige Eigenschaften von Funktionen der Logikalgebra (Russisch)* ⋄ B05 ⋄
POGORZELSKI, W.A. *A schema of deduction theorems for the propositional calculus (Polish) (Russian and English summaries)* ⋄ B05 ⋄
POGORZELSKI, W.A. *A survey of deduction theorems for the propositional calculi (Polish) (Russian and English summaries)* ⋄ B05 ⋄
ROSENBERG, I.G. *Detection et identification des fonctions Booleennes symetriques generalisees* ⋄ B05 ⋄
RUSSELL, L.J. *A layout for logical operations* ⋄ B05 ⋄
SAYRE, K.M. *Syllogistic inference within the propositional calculus* ⋄ A05 A10 B05 ⋄

TIRNOVEANU, M. *Elements of mathematical logic. Vol.1. Logic of bivalent propositions* ⋄ B05 ⋄
TROYANOVSKIJ, S.V. *On the equivalence of formulae with respect to a given function of the algebra of logic (Russian)* ⋄ B05 ⋄
TURQUETTE, A.R. *Peirce's icons for deductive logic* ⋄ A05 A10 B05 ⋄
VASIL'EV, YU.L. *On the "superposition" of reduced disjunctive normal forms (Russian)* ⋄ B05 ⋄
VASIL'EV, YU.L. *On the number of terminal and minimal disjunctive normal forms (Russian)* ⋄ B05 ⋄
ZHURAVLEV, YU.I. *Selection algorithms for sets of real variables of functions not everywhere defined in an algebra of logic (Russian)* ⋄ B05 ⋄
ZHURAVLEV, YU.I. *Set-theoretic methods in symbolic logic* ⋄ B05 E75 ⋄

1965
ANDERSON, D.E. & CLEAVER, F.L. *Venn-type diagrams for arguments of n terms* ⋄ B05 E20 ⋄
ARAI, Y. *On axiom systems of propositional calculi II,III,XII* ⋄ B05 B20 ⋄
ARAI, Y. & ISEKI, K. *On axiom systems of propositional calculi. VII* ⋄ B05 ⋄
BAUSCH, A.F. *Modus ponens under hypotheses* ⋄ B05 B20 ⋄
FRIDMAN, G.SH. & KAREV, G.P. & TRESKOV, S.A. *Estimate of the complexity of a function in the algebra of logic (Russian)* ⋄ B05 G05 ⋄
GEORGIEVA, N.V. *A problem in propositional calculus (Bulgarian) (Russian and English summaries)* ⋄ B05 ⋄
ISEKI, K. *On axiom systems of propositional calculi. IV* ⋄ B05 ⋄
ISEKI, K. & TANAKA, S. *On axiom systems of propositional calculi. V,X* ⋄ B05 ⋄
KOCHKAREV, B.S. *Estimation of the complexity of formulas for monotone functions of the algebra of logic in the class of disjunctive normal forms (d.n.f.) (Russian)* ⋄ B05 ⋄
KOROBKOV, V.K. *Monotone functions of the algebra of logic (Russian)* ⋄ B05 ⋄
KRNIC, L. *Types of bases of algebra of logic (Russian) (Serbo-Croatian summary)* ⋄ B05 ⋄
KUZ'MIN, V.A. *Realization of functions of the algebra of logic by means of automata, normal algorithms and Turing machines (Russian)* ⋄ B05 D03 D05 D10 ⋄
LOOMIS JR., H.H. & WYMAN JR., R.H. *On complete sets of logic primitives* ⋄ B05 ⋄
MANARA, C.F. *Un teorema di Beppo Levi riguardante la logica formale* ⋄ B05 ⋄
MIHAILESCU, E.G. *Properties of the Nicode functor in connection with eqivalence and disjunction* ⋄ B05 ⋄
MIHAILESCU, E.G. *Properties of the Sheffer functor in connection with reciprocity and conjunction* ⋄ B05 ⋄
MIHAILESCU, E.G. *Sur les formes normales par rapport a l'eqivalence, la reciprocite et la conjonction* ⋄ B05 ⋄
MIHAILESCU, E.G. *Sur quelques proprietes de la conjonction par rapport a la reciprocite* ⋄ B05 ⋄
MONTEIRO, A. *Generalisation d'un theoreme de R.Sikorski sur les algebres de Boole* ⋄ B05 ⋄

MOZHAROV, R.V. *One method for testing the completeness of systems of logical functions (Russian)* ◇ B05 ◇

RANDOLPH, J.F. *Cross-examining propositional calculus and set operations* ◇ B05 G05 ◇

REZNIKOFF, I. *Tout ensemble de formules de la logique classique est equivalent a un ensemble independant* ◇ B05 C40 ◇

RUDEANU, S. *Remarks on Motinonyi Goto's papers on boolean equations* ◇ B05 ◇

SCHARLE, T.W. *Axiomatization of porpositional calulus with Sheffer functors* ◇ B05 ◇

SIOSON, F.M. *A characterization of tautologies in the propositional calculus with three variables (Spanish)* ◇ B05 ◇

SZUKSZTA, W. *Die Definition benachbarter Konstituenten boolescher Funktionen mehrerer Veraenderlicher (Polish) (Russian and English summaries)* ◇ B05 ◇

TANAKA, K. *On axiom systems of propositional calculi. XI* ◇ B05 ◇

TIRNOVEANU, M. *Sur quelques proprietes des propositions generales* ◇ B05 B10 ◇

WANG, HAO *Note on rules of inference* ◇ B05 B22 ◇

WANG, XIANGHAO *Sequential Boolean equations (Chinese)* ◇ B05 ◇

YAJIMA, K. *On the decision method for rules and regulations by propositional logic (Japanese)* ◇ B05 ◇

1966

CALABRESE, P. *The Menger algebras of 2-place functions in the 2-valued logic* ◇ B05 G15 ◇

CARVALLO, M. *Equations de Boole* ◇ B05 ◇

CRESSWELL, M.J. *Functions of propositions* ◇ B05 ◇

FREJVALD, R.V. *Functional completeness for not everywhere defined functions of the algebra of logic (Russian)* ◇ B05 G15 ◇

GAVRILOV, G.P. & KUDRYAVTSEV, V.B. & YABLONSKIJ, S.V. *Boolesche Funktionen und Postsche Klassen (Russisch)* ◇ B05 B20 G05 G10 ◇

HEILWEIL, M.F. & HOERNES, G.E. *Introduction a l'algebre de Boole et aux dispositifs logiques* ◇ B05 G05 ◇

ISEKI, K. *An algebraic formulation of K − N propositional calculus* ◇ B05 ◇

ISEKI, K. *On axiom systems of propositional calculi. XV,XXI,XXIII,XXIV* ◇ B05 ◇

JASKOWSKI, S. *On formulas in which no individual variable occurs more than twice* ◇ B05 B20 ◇

KUDIELKA, V. & OLIVA, P. *Complete sets of functions of two and three binary variables* ◇ B05 ◇

LIBER, A.E. *Binaere boolesche Algebra und ihre Anwendungen (Russisch)* ◇ B05 B70 G05 ◇

NECHIPORUK, EH.I. *A boolean function (Russian)* ◇ B05 ◇

ONO, K. *A formalism for the classical sentence-logic* ◇ B05 ◇

PONASSE, D. *Structures et techniques de la logique mathematique* ◇ B05 ◇

QUINE, W.V.O. *On boolean functions* ◇ B05 ◇

ROSE, A. *Sur quelques resultats touchant la formalisation des calculs propositionnels a foncteurs variables* ◇ B05 B20 ◇

RUDEANU, S. *On solving boolean equations in the theory of graphs* ◇ B05 ◇

RUSSELL, L.J. *Note on a layout for logical operations* ◇ B05 ◇

SCHARLE, T.W. *Single axiom schemata for D and S* ◇ B05 ◇

SHOLOMOV, L.A. *Complexity criteria for Boolean functions (Russian)* ◇ B05 ◇

SKALA, H.L. *The irreducible generating sets of 2-place functions in the 2-valued logic* ◇ B05 ◇

SMULLYAN, R.M. *Finite nest structures and propositional logic* ◇ B05 ◇

SMULLYAN, R.M. *Trees and nest structures* ◇ B05 ◇

SURMA, S.J. *The termal completeness of sets of propositions (Polish) (English summary)* ◇ B05 ◇

TANAKA, S. *Axiom systems of B-algebra. VI* ◇ B05 G05 ◇

TANAKA, S. *On the propositional calculus with a variable functor $C\delta p C\delta N p\delta q$* ◇ B05 ◇

TOMS, R.M. *Systems of boolean equations* ◇ B05 ◇

VOJTISHEK, V.V. *On an approach to the classification of Boolean functions (Russian)* ◇ B05 ◇

1967

ANDON, F.I. *A method for simplifying the disjunctive normal form (DNF) of boolean functions (Russian)* ◇ B05 ◇

BLUDOV, V.S. *Algorithmus des Vergleichs der Wurzelformen von Funktionen der Logikalgebra (Russisch)* ◇ B05 ◇

CARVALLO, M. *Sur la resolution des equations de Post* ◇ B05 ◇

DOEHMANN, K. *Der Gruppencharakter der Transformationen der dyadischen Aussagenverknuepfungen* ◇ B05 G25 ◇

GLAGOLEV, V.V. *Certain estimates of disjunctive normal form of functions of the algebra of logic (Russian)* ◇ B05 ◇

HALKOWSKA, K. *A note on the system of propositional calculus with primitive rule of extensionality (Polish) (Russian and English summaries)* ◇ B05 ◇

HERNANDEZ S., J.L. *Teoremas de reversibilidad isomorfica de operaciones logicas multivariables* ◇ B05 ◇

KIPNIS, M.M. *A property of propositional formulas (Russian)* ◇ B05 ◇

KORNEV, YU.N. *Reduced disjunctive normal forms of partially defined boolean functions (Russian)* ◇ B05 ◇

MAL'TSEV, I.A. *A function of the algebra of logic (Russian)* ◇ B05 ◇

MARKOV, A.A. *Normal algorithms connected with computation of boolean functions (Russian)* ◇ B05 D03 ◇

MUCHNIK, B.A. *A criterion for the comparability of bases in the realization of functions of the algebra of logic by formulae (Russian)* ◇ B05 ◇

NIELAND, J.J.F. *Beth's tableau-method* ◇ B05 B25 B50 F07 ◇

PONASSE, D. *Algebrisation du calcul propositionnel au moyen d'anneaux booleens universels* ◇ B05 G05 ◇

RISER, J. *A Gentzen-type calculus for single-operator propositional logic* ◇ B05 ◇

SAPOZHENKO, A.A. *Estimate of the complexity of a function (Russian)* ◊ B05 ◊
STRAUSS, P. *Some systems of natural deduction* ◊ B05 B10 ◊
SURMA, S.J. *Indirect-deduction theorems (Polish) (Russian and English summaries)* ◊ B05 ◊
SVOBODA, A. *Ordering of implicants* ◊ B05 ◊
THOMSON, J. *Proof of the law of infinite conjunction using the perfect disjunctive normal form* ◊ B05 ◊
TISON, P.I. *Generalization of consensus theory and application to the minimization of boolean functions* ◊ B05 ◊
UMEZAWA, T. *A method of two-level simplification of boolean functions* ◊ B05 ◊
UMEZAWA, T. *An absolute simplification of boolean functions* ◊ B05 ◊
VASHCHENKO, V.P. *Mengentheoretischer Ansatz zur funktionalen Trennbarkeit (Russisch)* ◊ B05 ◊

1968

ADAM, ANDRAS *Truth functions and the problem of their realization by two-terminal graphs* ◊ B05 ◊
ANDON, F.I. *The minimization of the d.n.f.'s of the functions of the algebra of logic by the method of the sequential analysis of variants (Russian)* ◊ B05 ◊
CHERNYAEV, V.G. *Ueber einige Arten der Dekomposition unvollstaendig definierter boolescher Funktionen (Russisch)* ◊ B05 ◊
COLE, R. *Definitional boolean calculi* ◊ B05 ◊
GARDIES, J.-L. *Les deux tetraedres des liaisons logiques interpropositionnelles bivalentes* ◊ B05 ◊
ISEKI, K. *Symbolic logic (propositional logic) (Japanese)* ◊ B05 ◊
KECSKIC, J.D. *An axiomatization of the propositional caculus and the completeness theorem* ◊ B05 ◊
MUCHNIK, B.A. *Letter to the editors (Russian)* ◊ B05 ◊
PIECZKOWSKI, A. *Remarks on J.-L. Gardies "Les deux tetraedres des liaisons logiques interpropositionnelles bivalentes" (Polish)* ◊ B05 ◊
POGORZELSKI, W.A. *Some remarks on the concept of completeness of the propositional calculus I (Polish) (Russian and English summaries)* ◊ B05 ◊
RISTEA, T.G. *On propositional, truth and boolean functions* ◊ B05 ◊
ROBINSON, T.T. *Independence of two nice sets of axioms for the propositional calculus* ◊ B05 ◊
SETO, Y. *Proofs of some axioms by stroke function* ◊ B05 ◊
SHARONOV, V.I. & ZAMOV, N.K. *A certain algorithm for finding a full amplification of sequences in propositional calculus (Russian)* ◊ B05 ◊
TANAKA, S. *On proofs of some axioms with Sheffer functor 'D'* ◊ B05 ◊
WORKS, C. & YOURGRAU, W. *Note on duality in propositional calculus* ◊ B05 ◊

1969

BAER, G. *Ein Verfahren zur Umformung einer linearen Nebenbedingung in eine aequivalente alternative Normalform* ◊ B05 ◊
BLANCHE, R. *Sur le systeme des connecteurs interpropositionnels* ◊ B05 ◊
BOOLE, G. *L'analyse mathematique de la logique* ◊ A05 B05 ◊
BOTH, N. *Monotony in propositional calculus (Romanian) (French summary)* ◊ B05 ◊
BOTH, N. *On the completeness of an axiom system (Romanian) (French summary)* ◊ B05 ◊
BURLACU, E. *On the inverses of boolean matrices (Romanian) (Russian and French summaries)* ◊ B05 ◊
CARNES, R.D. *A reduction procedure for Sheffer stroke formulas* ◊ B05 ◊
KHASIN, L.S. *Complexity bounds for the realization of monotonic symmetrical functions by means of formulas in the basis $\vee, \wedge, \neg$ (Russian)* ◊ B05 D20 ◊
KHASIN, L.S. *The realization of monotone symmetric functions by formulae in the base $\vee, \wedge, \neg$ (Russian)* ◊ B05 B50 ◊
KOLPAKOV, V.I. *Estimate of the number of covers of the n-dimensional cube (Russian)* ◊ B05 ◊
KORSHUNOV, A.D. *Comparison of the complexity of the largest and shortest disjunctive normal forms and a lower estimate of the number of irredundant disjunctive normal forms for almost all Boolean functions (Russian)* ◊ B05 D20 ◊
KORSHUNOV, A.D. *The upper complexity bound of the shortest disjunctive normal forms of almost all boolean functions (Russian)* ◊ B05 ◊
LEVIN, A.A. *The relative complexity of the reduced disjunctive normal form (Russian)* ◊ B05 ◊
LUCKHARDT, H. *Kodifikation und Aussagenlogik* ◊ B05 B25 F50 ◊
MATEI, S. *Formulae for solving some boolean equations (Romanian) (Russian and English summaries)* ◊ B05 ◊
MEREDITH, C.A. & PRIOR, A.N. *Equational postulates for the Sheffer stroke* ◊ B05 G05 ◊
MIHAILESCU, E.G. *Les proprietes du foncteur Sheffer par rapport a l'equivalence et la disjonction* ◊ B05 ◊
MIKHEEVA, L. & SALUM, H. *Construction of abbreviated disjunctive normal forms of boolean functions by the method of masks (Russian)* ◊ B05 ◊
NECHIPORUK, EH.I. *On a boolean matrix (Russian)* ◊ B05 ◊
POEL VAN DER, W.L. *The foundation of the propositional calculus by means of one axiom (Dutch)* ◊ B05 ◊
POGORZELSKI, W.A. *Classical calculus of propositions. Outline of the theory (Polish)* ◊ B05 ◊
POGORZELSKI, W.A. *Non-creativity and translability of definitions in propositional calculus (Polish)* ◊ B05 ◊
POGORZELSKI, W.A. *Two-valued propositional calculus and deduction theorem (Polish)* ◊ B05 ◊
REICHBACH, J. *Propositional calculi and completeness theorem* ◊ B05 C07 ◊
SCHULZE, B. *Einfuehrung einer Halbordnung im Aussagenkalkuel* ◊ B05 ◊
SEDIVY, J. *Solution of simple logical problems by colouring graphs* ◊ B05 ◊
SHARONOV, V.I. & ZAMOV, N.K. *The strengthening of formulae that are provable in propositional calculus (Russian)* ◊ B05 ◊

TOMESCU, I. *Un algorithme pour la synthese des fonctions booleennes symetriques (Romanian) (French summary)* ◊ B05 ◊

VOLLRATH, H.J. *Einige neuere Beweismethoden in der Logik* ◊ B05 ◊

1970

ABIAN, A. *Completeness of the generalized propositional calculus* ◊ B05 ◊

ABIAN, A. *Generalized completeness theorem and solvability of systems of boolean polynomial equations* ◊ B05 ◊

ABIAN, A. *On the solvability of infinite systems of Boolean polynomial equations* ◊ B05 E25 ◊

BECCHIO, D. *Demonstration algebrique de l'independance des schemas d'axiomes du calcul propositionnel classique* ◊ B05 ◊

BECCHIO, D. *Demonstration algebrique du theoreme de Wajsberg et du theoreme methodologique de Jaskowski* ◊ B05 ◊

BILLERA, L.J. *Clutter decomposition and monotonic Boolean functions* ◊ B05 ◊

BLUM, A. *The missing premiss* ◊ B05 ◊

BRODSKIJ, I.N. *On certain variant of the calculus of rejected formulae of propositional logic (Russian)* ◊ B05 ◊

COWEN, R.H. *A new proof of the compactness theorem for propositional logic* ◊ B05 ◊

FEDINA, A.M. *The completeness of systems of logical inference (propositional calculus) (Russian)* ◊ B05 ◊

GLADSTONE, M.D. *On the number of variables in the axioms* ◊ B05 ◊

HENKIN, L. *Extending boolean operations* ◊ B05 ◊

KHASIN, L.S. *The use of negation for the realization of the monotone symmetric functions of the algebra of logic by formulae in the basis* $\vee, \wedge, \neg$ *(Russian)* ◊ B05 ◊

KOLPAKOV, V.I. *The correspondence between monotone functions and a set of irredundant tests for tables (Russian)* ◊ B05 ◊

LIEDL, R. *Harmonische Analysis bei Aussagenkalkuelen* ◊ B05 ◊

MENDES SILVA, A.J. *Geometrisierung der Aussagenlogik (Portugiesisch)* ◊ B05 ◊

MIHAILESCU, E.G. *Normal forms of the Nicod functors (Romanian)* ◊ B05 ◊

MULLER, D.E. & PREPARATA, F.P. *Generation of near optimal boolean functions* ◊ B05 ◊

PASTEL, A.M. *Caracterisation geometrique des fonctions logiques permutantes* ◊ B05 ◊

RUDEANU, S. *On reproductive solutions of boolean equations* ◊ B05 ◊

SADE, A. *Morphismes sur le systeme des operateurs propositionnels* ◊ B05 ◊

SETLUR, R.V. *A method to determine the expressive power of a set of connectives* ◊ B05 ◊

SETLUR, R.V. *On the equivalence of strong and weak validity of rule schemes in the two-valued propositional calculus* ◊ B05 ◊

SHOLOMOV, L.A. *On calculating the complexity of boolean functions on Turing machines (Russian)* ◊ B05 ◊

SIBAJIBAN *A remark on note on duality* ◊ B05 ◊

SOTIROV, V.KH. *A certain set of axioms for the propositional calculus (Russian)* ◊ B05 ◊

VARGAS, T. *Mathematische Logik fuer Anfaenger. Aussagenlogik* ◊ B05 ◊

WEBB, P. *A pair of primitive rules for the sentential calculus* ◊ B05 ◊

ZINOV'EV, A.A. *Classical and nonclassical propositional relations (Russian)* ◊ B05 ◊

1971

BANKOWSKI, J. *The reduction of boolean functions to the form of a mod 2 sum of products (Polish) (Russian and English summaries)* ◊ B05 ◊

CAVADIA, I.C. *An algorithm for the formal calculus of boolean expressions* ◊ B05 ◊

CLIMESCU, A. *Un calcul des applications (Romanian summary)* ◊ B05 G25 ◊

DENEV, J.D. *Evaluation du nombre des subfonctions des presque toutes fonctions de Boole (Bulgare)(Resume francais)* ◊ B05 ◊

DRABBE, J. *Sur la definissabilite equationelle* ◊ B05 ◊

GADSHIEV, M.M. *The maximal length of the abbreviated disjunctive normal form for boolean functions of five and six variables (Russian)* ◊ B05 ◊

GEORGIEVA, N.V. *Independence of system of axioms and rules of inference of propositional calculus (Russian) (Polish and English summaries)* ◊ B05 ◊

GEORGIEVA, N.V. *Independence of the axiom and rules of inference of one system of the extended propositional calculus* ◊ B05 ◊

GILEZAN, K. *Une generalisation du theoreme de Loewenheim sur les equations de Boole* ◊ B05 ◊

IBSEN, K. *Equivalent propositions (Danish) (English summary)* ◊ B05 ◊

KOSPANOV, EH.SH. *The product of the shortest disjunctive normal forms (Russian)* ◊ B05 ◊

LEVY, G. *Operations booleennes generalisees* ◊ B05 ◊

LIVOVSCHI, L. *Representation of boolean functions in arbitrary bases* ◊ B05 ◊

MUKHOPADHYAY, A. *Complete sets of logic primitives* ◊ B05 ◊

NEMESSZEGHY, E.A. & NEMESSZEGHY, E.Z. *Is $(p \supset q) = (\sim p \vee q)$ Df. a proper definition in the system of Principia Mathematica?* ◊ B05 ◊

PALMA DE, R. *L'algebre binaire de Boole et ses applications a l'informatique* ◊ B05 B70 G05 ◊

PINTER, C. *On simplifying truth functions: a preliminary reduction of coreless formulas* ◊ B05 ◊

POGORZELSKI, W.A. *Structural completeness of the propositional calculus (Russian summary)* ◊ B05 ◊

PRESIC, S.B. *Une methode de resolution des equations dont toutes les solutions appartiennent a un ensemble fini donne* ◊ B05 ◊

RAS, Z. *Deductive systems of computing machines* ◊ B05 D05 ◊

RAS, Z. *On a relationship between the propositional calculus and a grammar (Russian summary)* ◊ B05 D05 ◊

RICKEY, V.F. *On weak and strong validity of rules for the propositional calculus* ◊ B05 ◊

ROSE, A. *Tautologies sans constants* ◊ B05 ◊
SCHERER, D. *The form of reductio ad absurdum*
◊ A05 B05 ◊
SUSZKO, R. *Sentential variables versus arbitrary sentential constants (Polish summary)* ◊ A05 B05 ◊
WENDELIN, H. *Kurzer Weg zur Bestimmung der durch eine Aussagenverbindung dargestellte Wahrheitsfunktionen*
◊ B05 ◊

1972

BAER, G. *Zur linearen Darstellbarkeit von Ausdruecken des Aussagenkalkuels (English and Russian summaries)* ◊ B05 ◊
BAUM, J.D. *An arithmetic method in symbolic logic*
◊ B05 ◊
BUROSCH, G. & KUDRYAVTSEV, V.B. *Das Problem der Vollstaendigkeit fuer boolesche Funktionen ueber zwei Dualmengen* ◊ B05 ◊
CHIMEV, K.N. *Separable pairs and strongly essential variables of functions of three, four and five arguments (Bulgarian) (Russian and French summaries)* ◊ B05 ◊
CHIMEV, K.N. & KUTIRKOV, G.A. *The minimal number of separable pairs of functions of the algebra of logic that depend essentially on not more than fifteen variables (Bulgarian) (Russian and French summaries)*
◊ B05 G15 ◊
CHUBARYAN, A.A. & TSEJTIN, G.S. *Certain estimates of the length of logical deductions in classical propositional calculus (Russian) (Armenian summary)*
◊ B05 F07 F20 ◊
CORCORAN, J. *Strange arguments* ◊ B05 ◊
DELGADO, V.M. *Lecciones de logica I. Introduccion general. Logica de proposiciones* ◊ B05 B10 ◊
DENEV, J.D. *The complexity of the realization of almost all functions of the algebra of logic by the method of cascades (Russian)* ◊ B05 ◊
DESCHAMPS, J.-P. *Parametric solutions of boolean equations* ◊ B05 ◊
DOEPP, K. *Ueber die Kennzeichnung von Junktoren durch die Gestalt der herleitbaren Ausdruecke* ◊ B05 ◊
FADINI, A. *Algoritmo per la construzione de una funzione booleana composta mediante un'assegnata famiglia di funzioni booleana (English summary)* ◊ B05 ◊
FUJIKAWA, Y. *The construction of an implicational normal form* ◊ B05 ◊
FUJIMURA, T. *System of modern logic (Japanese)*
◊ B05 ◊
GOODSTEIN, R.L. *A new proof of completeness*
◊ B05 C07 ◊
GOODSTEIN, R.L. *The fundamental formula in the algebra of sets* ◊ B05 E20 G05 ◊
KIELKOPF, C.F. *Premisses are not axioms* ◊ B05 ◊
KRAVTSOV, S.S. *Certain topological properties of the functions of the algebra of logic (Russian)* ◊ B05 ◊
NAZARYAN, G.A. *Certain estimates of the realization of Boolean functions in algorithmic languages (Russian) (Armenian summmary)* ◊ B05 B75 ◊
POPOVICI, C.P. *Boolean expressions (Romanian)*
◊ B05 ◊
PRESIC, S.B. *Ein Satz ueber reproduktive Loesungen*
◊ B05 ◊
PRESIC, S.B. *Tautologies. I,II (Serbian)* ◊ B05 ◊
ROSENBERG, S. *A note on propositional calculus* ◊ B05 ◊
RUDEANU, S. *On boolean matrix equations* ◊ B05 ◊
SALLANTIN, J. *Systemes de propositions et informations*
◊ B05 ◊
SANCHEZ-MAZAS, M. *Calcul arithmetique des propositions* ◊ B05 ◊
SAPOZHENKO, A.A. *Ueber die Kompliziertheit der disjunktiven Normalformen, die sich mittels des Gradientenalgorithmus ergeben (Russisch)* ◊ B05 ◊
SERIKOV, YU.A. *An algebraic method of solving logic equations (Russian)* ◊ B05 ◊
STOJAKOVIC, M. *On the exponential properties of the implication* ◊ B05 ◊
SUSZKO, R. *Equational logic and theories in sentential language* ◊ B05 C05 C07 ◊
TOSIC, R. *S-bases of propositional algebra* ◊ B05 ◊
UHLIG, D. *Ueber eine Kompliziertheitscharakteristik boolescher Funktionen* ◊ B05 ◊
VAKARELOV, D. *Extensional logics (Russian)*
◊ B05 B25 ◊
WILLIAMSON, C. *Squares of opposition: comparisons between syllogistic and propositional logic*
◊ B05 B20 ◊

1973

ADAM, ANDRAS *Investigations on the superpositions of truth functions and on their repetition-free realizability by two-terminal graphs (Hungarian) (English summary)*
◊ B05 ◊
BAGINSKI, M. & GLAEWE, W. & GOLL, P. & LIST, G. & LOESCHAU, G. & MERTENS, A. & SCHWANITZ, G. & WALTER, M. *Einfuehrung in die mathematische Logik, Einfuehrung in die Mengenlehre, Aufbau der Zahlenbereiche* ◊ B05 B28 B98 E98 ◊
BLUM, A. *A correction in Copi's account of boolean normal forms* ◊ B05 ◊
CARVALLO, M. *Fonctions transitives generalisees*
◊ B05 ◊
CARVALLO, M. *Tautologies a operateur non specifie*
◊ B05 ◊
CAVADIA, I.C. *An algorithm for the formal calculus of boolean expressions. II* ◊ B05 G05 ◊
EHRICH, H.-D. *Minimale und m-minimale Variablenmengen fuer partielle Boole'sche Funktionen*
◊ B05 ◊
FERRO, R. *Un'osservazione sulle funzioni booleane simmetriche (English summary)* ◊ B05 ◊
FITZPATRICK, P.J. *An extension of Venn diagrams*
◊ B05 ◊
GLAGOLEV, V.V. *Ueber die Laenge einer verkuerzten disjunktiven Normalform fuer boolesche Funktionen der Dimension 1 (Russisch)* ◊ B05 ◊
GORELIK, E.S. *The complexity of the realization of elementary conjunctions and disjunctions in the base $\{x/y\}$ (Russian)* ◊ B05 ◊
HAMBLIN, C.L. *Language types and logical theorems*
◊ B05 D05 ◊
HEGENBERG, L. *Logica. O calculo sentential* ◊ B05 ◊
HIGHT, S.L. *Complex disjunctive decomposition of incompletely specified boolean functions* ◊ B05 ◊

IVANOV, V.A. *The noise immunity of Boolean functions (Russian)* ◊ B05 ◊

KABZINSKI, J.K. *On problems of definability of propositional connectives* ◊ A10 B05 ◊

KOEGST, M. *On the problem of the decomposition of Boolean functions* ◊ B05 ◊

LEE, J.M. *The form of reductio ad absurdum* ◊ A05 B05 ◊

MADUCH, M. *Lukasiewicz's rules of rejection (Polish) (English summary)* ◊ B05 ◊

MEREDITH, D. *On a property of certain propositional formulae* ◊ B05 ◊

MIHAILESCU, E.G. *Les proprietes du foncteur Nicod par rapport a la reciprocite et conjonction I* ◊ B05 ◊

MORGAN, C.G. *Proper definitions in Principia Mathematica* ◊ B05 ◊

MORGAN, C.G. *Sentential calculus for logical falsehoods* ◊ B05 ◊

PEKLO, B.T. *Logical inconsistencies* ◊ B05 ◊

PETER, R. *Mathematische Fassung der sogenannten "Entscheidungs-Tabellen" (Russian summary)* ◊ B05 ◊

PLUHACEK, A. *Ueber einige Eigenschaften der Loesungen boolescher Gleichungen* ◊ B05 ◊

POGORZELSKI, W.A. & PRUCNAL, T. *Some remarks on the notion of completeness of the propositional calculus. II* ◊ B05 ◊

POPOVICI, C.P. *Boolean functions (Romanian)* ◊ B05 ◊

RADOJEVIC, P. *Les tautologies dans les regles des deduction (Serbian)* ◊ B05 ◊

SEKIMOTO, T. *Propositional and implicational propositional calculus (Japanese)* ◊ B05 B20 ◊

SETTE, A.-M. *On the propositional calculus P^1* ◊ B05 ◊

STRAZDINJA, D.P. & STRAZDINS, I.E. *The possibilities of enlarging the Harvard classification of boolean functions (Russian)* ◊ B05 ◊

STRAZDINS, I.E. *Structure of the semigroup of endomorphisms of the algebra of two-place boolean functions. I (Russian) (Latvian and English summaries)* ◊ B05 ◊

TOMASZEWICZ, A. *Generators of propositional functions (Polish and Russian summaries)* ◊ B05 ◊

TVERDOKHLEBOVA, N.N. *A number-theoretic method of comparing Boolean functions (Russian)* ◊ B05 ◊

UHLIG, D. *On relations between the complexity of realisations of boolean functions by circuits and the number of their subfunctions (Russian)* ◊ B05 ◊

VASIL'EV, YU.L. *Ein Zusammenhang zwischen Abschaetzungen in der Theorie der disjunktiven Normalformen und in der kombinatorischen Analysis* ◊ B05 ◊

ZARNECKA-BIALY, E. *Wajsberg's algorithm for axiomatization of the classical propositional calculus* ◊ B05 ◊

ZUCKERMANN, M.M. *Formation sequences for propositional formulas* ◊ B05 ◊

1974

BAER, G. & MEILER, M. *Zur Umformung eines booleschen Ausdrucks in eine aequivalente Gleichung mit 0-1-Variablen* ◊ B05 ◊

BLUM, A. *A note on natural deduction* ◊ B05 ◊

BOTH, N. *Metric notions in propositional logic (Romanian) (French summary)* ◊ B05 ◊

COOK, S.A. & RECKHOW, R.A. *On the length of proofs in the propositional calculus: preliminary version* ◊ B05 D15 F20 ◊

DAVYDOV, G.V. & SUVOROV, P.YU. *Reduction of propositional tautology to graph coloring in three colors (Russian) (English summary)* ◊ B05 ◊

FREUND, H. & SORGER, P. *Aussagenlogik und Beweisverfahren* ◊ B05 ◊

GLEASON, G.G. *Normal and skew systems* ◊ B05 F07 ◊

JACUNOV, A.I. & ROMANKEVICH, A.M. *On a representation method of boolean functions (Russian)* ◊ B05 ◊

LADEGAILLERIE, Y. *Une propriete des consensus de monomes et une application* ◊ B05 ◊

LEVIN, A.A. *Complexity of the disjunctive normal form of a sum of terminal disjunctive normal forms with respect to the disjunctive normal form of a sum of minimal disjunctive normal forms (Russian)* ◊ B05 ◊

MARACCHIA, S. *Matematica antica e moderna in un caso di logica proposizionale* ◊ A10 B05 ◊

MARINO SARMIENTO, R. *Topology in logic: the compactness theorem (Spanish)* ◊ B05 ◊

MIHAILESCU, E.G. *Les proprietes du foncteur Nicod par rapport a la reciprocite et conjonction II* ◊ B05 ◊

MIJAJLOVIC, Z. *On decidability of one class of boolean formulas* ◊ B05 B25 ◊

MILICI, C. *A remark on axiom-system for the classical two-valued $\{\rightarrow, \neg\}$ propositional logic (Serbo-Croatian summary)* ◊ B05 ◊

MILICI, C. *On the propositional calculus defined with $\{\rightarrow, \&, \neg\}$* ◊ B05 ◊

MILLER, D.MICHAEL & MUZIO, J.C. *Two-place decomposition of binary functions* ◊ B05 ◊

MONTEIRO, A. *Matrices de Morgan caracteristiques pour le calcul propositionnel classique* ◊ B05 ◊

MONTELLA, E. *Il connettivo: "ne...ne..."* ◊ B05 ◊

POGORZELSKI, W.A. & PRUCNAL, T. *Equivalence of the structural completeness theorem for propositional calculus and the boolean representation theorem* ◊ B05 G05 ◊

POPOV, S.V. *An equivalence relation and a complete system of schemes of equivalent transformations of deductions in propositional calculus (Russian)* ◊ B05 F07 ◊

POSPESEL, H. *Introduction to logic. Propositional logic* ◊ B05 B98 ◊

ROBERTS, C. *Dale on material implication* ◊ B05 ◊

RUDEANU, S. *An algebraic approach to boolean equations* ◊ B05 ◊

SLAGHT, R.L. *A concise method for translating propositional formulae containing the standard truth-functional connectives into a Sheffer stroke equivalent; plus an extension of the method* ◊ B05 ◊

SMYTH, M.B. *Involution as a basis for propositional calculi* ◊ B05 F50 ◊

SOBOCINSKI, B. *A theorem concerning a restricted rule of substitution in the field of propositional calculi I & II* ◊ B05 B22 ◊

STRAZDINS, I.E. *A linear selfdual group over GF(2), and its action on the algebra of boolean functions (Russian) (English summary)* ◊ B05 ◊

STRAZDINS, I.E. *Structure of the semigroup of endomorphisms of the algebra of two-place boolean functions. II (Russian) (Latvian and English summaries)* ◊ B05 ◊

SUSZKO, R. *Equational logic and theories in sentential languages* ◊ B05 C05 C07 ◊

TUERKSEN, I.B. *Extensions on solutions of boolean equations via its conjunctions* ◊ B05 ◊

UHLIG, D. *About a family of classes of easily realizable Boolean functions (Russian)* ◊ B05 ◊

VIKULIN, A.P. *An estimate of the number of conjunctions in abbreviated disjunctive normal forms (Russian)* ◊ B05 ◊

WAESCHE, H. *Ueber den indirekten Beweis* ◊ A05 B05 ◊

WILDE, A.C. *A substitution property* ◊ B05 ◊

WILDE, A.C. *Generalizations of the distributive and associative laws* ◊ B05 ◊

YAKUBAJTIS, EH.A. *Subclasses and classes of boolean functions (Russian)* ◊ B05 ◊

ZAWADZKI, W. *Condition for the existence of the decomposition of a product function* ◊ B05 ◊

1975

ABDULLAEV, D.A. & YUNUSOV, D. *Ueber die Zerlegung symmetrischer Boolescher Funktionen (Russian)* ◊ B05 ◊

BAUM, J.D. *The tangled tale of Sheffer's stroke* ◊ A10 B05 ◊

CHUBARYAN, A.A. *A certain normal form, and a complexity characteristics of deduction in the classical propositional calculus (Russian) (Armenian and English summaries)* ◊ B05 ◊

CHUBARYAN, A.A. & TSEJTIN, G.S. *On some estimates of the length of derivation in classical propositional calculus (Russian)* ◊ B05 ◊

CHUBARYAN, A.A. & NGUEN VAN TIN' *Some estimates of the complexity characteristics of deductions in classical propositional calculus (Russian)* ◊ B05 F20 ◊

COOK, S.A. *Feasibly constructive proofs and the propositional calculus* ◊ B05 F20 ◊

COY, W. *Drei Komplexitaetsmasse zweistufiger Normalformen boolescher Funktionen* ◊ B05 ◊

DEMETROVICS, J. *The structural investigation of two-valued logic (Hungarian) (Russian summary)* ◊ B05 ◊

DESCHAMPS, J.-P. *Fermetures i-geneatrices - application aux fonctions booleennes permutantes* ◊ B05 G15 ◊

DOWNING, P. *Conditionals, impossibilities and material implications* ◊ A05 B05 ◊

FARRELL, R.J. *A note on the truth-table for $p \supset q$* ◊ B05 ◊

FISCHER, MICHAEL J. & MEYER, A.R. & PATERSON, M.S. *Lower bounds on the size of boolean formulars: Preliminary report* ◊ B05 ◊

FREGUGLIA, P. *Sull'evoluzione algebrica dei principi logici classici* ◊ A10 B05 ◊

HARPER, L.H. *A note on some classes of boolean functions* ◊ B05 D15 ◊

HOAK, D. *An all-rule axiomatization of the propositional calculus and the equivalence of some well-known axiomatizations* ◊ B05 ◊

HOTOMSKI, P. *A certain method of finding the values of formulas of the propositional calculus (Russian)* ◊ B05 ◊

JASKOWSKI, S. *Three contributions to the two-valued propositional calculus* ◊ B05 ◊

JOHNSON, L.E. *A perspective on propositions* ◊ B05 ◊

JUSSILA, E. *Functional logic. Part I. Sentences* ◊ B05 ◊

KASAHARA, S. *A remark on simplifications of boolean polynomials* ◊ B05 ◊

KERNTOPF, P. *Iterative compositions of boolean functions* ◊ B05 ◊

MANESSE, G. *Resolution d'une equation booleenne au moyen de l'arbre de developpement relatif a ses inconnues (English summary)* ◊ B05 ◊

MERLIER, T. *Applications du calcul propositionnel aux structures algebriques ordonnees* ◊ B05 ◊

MOSHCHENSKIJ, V.A. *Uniqueness of a certain normal form for propositional formulae (Russian)* ◊ B05 ◊

NAZARYAN, G.A. *On the complexity of frequency computations and extensions of boolean functions (Russian) (Armenian summary)* ◊ B05 B75 ◊

NAZARYAN, G.A. *On the realization of boolean functions in algorithmic languages (Russian)* ◊ B05 B75 D03 ◊

POPOV, S.V. *The complexity of deductions in certain propositional calculi (Russian)* ◊ B05 F20 ◊

PRATT, V.R. *The effect of basis on size of boolean expressions* ◊ B05 ◊

RICKEY, V.F. *Creative definitions in propositional calculi* ◊ B05 C40 ◊

ROSENBERG, Y. *In defense of Copi* ◊ B05 ◊

RYAN, R. *Basic digital electronics - understanding number systems, boolean algebra, & logic circuits* ◊ B05 B70 G05 ◊

SCHULTZ, P. *Semantic paradoxes* ◊ A05 B05 ◊

SEILER, E. *On the consistency of the theory of limits in topological unions. II (Romanian)* ◊ B05 E75 ◊

SHOLOMOV, L.A. *A sequence of complexly computable functions (Russian)* ◊ B05 ◊

SVOBODA, A. *The concept of term exclusiveness and its effect on the theory of boolean functions* ◊ B05 ◊

TARASOV, V.V. *The completeness problem for systems of functions of the algebra of logic with unreliable realization (Russian)* ◊ B05 ◊

TASEV, A.G. *Eine funktionale Abhaengigkeit boolescher Funktionen und ihre Anwendung in der Theorie der funktionalen Dekomposition (Bulgarian) (Russian and English summaries)* ◊ B05 ◊

THAYSE, A. *La detection des aleas dans les circuits logiques au moyen du calcul differentiel Booleen* ◊ B05 B70 ◊

UDALOV, V.I. *A method of determining whether boolean functions are monotypic (Russian)* ◊ B05 ◊

UTKIN, A.A. & ZAKREVSKIJ, A.D. *Ueber die Loesung logischer Differenzengleichungen (Russian)* ◊ B05 ◊

WEAVER, G.E. *Uniform compactness and interpolation theorems in sentential logic* ◊ B05 C40 ◊

WEBER, K. *Ueber verschiedene Kompliziertheitsmasse bei alternativen Normalformen* ◊ B05 ◊

1976

ARCHIE, L.C. *A simple defense of material implication* ◇ B05 ◇

ARMSTRONG, R.L. *A question about incompleteness* ◇ B05 ◇

BATOG, T. *On substitution for functorial variables* ◇ B05 ◇

BROWN, J.M. *Bernays's non-circular proof of the non-independence of the fourth axiom of "Principia Mathematica"* ◇ A05 B05 ◇

CERNY, E. *Comments on "Equational logic"* ◇ B05 ◇

CHIMEV, K.N. *Separable pairs of functions of six arguments (Bulgarian) (Russian and English summaries)* ◇ B05 ◇

CURTIS, H.A. *Simplified decomposition of boolean functions* ◇ B05 ◇

CZAJSNER, J. *Finitely axiomatizable sets on the basis of a system of the propositional calculus* ◇ B05 ◇

DANQUAH, J. *The circularity of the proof of the non-independence of the fourth axiom of "Principia Mathematica"* ◇ A05 B05 ◇

FARTOS MARTINEZ, M. *A system of rules for propositional calculus (Spanish)* ◇ A05 B05 ◇

FREUND, H. & SORGER, P. *Logik, Mengen, Relationen. Praxis des mathematischen Beweisens* ◇ B05 ◇

FUJITA, H. & NARUSHIMA, H. & NOJIMA, S. & ODAKA, A. *A new normal form of boolean functions* ◇ B05 ◇

GAGNON, L.S. *Nor logic: A system of natural deduction* ◇ B05 ◇

HAEUSSLER, A.F. *Polynomial beschraenkte nichtdeterministische Turingmaschinen und die Vollstaendigkeit des aussagelogischen Erfuellungsproblems* ◇ B05 D10 D15 ◇

HAYES, J.P. *Enumeration of fanout-free boolean functions* ◇ B05 ◇

HUGLY, P. & SAYWARD, C.W. *Prior on propositional identity* ◇ A05 B05 ◇

KHRAPCHENKO, V.M. *The complexity of the realization of symmetric functions of the algebra of logic by finitely based formulas (Russian)* ◇ B05 ◇

LAVIT, C. *Une generalisation de la notion de fonction Booleenne symetrique* ◇ B05 ◇

LORENZEN, P. *Die Vollstaendigkeit einer unverzweigten Variante des "analytischen" Entscheidungsverfahrens der klassischen Logik* ◇ B05 ◇

MAMATOV, YU.A. *Estimation of the number of functions of the algebra of logic realized by the plane d.n.f (Russian)* ◇ B05 B70 ◇

MCCOLL, W.F. *The depth of boolean functions* ◇ B05 ◇

MUZIO, J.C. *A complete classification of three-place functors in two valued logic* ◇ B05 ◇

NAZARYAN, G.A. *Complexity classes of sets of boolean functions (Russian) (Armenian summary)* ◇ B05 D15 ◇

NURMEEV, N.N. *Computation of boolean functions by Turing machines (Russian)* ◇ B05 D10 D15 ◇

PICCOLI, A. *Anelli nel calcolo delle proposizioni* ◇ B05 ◇

PICCOLI, A. *Gruppi del calcolo delle proposizioni* ◇ B05 ◇

POPOV, S.V. *On the complexity of derivations in classical propositional calculus (Russian)* ◇ B05 F20 ◇

PUPYREV, E.I. *Application of probability axioms to problems in the analysis of boolean functions (Russian)* ◇ B05 ◇

RODEK, I. *Normal forms as another method for testing whether rational expressions of the sentential calculus are tautologies* ◇ B05 ◇

RUDEANU, S. *Square roots and functional decomposition of boolean functions* ◇ B05 ◇

SCHMIDT, GUNTHER & STROEHLEIN, T. *A boolean matrix iteration in timetable construction* ◇ B05 ◇

SCHNORR, C.-P. *The combinational complexity of equivalence* ◇ B05 D15 ◇

SCHUSTER, P. *Probleme, die zum Erfuellungsproblem der Aussagenlogik polynomial aequivalent sind* ◇ B05 D15 ◇

STRAZDINS, I.E. *Self-complementary types of boolean functions (Russian)* ◇ B05 ◇

SURMA, S.J. *An axiomatization of the Sheffer function (Polish) (English summary)* ◇ B05 ◇

THIELE, H. *Ein graphentheoretisches Modell zur Beschreibung von Klassifizierungsprozessen bei nichtdisjunkten Systemen von Klassen* ◇ B05 B75 ◇

UGOL'NIKOV, A.B. *The realization of monotone functions by schemes of functional elements (Russian)* ◇ B05 ◇

WASSERMAN, H.C. *A note of evaluation mappings* ◇ B05 B45 ◇

1977

APRILE, G. *Calcolo proposizionale e calcolo inferenziale* ◇ B05 ◇

ARCHIE, L.C. & HURDLE JR., B.G.N. & THOMBLINSON, W.S. *A note on the truth table for "if P then Q"* ◇ B05 ◇

AVSARKISYAN, G.S. *A representation of boolean functions by the sum mod 2 of the implications of the arguments (Russian)* ◇ B05 ◇

BAUER, F.L. *Angstl's mechanism for checking wellformedness of parenthesis-free formulae* ◇ B05 ◇

BERGE, C. & RAMACHANDRA RAO, A. *A combinatorial problem in logic* ◇ B05 ◇

CAMPBELL, P.J. *An answer to Armstrong's question about incompleteness in Copi* ◇ A05 B05 ◇

CARVALLO, M. *Periodicite de fonctions logiques recurrentes (English summary)* ◇ B05 ◇

CERNY, E. *Unique and identity solutions of boolean equations* ◇ B05 ◇

CHIMEV, K.N. *On strongly essential variables of functions of six arguments (Bulgarian) (Russian and English summaries)* ◇ B05 ◇

COBURN, B. & MILLER, DAVID *Two comments on Lemmon's beginning logic* ◇ B05 ◇

COWEN, R.H. *Generalizing Koenig's infinity lemma* ◇ B05 E05 E07 ◇

DYWAN, Z. *On a certain condition of the finite structural axiomatization of the classical propositional calculus* ◇ B05 ◇

ECSEDI-TOTH, P. & MORICZ, F. & VARGA, A. *A note on symmetric boolean functions* ◇ B05 ◇

FREEMAN, J.B. *A caution on propositional identity* ◇ A05 B05 ◇

GAVRILOV, G.P. & SAPOZHENKO, A.A. *Collection of problems in discrete mathematics (Russian)* ⋄ B05 ⋄

GREENE, C. & TAKEUTI, G. *On the decomposition of boolean polynomials* ⋄ B05 ⋄

GRONAU, H.-D.O.F. *Erzeugung dualer Vektoren durch gewisse abgeschlossene Mengen boolescher Funktionen* ⋄ B05 ⋄

HUBIEN, H. *A new basis for classical propositional calculus* ⋄ B05 ⋄

KORSHUNOV, A.D. *Solution of Dedekind's problem on the number of monotonic Boolean functions (Russian)* ⋄ B05 ⋄

KROM, MELVEN R. *Complete rings of sets and sentential logic* ⋄ B05 E20 G05 ⋄

LEBLANC, H. & PAULOS, J.A. & WEAVER, G.E. *Rules of deduction and truth-tables* ⋄ B05 ⋄

LOPARIC, A. *Une etude semantique de quelques calculs propositionnels (English summary)* ⋄ B05 B25 ⋄

MARKOVA, V.P. *Calculation of the Fourier-spectrum of a system of boolean functions that are given by a disjunctive normal form (Russian)* ⋄ B05 ⋄

MCCOLL, W.F. & PATERSON, M.S. *The depth of all boolean functions* ⋄ B05 ⋄

MIJAJLOVIC, Z. *Some remarks on boolean terms - model theoretic approach* ⋄ B05 C50 G05 ⋄

PAUL, W.J. *A 2,5n-lower bound on the combinational complexity of boolean functions* ⋄ B05 B70 ⋄

POPOV, S.V. *Some complexity characteristics of derivations in propositional calculus (Russian) (English summary)* ⋄ B05 B70 F20 ⋄

PRIMENKO, EH.A. *On the number of types of invertible boolean functions (Russian)* ⋄ B05 ⋄

PULATOV, A.K. *On the influence of zero chains on the complexity of realization of boolean functions by contact schemes (Russian)* ⋄ B05 ⋄

PUPYREV, E.I. *Identification of redundant subformulae in boolean function formulae (Russian. Loose English translation)* ⋄ B05 ⋄

SCHNORR, C.-P. *The network complexity and the breadth of Boolean functions* ⋄ B05 B70 D15 ⋄

VASHCHENKO, V.P. *The general case of simple functional decomposition (Russian)* ⋄ B05 ⋄

WEBER, K. *Beziehungen zwischen verschiedenen Kompliziertheitsmassen bei alternativen Normalformen* ⋄ B05 ⋄

1978

ABDULLAEV, D.A. & YUNUSOV, D. *An algorithm for recognizing symmetric boolean functions (Russian)* ⋄ B05 ⋄

ADLEMAN, L.M. & BOOTH, K.S. & PREPARATA, F.P. & RUZZO, W.L. *Improved time and space bounds for boolean matrix multiplication* ⋄ B05 D15 ⋄

ALBRECHT, A. *Ueber die Kompliziertheit der Realisierung boolescher Funktionen durch Schemata aus Funktional- und Verbindungselementen* ⋄ B05 ⋄

BIBILO, P.N. & ENIN, S.V. *Nonessential nature af arguments and decomposition of boolean functions (Russian)* ⋄ B05 ⋄

BOJDAKOVA, V.N. & GLADKIJ, A.V. *On a system of axioms for propositional calculus (Russian)* ⋄ B05 ⋄

BOTH, N. *Measures and metrics in the two-valued logic* ⋄ B05 G05 ⋄

BREJTBART, YU.YA. *A family of Boolean functions with nonlinear formula size* ⋄ B05 ⋄

BUKHARAEVA, Z.K. & GOLUNKOV, YU.V. *Complexity and running time of normal algorithms computing boolean functions (Russian)* ⋄ B05 B35 D03 D15 ⋄

BUTOV, A.A. *A method of functional division of a completely defined boolean function (Russian)* ⋄ B05 ⋄

CAICEDO, X. *A formal system for the non-theorems of the propositional calculus* ⋄ B05 ⋄

CHANDRA, A.K. & MARKOWSKY, G. *On the number of prime implicants* ⋄ B05 ⋄

CHIRKOV, M.K. *Simple implicators of a system of generalized partial functions (Russian)* ⋄ B05 ⋄

COHEN, R.S. & DENEV, J.D. *Estimations of the number of subfunctions for almost all S-tuple boolean functions* ⋄ B05 ⋄

CZAJSNER, J. *Axiomatizable sets on the basis of a system of the propositional calculus* ⋄ B05 ⋄

DZEGELENOK, I.I. & MEDETOV, M.M. *Determination of the levels of complexity of a linear classifier (Russian)* ⋄ B05 ⋄

FANG, J. *The illogical in the logical* ⋄ A05 B05 ⋄

FLAGG, R.C. *On the independence of the Bigos-Kalmar axioms for sentential calculus* ⋄ B05 ⋄

GAITANIS, N. & HALATSIS, C. *Irredundant normal forms and minimal dependence sets of a boolean function* ⋄ B05 ⋄

GASHKOV, S.B. *Ueber die Tiefe der boolschen Funktionen (Russisch)* ⋄ B05 ⋄

GOESSEL, M. *A generalized principle of automata superposition relative to a pair of boolean functions (Russian) (English summary)* ⋄ B05 D05 ⋄

HAY, L. *Convex subsets of 2^n and bounded truth-table reducibility* ⋄ B05 D30 ⋄

KAMENOV, K. *Optimization of the special functional $\Theta(f)$ in factorization of boolean functions into factors (Russian) (English and Lithuanian summaries)* ⋄ B05 ⋄

KHRAPCHENKO, V.M. *The relation between the complexity and depth of formulas (Russian)* ⋄ B05 ⋄

KOEN, R.I. & MANEV, K.N. *Algebraic properties of the universal functions of the algebra of logic (Russian)* ⋄ B05 ⋄

KOSOVSKIJ, N.K. *The complexity of the solvability of boolean functional equations (Russian)* ⋄ B05 D15 G05 ⋄

KRNIC, L. *On generating systems and bases for F_2 (Serbo-Croatian) (English summary)* ⋄ B05 G05 ⋄

LEONTOVICH, A.M. & VIKTOROVA, I.I. *On the number of tests for the determination of significant variables of boolean functions (Russian)* ⋄ B05 ⋄

LEWIS, H.R. *Satisfiability problems for propositional calculi* ⋄ B05 ⋄

LOCKS, M.O. *Logical and probability analysis of systems* ⋄ A05 A10 B05 ⋄

MALATESTA, M. *Logistica II: Le tautologie. L'interdefinibilita dei funtori* ⋄ B05 ⋄

McColl, W.F. *Complexity hierarchies for boolean functions* ◇ B05 B70 ◇

McColl, W.F. *The circuit depth of symmetric boolean functions* ◇ B05 ◇

Medvedev, S.S. *Simplification of boolean functions on L-symmetrical matrices (Russian)* ◇ B05 ◇

Posthoff, C. *Die Loesung und Aufloesung binaerer Gleichungen mit Hilfe des booleschen Differentialkalkuels* ◇ B05 ◇

Rose, A. *A note on formalisation by the method of description of truth-tables* ◇ B05 ◇

Rose, A. *Sur certaines tautologies sans constantes qui possedent des variables de trois sortes (English summary)* ◇ B05 ◇

Rose, A. *Sur une extension du concept d'une tautologie sans constantes (English summary)* ◇ B05 ◇

Schaefer, T.J. *The complexity of satisfiability problems* ◇ B05 D15 ◇

Sholomov, L.A. *Informationseigenschaften von Kompliziertheits-Funktionalen fuer Systeme nicht total definierter boolscher Funktionen (Russisch)* ◇ B05 B70 ◇

Tarjan, R.E. *Complexity of monotone networks for computing conjunctions* ◇ B05 B70 D15 ◇

Tosic, R. *Some properties of monotone boolean functions over finite boolean algebras* ◇ B05 ◇

Ueda, T. *The fixing groups for the 2-asummable boolean functions* ◇ B05 ◇

Uhlig, D. *Boolean functions with linear combinational complexity* ◇ B05 ◇

Vashchenko, V.P. *Multiple separation of a function using a fixed adjoint function (Russian)* ◇ B05 ◇

Vashchenko, V.P. *Representation of functions of the algebra of logic by compositions of partial functions (Russian)* ◇ B05 ◇

Vasil'ev, Yu.L. *Massive classes of dense boolean functions. I (Russian)* ◇ B05 G05 ◇

Wasserman, H.C. *A second order axiomatic theory of strings* ◇ B05 ◇

Yaglom, I.M. *An unusual algebra (Russian)* ◇ B05 E20 G05 ◇

Zakrevskij, A.D. *Determination of minimal disjunctive base of a boolean matrix (Russian)* ◇ B05 ◇

Zuev, Yu.A. *Approximation of a partial Boolean function by a monotonic boolean function (Russian)* ◇ B05 ◇

1979

Aanderaa, S.O. & Boerger, E. *The Horn complexity of Boolean functions and Cook's problem* ◇ B05 D15 ◇

Albrecht, A. *Zur Kompliziertheit der Realisierung boolescher Funktionen durch F-V-Schemata* ◇ B05 ◇

Ambrosimov, A.S. & Sharov, N.N. *Some asymptotic expansions of the number of functions with non-trivial group of inertia (Russian)* ◇ B05 ◇

Aprile, G. *Breve teoria della equazioni logiche booleane (English summary)* ◇ B05 ◇

Aprile, G. *Costruzione automatica di tabelle di verita, mediante matrici pilota* ◇ B05 ◇

Aprile, G. *Risoluzione di eauazioni logiche mediante tavole di verita (English summary)* ◇ B05 ◇

Archie, L.C. *A simple defense of material implication* ◇ A05 B05 ◇

Asatryan, L.G. *Monotone boolean intersection functions (Russian)* ◇ B05 E07 ◇

Bibilo, P.N. & Enin, S.V. *Joint decomposition of a system of vector boolean functions (Russian)* ◇ B05 ◇

Bozoyan, Sh.E. & Torosyan, B.E. *Investigation of functions of the algebra of logic in terms of the activity of a set of variables (Russian) (English and Armenian summaries)* ◇ B05 ◇

Cippo, C.P. & Venini, E. *A neuro-model of the propositional calculus* ◇ B05 ◇

Commentz-Walter, B. *Size-depth tradeoff in monotone boolean formulae* ◇ B05 D15 ◇

Cook, S.A. & Reckhow, R.A. *The relative efficiency of propositional proof systems* ◇ B05 D15 F20 ◇

Danielsson, P.E. & Plavsic, V.M. *Sequential evaluation of boolean functions* ◇ B05 ◇

Denev, J.D. & Tonchev, V.D. *On the number of equivalence classes of boolean functions under a transformation group* ◇ B05 ◇

Detering, L. & Reusch, B. *On the generation of prime implicants* ◇ B05 ◇

Egiazaryan, Eh.V. *Quantitative Charakteristiken von Systemen von Gleichungen mit partiellen booleschen Funktionen (Russisch)* ◇ B05 ◇

Gibbins, P. *Material implication, the sufficiency condition, and conditional proof* ◇ A05 B05 ◇

Imielinski, T. *Functional dependencies and boolean forms. Deeper examination of connection* ◇ B05 ◇

Johnson, F.A. *Copi's method of deduction* ◇ B05 ◇

Karakhanyan, L.M. & Sapozhenko, A.A. *Estimates for parameters of DNFs of not everywhere defined (partial) functions of the algebra of logic (Russian)* ◇ B05 ◇

Koegst, M. & Oberst, E. *Zur graphischen Darstellung Boolescher Funktionen* ◇ B05 ◇

Koen, R.I. *Subfunctions and invariant classes of functions of the algebra of logic (Russian)* ◇ B05 ◇

Levitz, H. & Levitz, K. *Logic and Boolean algebra* ◇ B05 ◇

Lewis, H.R. *Satisfiability problems for propositional calculi* ◇ B05 D15 ◇

Lieberherr, K.J. & Specker, E. *Complexity of partial satisfaction* ◇ B05 D15 ◇

Mamatov, Yu.A. *Concerning a principle for obtaining lower bounds on the complexity of formulars (Russian)* ◇ B05 ◇

Mamatov, Yu.A. *On a principle for obtaining high (exponential for some parameter values) lower bounds for the complexity of disjunctive normal forms (Russian)* ◇ B05 ◇

Muzio, J.C. *Classes of universal decision elements using negative substitutions* ◇ B05 B70 ◇

Nazaryan, G.A. *Ueber disjunkte Zerlegungen von Mengen boolescher Funktionen (Russian)* ◇ B05 D15 ◇

Nurlybaev, A.N. *A boolean function (Russian)* ◇ B05 ◇

Okol'nishnikova, E.A. *On the role of negations in the realization of monotone boolean functions by means of formulas in the basis ($\vee, \&, \neg$) (Russian)* ◇ B05 ◇

PESCHEL, K. *Moeglichkeiten einer formallogischen Darstellung kommunikativer Situationen* ◇ A05 B05 ◇

RISER, J. *A simplification procedure for alternational normal schemata* ◇ B05 ◇

ROZENFEL'D, T.K. & SILAEV, V.N. *Boolean equations and decomposition of boolean functions (Russian)* ◇ B05 ◇

SALIMOV, F.I. *On the question of modeling Boolean random values by functions of the algebra of logic (Russian)* ◇ A05 B05 G05 ◇

SMARANDACHE, F.G. *Deducibility theorems in mathematical logic* ◇ B05 ◇

SNIR, M. *The covering problem of complete uniform hypergraphs* ◇ B05 ◇

STORK, H.-G. *Remarks on the satisfiability problem of propositional logic* ◇ B05 D15 ◇

TENNANT, N. *La barre de Scheffer dans la logique des sequents et des syllogismes* ◇ B05 ◇

TONOYAN, G.P. *The successive splitting of vertices of an n-dimensional unit cube into chains and decoding problems of monotonic Boolean functions (Russian)* ◇ B05 ◇

TOSIC, R. *An optimal identification algorithm for some subclasses of monotone boolean functions* ◇ B05 ◇

UEDA, T. *On the fixing group for a totally pre-ordered boolean function* ◇ B05 ◇

UHLIG, D. *A function of the algebra of logic with many subfunctions and a small complexity of realization (Russian)* ◇ B05 ◇

UHLIG, D. *Relations between the number of subfunctions and combinational complexity of boolean functions and vector functions* ◇ B05 ◇

VASHCHENKO, V.P. *On the computation of all nontrivial simple decompositions of a function of the algebra of logic (Russian)* ◇ B05 ◇

ZELICHENKO, A.I. *The correspondence of monotone boolean functions to systems of linear inequalities (Russian)* ◇ B05 ◇

1980

AROCHA, KH.L. *Minimization of a function of the algebra of logic by investigation of lower units and upper zeros (Russian)* ◇ B05 ◇

AVGUSTINOVICH, S.V. *On the approach of optaining lower estimates of the complexity for boolean functions (Russian)* ◇ B05 ◇

BENNETT, D.W. *Junctions* ◇ B05 G05 ◇

BIBILO, P.N. & ENIN, S.V. *Decomposition of a boolean function with minimum number of significant arguments of the subfunctions (Russian)* ◇ B05 ◇

BIBILO, P.N. & ENIN, S.V. *Joint decomposition of a system of boolean functions (Russian)* ◇ B05 ◇

COMERFORD, J.D. *Affine and general linear equivalences of boolean functions* ◇ B05 ◇

COMMENTZ-WALTER, B. & SATTLER, J. *Size-depth tradeoff in nonmonotone boolean formulae* ◇ B05 ◇

DAI, ZONGDUO & WAN, ZHEXIAN *Conditions for a Boolean function to be independent of some variable (Chinese)* ◇ B05 ◇

EGIAZARYAN, EH.V. *Estimates related to the number of solutions of systems of Boolean equations (Russian)* ◇ B05 ◇

ENGLEBRETSEN, G. *On propositional form* ◇ A05 A10 B05 ◇

JAIN, R.C. *On representation of boolean functions in simple form* ◇ B05 ◇

KARPOVA, N.A. *Complexity of realization of functions of the algebra of logic in certain infinite bases (Russian)* ◇ B05 ◇

LABORDE, J.-M. *A propos d'un probleme d'algebre de Boole* ◇ B05 ◇

LABORDE, J.-M. *Sur le cardinal maximum de la base complete d'une fonction Booleenne, en fonction du nombre de conjonctions de l'une de ses formes normales* ◇ B05 ◇

LEONG, Y. *Reductio ad absurdum (Proof by contradiction)* ◇ A10 B05 ◇

MAMATOV, YU.A. *A method for the synthesis of planar disjunctive normal forms for logic functions (Russian)* ◇ B05 ◇

MILICI, C. & RACHIN, N. *On the structures of formulas of propositional calculus defined with ($\vee, \neg$) (Romanian) (English summary)* ◇ B05 ◇

NAZARYAN, G.A. *The relations of some complexity characteristics of sets of boolean functions (Russian) (Armenian summary)* ◇ B05 D15 ◇

PALTANEA, R. *Example of a grammar that generates the formulas of propositional calculus* ◇ B05 D05 ◇

POVAROV, G.N. *Simple methods of comparing boolean functions (Russian)* ◇ B05 ◇

PRESIC, M.D. *On the embedding of propositional models* ◇ B05 ◇

SAPOZHENKO, A.A. *Estimation of the length and the number of dead-end disjunctive normal forms for almost all partial boolean functions (Russian)* ◇ B05 ◇

SILVER, C.L. *A simple strong-completeness proof for sentential logic* ◇ B05 ◇

THAYSE, A. *Boolean difference calculus* ◇ B05 ◇

THOM, R. *Predication et grammaire universelle (English summary)* ◇ B05 ◇

USTINOV, N.A. *The number of solutions of a system of logical equations (Russian)* ◇ B05 ◇

VASHCHENKO, V.P. *Investigation of simple decomposition of a Boolean function (Russian)* ◇ B05 ◇

WEGENER, I. *A new lower bound on the monotone network complexity of boolean sums* ◇ B05 B70 ◇

WOLNIEWICZ, B. *On the verifiers of disjunction* ◇ A05 B05 ◇

1981

AANDERAA, S.O. & BOERGER, E. *The equivalence of Horn and network complexity for boolean functions* ◇ B05 B70 D15 ◇

ALEKSANYAN, A.A. *Classes of functions of the algebra of logic (Russian)* ◇ B05 ◇

BIBILO, P.N. *On the probability of the existence of multiple decomposition of a completely defined boolean function (Russian) (English summary)* ◇ B05 ◇

BOTH, N. *Some remarks concerning Boolean functions* ◇ B05 ◇

BUTRICK, R. *Deduction and analysis* ◇ B05 ◇

CHIMEV, K.N. *On some properties of functions* ◇ B05 E20 ◇

CHUKHROV, I.P. *Estimates of the number of minimal disjunctive normal forms for a zonal function I (Russian)* ◇ B05 ◇

DZHAVADOV, R.M. *On ε-completeness of sets of functions in the algebra of logic (Russian)* ◇ B05 ◇

GABRIELIAN, A. *Pure grammars and pure languages* ◇ B05 D03 D05 D10 ◇

GOSTEV, YU.G. *Atomary languages and grammars. On the theory of data structure families (Russian)* ◇ B05 D05 ◇

GRONAU, H.-D.O.F. *Sperner type theorems and complexity of minimal disjunctive normal forms of monotone Boolean functions* ◇ B05 ◇

HENLEY, E.J. & OGUNBIYI, E.I. *Irredundant forms and prime implicants of a function with multistate variables* ◇ B05 ◇

KORSHUNOV, A.D. *Complexity of shortest disjunctive normal forms of random Boolean functions (Russian)* ◇ B05 ◇

KUTIRKOV, G.A. *One approach to the expression of functions of the algebra of logic* ◇ B05 ◇

LEVIN, A.A. *The comparative complexity of disjunctive normal forms (Russian)* ◇ B05 ◇

LIEBERHERR, K.J. & SPECKER, E. *Complexity of partial satisfaction* ◇ B05 D15 ◇

LOZHKIN, S.A. *The relation between depth and complexity of equivalent formulas and the depth of monotone functions of the algebra of logic (Russian)* ◇ B05 G05 ◇

MAREK, J.C. *Ueber weder notwendige noch hinreichende Bedingungen. Zur logischen Analyse einer gebraeuchlichen philosophischen Redewendung* ◇ A05 B05 ◇

STRAZDINS, I.E. *On fundamental transformation groups in the algebra of logic* ◇ B05 B50 ◇

TASEV, A.G. *Boolean differences and their properties (Bulgarian) (English and Russian summaries)* ◇ B05 ◇

THAYSE, A. *Boolean calculus of differences* ◇ B05 ◇

THAYSE, A. *Universal algorithms for evaluating boolean functions* ◇ B05 ◇

UHLIG, D. *On two classes of Boolean functions* ◇ B05 ◇

WEGENER, I. *An improved complexity hierarchy on the depth of boolean functions* ◇ B05 B70 D15 ◇

WEGENER, I. *Boolean functions whose monotone complexity is of size $n^2 \log n$* ◇ B05 D15 ◇

1982

BANKOVIC, D. *Some remarks on Boolean equations* ◇ B05 ◇

BIBILO, P.N. *Probability of nonessentialness of subsets of arguments of Boolean functions (Russian)* ◇ B05 ◇

CHUKHROV, I.P. *On the number of irredundant disjunctive normal forms (Russian)* ◇ B05 ◇

COWEN, R.H. *Solving algebraic problems in propositional logic by tableau* ◇ B05 ◇

DZHAVADOV, R.M. *On the complexity of approximate assignment of functions in the algebra of logic (Russian)* ◇ B05 ◇

FISCHER, MICHAEL J. & MEYER, A.R. & PATERSON, M.S. *$\Omega(n \log n)$ lower bounds on length of Boolean formulas* ◇ B05 D15 ◇

FISCHER, R. & UHLIG, D. *Fehlerkorrigierende Realisierungen zu Booleschen Funktionen und Automatenfunktionen mit linearer Kompliziertheit* ◇ B05 D05 ◇

HOTOMSKI, P. *Some characteristics of tables of Boolean functions, and the construction of expressions in Sheffer operators (Russian)* ◇ B05 B70 ◇

HYON, JONGRAK *A graph-theoretic method in the algebra of logic I (Korean) (English summary)* ◇ B05 ◇

KAMENSKIJ, M.I. & PETROVA, L.P. & SADOVSKIJ, B.N. *Mathematical logic. Methodical instructions (Russian)* ◇ B05 ◇

KATERINOCHKINA, N.N. *Search for a maximal upper zero for discrete monotone functions (Russian)* ◇ B05 ◇

KHACHATRYAN, M.A. *Complexity of derivation of certain propositional formulas (Russian) (Armenian summary)* ◇ B05 ◇

KIM, K.H. *Boolean matrix theory and applications* ◇ B05 ◇

LIEBERHERR, K.J. & VAVASIS, S.A. *Analysis of polynomial approximation algorithms for constant expressions* ◇ B05 D15 ◇

MILLER, DAVID *A geometry of logic* ◇ B05 ◇

NAZARYAN, G.A. *Realization of Boolean functions in algorithmic languages under constraints on the running time of the algorithms (Russian) (Armenian summary)* ◇ B05 D15 ◇

NOVIKOV, YA.A. *The algebraic construction of implicative normal form (Russian)* ◇ B05 ◇

OKOL'NISHNIKOVA, E.A. *The effect of negations on the complexity of realization of monotone Boolean functions by formulas of bounded depth (Russian)* ◇ B05 ◇

PETROSYAN, A.V. *Some differential properties of Boolean functions (Russian) (English summary)* ◇ B05 ◇

PLOTKIN, J.M. & ROSENTHAL, J.W. *The expected complexity of analytic tableaux analysis in propositional calculus* ◇ B05 B35 D15 F20 ◇

POLJAK, S. & TURZIK, D. *A polynomial algorithm for constructing a large bipartite subgraph, with an application to a satisfiability problem* ◇ B05 D15 ◇

REZUS, A. *On a theorem of Tarski* ◇ B05 B22 ◇

ROSE, A. *A generalisation of the concept of functional completeness and applications to modus ponens* ◇ B05 ◇

SEGERBERG, K. *Classical propositional operators. An exercise in the foundations of logic* ◇ B05 B22 ◇

SMITH, G.C. *The Boole-De Morgan correspondence, 1842-1864* ◇ A10 B05 ◇

STIEFEL, B. *Ueber Iterierte einer Booleschen Funktion* ◇ B05 ◇

SURMA, S.J. *On the axiomatic treatment of the Sheffer function* ◇ B05 ◇

WEBER, K. *The length of random Boolean functions (German and Russian summaries)* ◇ B05 ◇

WEGENER, I. *Boolean functions whose monotone complexity is of size $n^2/\log n$* ◇ B05 ◇

1983

ANDREEV, A.E. *On the synthesis of disjunctive normal forms close to minimum forms (Russian)* ◇ B05 ◇

AVGUSTINOVICH, S.V. *A lower bound for the length of a simultaneous covering by conjunctions of a system of boolean functions (Russian)* ⋄ B05 ⋄

BANKOVIC, D. *The general reproductive solution of Boolean equation* ⋄ B05 ⋄

BARRICELLI, N.A. *B-mathematics* ⋄ B05 ⋄

BENCIVENGA, E. *Dropping a few worlds* ⋄ B05 ⋄

DALE, A.J. *The non-independence of axioms in a propositional calculus formulated in terms of axiom schemata* ⋄ B05 ⋄

GAO, HENGSHAN *Comments on "The interpolation theorem for the propositional calculus $P(\kappa)$ when κ is a strongly inaccessible cardinal" by Luo, Libo (Chinese)* ⋄ B05 C40 C75 E55 ⋄

KORSHUNOV, A.D. *The maximal length of dead-end disjunctive normal forms for almost all Boolean functions (Russian)* ⋄ B05 ⋄

KUZNETSOV, S.E. *A lower bound for the length of the shortest d.n.f. of almost all Boolean functions (Russian)* ⋄ B05 ⋄

KUZNETSOV, S.E. *Complexity of realization of a sequence of Boolean functions by formulas of depth 3 in the basis $\{\vee, \&, \neg\}$ (Russian)* ⋄ B05 ⋄

LANGLEY, P. & LARSON, P. & SILAS, S. & WERTZ, S.K. *A proof of CN_qN_p from C_{pq} by the rule of detachment in Jeffrey's system 5.6* ⋄ B05 ⋄

LEVIN, A.A. *Gluing of Boolean functions and its application in estimates of the comparative complexity of DNF (Russian)* ⋄ B05 ⋄

LOZHKIN, S.A. *Depth of functions of the algebra of logic in certain bases (Russian)* ⋄ B05 ⋄

PAWLOWSKY, V. & TRILLAS, E. *On order and morphisms related to a Sheffer stroke* ⋄ B05 ⋄

POPOV, M.M. *Logical connectives as derivatives of the rules of inference (Russian)* ⋄ B05 ⋄

RUDEANU, S. *Linear Boolean equations and generalized minterms* ⋄ B05 ⋄

SCHUERFELD, U. *New lower bounds on the formula size of Boolean functions* ⋄ B05 ⋄

SEGERBERG, K. *Arbitrary truth-value functions and natural deduction* ⋄ B05 B20 ⋄

TERZILER, M. *La representation Booleienne via le calcul propositionnel* ⋄ B05 ⋄

WEBER, K. *Irredundant disjunctive normal forms of random Boolean functions (German and Russian summaries)* ⋄ B05 ⋄

1984

ADDISON, J.W. *Eloge: Alfred Tarski: 1901-1983* ⋄ A10 B05 ⋄

ANDREEV, A.E. *On the problem of minimization of disjunctive normal forms (Russian)* ⋄ B05 ⋄

ASLANSKI, M. & CHIMEV, K.N. *Structural characteristics of one class of Boolean functions* ⋄ B05 ⋄

CALL, R.L. *Constructing sequent rules for generalized propositional logics* ⋄ B05 F07 ⋄

CHANDRA, A.K. & STOCKMEYER, L.J. & VISHKIN, U. *Constant depth reducibility* ⋄ B05 ⋄

CHEN, JIYUAN *The satisfiability problem for simple boolean expressions belongs to P (Chinese) (English summary)* ⋄ B05 B25 D15 ⋄

CHUKHROV, I.P. *The number of minimal disjunctive normal forms (Russian)* ⋄ B05 ⋄

GAROCHE, F. & LEONARD, M. *On a class of Boolean functions with matroid property* ⋄ B05 ⋄

HEDTSTUECK, U. *On the argument complexity of multiply transitive Boolean functions* ⋄ B05 ⋄

HURST, S.L. *The relationship between the self-dualized classification of Boolean functions and spectral coefficient classification* ⋄ B05 ⋄

LUCKHARDT, H. *Obere Komplexitaetsschranken fuer TAUT- Entscheidungen* ⋄ B05 D15 ⋄

MARCHENKOV, S.S. *Existence of finite bases in closed classes of Boolean functions (Russian)* ⋄ B05 ⋄

MELTER, R.A. & RUDEANU, S. *Alternative definitions of Boolean functions and relations* ⋄ B05 ⋄

NURLYBAEV, A.N. *The shortest disjunctive normal functions of the algebra of logic (Russian) (Kazakh summary)* ⋄ B05 ⋄

PUDLAK, P. *Bounds for Hodes-Specker theorem* ⋄ B05 D15 ⋄

RADEV, S.R. *Propositional logics of formal languages* ⋄ B05 ⋄

REISCHER, C. & SIMOVICI, D.A. *Graph functions of Boolean functions* ⋄ B05 ⋄

SHOLOMOV, L.A. *Complexity of the composition representation of choice functions. I,II (Russian) (English summary)* ⋄ B05 ⋄

STEPANOV, V.A. *Propositional logic of reflexive sentences (Russian)* ⋄ B05 ⋄

WEGENER, I. *Optimal decision trees and one-time-only branching programs for symmetric Boolean functions* ⋄ B05 ⋄

WEGENER, I. *Proving lower bounds of the monotone complexity of Boolean functions* ⋄ B05 ⋄

1985

ANDREEV, A.E. *A method for obtaining lower bounds on the complexity of individual monotone functions (Russian)* ⋄ B05 ⋄

DYWAN, Z. *A new variant of the Goedel-Malcev theorem for the classical propositional calculus* ⋄ B05 ⋄

KOGAN, A.YU. & ZHURAVLEV, YU.I. *Realization of Boolean functions with a small number of zeros by disjunctive normal forms and related problems (Russian)* ⋄ B05 B75 ⋄

KRIEGEL, K. & WAACK, S. *Lower bounds for Boolean formulae of depth 3 and the topology of the n-cube* ⋄ B05 ⋄

LEONE, M. *On temporal invariance of menu systems* ⋄ B05 B75 C90 ⋄

MASON, I. *The metatheory of the classical propositional calculus is not axiomatizable* ⋄ B05 ⋄

MCGEE, V. *A counterexample to modus ponens* ⋄ B05 ⋄

RAUTENBERG, W. *Consequence relations of 2-element algebras* ⋄ B05 B22 ⋄

RAZBOROV, A.A. *Lower bounds on the monotone complexity of some Boolean functions (Russian)* ⋄ B05 ⋄

STEPANOV, V.A. *Propositional logic of reflexive sentences. II (Russian)* ⋄ B05 F30 ⋄

TANAKA, S. *A new axiom system of propositional calculus*
⋄ B05 ⋄
WEGENER, I. *On the complexity of slice functions*
⋄ B05 ⋄

WISNIEWSKI, A. *Propositional logic and erotetic inferences*
⋄ B05 ⋄

B10 Classical first-order logic

1879
FREGE, F.L.G. *Begriffsschrift, eine der arithmetischen nachgebildete Formelsprache des reinen Denkens*
⋄ A05 B03 B05 B10 ⋄

1900
PEANO, G. *Formules de logique mathematique*
⋄ B03 B10 ⋄

1905
RUSSELL, B. *The existential import of propositions*
⋄ A05 B10 ⋄

1910
WHITEHEAD, A.N. & RUSSELL, B. *Principia mathematica. Vol.I* ⋄ B05 B10 B15 B28 E30 E98 ⋄

1913
BURALI-FORTI, C. *Sur les lois generales de l'algorithme des symboles de fonction et d'operation* ⋄ B10 ⋄

1920
SKOLEM, T.A. *Logisch-kombinatorische Untersuchungen ueber die Erfuellbarkeit oder Beweisbarkeit mathematischer Saetze nebst einem Theorem ueber dichte Mengen* ⋄ B10 B25 C07 C65 C75 ⋄

1924
LUKASIEWICZ, J. *Demonstration de la compatibilite des axiomes de la theorie de la deduction* ⋄ A05 B10 ⋄
SCHOENFINKEL, M. *Ueber die Bausteine der mathematischen Logik* ⋄ A05 B10 B40 ⋄

1925
ACKERMANN, W. *Begruendung des "tertium non datur" mittels der Hilbertschen Theorie der Widerspruchsfreiheit* ⋄ A05 B10 F30 ⋄
LONDON, F. *Ueber die Irreversibilitaet deduktiver Schlussweisen* ⋄ A05 B10 F07 ⋄
ZARISKI, O. *Gli sviluppi piu recenti della teoria degli insiemi e il principio di Zermelo* ⋄ A05 B10 E25 F99 ⋄

1927
BERNAYS, P. *Probleme der theoretischen Logik*
⋄ A05 B10 ⋄
LANGFORD, C.H. *An analysis of some general propositions*
⋄ B10 ⋄

1928
DUBISLAV, W. *Zur kalkuelmaessigen Characterisierung der Definitionen* ⋄ A05 B10 ⋄
HERBRAND, J. *Sur la theorie de la demonstration*
⋄ A05 B10 F07 ⋄

1929
HERBRAND, J. *Sur le probleme fondamental des mathematiques* ⋄ A05 B10 F99 ⋄
HERBRAND, J. *Sur quelques proprietes des propositions vraies et leurs applications* ⋄ B10 F30 ⋄

1930
GOEDEL, K. *Die Vollstaendigkeit der Axiome des logischen Funktionenkalkuels* ⋄ B10 C07 ⋄
HAERLEN, H. *Die logische Grundlage eines mathematischen Beweisverfahrens*
⋄ A05 B10 E30 ⋄
HERBRAND, J. *Les bases de la logique Hilbertienne*
⋄ A05 B10 ⋄
HERBRAND, J. *Recherches sur la theorie de la demonstration* ⋄ B10 B25 F05 F07 F25 F30 ⋄
RAMSEY, F.P. *On a problem of formal logic*
⋄ B10 B25 E05 E20 E75 ⋄

1931
KURATOWSKI, K. & TARSKI, A. *Les operations logiques et les ensembles projectifs* ⋄ B10 D55 E15 ⋄

1932
BERNSTEIN, B.A. *Relation of Whitehead's and Russell's theory of deduction to the boolean logic of propositions*
⋄ A10 B10 ⋄
CHURCH, A. *A set of postulates for the foundation of logic*
⋄ B10 B40 D20 ⋄
HENLE, P. *The independence of the postulates of logic*
⋄ B10 ⋄
LESNIEWSKI, S. *Ueber die Definitionen in der sogenannten Theorie der Deduktion* ⋄ A05 B10 ⋄

1933
CHURCH, A. *A set of postulates for the foundation of logic (second paper)* ⋄ B10 B40 D20 ⋄
GOEDEL, K. *Zum Entscheidungsproblem des logischen Funktionenkalkuels* ⋄ B10 B20 B25 C13 D35 ⋄
WAJSBERG, M. *Beitrag zur Metamathematik*
⋄ A05 B10 B30 B96 C35 F99 ⋄
WAJSBERG, M. *Untersuchungen ueber den Funktionenkalkuel fuer endliche Individuenbereiche*
⋄ B10 C13 ⋄

1934
ACKERMANN, W. *Untersuchungen ueber das Eliminationsproblem der mathematischen Logik*
⋄ B10 ⋄
CHURCH, A. *The Richard paradox* ⋄ B10 ⋄
GOEDEL, K. *Ueber die Laenge von Beweisen*
⋄ B10 F20 F30 F35 ⋄
HUNTINGTON, E.V. *Independent postulates for an "informal Principia System with equality"*
⋄ A05 B10 ⋄

ROESSLER, K. *Beweis der Widerspruchsfreiheit des Funktionenkalkuels der mathematischen Logik (French summary)* ⋄ B10 F25 ⋄

TARSKI, A. *Some methodological investigations on the definability of terms (Polish)* ⋄ B10 C35 C40 ⋄

1935

ACKERMANN, W. *Zum Eliminationsproblem der mathematischen Logik* ⋄ B10 ⋄

GENTZEN, G. *Untersuchungen ueber das logische Schliessen I,II* ⋄ B10 B25 F05 F07 F30 F50 ⋄

ROSSER, J.B. *A mathematical logic without variables II* ⋄ A05 B10 ⋄

1936

BERNAYS, P. *Logical calculus* ⋄ B10 ⋄

MAL'TSEV, A.I. *Untersuchungen aus dem Gebiete der mathematischen Logik (Russian summary)* ⋄ B10 C07 ⋄

SKOLEM, T.A. *Ein Satz ueber die Erfuellbarkeit von einigen Zaehlausdruecken der Form* $(x)(\exists y_1,\ldots,y_n)K_1(x,y_1,\ldots,y_n) \& (x_1,x_2,x_3)K_2(x_1,x_2,x_3)$ ⋄ B10 B25 ⋄

TARSKI, A. *The establishment of scientific semantics (Polish)* ⋄ A05 B10 B30 C07 ⋄

1937

BERNSTEIN, B.A. *Remark on Nicod's reduction of "Principia Mathematica"* ⋄ B10 ⋄

HERMES, H. *Ein Axiomensystem fuer die Syntax des (klassischen) Logikkalkuels* ⋄ B10 ⋄

JOERGENSEN, J. *Outline of the recent development of the theory of deduction (Danish)* ⋄ B10 ⋄

QUINE, W.V.O. *On derivability* ⋄ B10 ⋄

WILKOSZ, W. *Remarque sur la notion de la definition conditionelle de Peano* ⋄ A05 B10 ⋄

1938

ACKERMANN, W. *Mengentheoretische Begruendung der Logik* ⋄ B10 E35 E70 ⋄

SCHMIDT, H.A. *Ueber deduktive Theorien mit mehreren Sorten von Grunddingen* ⋄ B10 ⋄

1939

ACKERMANN, W. *Bemerkungen zu den logisch-mathematischen Grundlagenproblemen* ⋄ A05 B10 ⋄

MAC LANE, S. *Symbolic logic* ⋄ B10 ⋄

TARSKI, A. *On undecidable statements in enlarged systems of logic and the concept of truth* ⋄ B10 F30 ⋄

1940

GOODMAN, NELSON & QUINE, W.V.O. *Elimination of extra-logical postulates* ⋄ A05 B10 ⋄

1941

BERRY, GEORGE D.W. *On Quine's axioms of quantification* ⋄ B10 ⋄

BETH, E.W. *Hauptstuecke aus der modernen formalen Logik. I (Dutch)* ⋄ B10 ⋄

FITCH, F.B. *Closure and Quine's *101* ⋄ B10 ⋄

KETONEN, O. *On the completeness of the predicate calculus (Finnish)* ⋄ B10 ⋄

1942

NEWMAN, M.H.A. *On theories with a combinatorial definition of "equivalence"* ⋄ B10 B40 ⋄

1944

KETONEN, O. *Untersuchungen zum Praedikatenkalkuel* ⋄ B10 ⋄

TARSKI, A. *The semantic conception of truth and the foundations of semantics* ⋄ A05 B10 B30 C07 ⋄

1945

QUINE, W.V.O. *On the logic of quantification* ⋄ B10 B25 ⋄

1946

FEYS, R. *Les methodes recentes de deduction naturelle* ⋄ B10 ⋄

MAUTNER, F.I. *An extension of Klein's Erlanger program: logic as invariant-theory* ⋄ A05 B10 ⋄

RIDDER, J. *Ueber den Aussagen- und den engeren Praedikatenkalkuel* ⋄ B10 ⋄

1947

AUBERT, K.E. *A group-theoretical remark of E.Hoff-Hansen concerning certain expressions in the quantification theory* ⋄ B10 ⋄

FEYS, R. *Note complementaire sur les methodes du deduction naturelle* ⋄ B10 ⋄

WANG, HAO *A note on Quine's principles of quantification* ⋄ B10 ⋄

1948

HALLDEN, S. *Certain problems connected with the definition of identity and of a definite description given in "Principia Mathematica"* ⋄ B10 ⋄

TARSKI, A. *A problem concerning the notion of definability* ⋄ B10 B15 C40 ⋄

1949

GOODMAN, NELSON *An improvement in the theory of simplicity* ⋄ A05 B10 ⋄

GOODMAN, NELSON *The logical simplicity of predicates* ⋄ A05 B10 ⋄

HENKIN, L. *The completeness of the first-order functional calculus* ⋄ B10 C07 ⋄

MOSTOWSKI, ANDRZEJ *Sur l'interpretation geometrique et topologique des notions logiques* ⋄ B10 C90 ⋄

NOVIKOV, P.S. *On regularity classes (Russian)* ⋄ B10 B45 ⋄

1950

CURRY, H.B. *A theory of formal deducibility* ⋄ B10 F50 F98 ⋄

FRAISSE, R. *Sur les types de polyrelation et sur une hypothese d'origine logistique* ⋄ B10 C07 E07 E10 ⋄

HASENJAEGER, G. *Ueber eine Art von Unvollstaendigkeit des Praedikatenkalkuels der ersten Stufe* ⋄ B10 ⋄

KREISEL, G. *Note on arithmetic models for consistent formulae of the predicate calculus I* ⋄ B10 C57 D45 D55 F30 ⋄

QUINE, W.V.O. *On natural deduction* ⋄ B10 ⋄

SAMPEI, Y. *Some remarks concerning identity* ⋄ B10 ⋄

SCHUETTE, K. *Schlussweisen-Kalkuele der Praedikatenlogik* ◇ B10 B25 F05 F07 F50 ◇

ZUBIETA RUSSI, G. *On the substitution of functional variables in the functional calculus of the first order* ◇ B10 ◇

1951

KALMAR, L. *Contributions to the reduction theory of the decision problem. III Prefix $(x_1)(Ex_2)...(Ex_{n-2})(x_{n-1})(x_n)$, a single binary predicate* ◇ B10 D35 ◇

MOSTOWSKI, ANDRZEJ *On the rules of proof in the pure functional calculus of the first order* ◇ B10 ◇

SCHMIDT, H.A. *Die Zulaessigkeit der Behandlung mehrsortiger Theorien mittels der ueblichen einsortigen Praedikatenlogik* ◇ B10 ◇

ZUBIETA RUSSI, G. *Sobre el calculo funcional de primer orden* ◇ B10 ◇

ZUBIETA RUSSI, G. *Some theorems in the theory of elementary quantification* ◇ B10 ◇

1952

ACKERMANN, W. *Widerspruchsfreier Aufbau einer typenfreien Logik I (erweitertes System)* ◇ B10 B15 E70 F25 F65 ◇

CRAIG, W. & QUINE, W.V.O. *On reduction to a symmetric relation* ◇ B10 C40 ◇

CURRY, H.B. *The permutability of rules in the classical inferential calculus* ◇ B10 ◇

CURRY, H.B. *The system LD* ◇ B10 F50 ◇

HASENJAEGER, G. *Topologische Untersuchungen zur Semantik und Syntax eines erweiterten Praedikatenkalkuels* ◇ B10 C07 G05 ◇

MO, SHAOKUI *A note on the theory of quantification* ◇ B10 ◇

ROSE, A. *Sur un ensemble de fonctions primitives pour le calcul des predicats du premier ordre lequel constitue son propre dual* ◇ B10 ◇

SCHOLZ, H. *Ein ungeloestes Problem in der symbolischen Logik (problem 1)* ◇ B10 C13 ◇

WANG, HAO *Logic of many-sorted theories* ◇ B10 ◇

1953

ACKERMANN, W. *Widerspruchsfreier Aufbau einer typenfreien Logik II* ◇ B10 B15 E35 E70 F30 F35 F65 ◇

BETH, E.W. *On Padoa's method in the theory of definition* ◇ B10 C40 ◇

DRESDEN, A. *Complete independence* ◇ B10 ◇

GOTTSCHALK, W.H. *The theory of quaternality* ◇ B10 ◇

HASENJAEGER, G. *Eine Bemerkung zu Henkin's Beweis fuer die Vollstaendigkeit des Praedikatenkalkuels der ersten Stufe* ◇ B10 C07 C57 D55 ◇

HENKIN, L. *Banishing the rule of substitution for functional variables* ◇ B10 B15 ◇

HINTIKKA, K.J.J. *Distributive normal forms in the calculus of predicates* ◇ B10 C07 ◇

KREISEL, G. *Note on arithmetic models for consistent formulae of the predicate calculus II* ◇ B10 C57 D45 D55 F30 ◇

MOSTOWSKI, ANDRZEJ & RASIOWA, H. *A geometric interpretation of logical formulae (Polish) (English and Russian summaries)* ◇ B10 F50 G05 G10 G15 ◇

RASIOWA, H. & SIKORSKI, R. *Algebraic treatment of the notion of satisfiability* ◇ B10 C90 G05 ◇

RASIOWA, H. & SIKORSKI, R. *On satisfiability and decidability in non-classical functional calculi* ◇ B10 B25 B45 C90 F50 ◇

STANLEY, R.L. *An extended procedure in quantification logic* ◇ B10 ◇

TRAKHTENBROT, B.A. *On recursively separability (Russian)* ◇ B10 C13 D35 ◇

1954

HAILPERIN, T. *Remarks on identity and description in first-order axiom systems* ◇ B10 ◇

HENKIN, L. *A generalization of the notion of ω-consistency* ◇ B10 C07 F30 ◇

IZUMI, Y. *Sur les formes normales* ◇ B10 ◇

PAP, A. *Propositions, sentences, and the semantic definition of truth* ◇ A05 B05 B10 ◇

QUINE, W.V.O. *Quantification and the empty domain* ◇ B10 B60 ◇

1955

AJDUKIEWICZ, K. *Concerning the plan of research in the field of logic (Polish) (Russian and English summaries)* ◇ A10 B10 ◇

BETH, E.W. *Remarks on natural deduction* ◇ A05 B10 F07 ◇

BETH, E.W. *Semantic entailment and formal derivability* ◇ B10 C07 ◇

FRAISSE, R. *La construction des γ-operateurs et leur application au calcul logique du premier ordre* ◇ B10 C07 E07 ◇

HASENJAEGER, G. *On definability and derivability* ◇ B10 B15 C07 D55 ◇

HINTIKKA, K.J.J. *Form and content in quantification theory* ◇ A05 B10 ◇

HINTIKKA, K.J.J. *Notes on quantification theory* ◇ B10 ◇

MAEHARA, S. *The predicate calculus with ε-symbol* ◇ B10 ◇

QUINE, W.V.O. *A proof procedure for quantification theory* ◇ B10 ◇

RASIOWA, H. *A proof of ε-theorems* ◇ B10 ◇

REICHBACH, J. *Completeness of the functional calculus of first order (Polish, Russian) (English summary)* ◇ B10 C07 ◇

RIEGER, L. *On a fundamental theorem of mathematical logic (Czech) (Russian and English summaries)* ◇ B10 C07 G05 G25 ◇

ROSSER, J.B. *Deux esquisses de logique* ◇ B10 B40 D20 E30 ◇

SCHROETER, K. *Methoden zur Axiomatisierung beliebiger Aussagen- und Praedikatenkalkuele* ◇ B10 ◇

SCHROETER, K. *Theorie des logischen Schliessens I* ◇ B10 F07 F50 ◇

STANLEY, R.L. *Simplified foundations for mathematical logic* ◇ B10 ◇

WANG, HAO *On denumerable bases of formal systems* ◇ B10 E25 E30 E35 E70 ◇

1956

ASSER, G. *Ueber die Ausdrucksfaehigkeit des Praedikatenkalkuels der ersten Stufe mit Funktionalen*
⋄ B10 ⋄

COBHAM, A. *Reduction to a symmetric predicate*
⋄ B10 D35 ⋄

COPI, I.M. *Another variant of natural deduction*
⋄ B10 ⋄

FRAISSE, R. *Application des γ-operateurs au calcul logique du premier echelon* ⋄ B10 C07 C65 ⋄

GUMIN, H. & HERMES, H. *Die Soundness des Praedikatenkalkuels auf der Basis der Quineschen Regel* ⋄ B10 ⋄

KALMAR, L. *Ein direkter Beweis fuer die allgemein-rekursive Unloesbarkeit des Entscheidungsproblems des Praedikatenkalkuels der ersten Stufe mit Identitaet* ⋄ B10 D35 ⋄

LOS, J. & MOSTOWSKI, ANDRZEJ & RASIOWA, H. *A proof of Herbrand's theorem* ⋄ B10 C07 F05 F50 ⋄

RASIOWA, H. *On the ε-theorems* ⋄ B10 G05 ⋄

SCHROETER, K. *Die Unabhaengigkeit der elementaren praedikatenlogischen Schlussregeln* ⋄ B10 ⋄

SCHUETTE, K. *Ein System des verknuepfenden Schliessens*
⋄ B10 F07 F30 ⋄

SOBOCINSKI, B. *On well-constructed axiom systems*
⋄ B10 ⋄

1957

ASSER, G. *Theorie der logischen Auswahlfunktionen*
⋄ B10 ⋄

BORKOWSKI, L. *Systems of the propositional and of the functional calculus based on one primitive term (Polish and Russian summaries)* ⋄ B10 ⋄

CRAIG, W. *Analysis of first-order implications*
⋄ B10 F05 ⋄

CRAIG, W. *Linear reasoning. A new form of the Herbrand-Gentzen theorem*
⋄ B10 C07 C40 F05 ⋄

CRAIG, W. *Three uses of the Herbrand-Gentzen theorem in relating model theory and proof theory*
⋄ B10 C07 C40 F05 ⋄

DREBEN, B. *Relation of m-valued quantificational logic to 2-valued quantificational logic* ⋄ B10 B50 ⋄

FEFERMAN, S. *Some recent work of Ehrenfeucht and Fraisse* ⋄ B10 C07 C65 E10 ⋄

HENKIN, L. *A generalization of the concept of ω-completeness* ⋄ B10 C07 F30 ⋄

HERMES, H. *Einfuehrung in die mathematische Logik. Klassische Praedikatenlogik* ⋄ B10 B98 ⋄

KALISH, D. & MONTAGUE, R. *Remarks on descriptions and natural deduction I,II* ⋄ B10 F07 ⋄

KANGER, S. *Provability in logic*
⋄ B10 F05 F07 F98 ⋄

LIGHTSTONE, A.H. & ROBINSON, A. *Syntactical transforms* ⋄ B10 C52 C60 ⋄

MAEHARA, S. *Equality axiom on Hilbert's ε-symbol*
⋄ B10 ⋄

PORTE, J. *Un systeme de postulats pour le calcul des predicats* ⋄ B10 ⋄

RASIOWA, H. *Sur la methode algebrique dans la methodologie des systemes deductifs elementaires*
⋄ B10 C07 G05 G10 ⋄

SMULLYAN, R.M. *Languages in which self reference is possible* ⋄ A05 B10 B28 F30 F40 ⋄

STAHL, G. *Les univers du discours et les calculs correspondants* ⋄ B10 ⋄

ZUBIETA RUSSI, G. *Clases aritmeticas definidas sin igualdad* ⋄ B10 B20 C07 C52 ⋄

1958

ACKERMANN, W. *Ein typenfreies System der Logik mit ausreichender mathematischer Anwendungsfaehigkeit I*
⋄ B10 B15 E70 ⋄

BERNAYS, P. *Remarques sur le probleme de la decision en logique elementaire* ⋄ B10 D35 ⋄

BORKOWSKI, L. & SLUPECKI, J. *A logical system based on rules and its application in teaching mathematical logic (Polish and Russian summaries)* ⋄ B10 ⋄

CRAIG, W. & VAUGHT, R.L. *Finite axiomatizability using additional predicates* ⋄ B10 C52 ⋄

NOLIN, L. *Sur l'algebre des predicats* ⋄ B10 G15 ⋄

NOLIN, L. *Sur un systeme de "deduction naturelle"*
⋄ B10 ⋄

RASIOWA, H. & SIKORSKI, R. *On the isomorphism of Lindenbaum algebras with fields of sets*
⋄ B10 G05 ⋄

SCHNEIDER, H.H. *Semantics of the predicate calculus with identity and the validity in the empty individual-domain*
⋄ B10 ⋄

SCHROETER, K. *Theorie des logischen Schliessens II*
⋄ B10 C07 F05 F07 ⋄

SIKORSKI, R. *On Herbrand's theorem*
⋄ B10 F05 G05 ⋄

STAHL, G. *Enfoque moderno de la logica clasica*
⋄ B05 B10 ⋄

TOERNEBOHM, H. *Outlines of a boolean tensor algebra with applications to the lower functional calculus*
⋄ B10 G05 G25 ⋄

1959

BEHMANN, H. *Der Praedikatenkalkuel mit limitierten Variablen. Grundlegung einer natuerlichen exakten Logik* ⋄ B10 ⋄

BETH, E.W. *Moderne Logik* ⋄ B10 ⋄

ROBINSON, A. *Outline of an introduction to mathematical logic IV* ⋄ B10 B98 C07 C98 G05 ⋄

STAHL, G. *Los universos del discurso y los sistemas correspondientes* ⋄ B10 ⋄

1960

BETH, E.W. *Completeness results for formal systems*
⋄ B10 C07 ⋄

BOCHVAR, D.A. *Antinomies based on sets of definitions of predicates, each individually consistent (Russian)*
⋄ B10 ⋄

CRAIG, W. *Bases for first-order theories and subtheories*
⋄ B10 D25 F05 ⋄

CURRY, H.B. *The inferential approach to logical calculus I*
⋄ B10 ⋄

GUILLAUME, M. *Certains aspects syntaxiques d'une notion de modele: Relativisation d'une fonction logique de choix* ⋄ B10 ⋄

GUILLAUME, M. *Sur une propriete remarquable du systeme de Bourbaki* ⋄ B10 ⋄

LIS, Z. *Logical consequence, semantic and formal*
  ◊ B10 C07 ◊
MAEHARA, S. *On the interpolation theorem of Craig (Japanese)* ◊ B10 C40 F05 ◊
RASIOWA, H. & SIKORSKI, R. *On the Gentzen theorem*
  ◊ B10 C07 F05 ◊

1961

CURRY, H.B. *The inferential approach to logical calculus II* ◊ B10 ◊
LEMMON, E.J. *Quantifier rules and natural deduction*
  ◊ B10 ◊
MCLAUGHLIN, T.G. *A muted variation on a theme of Mendelson* ◊ B10 ◊
POGORZELSKI, W.A. & SLUPECKI, J. *A variant of the proof of the completeness of the first order functional calculus (Polish and Russian summaries)* ◊ B10 C07 ◊
REICHBACH, J. *A note on my paper: on characterizations of the first-order functional calculus* ◊ B10 ◊
REICHBACH, J. *On characterizations of the first-order functional calculus* ◊ B10 ◊
REICHBACH, J. *On theses of the first-order functional calculus* ◊ B10 ◊
SCHNEIDER, H.H. *A syntactical characterization of the predicate calculus with identity and the validity in all individual-domains* ◊ B10 ◊
SCHOLZ, H. *Logik, Grammatik, Metaphysik*
  ◊ A05 B10 ◊
SMART, J.J.C. *Goedel's theorem, Church's theorem and mechanism* ◊ A05 B10 D35 F30 ◊

1962

BETH, E.W. *Umformung einer abgeschlossenen deduktiven oder semantischen Tafel in eine natuerliche Ableitung auf Grund der derivativen bzw. klassischen Implikationslogik* ◊ B10 F07 ◊
CHANG, C.C. & KEISLER, H.J. *An improved prenex normal form* ◊ B10 ◊
KUTSCHERA VON, F. *Zum Deduktionsbegriff der klassischen Praedikatenlogik erster Stufe* ◊ B10 ◊
MATULIS, V.A. *Two variantes of the classical predicate calculus without structural inference rules (Russian)*
  ◊ B10 ◊
OBERSCHELP, A. *Eine Bemerkung ueber den Kalkuel des natuerlichen Schliessens* ◊ B10 ◊
OBERSCHELP, A. *Untersuchungen zur mehrsortigen Quantorenlogik* ◊ B10 ◊
REICHBACH, J. *On generalization of the satisfiability definition and proof rules with remarks to my paper: On theses of the first-order functional calculus* ◊ B10 ◊
REICHBACH, J. *On the connection of the first-order functional calculus with many-valued propositional calculi* ◊ B10 B50 ◊
SUSZKO, R. *A note concerning the binary quantifiers*
  ◊ B10 ◊

1963

BORGERS, A. *Les lois et les regles de la logique symbolique elementaire des propositions et des predicats* ◊ B10 ◊
KANGER, S. *A simplified proof method for elementary logic*
  ◊ B10 F07 ◊
KROM, MELVEN R. *Separation principles in the hierarchy theory of pure first-order logic*
  ◊ B10 C40 C52 D55 ◊
LOS, J. *Semantic representation of the probability of formulas in formalized theories (Polish summary)*
  ◊ A05 B10 B48 B60 ◊
MASSEY, G.J. *Note on Copi's system* ◊ B10 ◊
MATULIS, V.A. *Variants of the classical predicate calculus with a single deduction tree (Russian)* ◊ B10 F07 ◊
PATTON, T.E. *A system of quantificational deduction*
  ◊ B10 ◊
REICHBACH, J. *About connection of the first-order functional calculus with many valued propositional calculi* ◊ B10 B50 ◊
REICHBACH, J. *Some characterizations of theses of the first-order functional calculus* ◊ B10 ◊
SAITO, S. *Truth value assignment in predicate calculus of first order* ◊ B10 ◊
SMIRNOV, V.A. *Bemerkungen zum System der Syllogistik und der allgemeinen Theorie der Deduktion (Russian)*
  ◊ B10 ◊
SMULLYAN, R.M. *A unifying principle in quantification theory* ◊ B10 F05 F07 ◊
WANG, HAO *Many-sorted predicate calculi* ◊ B10 ◊
WANG, HAO *Relative strength and reducibility*
  ◊ B10 E30 E70 F25 F30 ◊
WHOLEY, J.S. *Persistence and Herbrand expansions*
  ◊ B10 ◊

1964

ARRUDA, A.I. & COSTA DA, N.C.A. *Sur un theoreme de Hilbert et Bernays* ◊ B10 ◊
DUBIKAJTIS, L. *Decompositions of theories* ◊ B10 ◊
FARIS, J.A. *Quantification theory* ◊ B10 ◊
HINTIKKA, K.J.J. *Distributive normal forms and deductive interpolation* ◊ B10 C07 C40 ◊
HU, SHIHUA *Classic predicate calculus (Chinese)*
  ◊ B10 ◊
MO, SHAOKUI *Enumeration quantifiers and predicate calculus (Chinese)* ◊ B10 C07 C80 ◊
PATTON, T.E. *A liberalized system of quantificational deduction* ◊ B10 ◊
RAGGIO, A.R. *Direct consistency proof of Gentzen's system of natural deduction* ◊ B10 F07 ◊
REICHBACH, J. *A note about connection of the first-order functional calculus with many valued propositional calculi* ◊ B10 B50 ◊
RIEGER, L. *Zu den Strukturen der klassischen Praedikatenlogik* ◊ B10 C07 G05 ◊
ROBERTS, D.D. *The existential graphs and natural deduction* ◊ A05 B10 ◊
RUS, T. *Ueber ein formales System I,II (Rumaenisch)*
  ◊ B10 D20 ◊
SCHOCK, R. *Contribution to syntax, semantics, and the philosophy of science* ◊ A05 B10 ◊
SCHOCK, R. *On the logic of variable binders*
  ◊ B10 C80 ◊

1965

BACON, J. *An alternative contextual definition of descriptions* ◊ A05 B10 ◊

CHRISTIAN, C.C. *Die Interpretation logischer Formen* ⋄ B10 ⋄
HAILPERIN, T. *An incorrect theorem* ⋄ B10 ⋄
HERMES, H. *Eine Termlogik mit Auswahloperator* ⋄ B10 B60 ⋄
HINTIKKA, K.J.J. *Distributive normal forms in first-order logic* ⋄ B10 C07 ⋄
KALISH, D. & MONTAGUE, R. *On Tarski's formalisation of predicate logic with identity* ⋄ B10 ⋄
LEBLANC, H. *Marginalia on Gentzen's Sequenzen-Kalkule* ⋄ B10 F07 F50 ⋄
LEMMON, E.J. *A further note on natural deduction* ⋄ A05 B10 ⋄
PARRY, W.T. *Comments on a variant form of natural deduction* ⋄ B10 F07 ⋄
REICHBACH, J. *On the connection of the first-order functional calculus with $\aleph_0$ propositional calculus* ⋄ B10 B50 ⋄
SMULLYAN, R.M. *A unifying principle in quantification theory* ⋄ B10 F05 F07 ⋄
SMULLYAN, R.M. *Analytic natural deduction* ⋄ B10 F05 F07 ⋄
SUSZKO, R. *A note concerning the rules of inference for quantifiers* ⋄ B10 ⋄
TARSKI, A. *A simplified formalization of predicate logic with identity* ⋄ B10 ⋄
TIRNOVEANU, M. *Sur quelques proprietes des propositions generales* ⋄ B05 B10 ⋄
WESTON, K. *On predicate letter formulas which have no substitution instances provable in a first order language* ⋄ B10 ⋄

1966
CROSSLEY, J.N. *Some theorems in logic* ⋄ B10 C07 ⋄
GODDARD, L. *Predicates, relations and categories* ⋄ A05 B10 B50 ⋄
GOE, G. *A reconstruction of formal logic* ⋄ A05 B10 ⋄
HOSOI, T. *The separation theorem on the classical system* ⋄ B10 F50 ⋄
KEARNS, J.T. *Quantifiers and universal validity* ⋄ A05 B10 ⋄
LEBLANC, H. *Two separation theorems for natural deduction* ⋄ B10 F07 F50 ⋄
LEBLANC, H. *Two shortcomings of natural deduction* ⋄ B10 ⋄
MANGANI, P. *Sur certe algebre connesse con sistemi di logica elementare dotati dell' operator tau di Hilbert* ⋄ B10 G25 ⋄
MINTS, G.E. *Herbrand's theorem for the predicate calculus with equality and functional symbols (Russian)* ⋄ B10 F05 ⋄
PREVIALE, F. *Il "problema del prefisso" per i calcoli logici con piu tipi di variabili* ⋄ B10 ⋄
RESNIK, M.D. *A note on natural deduction* ⋄ B10 ⋄
SLATER, J.G. *The required correction to Copi's statement of UG* ⋄ B10 ⋄

1967
BINKLEY, R. & CLARK, R. *A cancellation algorithm for elementary logic* ⋄ B10 ⋄
CARRERAS MATAS, J.M. *Class of formulas of minimum content of an axiomatic system with respect to a given formula (Spanish)* ⋄ B10 ⋄
DAVYDOV, G.V. *A method of establishing deducibility in the classical predicate calculus (Russian)* ⋄ B10 B35 ⋄
DAVYDOV, G.V. *The correction of unprovable formulas (Russian)* ⋄ B10 B35 ⋄
DUMITRIU, A. *La negation des quantificateurs* ⋄ B10 ⋄
LEBLANC, H. & THOMASON, R.H. *All or none: a novel choice of primitives for elementary logic* ⋄ B10 ⋄
MINTS, G.E. *Herbrand's theorem (Russian)* ⋄ B10 F05 F07 ⋄
PRAWITZ, D. *A note on existential instantiation* ⋄ B10 ⋄
REICHBACH, J. *On generalizations of the satisfiability definition and Gentzen-Jaskowski's sequent proof rules* ⋄ B10 F07 ⋄
ROGAVA, M.G. *On sequential modifications of applied predicate calculi (Russian)* ⋄ B10 ⋄
SCOTT, D.S. *Existence and description in formal logic* ⋄ A05 B10 E30 ⋄
SHAFAAT, A. *Principle of localization for a more general type of languages* ⋄ B10 C55 C75 C90 ⋄
STRAUSS, P. *Some systems of natural deduction* ⋄ B05 B10 ⋄

1968
ALIMPIC, B.P. *On models of certain formulas of the predicate calculus of first order* ⋄ B10 ⋄
BELNAP JR., N.D. & DUNN, J.M. *The substitution interpretation of the quantifiers* ⋄ A05 B10 ⋄
BING, K. *Natural deduction with few restrictions on variables* ⋄ B10 F07 ⋄
BONDARENKO, A.S. *Use of semantic tableaux to establish the feasibility and general validity of certain formulae of predicate logic (Russian)* ⋄ B10 ⋄
BORKOWSKI, L. *Some remarks about the notion of definition (Polish) (Russian and English summaries)* ⋄ A05 B10 C40 ⋄
CURRY, H.B. *A deduction theorem for inferential predicate calculus* ⋄ B10 ⋄
CURRY, H.B. *The purposes of logical formalizations* ⋄ B10 ⋄
DELGADO, V.M. *Los lenguajes formalizados de la logica moderna* ⋄ B10 ⋄
FOLLESDAL, D. *Interpretation of quantifiers* ⋄ A05 B10 C90 ⋄
GUPTA, H.N. *On the rule of existential specification in systems of natural deduction* ⋄ A05 B10 F07 ⋄
HEINTZ, JOHN *Identity, quantification and predicables* ⋄ A05 B10 ⋄
HENKIN, L. *Relativization with respect to formulas and its use in proofs of independence* ⋄ B10 ⋄
HERMES, H. *Praedikatenlogik und Theorie der rekursiven Funktionen* ⋄ B10 D20 ⋄
LEBLANC, H. *A simplified account of validity and implication for quantificational logic* ⋄ B10 ⋄
LEBLANC, H. *Syntactically free, semantically bound (a note on variables)* ⋄ B10 ⋄
LEWANDOWSKI, A. & SUSZKO, R. *A note concerning the theory of descriptions (Polish and Russian summaries)* ⋄ B10 ⋄

LIGHTSTONE, A.H. *Group theory and the principle of duality* ◇ B10 ◇
LIS, Z. *An algebraic approach to traditional logic (Polish) (English and Russian summaries)* ◇ B10 ◇
MINTS, G.E. *Admissible and deductive rules (Russian)* ◇ B10 F07 F50 ◇
OREVKOV, V.P. *Glivenko classes of sequents (Russian)* ◇ B10 F50 ◇
OREVKOV, V.P. *Glivenko's sequence classes (Russian)* ◇ B10 B20 F50 ◇
PLIUSKEVICIUS, R. *Kanger's version of the predicate calculus with symbols for not everywhere defined functions (Russian)* ◇ B10 B60 ◇
POGORZELSKI, W.A. *On the scope of the classical deduction theorem* ◇ A05 B10 ◇
REICHBACH, J. *A note on theses of the first-order functional calculus* ◇ B10 ◇
SMULLYAN, R.M. *Analytic cut* ◇ B10 F05 F07 ◇
SMULLYAN, R.M. *Uniform Gentzen system* ◇ B10 ◇

1969

BENEJAM, J.-P. *Application du theoreme de Herbrand a la presentation de theses teratologiques de calcul des predicats elementaires* ◇ B10 ◇
BERLINSKI, D. & GALLIN, D. *Quine's definition of logical truth* ◇ A05 B10 ◇
BERTI, S.N. *Applications de la theorie des relations dans la logique mathematique (Russian)* ◇ A05 B10 ◇
EBERLE, R.A. *Denotationless terms and predicates expressive of positive qualities* ◇ B10 ◇
FELSCHER, W. *On the algebra of quantifiers (Russian summary)* ◇ B10 C05 C07 G15 ◇
GLEBSKIJ, YU.V. & KOGAN, D.I. & LIOGON'KIJ, M.I. & TALANOV, V.A. *Range and degree of realizability of formulas in the restricted predicate calculus (Russian)* ◇ B10 C07 C13 ◇
HAILPERIN, T. *A form of Herbrand's theorem* ◇ B10 F05 F07 ◇
ISSEL, W. *Semantische Untersuchungen ueber Quantoren I* ◇ B10 C55 C80 ◇
LEISENRING, A.C. *Mathematical logic and Hilbert's ε-symbol* ◇ B10 E25 F05 F98 ◇
LIGHTSTONE, A.H. *The notion of "consequence" in the predicate calculus* ◇ B10 ◇
LIOGON'KIJ, M.I. *A certain property of the equivalence axiom in first order predicate calculus (Russian)* ◇ B10 ◇
LIOGON'KIJ, M.I. *On the conditional satisfiability ratio of logical formulae (Russian)* ◇ B10 ◇
MOSCHOVAKIS, J.R. *A note on k-axiomatizations of identity* ◇ B10 ◇
REICHBACH, J. *Some examples of different methods of formal proofs with generalizations of the satisfiability definition* ◇ B10 ◇
RIBEIRO, H. *Elementary languages and mathematical structures (Portuguese)* ◇ B10 C07 ◇
ROUTLEY, R. *A simple natural deduction system* ◇ B10 ◇
SCHREIBER, P. *Ueber die Quantorenbasen der klassischen Praedikatenlogik* ◇ B10 ◇
SLOMSON, A. *An undecidable two sorted predicate calculus* ◇ B10 D35 ◇
STOPES-ROE, H.V. *An economy in the formation rules for quantification theory* ◇ B10 ◇

1970

ESSLER, W.K. *Ein nichtkonstruktiver Beweis des ersten ε-Theorems* ◇ B10 ◇
FRAISSE, R. *Aspects du theoreme de completude selon Herbrand* ◇ B10 C07 ◇
GOE, G. *Reconstructing formal logic: further developments and considerations* ◇ A05 B10 ◇
GREGG, J.R. *Axiomatic quasi-natural deduction* ◇ B10 F07 ◇
HAJEK, P. *Logische Kategorien* ◇ B10 C40 E40 F25 G30 ◇
REICHBACH, J. *On statistical tests in generalizations of Herbrand's theorems and asymptotic probabilistic models (main theorems of mathematics)* ◇ B10 C40 C90 ◇
SHARONOV, V.I. & ZAMOV, N.K. *Amplifications of formulae of predicate calculus (Russian)* ◇ B10 B20 D35 ◇
SHCHEGOL'KOVA, G.M. *Certain theorems of the theory of quantifiers (Russian)* ◇ B10 ◇
SMULLYAN, R.M. *Abstract quantification theory* ◇ B10 C07 F07 ◇
SZCZERBA, L.W. *Semantic method of proving theorems (Russian summary)* ◇ B10 ◇
WINNIE, J.A. *The completeness of Copi's system of natural deduction* ◇ B10 B45 ◇

1971

ALIMPIC, B.P. *On models of certain formulas with a predicate letter of length n* ◇ B10 ◇
BENEYTO, R. *Analytic labyrinths (Spanish)* ◇ B10 F07 ◇
BRADLEY, M.C. *Copi's method of deduction again* ◇ A05 B10 ◇
CORCORAN, J. & HERRING, J.M. *Notes on a semantic analysis of variable binding term operators* ◇ A05 A10 B10 C80 ◇
DREBEN, B. & GOLDFARB, W.D. *Note J* ◇ B10 F05 F07 ◇
DREBEN, B. & GOLDFARB, W.D. *Note N* ◇ A10 B10 F30 F35 ◇
DULAC, M.-H. *Decidabilite et operations entre theories* ◇ B10 B25 D35 ◇
FELSCHER, W. *An algebraic approach to first order logic* ◇ B10 C05 C07 C20 C90 G15 ◇
FLUM, J. *Eine Formulierung des Herbrandschen Satzes ohne Skolemfunktion* ◇ B10 ◇
FLUM, J. *Ganzgeschlossene und praedikatengeschlossene Logiken I,II* ◇ B10 C85 C95 ◇
GREGG, J.R. *Two modes of deductive inference* ◇ B10 ◇
HARRIS, J.H. *Ordinal theory in a conservative extension of predicate calculus* ◇ B10 E10 E30 ◇
HENKIN, L. & MONK, J.D. & TARSKI, A. *Cylindric algebras. Vol.I* ◇ B10 G15 ◇
HUNTER, G. *Metalogic: an introduction to the metatheory of standard firstorder logic* ◇ B10 B98 ◇
MARKWALD, W. *Praedikatenlogik mit partiell definierten Funktionen I* ◇ B10 B60 ◇

PARKS, R.Z. & RESCHER, N. *Restricted inference* ◊ B10 F07 ◊

PARSONS, C. *A plea for substitutional quantification* ◊ A05 B10 ◊

REICHBACH, J. *Some methods of formal proofs. III* ◊ B10 B15 ◊

RENNIE, M.K. *Completeness in the logic of predicate modifiers* ◊ B10 ◊

SANCHIS, L.E. *A generalization of the Gentzen Hauptsatz* ◊ B10 F05 ◊

SANCHIS, L.E. *Cut elimination, consistency and completeness in classical logic* ◊ B10 F05 ◊

TICHY, P. *On the vicious circle in definitions (Polish and Russian summaries)* ◊ A05 B10 ◊

WILCOX, W.C. *A mistake in Copi's discussion of completeness* ◊ B10 ◊

1972

ASSER, G. *Einfuehrung in die mathematische Logik, Teil II: Praedikatenkalkuel der ersten Stufe* ◊ B10 ◊

BOTH, N. *Monotonicity in the predicate calculus (Romanian) (French summary)* ◊ B10 ◊

BOWEN, K.A. *A note on cut elimination and completeness in first order theories* ◊ B10 C07 F05 ◊

BULLOCK, A.M. & SCHNEIDER, H.H. *A calculus for finitely satisfiable formulas with identity* ◊ B10 C07 C13 ◊

CALL, R.L. *The Goedel-Herbrand theorems* ◊ B10 C07 ◊

CORCORAN, J. & HATCHER, W.S. & HERRING, J.M. *Variable binding term operators* ◊ B10 C80 ◊

DELGADO, V.M. *Lecciones de logica I. Introduccion general. Logica de proposiciones* ◊ B05 B10 ◊

DRAGALIN, A.G. *On the use of classical calculi for establishing constructive truth (Russian) (English summary)* ◊ B10 F30 F50 ◊

FRAISSE, R. *Reflexions sur la completude selon Herbrand* ◊ B10 C07 ◊

HAUCK, J. & HERRE, H. & POSEGGA, M. *Zur Metatheorie formaler Systeme* ◊ B10 C07 F30 ◊

HOOKER, C.A. *Definite descriptions* ◊ A05 B10 ◊

LEBLANC, H. & SNYDER, D.P. *Duals of Smullyan trees* ◊ B10 F07 ◊

LEBLANC, H. & MEYER, R.K. *Matters of separation* ◊ B10 F07 F50 ◊

LOEB, M.H. *A reduction theorem for predicate logic* ◊ B10 ◊

MOISIL, G.C. *La logique formelle et son probleme actuel* ◊ B10 ◊

ROGAVA, M.G. *Permutability of the applications of rules in applied predicate calculi (Russian) (Georgian and English summaries)* ◊ B10 ◊

ROGAVA, M.G. *Sequential variants of applied predicate calculi without structural deductive rules (Russian)* ◊ B10 F07 ◊

SVENONIUS, L. *Translation and reduction* ◊ A05 B10 ◊

WESSEL, H. *Eine dialogische Begruendung logischer Gesetze* ◊ A05 B10 ◊

1973

BENEYTO, R. *Trees, logic and decision mechanisms (Spanish)* ◊ B10 ◊

BENNETT, D.W. *An elementary completeness proof for a system of natural deduction* ◊ B10 C07 F07 ◊

BULLOCK, A.M. & SCHNEIDER, H.H. *On generating the finitely satisfiable formulas* ◊ B10 C07 C13 D25 ◊

COSTA DA, N.C.A. *On the concept of transformation in the predicate calculus (Portuguese)* ◊ B10 ◊

HALKOWSKA, K. *Conditional definitions and the idea of meaningful expression with conditionally defined terms (Polish) (English summary)* ◊ B10 B20 ◊

HEGENBERG, L. *Logica. O calculo de predicados* ◊ B10 ◊

HINTIKKA, K.J.J. *Surface semantics: definition and its motivation* ◊ A05 B10 B45 ◊

ISEKI, K. *Symbolic logic. II: Predicate logic (Japanese)* ◊ B10 ◊

JERVELL, H.R. *Skolem and Herbrand theorems in first order logic* ◊ B10 C07 F05 ◊

KEISLER, H.J. & WALKOE JR., W.J. *The diversity of quantifier prefixes* ◊ B10 C07 C13 D55 ◊

MEYERS, L.F. *Simultaneous versus successive quantification* ◊ B10 ◊

MORGAN, C.G. *Truth, falsehood, and contigency in first-order predicate calculus* ◊ A05 B10 ◊

MOTOHASHI, N. *Two theorems on mix-relativization* ◊ B10 C40 C75 ◊

PAILLET, J.L. *La methode des systemes logiquement inductifs* ◊ B10 C07 C30 ◊

PREVIALE, F. *Diversificazione delle specie di individuo nella logica del I ordine (English summary)* ◊ B10 ◊

SCANLON, T.M. *The consistency of number theory via Herbrand's theorem* ◊ B10 F05 F30 ◊

SMIRNOV, V.A. *An absolute first order predicate calulus* ◊ B10 F05 ◊

SVENONIUS, L. *On the first-order logic of terms* ◊ B10 ◊

SVENONIUS, L. *Translation and reduction* ◊ A05 B10 ◊

WANG, JUENTIN *On the representation of generative grammars as first-order theories* ◊ B10 D05 ◊

1974

ANDREKA, H. & GERGELY, T. & NEMETI, I. *Sufficient and necessary condition for the completeness of a calculus* ◊ B10 C07 G05 ◊

BENTHEM VAN, J.F.A.K. *Semantic tableaus* ◊ B10 C07 F07 ◊

BIELA, A. *Structural incompleteness of the Quine's formalization of the classical predicate calculus* ◊ B10 ◊

BOCHVAR, D.A. & FUKSON, V.I. *Boolean operations over Russell cores (Russian)* ◊ B10 E30 G05 ◊

BOTH, N. *Considerations algebriques sur la theorie de la demonstration* ◊ B10 E20 G15 ◊

CHADADZE, O.S. & ROGAVA, M.G. *Sequential variants of predicate calculus with equality (Russian) (Georgian and English summaries)* ◊ B10 ◊

DELGADO, V.M. *Lecciones de logica II. Logica de la cuantificacion. Algebra de clases. Otros temas. Historia de la logica deductiva. Logica inductiva* ◊ A10 B10 ◊

EHDEL'MAN, G.S. *Certain radicals of metaideals (Russian)* ◊ B10 C07 C60 ◊

ENDO, H. *Quantification, games and existence* ◊ B10 ◊
FULTON, J.A. *Unary predicates* ◊ B10 ◊
GOODSTEIN, R.L. *Satisfiable in a larger domain* ◊ B10 C07 ◊
GRELL, B. *Un simple systeme de logique fonde sur regles* ◊ B10 ◊
HINTIKKA, K.J.J. *Quantifiers vs. quantification theory* ◊ A05 B10 C80 ◊
LINDSTROEM, P. *On characterizing elementary logic* ◊ B10 C95 ◊
MARKWALD, W. *Praedikatenlogik mit partiell definierten Funktionen. II* ◊ B10 B60 ◊
MINTS, G.E. *Functional form. Herbrand's theorem for non-prenex formulas (Russian)* ◊ B10 F05 ◊
MINTS, G.E. *Gentzen's formal systems (Russian)* ◊ B10 F05 F07 ◊
NESGOVOROVA, G.P. *Ueberfuehrung von Formeln des Praedikatenkalkuels erster Ordnung in die Praefixform (Russisch)* ◊ B10 ◊
PIECZKOWSKI, A. *Ueber Theorien im erweiterten Sinne* ◊ B10 ◊
POGORZELSKI, W.A. & PRUCNAL, T. *Structural completeness of the first-order predicate calculus* ◊ B10 ◊
RAGGIO, A.R. *A simple proof of Herbrand's theorem* ◊ B10 F05 F07 ◊
SAKAI, H. *On necessary but not-sufficient conditions* ◊ B10 F30 F35 G05 ◊
SMIRNOVA, E.D. *Consistency and eliminability in proof theory (Russian)* ◊ A05 B10 F99 ◊
VARSAVSKY, O. *Quantifiers and equivalence relations* ◊ B10 G15 ◊
WETTE, E. *Contradiction within pure number theory because of a system-internal "consistency"-deduction* ◊ B10 F30 ◊

1975

BAXTER, ROBERT & SEARBY, D. *Modus ponens eliminability in a formalisation of the predicate calculus* ◊ B10 ◊
BENEYTO, R. *A natural aspect of natural deduction* ◊ B10 ◊
CHANKVETADZE, O.E. *On some questions of the theory Ω_4 (Russian) (English summary)* ◊ B10 ◊
CHANKVETADZE, O.E. *On some questions of the theory Ω_5 (Russian) (English summary)* ◊ B10 ◊
COWEN, R.H. *A characterization of logical consequences in quantification theory* ◊ B10 ◊
DEUTSCH, M. *Zur Theorie der spektralen Darstellung von Praedikaten durch Ausdruecke der Praedikatenlogik 1.Stufe* ◊ B10 C13 D25 D35 ◊
GALLIE, R.D. *Substitutionalism and substitutional quantification* ◊ B10 ◊
HENDRY, H.E. *Another system of natural deduction* ◊ B10 F07 ◊
KENT, C.F. *Independence versus logical independence in the countable case* ◊ B10 ◊
KIRIN, V.G. *On free and independent terms of a formula (Serbo-Croatian summary)* ◊ B10 ◊

MARCHENKOV, S.S. *Elementary Skolem functions (Russian)* ◊ B10 ◊
MARINI, D. & MIGLIOLI, P.A. & ORNAGHI, M. *First order logic as a tool to solve and classify problems* ◊ B10 B25 F50 ◊
MELONI, G.C. *Introduzione semantica alla logica matematica* ◊ B10 ◊
ORLOWSKA, E. *On the Jaskowski's method of suppositions* ◊ B10 ◊
PAILLET, J.L. *Une characterisation de l'equivalence des formules modulo une theorie (English summary)* ◊ B10 ◊
POGORZELSKI, W.A. & PRUCNAL, T. *Structural completeness of the first-order predicate calculus* ◊ B10 ◊
POGORZELSKI, W.A. & PRUCNAL, T. *The substitution rule for predicate letters in the first-order predicate calculus* ◊ B10 ◊
PRAWITZ, D. *Comments on Gentzen-type procedures and the classical notion of truth* ◊ B10 F05 F07 F30 ◊
PRESIC, S.B. *Equational reformulation of formal theories* ◊ B10 ◊
PUDLAK, P. *The observational predicate calculus and complexity of computations* ◊ B10 D15 ◊
RANTALA, V. *Urn models: A new kind of non-standard model for first-order logic* ◊ A05 B10 C07 C90 ◊
REICHBACH, J. *Generalized models for classical and intuitionistic predicate calculi* ◊ B10 B25 C90 F50 ◊
SCHWARTZ, DIETRICH *Ultraprodukte in der Theorie der logischen Auswahlfunktionen* ◊ B10 C20 ◊
SLISENKO, A.O. *Finite approach to the problem of optimizing theorem-proving algorithms (Russian) (English summary)* ◊ B10 B35 ◊
STEVENSON, L. *A formal theory of sortal quantification* ◊ A05 B10 ◊
THARP, L.H. *Which logic is the right logic?* ◊ A05 B10 ◊
VENNERI, B.M. *Semantic implications of Herbrand's theory of fields* ◊ A05 B10 C07 ◊

1976

BUTH, M. *Logische Analyse eines mathematischen Lehrsatzes* ◊ A05 B10 ◊
CHANKVETADZE, O.E. *On the axiomatics of the Ω-calculus (Russian) (Georgian and English summaries)* ◊ B10 ◊
CHRISTEN, C. *Spektralproblem und Komplexitaetstheorie* ◊ B10 C13 D10 D15 D25 D35 ◊
CROSSLEY, J.N. *Introduction to mathematical logic: a new area of mathematics with ancient roots* ◊ B10 ◊
EMDEN VAN, M.H. & KOWALSKI, R. *The semantics of predicate logic as a programming language* ◊ B10 B75 ◊
IANNACCI, R. *Un teorema relativo alla estensione di Henkin del calcolo predicativo (English summary)* ◊ B10 ◊
LESISZ, W. & POGORZELSKI, W.A. *A simplified definition of the notion of similarity between formulas of the first-order predicate calculus* ◊ B10 ◊

LOEB, M.H. *Embedding first order predicate logic in fragments of intuitionistic logic* ⋄ B10 D35 F50 ⋄

MELZI, G. *I supporti fisici dell'inferenza formale. Vol. I* ⋄ B10 B75 D05 ⋄

NAGORNYJ, N.M. *Normal algorithms and first order languages (Russian)* ⋄ B10 D03 ⋄

POSPESEL, H. *Introduction to logic. Predicate logic* ⋄ B10 B98 ⋄

SCHNEIDER, H.H. *A deduction system for the full first-order predicate logic* ⋄ B10 ⋄

STEVENSON, L. *Freges zwei Definitionen der Quantifikation* ⋄ A05 A10 B10 ⋄

WALKOE JR., W.J. *A small step backwards* ⋄ B10 C80 ⋄

WELDING, S.O. *Logic as based on truth-value relations* ⋄ A05 B10 ⋄

1977

BARWISE, J. *An introduction to first-order logic* ⋄ B10 C07 C98 ⋄

BENNETT, D.W. *A note on the completeness proof for natural deduction* ⋄ B10 C07 F07 ⋄

BRADY, R.T. *Unspecified constants in predicate calculus and first-order theories* ⋄ B10 C07 ⋄

CHANKVETADZE, O.E. *Comparatively simple proofs of some criteria of N. Bourbaki (Russian) (Georgian and English summaries)* ⋄ B10 ⋄

CHUBARYAN, A.A. *The complexity of deductions in formal arithmetic and predicate calculus (Russian)* ⋄ B10 F20 F30 ⋄

KUZICHEV, A.S. *A system of λ-conversion with logical operators and an equality operator (Russian)* ⋄ B10 B40 F05 ⋄

LAMBERT JR., W.M. *Un tipo de solidez para un sistema de deduccion natural* ⋄ B10 F07 ⋄

NEDERPELT, R.P. *Presentation of natural deduction* ⋄ B10 F07 ⋄

OBERSCHELP, A. *Ein Ansatz zur Beruecksichtigung von Sprechhandlungen in der Praedikatenlogik* ⋄ B10 B65 ⋄

PUDLAK, P. *Generalized quantifiers and semisets* ⋄ B10 C80 E70 ⋄

RAGGIO, A.R. *Semi-formal Beth tableaux* ⋄ B10 F07 ⋄

RUKHAYA, KH.M. *A formal theory $\mathcal{T}$ (Russian) (Georgian and English summaries)* ⋄ B10 E30 ⋄

SCHWARTZ, DIETRICH *Sequenzenschliessen in der algebraischen Attributenlogik* ⋄ B10 G25 ⋄

SEMENOV, A.L. *Presburgerness of predicates regular in two number systems (Russian)* ⋄ B10 D20 F30 ⋄

SHAPIRO, S. *Incomplete translations of complete logics* ⋄ B10 ⋄

VOBACH, A.R. *The weak topology on logical calculi* ⋄ B10 ⋄

1978

ABIAN, A. *A proof of Rasiowa-Sikorski theorem via complete sequences* ⋄ B10 ⋄

BELLOMO, A. *Analisi di un'argomentazione del "Convivio" dantesco mediante il calcolo dei predicati* ⋄ A10 B10 ⋄

BENDALL, K. *Natural deduction, separation, and the meaning of logical operators* ⋄ A05 B10 ⋄

BOERGER, E. *Ein einfacher Beweis fuer die Unentscheidbarkeit der klassischen Praedikatenlogik* ⋄ B10 D03 D10 D35 ⋄

CHANKVETADZE, O.E. *Some fundamental questions concerning the theory Ω_6 (Russian) (Georgian and English Summary)* ⋄ B10 ⋄

ELLERMAN, D.P. & ROTA, G.-C. *A measure theoretic approach to logical quantification* ⋄ B10 C90 G25 ⋄

FILIPOIU, A. *Many-sorted polynomials and algebras of polynomials (Romanian) (English summary)* ⋄ B10 C05 ⋄

GUMB, R.D. *Metaphor theory* ⋄ B10 ⋄

HAFNER, I. *The deduction theorem (Slovenian) (English summary)* ⋄ B10 ⋄

ISAACS, R. *An example of a first-order language* ⋄ B10 ⋄

KIRIN, V.G. *A semantic characterization of terms free for some variable in a formula* ⋄ B10 ⋄

KIRIN, V.G. *Prolegomena to mathematics. Elements of mathematical logic (Serbo-Croatian)* ⋄ B10 ⋄

LABORDE, J.-M. *Un theoreme d'algebre de Boole et le theoreme d'Herbrand (English summary)* ⋄ B10 G05 G10 ⋄

MILNE, R. *Transforming predicate transformers* ⋄ B10 ⋄

MOTOHASHI, N. *An elimination theorem of uniqueness condition* ⋄ B10 C40 ⋄

PLISKO, V.E. *Some variants of the notion of realizability for predicate formulas (Russian)* ⋄ B10 F50 ⋄

RICKEY, V.F. *On creative definitions in first order functional calculi* ⋄ B10 C40 ⋄

RUKHAYA, KH.M. *A formal equality theory $\mathcal{T}_{eg}$ (Russian) (Georgian and English Summary)* ⋄ B10 ⋄

SCHUETTE, K. *Ein Ansatz zum Entscheidungsverfahren fuer eine Formelklasse der Praedikatenlogik mit Identitaet* ⋄ B10 B25 ⋄

SETTE, A.-M. & SETTE, J. *Functorialization of first-order language with finitely many predicates* ⋄ B10 G30 ⋄

THIEL, C. *Duality lost? Transforming Gentzen derivations into winning strategies for dialogue games* ⋄ B10 F07 ⋄

WOJTASIEWICZ, O. *The predicate calculus with extra-logical constants as an instrument of semantic description* ⋄ A05 B10 ⋄

ZUCKER, J.I. *The adequacy problem for classical logic* ⋄ A05 B10 ⋄

1979

BELLIN, G. *Herbrand's theorem for calculi of sequents LK and LJ* ⋄ B10 ⋄

BENDALL, K. *Negation as a sign of negative judgment* ⋄ B10 ⋄

BESSONOV, A.V. *Mixed interpretation of logical systems (Russian)* ⋄ B10 ⋄

BRANDOM, R.B. *A binary Sheffer operator which does the work of quantifiers and sentential connectives* ⋄ B10 ⋄

CARTWRIGHT, ROBERT & MCCARTHY, J. *Recursive programs as functions in a first order theory* ⋄ B10 B75 D20 ⋄

CIRULIS, J. *Theories with inclusion (Russian)* ⋄ A05 B10 ⋄

FUNAHASHI, S. & NAGATA, S. & OSHIBA, T. *A method for obtaining proof figures of valid formulas in the first order predicate calculus* ◇ B10 F07 ◇

GENRICH, H.J. & LAUTENBACH, K. *The analysis of distributed systems by means of predicate/transition-nets* ◇ B10 ◇

HADGOPOULOS, D.J. *The principle of the division into four figures in traditional logic* ◇ A05 A10 B10 ◇

HANAZAWA, M. *An interpretation of Skolem's paradox in the predicate calculus with ε-symbol*
◇ B10 C07 F05 F25 ◇

HANDSCHEL, G. *Eine graphische Veranschaulichung der logischen Operationen* ◇ B10 G15 ◇

KUZNETSOV, A.V. & RATSA, M.F. *A criterion for functional completeness in classical first-order predicate logic (Russian)* ◇ B10 ◇

LEBLANC, H. *Generalization in first order logic* ◇ B10 ◇

MO, SHAOKUI *Study on mathematical logic (Chinese) (English summary)* ◇ B10 ◇

NEDZYNSKI, T.G. *Quantification, domains of discourse, and existence* ◇ A05 B10 ◇

ROBINSON, JOHN ALAN *Logic: form and function. The mechanization of deductive reasoning*
◇ A10 B10 B35 B98 ◇

RUKHAYA, KH.M. *A variant of theory T extended by abbreviating symbols* ◇ B10 ◇

SCHREIBER, P. *Die allgemeinste Form der formalisierten Sprachen* ◇ B10 ◇

SCOTT, D.S. *A note on distributive normal forms*
◇ B10 C07 ◇

SMIRNOV, V.A. *Theory of quantification and ℰ-calculi*
◇ B10 F50 ◇

SUVOROV, P.YU. *Representation of proofs by colored graphs and the Hadwiger conjecture (Russian)*
◇ B10 F07 ◇

SWIGGART, P. *Domain restrictions in standard deductive logic* ◇ A05 B10 ◇

UMEZAWA, T. *A method for cut elimination in intuitionistic predicate logic and classical predicate logic*
◇ B10 F05 F50 ◇

VINOGRADOV, A.P. *A theorem on the local rank of one-place predicates (Russian)* ◇ B10 ◇

1980

ALCANTARA DE, L.P. & CORRADA, M. *Notes on many-sorted systems* ◇ B10 C07 ◇

CHANKVETADZE, O.E. *On some criteria of N. Bourbaki's quantified theory and the corresponding criteria of the Ω-calculus (Russian) (English and Georgian summaries)* ◇ B10 ◇

DAVIES, M.K. *A note on substitutional quantification*
◇ B10 ◇

HANAZAWA, M. *An extension of the notion of relativization to Hilbert's ε-symbol* ◇ B10 ◇

HUGLY, P. *Reflections on an extensionality theorem*
◇ B10 ◇

HYLTON, P. *Russell's substitutional theory*
◇ A05 A10 B10 ◇

JOHNSON WU, K. *On a tableau rule for identity* ◇ B10 ◇

LIFSCHITZ, V. *Semantical completeness theorems in logic and algebra* ◇ B10 F07 ◇

NERODE, A. & SHORE, R.A. *Second order logic and first order theories of reducibility orderings*
◇ B10 B15 D30 D35 F35 F40 ◇

NORGELA, S.A. *On approximation of some classes of classical predicate calculus by decidable classes. I,II (Russian) (English and Lithuanian summaries)*
◇ B10 B20 B25 ◇

PORTE, J. *Simplifying the axioms of the predicate calculus*
◇ B10 ◇

RUBCHINSKIJ, A.A. & VINOGRADSKAYA, T.M. *Logical forms of choice functions (Russian)* ◇ B10 E25 ◇

SCHNEIDER, H.H. *Substitutions for predicate variables and functional variables* ◇ B10 ◇

WEAVER, G.E. *A note on the compactness theorem in first order logic* ◇ B10 C07 ◇

WHALEY, T.P. & WILLIFORD, J. *Differentiability and permutations of quantifiers* ◇ B10 ◇

WILLIAMS, C.J.F. *Misinterpretations of quantifiers*
◇ A05 B10 ◇

1981

BENCIVENGA, E. *Semantic tableaux for a logic with identity* ◇ B10 ◇

BUNDER, M.W. *Predicate calculus and naive set theory in pure combinatory logic* ◇ B10 B40 E70 ◇

CASARI, E. *Positively omitting types* ◇ B10 C07 ◇

CHANKVETADZE, O.E. *Invariance of homogeneity with respect to homogeneous substitutions in N.Bourbaki's τ-calculus (Russian) (Georgian and English summaries)* ◇ B10 ◇

FERRARI, P.L. *Certe estensioni di teorie del I ordine mediante operatori logici di selezione (English summary)* ◇ B10 ◇

FINN, V.K. & SKVORTSOV, D.P. *A remark about an extension of the language of the many-sorted predicate logic (Russian)* ◇ B10 ◇

HAFNER, I. *First-order language with equality* ◇ B10 ◇

HINTIKKA, K.J.J. *Standard vs. nonstandard logic: higher order, modal, and first-order logics*
◇ B10 B15 B45 ◇

KLEINKNECHT, R. *Eliminability, noncreativity and explicit definability* ◇ B10 ◇

MATROSOV, V.M. & VASIL'EV, S.N. *Comparison principles in the dynamics of systems with distributed parameters (Russian)* ◇ B10 ◇

MUNDICI, D. *A group-theoretical invariant for elementary equivalence and its role in representations of elementary classes* ◇ B10 C07 C52 ◇

OSHIBA, T. *A method for obtaining proof figures of valid formulas in the first order predicate calculus* ◇ B10 ◇

PAULOS, J.A. *Probabilistic, truth-value, and standard semantics and the primacy of predicate logic*
◇ A05 B10 B48 ◇

QUINE, W.V.O. *Predicate functors revisited*
◇ B10 B40 ◇

RODRIGUEZ ARTALEJO, M. *Eine syntaktisch-algebraische Methode zur Konstruktion von Modellen*
◇ B10 C07 C30 C57 ◇

SCHMIDT, J. *Algebraic studies of first-order enlargements*
◇ B10 G15 G30 H20 ◇

TANG, TONGGAO *A note on Herbrand's theorem (Chinese) (English summary)* ◇ B10 F05 ◇

1982

BENCIVENGA, E. & LAMBERT, K. & MEYER, R.K. *The ineliminability of E! in free quantification theory without identity* ◇ B10 ◇

BERGSTRA, J.A. & TUCKER, J.V. *Two theorems about the completeness of Hoare's logic* ◇ B10 B75 ◇

CHANKVETADZE, O.E. *The invariance of uniformity with respect to uniform substitutions in N. Bourbaki's τ-calculus (Russian) (English and Georgian summaries)* ◇ B10 ◇

CHANKVETADZE, O.E. *The law of implication self-distributivity and the criterion for deduction in Bourbaki's τ-calculus (Russian) (English and Georgian summaries)* ◇ B10 ◇

CHANKVETADZE, O.E. *The rule of multiplication of implication and its application in τ-calculus (Russian) (English and Georgian summaries)* ◇ B10 ◇

CHUBARYAN, A.A. *Complexity characteristics of inferences in systems of predicate calculus and formal arithmetic (Russian) (Armenian summary)* ◇ B10 F20 F30 ◇

CVETKOVIC, D.M. & PEVAC, I. *Algorithms for transforming first order formulas in their natural form* ◇ B10 ◇

ECSEDI-TOTH, P. & TURI, L. *On enumerability of interpolants* ◇ B10 C40 ◇

FAUST, D.H. *The Boolean algebra of formulas of first-order logic* ◇ B10 G05 ◇

GLUBRECHT, J.-M. *Ein Vollstaendigkeitsbeweis fuer schnittfreie Kalkuele mit der Maximalisierungsmethode von Henkin* ◇ B10 C07 F05 ◇

HARRIS, J.H. *What's so logical about the "logical" axioms?* ◇ B10 ◇

HUGLY, P. & SAYWARD, C.W. *Indenumerability and substitutional quantification* ◇ A05 B10 C07 ◇

KFOURY, A.J. *Some connections between iterative programs, recursive programs, and first-order logic* ◇ B10 B75 ◇

KIRIN, V.G. *On decomposition of formulas of a first-order language* ◇ B10 ◇

KIRKERUD, B. *Completeness of Hoare-calculi revisited* ◇ B10 ◇

MA, XIWEN *A relational approach in semantics (Chinese) (English summary)* ◇ B10 ◇

MOTOHASHI, N. *ε-theorems and elimination theorems of uniqueness conditions* ◇ B10 C40 F05 ◇

MOTOHASHI, N. *An axiomatization theorem* ◇ B10 C07 C40 C75 ◇

MOTOHASHI, N. *Elimination theorems of uniqueness conditions* ◇ B10 ◇

PKHAKADZE, V.SH. *Substitutability of equivalences and equalities in the equality theory of Bourbaki (Russian) (English and Georgian summaries)* ◇ B10 ◇

RUKHAYA, KH.M. *Description of a derived formal mathematical theory $\mathcal{T}^+$ (Russian) (English and Georgian summaries)* ◇ B10 ◇

TUL'CHINSKIJ, G.L. *Some problems of the logical explication of "language games" (Russian)* ◇ B10 ◇

1983

BOCHAROV, V.A. *Subject-predicate calculus free from existential import* ◇ B10 ◇

COSTA DA, N.C.A. & MORTENSEN, C. *Notes on the theory of variable binding term operators* ◇ B10 ◇

ESSLER, W.K. & MARTINEZ CRUZADO, R.F. *Grundzuege der Logik I: Das logische Schliessen* ◇ B10 ◇

FROIDEVAUX, C. *La fonction logique ε de Hilbert a travers les "Grundlagen der Mathematik"* ◇ B10 ◇

GIOVANNETTI, E. *Une caracterisation des termes types dans un langage applicatif (English summary)* ◇ B10 B40 ◇

GLUBRECHT, J.-M. & OBERSCHELP, A. & TODT, G. *Klassenlogik* ◇ B10 B98 E30 ◇

GRUENBERG, T. *A tableau system of proof for predicate-functor logic with identity* ◇ B10 F07 ◇

HANF, W.P. & MYERS, D.L. *Boolean sentence algebras: isomorphism constructions* ◇ B10 C52 ◇

HELMAN, GLEN *An interpretation of classical proofs* ◇ A05 B10 ◇

HODGES, W. *Elementary predicate logic* ◇ B10 C98 ◇

HOSSAIN, M.F. *On Beth trees in mathematical logic (Bengali summary)* ◇ B10 ◇

KIRIN, V.G. *A note on the deduction theorem* ◇ B10 ◇

KUEHNRICH, M. *On the Hermes term logic* ◇ B10 C07 C20 ◇

KUHN, STEVEN T. *An axiomatization of predicate functor logic* ◇ B10 ◇

MARKUSZ, Z. *On first order many-sorted logic* ◇ B10 C07 ◇

RUBEL, L.A. *Conformal inequivalence of annuli and the first-order theory of subgroups of PSL(2,R)* ◇ B10 ◇

1984

BLASS, A.R. & GUREVICH, Y. & KOZEN, D. *A 0-1 law for logic with a fixed-point operator* ◇ B10 B75 C80 D70 ◇

BOOLOS, G. *Trees and finite satisfiability: proof of a conjecture of Burgess* ◇ B10 C13 F07 ◇

FITCH, F.B. *Correction to a definition of negation* ◇ B10 ◇

KIRIN, V.G. *On sentential tautologies within first order languages (Serb.-Croat. summary)* ◇ B10 ◇

LEBLANC, H. *A new semantics for first-order logic, multivalent and mostly intensional* ◇ B10 C30 ◇

LOBOVIKOV, V.O. *A non-logical interpretation of the mathematical apparatus of the classical first order predicate logic (Russian)* ◇ B10 ◇

MOTOHASHI, N. *A normal form theorem for first order formulas and its application to Gaifman's splitting theorem* ◇ B10 C07 C62 ◇

PHILLIPS, N.C.K. *Theorems ad infinitum* ◇ B10 ◇

PLAISTED, D.A. *Complete problems in the first-order predicate calculus* ◇ B10 D15 ◇

RENARDEL DE LAVALETTE, G.R. *Descriptions in mathematical logic* ◇ B10 F50 ◇

SHAROV, C.I. *On some generalizations of Beth's semantics (Russian)* ◇ B10 ◇

1985

BACON, J. *The completeness of a predicate-functor logic* ⋄ B10 ⋄

BELLOT, P. *A new proof for Craig's theorem* ⋄ B10 B40 ⋄

HENKIN, L. & MONK, J.D. & TARSKI, A. *Cylindric algebras. Part II* ⋄ B10 G15 ⋄

HOOK, J.L. *A note on interpretations of many-sorted theories* ⋄ B10 C07 F25 ⋄

KAPHENGST, H. *Zum Aufbau einer mehrsortigen elementaren Logik* ⋄ B10 ⋄

LEGENHAUSEN, G. *New semantics for the lower predicate calculus* ⋄ B10 ⋄

LEHMANN, G. *Modell- und rekursionstheoretische Grundlagen psychologischer Theorienbildung* ⋄ B10 C98 D80 D99 ⋄

MACKENZIE, J. *A pragmatic requirement for classically valid arguments* ⋄ B10 ⋄

MALINOWSKI, G. *Notes on sentential logic with identity* ⋄ B10 B60 ⋄

ONO, H. *Semantical analysis of predicate logics without the contraction rule* ⋄ B10 C90 F50 ⋄

ROEPER, P. *Generalisation of first-order logic to nonatomic domains* ⋄ B10 ⋄

B15 Higher-order logic and type theory

1908
RUSSELL, B. *Mathematical logic as based on the theory of types* ◊ A05 B15 ◊

1909
POINCARE, H. *La logique de l'infini* ◊ A05 B15 E30 ◊

1910
RUSSELL, B. *La theorie des types logiques* ◊ A05 B15 ◊
WHITEHEAD, A.N. & RUSSELL, B. *Principia mathematica. Vol.I* ◊ B05 B10 B15 B28 E30 E98 ◊

1912
CHWISTEK, L.B. *The principle of contradiction in the light of recent investigations of Bertrand Russell (Russian)* ◊ A05 B15 F35 ◊
WHITEHEAD, A.N. & RUSSELL, B. *Principia mathematica. Vol.II* ◊ B15 B28 E10 E30 E98 ◊

1913
WHITEHEAD, A.N. & RUSSELL, B. *Principia mathematica. Vol.III* ◊ B15 B28 E10 E30 E98 ◊

1922
CHWISTEK, L.B. *Grundlagen der reinen Typentheorie (Polish)* ◊ A05 B15 E70 ◊

1923
TAJTELBAUM, A. *On the primitive term of logistic (Polish)* ◊ B15 B20 ◊

1924
CHWISTEK, L.B. *The theory of constructive types (principles of logic and mathematics)* ◊ A05 B15 F65 ◊

1928
WEISS, P. *The theory of types* ◊ B15 ◊

1930
TARSKI, A. *Fundamentale Begriffe der Methodologie der deduktiven Wissenschaften I* ◊ A05 B15 B22 ◊

1933
TARSKI, A. *On the notion of truth in reference to formalized deductive sciences (Polish)* ◊ A05 B15 B30 C07 C40 F30 ◊

1934
QUINE, W.V.O. *A system of logistic* ◊ A05 B15 B98 ◊

1936
GENTZEN, G. *Die Widerspruchsfreiheit der Stufenlogik* ◊ B15 F05 F35 ◊
QUINE, W.V.O. *On the axiom of reducibility* ◊ B15 ◊

1937
BETH, E.W. *Une demonstration de la non-contradiction de la logique des types au point de vue fini* ◊ B15 F35 ◊
CHWISTEK, L.B. *Foundations of general theory of classes (Polish)* ◊ B15 E70 ◊
QUINE, W.V.O. *Logic based on inclusion and abstraction* ◊ B15 E30 ◊

1938
FITCH, F.B. *The consistency of the ramified Principia* ◊ B15 F25 ◊
QUINE, W.V.O. *On the theory of types* ◊ B15 E70 ◊

1939
CHURCH, A. *Schroeder's anticipation of the simple theory of types* ◊ A10 B15 ◊
LANGFORD, C.H. *A theorem on deducibility for second-order functions* ◊ B15 C35 C65 C85 ◊

1940
CHURCH, A. *A formulation of the simple theory of types* ◊ B15 ◊
SECKENDORFF VON, V. *Beweis und Definition durch verallgemeinerte transfinite Induktion. Ihr Aufbau und ihre Stellung in einem logischen Kalkuel* ◊ B15 E20 ◊

1941
ACKERMANN, W. *Ein System der typenfreien Logik I* ◊ B15 E70 ◊
SCHOLZ, H. *Metaphysik als strenge Wissenschaft* ◊ A05 B15 ◊

1942
NEWMAN, M.H.A. & TURING, A.M. *A formal theorem in Church's theory of types* ◊ B15 ◊

1943
BRUNER, F.G. *Mathematical logic with transfinite types* ◊ B15 ◊
NEWMAN, M.H.A. *Stratified systems of logic* ◊ B15 ◊

1945
BOCHVAR, D.A. *Some logical theorems on the normal sets and predicates (Russian) (English summary)* ◊ B15 E70 ◊

1947
BARCAN, R.C. *The identity of individuals in a strict functional calculus of second order* ◊ B15 ◊
MOSTOWSKI, ANDRZEJ *On absolute properties of relations* ◊ B15 C75 C85 E15 E47 ◊

1948
TARSKI, A. *A problem concerning the notion of definability*
⋄ B10 B15 C40 ⋄

TURING, A.M. *Practical forms of type theory* ⋄ B15 ⋄

1949
CHAUVIN, A. *Generalisation du theoreme de Goedel*
⋄ B15 F30 ⋄

SMART, J.J.C. *Whitehead and Russell's theory of types*
⋄ B15 ⋄

WANG, HAO *A theory of constructive types*
⋄ B15 F35 F65 ⋄

1950
ACKERMANN, W. *Widerspruchsfreier Aufbau der Logik I. Typenfreies System ohne tertium non datur*
⋄ B15 E70 F07 F25 F65 ⋄

COPI, I.M. *The inconsistency or redundancy of principia mathematica* ⋄ A05 B15 ⋄

HENKIN, L. *Completeness in the theory of types*
⋄ A05 B15 C85 F35 ⋄

NOVAK, I.L. *A construction for models of consistent systems* ⋄ B15 C07 E70 ⋄

ROSSER, J.B. & WANG, HAO *Non-standard models for formal logics*
⋄ B15 C62 E30 E70 F25 H15 H20 ⋄

SMART, J.J.C. *The theory of types again* ⋄ B15 ⋄

WANG, HAO *Remarks on the comparison of axiom systems* ⋄ B15 F25 F30 ⋄

1951
MYHILL, J.R. *Report on some investigations concerning the consistency of the axiom of reducibility* ⋄ B15 F35 ⋄

SMART, J.J.C. *The theory of types - a further note*
⋄ B15 ⋄

1952
ACKERMANN, W. *Widerspruchsfreier Aufbau einer typenfreien Logik I (erweitertes System)*
⋄ B10 B15 E70 F25 F65 ⋄

CHURCH, A. & QUINE, W.V.O. *Some theorems on definability and decidability* ⋄ B15 D35 ⋄

WANG, HAO *Negative types* ⋄ A05 B15 ⋄

1953
ACKERMANN, W. *Widerspruchsfreier Aufbau einer typenfreien Logik II*
⋄ B10 B15 E35 E70 F30 F35 F65 ⋄

BUECHI, J.R. *Investigation of the equivalence of the axiom of choice and Zorn's lemma from the viewpoint of the hierarchy of types* ⋄ B15 E25 ⋄

GROSS, M.W. & NORTHROP, F.S.C. *Alfred Whitehead: An anthology* ⋄ A05 B15 ⋄

HENKIN, L. *Banishing the rule of substitution for functional variables* ⋄ B10 B15 ⋄

L'ABBE, M. *Systems of transfinite types involving λ-conversion* ⋄ B15 B40 ⋄

ROSSER, J.B. *Logic for mathematicians*
⋄ B15 B98 E70 E98 ⋄

SCHUETTE, K. *Zur Widerspruchsfreiheit einer typenfreien Logik* ⋄ B15 B40 F05 F65 ⋄

SCHWABHAEUSER, W. *Zur Definition des geordneten Paares im einstelligen Stufenkalkuel* ⋄ B15 E20 ⋄

SEKI, S. *On the weakened type-logic (note on metamathematics I)* ⋄ B15 ⋄

SMART, J.J.C. *A note on categories* ⋄ B15 ⋄

TAKEUTI, G. *On a generalized logic calculus*
⋄ B15 F35 ⋄

ZYKOV, A.A. *The spectrum problem in the extended predicate calculus (Russian)* ⋄ B15 C13 ⋄

1954
GANDY, R.O. *On the possibility of proving the consistency of the simple theory of types* ⋄ B15 ⋄

KREISEL, G. *Remark on complete interpretations by models* ⋄ B15 F25 F30 ⋄

MAEHARA, S. *Gentzen's theorem on an extended predicate calculus* ⋄ B15 F05 ⋄

1955
GRZEGORCZYK, A. *The systems of Lesniewski in relation to contemporary logical research (Polish and Russian summaries)* ⋄ B15 ⋄

HASENJAEGER, G. *On definability and derivability*
⋄ B10 B15 C07 D55 ⋄

HINTIKKA, K.J.J. *Reductions in the theory of types*
⋄ B15 C85 ⋄

QUINE, W.V.O. *On Frege's way out* ⋄ B15 E30 ⋄

SLUPECKI, J. *A logical system without operators (Polish, Russian) (English summary)* ⋄ B15 ⋄

TAKEUTI, G. *On the fundamental conjecture of GLC I,II*
⋄ B15 F05 F35 ⋄

1956
GANDY, R.O. *On the axiom of extensionality I*
⋄ B15 E30 ⋄

KEMENY, J.G. *A new approach to semantics - part I,II*
⋄ A05 B15 ⋄

SIMAUTI, T. *Proof of a special case of the fundamental conjecture of Takeuti's GLC* ⋄ B15 F05 F35 ⋄

TAKEUTI, G. *A metamathematical theorem on functions*
⋄ B15 F05 F10 F35 ⋄

TAKEUTI, G. *Construction of ramified real numbers*
⋄ B15 F05 F35 F65 ⋄

TAKEUTI, G. *On the fundamental conjecture of GLC III,IV* ⋄ B15 F05 F35 ⋄

1957
BOCHVAR, D.A. *On paradoxes and the extended calculus of predicates (Russian)* ⋄ A05 B15 ⋄

GILMORE, P.C. *The monadic theory of types in the lower predicate calculus* ⋄ B15 ⋄

KOCHEN, S. *Completeness of algebraic systems in higher order calculi* ⋄ B15 C35 C60 C85 ⋄

OREY, S. *Strongly standard formulas* ⋄ B15 ⋄

THIELE, H. *Vollstaendigkeit im Stufenkalkul* ⋄ B15 ⋄

1958
ACKERMANN, W. *Ein typenfreies System der Logik mit ausreichender mathematischer Anwendungsfaehigkeit I*
⋄ B10 B15 E70 ⋄

ASSER, G. & SCHROETER, K. *Axiomatisierung der k-zahlig allgemeingueltigen Ausdruecke des Stufenkalkuels*
⋄ B15 ⋄

BORKOWSKI, L. *Reduction of arithmetic to logic based on the theory of types without the axiom of infinity and the typical ambiguity of arithmetical constant (Polish and Russian summaries)* ◇ B15 F35 ◇

FRAISSE, R. *Sur une extension de la polyrelation et des parentes tirant son origine du calcul logique du $k^{ème}$-echelon* ◇ B15 C07 C85 E07 ◇

GILMORE, P.C. *An addition to logic of many-sorted theories* ◇ B15 ◇

HASENJAEGER, G. *Ueber eine Interpretation der Praedikatenkalkuele hoeherer Stufe* ◇ B15 ◇

HASENJAEGER, G. *Zur Axiomatisierung der k-zahlig allgemeingueltigen Ausdruecke des Stufenkalkuels* ◇ B15 ◇

KURODA, S. *An investigation on the logical structure of mathematics. I. A logical system*
◇ B15 B28 E70 F07 ◇

KURODA, S. *An investigation on the logical structure of mathematics. III. Fundamental deductions. IV. Compendium for deductions*
◇ B15 B28 E70 F07 ◇

KURODA, S. *An investigation on the logical structure of mathematics XII. The principle of extensionality and of choice* ◇ B15 B28 E25 E70 ◇

STAHL, G. *Aspectos formales de algunas paradojas semanticas* ◇ B15 ◇

TAKEUTI, G. *On the fundamental conjecture of GLC V*
◇ B15 F05 F35 ◇

TAKEUTI, G. *Remark on the fundamental conjecture of GLC* ◇ B15 F05 F35 ◇

1959

KURODA, S. *An investigation on the logical struture of mathematics. IX. Deductions in the natural-number theory $T_1(N)$. X. Concepts and sets*
◇ B15 B28 B30 F30 ◇

MAL'TSEV, A.I. *Model correspondences (Russian)*
◇ B15 C05 C07 C52 C60 C85 ◇

OREY, S. *Model theory for the higher order predicate calculus* ◇ B15 C85 ◇

1960

BUECHI, J.R. *Weak second-order arithmetic and finite automata* ◇ B15 B25 C85 D05 F35 ◇

HASENJAEGER, G. *Unabhaengigkeitsbeweise in Mengenlehre und Stufenlogik durch Modelle*
◇ B15 E25 E35 ◇

MAEHARA, S. & NISHIMURA, T. & SEKI, S. *Non-constructive proofs of a metamathematical theorem concerning the consistency of analysis and its extension*
◇ B15 F05 F35 ◇

SCHUETTE, K. *Syntactical and semantical properties of simple type theory* ◇ B15 F05 F35 ◇

TAKEUTI, G. *An example on the fundamental conjecture of GLC* ◇ B15 F05 F35 ◇

VEITCH, E.W. *Logical equation minimization involving higher order solutions* ◇ B15 ◇

1961

MONTAGUE, R. *Semantical closure and non-finite axiomatizability I* ◇ B15 E30 F25 F30 ◇

MOSTOWSKI, ANDRZEJ *Concerning the problem of axiomatizability of the field of real numbers in the weak second order logic* ◇ B15 B28 C85 ◇

SKOLEM, T.A. *Interpretation of mathematical theories in the first order predicate calculus* ◇ B15 B30 ◇

SOBOCINSKI, B. *On the single axioms of prototothetic II,III*
◇ A05 B05 B15 ◇

TAKEUTI, G. *On the fundamental conjecture of GLC VI*
◇ B15 F05 F35 ◇

TAKEUTI, G. *On the inductive definition with quantifiers of second order* ◇ B15 F15 F35 ◇

1962

FREGE, F.L.G. *Funktion, Begriff, Bedeutung. Fuenf logische Studien*
◇ A05 B15 B28 B30 B96 E30 ◇

KAESBAUER, M. *Logisches System mit Praedikatquantoren* ◇ B15 ◇

MAEHARA, S. *Cut-elimination theorem concerning a formal system for ramified theory of types which admits quantification on types*
◇ B15 F05 F30 F35 F65 ◇

MOSTOWSKI, ANDRZEJ *On invariant, dual invariant and absolute formulas* ◇ B15 C40 C85 ◇

SPECKER, E. *Typical ambiguity* ◇ B15 E70 ◇

1963

ANDREWS, P.B. *A reduction of the axioms for the theory of propositional types* ◇ B15 ◇

HENKIN, L. *A theory of propositional types* ◇ B15 ◇

KUBINSKI, T. *A proof of consistency of Borkowski's logical system containing Peano's arithmetic (Polish and Russian summaries)* ◇ B15 F25 F30 ◇

STAHL, G. *A paratheory of type theory* ◇ A05 B15 ◇

WANG, HAO *Some formal details on predicative set theories* ◇ B15 E70 F65 ◇

1964

CARNAP, R. *The logicist foundations of mathematics*
◇ B15 ◇

GRZEGORCZYK, A. *A note on the theory of propositional types* ◇ B15 ◇

OREY, S. *New foundations and the axiom of counting*
◇ B15 E70 ◇

1965

ACKERMANN, W. *Der Aufbau einer hoeheren Logik*
◇ B15 E70 F30 F35 ◇

ANDREWS, P.B. *A transfinite type theory with type variables* ◇ B15 B40 ◇

BUECHI, J.R. *Decision methods in the theory of ordinals*
◇ B15 B25 C85 D05 E10 ◇

BUECHI, J.R. *Transfinite automata recursions and weak second order theory of ordinals*
◇ B15 B25 C85 D05 E10 ◇

CRAIG, W. *Satisfaction for n-th order languages defined in n-th order languages* ◇ B15 C40 ◇

FIEDLER, H. *Zur Stufenreduktion von Kalkuelen*
◇ B15 ◇

KAPLAN, DAVID & MONTAGUE, R. *Foundations of higher-order logic* ◇ B15 ◇

MONTAGUE, R. *Reductions of higher-order logic*
◇ B15 ◇

MONTAGUE, R. *Set theory and higher-order logic*
◇ B15 C62 E30 ◇

PRAWITZ, D. *Natural deduction. A proof-theoretical study*
⋄ B15 B98 F05 F07 F50 F98 ⋄

SANCHIS, L.E. *A predicative extension of elementary logic I* ⋄ B15 F05 F65 ⋄

SCHREIBER, P. *Untersuchungen ueber die Modelle der Typentheorie* ⋄ B15 C85 ⋄

SCHUETTE, K. *Eine Grenze fuer die Beweisbarkeit der transfiniten Induktion in der verzweigten Typenlogik*
⋄ B15 F15 F35 F65 ⋄

STANLEY, R.L. *Natural deduction, inference and consistency* ⋄ B15 ⋄

TANAKA, H. *On an arithmetical set of real numbers*
⋄ B15 ⋄

YANOVSKAYA, S.A. *Problems of introduction and elimination of abstractions of higher (than first) order (Russian)* ⋄ A05 B15 ⋄

1966

DRANGE, T. *Type crossings: sentential meaninglessness in the border area of linguistics and philosophy*
⋄ A05 B15 B65 ⋄

ELGOT, C.C. & RABIN, M.O. *Decidability and undecidability of extensions of second (first) order theory of (generalized) successor*
⋄ B15 B25 C85 D05 D35 F30 F35 ⋄

FEFERMAN, S. & KREISEL, G. *Persistent and invariant formulas relative to theories of higher order*
⋄ B15 C40 C52 C85 ⋄

KOGALOVSKIJ, S.R. *On higher order logic (Russian)*
⋄ B15 C40 C85 ⋄

KOGALOVSKIJ, S.R. *On the semantics of the theory of types (Russian)* ⋄ B15 C40 C85 ⋄

ONO, K. *Reinforced logics* ⋄ B15 ⋄

PARRY, W.T. *Quantification of the predicate and many-sorted logic* ⋄ B15 ⋄

TAIT, W.W. *A nonconstructive proof of Gentzen's Hauptsatz for second order predicate logic*
⋄ B15 F05 F35 ⋄

UESU, T. *On Zermelo's set-theory and the simple type-theory with the axiom of infinity*
⋄ B15 E30 E35 ⋄

1967

BAR-HILLEL, Y. *Theory of types* ⋄ A05 B15 ⋄

CHIKAWA, K. *On equivalences of laws in elementary protothetics. I* ⋄ B15 ⋄

COPI, I.M. *The theory of types (Introduction)* ⋄ B15 ⋄

HASENJAEGER, G. *On Loewenheim-Skolem-type insufficiencies of second order logic*
⋄ B15 C55 C85 ⋄

LOPEZ-ESCOBAR, E.G.K. *A complete, infinitary axiomatization of weak second-order logic*
⋄ B15 C75 C85 F35 ⋄

SCHWABHAEUSER, W. *Zur Axiomatisierbarkeit von Theorien in der schwachen Logik der zweiten Stufe*
⋄ B15 C85 ⋄

VENNE, M. *Ultraproduits de structures d'ordres superieurs*
⋄ B15 C20 ⋄

1968

CHIKAWA, K. *On equivalences of laws in elementary protothetics. II* ⋄ B15 ⋄

COCCHIARELLA, N.B. *Some remarks on second order logic with existence attributes* ⋄ A05 B15 ⋄

COPPOTELLI, F. *A first order type theory for the theory of sets* ⋄ B15 E30 E70 ⋄

GEISER, J.R. *Absolute properties in higher order structures*
⋄ B15 ⋄

KOGALOVSKIJ, S.R. *Some remarks on higher order logic (Russian)* ⋄ B15 C40 C85 F35 ⋄

LAEUCHLI, H. *A decision procedure for the weak second order theory of linear order* ⋄ B15 B25 C65 C85 ⋄

LOEB, M.H. *Die Vollstaendigkeit der verzweigten Typenlogik mit unendlicher Terminduktion*
⋄ B15 F35 ⋄

MOISIL, G.C. *Classical logic of propositions of higher type (Roumanian)* ⋄ B15 ⋄

MOORE VAN DER, W.A. *A weak ramified type theory*
⋄ B15 ⋄

PRAWITZ, D. *Hauptsatz for higher order logic*
⋄ B15 F05 F35 ⋄

RABIN, M.O. *Decidability of second-order theories and automata on infinite trees* ⋄ B15 B25 C85 D05 ⋄

RESNIK, M.D. *Professor Goddard and the simple theory of types* ⋄ B15 ⋄

SCHUETTE, K. *On simple type theory with extensionality*
⋄ B15 ⋄

SIEFKES, D. *Recursion theory and the theorem of Ramsey in one-place second order successor arithmetic*
⋄ B15 B25 C85 ⋄

TAKAHASHI, MOTO-O *On simple type theory (Japanese)*
⋄ B15 F05 F35 ⋄

TAKEUTI, G. *Formalization principle*
⋄ B15 E45 E55 E70 ⋄

THATCHER, J.W. & WRIGHT, J.B. *Generalized finite automata theory with an application to a decision problem of second order logic* ⋄ B15 B25 D05 ⋄

1969

BARWISE, J. & EKLOF, P.C. *Lefschetz's principle*
⋄ B15 C60 C75 C85 ⋄

BERNAYS, P. *Remark concerning the formalization of recursive definition in second order logic*
⋄ B15 D75 ⋄

BIBEL, W. *Schnittelimination in einem Teilsystem der einfachen Typenlogik* ⋄ B15 F05 F15 F35 ⋄

BOCHVAR, D.A. *Measures of kernels of reducibility axioms (Russian)* ⋄ B15 E70 ⋄

BUECHI, J.R. & LANDWEBER, L.H. *Definability in the monadic second-order theory of successor*
⋄ B15 B25 C40 C85 D05 F35 ⋄

COCCHIARELLA, N.B. *A second order logic of existence*
⋄ A05 B15 ⋄

COCCHIARELLA, N.B. *A substitution free axiom set for second order logic* ⋄ B15 ⋄

COCCHIARELLA, N.B. *Existence entailing attributes, modes of copulation, and modes of being in second order logic*
⋄ A05 B15 ⋄

COOPER, D.C. *Program scheme equivalences and second-order logic* ⋄ B15 D05 ⋄

JOHNSON, D.R. & THOMASON, R.H. *Predicate calculus with free quantifier variables* ⋄ B15 C80 ⋄

LEBLANC, H. *Three generalizations of a theorem of Beth's* ⋄ B15 C07 ⋄

LUCKHARDT, H. *Skolem-Normalformen (English summary)* ⋄ B15 ⋄

LUXEMBURG, W.A.J. *A general theory of monads* ⋄ B15 C20 H05 ⋄

MAEHARA, S. *A system of simple type theory with type variables* ⋄ B15 ⋄

PETER, R. *Ueber zweistufig definierte Sprachen* ⋄ B15 D20 ⋄

PREVIALE, F. *Rappresentabilita ed equipollenza di teorie assiomatiche I* ⋄ B15 ⋄

PRZELECKI, M. & WOJCICKI, R. *The problem of analyticity* ⋄ A05 B15 ⋄

RABIN, M.O. *Decidability of second order theories and automata on infinite trees* ⋄ B15 B25 C85 D05 ⋄

RESNIK, M.D. *A set theoretic approach to the simple theory of types* ⋄ B15 E70 ⋄

ROBINSON, JOHN ALAN *A note on mechanizing higher order logic* ⋄ B15 B35 ⋄

ROYSE, JAMES R. *Mathematical induction in ramified type theory* ⋄ B15 F35 F65 ⋄

1970

ENDERTON, H.B. *Finite partially-ordered quantifiers* ⋄ B15 C80 C85 ⋄

GASTEV, YU.A. *The expressible and deductive possibilities of logico-arithmetic calculi on the basis of the theory of types (Russian)* ⋄ B15 F35 ⋄

GOGUADZE, D.F. *The finite axiomatizability of a finite fragment of type theory (Russian) (Georgian and English summaries)* ⋄ B15 F35 ⋄

KOGALOVSKIJ, S.R. *Some reduction theorems for higher order logic (Russian)* ⋄ B15 F35 ⋄

LEBLANC, H. & MEYER, R.K. *Truth-value semantics for the theory of types* ⋄ B15 ⋄

POLLOCK, J.L. *Zermelo-Fraenkel set theory and cumulative type theory* ⋄ A05 B15 E30 ⋄

PREVIALE, F. *Rappresentabilita ed equipollenza di teorie assiomatiche II. Un 'applicazione alla geometria "senza punti"* ⋄ B15 ⋄

RABIN, M.O. *Weakly definable relations and special automata* ⋄ B15 B25 C40 C85 D05 ⋄

SEILER, E. *An application of higher order predicate logic (Romanian)* ⋄ B15 ⋄

SIEFKES, D. *Decidable theories I. Buechi's monadic second order successor arithmetic* ⋄ B15 B25 C85 D05 F35 F98 ⋄

SIEFKES, D. *Decidable extensions of monadic second order successor arithmetic* ⋄ B15 B25 C85 D05 F35 ⋄

TAUTS, A. *An analogue of Herbrand's theorem for second order predicate calculus (Russian) (Estonian and German summaries)* ⋄ B15 ⋄

WADLEIGH, H.J. *Expressibility in type theory* ⋄ B15 ⋄

WALKOE JR., W.J. *Finite partially-ordered quantification* ⋄ B15 C80 ⋄

1971

ANDREWS, P.B. *Resolution in type theory* ⋄ B15 F05 F35 ⋄

BATOG, T. *Is there a contradiction in the theory of types?* ⋄ A05 B15 ⋄

CASSEN, C.E. *Russell's distinction between the primary and secondary occurrence of definite descriptions* ⋄ A05 A10 B15 ⋄

COPI, I.M. *The theory of logical types* ⋄ B15 ⋄

DUMITRIU, A. *The antinomy of the theory of types* ⋄ A05 B15 ⋄

MAEHARA, S. & TAKEUTI, G. *Two interpolation theorems for a Π_1^1 predicate calculus* ⋄ B15 C40 F35 ⋄

MYHILL, J.R. *Embedding classical type theory in "intuitionistic" type theory* ⋄ B15 F35 F50 ⋄

PALYUTIN, E.A. *Boolean algebras with a categorical theory in a weak second order logic (Russian)* ⋄ B15 C35 C85 D35 G05 ⋄

RABIN, M.O. *Decidability and definability in second-order theories* ⋄ B15 B25 C40 C85 D05 ⋄

REICHBACH, J. *Some methods of formal proofs. III* ⋄ B10 B15 ⋄

SIEFKES, D. *Undecidable extensions of monadic second order successor arithmetic* ⋄ B15 B25 C85 D35 F35 ⋄

TICHY, P. *An approach to intensional analysis* ⋄ B15 ⋄

TICHY, P. *Synthetic components of infinite classes of postulates* ⋄ A05 B15 ⋄

UESU, T. *Simple type theory with constructive infinitely long expressions* ⋄ B15 C75 F05 F35 ⋄

1972

ANDREWS, P.B. *General models, descriptions, and choice in type theory* ⋄ B15 C85 E25 E35 F35 ⋄

ANDREWS, P.B. *General models and extensionality* ⋄ B15 C85 E35 F35 ⋄

ANIKEEV, A.S. *Classification of derivable propositional formulas (Russian)* ⋄ B15 D05 F20 F50 ⋄

ARMBRUST, M. & KAISER, KLAUS *On quasi-universal model classes* ⋄ B15 C05 C20 C52 C85 ⋄

BARWISE, J. *The Hanf number of second order logic* ⋄ B15 C55 C85 E30 ⋄

CHURCH, A. *Axioms for functional calculi of higher order* ⋄ B15 ⋄

COCCHIARELLA, N.B. *Properties as individuals in formal ontology* ⋄ A05 B15 ⋄

CRESSWELL, M.J. *Second-order intensional logic* ⋄ B15 B45 ⋄

DOEPP, K. *Bemerkung zu Henkins Beweis fuer die Nichtstandard-Vollstaendigkeit der Typentheorie* ⋄ B15 C62 C85 ⋄

FEHER, M. *Is there an antinomy in the theory of types* ⋄ B15 ⋄

HAUSCHILD, K. & HERRE, H. & RAUTENBERG, W. *Entscheidbarkeit der monadischen Theorie 2. Stufe der n-separierten Graphen (Russian, English and French summaries)* ⋄ B15 B25 ⋄

MOTOHASHI, N. *A new theorem on definability in a positive second order logic with countable conjunctions and disjunctions* ⋄ B15 C40 C75 C85 ⋄

RABIN, M.O. *Automata on infinite objects and Church's problem* ◇ B15 B25 C85 D05 ◇
ROEDDING, D. & SCHWICHTENBERG, H. *Bemerkungen zum Spektralproblem* ◇ B15 C13 D10 D15 ◇
WOLTER, H. *Ueber Mengen von Ausdruecken der Logik hoeherer Stufe, fuer die der Endlichkeitssatz und der Satz von Loewenheim-Skolem gelten*
◇ B15 C20 C85 ◇
WOLTER, H. *Untersuchung ueber Algebren und formalisierte Sprachen hoeherer Stufen (Russian, English and French summaries)*
◇ B15 C20 C30 C40 C85 ◇

1973

ANDREKA, H. & GERGELY, T. & NEMETI, I. *Some questions on languages of order n (Hungarian)*
◇ B15 C07 C85 ◇
BATLE, N. & CUFI, J. *Metamathematical justification of the substitution criteria of N.Bourbaki (Spanish)*
◇ B15 ◇
BOCHVAR, D.A. & FUKSON, V.I. *On logical approximation operators (Russian)* ◇ B15 E30 E35 ◇
BOWEN, K.A. *Cut elimination in transfinite type theory*
◇ B15 F05 F35 ◇
BUECHI, J.R. & SIEFKES, D. *Axiomatization of the monadic second order theory of ω_1* ◇ B15 B25 C85 E10 ◇
BUECHI, J.R. & SIEFKES, D. *The monadic second order theory of all countable ordinals*
◇ B15 B25 C85 E10 ◇
BUECHI, J.R. *The monadic second order theory of ω_1*
◇ B15 B25 C85 E10 ◇
COCCHIARELLA, N.B. *Whither Russell's paradox of predication?* ◇ A05 B15 ◇
FRIEDMAN, H.M. *The consistency of classical set theory relative to a set theory with intuitionistic logic*
◇ B15 E30 E35 F50 ◇
HAN, BYUNG HO *Proof without employment of higher order predicate calculus of Maehara's ε-theorem*
◇ B15 ◇
HUET, G. *The undecidability of unification on third order logic* ◇ B15 F35 ◇
KOGALOVSKIJ, S.R. & RORER, M.A. *On the question of the definability of the concept of definability (Russian)*
◇ B15 C40 C85 ◇
KREISEL, G. *Bertrand Arthur William Russell, Earl Russell 1872-1970* ◇ A10 B15 E99 ◇
LEBLANC, H. & WEAVER, G.E. *Truth-functionality and the ramified theory of types* ◇ B15 ◇
LINDSTROEM, P. *A note on weak second order logic with variables for elementarily definable relations*
◇ B15 C55 C62 ◇
MAREK, W. *Consistance d'une hypothese de Fraisse sur la definissabilite dans un langage du second ordre*
◇ B15 C65 E10 E35 E45 ◇
MAREK, W. *Sur la consistance d'une hypothese de Fraisse sur la definissabilite dans un langage du second ordre*
◇ B15 C65 E10 E35 E45 ◇
MARTIN-LOEF, P. *Hauptsatz for intuitionistic simple type theory* ◇ B15 F05 F35 F50 ◇

MOTOHASHI, N. *Model theory on a positive second order logic with countable conjunctions and disjunctions*
◇ B15 C40 C75 C85 ◇
PIETRZYKOWSKI, T. *A complete mechanization of second-order type theory* ◇ B15 B35 F35 ◇
PINUS, A.G. *A weak second order theory of fixed sets (Russian)* ◇ B15 D35 ◇
SEILER, E. *On the consistency of the theory of limits in topological unions. I (Romanian)* ◇ B15 E75 ◇
SHAFAAT, A. *On products of relational structures*
◇ B15 C30 ◇
SHELAH, S. *First order theory of permutation groups*
◇ B15 C55 C60 C85 ◇
SHELAH, S. *There are just four second-order quantifiers*
◇ B15 C80 C85 ◇
STAHL, G. *A ZF System combined with type theory (Spanish) (English summary)* ◇ B15 E70 ◇
STREET, R. & WALTERS, R.F.C. *The comprehensive factorization of a functor* ◇ B15 G30 ◇
THARP, L.H. *The characterization of monadic logic*
◇ B15 B20 C95 ◇
TITANI, S. *A proof of the cut-elimination theorem in simple type theory* ◇ B15 F05 F35 ◇
VERSHININ, K.P. *The connection of a formal language for writing down mathematical theories with the axiomatic systems of set theory (Russian) (English summary)*
◇ B15 E30 E70 ◇

1974

ANDREKA, H. & GERGELY, T. & NEMETI, I. *Some questions on languages of order n I,II (Russian) (English summaries)* ◇ B15 C07 C85 ◇
ANDREWS, P.B. *Provability in elementary type theory*
◇ B15 B25 D35 F35 ◇
ANDREWS, P.B. *Resolution and the consistency of analysis*
◇ B15 F05 F25 F35 ◇
BOWEN, K.A. *Systems of transfinite type theory based on intuitionistic and modal logics*
◇ B15 F05 F35 F50 ◇
CHURCH, A. *Russellian simple type theory*
◇ A10 B15 ◇
COCCHIARELLA, N.B. *A new formulation of predicative second order logic* ◇ A05 B15 F65 ◇
DUMITRIU, A. *The antinomy of the theory of types and the solution of logico-mathematical paradoxes*
◇ A05 B15 ◇
FRAISSE, R. *Isomorphisme local et equivalence associes a un ordinal; utilite en calcul des formules infinies a quanteurs finis* ◇ B15 C07 C75 C85 ◇
FROENING, R. *Maschinelles Beweisen in Struktur-Theorien* ◇ B15 B35 ◇
GARLAND, S.J. *Second-order cardinal characterizability*
◇ B15 C40 C55 C85 D55 E10 E47 E55 ◇
KAISER, KLAUS *Direct limits in quasi-universal model classes* ◇ B15 C30 C52 C85 ◇
KOGALOVSKIJ, S.R. *Certain criteria for localness for higher order formulas (Russian)* ◇ B15 C85 ◇
KOGALOVSKIJ, S.R. *Reductions in second-order logic (Russian)* ◇ B15 C85 ◇
MYHILL, J.R. *Embedding classical type theory in "intuitionistic" type theory, a correction*
◇ B15 F35 F50 ◇

RENNIE, M.K. *Some uses of type theory in the analysis of language* ◊ A05 B15 B65 ◊

STREET, R. *Elementary cosmoi I* ◊ B15 G30 ◊

VAJL', V.E. *Gentzen systems of postulates for set theory (Russian)* ◊ B15 E30 E70 ◊

WADLEIGH, H.J. *Translation of the simple theory of types into a first order language* ◊ B15 ◊

1975

ANDERSON, A.R. *Fitch on consistency*
◊ B15 F05 F25 ◊

ANDREKA, H. & GERGELY, T. & NEMETI, I. *Many-sorted languages and their connection with nth order languages (Russian) (English summary)*
◊ B15 C85 ◊

BOOLOS, G. *On second-order logic* ◊ A05 B15 ◊

BUCHHOLZ, W. *Ein ausgezeichnetes Modell fuer die intuitionistische Typenlogik* ◊ B15 F05 F35 F50 ◊

CHELLAS, B.F. *Quantity and quantification*
◊ A05 B15 ◊

CIRULIS, J. *Logic with inclusion (Russian)* ◊ B15 ◊

COCCHIARELLA, N.B. *A second order logic of variable-binding operators* ◊ A05 B15 C80 ◊

COCCHIARELLA, N.B. *Second order theories of predication: old and new foundations* ◊ A05 A10 B15 E70 ◊

COOLEY, J.E. *Theories of types and ordered pairs*
◊ B15 E20 ◊

CRABBE, M. *Types ambigus (English summary)*
◊ B15 E35 E70 ◊

DAVIS, CHARLES C. *An investigation concerning the Hilbert-Sierpinski logical form of the axiom of choice*
◊ B15 E25 ◊

FARTOS MARTINEZ, M. *On correspondences between monadic quantified formulas and products of nonquantified formulas (Spanish)* ◊ B15 ◊

HAUSCHILD, K. *Zur Uebertragbarkeit von Entscheidbarkeitsresultaten elementarer Theorien (Russian, English and French summaries)*
◊ B15 B25 C65 C85 ◊

INOUE, K. & NAKAMURA, A. *On the expressive power of logical metalanguages I^n and I_+* ◊ B15 F35 ◊

KAISER, KLAUS *Various remarks on quasi-universal model classes* ◊ B15 C05 C52 C85 ◊

LAKE, J. *Comparing type theory and set theory*
◊ B15 E30 E70 ◊

LEBLANC, H. *That "Principia Mathematica", first edition, has a predicative interpretation after all* ◊ B15 ◊

LOPEZ-ESCOBAR, E.G.K. & VELDMAN, W. *Intuitionistic completeness of a restricted second-order logic*
◊ B15 C90 F50 ◊

MOVSISYAN, YU.M. *Algebraic systems of the second level (Russian)* ◊ B15 C05 ◊

PERRIN, M.J. & ZALC, A. *Un theoreme de compacite en analyse. Application aux β_n-modeles et β_n-extensions (English summary)* ◊ B15 C62 ◊

RICKEY, V.F. *On creative definitions in the Principia Mathematica* ◊ B15 C40 ◊

ROUTLEY, R. *Universal semantics?* ◊ A05 B15 ◊

SEESE, D.G. *Zur Entscheidbarkeit der monadischen Theorie 2. Stufe baumartiger Graphen (Russian, English and French summaries)* ◊ B15 B25 ◊

TENNEY, R.L. *Second-order Ehrenfeucht games and the decidability of the second-order theory of an equivalence relation* ◊ B15 B25 C85 ◊

1976

AMER, M.A. *Typed boolean structures I (Arabic summary)*
◊ B15 B40 C90 E40 G05 ◊

CASTANEDA, H.-N. *Ontology and grammar. I: Russell's paradox and the general theory of properties in natural language* ◊ A05 B15 ◊

CHEPURNOV, B.A. & KOGALOVSKIJ, S.R. *Some criteria of heredity and locality for formulas of higher order (Russian)* ◊ B15 C52 C85 ◊

CHURCH, A. *Comparison of Russell's resolution of the semantical antinomies with that of Tarski*
◊ A05 A10 B15 ◊

COCCHIARELLA, N.B. *A note on the definition of identity in Quine's new foundations* ◊ A05 B15 E35 E70 ◊

FOSTER, C.C. & TENNEY, R.L. *Non-transitive dominance*
◊ B15 ◊

GRISHIN, V.N. *Reduction of comprehension axioms of a given depth to comprehension axioms of smaller depth (Russian)* ◊ B15 E70 ◊

KAISER, KLAUS *Quasi-axiomatic classes*
◊ B15 C20 C52 C75 C85 ◊

LEHMANN, S.K. *An interpretation of "finite" modal first-order languages in classical second-order languages* ◊ B15 B45 ◊

MIJOULE, R. *Theorie des types et modeles de la theorie des ensembles (English summary)* ◊ B15 C62 G30 ◊

THOMAS, WILLIAM J. *Consistency of n-order logics*
◊ B15 ◊

VANDERVEKEN, D.R. *The Lesniewski-Curry theory of syntactical categories and the categorially open functors*
◊ B15 D05 G30 ◊

1977

AMER, M.A. *Typed boolean structures II (Arabic summary)* ◊ B15 B40 C90 E40 G05 ◊

BOFFA, M. *A reduction of the theory of types*
◊ B15 E35 E70 ◊

BOFFA, M. *Modeles cumulatifs de la theorie des types*
◊ B15 C62 E30 ◊

BOFFA, M. *The consistency problem for NF*
◊ B15 E35 E70 ◊

BUECHI, J.R. *Using determinancy of games to eliminate quantifiers* ◊ B15 C10 C85 E60 ◊

COPPOTELLI, F. *On two first order type theories for the theory of sets* ◊ B15 E30 E70 ◊

DANIELS, C.B. & FREEMAN, J.B. *Classical second-order intensional logic with maximal propositions*
◊ B15 C85 C90 ◊

DEUTSCH, M. *Eine mengentheoretische Grundlegung der Theorie der Berechenbarkeit II*
◊ B15 D20 D35 E47 ◊

FEFERMAN, S. *Theories of finite type related to mathematical practice*
◊ B15 D55 D65 F10 F15 F35 F98 ◊

FRAISSE, R. *Deux relations denombrables, logiquement equivalentes pour le second ordre, sont isomorphes*
◊ B15 C15 C85 E07 E45 ◊

GACS, P. & LOVASZ, L. *Some remarks on generalized spectra* ◊ B15 C13 D15 ◊

GANDY, R.O. *The simple theory of types* ◊ A10 B15 ◊

GUREVICH, Y. *Monadic theory of order and topology I* ◊ B15 C65 C85 E07 E50 E75 ◊

HACKING, I. *Do-it-yourself semantics for classical sequent calculi including ramified type theory* ◊ B15 ◊

KRAWCZYK, A. & MAREK, W. *On the rules of proof generated by hierarchies* ◊ B15 C55 C85 E45 ◊

LADNER, R.E. *Application of model theoretic games to discrete linear orders and finite automata* ◊ B15 B25 C07 C13 C65 C85 D05 E07 E60 ◊

MIZUTANI, C. *Monadic second order logic with an added quantifier Q* ◊ B15 B25 C40 C55 C80 ◊

NABEBIN, A.A. *Expressibility in restricted second-order arithmetic (Russian)* ◊ B15 C40 D05 F35 ◊

SEESE, D.G. *Second order logic, generalized quantifiers and decidability* ◊ B15 B25 C55 C65 C80 C85 D35 ◊

SIMON, J. *Polynomially bounded quantification over higher types and a new hierarchy of the elementary sets* ◊ B15 D10 D15 D20 ◊

SURMA, S.J. (ED.) *On Lesniewski's systems. Proceedings of XXII conference on the history of logic* ◊ A10 B15 B97 ◊

VAEAENAENEN, J. *Remarks on generalized quantifiers and second order logics* ◊ B15 C40 C80 C85 C95 ◊

YUKAMI, T. *Transfinite type theory and provability of second order formulas* ◊ B15 ◊

1978

BAXTER, L.D. *The undecidability of the third order dyadic unification problem* ◊ B15 B40 D80 ◊

CIRULIS, J. *Logic with inclusion. II (Russian)* ◊ B15 ◊

CIRULIS, J. *Superinductive classes (Russian)* ◊ B15 E20 E25 ◊

CRABBE, M. *Ambiguity and stratification* ◊ B15 E35 E70 F25 ◊

CRABBE, M. *Ramification et predicativite* ◊ B15 ◊

DANIELS, C.B. & FREEMAN, J.B. *A logic of generalized quantification* ◊ B15 C80 C90 ◊

FERRO, R. *Interpolation theorem for $L^{2+}_{k,k}$* ◊ B15 C40 C75 C85 ◊

FRIEDMAN, H.M. *Classically and intuitionistically provably recursive functions* ◊ B15 D20 E70 F30 F50 ◊

MATERNA, P. *Theory of types and data description* ◊ B15 ◊

SALLE, P. *Une extension de la theorie des types en λ-calcul* ◊ B15 B40 ◊

SIEFKES, D. *An axiom system for the weak monadic second order theory of two successors* ◊ B15 B25 C85 D05 ◊

STREET, R. *A survey of topos theory* ◊ B15 G30 ◊

STREET, R. *Cosmoi of internal categories* ◊ B15 G30 ◊

STREET, R. & WALTERS, R.F.C. *Yoneda structures on 2-categories* ◊ B15 G30 ◊

TAUTS, A. *Ambiguity of the axiom of normal functions (Russian) (Estonian and German summaries)* ◊ B15 C75 E10 E55 E65 ◊

ZBIERSKI, P. *Axiomatizability of second order arithmetic with ω-rule* ◊ B15 C62 E45 E70 F35 ◊

1979

ASSER, G. *Ueber die Charakterisierbarkeit transfiniter Maechtigkeiten im Praedikatenkalkuel der zweiten Stufe* ◊ B15 C55 C80 C85 ◊

BOUDREAUX, J.C. *Defining general structures* ◊ B15 C85 ◊

CIRULIS, J. *Logic with inclusion. III (Russian)* ◊ B15 ◊

COCCHIARELLA, N.B. *The theory of homogeneous simple types as a second-order logic* ◊ B15 E70 ◊

DAVIS, L. *An alternate formulation of Kripke's theory of truth* ◊ A05 B15 ◊

FERRAZ DE ARAGON, D. *Finitary translation of mathematical theories* ◊ B15 E70 ◊

GUREVICH, Y. *Monadic theory of order and topology II* ◊ B15 C65 C85 E07 E45 E50 ◊

HAREL, D. *Characterizing second order logic with first order quantifiers* ◊ B15 C80 C85 ◊

HINNION, R. *Modele constructible de la theorie des ensembles de Zermelo dans la theorie des types* ◊ B15 C62 E30 E35 E45 ◊

KISELEV, A.A. & KOGALOVSKIJ, S.R. *Some remarks on higher order monadic languages (Russian)* ◊ B15 C85 ◊

KRYNICKI, M. *On the expressive power of the language using the Henkin quantifier* ◊ B15 B25 C80 C85 ◊

KRYNICKI, M. & LACHLAN, A.H. *On the semantics of the Henkin quantifier* ◊ B15 B25 C40 C55 C80 C85 ◊

MIKELADZE, Z.N. *A class of logical concepts (Russian)* ◊ B15 ◊

MOSTOWSKI, A.WLODZIMIERZ *A note concerning the complexity of a decision problem for positive formulas in SkS* ◊ B15 B25 D15 ◊

OIKKONEN, J. *Hierarchies of model-theoretic definability - an approach to second-order logics* ◊ B15 C40 C70 C75 C80 C85 ◊

SCHWABHAEUSER, W. *Non-finitizability of a weak second-order theory* ◊ B15 C62 C65 C85 F35 ◊

SEESE, D.G. *Some graph-theoretical operations and decidability* ◊ B15 B25 C65 C85 ◊

SHVARTS, G.F. *Some extensions of intuitionistic type theory (Russian) (English summary)* ◊ B15 F35 F50 ◊

STARK, W.R. *Type hierarchies and type-level reduction* ◊ B15 ◊

TAJTSLIN, M.A. *A description of algebraic systems in weak logic of order ω and in program logic (Russian)* ◊ B15 B75 C35 C85 ◊

ZHEZHERUN, A.P. *Decidability of the unification problem for second-order languages with unary functional symbols (Russian) (English summary)* ◊ B15 B35 ◊

1980

ANDERSON, C.A. *Some new axioms for the logic of sense and denotation: alternative (0)* ◊ A05 B15 ◊

BOSTOCK, D. *A study of type-neutrality I,II (II: Relations)* ◊ A05 B15 ◊

BOUDREAUX, J.C. *Frames versus minimally restricted structures* ◊ B15 C85 ◊

BUNDER, M.W. *The consistency of a higher order predicate calculus and set theory based on combinatory logic* ⋄ B15 B40 F05 ⋄

CHLEBUS, B.S. *Decidability and definability results concerning well-orderings and some extensions of first order logic*
⋄ B15 B25 C40 C65 C80 D35 E07 ⋄

COCCHIARELLA, N.B. *Nominalism and conceptualism as predicative second-order theories of predication*
⋄ A05 B15 C85 ⋄

CORCORAN, J. *Categoricity* ⋄ A05 B15 C35 C85 ⋄

DRAGALIN, A.G. *Higher-order predicate logic in the form of calculus realization (Russian)*
⋄ B15 F05 F35 F50 ⋄

HERRMANN, E. & WOLTER, H. *Untersuchungen zu schwachen Logiken der zweiten Stufe*
⋄ B15 C15 C85 ⋄

MALITZ, J. & RUBIN, M. *Compact fragments of higher order logic* ⋄ B15 C55 C80 C85 ⋄

MO, SHAOKUI *On the nature of higher order functions and operators (Chinese) (English summary)* ⋄ B15 ⋄

NERODE, A. & SHORE, R.A. *Second order logic and first order theories of reducibility orderings*
⋄ B10 B15 D30 D35 F35 F40 ⋄

PABION, J.F. TT_3I *est equivalent a l'arithmetique du second ordre (English summary)* ⋄ B15 F25 F35 ⋄

PLOTKIN, G.D. λ *-definability in the full type hierarchy*
⋄ B15 B40 ⋄

SALLE, P. *Une generalisation de la theorie des types en* λ *-calcul I,II* ⋄ B15 B40 ⋄

THOMAS, WOLFGANG *On the bounded monadic theory of well-ordered structures* ⋄ B15 B25 C85 E07 ⋄

1981

ASSER, G. *Einfuehrung in die mathematische Logik. Teil III: Praedikatenlogik hoeherer Stufe* ⋄ B15 B98 ⋄

BOFFA, M. *La theorie des types et NF* ⋄ B15 E70 ⋄

FOKKINGA, M.M. *On the notion of strong typing*
⋄ B15 ⋄

GEACH, P.T. *Second-order quantification in Frege (Spanish)* ⋄ A10 B15 ⋄

GOLDFARB, W.D. *The undecidability of the second-order unification problem* ⋄ B15 D35 F20 ⋄

GRABOWSKI, M. *Full weak second-order logic versus algorithmic logic* ⋄ B15 ⋄

HINTIKKA, K.J.J. *Standard vs. nonstandard logic: higher order, modal, and first-order logics*
⋄ B10 B15 B45 ⋄

KUEHNRICH, M. *Zur Praedikatentheorie der ersten und zweiten Stufe* ⋄ B15 ⋄

MOSTOWSKI, A.WLODZIMIERZ *The complexity of automata and subtheories of monadic second order arithmetics* ⋄ B15 B25 D05 D15 F35 ⋄

MULLER, D.E. & SCHUPP, P.E. *Context-free languages, groups, the theory of ends, second-order logic, tiling problems, cellular automata, and vector addition systems* ⋄ B15 B25 D05 D80 ⋄

NAKAMURA, A. & ONO, H. *Undecidability of extensions of the monadic first-order theory of successor and two-dimensional finite automata*
⋄ B15 B25 D05 D35 ⋄

YAKUBOVICH, A.M. *On the consistency of the theory of types with the axiom of choice relative to type theory (Russian)* ⋄ B15 E25 E35 F35 ⋄

YAKUBOVICH, A.M. *Variants of the axiom of choice in the simple theory of types (Russian)* ⋄ B15 E25 E35 ⋄

1982

BACON, J. *First-order logic based on inclusion and abstraction* ⋄ B15 ⋄

BOFFA, M. *Algebres de Boole atomiques et modeles de la theorie des types* ⋄ B15 C62 E70 ⋄

BUNDER, M.W. *Some results in Aczel-Feferman logic and set theory* ⋄ B15 E70 ⋄

CIRULIS, J. *Axioms of protothetics with a strengthened capacity rule (Russian)* ⋄ B15 B60 ⋄

JANSOHN, H.-S. & LANDWEHR, R. & WRIGHTSON, G. *An interactive proof system for higher order logic*
⋄ B15 B35 F35 ⋄

SEELY, R.A.G. *Locally Cartesian closed categories and type theory 1* ⋄ B15 F35 F50 G30 ⋄

TICHY, P. *Foundations of partial type theory* ⋄ B15 ⋄

ZALTA, E.N. *Meinongian type theory and its applications*
⋄ A05 B15 ⋄

1983

BEALER, G. *Completeness in the theory of properties, relations, and propositions* ⋄ B15 ⋄

BELYAKIN, N.V. *A means of modeling a classical second-order arithmetic (Russian)*
⋄ B15 C62 D55 F35 ⋄

BENCIVENGA, E. *Compactness of a supervaluation language* ⋄ B15 ⋄

BOFFA, M. *Arithmetic and theory of types*
⋄ B15 C62 E70 F35 ⋄

BUECHI, J.R. & ZAIONTZ, C. *Deterministic automata and the monadic theory of ordinals* $< \omega_2$
⋄ B15 B25 C85 D05 E10 ⋄

BUECHI, J.R. & SIEFKES, D. *The complete extensions of the monadic second order theory of countable ordinals*
⋄ B15 C35 C85 D05 E10 ⋄

BUNDER, M.W. *A one axiom set theory based on higher order predicate calculus* ⋄ B15 B40 E70 ⋄

BUNDER, M.W. *Predicate calculus of arbitrarily high finite order* ⋄ B15 ⋄

COUTURE, J. *Les classes dans les Principia mathematica sont-elles des expressions incompletes? (English and German summaries)* ⋄ A05 B15 E30 ⋄

CRABBE, M. *On the reduction of type theory* ⋄ B15 ⋄

CRABBE, M. *The axiom of choice in type theory with ambiguity axioms* ⋄ B15 E25 E70 ⋄

DEGEN, J.W. *Systeme der kumulativen Logik* ⋄ B15 ⋄

FORTUNE, S. & LEIVANT, D. & O'DONNELL, M.J. *The expressiveness of simple and second-order type structures* ⋄ B15 ⋄

GUREVICH, Y. & SHELAH, S. *Interpreting second-order logic in the monadic theory of order*
⋄ B15 C65 C85 D35 E07 E50 F25 ⋄

GUREVICH, Y. & SHELAH, S. *Rabin's uniformization problem* ⋄ B15 C40 C65 C85 E40 ⋄

GUREVICH, Y. & MAGIDOR, M. & SHELAH, S. *The monadic theory of* ω_2
⋄ B15 B25 C65 C85 D35 E10 E35 ⋄

KANOVEJ, V.G. *Some problems of descriptive set theory and definability in the theory of types (Russian)*
⋄ B15 E15 ⋄

KUEHNRICH, M. *Eine aequivalente Formalisierung der Logik von Feferman und Aczel* ⋄ B15 E70 ⋄

PAEPPINGHAUS, P. *Completeness properties of classical theories of finite type and the normal form theorem*
⋄ B15 F05 F35 ⋄

SCHWICHTENBERG, H. *On Martin-Loef's theory of types*
⋄ B15 F35 F50 ⋄

SKVORTSOV, D.P. *On some ways of construction logical languages with quantifiers over finite sequences (Russian)* ⋄ B15 C80 ⋄

SMIRNOV, V.A. *Embedding the elementary ontology of Stanislaw Lesniewski into the monadic second-order calculus of predicates* ⋄ B15 E70 ⋄

1984

BOFFA, M. *Arithmetic and the theory of types*
⋄ B15 F30 F35 ⋄

CHIHARA, C.S. *A simple type theory without Platonic domains* ⋄ B15 ⋄

CONSTABLE, R.L. & ZLATIN, D.R. *The type theory of PL/CV3* ⋄ B15 B75 ⋄

GIRARD, J.-Y. *The Ω-rule* ⋄ B15 C62 C75 F35 ⋄

GRAYSON, R.J. *Heyting-valued semantics*
⋄ B15 C90 F35 F50 G30 ⋄

HORT, C. & OSSWALD, H. *On nonstandard models in higher order logic*
⋄ B15 C55 C85 E55 H05 H20 ⋄

LAMBEK, J. & SCOTT, P.J. *Aspects of higher order categorical logic* ⋄ B15 F35 F50 G30 ⋄

LISOVIK, L.P. *Monadic second-order theories of two successor functions with an additional predicate (Russian)* ⋄ B15 B25 C10 C85 D05 ⋄

NEUMANN, O. *On the definition of ordered n-tuples*
⋄ B15 E20 ⋄

RONCHI DELLA ROCCA, S. & VENNERI, B.M. *Principal type schemes for an extended type theory* ⋄ B15 ⋄

SCARPELLINI, B. *Complete second order spectra*
⋄ B15 C13 ⋄

SCARPELLINI, B. *Second order spectra* ⋄ B15 C13 ⋄

SMITH, JAN *An interpretation of Martin-Loef's type theory in a type-free theory of propositions* ⋄ B15 ⋄

1985

COCCHIARELLA, N.B. *Two λ-extensions of the theory of homogeneous simple types as a second-order logic*
⋄ B15 B40 ⋄

CONSTABLE, R.L. & MENDLER, N.P. *Recursive definitions in type theory* ⋄ B15 F35 ⋄

FRAISSE, R. *Deux relations denombrables, logiquement equivalentes pour le second ordre, sont isomorphes (modulo un axiome de constructibilite)*
⋄ B15 C15 C85 E45 ⋄

HUANG, QIEYUAN *Type structures (Chinese)* ⋄ B15 ⋄

MEYER, A.R. & MITCHELL, J.C. *Second-order logical relations* ⋄ B15 B22 ⋄

MUCHNIK, A.A. *Games on infinite trees and automata with dead ends. A new proof of decidability for the monadic theory with two successor functions (Russian)*
⋄ B15 B25 C85 D05 ⋄

MULLER, D.E. & SCHUPP, P.E. *The theory of ends, pushdown automata, and second-order logic*
⋄ B15 B25 D10 D80 ⋄

SCARPELLINI, B. *Lower bound results on lengths of second-order formulas*
⋄ B15 C13 D10 F20 F35 ⋄

SCEDROV, A. *Extending Goedel's modal interpretation to type theory and set theory*
⋄ B15 B45 E70 F35 F50 ⋄

WASILEWSKA, A. *Monadic second-order definability as a common characterization of finite automata, certain classes of programs and logics*
⋄ B15 B75 C40 D05 ⋄

B20 Fragments of classical logic

1880
PEIRCE, C.S. *On the algebra of logic, chapter I.-Syllogistic. chapter II.-The logic of non-relative terms, chapter III. The logic of relatives* ⋄ A05 A10 B20 ⋄

1881
VENN, J. *Symbolic logic* ⋄ A10 B20 B98 ⋄

1885
PEIRCE, C.S. *On the algebra of logic: A contribution to the philosophy of notation* ⋄ A05 A10 B20 ⋄

1890
SCHROEDER, F.W.K.E. *Vorlesungen ueber die Algebra der Logik (exakte Logik) Vol.1*
⋄ A05 A10 B20 B98 G98 ⋄

1891
PEANO, G. *Formole di logica matematica* ⋄ B20 ⋄
PEANO, G. *Principii di logica matematica* ⋄ B20 ⋄
SCHROEDER, F.W.K.E. *Vorlesungen ueber die Algebra der Logik (exakte Logik) Vol.2*
⋄ A05 A10 B20 B98 G05 G15 G98 ⋄

1895
DODGSON, C.L. *Symbolic logic. Part I. Elementary*
⋄ B20 B98 ⋄
FREGE, F.L.G. *Kritische Beleuchtung einiger Punkte in E.Schroeder's Vorlesungen ueber die Algebra der Logik*
⋄ A05 B20 ⋄
SCHROEDER, F.W.K.E. *Vorlesungen ueber die Algebra der Logik (exakte Logik) Vol.3 Algebra und Logik der Relative* ⋄ A05 A10 B20 B98 G98 ⋄

1896
MACCOLL, H. *The calculus of equivalent statements*
⋄ B20 ⋄

1897
PEANO, G. *Formulaire de mathematiques Vol.2, chap.1: Logique mathematique* ⋄ B20 ⋄
PEANO, G. *Studii di logica matematica*
⋄ A10 B20 B28 ⋄

1906
RUSSELL, B. *The theory of implication* ⋄ A05 B20 ⋄

1908
LOEWENHEIM, L. *Ueber das Aufloesungsproblem im logischen Klassenkalkuel* ⋄ B20 B25 G05 ⋄

1909
SCHROEDER, F.W.K.E. *Abriss der Algebra der Logik 1: Elementarlehre* ⋄ B20 G05 ⋄

1910
SCHROEDER, F.W.K.E. *Abriss der Algebra der Logik 2: Aussagentheorie, Funktionen, Gleichungen und Ungleichungen* ⋄ B20 G05 ⋄

1912
WIENER, N. *A simplification of the logic of relations*
⋄ B20 E07 ⋄

1918
DANIELL, P.J. *Independence proofs and the theory of implication* ⋄ A05 B20 ⋄

1919
SKOLEM, T.A. *Untersuchungen ueber die Axiome des Klassenkalkuels und ueber Produktions- und Summationsprobleme, welche gewisse Klassen von Aussagen betreffen* ⋄ B20 B25 C10 D35 G05 ⋄

1921
POST, E.L. *Introduction to a general theory of elementary propositions* ⋄ B05 B20 B50 ⋄

1922
BEHMANN, H. *Beitraege zur Algebra der Logik, insbesondere zum Entscheidungsproblem*
⋄ B20 B25 D35 ⋄

1923
BEHMANN, H. *Algebra der Logik und Entscheidungsproblem* ⋄ B20 B25 D35 ⋄
TAJTELBAUM, A. *On the primitive term of logistic (Polish)*
⋄ B15 B20 ⋄

1926
LANGFORD, C.H. *Analytic completeness of postulate sets*
⋄ B20 C35 C65 E07 ⋄
YULE, D. *Zur Grundlegung des Klassenkalkuels*
⋄ B20 E30 G05 ⋄

1928
BERNAYS, P. & SCHOENFINKEL, M. *Zum Entscheidungsproblem der mathematischen Logik*
⋄ B20 B25 D35 ⋄
KALMAR, L. *Eine Bemerkung zur Entscheidungstheorie*
⋄ B20 B25 F30 ⋄

1929
GOEDEL, K. *Ein Spezialfall des Entscheidungsproblems der theoretischen Logik* ⋄ B20 B25 ⋄

1931
HERTZ, P. *Vom Wesen des Logischen, insbesondere der Bedeutung des modus barbara* ⋄ A05 B20 ⋄

1932
KALMAR, L. *Zum Entscheidungsproblem der mathematischen Logik* ◊ B20 B25 D35 ◊

SKOLEM, T.A. *Ueber die symmetrisch allgemeinen Loesungen im Klassenkalkuel* ◊ B20 ◊

1933
GOEDEL, K. *Zum Entscheidungsproblem des logischen Funktionenkalkuels* ◊ B10 B20 B25 C13 D35 ◊

GONSETH, F. *Sur l'axiomatique de la theorie des ensembles et sur la logique des relations*
◊ A05 B20 E07 E30 ◊

KALMAR, L. *Ueber die Erfuellbarkeit derjenigen Zaehlausdruecke, welche in der Normalform zwei benachbarte Allzeichen enthalten* ◊ B20 B25 ◊

SCHUETTE, K. *Untersuchungen zum Entscheidungsproblem der mathematischen Logik*
◊ B20 B25 D35 ◊

SKOLEM, T.A. *Ein kombinatorischer Satz mit Anwendung auf ein logisches Entscheidungsproblem*
◊ B20 B25 E05 ◊

WAJSBERG, M. *Ein erweiterter Klassenkalkuel*
◊ B20 E20 ◊

1934
HIRANO, T. *Die kontradiktorische Logik* ◊ B20 ◊

JASKOWSKI, S. *On the rules of suppositions in formal logic*
◊ A05 B20 ◊

SCHUETTE, K. *Ueber die Erfuellbarkeit einer Klasse von logischen Formeln* ◊ B20 B25 ◊

1935
HENLE, P. & SMITH, HENRY BRADFORD *A note on the validity of aristotelian logic* ◊ A05 A10 B20 ◊

SOBOCINSKI, B. *Die Axiomatisierung der implikativkonjunktiven Deduktionstheorie (Polish)*
◊ B20 ◊

VAIDYANATHASWAMY, R. *On the arithmetico-logical symmetric functions of n attributes* ◊ B20 ◊

1936
CHURCH, A. *A note on the "Entscheidungsproblem"*
◊ B20 D20 D35 ◊

CURRY, H.B. *A mathematical treatment of the rules of the syllogism* ◊ B20 ◊

PEPIS, J. *Beitraege zur Reduktionstheorie des logischen Entscheidungsproblemes* ◊ B20 D35 ◊

SKOLEM, T.A. *Einige Reduktionen des Entscheidungsproblems* ◊ B20 D35 ◊

1937
FEYS, R. *Les logiques nouvelles des modalites*
◊ B20 B48 ◊

HIRANO, J. *Zum Zerlegungssatz im erweiterten einstelligen Praedikatenkalkuel* ◊ B20 ◊

HUNTINGTON, E.V. *Postulates for assertion, conjunction, negation, and equality* ◊ A10 B20 ◊

KALMAR, L. *Zur Reduktion des Entscheidungsproblems*
◊ B20 D35 ◊

KALMAR, L. *Zurueckfuehrung des Entscheidungsproblems auf den Fall von Formeln mit einer einzigen, binaeren Funktionsvariablen* ◊ B20 D35 ◊

MIHAILESCU, E.G. *Recherches sur un sous-systeme du calcul de propositions* ◊ B20 ◊

MIHAILESCU, E.G. *Recherches sur la negation et l'equivalence dans le calcul des propositions* ◊ B20 ◊

MIHAILESCU, E.G. *Sur le principe de contradiction*
◊ B20 ◊

MOISIL, G.C. *Le principe d'identite et le principe du syllogisme* ◊ B20 ◊

PEPIS, J. *Ueber das Entscheidungsproblem des engeren logischen Funktionskalkuels (Polish) (German summary)* ◊ B20 B25 D35 ◊

SKOLEM, T.A. *Eine Bemerkung zum Entscheidungsproblem* ◊ B20 B25 D35 ◊

WAJSBERG, M. *Metalogische Beitraege I* ◊ A05 B20 ◊

1938
MIHAILESCU, E.G. *Sur certains sous-systemes de la logique positive classique* ◊ B20 ◊

MILLER, JAMES WILKIN *The structure of Aristotelian logic*
◊ A10 B20 ◊

MORDUKHAJ-BOLTOVSKOJ, D. *Sur les syllogismes en logique et les hypersyllogismes en metalogique*
◊ B20 ◊

PEPIS, J. *Ein Verfahren der mathematischen Logik*
◊ B20 D35 ◊

PEPIS, J. *Untersuchungen ueber das Entscheidungsproblem der mathematischen Logik* ◊ B20 D35 ◊

1939
KALMAR, L. *On the reduction of the decision problem I: Ackermann prefix, a single binary predicate*
◊ B20 D35 ◊

LUKASIEWICZ, J. *The equivalential calculus (Polish)*
◊ A05 B20 ◊

MIHAILESCU, E.G. *Recherches sur l'equivalence, la negation et la reciprocite dans le calcul des propositions*
◊ B20 ◊

MIHAILESCU, E.G. *Recherches sur les formes normales par rapport a l'equivalence et la disjonction, dans le calcul des propositions* ◊ B20 ◊

MOISIL, G.C. *Recherches sur le syllogisme* ◊ B20 ◊

ROSSER, J.B. *An informal exposition of proofs of Goedel's theorems and Church's theorem*
◊ A05 B20 D35 F30 ◊

WAJSBERG, M. *Metalogische Beitraege II* ◊ A05 B20 ◊

ZHEGALKIN, I.I. *Sur l'Entscheidungsproblem (Russian) (French summary)* ◊ B20 B25 D35 ◊

1941
DOTTERER, R.H. *A generalization of the antilogism*
◊ A05 B20 ◊

POST, E.L. *The two-valued iterative systems of mathematical logic* ◊ B05 B20 ◊

TARSKI, A. *On the calculus of relations*
◊ B20 E07 G15 ◊

1943
BERGMANN, G. *Notes on identity* ◊ A05 B20 ◊

DOTTERER, R.H. *A supplementary note on the rules of the antilogism* ◊ B20 ◊

MCKINSEY, J.C.C. *The decision problem for some classes of sentences without quantifiers*
◊ B20 B25 C05 C60 G10 ◊

SURANYI, J. *Zur Reduktion des Entscheidungsproblems des logischen Funktionskalkuels (Hungarian) (German summary)* ⋄ B20 D35 ⋄

1944
FITCH, F.B. *A minimum calculus for logic* ⋄ B20 B40 ⋄
FITCH, F.B. *Representation of calculi* ⋄ B20 B40 ⋄

1946
LOS, J. *Une preuve d'axiomatisation de la logique traditionelle* ⋄ B20 ⋄
SLUPECKI, J. *Les remarques sur la syllogistique d'Aristotle (Polish)* ⋄ A10 B20 ⋄
TOLEDO TOLEDO, R. *Mathematische Grundlagen einer strukturellen Logik (Spanisch)* ⋄ B20 ⋄
ZHEGALKIN, I.I. *Sur le probleme de la resolubilite pour les classes finies (Russian) (French summary)* ⋄ B20 B25 C13 ⋄

1947
HOENEN, P. *Recherches de logique formelle. La structure du systeme des syllogismes et des sorites – la logique des notions "au moins" et "tout au plus"* ⋄ B20 ⋄
KALMAR, L. & SURANYI, J. *On the reduction of the decision problem II: Goedel prefix, a single binary predicate* ⋄ B20 D35 ⋄
POPPER, K.R. *Functional logic without axioms or primitive rules of inference* ⋄ A05 B20 F50 ⋄
RASIOWA, H. *Axiomatisation d'un systeme partiel de la theorie de la deduction* ⋄ B20 ⋄
RIDDER, J. *Ueber den Aussagen- und den engeren Praedikatenkalkuel. II* ⋄ B20 ⋄

1948
FITCH, F.B. *An extension of basic logic* ⋄ B20 B40 F35 ⋄
LUKASIEWICZ, J. *The shortest axiom of the implicational calculus of proposition* ⋄ B20 ⋄
POPPER, K.R. *On the theory of deduction I: Derivation and its generalizations. II: The definitions of classical and intuitionistic negation* ⋄ B20 ⋄
SUSZKO, R. *Concerning logic without axioms (Russian)* ⋄ B20 ⋄
WRIGHT VON, G.H. *On the idea of logical truth I* ⋄ A05 B05 B20 B25 ⋄

1949
FITCH, F.B. *A further consistent extension of basic logic* ⋄ B20 B40 ⋄
FRAENKEL, A.A. *The relation of equality in deduction systems* ⋄ B20 ⋄
HENKIN, L. *Fragments of propositional calculus* ⋄ B20 ⋄
SURANYI, J. *Reduction of the decision problem to formulas containing a bounded number of quantifiers only* ⋄ B20 D35 ⋄

1950
BAR-HILLEL, Y. *Bolzano's propositional logic* ⋄ A05 A10 B20 ⋄
BEHMANN, H. *Das Aufloesungsproblem in der Klassenlogik* ⋄ A05 B20 B25 ⋄
BETH, E.W. *Decision problems of logic and mathematics* ⋄ B20 B25 C10 D35 ⋄
HENKIN, L. *An algebraic characterization of quantifiers* ⋄ B20 ⋄
KALMAR, L. *Contribution to the reduction theory of the decision problem I. Prefix $(x_1)(x_2)(Ex_3)\ldots(Ex_{n-1})(x_n)$, a single binary predicate* ⋄ B20 D35 ⋄
KALMAR, L. & SURANYI, J. *On the reduction of the decision problem III: Pepis prefix, a single binary predicate* ⋄ B20 D35 ⋄
LUKASIEWICZ, J. *Concerning an axiom system of the implicational propositional calculus (Polish)* ⋄ B20 ⋄
MO, SHAOKUI *The deduction theorems and two new logical systems* ⋄ B20 B45 ⋄
SLUPECKI, J. *On Aristotelian syllogistic* ⋄ A10 B20 ⋄
SURANYI, J. *Contributions to the reduction theory of the decision problem II: Three universal, one existential quantifier* ⋄ B20 D35 ⋄
WRIGHT VON, G.H. *On the idea of logical truth II* ⋄ A05 B05 B20 B25 ⋄

1951
CHURCH, A. *The weak theory of implication* ⋄ B20 ⋄
FRAENKEL, A.A. *On the crisis of the principle of the excluded middle* ⋄ A05 B20 F99 ⋄
L'ABBE, M. *On the independence of Henkin's Axiom for fragments of the propositional calculus* ⋄ B20 ⋄
LYNDON, R.C. *Identities in two-valued calculi* ⋄ B05 B20 C05 ⋄
MIHAILESCU, E.G. *Researches on sub-systems of the propositional calculus (Roumanian)* ⋄ B20 ⋄
ROSE, A. *Strong completeness of fragments of the propositional calculus* ⋄ B20 ⋄
ROSE, A. *The degree of completeness of some Lukasiewicz-Tarski propositional calculi* ⋄ B20 B50 ⋄
ROSE, A. *The degrees of completeness of a partial system of the 2-valued propositional calculus* ⋄ B20 ⋄
SURANYI, J. *Contributions to the reduction theory of the decision problem V: Ackermann prefix with three universal quantifiers* ⋄ B20 D35 ⋄

1952
CURRY, H.B. *On the definition of negation by a fixed proposition in inferential calculus* ⋄ B20 ⋄
CURRY, H.B. *The inferential theory of negation* ⋄ A05 B20 F50 ⋄
GEACH, P.T. & WRIGHT VON, G.H. *On an extended logic of relations* ⋄ B20 B25 ⋄
GERICKE, H. *Algebraische Betrachtungen zu den Aristotelischen Syllogismen* ⋄ A05 B20 ⋄
REICHENBACH, H. *The syllogism revised* ⋄ B20 ⋄
THOMAS, I. *A new decision procedure for Aristotle's syllogistic* ⋄ A10 B20 B25 ⋄
VACCARINO, G. *La sillogistica. I,II* ⋄ B20 ⋄
WRIGHT VON, G.H. *On double quantification* ⋄ A05 B20 B25 ⋄

1953
CUESTA DUTARI, N. *Ordinal deductive models (Spanish)* ⋄ B20 E07 ⋄
HAERTIG, K. *Axiomatische Probleme in der klassischen Syllogistik* ⋄ B20 ⋄

HAILPERIN, T. *Quantification theory and empty individual domains* ⋄ B20 ⋄
JOHANSSON, I. *Sur le concept de "le" (ou de "ce qui") dans le calcul affirmatif et dans les calculs intuitionnistes* ⋄ B20 F50 ⋄
LUSZCZEWSKA-ROMAHNOWA, S. *An analysis and generalization of Venn's diagrammatic decision procedure (Polish, Russian) (English summary)* ⋄ B20 ⋄
MEREDITH, C.A. *A single axiom of positive logic* ⋄ B20 F50 ⋄
MEREDITH, C.A. *The figures and moods of the n-term aristotelian syllogism* ⋄ A10 B20 ⋄
MYHILL, J.R. *On the interpretation of the sign " → "* ⋄ B20 ⋄
OMORI, S. *Formalization of an intensional logic (Japanese)* ⋄ B20 ⋄

1954

HARROP, R. *An investigation of the propositional calculus used in a particular system of logic* ⋄ B20 B25 ⋄
SOBOCINSKI, B. *Axiomatization of a conjunctive-negative calculus of propositions* ⋄ B20 ⋄
VAUGHT, R.L. *On sentences holding in direct products of relational systems* ⋄ B20 B25 C30 ⋄

1955

MESERVE, B.E. *Decision methods for elementary algebra* ⋄ B20 ⋄
RASIOWA, H. *On a fragment of the implicative propositional calculus (Polish) (Russian and English summaries)* ⋄ B20 G25 ⋄
ROSE, A. *A single axiom for a partial system of the propositional calculus* ⋄ B20 ⋄
SURANYI, J. *On the reduction theory of the decision problem of symbolic logic (Hungarian) (Russian and English summaries)* ⋄ B20 D35 ⋄
VALPOLA, V. *Ein System der negationslosen Logik mit ausschliesslich realisierbaren Praedikaten* ⋄ B20 F50 ⋄
VALPOLA, V. *Eine Eigenschaft gewoehnlicher negationsloser Kalkuele der Propositionen- und Praedikatenlogik* ⋄ B20 F50 ⋄

1956

HINTIKKA, K.J.J. *Identity, variables and impredicative definitions* ⋄ A05 B20 F65 ⋄
KALMAR, L. *A direct proof of the unsolvability of the decision problem by means of a general recursive algorithm (Hungarian)* ⋄ B20 D35 ⋄
KEMPSKI VON, J. *Relationen- und praedikatenlogische Untersuchungen zur Syllogistik* ⋄ B20 ⋄
LORENZEN, P. *Zur Interpretation der Syllogistik* ⋄ A05 B20 ⋄
ROGERS JR., H. *Certain logical reduction and decision problems* ⋄ B20 D35 ⋄
SCHROETER, K. *Ueber den Zusammenhang der in den Implikationsaxiomen vollstaendigen Axiomensysteme des zweiwertigen mit denen des intuitionistischen Aussagenkalkuel* ⋄ B20 F50 ⋄
SHEPHERDSON, J.C. *On the interpretation of Aristotelian syllogistic* ⋄ A10 B20 ⋄

THIELE, H. *Eine Axiomatisierung der zweiwertigen Praedikatenkalkuele der ersten Stufe, welche die Implikation erhalten* ⋄ B20 ⋄

1957

BORKOWSKI, L. *The first modern monography of Aristotle's syllogistic (Polish)* ⋄ B20 ⋄
DELGADO, V.M. *La interpretacion formalista de la silogistica de Aristoteles* ⋄ A10 B20 ⋄
DREBEN, B. *Systematic treatment of the decision problem* ⋄ B20 B25 D35 ⋄
HAILPERIN, T. *A theory of restricted quantification I,II* ⋄ B20 ⋄
LORENZEN, P. *Ueber die Syllogismen als Relationenmultiplikationen* ⋄ A05 B20 ⋄
SCHROETER, K. *Die Vollstaendigkeit der die Implikation enthaltenden zweiwertigen Aussagenkalkuele und Praedikatenkalkuele der ersten Stufe* ⋄ B20 ⋄
THOMAS, I. *Eulerian syllogistic* ⋄ A10 B20 ⋄
ZUBIETA RUSSI, G. *Clases aritmeticas definidas sin igualdad* ⋄ B10 B20 C07 C52 ⋄

1958

BIERMANN, K.R. & MAU, J. *Ueberpruefung einer fruehen Anwendung der Kombinatorik in der Logik* ⋄ A10 B20 ⋄
BORKOWSKI, L. *On proper quantifiers I (Polish and Russian summaries)* ⋄ B20 B25 ⋄
GLAZOWSKA, K. *The structure of valid n-term syllogisms (Polish) (Russian and English summaries)* ⋄ B20 ⋄
GUILLAUME, M. *Les tableaux semantiques du calcul des predicats restreint* ⋄ B20 F07 F98 ⋄
KAMINSKI, S. *On the number of concluding syllogistic moods (Polish) (Russian and English summaries)* ⋄ B20 ⋄
KORCIK, A. *Verification of Aristotle's theory of syllogism by means of Gergonne's method (Polish) (English summary)* ⋄ B20 ⋄
LEJEWSKI, C. *On implicational definitions (Polish and Russian summaries)* ⋄ B05 B20 ⋄
MIHAILESCU, E.G. *Recherches sur quelques systemes du calcul des propositions* ⋄ B20 ⋄
MOISIL, G.C. *Sur la logique positive (Russian and English summaries)* ⋄ B20 ⋄
OBERSCHELP, A. *Ueber die Axiome produkt-abgeschlossener arithmetischer Klassen* ⋄ B20 B25 C30 C40 C52 ⋄
SLUPECKI, J. *On some partial systems of the propositional calculus (Polish) (Russian and English summaries)* ⋄ B20 ⋄

1959

LYNDON, R.C. *Existential Horn sentences* ⋄ B20 C30 C40 C52 ⋄
PATZIG, G. *Die Aristotelische Syllogistik. Logisch-Philosophische Untersuchungen ueber Buch "A" der "Ersten Analytiken"* ⋄ A05 B20 ⋄
SLUPECKI, J. *The Lukasiewicz function (Polish)* ⋄ B05 B20 ⋄
SURANYI, J. *Reduktionstheorie des Entscheidungsproblems im Praedikatenkalkuel der ersten Stufe* ⋄ B20 D35 ⋄

UEMOV, A.I. *Leere Klassen und aristotelische Logik (Russisch)* ⋄ B20 E20 ⋄

ZYKOV, A.A. *Remarks in connection with the reduction theorem for logical calculi (Russian)* ⋄ B20 D35 ⋄

1960

BIRD, O. *The formalizing of the topics in mediaeval logic* ⋄ A05 A10 B20 ⋄

BORKOWSKI, L. *On proper quantifiers II (Polish and Russian summaries)* ⋄ B20 B25 ⋄

GODDARD, L. *The exclusive "or"* ⋄ B20 ⋄

HAILPERIN, T. *Corrections to a theory of restricted quantification* ⋄ B20 ⋄

STAAL, J.F. *Formal structures in Indian logic* ⋄ A05 A10 B20 ⋄

THOMAS, I. *Functional completeness of Henkin's propositional fragments* ⋄ B20 ⋄

THOMAS, I. *Independence of faris-rejection-axioms* ⋄ A10 B20 ⋄

THOMAS, I. *Independence of Tarski's law in Henkin's propositional fragments* ⋄ B20 ⋄

1961

BIRD, O. *Topic and consequence in Ockham's logic* ⋄ A05 A10 B20 ⋄

BORKOWSKI, L. *A didactical approach to the zero-one decision procedure of the expressions of the first order monadic predicate calculus (Polish) (Russian and English summaries)* ⋄ B20 B25 C10 ⋄

HAILPERIN, T. *A complete set of axioms for logical formulas invalid in some finite domain* ⋄ B20 C13 ⋄

KAMINSKI, S. *Traditional theory of immediate inference as a fragment of two-valued propositional calculus (Polish) (Russian and English summaries)* ⋄ B20 ⋄

KREISEL, G. & TAIT, W.W. *Finite definability of number-theoretic functions and parametric completeness of equational calculi* ⋄ B20 D20 ⋄

LEBLANC, H. *An extension of the equivalence calculus* ⋄ B20 ⋄

LEJEWSKI, C. *On prosleptic syllogisms* ⋄ A05 A10 B20 ⋄

PLYMEN, R.J. *A model of the arithmetic of alephs in the equation calculus* ⋄ B20 E10 ⋄

SIKORSKI, R. *A topological characterization of open theories* ⋄ B20 C07 G05 ⋄

TRAKHTENBROT, B.A. *Certain constructions in the logic of one-place predicates (Russian)* ⋄ B20 F35 ⋄

TRAKHTENBROT, B.A. *Finite automata and the logic of one-place predicates (Russian)* ⋄ B20 D05 F35 ⋄

1962

BETH, E.W. *Logique inférentielle et logique bivalente de l'implication* ⋄ B20 ⋄

BUECHI, J.R. *Turing machines and the "Entscheidungsproblem"* ⋄ B20 D10 D35 ⋄

BURKS, A.W. & WRIGHT, J.B. *Sequence generators, graphs and formal languages* ⋄ B20 D05 ⋄

BURKS, A.W. & WRIGHT, J.B. *Sequence generators and digital computers* ⋄ B20 D05 ⋄

CUESTA DUTARI, N. *Implication structures* ⋄ B20 ⋄

DREBEN, B. & KAHR, A.S. & WANG, HAO *Classification of AEA formulas by letter atoms* ⋄ B20 B25 D35 ⋄

DREBEN, B. *Solvable Suranyi subclasses: an introduction to the Herbrand theory* ⋄ B20 B25 ⋄

FORDER, H.G. & KALMAN, J.A. *Implication in equational logic* ⋄ B20 ⋄

FRAYNE, T.E. & MOREL, A.C. & SCOTT, D.S. *Reduced direct products* ⋄ B20 C20 C55 ⋄

KAHR, A.S. & MOORE, E.F. & WANG, HAO *Entscheidungsproblem reduced to the $\forall\exists\forall$ case* ⋄ B20 D35 ⋄

KRASZEWSKI, Z. *Modification of a decision method of theorems of 1-place functional calculus (Polish)* ⋄ B20 ⋄

MENNE, A. *Some results of investigation of the syllogism and their philosophical consequences* ⋄ A05 A10 B20 ⋄

MIHAILESCU, E.G. *Sur les proprietes de l'implication par rapport a l'equivalence et la disjonction* ⋄ B20 ⋄

MIHAILESCU, E.G. *Systeme d'equivalence, conjonction et reciprocite (Romanian) (Russian and French summaries)* ⋄ B20 ⋄

PAGER, D. *An emendation of the axiom system of Hilbert and Ackermann for the restricted calculus of predicates* ⋄ B20 ⋄

SIKORSKI, R. *On open theories* ⋄ B20 C07 G05 ⋄

SMILEY, T.J. *Syllogism and quantification* ⋄ A05 A10 B20 ⋄

STANDLEY, G.B. *Two arithmetical techniques with numbered classes* ⋄ B20 ⋄

THOMAS, I. *On the infinity of positive logic* ⋄ A05 B20 ⋄

THOMAS, I. *The rule of excision in positive implication* ⋄ A05 B20 ⋄

TRAKHTENBROT, B.A. *Finite automata and the logic of one-place predicates (Russian)* ⋄ B20 D05 F35 ⋄

1963

GOODSTEIN, R.L. *A decidable fragment of recursive arithmetic* ⋄ B20 B25 F30 ⋄

PORTA, HORACIO *Sur un theoreme de Skolem* ⋄ B20 G25 ⋄

STOICHITA, R. *La transcription du carre logique en calcul propositionnel* ⋄ B20 ⋄

WANG, HAO *Dominoes and the AEA case of the decision problem* ⋄ B20 B25 D05 D35 ⋄

ZINOV'EV, A.A. *Eine Verallgemeinerung der Syllogistik (Russian)* ⋄ B20 ⋄

1964

BIRD, O. *Syllogistic and its extensions* ⋄ B20 ⋄

CHIMEV, K.N. *Some functions of P_2 (Bulgarian) (Russian summary)* ⋄ B20 ⋄

GENENZ, J. *Reduktionstheorie nach der Methode von Kahr-Moore-Wang* ⋄ B20 D35 D98 ⋄

JOHNSTONE JR., H.W. & PRICE, ROBERT *Axioms for the implicational calculus with one variable* ⋄ B20 ⋄

KOSTYRKO, V.F. *The reduction class $\forall\exists^n\forall$ (Russian)* ⋄ B20 D35 ⋄

KROM, MELVEN R. *A decision procedure for a class of formulas of first order predicate calculus* ⋄ B20 B25 ⋄

LEJEWSKI, C. *Aristotle's syllogistic and its extensions* ⋄ A10 B20 ⋄

PIECZKOWSKI, A. *On the equivalence of the calculus of dependent sentential variables and the cylindrical algebra without diagonal elements* ⋄ B20 G15 ⋄

PRIOR, A.N. *Two additions to positive implications* ⋄ B20 B45 ⋄

RASIOWA, H. *A generalization of a formalized theory of fields of sets on nonclassical logics* ⋄ B20 E70 G10 G25 ⋄

RINON, S. & YOELI, M. *Application of ternary algebra to the study of static hazards* ⋄ B20 ⋄

ROSE, A. *Fragments du calcul propositionnel bivalent a foncteurs variables* ⋄ B20 ⋄

SINGLETARY, W.E. *A complex of problems proposed by Post* ⋄ B20 D03 D25 D30 D35 ⋄

SOBOCINSKI, B. *On the propositional system A of Vuchkovich and its extension I,II* ⋄ B20 B60 ⋄

SUBBOTIN, A.L. *Aristotelian syllogistics from an algebraic point of view (Russian)* ⋄ B20 G05 ⋄

TAUTS, A. *Solution of logical equations in the first-order one-place predicate calculus (Russian) (English summary)* ⋄ B20 B25 ⋄

THOMAS, I. *Universal variable non-Tarskian functors* ⋄ A05 B20 ⋄

YNTEMA, M.K. *A detailed argument for the Post-Linial theorems* ⋄ B20 D35 ⋄

1965

ARAI, Y. *On axiom systems of propositional calculi II,III,XII* ⋄ B05 B20 ⋄

BAUSCH, A.F. *Modus ponens under hypotheses* ⋄ B05 B20 ⋄

BERNAYS, P. *Betrachtungen zum Sequenzen-Kalkuel* ⋄ B20 F05 F07 ⋄

BETH, E.W. *Semantische Begruendung der derivativen Implikationslogik* ⋄ B20 ⋄

CONSTANTINESCU, M. *Das Logische und Historische in der Interpretation der aristotelischen Syllogistik* ⋄ A10 B20 ⋄

FISK, M. *The logic of either-or* ⋄ B20 ⋄

GLADSTONE, M.D. *Some ways of constructing a propositional calculus of any required degree of unsolvability* ⋄ B20 D25 D35 ⋄

GUREVICH, Y. *Existential interpretation (Russian)* ⋄ B20 B25 C60 D35 F25 ⋄

IMAI, Y. & ISEKI, K. *On axiom systems of propositional calculi. I* ⋄ B20 ⋄

ISEKI, K. *Algebraic formulations of propositional calculi* ⋄ B20 G05 G25 ⋄

KAMINSKI, S. *Rules for syllogisms with the consideration of schemata with negated subject terms (Polish) (Russian and English summaries)* ⋄ B20 ⋄

KEISLER, H.J. *Reduced products and Horn classes* ⋄ B20 C20 C52 C55 ⋄

MASLOV, S.YU. & MINTS, G.E. & OREVKOV, V.P. *Unsolvability in the constructive predicate calculus of certain classes of formulas containing only monadic predicate variables (Russian)* ⋄ B20 D35 F50 ⋄

MOISIL, G.C. *Sur la logique strictement positive (Romanian and Russian summaries)* ⋄ B20 ⋄

MOISIL, G.C. *Sur les lois distributives dans le calcul des propositions* ⋄ B20 ⋄

MONK, J.D. *Substitutionless predicate logic with identity* ⋄ B20 ⋄

ONO, K. *A certain kind of formal theories* ⋄ B20 ⋄

RASIOWA, H. *On non-classical calculi of classes* ⋄ B20 B50 E70 G10 G25 ⋄

ROBITASHVILI, N.G. *On the problem of reducing a pp-formula of the pure first-order functional calculus to some special case of the decision problem (Georgian) (Russian summary)* ⋄ B20 B25 ⋄

ROSE, A. *Two non-henkinian fragments of the 2-valued propositional calculus with variable functors* ⋄ B20 ⋄

SHUKLA, A. *A set of axioms for the propositional calculus with implication and converse nonimplication* ⋄ B20 ⋄

TANAKA, S. *On axiom systems of propositional calculi. VI,VIII,IX,XIII* ⋄ B20 ⋄

WANG, HAO *Remarks on machines, sets, and the decision problem* ⋄ B20 B25 D03 D05 D10 D35 E30 ⋄

1966

APRILE, G. *Su talune restrizioni nell'uso dei simboli per la construzione dei grafi semantici* ⋄ B20 ⋄

ARAI, Y. *On axiom systems of propositional calculi. XVII* ⋄ B20 ⋄

ARAI, Y. & TANAKA, S. *On axiom systems of propositional calculi. XIX* ⋄ B20 ⋄

ARRUDA, A.I. & COSTA DA, N.C.A. *Transformations in restricted predicate calculus (Portuguese)* ⋄ B20 ⋄

BAKER, A.J. *Non-empty complex terms* ⋄ A05 B20 ⋄

BORGERS, A. *La logique des classes* ⋄ B20 ⋄

GAVRILOV, G.P. & KUDRYAVTSEV, V.B. & YABLONSKIJ, S.V. *Boolesche Funktionen und Postsche Klassen (Russisch)* ⋄ B05 B20 G05 G10 ⋄

GOODSTEIN, R.L. & LEE, R.D. *A decidable class of equations in recursive arithmetic* ⋄ B20 B25 F30 ⋄

GUREVICH, Y. *Effective recognition of satisfiability of formulae of the restricted predicate calculus (Russian)* ⋄ B20 B25 D35 ⋄

GUREVICH, Y. *On the decision problem for pure restricted predicate logic (Russian)* ⋄ B20 C13 D35 ⋄

GUREVICH, Y. *The decision problem for the restricted predicate calculus (Russian)* ⋄ B20 C13 D35 ⋄

HATCHER, W.S. *Logical truth and logical implication* ⋄ A05 B20 ⋄

HOSOI, T. *Algebraic proof of the separation theorem on classical propositional calculus* ⋄ B20 ⋄

IMAI, Y. & ISEKI, K. *On axiom systems of propositional calculi. XIV* ⋄ B20 ⋄

ISEKI, K. *A characterization of the NB-system* ⋄ B20 ⋄

ISEKI, K. *Algebraic formulations of propositional calculi with variable forming functors* ⋄ B20 ⋄

ITO, J. & ONO, K. *On a characteristic feature of the positive logics* ⋄ B20 ⋄

JASKOWSKI, S. *On formulas in which no individual variable occurs more than twice* ⋄ B05 B20 ⋄

JOHNSTONE JR., H.W. *A definition of conjunction in the pure implicational calculus with one variable* ⋄ B20 ⋄

KOSTYRKO, V.F. *On the AEA reduction class (Russian) (English summary)* ⋄ B20 D35 ⋄

MASLOV, S.YU. *Application of the inverse method for establishing deducibility to the theory of decidable fragments in the classical predicate calculus (Russian)* ⋄ B20 B25 ⋄

MEREDITH, C.A. *Postulates for implicational calculi*
⋄ B20 ⋄

MIHAILESCU, E.G. *The properties of the converse implication in respect to the equivalence and disjunction*
⋄ B20 ⋄

MIURA, S. *A remark on the intersection of two logics*
⋄ B20 ⋄

MOISIL, G.C. *Remarques sur les modeles de la logique elementaire* ⋄ B20 ⋄

MORGAN DE, A. *Logic* ⋄ B20 ⋄

MORGAN DE, A. *On the syllogism I: On the structure of syllogism. II: On the symbols of logic, the theory of syllogism, and in particular the copula. III: And on logic in general. IV: And on the logic of relations. V: And on various points of the onymatic system* ⋄ B20 ⋄

MORGAN DE, A. *Syllabus of a proposed system of logic*
⋄ B20 ⋄

NELSON, D. *Non-null implication* ⋄ B20 B28 F50 ⋄

ONO, K. *A lemma which distinguishes minimal logics from other logics* ⋄ B20 ⋄

ONO, K. *Formal system having just one primitive notion*
⋄ B20 ⋄

ONO, K. *On development of formal systems starting from primitive logic* ⋄ B20 ⋄

ONO, K. *On universal character on the primitive logic*
⋄ B20 ⋄

ROSE, A. *Sur quelques resultats touchant la formalisation des calculs propositionnels a foncteurs variables*
⋄ B05 B20 ⋄

ROUSSEAU, G. *A decidable class of number theoretic equations* ⋄ B20 B25 F30 ⋄

SCHROEDER, F.W.K.E. *Der Operationskreis des Logikkalkuels* ⋄ A05 B20 ⋄

SHESTOPAL, G.A. *Simple basis in closed classes of functions of the algebra of logic (Russian)* ⋄ B20 ⋄

SHUKLA, A. *A set of axioms for the propositional calculus with implication and non-equivalence* ⋄ B20 ⋄

TAJMANOV, A.D. *On formulae of Horn-type (Russian)*
⋄ B20 D35 ⋄

TANAKA, S. *On axiom systems of propositional calculi. XVI,XVIII,XX,XXII* ⋄ B20 ⋄

TANAKA, S. *On variants of axiom systems of propositional calculus I,II* ⋄ B20 ⋄

TURQUETTE, A.R. *A method for constructing implication logics* ⋄ B20 ⋄

1967

APRILE, G. *Grafi inferenziali per il sillogismo categorico*
⋄ B20 ⋄

APRILE, G. *Risoluzione de equazioni logiche con l'ausilio di grafi inferenziali* ⋄ B20 ⋄

BACON, J. *Syllogistic without existence* ⋄ B20 ⋄

CRAIG, W. *Modus ponens and derivations from Horn formulas* ⋄ B20 F05 F50 ⋄

GALVIN, F. *Reduced products, Horn sentences, and decision problems*
⋄ B20 B25 C20 C30 C40 D35 ⋄

HACKER, E.A. *Number system for the immediate inferences and the syllogism in aristotelian logic*
⋄ B20 ⋄

HACKER, E.A. & PARRY, W.T. *Pure numerical boolean syllogisms* ⋄ A05 B20 ⋄

HENKIN, L. *Logical systems containing only a finite number of symbols* ⋄ B20 ⋄

ISEKI, K. *Algebraic formulation of propositional calculi with general detachment rule* ⋄ B20 G25 ⋄

ISEKI, K. *Axiom systems of aristotle traditional logic*
⋄ B20 ⋄

KROM, MELVEN R. *The decision problem for segregated formulas in first-order logic* ⋄ B20 B25 D35 ⋄

KROM, MELVEN R. *The decision problem for a class of first-order formulas in which all disjunctions are binary*
⋄ B20 ⋄

LAMBERT, K. & SCHARLE, T.W. *A translation theorem for two systems of free logic* ⋄ B20 B60 F25 ⋄

LIFSCHITZ, V. *Deductive general validity and reduction classes (Russian)* ⋄ B20 D35 ⋄

ONO, K. *Reduction of logics to the primitive logic*
⋄ B20 ⋄

SINGLETARY, W.E. *A note on finite axiomatization of partial propositional calculi* ⋄ B20 ⋄

SMIRNOV, V.A. *Imbedding syllogistics into predicate calculus (Russian)* ⋄ B20 ⋄

TAJMANOV, A.D. *Systems with a solvable universal theory (Russian)* ⋄ B20 B25 C20 ⋄

TANAKA, S. *Axiom systems of aristotle traditional logic. II*
⋄ B20 ⋄

TANAKA, S. *On axiom systems of propositional calculi. XXV* ⋄ B20 ⋄

TUGUE, T. *A Lemma for negationless propositional logics and its applications* ⋄ B20 ⋄

VILLARS, R. *Eine semantische Charakterisierung der durch die Implikation allein darstellbaren Wahrheitsfunktionen* ⋄ B20 ⋄

1968

GLADSTONE, M.D. *A single-axiom implicational calculus of given unsolvability* ⋄ B20 D25 D30 D35 ⋄

KROM, MELVEN R. *Some interpolation theorems for first-order formulas in which all disjunctions are binary*
⋄ B20 C40 ⋄

LEBLANC, H. & THOMASON, R.H. *Completeness theorem for some presupposition-free logics* ⋄ B20 ⋄

MEREDITH, C.A. & PRIOR, A.N. *Equational logic*
⋄ B20 F50 G05 ⋄

MONTGOMERY, H. & ROUTLEY, R. *On systems containing Aristotle's thesis* ⋄ B20 ⋄

ONO, K. *A study on formal deductions in the primitive logic* ⋄ B20 ⋄

ONO, K. *On a class of truth-value evaluations of the primitive logic* ⋄ B20 ⋄

ONO, K. *On formal theories* ⋄ B20 ⋄

OREVKOV, V.P. *Glivenko's sequence classes (Russian)*
⋄ B10 B20 F50 ⋄

OREVKOV, V.P. *Two undecidable classes of formulas in classical predicate calculus (Russian)* ⋄ B20 D35 ⋄

PACHOLSKI, L. & WEGLORZ, B. *Topologically compact structures and positive formulas* ⋄ B20 C40 C50 ⋄

PIECZKOWSKI, A. *Undecidability of the homogeneous formulas of degree 3 of the predicate calculus (Polish and Russian summaries)* ⋄ B20 D35 ⋄

RAUTENBERG, W. *Unterscheidbarkeit endlicher geordneter Mengen mit gegebener Anzahl von Quantoren* ◊ B20 C07 C13 C65 D35 ◊

ROSE, L.E. *Aristotle's syllogistic* ◊ A10 B20 ◊

ROUSSEAU, G. *Note on the decidability of a certain class of number theoretic equations* ◊ B20 B25 F30 ◊

SINGLETARY, W.E. *Results regarding the axiomatization of partial propositional calculi* ◊ B20 ◊

SURMA, S.J. *Theorems on deduction for descending implications (Polish) (English and Russian summaries)* ◊ B20 ◊

TANAKA, S. *An algebraic formulation of K − N propositional calculus. IV* ◊ B20 ◊

TANAKA, S. *Axiom systems of Aristotle traditional logic. IV* ◊ A10 B20 ◊

TURSMAN, R. *The shortest axioms of the implicational calculus* ◊ B20 ◊

1969

BACON, J. *Ontological commitment and free logic* ◊ A05 B20 ◊

HAUSCHILD, K. *Equivalence in respect to special classes of sentences* ◊ B20 ◊

IWANUS, B. *An extension of the traditional logic containing the elementary ontology and the algebra of classes* ◊ B20 G05 ◊

IWANUS, B. *Remarks about syllogistic with negative terms (Polish and Russian summaries)* ◊ B20 ◊

JASKOWSKI, S. *On the interpretations of aristotelian categorical propositions in the predicate calculus* ◊ A10 B20 ◊

KEENE, G.B. *The relational syllogism. A systematic approach to relational logic* ◊ B20 ◊

LEBLANC, H. & MEYER, R.K. *Open formulas and the empty domain* ◊ B20 ◊

MAKINSON, D. *Remarks on the concept of distribution in traditional logic* ◊ A05 A10 B20 ◊

MEREDITH, C.A. *Terminal functors permissible with syllogistic* ◊ A05 B20 ◊

MIHAILESCU, E.G. *L'ordre d'incomplétitude pour le systeme d'equivalence la negation et la reciprocite* ◊ B20 ◊

NADIU, G.S. *The interpolation theorem in strict positive logic* ◊ B20 C40 ◊

OEFFENBERGER, G.N. *Zur Frage der Bestimmbarkeit des Wahrheitswertes der schlusskraeftigen syllogistischen Modi im Falle falscher Praemissenkonjunktionen (Polish summary)* ◊ B20 ◊

ONO, K. *On a class of set-theoretical interpretations of the primitive logic* ◊ B20 E75 ◊

PRIOR, A.N. *Propositional calculus in implication and non-equivalence* ◊ B20 ◊

ROGAVA, M.G. *A sequential version of the calculus of equivalences (Russian)* ◊ B20 ◊

ROGAVA, M.G. *A specialization of certain logical rules in applied predicate calculi (Russian) (Georgian and English summaries)* ◊ B20 ◊

ROSE, A. *Calcul implicatif de non-contradiction* ◊ B20 ◊

ROSE, A. *Single generators for Henkinian fragments of the 2-valued propositional calculus* ◊ B20 ◊

SAITO, S. *A theory of categorical syllogism* ◊ B20 ◊

SHUKLA, A. *A note on independence* ◊ B20 ◊

STEINER, H.-G. *Representation de la syllogistique aristotelicienne sur la base de la theorie des relations* ◊ A10 B20 ◊

STRAZDINS, I.E. *Automorphism groups of an algebra of boolean functions and Post classes (Russian)* ◊ B20 ◊

TABATA, H. *Free structures and universal Horn sentences* ◊ B20 C05 C52 ◊

TANAKA, S. *On the proposition $C\delta CpqC\delta p\delta q$ with a variable functor* ◊ B20 ◊

1970

APRILE, G. *Grafi inferenziali per le relazioni diadiche* ◊ B20 ◊

BLACKHURST, J.H. *Syllogistic and non-syllogistic aspects of the comparative argument* ◊ A05 B20 ◊

FEDINA, A.M. *The syllogistics of classes (Russian)* ◊ B20 ◊

GALVIN, F. *Horn sentences* ◊ B20 C20 C30 C40 D35 ◊

GUREVICH, Y. *Minsky machines and the case $\forall\exists\forall$ & $\exists^\infty$ of the decision problem (Russian)* ◊ B20 D05 D35 ◊

IRVIN, A.A. *Connective implication (Russian)* ◊ B20 ◊

ITO, J. *On reducibility of provability in the primitive logic* [LO] ◊ B20 ◊

KROM, MELVEN R. *The decision problem for formulas in prenex conjunctive normal form with binary disjunctions* ◊ B20 D03 D35 ◊

KUZICHEV, A.S. *Solution of certain problems of mathematical logic by means of Venn diagrams (Russian)* ◊ B20 ◊

LEBLANC, H. & MEYER, R.K. *On prefacing $(\forall X)A \supset A(Y/X)$ with $(\forall Y)$ − A free quantification theory without identity* ◊ B20 ◊

LIOGON'KIJ, M.I. *On the question of quantitative characteristics of logical formulae (Russian) (English summary)* ◊ B20 B25 C13 ◊

LONDEY, D. *On the strong completeness of some equivalential systems* ◊ B20 ◊

SETLUR, R.V. *The product of implication and counter-implication systems* ◊ B20 ◊

SHARONOV, V.I. & ZAMOV, N.K. *Amplifications of formulae of predicate calculus (Russian)* ◊ B10 B20 D35 ◊

SHEJNBERGAS, I.M. *Simple bases and the number of functions in certain Post classes (Russian)* ◊ B20 ◊

SZYNDLER, J. *A note on a certain property of Hilbert's implicative sentential calculus (Polish) (English summary)* ◊ B20 ◊

THOMAS, I. *Final word on a shortest implicational axiom* ◊ B20 ◊

TREW, A. *Nonstandard theories of quantifications and identity* ◊ B20 B60 ◊

UEMOV, A.I. *A construction of propositional logic without the affirmation principle (Russian)* ◊ B20 ◊

1971

AANDERAA, S.O. *On the decision problem for formulas in which all disjunctions are binary* ◇ B20 D35 ◇

BRUSENTSOV, N.P. *A mathematical theory of syllogistics (Russian)* ◇ B20 ◇

BUNDER, M.W. *Predicate calculus without free variables* ◇ B20 ◇

CHAPIN JR., E.W. *Gentzen-like systems for partial propositional calculi I,II* ◇ B20 ◇

CHAPIN JR., E.W. *The strong decidability of cut-logics I: Partial propositional calculi. II: Generalizations* ◇ B20 B25 ◇

CHEVALLARD, Y. *Etude de compacite sur les calculs propositionnels incomplets* ◇ B20 E25 ◇

FERRO, R. *Le formule positive e negative, un teorema debole di sostituzione (English summary)* ◇ B20 ◇

FITTING, M. *A tableau proof method admitting the empty domain* ◇ B20 ◇

FUJIWARA, T. *On the construction of the least universal Horn class containing a given class* ◇ B20 C05 C52 ◇

GEORGIEVA, N.V. *A logical system which has ≡ and ∨ as primitive connectives (Russian) (Polish and English summaries)* ◇ B20 ◇

HERMES, H. *A simplified proof for the unsolvability of the decision problem in the case ∀∃∀* ◇ B20 D35 ◇

KOSTYRKO, V.F. *The reduction class $\forall x \forall y \exists z F(x,y,z) \wedge \forall^m \mathfrak{A}(F)$ (Russian) (English summary)* ◇ B20 C13 D35 ◇

LEBLANC, H. *Truth-value semantics for a logic of existence* ◇ B20 ◇

MCKENZIE, R. & MYCIELSKI, J. & THOMPSON, D. *On boolean functions and connected sets* ◇ B20 ◇

MINTS, G.E. *Quantifier-free and one-quantifier systems (Russian) (English summary)* ◇ B20 D20 F30 ◇

MOISIL, G.C. *La logique elementaire* ◇ B20 ◇

MONK, J.D. *Provability with finitely many variables* ◇ B20 ◇

MONTALI, T. *Pseudolimiti e formule SH* ◇ B20 C30 G30 ◇

NOLIN, L. *Types a deux variables (methode de Lindenbaum)* ◇ B20 ◇

OREVKOV, V.P. *On biconjunctive reduction classes (Russian) (English summary)* ◇ B20 D35 ◇

PAHI, B. *Full models and restricted extensions of propositional calculi* ◇ B20 ◇

PAHI, B. *Restricted extensions of implicational calculi* ◇ B20 ◇

PAOLA DI, R.A. *The relational data file and the decision problem for classes of proper formulas* ◇ B20 B75 D35 ◇

PASTEL, A.M. *Sur les fonctions logiques permutantes* ◇ B20 ◇

POLLOCK, J.L. *Henkin style completeness proofs in theories lacking negation* ◇ B20 ◇

SANCHEZ-MAZAS, M. *Arithmetical calculus of propositions (Spanish)* ◇ B20 ◇

SMYTH, M.B. *A diagrammatic treatment of syllogistic* ◇ A05 B20 ◇

STASZEK, W. *On proofs of rejection (Polish and Russian summaries)* ◇ B20 ◇

SURANYI, J. *Reduction of the decision problem of the first order predicate calculus to reflexive and symmetrical binary predicates* ◇ B20 D35 ◇

SURDU, A. *Aristotle's definition of the syllogism (Romanian)* ◇ A10 B20 ◇

SURMA, S.J. *Method of natural deduction in the equivalential and equivalential-negational propositional calculus (Polish summary)* ◇ B20 ◇

TABATA, H. *A generalized free structure and several properties of universal Horn classes* ◇ B20 C05 C52 ◇

1972

BAKER, A.J. *Syllogistic with complex terms* ◇ B20 ◇

BANACHOWSKI, L. *On a certain subset of the set of predicate tautologies (Russian summary)* ◇ B20 ◇

BOLLMAN, D.A. & TAPIA, M. *On the recursive unsolvability of the provability of the deduction theorem in partial propositional calculi* ◇ B20 D35 ◇

CHAPIN JR., E.W. *Measures of centrality and complexity for partial propositional calculi* ◇ B20 ◇

CORCORAN, J. *Completeness of an ancient logic* ◇ A10 B20 ◇

FRIDMAN, EH.I. & PENZIN, YU.G. *The elementary and the universal theory of the ordered group of integers with maximal subgroups (Russian)* ◇ B20 B25 D35 ◇

GROSJEAN, P.V. *Theorie algebrique du syllogisme categorique* ◇ B20 ◇

HAGIWARA, Y. *A system for recognition of the implication relation between formulas in the logic of predicates* ◇ B20 B35 ◇

LAMBERT, K. *Notes on free description theory: some philosophical issues and consequences* ◇ A05 B20 ◇

LOPEZ-ESCOBAR, E.G.K. *Constructions and negationless logic (Polish and Russian summaries)* ◇ B20 E70 F50 ◇

MASLOV, S.YU. & OREVKOV, V.P. *Decidable classes that reduce to a single quantifier class (Russian)* ◇ B20 B25 ◇

RAS, Z. *On a relationship between certain grammars and enumerable first order predicate calculi (Russian summary)* ◇ B20 D05 ◇

SEKIMOTO, T. *On the uniqueness of the shortest single axiom for the implicational calculus of propositions* ◇ B20 ◇

SLUPECKI, J. *Proof of the completeness of a fragmentary system of the propositional calculus (Polish) (English summary)* ◇ B20 ◇

STASZEK, W. *A certain Interpretation of the theory of rejected propositions (Polish and Russian summaries)* ◇ B20 ◇

SURMA, S.J. *A method of axiomatization of two-valued propositional connectives* ◇ B20 ◇

SURMA, S.J. *A survey of the results and methods of investigations of the equivalential propositional calculus (Polish summary)* ◇ B20 ◇

SURMA, S.J. *A uniform method of proof of the completeness theorem for the equivalential propositional calculus and for some of its extensions (Polish summary)* ◇ B20 ◇

WILLIAMSON, C. *Squares of opposition: comparisons between syllogistic and propositional logic* ⋄ B05 B20 ⋄

ZARNECKA-BIALY, E. *The deduction theorems for propositional calculi when implication and falsum is present (Polish summary)* ⋄ B20 F50 ⋄

1973

AANDERAA, S.O. & LEWIS, H.R. *Prefix classes of Krom formulas* ⋄ B20 B25 D35 ⋄

CORCORAN, J. *A mathematical model of Aristotle's syllogistic* ⋄ A05 A10 B20 ⋄

DENISE, T.C. *The two logics: traditional and modern* ⋄ A05 B20 ⋄

ERMOLAEVA, N.M. *Logics akin to Hao Wang's calculus (Russian)* ⋄ B20 ⋄

FUJIWARA, T. *Note on generalized atomic sets of formulas* ⋄ B20 C05 C40 ⋄

GENSLER, H.J. *A simplified decision procedure for categorical syllogisms* ⋄ B20 B25 ⋄

GOLDFARB, W.D. & LEWIS, H.R. *The decision problem for formulas with a small number of atomic subformulas* ⋄ B20 D35 ⋄

GUREVICH, Y. & TURASHVILI, T.V. *A strengthening of a certain result of J. Suranyi (Russian) (Georgian and English summaries)* ⋄ B20 D35 ⋄

GUREVICH, Y. *Formulas with a single ∀ (Russian)* ⋄ B20 B25 C13 ⋄

HALKOWSKA, K. *Conditional definitions and the idea of meaningful expression with conditionally defined terms (Polish) (English summary)* ⋄ B10 B20 ⋄

HAMBLIN, C.L. *A felicitous fragment of the predicate calculus* ⋄ B20 B25 ⋄

HIZ, H. *A completeness proof for C-calculus* ⋄ B20 ⋄

JOHNSON, J.S. *Axiom systems for first order logic with finitely many variables* ⋄ B20 C05 G15 ⋄

LEHMANN, A. *Two sets of perfect syllogisms* ⋄ B20 ⋄

MAKSIMOVA, L.L. *Implication lattices (Russian)* ⋄ B20 G10 ⋄

MEYER, R.K. *On conserving positive logics* ⋄ B20 ⋄

MONTEVERDI, D. *SH a piu sorte di variabili* ⋄ B20 C30 G30 ⋄

NEGRU, I.S. *Certain cardinality properties of the lattice of primitive logics (Russian)* ⋄ B20 ⋄

PAOLA DI, R.A. *The solvability of the decision problem for classes of proper formulas and related results* ⋄ B20 B25 ⋄

SANMARTIN ESPLUGUES, J. *Syllogistics, many-valued logic and model theory (Spanish)* ⋄ A10 B20 B50 C90 ⋄

SEKIMOTO, T. *Propositional and implicational propositional calculus (Japanese)* ⋄ B05 B20 ⋄

SMILEY, T.J. *What is a syllogism?* ⋄ A05 B20 ⋄

STEPIEN, T. *A survey of minor Wajsberg's results concerning fragmentary systems of the classical propositional calculus* ⋄ B20 ⋄

STIHI, T. *A case of decidability in the first order predicate calculus (Romanian)* ⋄ B20 B25 ⋄

STIHI, T. *Une generalisation du carre logique* ⋄ A10 B20 ⋄

SURMA, S.J. *A method of axiomatization of two-valued propositional connectives* ⋄ B20 ⋄

SZCZECH, W. *On one of Wajsberg's theorem* ⋄ B20 ⋄

THARP, L.H. *The characterization of monadic logic* ⋄ B15 B20 C95 ⋄

ZARNECKA-BIALY, E. *Negation in Ch.S.Peirce's propositional calculus* ⋄ A10 B20 F50 ⋄

1974

AANDERAA, S.O. & LEWIS, H.R. *Linear sampling and the ∀∃∀ case of the decision problem* ⋄ B20 D03 D05 D35 D80 ⋄

BALAN, T. *Quaternary relations* ⋄ A05 B20 E07 ⋄

BARTHELEMY, J.-P. *Sur la refutabilite* ⋄ B20 F50 G30 ⋄

BOERGER, E. *Σ_3-completude de l'ensemble des types de reduction* ⋄ B20 D35 D55 ⋄

BOERGER, E. *Beitrag zur Reduktion des Entscheidungsproblems auf Klassen von Hornformeln mit kurzen Alternationen* ⋄ B20 B25 D35 ⋄

CHAPIN JR., E.W. *Translations and structures for partial propositional calculi* ⋄ B20 ⋄

CHYTIL, M.K. *The decomposition calculus and its semantics* ⋄ B20 ⋄

CORCORAN, J. *Aristotle's natural deduction system* ⋄ A05 A10 B20 ⋄

DUTHIE, G.D. *Logic of terms* ⋄ B20 ⋄

EPSTEIN, G. & HORN, A. *Finite limitations on a propositional calculus for affirmation and negation* ⋄ B20 G10 ⋄

FLUM, J. *On Horn theories* ⋄ B20 C20 C80 ⋄

MARCJA, A. & TULIPANI, S. *Questioni di teoria dei modelli per linguaggi universali positivi (English summary)* ⋄ B20 C07 C10 ⋄

MARSDEN, E.L. *Reducible implicative models* ⋄ B20 B45 C05 G25 ⋄

RICKEY, V.F. *The one variable implicational calculus* ⋄ B20 ⋄

SIMONS, L. *Logic without tautologies* ⋄ B20 ⋄

SINGLETARY, W.E. *Many-one degrees associated with partial propositional calculi* ⋄ B20 D25 ⋄

TAMAKI, I. *Syllogistic and calculus of classes* ⋄ B20 ⋄

THOMAS, I. *On Meredith's sole positive axiom* ⋄ B20 ⋄

WEAVER, G.E. *Finite partitions and their generators* ⋄ B20 B25 C07 C40 ⋄

WIEDMER, E. *Ein neuer negationsloser Beweis eines Satzes von G.F.C.Griss* ⋄ B20 F55 ⋄

WRONSKI, A. *The degree of completeness of some fragments of the intuitionistic propositional logic* ⋄ B20 F50 ⋄

ZAKREVSKIJ, A.D. *On the theory of polysyllogisms (Russian)* ⋄ B20 ⋄

1975

BACSICH, P.D. *Amalgamation properties and interpolation theorems for equational theories* ⋄ B20 C05 C40 C52 ⋄

BAKER, J.R. *On two immediate inferences by limitation* ⋄ B20 ⋄

DANCY, R.M. *Sense and contradiction: A study in Aristotle* ⋄ A10 B20 ⋄

EDELSTEIN, R. *An interpolation lemma for the pure implicational calculus* ⋄ B20 ⋄

FUJIWARA, T. *Universal sentences preserved under certain extensions* ⋄ B20 C40 ⋄

GOLDFARB, W.D. & LEWIS, H.R. *Skolem reduction classes* ⋄ B20 D35 ⋄

GRACIA, J.J.E. *Propositions as premises of syllogisms in medieval logic* ⋄ A10 B20 ⋄

HACKER, E.A. *The octagon of opposition* ⋄ A10 B20 ⋄

HALKOWSKA, K. & PIROG-RZEPECKA, K. *On proofs in the theories containig conditional definition (Polish) (English summary)* ⋄ B20 C07 ⋄

JORDAN, I.B. & MOUFTAH, H.T. *A design technique for an integrable ternary arithmetic unit* ⋄ B20 B70 ⋄

KABZINSKI, J.K. *Some problems connected with equivalential formalization of classical sentential calculus (Polish and English)* ⋄ B20 ⋄

KANOVICH, M.I. *A hierachical semantic system with set variables (Russian)* ⋄ B20 B65 C40 D05 F50 ⋄

KEANE, O. *Abstract Horn theories* ⋄ B20 G30 ⋄

KHOMICH, V.I. *Weakly and strongly nonrealizable propositional formulae (Russian)* ⋄ B20 D15 D20 F50 ⋄

KOPIEKI, R. & SARALSKI, B. & WALIGORA, G. *The research of Jaskowski on decidability theory of first order sentences I* ⋄ A10 B20 B25 D35 ⋄

McKENZIE, R. *On spectra, and the negative solution of the decision problem for identities having a finite nontrivial model* ⋄ B20 C05 C13 D35 ⋄

MEZHLUMBEKOVA, V.F. *Cut-elimination in a system of negationless arithmetic (Russian) (English summary)* ⋄ B20 F05 F30 F50 ⋄

MEZHLUMBEKOVA, V.F. *On systems of the negationless calculus of predicates (Russian) (English summary)* ⋄ B20 F50 ⋄

MILICI, C. *Note on the C-calculus* ⋄ B20 ⋄

MORTIMER, M. *On languages with two variables* ⋄ B20 ⋄

POIZAT, B. *Theoremes globaux (English summary)* ⋄ B20 C40 ⋄

SCHUMM, G.F. *A Henkin-style completeness proof for the pure implicational calculus* ⋄ B20 ⋄

SMOLIN, V.P. *Monadic algebras and the one-place predicate calculus* ⋄ B20 C07 G15 ⋄

STRAZDINS, I.E. *Affine classification of Boolean functions of five variables (Russian)* ⋄ B20 ⋄

THOMAS, I. *Nice implicational axioms* ⋄ B20 ⋄

THOMAS, I. *Shorter development of an axiom* ⋄ B20 ⋄

THOMAS, I. *Simple implicational development* ⋄ B20 ⋄

TULIPANI, S. *Questioni di teoria dei modelli per linguaggi universali positivi II: Metodi di "back and forth" (English summary)* ⋄ B20 C07 C40 C50 ⋄

TURASHVILI, T.V. *The decidability problem of first order predicate logic (Russian)* ⋄ B20 D35 ⋄

TUREK, C. *Lehmann on the rules of the invalid syllogisms* ⋄ B20 ⋄

WYBRANIEC-SKARDOWSKA, U. *Mutual definability of the notions of entailment and inconsistency (Polish) (English summary)* ⋄ B20 ⋄

1976

BERNHARDT, K. *Semantics for finitary predicate calculi* ⋄ B20 C90 ⋄

EVANS, M.G. *A truth-functional analysis of Aristotelian logic* ⋄ A10 B20 ⋄

GRZEGOREK, E. *On axial maps of direct products, II* ⋄ B20 E20 ⋄

GUREVICH, Y. *Semi-conservative reduction* ⋄ B20 B25 D35 ⋄

GUREVICH, Y. *The decision problem for standard classes* ⋄ B20 B25 D35 ⋄

HENSCHEN, L.J. *Semantic resolution for Horn sets* ⋄ B20 B35 ⋄

HERRING, J.M. *Equivalence of several notions of theory completeness in a free logic* ⋄ B20 C35 ⋄

HUGHES, C.E. *A reduction class containing formulas with one monadic predicate and one binary function symbol* ⋄ B20 D35 ⋄

HUGHES, C.E. *Two variable implicational calculi of prescribed many-one degrees of unsolvability* ⋄ B20 D25 ⋄

LEWIS, H.R. *Krom formulas with one dyadic predicate letter* ⋄ B20 D05 D35 ⋄

MAL'TSEV, A.I. *Iterative Post algebras (Russian)* ⋄ B20 ⋄

NEMESSZEGHY, E.A. *Note on an independence proof of Johansson* ⋄ B20 ⋄

NORGELA, S.A. *On approximating reduction classes of CPC by decidable classes (Russian) (English summary)* ⋄ B20 B25 D35 ⋄

PETERSON, J.G. *Shortest single axiom for the classical equivalential calculus* ⋄ B20 ⋄

PRULLAGE, M.M. *A theory of restricted variables without existence assumptions* ⋄ B20 ⋄

ROSE, A. *Formalisations with non-standard degrees of completeness* ⋄ B20 ⋄

RYBAK, JANET & RYBAK, JOHN *Venn diagrams extended: Map logic* ⋄ B20 E20 ⋄

SAURIOL, P. *La structure tetrahexaedrique du systeme complet des propositions categoriques* ⋄ A05 B20 ⋄

SURMA, S.J. *Every reduct of two-element boolean algebra can be finitely axiomatized by modus ponens or substitution rule* ⋄ B20 ⋄

TAMURA, S. *On NBN-systems* ⋄ B20 ⋄

THOMAS, I. *One dimension in PS and PSI* ⋄ B20 ⋄

ULRICH, D. *On a property of matrices for subsystems of IC^+* ⋄ B20 F50 ⋄

1977

BAKER, A.J. *Classical logical relations* ⋄ A10 B20 ⋄

BANASCHEWSKI, B. *On G. Spencer Brown's laws of form* ⋄ B20 ⋄

BRUSENTSOV, N.P. *Lewis Carrol diagrams and the Aristotelian syllogistics (Russian)* ⋄ B20 ⋄

CHYTIL, M.K. *Semantique de formules logiques en forme d'equivalence n-aire (demi-modeles)* ⋄ B20 ⋄

CZERMAK, J. *A remark on Gentzen's calculus of sequents* ⋄ B20 F07 ⋄

DZHIDZHYAN, R.Z. *Extended syllogistics (Russian)* ⋄ B20 ⋄

HASENJAEGER, G. *Von der Syllogistik zur Mengentheorie* ⋄ A10 B20 E30 ⋄

HENKIN, L. *The logic of equality* ⋄ B20 C05 ⋄

ISHIMOTO, A. *A propositional fragment of Lesniewski's ontology* ⋄ A05 B20 ⋄

McNULTY, G.F. *Fragments of first order logic, I: Universal Horn logic* ⋄ B20 ⋄

NELSON, EVELYN *Classes defined by implications* ⋄ B20 C05 C52 ⋄

PETERSON, J.G. *Single axioms for the classical equivalential calculus* ⋄ B20 B30 ⋄

ROSE, A. *Simplified formalizations of fragments of the propositional calculus* ⋄ B20 ⋄

RUKHAYA, KH.M. *A generalization of the theory $\mathcal{T}$ (Russian) (Georgian and English summaries)* ⋄ B20 E30 ⋄

SHELAH, S. *Decidability of a portion of the predicate calculus* ⋄ B20 B25 ⋄

TAMURA, S. *The separation theorem of NBN-system* ⋄ B20 ⋄

TEMPLE, G. *Inference without axiom or paradoxes* ⋄ A10 B20 ⋄

TURASHVILI, T.V. *On the undecidable minimal classes of first order predicate logic (Russian) (English summary)* ⋄ B20 D35 ⋄

WIRSING, M. *Das Entscheidungsproblem der Klasse von Formeln, die hoechstens zwei Primformeln enthalten* ⋄ B20 B25 D35 ⋄

1978

APRILE, G. *Grafi inferenziali per i problemi logici con l'identita* ⋄ B20 ⋄

ASHWORTH, E.J. *Multiple quantification and the use of special quantifiers in early sixteenth century logic* ⋄ A10 B20 ⋄

BARJA PEREZ, J. *Algebras universales en el calculo de proposiciones* ⋄ B20 C05 G99 ⋄

BOERGER, E. *Bemerkung zu Gurevich's Arbeit ueber das Entscheidungsproblem fuer Standardklassen* ⋄ B20 D10 D35 ⋄

FERRO, A. & OMODEO, E.G. *An efficient validity test for formulae in extensional two-level syllogistic* ⋄ B20 B25 B35 ⋄

FREGUGLIA, P. *Su una interpretazione algebrico-categoriale del sillogismo* ⋄ A05 B20 ⋄

GIVANT, S. *Universal Horn classes categorical or free in power* ⋄ B20 C05 C35 ⋄

HERRERA MIRANDA, J. *Les theories convexes de Horn (English summary)* ⋄ B20 C05 C52 G30 ⋄

KALMAN, J.A. *A shortest single axiom for the classical equivalential calculus* ⋄ B20 ⋄

LEWIS, H.R. *Complexity of solvable cases of the decision problem for the predicate calculus* ⋄ B20 B25 D15 ⋄

LEWIS, H.R. *Renaming a set of clauses as a Horn set* ⋄ B20 D15 ⋄

MEREDITH, D. *Positive logic and λ-constants* ⋄ B20 B25 B40 ⋄

MUZIO, J.C. *A note concerning a sole sufficient operator* ⋄ B20 ⋄

SIMONS, L. *More logics without tautologies* ⋄ B20 ⋄

SLATER, B.H. *A fragment of new propositional logic* ⋄ B20 ⋄

SMITH, ROBIN *The mathematical origins of Aristotle's syllogistic* ⋄ A10 B20 ⋄

SOBOCINSKI, B. *Note about Lukasiewicz's theorem concerning the system of axioms of the implicational propositional calculus* ⋄ B20 ⋄

WIRSING, M. *Kleine unentscheidbare Klassen der Praedikatenlogik mit Identitaet und Funktionszeichen* ⋄ B20 D35 ⋄

WOZNIAKOWSKA, B. *The representation theorem for the algebras determined by the fragments of infinite-valued logic of Lukasiewicz* ⋄ B20 B50 G20 G25 ⋄

1979

ANGELELLI, I. *Aristotelian modal syllogistic* ⋄ A05 B20 ⋄

BERKA, K. *A reinterpretation of Aristotle's syllogistic* ⋄ A10 B20 ⋄

CORCORAN, J. & ZIEWACZ, S. *Identity logics* ⋄ B20 ⋄

DALE, T. *A natural deduction system for "if then"* ⋄ B20 ⋄

GIVANT, S. *A representation theorem for universal Horn classes categorical in power* ⋄ B20 C05 C35 ⋄

GLADSTONE, M.D. *The decidability of one-variable propositional calculi* ⋄ B20 B25 ⋄

GOTTLIEB, D. & McCARTHY, T. *Substitutional quantification and set theory* ⋄ A05 B20 ⋄

HAFNER, I. *On proof length in the equivalential calculus* ⋄ B20 F20 ⋄

HUGLY, P. & SAYWARD, C.W. *A semantical account of the vicious circle principle* ⋄ A05 B20 ⋄

KHOMICH, V.I. *Separability of superintuitionistic propositional calculi (Russian)* ⋄ B20 B55 ⋄

KUZNETSOV, A.V. *Means for detection of nondeducibility and inexpressibility (Russian)* ⋄ B20 ⋄

LEWIS, H.R. *Unsolvable classes of quantificational formulas* ⋄ B20 B25 D35 D98 ⋄

MANDERS, K.L. *The theory of all substructures of a structure: Characterisation and decision problems* ⋄ B20 B25 C57 C60 ⋄

MEREDITH, D. *Axiomatics for implication* ⋄ B20 F50 ⋄

MUELLER, I. *The completeness of Stoic propositional logic* ⋄ A10 B20 ⋄

PINUS, A.G. *A calculus of one-place predicates (Russian)* ⋄ B20 B25 C55 C80 C85 ⋄

POUZET, M. *Chaines de theories universelles* ⋄ B20 C52 ⋄

ROZENBLAT, B.V. *Positive theories of free inverse semigroups (Russian)* ⋄ B20 C05 D35 D40 ⋄

SLATER, B.H. *Aristotle's propositional logic* ⋄ A05 A10 B20 ⋄

TABAKOV, M. *A formal system without well-formed formulas* ⋄ B20 ⋄

THOM, P. *Aristotle's syllogistic* ⋄ A05 A10 B20 ⋄

TSITKIN, A.I. *On the question of an error in a famous work due to Wajsberg (Russian)* ⋄ A10 B20 F50 ⋄

UGOL'NIKOV, A.B. *The synthesis of schemes and formulas in incomplete bases (Russian)* ⋄ B20 ⋄

ZAKREVSKIJ, A.D. *On a formalization of polysyllogistics (Russian)* ⋄ B20 ⋄

1980

BALOWITZ, V. *The rules of the syllogism without distribution* ◊ B20 ◊

BATENS, D. *A completeness-proof method for extensions of the implicational fragment of the propositional calculus* ◊ B20 ◊

BOCHAROV, V.A. *Algebraic reconstructions of syllogistics (Russian)* ◊ B20 ◊

BUNDER, M.W. *Modus ponens free (implicational) logics* ◊ B20 ◊

DYWAN, Z. & STEPIEN, T. *Every two-valued propositional calculus has the interpolation property* ◊ B20 C40 ◊

ENGLEBRETSEN, G. *Noncategorical syllogisms in the Analytics* ◊ A05 A10 B20 ◊

FRIEDMAN, W.H. *Calculemus* ◊ B20 ◊

GRIEDER, A. *On the logic of relations* ◊ B20 G15 ◊

JACOBS, W. *The existential presuppositions of Aristotle's logic* ◊ A05 B20 ◊

KODERA, H. *Remark on classical logic and intuitionistic logic* ◊ B20 F05 F50 ◊

LEWIS, H.R. *Complexity results for classes of quantificational formulas* ◊ B20 B25 D10 D15 ◊

LOPES DOS SANTOS, L.H. *Constructive completeness proofs for positive propositional calculi* ◊ B20 ◊

MEREDITH, D. *A positive logic proof procedure* ◊ B20 B35 B40 F50 ◊

MISERCQUE, D. *Sur le treillis distributif des $\forall_1$-formules fermees de l'arithmetique de Peano (English summary)* ◊ B20 F30 G10 ◊

NORGELA, S.A. *On approximation of some classes of classical predicate calculus by decidable classes. I,II (Russian) (English and Lithuanian summaries)* ◊ B10 B20 B25 ◊

NOVAK, JOSEPH A. *Some recent work on the assertoric syllogistic* ◊ A10 B20 ◊

SHABANOV-KUSHNARENKO, YU.P. *Formulas of the algebra of finite predicates (Russian)* ◊ B20 ◊

SMIRNOV, V.A. *Adequate conversion of assertions of syllogistics in predicate calculus (Russian)* ◊ B20 ◊

SURMA, S.J. *On the axiomatic treatment of the Sheffer stroke and of the dual Sheffer stroke* ◊ B20 ◊

1981

BRANDOM, R.B. *Semantic paradox of material implication* ◊ A05 B20 ◊

CIRULIS, J. *On the EA-fragment of classical propositional logic* ◊ B20 ◊

CIRULIS, J. *On the EK-fragments of positive and classical propositional logics* ◊ B20 ◊

DELOBEL, C. & FAGIN, R. & PARKER JR., D.S. & SAGIV, Y. *An equivalence between relational database dependencies and a fragment of propositional logic* ◊ B20 ◊

DEUTSCH, M. *Zur Reduktionstheorie des Entscheidungsproblems* ◊ B20 D35 ◊

FUERER, M. *Alternation and the Ackermann case of the decision problem* ◊ B20 B25 D15 ◊

GOLDFARB, W.D. *On the Goedel class with identity* ◊ B20 B25 D35 ◊

KIRK, R.E. *A complete semantics for implicational logics* ◊ B20 ◊

KLEINE BUENING, H. *Some undecidable theories with monadic predicates and without equality* ◊ B20 D35 ◊

MARKOVIC, Z. *On axiomatizability and preservation in Kripke models* ◊ B20 ◊

RAUTENBERG, W. *2-element matrices* ◊ B20 B22 ◊

RENARDEL DE LAVALETTE, G.R. *The interpolation theorem in fragments of logics* ◊ B20 B55 C40 F50 ◊

ROGAVA, M.G. *On the sequential version of the propositional fragment of the first-order calculus of terms (Russian) (Georgian and English summaries)* ◊ B20 ◊

SHEN, YOUDING *The part of the first-order logical calculus (with identity) which is independent of quantifiers (Chinese)* ◊ B20 ◊

THOM, P. *The syllogism* ◊ A05 B20 ◊

1982

AANDERAA, S.O. & BOERGER, E. & LEWIS, H.R. *Conservative reduction classes of Krom formulas* ◊ B20 C13 D35 ◊

AANDERAA, S.O. & BOERGER, E. & GUREVICH, Y. *Prefix classes of Krom formulae with identity (German summary)* ◊ B20 B25 D35 ◊

BRUSENTSOV, N.P. *Complete system of categorical syllogisms of Aristotle (Russian)* ◊ B20 ◊

BRUSENTSOV, N.P. *Representation of syllogistics in predicate logic (Russian)* ◊ B20 ◊

GUREVICH, Y. *Existential interpretation II* ◊ B20 D35 F25 ◊

HUSAIN, M. *"If...then" and Aristotle's syllogism* ◊ A10 B20 ◊

ISHIMOTO, A. & KOBAYASHI, M. *A propositional fragment of Lesniewski's ontology and its formulation by the tableau method* ◊ B20 ◊

JOHNSTONE JR., H.W. *The syllogism on the negative-entailing interpretation of affirmative propositions* ◊ B20 ◊

KOMORI, Y. & ONO, H. *Logics without the contradiction rule* ◊ B20 B53 ◊

KREJNOVICH, V.YA. & OSWALD, U. *A decision method for the universal theorems of Quine's new foundations* ◊ B20 B25 E70 ◊

MARCHENKOV, S.S. *Undecidability of the positive $\forall\exists$-theory of a free semigroup (Russian)* ◊ B20 D10 D35 ◊

MINTS, G.E. & TYUGU, E.KH. *The completeness of structural synthesis rules (Russian)* ◊ B20 B75 D20 F50 ◊

POIZAT, B. *Deux ou trois choses que je sais de L_n* ◊ B20 C07 C13 C50 ◊

POPOV, S.V. *Diagrams of natural deductions (Russian)* ◊ B20 ◊

POPOV, V.M. *Solvability of the syllogistic with negative terms (Russian)* ◊ B20 ◊

PRAWITZ, D. *Beweise und die Bedeutung und Vollstaendigkeit der logischen Konstanten* ◊ A05 B20 F07 ◊

SANCHEZ-MAZAS, M. *Un modele arithmetique de la syllogistique et ses extensions* ◊ B20 ◊

1983

BONOTTO, C. & BRESSAN, A. *On a synonymy relation for extensional first order theories I: A notion of synonymy. II: A sufficient criterion for non synonymy. Applications* ⋄ B20 ⋄

DOSHITA, S. & YAMASAKI, S. *The satisfiability problem for a class consisting of Horn sentences and some non-Horn sentences in propositional logic* ⋄ B20 B25 D15 ⋄

GUESSARIAN, I. *Survey on classes of interpretations and some of their applications* ⋄ B20 ⋄

GUREVICH, Y. & SHELAH, S. *Random models and the Goedel case of the decision problem* ⋄ B20 B25 C13 D35 ⋄

KAPETANOVIC, M. *Some simple decidability proofs* ⋄ B20 B25 ⋄

MARKOVIC, Z. *Some preservation results for classical and intuitionistic satisfiability in Kripke models* ⋄ B20 C25 C90 F50 ⋄

MUNDICI, D. *A lower bound for the complexity of Craig's interpolants in sentential logic* ⋄ B20 C40 D15 ⋄

ROZENBLAT, B.V. & VAZHENIN, YU.M. *On positive theories of free algebraic systems (Russian)* ⋄ B20 B25 C05 ⋄

SEGERBERG, K. *Arbitrary truth-value functions and natural deduction* ⋄ B05 B20 ⋄

1984

AANDERAA, S.O. *On the solvability of the extended $\forall \exists \wedge \exists \forall^*$-Ackermann class with identity* ⋄ B20 B25 ⋄

BEZHANISHVILI, M.N. *Model-theoretic interpretation of Aristotle's apodictic syllogistics (Russian) (English summary)* ⋄ A10 B20 ⋄

BLASS, A.R. *There are not exactly five objects* ⋄ B20 C07 ⋄

BOCHAROV, V.A. *Aristotle and traditional logic* ⋄ A10 B20 ⋄

BOERGER, E. *Decision problems in predicate logic* ⋄ B20 B25 D35 ⋄

BONEVAC, D. *Skolem fragments* ⋄ B20 ⋄

BONOTTO, C. & BRESSAN, A. *On a synonymy relation for extensional first order theories. III: A necessary and sufficient condition for synonymy* ⋄ B20 ⋄

CHIKHACHEV, S.A. *An example of a theory without weakly (Σ, Σ)-atomic models (Russian)* ⋄ B20 C50 ⋄

DENENBERG, L. & LEWIS, H.R. *The complexity of the satisfiability problem for Krom formulas* ⋄ B20 B25 D15 ⋄

DEUTSCH, M. *Reductions for the satisfiability with a simple interpretation of the predicate variable* ⋄ B20 D35 ⋄

DYWAN, Z. *An interpretation of Aristotle's syllogistic and a certain fragment of set theory in propositional calculi* ⋄ B20 E20 ⋄

DYWAN, Z. *An interpretation of a certain fragment of arithmetic in some propositional calculus* ⋄ B20 F30 ⋄

FUERER, M. *The computational complexity of the unconstrained limited domino problem (with implications for logical decision problems)* ⋄ B20 D15 D80 ⋄

GOLDFARB, W.D. & GUREVICH, Y. & SHELAH, S. *A decidable subclass of the minimal Goedel class with identity* ⋄ B20 B25 ⋄

GOLDFARB, W.D. *The unsolvability of the Goedel class with identity* ⋄ B20 D35 ⋄

GOLDFARB, W.D. *The Goedel class with identity is unsolvable* ⋄ B20 D35 ⋄

HENSON, C.W. & RUBEL, L.A. *Some applications of Nevanlinna theory to mathematical logic: identities of exponential functions* ⋄ B20 C65 F30 ⋄

KETONEN, J. & WEYHRAUCH, R.W. *A decidable fragment of predicate calculus* ⋄ B20 B25 ⋄

KRIVTSOV, V.N. *A type of formal negationfree systems (Russian)* ⋄ B20 ⋄

KRIVTSOV, V.N. *Deductive potentialities of the negationless predicate calculus (Russian)* ⋄ B20 ⋄

MOTOHASHI, N. *Equality and Lyndon's interpolation theorem* ⋄ B20 C40 C75 ⋄

STENGER, H.-J. *Algebraic characterisations of NTIME(F) and NTIME(F,A)* ⋄ B20 B35 D15 D20 ⋄

1985

DASSOW, J. & WOLTER, U. *Remarks on equality languages* ⋄ A05 B20 D05 ⋄

GOGUEN, J.A. & MESEGUER, J. *Completeness of many-sorted equational logic* ⋄ B20 C05 C07 ⋄

KOMORI, Y. & ONO, H. *Logics without the contraction rule* ⋄ B20 ⋄

PAUL, E. *On solving the equality problem in theories defined by Horn clauses* ⋄ B20 B35 ⋄

B22 Abstract deductive systems

1929
TARSKI, A. *Remarques sur les notions fondamentales de la methodologie des mathematiques* ⋄ A05 B22 ⋄

1930
TARSKI, A. *Fundamentale Begriffe der Methodologie der deduktiven Wissenschaften I* ⋄ A05 B15 B22 ⋄
TARSKI, A. *Ueber einige fundamentale Begriffe der Metamathematik* ⋄ A05 B22 ⋄

1935
HENLE, R. *A definition of abstract systems* ⋄ B22 ⋄
TARSKI, A. *Grundzuege des Systemkalkuels. Teil I* ⋄ B22 C07 ⋄
TARSKI, A. *Ueber die Erweiterungen der unvollstaendigen Systeme des Aussagenkalkuels* ⋄ B05 B22 ⋄

1936
TARSKI, A. *Grundzuege des Systemkalkuels. Teil II* ⋄ A05 B22 C07 C35 G15 ⋄

1937
TARSKI, A. *Ideale in den Mengenkoerpern* ⋄ B22 G05 ⋄

1946
HIZ, H. *Remarques sur le degre de completude* ⋄ B22 F50 ⋄

1948
KEMENY, J.G. *Models of logical systems* ⋄ B22 ⋄

1949
LOS, J. *On logical matrices (Polish)* ⋄ B22 B50 ⋄

1950
SUSZKO, R. *Canonic axiomatic systems* ⋄ B22 E30 ⋄

1951
HERMES, H. *Zur Theorie der aussagenlogischen Matrizen* ⋄ B22 ⋄

1952
SCHROETER, K. *Deduktiv abgeschlossene Mengen ohne Basis* ⋄ B22 ⋄

1953
SCHMIDT, J. *Einige grundlegende Begriffe und Saetze aus der Theorie der Huellenoperatoren* ⋄ B22 C05 E20 ⋄

1954
CUESTA DUTARI, N. *Deductive structures (Spanish)* ⋄ B22 ⋄

1955
LOS, J. *The algebraic treatment of the methodology of elementary deductive system(Polish and Russian summaries)* ⋄ B22 C07 ⋄

1956
HENKIN, L. & MONTAGUE, R. *On the definition of "formal deduction"* ⋄ B22 ⋄

1957
HIZ, H. *Complete sentential calculus admitting extensions* ⋄ B22 ⋄
MO, SHAOKUI *On the rules of logical systems (Chinese)* ⋄ B22 ⋄

1958
CURRY, H.B. *Calculuses and formal systems* ⋄ A05 B22 ⋄
CURRY, H.B. *On definitions in formal systems* ⋄ B22 ⋄
HARROP, R. *On the existence of finite models and decision procedures for propositional calculi* ⋄ B22 B25 ⋄
LOS, J. & SUSZKO, R. *Remarks on sentential logics* ⋄ B22 ⋄
NOLIN, L. *Sur les proprietes communes aux calculs des propositions et a leurs modeles* ⋄ B22 G25 ⋄

1959
HARROP, R. *The finite model property and subsystems of classical propositional calculus* ⋄ B22 ⋄
HIZ, H. *Extendible sentential calculus* ⋄ B22 ⋄

1960
POGORZELSKI, W.A. & SLUPECKI, J. *Basic properties of deductive systems based on nonclassical logics I,II (Polish) (Russian and English summaries)* ⋄ B22 F50 ⋄
SMETANICH, YA.S. *On the completeness of the propositional calculus with additional operations in one argument (Russian)* ⋄ B22 ⋄
VUCKOVIC, V. *Rekursive Modelle einiger nichtklassischer Aussagenkalkuele (Serbo-Croatian)* ⋄ B22 F30 ⋄

1961
KUBINSKI, T. *On the scope of Lindenbaum's theorem on complete supersystems (Polish) (Russian and English summaries)* ⋄ B22 ⋄
SMETANICH, YA.S. *On propositional calculi with an additional operation (Russian)* ⋄ B22 ⋄
SUSZKO, R. *Concerning the method of logical schemes, the notion of logical calculus and the role of consequence relations (Polish and Russian summaries)* ⋄ A05 B22 ⋄

1962

POGORZELSKI, W.A. *The adequacy of theory of deductive systems with respect to sentential calculi (Polish) (Russian and English summaries)* ⋄ B22 ⋄

SMILEY, T.J. *The independence of connectives* ⋄ B22 ⋄

1963

BELNAP JR., N.D. & THOMASON, R.H. *A rule-completeness theorem* ⋄ B22 F07 F50 ⋄

KUZNETSOV, A.V. *Undecidability of the general problems of completeness solvability and equivalence for propositional calculi (Russian)* ⋄ B22 D35 ⋄

1964

DUBIKAJTIS, L. *Operations on deductive systems* ⋄ B22 ⋄

HARROP, R. *A relativization procedure for propositional calculi, with an application to a generalized form of Post's theorem* ⋄ B22 D35 ⋄

HATCHER, W.S. *La notion d'equivalence entre systemes formels* ⋄ B22 F25 ⋄

UEMOV, A.I. *The problem of equivalence of logical structures (Russian)* ⋄ B22 ⋄

1965

HARROP, R. *Some generalizations and applications of a relativization procedure for propositional calculi* ⋄ B22 B25 D35 ⋄

HARROP, R. *Some structure results for propositional calculi* ⋄ B22 ⋄

MURSKIJ, V.L. *The existence in three-valued logic of a closed class with finite basis, not having a finite complete system of identities (Russian)* ⋄ B22 B50 ⋄

PORTE, J. *Recherches sur la theorie generale des systemes formels et sur les systemes connectifs* ⋄ B22 ⋄

SURMA, S.J. *On the relation of formal inference and some related concepts (Polish) (English and Russian summaries)* ⋄ B22 ⋄

WANG, HAO *Note on rules of inference* ⋄ B05 B22 ⋄

ZARNECKA-BIALY, E. *Certain non-standard concept of consequence (Polish) (English and Russian summaries)* ⋄ B22 ⋄

1966

HERMES, H. *Zum Folgerungsbegriff* ⋄ A05 B22 ⋄

MAGARI, R. *Calcoli generali e spazi V_α* ⋄ B22 B60 ⋄

MAGARI, R. *Calcoli generali con piu di un operatore deduzione* ⋄ B22 ⋄

MLEZIVA, M. *Ueber das Enthaltensein des klassischen Aussagenkalkuels in den nicht-klassischen Aussagenkalkuelen (Czech summary)* ⋄ B22 ⋄

ZANDAROWSKA, W. *On certain connections between consequence, inconsistency and completeness (Polish) (Russian and English summaries)* ⋄ B22 ⋄

ZARNECKA-BIALY, E. *On the equivalence of logical systems (Polish) (English summary)* ⋄ B22 ⋄

1967

MIURA, S. & ONO, K. *On pairs of very-close formal systems* ⋄ B22 ⋄

1968

ANDERSON, J.G. *Concerning the finite model property for propositional calculi* ⋄ B22 ⋄

BAMMERT, J. *Quasideduktive Systeme und S-Algebren I,II* ⋄ B22 G12 G25 ⋄

CURRY, H.B. *The equivalence of two definitons of elementary formal system* ⋄ B22 ⋄

HARROP, R. *Some forms of models of propositional calculi* ⋄ B22 ⋄

LEISENRING, A.C. *An abstract property of formalized languages which contain Hilbert's ε-symbol* ⋄ B22 ⋄

LEJEWSKI, C. *A propositional calculus in which three mutually undefinable functors are used as primitive terms (Polish and Russian summaries)* ⋄ B22 ⋄

MESAROVIC, M.D. *On some metamathematical results as properties of general systems* ⋄ B22 ⋄

SURMA, S.J. *A simplified axiom system for Tarski's consequence theory (Polish) (English summary)* ⋄ B22 ⋄

SURMA, S.J. *Four studies in metamathematics (Polish) (Russian and English summaries)* ⋄ B22 C07 C35 E25 ⋄

SURMA, S.J. *Investigations into the theory of deduction (Polish) (English summary)* ⋄ B22 ⋄

SURMA, S.J. *Kinds of completeness of deductive theories (Polish) (English summary)* ⋄ B22 ⋄

SURMA, S.J. *On metamathematical finite character properties* ⋄ B22 ⋄

TOTI RIGATELLI, L. *Sullo spazio dei modelli di un calcolo generale (French summary)* ⋄ B22 ⋄

1969

MAZURKIEWICZ, A.W. *On the simple form of the deduction systems* ⋄ B22 ⋄

SURMA, S.J. *On the axiomatic treatment of the theory of models I: Theory of models as an extension of Tarski's consequence theory (Polish summary)* ⋄ B22 C07 ⋄

ULRICH, D. *Solution to a problem posed by Kalicki* ⋄ B22 ⋄

WOJCICKI, R. *Logical matrices strongly adequate for structural sentential calculi (Russian summary)* ⋄ B22 ⋄

1970

APPLEBEE, R.C. & PAHI, B. *An unsolvable problem concerning implicational calculi* ⋄ B22 D35 ⋄

BLOOM, S.L. & BROWN, D.J. & SUSZKO, R. *A note on abstract logics (Russian summary)* ⋄ B22 ⋄

BLOOM, S.L. & BROWN, D.J. & SUSZKO, R. *Some theorems on abstract logics (Russian)* ⋄ B22 ⋄

BRYLL, G. & SLUPECKI, J. & WYBRANIEC-SKARDOWSKA, U. *A certain theory that is equivalent to Tarski's theory of deductive systems (Polish) (English summary)* ⋄ B22 ⋄

KOSSOWSKI, P. *On the axiomatic treatment of the theory of models IV: Independence of some axiom system of model theory (Polish summary)* ⋄ B22 C07 ⋄

KOTAS, J. & PIECZKOWSKI, A. *Allgemeine logische und mathematische Theorien* ⋄ B22 ⋄

LOKTIONOV, V.I. *A certain theorem on logical semantics (Russian)* ⋄ B22 ⋄

MADUCH, M. *Bases of classical consequences* ⋄ B22 ⋄
NAKAMURA, A. *On a propositional calculus whose decision problem is recursively unsolvable* ⋄ B22 D35 ⋄
PAGLI, P. *Ammissibilita di connetivi in un calcolo generale (English summary)* ⋄ B22 ⋄
SURMA, S.J. *On the axiomatic treatment of the theory of models II: Syntactical characterization of a fragment of the theory of models (Polish summary)* ⋄ B22 C07 C75 ⋄
SURMA, S.J. *On the axiomatic treatment of the theory of models III: The function of model content (Polish summary)* ⋄ B22 C07 ⋄
ULRICH, D. *Decidability results for some classes of propositional calculi* ⋄ B22 ⋄
WOJCICKI, R. *Some remarks on the consequence operation in sentential logics* ⋄ B22 ⋄

1971

BRYLL, G. & SLUPECKI, J. & WYBRANIEC-SKARDOWSKA, U. *Theory of rejected propositions I (Polish and Russian summaries)* ⋄ A10 B22 ⋄
CHAPIN JR., E.W. *On axioms and their corresponding deduction rules: A survey* ⋄ B22 ⋄
HARROP, R. *On the equivalence for nonderivability testing of finite Smiley models and finite modified Smiley models* ⋄ B22 ⋄
KOTAS, J. *Logical systems with implications (Polish and Russian summaries)* ⋄ B22 G25 ⋄
RISTEA, T.G. *La theorie des systemes deductifs chez A. Tarski (Romanian) (French summary)* ⋄ B22 ⋄
TOKARZ, M. & WOJCICKI, R. *The problem of reconstructability of propositional calculi (Polish and Russian summaries)* ⋄ B22 F50 ⋄

1972

BLOOM, S.L. *A note on regular congruences* ⋄ B22 C05 ⋄
BLOOM, S.L. *Model-consequence operations and quasi-complete theories* ⋄ B22 ⋄
BOTH, N. *La caracterisation algebrique des systemes et theories deductives (Romanian) (French summary)* ⋄ B22 C05 G25 ⋄
BRYLL, G. & SLUPECKI, J. & WYBRANIEC-SKARDOWSKA, U. *The theory of rejected propositions II (Polish and Russian summaries)* ⋄ A05 B22 ⋄
CORCORAN, J. *Weak and strong completeness in sentential logics* ⋄ B22 ⋄
GRZEGORCZYK, A. *An approach to logical calculus (Polish and Russian summaries)* ⋄ B22 F50 ⋄
NEGRU, I.S. *Certain sublattices of the lattice of all propositional logics with the ordinary concept of formula (Russian)* ⋄ B22 ⋄
PAHI, B. *A method for proving the nonexistence of finite characteristic models for implicational calculi* ⋄ B22 ⋄
PAHI, B. *A theorem on the interrelationship of axiom systems for implicational calculi* ⋄ B22 ⋄
SIDORENKO, J.A. *Die logische Folgebeziehung und ihre Formalisierung* ⋄ A05 B22 ⋄
SLUPECKI, J. *L-decidability and decidability* ⋄ B22 B25 ⋄

TOKARZ, M. *Connections between some notions of completeness of structural propositional calculi* ⋄ B22 ⋄
ULRICH, D. *Some results concerning finite models for sentential calculi* ⋄ B22 ⋄
WAGNER, K. *Zur Axiomatisierung eines sequentiellen Aussagenkalkuels (Russian, English and French summaries)* ⋄ B22 D05 ⋄
WOJCICKI, R. *Some properties of strongly finite propositional calculi* ⋄ B22 ⋄
ZYGMUNT, J. *Direct products of consequences operations* ⋄ A10 B22 ⋄

1973

BLOOM, S.L. *A theorem on well-finite standard consequence operations* ⋄ B22 ⋄
BLOOM, S.L. & BROWN, D.J. *Classical abstract logics* ⋄ B22 G05 ⋄
BROWN, D.J. & SUSZKO, R. *Abstract logics* ⋄ B22 ⋄
GABBAY, D.M. *Applications of Scott's notion of consequence to the study of general binary intensional connectives and entailment* ⋄ B22 B46 ⋄
HARROP, R. *On simple, weak and strong models of propositional calculi* ⋄ B22 ⋄
MADUCH, M. *Consequence operation defined by cartesian product of matrices (Polish) (English summary)* ⋄ B22 ⋄
PERZANOWSKI, J. *A linguistic criterion of structural incompleteness* ⋄ B22 ⋄
SEGERBERG, K. *Hallden's theorem on Post completeness* ⋄ B22 ⋄
SPASOWSKI, M. *Some connections between Cn, Cn^{-1} and dCn* ⋄ B22 ⋄
TAVANETS, P.V. (ED.) *Theorie der logischen Folgerungen (Russian)* ⋄ B22 B97 ⋄
TOKARZ, M. *Connections between some notions of completeness of structural propositional calculi (Polish and Russian summaries)* ⋄ B22 ⋄
WOJCICKI, R. *Dual counterparts of consequence operations* ⋄ B22 ⋄
WOJCICKI, R. *Matrix approach in methodology of sentential calculi (Polish and Russian summaries)* ⋄ B22 ⋄

1974

AHMAD, S. *On implicational completeness* ⋄ B22 C05 ⋄
BIELA, A. *On Lindenbaum's extensions* ⋄ B22 ⋄
BIELA, A. & POGORZELSKI, W.A. *The power of the class of Lindenbaum-Asser extensions of consistent sets of formulas* ⋄ B22 ⋄
BLOOM, S.L. *On "generalized logics"* ⋄ B22 ⋄
BOTH, N. *L'independance (Roumain) (Resume francais)* ⋄ B22 ⋄
CZELAKOWSKI, J. *Logics based on partial boolean σ-algebras. I* ⋄ B22 G05 ⋄
GABBAY, D.M. & JONGH DE, D.H.J. *A sequence of decidable finitely axiomatizable intermediate logics with the disjunction property* ⋄ B22 B25 B55 ⋄
MAKINSON, D. & SEGERBERG, K. *Post completeness and ultrafilters* ⋄ B22 B45 G20 ⋄

MOSHCHENSKIJ, V.A. *Abnehmende Folge von Aussagenkalkuelen mit kompliziert werdender Deduktion* ⋄ B22 ⋄

NEGRU, I.S. *On algebraic properties of the totality of propositional logics (Russian)* ⋄ B22 ⋄

PAHI, B. *On the non-existence of finite characteristic models for some classes of implicational calculi* ⋄ B22 ⋄

POGORZELSKI, W.A. *Concerning the notion of completeness of invariant propositional calculi* ⋄ B22 ⋄

PRUCNAL, T. & WRONSKI, A. *An algebraic characterization of the notion of structural completeness* ⋄ B22 ⋄

SCHREIBER, P. *Durch Schlussregeln definierte logische Kalkuele* ⋄ B22 ⋄

SCOTT, D.S. *Rules and derived rules* ⋄ B22 ⋄

SOBOCINSKI, B. *A theorem concerning a restricted rule of substitution in the field of propositional calculi I & II* ⋄ B05 B22 ⋄

SPASOWSKI, M. *Some properties of the operation d* ⋄ B22 ⋄

SURMA, S.J. *An algorithm for axiomatizing every finite logic* ⋄ B22 ⋄

SURMA, S.J. *On axiomatization of finite logics* ⋄ B22 ⋄

URQUHART, A.I.F. *Proofs, snakes and ladders* ⋄ B22 F07 ⋄

WASSERMAN, H.C. *Admissible rules, derivable rules, and extendible logistic systems* ⋄ B22 ⋄

WOJCICKI, R. *Degree of maximality versus degree of completeness* ⋄ B22 ⋄

1975

BIELA, A. *On Lindenbaum's extensions II* ⋄ B22 ⋄

BLOOM, S.L. *A representation theorem for the lattice of standard consequence operations* ⋄ B22 ⋄

BLOOM, S.L. *Some theorems on structural consequence operations* ⋄ B22 C05 C20 ⋄

BRYLL, G. & SLUPECKI, J. *Contributions to the general theory of deductive systems (Polish) (English summary)* ⋄ B22 ⋄

DZIK, W. *On structural completeness of some nonclassical predicate calculi* ⋄ B22 F50 ⋄

GRANT, J. *Inconsistent and incomplete logics* ⋄ B22 B53 ⋄

POGORZELSKI, W.A. *On the notion of the rule of inference and completeness of systems: some comments on H.C. Wasserman's remarks "Admissible rules, derivable rules, and extendible logistic systems"* ⋄ B22 ⋄

WOJCICKI, R. *A theorem on strongly finite propositional calculi* ⋄ B22 ⋄

1976

BIELA, A. *On the so-called Tarski's property in the theory of Lindenbaum's oversystems* ⋄ B22 ⋄

BIELA, A. *On Lindenbaum's extensions. III* ⋄ B22 ⋄

BLOOM, S.L. *Projective and inductive generation of abstract logics* ⋄ B22 ⋄

DALLA CHIARA SCABIA, M.L. *A general approach to non-distributive logics* ⋄ B22 ⋄

FRANK, W. *A note on the adequacy of translations* ⋄ B22 ⋄

GOL'DSHTEJN, M.SH. *Ueber die Darstellung abstrakter Logiken (Russisch)* ⋄ B22 G12 ⋄

HARROP, R. *Some results concerning finite model separability of propositional calculi* ⋄ B22 B25 ⋄

LOGAN, G.J. *Note on the Frattini-Neumann-Schmidt intersection theorem* ⋄ B22 E20 ⋄

MAKINSON, D. *A characterization of structural completeness of a structural consequence operation* ⋄ B22 ⋄

MAKSIMOVA, L.L. *The principle of separation of variables in propositional logics (Russian)* ⋄ B22 ⋄

TOKARZ, M. *A strongly finite logic with infinite degree of maximality* ⋄ B22 ⋄

WRONSKI, A. *On finitely based consequence operations* ⋄ B22 ⋄

1977

BLOOM, S.L. *A note on Ψ-consequences* ⋄ B22 F50 ⋄

MAKSIMOVA, L.L. *Craig's interpolation theorem and amalgamable varieties (Russian)* ⋄ B22 B55 C05 C40 C90 G10 ⋄

SURMA, S.J. *A uniform method for axiomatization of finite logics* ⋄ B22 ⋄

URQUHART, A.I.F. *A finite matrix whose consequence relation is not finitely axiomatizable* ⋄ B22 ⋄

WOJCICKI, R. & ZYGMUNT, J. *Results and open problems in the theory of strongly finite sentential calculi* ⋄ B22 ⋄

WOJCICKI, R. *Strongly finite sentential calculi* ⋄ B22 ⋄

WRONSKI, A. *A method of axiomatizing an intersection of propositional logics* ⋄ B22 ⋄

WRONSKI, A. *On the depth of a consequence operation* ⋄ B22 ⋄

1978

DYWAN, Z. *Decidability of structural completeness for strongly finite propositional calculi* ⋄ B22 ⋄

DYWAN, Z. *Dual counterparts of strongly finite consequences* ⋄ B22 ⋄

GABBAY, D.M. *What is a classical connective?* ⋄ B22 F07 ⋄

MAGGIOLO-SCHETTINI, A. & WINKOWSKI, J. *An algebraic characterization of derivability relations* ⋄ B22 ⋄

MALINOWSKI, G. *A proof of Ryszard Wojcicki's conjecture* ⋄ B22 ⋄

SHOESMITH, D.J. & SMILEY, T.J. *Multiple-conclusion logic* ⋄ B22 F07 F98 ⋄

SOBOCINSKI, B. *Awkward axiom - systems* ⋄ B22 ⋄

STACHNIAK, Z. *Some notes on characteristic consequence operations* ⋄ B22 G25 ⋄

1979

BUNDER, M.W. *Alternative forms of propositional calculus for a given deduction theorem* ⋄ B22 ⋄

CZELAKOWSKI, J. *A characterization of Matr(C)* ⋄ B22 C05 ⋄

CZELAKOWSKI, J. *Large matrices which induce finite consequence operations* ⋄ B22 C20 ⋄

DYWAN, Z. *Finite structural axiomatization of every finite-valued propositional calculus* ⋄ B22 ⋄

DZIOBIAK, W. *An example concerning the lattice of the structural consequence operations* ⋄ B22 ⋄

DZIOBIAK, W. *On strongly finite consequence operations*
⋄ B22 ⋄
DZIOBIAK, W. & SACHWANOWICZ, W. *On two notions concerning the structural sentential calculi* ⋄ B22 ⋄
LAMBROS, C.H. *A generalized theorem concerning a restricted rule of substitution in the field of propositional calculi* ⋄ B22 ⋄
LAMBROS, C.H. *A shortened proof of Sobocinski's theorem concerning a restricted rule of substitution in the field of propositional calculi* ⋄ B22 ⋄
MAKINSON, D. *Qu'est-ce la completude structurale?*
⋄ B22 ⋄
MALINOWSKI, G. *Topics in the theory of strengthenings of sentential calculi* ⋄ B22 G20 ⋄
SUSZKO, R. & WEINFELD, T. *Filters and natural extensions of closure systems* ⋄ B22 ⋄
VERDU I SOLANS, V. *Distributive and boolean logics (Catalan) (English summary)* ⋄ B22 G10 ⋄
VUJOSEVIC, S.T. *On the limits of the families of Lindenbaum algebras* ⋄ B22 G05 ⋄
WOJCICKI, R. *A theorem on strictly finite propositional calculi (Russian)* ⋄ B22 ⋄
WOJTYLAK, P. *Matrix representations for structural strengthenings of a propositional logic* ⋄ B22 G25 ⋄
WOJTYLAK, P. *Strongly finite logics: Finite axiomatizability and the problem of supremum* ⋄ B22 ⋄
WRONSKI, A. *A three element matrix whose consequence operation is not finitely based* ⋄ B22 ⋄

1980

BERNERT, J. *The strongly adequate matrices of the form of product of Lindenbaum's matrices* ⋄ B22 ⋄
BIELECKA-HOLDA, M. *Independent basis for the consequence determined by nondegenerated distributive lattices* ⋄ B22 G10 G25 ⋄
CZELAKOWSKI, J. *Model-theoretic methods in methodology of propositional calculi*
⋄ B22 C05 G99 ⋄
CZELAKOWSKI, J. *Reduced products of logical matrices*
⋄ B22 C05 C20 ⋄
DYRDA, K. & PRUCNAL, T. *On finitely based consequence determined by a distributive lattice* ⋄ B22 G25 ⋄
DYWAN, Z. *Finite structural axiomatization of every finite-valued propositional calculus* ⋄ B22 ⋄
DYWAN, Z. *Quasi-strongly finite sentential calculi*
⋄ B22 ⋄
DZIOBIAK, W. *An example of strongly finite consequence operation with $2^{\aleph_0}$ standard strengthenings* ⋄ B22 ⋄
GUCCIONE, S. & TORTORA, R. *A general deduction theorem* ⋄ B22 ⋄
HAWRANEK, J. & ZYGMUNT, J. *A theorem on the degree of complexity of some sentential logics* ⋄ B22 ⋄
HAWRANEK, J. *Some examples concerning uniformity and complexity of sentential logics* ⋄ B22 ⋄

1981

BIELA, A. *On the so-called Tarski's property in the theory of Lindenbaum's oversystem. II: Axiomatizable systems*
⋄ B22 ⋄
CHUBARYAN, A.A. *On the complexity of conclusions in different systems of propositional calculus (Russian)*
⋄ B22 ⋄

CZELAKOWSKI, J. *Equivalential logics. I,II* ⋄ B22 ⋄
DYWAN, Z. *Finite unaxiomatizability of propositional cacluli with one variable* ⋄ B22 ⋄
DZIK, W. *On the content of lattices of logics I. The representation theorem for lattices of logics* ⋄ B22 ⋄
DZIK, W. *The existence of Lindenbaum's extensions is equivalent to the axiom of choice* ⋄ B22 E25 ⋄
DZIOBIAK, W. *The lattice of strengthenings of a strongly finite consequence operation* ⋄ B22 ⋄
HAWRANEK, J. & ZYGMUNT, J. *Another proof of Wojtylak's theorem* ⋄ B22 ⋄
HAWRANEK, J. & ZYGMUNT, J. *On the degree of complexity of sentential logics. A couple of examples*
⋄ B22 ⋄
KABZINSKI, J.K. *On equivalential fragment of implicative extensional consequence* ⋄ B22 G25 ⋄
MALINOWSKI, G. *A characterization of strengthenings of a two-valued nonuniform sentential calculus* ⋄ B22 ⋄
POGORZELSKI, W.A. *On Hilbert's operation on logical rules I* ⋄ B22 ⋄
RAUTENBERG, W. *2-element matrices* ⋄ B20 B22 ⋄
STEPIEN, T. *Craig-Goedel-Lindenbaum's property and Sobocinski-Tarski's property in propositional calculi*
⋄ B22 C40 ⋄
THOMPSON, P.B. *Bolzano's deducibility and Tarski's logical consequence* ⋄ A10 B22 ⋄
URQUHART, A.I.F. *Decidability and the finite model property* ⋄ B22 B25 D35 ⋄
WOJTYLAK, P. *Mutual interpretability of sentential logics. I,II* ⋄ B22 F25 ⋄
ZYGMUNT, J. *Notes on decidability and finite approximability of sentential logics*
⋄ A10 B22 B25 ⋄

1982

BERNERT, J. *A note on existence of matrices strongly adequate for some positive logics* ⋄ B22 ⋄
BIELA, A. & DZIOBIAK, W. *On two properties of structurally complete logics* ⋄ B22 ⋄
BUNDER, M.W. *Deduction theorems for weak implicational logics* ⋄ B22 ⋄
CZELAKOWSKI, J. *Logical matrices and the amalgamation property* ⋄ B22 C40 C52 C90 ⋄
FURMANOWSKI, T. *Adjoint interpretations of sentential calculi* ⋄ B22 F25 G30 ⋄
KIRK, R.E. *A result on propositional logics having the disjunction property* ⋄ B22 ⋄
POGORZELSKI, W.A. & WOJTYLAK, P. *Elements of the theory of completeness in propositional logic (Polish summary)* ⋄ B22 ⋄
REZUS, A. *On a theorem of Tarski* ⋄ B05 B22 ⋄
SCOTT, D.S. *Domains for denotational semantics*
⋄ B22 B40 ⋄
SEGERBERG, K. *Classical propositional operators. An exercise in the foundations of logic* ⋄ B05 B22 ⋄
SHEKHTMAN, V.B. *Undecidable propositional calculi (Russian)* ⋄ B22 B55 D35 ⋄
TORRENS TORRELL, A. *The general deduction principle (Catalan) (English summary)* ⋄ B22 ⋄
ULRICH, D. *Answer to a question raised by Harrop*
⋄ B22 ⋄

VERDU I SOLANS, V. *Algebras characterized by means of inference rules (Catalan)* ⋄ B22 C05 G05 G10 ⋄

WOJCICKI, R. *Referential matrix semantics for propositional calculi* ⋄ B22 B50 ⋄

1983

BIELA, A. & STEPIEN, T. *Lindenbaum's extensions* ⋄ B22 ⋄

BIELA, A. & DZIOBIAK, W. *On two properties of structurally complete logics* ⋄ B22 ⋄

CZELAKOWSKI, J. *Algebraic aspects of deduction theorems* ⋄ B22 ⋄

CZELAKOWSKI, J. *Matrices, primitive satisfaction and finitely based logics* ⋄ B22 ⋄

HAWRANEK, J. & ZYGMUNT, J. *On normal extensions of sentential logics (Polish) (English summary)* ⋄ B22 ⋄

IDZIAK, P.M. *A finite base for the consequence operation determined by the ring Z_p in the language without constants* ⋄ B22 ⋄

MALINOWSKI, G. *Pseudo-referential matrix semantics for propositional logics* ⋄ B22 ⋄

PIOCHI, B. *Logical matrices and non-structural consequence operators* ⋄ B22 ⋄

SPASOWSKI, M. *Some properties of dual counterparts of consequence operations (Polish) (English summary)* ⋄ B22 ⋄

STEPIEN, T. *System $\overline{S}$* ⋄ B22 ⋄

ULRICH, D. *The finite model property and recursive bounds on the size of countermodels* ⋄ B22 B25 C13 ⋄

WOJTYLAK, P. *Corrections to the paper: "Structural completeness of Lewis's system S5" by T. Prucnal* ⋄ B22 ⋄

ZYGMUNT, J. *An application of the Lindenbaum method in the domain of strongly finite sentential calculi* ⋄ B22 ⋄

ZYGMUNT, J. *On decidability and finite approximability of sentential logics* ⋄ B22 B25 ⋄

ZYGMUNT, J. *Some remarks on matrix consequence operations* ⋄ B22 ⋄

1984

BULL, R. *The classical propositional calculus of arguments* ⋄ B22 ⋄

CZELAKOWSKI, J. *Filter distributive logics* ⋄ B22 ⋄

CZELAKOWSKI, J. *Remarks on finitely based logics* ⋄ B22 C90 ⋄

DZIK, W. & TOKARZ, M. *Invariant matrix consequences* ⋄ B22 ⋄

FELSCHER, W. & SCHULTE-MOENTING, J. *Algebraic and deductive consequence operations* ⋄ B22 ⋄

HAWRANEK, J. & ZYGMUNT, J. *On the degree of complexity of sentential logics. II. An example of the logic with semi-negation* ⋄ B22 ⋄

LESISZ, W. *On propositional calculus with a variable functor* ⋄ B22 ⋄

MALINOWSKI, G. *Matrix referentiality* ⋄ B22 ⋄

MAREK, I. *Consequence operations defined by partial matrices* ⋄ B22 ⋄

POGORZELSKI, W.A. *On Hilbert's operation on logical rules. II* ⋄ B22 ⋄

SCHROEDER-HEISTER, P. *A natural extension of natural deduction* ⋄ B22 F07 F50 ⋄

STEPIEN, T. *A sufficient and necessary condition for Tarski's property in Lindenbaum's extensions* ⋄ B22 ⋄

STEPIEN, T. *The sufficient and necessary condition for Tarski's property in Lindenbaum's extensions* ⋄ B22 ⋄

WOJTYLAK, P. *An example of a finite though finitely non-axiomatizable matrix* ⋄ B22 ⋄

ZYGMUNT, J. *An essay in matrix semantics for consequence relations* ⋄ B22 ⋄

1985

CZELAKOWSKI, J. *Sentential logics and Maehara interpolation property* ⋄ B22 C05 C40 C52 ⋄

HOARE, C.A.R. *A couple of novelties in the propositional calculus* ⋄ B22 ⋄

JANKOWSKI, A.W. *Galois structures* ⋄ B22 C95 ⋄

JONES, JOHN *Formalisations of many valued propositional calculi with variable functors* ⋄ B22 ⋄

MALINOWSKI, G. *The problem of degrees of maximality (a survey) (Polish summary)* ⋄ B22 ⋄

MEYER, A.R. & MITCHELL, J.C. *Second-order logical relations* ⋄ B15 B22 ⋄

PRAWITZ, D. *Remarks on some approaches to the concept of logical consequence* ⋄ A05 B22 F07 ⋄

PRUCNAL, T. *On finitely e-based consequence determined by Wronski's matrix* ⋄ B22 ⋄

RAUTENBERG, W. *Consequence relations of 2-element algebras* ⋄ B05 B22 ⋄

STACHNIAK, Z. *Note on structural logics* ⋄ B22 ⋄

STEPIEN, T. *Logic based on atomic entailment* ⋄ B22 B60 ⋄

STEPIEN, T. *On number of Lindenbaum's oversystems of propositional and predicate calculi* ⋄ B22 ⋄

B28 Classical foundations of number systems

1884
FREGE, F.L.G. *Die Grundlagen der Arithmetik, eine logisch mathematische Untersuchung ueber den Begriff der Zahl* ⋄ A05 B28 E75 ⋄

1885
FREGE, F.L.G. *Ueber formale Theorien der Arithmetik* ⋄ A05 B28 ⋄

STOLZ, O. *Vorlesungen ueber allgemeine Arithmetik. Vol. I* ⋄ B28 ⋄

1886
STOLZ, O. *Vorlesungen ueber allgemeine Arithmetik. Vol. II* ⋄ B28 ⋄

1887
KRONECKER, L. *Ueber den Zahlbegriff* ⋄ B28 F55 F65 ⋄

1888
DEDEKIND, R. *Was sind und was sollen die Zahlen?* ⋄ B28 D20 E75 ⋄

1889
PEANO, G. *The principles of arithmetic, presented by a new method (Latin)* ⋄ B03 B28 E75 F30 ⋄

1891
STOLZ, O. *Groessen und Zahlen* ⋄ B28 ⋄

1893
FREGE, F.L.G. *Grundgesetze der Arithmetik, Begriffsschriftlich abgeleitet. Vol.1* ⋄ A05 B28 E30 E75 ⋄

1895
FREGE, F.L.G. *Le nombre entier* ⋄ A05 B28 ⋄

PEANO, G. *Review of : Frege : Grundgesetze der Arithmetik, begriffsschriftlich abgeleitet, Vol.1* ⋄ B28 ⋄

1896
CONANT, L.L. *The number concept* ⋄ A10 B28 ⋄

1897
PEANO, G. *Studii di logica matematica* ⋄ A10 B20 B28 ⋄

1898
PEANO, G. *Formulaire de mathematiques Vol.2, chap.2: Arithmetique* ⋄ B28 ⋄

1902
HUNTINGTON, E.V. *A complete set of postulates for the theory of absolute continuous magnitude* ⋄ B28 ⋄

KEYSER, C.J. *Concerning the axiom of infinity and mathematical induction* ⋄ A05 B28 E30 ⋄

1903
FREGE, F.L.G. *Grundgesetze der Arithmetik, Begriffsschriftlich abgeleitet. Vol.2* ⋄ A05 B28 E30 E75 ⋄

PEANO, G. *Formulaire mathematique Vol.4* ⋄ B28 ⋄

RUSSELL, B. *The principles of mathematics. Vol. 1* ⋄ A05 B28 B30 E30 ⋄

1904
FREGE, F.L.G. *Was ist eine Funktion ?* ⋄ A05 B28 E20 ⋄

SCHUBERT, H. *Principes fondamentaux de l'arithmetique* ⋄ B28 ⋄

1905
HILBERT, D. *Ueber die Grundlagen der Logik und der Arithmetik* ⋄ A05 B28 ⋄

HUNTINGTON, E.V. *The continuum as a type of order: an exposition of the modern theory* ⋄ B28 E07 E10 E75 ⋄

1906
WEBER, H. *Elementare Mengenlehre* ⋄ B28 E10 ⋄

1907
MOLLERUP, J. *Die Definition des Mengenbegriffs* ⋄ A05 B28 E30 ⋄

MOLLERUP, J. *Sur la theorie des ensembles et le concept de nombre* ⋄ B28 E20 ⋄

1908
FREGE, F.L.G. *Die Unmoeglichkeit der Thomaeschen formalen Arithmetik aufs neue nachgewiesen* ⋄ A05 B28 ⋄

PEANO, G. *Formulario mathematice Vol.5* ⋄ B28 ⋄

1909
PASCH, M. *Grundlagen der Analysis* ⋄ B28 ⋄

ZERMELO, E. *Ueber die Grundlagen der Arithmetik* ⋄ B28 E20 ⋄

1910
WHITEHEAD, A.N. & RUSSELL, B. *Principia mathematica. Vol.I* ⋄ B05 B10 B15 B28 E30 E98 ⋄

1911
BOEHM, K. *Axiome der Arithmetik* ⋄ B28 ⋄

ENRIQUES, F. *Sui numeri non archimedei e su alcune loro interpretazioni* ⋄ B28 E10 E20 ⋄

1912
WHITEHEAD, A.N. & RUSSELL, B. *Principia mathematica. Vol.II* ⋄ B15 B28 E10 E30 E98 ⋄

1913

BLUMBERG, H. *A set of postulates for arithmetic and algebra* ◊ B28 ◊

PEANO, G. *Review of : A.N.Whitehead and B.Russell : Principia Mathematica I,II* ◊ A10 B28 ◊

PUCCIANO, G. *I principii dell' ordinamento naturale e della continuita* ◊ B28 ◊

WHITEHEAD, A.N. & RUSSELL, B. *Principia mathematica. Vol.III* ◊ B15 B28 E10 E30 E98 ◊

1914

HOELDER, O. *Die Arithmetik in strenger Begruendung* ◊ B28 ◊

KORSELT, A. *Allgemeinste vollstaendige Induktion* ◊ B28 E07 ◊

PUCCIANO, G. *Il continuo lineare aperto ed omogeneo e la geometria archimedea della retta* ◊ B28 E75 ◊

1915

FREGE, F.L.G. *The fundamental laws of arithmetic* ◊ A05 B28 E30 E75 ◊

1916

FREGE, F.L.G. *The fundamental laws of arithmetic: Psychological logic* ◊ A05 B28 ◊

1917

FREGE, F.L.G. *Class, function, concept, relation* ◊ A05 B28 E30 ◊

1918

WEYL, H. *Das Kontinuum* ◊ A05 B28 E30 F65 ◊

1919

WEYL, H. *Der "Circulus vitiosus" in der heutigen Begruendung der Analysis* ◊ A05 B28 F65 ◊

1922

BERNAYS, P. *Ueber Hilberts Gedanken zur Grundlegung der Arithmetik* ◊ A05 B28 ◊

1923

BERNAYS, P. *Erwiderung auf die Note von Herrn Alois Mueller: "Ueber Zahlen als Zeichen"* ◊ A05 B28 ◊

KHINCHIN, A.YA. *Das Stetigkeitsaxiom des Linearkontinuums als Induktionsprinzip betrachtet* ◊ B28 E75 ◊

SKOLEM, T.A. *Begruendung der elementaren Arithmetik durch die rekurrierende Denkweise ohne Anwendung scheinbarer Veraenderlichen mit unendlichem Ausdehnungsbereich* ◊ A05 B28 D20 F30 ◊

1924

TARSKI, A. *Sur les ensembles finis* ◊ B28 E10 E20 E25 ◊

1925

FRAENKEL, A.A. *Die neueren Ideen zur Grundlegung der Analysis und Mengenlehre* ◊ A05 B28 B30 E30 E75 ◊

TARSKI, A. *Une remarque concernant les principes d'arithmetique theorique* ◊ B28 ◊

1926

HILBERT, D. *Ueber das Unendliche* ◊ A05 B28 D20 E10 E30 ◊

WAESCHE, H. *Grundzuege zu einer Logik der Arithmetik* ◊ B28 ◊

1927

BELL, E.T. *Algebraic arithmetic* ◊ B28 ◊

GRUZINTSEV, G.A. *The concept of relation and the axiomatical definition of number (Russian) (French summary)* ◊ B28 F99 ◊

LANGFORD, C.H. *On inductive relations* ◊ A05 B28 ◊

MORDUKHAJ-BOLTOVSKOJ, D. *Lobachevskij and fundamental logical problems in mathematics (Russian)* ◊ B28 ◊

1928

ACKERMANN, W. *Zum Hilbertschen Aufbau der reellen Zahlen* ◊ B28 D20 F15 ◊

NEUMANN VON, J. *Die Axiomatisierung der Mengenlehre* ◊ B28 E30 E70 ◊

1929

BAER, REINHOLD *Zur Axiomatik der Kardinalzahlarithmetik* ◊ B28 E10 E30 E35 ◊

KHINCHIN, A.YA. *The role and the character of induction in mathematics (Russian)* ◊ B28 ◊

1930

HILBERT, D. *Die Grundlegung der elementaren Zahlenlehre* ◊ A05 B28 ◊

LANDAU, E. *Grundlagen der Analysis* ◊ B28 E75 E98 ◊

PRESBURGER, M. *Ueber die Vollstaendigkeit eines gewissen Systems der Arithmetik ganzer Zahlen, in welchem die Addition als einzige Operation hervortritt* ◊ B25 B28 C10 C35 F30 ◊

1931

GOEDEL, K. *Ueber formal unentscheidbare Saetze der "Principia Mathematica" und verwandter Systeme I* ◊ B28 B30 D20 D35 F25 F30 F35 ◊

HILBERT, D. *Beweis des "tertium non datur"* ◊ A05 B28 F50 ◊

SKOLEM, T.A. *Ueber einige Satzfunktionen in der Arithmetik* ◊ B25 B28 C10 F30 ◊

TARSKI, A. *Sur les ensembles definissables de nombres reels I*
◊ B25 B28 C10 C40 C60 C65 D55 E15 E47 F35 ◊

1932

DEDEKIND, R. *Gesammelte mathematische Werke. Vol.III* ◊ B28 B96 E75 E96 ◊

KACZMARZ, S. *Axioms for arithmetic* ◊ B28 ◊

ZERMELO, E. *Anmerkung* ◊ A05 A10 B28 ◊

1933

BACHMANN, F. & SCHOLZ, H. *Die logischen Grundlagen der Arithmetik im Anschluss an Frege und Dedekind* ◊ A05 A10 B28 ◊

FINSLER, P. *Die Existenz der Zahlenreihe und des Kontinuums* ◊ A05 B28 E20 E50 ◊

SKOLEM, T.A. *Ueber die Unmoeglichkeit einer Charakterisierung der Zahlenreihe mittels eines endlichen Axiomensystems*
⋄ B28 C20 C62 F30 H15 ⋄

TARSKI, A. *Einige Betrachtungen ueber die Begriffe der ω-Widerspruchsfreiheit und der ω-Vollstaendigkeit*
⋄ B28 C62 F30 ⋄

1934

BACHMANN, F. *Untersuchungen zur Grundlegung der Arithmetik mit besonderer Beziehung auf Dedekind, Frege und Russell* ⋄ A05 A10 B28 ⋄

HETPER, W. *Semantische Arithmetik* ⋄ B28 F30 ⋄

QUINE, W.V.O. *A method of generating part of arithmetic without use of intuitive logic* ⋄ B28 ⋄

SKOLEM, T.A. *Ein Satz ueber Zaehlausdruecke* ⋄ B28 ⋄

SKOLEM, T.A. *Ueber die Nichtcharakterisierbarkeit der Zahlenreihe mittels endlich oder abzaehlbar unendlich vieler Aussagen mit ausschliesslich Zahlenvariablen*
⋄ B28 C20 C62 F30 H15 ⋄

WHITEHEAD, A.N. *Logical definitions of extension, class, and number* ⋄ A05 B28 E30 ⋄

1935

BERNAYS, P. *Hilberts Untersuchungen ueber die Grundlagen der Arithmetik* ⋄ A05 B28 ⋄

KLEENE, S.C. *A theory of positive integers in formal logic I,II* ⋄ B28 B40 ⋄

SKOLEM, T.A. *Ueber die Erfuellbarkeit gewisser Zaehlausdruecke* ⋄ B25 B28 ⋄

1936

BACHMANN, F. *Die Fragen der Abhaengigkeit und der Entbehrlichkeit von Axiomen in Axiomensystemen, in denen ein Extremalaxiom auftritt* ⋄ A05 B28 B30 ⋄

BACHMANN, F. & CARNAP, R. *Ueber Extremalaxiome*
⋄ B28 B30 ⋄

GENTZEN, G. *Die Widerspruchsfreiheit der reinen Zahlentheorie* ⋄ B28 F05 F07 F30 ⋄

GLIVENKO, V.I. *The crisis in the foundations of mathematics in the present state of their development (Russian)* ⋄ B28 ⋄

PADOA, A. *Les extensions successives de l'ensemble des nombres au point de vue deductif* ⋄ B28 ⋄

SKOLEM, T.A. *Ueber die Zurueckfuehrbarkeit einiger durch Rekursionen definierter Relationen auf "arithmetische"*
⋄ B28 F30 ⋄

1937

SECKENDORFF VON, V. *Beweis des Induktionsschlusses der natuerlichen Zahlen aus der Dedekindschen Definition endlicher Mengen* ⋄ B28 E10 F30 ⋄

STONE, M.H. *Note on formal logic* ⋄ B28 G05 G15 ⋄

1938

BURCKHARDT, J.J. *Zur Neubegruendung der Mengenlehre*
⋄ B28 E70 ⋄

DIENES, P. *Logic of algebra* ⋄ A05 B28 F55 G05 ⋄

1939

BACHMANN, F. *Aufbau des Zahlensystems* ⋄ B28 ⋄

KNOPP, K. *Darstellung der reellen Zahlen durch Grenzprozesse* ⋄ B28 ⋄

1940

KALMAR, L. *On the possibility of definition by recursion*
⋄ B28 ⋄

1941

CURRY, H.B. *A formalization of recursive arithmetic*
⋄ B28 D20 F30 ⋄

1943

MARTIN, R.M. *A homogeneous system for formal logic*
⋄ A05 B28 G05 ⋄

NOVIKOV, P.S. *On the consistency of certain logical calculus (Russian summary)* ⋄ B05 B28 C75 ⋄

WANG, XIANGHAO *A system of completely independent axioms for the sequence of natural numbers* ⋄ B28 ⋄

1944

KONDO, M. *Une methode operationelle dans la theorie des nombres naturels* ⋄ B28 ⋄

1945

GOODSTEIN, R.L. *Function theory in an axiom-free equation calculus* ⋄ B28 F30 F60 ⋄

KLEENE, S.C. *On the interpretation of intuitionistic number theory* ⋄ B28 F50 ⋄

TAKAGI, T. *Zur Axiomatik der ganzen und der reellen Zahlen* ⋄ B28 ⋄

1946

BING, K. *On B. Germansky's systems for the foundations of natural numbers. I,II (Hebrew)* ⋄ B28 ⋄

KOLMOGOROV, A.N. *On the foundations of the theory of real numbers (Russian)* ⋄ B28 ⋄

QUINE, W.V.O. *Concatenation as a basis for arithmetic*
⋄ B28 ⋄

1947

KAVUN, N.I. *A development of the theory of real numbers by the method of A.N. Kolmogorov (Russian)* ⋄ B28 ⋄

VALPOLA, V. *The position of negation in a language which expresses knowledge (Finnish)*
⋄ A05 B28 F30 F50 ⋄

1948

DEDEKIND, R. *Essays on the theory of numbers. I. Continuity and irrational numbers. II. The nature and meaning of numbers* ⋄ B28 E75 ⋄

TARSKI, A. *Axiomatic and algebraic aspects of two theorems on sums of cardinals*
⋄ B28 C15 E10 E25 G05 ⋄

WANG, HAO *A new theory of element and number*
⋄ B28 E70 ⋄

ZICH, O.V. *Sur la notion du nombre entier*
⋄ A05 B28 ⋄

1949

BORGERS, A. *The natural number (Dutch)* ⋄ B28 ⋄

FITCH, F.B. *On natural numbers, integers, and rationals*
⋄ B28 ⋄

FITCH, F.B. *The Heine-Borel theorem in extended basic logic* ⋄ B28 B40 E70 F35 ⋄

KHINCHIN, A.YA. *The simplest linear continuum (Russian)* ⋄ B28 E07 ⋄

NOVIKOV, P.S. *On the axiom of complete induction (Russian)* ⋄ B25 B28 D35 F30 ⋄

1950

FOGELIS, E. *On finite proofs of arithmetical theorems*
  ⋄ B28 F30 ⋄
GONZALEZ, M.O. & MANCILL, J.D. *On the system of natural numbers* ⋄ B28 ⋄
KALMAR, L. *Ueber die Cantorsche Theorie der reellen Zahlen* ⋄ B28 ⋄
KUREPA, D. *Demonstration du principe de l'induction totale* ⋄ B28 E20 ⋄
MARTIN, R.M. *On virtual classes and real numbers*
  ⋄ B28 E70 ⋄
MYHILL, J.R. *A complete theory of natural, rational, and real numbers* ⋄ B28 ⋄
MYHILL, J.R. *A reduction in the number of primitive ideas of arithmetic* ⋄ B28 ⋄
WANG, HAO *Set-theoretical basis for real numbers*
  ⋄ B28 E75 ⋄

1951

QUINE, W.V.O. *The ordered pair in number theory*
  ⋄ B28 ⋄
ROHRBACH, H. *Das Axiomensystem von Erhard Schmidt fuer die Menge der natuerlichen Zahlen* ⋄ B28 ⋄
SUPPES, P. *A set of independent axioms for extensive quantities (Portuguese)* ⋄ B28 ⋄

1952

BAR-HILLEL, Y. *Rational numbers as triplets of natural numbers (Hebrew) (English summary)* ⋄ B28 ⋄
MYHILL, J.R. *A derivation of number theory from ancestral theory* ⋄ B28 ⋄
ROSENBAUM, I. *A logistic proof of a theorem related to Landau's theorem 4* ⋄ B28 ⋄
RYLL-NARDZEWSKI, C. *The role of the axiom of induction in elementary arithmetic* ⋄ B28 ⋄

1953

KREISEL, G. *A variant to Hilbert's theory of the foundations of arithmetic* ⋄ A05 B28 D30 D55 ⋄
SKOLEM, T.A. *The logical background of arithmetic*
  ⋄ A05 B28 ⋄
WANG, HAO *Between number theory and set theory*
  ⋄ B28 E30 F30 ⋄

1954

GENTZEN, G. *Zusammenfassung von mehreren vollstaendigen Induktionen zu einer einzigen*
  ⋄ B28 F30 ⋄
HADAMARD, J. *Sur l'impossibilite de demontrer la compatibilite des axiomes de l'arithmetique*
  ⋄ A05 B28 F99 ⋄
MCNAUGHTON, R. *A non-standard truth definition*
  ⋄ B28 C62 E30 H10 H20 ⋄

1955

BERNAYS, P. *Betrachtungen ueber das Vollstaendigkeitsaxiom und verwandte Axiome*
  ⋄ B28 C60 C65 ⋄
DEVIDE, V. *Ein Axiomensystem fuer die natuerlichen Zahlen* ⋄ B28 ⋄
GOODSTEIN, R.L. *On non-constructive theorems of analysis and the decision problem*
  ⋄ B25 B28 F30 F60 ⋄
SKOLEM, T.A. *Peano's axioms and models of arithmetic*
  ⋄ B28 C20 C62 F30 H15 ⋄
WANG, HAO *Undecidable sentences generated by semantic paradoxes* ⋄ A05 B28 F30 ⋄

1956

KONDO, M. *Sur les nombres reels et nommables*
  ⋄ B28 D55 E15 F65 ⋄
SHEPHERDSON, J.C. *Note on a system of Myhill* ⋄ B28 ⋄
SKOLEM, T.A. *A version of the proof of equivalence between complete induction and the uniqueness of primitive recursion* ⋄ B28 F30 ⋄
TRAKHTENBROT, B.A. *The definition of a finite set and the deductive incompleteness of the theory of sets (Russian)*
  ⋄ B28 D35 E30 ⋄

1957

BING, K. *On the axiom of order and succession* ⋄ B28 ⋄
GOODSTEIN, R.L. *Recursive number theory. A development of recursive arithmetic in a logic-free equation calculus*
  ⋄ B28 F30 F50 F60 F98 ⋄
MAEHARA, S. *Remark on Skolem's theorem concerning the impossibility of characterization of the natural number sequence* ⋄ B28 C62 F30 ⋄
MAEHARA, S. *Ueber die rekursive Einfuehrung der Funktionen in der reinen Zahlentheorie* ⋄ B28 ⋄
ROBINSON, R.M. *Restricted set-theoretical definitions in arithmetic* ⋄ B28 F30 F35 ⋄
SMULLYAN, R.M. *Languages in which self reference is possible* ⋄ A05 B10 B28 F30 F40 ⋄
TAKEUTI, G. *On the theory of ordinal numbers*
  ⋄ B28 E10 E30 E35 E45 F05 ⋄
WANG, HAO *The axiomatization of arithmetic*
  ⋄ A05 B28 F30 ⋄

1958

FELSCHER, W. & SCHMIDT, J. *Natuerliche Zahlen, Ordnung, Nachfolge* ⋄ B28 ⋄
KONDO, M. *Sur les ensembles nommables et le fondement de l'analyse mathematique I* ⋄ B28 E15 ⋄
KURODA, S. *An investigation on the logical structure of mathematics. I. A logical system*
  ⋄ B15 B28 E70 F07 ⋄
KURODA, S. *An investigation on the logical structure of mathematics. III. Fundamental deductions. IV. Compendium for deductions*
  ⋄ B15 B28 E70 F07 ⋄
KURODA, S. *An investigation on the logical structure of mathematics XII. The principle of extensionality and of choice* ⋄ B15 B28 E25 E70 ⋄
LAUGWITZ, D. & SCHMIEDEN, C. *Eine Erweiterung der Infinitesimalrechnung* ⋄ B28 C65 E20 H05 ⋄
LENZ, H. *Zur Axiomatik der Zahlen* ⋄ B28 ⋄
NAGEL, E. & NEWMAN, J.R. *Goedel's proof*
  ⋄ B28 D35 F30 F98 ⋄
ROBINSON, R.M. *Restricted set-theoretical definitions in arithmetic* ⋄ B28 F30 F35 ⋄

1959

GASTEV, YU.A. *Ueber die Konstruktion der Analysis auf Grund einer axiomatisierten Geometrie der Geraden I (Russisch)* ⋄ B28 ⋄

HAERTIG, K. *Einstellige Funktionen als Grundbegriffe der elementaren Zahlentheorie* ◊ B28 ◊

KURODA, S. *An investigation on the logical structure of mathematics. II. Transformation of the proof* ◊ B28 B30 F05 ◊

KURODA, S. *An investigation on the logical structure of mathematics. VI. Consistent V-systems T(V) . VII. Set-theoretical contradictions* ◊ B28 B30 E70 F30 ◊

KURODA, S. *An investigation on the logical structure of mathematics. VIII. Consistency of the natural-number theory $T_1(N)$* ◊ B28 B30 F30 ◊

KURODA, S. *An investigation on the logical struture of mathematics. IX. Deductions in the natural-number theory $T_1(N)$. X. Concepts and sets* ◊ B15 B28 B30 F30 ◊

1960

HAJEK, O. *Three principles of induction in mathematics (Czech)* ◊ B28 E20 ◊

HENKIN, L. *On mathematical induction* ◊ A05 B28 ◊

SOUDIEUX, C. *De l'infini arithmetique* ◊ B28 F30 F98 ◊

1961

ACKERMANN, W. *Grundgedanken einer typenfreien Logik* ◊ B28 E70 ◊

GERMANSKY, B. *The induction axiom and the axiom of choice* ◊ B28 E25 ◊

HAMILTON, N.T. & LANDIN, J. *Set theory: the structure of arithmetic* ◊ B28 E75 E98 ◊

KLAUA, D. *Konstruktion ganzer, rationaler und reeller Ordinalzahlen und die diskontinuierliche Struktur der transfiniten reellen Zahlenraeume* ◊ B28 E10 ◊

MACDOWELL, R. & SPECKER, E. *Modelle der Arithmetik* ◊ B28 C20 C55 C62 ◊

MARGARIS, A. *Successor axioms for the integers* ◊ B28 ◊

MOSTOWSKI, ANDRZEJ *Concerning the problem of axiomatizability of the field of real numbers in the weak second order logic* ◊ B15 B28 C85 ◊

NAGEL, E. & NEWMAN, J.R. *Discussion: Putnam's review of Goedel's proof* ◊ B28 F30 ◊

QUINE, W.V.O. *A basis for number theory in finite classes* ◊ B28 E75 ◊

RABIN, M.O. *Non-standard models and independence of the induction axiom* ◊ B28 C62 F30 H15 ◊

1962

BARROS DE, C.M. *Sur l'existence de fonctions recurrentes* ◊ B28 E20 ◊

BETH, E.W. *Formal methods. An introduction to symbolic logic and to the study of effective operations in arithmetic and logic* ◊ B28 B98 F30 F50 F98 ◊

FREGE, F.L.G. *Funktion, Begriff, Bedeutung. Fuenf logische Studien* ◊ A05 B15 B28 B30 B96 E30 ◊

HENKIN, L. & SMITH, N.W. & VARINEAU, V.J. & WALSH, M.J. *Retracing elementary mathematics* ◊ B28 E98 ◊

LINES ESCARDO, E. *Ueber die Struktur der Menge der natuerlichen Zahlen (Spanisch)* ◊ B28 E20 ◊

1963

ASENJO, F.G. *Relations irreducible to classes* ◊ B28 E70 ◊

BETH, E.W. *La logique de quanteurs. Axiomatisations faibles pour l'arithmetique* ◊ B28 ◊

BUCK, R.C. *Mathematical induction and recursive definitions* ◊ B28 D20 ◊

GASTEV, YU.A. *Ueber die Konstruktion der Analysis auf Grund einer axiomatisierten Geometrie der Geraden II (Russisch)* ◊ B28 ◊

LEWANDOWSKI, A. *A modification of Hilbert's axiom of order (Polish) (Russian and English summaries)* ◊ B28 ◊

WANG, HAO *Some partial systems* ◊ B28 E70 ◊

1964

FEFERMAN, S. *The number systems. Foundations of algebra and analysis* ◊ B28 E98 ◊

FREGE, F.L.G. *The basic laws of arithmetic. Exposition of the system* ◊ A05 A10 B28 ◊

FREGE, F.L.G. *The concept of number* ◊ B28 ◊

KATZ, R. *Axiomatic analysis: an introduction to logic and the real number system* ◊ B28 ◊

1965

ASENJO, F.G. *The arithmetic of the term-relation number theory* ◊ B28 ◊

HAYASAKA, S. *On the axiom of mathematical induction (Japanese) (English summary)* ◊ B28 ◊

LORENZEN, P. *Die klassische Analysis als eine konstruktive Theorie* ◊ B28 F65 ◊

MCNAUGHTON, R. *Undefinability of addition from one unary operator* ◊ B28 C10 ◊

RITCHIE, R.W. *A rudimentary definition of addition* ◊ A05 B28 ◊

ROUTLEY, R. *What numbers are* ◊ A05 B28 E10 ◊

1966

APELT, H. *Axiomatische Untersuchungen ueber einige mit der Presburgerschen Arithmetik verwandte Systeme* ◊ B28 C35 C80 F30 ◊

LIGHTSTONE, A.H. *Symbolic logic and the real number system: An introduction to the foundations of number systems* ◊ B28 ◊

NELSON, D. *Non-null implication* ◊ B20 B28 F50 ◊

ROHRBACH, H. *Was sind und was sollen Zahlen ?* ◊ B28 ◊

VACALIUC, I. *Demonstration de quelques theoremes concernant l'addition et la multiplication dans l'arithmetique a logique elementaire* ◊ B28 ◊

1967

BOLLMAN, D.A. *Formal nonassociative number theory* ◊ B28 D35 ◊

BOTH, N. *On the bases of complete induction. Simultaneous induction (Romanian)* ◊ B28 F30 ◊

DEDEKIND, R. *Letter to Keferstein* ◊ B28 E75 ◊

FREGE, F.L.G. *Kleine Schriften* ◊ A05 B28 B30 B96 ◊

SINGH, S. *The natural number arithmetic in Goedel's axiomatic set theory* ◊ B28 F30 ◊

TSINMAN, L.L. *On the complete induction axiom (Russian)* ◊ B28 F30 ◊

TYTUS, F.J. *An elementary construction of the natural numbers* ◊ B28 ◊

ZAHN, P. *Eine Einfuehrung der reellen Zahlen in der operativen Mathematik ohne die Unterscheidung von Sprachschichten* ◊ B28 F35 F65 ◊

1968

CHENG, CHUNGYING *On explanation of number progression* ◊ A05 B28 ◊

CLAY, R.E. *The consistency of Lesniewski's mereology relative to the real number system* ◊ B28 E35 E70 F25 ◊

HAUSCHILD, K. *Metatheoretische Eigenschaften gewisser Klassen von elementaren Theorien* ◊ B28 C15 C35 E30 E70 ◊

ISAACS, G.L. *Real numbers: a development of the real numbers in an axiomatic set theory* ◊ B28 E75 ◊

LENSTRA JR., H.W. *A definition of the system of natural numbers, equivalent to that of Peano* ◊ B28 ◊

OBERSCHELP, A. *Aufbau des Zahlensystems* ◊ B28 ◊

PANCZAKIEWICZ, M. *Set-theoretic foundations of the arithmetic of natural numbers (Polish) (English summary)* ◊ B28 ◊

TSINMAN, L.L. *The role of the principle of induction in a formal arithmetical system (Russian)* ◊ B28 F30 ◊

1969

DEJON, B. & HENRICI, P. (EDS.) *Constructive aspects of the fundamental theorem of algebra: proceedings of a symposium, IBM* ◊ B28 C97 F97 ◊

POPESCU, E. *A construction of the set of rational numbers starting from the cartesian product* $N \times N \times (N \setminus \{0\})$ ◊ B28 ◊

STERN, S.T. *A number theory for the seminaturals* ◊ B28 ◊

1970

BERNAYS, P. *The original Gentzen consistency proof for number theory* ◊ A10 B28 F30 F50 ◊

BOSCH, W. & KRAJKIEWICZ, P. *A categorical system of axioms for the complex numbers* ◊ B28 ◊

COLLINS, G.E. & HALPERN, J.D. *On the interpretability of arithmetic in set theory* ◊ B28 D35 E30 F25 F30 ◊

KIRSCH, A. *Die Einfuehrung der natuerlichen Zahlen als Operatoren* ◊ B28 ◊

MATIYASEVICH, YU.V. *Solution of the tenth problem of Hilbert* ◊ B28 D25 D35 ◊

METZLER, W. *Eine Einfuehrung der positiven rationalen und positiven reellen Zahlen auf Grund von Messvorgaengen* ◊ B28 ◊

NAZAROV, A.G. *Certain properties of transfinite real numbers (Russian) (Armenian summary)* ◊ B28 ◊

NAZAROV, A.G. *Transfinite numbers, endowed with properties of real numbers (Russian) (Armenian summary)* ◊ B28 ◊

WETTE, E. *Vom Unendlichen zum Endlichen* ◊ A05 B28 E30 F30 ◊

1971

BAKER, A. *Effective methods in the theory of numbers* ◊ B25 B28 ◊

BOTH, N. *Fonctions de succession* ◊ B28 ◊

FREGE, F.L.G. *On the foundations of geometry and formal theories of arithmetic* ◊ B28 B30 ◊

GOTTWALD, S. *Eine Konstruktion der reellen Zahlen "Unendlich"* ◊ B28 ◊

GRZEGORCZYK, A. *Outline of theoretical arithmetic (Polish)* ◊ B28 C62 F30 F35 F98 ◊

HAUSCHILD, K. *Nichtaxiomatisierbarkeit von Satzmengen durch Ausdruecke spezieller Gestalt* ◊ B28 C20 C40 C52 F30 ◊

NAZAROV, A.G. *A transfinite number line (Russian) (Armenian summary)* ◊ B28 ◊

PAPIC, P. *Sur le continu de Souslin (Serbo-Croatian summary)* ◊ B28 E65 ◊

WEIDIG, I. *Einfuehrung der ganzen Zahlen durch eine Ordnungsstruktur* ◊ B28 E07 ◊

1972

CHRISTIAN, C.C. *Konsistenzkriterien fuer formale Theorien und ihre Anwendung auf Zahlen- und Mengentheorie* ◊ B28 E30 F30 ◊

FRANEK, M. *Ueber eine Konstruktion reeller Zahlen. I* ◊ B28 ◊

FREYD, P. *Aspects of topoi* ◊ B28 E70 E98 G30 ◊

GEL'FOND, M.G. *Relationship between the classical and constructive developments of mathematical analysis (Russian) (English summary)* ◊ B28 F35 F60 ◊

HAUSCHILD, K. *Ueber die Universalitaet axiomatisierbarer Fragmente der Zahlentheorie* ◊ B28 ◊

VILLA, G. *Su di una definizione dei numeri naturali* ◊ B28 ◊

1973

BAGINSKI, M. & GLAEWE, W. & GOLL, P. & LIST, G. & LOESCHAU, G. & MERTENS, A. & SCHWANITZ, G. & WALTER, M. *Einfuehrung in die mathematische Logik, Einfuehrung in die Mengenlehre, Aufbau der Zahlenbereiche* ◊ B05 B28 B98 E98 ◊

BOLLMAN, D.A. & LAPLAZA, M.L. *A set-theoretic model for nonassociative number theory* ◊ B28 E75 F30 ◊

MENDELSON, E. *Number systems of the foundations of analysis* ◊ B28 ◊

WOLTER, H. *Eine Erweiterung der elementaren Praedikatenlogik: Anwendungen in der Arithmetik und anderen mathematischen Theorien* ◊ B28 C10 C80 F30 ◊

1974

BOSTOCK, D. *Logic and arithmetic I: Natural numbers* ◊ A05 B28 ◊

EBBINGHAUS, H.-D. *Zur mengentheoretischen Begruendung der natuerlichen Zahlen* ◊ B28 E75 ◊

MAKEEV, A.I. *On the Dedekind construction of the field of real numbers (Russian)* ◊ B28 ◊

MCMINN, T.J. *A formal number-termed number system based on recursion* ◊ B28 F30 ◊

MILEV, L.K. *Die Konstruktion rationaler Zahlen in* $\mathbb{N} \times \{1,2\} \times \mathbb{N}$ *(Bulgarian) (Russian and French summaries)* ◊ B28 ◊

TAKAHASHI, S. *Methodes logiques en geometrie diophantienne* ⋄ B28 C98 D98 G30 H98 ⋄

1975

CARILLO GALLEGO, D. & MIRA ROS, J.M. *Ueber verschiedene Definitionen der "natuerlichen Zahl" (Spanisch)* ⋄ B28 ⋄

DEUTSCH, M. *Zur Benutzung der Verkettung als Basis fuer die Arithmetik* ⋄ B28 F30 ⋄

JASKOWSKI, S. *About certain groups of classes of sets and their application to the definitions of numbers* ⋄ B28 ⋄

MUELLER, D.W. *Endliche Mengen und natuerliche Zahlen* ⋄ A10 B28 E10 ⋄

RAUTENBERG, W. *Eine Synthese der axiomatischen und der kardinalen Definition der natuerlichen Zahlen* ⋄ B28 ⋄

SELDIN, J.P. *Arithmetic as a study of formal systems* ⋄ A05 B28 ⋄

SOLS, I. *Bon ordre dans l'objet des nombres naturels d'un topos booleen (English summary)* ⋄ B28 E70 G30 ⋄

STOJAKOVIC, M. *Inductive and Peano models (Serbo-Croatian) (English summary)* ⋄ B28 C62 ⋄

TOMAS, F. *A test for consistency and its application to recursive arithmetic (Spanish)* ⋄ B28 F30 ⋄

1976

BEL'TYUKOV, A.P. *Decidability of the universal theory of natural numbers with addition and divisibility (Russian)* ⋄ B25 B28 F30 ⋄

BRUIJN DE, N.G. *Defining reals without the use of rationals* ⋄ B28 ⋄

CHRISTIAN, C.C. *Peano-Systeme* ⋄ B28 F30 F35 ⋄

CONWAY, J.H. *On numbers and games* ⋄ B28 E10 E70 E75 H15 H20 ⋄

FLORES, A. *Sulla rappresentazione dei numeri razionali mediante terne di numeri naturali* ⋄ B28 ⋄

GOODSTEIN, R.L. *Arithmetic without sets* ⋄ B28 ⋄

HISCHER, HORST & LUCHT, L. *Zum Verstaendnis des Induktionsaxioms* ⋄ B28 F30 ⋄

MAINZER, K. *Das Begruendungsproblem des mathematischen Kontinuums in der neuzeitlichen Entwicklung der Grundlagenforschung* ⋄ A10 B28 ⋄

MIKUSINSKI, J. & MIKUSINSKI, P. *Natuerliche Zahlen und ihre Axiomatik (Polnisch)* ⋄ B28 ⋄

MOZZOCHI, C.J. *Another approach to some recursion theorems of Landau* ⋄ B28 ⋄

OBERSCHELP, A. *On Frege's numbers in set theory* ⋄ B28 E70 ⋄

VINNER, S. *Implicit axioms, ω-rule and the axiom of induction in high school mathematics* ⋄ B28 C75 ⋄

1977

BRUIJN DE, N.G. *Construction of the system of real numbers* ⋄ B28 ⋄

DEUTSCH, M. *Zum Begriff der Wortmischung als Basis fuer die Arithmetik* ⋄ B28 D05 D75 ⋄

HAMBOURGER, R. *A difficulty with the Frege-Russell definition of number* ⋄ A05 B28 ⋄

MINOGUE, B.P. *Numbers, properties, and Frege* ⋄ A05 B28 ⋄

NAGUMO, M. *Quantities and real numbers* ⋄ B28 ⋄

PTAK, V. *Nondiscrete mathematical induction* ⋄ B28 E75 ⋄

1978

FELSCHER, W. *Naive Mengen und abstrakte Zahlen I,II (II: Algebraische und reelle Zahlen)* ⋄ A10 B28 E98 ⋄

GABRIEL, G. *Implizite Definitionen -- eine Verwechslungsgeschichte* ⋄ A10 B28 ⋄

HASENJAEGER, G. *Praedikatenvariablen in der Zahlentheorie* ⋄ B28 ⋄

MEDVEDEV, F.A. *Cantor's theory of real numbers (Russian)* ⋄ A10 B28 E30 ⋄

PARIS, J.B. *Note on an induction axiom* ⋄ B28 C62 F30 ⋄

VOLKOV, V.A. *Elements of set theory and the development of the concept of number (Russian)* ⋄ A05 B28 E98 ⋄

1979

BECK, JON M. *Simplicial sets and the foundations of analysis* ⋄ B28 F65 ⋄

BOOLOS, G. *The unprovability of consistency. An essay in modal logic* ⋄ B25 B28 B45 B98 F30 F98 ⋄

BORISOV, YU.F. *Mechanical determinism and the structure of the number line (Russian)* ⋄ A05 B28 ⋄

BOSTOCK, D. *Logic and arithmetic. Vol. 2: Rational and irrational numbers* ⋄ A05 B28 ⋄

DETLEFSEN, M. *On interpreting Goedel's second theorem* ⋄ A05 B28 ⋄

FELSCHER, W. *Naive Mengen und abstrakte Zahlen III* ⋄ B28 E98 ⋄

GUTIERREZ-NOVOA, L. *Non-Euclidean real numbers* ⋄ B28 C65 ⋄

LEVIN, A.M. *On an interesting axiomatic theory (Russian)* ⋄ B28 F35 ⋄

LILLO DE, N.J. *Models of an extension of the theory ORD* ⋄ B28 C40 C65 E10 E30 ⋄

RAUTENBERG, W. *Reelle Zahlen in elementarer Darstellung* ⋄ B28 ⋄

SCIENZA, G. *Elementary mathematics from an information-theoretic viewpoint: the real numbers* ⋄ B28 F60 ⋄

THIEL, C. *Zur Bestimmung der Arithmetik* ⋄ B28 ⋄

1980

BOOLOS, G. *ω-consistency and the diamond* ⋄ B28 F30 ⋄

EDWARDS, R.E. *A formal background to mathematics IIa,IIb: A critical approach to elementary analysis* ⋄ B28 ⋄

FRIEDMAN, H.M. *A strong conservative extension of Peano arithmetic* ⋄ B28 F30 F35 ⋄

HANSEN, R.T. & SWANSON, L.G. *Placing the pigeonhole principle within the defining axioms of the integers* ⋄ B28 ⋄

YUAN, XIANGWAN & ZHENG, YUXIN & ZHU, WUJIA *Concerning sets and point-set spaces (Chinese) (English summary)* ⋄ A05 B28 ⋄

1981

CEGIELSKI, P. *La theorie elementaire de la multiplication est consequence d'un nombre fini d'axiomes de $I\Sigma_0$*
⋄ B28 ⋄

HAJEK, P. *On interpretability in theories containing arithmetics II* ⋄ B28 B45 F25 F30 ⋄

MALINA, S. *Notes on a method of introducing real numbers (Slovak) (English and Russian summaries)* ⋄ B28 ⋄

MUELLER, GERT H. *Framing Mathematics*
⋄ A05 B28 B30 E30 ⋄

MYCIELSKI, J. *Finitistic real analysis* ⋄ B28 F65 ⋄

PABION, J.F. & RICHARD, D. *Synonymy and re-interpretation for some sublanguages of Peano arithmetic* ⋄ B28 C62 F30 ⋄

SCHMIDT, GARFIELD C. *Nonreflexive, well founded sets and natural numbers* ⋄ B28 E20 ⋄

STEIN, M. *Eine Bemerkung zur Wohlordungseigenschaft der natuerlichen Zahlen* ⋄ B28 ⋄

1982

DAPUETO, C. *Sull' aritmetizzazione della sintassi del primo ordine (French summary)* ⋄ B28 F30 F40 ⋄

GAIFMAN, H. & SNIR, M. *Probabilities over rich languages, testing and randomness* ⋄ B28 ⋄

GILLIES, D.A. *Frege, Dedekind and Peano on the foundations of arithmetic* ⋄ A10 B28 ⋄

HAMILTON, A.G. *Numbers, sets and axioms* ⋄ B28 ⋄

POGORZELSKI, H.A. & RYAN, W.J. *Foundations of semilogical theory of numbers. Vol.1* ⋄ B28 ⋄

SCHMERL, J.H. & SIMPSON, S.G. *On the role of Ramsey quantifiers in first order arithmetic*
⋄ B28 C10 C62 C80 ⋄

1983

ARTMANN, B. *Der Zahlbegriff* ⋄ B28 E98 ⋄

BRANZEI, D. & MIRON, R. *The foundations of arithmetic and geometry (Romanian)* ⋄ B28 ⋄

CLOTE, P. & McALOON, K. *Two further combinatorial theorems equivalent to the 1-consistency of Peano arithmetic* ⋄ B28 C62 F30 ⋄

HAJEK, P. *Arithmetical interpretations of dynamic logic*
⋄ B28 B45 ⋄

MIKOLAJEWICZ, B. *Some classes of models for Peano arithmetic and for some related theories (Polish) (English summary)* ⋄ B28 F30 ⋄

PUDLAK, P. *A definition of exponentiation by a bounded arithmetical formula* ⋄ B28 C62 F30 ⋄

WRIGHT, C. *Frege's conception of numbers as objects*
⋄ A05 B28 ⋄

ZAIONTZ, C. *Axiomatization of the monadic theory of ordinals $<\omega_2$* ⋄ B25 B28 C65 C85 ⋄

1984

CEGIELSKI, P. *La theorie elementaire de la divisibilite est finiment axiomatisable (English summary)*
⋄ B25 B28 F30 ⋄

DING, DECHENG & MO, SHAOKUI *Simplification of the axioms of recursive arithmetic (Chinese)*
⋄ B28 F30 ⋄

GUREVICH, R. *Decidability of the equational theory of positive numbers with raising to a power (Russian)*
⋄ B25 B28 C65 ⋄

MANSFIELD, R. *A complete axiomatization of computer arithmetic* ⋄ B28 ⋄

RICHARD, D. *Les relations arithmetiques sur les entiers primaires sont definissables au premier ordre par successeur et coprimarite (English summary)*
⋄ B28 F30 ⋄

RICHARD, D. *The arithmetics as theories of two orders (English and French summaries)*
⋄ B28 C62 D35 F30 ⋄

1985

BOSCH, J.E. *Sur les nombres natureles et les fondements de l'arithmetique et la geometrie* ⋄ B28 F30 ⋄

COERS, H. *Die Unabhaengigkeit des archimedischen Axioms vom starken Intervallschachtelungsaxiom*
⋄ B28 ⋄

COMER, S.D. *The elementary theory of interval real numbers* ⋄ B25 B28 C65 ⋄

GUREVICH, R. *Equational theory of positive numbers with exponentiation* ⋄ B25 B28 C65 ⋄

LIFSCHITZ, V. *Calculable natural numbers*
⋄ B28 F30 F50 ⋄

MALIAUKIENE, L. *Elimination of the induction axiom in the multiplicative arithmetic with restricted difference (Russian) (English and Lithuanian summaries)*
⋄ B28 F30 ⋄

PARIS, J.B. & WILKIE, A.J. *Counting problems in bounded arithmetic* ⋄ B28 C62 F30 ⋄

POGORZELSKI, H.A. & RYAN, W.J. *Foundations of semilogical theory of numbers II* ⋄ B28 ⋄

RICHARD, D. *All arithmetical sets of powers of primes are first-order definable in terms of the successor function and the coprimeness predicate* ⋄ B28 F30 ⋄

RICHARD, D. *Definissabilite de l'arithmetique par successeur, coprimarite et puissance* ⋄ B28 F30 ⋄

SHAPIRO, S. *Epistemic and intuitionistic arithmetic*
⋄ B28 F30 F50 ⋄

SKVORTSOV, D.P. *The question of "how many": definition of the notion of cardinality for finite sets in some arithmetic systems (Russian)* ⋄ B28 E30 ⋄

WANG, SHUTANG *Generalized number system and its applications I* ⋄ B28 ⋄

B30 Logical foundations of other classical theories; axiomatics

1882

FREGE, F.L.G. *Ueber den Zweck der Begriffsschrift*
⋄ A05 A10 B30 ⋄

FREGE, F.L.G. *Ueber die wissenschaftliche Berechtigung einer Begriffsschrift* ⋄ A05 A10 B30 ⋄

PASCH, M. *Vorlesungen ueber neuere Geometrie*
⋄ B30 ⋄

1897

RUSSELL, B. *An essay on the foundations of geometry*
⋄ A05 B30 ⋄

1898

SCHROEDER, F.W.K.E. *Ueber zwei Definitionen der Endlichkeit und G.Cantor'sche Saetze*
⋄ B30 E20 E25 ⋄

1899

HILBERT, D. *Grundlagen der Geometrie*
⋄ A05 B30 F99 ⋄

1900

HOELDER, O. *Anschauung und Denken in der Geometrie (Antrittsvorlesung)* ⋄ B30 ⋄

1902

PADOA, A. *Essai d'une theorie algebrique des nombres entiers, precede d'une introduction logique a une theorie deductive quelconque* ⋄ A10 B30 C40 ⋄

PEANO, G. *Les definitions mathematiques* ⋄ B30 ⋄

1903

FREGE, F.L.G. *Ueber die Grundlagen der Geometrie*
⋄ A05 B30 ⋄

POINCARE, H. *La science et l'hypothese* ⋄ A05 B30 ⋄

RUSSELL, B. *The principles of mathematics. Vol. 1*
⋄ A05 B28 B30 E30 ⋄

1904

HUNTINGTON, E.V. *Sets of independent postulates for the algebra of logic* ⋄ B05 B30 G05 ⋄

VEBLEN, O. *A system of axioms for geometry* ⋄ B30 ⋄

1905

COUTURAT, L. *Les principes des mathematiques. Avec un appendice sur la philosophie des mathematiques de Kant* ⋄ A05 A10 B30 ⋄

1906

FREGE, F.L.G. *Ueber die Grundlagen der Geometrie*
⋄ A05 B30 ⋄

1907

WHITEHEAD, A.N. *The axioms of descriptive geometry*
⋄ B30 ⋄

1908

POINCARE, H. *Science et methode* ⋄ A05 B30 ⋄

1910

NATORP, P. *Die logischen Grundlagen der exakten Wissenschaften* ⋄ A05 B30 ⋄

WEYL, H. *Ueber die Definition der mathematischen Grundbegriffe* ⋄ A05 B30 F99 ⋄

1911

SCHOENFLIES, A. *Ueber die Stellung der Definition in der Axiomatik* ⋄ A05 B30 ⋄

1912

TAYLOR, J.S. *A set of five postulates for boolean algebras in terms of the operation "exception"*
⋄ A05 B30 G05 ⋄

1913

HUNTINGTON, E.V. *A set of postulates for abstract geometry, expressed in terms of the simple relation of inclusion* ⋄ B30 ⋄

1914

KOENIG, J. *Neue Grundlagen der Logik, Arithmetik und Mengenlehre* ⋄ A05 B30 E10 E98 ⋄

1916

LANDIS, E.H. & RICHARDSON, R.P. *Fundamental conceptions of modern mathematics. Variables and quantifiers, with a discussion of the general conception of the functional relation* ⋄ A05 B30 ⋄

1917

BERNSHTEJN, S.N. *Essay of axiomatical foundation of probability theory (Russian)* ⋄ B30 ⋄

HUNTINGTON, E.V. *Complete existential theory of the postulates for serial order* ⋄ B30 C13 C35 E07 ⋄

HUNTINGTON, E.V. *Complete existential theory of the postulates for well ordered sets*
⋄ B30 C13 C35 E07 ⋄

TAYLOR, J.S. *Complete existential theory of Bernstein's set of four postulates for boolean algebras* ⋄ B30 G05 ⋄

1918

HILBERT, D. *Axiomatisches Denken* ⋄ A05 B30 ⋄

1922

HERTZ, P. *Ueber Axiomensysteme fuer beliebige Satzsysteme I* ⋄ A05 B30 ⋄

1923

HERTZ, P. *Ueber Axiomensysteme fuer beliebige Satzsysteme II* ⋄ A05 B30 ⋄

HILBERT, D. *Die logischen Grundlagen der Mathematik*
⋄ A05 B30 ⋄

SKOLEM, T.A. *Einige Bemerkungen zur axiomatischen Begruendung der Mengenlehre*
⋄ A05 B30 C07 C62 E30 ⋄

1924

HOELDER, O. *Die mathematische Methode. Logisch-erkenntnistheoretische Untersuchungen im Gebiet der Mathematik, Mechanik und Physik*
⋄ A05 B30 ⋄

WEYL, H. *Randbemerkungen zu Hauptproblemen der Mathematik* ⋄ B30 F55 ⋄

1925

FRAENKEL, A.A. *Die neueren Ideen zur Grundlegung der Analysis und Mengenlehre*
⋄ A05 B28 B30 E30 E75 ⋄

LYUSTERNIK, L.A. *To the question of the foundation of analysis and the founding of geometry without set theory (Russian)* ⋄ B30 ⋄

1926

CHWISTEK, L.B. *Ueber die Hypothesen der Mengenlehre*
⋄ A05 B30 E30 ⋄

LANGFORD, C.H. *Some theorems on deducibility*
⋄ B30 C10 C35 C65 ⋄

LANGFORD, C.H. *Theorems on deducibility (second paper)*
⋄ B30 C10 C35 C65 ⋄

1927

LINDENBAUM, A. & TARSKI, A. *Sur l'independance des notions primitives dans les systemes mathematiques*
⋄ B30 C40 E75 ⋄

1928

HILBERT, D. *Die Grundlagen der Mathematik*
⋄ A05 B30 E50 ⋄

MENGER, K. *Bemerkungen zu Grundlagenfragen I*
⋄ A05 B30 E30 F99 ⋄

MENGER, K. *Bemerkungen zu Grundlagenfragen II*
⋄ A05 B30 E30 ⋄

MENGER, K. *Bemerkungen zu Grundlagenfragen, IV. Axiomatik der endlichen Mengen und der elementargeometrischen Verknuepfungsbeziehungen*
⋄ B30 E30 ⋄

MISES VON, R. *Wahrscheinlichkeit, Statistik und Wahrheit*
⋄ B30 ⋄

1929

CHWISTEK, L.B. *Neue Grundlagen der Logik und Mathematik* ⋄ A05 B30 F99 ⋄

HERTZ, P. *Ueber Axiomensysteme von Satzsystemen*
⋄ A05 B30 ⋄

HERTZ, P. *Ueber Axiomensysteme fuer beliebige Satzsysteme* ⋄ A05 B30 ⋄

HILBERT, D. *Probleme der Grundlegung der Mathematik*
⋄ A05 B30 ⋄

SKOLEM, T.A. *Ueber einige Grundlagenfragen der Mathematik* ⋄ A05 A10 B30 C07 E30 ⋄

TARSKI, A. *Les fondements de la geometrie des corps*
⋄ B30 G05 ⋄

1930

HILBERT, D. *Naturerkennen und Logik* ⋄ A05 B30 ⋄

LINDENBAUM, A. *Remarques sur une question de la methode axiomatique* ⋄ A05 B30 ⋄

NICOD, J. *Geometry and induction, containing "Geometry in the sensible world", and "The logical problem of induction"* ⋄ A05 B30 F98 ⋄

PADOA, A. *Proposizioni assiomatiche* ⋄ B30 ⋄

SCHOLZ, H. *Die Axiomatik der Alten* ⋄ A10 B30 ⋄

1931

GOEDEL, K. *Diskussion zur Grundlegung der Mathematik*
⋄ A05 B30 ⋄

GOEDEL, K. *Ueber formal unentscheidbare Saetze der "Principia Mathematica" und verwandter Systeme I*
⋄ B28 B30 D20 D35 F25 F30 F35 ⋄

HERBRAND, J. *Sur le probleme fondamental de la logique mathematique* ⋄ A05 B25 B30 F30 F99 ⋄

1932

BERNAYS, P. *Methoden des Nachweises von Widerspruchsfreiheit und ihre Grenzen*
⋄ A05 B30 F99 ⋄

CHWISTEK, L.B. *Neue Grundlagen der Logik und Mathematik. Zweite Mitteilung* ⋄ A05 B30 E70 ⋄

GENTZEN, G. *Ueber die Existenz unabhaengiger Axiomensysteme zu unendlichen Satzsystemen*
⋄ B30 ⋄

REICHENBACH, H. *Axiomatik der Wahrscheinlichkeitsrechnung* ⋄ B30 ⋄

REICHENBACH, H. *Wahrscheinlichkeitslogik*
⋄ A05 B30 B48 ⋄

TARSKI, A. *Der Wahrheitsbegriff in den Sprachen der deduktiven Disziplinen*
⋄ A05 B30 C07 C40 F30 ⋄

1933

KLEIN-BARMEN, F. *Ueber gekoppelte Axiomensysteme in der Theorie der abstrakten Verknuepfungen* ⋄ B30 ⋄

REICHENBACH, H. *Die logischen Grundlagen des Wahrscheinlichkeitsbegriffs* ⋄ A05 B30 ⋄

TARSKI, A. *On the notion of truth in reference to formalized deductive sciences (Polish)*
⋄ A05 B15 B30 C07 C40 F30 ⋄

WAJSBERG, M. *Beitrag zur Metamathematik*
⋄ A05 B10 B30 B96 C35 F99 ⋄

1934

HUNTINGTON, E.V. *Independent postulates for the "informal" part of "Principia Mathematica"*
⋄ A05 B30 ⋄

HUNTINGTON, E.V. *The postulational method in mathematics* ⋄ A05 B30 ⋄

ITO, MAKOTO *Einige Anwendungen der Theorie des Entscheidungsproblems zur Axiomatik I*
⋄ B30 C75 F25 ⋄

1935

BARZIN, M. *Sur les demonstrations de non-contradiction des axiomes* ⋄ A05 B30 F99 ⋄

CARNAP, R. *Ein Gueltigkeitskriterium fuer die Saetze der klassischen Mathematik* ⋄ B30 F07 ⋄

HELMER, O. *Axiomatischer Aufbau der Geometrie in formalisierter Darstellung* ◊ B30 ◊
HUNTINGTON, E.V. *The inter-deducibility of the new Hilbert-Bernays theory and "Principia Mathematica"* ◊ B30 E30 ◊
ITO, MAKOTO *Einige Anwendungen der Theorie des Entscheidungsproblems zur Axiomatik. II* ◊ B30 F25 ◊
LINDENBAUM, A. & TARSKI, A. *Ueber die Beschraenktheit der Ausdrucksmittel deduktiver Theorien* ◊ B30 C07 C35 C65 ◊
MOORE, R.L. *A set of axioms for plane analysis situs* ◊ B30 ◊
SCHOLZ, H. & SCHWEITZER, H. *Die sogenannten Definitionen durch Abstraktion. Eine Theorie der Definitionen durch Bildung von Gleichheitsverwandtschaften* ◊ B30 ◊

1936

BACHMANN, F. *Die Fragen der Abhaengigkeit und der Entbehrlichkeit von Axiomen in Axiomensystemen, in denen ein Extremalaxiom auftritt* ◊ A05 B28 B30 ◊
BACHMANN, F. & CARNAP, R. *Ueber Extremalaxiome* ◊ B28 B30 ◊
BERNSTEIN, B.A. *Postulates for boolean algebra involving the operation of complete disjunction* ◊ B30 G05 ◊
LINDENBAUM, A. *Sur la simplicite formelle des notions* ◊ B30 ◊
TARSKI, A. *The establishment of scientific semantics (Polish)* ◊ A05 B10 B30 C07 ◊

1937

TARSKI, A. *Einfuehrung in die mathematische Logik und in die Methodologie der Mathematik* ◊ A05 B30 B98 ◊
VASILIOU, P. *Ueber den gegenwaertigen Stand der axiomatischen Methode* ◊ B30 ◊
WOODGER, J.-H. *The axiomatic method in biology* ◊ B30 B80 ◊

1938

CAVAILLES, J. *Methode axiomatique et formalisme. Essai sur le probleme du fondement des mathematiques* ◊ A05 B30 ◊
CHWISTEK, L.B. & HETPER, W. *New foundation of formal metamathematics* ◊ B30 F35 ◊
GOEDEL, K. *The consistency of the axiom of choice and of the generalized continuum-hypothesis* ◊ B30 E25 E35 E45 E50 ◊
HERMES, H. *Eine Axiomatisierung der allgemeinen Mechanik* ◊ B30 ◊
HETPER, W. *Le role des schemas independants dans le systeme de la semantique elementaire (Polish) (French summary)* ◊ B30 ◊
HETPER, W. *Relations ancestrales dans le systeme de la semantique (Polish) (French summary)* ◊ B30 E07 ◊
MENGER, K. *A new foundation of non-euclidean, affine, real projective and euclidean geometry* ◊ B30 ◊
STRAUSS, M. *Mathematics as logical syntax - a method to formalize the language of a physical theory* ◊ B30 ◊
TARSKI, A. *Ein Beitrag zur Axiomatik der abelschen Gruppen* ◊ B30 C60 ◊

1939

BERNAYS, P. *Bemerkungen zur Grundlagenfrage* ◊ A05 B30 ◊
DIEUDONNE, J. *Les methodes axiomatiques modernes et les fondements des mathematiques* ◊ A05 B30 ◊
EVANS, H.P. & KLEENE, S.C. *A postulational basis for probability* ◊ B30 B48 ◊
REICHENBACH, H. *Ueber die semantische und die Objekt-Auffassung von Wahrscheinlichkeitsausdruecken* ◊ B30 B48 ◊
SCHOLZ, H. *Was ist eine formalisierte Theorie?* ◊ B30 ◊

1940

LEVI, B. *La nocion de "dominio deductivo" como elemento de orientacion en las cuestiones de fundamentos de las teorias matematicas* ◊ A05 B30 ◊
LOEWENHEIM, L. *Einkleidung der Mathematik in Schroederschen Relativkalkuel* ◊ B30 E30 ◊

1941

COPELAND SR., A.H. *Postulates for the theory of probability* ◊ B30 ◊
LEBESGUE, H. *Les controverses sur la theorie des ensembles et la question des fondements* ◊ A05 B30 E30 ◊
POLYA, G. *Heuristic reasoning and the theory of probability* ◊ B30 ◊
REICHENBACH, H. *Note on probability implication* ◊ B30 ◊
SCHOLZ, H. *Eine neue Gestalt der Grundlagenforschung* ◊ A05 A10 B30 ◊
STECK, M. *Unbekannte Briefe Frege's ueber die Grundlagen der Geometrie und Antwortbrief Hilbert's an Frege* ◊ A10 B30 ◊

1944

TARSKI, A. *The semantic conception of truth and the foundations of semantics* ◊ A05 B10 B30 C07 ◊

1945

MARGENAU, H. *On the frequency theory of probability* ◊ A05 B30 ◊

1946

WEYL, H. *Mathematics and logic: a brief survey serving as a review of "The philosophy of Bertrand Russell"* ◊ A05 B30 ◊

1948

TARSKI, A. *A decision method for elementary algebra and geometry* ◊ B25 B30 C10 C35 C60 ◊

1950

BERNAYS, P. *Mathematische Existenz und Widerspruchsfreiheit* ◊ A05 B30 F99 ◊
FITCH, F.B. *A demonstrably consistent mathematics. Part I* ◊ B30 B40 E70 ◊
LOS, J. *Sur la notion independance dans la metamathematique (Polish summary)* ◊ B30 ◊
MYHILL, J.R. *A system which can define its own truth* ◊ B30 D20 F25 F30 ◊

1951

FITCH, F.B. *A demonstrably consistent mathematics. Part II* ◊ B30 B40 E70 ◊

FREUDENTHAL, H. *On the foundations of geometry to be sought in intuition and abstraction (Latin)*
◊ A05 B30 ◊
HERMES, H. *Zum Begriff der Axiomatisierbarkeit*
◊ A05 B30 D25 ◊

1952
GOETLIND, E. *A note on Chwistek and Hetper's foundation of formal metamathematics* ◊ B30 F35 ◊
HERMES, H. *Ueber den Begriff der Grenze in der Mathematik* ◊ B30 ◊
RICHTER, H. *Zur Grundlegung der Wahrscheinlichkeitstheorie I,II,III* ◊ B30 ◊
WOODGER, J.-H. *From biology to mathematics*
◊ B30 B80 ◊

1953
BERNAYS, P. *Existence et non-contradiction en mathematiques. Avec une note de G.Bouligand*
◊ A05 B30 F99 ◊
DESTOUCHES, J.-L. & FEVRIER, P. & JOFFRE, D. (EDS.) *Les methodes formelles en axiomatiques* ◊ B30 ◊
HERMES, H. *Sur le concept d'axiomatisabilite* ◊ B30 ◊
LUKASIEWICZ, J. *Sur la formalisation des theories mathematiques* ◊ B30 ◊
SCHOLZ, H. *Der klassische und der moderne Begriff einer mathematischen Theorie* ◊ A05 A10 B30 ◊
WANG, HAO *Quelques notions d'axiomatique*
◊ B25 B30 C07 C35 E30 F30 ◊

1954
CARNAP, R. *Einfuehrung in die symbolische Logik mit besonderer Beruecksichtigung ihrer Anwendungen*
◊ A05 B30 B98 ◊
DESTOUCHES, J.-L. *La logique et les theories physiques*
◊ B30 ◊
HADAMARD, J. *La geometrie non euclidienne et les definitions axiomatiques* ◊ B30 ◊
HADAMARD, J. *La Geometrie non-euclidienne et les definitions axiomatiques* ◊ A05 B30 ◊
ROBINSON, A. *L'application de la logique formelle aux mathematiques* ◊ B30 C60 ◊
SHOENFIELD, J.R. *A relative consistency proof*
◊ B30 E30 E35 E70 F25 ◊
VUYSJE, D. *L'importance du point de depart psycho-linguistique pour l'application de la logistique aux sciences non mathematiques* ◊ B30 ◊
WANG, HAO *The formalization of mathematics*
◊ A05 B30 E70 F35 F65 ◊
WRIGHT, J.B. *Quasi-projective geometry of two dimensions* ◊ B30 ◊

1955
FREUDENTHAL, H. *The concepts axiom and axiomatics in mathematics and physics (Dutch)* ◊ A05 B30 ◊
LOS, J. *On the axiomatic treatment of probability*
◊ B30 ◊
MCKINSEY, J.C.C. & SUPPES, P. *On the notion of invariance in classical mechanics* ◊ B30 ◊
RASIOWA, H. *Algebraic models of axiomatic theories*
◊ B30 C90 F50 G05 G10 G25 ◊

1956
BETH, E.W. & TARSKI, A. *Equilaterality as the only primitive notion of Euclidean geometry* ◊ B30 ◊
LEBLANC, H. *Two probability concepts* ◊ B30 ◊
SCHWABHAEUSER, W. *Ueber die Vollstaendigkeit der elementaren euklidischen Geometrie*
◊ B30 C35 C65 ◊
SCOTT, D.S. *A symmetric primitive notion for euclidean geometry* ◊ B30 ◊
TARSKI, A. *A general theorem concerning primitive notions of Euclidean geometry* ◊ B30 ◊
WITTGENSTEIN, L. *Remarks on the foundations of mathematics* ◊ A05 B30 ◊

1957
DELGADO, V.M. *Un nuevo modo de presentar la formalizacion de la cienca* ◊ B30 ◊
MONTAGUE, R. & TARSKI, A. *Independent recursive axiomatizability* ◊ B30 D25 ◊
MONTAGUE, R. *Non-finite axiomatizability*
◊ B30 F25 F30 ◊
MUELLER, GERT H. *Zur operativen Begruendung von Logik und Mathematik* ◊ B30 B60 ◊
STEGMUELLER, W. *Das Wahrheitsproblem und die Idee der Semantik. Eine Einfuehrung in die Theorien von A.Tarski und R.Carnap* ◊ A05 B30 ◊

1958
KIESOW, H. *Einige Bemerkungen zur Grundlegung der Wahrscheinlichkeitstheorie in elementaren Sprachen*
◊ B30 ◊
LAMBEK, J. *The mathematics of sentence structures*
◊ B30 B65 F07 ◊
PICKERT, G. *Ebene Inzidenzgeometrie: Beispiele zur Axiomatik mit einer Einfuehrung in die formale Logik*
◊ B30 ◊
SCOTT, D.S. & SUPPES, P. *Foundational aspects of theories of measurement* ◊ B30 C13 C52 ◊
SKOLEM, T.A. *Reduction of axiom systems with axiom schemes to systems with only simple axioms* ◊ B30 ◊
SKOLEM, T.A. *Une relativisation des notions mathematiques fondamentales* ◊ A05 B30 ◊
SPECKER, E. *Dualitaet* ◊ B30 C07 E70 ◊

1959
BERNAYS, P. *Die Mannigfaltigkeiten der Direktiven fuer die Gestaltung geometrischer Axiomensysteme*
◊ A05 B30 ◊
BLUMENTHAL, L.M. *New metric postulates for elliptic n-space* ◊ B30 ◊
BRAITHWAITE, R.B. *Axiomatizing a scietific system by axioms in the form of identifications* ◊ B30 ◊
FEVRIER, P. *Logical structure of physical theories*
◊ B30 ◊
FREUDENTHAL, H. *Logic as method and subject (Dutch)*
◊ A05 B30 ◊
HENKIN, L. & SUPPES, P. & TARSKI, A. (EDS.) *The axiomatic method, with special reference to geometry and physics* ◊ B30 B97 C65 D35 ◊
HERMES, H. *Zur Axiomatisierung der Mechanik*
◊ B30 ◊

HEYTING, A. *Axioms for intuitionistic plane affine geometry* ⋄ B30 F55 ⋄

KURODA, S. *An investigation on the logical structure of mathematics. II. Transformation of the proof*
⋄ B28 B30 F05 ⋄

KURODA, S. *An investigation on the logical structure of mathematics. VI. Consistent V-systems $T(V)$. VII. Set-theoretical contradictions*
⋄ B28 B30 E70 F30 ⋄

KURODA, S. *An investigation on the logical structure of mathematics. VIII. Consistency of the natural-number theory $T_1(N)$* ⋄ B28 B30 F30 ⋄

KURODA, S. *An investigation on the logical struture of mathematics. IX. Deductions in the natural-number theory $T_1(N)$. X. Concepts and sets*
⋄ B15 B28 B30 F30 ⋄

MENGER, K. *An axiomatic theory of functions and fluents*
⋄ A05 B30 ⋄

MUCHNIK, A.A. & YANOV, YU.I. *Existence of k-valued closed classes without a finite basis (Russian)*
⋄ B30 B50 ⋄

OBERSCHELP, W. *Ueber Einfachheitsprinzipien in der Wahrscheinlichkeitstheorie I* ⋄ B30 ⋄

RUDEANU, S. *Independent systems of axioms in lattice theory* ⋄ B30 G10 ⋄

SCHWABHAEUSER, W. *Entscheidbarkeit und Vollstaendigkeit der elementaren hyperbolischen Geometrie* ⋄ B25 B30 C35 C65 ⋄

SCOTT, D.S. *Dimension in elementary euclidean geometry*
⋄ B30 ⋄

SIMON, H.A. *Definable terms and primitives in axiom systems* ⋄ B30 ⋄

SZMIELEW, W. *Absolute calculus of segments and its metamathematical implications* ⋄ B30 ⋄

SZMIELEW, W. *Some metamathematical problems concerning elementary hyperbolic geometry* ⋄ B30 ⋄

TARSKI, A. *What is elementary geometry ?* ⋄ B30 ⋄

WOODGER, J.-H. *Studies in the foundations of genetics*
⋄ B30 B80 ⋄

1960

AJDUKIEWICZ, K. *The axiomatic systems from the methodological point of view (Polish and Russian summaries)* ⋄ A05 B30 ⋄

BORGERS, A. *La methode axiomatique et la logique symbolique* ⋄ A10 B30 ⋄

LEBLANC, H. *On requirements for conditional probability functions* ⋄ B05 B30 B48 ⋄

LOS, J. *Fields of events and their definition in the axiomatic treatment of probability (Polish) (Russian and English summaries)* ⋄ B30 ⋄

1961

ELLIS, J.W. *Another very independent axiom system*
⋄ B30 ⋄

GRZEGORCZYK, A. *Axiomatizability of geometry without points* ⋄ B30 ⋄

HARARY, F. *A very independent axiom system*
⋄ B30 E07 ⋄

HEYTING, A. *Axiomatic method and intuitionism*
⋄ B30 F55 ⋄

MUNOZ, V. *De la axiomatica a los sistemas formales*
⋄ B30 B98 ⋄

SKOLEM, T.A. *Interpretation of mathematical theories in the first order predicate calculus* ⋄ B15 B30 ⋄

1962

BLANCHE, R. *Axiomatics* ⋄ A10 B25 B30 ⋄

FREGE, F.L.G. *Funktion, Begriff, Bedeutung. Fuenf logische Studien*
⋄ A05 B15 B28 B30 B96 E30 ⋄

GRZEGORCZYK, A. *On the validation of the sets of axioms in mathematical theories (Polish) (Russian and English summaries)* ⋄ A05 B30 ⋄

OBERSCHELP, W. *Ueber die Begruendung wahrscheinlichkeitstheoretischer Axiome durch Wetten*
⋄ B30 ⋄

RAUTENBERG, W. *Ueber metatheoretische Eigenschaften einiger geometrischer Theorien* ⋄ B30 C65 D35 ⋄

SIKORSKI, R. *Applications of topology to foundations of mathematics* ⋄ B30 E75 F50 ⋄

SZMIELEW, W. *New foundations of absolute geometry*
⋄ B30 C65 ⋄

1963

HARARY, F. *A measure of axiomatic independence*
⋄ B30 ⋄

LOS, J. *Remarks on foundations of probability. Semantical interpretation of the probability of formulas* ⋄ B30 ⋄

SUPPES, P. & ZINNES, J.L. *Basic measurement theory*
⋄ B30 ⋄

1964

MAGARI, R. *Sui sistemi di assiomi "minimali" per una data teoria* ⋄ B30 ⋄

NEUMANN VON, J. *The formalist foundations of mathematics* ⋄ B30 ⋄

1965

BOUVERE DE, K.L. *Synonymous theories* ⋄ B30 C40 ⋄

EMCH, G. & JAUCH, J.M. *Structures logiques et mathematiques en physique quantique* ⋄ B30 ⋄

MAKOWIECKA, H. *On a primitive notion of one-dimensional geometries over some fields*
⋄ B30 C60 ⋄

ROBERT, P. *Methodes fonctorielles sur l'axiomatique des systemes de generateurs, des rangs, etc*
⋄ B30 E20 G10 G30 ⋄

SCHWABHAEUSER, W. *Metamathematical methods in foundations of geometry* ⋄ B30 ⋄

SUPPES, P. *Logics appropriate to empirical theories*
⋄ A05 B30 ⋄

SZCZERBA, L.W. & TARSKI, A. *Metamathematical properties of some affine geometries*
⋄ B25 B30 C52 C65 D35 ⋄

1966

AJDUKIEWICZ, K. *From the methodology of the deductive sciences* ⋄ B30 ⋄

MUELLER, GERT H. *Definitionen und Postulatensysteme*
⋄ B30 ⋄

RAUTENBERG, W. *Ueber Hilberts Schnittpunktsaetze*
⋄ B30 C65 ⋄

SCHMIDT, J. *Der baryzentrische Kalkuel als axiomatische Grundlage der affinen Geometrie* ◊ B30 ◊
SURMA, S.J. *The concept of measurable magnitude in set-theoretical treatment (Polish) (English summary)* ◊ B30 ◊
ZELINKA, B. *Independence of Birkhoff's postulate system for distributive lattices with a unit element (Czech) (Russian and English summaries)* ◊ B30 G10 ◊

1967

FREGE, F.L.G. *Kleine Schriften* ◊ A05 B28 B30 B96 ◊
HERMES, H. *Die Rolle der Logik beim Aufbau naturwissenschaftlicher Theorien* ◊ B30 ◊
KREISEL, G. *Informal rigour and completeness proofs* ◊ A05 B30 C07 E30 E50 F50 ◊
KRIPKE, S.A. & POUR-EL, M.B. *Deduction-preserving "recursive isomorphisms" between theories* ◊ B30 D25 D35 F30 ◊
ROBERT, P. *Sur l'axiomatique des systemes generateurs, des rangs, etc* ◊ B30 ◊
RUSSELL, B. & WHITEHEAD, A.N. *Incomplete symbols: descriptions* ◊ B30 ◊
TARSKI, A. *The completeness of elementary algebra and geometry* ◊ B25 B30 C10 C35 C60 C65 ◊

1968

AX, J. *The elementary theory of finite fields* ◊ B25 B30 C13 C20 C60 ◊
LAROUCHE, L. *Examination of the axiomatic foundations of a theory of change I* ◊ B30 ◊
MOLER, N. & SUPPES, P. *Quantifier-free axioms for constructive plane geometry* ◊ B30 ◊
PFANZAGL, J. *Theory of measurement* ◊ A05 B30 E75 ◊
POUR-EL, M.B. *Effectively extensible theories* ◊ B30 D25 ◊
PREVIALE, F. *Introduzione a una teoria generale della rappresentazione per le teorie assiomatiche* ◊ B30 ◊
ROEDDING, W. *Eine Art von Gleichgewicht zahlentheoretischer und mengentheoretischer Axiomensysteme* ◊ B30 E30 F30 ◊

1969

BOTH, N. *A syntactic characterization of the independence of a system of axioms (Romanian) (French summary)* ◊ B30 ◊
KOLMOGOROV, A.N. *Logical basis of information theory and probability theory (Russian)* ◊ B30 D15 D20 D80 ◊
KORDOS, M. & SZCZERBA, L.W. *On the ΠΣ-axiom systems of hyperbolic and some related geometries (Russian summary)* ◊ B30 ◊
KREISEL, G. *Axiomatisations of nonstandard analysis that are conservative extensions of formal systems for classical standard analysis* ◊ B30 F35 H05 ◊
LAMBERT, K. *Logical truth and microphysics* ◊ B30 ◊
LAROUCHE, L. *Examination of the axiomatic foundations of a theory of change II,III* ◊ B30 ◊
RAUTENBERG, W. *Euklidische und Minkowskische Orthogonalitaetsrelationen* ◊ B30 ◊
RUBIN, H. *A new approach to the foundations of probability* ◊ B30 ◊

1970

BENIOFF, P.A. *Some aspects of the relationship between mathematical logic and physics I* ◊ B30 ◊
ENGELER, E. *Geometry and language* ◊ B30 C75 ◊
FRAASSEN VAN, B.C. *On the extension of Beth's semantics of physical theories* ◊ B30 ◊
GERMANO, G. *Metamathematische Begriffe in Standardtheorien* ◊ A05 B30 F30 ◊
HANSSON, B. *Transitivity and topological structure of the preference space* ◊ B30 ◊
MARTIN-LOEF, P. *On the notion of randomness* ◊ B30 D55 D80 E15 ◊
STEGMUELLER, W. *Probleme und Resultate der Wissenschaftstheorie und Analytischen Philosophie. Band II: Theorie und Erfahrung. 1. Halbband: Begriffsformen, Wissenschaftssprache, empirische Signifikanz und theoretische Begriffe* ◊ A05 B30 ◊
SZCZERBA, L.W. *Independence of Pasch's axiom* ◊ B30 ◊
SZCZERBA, L.W. *On the euclidean geometry without the Pasch axiom* ◊ B30 ◊

1971

BENIOFF, P.A. *Some aspects of the relationship between mathematical logic and physics II* ◊ B30 ◊
CARNAP, R. & JEFFREY, R.C. (EDS.) *Studies in inductive logic and probability I* ◊ B30 B48 ◊
CAZANESCU, V.E. *Remarques sur les theories deductives* ◊ B30 ◊
FREGE, F.L.G. *On the foundations of geometry and formal theories of arithmetic* ◊ B28 B30 ◊
HIRST, K.E. & RHODES, F. *Conceptual models in mathematics* ◊ B30 ◊
LAROUCHE, L. *Examination of the axiomatic foundations of a theory of change IV* ◊ A05 B30 ◊
MARTIN-LOEF, P. *Complexity oscillations in infinite binary sequences* ◊ B30 ◊
MOROZ, B.Z. *Formal systems arising in analysis of physical theories (Russian)* ◊ B30 ◊
RASIOWA, H. *Introduction to modern mathematics (Polish)* ◊ B30 B98 C05 E98 ◊
SCHNORR, C.-P. *Zufaelligkeit und Wahrscheinlichkeit. Eine algorithmische Begruendung der Wahrscheinlichkeitstheorie* ◊ B30 B75 D80 ◊
SNEED, J.D. *The logical structure of mathematical physics* ◊ A05 B30 G12 ◊
SZCZERBA, L.W. *Undefinability of order in Pasch-free geometry* ◊ B30 ◊
UEMOV, A.I. *Logical foundations of the method of models (Russian)* ◊ A05 B30 ◊

1972

HATCHER, W.S. *Foundations as a branch of mathematics* ◊ A05 B30 ◊
KENNEDY, H.C. *The origins of modern axiomatics: Pasch to Peano* ◊ A10 B30 ◊
KIRSCHENMANN, P. *Concepts of randomness* ◊ B30 ◊
LAROUCHE, L. *Examination of the axiomatic foundations of a theory of change V* ◊ B30 ◊

MARINO SARMIENTO, R. *Konsistenz und Inkonsistenz in der Mathematik (aus einem deduktiven Axiomensystem)* ◊ B30 ◊

SZCZERBA, L.W. *A paradoxical model of euclidean affine geometry* ◊ B30 ◊

1973

FINE, T.L. *Theories of probability: An examination of foundations* ◊ A05 B30 ◊

KINOKUNIYA, Y. *Relativities between sets and measurements* ◊ B30 E50 E75 ◊

LAUGWITZ, D. *A new theory of contact angles* ◊ B30 H05 ◊

QUAISSER, E. *Winkelmetrik in affin-metrischen Ebenen* ◊ B30 ◊

ROBINSON, A. *Metamathematical problems* ◊ A05 B30 C60 C98 H05 ◊

WHITELEY, W. *Logic and invariant theory I. Invariant theory of projective properties* ◊ B30 C60 ◊

WOJCICKI, R. *Basic concepts of formal methodology of empirical sciences* ◊ A05 B30 ◊

1974

BOUVERE DE, K.L. *Some remarks concerning logical and ontological theories* ◊ A05 B30 ◊

GENTZEN, G. *Ueber das Verhaeltnis zwischen intuitionistischer und klassischer Arithmetik* ◊ B30 F30 F50 ◊

KREISEL, G. *A notion of mechanistic theory* ◊ A05 B30 D10 ◊

LORENZEN, P. *Justification of the deductive method (Russian)* ◊ A05 B30 ◊

MARLOW, A.R. *Physical axiomatics* ◊ B30 ◊

MILLER, DAVID *On the comparison of false theories by their bases* ◊ A05 B30 ◊

PABION, J.F. *L'axiomatisation de la syntaxe et le second theoreme de Goedel* ◊ B30 F30 ◊

PRZELECKI, M. *A set theoretic versus a model thoretic approach to the logical structure of physical theories* ◊ B30 ◊

SUPPES, P. *The axiomatic method in the empirical sciences* ◊ A05 A10 B30 ◊

SZMIELEW, W. *The order and the semi-order of n-dimensional euclidean space in the axiomatic and model-theoretic aspects* ◊ B30 C65 ◊

WILLIAMS, P.M. *Certain classes of models for empirical systems* ◊ A05 B30 ◊

WOJCICKI, R. *Formal methods and epistemological problems (Russian)* ◊ A05 B30 ◊

WOJCICKI, R. *Set theoretic representations of empirical phenomena* ◊ A05 B30 E75 ◊

1975

FALMAGNE, J.-C. *A set of independent axioms for positive Holder systems* ◊ B30 ◊

LEVIN, A.M. *The axiom of choice in classical analysis (Russian) (English summary)* ◊ B30 E15 E25 E35 E75 F35 ◊

MAKOWIECKA, H. *An elementary geometry in connection with decomposition of a plane* ◊ B30 C60 ◊

MAKOWIECKA, H. *On primitive notions in n-dimensional elementary geometries* ◊ B30 C60 ◊

MAKOWIECKA, H. *Quaternary relations in weak euclidean geometries (Russian summary)* ◊ B30 C60 ◊

MAKOWIECKA, H. *The norm relation in weak euclidean geometry with the axiom on two circles (Russian summary)* ◊ B30 C60 ◊

MAKOWIECKA, H. *The theory of bi-proportionality as a geometry* ◊ B30 C60 ◊

SCHREIBER, A. *Theorie und Rechtfertigung. Untersuchungen zum Rechtfertigungsproblem axiomatischer Theorien in der Wissenschaftstheorie* ◊ A05 B30 ◊

1976

DALLA CHIARA SCABIA, M.L. & TORALDO DI FRANCIA, G.G. *The logical dividing line between deterministic and indeterministic theories* ◊ A05 B30 ◊

GILES, R. *A pragmatic approach to the formalization of empirical theories* ◊ B30 ◊

HOWSON, C. *The development of logical probability* ◊ A05 B30 B48 ◊

KUIPERS, T.A.F. *A two-dimensional continuum of a priori probability distributions on constituents* ◊ A05 B30 ◊

LEVEILLE, J.P. & ROMAN, P. *Group representations in certain lattices of propositions* ◊ B30 ◊

MAKOWIECKA, H. *A general property of ternary relations in elementary geometry (Russian summary)* ◊ B30 C60 ◊

MALINOWSKI, G. & PRZELECKI, M. & SZANIAWSKI, K. & WOJCICKI, R. (EDS.) *Formal methods in the methodology of empirical sciences* ◊ B30 ◊

SKYRMS, B. *Possible worlds physics and metaphysics* ◊ B30 ◊

SOCHOR, A. *The alternative set theory* ◊ B30 E70 E75 H20 ◊

TOTH, I. *Un probleme de logique et de linguistique concernant le rapport entre geometrie euclidienne (GE) et geometrie non-euclidienne (GNE)* ◊ B30 ◊

WILLIAMS, P.M. *Indeterminate probabilities* ◊ B30 ◊

WITTGENSTEIN, L. *Wittgenstein's lectures on the foundations of mathematics, Cambridge, 1939* ◊ A05 B30 B96 ◊

1977

CARNAP, R. *Two essays on entropy.* ◊ A05 B30 ◊

CHUAQUI, R.B. *A model theoretical definition of probability* ◊ B30 ◊

HILPINEN, R. & NIINILUOTO, I. & SAARINEN, E. (EDS.) *Foundations of probability and statistics. I,II,III* ◊ A05 B30 ◊

MAKOWIECKA, H. *On minimal systems of primitives in elementary euclidean geometry* ◊ B30 ◊

MATIYASEVICH, YU.V. *Some purely mathematical results inspired by mathematical logic* ◊ B30 D25 D35 D80 ◊

OMELYANOVSKIJ, M. *Axiomatics and the search for the foundations of physics* ◊ B30 ◊

PETERSON, J.G. *Single axioms for the classical equivalential calculus* ◊ B20 B30 ◊

PKHAKADZE, SH.S. *Some questions of notation theory (Russian)* ◊ B30 ◊

PRIMAS, H. *Theory reduction and non-boolean theories* ◊ B30 G12 ◊

SCHNORR, C.-P. *A survey of the theory of random sequences* ◊ B30 ◊

SCHWEIZER, B. & SKLAR, A. *The axiomatic characterization of functions* ◊ B30 E20 E30 ◊

SZCZERBA, L.W. *Interpretability of elementary theories* ◊ B30 F25 ◊

WHITELEY, W. *Logic and invariant theory III. Axiom systems and basic syzygies* ◊ B30 C60 ◊

1978

BORGA, M. *Il problema dell'indipendenza dei simboli primitivi di una teoria assiomatica* ◊ B30 ◊

FRAISSE, R. *Les axiomatiques sont-elles un jeu?* ◊ A05 B30 ◊

KAUFMANN, F. *The infinite in mathematics. Logico-mathmatical writings* ◊ A05 A10 B30 ◊

KYBURG JR., H.E. *Subjective probability: Criticisms, reflections, and problems* ◊ A05 B30 ◊

LAKATOS, I. *Mathematics, science and epistemology. Philosophical papers. Vol. 2* A05 B30 ◊

LORENZEN, P. *Eine konstruktive Deutung des Dualismus in der Wahrscheinlichkeitstheorie* ◊ A05 B30 ◊

SAMOKHVALOV, K.F. *On the axiomatic representation of empirical theories (Russian)* ◊ A05 B30 ◊

SEELAND, H. *Algorithmische Theorien und konstruktive Geometrie* ◊ B30 B75 ◊

SOLOMONOFF, R.J. *Complexity-based induction systems: Comparisons and convergence theorems* ◊ B30 D10 ◊

URBANIK, K. *An axiomatic definition of information* ◊ B30 G10 ◊

WHITELEY, W. *Logic and invariant theory II: Homogeneous coordinates, the introduction of higher quantities, and structural geometry* ◊ B30 C60 ◊

1979

CANTINI, A. *I fondamenti della matematica da Dedekind a Tarski* ◊ A05 B30 ◊

DALLA CHIARA SCABIA, M.L. & TORALDO DI FRANCIA, G.G. *Formal analysis of physical theories* ◊ A05 B30 B51 ◊

FERENCZI, M. *On the foundations of the axiomatic theory of probability in symbolic logic* ◊ A05 B30 ◊

GOOSENS, W.K. *Alternative axiomatizations of elementary probability theory* ◊ A05 B30 ◊

KIRUTA, A.YA. *The theory of expected utility for intransitive preferences (Russian)* ◊ B30 ◊

LEVI, I. *Serious possibility* ◊ A05 B30 ◊

PASINI, A. *Model theoretic approach to geometries* ◊ B30 ◊

PKHAKADZE, SH.S. *Some questions of notation theory (Russian)* ◊ B30 ◊

PRESTEL, A. & SZCZERBA, L.W. *Nonaxiomatizability of real general affine geometry* ◊ B30 ◊

SIMANOV, A.L. & TAJMANOV, A.D. *On the problem of the logic of the foundation of scientific theories (Russian)* ◊ B30 ◊

STERNFELD, R. *Frege's achievements and literal scientific discourse* ◊ A05 A10 B30 ◊

SZCZERBA, L.W. *Interpretability and axiomatizability* ◊ B30 F25 ◊

SZCZERBA, L.W. & TARSKI, A. *Metamathematical discussion of some affine geometries* ◊ B25 B30 C35 C52 C65 D35 ◊

TOMAS, F. *The absolute consistency of the geometry of the ruler and transporter of segments (Spanish)* ◊ B30 ◊

WHEELER, W.H. *The first order theory of N-colorable graphs* ◊ B30 C10 C25 C35 C65 ◊

1980

GUZ, W. *A non-symmetric transition probability in quantum mechanics* ◊ B30 ◊

LUCAS, J.R. *Truth, probability and set theory* ◊ A05 B30 ◊

MUELLER, GERT H. *Mathematisierung* ◊ B30 ◊

ROEPER, P. *Intervals and tenses* ◊ A05 B30 B45 ◊

SMIRNOVA, E.D. *Analysis of expressive possibilities of languages and theories (Russian)* ◊ B30 ◊

1981

BUGAJSKI, S. & MOTYKA, Z. *Generalized Borel law and quantum probabilities* ◊ B30 G12 ◊

ERNE, M. *Isomorphismen und Identifikationen in der Ordnungstheorie* ◊ B30 E07 ◊

GINSBERG, A. *Quantum theory and the identity of indiscernibles revisited* ◊ A05 B30 ◊

LUDWIG, G. & NEUMANN, H. *Connections between different approaches to the foundations of quantum mechanics* ◊ B30 ◊

MUELLER, GERT H. *Framing Mathematics* ◊ A05 B28 B30 E30 ◊

PARKER-RHODES, A.F. *The theory of indistinguishables. A search for explanatory principles below the level of physics* ◊ B30 ◊

SAMOKHVALOV, K.F. *Ways of measuring the simplicity of empirical theories (Russian)* ◊ B30 ◊

TSUKADA, H. *Category theory not based upon set theory* ◊ B30 G30 ◊

ZIL'BER, B.I. *Solution of the problem of finite axiomatizability for theories that are categorical in all infinite powers (Russian)* ◊ B30 C35 C45 C52 ◊

1982

FINKELSTEIN, D. *Quantum sets and Clifford algebras* ◊ B30 ◊

JUZA, M. *About the sixth Hilbert's problem* ◊ B30 ◊

MANDERS, K.L. *On the space-time ontology of physical theories* ◊ B30 ◊

SNEED, J.D. *The logical structure of Bayesian decision theory* ◊ B30 ◊

WINTER, D.J. *Axiomatic game thoery* ◊ B30 E30 ◊

1983

GEERTSEMA, J.C. *Recent views in the foundational controversy in statistics* ◊ B30 ◊

MUNDY, B. *Relational theories of Euclidean space and Minkowski spacetime* ◊ A05 B30 ◊

MYCIELSKI, J. *The meaning of the conjecture $P \neq NP$ for mathematical logic* ◊ B30 D15 ◊

SCHEIBE, E. *Two types of successor relations between theories* ◊ B30 ◊

SCHWABHAEUSER, W. & SZMIELEW, W. & TARSKI, A. *Metamathematische Methoden in der Geometrie* ◇ B30 C65 C98 D35 ◇

SCHWABHAEUSER, W. *Metamathematische Methoden in der Geometrie. Teil II: Metamathematische Betrachtungen* ◇ B30 C65 C98 ◇

SZCZERBA, L.W. *Interpretations* ◇ B30 C07 F25 ◇

1984

AUGENSTEIN, B.W. *Hadron physics and transfinite set theory* ◇ B30 E75 ◇

GUDDER, S.P. *An extension of classical measure theory* ◇ B30 ◇

JOZSA, R. *Sheaf models and massless fields* ◇ B30 C90 F50 G30 ◇

KANOVICH, M.I. *On the independence of invariant propositions (Russian)* ◇ B30 D20 ◇

KYBURG JR., H.E. *Theory and measurement* ◇ B30 ◇

MCKEE, T.A. *Logical aspects of combinatorial duality* ◇ B30 ◇

STRUVE, H. *Affine Ebenen mit Orthogonalitaetsrelation* ◇ B30 ◇

1985

JONGELING, T.B. *On an axiomatization of evolutionary theory* ◇ A05 B30 ◇

VERDU I SOLANS, V. *Some algebraic structures determined by closure operators* ◇ B30 C05 G05 G10 ◇

WEYL, H. *Axiomatic versus constructive procedures in mathematics* ◇ A05 B30 F99 ◇

B35 Mechanization of proofs and logical operations

1950

MAYS, W. & PRINZ, D.G. *A relay machine for the demonstration of symbolic logic* ⋄ B35 ⋄

1951

ABBOTT, W.R. *Computing logical truth with the California digital computer* ⋄ B35 ⋄

HANSEL, C.E.M. & HENRY, DESMOND PAUL & MAYS, W. *Note on the exhibition of logical machines at the joint session, July 1950* ⋄ B35 ⋄

MCCALLUM, D.B. & SMITH, J.B. *Mechanized reasoning – logical computers and their design* ⋄ B35 ⋄

1952

ABRAMOV, A.A. & LYUSTERNIK, L.A. & SHESTAKOV, V.I. & SHURA-BURA, M.R. *Solution of mathematical problems by automatic computers. Programming of electronic computers (Russian)* ⋄ B35 ⋄

HERMES, H. *Maschinen zur Entscheidung von mathematischen Problemen* ⋄ B35 ⋄

ROSE, A. *An extension of computational logic* ⋄ B35 ⋄

1953

CARTER, W.C. & RETTIG, A.S. *Analytic minimization methods I. Conjunctive forms* ⋄ B35 ⋄

1956

MCCLUSKEY JR., E.J. *Detection of group invariance or total symmetry of a boolean function* ⋄ B35 ⋄

MCCLUSKEY JR., E.J. *Minimization of boolean functions* ⋄ B35 ⋄

ROHLEDER, H. *Zur Umformung logischer Ausdruecke mit Hilfe programmgesteuerter Rechenanlagen* ⋄ B35 ⋄

ROUTLEDGE, N.A. *Logic on electronic computers: a practical method for reducing expressions to conjunctive normal form* ⋄ B35 ⋄

TRAKHTENBROT, B.A. *Algorithms and mechanical solution of problems I,II (Russian)* ⋄ B35 ⋄

1957

BRENDER, D.M. *The logical procedures needed for finding the minimals of a boolean function on a digital computer* ⋄ B35 ⋄

DAVIS, MARTIN D. *A program for Presburger's algorithm* ⋄ B35 ⋄

FRIDSHAL, R. *The Quine algorithm* ⋄ B35 ⋄

GAZALE, M.J. *Irredundant disjunctive and conjunctive forms of a boolean function* ⋄ B35 ⋄

GELERNTER, H.L. *Theorem proving by machine* ⋄ B35 ⋄

NEWELL, A. & SHAW, J.C. & SIMON, H.A. *Empirical explorations of the logic theory machine: a case study in heuristic* ⋄ B35 ⋄

NORTH, J.H. *A machine evaluation of logical building blocks* ⋄ B35 ⋄

ROBINSON, A. *Proving a theorem (as done by man, logician or machine)* ⋄ A05 B35 ⋄

SHAW, J.C. *Programming the logic theory machine* ⋄ B35 ⋄

ZHURAVLEV, YU.I. *On the separability of subsets of the vertices of an n-dimensional unit cube (Russian)* ⋄ B35 ⋄

1958

BETH, E.W. *On machines which prove theorems* ⋄ B35 ⋄

LYUBCHENKO, G.G. *Methods for determining the identical truth or falsity of the calculus of assertion formulae of bivalent logic (Ukrainian) (Russian and English summaries)* ⋄ B35 ⋄

1959

GELERNTER, H.L. *A note on syntactic symmetry and the manipulation of formal systems by machine* ⋄ B35 ⋄

PORTER, A. & VASWANI, P.K.T. *The optimization of logical goal-seeking procedures* ⋄ B35 ⋄

ROHLEDER, H. *Ein Verfahren zum Aufstellen optimaler Normalformen bei gegebenen Primimplikanden* ⋄ B35 ⋄

ZHURAVLEV, YU.I. *Construction of minimal disjunctive normal forms for functions of the algebra of logic (Russian)* ⋄ B35 ⋄

1960

BAUER, F.L. *The formula-controlled logical computer "Stanislaus"* ⋄ B35 ⋄

DAVIS, MARTIN D. & PUTNAM, H. *A computing procedure for quantification theory* ⋄ B35 ⋄

DUNHAM, B. & FRIDSHAL, R. & SWARD, G.L. *A non-heuristic program for proving elementary logical theorems* ⋄ B35 ⋄

GELERNTER, H.L. *Realization of a geometry theorem proving machine* ⋄ B35 ⋄

GILMORE, P.C. *A program for the production from axioms, of proofs for theorems derivable within the first order predicate calculus* ⋄ B35 ⋄

GILMORE, P.C. *A proof method for quantification theory: Its justification and realization* ⋄ B35 ⋄

KURODA, S. *An investigation on the logical structure of mathematics. XIII. A method of programming of proofs in mathematics for electronic computing machines* ⋄ B35 ⋄

MOTT JR., T.H. *Determination of the irredundant normal forms of a truth function by iterated consensus of the prime implicants* ⋄ B35 ⋄

NEWELL, A. & SHAW, J.C. & SIMON, H.A. *Report on general problem-solving program* ⋄ B35 ⋄

PETRICK, S.R. *On the minimization of boolean functions* ⋄ B35 ⋄

PRATHER, R.E. *Computational aids for determining the minimal form of a truth function* ◇ B35 ◇

PRAWITZ, D. & PRAWITZ, H. & VOGHERA, N. *A mechanical proof procedure and its realization in an electronic computer* ◇ B35 ◇

PRAWITZ, D. *An improved proof procedure* ◇ B35 ◇

ROBINSON, A. *On the mechanization of the theory of equations* ◇ B35 ◇

SEKI, S. *Programs for mathematical proofs by machines-present status of inferential analysis (Japanese)* ◇ B35 ◇

TRAKHTENBROT, B.A. *Algorithms and machine solutions of problems (Russian)* ◇ B25 B35 D20 ◇

WANG, HAO *Proving theorems by pattern recognition I* ◇ B35 ◇

WANG, HAO *Toward mechanical mathematics* ◇ A05 B35 ◇

ZHURAVLEV, YU.I. *On simplification algorithms for disjunctive normal forms (Russian)* ◇ B35 ◇

ZHURAVLEV, YU.I. *On the impossibility of constructing minimal disjunctive normal forms for functions of the algebra of logic in a single class of algorithms (Russian)* ◇ B35 ◇

ZHURAVLEV, YU.I. *Various notions of minimality of disjunctive normal forms (Russian)* ◇ B35 ◇

1961

CHOW, C.K. *Boolean functions realizable with single threshold devices* ◇ B35 ◇

COBHAM, A. & FRIDSHAL, R. & NORTH, J.H. *An application of linear programming to the minimization of boolean functions* ◇ B35 ◇

FOXLEY, E. *Testing the independence of a system of axioms, using a logical computer* ◇ B35 ◇

MCCLUSKEY JR., E.J. *An essay on prime implicant tables* ◇ B35 ◇

MCCLUSKEY JR., E.J. *Minimal sums for boolean functions having many unspecified fundamental products* ◇ B35 ◇

ROHLEDER, H. *Zum Ausmultiplizieren der Klammern beim Verfahren von Nelson* ◇ B35 D05 ◇

SUBBOTOVSKAYA, B.A. *Realizations of linear functions by formulas using* $\vee, \&, -$ *(Russian)* ◇ B35 ◇

WANG, HAO *Proving theorems by pattern recognition II* ◇ B35 ◇

ZHURAVLEV, YU.I. *Finite index algorithms for the simplification of disjunctive normal forms (Russian)* ◇ B35 ◇

1962

BRUMM, G.L. *The method of possibility diagrams for testing the validity of certain types of inferences, based on Jevons' logical alphabet* ◇ B35 ◇

CANNONITO, F.B. *The Goedel incompleteness theorem and intelligent machines* ◇ B35 D10 D25 F30 ◇

DAVIS, MARTIN D. & LOGEMANN, G. & LOVELAND, D.W. *A machine program for theorem-proving* ◇ B35 ◇

DUNHAM, B. & FRIDSHAL, R. & NORTH, J.H. *Exploratory mathematics by machine* ◇ B35 ◇

FOXLEY, E. *The determination of all Sheffer functions in 3-valued logic, using a logical computer* ◇ B35 B50 ◇

HOCKNEY, R. *An intersection algorithm giving all irredundant normal forms from a prime implicant list* ◇ B05 B35 ◇

MCCARTHY, J. *Computer programs for checking mathematical proofs* ◇ B35 ◇

MCCLUSKEY JR., E.J. & PYNE, I.B. *The reduction of redundancy in solving prime implicant tables* ◇ B35 ◇

MEO, A.R. *On the minimal third order expression of a boolean function* ◇ B35 ◇

RADO, T. *Comments on the presence function of gazale* ◇ B35 ◇

SUMMERSBEE, S. & WALTERS, A. *Programming the functions of formal logic* ◇ B35 ◇

ZELEZNIKAR, A. *Behandlung logistischer Probleme mit Ziffernrechner (Serbo-Croatian summary)* ◇ B35 ◇

ZHURAVLEV, YU.I. *Set theoretical methods in the algebra of logic (Russian)* ◇ B35 ◇

1963

BETH, E.W. *Observations concerning computing, deduction and heuristics* ◇ B35 B80 ◇

CHANG, D.M.Y. & MOTT JR., T.H. *Computing irredundant normal forms from abbreviated presence functions* ◇ B35 ◇

DAVIS, MARTIN D. *Eliminating the irrelevant from mechanical proofs* ◇ B35 ◇

DEUSSEN, P. *Bericht ueber ein Programm zur Uebersetzung boolescher Ausdruecke in disjunktive Normalform* ◇ B35 ◇

DUNHAM, B. & NORTH, J.H. *Theorem testing by computer* ◇ B35 ◇

FRIEDMAN, JOYCE *A computer program for a solvable case of the decision problem* ◇ B35 ◇

FRIEDMAN, JOYCE *A semi-decision procedure for the functional calculus* ◇ B25 B35 ◇

GELERNTER, H.L. & HANSEN, J.R. & LOVELAND, D.W. *Empirical explorations of the geometry-theorem proving machine* ◇ B35 ◇

KUDIELKA, V. *Programmierung von Minimisierungsverfahren fuer zweistufige Logik* ◇ B35 ◇

LEBLANC, H. *Proof routines for the propositional calculus* ◇ B35 ◇

LEHMER, D.H. *Some high-speed logic* ◇ B35 ◇

MCCLUSKEY JR., E.J. & SCHORR, H. *Essential multiple-output prime implicants* ◇ B35 ◇

ROBINSON, JOHN ALAN *Theorem-proving on the computer* ◇ B35 ◇

ROHLEDER, H. *Ein Verfahren zum Aufsuchen minimaler Ausdruecke* ◇ B35 ◇

SIMAUTI, T. *Mechanization of mathematics* ◇ B35 ◇

SUMMERSBEE, S. & WALTERS, A. *Programming the functions of formal logic II (multi-valued logics)* ◇ B35 ◇

WANG, HAO *Mechanical mathematics and inferential analysis* ◇ B35 ◇

WANG, HAO *The mechanization of mathematical arguments* ◇ B35 ◇

WANG, HAO *Toward mechanical mathematics* ◇ B35 ◇

WILHELMY, A. *Minimisierung mit nichtkombinatorischen Methoden* ◇ B35 ◇

ZHURAVLEV, YU.I. *Algorithms with finite memory on disjunctive forms (Russian)* ◇ B35 ◇

ZHURAVLEV, YU.I. *On a class of algorithms over finite sets (Russian)* ◇ B35 ◇

1964

CARSON, D.F. & ROBINSON, G.A. & WOS, L. *The unit preference strategy in theorem proving* ◇ B35 ◇

DREBEN, B. & WANG, HAO *A refutation procedure and its model-theoretic justification* ◇ B35 ◇

FOXLEY, E. *Determination of the set of all four-variable formulae corresponding to universal decision elements using a logical computer* ◇ B35 ◇

KRAKOVSKAYA, O.S. & MAJSTROVA, T.D. *On a method of minimization of normal forms of boolean functions (Russian)* ◇ B35 ◇

LAWLER, E.L. *An approach to multilevel boolean minimization* ◇ B35 ◇

MASLOV, S.YU. *An inverse method of establishing deducibilities in the classical predicate calculus (Russian)* ◇ B35 ◇

PSHENICHNIKOVA, S.V. *On an algorithm for the automatic proof of certain theorems in analysis (Russian) (Azerbaijani summary)* ◇ B35 ◇

ROBINSON, JOHN ALAN *On automatic deduction* ◇ B35 ◇

TAUTS, A. *Solution of logical equations by an iteration method in the first-order predicate calculus (Russian) (English summary)* ◇ B35 ◇

TRAVIS, L.E. *Experiments with a theorem-utilizing program* ◇ B35 ◇

ZHURAVLEV, YU.I. *Appraisals of the complexity of algorithms for the construction of minimal disjunctive normal forms for the functions of the algebra of logic (Russian)* ◇ B35 ◇

1965

CARSON, D.F. & ROBINSON, G.A. & WOS, L. *Efficiency and completeness of the set of support strategy in theorem proving* ◇ B35 ◇

CHANG, D.M.Y. & MOTT JR., T.H. *Computing irredundant normal forms from abbreviated presence functions* ◇ B35 ◇

DAVYDOV, G.V. & MASLOV, S.YU. & MINTS, G.E. & OREVKOV, V.P. & SHANIN, N.A. & SLISENKO, A.O. *An algorithm of computer searching for natural logical proof in propositional calculus (Russian)* ◇ B35 ◇

GELERNTER, H.L. *Machine-generated problem-solving graphs* ◇ B35 ◇

HOUSE, R.W. & RADO, T. *A generalization of Nelson's algorithm for obtaining prime implicants* ◇ B35 ◇

KONDO, M. & MURATA, H. *Problem-solving machines I,II,III* ◇ B35 ◇

LIN, JINHUO & MENG, ZHANGRONG *To compute Boolean expression with best order (Chinese)* ◇ B35 ◇

ROBINSON, JOHN ALAN *A machine-oriented logic based on the resolution principle* ◇ B35 ◇

ROBINSON, JOHN ALAN *Automatic deduction with hyper-resolution* ◇ B35 ◇

ROHLEDER, H. *Eine Bemerkung zur Zeilendominanz beim Aufsuchen optimaler Normalformen* ◇ B35 ◇

ROWICKI, A. *On minimal programming of formulas in Lukasiewicz's notation* ◇ B35 ◇

SHENG, C.L. *Detection of totally symmetric boolean functions* ◇ B35 ◇

SLAGLE, J.R. *Experiments with a deductive question-answering program* ◇ B35 ◇

TIMOFEEVA, L.M. *A machine for proving geometrical functions (Russian)* ◇ B35 ◇

WANG, HAO *Formalization and automatic theorem-proving* ◇ B35 ◇

WANG, HAO *Logic and computers* ◇ A05 B35 D03 D10 ◇

ZAKREVSKIJ, A.D. *Algorithms of minimalization of weakly defined Boolean functions (Russian) (English summary)* ◇ B35 D20 ◇

1966

ABRAHAMS, P.W. *Machine verification of mathematical proof* ◇ B35 ◇

ANDON, F.I. *A simplification algorithm of a disjunctive normal form of the boolean functions (Russian)* ◇ B35 ◇

ANDON, F.I. *An approach to the minimization of systems of boolean functions (Russian)* ◇ B35 ◇

ANUFRIEV, F.V. & ASEL'DEROV, Z.M. & DIDUKH, I.I. & FEDYURKO, V.V. & LETICHEVS'KIJ, O.A. *On a certain algorithm for search of proofs of theorems in the theory of groups (Russian) (English summary)* ◇ B35 ◇

BENZAKEN, C. *Algorithmes de dualisation d'une fonction booleenne* ◇ B35 ◇

CARSON, D.F. & ROBINSON, G.A. & WOS, L. *Automatic generation of proofs in the language of mathematics* ◇ B35 ◇

CHOUDHURY, A.K. & DAS, S.R. *Computing irredundant normal forms from abbreviated presence functions* ◇ B35 ◇

CHYTIL, M.K. & HAJEK, P. & HAVEL, I.M. *GUHA - the method of systematical hypotheses searching (Czech) (English summary)* ◇ B35 B75 ◇

COOPER, D.C. *Theorem-proving in computers* ◇ B35 ◇

DARLINGTON, J.L. *Machine methods for proving logical arguments expressed in English* ◇ B35 ◇

FRIEDMAN, JOEL I. *The decision method for real algebra: Is it practical?* ◇ B35 ◇

FRIEDMAN, JOYCE *Computer realization of a decision procedure in logic* ◇ B25 B35 ◇

KOLKER, R.J. *Computing boolean functions with finite state machines* ◇ B35 ◇

MELTZER, B. *Theorem-proving for computers: Some results on resolution and renaming* ◇ B35 ◇

OBERST, E. *Ueber eine Moeglichkeit zur Bestimmung der Primkonjunktionen einer kanonischen alternativen Normalform* ◇ B35 ◇

ONO, K. *A formalism for primitive logic and mechanical proof-checking* ◇ B35 ◇

PITRAT, J. *Realization of a program which chooses the theorems it proves* ◇ B35 ◇

ROSENBERG, I.G. *Ein anschauliches Modell fuer die Minimisierung boolescher Funktionen* ◇ B35 ◇

SLAGLE, J.R. *A multipurpose theorem proving heuristic program that learns* ◇ B35 ◇

1967

ABRAHAMS, P.W. *Machine verification of mathematical proof* ◊ B35 ◊

AMAREL, S. *An approach to heuristic problem solving and theorem proving in the propositional calculus* ◊ B35 ◊

ANKERLIN, R.A. & JACKSON, C.L. *A rapid method for the identification of the type of a four-variable boolean function* ◊ B35 ◊

CARSON, D.F. & ROBINSON, G.A. & SHALLA, L. & WOS, L. *The concept of demodulation in theorem proving* ◊ B35 ◊

CHERNYAVSKIJ, A.L. *Computer simulation of the process of solving complex logical problems (heuristic programming) (Russian)* ◊ B35 ◊

DAVYDOV, G.V. *A method of establishing deducibility in the classical predicate calculus (Russian)* ◊ B10 B35 ◊

DAVYDOV, G.V. *The correction of unprovable formulas (Russian)* ◊ B10 B35 ◊

EHRENFEUCHT, A. & ORLOWSKA, E. *Mechanical proof procedure for propositional calculus* ◊ B35 ◊

FENSKE, C. *Beweisprogramme fuer die Praedikatenlogik und der Vollstaendigkeitssatz von Beth* ◊ B35 ◊

FOXLEY, E. *The construction of scale-of-two mechanisms from universal decision elements* ◊ B35 ◊

GREEN, C.C. & RAPHAEL, B. *The use of theorem-proving techniques in question-answering systems* ◊ B35 ◊

GRIGOR'YAN, YU.G. *The variational problem of functions of logical algebra and a method for its computer realization (Russian)* ◊ B35 ◊

KORNEV, YU.N. *Prinzip der Erhoehung des Unbestimmtheitsgrades boolescher Funktionen. Algorithmus zur Bestimmung der kuerzesten disjunktiven Normalformen (Russian)* ◊ B35 ◊

LIGMANOWSKI, M. & SORDYL, E. *Eine Methode zur bestimmung der Primimplikanten einer Menge logischer Funktionen (Polish) (English and Russian summaries)* ◊ B35 ◊

LUCKHAM, D.C. *The resolution principle in theorem-proving* ◊ B35 ◊

MASLOV, S.YU. *An inverse method for establishing deducibility of nonprenex formulas of predicate calculus (Russian)* ◊ B35 ◊

MINTS, G.E. *Variation in the deduction search tactics in sequential calculi (Russian)* ◊ B35 F07 F50 ◊

MORREALE, E. *Partitioned list algorithms for prime implicant determination from canonical forms* ◊ B35 ◊

NECULA, N.N. *A numerical procedure for determination of the prime implicants of a boolean function* ◊ B35 ◊

ORLOWSKA, E. *Mechanical proof procedure for the n-valued propositional calculus (Russian summary)* ◊ B35 ◊

POPPLESTONE, R.J. *Beth tree methods in theorem proving* ◊ B35 ◊

ROBINSON, JOHN ALAN *A review of automatic theorem-proving* ◊ B35 ◊

ROBINSON, JOHN ALAN *Heuristic and complete processes in the mech. of theorem proving* ◊ B35 ◊

RUS, T. *Some observations concerning the application of the electronic computers in order to solve nonarithmetical Problems* ◊ B35 ◊

SCHWARTZ, J.T. (ED.) *Mathematical aspects of computer science* ◊ B35 B65 B75 B97 D10 D80 D97 ◊

SLAGLE, J.R. *Automatic theorem proving with renamable and semantic resolution* ◊ B35 ◊

TORTORICI, M. *Calcolo inferenziale automatico* ◊ B35 ◊

VEENKER, G. *Beweisalgorithmen fuer die Praedikatenlogik* ◊ B35 ◊

WEINBERG, G.M. *Computing machines* ◊ B35 D10 ◊

1968

ALLEN, C.M. & GIVONE, D.D. *A minimization technique for multiple-valued logic systems* ◊ B35 ◊

ANDREWS, P.B. *On simplifying the matrix of a wff* ◊ B35 ◊

ANDREWS, P.B. *Resolution with merging* ◊ B35 ◊

BRAFFORT, P. & SCHEEPEN VAN, F. (EDS.) *Automation in language translation and theorem proving* ◊ B35 ◊

BURSKY, P. & SLAGLE, J.R. *Experiments with a multipurpose, theorem-proving heuristic program* ◊ B35 ◊

DARLINGTON, J.L. *Automatic theorem-proving with equality substitutions and mathematical induction* ◊ B35 ◊

DARLINGTON, J.L. *Some theorem-proving strategies based on the resolution principle* ◊ B35 ◊

DAVYDOV, G.V. *Some remarks on the search for a deduction in the predicate calculus (Russian)* ◊ B35 ◊

DRAGAN, I. *Un algorithme lexicographique pour la resolution des programmes polynomiaux en variables entieres* ◊ B35 ◊

HODES, L. & SPECKER, E. *Lengths of formulas and elimination of quantifiers I* ◊ B35 C10 C40 ◊

HUNT, E.B. & QUINLAND, J.R. *A formal deductive problem-solving system* ◊ B35 ◊

HURST, S.L. *An extension of binary minimisation techniques to ternary equations* ◊ B35 ◊

LOVELAND, D.W. *Mechanical theorem-proving by model elimination* ◊ B35 ◊

LUCKHAM, D.C. *Some tree-paring strategies for theorem-proving* ◊ B35 ◊

MASLOV, S.YU. *An inverse method of establishing deducibility for logical calculi (Russian)* ◊ B35 ◊

MELTZER, B. *A new look at mathematics and its mechanization* ◊ B35 ◊

MELTZER, B. *Some notes on resolution strategies* ◊ B35 ◊

MELTZER, B. *Some recent developments in complete strategies for theorem-proving by computer* ◊ B35 ◊

ROBINSON, JOHN ALAN *The generalized resolution principle* ◊ B35 ◊

WESTRHENEN VAN, S.C. *A computer programme for the first order predicate calculus without identity* ◊ B35 ◊

WESTRHENEN VAN, S.C. *A random generator for sentential calculus* ◊ B35 ◊

WESTRHENEN VAN, S.C. *A simple application of discrete Markov chains to mathematical logic* ◊ B35 ◊

WESTRHENEN VAN, S.C. *A simple stochastic model of the reduction of a semantic tableau* ◊ B35 ◊

WESTRHENEN VAN, S.C. *Some remarks on the statistical estimation of probability in first-order predicate calculus* ◊ B35 ◊

WESTRHENEN VAN, S.C. *Statistical estimation of definability* ◊ B35 C40 ◊

WESTRHENEN VAN, S.C. *Two programmes for the calculus of propositional logic* ◊ B35 ◊

1969

ALLEN, JOHN & LUCKHAM, D.C. *An interactive theorem-proving program* ◊ B35 ◊

BALLANTYNE, M. & BENNETT, J.H. *Semi-automated mathematics* ◊ B35 ◊

BENNETT, J.H. & GUARD, J.R. & OGLESBY, F.C. & SETTLE, L.G. *Semi-automated mathematics* ◊ B35 B75 ◊

CERUTTI, E. & DAVIS, P.J. *Formac meets Pappus: Some observations on elementary analytic geometry by computer* ◊ B35 ◊

CHANG, C.L. & LEE, R.C.T. & SLAGLE, J.R. *Completeness theorems for semantic resolution in consequence finding* ◊ B35 ◊

CULIK, K. *Problems in the theory of minimization of boolean expressions (Czech) (English summary)* ◊ B35 ◊

DARLINGTON, J.L. *Theorem proving and information retrieval* ◊ B35 ◊

DARLINGTON, J.L. *Theorem provers as question answerers* ◊ B35 ◊

DAVYDOV, G.V. & MASLOV, S.YU. & MINTS, G.E. & OREVKOV, V.P. & SLISENKO, A.O. *A computer algorithm for establishing deducibility based on inverse method (Russian)* ◊ B35 ◊

GREEN, C.C. *Application of theorem proving to problem solving* ◊ B35 ◊

GREEN, C.C. *Theorem-proving by resolution as a basis for question-answering systems* ◊ B35 ◊

HAYES, P.J. & KOWALSKI, R. *Semantic trees in automatic theorem-proving* ◊ B35 ◊

HEWITT, C. *PLANNER: A language for proving theorems in robots* ◊ B35 ◊

KALLICK, B. *A decision procedure based on the resolution method* ◊ B25 B35 ◊

KOWALSKI, R. *Search strategies for theorem-proving* ◊ B35 ◊

LOVELAND, D.W. *Theorem-provers combining model elimination and resolution* ◊ B35 ◊

MASLOV, S.YU. *The connection between tactics of the inverse method and the resolution method (Russian)* ◊ B35 ◊

MCKELLAR, A.C. & SHEN, V.Y. & WEINER, P. *A fast algorithm for testing switching functions for disjunctive decompositions* ◊ B35 B70 ◊

MELTZER, B. *Power amplification for theorem provers* ◊ B35 ◊

MELTZER, B. *The use of symbolic logic in proving mathematical theorems by means of a digital computer* ◊ B35 ◊

MORRIS, J.B. *E-resolution: Extensions of resolution to include the equality relation* ◊ B35 ◊

NIETHAMMER, W. & VEENKER, G. *Maschinen und mathematische Beweise* ◊ B35 ◊

OREVKOV, V.P. *On nonlengthening applications of equality rules (Russian)* ◊ B25 B35 F05 F07 ◊

ORLOWSKA, E. *Automatic theorem proving in a certain class of formulae of predicate calculus* ◊ B35 ◊

ORLOWSKA, E. *Mechanical theorem proving in a certain class of formulae of the predicate calculus (Polish) (Russian summary)* ◊ B35 ◊

PETRI, N.V. *Algorithms connected with predicates and Boolean functions (Russian)* ◊ B35 D15 ◊

PETRI, N.V. *The complexity of algorithms and their operating time (Russian)* ◊ B35 D15 D20 ◊

PLOTKIN, G.D. *A note on inductive generalization* ◊ B35 ◊

PRAWITZ, D. *Advances and problems in mechanical proof procedures* ◊ B35 ◊

PURTILL, R.L. *Doing logic by computer* ◊ B35 ◊

REYNOLDS, J.C. *A generalized resolution principle based upon context-free grammars* ◊ B35 D05 ◊

REYNOLDS, J.C. *Transformational systems and the algebraic structure of atomic formulas* ◊ B35 G10 ◊

ROBINSON, G.A. & WOS, L. *Paramodulation and theorem-proving in first order theories with equality* ◊ B35 ◊

ROBINSON, JOHN ALAN *A note on mechanizing higher order logic* ◊ B15 B35 ◊

ROBINSON, JOHN ALAN *Mechanizing higher-order logic* ◊ B35 B40 F35 ◊

ROBINSON, JOHN ALAN *New directions in mechanical theorem proving* ◊ B35 ◊

SHARONOV, V.I. & ZAMOV, N.K. *A certain class of strategies used in theorem proving by the resolution method (Russian)* ◊ B35 ◊

SHARONOV, V.I. & ZAMOV, N.K. *A class of strategies for the determination of provability by the resolution method (Russian)* ◊ B35 ◊

SIBERT JR., E.E. *A machine-oriented logic incorporating the equality relation* ◊ B35 ◊

SILTERE, M.Y. *Mechanical deduction of arithmetical identities* ◊ B35 ◊

VEENKER, G. *Ein Entscheidungsverfahren fuer den Aussagenkalkuel und seine Realisierung in einem Rechenautomaten* ◊ B35 ◊

WESTRHENEN VAN, S.C. *A probabilistic machine for the estimation of provability in the first order predicate calculus* ◊ B35 ◊

ZAKREVSKIJ, A.D. *New algorithm for minimizing weakly defined boolean functions* ◊ B35 ◊

1970

ANDERSON, ROBERT M. & BLEDSOE, W.W. *A linear format for resolution with merging and a new technique for establishing completeness* ◊ B35 ◊

ANDERSON, ROBERT M. *Completeness results for E-resolution* ◊ B35 ◊

ARNOLD, A. *Presentation d'un langage de formalisation des demonstrations mathematiques naturelles* ◊ B35 ◊

CAVINESS, B.F. *On canonical forms and simplification* ◊ B35 D35 ◊

CHANG, C.L. & LEE, R.C.T. & SLAGLE, J.R. *A new algorithm for generating prime implicants* ◊ B35 ◊

CHANG, C.L. *Renamable paramodulation for automatic theorem proving with equality* ◊ B35 ◊

CHANG, C.L. *The unit and the input proof in theorem proving* ⋄ B35 ⋄

FLOYD, R.W. & KING, J. *An interpretation-oriented theorem prover over integers* ⋄ B35 ⋄

GILMORE, P.C. *An examination of the geometry theorem machine* ⋄ B35 ⋄

HODES, L. *The logical complexity of geometric properties on the plane* ⋄ B35 C40 ⋄

KANGER, S. *Equational calculi and automatic demonstration* ⋄ B25 B35 C05 ⋄

KHOMICH, V.I. *On the complexity of realization of propositional formulae (Russian)* ⋄ B35 F50 ⋄

KONDO, M. & MURATA, H. *Problem-solving machines for Euclidean goemetry* ⋄ B35 ⋄

KOWALSKI, R. & KUEHNER, D. *Linear resolution with selection function* ⋄ B35 ⋄

KOWALSKI, R. *The case for using equality axioms in automatic demonstration* ⋄ B35 ⋄

KREISEL, G. *Hilbert's programme and the search for automatic proof procedures* ⋄ A05 B35 F99 ⋄

LACOMBE, D. & LAUDET, M. & NOLIN, L. & SCHUETZENBERGER, M.-P. (EDS.) *Symposium on automatic demonstration* ⋄ B35 B97 F97 ⋄

LOVELAND, D.W. *A linear format for resolution* ⋄ B35 ⋄

LUCKHAM, D.C. *Refinement theorems in resolution theory* ⋄ B35 ⋄

NEUBUESER, J. *Investigations of groups on computers* ⋄ B35 ⋄

PAWLAK, Z. *Definitional approach to automatic demonstration* ⋄ B35 ⋄

PRAWITZ, D. *A proof procedure with matrix reduction* ⋄ B35 ⋄

REITER, R. *The predicate elimination strategy in theorem-proving* ⋄ B35 ⋄

ROBINSON, G.A. & WOS, L. *Axiom systems in automatic theorem proving* ⋄ B35 ⋄

ROBINSON, G.A. & WOS, L. *Paramodulation and set of support* ⋄ B35 ⋄

ROBINSON, JOHN ALAN *An overview of mechanical theorem proving* ⋄ B35 ⋄

ROBINSON, JOHN ALAN *Computational logic: The unification computation* ⋄ B35 ⋄

ROBINSON, JOHN ALAN *The present state of mechanical theorem proving* ⋄ B35 ⋄

SLAGLE, J.R. *Interpolation theorems for resolution in lower predicate calculus* ⋄ B35 C40 ⋄

TORTORICI, M. *Il calcolo inferenziale automatico nella trattazione delle equazioni logiche* ⋄ B35 C05 ⋄

WANG, HAO *On the long-range prospects of automatic theorem-proving* ⋄ B35 ⋄

1971

AUSIELLO, G. *Automatic reduction of CUCH expression by means of the value method* ⋄ B35 ⋄

BISWAS, N.N. *Minimization of boolean functions* ⋄ B35 ⋄

BLEDSOE, W.W. *Splitting and reduction heuristics in automatic theorem proving* ⋄ B35 ⋄

BLISS, K. & CHIEN, R. & STOHL, F. *R2 a natural language question-answering system* ⋄ B35 ⋄

CHANG, C.L. & SLAGLE, J.R. *Completeness of linear refutation for theories with equality* ⋄ B35 ⋄

DARLINGTON, J.L. *Deductive question answering and plan formation in second order logic* ⋄ B35 ⋄

DAVYDOV, G.V. *Synthesis of the resolution method with the inverse method (Russian) (English summary)* ⋄ B35 ⋄

ERNST, G.W. *The utility of independent subgoals in theorem proving* ⋄ B35 ⋄

FIKES, R.E. & NILSSON, N.J. *STRIPS: A new approach to the application of theorem proving to problem solving* ⋄ B35 ⋄

HOFFMANN, GEERD-RUEDIGER & VEENKER, G. *The unit-clause proof procedure with equality (German summary)* ⋄ B35 ⋄

KANOVICH, M.I. *On complexity of Boolean function minimzation (Russian)* ⋄ B35 D15 ⋄

LUCKHAM, D.C. & NILSSON, N.J. *Extracting information from resolution proof trees* ⋄ B35 ⋄

MASLOV, S.YU. *Extension of the inverse method to calculus with an equality (Russian) (English summary)* ⋄ B35 ⋄

MASLOV, S.YU. & MINTS, G.E. & OREVKOV, V.P. *Mechanical proof-search and the theory of logical deduction in the USSR* ⋄ B35 ⋄

MASLOV, S.YU. *Proof-search strategies for methods of the resolution type* ⋄ B35 ⋄

MELTZER, B. *Prolegomena to a theory of efficiency of proof procedures* ⋄ B35 ⋄

MORE JR., T. *An interactive method for algebraic proofs* ⋄ B35 ⋄

MORGAN, C.G. *Hypothesis generation by machine* ⋄ B35 ⋄

PLIUSKEVICIENE, A. *Specialization of the use of axioms for deduction search in axiomatic theories with equality (Russian) (English summary)* ⋄ B35 ⋄

REITER, R. *Two results on ordering for resolution with merging and linear format* ⋄ B35 ⋄

ROBINSON, JOHN ALAN *Building deduction machines. Art. int. and heuristic programming* ⋄ B35 ⋄

ROBITASHVILI, N.G. *A combination of the inverse method and the method of resolution (Russian) (Georgian and English summaries)* ⋄ B35 F07 ⋄

VEENKER, G. *Maschinelles Beweisen* ⋄ B35 ⋄

1972

ANUFRIEV, F.V. & KOSTYAKOV, V.M. & MALASHONOK, A.I. *Algorithm and machine trial in searches for proofs of theorems in propositional calculus (Russian) (English summary)* ⋄ B35 ⋄

ANUFRIEV, F.V. & ASEL'DEROV, Z.M. & KAPITONOVA, YU.V. *Organization of data for the evidence algorithm in the search for proofs of theorems in formalized theories (Russian) (English summary)* ⋄ B35 ⋄

ANUFRIEV, F.V. & ASEL'DEROV, Z.M. *The evidence algorithm (Russian) (English summary)* ⋄ B35 ⋄

BOBROV, A.E. & ZHURKIN, V.A. *Synthesis of many-valued logic circuits in the system: maximum, minimum, cycle and special one-place functions (Russian) (English summary)* ⋄ B35 ⋄

BOEHM, C. & DEZANI, M. *A CUCH-machine: The automatic treatment of bound variables* ⋄ B35 B40 ⋄

COOPER, D.C. *Theorem proving in arithmetic without multiplication* ⋄ B25 B35 F30 ⋄

GLUSHKOV, V.M. & KAPITONOVA, YU.V. & LETICHEVS'KIJ, O.A. & MALEVANYI, N.P. & VERSHININ, K.P. *On the construction of a practical formal language for the transcription of mathematical theories (Russian) (English summary)* ⋄ B35 ⋄

HAGIWARA, Y. *A system for recognition of the implication relation between formulas in the logic of predicates* ⋄ B20 B35 ⋄

KANTARIYA, G.V. *Eine optimale Auswahl mit Vertraeglichkeit alternativer Hypothesen (Russian) (Georgian and English summary)* ⋄ B35 ⋄

K"NEV, P. *An algorithm for the determination of the absolute minimum form of weakly determined functions* ⋄ B35 ⋄

KRAMOSIL, I. *A method for random sampling of well-formed formulas (A method for random sampling of formulas of an elementary theory and statistical estimation of their deducibility equipped by a program. I)* ⋄ B35 ⋄

LEE, R.C.T. *An algorithm to generate prime implicants and its application to the selection problem* ⋄ B35 ⋄

LEE, R.C.T. *Fuzzy logic and the resolution principle* ⋄ B35 B52 ⋄

LOVELAND, D.W. *A unifying view of some linear Herbrand procedures* ⋄ B35 ⋄

MASLOV, S.YU. *Deduction search in calculi of general type (Russian) (English summary)* ⋄ B35 D03 F07 ⋄

MASLOV, S.YU. *Search for deduction as a model of a heuristic process (Russian) (English summary)* ⋄ B35 ⋄

MASLOV, S.YU. *The inverse method, and tactics for establishing deducibility for a calculus with functional symbols (Russian)* ⋄ B35 ⋄

MINICOZZI, E. & REITER, R. *A note on linear resolution in consequence-finding* ⋄ B35 ⋄

NOVOSELOV, V.G. *Statistical estimation of the efficiency of methods for minimization of boolean functions (Russian)* ⋄ B35 ⋄

PAPP, B. *Procede pour determiner les formes normales minimales des fonctions booleennes, en utilisant les regles de minimisation de la fonction de Cout* ⋄ B35 ⋄

PASZTOR-VARGA, K. *On some minimizing algorithms of boolean functions* ⋄ B35 ⋄

PLIUSKEVICIENE, A. *Extension of the inverse method to axiomatic theories with equality (Russian) (English summary)* ⋄ B35 ⋄

PLOTKIN, G.D. *Building-in equational theories* ⋄ B35 ⋄

SZCZERBA, L.W. *Automatic theorem proving* ⋄ B35 ⋄

WESTRHENEN VAN, S.C. *Statistical studies of theoremhood in classical propositional and first order predicate calculus* ⋄ B35 ⋄

WOJCIK, A.S. *The minimization of higher-order Boolean functions* ⋄ B35 ⋄

ZAMOV, N.K. *On a bound for the complexity of terms in the resolution method (Russian)* ⋄ B35 ⋄

1973

BHOWMIK, BIMAL & BHOWMIK, KANTI BHUSAN *The P-table and its application to second and third-order simplification of boolean functions* ⋄ B35 ⋄

BOBROV, A.E. & KOTS, V.S. *Minimierung von Funktionen einer mehrwertigen Logik, die durch das Maximum, das Minimum und alle einstelligen Funktionen bestimmt sind (Russisch)* ⋄ B35 ⋄

CHANG, C.L. & LEE, R.C.T. *Symbolic logic and mechanical theorem proving* ⋄ B35 B98 ⋄

DEGTYAREV, A.I. & KAPITONOVA, YU.V. & LYALETSKIJ, A.V. *Use of heuristic procedures in programs for the search for proofs of theorems (a survey) (Russian) (English summary)* ⋄ B35 ⋄

DELL'ORCO, P. *A syntactic approach to automatic theorem proving* ⋄ B35 ⋄

HAJEK, P. *Automatic listing of important observational statements. I,II* ⋄ B35 C13 C80 C90 ⋄

HAJEK, P. *Some logical problems of automated research* ⋄ B35 ⋄

HINTIKKA, K.J.J. & NIINILUOTO, I. *On the surface semantics of quantificational proof procedures* ⋄ B35 C07 F07 ⋄

JOYNER JR., W.M. *Automatic theorem-proving and the decision problem* ⋄ B25 B35 ⋄

KRAMOSIL, I. *A method for statistical testing of an at random sampled formula* ⋄ B35 ⋄

KRAMOSIL, I. *Statistical estimation of deducibility in a random sequence of formulas* ⋄ A05 B35 ⋄

KRAMOSIL, I. *Statistical estimation of deducibility in formalized theories* ⋄ B35 ⋄

LOSEV, G.F. *Best local algorithm of index 1 for constructing a sum of minimal disjunctive normal forms of a boolean function (Russian)* ⋄ B35 ⋄

LYSENKO, E.V. & MOKLYAK, N.G. & POPOV, V.A. & SKIBENKO, I.T. *Ueber einen Algorithmus der Suche von Arten unbestimmter boolescher Funktionen (Russian)* ⋄ B35 ⋄

MORGAN, C.G. *On the algorithmic generation of hypotheses* ⋄ B35 ⋄

ORLOWSKA, E. *Theorem-proving systems* ⋄ B35 ⋄

PIETRZYKOWSKI, T. *A complete mechanization of second-order type theory* ⋄ B15 B35 F35 ⋄

ROBINSON, G.A. & WOS, L. *Maximal models and refutation completeness: semidecision procedures in automatic theorem proving* ⋄ B35 ⋄

SHARONOV, V.I. & ZAMOV, N.K. *Anwendung von Isogrammen bei der Ermittlung des Schlusses (Russisch)* ⋄ B35 ⋄

TERENKOV, V.I. *The accuracy of algorithms for the computation of estimates for tables that are generated by monotone boolean functions (Russian)* ⋄ B35 ⋄

1974

BIBEL, W. *An approach to a systematic theorem proving procedure in first-order logic (German summary)* ⋄ B35 ⋄

BISHOP, P. & GREIF, I. & HALE, R.L.V. & HEWITT, C. & MATSON, T. & SMITH, BRIAN & STEIGER, R. *Behavioral semantics of nonrecursive control structures* ⋄ B35 B40 ⋄

CAIN, R.G. & HONG, SEJUNE & OSTAPKO, D.L. *Mini: a heuristic approach for logic minimization* ⋄ B35 G05 ⋄

FROENING, R. *Maschinelles Beweisen in Struktur-Theorien* ⋄ B15 B35 ⋄

GALDA, K. & PASSOS, E.P. *Application of quantifier elimination to mechanical theorem proving* ⋄ B35 C10 ⋄

GODLEVSKIJ, A.B. & LETICHEVS'KIJ, O.A. (EDS.) *Automata and algorithm theory, and mathematical logic (Russian)* ⋄ B35 B97 D05 D20 D97 ⋄

GUREVICH, I.B. & ZHURAVLEV, YU.I. *Minimierung boolescher Funktionen und effektive Erkennungsalgorithmen (Russisch)* ⋄ B35 ⋄

GUREVICH, I.B. *The noncomputability within a class of local algorithms of certain predicates connected with the minimization of Boolean functions (Russian)* ⋄ B35 D20 ⋄

HAJEK, P. *Automatic listing of important observational statements. III* ⋄ B35 C13 C80 C90 ⋄

HONG, SEJUNE & OSTAPKO, D.L. *Generating test examples for heuristic boolean minimization* ⋄ B35 ⋄

IWAMARU, Y. & NAGATA, M. & NAKANISHI, M. & NISHIMURA, T. *Implementation of Gentzen-type formal system representing properties of functions* ⋄ B35 ⋄

KANOVICH, M.I. *A theorem on speeding up in formal systems (Russian)* ⋄ B35 D15 ⋄

LONGO, G. *I problemi di decisione e la loro complessita* ⋄ B25 B35 D15 F20 ⋄

LOSEV, G.F. *The best local algorithm for constructing the sum of minimal-essential disjunctive normal forms of a boolean function, using neighbourhoods of minimal order (Russian)* ⋄ B35 ⋄

MAYEGA, J. *Statistical decidability of theorems* ⋄ B35 ⋄

MAYOH, B.H. *Extracting information from logical proofs* ⋄ B35 F07 ⋄

MENDES DES SANTOS, S. & MILLAN, M.R. *Automatic proofs for theorems on predicate calculus* ⋄ B35 ⋄

REITER, R. *On self-modifying programs* ⋄ B35 B75 ⋄

SANCHIS, L.E. *Logical completeness of directed resolution* ⋄ B35 ⋄

SCHNORR, C.-P. *On maximal merging of information in boolean computations* ⋄ B35 ⋄

SHEPPARD, D.A. & VRANESIC, Z.G. *Fault detection of binary sequential machines using R-valued test machines* ⋄ B35 ⋄

SHIMADA, K. *A theorem prover for intuitionistic propositional logic* ⋄ B35 F50 ⋄

VENTURINI ZILLI, M. *Su alcune strategie della risoluzione* ⋄ B35 ⋄

WEHRFRITZ, H. *Techniques for the transformation of logic equations* ⋄ B35 ⋄

ZAMOV, N.K. *Beschraenkung der Kompliziertheit der Terme in Ableitungen und aufloesbare Fragmente der Praedikatenkalkuele (Russisch)* ⋄ B35 F20 ⋄

1975

AZRIEL, A. *On the minimization of boolean functions (Bulgarian) (Russian and English summaries)* ⋄ B35 ⋄

BECVAR, J. (ED.) *Mathematical foundations of computer science 1975* ⋄ B35 B75 ⋄

BIBEL, W. *Die systematische Beweismethode und ihre Anwendungen* ⋄ B35 ⋄

BIBEL, W. *Effizienzvergleiche von Beweisprozeduren* ⋄ B35 ⋄

BIBEL, W. & SCHREIBER, J. *Proof search in a Gentzen-like system of first-order logic* ⋄ B35 F07 ⋄

BRAND, D. *Proving theorems with the modification method* ⋄ B35 ⋄

BREJTBART, YU.YA. & REITER, A. *A branch-and-bound algorithm to obtain an optimal evaluation tree for monotonic boolean functions* ⋄ B35 ⋄

CHIRKOV, M.K. & EGLITIS, L.V. *Generalized partial functions of the logic algebra and their minimization (Russian)* ⋄ B35 ⋄

KIREMITDJIAN, G. *Verification interactive de demonstrations mathematiques* ⋄ B35 ⋄

KUL'YANOV, E.G. *An algorithm for finding the maximal upper zero of an arbitrary monotonic function of the algebra of logic (Russian)* ⋄ B35 ⋄

LASKI, J. *An algorithm for generating orthogonal expansions of logical functions* ⋄ B35 ⋄

LYALETSKIJ, A.V. & MALASHONOK, A.I. *A calculus of k-disjuncts with the rule of a latent clash-resolution (Russian)* ⋄ B35 F07 ⋄

LYALETSKIJ, A.V. *A k-disjunct calculus (Russian)* ⋄ B35 F07 ⋄

MASLOV, S.YU. *The theory of deduction search, and some of its applications (Russian)* ⋄ B35 ⋄

MITEW, W. & POSTHOFF, C. *Ein Verfahren zur Loesung boolescher Gleichungen* ⋄ B35 ⋄

NAGATA, M. & NAKANISHI, M. & NISHIMURA, T. *TKP1.2 - the extension of TKP1 by adding some facilities* ⋄ B35 F35 ⋄

NEVINS, A.J. *Plane geometry theorem proving using forward chaining* ⋄ B35 ⋄

PUDLAK, P. *Polynomially complete problems in the logic of automated discovery* ⋄ B35 D15 ⋄

RICHTER, M.M. *Automatisches Beweisen und Gleichheitslogik* ⋄ B35 ⋄

SLISENKO, A.O. *Finite approach to the problem of optimizing theorem-proving algorithms (Russian) (English summary)* ⋄ B10 B35 ⋄

ZUCKER, J.I. *Formalization of classical mathematics in automath* ⋄ B35 ⋄

1976

ANDREWS, P.B. *Refutations by mapings* ⋄ B35 ⋄

AUSIELLO, G. *Difficult logical theories and their computer approximations* ⋄ B25 B35 D15 ⋄

BIBEL, W. *Maschinelles Beweisen* ⋄ B35 ⋄

BRADY, R.T. *A computer program for determining matrix models of propositional calculi* ⋄ B35 ⋄

CHAN, S.P. & PICHAI, V. *A minimal reduction of cyclic prime implicant tables by a graph theoretic method* ⋄ B35 ⋄

COLOMBETTI, M. & PAGELLO, E. *Programs, computations and temporal features* ⋄ B35 B75 ⋄

DARLINGTON, J.L. *Automatic synthesis of SNOBOL programs* ⋄ B35 ⋄

DUNHAM, B. & WANG, HAO *Towards feasible solutions of the tautology problem* ◊ B35 D15 ◊

ERNST, G.W. *A definition-driven theorem prover* ◊ B35 ◊

GILBERT, M.A. *A heuristic procedure for natural deduction derivations using reductio ad absurdum* ◊ B35 F07 ◊

GILL, A. *Applied algebra for the computer sciences* ◊ B35 ◊

HARTMANIS, J. *On effective speed-up and long proofs of trivial theorems in formal theories (French summary)* ◊ B25 B35 D15 D20 F20 ◊

HAVEL, I.M. & STEPANKOVA, O. *A logical theory of robot problem solving* ◊ B35 ◊

HENSCHEN, L.J. *Semantic resolution for Horn sets* ◊ B20 B35 ◊

JORGENSEN, P. & STRUNZ, H. *Anwendung graphentheoretischer Verfahren in der Entscheidungstabellentechnik* ◊ B35 ◊

JOYNER JR., W.M. *Resolution strategies as decision procedures* ◊ B25 B35 ◊

KALMAN, J.A. *Computer studies of $T_\to - W - I$* ◊ B35 B46 ◊

LOCKS, M.O. *Reversal in boolean minimalization* ◊ B35 ◊

LOVELAND, D.W. & STICKEL, M.E. *A hole in goal trees: some quidance from resolution theory* ◊ B35 ◊

MILNER, R. *Program semantics and mechanized proof* ◊ B35 ◊

MORGAN, C.G. *Methods for automated theorem proving in nonclassical logics* ◊ B35 F50 ◊

NAZARYAN, G.A. *Some comparative complexity characteristics of computations of boolean functions (Russian) (English and Armenian summaries)* ◊ B35 ◊

NIGMATULLIN, R.G. *Computation of boolean functions on digital machines (Russian)* ◊ B35 ◊

ORLOWSKA, E. *Herbrand systems for proving theorems of predicate calculus (Polish) (English and Russian summaries)* ◊ B35 ◊

PETERSON, G.E. *Theorem proving with lemmas* ◊ B35 ◊

REITER, R. *A semantically guided deductive system for automatic theorem proving* ◊ B35 ◊

SANDEWALL, E. *Conversion of predicate-calculus axioms to corresponding deterministic programs* ◊ B35 B75 ◊

SHOSTAK, R.E. *Refutation graphs* ◊ B35 ◊

SUVOROV, P.YU. *On the recognition of the tautological nature of propositional formulas (Russian) (English summary)* ◊ B35 ◊

VENTURINI ZILLI, M. *Complexity of the unification algorithm for first-order expressions* ◊ B35 ◊

WINKER, S.K. *Complete demodulation for automatic theorem proving* ◊ B35 ◊

1977

ANDREWS, P.B. & LONGINI COHEN, E. *Theorem proving in type theory* ◊ B35 ◊

BALLANTYNE, A.M. & BLEDSOE, W.W. *Automatic proofs of theorems in analysis using nonstandard techniques* ◊ B35 H05 ◊

BENTHEM JUTTING VAN, L.S. *Checking Landau's "Grundlagen" in the AUTOMATH system* ◊ B35 ◊

BERMAN, G. & COLIGN, A.W. *A computer representation of boolean and pseudo-boolean functions* ◊ B35 ◊

BIBEL, W. *A syntactic connection between proof procedures and refutation procedures* ◊ B35 ◊

BRADY, J.M. *Hints on proofs by recursion induction* ◊ B35 D20 ◊

CHEREDNICHENKO, N.D. & KOROTKOVA, M.A. & POPOV, S.V. *A program for finding proofs (Russian) (English summary)* ◊ B35 ◊

CHERNYSHEV, YU.O. *Optimization methods for combinatorial devices (Russian)* ◊ B35 ◊

CHIRKOV, M.K. & EGLITIS, L.V. *The minimization of systems of generalized partial functions of the logic algebra (Russian)* ◊ B35 ◊

COX, P.T. *A graphical proof procedure for first-order logic* ◊ B35 ◊

DALEY, R.P. *On the inference of optimal descriptions* ◊ B35 D15 ◊

DUDEK, W. *Minimization of weakly defined multiargument logic functions by means of the reduction of quasi-implicants* ◊ B35 ◊

ECSEDI-TOTH, P. & VARGA, A. *An effective theorem proving algorithm* ◊ B35 ◊

GALIL, Z. *On resolution with clauses of bounded size* ◊ B35 D15 ◊

GALIL, Z. *On the complexity of regular resolution and the Davis-Putnam procedure* ◊ B35 D15 ◊

GOTO, M. & KAO, S. & NINOMIYA, T. *Determination of many-valued truth tables for undefined operators in axioms by a computer and their applications* ◊ B35 ◊

GUBA, W. *Ein maximaler lokaler Algorithmus fuer Klassen unverkuerzbarer Ueberdeckungen* ◊ B35 ◊

HAJEK, P. & HAVRANEK, T. *On generation of inductive hypotheses* ◊ B35 C90 ◊

HARMS, S. & KLIX, W.-D. *Ein Algorithmus zur automatischen Loesung konstruktiver Problemstellungen* ◊ B35 F65 ◊

HAVEL, I.M. & STEPANKOVA, O. *Incidental and state-dependent phenomena in robot problem solving* ◊ B35 ◊

HELBIG, H. *A new method for deductive answer finding in a question-answering system* ◊ B35 ◊

IVANOV, P.M. *On an approach to the simplification of boolean functions (Russian)* ◊ B35 ◊

JENSEN, D.C. & PIETRZYKOWSKI, T. *Mechanizing ω-order type theory through unification* ◊ B35 ◊

KARAPETYAN, B.K. & POGOSYAN, EH.M. *Inductors and their connection with the method of empirical prediction (Russian)* ◊ B35 D70 ◊

KITAHASHI, T. & OGAWA, H. & TANAKA, K. *Formal deduction based on a directed AND/OR graph and its completeness (Japanese)* ◊ B35 ◊

KITAHASHI, T. & OGAWA, H. & TANAKA, K. *Minimum deduction process by D graph deduction (Japanese)* ◊ B35 ◊

KOLODZIEJ, R. *Algorithms for the validity of formulas of predicate calculus through hypergraphs (Polish) (English and Russian summaries)* ◊ B35 ◊

KREISEL, G. *From foundations to science: Justifying and unwinding proofs* ⋄ A05 B35 F07 ⋄

KRUPKO, N.A. *On one automatic system of logical conclusion (Russian)* ⋄ B35 ⋄

MACALUSO, A.T. & TORTORICI, M. *Un processo inferenziale per la risolutione, in calcolo automatico, di equazioni e sistemi di equazioni logiche* ⋄ B35 ⋄

MASLOV, S.YU. & NORGELA, S.A. *Herbrand strategies and the "greater deducibility" relation (Russian) (English summary)* ⋄ B25 B35 F05 ⋄

NORGELA, S.A. *On recursive nonseparability of the strategies of deduction-search in the classical predicate calculus (Russian)* ⋄ B35 D35 ⋄

SHOSTAK, R.E. *On the role of unification in mechanical theorem proving* ⋄ B35 ⋄

SHOSTAK, R.E. *On the SUP-INF method for proving Presburger formulas* ⋄ B25 B35 F30 ⋄

STATMAN, R. *Complexity and derivations from quantifier-free Horn formulae, mechanical introduction of explicit definitions, and refinement of completeness theorems* ⋄ B35 F20 ⋄

TAKAMATSU, T. *Computer experiment on proof of chemical reaction* ⋄ B35 ⋄

WINTERSTEIN, G. *Unification in second order logic* ⋄ B35 ⋄

1978

ABDUL-KARIM, M.A.H. & BERBAT, N.E. *A simultaneous analog-ternary converter* ⋄ B35 ⋄

ANDRE, C. & AUGUIN, M. & BOERI, F. *An algorithm for designing multiple boolean functions: application to PLAs* ⋄ B35 G05 ⋄

AREVALO, Z. & BREDESON, J.G. *A method to simplify a boolean function into a near minimal sum-of-products for programmable logic arrays* ⋄ B35 ⋄

BREJTBART, YU.YA. & GAL, S. *Analysis of algorithms for the evaluation of monotonic boolean functions* ⋄ B35 ⋄

BROWN, F.M. *Towards the automation of set theory and its logic* ⋄ B35 ⋄

BRUIJN DE, N.G. *Lambda-calculus with namefree formulas involving symbols that represent reference transforming mappings* ⋄ B35 B40 ⋄

BUKHARAEVA, Z.K. & GOLUNKOV, YU.V. *Complexity and running time of normal algorithms computing boolean functions (Russian)* ⋄ B05 B35 D03 D15 ⋄

CARTON, M. *La logique informationnelle recursive* ⋄ B35 B75 ⋄

CASE, J. & SMITH, C.H. *Anomaly hierarchies of mechanized inductive inference* ⋄ B35 D20 ⋄

COMMENTZ-WALTER, B. *Size-depth tradeoff in boolean formulas* ⋄ B35 ⋄

DAVIO, M. & DESCHAMPS, J.-P. & THAYSE, A. *Optimization of multivalued decision algorithms* ⋄ B35 ⋄

FERRO, A. & OMODEO, E.G. *An efficient validity test for formulae in extensional two-level syllogistic* ⋄ B20 B25 B35 ⋄

GERGELY, T. & VERSHININ, K.P. *Model theoretical investigation of theorem proving methods* ⋄ B35 C07 ⋄

GOTO, M. & KAO, S. & NINOMIYA, T. *Determination of the fittest number of truth-values and canonical forms of logical functions for a many-valued axiom set by a computer* ⋄ B35 ⋄

HAJEK, P. & HAVRANEK, T. *Mechanizing hypothesis formation. Mathematical foundations for a general theory* ⋄ A05 B35 C13 C80 C90 D80 ⋄

HAJEK, P. & HAVRANEK, T. *The GUHA method – its aims and techniques (twenty-four questions and answers)* ⋄ A05 B35 ⋄

HAVRANEK, T. & VOSAHLO, J. *A GUHA procedure with correlational quantifiers* ⋄ B35 C80 ⋄

KABULOV, A.V. & LOSEV, G.F. *On local simplification algorithms for disjunctive normal forms of boolean functions (Russian)* ⋄ B35 ⋄

KOSTYUKOVICH, A.I. *On recognizing the equivalence of boolean functions with respect to the inversion of variables (Russian) (English summary)* ⋄ B35 ⋄

KOSTYUKOVICH, A.I. *On recognizing the membership of boolean functions to one class and one type (Russian)* ⋄ B35 ⋄

LEVIN, A.G. *A method for the solution of difference logic equations (Russian)* ⋄ B35 ⋄

LOCKS, M.O. *Minimization of boolean polynomials, truth functions, and lattices* ⋄ B35 G10 ⋄

LOVELAND, D.W. *Automated theorem proving: a logical Basis* ⋄ B35 ⋄

MITANI, S. *Completion theory performed on the basis of LJ* ⋄ B35 ⋄

MOGILEVSKIJ, G.L. & OSTROUKHOV, D.A. *On a mechanical propositional calculus using Smullyan's analytical tableaux (Russian)* ⋄ B35 ⋄

NAZARYAN, G.A. *Ueber eine Synthese von Algorithmen approximativ berechenbarer boolescher Funktionen (Russian)* ⋄ B35 D15 D20 ⋄

NORGELA, S.A. *Herbrand strategies of deduction-search in predicate calculus I (Russian) (English and Lithuanian summaries)* ⋄ B25 B35 D35 F07 ⋄

ORLOWSKA, E. *The resolution principle for ω^+-valued logic* ⋄ B35 ⋄

PANKOV, P.S. *A combined method for proving certain theorems of mathematical analysis via a computer (Russian) (English summary)* ⋄ B35 ⋄

PARKIN, I.A. *On determining the prime implicants of a boolean function without recourse to its minterm form* ⋄ B35 ⋄

PASTRE, D. *Automatic theorem proving in set theory* ⋄ B35 ⋄

PETERSON, J.G. *An automatic theorem prover for substitution and detachment systems* ⋄ B35 ⋄

POPOV, S.V. & ZAKHAR'YASHCHEV, M.V. *Deduction procedure based on the syntactic tree method (Russian)* ⋄ B35 ⋄

RICHTER, M.M. *Logikkalkuele* ⋄ B35 B98 F05 G10 ⋄

ROMANOV, A.M. *The shortest disjunctive normal forms for products of boolean functions (Russian)* ⋄ B35 ⋄

SAMET, H. *A canonical form algorithm for proving equivalence of conditional forms* ⋄ B35 ⋄

SHOSTAK, R.E. *An algorithm for reasoning about equality* ⋄ B35 ⋄

SOCHILINA, A.V. *Algorithm and program for establishing derivability resolving wide classes of formulas (Russian) (English summary)* ⋄ B35 ⋄

STATMAN, R. *Bounds for proof-search and speed-up in the predicate calculus* ⋄ B35 F05 F20 ⋄

SZABO, P. *The undecidability of the D_A-unification problem* ⋄ B35 D35 ⋄

SZABO, P. & UNVERICHT, E. *The unification problem for distributive terms* ⋄ B35 ⋄

VASHCHENKO, V.P. *A minimization of boolean functions by the synthesis of module schemes (Russian)* ⋄ B35 ⋄

WASILEWSKA, A. *Machines, logics and decidability* ⋄ B25 B35 ⋄

WU, WENJUN *Mechanical theorem proving in elementary differential geometry (Chinese)* ⋄ B35 ⋄

WU, WENJUN *On the decision problem and the mechanization of theorem-proving in elementary geometry (Chinese)* ⋄ B35 ⋄

ZHENG, XIANCHANG *Mechanical theorem proving and Dedekind's problem (Chinese)* ⋄ B35 ⋄

1979

ABDUGALIEV, U.A. *An extremal problem connected with the distribution of work between collectives (Russian)* ⋄ B35 ⋄

ASPVALL, B. & PLASS, M. & TARJAN, R.E. *A linear-time algorithm for testing the truth of certain quantified boolean formulas* ⋄ B35 D15 ⋄

AUBIN, R. *Mechanizing structural induction I. Formal system. II Strategies* ⋄ B35 ⋄

BERESNEV, V.L. *Algorithms for the minimization of polynomials of boolean variables (Russian)* ⋄ B35 ⋄

BIBEL, W. *Tautology testing with a generalized matrix reduction method* ⋄ B35 D15 F05 ⋄

BLEDSOE, W.W. *A maximal method for set variables in automatic theorem-proving* ⋄ B35 ⋄

BREJTBART, YU.YA. & VAIRAVAN, K. *The computational complexity of a class of minimization algorithms for switching functions* ⋄ B35 ⋄

BROWN, F.M. & TAERNLUND, S.-A. *Inductive reasoning on recursive equations* ⋄ B35 ⋄

BROWN, F.M. *On a convenient division of labor in the generation of prime implicants* ⋄ B35 ⋄

CHAKRAPANI, N. & RANGASWAMY, S.V. & TIKEKAR, V. *Development of a structured program for conversion to prenex normal form* ⋄ B35 ⋄

CHANG, C.L. & SLAGLE, J.R. *Using rewriting rules for connection graphs to prove theorems* ⋄ B35 ⋄

COHEN, JACQUES & RUBIN, A.L. *An interactive system for proving theorems in the predicate calculus* ⋄ B35 ⋄

CURTIS, H.A. *Short-cut method of deriving nearly optimal arrays of NAND trees* ⋄ B35 ⋄

DANTSIN, E.YA. *Parameters defining the time of tautology recognition by the splitting method (Russian)* ⋄ B35 D15 ⋄

DARDZHANIYA, G.K. *Polynomial complexities of deduction of some logical calculi (Russian) (English summary)* ⋄ B35 F20 ⋄

DAVIS, MARTIN D. & SCHWARTZ, J.T. *Metamathematical extensibility for theorem verifiers and proof-checkers* ⋄ B35 ⋄

DEGANO, P. & SIROVICH, F. *Inductive generalization and proofs of function properties* ⋄ B35 B75 D20 D80 ⋄

DELIYANNI, A. & KOWALSKI, R. *Logic and semantic networks* ⋄ B35 ⋄

DESCHAMPS, J.-P. & LAPSCHER, F. *Presentation et optimisation d'un programme de recherche des decompositions disjointes d'une fonction booleenne. Extension aux fonctions booleennes generales* ⋄ B35 ⋄

DUDEK, W. *Minimization of multiargument weakly-defined logic functions with multiargument bindings* ⋄ B35 ⋄

EMDEN VAN, M.H. *Relational programming illustrated by a program for the game of mastermind* ⋄ B35 ⋄

FIBY, R. & SOKOL, J. & SUDOLSKY, M. *Efficient resolution theorem proving in the propositional logic* ⋄ B35 ⋄

GOTO, M. & KAO, S. & NINOMIYA, T. *Axiomatization of Kleene's three-valued logic by a computer* ⋄ B35 ⋄

GRIGORYAN, A.K. & SHIPILINA, L.B. *Effective algorithm for minimization of circuit complexity of underdetermined automata specified by nonbranching graphs (Russian)* ⋄ B35 ⋄

HENSCHEN, L.J. *Theorem proving by covering expressions* ⋄ B35 ⋄

JEANROND, H. *A unique termination theorem for a theory with generalized commutative axioms* ⋄ B35 ⋄

KARATUEV, V.G. & MATROSOV, V.M. & NOVIKOV, M.A. & SUMENKOV, E.A. & VASIL'EV, S.N. & YADYKIN, S.A. *Machine deduction of theorems on dynamic properties with vector functions of Lyapunov (Russian)* ⋄ B35 ⋄

KIRSANOV, G.M. & TSEJTLIN, G.E. & YUSHCHENKO, E.L. *ANALYST: A program package for proving identities (theorems) in axiomatic algorithmic-algebra systems (Russian)* ⋄ B35 ⋄

KLETTE, R. & LINDNER, R. *Zweidimensional arbeitende Vektormaschinen und ihr Leistungsvermoegen bei der Loesung von Entscheidungsproblemen der Aussagenlogik* ⋄ B35 D05 D15 ⋄

KOENIG, E.C. *Establishing valid arguments by computer and storing their meanings: A premiss of form a of Aristotelian logic* ⋄ B35 ⋄

KOWALSKI, R. *Logic for problem solving* ⋄ B35 ⋄

KRAMOSIL, I. *A note on computational complexity of a statistical deducibility testing procedure* ⋄ B35 D15 ⋄

KRAMOSIL, I. *Statistical approach to proof theory* ⋄ B35 ⋄

KUR'EROV, YU.N. *Normal form of mutual-absorption tactics (Russian)* ⋄ B35 F05 ⋄

MASLOV, S.YU. *Theory of inference search and questions of the psychology of creation (Russian)* ⋄ A05 B35 ⋄

MILNER, R. *ICF: A way of doing proofs with a machine* ⋄ B35 B40 ⋄

MOORE, J.S. *A mechanical proof of the termination of Takeuchi's function* ⋄ B35 D20 ⋄

NILSSON, N.J. *A production system for automatic deduction* ⋄ B35 ⋄

NORGELA, S.A. *Herbrand's strategies of deduction-search in predicate calculus II (Russian) (Lithuanian summary)* ◊ B35 ◊

NURLYBAEV, A.N. *On the construction of a reduced disjunctive normal form by the Nelson method (Russian)* ◊ B35 ◊

ORLOWSKA, E. *A generalization of the resolution principle* ◊ B35 ◊

PAPAKONSTANTINOU, G. *Minimization of modulo-2 sum of products* ◊ B35 ◊

PUDLAK, P. & SPRINGSTEEL, F. *Complexity in mechanized hypothesis formation* ◊ B35 D15 ◊

ROBINSON, JOHN ALAN *Logic: form and function. The mechanization of deductive reasoning* ◊ A10 B10 B35 B98 ◊

SHOSTAK, R.E. *A practical decision procedure for arithmetic with function symbols* ◊ B25 B35 F30 ◊

SHVARTSMAN, M.I. *Generalized approach to minimization of boolean functions (Russian)* ◊ B35 ◊

SIEKMANN, J. & STEPHAN, W. *Completeness and soundness of the connection graph proof procedure* ◊ B35 ◊

TLAS-YAKO, EH. *Minimization of boolean functions using a new form of canonical decomposition (Russian)* ◊ B35 ◊

UDALOV, V.I. *The diagnosis of a programmed logical matrix with the use of algorithms for the minimization of boolean functions (Russian)* ◊ B35 ◊

VERSHININ, K.P. *About using of auxiliary statements during search of proof (Russian)* ◊ B35 ◊

WOJCIECHOWSKI, W. & WOJCIK, A.S. *Multiple-valued logic design by theorem proving* ◊ B35 ◊

WU, WENJUN *On the mechanization of theorem-proving in elementary differential geometry (Chinese) (English summary)* ◊ B35 ◊

ZHEZHERUN, A.P. *Decidability of the unification problem for second-order languages with unary functional symbols (Russian) (English summary)* ◊ B15 B35 ◊

ZHURAVLEV, YU.I. *Local algorithms over disjunctive normal forms (Russian)* ◊ B35 G05 ◊

1980

AIELLO, L. *Using meta-theoretic reasoning to do algebra* ◊ B35 ◊

ALFONSECA, M. *Automatic solution of sorites* ◊ B35 ◊

ANDREWS, P.B. *Transforming matings into natural deduction proofs* ◊ B35 ◊

ASPVALL, B. *Recognizing disguised NR(1) instances of the satisfiability problem* ◊ B35 D15 ◊

BEN-ARI, M. *A simplified proof that regular resolution is exponential* ◊ B35 D15 F20 ◊

BEN-ARI, M. *Comments on "Tautology testing with a generalized matrix reduction method" by W. Bibel* ◊ B35 ◊

BIBEL, W. *A theoretical basis for the systematic proof method* ◊ B35 ◊

BIBEL, W. & KOWALSKI, R. (EDS.) *5th Conference on Automated Deduction* ◊ B35 B97 ◊

BLEDSOE, W.W. & HINES, L.M. *Variable elemination and chaining in a resolution-based prover for inequalities* ◊ B35 ◊

BROWN, F.M. *An investigation into the goals of research in automatic theorem proving as related to mathematical reasoning* ◊ B35 ◊

BRUIJN DE, N.G. *A survey of the project AUTOMATH* ◊ B35 B40 ◊

BULLERS, W.I. & NOF, S.Y. & WHINSTON, A.B. *Artificial intelligence in manufacturing planning and control* ◊ B35 B75 ◊

CHAKRAPANI, N. & RANGASWAMY, S.V. & TIKEKAR, V. *A computational algorithm for the verification of tautologies in propositional calculus* ◊ B35 ◊

COHEN, JACK K. & WATTS, D.E. *Computer-implemented set theory* ◊ B35 E20 ◊

COX, P.T. & PIETRZYKOWSKI, T. *A complete, nonredundant algorithm for reversed Skolemization* ◊ B35 ◊

CUTLER, R.B. & MUROGA, S. *Useless prime implicants of incompletely specified multiple-output switching functions* ◊ B35 ◊

DABIJA, V. *On a method for automated minimization of boolean functions based on if-then-else trees* ◊ B35 ◊

DOSHITA, S. & ISHIBASHI, T. & YAMASAKI, S. *Unit resolution for a subclass of the Ackermann class* ◊ B35 D15 ◊

ECSEDI-TOTH, P. & MORICZ, F. & VARGA, A. *A heuristic method for speeding up manual optimization of boolean functions* ◊ B35 ◊

ERICKSON, R.W. & MUSSER, D.R. *The AFFIRM theorem prover: Proof forests and management of large proofs* ◊ B35 ◊

FROLOV, A.B. *Extension of the notion of resolvent deduction and the resolution principle (Russian)* ◊ B35 ◊

GASHKOV, S.B. *Complexity of realization of boolean functions by schemes made up of functional elements and by formulas in bases whose elements realize continuous functions (Russian)* ◊ B35 ◊

GLAZUNOV, N.M. & KALUZHNIN, L.A. & SUSHCHANSKIJ, V.I. *A programming system for solution of combinatorial problems of contemporary algebra (Russian)* ◊ B35 ◊

GOAD, C.A. *Proofs as descriptions of computation* ◊ B35 F07 F50 ◊

HENNESSY, M. *A proof system for the first-order relational calculus* ◊ B35 C07 ◊

HULLOT, J.-M. *Canonical forms and unification* ◊ B35 ◊

IGARASHI, Y. *The size of arrays for a prime implicant generating algorithm* ◊ B35 ◊

KRAKOWSKI, I. *The four color problem reconsidered* ◊ A05 B35 ◊

KRAMOSIL, I. *Computational complexity of a statistical verification procedure for propositional calculus* ◊ B35 D15 ◊

KRAMOSIL, I. *Gentzen-like random axiomatic systems* ◊ B35 ◊

KRAMOSIL, I. *Statistical testing procedure for lengths of formalized proofs* ◊ B35 F20 ◊

LIU, CHUCHANG *The mechanical algorithm for solving Dedekind's problem (Chinese) (English summary)* ◊ B35 ◊

LOVASZ, L. *Efficient algorithms: an approach by formal logic* ⋄ B35 C13 D15 ⋄

LUSK, E.L. & OVERBEEK, R.A. *Data structures and control architecture for implementation of theorem-proving programs* ⋄ B35 ⋄

MCROBBIE, M.A. & MEYER, R.K. & THISTLEWAITE, P.B. *A mechanized decision procedure for non-classical logics: The program KRIPKE* ⋄ B25 B35 ⋄

MEREDITH, D. *A positive logic proof procedure* ⋄ B20 B35 B40 F50 ⋄

MIKELADZE, L.L. *A system for drawing relevant conclusions from a large volume of data (Russian)* ⋄ B35 ⋄

NEDERPELT, R.P. *An approach to theorem proving on the basis of a typed lambda-calculus* ⋄ B35 B40 F60 ⋄

NILSSON, N.J. *Principles of artificial intelligence* ⋄ B35 ⋄

NOLL, H. *A note on resolution: How to get rid of factoring without loosing completeness* ⋄ B35 ⋄

OPPEN, D.C. *Complexity, convexity and combinations of theories* ⋄ B25 B35 D15 ⋄

ORLOWSKA, E. *Resolution systems and their applications. I,II* ⋄ B35 ⋄

PLAISTED, D.A. *Abstraction mappings in mechanical theorem proving* ⋄ B35 ⋄

PLAISTED, D.A. *The application of multivariate polynomials to inference rules and partial tests for unsatisfiability* ⋄ B35 D15 F20 ⋄

REITER, R. *A logic for default reasoning* ⋄ B35 ⋄

ROMANSKI, R. *Algorithms for functional decomposition of boolean function (Bulgarian) (Russian and English summaries)* ⋄ B35 ⋄

SANDFORD, D.M. *Using sophisticated models in resolution theorem proving* ⋄ B35 ⋄

SCHWARTZ, J.T. *Fast probabilistic algorithms for verification of polynomial identities* ⋄ B35 ⋄

SCHWIND, C.B. *Natural language analysis by theorem proving methods: disambiguating pronouns in natural language texts* ⋄ B35 B45 B65 ⋄

SETHI, I.K. *Fast sequential evaluation of monotonic boolean functions* ⋄ B35 ⋄

SIEKMANN, J. & WRIGHTSON, G. *Paramodulated connection graphs* ⋄ B35 ⋄

STARK, W.R. *Automatic model construction* ⋄ B35 ⋄

STATMAN, R. *Solution to a problem of Chang and Lee* ⋄ B35 ⋄

TOWNLEY, J.A. *A pragmatic approach to resolution-based theorem proving* ⋄ B35 ⋄

WEYHRAUCH, R.W. *Prolegomena to a theory of mechanized formal reasoning* ⋄ B35 ⋄

WON, CHANGSA *On the computerization of minimizing the logic circuit by the MASK and the cost table method* ⋄ B35 ⋄

ZAKREVSKIJ, A.D. *Some combinatorial problems of artificial intelligence (Russian)* ⋄ B35 ⋄

1981

ABDULLAEVA, M. & BUZURKHANOV, V. & KASYMOV, N.KH. *On an approach to the realization of the resolution method (Russian)* ⋄ B35 ⋄

ANDREWS, P.B. *Theorem proving via general matings* ⋄ B35 ⋄

BHATTACHARJEE, P.R. & BHOWMIK, KANTI BHUSAN *A simpler method for minimization of switching functions* ⋄ B35 ⋄

BIBEL, W. *On matrices with connections* ⋄ B35 ⋄

BILLARD, B. *Polynomial manipulation with APL* ⋄ B35 ⋄

BUNDY, A. & WELHAM, B. *Using meta-level inference for selective application of multiple rewrite rule sets in algebraic manipulation* ⋄ B35 ⋄

CAI, JINGQIU *Negative hyper ordered resolution principle (Chinese) (English summary)* ⋄ B35 ⋄

CHAMPEAUX DE, D. *Other directions for automatic theorem proving* ⋄ B35 ⋄

COX, P.T. & PIETRZYKOWSKI, T. *Deduction plans: A basis for intelligent backtracking* ⋄ B35 ⋄

DABIJA, V. *On an IF-THEN-ELSE technique for automated theorem proving in propositional calculus* ⋄ B35 ⋄

DANTSIN, E.YA. *Two systems for proving tautologies based on the splitting method (Russian) (English summary)* ⋄ B35 D15 F20 ⋄

ERNI, W. & LAPSIEN, R. *On the time and tape complexity of weak unification* ⋄ B35 D15 ⋄

FUNAHASHI, S. & MIYAWAKI, M. & NAGATA, S. & OSHIBA, T. *A program for checking the validities of formulas in the first-order predicate calculus (Japanese) (English summary)* ⋄ B35 ⋄

GERGELY, T. & VERSHININ, K.P. *Concept sensitive formal language for task specification* ⋄ B35 ⋄

GOTO, M. & KAO, S. & NINOMIYA, T. *Axiomatization of Bochvar's three-valued logic by a computer* ⋄ B35 ⋄

HE, YANXIANG & HU, JIUQING & LI, WEIHUA *A way of mechanization of programs correctness proofs (Chinese) (English summary)* ⋄ B35 B75 ⋄

IBRAHIM, R.L.R. *A computer system for axiomatic investigation* ⋄ B35 ⋄

JIRKU, P. *Logical and linguistic aspects of computer-based inference processes* ⋄ B35 ⋄

JUSTEN, K. *A note on regular resolution* ⋄ B35 F20 ⋄

KUR'EROV, YU.N. *Equiprobable canonical calculi (Russian)* ⋄ B35 D03 ⋄

LI, DAFA *A method for determining the unsatisfiability of a set of ground clauses (Chinese) (English summary)* ⋄ B35 ⋄

LI, WEIHUA *Three fast algorithms for theorem proving (Chinese) (English summary)* ⋄ B35 ⋄

LIU, ZUNQUAN & QIN, CHAOBIN *Mechanical deduction of formulas of differential equations. I (Chinese)* ⋄ B35 ⋄

LIU, ZUNQUAN & QIN, CHAOBIN *Mechanical deduction of formulas of differential equations. II (Chinese)* ⋄ B35 ⋄

LOVELAND, D.W. & REDDY, C.R. *Deleting repeated goals in the problem reduction format* ⋄ B35 ⋄

LU, ZHONGWAN *Mathematical logic and mechanical proofs (Chinese)* ⋄ B35 ⋄

LUSK, E.L. & WINKER, S.K. & WOS, L. *Semigroups, antiautomorphisms, and involutions: a computer solution to an open problem. I* ⋄ B35 ⋄

LYALETSKIJ, A.V. *A variant of Herbrand's theorem for formulas in prenex form (Russian)* ⋄ B35 F05 ⋄

MANNA, Z. & WALDINGER, R. *Deductive synthesis of the unification algorithm* ◊ B35 ◊

MARTEM'JANOV, YU.S. *Controlled derivation of sentences (Russian) (English summary)* ◊ B35 ◊

PLAISTED, D.A. *Theorem proving with abstraction* ◊ B35 ◊

SCHMIDT, K. *Ein Rechenverfahren fuer die elementare Logik (English summary)* ◊ B25 B35 ◊

SCHOENFELD, W. *Gleichungen in der Algebra der binaeren Relationen* ◊ B35 E07 ◊

SILVESTRINI, D. *Alcune considerazioni sul metodo delle supervalutazioni e una semantica a supervalutazioni per il calcolo predicativo classico* ◊ B35 B60 ◊

SPRINGSTEEL, F. *Complexity of hypothesis formation problems* ◊ B35 D15 ◊

STICKEL, M.E. *A unification algorithm for associative-commutative functions* ◊ B35 ◊

TORBASOVA, V.P. *Decomposable machines over algebraic systems (Russian)* ◊ B35 C05 ◊

WANG, SHITIE *The HOI resolution and HOL resolution in Horn sets (Chinese) (English summary)* ◊ B35 ◊

WESTRHENEN VAN, S.C. *Note on probabilistic proof procedure for the first order predicate calculus* ◊ B35 ◊

WIETLISBACH, M.N. *Zur Komplexitaet von Entscheidungsalgorithmen, die auf dem Herbrand'schen Satz und regulaerer Resolution beruhen* ◊ B35 D15 ◊

1982

ANDREWS, P.B. & LONGINI COHEN, E. & MILLER, DALE A. *A look at TPS* ◊ B35 ◊

APOLLONI, B. & GREGORIO DI, S. *A probabilistic analysis of a new satisfiability algorithm (French summary)* ◊ B35 D15 ◊

APT, K.R. & EMDEN VAN, M.H. *Contributions to the theory of logic programming* ◊ B35 B75 ◊

ASLANYAN, L.A. *On the question of minimization of systems of poorly defined Boolean functions (Russian)* ◊ B35 ◊

BAJKHUMANOV, A.A. & KABULOV, A.V. *Solution of systems of nonlinear Boolean equations (Russian)* ◊ B35 ◊

BAYACHOROVA, B.D. & PANKOV, P.S. & YUGAI, S.A. *Numerical theorem proving by electronic computers and its applications in various branches of mathematics* ◊ B35 ◊

BERRETA PION, M.M. & CARVALHO DE, R.L. & LOPES PASSOS, E.P. *Interactive system to construct minimal model on the Herbrand universe* ◊ B35 ◊

BIBEL, W. *A comparative study of several proof procedures* ◊ B35 F07 ◊

BIBEL, W. *Automated theorem proving* ◊ B35 ◊

BIBEL, W. *Computationally improved versions of Herbrand's theorem* ◊ B35 ◊

BIBEL, W. *Deduktionsverfahren* ◊ B35 F07 ◊

BIBEL, W. & HOERNIG, K.M. *Improvements of a tautology-testing algorithm* ◊ B35 ◊

BIBEL, W. & SIEKMANN, J. (EDS.) *Kuenstliche Intelligenz* ◊ B35 ◊

BROWN, CYNTHIA A. & GOLDBERG, ALLEN & PURDOM JR., P.W. *Average time analyses of simplified Davis-Putnam procedures* ◊ B35 D15 ◊

CAFERRA, R. *Proof by matrix reduction as plan + validation* ◊ B35 ◊

COSTA, E.J.F. *Automatic program transformation viewed as theorem proving* ◊ B35 ◊

DEGTYAREV, A.I. *On the forms of deduction in calculi with equality and the paramodulation rule (Russian)* ◊ B35 ◊

DERSHOWITZ, N. *Orderings for term-rewriting systems* ◊ B35 ◊

EFIMOV, E.I. *Intellectual problem solvers (Russian)* ◊ B35 ◊

FARINAS DEL CERRO, L. *A simple deduction method for modal logic* ◊ B35 B45 ◊

GREENBAUM, S. & NAGASAKA, A. & O'RORKE, P. & PLAISTED, D.A. *Comparison of natural deduction and locking resolution implementations* ◊ B35 ◊

HENSCHEN, L.J. & NAQVI, S.A. *Representing infinite sequences of resolvents in recursive first-order Horn databases* ◊ B35 ◊

HOTOMSKI, P. *An induction law in proofs by contradiction with an application to automatic theorem proving (Russian)* ◊ B35 F30 ◊

JANSOHN, H.-S. & LANDWEHR, R. & WRIGHTSON, G. *An interactive proof system for higher order logic* ◊ B15 B35 F35 ◊

KOSSEJ, I.P. *Four fast algorithms in the method of resolutions (Russian)* ◊ B35 ◊

KRAMOSIL, I. *Three semantical interpretations of a statistical theoremhood testing procedure* ◊ B35 ◊

LI, DAFA *Some results on unit proofs (Chinese) (English summary)* ◊ B35 ◊

LIU, XUHUA *Delection strategy in the resolution principle (Chinese) (English summary)* ◊ B35 ◊

LIU, XUHUA & WANG, XIANGHAO *Generalized resolution (Chinese) (English summary)* ◊ B35 ◊

LOSEV, G.F. *Local algorithms for information calculation and the minimum covering problem (Russian)* ◊ B35 ◊

LOVELAND, D.W. (ED.) *6th conference on automated deduction* ◊ B35 ◊

LU, RUQIAN *A resolution theorem on Horn sets (Chinese)* ◊ B35 ◊

LU, RUQIAN *Semantic mappings on Herbrand base (Chinese)* ◊ B35 ◊

LU, RUQIAN *The principle of strong ordered input resolution (Chinese)* ◊ B35 ◊

LUSK, E.L. & OVERBEEK, R.A. *Experiments with resolution-based theorem-proving algorithms* ◊ B35 ◊

LUSK, E.L. & MCCUNE, W. & OVERBEEK, R.A. *Logic machine architecture: inference mechanisms* ◊ B35 ◊

LYALETSKIJ, A.V. *Testing the admissibility of substitutions (Russian)* ◊ B35 ◊

MARTELLI, A. & MONTANARI, U. *An efficient unification algorithm* ◊ B35 ◊

MATWIN, S. & PIETRZYKOWSKI, T. *Exponential improvement of efficient backtracking: a strategy for plan-based deduction* ◊ B35 ◊

MELLIAR-SMITH, P.M. & SCHWARTZ, R.L. & SHOSTAK, R.E. *STP: a mechanized logic for specification and verification* ◇ B35 ◇

MINKER, J. & ZANON, G. *An extension to linear resolution with selection function* ◇ B35 ◇

MINKER, J. *On closed databases and the closed world assumption* ◇ B35 ◇

MURRAY, N.V. *Completely non-clausal theorem proving* ◇ B35 ◇

NORGELA, S.A. *Some applications of the inverse method of proof search (Russian) (French and Lithuanian summaries)* ◇ B35 ◇

NORWOOD, F.H. *Long proofs* ◇ B35 F20 ◇

OMODEO, E.G. *The linked conjunct method for automatic deduction and related search techniques* ◇ B35 ◇

PLAISTED, D.A. *A simplified problem reduction format* ◇ B35 ◇

PLOTKIN, J.M. & ROSENTHAL, J.W. *The expected complexity of analytic tableaux analysis in propositional calculus* ◇ B05 B35 D15 F20 ◇

SCHOENFELD, W. *Upper bounds for proof-search in a sequent calculus for relational equations* ◇ B35 F20 ◇

SCHWARTZ, T. *No minimally reasonable collective-choice process can be strategy-proof* ◇ B35 ◇

SHOSTAK, R.E. *Deciding combinations of theories* ◇ B25 B35 ◇

SIEKMANN, J. & SZABO, P. *Universal unification and a classification of equational theories* ◇ B25 B35 C05 ◇

SILVER, B. *The application of homogenization to simultaneous equations* ◇ B35 ◇

SMITH, D.R. *Derived preconditions and their use in program synthesis* ◇ B35 ◇

WEYHRAUCH, R.W. *An example of FOL using metatheory: formalizing reasoning systems and introducing derived inference rules* ◇ B35 ◇

WINKER, S.K. *Generation and verification of finite models and counterexamples using an automated theorem prover answering two open questions* ◇ B35 ◇

WINKER, S.K. & WOS, L. *Procedure implementation through demodulation and related tricks* ◇ B35 ◇

WOS, L. *Solving open questions with an automated theorem-proving program* ◇ B35 ◇

WU, WENJUN *Mechanical theorem proving in elementary geometry and differential geometry* ◇ B35 ◇

WU, WENJUN *Toward mechanization of geometry – some comments on Hilbert's "Grundlagen der Geometrie"* ◇ B35 ◇

ZHENG, XIANCHANG *On the algorithm to realize theorem proving and solving problem (Chinese)* ◇ B35 ◇

ZHU, CHANGHONG *Subsemantic resolution (Chinese)* ◇ B35 ◇

1983

AMLIEN, J. & BARRICELLI, N.A. *The use of data processing machines in the recognition and proof of arithmetical theorems translated into B-mathematical language* ◇ B35 B80 ◇

ARNON, D.S. & SMITH, S.F. *Towards mechanical solution of the Kahan ellipse problem* ◇ B35 ◇

BARRICELLI, N.A. & COLBAN, E.A. & HANSEN, B.B. *Arithmetical applications of B-mathematics* ◇ B35 ◇

BENDA, W. & HORNUNG, G. & RAULEFS, P. & VOLLMANN, F. *Der META-Beweiser fuer die Zahlentheorie* ◇ B35 ◇

BUNDY, A. *The computer modelling of mathematical reasoning* ◇ B35 ◇

CASE, J. & SMITH, C.H. *Comparison of identification criteria for machine inductive inference* ◇ B35 B48 D20 ◇

CHYTIL, M.K. & HAJEK, P. & HAVRANEK, T. *The GUHA method (Czech)* ◇ B35 ◇

CVETKOVIC, D.M. *Discussing graph theory with a computer II. Theorems suggested by the computer* ◇ B35 ◇

CVETKOVIC, D.M. & PEVAC, I. *Discussing graph theory with a computer III. Man-machine theorem proving* ◇ B35 ◇

DERSHOWITZ, N. & HSIANG, JIEH *Rewrite methods for clausal and non-clausal theorem proving* ◇ B35 ◇

DEWDNEY, M. & WALSH, T.R.S. & WEHLAU, D.L. *Average-time testing of satisfiability algorithms* ◇ B35 ◇

DILGER, W. & JANSON, A. *Unifikationsgraphen fuer intelligentes Backtracking in Deduktionssystemen* ◇ B35 ◇

FINN, V.K. *Machine-oriented formalization of plausible arguments in the style of F.Bacon and J.S.Mill (Russian)* ◇ B35 ◇

FRANCO, J. & PAULL, M. *Probabilistic analysis of the Davis-Putnam procedure for solving the satisfiability problem* ◇ B35 D15 ◇

HAYASHI, S. *Extracting Lisp programs from constructive proofs: a formal theory of constructive mathematics based on Lisp* ◇ B35 F35 F50 ◇

HENSCHEN, L.J. & WOS, L. *Automated theorem proving 1965-1970* ◇ B35 ◇

HENSCHEN, L.J. & SMITH, BRIAN & VEROFF, R. & WINKER, S.K. & WOS, L. *Questions concerning possible shortest single axioms for the equivalential calculus: an application of automated theorem proving to infinite domains* ◇ B35 ◇

HEROLD, A. *Universal unification and a class of equational theories* ◇ B35 C05 ◇

HOTOMSKI, P. *A way of incorporating an induction rule in an automatic theorem-proving procedure with resolution (Russian)* ◇ B35 ◇

JOUANNAUD, J.-P. & KIRCHNER, C. & KIRCHNER, H. *Incremental construction of unification algorithms in equational theories* ◇ B35 ◇

KALMAN, J.A. *Condensed detachment as a rule of inference* ◇ B35 ◇

KRAMOSIL, I. *Statistical verification procedures for propositional calculus* ◇ B35 ◇

LI, DAFA *Comparing the efficiency of input and unit resolution (Chinese) (English summary)* ◇ B35 ◇

LI, DAFA *Proof of the conjecture of Henschen and Wos (Chinese) (English summary)* ◇ B35 ◇

LIU, XUHUA & WANG, XIANGHAO *Factoring problem in resolution (Chinese)* ◇ B35 ◇

MITANI, S. *Dimension theory constructed on the basis of LJ* ◊ B35 F55 ◊

MOROKHOVETS, M.K. & VERSHININ, K.P. *Strategies of the search for derivation of statements with restricted quantifiers (Russian)* ◊ B35 ◊

OHLBACH, H.-J. *Ein regelbasiertes Klauselgraph-Beweisverfahren* ◊ B35 ◊

PAULSON, L. *A higher-order implementation of rewriting* ◊ B35 ◊

PETERSON, G.E. *A technique for establishing completeness results in theorem proving with equality* ◊ B35 ◊

REZUS, A. *Abstract AUTOMATH* ◊ B35 ◊

SAMITOV, R.K. *On the complexity of linear deductions in resolution theory (Russian)* ◊ B35 ◊

SCHOENFELD, W. *Proof search for unprovable formulas* ◊ B35 F07 F20 ◊

SEN, M. *Minimization of Boolean functions of any number of variables using decimal labels* ◊ B35 ◊

SIEKMANN, J. & WRIGHTSON, G. (EDS.) *Automation of reasoning I,II* ◊ B35 ◊

SMOLKA, G. *Completeness of the connection graph proof procedure for unit-refutable clause sets* ◊ B35 ◊

STAPLES, J. *Two-level expression representation for faster evaluation* ◊ B35 B40 D03 ◊

WU, WENJUN *Some remarks on mechanical theoremproving in elementary geometry* ◊ B35 ◊

YUKAMI, T. *A theorem on lengths of proof of Presburger formulas* ◊ B35 F20 ◊

ZHALDOKAS, R. *An approach to the construction of complete term rewriting systems (Russian) (English and Lithuanian summaries)* ◊ B25 B35 B75 C05 ◊

1984

ANDREWS, P.B. & LONGINI COHEN, E. & MILLER, DALE A. & PFENNING, F. *Automating higher-order logic* ◊ B35 ◊

BIBEL, W. *Automatische Inferenz* ◊ B35 ◊

BIUNDO, S. & ZBORAY, F. *Automated induction proofs using methods of program synthesis* ◊ B35 ◊

BLEDSOE, W.W. & LOVELAND, D.W. (EDS.) *Automated theorem proving* ◊ B35 ◊

BLEDSOE, W.W. *Some automatic proofs in analysis* ◊ B35 ◊

BOERGER, E. & HASENJAEGER, G. & ROEDDING, D. (EDS.) *Logic and machines: decision problems and complexity* ◊ B35 D97 ◊

BOYER, R.S. & MOORE, J.S. *A mechanical proof of the Turing completeness of pure LISP* ◊ B35 D20 ◊

BOYER, R.S. & MOORE, J.S. *Proof checking the RSA public key encryption algorithm* ◊ B35 ◊

BOYER, R.S. & MOORE, J.S. *Proof-checking, theorem-proving, and program verification* ◊ B35 ◊

BRADY, B. & BUTLER, G.A. & LANKFORD, D.S. *Abelian group unification algorithms for elementary terms* ◊ B35 ◊

BROY, M. *On the Herbrand Kleene universe for nondeterministic computations* ◊ B35 ◊

BURGINA, E.S. *A graduated semantic system for set theory (Russian)* ◊ B35 E70 ◊

CHENADEC LE, P. *Canonical forms in finitely presented algebras* ◊ B35 C05 ◊

CHOU, SHANGCHING *Proving elementary geometry theorems using Wu's algorithm* ◊ B35 ◊

COX, P.T. & PIETRZYKOWSKI, T. *A complete, nonredundant algorithm for reversed Skolemization* ◊ B35 ◊

CVETKOVIC, D.M. *Discussing graph theory with a computer. IV. Knowledge organization and examples of theorem proving* ◊ B35 ◊

CVETKOVIC, D.M. & PEVAC, I. *Some heuristic in automatic theorem proving* ◊ B35 ◊

DILGER, W. & MUELLER, J. *An associative processor for theorem proving* ◊ B35 ◊

DOSHITA, S. & HIRATA, M. & YAMASAKI, S. & YOSHIDA, M. *A new combination of input and unit deductions for Horn sentences* ◊ B35 ◊

DOWLING, W.F. & GALLIER, J.H. *Linear-time algorithms for testing the satisfiability of propositional Horn formuae* ◊ B35 ◊

EFIMOV, E.I. *Automatic construction of axioms of the applied propositional calculus (Russian)* ◊ B35 ◊

EFSTATHIOU, J. & MAMDANI, E.H. *An analysis of formal logics are inference mechanisms in expert systems* ◊ B35 ◊

FAGES, F. *Associative-commutative unification* ◊ B35 ◊

FORSYTHE, K. & MATWIN, S. *Implementation strategies for plan-based deduction* ◊ B35 ◊

FRIBOURG, L. *A narrowing procedure for theories with constructors* ◊ B35 ◊

FRIBOURG, L. *Oriented equational clauses as a programming language* ◊ B35 ◊

FUNAHASHI, S. & NAGATA, S. & OSHIBA, T. *A procedure for checking validities of formulas by using adjoint formulas (Japanese) (English summary)* ◊ B35 ◊

GELDER VAN, A. *A satisfiability tester for non-clausal propositional calculus* ◊ B35 ◊

KAPUR, D. & KRISHNAMURTHY, B. *A natural proof sytem based on rewriting techniques* ◊ B35 F07 ◊

KETONEN, J. *EKL-a mathematically oriented proof checker* ◊ B35 ◊

LENAT, D.B. *Automated theory formation in mathematics* ◊ B35 ◊

LESCANNE, P. *Term rewriting systems and algebra* ◊ B35 ◊

MART'YANOV, V.I. *Invariant transformations of formulas (Russian)* ◊ B35 ◊

MCCUNE, W. & SMITH, BRIAN & VEROFF, R. & WOS, L. *The linked inference principle. II. The user's viewpoint* ◊ B35 ◊

MCDONALD, J.L. & SUPPES, P. *Student use of an interactive theorem prover* ◊ B35 ◊

MILLER, DALE A. *Expansion tree proofs and their conversion to natural deduction proofs* ◊ B35 F07 ◊

MILNER, R. *The use of machines to assist in rigorous proof. Mathematical logic and programming languages (with discussion)* ◊ B35 ◊

MULMULEY, K. *The mechanization of existence proofs of recursive predicates* ◊ B35 ◊

NELSON, GREG *Combining satisfiability procedures by equality-sharing* ◊ B25 B35 ◊

NOGUCHI, S. & TOGASHI, A. *A finite termination problem for term rewriting systems (Japanese) (English summary)* ◊ B35 ◊

OHLBACH, H.-J. & WRIGHTSON, G. *Solving a problem in relevance logic with an automated theorem prover* ◊ B35 B46 ◊

PAUL, E. *A new interpretation of the resolution principle* ◊ B35 ◊

PFENNING, F. *Analytic and non-analytic proofs* ◊ B35 ◊

PLAISTED, D.A. *Using examples, case analysis, and dependency graphs in theorem proving* ◊ B35 ◊

RYBAK, JANET & RYBAK, JOHN *Mechanizing logic I: Map logic extended formally to relational arguments. II: Automated map logic method for relational arguments on paper and by computer* ◊ B35 ◊

SCHMIDT, DAVID A. *A programming notation for tactical reasoning* ◊ B35 ◊

SHOSTAK, R.E. *Deciding combinations of theories* ◊ B35 ◊

SHOSTAK, R.E. (ED.) *7th international conference on automated deduction* ◊ B35 ◊

SIEKMANN, J. *Universal unification* ◊ B35 ◊

SIMON, D. *A linear time algorithm for a subcase of second order instantiation* ◊ B35 ◊

SLISENKO, A.O. *Linguistic considerations in devising effective algorithms* ◊ B35 D20 ◊

SMITH, BRIAN & WINKER, S.K. & WOS, L. *A new use of an automated reasoning assistant: Open questions in equivalential calculus and the study of infinite domains* ◊ B35 ◊

STENGER, H.-J. *Algebraic characterisations of NTIME(F) and NTIME(F,A)* ◊ B20 B35 D15 D20 ◊

STICKEL, M.E. *A case study of theorem proving by the Knuth-Bendix method: discovering that $x^3 = x$ implies ring commutativity* ◊ B35 ◊

SUPPES, P. *The next generation of interactive theorem provers* ◊ B35 ◊

WALTHER, C. *Ein mehrsortiger Resolutionskalkuel mit Paramodulation* ◊ B35 B75 ◊

WANG, HAO *Computer theorem proving and artificial intelligence* ◊ B35 ◊

WINKER, S.K. & WOS, L. *Open questions solved with the assistance of AURA* ◊ B35 ◊

WU, WENJUN *On the decision problem and the mechanization of theorem-proving in elementary geometry* ◊ B35 ◊

WU, WENJUN *Some recent advances in mechanical theorem-proving of geometries* ◊ B35 ◊

1985

ABADI, M. & MANNA, Z. *Nonclausal temporal deduction* ◊ B35 ◊

AN, ZHI & LIU, XUHUA *Generalized resolution using an equational substitution strategy (Chinese)* ◊ B35 ◊

ASPETSBERGER, K. *Substitution expressions: extacting solutions of non-Horn clause proofs* ◊ B35 ◊

BACK, R.J.R. & MANNILA, H. *On the suitability of trace semantics for modular proofs of communicating processes* ◊ B35 ◊

BIBEL, W. *Towards a connection machine for logical inference* ◊ B35 ◊

BLAIR, H.A. & TRYBULEC, A. *Computer aided reasoning* ◊ B35 ◊

BLEDSOE, W.W. & KUNEN, K. & SHOSTAK, R.E. *Completeness results for inequality provers* ◊ B35 ◊

BOETTGER, H. *Automatisches Theorembeweisen mit Konfigurationen (English and Russian summaries)* ◊ B35 ◊

BUCHBERGER, B. *Basic features and development of the critical-pair/completion procedure* ◊ B35 ◊

CARNIELLI, W.A. *An algorithm for axiomatizing and theorem proving in finite many-valued logics* ◊ B35 ◊

CARVALLO, M. *Sur la minimisation d'une expression representant une famille de fonctions booleennes (English summary)* ◊ B35 ◊

COQUAND, T. & HUET, G. *Constructions: a higher order proof system for mechanizing mathematics* ◊ B35 ◊

CUNNINGHAM, R.J. & DICK, A.J.J. *Rewrite systems on a lattice of types* ◊ B35 ◊

EDER, E. *Properties of substitutions and unifications* ◊ B35 ◊

FARINAS DEL CERRO, L. *Resolution modal logics* ◊ B35 ◊

FORTENBACHER, A. *An algebraic approach to unification under associativity and commutativity* ◊ B35 ◊

FRIBOURG, L. *A superposition oriented theorem prover* ◊ B35 ◊

GHELFO, S. & OMODEO, E.G. *Towards practical implementations of syllogistic* ◊ B35 ◊

GOTTLOB, G. & LEITSCH, A. *Fast subsumption algorithms* ◊ B35 ◊

HAKEN, A. *The intractability of resolution* ◊ B35 D15 D35 ◊

HOTOMSKI, P. *An automatic theorem-proving system with resolution, induction and symmetry (Russian)* ◊ B35 ◊

HSIANG, JIEH & SRIVAS, M. *PROLOG-based inductive theorem proving* ◊ B35 ◊

HSIANG, JIEH *Two results in term rewriting theorem proving* ◊ B35 ◊

HUSSMANN, H. *Unification in conditional-equational theories* ◊ B35 ◊

KANOVICH, M.I. *Efficient logical algorithms for analysis and syntheis of dependencies (Russian)* ◊ B35 ◊

KRAMOSIL, I. & SINDELAR, J. *Computational complexity of probabilistic searching algorithms over Herbrand universes* ◊ B35 ◊

KRISHNAMURTHY, B. *Short proofs for tricky formulas* ◊ B35 ◊

KROEGER, F. *On termporal program verification rules (French summary)* ◊ B35 ◊

LINIAL, N. & TARSI, M. *Deciding hypergraph 2-colourability by H-resolution* ◊ B35 ◊

LIU, XUHUA *The input semicancellation resolution principle on Horn sets (Chinese)* ◊ B35 ◊

MANNA, Z. & WALDINGER, R. *Deduction with relation matching* ◊ B35 ◊

MANNA, Z. & WALDINGER, R. *Special relations in automated deduction* ◊ B35 ◊

MOROKHOVETS, M.K. *Procedures for inference search, and transitive relations (Russian) (English summary)* ◊ B35 ◊

MURRAY, N.V. & ROSENTHAL, E. *Path resolution and semantic graphs* ◊ B35 ◊

NOSSUM, R. *Automated theorem proving methods* ⋄ B35 ⋄

PAUL, E. *On solving the equality problem in theories defined by Horn clauses* ⋄ B20 B35 ⋄

PLIUSKEVICIENE, A. *Generalized disjunction and existence properties for the logic of provability (Russian)* ⋄ B35 B45 ⋄

PNUELI, A. *In transition from global to modular temporal reasoning about programs* ⋄ B35 B45 B75 ⋄

PNUELI, A. *Linear and branching structures in the semantics and logics of reactive systems* ⋄ B35 B45 B75 ⋄

PRADE, H. *A computational approach to approximate and plausible reasoning with applications to expert systems* ⋄ B35 ⋄

RUSINOWITCH, M. *Path of subterms ordering and recursive decomposition ordering revisited* ⋄ B35 ⋄

SIEBER, K. *A partial correctness logic for procedures (in an ALGOL-like language)* ⋄ B35 B75 ⋄

TAUTS, A. *Extraction of a program from a derivation and its regularity I (Russian) (English summary)* ⋄ B35 ⋄

TULIPANI, S. *An algorithm to determine for any prime p, a polynomial-sized Horn sentence which expresses "the cardinality is not p"* ⋄ B35 D15 ⋄

VENKATESH, G. *A decision method for temporal logic based on resolution* ⋄ B35 B45 ⋄

WALTHER, C. *A mechanical solution of Schubert's steamroller by many-sorted resolution* ⋄ B35 ⋄

WASILEWSKA, A. *Some remarks on theorem proving systems and Mazurkiewics algorithms associated with them* ⋄ B35 ⋄

WOS, L. *Automated reasoning* ⋄ B35 ⋄

YELICK, K. *Combining unification algorithms for confined regular equational theories* ⋄ B35 ⋄

ZALDOKAS, R. *On the question of the construction of complete term rewriting systems (Russian) (English and Lithuanian summaries)* ⋄ B35 ⋄

H05 Infinitesimal analysis in pure mathematics

1939
LOEWNER, K. *Grundzuege einer Inhaltslehre im Hilbertschen Raume* ⋄ H05 ⋄

1958
LAUGWITZ, D. & SCHMIEDEN, C. *Eine Erweiterung der Infinitesimalrechnung* ⋄ B28 C65 E20 H05 ⋄

1961
ROBINSON, A. *Non-standard analysis* ⋄ H05 H15 ⋄
ROBINSON, A. *On the d-calculus for linear differential equations with constant coefficients* ⋄ H05 ⋄

1962
LUXEMBURG, W.A.J. *Nonstandard analysis. Lectures on A. Robinson's theory of infinitesimals and infinitely large numbers* ⋄ C20 H05 H98 ⋄
LUXEMBURG, W.A.J. *Two applications of the method of construction by ultrapowers to analysis* ⋄ C20 E05 E75 H05 ⋄
ROBINSON, A. *Complex function theory over non-archimedean fields* ⋄ H05 ⋄

1963
ROBINSON, A. *On languages which are based on non-standard arithmetic* ⋄ C75 H05 H15 ⋄

1964
ROBINSON, A. *On generalized limits and linear functionals* ⋄ C20 H05 ⋄

1965
AMEMIYA, I. *On non-standard analysis* ⋄ H05 ⋄
ROBINSON, A. *On the theory of normal families* ⋄ H05 ⋄
ROBINSON, A. *Topics in non-archimedean mathematics* ⋄ H05 H15 H20 ⋄
STONE, A.L. *Extensive ultraproducts and Haar measures* ⋄ C20 C65 E75 H05 ⋄

1966
BERNSTEIN, A.R. & ROBINSON, A. *Solution of an invariant subspace problem of K.T.Smith and P.R.Halmos* ⋄ H05 ⋄
HALMOS, P.R. *Invariant subspaces of polynomially compact operators* ⋄ H05 ⋄
ROBINSON, A. *Non-standard analysis* ⋄ A10 B98 H05 H10 H15 H98 ⋄
ROBINSON, A. *On some applications of model theory to algebra and analysis* ⋄ C60 C98 H05 H98 ⋄

1967
BERNSTEIN, A.R. *Invariant subspaces of polynomially compact operators on Banach space* ⋄ H05 ⋄
BUNYATOV, M.R. *Dirac spaces (Russian)* ⋄ C20 C60 H05 H10 ⋄
DEMUTH, O. *Necessary and sufficient conditions for Riemann integrability of constructive functions (Russian)* ⋄ F60 H05 ⋄
FENSTAD, J.E. *A note on "standard" versus "non-standard" topology* ⋄ E75 H05 ⋄
FLEISCHER, I. *Infinitesimals* ⋄ H05 ⋄
LUXEMBURG, W.A.J. *A new approach to the theory of monads* ⋄ H05 ⋄

1968
DEMUTH, O. *The connection between Riemann and Lebesgue integrability of constructive functions (Russian)* ⋄ F60 H05 ⋄
DEMUTH, O. *The Lebesgue integral and the concept of measureability of functions in constructive analysis (Russian)* ⋄ F60 H05 ⋄
WOLTER, H. *Eine Erweiterung der klassischen Analysis* ⋄ C20 H05 ⋄

1969
BERNSTEIN, A.R. & WATTENBERG, F. *Nonstandard measure theory* ⋄ H05 ⋄
CHUDNOVSKY, D.V. *Nonstandard analysis and homeomorphisms of B-spaces (Russian)* ⋄ H05 ⋄
COOPER, J.L.B. *k-fold preordered sets* ⋄ H05 ⋄
DNEPROVSKAYA, N.V. *Generalization of Kolmogorov's criterion in a linear normed space over an enlarged field of real numbers (Russian)* ⋄ H05 ⋄
DNEPROVSKAYA, N.V. *The Weierstrass theorems in the nonstandard space of continuous functions (Russian)* ⋄ H05 ⋄
DRESS, F. *Logique mathématique et analyse non-standard* ⋄ H05 H98 ⋄
HIRSCHFELD, J. & MACHOVER, M. *Lectures on non-standard analysis* ⋄ B98 H05 ⋄
JUHASZ, I. & MACHOVER, M. *A note on non-standard topology* ⋄ H05 ⋄
KREISEL, G. *Axiomatisations of nonstandard analysis that are conservative extensions of formal systems for classical standard analysis* ⋄ B30 F35 H05 ⋄
KUGLER, L.D. *A nonstandard approach to linear functions* ⋄ H05 ⋄
KUGLER, L.D. *Nonstandard almost periodic functions on a group* ⋄ H05 ⋄
KUGLER, L.D. *Nonstandard analysis of almost periodic functions* ⋄ H05 ⋄
LUXEMBURG, W.A.J. *A general theory of monads* ⋄ B15 C20 H05 ⋄
LUXEMBURG, W.A.J. *Reduced powers of the real number system and equivalents of the Hahn-Banach extension theorem* ⋄ C20 E25 E75 H05 ⋄
MUELLER, D.W. *Nonstandard proofs of invariance principles in probability theory* ⋄ H05 ⋄

PARIKH, R. *A nonstandard theory of topological groups* ◊ H05 ◊

PHILLIPS, R.G. *Liouville's theorem* ◊ C65 H05 ◊

ROBINSON, A. & ZAKON, E. *A set-theoretical characterization of enlargements* ◊ C20 C62 E75 H05 H20 ◊

ROBINSON, A. *Compactification of groups and rings and non-standard analysis* ◊ C40 C60 H05 H20 ◊

ROBINSON, A. *Germs* ◊ H05 ◊

ROBINSON, A. *Topics in nonstandard algebraic number theory* ◊ C60 H05 H15 H20 ◊

SCOTT, D.S. *Boolean models and nonstandard analysis* ◊ E35 E40 E50 H05 ◊

STONE, A.L. *Nonstandard analysis in topological algebra* ◊ H05 ◊

TAKAHASHI, S. *Analysis in categories* ◊ G30 H05 ◊

TAYLOR, R.F. *On some properties of bounded internal functions* ◊ H05 ◊

ZAKON, E. *Remarks on the nonstandard real axis* ◊ C65 H05 ◊

1970

ASENJO, F.G. *Generalized reals* ◊ H05 ◊

BERNSTEIN, A.R. *A new kind of compactness for topological spaces* ◊ E75 H05 ◊

CONNES, A. *Determination de modeles minimaux en analyse non standard et application* ◊ H05 H15 ◊

CONNES, A. *Ultrapuissances et applications dans le cadre de l'analyse non standard* ◊ C20 H05 ◊

FENSTAD, J.E. *Non-standard models for arithmetic and analysis* ◊ H05 H15 ◊

GEISER, J.R. *Nonstandard analysis* ◊ H05 ◊

JENSEN, A. *The possible influence of non-standard analysis on elementary mathematics* ◊ H05 ◊

LUXEMBURG, W.A.J. & TAYLOR, R.F. *Almost commuting matrices are near commuting matrices* ◊ C60 H05 ◊

PRASAD, S.N. *An extension of Bolzano-Weierstrass theorem to the field of ultra real products* ◊ H05 ◊

SMIRNOV, S.V. *Chebyshev approximations and the theory of models (Russian)* ◊ H05 ◊

1971

CLEAVE, J.P. *Cauchy, convergence and continuity* ◊ A10 H05 ◊

FENSTAD, J.E. & NYBERG, A.M. *Standard and non-standard methods in uniform topology* ◊ E75 H05 ◊

GONSHOR, H. *The ring of finite elements in a non-standard model of the reals* ◊ H05 ◊

LOEB, P.A. *A nonstandard representation of measurable spaces and L_∞* ◊ H05 ◊

MELONI, G.C. *Il reticolo dei filtri dal punto di vista dell'analisi non standard* ◊ H05 ◊

NARENS, L. *A nonstandard proof of the Jordan curve theorem* ◊ H05 ◊

WATTENBERG, F. *Nonstandard topology and extensions of monad systems to infinite points* ◊ H05 ◊

1972

ASENJO, F.G. & MCKEAN, J.M. *Weierstrass's final theorem of arithmetic is not final* ◊ H05 ◊

BACSICH, P.D. *Compact injectives and non-standard analysis* ◊ H05 ◊

BERNSTEIN, A.R. *Invariant subspaces for certain commuting operators on Hilbert space* ◊ H05 ◊

BERNSTEIN, A.R. *The spectral theorem - a non-standard approach* ◊ H05 ◊

BRACE, J.W. & KNEECE, R.R. *Approximation of strictly singular and strictly cosingular operators using nonstandard analysis* ◊ H05 ◊

CHADWICK, J.J.M. & CROSS, R.W. *A characterization of pre-near-standardness in locally convex linear topological spaces* ◊ H05 ◊

CHERLIN, G.L. & HIRSCHFELD, J. *Ultrafilters and ultraproducts in non-standard analysis* ◊ C20 E05 H05 ◊

EIFRIG, B. *Ein nicht-standard Beweis fuer die Existenz eines starken Liftings in $\mathscr{L}_\infty(0,1]$* ◊ E05 E50 E75 H05 ◊

EIFRIG, B. *Ein Nicht-Standard-Beweis fuer die Existenz eines liftings* ◊ H05 ◊

FLETCHER, P. & LINDGREN, W.F. *Transitive quasi-uniformities* ◊ E75 H05 ◊

GIORELLO, G. *Strutture non-standard della teoria dei numeri reali* ◊ H05 ◊

HAUSNER, M. *On a non-standard construction of Haar measure* ◊ E75 H05 ◊

HECHLER, S.H. *On monads in saturated enlargements* ◊ H05 ◊

HENSON, C.W. *On the nonstandard representation of measures* ◊ H05 ◊

HENSON, C.W. *The nonstandard hulls of a uniform space* ◊ C65 H05 ◊

HENSON, C.W. & MOORE JR., L.C. *The nonstandard theory of topological vector spaces* ◊ C65 H05 ◊

JANSSEN, G. *Restricted ultraproducts of finite von Neumann algebras* ◊ C20 C60 H05 ◊

JENSEN, A. *A computer oriented version of "non-standard analysis"* ◊ H05 ◊

JUHASZ, I. *Non-standard notes on the hyperspace* ◊ E75 H05 ◊

KELEMEN, P.J. & ROBINSON, A. *The nonstandard $\lambda:\varphi_2^4(x)$: model. I: The technique of nonstandard analysis in theoretical physics. II: The standard model from a nonstandard point of view* ◊ H05 H10 ◊

LIGHTSTONE, A.H. *Infinitesimals* ◊ H05 ◊

LOEB, P.A. *A non-standard representation of measurable spaces, L_∞, and L_∞^** ◊ E75 H05 ◊

LUXEMBURG, W.A.J. *A nonstandard analysis approach to Fourier analysis* ◊ H05 ◊

LUXEMBURG, W.A.J. *A remark on the Cantor-Lebesque lemma* ◊ H05 ◊

LUXEMBURG, W.A.J. & ROBINSON, A. (EDS.) *Contributions to non-standard analysis* ◊ B97 H05 H97 ◊

LUXEMBURG, W.A.J. *On some concurrent binary relations occuring in analysis* ◊ H05 ◊

MCCORD, M.C. *Non-standard analysis and homology* ◊ H05 ◊

NARENS, L. *Topologies of closed subsets* ◊ H05 ◊

OSDOL VAN, D.H. *Truth with respect to an ultrafilter or how to make intuition rigorous* ◊ C20 H05 ◊

PARIKH, R. & PARNES, M. *Conditional probability can be defined for all pairs of sets of reals* ◊ H05 ◊

POTTHOFF, K. *Ordnungseigenschaften von Nichtstandardmodellen* ◊ C20 C62 H05 H15 ◊

PURITZ, C.W. *Almost perpendicular vectors* ◊ H05 ◊

STROYAN, K.D. *Additional remarks on the theory of monads* ◊ H05 ◊

STROYAN, K.D. *Uniform continuity and rates of growth of meromorphic functions* ◊ H05 ◊

TACON, D.G. *Weak compactness in normed linear spaces* ◊ H05 ◊

WOLFF, M. *Nonstandard-Komplettierung von Cauchy-Algebren* ◊ C65 H05 ◊

YOUNG, L. *Functional analysis - a non-standard treatment with semifields* ◊ C65 H05 ◊

1973

ADLER, A. *F-planar graphs* ◊ E05 H05 ◊

ADLER, A. & HAMILTON, J. *Invariant means via the ultrapower* ◊ C20 H05 ◊

BERNSTEIN, A.R. *Non-standard analysis* ◊ C60 C65 H05 ◊

CHAPIN JR., E.W. & WEBB, S.M. *A non-standard proof in the theory of integration* ◊ H05 ◊

CHARRETTON, C. & RICHARD, D. *Elements d'une theorie non standard des groupes topologiques* ◊ C60 H05 ◊

EIFRIG, B. *Zur Existenz eines Spielwertes bei Spielen auf kompakten Raeumen mit stetiger Auszahlungsfunktion* ◊ H05 H10 ◊

GIORELLO, G. *Una rappresentazione non-standard delle distribuzioni functoriali* ◊ H05 ◊

GIORELLO, G. *Una rappresentazione nonstandard delle distribuzioni temperate e la trasformazione di Fourier* ◊ H05 ◊

GONSHOR, H. *Projective covers as subquotients of enlargements* ◊ H05 H20 ◊

HENSON, C.W. & MOORE JR., L.C. *Invariance of the nonstandard hulls of locally convex spaces* ◊ C65 H05 ◊

KASAHARA, S. *A characterization of nonstandard real fields* ◊ C20 C60 H05 ◊

KHRISTOV, KH.YA. & TODOROV, T.D. *Asymptotic numbers - a generalization of the notion of number* ◊ H05 H20 ◊

KOMKOV, V. & WAID, C.C. *Asymptotic behavior of non-linear inhomogeneous equations via non-standard analysis I: Second order equations* ◊ H05 ◊

LAUGWITZ, D. *A new theory of contact angles* ◊ B30 H05 ◊

LIGHTSTONE, A.H. *Infinitesimals and integration* ◊ H05 ◊

LOEB, P.A. *A combinatorial analog of Lyapunov's theorem for infinitesimal generated atomic vector measures* ◊ H05 ◊

LUCAS, T. *L'analyse non-standard* ◊ H05 ◊

LUXEMBURG, W.A.J. *Non-standard analysis* ◊ H05 ◊

MAHE, L. *Topos infinitesimal* ◊ G30 H05 ◊

MEISTERS, G.H. & MONK, J.D. *Construction of the reals via ultrapowers* ◊ C20 H05 ◊

MELONI, G.C. *Analisi non standard di anelli e corpi topologici (English summary)* ◊ C60 H05 H20 ◊

NELSON, GEORGE C. *Nonconstructivity of models of the reals (Russian summary)* ◊ C62 C65 D55 H05 H15 ◊

ROBINSON, A. *Function theory on some nonarchimedian fields* ◊ C60 C65 H05 ◊

ROBINSON, A. *Metamathematical problems* ◊ A05 B30 C60 C98 H05 ◊

ROBINSON, A. *Standard and nonstandard number systems* ◊ C25 C60 C98 H05 H15 H20 H98 ◊

SMIRNOVA, O.S. *Some examples of nonstandard extensions (Russian)* ◊ H05 ◊

STROYAN, K.D. *A characterization of the Mackey uniformity $m(L^\infty, L^1)$ for finite measures* ◊ H05 ◊

TACON, D.G. *Weak compactness in locally convex spaces* ◊ H05 ◊

VOROS, A. *Introduction to non-standard analysis* ◊ H05 ◊

WATTENBERG, F. *Monads of infinite points and finite product spaces* ◊ H05 ◊

1974

ABIAN, A. *Nonstandard models for arithmetics and analysis* ◊ C20 H05 H15 ◊

BEHRENS, M.F. *A local inverse function theorem* ◊ H05 ◊

BEHRENS, M.F. *Analytic sets in $\mathcal{M}(D)$* ◊ H05 ◊

BEHRENS, M.F. *Boundary values for meromorphic functions defined in the open unit disk* ◊ H05 ◊

BELLENOT, S.F. *Nonstandard topological vector spaces* ◊ H05 ◊

BERNSTEIN, A.R. *A non-standard integration theory for unbounded functions* ◊ H05 ◊

BERNSTEIN, A.R. & LOEB, P.A. *A nonstandard integration theory for unbounded functions* ◊ H05 ◊

BERNSTEIN, A.R. & WATTENBERG, F. *Cardinality-dependent properties of topological spaces* ◊ H05 ◊

BONACINI, R. & MELONI, G.C. *Teoria non-standard delle probabilita* ◊ H05 ◊

BORODYANSKIJ, B.M. *Nonstandard analysis and enlargements of Hilbert spaces (Russian)* ◊ H05 ◊

BORODYANSKIJ, B.M. *Nonstandard analysis and group representations (Russian)* ◊ H05 ◊

BORODYANSKIJ, B.M. *Outline of the basic concepts of nonstandard analysis (Russian)* ◊ H05 ◊

CHARRETTON, C. & RICHARD, D. *Preuves non standards de resultats classiques des groupes topologiques et quelques resultats non standard* ◊ H05 ◊

COZART, D. & MOORE JR., L.C. *The nonstandard hull of a Riesz space* ◊ C60 H05 ◊

GONSHOR, H. *Enlargements contain various kinds of completions* ◊ H05 H20 ◊

HADDAD, L. *Introduction a l'analyse nonstandard* ◊ H05 ◊

HENSON, C.W. & MOORE JR., L.C. *Invariance of the nonstandard hulls of a uniform space* ◊ C65 H05 ◊

HENSON, C.W. & MOORE JR., L.C. *Nonstandard hulls of the classical Banach spaces* ◊ C65 H05 ◊

HENSON, C.W. & MOORE JR., L.C. *Semi-reflexivity of the nonstandard hulls of a locally convex space*
⋄ C65 H05 ⋄

HENSON, C.W. & MOORE JR., L.C. *Subspaces of the nonstandard hull of a normed space* ⋄ C65 H05 ⋄

HENSON, C.W. *The isomorphism property in nonstandard analysis and its use in the theory of Banach spaces*
⋄ C65 H05 ⋄

HIRSCHFELDER, J.J. *Nonstandard analysis in a nutshell*
⋄ H05 ⋄

HURD, A.E. *Near periods and Bohr compactifications*
⋄ H05 ⋄

KASAHARA, S. *A remark on nonstandard real fields*
⋄ H05 ⋄

KEISLER, H.J. *Monotone complete fields*
⋄ C60 C65 H05 ⋄

KHAN, M.A. *Approximately convex average sums of unbounded sets* ⋄ H05 ⋄

KHRISTOV, KH.YA. *Eine neue Art von verallgemeinerten Funktionen – die asymptotischen Funktionen* ⋄ H05 ⋄

KOCK, A. & MIKKELSEN, C.J. *Topos-theoretic factorization of non-standard extensions* ⋄ G30 H05 ⋄

KOMKOV, V. *Asymptotic behavior of non-linear, inhomogeneous differential equations via non-standard analysis II: Some applications to higher order equation*
⋄ H05 ⋄

LEVITZ, H. *Non-standard analysis: an exposition*
⋄ H05 ⋄

LOEB, P.A. *A nonstandard representation of Borel measures and σ-finite measures* ⋄ H05 ⋄

LOEB, P.A. *A note on continuity for Robinson's predistribution* ⋄ H05 ⋄

LUXEMBURG, W.A.J. *On a theorem of helly and a theorem about liftings* ⋄ H05 ⋄

NARENS, L. *Field embeddings of generalized metric spaces*
⋄ H05 ⋄

NARENS, L. *Homeomorphism types of generalized metric spaces* ⋄ H05 ⋄

NARENS, L. *Measurement without Archimedean axioms*
⋄ H05 ⋄

PARIKH, R. & PARNES, M. *Conditional probabilities and uniform sets* ⋄ H05 ⋄

PINCUS, D. *The strength of the Hahn-Banach theorem*
⋄ E25 E75 H05 ⋄

ROBINSON, A. *Enlarged sheaves* ⋄ H05 ⋄

SAENDIG, A.-M. *Integraltransformationen von Predistributionen (Italian summary)* ⋄ H05 ⋄

SAENDIG, A.-M. *Lokale Werte und Grenzwerte von Predistributionen* ⋄ H05 ⋄

SAENDIG, A.-M. *Ueber die Regularisierung und Ordnung von Predistributionen* ⋄ H05 ⋄

SPIVAKOV, YU.L. *Algebraic extensions of the field of formal power series $F_D((t))$ (Russian) (Uzbek summary)* ⋄ C20 C60 H05 ⋄

STROYAN, K.D. *A nonstandard characterization of mixed topologies* ⋄ H05 ⋄

STROYAN, K.D. *Infinitesimal relations on the space of bounded holomorphic functions* ⋄ H05 ⋄

WATTENBERG, F. *Two topologies with the same monads*
⋄ H05 ⋄

ZAKON, E. *A new variant of non-standard analysis*
⋄ C20 H05 ⋄

1975

CHARRETTON, C. & RICHARD, D. *Theoreme d'Ascoli et application aux groupes topologiques localement compacts en analyse non standard* ⋄ H05 ⋄

DACUNHA-CASTELLE, D. & KRIVINE, J.-L. *Sous-espaces de L^1 (English summary)* ⋄ H05 ⋄

EARMAN, J. *Infinities, infinitesimals, and indivisibles: The Leibnizian labyrinth* ⋄ A05 A10 H05 ⋄

GRAINGER, A.D. *Invariant subspaces of compact operators on topological vector spaces* ⋄ C60 C65 H05 ⋄

GREENWOOD, P. & HERSH, R. *Stochastic differential and quasi-standard random variables* ⋄ H05 ⋄

HENSON, C.W. *The monad system of the finest compatible uniform structure* ⋄ C65 H05 ⋄

HENSON, C.W. *When do two Banach spaces have isometrically isomorphic nonstandard hulls?*
⋄ C65 H05 ⋄

HERRMANN, R.A. *Nonstandard topological extensions*
⋄ H05 ⋄

JOHNSON, D.R. & MATTSON, D.A. *Some applications of non-standard analysis to proximity spaces* ⋄ H05 ⋄

KIRK, R.B. *The Haar integral via non-standard analysis*
⋄ E75 H05 ⋄

LIGHTSTONE, A.H. & WONG, KAM *Dirac delta functions via nonstandard analysis* ⋄ H05 ⋄

LIGHTSTONE, A.H. & ROBINSON, A. *Nonarchimedean fields and asymptotic expansions* ⋄ H05 ⋄

LOEB, P.A. *Conversion from nonstandard to standard measure spaces and applications in probability theory*
⋄ H05 ⋄

LUXEMBURG, W.A.J. *Nichtstandard-Zahlsysteme und die Begruendung des Leibnizschen Infinitesimalkalkuels*
⋄ H05 ⋄

POPOWICZ, Z. *Remarks on dual structures in a Dirac space (Russian summary)* ⋄ H05 ⋄

ROBINSON, A. & ROQUETTE, P. *On the finiteness theorem of Siegel and Mahler concerning diophantine equations*
⋄ C60 H05 H15 H20 ⋄

SHORB, A.M. *Completely additive measure and integration* ⋄ H05 ⋄

WEBB, S.M. *Non-standard probability* ⋄ H05 ⋄

1976

BELLENOT, S.F. *On nonstandard hulls on convex spaces*
⋄ H05 ⋄

BROWN, D.J. & LOEB, P.A. *The values of nonstandard exchange economies* ⋄ H05 H10 ⋄

BUTTON, R.W. *A non-standard characterization of perfect mappings* ⋄ H05 ⋄

BUTTON, R.W. *Monads for regular and normal spaces*
⋄ H05 ⋄

EIFRIG, B. *Ein Nicht-Standard-Beweis fur die Existenz eines Liftings* ⋄ H05 ⋄

FLUM, J. *Non-standard analysis (Spanish)*
⋄ H05 H98 ⋄

HENSON, C.W. *Nonstandard hulls of Banach spaces*
⋄ C65 H05 ⋄

HENSON, C.W. *Ultraproducts of Banach spaces*
 ⋄ C20 C65 H05 ⋄
HIRSCHFELD, J. *Non standard analysis and the compactification of groups* ⋄ C60 H05 H20 ⋄
JOCKUSCH JR., C.G. & SIMPSON, S.G. *A degree theoretic definition of the ramified analytical hierarchy*
 ⋄ D30 D55 E40 E45 H05 ⋄
KEISLER, H.J. *Elementary calculus* ⋄ H05 H98 ⋄
KEISLER, H.J. *Foundations of infinitesimal calculus*
 ⋄ C20 H05 H98 ⋄
KHRISTOV, KH.YA. & TODOROV, T.D. *Asymptotic numbers - algebraic operations with them* ⋄ H05 ⋄
LOEB, P.A. *Applications of nonstandard analysis to ideal boundaries in potential theory* ⋄ H05 ⋄
MACDONALD, A.L. *Sturm-Liouville theory via nonstandard analysis* ⋄ H05 ⋄
MOORE JR., L.C. *Hyperfinite extensions of bounded operators on a Hilbert space* ⋄ H05 ⋄
O'BRIAN, N.R. *Local properties of analytic functions and non-standard analysis* ⋄ H05 ⋄
PATTEN, P.R. *A nonstandard model of the real numbers with applications to limits and continuity* ⋄ H05 ⋄
PURITZ, C.W. *Quasimonad spaces: a nonstandard approach to convergence* ⋄ H05 ⋄
RICHARD, D. *Mesure de Haar en analyse non-standard*
 ⋄ H05 ⋄
RICHTER, M.M. *Ueber die unendlich kleinen Groessen in der Analysis* ⋄ H05 ⋄
RYABTSEV, I.I. *The local definition of a generalized function by means of nonstandard analysis (Russian)*
 ⋄ H05 ⋄
SAITO, M. *Ultraproducts and nonstandard analysis (Japanese)* ⋄ C20 H05 ⋄
SINGER, M.F. *One parameter subgroups and nonstandard analysis* ⋄ H05 ⋄
TAKEUCHI, Y. *Representation of nonstandard numbers by means of hermitian operators (Spanish) (English summary)* ⋄ H05 ⋄
WINKLER, R. *Ueber moegliche Erweiterungen der Analysis und ihre praktische Bedeutung* ⋄ H05 ⋄

1977

BALLANTYNE, A.M. & BLEDSOE, W.W. *Automatic proofs of theorems in analysis using nonstandard techniques*
 ⋄ B35 H05 ⋄
BUTTON, R.W. *When do *continuous extensions exist?*
 ⋄ H05 H20 ⋄
CAPINSKI, M. *Ultraproducts of higher-order models and non-standard analysis* ⋄ H05 ⋄
CHADWICK, J.J.M. & WICKSTEAD, A.W. *A quotient of ultrapowers of Banach spaces and semi-Fredholm operators* ⋄ C20 C65 H05 ⋄
CHADWICK, J.J.M. *Standard biorthogonal systems in the enlargement of a normed space* ⋄ H05 ⋄
CHUDACEK, J. *Topological problems in alternative set theory* ⋄ E70 E75 H05 H20 ⋄
CUDA, K. *The relation between ε-δ procedures and the infinitely small in nonstandard methods* ⋄ H05 ⋄
DAVIS, MARTIN D. *Applied nonstandard analysis*
 ⋄ H05 H98 ⋄

HENSON, C.W. & JOCKUSCH JR., C.G. & RUBEL, L.A. & TAKEUTI, G. *First-order topology*
 ⋄ B25 C65 C75 D35 H05 ⋄
HERRMANN, R.A. *A nonstandard generalization for perfect maps* ⋄ H05 H20 ⋄
HERRMANN, R.A. *Nonstandard quasi-Hausdorff, Urysohn and regular-closed extensions* ⋄ H05 ⋄
ISHIKAWA, S. *The weak law of large numbers by counting probability* ⋄ H05 ⋄
JHA, S.N. *A non-standard treatment of some aspects of two variables function theory* ⋄ H05 ⋄
LUXEMBURG, W.A.J. *Nichtstandard-Zahlsysteme und die Begruendung des Leibnizschen Infinitesimalkalkuels (Slowakisch)* ⋄ H05 ⋄
LUXEMBURG, W.A.J. *Non-standard analysis*
 ⋄ A10 H05 ⋄
MURAKAMI, H. & NAKAGIRI, S. & YEH, CHEHCHIH *Asymptotic behavior of solutions of nonlinear functional equations via nonstandard analysis (Italian summary)*
 ⋄ H05 ⋄
MUSES, C. *Applied hypernumbers. Computational concepts* ⋄ B75 H05 ⋄
NELSON, EDWARD *Internal set theory: A new approach to nonstandard analysis* ⋄ E70 H05 ⋄
REEB, G. & STERN, J. *Seance debat sur l'analyse nonstandard* ⋄ H05 ⋄
SAITO, M. *On the non-standard representation of linear mappings from a function space* ⋄ H05 ⋄
SOCHOR, A. *Differential calculus in the alternative set theory* ⋄ E70 E75 H05 ⋄
STROYAN, K.D. *Infinitesimal analysis of curves and surfaces* ⋄ H05 ⋄
TAKEUCHI, Y. *Construction of nonstandard numbers from the rationals (Spanish)* ⋄ C20 H05 ⋄
WATTENBERG, F. *Nonstandard measure theory-Hausdorff measure* ⋄ H05 ⋄
WATTENBERG, F. *Topologies on the set of closed subsets*
 ⋄ H05 ⋄

1978

ABIAN, A. *Passages between finite and infinite*
 ⋄ A05 H05 ⋄
ANDERSON, ROBERT M. & RASHID, S. *A nonstandard characterization of weak convergence* ⋄ H05 ⋄
BUTTON, R.W. *A note on the Q-topology* ⋄ H05 H20 ⋄
CALLOT, J.-L. & DIENER, F. & DIENER, M. *Le probleme de la "chasse au canard" (English summary)* ⋄ H05 ⋄
CHONG, C.T. *Non-standard analysis* ⋄ H05 ⋄
GIANNONE, A. *An introduction to nonstandard methods via simply additive measures* ⋄ H05 ⋄
HADDAD, L. *Comments on nonstandard topology*
 ⋄ H05 ⋄
HASHIMOTO, N. *Writing a textbook on proof theory using nonstandard analysis (Japanese)* ⋄ H05 ⋄
HERRMANN, R.A. *Perfect maps and remoteness* ⋄ H05 ⋄
HERRMANN, R.A. *The nonstandard theory of semi-uniform spaces* ⋄ C65 H05 H20 ⋄
HRBACEK, K. *Axiomatic foundations for nonstandard analysis* ⋄ E35 E70 H05 H20 ⋄
LAKATOS, I. *Cauchy and the continuum: the significance of non-standard analysis for the history and philosophy of mathematics* ⋄ H05 ⋄

LAUGWITZ, D. *Infinitesimalkalkuel: Kontinuum und Zahlen - eine elementare Einfuehrung in die Nichtstandard-Analysis* ◊ H05 ◊

LI, BANGHE *Nonstandard analysis and multiplication of distributions (Chinese)* ◊ H05 ◊

LIGHTSTONE, A.H. *Mathematical logic. An introduction to model theory. Edited by H. B. Enderton*
◊ B98 C98 H05 ◊

LOLLI, G. *Alcune applicazioni della compattezza*
◊ C20 C60 H05 ◊

MUSES, C. *Hypernumbers. II. Further concepts and computational applications* ◊ A05 H05 ◊

ROUSSEAU, C. *Topos theory and complex analysis*
◊ F50 G30 H05 ◊

SAITO, M. *Introduction of a few set theories within nonstandard analysis (Japanese)* ◊ E70 H05 ◊

SCHMID, J. **-compactifications* ◊ H05 ◊

SLOAN, A. *A note on exponentials of distributions*
◊ H05 ◊

SMIRNOV, S.V. *Approximations of mappings into a group (Russian)* ◊ H05 ◊

STROYAN, K.D. *Infinitesimal calculus on locally convex spaces I: Fundamentals* ◊ H05 ◊

TAKEUCHI, Y. *Representation of non-standard numbers by asymptotic behaviour of real functions (Spanish)*
◊ H05 ◊

TROESCH, A. & URLACHER, E. *Perturbations singulieres et analyse non standard (English summary)* ◊ H05 ◊

TULIPANI, S. *On solutions of algebraic equations whose coefficients are germs of continuous functions*
◊ H05 ◊

WATTENBERG, F. *Nonstandard analysis and the theory of Shape* ◊ H05 ◊

WINKLER, R. *Ein Existenzbeweis fur gewoehnliche Differentialgleichungen mit Methoden der non-standard-Analysis* ◊ H05 ◊

1979

BELYAKIN, N.V. *Nonstandardly finite sets (Russian)*
◊ E30 H05 H15 ◊

BUTTON, R.W. *When do two topologies have the same monads?* ◊ C20 H05 ◊

COLETTI, G. & REGOLI, G. & VINCENTI, R. *Algebraic and topological structure of nonstandard models of the reals*
◊ H05 H20 ◊

DIENER, F. *Famille d'equations a cycle limite unique (English summary)* ◊ H05 ◊

FENSTAD, J.E. *On the metaphysics of the real line*
◊ A05 H05 ◊

GONSHOR, H. *An application of nonstandard analysis to category theory* ◊ H05 ◊

GRAINGER, A.D. *Finite points of filters in infinite-dimensional vector spaces* ◊ H05 ◊

HENLE, J.M. & KLEINBERG, E.M. *Infinitesimal calculus*
◊ H05 ◊

HENSON, C.W. *Analytic sets, Baire sets and the standard part map* ◊ H05 ◊

HENSON, C.W. *Unbounded Loeb measures*
◊ H05 H10 ◊

HERRMANN, R.A. *A nonstandard approach to S-closed spaces* ◊ H05 ◊

HERRMANN, R.A. *Convergence spaces and nonstandard compactifications* ◊ H05 ◊

HERRMANN, R.A. *Nonstandard implication algebras*
◊ G10 G25 H05 H20 ◊

HERRMANN, R.A. *Point monads and p-closed spaces*
◊ H05 H20 ◊

HRBACEK, K. *Nonstandard set theory*
◊ E70 H05 H20 ◊

HUANG, CHENGGUI *Two-phase calculus (Chinese) (English summary)* ◊ H05 H10 ◊

KAWAI, T. *An axiom system for nonstandard set theory*
◊ E70 H05 H20 ◊

KOMKOV, V. *A note on a formal manipulation of divergent series and integrals* ◊ H05 ◊

LAUGWITZ, D. *A nonstandard approach to distributions and operational calculus* ◊ H05 ◊

LI, BANGHE *Differential and integral calculus on a non-Archimedean field (Chinese) (English summary)*
◊ H05 H10 ◊

LOEB, P.A. *An introduction to nonstandard analysis and hyperfinite probability theory* ◊ H05 ◊

LOEB, P.A. *Weak limits of measures and the standard part map* ◊ H05 ◊

MLCEK, J. *Valuations of structures* ◊ E70 H05 ◊

NEPEJVODA, N.N. *Some remarks about constructive non-standard analysis (Russian)* ◊ F99 H05 ◊

RASKOVIC, M. *On existence of expansion of a complex function* ◊ H05 ◊

REEB, G. *Equations differentielles et analyse non classique (d'apres J.-L. Callot (Oran))* ◊ H05 ◊

ROBINSON, A. *Selected papers of Abraham Robinson. Vol.II: Nonstandard analysis and philosophy*
◊ A05 B96 C96 H05 H10 H96 ◊

ROUSSEAU, C. *Topos theory and complex analysis*
◊ F50 F55 G30 H05 ◊

RUDIN, M.E. & SHELAH, S. *Unordered types of ultrafilters*
◊ E05 E75 H05 ◊

SOCHOR, A. *Some remarks to the connection between the alternative set theory and nonstandard methods*
◊ E70 H05 ◊

TACON, D.G. *Two characterizations of power compact operators* ◊ H05 ◊

TALL, D. *The calculus of Leibniz - an alternative modern approach* ◊ A05 H05 ◊

WANG, SHUTANG *A generalized number system and its application I (Chinese) (English summary)* ◊ H05 ◊

WATTENBERG, F. *Nonstandard measure theory - avoiding pathological sets* ◊ H05 ◊

YANG, ANZHOU *Problem of cardinal number and other problems of non-standard model *R (Chinese)*
◊ E10 H05 ◊

YOHE, J. *Implementing nonstandard arithmetics*
◊ H05 H20 ◊

1980

BANTEA, R. *Study of the completion of linear topological spaces by nonstandard methods* ◊ H05 ◊

BARONE, E. & GIANNONE, A. & SCOZZAFAVA, R. *On some aspects of the theory and applications of finitely additive probability measures* ◊ C20 H05 ◊

BECK, JON M. *On the relationship between algebra and analysis* ◊ H05 ◊

BELLENOT, S.F. *Basic sequences in non-Schwartz Frechet spaces* ◊ H05 ◊

BIRKELAND, B. *A singular Sturm-Liouville problem treated by nonstandard analysis* ◊ H05 ◊

CSIRMAZ, L. *Structure of program runs of nonstandard time* ◊ H05 ◊

CUDA, K. *An elimination of infinitely small quantities and infinitely large numbers (within the framework of AST)* ◊ E70 H05 H15 ◊

ESTERLE, J. *Homomorphismes discontinus des algebres de Banach commutatives separables* ◊ H05 ◊

GRAINGER, A.D. *On the nonstandard duality theory of locally convex spaces* ◊ H05 ◊

HEINRICH, S. *The isomorphic problem of envelopes* ◊ C20 C65 H05 ◊

HEINRICH, S. *Ultraproducts in Banach space theory* ◊ C20 C65 C98 H05 ◊

HERRMANN, R.A. *A nonstandard approach to pseudotopological compactifications* ◊ H05 ◊

HU, SHIGENG *The general form of separation axiom (Chinese) (English summary)* ◊ H05 ◊

HUANG, CHENGGUI & SHI, ZUIJIAN *The problem of singularity of the δ-function (Chinese English summary)* ◊ H05 ◊

JANZ, A. *Eine neue Variante der Nichtstandart-Analysis und einige ihrer Anwendungen in der allgemeinen Topologie* ◊ H05 ◊

JANZ, A. *Konstruktion von Kompaktifizierungen topologischer Raeume mit Hilfe von Nichtstandard-Modellen* ◊ H05 H20 ◊

JANZ, A. *Zwei Nichtstandard-Metrisationstheoreme* ◊ H05 H20 ◊

KAKUDA, Y. *Non-standard analysis in boolean-valued models* ◊ E40 H05 ◊

KOMKOV, V. *Asymptotic behavior of nonlinear differential equations via nonstandard analysis III: Boundedness and monotone behavior of the equation $(a(t)\varphi(x)x')' + c(t)f(x) = q(t)$* ◊ H05 ◊

LI, BANGHE *Integral mean value theorems on the field of formal power series (Chinese)* ◊ H05 H10 ◊

LI, BANGHE *The differential and integral calculus on the field of formal power series (Chinese)* ◊ H05 H10 ◊

LIU, SHICHAO *A proof-theoretic approach to nonstandard analysis with emphasis on distinguishing between constructive and nonconstructive results* ◊ E70 F99 H05 H20 ◊

MCKEE, T.A. *Monadic characterizations in nonstandard topology* ◊ C10 H05 H20 ◊

MOORE, S.M. *Nonstandard analysis and generalized functions* ◊ H05 ◊

MOORE JR., L.C. *Hyperfinite-dimensional subspaces of the nonstandard hull of c_0* ◊ H05 ◊

PLA I CARRERA, J. *On Keisler's axiomatization of nonstandard analysis (Catalan) (English summary)* ◊ H05 ◊

TACON, D.G. *Nonstandard extensions of transformations between Banach spaces* ◊ H05 ◊

TALL, D. *Looking at graphs through infinitesimal microscopes, windows and telescopes* ◊ A10 H05 ◊

TODOROV, T.D. *Asymptotic numbers. I: Algebraic properties. II Order relations, infinitesimals and interval topology (Russian summaries)* ◊ H05 H20 ◊

TZOUVARAS, A.D. *A non-standard characterization of the norm of free ultrafilters* ◊ C55 E05 H05 ◊

URSINI, A. *Dai numeri razionali ai numeri iperreali* ◊ H05 ◊

1981

BANTEA, R. *On the extension of vector measures by nonstandard methods (Romanian) (English summary)* ◊ H05 ◊

BENOIT, E. & CALLOT, J.-L. *Chasse au canard I,IV* ◊ H05 ◊

BENOIT, E. *Equations differentielles: relation entree-sortie (English summary)* ◊ H05 ◊

BENOIT, E. *Tunnels et entonnoirs (English summary)* ◊ H05 ◊

BERG VAN DEN, I. & DIENER, M. *Diverses applications du lemme de Robinson en analyse non standard (English summary)* ◊ H05 H10 ◊

BERG VAN DEN, I. & DIENER, M. *Halos et galaxies: une extension du lemme de Robinson (English summary)* ◊ H05 ◊

DAMYANOV, B.P. & KHRISTOV, KH.YA. *Changing of the independent variable in the theory of asymptotic functions* ◊ H05 ◊

GHITA, A. *A class of fields of hyperreal numbers* ◊ H05 ◊

GOZE, M. & LUTZ, R. *Nonstandard analysis. A pratical guide with applications* ◊ B98 H05 ◊

HEINRICH, S. *Ultraproducts of L_1-predual spaces* ◊ C20 C65 H05 ◊

HENSON, C.W. & WATTENBERG, F. *Egoroff's theorem and the distribution of standard points in a nonstandard model* ◊ H05 ◊

JI, ZHERUI *On the axiomatics of hyperreal number I (Chinese) (English summary)* ◊ H05 ◊

JONES, C.K.R.T. & KELEMEN, P.J. *The ϱ-calculus* ◊ H05 ◊

KAWAI, T. *Axiom systems of nonstandard set theory* ◊ E70 H05 H20 ◊

KAWAI, T. *Nonstandardization of Feferman's set theory and a conservation theorem (Japanese)* ◊ E35 E70 H05 H20 ◊

LAUGWITZ, D. *Verallgemeinerte Grenzwerte beschraenkter Zahlfolgen* ◊ H05 ◊

LUTZ, R. & SARI, T. *Sur le comportement asymptotique des solutions dans un probleme aux limites non lineaires (English summary)* ◊ H05 ◊

MAZZANTI, G. & MIROLLI, M. *Loeb operators and interior operators* ◊ H05 ◊

MEJLBO, L.C. *On nonstandard analysis* ◊ H05 ◊

MOORE JR., L.C. *Approximately finite-dimensional Banach spaces* ◊ H05 ◊

RUBIO DE FRANCIA, J.L. *Contribuciones al analisis funcional no-standard* ◊ H05 ◊

SARI, T. *Sur le comportement asymptotique des solutions dans un probleme aux limites semi-lineaire (English summary)* ◊ H05 ◊

SOLON, B.YA. *PC-degrees inside an e-degree of a hyperimmune retraceable set (Russian)*
⋄ D25 D30 D50 H05 ⋄

TODOROV, T.D. *Asymptotic functions and the problem of multiplication of distributions* ⋄ H05 ⋄

TODOROV, T.D. *Extended asymptotic functions - some examples (Russian summary)* ⋄ H05 ⋄

USPENSKIJ, V.A. *Nonstandard analysis (Bulgarian)*
⋄ H05 H98 ⋄

VESLEY, R.E. *An intuitionistic infinitesimal calculus*
⋄ H05 ⋄

WALLET, G. *Holonomie et cycle evanouissant* ⋄ H05 ⋄

1982

BENOIT, E. *Les canards de R^3* ⋄ H05 ⋄

CARBONI, A. *Analisi non-standard e topos*
⋄ H05 H10 ⋄

FISHER, D. *Extending functions to infinitesimals of finite order* ⋄ H05 ⋄

HATCHER, W.S. *Clone embeddings and the hyperreals*
⋄ H05 ⋄

HEINRICH, S. & MANKIEWICZ, P. *Applications of ultrapowers to the uniform and Lipschitz classification of Banach spaces* ⋄ C20 C65 H05 ⋄

HEINRICH, S. *The isomorphic problem of envelopes*
⋄ C20 C65 H05 ⋄

HELMS, L.L. & LOEB, P.A. *A nonstandard proof of the martingale convergence theorem* ⋄ H05 H10 ⋄

HINTIKKA, K.J.J. & HINTIKKA, M.P. *Sherlock Holmes confronts modern logic: toward a theory of information-seeking through questioning*
⋄ A05 H05 ⋄

HOSKINS, R.F. *Infinitesimals, nonstandard analysis and generalised functions* ⋄ H05 ⋄

HOSKINS, R.F. *Superreals and superfunctions* ⋄ H05 ⋄

INGLETON, A.W. *An introduction to nonstandard analysis*
⋄ H05 ⋄

LEVITZ, H. *Calculation of an order type: an application of nonstandard methods* ⋄ H05 ⋄

LINDSTROEM, T.L. *A Loeb-measure approach to theorems by Prohorov, Sazonov and Gross* ⋄ B50 H05 ⋄

PECORA, L.M. *A nonstandard infinite-dimensional vector space approach to Gaussian functional measures*
⋄ H05 ⋄

PRIDA, J.F. *A non-standard study of the theory of relative recursivity (Spanish)* ⋄ D20 D30 H05 ⋄

RICHTER, M.M. *Ideale Punkte, Monaden und Nichtstandard-Methoden*
⋄ C20 C60 E70 H05 H98 ⋄

ROGERS, L. *Infinitesimal, continuity and meaning*
⋄ H05 ⋄

ROITMAN, J. *Non-isomorphic hyper-real fields from non-isomorphic ultrapowers* ⋄ C20 H05 H20 ⋄

SIMS, B. *"Ultra"-techniques in Banach space theory*
⋄ H05 ⋄

TALL, D. *Elementary axioms and pictures for infinitesimal calculus* ⋄ H05 ⋄

TROESCH, A. *Etude macroscopique de l'equation de van der Pol* ⋄ H05 ⋄

XU, LIZHI & YUAN, XIANGWAN & ZHENG, YUXIN & ZHU, WUJIA *Antinomies and the foundational problem of mathematics I (Chinese) (English summary)*
⋄ A05 H05 ⋄

1983

BANKSTON, P. *Coarse topologies in nonstandard extensions via separative ultrafilters*
⋄ E50 E75 H05 ⋄

BENOIT, E. *Canards et chaos dans R^3* ⋄ H05 ⋄

BENOIT, E. *Systemes lents-rapides dans $\mathbf{R}^3$ et leurs canards* ⋄ H05 ⋄

BERG VAN DEN, I. *Un point de vue nonstandard sur les developpements en serie de Taylor* ⋄ H05 ⋄

BERG VAN DEN, I. *Un principe de permanence general*
⋄ E70 H05 ⋄

BERGER, M. & SLOAN, A. *Explicit solutions of partial differential equations* ⋄ H05 ⋄

CALLOT, J.-L. & SARI, T. *Stroboscopie infinitesimale et moyennisation dans les systemes d'equations differentielles a solution rapidement oscillantes*
⋄ H05 ⋄

CHRISTIAN, C.C. *Der Beitrag Goedels fuer die Rechtfertigung der Leibnizschen Idee von den Infinitesimalien* ⋄ A05 A10 H05 ⋄

CRISMA, L. & HOLZER, S. *Starconcepts* ⋄ H05 ⋄

CUDA, K. & KUSSOVA, B. *Monads in basic equivalences*
⋄ E70 H05 ⋄

DEVITO, C.L. *Compactlike operators on locally convex spaces* ⋄ H05 ⋄

DIENER, M. *Canard et bifurcations* ⋄ H05 ⋄

GOL'DSHTEJN, B.G. *Decomposition of logics (Russian)*
⋄ C95 H05 ⋄

HARTHONG, J. *Elements pour une theorie du continu*
⋄ H05 ⋄

HATCHER, W.S. & LAFLAMME, C. *On the order structure of the hyperreal line* ⋄ H05 ⋄

HENSON, C.W. & MOORE JR., L.C. *Nonstandard analysis and the theory of Banach spaces* ⋄ C65 H05 ⋄

HOOVER, D.N. & PERKINS, E. *Nonstandard construction of the stochastic integral and applications to stochastic differential equations I,II* ⋄ H05 ⋄

HURD, A.E. (ED.) *Nonstandard analysis - recent developments* ⋄ H05 ⋄

KAWAI, T. *Nonstandard analysis by axiomatic method*
⋄ H05 ⋄

KOSCIUK, S.A. *Stochastic solutions to partial differential equations* ⋄ H05 ⋄

LAUGWITZ, D. *Ω-calculus as a generalization of field extension; an alternative approach to nonstandard analysis* ⋄ H05 ⋄

LAUGWITZ, D. *Nichtstandard-Mathematik, begruendet durch eine Verallgemeinerung der Koerpererweiterung*
⋄ A10 H05 ⋄

LUTZ, R. *L'intrusion de l'analyse nonstandard dans l'etude des perturbations singulieres* ⋄ H05 ⋄

LYANTSE, V.E. *Can nonstandard analysis be ignored? (Jordan form of an operator in an infinite-dimensional space) (Russian)* ⋄ H05 ⋄

MAGAJNA, B. *Infinitesimals (Slovenian) (English summary)* ⋄ C20 H05 ⋄

MAGIDOR, M. & SHELAH, S. & STAVI, J. *On the standard part of nonstandard models of set theory*
⋄ C62 E45 H05 ⋄

NORMANN, D. *Characterizing the continuous functionals*
⋄ D65 H05 ⋄

POTTHOFF, K. *Quelques applications des methodes non-standard a la theorie des groupes*
⋄ C60 H05 H20 ⋄

STROYAN, K.D. *Infinitesimal analysis of l^∞ in its Mackey topology* ⋄ H05 H10 ⋄

STROYAN, K.D. *Locally convex infinitesimal calculus II. Computations on Mackey (l^∞)* ⋄ H05 H10 ⋄

TACON, D.G. *Generalized semi-Fredholm transformations*
⋄ H05 ⋄

XU, LIZHI *Generalized Moebius-Rota inversion theory associated with nonstandard analysis (Chinese summary)* ⋄ H05 ⋄

YASUMOTO, M. *Nonstandard arithmetic of function fields over H-convex subfields of $^*Q$ (Japanese)*
⋄ H05 H15 ⋄

YASUMOTO, M. *Nonstandard arithmetic of function fields over H-convex subfields of $^*Q$* ⋄ H05 H15 ⋄

ZIVALJEVIC, R. *The notions of w-net and Y-compact space viewed under infinitesimal microscope* ⋄ H05 ⋄

1984

BEHRENS, M.F. *Interpolation and Gleason parts in L-domains* ⋄ H05 ⋄

BENNINGHOFEN, B. *Superinfinitesimals and the calculus of the generalized Rieman integral* ⋄ H05 ⋄

BESHENKOV, S.A. *Formal consistency as a positive principle of non-standard arithmetic (Russian)* ⋄ H05 ⋄

BUFF, H.W. *ω-Konservativitaet der Nonstandardmengenlehre von Nelson bezueglich ZF+Kompaktheitssatz* ⋄ E25 E35 E70 H05 ⋄

CUDA, K. *Translation of nonstandard definitions to standard ones* ⋄ H05 ⋄

CUTLAND, N.J. *A question of Borel hyperdeterminacy*
⋄ E60 H05 ⋄

DIENER, M. *The canard unchained or how fast/slow dynamical systems bifurcate* ⋄ H05 ⋄

DOEPP, K. *Filterkonvergenz in der Nichtstandard-Analysis*
⋄ H05 ⋄

DOEPP, K. *Filterkonvergenz in der Nichtstandard-Analysis bei nichtelementaren Funktionen* ⋄ H05 ⋄

FACENDA AGUIRRE, J.A. *(HM)-spaces and measurable cardinals* ⋄ E55 E75 H05 ⋄

FARKAS, E.J. & SZABO, M.E. *On the plausibility of nonstandard proofs in analysis* ⋄ H05 ⋄

FERRO, R. *Una nota sulla nozione di molto maggiore (English summary)* ⋄ H05 ⋄

GOODYEAR, P. *Double enlargements of topological spaces*
⋄ H05 ⋄

HENLE, J.M. *Tangent planes with infinitesimals* ⋄ H05 ⋄

HENSON, C.W. & KAUFMANN, M. & KEISLER, H.J. *The strength of nonstandard methods in arithmetic*
⋄ C62 E30 F30 F35 H05 H15 ⋄

HORT, C. & OSSWALD, H. *On nonstandard models in higher order logic*
⋄ B15 C55 C85 E55 H05 H20 ⋄

KAKUDA, Y. *Nonstandard analysis without nonstandard models* ⋄ H05 ⋄

KEISLER, H.J. *An infinitesimal approach to stochastic analysis* ⋄ H05 ⋄

KOMKOV, V. & MCLAUGHLIN, T.G. *Local analysis of nonstandard C^∞ functions of predistributional type*
⋄ H05 ⋄

LIU, SHICHAO *A proof-theoretic approach to nonstandard analysis (continued)* ⋄ E70 F50 H05 ⋄

LOEB, P.A. *A functional approach to nonstandard measure theory* ⋄ H05 ⋄

MILLER, H.I. & ZIVALJEVIC, B. *Remarks on the zero-one law (Russian summary)* ⋄ H05 ⋄

NEUBRUNN, T. & RIECANOVA, Z. *Elementary nonstandard approach to metric spaces* ⋄ H05 ⋄

RICHTER, M.M. *Some aspects of nonstandard methods in general algebra* ⋄ C05 C30 H05 ⋄

ROBERT, A. *L'analyse non standard* ⋄ H05 ⋄

ROBERT, A. *Une approche naive de l'analyse non-standard*
⋄ H05 ⋄

TROESCH, A. *Etude macroscopique de systemes differentiels* ⋄ H05 ⋄

1985

BARBANCON, G. *On the analytical and approximate solutions of $\in y'' = yy'$* ⋄ H05 ⋄

BENIS-SINACEUR, H. *La theorie d'Artin et Schreier et l'analyse non-standard d'Abraham Robinson*
⋄ H05 ⋄

BERTOSSI, L. & CHUAQUI, R.B. *Approximation to truth and theory of errors* ⋄ C90 H05 ⋄

CHEN, GUANGYI *Toeplitz generalized summation on the field of hyperreal numbers (Chinese) (English summary)*
⋄ H05 ⋄

FENSTAD, J.E. *Is nonstandard analysis relevant for the philosophy of mathematics?* ⋄ A05 H05 ⋄

GONSHOR, H. *Remarks on the Dedekind completion of a nonstandard model of the reals* ⋄ H05 ⋄

HATCHER, W.S. *Elementary extension and the hyperreal numbers* ⋄ C65 H05 ⋄

HERRMANN, R.A. *Supernear functions* ⋄ H05 ⋄

HURD, A.E. & LOEB, P.A. *An introduction to nonstandard real analysis* ⋄ H05 ⋄

KUTATELADZE, S.S. *Nonstandard analysis of tangent cones (Russian)* ⋄ H05 ⋄

LEWIS, A.A. *Hyperfinite von Neumann games* ⋄ H05 ⋄

LEWIS, A.A. *Loeb-measurable solutions to *finite games*
⋄ H05 ⋄

LIN, PEIKEE *Unconditional bases and fixed points of nonexpansive mappings* ⋄ C20 C65 H05 ⋄

LOEB, P.A. *A nonstandard functional approach to Fubini's theorem* ⋄ C60 C65 H05 H20 ⋄

LYUBETSKIJ, V.A. *Some algebraic questions of nonstandard analysis (Russian)* ⋄ C60 C90 H05 ⋄

MYCIELSKI, J. *Sullivan's lamination of a planar region*
⋄ H05 ⋄

RASKOVIC, M. *An application of nonstandard analysis to functional equations* ⋄ H05 ⋄

ROBERT, A. *Analyse non standard* ⋄ B98 H05 ⋄

SAMBORSKIJ, S.N. *Limit trajectories of singulary pertubed differential equations (Russian) (English summary)* ◊ H05 ◊

THOMPSON, C. *An introduction to nonstandard analysis* ◊ H05 ◊

ZHANG, XIANG *An initial study of the use of hypernets to construct nonstandard models (Chinese)* ◊ H05 ◊

ZIVALJEVIC, R. *Loeb completion of internal vector-valued measures* ◊ H05 ◊

H10 Other applications of infinitesimal analysis

1954
MCNAUGHTON, R. *A non-standard truth definition*
⋄ B28 C62 E30 H10 H20 ⋄

1962
TAKEUTI, G. *Dirac space* ⋄ H10 ⋄

1966
ROBINSON, A. *Non-standard analysis*
⋄ A10 B98 H05 H10 H15 H98 ⋄

1967
ASENJO, F.G. *Rings of term-relation numbers as non-standard models* ⋄ H10 H20 ⋄
BUNYATOV, M.R. *Dirac spaces (Russian)*
⋄ C20 C60 H05 H10 ⋄

1968
GEISER, J.R. *Nonstandard logic* ⋄ C20 H10 H15 ⋄
HURD, A.E. & ROBINSON, A. *On flexural wave propagation on nonhomogeneous elastic plates*
⋄ H10 ⋄

1969
ANONYMOUS *Les activites du centre national de recherche de logique en 1968* ⋄ A10 H10 ⋄

1971
HURD, A.E. *Local conditions for equivalence of compact dynamical systems* ⋄ H10 ⋄
HURD, A.E. *Nonstandard analysis of dynamical systems. I: Limit motions, stability* ⋄ H10 ⋄
WESLEY, E. *An application of non-standard analysis to game theory* ⋄ H10 ⋄

1972
BROWN, D.J. & ROBINSON, A. *A limit theorem on the cores of large standard exchange economies* ⋄ H10 ⋄
KELEMEN, P.J. & ROBINSON, A. *The nonstandard* λ: $\varphi_2^4(x)$: *model. I: The technique of nonstandard analysis in theoretical physics. II: The standard model from a nonstandard point of view* ⋄ H05 H10 ⋄

1973
EIFRIG, B. *Zur Existenz eines Spielwertes bei Spielen auf kompakten Raeumen mit stetiger Auszahlungsfunktion*
⋄ H05 H10 ⋄

1974
ALAGIC, M. *A monadic approach to k-spaces* ⋄ H10 ⋄
GEISER, J.R. *A formalization of Essenin-Volpin's proof theoretical studies by means of nonstandard analysis*
⋄ C75 F50 H10 H15 ⋄
HURD, A.E. *Nonstandard dynamical systems* ⋄ H10 ⋄

KATZ, JOSE & THURBER, J.K. *Applications of fractional powers of delta functions* ⋄ H10 ⋄
KELEMEN, P.J. *Quantum mechanics, quantum field theory, hyper-quantum mechanics* ⋄ H10 ⋄

1975
FARRUKH, M.O. *Application of nonstandard analysis to quantum mechanics* ⋄ H10 ⋄

1976
ANDERSON, ROBERT M. *A non-standard representation for Brownian motion and Ito integration* ⋄ H10 ⋄
ANDERSON, ROBERT M. *A nonstandard representation for Brownian motion and Ito integration* ⋄ H10 ⋄
BROWN, D.J. & LOEB, P.A. *The values of nonstandard exchange economies* ⋄ H05 H10 ⋄
KHAN, M.A. *Oligopoly in markets with a continuum of traders: an asymptotic interpretation* ⋄ H10 ⋄

1977
KHAN, M.A. *Some remarks on sets with unbounded non-convexities* ⋄ H10 ⋄

1978
BLANCHARD, P. & TARSKI, J. *Renormalizable interactions in two dimensions and sharp-time fields* ⋄ H10 ⋄
DOMOTOR, Z. *Axiomatization of Jeffrey utilities*
⋄ H10 ⋄
JHA, M.N. *Generalised metric via ultra-real numbers*
⋄ H10 ⋄
KELLER, J.P. *The intermittent server* ⋄ H10 ⋄
RANTALA, V. *Correspondence and non-standard models: A case study* ⋄ A05 H10 ⋄
RASHID, S. *Existence of equilibrium in infinite economies with production* ⋄ H10 ⋄
SHELAH, S. & STERN, J. *The Hanf number of the first order theory of Banach spaces* ⋄ C55 C65 H10 ⋄
TARSKI, J. *Short introduction to nonstandard analysis and its physical applications* ⋄ H10 ⋄

1979
ALBEVERIO, S. & FENSTAD, J.E. & HOEEGH-KROHN, R. *Singular perturbations and nonstandard analysis* ⋄ H10 ⋄
BOPP, F. *Ueber den Zustandsraum der Quantenphysik*
⋄ H10 ⋄
HELMS, L.L. & LOEB, P.A. *Applications of nonstandard analysis to spin models* ⋄ H10 ⋄
HENSON, C.W. *Unbounded Loeb measures*
⋄ H05 H10 ⋄
HUANG, CHENGGUI *Two-phase calculus (Chinese) (English summary)* ⋄ H05 H10 ⋄
KAWAI, T. *An application of nonstandard analysis to characters of groups of continuous functions*
⋄ C65 H10 ⋄

LI, BANGHE *Differential and integral calculus on a non-Archimedean field (Chinese) (English summary)* ⋄ H05 H10 ⋄

ROBINSON, A. *Selected papers of Abraham Robinson. Vol.II: Nonstandard analysis and philosophy* ⋄ A05 B96 C96 H05 H10 H96 ⋄

WOZNIAK, C. *Non-standard analysis and its application to mechanics* ⋄ H10 ⋄

1980

BROWN, D.J. & KHAN, M.A. *An extension of the Brown-Robinson equivalence theorem* ⋄ H10 ⋄

FENSTAD, J.E. *Nonstandard methods in stochastic analysis and mathematical physics* ⋄ H10 ⋄

HARTHONG, J. *Le moire (English summary)* ⋄ H10 ⋄

LEROY, S.F. *Entry and equilibrium under adjustment costs* ⋄ H10 ⋄

LI, BANGHE *Integral mean value theorems on the field of formal power series (Chinese)* ⋄ H05 H10 ⋄

LI, BANGHE *The differential and integral calculus on the field of formal power series (Chinese)* ⋄ H05 H10 ⋄

LI, YACHING *Some results on products of distributions (Chinese)* ⋄ H10 ⋄

LINDSTROEM, T.L. *Hyperfinite stochastic integration. I: The nonstandard theory. II: Comparison with the standard-theory. III: Hyperfinite representations of standard martingales* ⋄ H10 ⋄

LUE, QICI *Concurrent relations and nonstandard models (Chinese) (English summary)* ⋄ H10 ⋄

MOORE, S.M. *Stochastic fields from stochastic mechanics* ⋄ H10 ⋄

WOZNIAK, C. *Nonstandard analysis and material systems in mechanics. I,II* ⋄ H10 ⋄

1981

ARKERYD, L. *Intermolecular forces of infinite range and the Boltzmann equation* ⋄ H10 ⋄

BANTEA, R. *Nonstandard vector measures (Romanian) (English summary)* ⋄ H10 ⋄

BERG VAN DEN, I. & DIENER, M. *Diverses applications du lemme de Robinson en analyse non standard (English summary)* ⋄ H05 H10 ⋄

CSIRMAZ, L. *Nonstandard runs and program verification* ⋄ B75 H10 ⋄

FRANCIS, C.E. *Applications of nonstandard analysis to relativistic quantum mechanics. I* ⋄ H10 ⋄

GORDON, E.I. & LYUBETSKIJ, V.A. *Some applications of nonstandard analysis in the theory of Boolean-valued measures (Russian)* ⋄ H10 ⋄

HARTHONG, J. *La propagation des ondes (English summary)* ⋄ H10 ⋄

HARTHONG, J. *Le moire* ⋄ H10 ⋄

HERRMANN, R.A. *Rigorous infinitesimal modelling* ⋄ H10 ⋄

HURD, A.E. *Nonstandard analysis and lattice statistical mechanics: A variational principle* ⋄ H10 ⋄

LI, BANGHE & LI, YACHING *On multiplication of distributions (Chinese) (English summary)* ⋄ H10 ⋄

MARAVALL CASESNOVES, D. *Attempt at a theory of the infinitely large and the infinitely small, certain and random, and the laws of large numbers (Spanish) (English summary)* ⋄ H10 ⋄

MAZZANTI, S. & SCHMITT, B.V. *Solutions periodiques symetriques de l'equation de Duffing sans dissipation* ⋄ H10 ⋄

PERKINS, E. *A global intrinsic characterization of Brownian local time* ⋄ H10 ⋄

SLOAN, A. *The strong convergence of Schroedinger propagators* ⋄ H10 ⋄

1982

ALBEVERIO, S. & BLANCHARD, P. & HOEEGH-KROHN, R. *Some applications of functional integration* ⋄ H10 ⋄

ANDERSON, ROBERT M. *Star-finite representations of measures spaces* ⋄ H10 ⋄

ANDREKA, H. & NEMETI, I. & SAIN, I. *A complete logic for reasoning about programs via nonstandard model theory. I,II* ⋄ B75 C90 H10 H20 ⋄

ARKERYD, L. *Asymptotic behaviour of the Boltzmann equation with infinite range forces* ⋄ H10 ⋄

BAYOD, J.M. *Spherical completeness with infinitesimals* ⋄ H10 ⋄

CARBONI, A. *Analisi non-standard e topos* ⋄ H05 H10 ⋄

CARTIER, P. *Perturbations singulieres des equations differentielles ordinaires et analyse non-standard* ⋄ H10 ⋄

CSIRMAZ, L. *Nonstandard program runnings and correctness of programs* ⋄ B75 H10 ⋄

CUTLAND, N.J. *Infinitesimal methods in measure theory, probability theory and stochastic analysis* ⋄ H10 ⋄

CUTLAND, N.J. *On the existence of solutions to stochastic differential equations on Loeb spaces* ⋄ H10 ⋄

CUTLAND, N.J. *Optimal controls for partially observed stochastic systems using nonstandard analysis* ⋄ H10 ⋄

DYRE, J.C. *Nonstandard characterizations of ideals in $C(X)$* ⋄ H10 ⋄

GIANNONE, A. *Probabilita non σ-additive e analisi non-standard (English summary)* ⋄ C20 H10 ⋄

HELMS, L.L. & LOEB, P.A. *A nonstandard proof of the martingale convergence theorem* ⋄ H05 H10 ⋄

KAMAE, T. *A simple proof of the ergodic theorem using nonstandard analysis* ⋄ H10 ⋄

LUTZ, R. & SARI, T. *Applications of nonstandard analysis to boundary value problems in singular perturbation theory* ⋄ H10 ⋄

MOORE, S.M. *Nonstandard analysis applied to path integrals and generalized functions (Italian and Russian summaries)* ⋄ C65 H10 H20 ⋄

NOBIS, K. & WOZNIAK, C. *Nonstandard analysis and balance equations in the theory of porous media (Russian summary)* ⋄ H10 ⋄

PERKINS, E. *On the construction and distribution of a local martingale with a given absolute value* ⋄ H10 ⋄

PERKINS, E. *Weak invariance principles for local time* ⋄ H10 ⋄

REEB, G. *Analyse non standard et theorie du moire* ⋄ H10 ⋄

RODENHAUSEN, H. *A characterization of nonstandard liftings of measurable functions and stochastic processes* ⋄ H10 ⋄

WIERZBICKI, E. & WOZNIAK, C. *On the formation of implicit constraints and free-boundary problems for elastodynamics by the nonstandard analysis technique (Russian summary)* ◊ H10 ◊

WOZNIAK, C. *On the nonstandard interrelation between mass-point mechanics and continuum mechanics (Russian summary)* ◊ H10 ◊

WOZNIAK, C. *On the nonstandard model of the theory of elasticity (Russian summary)* ◊ H10 ◊

ZIVALJEVIC, R. *A Loeb measure approach to the Riesz representation theorem* ◊ H10 ◊

1983

BASARAB, S.A. *Roth's theorem: non-standard aspects (Romanian)* ◊ H10 H15 ◊

CERRUTI, U. *"Nonstandard" concepts in fuzzy topology* ◊ E72 E75 H10 ◊

CHUAQUI, R.B. & MALITZ, J. *Preorderings compatible with probability measures* ◊ E75 H10 ◊

CUTLAND, N.J. *Internal controls and relaxed controls* ◊ H10 ◊

CUTLAND, N.J. *Nonstandard measure theory and its applications* ◊ H10 H98 ◊

CUTLAND, N.J. *Optimal controls for partially observed stochastic systems: an infinitesimal approach* ◊ H10 ◊

ECKHAUS, W. *Relaxation oscillations including a standard chase on French ducks* ◊ H10 ◊

FREY, G. *Nonstandard arithmetic and application to height functions* ◊ C60 H10 H15 H20 ◊

GOZE, M. *Etude locale des courbes algebriques planes* ◊ H10 ◊

HEINRICH, S. *Ultrapowers of locally convex spaces and applications I* ◊ C20 C65 H10 ◊

HEINRICH, S. *Ultrapowers of locally convex spaces and applications II* ◊ C20 C65 H10 ◊

HELMS, L.L. *Hyperfinite spin models* ◊ H10 ◊

KAYUNOV, O.N. *Nonstandard solutions of equations of mathematical physics (Russian)* ◊ H10 ◊

LAMBALGEN VAN, M. *Quantum set theory* ◊ E70 G12 H10 ◊

MURAKAMI, H. & NAKAGIRI, S. & YEH, CHEHCHIH *Asymptotic behavior of solutions of nonlinear differential equations with deviating arguments via nonstandard analysis* ◊ H10 ◊

PERKINS, E. *Stochastic processes and nonstandard analysis* ◊ H10 ◊

RICHTER, M.M. & SZABO, M.E. *Towards a nonstandard analysis of programs* ◊ B75 H10 ◊

RICHTER, M.M. *Variationen des Endlichkeitsbegriffes und Perspektiven fuer moegliche Anwendungen* ◊ E75 H10 ◊

SAIN, I. *Total correctness in nonstandard dynamic logic* ◊ B75 H10 ◊

STROYAN, K.D. *Infinitesimal analysis of l^∞ in its Mackey topology* ◊ H05 H10 ◊

STROYAN, K.D. *Locally convex infinitesimal calculus II. Computations on Mackey (l^∞)* ◊ H05 H10 ◊

TAKEUCHI, Y. *Nonstandard functions and distribution theory (Spanish) (English summary)* ◊ H10 ◊

WOZNIAK, C. *Nonstandard analysis in Newtonian mechanics of a particle (Polish) (English and Russian summaries)* ◊ H10 ◊

WOZNIAK, C. *On the nonstandard analysis and the interrelation between mechanics of mass-point systems and continuum mechanics (Russian and Polish summaries)* ◊ H10 ◊

XU, LIZHI *Generalized Moebius inversion theory associated with non-standard analysis (Chinese)* ◊ H10 ◊

1984

ALBEVERIO, S. & FENSTAD, J.E. & HOEEGH-KROHN, R. & KARWOWSKI, W. & LINDSTROEM, T.L. *Perturbations of the Laplacian supported by null sets, with applications to Polymer measures and quantum fields* ◊ H10 ◊

ARKERYD, L. *Loeb solutions of the Boltzmann equation* ◊ H10 ◊

BAYOD, J.M. *A spherical completion within the infinitesimal hull of an ultrametric space* ◊ H10 ◊

DRIES VAN DEN, L. & SCHMIDT, K. *Bounds in the theory of polynomial rings over fields. A nonstandard approach* ◊ C60 H10 H20 ◊

EMMONS, D.W. *Existence of Lindahl equilibria in measure theoretic economies without ordered preferences* ◊ H10 ◊

FARKAS, E.J. & SZABO, M.E. *A star-finite relational semantics for parallel programs* ◊ B75 H10 ◊

FITTLER, R. *Some nonstandard quantum electrodynamics* ◊ H10 ◊

HOOVER, D.N. & KEISLER, H.J. *Adapted probability distributions* ◊ C50 C65 H10 ◊

KRUPA, A. & ZAWISZA, B. *Applications of ultrapowers in analysis of unbounded selfadjoint operators* ◊ C20 C65 H10 ◊

LAUGWITZ, D. *Infinitesimals in physics (an introduction to the application of nonstandard methods)* ◊ H10 ◊

LEVY, M. & RAYNAUD, Y. *Ultrapuissances des espaces $L^p(L^q)$ (English summary)* ◊ C20 C65 H10 H20 ◊

LOEB, P.A. *Measure spaces in nonstandard models underlying standard stochastic processes* ◊ H10 ◊

NEUBRUNN, T. & RIECAN, B. & RIECANOVA, Z. *An elementary approach to some applications of nonstandard analysis* ◊ H10 ◊

NOBIS, K. *On the application of nonstandard analysis in mechanics of porous media* ◊ H10 ◊

NOBIS, K. & WIERZBICKI, E. & WOZNIAK, C. *On the physical interpretation of nonstandard methods in mechanics* ◊ H10 ◊

NOTTALE, L. & SCHNEIDER, J. *Fractals and nonstandard analysis* ◊ H10 ◊

RADINOVIC, S. *A hereditary property of HM-spaces* ◊ H10 ◊

SHUBIN, M.A. & ZVONKIN, A.K. *Nonstandard analysis and singular perturbations of the ordinary differential equations (Russian)* ◊ H10 ◊

WAKITA, H. *Mathematical framework of quantum electrodynamics* ◊ H10 ◊

WAKITA, H. *Solutions of the renormalized Tomonaga-Schwinger equation* ◊ H10 ◊

WIERZBICKI, E. *On the formation of internal constraints by the technique of nonstandard analysis* ⋄ H10 ⋄

WOLFF, M. *Spectral theory of group representations and their nonstandard hull* ⋄ C65 H10 ⋄

1985

ALBEVERIO, S. *Nonstandard analysis; polymer models, quantum fields* ⋄ H10 ⋄

ALLING, N.L. *Conway's field of surreal numbers* ⋄ H10 H15 H20 ⋄

CUTLAND, N.J. *Simplified existence for solutions to stochastic differential equations* ⋄ H10 ⋄

HORTALA-GONZALEZ, M.T. & RODRIGUEZ ARTALEJO, M. *Hoare's logic for nondeterministic regular programs: a nonstandard completeness theorem* ⋄ B75 H10 ⋄

LI, BANGHE & LI, YACHING *Nonstandard analysis and multiplication of distributions in any dimension (Chinese)* ⋄ H10 ⋄

RASHID, S. *Nonstandard analysis and infinite economies: The Cournot-Nash solution* ⋄ H10 ⋄

RICHTER, M.M. & SZABO, M.E. *Nonstandard computation theory* ⋄ D20 D75 H10 ⋄

B80 Other applications of logic

1879
FREGE, F.L.G. *Ueber Anwendungen der Begriffsschrift*
◇ A05 A10 B80 ◇

1924
NICOD, J. *Geometry in the sensible world (French)*
◇ A05 B80 F99 ◇

1929
LANGER, S.K. *A set of postulates for the logical structure of music* ◇ B80 ◇

1937
WOODGER, J.-H. *The axiomatic method in biology*
◇ B30 B80 ◇

1942
BERKELEY, E.C. *Conditions affecting the application of symbolic logic* ◇ B80 ◇

1948
SEARLES, H.L. *Logic and scientific methods. An introductory course* ◇ A05 B80 ◇

1951
KLUG, U. *Juristische Logik* ◇ B80 ◇

1952
STABLER, E.R. *Applied logic and modern problems*
◇ A05 B80 ◇
WOODGER, J.-H. *From biology to mathematics*
◇ B30 B80 ◇

1953
GRENIEWSKI, H. *An attempt at "rejuvenation" of the square of opposition (Polish, Russian) (English summary)* ◇ A05 B80 ◇

1954
DESTOUCHES, J.-L. & FEVRIER, P. (EDS.) *Applications scientifiques de la logique mathematique. Actes du 2eme colloque international de logique mathematique*
◇ B80 ◇
HINTIKKA, K.J.J. *An application of logic to algebra*
◇ B80 G10 ◇
TAMARI, D. *Some mutual applications of logic and mathematics* ◇ B80 ◇
WOODGER, J.-H. *Problems arising from the application of mathematical logic to biology* ◇ B80 ◇

1955
GURMUND, L. *The problem of correct symbolism as related to some problems of social psychology* ◇ B80 ◇

1956
KOKOSZYNSKA, M. & KUBINSKI, T. & SLUPECKI, J. *The application of logistic concepts to the explication of some concepts on natural sciences (Polish) (Russian and English summaries)* ◇ B80 ◇

1958
SALAMUCHA, J. *The proof "ex motu" for the existence of God: Logical analysis of St. Thomas' arguments*
◇ A05 B80 ◇

1959
FENSTAD, J.E. *Notes on the application of formal methods in the soft sciences* ◇ B80 ◇
FEYS, R. & MOTTE, M.-T. *Logique juridique, systemes juridiques* ◇ B80 ◇
WOODGER, J.-H. *Studies in the foundations of genetics*
◇ B30 B80 ◇

1960
FREUDENTHAL, H. *Lincos: Design of a language for cosmic intercourse. Part I* ◇ B80 ◇
POHM, A. & REID, A. & SCHAUER, R. & STEWART, R. *Some applications of magnetic film parametrons as logical devices* ◇ B80 ◇

1961
FEYS, R. *Logique formalisee et raissonement juridique*
◇ B80 ◇
LYUBCHENKO, G.G. *Binary codes for logical machines (Ukrainian) (Russian and English summaries)*
◇ B80 ◇

1962
KLUG, U. *Bemerkungen zur logischen Analyse einiger rechtstheoretischer Begriffe und Behauptungen*
◇ B80 ◇
SCHREIBER, R. *Logik des Rechts* ◇ B80 ◇

1963
BETH, E.W. *Observations concerning computing, deduction and heuristics* ◇ B35 B80 ◇
KASSLER, M. *A sketch of the use of formalized languages for the assertion of music* ◇ B80 ◇
SCARPELLINI, B. *Eine Anwendung der unendlichwertigen Logik auf topologische Raeume* ◇ B80 C65 ◇

1965
GIZOWA, H. *From Zygmunt Kramsztyk's work in the field of applied logic in medicine (Polish) (English and Russian summaries)* ◇ B80 ◇

1967
BAER, G. & ROHLEDER, H. *Ueber einen arithmetisch-aussagenlogischen Kalkuel und seine Anwendung auf ganzzahlige Optimierungsprobleme*
◇ B80 ◇

HOROVITZ, J. *La logique et le droit* ⋄ B80 ⋄
KASSLER, M. *Towards a theory that is the twelve-note-class system* ⋄ B80 ⋄

1969

BORETZ, B. *Meta-variations: studies in the foundations of musical thought (I)* ⋄ B80 ⋄
DICKOFF, J.W. & JAMES, P.A. *Principles of symbolic logic for pattern recognition* ⋄ B80 ⋄
LIPCZYNSKA, M. & WOLTER, H. *Elements of logic. An exposition for law* ⋄ B80 ⋄

1970

BORETZ, B. *Sketch of a musical system (meta-variations, part II)* ⋄ B80 ⋄
BORETZ, B. *The construction of musical syntax I* ⋄ B80 ⋄
MARTIN, R.M. *On the proto-theory of musical structure* ⋄ B80 ⋄
ZIEMBINSKI, Z. *Conditions preliminaires de l'application de la logique deontique dans les raisonnements juridique* ⋄ B80 ⋄

1971

BERGMANN, H. *Eine Faerbungstheorie fuer endliche Graphen* ⋄ B80 ⋄
BIRKHOFF, GARRETT *Mathematics and psychology I,II (Bulgarian)* ⋄ A05 B80 ⋄
BORETZ, B. *Musical syntax (II)* ⋄ B80 ⋄
JARDINE, C.J. & JARDINE, N. *The matching of parts of things (Polish and Russian summaries)* ⋄ A05 B80 E70 ⋄
PEKLO, B.T. *Recht und Logik* ⋄ B80 ⋄
PINKAVA, V. *Logical models of sexual deviations* ⋄ B80 ⋄
SCHLINK, B. *On a principle of contradiction in normative logic and jurisprudence* ⋄ B80 ⋄

1972

HOROVITZ, J. *Law and logic. A critical account of legal argument* ⋄ B80 ⋄
KANGER, S. *Law and logic* ⋄ A05 B80 ⋄
MATIYASEVICH, YU.V. *The application of the methods of the theory of logical derivation to graph theory (Russian)* ⋄ B80 ⋄
PEKLO, B.T. *Mancherlei ueber rechtslogische Fragen (juristisch-logisches quodlibet.)* ⋄ B80 ⋄

1973

ALLEN, C.M. & GIVONE, D.D. & MATTREY, R.F. *Applying multiple-valued algebra concepts to neural modeling* ⋄ B80 ⋄
BREUER, M.A. & EPSTEIN, G. *The smallest many-valued logic for treatment of complemented and uncomplemented error signal* ⋄ B80 ⋄
HUSSON, L. *L'infrastructure du raisonnement juridique* ⋄ B80 ⋄
RVACHEV, V.L. *The concept of solution structure of a boundary value problem (Russian)* ⋄ B80 ⋄
SENYUKOVA, A.G. *Methode der logischen Gleichungen in Anwendung auf das Problem der Vergleichung nichtorientierter Graphen, die durch Nachbarschaftmatrizen vorgegeben sind (Russian)* ⋄ B80 ⋄

1974

BRESSAN, A. *On the usefulness of modal logic in axiomatizations of physics* ⋄ B45 B80 ⋄
GOLDBERG, ADELE *Design of a computer-tutor for elementary mathematical logic* ⋄ B80 ⋄
HAVRANEK, T. *An application of logical-probabilistic expressions to the realization of stochastic automata* ⋄ B70 B80 D05 ⋄
MATIYASEVICH, YU.V. *A proof-scheme in discrete mathematics (Russian) (English summary)* ⋄ B80 ⋄

1975

HANSEN, J.C. *Some applications of a general theory of digraph measures* ⋄ B80 ⋄
LEKOMTSEV, YU.K. *The algebraic approach to the syntax of colors in painting (Russian)* ⋄ B80 ⋄
MARTIN, J.N. *A many-valued semantics for category mistakes* ⋄ A05 B80 ⋄
MATIYASEVICH, YU.V. *Metamathematical approach to proving theorems in discrete mathematics (Russian) (English summary)* ⋄ B80 ⋄
PALLASCHKE, D. *Eine algebraische Formulierung des Beobachtbarkeitsbegriffes in der Kontrolltheorie* ⋄ B80 G05 ⋄
RINE, D.C. *Associative and multi-valued logic for possible improvements in some X-ray image processing* ⋄ B80 ⋄
SCHEPERS, H. *Leibniz Disputation "de conditionibus". Ansaetze zu einer juristischen Aussagenlogik* ⋄ A10 B80 ⋄
WECHSLER, H. *Applications of fuzzy logic to medical diagnosis* ⋄ B80 ⋄

1976

ALBRECHT-BUEHLER, G. *Numerical evaluation of the validity of experimental proofs in biology* ⋄ B80 ⋄
BECKER, J.A. & LIPSHITZ, L. *An application of logic to analysis* ⋄ B80 C65 ⋄
CHILAUSKY, R. & JACOBSEN, B. & MICHALSKI, R.S. *An application of variable-valued logic to inductive learning of plant disease diagnostic rules* ⋄ B80 ⋄
GARDIES, J.-L. *Modalites et normes* ⋄ B80 ⋄
GOLDBERG, ADELE & SUPPES, P. *Computer-assisted instruction in elementary logic at the university level* ⋄ B80 ⋄
KASSLER, M. *The decidability of languages that assert music* ⋄ B25 B80 ⋄
KOHOUT, L.J. *Application of multi-valued logics to the study of human movement control and of movement disorders* ⋄ B80 ⋄
LEE, S.C. *Vector boolean algebra and calculus* ⋄ B80 G05 ⋄
RVACHEV, V.L. & SLESARENKO, A.P. *Algebra of logic and integral transforms in boundary value problems (Russian)* ⋄ B80 G05 ⋄

1977

GEORGESCU, G. *Information storage and retrieval systems based on the θ-valued logic (Romanian) (English summary)* ⋄ B75 B80 ⋄
LUEKOE, G. *Truth functions and problems in graph colouring* ⋄ B80 ⋄

MORSCHER, E. *Betrachtungen zur Praedikatenlogik*
◇ B80 ◇

WEISS, M. *Bemerkungen zu Bewertungsmoeglichkeiten von Erzeugnissen, deren Funktionsweise mit Hilfe boolescher Funktionen beschrieben werden kann*
◇ B80 ◇

1978

BAER, G. *Eine theoretische Fundierung der diskreten Optimierung mit logischen Mitteln* ◇ B80 ◇

HOOKER, C.A. & LEACH, J.J. & MCCLENNEN, E. (EDS.) *Foundations and applications of decision theory. Papers resulting from a workshop* ◇ B80 ◇

SCHWIND, C.B. *A formalism for the description of question answering systems* ◇ B65 B80 ◇

1979

BELLACICCO, A. & FREGUGLIA, P. *Caratterizzazione semantica di tavole oggetto-predicato soggette a procedure di clustering* ◇ B80 ◇

BELYANIN, P.N. & BRUEVICH, N.G. & CHELISHCHEV, B.E. & GONSALES-SABATER, A. *Mathematical theory of assembly technology (Russian)* ◇ B80 ◇

ERSHOV, YU.L. *Some questions of the application of formalized languages for the investigation of philosophical problems (Russian)* ◇ A05 B80 ◇

FIESCHI, M. & JOUBERT, M. *Application d'une methode logique a l'aide a la decision en medecine. La methode de Davis et Putnam en logique propositionnelle* ◇ B80 ◇

GAINES, B.R. *Logical foundations for database systems* ◇ B80 ◇

KENTON, S.A. *Mathematical foundations of constitutional law* ◇ A05 B80 ◇

SANCHEZ, E. *Inverses of fuzzy relations. Application to possibility distributions and medical diagnosis* ◇ B80 E07 E72 ◇

STEINAUER, P.H. *La logique au service du droit. Etude de logique contemporaine pour une meilleure communication de la pensee juridique* ◇ B80 ◇

STEPANKOVA, O. *Planning in uncertain environments through situation calculus* ◇ B80 ◇

STEPANKOVA, O. *Uniform formulas and their proof in a situation theory (Russian and Czech summaries)* ◇ B75 B80 ◇

1980

BUCUR, I. *Special chapters of algebra (Romanian)* ◇ B80 C98 D98 ◇

JANSSEN, T.M.V. *Logical investigations on PTQ arising from programming requirements* ◇ B80 ◇

KIRIN, V.G. *On some applications of a tautology* ◇ B80 ◇

KONDO, K. *Hering's opponent colour scheme as tristimulus projective - geometrical correlation and a boolean logical analysis of colour naming* ◇ B80 ◇

LEVIN, V.A. & PEREL'ROIZEN, E.Z. *Logical methods in the field of computer vision (Russian)* ◇ B80 ◇

LEVIN, V.A. *Qualitative analysis of the problem of combinatorial optimization by the method of logical determinants (Russian)* ◇ B80 ◇

SANCHEZ, E. *Fuzzy logics with application to medical diagnosis* ◇ B80 ◇

1981

AGAZZI, E. *Logic and the methodology of empirical sciences* ◇ B80 ◇

ALEXANDER, P. *The case of the lonely corpuscle: reductive explanation and primitive expressions* ◇ B80 ◇

BIELTZ, P. *The logic of elections and social decision (Romanian)* ◇ B80 ◇

CZOGALA, E. & PEDRYCZ, W. *Methods of multivalued logics and fuzzy set theory in computer-aided decision making in diagnostic and therapeutic processes* ◇ B80 ◇

EMEL'YANOV, A.M. *A method of analysis of human managerial work by frames and special modal logic (Russian)* ◇ B80 ◇

GRIGOR'EV, D.YU. & OREVKOV, V.P. (EDS.) *Theoretical applications of methods of mathematical logic. III. Work collection (Russian)* ◇ B80 ◇

HUTCHINS, G.M. & MOORE, G.W. *A Hintikka possible worlds model for certainty levels in medical decision making* ◇ B80 ◇

KREISEL, G. *Aussagenquantoren: Ein Beitrag der mathematischen Logik zur Erkenntnistheorie* ◇ A05 B80 ◇

STEPANKOVA, O. *Normal form of proof of certain formulas of situation calculus* ◇ B80 ◇

VESELY, A. *Logically orientated Cluster analysis* ◇ B80 ◇

1982

ALBIN, P.S. *The metalogic of economic predictions, calculations and propositions* ◇ B80 ◇

BARRICELLI, N.A. *B-mathematics as a language* ◇ B80 ◇

BELLIA, M. & DAMERI, E. & DEGANO, P. & LEVI, G. & MARTELLI, M. *Applicative communicating processes in first order logic* ◇ B80 ◇

COHEN, Y. *Logical polygraph* ◇ B80 ◇

DOLBY, G.R. *The role of statistics in the methodology of the life sciences* ◇ B80 ◇

ERNST, C.J. *An approach to management expert systems using fuzzy logic* ◇ B80 ◇

EVANGELIST, M. *Non-standard propositional logics and their application to complexity theory*
◇ B80 D15 F20 ◇

FARINAS DEL CERRO, L. *Logique modal et processus communicants* ◇ B80 ◇

HIGUCHI, T. & KAMEYAMA, M. *Construction of a processor exclusively for picture processing using many-valued logic (Japanese)* ◇ B80 ◇

HINTIKKA, K.J.J. *A dialogical model of teaching* ◇ B80 ◇

LOBOVIKOV, V.O. *Some applications of modal logic to ethical categories (Russian)* ◇ B80 ◇

MA, HUAXIAO *A simplified algorithm for finding Boolean differences and its applications in engineering (Chinese) (English summary)* ◇ B80 ◇

SANCHEZ, E. & SOULA, G. *Soft deduction rules in medical diagnostic processes* ◇ B80 ◇

SCHREINER, H. *Information systems and artificial intelligence in law. Logical procedures for the application of technical intelligence in juridical decisions* ◇ B80 ◇

1983

AMLIEN, J. & BARRICELLI, N.A. *The use of data processing machines in the recognition and proof of arithmetical theorems translated into B-mathematical language* ◇ B35 B80 ◇

BARONE, F. & GRASSINI, R. *Logico algebraic approach to Lagrangian systems* ◇ B80 ◇

ELLETT JR., F.S. & ERICSON, D.P. *The logic of causal methods in social science* ◇ B80 ◇

FISHBURN, P.C. *A generalization of comparative probability on finite sets* ◇ B80 ◇

MASLOV, S.YU. *Deductive systems and their economic applications (Russian)* ◇ B80 ◇

PARSONS, C. *Mathematics in philosophy. Selected essays* ◇ A05 B80 ◇

WARNER, M.W. *Lattices and lattice-valued relations in biology* ◇ B80 G10 ◇

ZABSKI, E. *A formalization of the Weber-Fechner theory (Polish) (English summary)* ◇ B80 ◇

ZABSKI, E. *Some elementary theories of empirical precedence (Polish) (English summary)* ◇ B80 ◇

1984

ROBERTS, F.S. *Applications of the theory of meaningfulness to order and matching experiments* ◇ B80 ◇

RUBINSTEIN, A. *The single profile analogues to multiprofile theorems: mathematical logic's approach* ◇ B80 ◇

SHAPIRO, S.I.M. *Solution of logical and game-theoretic problems. Logical-psychological studies (Russian)* ◇ B80 ◇

1985

CLARKE, E.M. *Using temporal logic for automatic verification of finite state systems* ◇ B80 ◇

B97 Proceedings

1883
PEIRCE, C.S. (ED.) *Studies in logic, by members of the Johns Hopkins University* ◊ B97 ◊

1941
GONSETH, F. (ED.) *Les entretiens de Zuerich sur les fondements et la methode des sciences mathematiques: Exposes et discussions* ◊ B97 E97 F97 ◊

1959
HENKIN, L. & SUPPES, P. & TARSKI, A. (EDS.) *The axiomatic method, with special reference to geometry and physics* ◊ B30 B97 C65 D35 ◊

1962
NAGEL, E. & SUPPES, P. & TARSKI, A. (EDS.) *Logic, methodology and philosophy of science* ◊ B97 ◊

1963
BRAFFORT, P. & HIRSCHBERG, D. (EDS.) *Computer programming and formal systems* ◊ B97 D05 D97 ◊

1965
CROSSLEY, J.N. & DUMMETT, M. (EDS.) *Formal systems and recursive functions* ◊ B97 D20 D97 ◊
KALMAR, L. (ED.) *Colloque sur les fondements des mathematiques, les machines mathematiques, et leurs applications* ◊ B97 D05 D10 D97 ◊
TYMIENIECKA, A.-T. (ED.) *Contributions to logic and methodology, in honor of I.M.Bochenski* ◊ A05 B97 ◊

1967
COPI, I.M. & GOULD, JAMES A. (EDS.) *Contemporary readings in logical theory* ◊ A05 A10 B97 ◊
CROSSLEY, J.N. (ED.) *Sets, models and recursion theory* ◊ B97 C97 D97 ◊
DESTOUCHES, J.-L. (ED.) *Logic & foundations of science. E.W.Beth memorial colloquium* ◊ B97 ◊
HEIJENOORT VAN, J. (ED.) *From Frege to Goedel: a source book in mathematical logic, 1879-1931* ◊ A10 B97 F97 ◊

1968
OREVKOV, V.P. (ED.) *The calculi of symbolic logic I (Russian)* ◊ B97 F97 ◊
SCHMIDT, H.A. & SCHUETTE, K. & THIELE, H. (EDS.) *Contribtions to mathematical logic. Proc. of the logic colloq., Hannover 1966* ◊ B97 ◊

1969
DAVIDSON, D. & HINTIKKA, K.J.J. (EDS.) *Words and objections: Essays on the work of W.V.Quine* ◊ A05 B97 ◊

1970
LACOMBE, D. & LAUDET, M. & NOLIN, L. & SCHUETZENBERGER, M.-P. (EDS.) *Symposium on automatic demonstration* ◊ B35 B97 F97 ◊
TAVANETS, P.V. (ED.) *Studies in systems of logic (Russian)* ◊ B97 ◊

1971
FENSTAD, J.E. (ED.) *Proceedings of the Second Scandinavian Logic Symposium* ◊ B97 ◊

1972
HODGES, W. (ED.) *Conference in Mathematical Logic -- London '70* ◊ B97 C97 ◊
LUXEMBURG, W.A.J. & ROBINSON, A. (EDS.) *Contributions to non-standard analysis* ◊ B97 H05 H97 ◊
OREVKOV, V.P. (ED.) *Logical and logico-mathematical calculi 2 (Russian)* ◊ B97 ◊

1973
BELL, J.L. & COLE, J.C. & PRIEST, G. & SLOMSON, A. (EDS.) *The Proceedings of the Bertrand Russell Memorial Logic Conference (Uldum, 4 - 16 August 1971)* ◊ B97 ◊
ERSHOV, YU.L. & KARGAPOLOV, M.I. & MERZLYAKOV, YU.I. & SHIRSHOV, A.I. & SMIRNOV, D.M. (EDS.) *Selected questions of algebra and logic (Russian,English)* ◊ B97 ◊
MATHIAS, A.R.D. & ROGERS JR., H. (EDS.) *Cambridge summer school in mathematical logic (Held in Cambridge, England, August 1 - 21, 1971)* ◊ B97 ◊
TAVANETS, P.V. (ED.) *Theorie der logischen Folgerungen (Russian)* ◊ B22 B97 ◊

1974
ADDISON, J.W. & CHANG, C.C. & CRAIG, W. & HENKIN, L. & SCOTT, D.S. & VAUGHT, R.L. (EDS.) *Proceedings of the Tarski Symposium* ◊ B97 C97 ◊
GODLEVSKIJ, A.B. & LETICHEVS'KIJ, O.A. (EDS.) *Automata and algorithm theory, and mathematical logic (Russian)* ◊ B35 B97 D05 D20 D97 ◊
SMIRNOV, V.A. & TAVANETS, P.V. (EDS.) *Philosophy and logic (Russian)* ◊ A05 B97 ◊

1975
BOKSHTEJN, M.F. & PETRI, N.V. & SHCHEGOL'KOV, E.A. (EDS.) *Current questions in mathematical logic and set theory (Russian)* ◊ B97 E97 ◊
CROSSLEY, J.N. (ED.) *Algebra and logic* ◊ B97 C97 G97 ◊
HENKIN, L. & JOJA, A. & MOISIL, G.C. & SUPPES, P. (EDS.) *Logic, methodology and philosophy of science, IV. Proceedings of the fourth international*

congress for logic, methodology and philosophy of science ⋄ B97 ⋄

KOTAS, J. (ED.) *Stanislaw Jaskowski's achievements in mathematical logic* ⋄ A10 B97 ⋄

MAGNARADZE, D.G. & PKHAKADZE, SH.S. (EDS.) *Studies in mathematical logic and the theory of algorithms (Russian)* ⋄ B97 D97 ⋄

MASLOV, S.YU. & MINTS, G.E. (EDS.) *Theoretical applications of the methods of mathematical logic I (Russian)* ⋄ B97 ⋄

MUELLER, GERT H. & OBERSCHELP, A. & POTTHOFF, K. (EDS.) *ISILC Logic Conference* ⋄ B97 ⋄

PARIKH, R. (ED.) *Logic Colloquium* ⋄ B97 ⋄

ROSE, H.E. & SHEPHERDSON, J.C. (EDS.) *Proceedings of the Logic Colloquium (Bristol, July, 1973)* ⋄ B97 ⋄

1976

KASHER, A. (ED.) *Language in focus: foundations, methods and systems. Essays in memory of Yehoshua Bar-Hillel* ⋄ B97 ⋄

KUZNETSOV, A.V. (ED.) *Fourth All-Union Conference on Mathematical Logic (Russian)* ⋄ B97 ⋄

SCHIRN, M. (ED.) *Studien zu Frege. I,II,III* ⋄ A05 A10 B97 ⋄

1977

BUTTS, R.E. & HINTIKKA, K.J.J. (EDS.) *Proceedings of the fifth international congress of logic, methodology and philosophy of science. I,II,III,IV* ⋄ A05 A10 B97 ⋄

GANDY, R.O. & HYLAND, J.M.E. (EDS.) *Logic Colloquium 76. Proceedings of a conference held in Oxford in July 1976* ⋄ B97 C97 D97 ⋄

HAKAMIES, A. (ED.) *Logik, Mathematik und Philosophie des Transzendenten* ⋄ A05 B97 ⋄

MAGNARADZE, L.G. & PKHAKADZE, SH.S. (EDS.) *Studies in mathematical logic and the theory of algorithms. No. II (Russian)* ⋄ B97 D97 ⋄

MIHALJINEC, M. (ED.) *Colloque international de logique, Clermont-Ferrand, 18-25 juillet 1975* ⋄ B97 ⋄

1978

BIRYUKOV, B.V. & SPIRKIN, A.G. (EDS.) *Cybernetics and logic. The growth of cybernetical ideas and the development of computing machinery in their mathematical-logical aspects (Russian) (English summary)* ⋄ B97 ⋄

MACINTYRE, A. & PACHOLSKI, L. & PARIS, J.B. (EDS.) *Logic colloquium '77. Proceedings of the colloquium held in Wroclaw, august 1977* ⋄ B97 C97 ⋄

MAGNARADZE, L.G. & PKHAKADZE, SH.S. (EDS.) *Studies in mathematical logic and the theory of algorithms (Russian)* ⋄ A05 B97 ⋄

1979

BOFFA, M. & DALEN VAN, D. & MCALOON, K. (EDS.) *Logic colloquium '78. Proceedings of the colloquium held in Mons, August 1978* ⋄ B97 ⋄

BOLCK, F. (ED.) *Begriffsschrift -- Jenaer Frege-Konferenz* ⋄ A10 B97 ⋄

BORISOV, YU.F. (ED.) *Methodological problems of mathematics (Russian)* ⋄ B97 ⋄

ERSHOV, YU.L. (ED.) *Fifth All-Union conference on mathematical logic. Dedicated to Academician A. I. Mal'tsev. Abstracts of papers (Russian)* ⋄ B97 ⋄

JENSEN, F.V. & MAYOH, B.H. & MOELLER, K.K. (EDS.) *Proceedings from 5th Scandinavian Logic Symposium, Aalborg, 17-19, January 1979* ⋄ B97 ⋄

LOURENCO, M. (ED.) *Goedel's theorem and the continuum hypothesis (Portuguese)* ⋄ B97 E50 F30 F97 ⋄

MATIYASEVICH, YU.V. & SLISENKO, A.O. (EDS.) *Studies in constructive mathematics and mathematical logic VIII (Russian)* ⋄ B97 F97 ⋄

SMIRNOV, V.A. (ED.) *Logical inference (Russian)* ⋄ B97 ⋄

1980

BARWISE, J. & KEISLER, H.J. & KUNEN, K. (EDS.) *The Kleene Symposium* ⋄ B97 ⋄

BIBEL, W. & KOWALSKI, R. (EDS.) *5th Conference on Automated Deduction* ⋄ B35 B97 ⋄

BORGA, M. & FREGUGLIA, P. & PALLADINO, D. (EDS.) *Rassegna di matematica. Logica matematica. Matematica applicata. Didattica della matematica* ⋄ B97 D97 ⋄

GRASS, W. & HALLER, R. (EDS.) *Language, logic, and philosophy* ⋄ A05 B65 B97 ⋄

NICKLES, T. (ED.) *Scientific discovery, logic, and rationality* ⋄ A05 B97 ⋄

POPOVIC, M.V. (ED.) *Actual problems of logic and methodology of science (Russian)* ⋄ A05 B97 ⋄

1981

AGAZZI, E. & GRUENDER, D. & HINTIKKA, K.J.J. (EDS.) *Proceedings of the 1978 Pisa conference on the history and philosophy of science. 2nd international conference I,II* ⋄ A10 B97 ⋄

DOEMOELKI, B. & GERGELY, T. (EDS.) *Mathematical logic in computer science (Colloquium held in Salgotarjan, Hungary, September 10-15, 1978)* ⋄ B97 ⋄

LERMAN, M. & SCHMERL, J.H. & SOARE, R.I. (EDS.) *Logic year 1979-80, The University of Connecticut, USA* ⋄ B97 C97 ⋄

MUELLER, GERT H. & TAKEUTI, G. & TUGUE, T. (EDS.) *Logic symposia, Hakone 1979, 1980. Proceedings* ⋄ B97 E97 ⋄

1982

COHEN, L.J. & LOS, J. & PFEIFFER, H. & PODEWSKI, K.-P. (EDS.) *Logic, methodology and philosophy of science VI* ⋄ B97 D98 ⋄

CRESSWELL, M.J. & GOLDBLATT, R.I. & MEYERHOFF CRESSWELL, M. & SEGERBERG, K. *Symbolic logic. Proceedings of the 1981 Annual Conference of the Australian Association of Symbolic Logic held in Wellington, New Zealand, from 2-5 July 1981* ⋄ B97 ⋄

DALEN VAN, D. & LASCAR, D. & SMILEY, T.J. (EDS.) *Logic colloquium '80. Eur. Summer Meet., Prague 1980* ⋄ B97 D97 ⋄

HORECKY, J. *COLING 82. Proceedings of the Ninth International Conference on Computational Linguistics, Prague, July 5-10, 1982* ⋄ B97 ⋄

Proceedings

METAKIDES, G. (ED.) *Patras logic symposium*
⋄ B97 C97 ⋄
STERN, J. (ED.) *Proceedings of the Herbrand Symposium. Logic colloquium '81, held in Marseille, France, July 1981* ⋄ B97 D97 ⋄
TROELSTRA, A.S. & DALEN VAN, D. (EDS.) *The L.E.J. Brouwer centenary symposium*
⋄ A05 A10 B97 F50 F55 F97 ⋄
WAHLSTER, W. (ED.) *GWAI-82* ⋄ B97 ⋄

1983
CHONG, C.T. & WICKS, M.J. (EDS.) *Southeast Asian conference on logic* ⋄ B97 ⋄
COHEN, R.S. & WARTOFSKY, M.W. (EDS.) *Language, logic and method* ⋄ B97 ⋄
GLADKIJ, A.V. (ED.) *Mathematical logic, mathematical linguistics and theory of algorithms (Russian)*
⋄ B97 D97 ⋄
MEULEN TER, A.G.B. (ED.) *Studies in modeltheoretic semantics* ⋄ B97 C97 ⋄

1984
BOERGER, E. & OBERSCHELP, W. & RICHTER, M.M. & SCHINZEL, B. & THOMAS, WOLFGANG (EDS.) *Computation and proof theory* ⋄ B97 D97 F97 ⋄
DELON, F. & LASCAR, D. & PARIGOT, M. & SABBAGH, G. (EDS.) *Logique, Octobre 1983, Paris. Compte rendu de la table ronde de logique des 15 et 16 octobre 1983, Paris* ⋄ B97 C97 ⋄
ERSHOV, YU.L. & LAVROV, I.A. & PAVILENIS, R.I. & PETROV, V.V. & SMIRNOV, V.A. *Logic, the foundations of mathematics and linguistics (Russian) (English summary)* ⋄ B97 ⋄

LANDMAN, F. & VELTMAN, F. (EDS.) *Varieties of formal semantics* ⋄ B97 ⋄
LOLLI, G. & LONGO, G. & MARCJA, A. (EDS.) *Logic colloquium '82. Proceedings of the colloquium held in Florence, 23-28 August, 1982*
⋄ B97 C97 D97 F97 ⋄
MUELLER, GERT H. & RICHTER, M.M. (EDS.) *Models and sets* ⋄ B97 C97 D97 H97 ⋄
WECHSUNG, G. (ED.) *Frege conference 1984. Proceedings of the International Conference held at Schwerin, Sept.10-14, 1984* ⋄ A05 A10 B97 C97 ⋄

1985
ALCANTARA DE, L.P. (ED.) *Mathematical logic and formal systems. A collection of papers in honor of Professor Newton C.A. da Costa* ⋄ B97 ⋄
DELON, F. & LASCAR, D. & LOUVEAU, A. & SABBAGH, G. (EDS.) *Seminaire general de logique 1982-83* ⋄ B97 C98 D97 ⋄
DORN, G. & WEINGARTNER, P. (EDS.) *Foundations of logic and linguistics. Problems and their solutions. Papers from the seventh international congress on logic, methodology and philosophy of science* ⋄ B97 ⋄
HARRINGTON, L.A. & MORLEY, M.D. & SCEDROV, A. & SIMPSON, S.G. *Harvey Friedman's research on the foundations of mathematics*
⋄ B97 C97 D97 E97 F97 ⋄
PRISCO DI, C.A. (ED.) *Methods in mathematical logic. Proceedings of the 6th Latin American Symposium on Mathematical Logic held in Caracas, Venezuela, Aug. 1-6, 1983* ⋄ B97 C97 ⋄

B98 Textbooks, surveys

1880
JEVONS, W.S. *Studies in deductive logic: a manual for students* ⋄ B98 ⋄

1881
VENN, J. *Symbolic logic* ⋄ A10 B20 B98 ⋄

1884
KEYNES, J.N. *Studies and exercises in formal logic, including a generalisation of logical processes in their application to complex inferences* ⋄ B98 ⋄

1887
DODGSON, C.L. *The game of logic* ⋄ B98 ⋄

1889
BAIN, A. *Logic: deductive and inductive* ⋄ A05 B98 ⋄

1890
SCHROEDER, F.W.K.E. *Vorlesungen ueber die Algebra der Logik (exakte Logik) Vol.1*
⋄ A05 A10 B20 B98 G98 ⋄

1891
SCHROEDER, F.W.K.E. *Vorlesungen ueber die Algebra der Logik (exakte Logik) Vol.2*
⋄ A05 A10 B20 B98 G05 G15 G98 ⋄

1894
BURALI-FORTI, C. *Logica matematica* ⋄ B98 ⋄

1895
DODGSON, C.L. *Symbolic logic. Part I. Elementary*
⋄ B20 B98 ⋄
SCHROEDER, F.W.K.E. *Vorlesungen ueber die Algebra der Logik (exakte Logik) Vol.3 Algebra und Logik der Relative* ⋄ A05 A10 B20 B98 G98 ⋄
SIGWART, C. *Logic. Vol.II: logical methods* ⋄ B98 ⋄

1896
SCHUBERT, H. *Mathematical essays and recreations*
⋄ B98 ⋄

1902
POINCARE, H. *Du role de l'intuition et de la logique en mathematiques* ⋄ A05 B98 ⋄

1904
COUTURAT, L. *L'algebre de la logique* ⋄ B98 G05 ⋄
NATORP, P. *Logik. Grundlegung und logischer Aufbau der Mathematik und mathematischen Naturwissenschaft*
⋄ B98 ⋄
SIGWART, C. *Logik Vol.1* ⋄ B98 ⋄

1906
MACCOLL, H. *Symbolic logic and its applications*
⋄ B05 B98 ⋄

1911
BOSANQUET, B. *Logic. Vol. II* ⋄ B98 ⋄
WHITEHEAD, A.N. *Introduction to mathematics* ⋄ B98 ⋄

1912
PADOA, A. *La logique deductive dans sa derniere phase de developpement* ⋄ A10 B98 ⋄

1918
LEWIS, C.I. *A survey of symbolic logic*
⋄ B45 B46 B98 ⋄
RIEBER, C.H. *Footnotes to formal logic* ⋄ A05 B98 ⋄

1926
RAMSEY, F.P. *The foundations of mathematics*
⋄ A05 B98 ⋄

1927
BEHMANN, H. *Mathematik und Logik* ⋄ A05 B98 ⋄

1928
HILBERT, D. & ACKERMANN, W. *Grundzuege der theoretischen Logik* ⋄ B25 B98 D35 ⋄
SKOLEM, T.A. *Ueber die mathematische Logik*
⋄ A05 B25 B98 C07 C10 C35 ⋄

1929
CARNAP, R. *Abriss der Logistik, mit besonderer Beruecksichtigung der Relationstheorie und ihrer Anwendungen* ⋄ A05 B98 ⋄
KOTARBINSKI, T. *Elements of epistemology, formal logic, and methodology of the sciences (Polish)*
⋄ A05 B98 ⋄
LUKASIEWICZ, J. *Elements of mathematical logic (Polish)*
⋄ B98 ⋄

1930
STEBBING, L.S. *A modern introduction to logic* ⋄ B98 ⋄

1931
EATON, R.M. *General logic: an introductory survey*
⋄ B98 ⋄
JEFFREYS, H. *Scientific inference* ⋄ A05 B98 ⋄
JORGENSEN, J. *A treatise of formal logic: its evolution and main branches, with its relations to mathematics and philosophy. I,II,III* ⋄ A05 A10 B98 ⋄

1932
BURKAMP, W. *Logik* ⋄ B98 ⋄
LANGFORD, C.H. & LEWIS, C.I. *Symbolic logic*
⋄ B25 B98 C10 C98 ⋄
WHITEHEAD, A.N. & RUSSELL, B. *Einfuehrung in die mathematische Logik* ⋄ A05 B98 ⋄

1933

CHAPMAN, F.M. & HENLE, P. *The fundamentals of logic*
◊ B98 ◊

MACE, C.A. *The principles of logic: an introductory survey*
◊ B98 ◊

1934

COHEN, M.R. & NAGEL, E. *An introduction to logic and scientific method* ◊ B98 ◊

HILBERT, D. & BERNAYS, P. *Grundlagen der Mathematik I* ◊ A05 B98 F05 F30 F98 ◊

QUINE, W.V.O. *A system of logistic* ◊ A05 B15 B98 ◊

ZHU, YANJUN *A summary on mathematical logic (Chinese)* ◊ B98 ◊

1936

CHURCH, A. *A bibliography of symbolic logic*
◊ A10 B98 ◊

CHURCH, A. *Mathematical logic (mimeographed notes)*
◊ B40 B98 D20 ◊

TARSKI, A. *On mathematical logic and the deductive method (Polish)* ◊ B98 ◊

USHENKO, A.P. *The theory of logic* ◊ B98 ◊

WAISMANN, F. *Einfuehrung in das mathematische Denken. Die Begriffsbildung der modernen Mathematik* ◊ A05 B98 ◊

ZHU, YANJUN *Introduction to mathematical logic*
◊ B98 ◊

1937

BOULIGAND, G. *Structure des theories. Problemes infinis*
◊ A05 B98 ◊

GONSETH, F. *Qu'est-ce que la logique* ◊ B98 ◊

LANGER, S.K. *An introduction to symbolic logic* ◊ B98 ◊

LUKASIEWICZ, J. *In defense of logistic (Polish)*
◊ A05 B98 ◊

TARSKI, A. *Einfuehrung in die mathematische Logik und in die Methodologie der Mathematik*
◊ A05 B30 B98 ◊

1938

BOCHENSKI, I.M. *Nove lezioni di logica simbolica*
◊ A05 B98 ◊

1939

BAYLIS, C.A. & BENNETT, A.A. *Formal logic: A modern introduction* ◊ B98 ◊

CARNAP, R. *Foundations of logic and mathematics*
◊ A05 B98 ◊

FEYS, R. *Principes de logistique, premier volume*
◊ A05 B98 ◊

HILBERT, D. & BERNAYS, P. *Grundlagen der Mathematik II* ◊ A05 B98 F05 F15 F30 F40 F98 ◊

REICHENBACH, H. *Introduction a la logistique* ◊ B98 ◊

1940

BETH, E.W. *Einfuehrung in die Philosophie der Mathematik (Dutch)* ◊ A05 B98 ◊

CHURCH, A. *Elementary topics in mathematical logic*
◊ B98 ◊

CHURCHMAN, C.W. *Elements of logic and formal science*
◊ B98 ◊

QUINE, W.V.O. *Mathematical logic* ◊ A05 B98 E70 ◊

1941

LUKASIEWICZ, J. *Die Logik und das Grundlagenproblem*
◊ A05 B98 ◊

QUINE, W.V.O. *Elementary logic* ◊ B98 ◊

USHENKO, A.P. *The problems of logic* ◊ A05 B98 ◊

1942

BETH, E.W. *Summulae logicales. Supplement to formal logic (Dutch)* ◊ A05 B98 ◊

BOLL, M. *Elements de logique scientifique*
◊ A05 B98 ◊

COOLEY, J.C. *A primer of formal logic* ◊ B98 ◊

1943

CARNAP, R. *Formalization of logic* ◊ A05 B98 ◊

STEBBING, L.S. *A modern elementary logic* ◊ B98 ◊

1944

CHURCH, A. *Introduction to mathematical logic. Part I*
◊ B98 ◊

FEYS, R. *Logistic. Formal logic I. General survey. Logic of propositions and classes (Dutch)* ◊ B98 ◊

GOEDEL, K. *Russell's mathematical logic* ◊ A10 B98 ◊

QUINE, W.V.O. *O sentido da nova logica* ◊ B98 ◊

1945

DAVAL, R. & GUILBAUD, G.T. *Le raisonnement mathematique* ◊ A05 B98 ◊

SERRUS, C. *Traite de logique* ◊ A05 B98 ◊

1946

QUINE, W.V.O. *A short course in logic* ◊ B98 ◊

1947

BOOLE, G. *The mathematical analysis of logic, being an essay toward a calculus of deductive reasoning*
◊ A05 A10 B98 G05 ◊

GOEDEL, K. *What is Cantor's continuum problem?*
◊ A05 B98 E50 E98 ◊

LIEBER, L.R. *Mits, wits, and logic* ◊ B60 B98 ◊

REICHENBACH, H. *Elements of symbolic logic*
◊ A05 B98 ◊

1948

AMBROSE, A. & LAZEROWITZ, M. *Fundamentals of symbolic logic* ◊ B98 ◊

LORENZEN, P. *Einfuehrung in die Logik* ◊ B98 ◊

MOSTOWSKI, ANDRZEJ *Logique mathematique. Cours donne a l'universite (Polish)* ◊ B98 ◊

QUINE, W.V.O. *Theory of deduction. Parts I-IV* ◊ B98 ◊

SCHOLZ, H. *Vorlesungen ueber Grundzuege der mathematischen Logik I* ◊ B98 ◊

1949

BLACK, M. *Language and philosophy: studies in method*
◊ A05 B98 ◊

BOCHENSKI, I.M. *Precis de logique mathematique*
◊ B98 ◊

BOURBAKI, N. *Foundations of mathematics for the working mathematician* ◊ A05 B98 ◊

REICHENBACH, H. *The theory of probability*
◊ A05 B98 ◊

SCHOLZ, H. *Vorlesungen ueber Grundzuege der mathematischen Logik II* ◊ B98 ◊

1950

BETH, E.W. *Les fondements logiques des mathématiques*
⋄ A05 B98 ⋄

MARC-WOGAU, K. *Modern logic. An elementary text-book (Swedish)* ⋄ B98 ⋄

QUINE, W.V.O. *Methods of logic* ⋄ B98 ⋄

ROSENBAUM, I. *Introduction to mathematical logic and its applications* ⋄ B98 ⋄

ROSENBLOOM, P.C. *The elements of mathematical logic*
⋄ B98 ⋄

SCHMIDT, H.A. *Mathematische Grundlagenforschung*
⋄ B98 ⋄

1951

BECKER, O. *Einfuehrung in die Logistik, vorzueglich in den Modalkalkuel* ⋄ B98 ⋄

FREYTAG-LOERINGHOFF BARON VON, B. *Logik I. Das System der reinen Logik und ihr Verhaeltnis zur Logistik* ⋄ A05 B98 ⋄

GOODSTEIN, R.L. *Constructive formalism: Essays on the foundations of mathematics*
⋄ A05 B98 F30 F60 F98 ⋄

HENLE, P. & KALLEN, H.M. &
LANGER, S.K. (EDS.)*Structure, method, and meaning. Essays in honor of Henry M. Scheffer* ⋄ B98 ⋄

SESMAT, A. *Logique. vol.II: les raisonnements, la logistique* ⋄ A05 B98 ⋄

STENIUS, E. *Modern logic (Swedish)* ⋄ B98 ⋄

WAISMANN, F. *Introduction to mathematical thinking. The formation of concepts in modern mathematics*
⋄ B98 ⋄

1952

BERKELEY, E.C. *Symbolic logic. Twenty problems and solutions* ⋄ B98 ⋄

BOOLE, G. *Studies in logic and probability* ⋄ A05 B98 ⋄

CHURCH, A. *Brief bibliography of formal logic*
⋄ A10 B98 ⋄

CURRY, H.B. *Lecons de logique algebrique*
⋄ B98 G15 G98 ⋄

FITCH, F.B. *Symbolic logic. An introduction*
⋄ B40 B98 ⋄

HERMES, H. & SCHOLZ, H. *Mathematische Logik*
⋄ B98 ⋄

KLEENE, S.C. *Introduction to metamathematics*
⋄ A05 B98 D98 F30 F50 F98 ⋄

SKOLEM, T.A. *Consideraciones sobre los fundamentos de la matematica I* ⋄ B98 F30 F50 ⋄

STRAWSON, P.F. *Introduction to logical theory*
⋄ A05 B98 ⋄

WILDER, R.L. *Introduction to the foundations of mathematics* ⋄ A05 B98 ⋄

1953

BASSON, A.H. & O'CONNOR, D.J. *Introduction to symbolic logic* ⋄ B98 ⋄

COPI, I.M. *Introduction to logic* ⋄ A10 B98 ⋄

ROSSER, J.B. *Logic for mathematicians*
⋄ B15 B98 E70 E98 ⋄

SKOLEM, T.A. *Consideraciones sobre los fundamentos de la matematica II* ⋄ B98 F30 F50 ⋄

STABLER, E.R. *An introduction to mathematical thought*
⋄ A05 B98 ⋄

1954

ALBRECHT, W. *Die Logik der Logistik* ⋄ B98 ⋄

BOCHENSKI, I.M. & MENNE, A. *Grundriss der Logistik*
⋄ B98 ⋄

CARNAP, R. *Einfuehrung in die symbolische Logik mit besonderer Beruecksichtigung ihrer Anwendungen*
⋄ A05 B30 B98 ⋄

COPI, I.M. *Symbolic logic* ⋄ B98 ⋄

CURRY, H.B. *Remarks on the definition and nature of mathematics* ⋄ A05 B98 ⋄

DUERR, K. *Lehrbuch der Logistik* ⋄ B98 ⋄

JOHNSTONE JR., H.W. *Elementary deductive logic*
⋄ B98 ⋄

JUHOS VON, B. *Elemente der neuen Logik* ⋄ B98 ⋄

KREISEL, G. *Applications of mathematical logic to various branches of mathematics* ⋄ B98 C65 ⋄

KREYCHE, R.J. *Logic for undergraduates* ⋄ B98 ⋄

SKOLEM, T.A. *Results in investigations in the foundations (Norwegian)* ⋄ A05 B98 D03 F55 ⋄

1955

BETH, E.W. *Die Stellung der Logik im Gebaeude der heutigen Wissenschaft* ⋄ B98 ⋄

BLANCHE, R. *L'axiomatique* ⋄ B98 ⋄

GRENIEWSKI, H. *Elements of formal logic (Polish)*
⋄ B98 ⋄

GRZEGORCZYK, A. & JASKOWSKI, S. & LOS, J. &
MAZUR, S. & MOSTOWSKI, ANDRZEJ & RASIOWA, H. &
SIKORSKI, R. *The present state of investigations on the foundations of mathematics (Polish)*
⋄ A10 B98 E98 ⋄

LEBLANC, H. *An introduction to deductive logic* ⋄ B98 ⋄

LORENZEN, P. *Einfuehrung in die operative Logik und Mathematik* ⋄ A05 B98 F50 F65 F98 ⋄

MILLER, JAMES WILKIN *Exercises in introductory symbolic logic* ⋄ B98 ⋄

MOSAHEB, G.-H. *Introduction to formal logic (Iranian)*
⋄ B98 ⋄

PRIOR, A.N. *Formal logic* ⋄ A05 A10 B98 E30 ⋄

WANG, HAO *On formalization* ⋄ A05 B98 ⋄

1956

BETH, E.W. *L'existence en mathematiques*
⋄ A05 B98 F55 ⋄

BOCHENSKI, I.M. *Formale Logik* ⋄ A10 B98 ⋄

KEMENY, J.G. *Semantics as a branch of logic* ⋄ B98 ⋄

STAHL, G. *Introduccion a la logica simbolica* ⋄ B98 ⋄

1957

BLANCHE, R. *Introduction a la logique contemporaine*
⋄ A05 B98 F50 ⋄

BLYTH, J.W. *A modern introduction to logic* ⋄ B98 ⋄

BRENNAN, J.G. *A handbook of logic* ⋄ B98 ⋄

CHAUVINEAU, J. *La logique moderne* ⋄ B98 ⋄

CURRY, H.B. *Combinatory logic* ⋄ B40 B98 ⋄

DUBARLE, H.D. *Initiation a la logique* ⋄ B98 ⋄

GOODSTEIN, R.L. *Mathematical logic* ⋄ B98 ⋄

HERMES, H. *Einfuehrung in die mathematische Logik. Klassische Praedikatenlogik* ⋄ B10 B98 ⋄

Hu, Shihua *Mathematical logic, its fundamental properties and scientific significance (Chinese)*
⋄ B98 ⋄

Ladriere, J. *Les limitations internes des formalismes. Etude sur la signification du theoreme de Goedel et des theoremes apparentes dans la theorie des fondements des mathematiques* ⋄ A05 B98 F30 F98 ⋄

Nidditch, P.H. *Introductory formal logic of mathematics*
⋄ B98 ⋄

Pasquinelli, A. *Introduzione alla logica simbolica*
⋄ B98 ⋄

Suppes, P. *Introduction to logic* ⋄ B98 ⋄

1958

Ajdukiewicz, K. *Abriss der Logik* ⋄ B98 ⋄

Bar-Hillel, Y. & Dalen van, D. & Fraenkel, A.A. & Levy, A. *Foundations of set theory*
⋄ A05 B98 E30 E70 E98 F98 ⋄

Bernays, P. & Fraenkel, A.A. *Axiomatic set theory*
⋄ B98 E98 ⋄

Boole, G. *An investigation of the laws of thought, on which are founded the mathematical theories of logic and probabilities* ⋄ A05 B98 G05 ⋄

Christian, R.R. *Introduction to logic and sets*
⋄ B98 E98 ⋄

Culbertson, J.T. *Mathematics and logic for digital devices* ⋄ B98 G05 ⋄

Curry, H.B. & Feys, R. *Combinatory logic. Volume I*
⋄ B40 B98 ⋄

Heyting, A. *Intuitionism in mathematics*
⋄ A05 B98 F50 F55 F98 ⋄

Miller, James Wilkin *Logic workbook* ⋄ B98 ⋄

Robinson, A. *Outline of an introduction to mathematical logic I,II,III* ⋄ B98 C07 G05 ⋄

Wang, Hao *Eighty years of foundational studies (German and French summaries)* ⋄ A05 A10 B98 ⋄

1959

Beth, E.W. *The foundation of mathematics. A study in the philosophy of science* ⋄ A05 B98 ⋄

Bochenski, I.M. *A precis of mathematical logic*
⋄ B98 ⋄

Davis, Martin D. *Lecture notes on mathematical logic*
⋄ B98 ⋄

Exner, R.M. & Rosskopf, M.F. *Logic in elementary mathematics* ⋄ B98 ⋄

Fraenkel, A.A. *Mengenlehre und Logik* ⋄ B98 E98 ⋄

Novikov, P.S. *Elements of mathematical logic (Russian)*
⋄ B98 ⋄

Popov, A.I. *Introduction to mathematical logic (Russian)*
⋄ B98 ⋄

Robinson, A. *Outline of an introduction to mathematical logic IV* ⋄ B10 B98 C07 C98 G05 ⋄

Schipper, E.W. & Schuh, E. *A first course in modern logic* ⋄ B98 ⋄

1960

Casari, E. *Lineamenti di logica matematica* ⋄ B98 ⋄

Halberstadt, W.H. *Introduction to modern logic: an elementary textbook of symbolic logic* ⋄ B98 ⋄

Kleene, S.C. *Mathematical logic: constructive and non-constructive operations* ⋄ B98 D98 ⋄

Lewis, C.I. *A survey of symbolic logic (Reprinted edition)*
⋄ B45 B98 ⋄

Lu, Yangci & Lu, Zhongwan & Tang, Zhisong & Wan, Zhexian *Mathematical logic and its application (Chinese)* ⋄ B98 ⋄

Prijatelj, N. *Introduction to mathematical logic (Slovenian)* ⋄ B98 ⋄

Ruby, L. *Logic: An introduction* ⋄ B98 ⋄

Schmidt, H.A. *Mathematische Gesetze der Logik. I. Vorlesungen ueber Aussagenlogik*
⋄ B05 B98 F50 F98 ⋄

Schuette, K. *Beweistheorie*
⋄ B98 F05 F15 F30 F35 F98 ⋄

Sugihara, T. *Modern logic (Japanese)* ⋄ B98 ⋄

1961

Allen, L.E. & Brooks, R.B.S. & Dickoff, J.W. & James, P.A. *The ALL project (accelerated learning of logic)* ⋄ B98 ⋄

Ambrose, A. & Lazerowitz, M. *Logic: the theory of formal inference* ⋄ B98 ⋄

Freudenthal, H. *Exact logic (Dutch)* ⋄ A05 B98 ⋄

Grzegorczyk, A. *An outline of mathematical logic (Polish)* ⋄ B98 ⋄

Grzegorczyk, A. *Logika popularna* ⋄ B98 ⋄

Lee, H.N. *Symbolic logic* ⋄ B98 ⋄

Lukasiewicz, J. *Logistic and philosophy (Polish)*
⋄ A05 B98 ⋄

Munoz, V. *De la axiomatica a los sistemas formales*
⋄ B30 B98 ⋄

Scholz, H. & Hasenjaeger, G. *Grundzuege der mathematischen Logik* ⋄ B98 ⋄

Smullyan, R.M. *Theory of formal systems*
⋄ B98 D03 D05 D20 D25 D98 ⋄

Stoll, R.R. *Sets, logic and axiomatic theories*
⋄ B98 E98 ⋄

1962

Anderson, J.M. & Johnstone Jr., H.W. *Natural deduction. The logical basis of axiom systems*
⋄ B98 F07 F98 ⋄

Arnold, B.H. *Logic and boolean algebra*
⋄ B05 B98 G05 ⋄

Badawi, A.R. *Formale und mathematische Logik (Arabic)* ⋄ B98 ⋄

Beth, E.W. *Formal methods. An introduction to symbolic logic and to the study of effective operations in arithmetic and logic* ⋄ B28 B98 F30 F50 F98 ⋄

Clark, R. & Welsh Jr., P.J. *Introduction to logic*
⋄ B98 ⋄

Dopp, J. *Logiques construites par une methode de deduction naturelle* ⋄ B98 F05 F07 F50 F98 ⋄

Hasenjaeger, G. *Einfuehrung in die Grundbegriffe und Probleme der modernen Logik* ⋄ A05 A10 B98 ⋄

Kaesbauer, M. & Kutschera von, F. (eds.) *Logik und Logikkalkuel* ⋄ B98 ⋄

Lorenzen, P. *Metamathematik*
⋄ A05 B98 C60 D20 D35 D98 F98 ⋄

Mitchell, D. *An introduction to logic* ⋄ B98 ⋄

Pogorzelski, W.A. & Slupecki, J. *On mathematical proof (Polish)* ⋄ B98 ⋄

SCHUETTE, K. *Lecture notes in mathematical logic. Vol.I*
  ◇ B98 F15 ◇
SMULLYAN, A. *Fundamentals of logic* ◇ B98 ◇
ZINOV'EV, A.A. *Propositional logic and theory of deduction (Russian)* ◇ B98 ◇

1963

AJZERMAN, M.A. & GUSEV, L.A. & ROZONOEHR, L.I. & SMIRNOVA, I.M. & TAL', A.A. *Logic, automata and algorithms (Russian)* ◇ B98 D05 D10 D98 ◇
BLYTH, J.W. & JACOBSON JR., J.H. *Class logic. A programmed text* ◇ B98 ◇
BORKOWSKI, L. & SLUPECKI, J. *Elements of mathematical logic and set theory (Polish)* ◇ A05 B98 E98 ◇
BRAUN, E.L. *Digital computer design: logic, circuitry, and synthesis* ◇ B70 B98 ◇
CURRY, H.B. *Foundations of mathematical logic* ◇ B98 ◇
HARBECK, G. *Einfuehrung in die formale Logik* ◇ B98 ◇
HILTON, A.M. *Logic, computing machines, and automation* ◇ A05 B98 D05 ◇
KHARIN, N.N. *Mathematical logic and the theory of sets (Russian)* ◇ A05 B98 ◇
KNEEBONE, G.T. *Mathematical logic and the foundations of mathematics: an introduction* ◇ A05 A10 B98 ◇
MO, SHAOKUI *Notes on mathematical logic (Chinese)* ◇ B98 ◇
MOURANT, J.A. *Formal logic: an introductory text book* ◇ B98 ◇
QUINE, W.V.O. *Set theory and its logic*
  ◇ A10 B98 E30 E98 ◇
RASIOWA, H. & SIKORSKI, R. *The mathematics of metamathematics*
  ◇ B98 C98 F50 F98 G05 G10 G98 ◇
ROBINSON, A. *Introduction to model theory and to the metamathematics of algebra*
  ◇ B98 C60 C98 H15 ◇
SALMON, W.C. *Logic* ◇ B98 ◇
SCHUETTE, K. *Lecture notes in mathematical logic. Vol.II*
  ◇ B98 F15 ◇
SMULLYAN, R.M. *First order logic* ◇ B98 F07 F98 ◇
STOLL, R.R. *Set theory and logic* ◇ B98 E98 ◇
SULLIVAN, D.J. *Fundamentals of logic* ◇ B98 ◇
WANG, HAO *The predicate calculus* ◇ B98 ◇

1964

AGAZZI, E. *La logica simbolica* ◇ A05 B98 ◇
ANGELL, R.B. *Reasoning and logic* ◇ B98 ◇
BAR-HILLEL, Y. *Language and information: selected essays on their theory and application*
  ◇ A05 B65 B98 ◇
BLACK, M. *Critical thinking: an introduction to logic and scientific method* ◇ A05 B98 ◇
BROWNE, S.S.S. *Fundamentals of deductive logic*
  ◇ B98 ◇
CARNEY, J.D. & SCHEER, R.K. *Fundamentals of logic*
  ◇ B98 ◇
DEVIDE, V. *Mathematical logic. Part I (The classical logic of propositions) (Croatian) (English summary)*
  ◇ B05 B98 ◇
DINKINES, F. *Elementary concepts of modern mathematics. Part 1. Elementary theory of sets. Part 2. Introduction to mathematical logic* ◇ B98 E98 ◇

FISK, M. *A modern formal logic* ◇ B98 ◇
HILL, S. & SUPPES, P. *First course in mathematical logic*
  ◇ B98 ◇
KALISH, D. & MONTAGUE, R. *Logic. Techniques of formal reasoning* ◇ B98 ◇
KEENE, G.B. *First-order functional calculus* ◇ B98 ◇
KLAUS, G. *Moderne Logik. Abriss der formalen Logik*
  ◇ B98 ◇
LIGHTSTONE, A.H. *The axiomatic method: an introduction to mathematical logic* ◇ B98 ◇
MANGIONE, C. *Elementi di logica matematica* ◇ B98 ◇
MENDELSON, E. *Introduction to mathematical logic*
  ◇ B98 ◇
RESCHER, N. *Introduction to logic* ◇ A05 B98 ◇
STAHL, G. *Elementos de la metalogica y metamatematica*
  ◇ B98 ◇

1965

AAQVIST, L. *A new approach to the logical theory of interrogatives. Part I. Analysis* ◇ A05 B98 ◇
AJDUKIEWICZ, K. *Pragmatic logic (Polish)*
  ◇ A05 B98 ◇
BARKER, S.F. *The elements of logic* ◇ B98 ◇
BARLINGAY, S.S. *A modern introduction to Indian logic*
  ◇ A10 B98 ◇
BINFORD, F. *Solution to the exercises in: "First course in mathematical logic"* ◇ B98 ◇
BROWN, P.L. & STUBERMAN, W.E. *Elementary modern logic* ◇ B98 ◇
CHOMSKY, N. *Aspects of the theory of syntax* ◇ B98 ◇
DICKOFF, J.W. & JAMES, P.A. *Symbolic logic and language: a programmed text* ◇ B98 ◇
EVES, H.W. *An introduction to the foundations and fundamental concepts of mathematics* ◇ B98 ◇
HU, SZETIEN *Threshold logic* ◇ B70 B98 ◇
HUGHES, G.E. & LONDEY, D.G. *The elements of formal logic* ◇. B98 ◇
JEFFREY, R.C. *The logic of decision* ◇ B48 B98 ◇
KREISEL, G. *Mathematical logic*
  ◇ B98 F10 F35 F50 F98 ◇
KRETZMANN, N. *Elements of formal logic* ◇ B98 ◇
LEMMON, E.J. *Beginning logic* ◇ B98 ◇
LUCHINS, A.S. & LUCHINS, E.H. *Logical foundation of mathematics for behavioral scientists* ◇ B98 ◇
MATES, B. *Elementary logic* ◇ A10 B98 ◇
MO, SHAOKUI *Introduction to mathematical logic (Chinese)* ◇ B98 ◇
MOSTOWSKI, ANDRZEJ *Thirty years of foundational studies. Lectures on the development of mathematical logic and the studies of the foundations of mathematics in 1930-1964* ◇ A10 B98 C98 D98 E98 F98 ◇
PRAWITZ, D. *Natural deduction. A proof-theoretical study*
  ◇ B15 B98 F05 F07 F50 F98 ◇

1966

ACKERMANN, R.J. *Non-deductive inference* ◇ B98 ◇
CAPALDI, N. *Introduction to deductive logic* ◇ B98 ◇
CORNMAN, J.W. *Metaphysics, reference and language*
  ◇ A05 B98 ◇
DUTTON, J.D. (ED.) *Logics: an introduction with exercises*
  ◇ B98 ◇

Textbooks, surveys

ESSLER, W.K. *Einfuehrung in die Logik* ◊ B98 ◊
GRASSMANN, G. *Die Formenlehre oder Mathematik. 1.die Groessenlehre. 2.die Begriffslehre der Logik. 3.die Bindelehre oder Kombinationslehre. 4.die Zahlenlehre oder Arithmetik. 5.die Aussenlehre oder Ausdehnungslehre (reprinted)* ◊ B98 ◊
HACKSTAFF, L.H. *Systems of formal logic* ◊ B98 ◊
HASSE, M. *Grundbegriffe der Mengenlehre und Logik* ◊ B98 E98 ◊
KORFHAGE, R. *Logic and algorithms with applications to the computer and information sciences* ◊ B98 D98 ◊
KUPPERMAN, J. & MCGRADE, A.S. *Fundamentals of logic* ◊ B98 ◊
LEBLANC, H. *Techniques of deductive inference* ◊ B98 ◊
MAEHARA, S. *Introduction to mathematical logic (Japanese)* ◊ B98 ◊
MENNE, A. *Einfuehrung in die Logik* ◊ B98 ◊
ROBINSON, A. *Non-standard analysis* ◊ A10 B98 H05 H10 H15 H98 ◊
SKYRMS, B. *Choice and chance* ◊ A05 B48 B98 ◊
THOMAS, N.L. *Modern logic. An introduction* ◊ B98 ◊

1967

ACKERMANN, R.J. *An introduction to many-valued logics* ◊ B50 B98 ◊
ACKERMANN, R.J. *Introduction to many-valued logics* ◊ B50 B98 ◊
BLUMBERG, A.E. *Modern logic* ◊ A05 B98 ◊
CANGELOSI, V.E. *Compound statements and mathematical logic* ◊ B98 ◊
COATES, C.L. & LEWIS II, P.M. *Threshold logic* ◊ B70 B98 ◊
COPI, I.M. & GOULD, JAMES A. *Deontic logic, introduction* ◊ B45 B98 ◊
CURRY, H.B. *Combinatory logic* ◊ A05 B40 B98 ◊
DOPP, J. *Notions de logique formelle* ◊ B98 ◊
FRAISSE, R. *Cours de logique mathematique. Tome I. Relation, formule logique, compacite, completude* ◊ B98 C98 ◊
HAMBLIN, C.L. *Elementary formal logic* ◊ B98 ◊
JEFFREY, R.C. *Formal logic: Its scope and limits* ◊ B98 ◊
KENELLY, J.W. *Informal logic* ◊ B98 ◊
KILMISTER, C.W. *Language, logic and mathematics* ◊ A05 B98 D35 ◊
KLEENE, S.C. *Mathematical logic* ◊ B98 D20 D25 D55 D98 F30 F98 ◊
KREISEL, G. & KRIVINE, J.-L. *Elements de logique mathematique. Theorie des modeles* ◊ B98 C98 ◊
KUTSCHERA VON, F. *Elementare Logik* ◊ B98 ◊
LEONARD, H.S. *Principles of reasoning: an introduction to logic, methodology, and the theory of signs* ◊ A05 B98 ◊
MARGARIS, A. *First order mathematical logic* ◊ B98 ◊
NEIDORF, R. *Deductive forms: An elementary logic* ◊ B98 ◊
PONASSE, D. *Logique mathematique. Elements de base: calcul propositionnel, calcul des predicats* ◊ B98 ◊
RIEGER, L. *Algebraic methods of mathematical logic* ◊ A05 B98 G05 G15 ◊
ROURE, M.-L. *Elements de logique contemporaine* ◊ B98 ◊

SCHOCK, R. *Logik* ◊ B98 ◊
SHOENFIELD, J.R. *Mathematical logic* ◊ B98 C98 D98 E98 F98 ◊
SMIRNOVA, E.D. & TAVANETS, P.V. *Semantics in logic (Russian)* ◊ A05 B98 ◊
TERRELL, D.B. *Logic: A modern introduction to deductive reasoning* ◊ B98 ◊

1968

CHOMSKY, N. *Language and mind* ◊ A05 B65 B98 ◊
CRESSWELL, M.J. & HUGHES, G.E. *An introduction to modal logic* ◊ B45 B98 ◊
EHLERS, F. *Logic by way of set theory* ◊ B98 E98 ◊
HATCHER, W.S. *Foundations of mathematics* ◊ B98 E30 E70 E98 G30 ◊
HERMES, H. *Methodik der Mathematik und Logik* ◊ A05 A10 B98 ◊
ISEMINGER, G. *An introduction to deductive logic* ◊ B98 ◊
KAC, M. & ULAM, S.M. *Mathematics and logic: Retrospect and prospects* ◊ A05 A10 B98 ◊
KILGORE, W.J. *An introductory logic* ◊ B98 ◊
KRUGER, A.N. & MANICAS, P.T. *Essentials of logic* ◊ B98 ◊
PENZOV, YU.E. *Elements of mathematical logic and set theory (Russian)* ◊ B98 E98 ◊
ROOTSELAAR VAN, B. & STAAL, J.F. (EDS.) *Logic, methodology and philosophy of science III* ◊ B98 ◊
SCHAGRIN, M.L. *The language of logic. A programmed text* ◊ B98 ◊
SUMMERS, G.J. *New puzzles in logical deduction* ◊ B98 ◊
ZOLL, E.J. *Logic: A programmed text for 2-valued and 3-valued logics* ◊ B50 B98 ◊

1969

CHELLAS, B.F. *The logical form of imperatives* ◊ A05 B98 ◊
DODGE, C.W. *Sets, logic and numbers* ◊ B98 E98 ◊
ENNIS, R.H. *Ordinary logic* ◊ B98 ◊
FEYS, R. & FITCH, F.B. *Dictionary of symbols of mathematical logic* ◊ B98 ◊
FITTING, M. *Intuitionistic logic, model theory, and forcing* ◊ B98 C90 C98 E25 E35 E45 E50 F50 F98 ◊
FOULIS, D.J. *Fundamental concepts of mathematics* ◊ B98 ◊
GRIZE, J.-B. *Logique moderne. fasc.I: Logique des propositions et des predicats, deduction naturelle* ◊ B98 ◊
HIRSCHFELD, J. & MACHOVER, M. *Lectures on non-standard analysis* ◊ B98 H05 ◊
KAHANE, H. *Logic and philosophy. A modern introduction* ◊ B98 ◊
KEARNS, J.T. *Deductive logic: a programmed introduction* ◊ B98 ◊
KLEENE, S.C. *The new logic* ◊ B98 ◊
MICHALOS, A.C. *Principles of logic* ◊ B98 ◊
MIHAILESCU, E.G. *Mathematical logic. Elements of propositional and predicate calculus (Romanian)* ◊ B98 ◊
NAESS, A. *Introduction to logic and scientific method* ◊ B98 ◊

POLLOCK, J.L. *Introduction to symbolic logic* ⋄ B98 ⋄
RESCHER, N. *Topics in philosophical logic*
⋄ A05 B45 B98 C90 ⋄
ROBBIN, J.W. *Mathematical logic: a first course* ⋄ B98 ⋄
ROBISON, G.B. *An introduction to mathematical logic*
⋄ B98 ⋄
SHENG, C.L. *Threshold logic* ⋄ B75 B98 ⋄

1970

ACKERMANN, R.J. *Modern deductive logic* ⋄ B98 ⋄
AVENOSO, F.J. & CHEIFETZ, P.M. *Logic and set theory*
⋄ B98 E98 ⋄
BEACH, J. *Introduction to logic* ⋄ B98 ⋄
BETH, E.W. *Aspects of modern logic* ⋄ B98 ⋄
BITTINGER, M.L. *Logic and proof* ⋄ B98 ⋄
CARNEY, J.D. *Introduction to symbolic logic* ⋄ B98 ⋄
CASEY, H. & CLARK, M. *Logic: a practical approach*
⋄ B98 ⋄
COHEN, L.J. *The implications of induction*
⋄ A05 B48 B98 ⋄
DELONG, H. *A profile of mathematical logic*
⋄ A05 A10 B98 ⋄
DETLOVS, V.K. *Elements of mathematical logic and set theory (Latvian)* ⋄ B98 E98 ⋄
FIORENTINI, M. & MARRUCCELLI, A. *Complementi di matematiche moderne; Logica matematicea, teorie degli insiemi, strutture algebriche* ⋄ B98 E98 ⋄
LAVROV, I.A. *Logic and algorithms (Russian)*
⋄ B98 D98 ⋄
LAVROV, I.A. & MAKSIMOVA, L.L. *Problems in logic (Russian)* ⋄ B98 E98 ⋄
LORENZEN, P. *Logica formal* ⋄ B98 ⋄
MASSEY, G.J. *Understanding symbolic logic*
⋄ A05 B98 ⋄
MATSUMOTO, K. *Mathematical logic (Japanese)*
⋄ B98 ⋄
MINTS, G.E. (ED.) *Mathematical logic (Russian)*
⋄ A10 B98 ⋄
MITCHELL, D. *An introduction to logic* ⋄ B98 ⋄
MOSTOWSKI, A.WLODZIMIERZ & PAWLAK, Z. *Logik fuer Ingenieure (Polnisch)* ⋄ B98 ⋄
OBERSCHELP, A. *Mengenlehre und Logik* ⋄ B98 E98 ⋄
PANCZAKIEWICZ, M. *Mathematical logic (Polish)*
⋄ B98 ⋄
RENYI, A. *Foundations of probability* ⋄ B98 ⋄
RESNIK, M.D. *Elementary logic* ⋄ B98 ⋄
STOLYAR, A.A. *Introduction to elementary mathematical logic* ⋄ B98 ⋄
TAJTSLIN, M.A. *Model theory (Russian)* ⋄ B98 C98 ⋄
THOMASON, R.H. *Symbolic logic. An introduction*
⋄ B98 ⋄

1971

ALCHOURRON, C.E. & BULYGIN, E. *Normative systems*
⋄ B45 B98 ⋄
ALTHAM, J.E.J. *The logic of plurality*
⋄ A05 B98 C80 ⋄
BIZAM, G. & HERCZEG, J. *Play and logic (Hungarian)*
⋄ B98 ⋄
BREITKOPF, A. & KUTSCHERA VON, F. *Einfuehrung in die moderne Logik* ⋄ B98 ⋄

COMBES, M. *Fondements des mathematiques*
⋄ A05 A10 B98 E98 ⋄
FRAASSEN VAN, B.C. *Formal semantics and logic*
⋄ B98 ⋄
GEORGESCU, G. *Praedikatenkalkuel I* ⋄ B98 ⋄
GLUSHKOV, V.M. *Einfuehrung in die technische Kybernetik Band 1, 2* ⋄ B98 ⋄
GRIZE, J.-B. *Logique moderne. fasc.II: Logique des propositions et des predicats, tables de verite et axiomatisation* ⋄ B98 ⋄
HANNA, S.C. & SABER, J.C. *Sets and logic* ⋄ B98 E98 ⋄
HUNTER, G. *Metalogic: an introduction to the metatheory of standard firstorder logic* ⋄ B10 B98 ⋄
KNEEBONE, G.T. *Mathematical logic in relation to ordinary mathematics* ⋄ A05 B98 ⋄
ONICESCU, O. *Principes de logique et de philosophie mathematique* ⋄ A05 B98 ⋄
POSPESEL, H. *Arguments: deductive logic exercises*
⋄ B98 ⋄
PURTILL, R.L. *Logic for philosophers* ⋄ B98 ⋄
RASIOWA, H. *Introduction to modern mathematics (Polish)* ⋄ B30 B98 C05 E98 ⋄
RESCHER, N. & URQUHART, A.I.F. *Temporal logic*
⋄ A05 B45 B98 ⋄
ROGERS, R. *Mathematical logic and formalized theories. A survey of basic concepts and results* ⋄ B98 C98 ⋄
SABBAGH, G. *Logique mathematique I. Generalites*
⋄ B98 ⋄
STANDLEY, G.B. *New methods in symbolic logic* ⋄ B98 ⋄
YASUHARA, A. *Recursive function theory and logic*
⋄ B98 D10 D20 D98 ⋄

1972

ASH, C.J. & BRICKHILL, C.J. & CROSSLEY, J.N. & STILLWELL, J.C. & WILLIAMS, N.H. *What is mathematical logic ?* ⋄ B98 ⋄
BALLARD, K.E. *Study guide for Copi: Introduction to logic. A self-instructional supplement* ⋄ B98 ⋄
BERNARDO DI, G. *Introduzione alla logica dei sistemi normativi* ⋄ B98 ⋄
BRODSKIJ, I.N. *Elementare Einfuehrung in die Logik. 2.ueberarb. Aufl (Russian)* ⋄ B98 ⋄
BULL, R.A. *Mathematical logic* ⋄ B98 ⋄
CHOMSKY, N. *Studies on semantics in generative grammars* ⋄ B65 B98 D05 ⋄
DALEN VAN, D. *Logik und formale Theorien (Niederlaendisch)* ⋄ B98 ⋄
EMDE, H. & REYERSBACH, W. & STROMBACH, W. *Mathematische Logik. Ihre Grundprobleme in Theorie und Anwendung* ⋄ B98 G05 G10 ⋄
ENDERTON, H.B. *A mathematical introduction to logic*
⋄ B98 C98 F30 ⋄
FRAISSE, R. *Cours de logique mathematique. Tome 2: Theorie des modeles* ⋄ B98 C98 ⋄
GRADSHTEJN, I.S. *Direct and inverse theorems (Russian)*
⋄ B98 ⋄
GUTTENPLAN, S.D. & TAMNY, M. *Logic: a comprehensive introduction* ⋄ A05 B98 ⋄
HACKING, I. *A concise introduction to logic* ⋄ B98 ⋄
HERINGER, H.J. *Formale Logik und Grammatik*
⋄ B65 B98 D05 ⋄

HINDLEY, J.R. & LERCHER, B. & SELDIN, J.P. *Introduction to combinatory logic* ◊ B40 B98 F98 ◊

KLOETZER, G. & RAUTENBERG, W. *Im Grenzbereich Algebra, Logik, Maschinen. 10 Jahre Forschungsarbeit der Nowosibirsker Schule A.I. Malcev's (eine Studie)* ◊ A10 B98 C05 D05 D98 ◊

LEBLANC, H. & WISDOM, W.A. *Deductive logic* ◊ B98 ◊

MAREK, W. & ONYSZKIEWICZ, J. *Elements of logic and foundations of mathematics in problems (Polish)* ◊ B98 C98 D98 E98 ◊

MARKWALD, W. *Einfuehrung in die formale Logik und Metamathematik* ◊ B98 ◊

OTEPANOV, V.I. & VERKHOZIN, O.M. *Problems in logic (Russian)* ◊ B98 E98 ◊

PORTE, J. *La logique mathematique et le calcul mecanique* ◊ B98 D98 ◊

PURTILL, R.L. *Logical thinking* ◊ B98 ◊

RASIOWA, H. *Introduction to set theory and mathematical logic (Bulgarian)* ◊ B98 C05 E98 ◊

RAUTENBERG, W. *Zur praktischen und theoretischen Wirksamkeit der mathematischen Logik* ◊ B98 ◊

SEREBRYANNIKOV, O.F. *Heuristic principles and logical calculi* ◊ B98 ◊

SMIRNOV, V.A. *Formal derivation and logical calculi (Russian)* ◊ B98 C07 C40 C98 F05 F07 F98 ◊

STEEN, S.W.P. *Mathematical logic, with special reference to the natural numbers* ◊ B98 F30 F98 ◊

1973

BAGINSKI, M. & GLAEWE, W. & GOLL, P. & LIST, G. & LOESCHAU, G. & MERTENS, A. & SCHWANITZ, G. & WALTER, M. *Einfuehrung in die mathematische Logik, Einfuehrung in die Mengenlehre, Aufbau der Zahlenbereiche* ◊ B05 B28 B98 E98 ◊

BRODY, B.A. *Logic, theoretical and applied* ◊ B98 ◊

BYERLY, H.C. *A primer of logic* ◊ B98 ◊

CARNAP, R. *Grundlagen der Logik und Mathematik* ◊ A05 B98 ◊

CHANG, C.L. & LEE, R.C.T. *Symbolic logic and mechanical theorem proving* ◊ B35 B98 ◊

CRESSWELL, M.J. *Logics and languages* ◊ A05 B98 ◊

EARLE, J.N.F. *Logic* ◊ B98 ◊

ERSHOV, YU.L. & PALYUTIN, E.A. & TAJTSLIN, M.A. *Mathematical logic (Russian)* ◊ B98 C98 ◊

GEORGESCU, G. *The theory of categories and mathematical logic (Romanian)* ◊ B98 ◊

GIRLING, B. & MORING, H. *Logic and logic design* ◊ B70 B98 ◊

MAKINSON, D. *Topics in modern logic* ◊ B98 ◊

MOSHCHENSKIJ, V.A. *Lectures on mathematical logic (Russian)* ◊ B98 D10 D20 ◊

NIINILUOTO, I. *Mathematical logic (Finnish)* ◊ B98 ◊

PONASSE, D. *Mathematical logic* ◊ B98 ◊

RUTZ, P. *Zweiwertige und mehrwertige Logik - ein Beitrag zur Geschichte und Einheit der Logik* ◊ A10 B98 ◊

SEIFFERT, H. *Einfuehrung in die Logik. Logische Propaedeutik und formale Logik* ◊ B98 ◊

STAHL, G. *Elementos de metamatematica* ◊ B98 ◊

1974

ASSER, G. *Mathematische Logik und Grundlagen der Mathematik* ◊ B98 ◊

BOOLOS, G. & JEFFREY, R.C. *Computability and logic* ◊ B98 D98 ◊

CHENIQUE, F. *Comprendre la logique moderne. Tome I: Classes, propositions et predicats* ◊ B98 E98 ◊

CHENIQUE, F. *Comprendre la logique moderne. Tome II: Logiques non classiques, relations et structures* ◊ B98 ◊

COOPER, W.S. *Set theory and syntactic description* ◊ B98 E75 ◊

CORNIDES, T. *Ordinale Deontik* ◊ B45 B98 ◊

DAVIS, MARTIN D. *Computability* ◊ B98 D98 ◊

DETLOVS, V.K. *Mathematical logic (Latvian)* ◊ B98 ◊

FITCH, F.B. *Elements of combinatory logic* ◊ B40 B98 ◊

GLADKIJ, A.V. *Lecture notes on mathematical logic and set theory (Russian)* ◊ B98 E98 ◊

HALLERBERG, A.E. *Logic in mathematics: An elementary approach* ◊ B98 ◊

HALLERBERG, A.E. *Mathematical proof: An elementary approach* ◊ B98 ◊

JACOBSON, A. & KLEMKE, E.D. & ZABECH, F. (EDS.) *Readings in semantics* ◊ A05 B98 ◊

OBERSCHELP, A. *Elementare Logik und Mengenlehre I* ◊ B98 E98 ◊

PETERS, F.E. *Einfuehrung in mathematische Methoden der Informatik* ◊ B70 B98 D80 D98 ◊

POGORZELSKI, W.A. & PRUCNAL, T. *Introduction to mathematical logic. Part I. Elements of the algebra of propositional logic (Polish)* ◊ B98 G05 ◊

POSPESEL, H. *Introduction to logic. Propositional logic* ◊ B05 B98 ◊

RASIOWA, H. *An algebraic approach to non-classical logics* ◊ B98 F50 G05 G10 G20 G25 G98 ◊

SCHENK, G. *Die Logik* ◊ A10 B98 ◊

STYAZHKIN, N.I. *Logic with the elements of mathematical logic (Russian)* ◊ B98 ◊

1975

ADAMS, E.W. *The logic of conditionals. An application of probability to deductive logic* ◊ A05 B48 B98 ◊

ANDERSON, A.R. & BELNAP JR., N.D. *Entailment. The logic of relevance and necessity I* ◊ A05 B46 B98 ◊

ANDERSON, A.R. & BARCAN MARCUS, R. & MARTIN, R.M. (EDS.) *The logical enterprise* ◊ A05 A10 B98 ◊

BARNES, D.W. & MACK, J.M. *An algebraic introduction to mathematical logic* ◊ B98 G98 ◊

BARWISE, J. *Admissible sets and structures. An approach to definability theory* ◊ B98 C40 C70 C98 D60 D98 E30 E98 ◊

BISWAS, N.N. *Introduction to logic and switching theory* ◊ B98 ◊

BODDENBERG, E. *Logik. I* ◊ B98 ◊

CIRULIS, J. *Lectures on mathematical logic and set theory. Part I: Mathematical logic (Russian)* ◊ B98 ◊

CROSSLEY, J.N. *What is mathematical logic ?* ◊ B98 F30 ◊

EDEL'MAN, S.L. *Mathematical logic (Russian)* ⋄ B98 ⋄
FRIEDMAN, H.M. *One hundred and two problems in mathematical logic* ⋄ B98 C98 D98 E98 F98 ⋄
HEGENBERG, L. *Logica: Simbolizacao e deducao* ⋄ B98 ⋄
KREISEL, G. *Was hat die Logik in den letzten 25 Jahren fur die Mathematik geleistet?* ⋄ A05 B98 C75 D40 ⋄
LAVROV, I.A. & MAKSIMOVA, L.L. *Problems in set theory, mathematical logic and the theory of algorithms (Russian)* ⋄ B98 D98 E98 ⋄
MANASTER, A.B. *Completeness, compactness, and undecidability: an introduction to mathematical logic* ⋄ B98 C07 D35 D98 ⋄
MOSS, J.M.B. & SCOTT, D.S. *Bibliography of logic books* ⋄ A10 B98 ⋄
RESCHER, N. *A theory of possibility* ⋄ A05 B98 ⋄
WESSEL, H. & ZINOV'EV, A.A. *Logische Sprachregeln. Eine Einfuehrung in die Logik* ⋄ A05 B98 ⋄

1976

BLUMBERG, A.E. *Logic: A first course* ⋄ B98 ⋄
BORKOWSKI, L. *Formale Logik. Logische Systeme. Einfuehrung in die Metalogik* ⋄ A05 B98 ⋄
CHRISTIAN, C.C. *Die Bedeutung der Mengentheorie als Grundlagenwissenschaft* ⋄ A05 B98 E10 E30 ⋄
CHUAQUI, R.B. *Introduccion a la metamatematica y sus aplicaciones* ⋄ B98 ⋄
COURVOISIER, M. & LAGASSE, J. & RICHARD, J. *Logique combinatoire. 3e ed* ⋄ B40 B98 ⋄
HALKOWSKA, K. & PIROG-RZEPECKA, K. & SLUPECKI, J. *Mathematical logic (Polish)* ⋄ B98 C07 C98 E98 ⋄
LEBLANC, H. *Truth-value semantics* ⋄ B98 ⋄
MALATESTA, M. *Logistica I: Introduzione. La logica degli enunciati* ⋄ B98 ⋄
MONK, J.D. *Mathematical logic* ⋄ B98 ⋄
MOSTERIN, J. *Logica de primer order* ⋄ B98 ⋄
PABION, J.F. *Logique mathematique* ⋄ B98 ⋄
POSPESEL, H. *Introduction to logic. Predicate logic* ⋄ B10 B98 ⋄
RICHTER, M.M. *Mathematische Logik I* ⋄ B98 ⋄
TAPSCOTT, B.L. *Elementary applied symbolic logic* ⋄ B98 ⋄
ZIEMBINSKI, Z. *Practical logic (Polish)* ⋄ A05 B45 B98 ⋄

1977

BARWISE, J. (ED.) *Handbook of mathematical logic* ⋄ B98 C98 D98 E98 F98 H98 ⋄
BELL, J.L. & MACHOVER, M. *A course in mathematical logic* ⋄ B98 ⋄
BERGMANN, E. & NOLL, H. *Mathematische Logik mit Informatik-Anwendungen* ⋄ A05 A10 B98 ⋄
BURKS, A.W. *Chance, cause, reason. An inquiry into the nature of scientific evidence* ⋄ A05 B98 ⋄
GEORGE, F.H. *Precision, language and logic* ⋄ A05 B98 ⋄
GLADKIJ, A.V. *The language of mathematical logic* ⋄ B98 ⋄
GRANDY, R.E. *Advanced logic for applications* ⋄ B98 ⋄
HODGES, W. *Logic* ⋄ B98 ⋄
KAAZ, M.A. *Elemente der mathematischen Logik fuer den Gebrauch in Physik und Technik* ⋄ B98 ⋄

KARPOV, V.G. & MOSHCHENSKIJ, V.A. *Mathematical logic and discrete mathematics (Russian)* ⋄ B98 E98 ⋄
KREMPA, J. & MAZBIC-KULMA, B. *Elements of logic, set theory and algebra (Elementy logiki, teorii mnogosci i algebry) (Polish)* ⋄ B98 E98 ⋄
LEMMON, E.J. *An introduction to modal logic* ⋄ A10 B45 B98 ⋄
MANIN, YU.I. *A course in mathematical logic* ⋄ B98 C07 D98 E35 E50 F30 G12 ⋄
SCHREIBER, P. *Grundlagen der Mathematik* ⋄ B98 ⋄
STILLWELL, J.C. *Concise survey of mathematical logic* ⋄ B98 C98 D10 D35 ⋄
VASILACHE, S. *Ensembles, structures, categories, faisceaux* ⋄ B98 E98 G30 ⋄

1978

BAXANDALL, P.R. & BROWN, W.S. & ROSE, G.S.C. & WATSON, F.R. *Proof in mathematics ("If", "then" and "perhaps"). A collection of material illustrating the nature and variety of the idea of proof in mathematics* ⋄ B98 ⋄
COOPER, W.S. *Foundations of logico-linguistics: A unified theory of information, language, and logic* ⋄ A05 B98 ⋄
DENNING, P.J. & DENNIS, J.B. & QUALITZ, J.E. *Machines, languages, and computation* ⋄ B98 D05 D10 D98 ⋄
DILLER, J. *Klassische Praedikatenlogik* ⋄ B98 ⋄
DURNEV, V.G. *Elements of set theory and mathematical logic (Russian)* ⋄ B98 E98 ⋄
EBBINGHAUS, H.-D. & FLUM, J. & THOMAS, WOLFGANG *Einfuehrung in die mathematische Logik* ⋄ B98 C95 C98 ⋄
FACIONE, P.A. & SCHERER, D. *Logic and logical thinking: a modular approach* ⋄ A05 B98 ⋄
GAUTHIER, Y. *Methodes et concepts de la logique formelle* ⋄ A05 B98 ⋄
HALKOWSKA, K. & PIROG-RZEPECKA, K. & SLUPECKI, J. *Logic and set theory (Polish)* ⋄ B98 E98 ⋄
HAMILTON, A.G. *Logic for mathematicians* ⋄ B98 ⋄
KHROMOJ, YA.V. *A collection of exercises and problems in mathematical logic (Ukrainian)* ⋄ B98 ⋄
KONDAKOV, N.I. *Woerterbuch der Logik* ⋄ A05 A10 B98 ⋄
LEMMON, E.J. *Beginning logic* ⋄ B98 ⋄
LIGHTSTONE, A.H. *Mathematical logic. An introduction to model theory. Edited by H. B. Enderton* ⋄ B98 C98 H05 ⋄
LOLLI, G. *Lezioni di logica matematica* ⋄ B98 ⋄
OBERSCHELP, A. *Elementare Logik und Mengenlehre II* ⋄ B98 E98 ⋄
RICHTER, M.M. *Logikkalkuele* ⋄ B35 B98 F05 G10 ⋄
RUTKOWSKI, A. *Elements of mathematical logic (Polish)* ⋄ B98 ⋄
SMULLYAN, R.M. *What is the name of this book? The riddle of Dracula and other logical puzzles* ⋄ B98 ⋄
TAKEUTI, G. *Two applications of logic to mathematics* ⋄ B98 C90 E40 E75 F05 F30 F35 G12 ⋄
TENNANT, N. *Natural logic* ⋄ B98 ⋄

1979

APT, K.R. *Ten years of Hoare's logic, a survey* ⋄ B98 ⋄

BOOLOS, G. *The unprovability of consistency. An essay in modal logic* ⋄ B25 B28 B45 B98 F30 F98 ⋄

BOYER, R.S. & MOORE, J.S. *A computational logic* ⋄ B98 D70 ⋄

CARDWELL, C.E. *Arguments and inference. An introduction to symbolic logic* ⋄ B98 ⋄

EDWARDS, R.E. *A formal background to mathematics. Ia, Ib: Logic, sets and numbers* ⋄ B98 E98 ⋄

ERSHOV, YU.L. & PALYUTIN, E.A. *Mathematical logic (Russian)* ⋄ B98 ⋄

GOLDBLATT, R.I. *Topoi. The categorial analysis of logic* ⋄ B98 C98 E98 F35 F50 F98 G30 ⋄

GOMEZ CALDERON, J. *Logica simbolica* ⋄ B98 ⋄

HOFSTADTER, D.R. *Goedel, Escher, Bach: an eternal golden braid* ⋄ A05 B98 D99 ⋄

MALITZ, J. *Introduction to mathematical logic. Set theory, computable functions, model theory* ⋄ B98 C98 D98 ⋄

MANIN, YU.I. *Provable and unprovable (Russian)* ⋄ A05 B98 E35 E50 F30 F98 G12 ⋄

MOSTOWSKI, ANDRZEJ *An excerpt from the book Mathematical logic* ⋄ B98 ⋄

PURTILL, R.L. *Logic. Argument, refutation, and proof* ⋄ B98 ⋄

RAUTENBERG, W. *Klassische und nichtklassische Aussagenlogik* ⋄ B45 B50 B98 C90 ⋄

ROBINSON, JOHN ALAN *Logic: form and function. The mechanization of deductive reasoning* ⋄ A10 B10 B35 B98 ⋄

THIELE, R. *Mathematische Beweise* ⋄ B98 ⋄

WOJCIECHOWSKA, A. *Elements of logic and set theory* ⋄ B98 E98 ⋄

WU, YUNZENG *Two problems of modern mathematical logic and its philosophic significance (Chinese)* ⋄ A05 B98 ⋄

1980

ARRUDA, A.I. *A survey of paraconsistent logic* ⋄ A05 B53 B98 E70 ⋄

BARTNICK, J. *Predicate logic without bound variables* ⋄ B98 ⋄

BAUDISCH, A. & SEESE, D.G. & TUSCHIK, H.-P. & WEESE, M. *Decidability and generalized quantifiers* ⋄ B25 B98 C10 C60 C80 C98 ⋄

BERGMANN, MERRIE & MOOR, J. & NELSON, JACK *The logic book* ⋄ B98 ⋄

CHELLAS, B.F. *Modal logic. An introduction* ⋄ B45 B98 ⋄

DALEN VAN, D. *Logic and structure* ⋄ B98 C98 ⋄

LYNCH, E.P. *Applied symbolic logic* ⋄ B98 ⋄

MO, SHAOKUI *Introduction to mathematical logic (Chinese)* ⋄ B98 ⋄

PUTNAM, H. *Models and reality* ⋄ A05 B98 C99 ⋄

SKORUBSKIJ, V.I. *Arithmetical and logical foundations of digital machines. Textbook (Russian)* ⋄ B70 B98 ⋄

1981

ASSER, G. *Einfuehrung in die mathematische Logik. Teil III: Praedikatenlogik hoeherer Stufe* ⋄ B15 B98 ⋄

BARENDREGT, H.P. *The lambda calculus, its syntax and semantics* ⋄ B40 B98 ⋄

BECKMAN, F.S. *Mathematical foundations of programming* ⋄ B98 ⋄

BOEHME, G. *Einstieg in die mathematische Logik* ⋄ B98 ⋄

GAINES, B.R. & MAMDANI, E.H. (EDS.) *Fuzzy reasoning and its applications* ⋄ B52 B98 ⋄

GLADKIJ, M. *Mathematical logic and mathematical linguistics (Russian)* ⋄ B98 ⋄

GOZE, M. & LUTZ, R. *Nonstandard analysis. A pratical guide with applications* ⋄ B98 H05 ⋄

HARPER, W.L. *A sketch of some recent developments in the theory of conditionals* ⋄ B48 B98 ⋄

HU, SHIHUA & LU, ZHONGWAN *Foundation of mathematical logic I (Chinese)* ⋄ B98 ⋄

KOSOVSKIJ, N.K. *Elements of mathematical logic and its applications to the theory of subrecursive algorithms (Russian)* ⋄ B98 D20 D98 F30 F60 F98 ⋄

MCCAWLEY, J.D. *Everything that linguists have always wanted to know about logic, but were ashamed to ask* ⋄ B98 ⋄

NUTE, D.E. *Essential formal semantics* ⋄ B98 ⋄

POGORZELSKI, W.A. *Classical calculus of quantifiers. Outline of the theory (Polish)* ⋄ B98 ⋄

TAKEUTI, G. *Logic and set theory* ⋄ B51 B98 E40 E70 F50 ⋄

WANG, HAO *Popular lectures on mathematical logic (Chinese)* ⋄ A05 B98 C98 E98 F30 F98 ⋄

1982

BITTINGER, M.L. *Logic, proof, and sets* ⋄ B98 ⋄

FISHER, A. *Formal number theory and computability* ⋄ B98 D20 F30 F98 ⋄

HATCHER, W.S. *The logical foundations of mathematics* ⋄ B98 ⋄

HU, SHIHUA & LU, ZHONGWAN *Foundation of mathematical logic II (Chinese)* ⋄ B98 ⋄

PRIJATELJ, N. *Foundations of mathematical logic I (Slovenian)* ⋄ B98 ⋄

SCOTT, D.S. *Lectures on a mathematical theory of computation* ⋄ B40 B98 D75 ⋄

TSELISHCHEV, V.V. (ED.) *Problems of logic and methodology of science (Russian)* ⋄ B98 ⋄

WANG, XIANJUN *Introduction to mathematical logic (Chinese)* ⋄ B98 ⋄

XU, LIZHI & YUAN, XIANGWAN & ZHENG, YUXIN & ZHU, WUJIA *Antinomies and the foundational problem of mathematics II (Chinese) (English summary)* ⋄ A05 B98 ⋄

ZAHN, P. *Ein argumentativer Weg zur Logik* ⋄ B98 ⋄

1983

BACHMANN, H. *Der Weg der mathematischen Grundlagenforschung* ⋄ A10 B98 ⋄

EICHHORN, H. *Conceptual and conventional definitions in the mathematical sciences* ⋄ A05 B98 ⋄

ENGELER, E. *Metamathematik der Elementarmathematik* ⋄ B98 ⋄

GABBAY, D.M. & GUENTHNER, F. (EDS.) *Handbook of philosophical logic Vol. I: Elements of classical logic* ⋄ B98 C98 ⋄

GLUBRECHT, J.-M. & OBERSCHELP, A. & TODT, G. *Klassenlogik* ◊ B10 B98 E30 ◊

GOE, G. *Lezioni di logica* ◊ B98 ◊

SMULLYAN, R.M. *Dame oder Tiger? Logische Denkspiele und eine mathematische Novelle ueber Goedels grosse Entdeckung* ◊ B98 F99 ◊

TAMAS, G. (ED.) *Studien zur Logik* ◊ B98 ◊

1984

ATIYAH, M. & HOARE, C.A.R. & SHEPHERDSON, J.C. (EDS.) *Mathematical logic and programming languages* ◊ B98 ◊

BENCIVENGA, E. *Il primo libro di logica. Introduzione ai metodi della logica contemporanea* ◊ B98 ◊

GABBAY, D.M. & GUENTHNER, F. (EDS.) *Handbook of philosophical logic Vol II. Extensions of classical logic* ◊ B98 C90 C98 ◊

HAAS, G. *Konstruktive Einfuehrung in die formale Logik* ◊ B98 F98 ◊

MARKOV, A.A. *Elements of mathematical logic (Russian)* ◊ B98 ◊

MO, SHAOKUI & SHEN, BAIYING & XU, YONGSHEN *Mathematical logic (Chinese)* ◊ B98 ◊

SCHMUCKER, K.J. *Fuzzy sets, natural language computations, and risk analysis, Foreword by Lotfi A. Zadeh* ◊ B52 B98 ◊

WOJCICKI, R. *Lectures on propositional calculi* ◊ B98 ◊

1985

DALLA CHIARA SCABIA, M.L. (ED.) *Present state of the problem of the foundations of mathematics* ◊ B98 ◊

DRAKE, F.R. *How recent work in mathematical logic relates to the foundations of mathematics* ◊ A05 B98 E98 ◊

GINDIKIN, S.G. *Algebraic logic* ◊ B98 ◊

KREISEL, G. *Mathematical logic: tool and object lesson for science* ◊ A05 B98 ◊

ROBERT, A. *Analyse non standard* ◊ B98 H05 ◊

Author Index

AANDERAA, S.O. [1971] *On the decision problem for formulas in which all disjunctions are binary* (P 0604) Scand Logic Symp (2);1970 Oslo 1-18
⋄ B20 D35 ⋄ REV MR 50#1864 Zbl 232#02034 JSL 40.503 • ID 00087

AANDERAA, S.O. & LEWIS, H.R. [1973] *Prefix classes of Krom formulas* (J 0036) J Symb Logic 38*628-642
⋄ B20 B25 D35 ⋄ REV MR 49#2326 Zbl 326#02035 • ID 00010

AANDERAA, S.O. & LEWIS, H.R. [1974] *Linear sampling and the ∀∃∀ case of the decision problem* (J 0036) J Symb Logic 39*519-548
⋄ B20 D03 D05 D35 D80 ⋄ REV MR 51#114 Zbl 301#02042 • ID 03803

AANDERAA, S.O. & BOERGER, E. [1979] *The Horn complexity of Boolean functions and Cook's problem* (P 2615) Scand Logic Symp (5);1979 Aalborg 231-256
⋄ B05 D15 ⋄ REV MR 83b:03048b Zbl 429#03022 • ID 53853

AANDERAA, S.O. & BOERGER, E. [1981] *The equivalence of Horn and network complexity for boolean functions* (J 1431) Acta Inf 15*303-307
⋄ B05 B70 D15 ⋄ REV MR 83b:03048a Zbl 477#94034 • ID 55613

AANDERAA, S.O. & BOERGER, E. & LEWIS, H.R. [1982] *Conservative reduction classes of Krom formulas* (J 0036) J Symb Logic 47*110-130
⋄ B20 C13 D35 ⋄ REV MR 83e:03021 Zbl 487#03005 • ID 35210

AANDERAA, S.O. & BOERGER, E. & GUREVICH, Y. [1982] *Prefix classes of Krom formulae with identity (German summary)* (J 0009) Arch Math Logik Grundlagenforsch 22*43-49
⋄ B20 B25 D35 ⋄ REV MR 83m:03019 Zbl 494#03007 • ID 33756

AANDERAA, S.O. [1984] *On the solvability of the extended ∀∃ ∧ ∃∀*-Ackermann class with identity* (P 2342) Symp Rek Kombin;1983 Muenster 270-284
⋄ B20 B25 ⋄ REV MR 86h:03016 Zbl 575#03005 • ID 41791

AANDERAA, S.O. see Vol. III, IV, VI for further entries

AAQVIST, L. [1965] *A new approach to the logical theory of interrogatives. Part I. Analysis* (X 0882) Univ Filos Foeren: Uppsala iv+174pp
⋄ A05 B98 ⋄ REV JSL 32.403 • ID 16716

AAQVIST, L. see Vol. II for further entries

ABADI, M. & MANNA, Z. [1985] *Nonclausal temporal deduction* (P 4571) Log of Progr;1985 Brooklyn 1-15
⋄ B35 ⋄ REV Zbl 567#03004 • ID 49209

ABBOTT, W.R. [1951] *Computing logical truth with the California digital computer* (J 0235) Math Tables Other Aids Comp 5*120-128
⋄ B35 ⋄ REV MR 14.211 Zbl 45.400 JSL 17.280 • ID 00013

ABDUGALIEV, U.A. [1979] *An extremal problem connected with the distribution of work between collectives (Russian)* (C 2065) Teor Nereg Kriv Raz Geom Post 3-8
⋄ B35 ⋄ REV MR 81h:03047 • ID 70253

ABDUGALIEV, U.A. see Vol. II for further entries

ABDUL-KARIM, M.A.H. & BERBAT, N.E. [1978] *A simultaneous analog-ternary converter* (P 2014) Int Symp Multi-Val Log (8);1978 Rosemont 73-75
⋄ B35 ⋄ ID 35915

ABDULLAEV, D.A. & YUNUSOV, D. [1975] *Ueber die Zerlegung symmetrischer Boolescher Funktionen (Russian)* (J 0474) Avtom Vychis Tekh, Akad Nauk Latv SSR 1975/2*12-13
⋄ B05 ⋄ REV Zbl 298#94050 • ID 60011

ABDULLAEV, D.A. & YUNUSOV, D. [1978] *An algorithm for recognizing symmetric boolean functions (Russian)* (J 0977) Izv Akad Nauk SSSR, Tekh Kibern 1978/5*214-218
• TRANSL [1978] (J 0522) Engin Cybern 16/5*164-168
⋄ B05 ⋄ REV Zbl 435#94030 • ID 69029

ABDULLAEVA, M. & BUZURKHANOV, V. & KASYMOV, N.KH. [1981] *On an approach to the realization of the resolution method (Russian)* (J 3287) Vopr Vychisl Prikl Mat (Tashkent) 65*149-157
⋄ B35 ⋄ REV Zbl 473#68084 • ID 55401

ABHYANKAR, S. [1958] *Minimal "sum of products of sums" expressions of boolean functions* (J 0072) IRE Trans Electr Comp EC-7*268-276
⋄ B05 ⋄ REV JSL 24.254 • ID 00011

ABHYANKAR, S. [1959] *Absolute minimal expressions of boolean functions* (J 0072) IRE Trans Electr Comp EC-8*3-8
⋄ B05 ⋄ REV JSL 24.255 • ID 00012

ABIAN, A. [1970] *Completeness of the generalized propositional calculus* (J 0047) Notre Dame J Formal Log 11*449-452
⋄ B05 ⋄ REV MR 47#39 Zbl 169.305 • ID 00054

ABIAN, A. [1970] *Generalized completeness theorem and solvability of systems of boolean polynomial equations* (J 0068) Z Math Logik Grundlagen Math 16*263-264
⋄ B05 ⋄ REV MR 43#3183 Zbl 202.7 JSL 40.88 • ID 00055

ABIAN, A. [1970] *On the solvability of infinite systems of Boolean polynomial equations* (S 0019) Colloq Math (Warsaw) 21*27-30
⋄ B05 E25 ⋄ REV MR 41#3343 Zbl 192.336 JSL 40.88 • ID 00051

ABIAN, A. [1974] *Nonstandard models for arithmetics and analysis* (J 0063) Studia Logica 33*11-22
⋄ C20 H05 H15 ⋄ REV MR 50#92 Zbl 287#02038 • ID 03810

ABIAN, A. [1978] *A proof of Rasiowa-Sikorski theorem via complete sequences* (J 0050) Port Math 37*53-54
⋄ B10 ⋄ REV MR 82f:03056 Zbl 454#03029 • ID 54241

ABIAN, A. [1978] *Passages between finite and infinite* (J 0047) Notre Dame J Formal Log 19*452-456
⋄ A05 H05 ⋄ REV MR 58#14 Zbl 305#02067 • ID 31604

ABIAN, A. also published under the name ABIAN, S.

ABIAN, A. see Vol. II, III, IV, V for further entries

ABRAHAMS, P.W. [1966] *Machine verification of mathematical proof* (J 0074) Math Algor 1*11-32
⋄ B35 ⋄ REV JSL 37.411 • ID 00094

ABRAHAMS, P.W. [1967] *Machine verification of mathematical proof* (J 0074) Math Algor 2*28-79
⋄ B35 ⋄ REV JSL 37.411 • ID 00103

ABRAMOV, A.A. & LYUSTERNIK, L.A. & SHESTAKOV, V.I. & SHURA-BURA, M.R. [1952] *Solution of mathematical problems by automatic computers. Programming of electronic computers (Russian)* (X 3333) Unknown Publisher: See Remarks 327pp
⋄ B35 ⋄ REM Moskva-Leningrad • ID 45703

ACKERMANN, R.J. [1966] *Non-deductive inference* (X 0866) Routledge & Kegan Paul: Henley on Thames v+130pp
⋄ B98 ⋄ ID 25397

ACKERMANN, R.J. [1967] *An introduction to many-valued logics* (X 0813) Dover: New York 50pp
⋄ B50 B98 ⋄ ID 00099

ACKERMANN, R.J. [1967] *Introduction to many-valued logics* (X 0866) Routledge & Kegan Paul: Henley on Thames v+90pp
⋄ B50 B98 ⋄ ID 25260

ACKERMANN, R.J. [1970] *Modern deductive logic* (X 0878) Doubleday: London ix+261pp
⋄ B98 ⋄ REV MR 50#12628 • ID 04128

ACKERMANN, R.J. see Vol. II for further entries

ACKERMANN, W. [1925] *Begruendung des "tertium non datur" mittels der Hilbertschen Theorie der Widerspruchsfreiheit* (J 0043) Math Ann 93*1-36
⋄ A05 B10 F30 ⋄ ID 00104

ACKERMANN, W. [1928] see HILBERT, D.

ACKERMANN, W. [1928] *Zum Hilbertschen Aufbau der reellen Zahlen* (J 0043) Math Ann 99*118-133
• TRANSL [1967] (C 0675) From Frege to Goedel 493-507
⋄ B28 D20 F15 ⋄ REV FdM 54.56 • ID 00106

ACKERMANN, W. [1934] *Untersuchungen ueber das Eliminationsproblem der mathematischen Logik* (J 0043) Math Ann 110*390-413
⋄ B10 ⋄ REV Zbl 9.386 FdM 60.22 • ID 00110

ACKERMANN, W. [1935] *Zum Eliminationsproblem der mathematischen Logik* (J 0043) Math Ann 111*61-63
⋄ B10 ⋄ REV Zbl 11.3 FdM 61.51 • ID 00111

ACKERMANN, W. [1938] *Mengentheoretische Begruendung der Logik* (J 0043) Math Ann 115*1-22
⋄ B10 E35 E70 ⋄ REV Zbl 17.242 JSL 3.85 • ID 00114

ACKERMANN, W. [1939] *Bemerkungen zu den logisch-mathematischen Grundlagenproblemen* (C 0656) Phil Mathematique 76-82
⋄ A05 B10 ⋄ REV JSL 5.78 • ID 16703

ACKERMANN, W. [1941] *Ein System der typenfreien Logik I* (J 0956) Forsch Logik Grundl exakt Wiss 7*29pp
⋄ B15 E70 ⋄ REV Zbl 28.100 JSL 7.93 • ID 16704

ACKERMANN, W. [1950] *Widerspruchsfreier Aufbau der Logik I. Typenfreies System ohne tertium non datur* (J 0036) J Symb Logic 15*33-57
⋄ B15 E70 F07 F25 F65 ⋄ REV MR 12.384 Zbl 36.147 JSL 16.72 • ID 00122

ACKERMANN, W. [1952] *Widerspruchsfreier Aufbau einer typenfreien Logik I (erweitertes System)* (J 0044) Math Z 55*364-384
⋄ B10 B15 E70 F25 F65 ⋄ REV MR 14.344 Zbl 50.245 JSL 19.295 • REM Part II 1953 • ID 00124

ACKERMANN, W. [1953] *Widerspruchsfreier Aufbau einer typenfreien Logik II* (J 0044) Math Z 57*155-166
⋄ B10 B15 E35 E70 F30 F35 F65 ⋄ REV MR 14.834 Zbl 50.245 JSL 19.295 • REM Part I 1952 • ID 00125

ACKERMANN, W. [1958] *Ein typenfreies System der Logik mit ausreichender mathematischer Anwendungsfaehigkeit I* (J 0009) Arch Math Logik Grundlagenforsch 4*3-26
⋄ B10 B15 E70 ⋄ REV MR 20#4477 Zbl 83.1 JSL 32.259 • REM Part II 1959 • ID 00131

ACKERMANN, W. [1961] *Grundgedanken einer typenfreien Logik* (C 0622) Essays Found of Math (Fraenkel) 143-155
⋄ B28 E70 ⋄ REV MR 29#3357 JSL 32.259 • ID 00192

ACKERMANN, W. [1965] *Der Aufbau einer hoeheren Logik* (J 0009) Arch Math Logik Grundlagenforsch 7*5-22
⋄ B15 E70 F30 F35 ⋄ REV MR 34#26 Zbl 154.255 JSL 40.458 • ID 00135

ACKERMANN, W. see Vol. II, III, IV, V, VI for further entries

ADAM, ANDRAS [1960] *Zur Theorie der Wahrheitsfunktionen* (J 0002) Acta Sci Math (Szeged) 21*47-52
⋄ B05 ⋄ REV MR 22#9439 Zbl 91.8 • ID 00149

ADAM, ANDRAS [1962] *Ueber die monotone Superposition der Wahrheitsfunktionen* (J 0002) Acta Sci Math (Szeged) 23*18-37
⋄ B05 ⋄ REV MR 26#3584 Zbl 106.237 • ID 00150

ADAM, ANDRAS [1968] *Truth functions and the problem of their realization by two-terminal graphs* (X 0928) Akad Kiado: Budapest 206pp
⋄ B05 ⋄ REV MR 39#3903 • ID 19517

ADAM, ANDRAS [1973] *Investigations on the superpositions of truth functions and on their repetition-free realizability by two-terminal graphs (Hungarian) (English summary)* (J 0462) Mat Fiz Oszt Koezlem, Acad Sci Hung 21*1-42
⋄ B05 ⋄ REV MR 47#3110 Zbl 273#94037 • ID 60053

ADAM, ANDRAS see Vol. IV, V for further entries

ADAMS, E.W. [1975] *The logic of conditionals. An application of probability to deductive logic* (S 3307) Synth Libr 86*xiii+155pp
⋄ A05 B48 B98 ⋄ REV MR 58#5043 Zbl 324#02002 • ID 60060

ADAMS, E.W. see Vol. II, III for further entries

ADDISON, J.W. & CHANG, C.C. & CRAIG, W. & HENKIN, L. & SCOTT, D.S. & VAUGHT, R.L. (EDS.) [1974] *Proceedings of the Tarski Symposium* (S 3304) Proc Symp Pure Math 25*xxi+498pp
  ◇ B97 C97 ◇ REV MR 50#1829 • REM Corr. ed. 1979; xx+498pp. Contains bibliography of A.Tarski with a supplement in the corr. ed. • ID 70206

ADDISON, J.W. [1984] *Eloge: Alfred Tarski: 1901-1983* (J 3789) Ann Hist of Comp 6*335-336
  ◇ A10 B05 ◇ REV MR 85k:01041 • ID 43948

ADDISON, J.W. see Vol. III, IV, V for further entries

ADLEMAN, L.M. & BOOTH, K.S. & PREPARATA, F.P. & RUZZO, W.L. [1978] *Improved time and space bounds for boolean matrix multiplication* (J 1431) Acta Inf 11*61-70
  ◇ B05 D15 ◇ REV MR 80f:68038 Zbl 389#68016 • ID 69046

ADLEMAN, L.M. see Vol. IV for further entries

ADLER, A. [1973] *F-planar graphs* (J 0033) J Comb Th, Ser B 15*207-210
  ◇ E05 H05 ◇ REV MR 48#174 Zbl 266#05104 • ID 00189

ADLER, A. & HAMILTON, J. [1973] *Invariant means via the ultrapower* (J 0043) Math Ann 202*71-76
  ◇ C20 H05 ◇ REV MR 48#2664 Zbl 235#43001 • ID 00190

ADLER, A. see Vol. III, IV, V for further entries

AGAZZI, E. [1964] *La logica simbolica* (X 1348) Scuola: Brescia 396pp
  ◇ A05 B98 ◇ REV Zbl 166.248 JSL 39.327 • ID 16707

AGAZZI, E. [1981] *Logic and the methodology of empirical sciences* (C 2617) Modern Log Survey 255-282
  ◇ B80 ◇ REV MR 82f:03002 Zbl 464#03001 • ID 42771

AGAZZI, E. & GRUENDER, D. & HINTIKKA, K.J.J. (EDS.) [1981] *Proceedings of the 1978 Pisa conference on the history and philosophy of science. 2nd international conference I,II* (S 3307) Synth Libr 145*xiv+354pp,146*xiv+326pp
  ◇ A10 B97 ◇ REV MR 82f:01001 • ID 48651

AGAZZI, E. see Vol. II, VI for further entries

AHMAD, S. [1974] *On implicational completeness* (J 0017) Canad J Math 26*761-768
  ◇ B22 C05 ◇ REV MR 50#2024 Zbl 295#08003 • ID 03819

AIELLO, L. [1980] *Using meta-theoretic reasoning to do algebra* (P 3063) Autom Deduct (5);1980 Les Arcs 1-13
  ◇ B35 ◇ REV Zbl 438#68040 • ID 69057

AIELLO, L. see Vol. VI for further entries

AJDUKIEWICZ, K. [1955] *Concerning the plan of research in the field of logic (Polish) (Russian and English summaries)* (J 0063) Studia Logica 2*267-277,328
  ◇ A10 B10 ◇ REV Zbl 67.248 • ID 33065

AJDUKIEWICZ, K. [1958] *Abriss der Logik* (X 1230) Aufbau: Berlin 204pp
  ◇ B98 ◇ REV JSL 19.235 • ID 22390

AJDUKIEWICZ, K. [1960] *The axiomatic systems from the methodological point of view (Polish and Russian summaries)* (J 0063) Studia Logica 9*205-220
  • REPR [1977] (C 3174) Twenty-five Years Log Meth Poland 49-63
  ◇ A05 B30 ◇ REV MR 24#A1184 Zbl 373#02006 • ID 33091

AJDUKIEWICZ, K. [1965] *Pragmatic logic (Polish)* (X 1034) PWN: Warsaw
  • TRANSL [1974] (S 3307) Synth Libr 62*xv+460pp
  ◇ A05 B98 ◇ REV Zbl 307#02005 • ID 60083

AJDUKIEWICZ, K. [1966] *From the methodology of the deductive sciences* (J 0063) Studia Logica 19*9-45
  ◇ B30 ◇ REV MR 33#5447 Zbl 301#02040 • ID 00233

AJDUKIEWICZ, K. see Vol. II for further entries

AJZERMAN, M.A. & GUSEV, L.A. & ROZONOEHR, L.I. & SMIRNOVA, I.M. & TAL', A.A. [1963] *Logic, automata and algorithms (Russian)* (X 3709) Izdat Fiz-Mat Lit: Moskva 556pp
  • TRANSL [1971] (X 0801) Academic Pr: New York xii+433pp [1967] (X 0814) Oldenbourg: Muenchen x+431pp (German) [1971] (X 1226) Academia: Prague 408pp (Czech)
  ◇ B98 D05 D10 D98 ◇ REV MR 29#5690 MR 35#1411 MR 43#3044 MR 48#1818 Zbl 131.8 Zbl 216.7 JSL 31.109 JSL 37.625 • ID 00220

AJZERMAN, M.A. see Vol. II, IV for further entries

AKERS JR., S.B. [1961] *A truth table method for the synthesis of combinational logic* (J 0072) IRE Trans Electr Comp EC-10*604-615
  ◇ B05 ◇ REV MR 26#7172 JSL 28.290 • ID 00240

ALAGIC, M. [1974] *A monadic approach to k-spaces* (J 0042) Mat Vesn, Drust Mat Fiz Astron Serb 11(26)*239-243
  ◇ H10 ◇ REV MR 53#1569 Zbl 293#54021 • ID 26609

ALBEVERIO, S. & FENSTAD, J.E. & HOEEGH-KROHN, R. [1979] *Singular perturbations and nonstandard analysis* (J 0064) Trans Amer Math Soc 252*275-295
  ◇ H10 ◇ REV MR 80k:34029 Zbl 424#35014 • ID 80528

ALBEVERIO, S. & BLANCHARD, P. & HOEEGH-KROHN, R. [1982] *Some applications of functional integration* (P 4264) Math Probl in Th Phys;1981 Berlin 265-275
  ◇ H10 ◇ REV MR 84j:81006 • ID 44886

ALBEVERIO, S. & FENSTAD, J.E. & HOEEGH-KROHN, R. & KARWOWSKI, W. & LINDSTROEM, T.L. [1984] *Perturbations of the Laplacian supported by null sets, with applications to Polymer measures and quantum fields* (J 2813) Phys Lett A 104*396-400
  ◇ H10 ◇ REV MR 86f:81030 • ID 39337

ALBEVERIO, S. [1985] *Nonstandard analysis; polymer models, quantum fields* (P 4300) Stoch Method & Comput Techn in Quant Dynam;1984 Graz 233-254
  ◇ H10 ◇ REV MR 85m:81005 Zbl 541#60031 • ID 45641

ALBIN, P.S. [1982] *The metalogic of economic predictions, calculations and propositions* (J 3914) Math Soc Sci 3*329-358
  ◇ B80 ◇ REV MR 84c:90003 Zbl 503#90028 • ID 36675

ALBRECHT, A. [1978] *Ueber die Kompliziertheit der Realisierung boolescher Funktionen durch Schemata aus Funktional- und Verbindungselementen* (S 2829) Rostocker Math Kolloq 10*7-10
⋄ B05 ⋄ REV Zbl 412#94029 • ID 69072

ALBRECHT, A. [1979] *Zur Kompliziertheit der Realisierung boolescher Funktionen durch F-V-Schemata* (J 0129) Elektr Informationsverarbeitung & Kybern 15*159-171
⋄ B05 ⋄ REV MR 80m:94082 Zbl 408#68040 • ID 69071

ALBRECHT, A. see Vol. IV for further entries

ALBRECHT, W. [1954] *Die Logik der Logistik* (X 1258) Duncker & Humblot: Berlin 60pp
⋄ B98 ⋄ ID 22391

ALBRECHT-BUEHLER, G. [1976] *Numerical evaluation of the validity of experimental proofs in biology* (J 0154) Synthese 33*283-312
⋄ B80 ⋄ REV MR 57#9477 Zbl 344#02021 • ID 60090

ALCANTARA DE, L.P. & CORRADA, M. [1980] *Notes on many-sorted systems* (P 3006) Brazil Conf Math Log (3);1979 Recife 83-108
⋄ B10 C07 ⋄ REV MR 83k:03015 Zbl 455#03005 • ID 54266

ALCANTARA DE, L.P. (ED.) [1985] *Mathematical logic and formal systems. A collection of papers in honor of Professor Newton C.A. da Costa* (S 3310) Lect Notes Pure Appl Math 94*xiv+297pp
⋄ B97 ⋄ REV MR 86e:03001 Zbl 563#00002 • ID 47463

ALCANTARA DE, L.P. see Vol. II, V for further entries

ALCHOURRON, C.E. & BULYGIN, E. [1971] *Normative systems* (X 0902) Springer: Wien xviii+208pp
⋄ B45 B98 ⋄ REV Zbl 231#02006 JSL 38.326 • ID 15012

ALCHOURRON, C.E. see Vol. II for further entries

ALEKSANYAN, A.A. [1981] *Classes of functions of the algebra of logic (Russian)* (X 2265) Akad Nauk Vychis Tsentr: Moskva 14pp
⋄ B05 ⋄ REV MR 83h:06018 • ID 39250

ALEXANDER, P. [1981] *The case of the lonely corpuscle: reductive explanation and primitive expressions* (C 4283) Reduct, Time & Reality 17-35
⋄ B80 ⋄ REV MR 84g:03008 • ID 45727

ALFONSECA, M. [1980] *Automatic solution of sorites* (J 1429) Kybernetes 9*37-44
⋄ B35 ⋄ REV Zbl 427#03006 • ID 53693

ALIMPIC, B.P. [1968] *On models of certain formulas of the predicate calculus of first order* (J 0042) Mat Vesn, Drust Mat Fiz Astron Serb 5(20)*347-351
⋄ B10 ⋄ REV MR 39#6754 Zbl 176.274 • ID 00268

ALIMPIC, B.P. [1971] *On models of certain formulas with a predicate letter of length n* (J 0042) Mat Vesn, Drust Mat Fiz Astron Serb 8(23)*287-291
⋄ B10 ⋄ REV MR 45#4943 Zbl 227#02006 • ID 00269

ALLEN, C.M. & GIVONE, D.D. [1968] *A minimization technique for multiple-valued logic systems* (J 0187) IEEE Trans Comp C-17*182-184
⋄ B35 ⋄ REV Zbl 157.338 • ID 41932

ALLEN, C.M. & GIVONE, D.D. & MATTREY, R.F. [1973] *Applying multiple-valued algebra concepts to neural modeling* (P 2009) Int Symp Multi-Val Log (3);1973 Toronto 127-136
⋄ B80 ⋄ REV MR 50#4244 • ID 42063

ALLEN, C.M. see Vol. II for further entries

ALLEN, JOHN & LUCKHAM, D.C. [1969] *An interactive theorem-proving program* (J 0508) Machine Intelligence 5*321-326
• REPR [1983] (C 4659) Autom of Reasoning 2*417-434
⋄ B35 ⋄ REV Zbl 219#68050 • ID 47285

ALLEN, L.E. & BROOKS, R.B.S. & DICKOFF, J.W. & JAMES, P.A. [1961] *The ALL project (accelerated learning of logic)* (J 0005) Amer Math Mon 68*497-500
⋄ B98 ⋄ REV JSL 35.484 • ID 00270

ALLEN, L.E. [1962] *Wff'n proof: the game of modern logic* (X 1371) Wff'n Proof: Ann Arbor vii+224pp
⋄ B05 ⋄ REV JSL 30.105 • ID 22452

ALLEN, L.E. [1963] *Wff: the beginner's game of modern logic* (X 1371) Wff'n Proof: Ann Arbor 78pp
⋄ B05 ⋄ REV JSL 30.105 • ID 22453

ALLING, N.L. [1985] *Conway's field of surreal numbers* (J 0064) Trans Amer Math Soc 287*365-386
⋄ H10 H15 H20 ⋄ REV MR 86f:04002 • ID 44331

ALLING, N.L. see Vol. III, V for further entries

ALTHAM, J.E.J. [1971] *The logic of plurality* (X 0816) Methuen: London & New York ix+84pp
⋄ A05 B98 C80 ⋄ REV MR 47#12 Zbl 276#02008 • ID 00285

ALTHAM, J.E.J. see Vol. III for further entries

ALVES, M.T. [1951] *A theorem of metamathematics* (J 0084) Gaz Mat (Lisboa) 12/49*6-8
⋄ B05 ⋄ REV MR 13.309 Zbl 43.247 • ID 00292

AMAREL, S. [1967] *An approach to heuristic problem solving and theorem proving in the propositional calculus* (P 1390) Syst & Comput Sci;1965 London ON 125-220
⋄ B35 ⋄ REV MR 41#31 • ID 16710

AMBROSE, A. & LAZEROWITZ, M. [1948] *Fundamentals of symbolic logic* (X 0818) Holt Rinehart & Winston: New York ix+310pp
⋄ B98 ⋄ REV JSL 14.191 JSL 40.607 • ID 14474

AMBROSE, A. & LAZEROWITZ, M. [1961] *Logic: the theory of formal inference* (X 0818) Holt Rinehart & Winston: New York vi+78pp
⋄ B98 ⋄ REV Zbl 98.241 JSL 28.169 • ID 07900

AMBROSE, A. see Vol. II, VI for further entries

AMBROSIMOV, A.S. & SHAROV, N.N. [1979] *Some asymptotic expansions of the number of functions with non-trivial group of inertia (Russian)* (J 0052) Probl Kibern 36*65-84,279
⋄ B05 ⋄ REV MR 81j:05008 Zbl 447#94027 • ID 69078

AMEMIYA, I. [1965] *On non-standard analysis* (J 0091) Sugaku 16*158-161
⋄ H05 ⋄ REV MR 35#5304 • ID 00300

AMER, M.A. [1976] *Typed boolean structures I (Arabic summary)* (J 0397) Proc Math Phys Soc Egypt 41*15-22
⋄ B15 B40 C90 E40 G05 ⋄ REV MR 52#7860 Zbl 418#03009 • REM Part II 1977 • ID 17946

AMER, M.A. [1977] *Typed boolean structures II (Arabic summary)* (J 0397) Proc Math Phys Soc Egypt 44*11-15
⋄ B15 B40 C90 E40 G05 ⋄ REV MR 56#2784 Zbl 427#03011 • REM Part I 1976 • ID 30619

AMER, M.A. see Vol. II, III, VI for further entries

AMLIEN, J. & BARRICELLI, N.A. [1983] *The use of data processing machines in the recognition and proof of arithmetical theorems translated into B-mathematical language* (X 3805) Blindern Theoretic Res Team: Oslo 16pp
⋄ B35 B80 ⋄ REV MR 85b:03018 • ID 40574

AN, ZHI & LIU, XUHUA [1985] *Generalized resolution using an equational substitution strategy (Chinese)* (J 2771) Kexue Tongbao 30*1601-1603
⋄ B35 ⋄ ID 49708

ANDERSON, A.R. & BELNAP JR., N.D. [1959] *A simple treatment of truth functions* (J 0036) J Symb Logic 24*301-302
⋄ B05 ⋄ REV Zbl 96.242 JSL 28.291 • ID 00323

ANDERSON, A.R. & BELNAP JR., N.D. [1975] *Entailment. The logic of relevance and necessity I* (X 0857) Princeton Univ Pr: Princeton xxxii+543pp
⋄ A05 B46 B98 ⋄ REV MR 53#10542 Zbl 323#02030 JSL 42.311 • ID 23056

ANDERSON, A.R. [1975] *Fitch on consistency* (C 1856) Log Enterprise 123-141
⋄ B15 F05 F25 ⋄ REV Zbl 361#02040 • ID 50675

ANDERSON, A.R. & BARCAN MARCUS, R. & MARTIN, R.M. (EDS.) [1975] *The logical enterprise* (X 0875) Yale Univ Pr: New Haven x+261pp
⋄ A05 A10 B98 ⋄ REV Zbl 344#00006 • ID 48638

ANDERSON, A.R. see Vol. II, III for further entries

ANDERSON, C.A. [1980] *Some new axioms for the logic of sense and denotation: alternative (0)* (J 0097) Nous, Quart J Phil 14*217-234
⋄ A05 B15 ⋄ REV MR 81m:03012 • ID 70406

ANDERSON, D.E. & CLEAVER, F.L. [1965] *Venn-type diagrams for arguments of n terms* (J 0036) J Symb Logic 30*113-118
⋄ B05 E20 ⋄ REV MR 34#2430 Zbl 131.5 • ID 00336

ANDERSON, J.G. [1968] *Concerning the finite model property for propositional calculi* (J 0053) Proc Amer Math Soc 19*1207-1210
⋄ B22 ⋄ REV MR 38#25 Zbl 165.310 • ID 00343

ANDERSON, J.G. see Vol. II, III for further entries

ANDERSON, J.M. & JOHNSTONE JR., H.W. [1962] *Natural deduction. The logical basis of axiom systems* (X 0821) Wadsworth Publ: Belmont xii+418pp
⋄ B98 F07 F98 ⋄ REV MR 25#4986 JSL 29.93 • ID 00341

ANDERSON, ROBERT M. & BLEDSOE, W.W. [1970] *A linear format for resolution with merging and a new technique for establishing completeness* (J 0037) ACM J 17*525-534
• REPR [1983] (C 4659) Autom of Reasoning 2*321-330
⋄ B35 ⋄ REV MR 47#1335 Zbl 199.315 • ID 47288

ANDERSON, ROBERT M. [1970] *Completeness results for E-resolution* (P 4202) AFIPS Spring Jt Computer Conf;1970 652-656
• REPR [1983] (C 4659) Autom of Reasoning 2*317-320
⋄ B35 ⋄ ID 47287

ANDERSON, ROBERT M. [1976] *A non-standard representation for Brownian motion and Ito integration* (J 0029) Israel J Math 25*15-46
⋄ H10 ⋄ REV MR 57#4311 Zbl 353#60052 JSL 50.243 • ID 26071

ANDERSON, ROBERT M. [1976] *A nonstandard representation for Brownian motion and Ito integration* (J 0015) Bull Amer Math Soc 82*99-101
⋄ H10 ⋄ REV MR 53#9374 Zbl 327#60039 • ID 23047

ANDERSON, ROBERT M. & RASHID, S. [1978] *A nonstandard characterization of weak convergence* (J 0053) Proc Amer Math Soc 69*327-332
⋄ H05 ⋄ REV MR 58#1073 Zbl 393#03047 • ID 52467

ANDERSON, ROBERT M. [1982] *Star-finite representations of measures spaces* (J 0064) Trans Amer Math Soc 271*667-687
⋄ H10 ⋄ REV MR 83m:03077 Zbl 494#28005 • ID 35473

ANDON, F.I. [1966] *A simplification algorithm of a disjunctive normal form of the boolean functions (Russian)* (J 0040) Kibernetika, Akad Nauk Ukr SSR 1966/6*12-14
⋄ B35 ⋄ REV Zbl 217.293 JSL 35.330 • ID 00355

ANDON, F.I. [1966] *An approach to the minimization of systems of boolean functions (Russian)* (J 0040) Kibernetika, Akad Nauk Ukr SSR 1966/5*44-48
• TRANSL [1966] (J 0021) Cybernetics 2/5*34-37
⋄ B35 ⋄ REV Zbl 192.86 JSL 35.330 • ID 00354

ANDON, F.I. [1967] *A method for simplifying the disjunctive normal form (DNF) of boolean functions (Russian)* (J 0040) Kibernetika, Akad Nauk Ukr SSR 1967/1*21-25
• TRANSL [1967] (J 0021) Cybernetics 3/1*17-20
⋄ B05 ⋄ REV MR 45#130 Zbl 153.12 • ID 00357

ANDON, F.I. [1968] *The minimization of the d.n.f.'s of the functions of the algebra of logic by the method of the sequential analysis of variants (Russian)* (C 4203) Teor Avtom & Met Formal Sint Vychisl Mash & Sist Kiev 1968 3*18-41
⋄ B05 ⋄ REV MR 39#6730 • ID 48085

ANDRE, C. & AUGUIN, M. & BOERI, F. [1978] *An algorithm for designing multiple boolean functions: application to PLAs* (J 2701) Digit Processes 4*215-230
⋄ B35 G05 ⋄ REV MR 80g:94079 Zbl 403#94027 • ID 69106

ANDREEV, A.E. [1983] *On the synthesis of disjunctive normal forms close to minimum forms (Russian)* (J 0023) Dokl Akad Nauk SSSR 269*11-15
• TRANSL [1983] (J 0062) Sov Math, Dokl 27*265-269
⋄ B05 ⋄ REV MR 85a:94041 Zbl 556#94015 • ID 38926

ANDREEV, A.E. [1984] *On the problem of minimization of disjunctive normal forms (Russian)* (J 0023) Dokl Akad Nauk SSSR 274*265-269
• TRANSL [1984] (J 0062) Sov Math, Dokl 29*32-36
⋄ B05 ⋄ REV MR 86f:03103 • ID 45319

ANDREEV, A.E. [1985] *A method for obtaining lower bounds on the complexity of individual monotone functions (Russian)* (J 0023) Dokl Akad Nauk SSSR 282*1033-1037
• TRANSL [1985] (J 0062) Sov Math, Dokl 31*530-534
⋄ B05 ⋄ ID 47696

ANDREKA, H. & GERGELY, T. & NEMETI, I. [1973] *Some questions on languages of order n (Hungarian)* (J 0396) Mat Lapok 24*63-94
⋄ B15 C07 C85 ⋄ REV MR 53#5249 • ID 22879

ANDREKA, H. & GERGELY, T. & NEMETI, I. [1974] *Some questions on languages of order n I,II (Russian) (English summaries)* (J 0040) Kibernetika, Akad Nauk Ukr SSR 1974/5*61-67,1974/6*79-83
- TRANSL [1974] (J 0021) Cybernetics 10*804-812,1003-1008
- ◊ B15 C07 C85 ◊ REV MR 56#1811 MR 56#1812 Zbl 316#68048 Zbl 316#68056 • ID 60135

ANDREKA, H. & GERGELY, T. & NEMETI, I. [1974] *Sufficient and necessary condition for the completeness of a calculus* (J 0068) Z Math Logik Grundlagen Math 20*433-434
- ◊ B10 C07 G05 ◊ REV MR 51#10023 Zbl 305#02063
- • ID 03829

ANDREKA, H. & GERGELY, T. & NEMETI, I. [1975] *Many-sorted languages and their connection with nth order languages (Russian) (English summary)* (J 0040) Kibernetika, Akad Nauk Ukr SSR 1975/4*86-92
- TRANSL [1975] (J 0021) Cybernetics 11*605-612
- ◊ B15 C85 ◊ REV MR 56#11742 Zbl 436#03004
- • ID 55844

ANDREKA, H. & NEMETI, I. & SAIN, I. [1982] *A complete logic for reasoning about programs via nonstandard model theory. I,II* (J 1426) Theor Comput Sci 17*193-212,259-278
- ◊ B75 C90 H10 H20 ◊ REV MR 84d:68007 Zbl 475#68009 Zbl 475#68010 • ID 55505

ANDREKA, H. see Vol. III, V for further entries

ANDREOLI, G. [1959] *Proprieta delle funzioni simmetriche elementari nelle algebre di Boole e nelle algebre dei livelli* (J 0099) Ricerca, Riv Mat Pure & Appl 10/3*1-10
- ◊ B05 G05 ◊ REV MR 23#A2347 Zbl 90.23 JSL 38.153
- • ID 44075

ANDREOLI, G. see Vol. V for further entries

ANDREWS, P.B. [1963] *A reduction of the axioms for the theory of propositional types* (J 0027) Fund Math 52*345-350
- ◊ B15 ◊ REV MR 27#3498 Zbl 127.7 JSL 30.385
- • ID 00362

ANDREWS, P.B. [1965] *A transfinite type theory with type variables* (X 0809) North Holland: Amsterdam xv+143pp
- ◊ B15 B40 ◊ REV MR 37#51 Zbl 132.245 JSL 33.112
- • ID 00402

ANDREWS, P.B. [1968] *On simplifying the matrix of a wff* (J 0036) J Symb Logic 33*180-192
- REPR [1983] (C 4659) Autom of Reasoning 2*102-116
- ◊ B35 ◊ REV MR 41#33 Zbl 157.335 • ID 00363

ANDREWS, P.B. [1968] *Resolution with merging* (J 0037) ACM J 15*367-381 • ERR/ADD ibid 15*720
- REPR [1983] (C 4659) Autom of Reasoning 2*85-101
- ◊ B35 ◊ REV Zbl 182.25 JSL 35.159 • ID 00364

ANDREWS, P.B. [1971] *Resolution in type theory* (J 0036) J Symb Logic 36*414-432
- REPR [1983] (C 4659) Autom of Reasoning 2*487-507
- ◊ B15 F05 F35 ◊ REV MR 46#1551 Zbl 231#02038
- • ID 00365

ANDREWS, P.B. [1972] *General models, descriptions, and choice in type theory* (J 0036) J Symb Logic 37*385-394
- ◊ B15 C85 E25 E35 F35 ◊ REV MR 47#6433 Zbl 264#02049 • ID 00367

ANDREWS, P.B. [1972] *General models and extensionality* (J 0036) J Symb Logic 37*395-397
- ◊ B15 C85 E35 F35 ◊ REV MR 47#6434 Zbl 264#02050 • ID 00366

ANDREWS, P.B. [1974] *Provability in elementary type theory* (J 0068) Z Math Logik Grundlagen Math 20*411-418
- ◊ B15 B25 D35 F35 ◊ REV MR 52#7867 Zbl 306#02017 • ID 03832

ANDREWS, P.B. [1974] *Resolution and the consistency of analysis* (J 0047) Notre Dame J Formal Log 15*73-84
- ◊ B15 F05 F25 F35 ◊ REV MR 52#7886 Zbl 226#02042 • ID 00368

ANDREWS, P.B. [1976] *Refutations by mapings* (J 0187) IEEE Trans Comp C-25*801-807
- ◊ B35 ◊ REV MR 58#13990 Zbl 331#68050 • ID 30630

ANDREWS, P.B. & LONGINI COHEN, E. [1977] *Theorem proving in type theory* (P 1676) Int Joint Conf Artif Intell (5);1977 Cambridge MA 566
- ◊ B35 ◊ ID 30631

ANDREWS, P.B. [1980] *Transforming matings into natural deduction proofs* (P 3063) Autom Deduct (5);1980 Les Arcs 281-292
- ◊ B35 ◊ REV MR 81i:68006 Zbl 438#68047 • ID 69080

ANDREWS, P.B. [1981] *Theorem proving via general matings* (J 0037) ACM J 28*193-214
- ◊ B35 ◊ REV MR 82m:03013 Zbl 456#68119 • ID 54324

ANDREWS, P.B. & LONGINI COHEN, E. & MILLER, DALE A. [1982] *A look at TPS* (P 3840) Autom Deduct (6);1982 New York 50-69
- ◊ B35 ◊ REV MR 85g:03024 • ID 43497

ANDREWS, P.B. & LONGINI COHEN, E. & MILLER, DALE A. & PFENNING, F. [1984] *Automating higher-order logic* (P 3084) Autom Theor Prov After 25 Yea;1983 Denver 169-192
- ◊ B35 ◊ REV MR 85d:68005 Zbl 551#68075 • ID 43999

ANDREWS, P.B. see Vol. VI for further entries

ANGELELLI, I. [1979] *Aristotelian modal syllogistic* (J 0162) Teorema (Valencia) 9*165-182
- ◊ A05 B20 ◊ REV MR 81c:03012 • ID 70436

ANGELL, R.B. [1960] *Note on a less restricted type of rule of inference* (J 0094) Mind 69*253-255
- ◊ B05 ◊ REV MR 22#667 JSL 40.602 • ID 15089

ANGELL, R.B. [1960] *The sentential calculus using rule of inference R_e* (J 0036) J Symb Logic 25*143
- ◊ B05 ◊ REV MR 27#1354 Zbl 105.247 JSL 40.602
- • ID 00372

ANGELL, R.B. [1964] *Reasoning and logic* (X 1228) Appleton-Century-Crofts: New York xiv+625pp
- ◊ B98 ◊ REV JSL 31.674 • ID 22456

ANGELL, R.B. see Vol. II for further entries

ANGSTL, H. & MENNE, A. & WILHELMY, A. (EDS.) [1951] *Kontrolliertes Denken. Untersuchungen zum Logikkalkuel und zur Logik der Einzelwissenschaften. Festschrift fuer Wilhelm Britzelmayr* (X 0826) Alber: Freiburg 122pp
- ◊ A05 B96 ◊ ID 48617

ANGSTL, H. [1951] *Ueber Gleichungen im Aussagenkalkuel* (C 0621) Kontrolliertes Denken (Britzelmayr) 8-9
- ◊ B05 ◊ REV JSL 18.329 • ID 00374

ANGSTL, H. see Vol. II for further entries

ANIKEEV, A.S. [1972] *Classification of derivable propositional formulas (Russian)* (J 0087) Mat Zametki (Akad Nauk SSSR) 11*165-174
- TRANSL [1972] (J 1044) Math Notes, Acad Sci USSR 11*106-110
⋄ B15 D05 F20 F50 ⋄ REV MR 45 #6568 Zbl 239 #02003 • ID 00375

ANKERLIN, R.A. & JACKSON, C.L. [1967] *A rapid method for the identification of the type of a four-variable boolean function* (J 4305) IEEE Trans Electr Comp EC-16*870-871
⋄ B35 ⋄ REV Zbl 159.17 • ID 06484

ANONYMOUS [1969] *Les activites du centre national de recherche de logique en 1968* (J 0079) Logique & Anal, NS 12*199-202
⋄ A10 H10 ⋄ REV JSL 35.484 • ID 00380

ANUFRIEV, F.V. & ASEL'DEROV, Z.M. & DIDUKH, I.I. & FEDYURKO, V.V. & LETICHEVS'KIJ, O.A. [1966] *On a certain algorithm for search of proofs of theorems in the theory of groups (Russian) (English summary)* (J 0040) Kibernetika, Akad Nauk Ukr SSR 1966/1*23-29
⋄ B35 ⋄ REV MR 33 #7200 Zbl 259 #68039 • ID 60152

ANUFRIEV, F.V. & KOSTYAKOV, V.M. & MALASHONOK, A.I. [1972] *Algorithm and machine trial in searches for proofs of theorems in propositional calculus (Russian) (English summary)* (J 0040) Kibernetika, Akad Nauk Ukr SSR 1972/5*68-73
⋄ B35 ⋄ REV MR 47 #6148 Zbl 256 #68039 • ID 00391

ANUFRIEV, F.V. & ASEL'DEROV, Z.M. & KAPITONOVA, YU.V. [1972] *Organization of data for the evidence algorithm in the search for proofs of theorems in formalized theories (Russian) (English summary)* (J 0040) Kibernetika, Akad Nauk Ukr SSR 1972/5*61-67
⋄ B35 ⋄ REV MR 47 #6147 • ID 00510

ANUFRIEV, F.V. & ASEL'DEROV, Z.M. [1972] *The evidence algorithm (Russian) (English summary)* (J 0040) Kibernetika, Akad Nauk Ukr SSR 1972/5*29-60
⋄ B35 ⋄ REV MR 47 #6146 • ID 00389

APELT, H. [1966] *Axiomatische Untersuchungen ueber einige mit der Presburgerschen Arithmetik verwandte Systeme* (J 0068) Z Math Logik Grundlagen Math 12*131-168
⋄ B28 C35 C80 F30 ⋄ REV MR 36 #6279 Zbl 149.7 • ID 00403

APOLLONI, B. & GREGORIO DI, S. [1982] *A probabilistic analysis of a new satisfiability algorithm (French summary)* (J 3441) RAIRO Inform Theor 16*201-223
⋄ B35 D15 ⋄ REV MR 84a:68032 Zbl 489 #68038 • ID 38805

APPLEBEE, R.C. & PAHI, B. [1970] *An unsolvable problem concerning implicational calculi* (J 0047) Notre Dame J Formal Log 11*200-202
⋄ B22 D35 ⋄ REV MR 44 #74 Zbl 169.305 JSL 37.417 • ID 43922

APPLEBEE, R.C. see Vol. II for further entries

APRILE, G. [1966] *Su talune restrizioni nell'uso dei simboli per la construzione dei grafi semantici* (J 3522) Rend Circ Mat Palermo, Ser 2 15*129-132
⋄ B20 ⋄ REV MR 37 #3909 • ID 00424

APRILE, G. [1967] *Grafi inferenziali per il sillogismo categorico* (J 0104) Atti Accad Sci Lett Arti Palermo, Ser 4/I 28*255-260
⋄ B20 ⋄ REV MR 41 #6661 Zbl 235 #02007 • ID 27843

APRILE, G. [1967] *Risoluzione de equazioni logiche con l'ausilio di grafi inferenziali* (J 0104) Atti Accad Sci Lett Arti Palermo, Ser 4/I 28*441-448
⋄ B20 ⋄ REV MR 41 #8232 Zbl 235 #02009 • ID 27845

APRILE, G. [1970] *Grafi inferenziali per le relazioni diadiche* (J 0104) Atti Accad Sci Lett Arti Palermo, Ser 4/I 29*311-321
⋄ B20 ⋄ REV MR 45 #6569 Zbl 235 #02010 • ID 27846

APRILE, G. [1973] *Sulla sintassi della notazione polacca* (J 0104) Atti Accad Sci Lett Arti Palermo, Ser 4/I 33*435-445
⋄ B03 ⋄ REV MR 51 #10052 • ID 17546

APRILE, G. [1977] *Calcolo proposizionale e calcolo inferenziale* (J 0104) Atti Accad Sci Lett Arti Palermo, Ser 4/I 36*599-603
⋄ B05 ⋄ REV MR 81e:03002 Zbl 464 #03012 • ID 54602

APRILE, G. [1978] *Grafi inferenziali per i problemi logici con l'identita* (J 0104) Atti Accad Sci Lett Arti Palermo, Ser 4/I 37*71-77
⋄ B20 ⋄ REV MR 82m:03042a Zbl 449 #03014 • ID 56681

APRILE, G. [1979] *Breve teoria della equazioni logiche booleane (English summary)* (J 0104) Atti Accad Sci Lett Arti Palermo, Ser 4/I 38*69-75
⋄ B05 ⋄ REV MR 83g:03065 Zbl 492 #03004 • ID 36021

APRILE, G. [1979] *Costruzione automatica di tabelle di verita, mediante matrici pilota* (J 0104) Atti Accad Sci Lett Arti Palermo, Ser 4/I 38*61-68
⋄ B05 ⋄ REV Zbl 492 #03003 • ID 38093

APRILE, G. [1979] *Risoluzione di eauazioni logiche mediante tavole di verita (English summary)* (J 0104) Atti Accad Sci Lett Arti Palermo, Ser 4/I 38*349-355
⋄ B05 ⋄ REV MR 83g:03066 Zbl 492 #03005 • ID 36022

APRILE, G. see Vol. II, IV, V for further entries

APT, K.R. [1979] *Ten years of Hoare's logic, a survey* (P 2615) Scand Logic Symp (5);1979 Aalborg 1-44
⋄ B98 ⋄ REV MR 82m:68018 Zbl 426 #68004 • REM Part I. Part II 1983 • ID 53676

APT, K.R. & EMDEN VAN, M.H. [1982] *Contributions to the theory of logic programming* (J 0037) ACM J 29*841-862
⋄ B35 B75 ⋄ REV MR 83h:68016 Zbl 483 #68004 • ID 39187

APT, K.R. see Vol. III, IV, V, VI for further entries

ARAI, Y. [1965] *On axiom systems of propositional calculi II,III,XII* (J 0081) Proc Japan Acad 41*440-442,570-574,901-903
⋄ B05 B20 ⋄ REV MR 32 #2314 MR 33 #3901 MR 33 #5452 Zbl 156.248 Zbl 156.249 Zbl 223 #02008 JSL 34.122 • REM Part I 1965 by Imai,Y. & Iseki,K. Part IV 1965 by Iseki,K. Part XI 1965 by Tanaka,K. Part XIII 1965 by Tanaka,S. • ID 00442

ARAI, Y. & ISEKI, K. [1965] *On axiom systems of propositional calculi. VII* (J 0081) Proc Japan Acad 41*667-669
⋄ B05 ⋄ REV MR 33 #5456 Zbl 156.248 JSL 34.122 • REM Parts VI,VIII 1965 by Tanaka,S. • ID 00445

ARAI, Y. [1966] *On axiom systems of propositional calculi. XVII* (J 0081) Proc Japan Acad 42*351-354
◇ B20 ◇ REV MR 34#2431a Zbl 156.249 JSL 38.521
• REM Parts XVI,XVIII 1966 by Tanaka,S. • ID 48363

ARAI, Y. & TANAKA, S. [1966] *On axiom systems of propositional calculi. XIX* (J 0081) Proc Japan Acad 42*358-360
◇ B20 ◇ REV MR 34#2431c Zbl 156.249 JSL 38.521
• REM Parts XVIII,XX 1966 by Tanaka,S. • ID 00447

ARANGO, H. & SANTOS, J. [1957] *La operacion puente en las algebras de Boole* (P 1412) Union Mat Argentina Jorn Cientif (10);1957 Bahia Blanca 25-28
◇ B05 G05 ◇ REV JSL 34.512 • ID 43325

ARCHIE, L.C. [1976] *A simple defense of material implication* (J 1893) Relevance Logic Newslett 1*119-122
◇ B05 ◇ REV Zbl 343#02007 • ID 60177

ARCHIE, L.C. & HURDLE JR., B.G.N. & THOMBLINSON, W.S. [1977] *A note on the truth table for "if P then Q"* (J 0047) Notre Dame J Formal Log 18*596-598
◇ B05 ◇ REV MR 58#115 Zbl 315#02004 • ID 24300

ARCHIE, L.C. [1979] *A simple defense of material implication* (J 0047) Notre Dame J Formal Log 20*412-414
◇ A05 B05 ◇ REV MR 81i:03011 Zbl 394#03009
• ID 54717

AREVALO, Z. & BREDESON, J.G. [1978] *A method to simplify a boolean function into a near minimal sum-of-products for programmable logic arrays* (J 0187) IEEE Trans Comp C-27*1028-1039
◇ B35 ◇ REV MR 82e:94068 Zbl 388#94021 • ID 69092

ARKERYD, L. [1981] *Intermolecular forces of infinite range and the Boltzmann equation* (J 3596) Arch Rational Mech Anal 77*11-21
◇ H10 ◇ REV MR 83k:76057 Zbl 547.76085 • ID 40340

ARKERYD, L. [1982] *Asymptotic behaviour of the Boltzmann equation with infinite range forces* (J 1113) Commun Math Phys 86*475-484
◇ H10 ◇ REV MR 85d:76020 Zbl 514#35075 • ID 39092

ARKERYD, L. [1984] *Loeb solutions of the Boltzmann equation* (J 3596) Arch Rational Mech Anal 86*85-97
◇ H10 ◇ REV MR 85m:76047 • ID 48181

ARMBRUST, M. & KAISER, KLAUS [1972] *On quasi-universal model classes* (J 0068) Z Math Logik Grundlagen Math 18*403-406
◇ B15 C05 C20 C52 C85 ◇ REV MR 47#4784 Zbl 249#02025 • ID 00468

ARMBRUST, M. see Vol. III, V for further entries

ARMSTRONG, R.L. [1976] *A question about incompleteness* (J 0047) Notre Dame J Formal Log 17*295-296
◇ B05 ◇ REV MR 54#9968 Zbl 262#02048 • ID 25594

ARNOLD, A. [1970] *Presentation d'un langage de formalisation des demonstrations mathematiques naturelles* (P 0625) Symp Autom Demonst;1968 Versailles 6-28
◇ B35 ◇ REV MR 43#1770 Zbl 214.24 • ID 00474

ARNOLD, A. see Vol. IV for further entries

ARNOLD, B.H. [1962] *Logic and boolean algebra* (X 0819) Prentice Hall: Englewood Cliffs viii+144pp
◇ B05 B98 G05 ◇ REV MR 27#65 Zbl 121.27 JSL 29.95
• ID 23577

ARNOLD, R.F. & HARRISON, M.A. [1963] *Algebraic properties of symmetric and partially symmetric boolean functions* (J 4305) IEEE Trans Electr Comp EC-12*244-251
◇ B05 ◇ REV MR 27#3481 Zbl 121.28 • ID 04243

ARNON, D.S. & SMITH, S.F. [1983] *Towards mechanical solution of the Kahan ellipse problem* (P 4009) Europ Comput Algeb Conf;1983 London 36-44
◇ B35 ◇ REV Zbl 553#68031 • ID 43433

ARNON, D.S. see Vol. III for further entries

AROCHA, KH.L. [1980] *Minimization of a function of the algebra of logic by investigation of lower units and upper zeros (Russian)* (J 1516) Vest Ser Fiz Mat Mekh, Univ Minsk 1980/2*47-49,64
◇ B05 ◇ REV MR 81k:94049 • ID 80573

ARRUDA, A.I. & COSTA DA, N.C.A. [1964] *Sur un theoreme de Hilbert et Bernays* (J 0109) C R Acad Sci, Paris 258*6311-6312
◇ B10 ◇ REV MR 29#2166 Zbl 154.254 • ID 00481

ARRUDA, A.I. & COSTA DA, N.C.A. [1966] *Transformations in restricted predicate calculus (Portuguese)* (J 0110) Anais Acad Bras Cienc 38*385-390
◇ B20 ◇ REV MR 36#2465 Zbl 191.285 • ID 00485

ARRUDA, A.I. [1980] *A survey of paraconsistent logic* (P 2958) Latin Amer Symp Math Log (4);1978 Santiago 1-41
◇ A05 B53 B98 E70 ◇ REV MR 81i:03033 Zbl 426#03031 • ID 53627

ARRUDA, A.I. see Vol. II, III, V, VI for further entries

ARTMANN, B. [1983] *Der Zahlbegriff* (X 0903) Vandenhoeck & Ruprecht: Goettingen viii+265pp
◇ B28 E98 ◇ REV MR 84f:04001 Zbl 506#10001
• ID 34479

ASATRYAN, L.G. [1979] *Monotone boolean intersection functions (Russian)* (J 0346) Dokl Akad Nauk Armyan SSR 68*10-13
◇ B05 E07 ◇ REV MR 80k:94038 Zbl 462#05060
• ID 54526

ASEL'DEROV, Z.M. [1966] see ANUFRIEV, F.V.

ASEL'DEROV, Z.M. [1972] see ANUFRIEV, F.V.

ASENJO, F.G. [1963] *Relations irreducible to classes* (J 0047) Notre Dame J Formal Log 4*193-200
◇ B28 E70 ◇ REV MR 29#2180 Zbl 126.12 • ID 00514

ASENJO, F.G. [1965] *The arithmetic of the term-relation number theory* (J 0047) Notre Dame J Formal Log 6*223-228
◇ B28 ◇ REV MR 33#2538 • ID 00515

ASENJO, F.G. [1967] *Rings of term-relation numbers as non-standard models* (J 0047) Notre Dame J Formal Log 8*24-26
◇ H10 H20 ◇ REV MR 38#4299 Zbl 189.285 • ID 00517

ASENJO, F.G. [1970] *Generalized reals* (J 0047) Notre Dame J Formal Log 11*473-476
◇ H05 ◇ REV MR 45#3190 Zbl 177.11 • ID 00519

ASENJO, F.G. & MCKEAN, J.M. [1972] *Weierstrass's final theorem of arithmetic is not final* (J 0047) Notre Dame J Formal Log 13*91-94
◇ H05 ◇ REV MR 45#6611 Zbl 227#02035 • ID 00520

ASENJO, F.G. see Vol. II, V for further entries

ASH, C.J. & BRICKHILL, C.J. & CROSSLEY, J.N. & STILLWELL, J.C. & WILLIAMS, N.H. [1972] *What is mathematical logic?* (**X** 0894) Oxford Univ Pr: Oxford ix+82pp
- ◇ B98 ◇ REV MR 54#2411 Zbl 251#02001 JSL 40.241 • ID 17019

ASH, C.J. see Vol. III, IV, V for further entries

ASHENHURST, R.L. [1952] *The application of counting techniques* (**P** 0700) ACM Proc Conf;1952 Pittsburgh 293-305
- ◇ B05 ◇ REV MR 15.93 JSL 19.56 • ID 42242

ASHWORTH, E.J. [1978] *Multiple quantification and the use of special quantifiers in early sixteenth century logic* (**J** 0047) Notre Dame J Formal Log 19*599-613
- ◇ A10 B20 ◇ REV MR 81k:01016 Zbl 283#02002 Zbl 402#03002 • ID 54649

ASHWORTH, E.J. see Vol. II for further entries

ASLANSKI, M. & CHIMEV, K.N. [1984] *Structural characteristics of one class of Boolean functions* (**J** 2774) Koezlem MTA Szam & Autom: Kutat Intez 31*23-31
- ◇ B05 ◇ ID 47765

ASLANYAN, L.A. [1982] *On the question of minimization of systems of poorly defined Boolean functions (Russian)* (**J** 2845) Tanulmanyok 135*51-72
- ◇ B35 ◇ REV MR 84d:94026 • ID 39738

ASPETSBERGER, K. [1985] *Substitution expressions: extacting solutions of non-Horn clause proofs* (**P** 4601) EUROCAL;1985 Linz 2*78-86
- ◇ B35 ◇ ID 49647

ASPVALL, B. & PLASS, M. & TARJAN, R.E. [1979] *A linear-time algorithm for testing the truth of certain quantified boolean formulas* (**J** 0232) Inform Process Lett 8*121-123
- ◇ B35 D15 ◇ REV MR 80b:68050 Zbl 398#68042 • ID 52791

ASPVALL, B. [1980] *Recognizing disguised NR(1) instances of the satisfiability problem* (**J** 2746) J Algor 1*97-103
- ◇ B35 D15 ◇ REV MR 82a:68051 Zbl 451#68037 • ID 69096

ASSER, G. [1956] *Einfuehrung in die mathematische Logik. Teil I: Aussagenkalkuel* (**X** 1079) Teubner: Leipzig vi+184pp
- • LAST ED [1982] (**X** 1054) Harri Deutsch: Frankfurt
- ◇ B05 ◇ REV MR 26#4885 MR 31#5793 Zbl 88.9 JSL 25.276 JSL 33.304 • REM Part II 1972 • ID 22420

ASSER, G. [1956] *Ueber die Ausdrucksfaehigkeit des Praedikatenkalkuels der ersten Stufe mit Funktionalen* (**J** 0068) Z Math Logik Grundlagen Math 2*250-264
- ◇ B10 ◇ REV MR 18.866 Zbl 74.248 JSL 23.39 • ID 00542

ASSER, G. [1957] *Theorie der logischen Auswahlfunktionen* (**J** 0068) Z Math Logik Grundlagen Math 3*30-68
- ◇ B10 ◇ REV MR 20#3063 Zbl 88.247 JSL 23.39 • ID 00543

ASSER, G. & SCHROETER, K. [1958] *Axiomatisierung der k-zahlig allgemeingueltigen Ausdruecke des Stufenkalkuels* (**J** 0114) Math Nachr 19*73-86
- ◇ B15 ◇ REV MR 21#2581 JSL 25.176 • ID 00545

ASSER, G. & RAUTENBERG, W. [1960] *Ein Verfahren zur Axiomatisierung der Kontradiktionen gewisser zweiwertiger Aussagenkalkuele* (**J** 0068) Z Math Logik Grundlagen Math 6*309-318
- ◇ B05 ◇ REV MR 23#A1513 Zbl 95.7 • ID 00550

ASSER, G. [1972] *Einfuehrung in die mathematische Logik, Teil II: Praedikatenkalkuel der ersten Stufe* (**X** 1079) Teubner: Leipzig v+190pp
- ◇ B10 ◇ REV MR 50#9521 Zbl 252#02002 • REM 2nd ed. 1975. Part I 1956. Part III 1981 • ID 03842

ASSER, G. [1974] *Mathematische Logik und Grundlagen der Mathematik* (**C** 2318) Entwicklung Math in DDR 3-17
- ◇ B98 ◇ REV Zbl 311#02006 • ID 29592

ASSER, G. [1979] *Ueber die Charakterisierbarkeit transfiniter Maechtigkeiten im Praedikatenkalkuel der zweiten Stufe* (**P** 2539) Frege Konferenz (1);1979 Jena 33-42
- ◇ B15 C55 C80 C85 ◇ REV MR 82d:03017 • ID 70532

ASSER, G. [1981] *Einfuehrung in die mathematische Logik. Teil III: Praedikatenlogik hoeherer Stufe* (**X** 1079) Teubner: Leipzig 164pp
- • REPR [1981] (**X** 1054) Harri Deutsch: Frankfurt 164pp
- ◇ B15 B98 ◇ REV MR 84k:03032 Zbl 471#03001 Zbl 471#03002 • ID 55197

ASSER, G. see Vol. II, III, IV, V for further entries

ATAKA, H. [1963] *Some theorems on weighted-majority functions* (**J** 0116) Electr & Comm Japan 46/9*77-79
- ◇ B05 ◇ ID 00561

ATIYAH, M. & HOARE, C.A.R. & SHEPHERDSON, J.C. (EDS.) [1984] *Mathematical logic and programming languages* (**J** 0354) Phil Trans Roy Soc London, Ser A 312*343-518
- • REPR [1985] (**X** 0819) Prentice Hall: Englewood Cliffs 184pp
- ◇ B98 ◇ REV MR 85k:68005 • ID 47111

AUBERT, K.E. [1947] *A group-theoretical remark of E.Hoff-Hansen concerning certain expressions in the quantification theory* (**J** 0117) Arch Math Naturvid 49/7*151-156
- ◇ B10 ◇ REV MR 9.403 Zbl 33.6 • ID 00562

AUBERT, K.E. see Vol. V for further entries

AUBIN, R. [1979] *Mechanizing structural induction I. Formal system. II Strategies* (**J** 1426) Theor Comput Sci 9*329-345,347-362
- ◇ B35 ◇ REV MR 81h:68091 Zbl 423#68050 Zbl 423#68051 • ID 80583

AUGENSTEIN, B.W. [1984] *Hadron physics and transfinite set theory* (**J** 2736) Int J Theor Phys 23*1197-1205
- ◇ B30 E75 ◇ ID 44872

AUGUIN, M. [1978] see ANDRE, C.

AUSIELLO, G. [1971] *Automatic reduction of CUCH expression by means of the value method* (**J** 3434) Pubbl Ist Appl Calcolo, Ser 3 65*5-19
- ◇ B35 ◇ REV Zbl 302#68025 • ID 60214

AUSIELLO, G. [1976] *Difficult logical theories and their computer approximations* (**J** 1620) Asterisque 38-39*3-21
- ◇ B25 B35 D15 ◇ REV MR 57#16032 Zbl 365#02022 • ID 51023

AUSIELLO, G. see Vol. IV, VI for further entries

AVENOSO, F.J. & CHEIFETZ, P.M. [1970] *Logic and set theory* (**X** 0821) Wadsworth Publ: Belmont 196pp
- ◇ B98 E98 ◇ ID 22631

AVGUSTINOVICH, S.V. [1980] *On the approach of optaining lower estimates of the complexity for boolean functions (Russian)* (J 0071) Met Diskr Analiz (Novosibirsk) 35*3-8
⋄ B05 ⋄ REV MR 83e:68030 Zbl 473 # 68037 • ID 69124

AVGUSTINOVICH, S.V. [1983] *A lower bound for the length of a simultaneous covering by conjunctions of a system of boolean functions (Russian)* (J 0071) Met Diskr Analiz (Novosibirsk) 39*3-6
⋄ B05 ⋄ REV MR 85k:94060 Zbl 529 # 94017 • ID 45497

AVSARKISYAN, G.S. [1977] *A representation of boolean functions by the sum mod 2 of the implications of the arguments (Russian)* (J 0474) Avtom Vychis Tekh, Akad Nauk Latv SSR 1977/1*8-11,91
• TRANSL [1977] (J 2666) Autom Control Comput Sci 11/1*8-11
⋄ B05 ⋄ REV MR 58 # 10432 Zbl 348 # 94044 • ID 70556

AVSARKISYAN, G.S. see Vol. II for further entries

AX, J. [1968] *The elementary theory of finite fields* (J 0120) Ann of Math, Ser 2 88*239-271
⋄ B25 B30 C13 C20 C60 ⋄ REV MR 37 # 5187 Zbl 195.57 JSL 38.162 • ID 00580

Ax, J. see Vol. III, IV for further entries

AZRIEL, A. [1975] *On the minimization of boolean functions (Bulgarian) (Russian and English summaries)* (J 3171) God Vissh Ucheb Zaved, Prilozhna Mat, Sofiya 10/1*123-136
⋄ B35 ⋄ REV Zbl 339 # 02005 • ID 60223

BACHMANN, F. & SCHOLZ, H. [1933] *Die logischen Grundlagen der Arithmetik im Anschluss an Frege und Dedekind* (1111) Preprints, Manuscr., Techn. Reports etc. xi+235pp
⋄ A05 A10 B28 ⋄ REM Lithographed, Muenster, I.W., X • ID 16722

BACHMANN, F. [1934] *Untersuchungen zur Grundlegung der Arithmetik mit besonderer Beziehung auf Dedekind, Frege und Russell* (X 1282) Hirzel: Stuttgart 78pp
⋄ A05 A10 B28 ⋄ REV FdM 60.853 • ID 16723

BACHMANN, F. [1936] *Die Fragen der Abhaengigkeit und der Entbehrlichkeit von Axiomen in Axiomensystemen, in denen ein Extremalaxiom auftritt* (P 0632) Congr Int Phil des Sci;1935 Paris 7*39-52
⋄ A05 B28 B30 ⋄ REV Zbl 15.50 JSL 2.56 FdM 62.1057 • ID 00607

BACHMANN, F. & CARNAP, R. [1936] *Ueber Extremalaxiome* (J 0748) Erkenntnis (Leipzig) 6*166-188
• TRANSL [1981] (J 2028) Hist & Phil Log 2*67-85 (English)
⋄ B28 B30 ⋄ REV MR 83i:03024 Zbl 15.49 Zbl 512 # 03003 JSL 2.42 FdM 62.1056 • ID 34848

BACHMANN, F. [1939] *Aufbau des Zahlensystems* (X 0823) Teubner: Stuttgart 28pp
⋄ B28 ⋄ REV Zbl 21.205 FdM 65.33 • ID 23320

BACHMANN, H. [1983] *Der Weg der mathematischen Grundlagenforschung* (X 2865) Lang: Frankfurt 240pp
⋄ A10 B98 ⋄ REV MR 85a:01002 • ID 38832

BACHMANN, H. see Vol. V, VI for further entries

BACK, R.J.R. & MANNILA, H. [1985] *On the suitability of trace semantics for modular proofs of communicating processes* (J 1426) Theor Comput Sci 39*47-68
⋄ B35 ⋄ REV Zbl 567 # 68018 • ID 49183

BACK, R.J.R. see Vol. III, VI for further entries

BACKUS, J.W. [1958] *The syntax and semantics of the proposed international algebraic language of the Zurich ACM-GAMM conference* (X 2471) IBM: Armonk
⋄ B03 ⋄ REV Zbl 112.83 • ID 41723

BACON, J. [1965] *An alternative contextual definition of descriptions* (J 0095) Philos Stud 16*75-76
⋄ A05 B10 ⋄ ID 31333

BACON, J. [1967] *Syllogistic without existence* (J 0047) Notre Dame J Formal Log 8*195-219
⋄ B20 ⋄ REV MR 38 # 3127 Zbl 174.8 • ID 00616

BACON, J. [1969] *Ontological commitment and free logic* (J 0320) Monist 53*310-319
⋄ A05 B20 ⋄ ID 31334

BACON, J. [1982] *First-order logic based on inclusion and abstraction* (J 0036) J Symb Logic 47*793-808
⋄ B15 ⋄ REV MR 84b:03020 Zbl 501 # 03004 • ID 35618

BACON, J. [1985] *The completeness of a predicate-functor logic* (J 0036) J Symb Logic 50*903-926
⋄ B10 ⋄ ID 49323

BACON, J. see Vol. II for further entries

BACSICH, P.D. [1972] *Compact injectives and non-standard analysis* (P 2080) Conf Math Log;1970 London 255*10-28
⋄ H05 ⋄ REV MR 49 # 2875 Zbl 251 # 02054 • ID 28912

BACSICH, P.D. [1975] *Amalgamation properties and interpolation theorems for equational theories* (J 0004) Algeb Universalis 5*45-55
⋄ B20 C05 C40 C52 ⋄ REV MR 52 # 2873 Zbl 324 # 02036 • ID 17658

BACSICH, P.D. see Vol. III, V for further entries

BADAWI, A.R. [1962] *Formale und mathematische Logik (Arabic)* (X 3333) Unknown Publisher: See Remarks 330pp
⋄ B98 ⋄ REM 4th ed. 1976, Kuwait • ID 43354

BAER, G. & ROHLEDER, H. [1967] *Ueber einen arithmetisch-aussagenlogischen Kalkuel und seine Anwendung auf ganzzahlige Optimierungsprobleme* (J 0129) Elektr Informationsverarbeitung & Kybern 3*171-195
⋄ B80 ⋄ REV MR 35 # 6451 • ID 16925

BAER, G. [1969] *Ein Verfahren zur Umformung einer linearen Nebenbedingung in eine aequivalente alternative Normalform* (J 0068) Z Math Logik Grundlagen Math 15*163-180
⋄ B05 ⋄ REV MR 39 # 3816 • ID 00629

BAER, G. [1972] *Zur linearen Darstellbarkeit von Ausdruecken des Aussagenkalkuels (English and Russian summaries)* (J 0129) Elektr Informationsverarbeitung & Kybern 8*353-378
⋄ B05 ⋄ REV MR 47 # 4756 Zbl 261 # 94044 • ID 00630

BAER, G. & MEILER, M. [1974] *Zur Umformung eines booleschen Ausdrucks in eine aequivalente Gleichung mit 0-1-Variablen* (J 0129) Elektr Informationsverarbeitung & Kybern 10*341-353
⋄ B05 ⋄ REV MR 50 # 9423 Zbl 291 # 94016 • ID 63854

BAER, G. [1978] *Eine theoretische Fundierung der diskreten Optimierung mit logischen Mitteln* (J 0129) Elektr Informationsverarbeitung & Kybern 14*305-319
⋄ B80 ⋄ REV MR 80f:90088 Zbl 387 # 90076 • ID 52254

BAER, REINHOLD [1929] *Zur Axiomatik der Kardinalzahlarithmetik* (J 0044) Math Z 29*381-396
⋄ B28 E10 E30 E35 ⋄ REV FdM 54.90 • ID 00632

BAER, REINHOLD see Vol. III, V for further entries

BAGINSKI, M. & GLAEWE, W. & GOLL, P. & LIST, G. & LOESCHAU, G. & MERTENS, A. & SCHWANITZ, G. & WALTER, M. [1973] *Einfuehrung in die mathematische Logik, Einfuehrung in die Mengenlehre, Aufbau der Zahlenbereiche* (X 1036) Volk & Wissen: Berlin 327pp
  ◊ B05 B28 B98 E98 ◊ REV MR 50 # 12630 • ID 75585

BAIN, A. [1889] *Logic: deductive and inductive* (X 1228) Appleton-Century-Crofts: New York 731pp
  ◊ A05 B98 ◊ REM revised edition of "Logic, deduction" and "Logic, induction" (1870) • ID 22445

BAJKHUMANOV, A.A. & KABULOV, A.V. [1982] *Solution of systems of nonlinear Boolean equations (Russian)* (J 3287) Vopr Vychisl Prikl Mat (Tashkent) 69*161-172
  ◊ B35 ◊ REV Zbl 513 # 03032 • ID 37230

BAKER, A. [1971] *Effective methods in the theory of numbers* (P 0743) Int Congr Math (II,11,Proc);1970 Nice 1*19-26
  ◊ B25 B28 ◊ REV MR 54 # 10162 Zbl 222 # 10001 JSL 37.606 • ID 42680

BAKER, A. see Vol. III for further entries

BAKER, A.J. [1966] *Non-empty complex terms* (J 0047) Notre Dame J Formal Log 7*48-56
  ◊ A05 B20 ◊ ID 00680

BAKER, A.J. [1972] *Syllogistic with complex terms* (J 0047) Notre Dame J Formal Log 13*69-87
  ◊ B20 ◊ REV MR 45 # 6580 Zbl 228 # 02006 • ID 29113

BAKER, A.J. [1977] *Classical logical relations* (J 0047) Notre Dame J Formal Log 18*164-168
  ◊ A10 B20 ◊ REV Zbl 271 # 02005 • ID 60254

BAKER, J.R. [1975] *On two immediate inferences by limitation* (J 0047) Notre Dame J Formal Log 16*496-500
  ◊ B20 ◊ REV MR 52 # 5339 Zbl 262 # 02002 • ID 17964

BAKER, J.R. see Vol. II for further entries

BALAN, T. [1974] *Quaternary relations* (J 0060) Rev Roumaine Math Pures Appl 19*841-848
  ◊ A05 B20 E07 ◊ REV MR 50 # 12728 Zbl 293 # 54002 • ID 70609

BALLANTYNE, A.M. & BLEDSOE, W.W. [1977] *Automatic proofs of theorems in analysis using nonstandard techniques* (J 0037) ACM J 24*353-374
  ◊ B35 H05 ◊ REV MR 56 # 1840 Zbl 359 # 68109 • ID 50632

BALLANTYNE, A.M. see Vol. IV for further entries

BALLANTYNE, M. & BENNETT, J.H. [1969] *Semi-automated mathematics* (J 0037) ACM J 16*1
  ◊ B35 ◊ ID 47289

BALLARD, K.E. [1972] *Study guide for Copi: Introduction to logic. A self-instructional supplement* (X 0843) Macmillan : New York & London 274pp
  ◊ B98 ◊ ID 22457

BALOWITZ, V. [1980] *The rules of the syllogism without distribution* (J 0286) Int Logic Rev 22*99-104
  ◊ B20 ◊ REV MR 83i:03023 • ID 46679

BAMMERT, J. [1968] *Quasideduktive Systeme und S-Algebren I,II* (J 0009) Arch Math Logik Grundlagenforsch 11*56-67,101-112
  ◊ B22 G12 G25 ◊ REV MR 40 # 2527 MR 40 # 2528 Zbl 162.312 • ID 00749

BANACHOWSKI, L. [1972] *On a certain subset of the set of predicate tautologies (Russian summary)* (J 0014) Bull Acad Pol Sci, Ser Math Astron Phys 20*811-817
  ◊ B20 ◊ REV MR 47 # 11 Zbl 249 # 02007 • ID 00753

BANACHOWSKI, L. see Vol. II, IV for further entries

BANASCHEWSKI, B. [1977] *On G. Spencer Brown's laws of form* (J 0047) Notre Dame J Formal Log 18*507-509
  ◊ B20 ◊ REV MR 58 # 4749 Zbl 336 # 02008 • ID 50122

BANASCHEWSKI, B. see Vol. III, IV, V for further entries

BANKOVIC, D. [1982] *Some remarks on Boolean equations* (P 3758) Algeb Conf (2);1981 Novi Sad 47-54
  ◊ B05 ◊ REV MR 84c:06017 Zbl 502 # 06004 • ID 39643

BANKOVIC, D. [1983] *The general reproductive solution of Boolean equation* (J 0400) Publ Inst Math, NS (Belgrade) 34(48)*7-11
  ◊ B05 ◊ REV MR 86b:06008 Zbl 554 # 03031 • ID 44945

BANKOVIC, D. see Vol. V for further entries

BANKOWSKI, J. [1971] *The reduction of boolean functions to the form of a mod 2 sum of products (Polish) (Russian and English summaries)* (J 0141) Arch Autom & Telemech 16*449-453
  ◊ B05 ◊ REV MR 45 # 6601 • ID 00768

BANKSTON, P. [1983] *Coarse topologies in nonstandard extensions via separative ultrafilters* (J 0316) Illinois J Math 27*459-466
  ◊ E50 E75 H05 ◊ REV MR 85c:54008 Zbl 504 # 54048 • ID 38418

BANKSTON, P. see Vol. III, V for further entries

BANTEA, R. [1980] *Study of the completion of linear topological spaces by nonstandard methods* (J 0197) Stud Cercet Mat Acad Romana 32*119-134
  ◊ H05 ◊ REV MR 81g:03082 Zbl 454 # 46014 • ID 70654

BANTEA, R. [1981] *Nonstandard vector measures (Romanian) (English summary)* (J 0197) Stud Cercet Mat Acad Romana 33*503-514
  ◊ H10 ◊ REV MR 83i:28011a Zbl 488 # 28007 • ID 39392

BANTEA, R. [1981] *On the extension of vector measures by nonstandard methods (Romanian) (English summary)* (J 0197) Stud Cercet Mat Acad Romana 33*573-600
  ◊ H05 ◊ REV MR 83i:28011b Zbl 488 # 28008 • ID 39393

BAR-HILLEL, Y. [1950] *Bolzano's propositional logic* (J 0009) Arch Math Logik Grundlagenforsch 1*65-98
  ◊ A05 A10 B20 ◊ REV MR 14.121 Zbl 47.12 JSL 21.386 • ID 00903

BAR-HILLEL, Y. [1952] *Rational numbers as triplets of natural numbers (Hebrew) (English summary)* (J 0173) Riveon Lematemat 5*53-54
  ◊ B28 ◊ REV MR 13.812 • ID 00904

BAR-HILLEL, Y. [1953] *A quasi-arithmetical notation for syntactic description* (J 0242) Language (Baltimore) 29*47-58
  ◊ B03 ◊ REV JSL 20.193 • ID 37302

BAR-HILLEL, Y. [1954] *Can translation be mechanized?* (J 1843) Amer Sci 42*248-260
  ◊ B03 B65 ◊ REV JSL 20.193 • ID 42480

BAR-HILLEL, Y. & DALEN VAN, D. & FRAENKEL, A.A. & LEVY, A. [1958] *Foundations of set theory* (X 0809) North Holland: Amsterdam x+415pp
- ◇ A05 B98 E30 E70 E98 F98 ◇ REV MR 21 #648 MR 49 #10546 Zbl 248 #02071 Zbl 82.262 JSL 29.141
- REM 2nd ed. 1973; x+404pp; 1st ed. by Fraenkel,A.A. & Bar-Hillel,Y. • ID 00805

BAR-HILLEL, Y. [1964] *Language and information: selected essays on their theory and application* (X 0833) Acad Pr: Jerusalem x+388pp
- LAST ED [1964] (X 0832) Addison-Wesley: Reading x+388pp
- ◇ A05 B65 B98 ◇ REV MR 29 #6959 Zbl 158.241 JSL 30.382 • ID 37308

BAR-HILLEL, Y. [1967] *Theory of types* (C 0601) Encycl of Philos 8*168-172
- ◇ A05 B15 ◇ REV JSL 35.301 • ID 00807

BAR-HILLEL, Y. see Vol. II, III, IV, V for further entries

BARBANCON, G. [1985] *On the analytical and approximate solutions of* $\in y'' = yy'$ (J 0034) J Math Anal & Appl 111*637-642
- ◇ H05 ◇ ID 49585

BARCAN, R.C. [1947] *The identity of individuals in a strict functional calculus of second order* (J 0036) J Symb Logic 12*12-15
- ◇ B15 ◇ REV MR 8.429 JSL 12.95 JSL 23.342 • ID 00793

BARCAN, R.C. also published under the name BARCAN MARCUS, R.

BARCAN, R.C. see Vol. II for further entries

BARCAN MARCUS, R. [1975] see ANDERSON, A.R.

BARCAN MARCUS, R. also published under the name BARCAN, R.C.

BARCAN MARCUS, R. see Vol. II for further entries

BARENDREGT, H.P. [1981] *The lambda calculus, its syntax and semantics* (S 3303) Stud Logic Found Math 103*xiv+615pp
- TRANSL [1985] (X 0885) Mir: Moskva 606pp
- ◇ B40 B98 ◇ REV MR 83b:03016 MR 86a:03012 Zbl 467 #03010 Zbl 551 #03007 JSL 49.301 • REM Revised ed.1984, xv+621pp • ID 55008

BARENDREGT, H.P. see Vol. IV, V, VI for further entries

BARJA PEREZ, J. [1978] *Algebras universales en el calculo de proposiciones* (J 3109) Alxebra 22*70pp
- ◇ B20 C05 G99 ◇ REV MR 80b:03032 Zbl 383 #03007
- ID 51989

BARKER, S.F. [1965] *The elements of logic* (X 0822) McGraw-Hill: New York xii+336pp
- ◇ B98 ◇ REV Zbl 168.3 • ID 22458

BARKER, S.F. see Vol. II for further entries

BARLINGAY, S.S. [1965] *A modern introduction to Indian logic* (X 4322) Nat Publ House: New Delhi xv+238pp
- ◇ A10 B98 ◇ REV JSL 33.603 • ID 43203

BARNES, D.W. & MACK, J.M. [1975] *An algebraic introduction to mathematical logic* (X 0811) Springer: Heidelberg & New York viii+121pp
- ◇ B98 G98 ◇ REV MR 52 #10362 Zbl 311 #02001
- ID 21669

BARNES, D.W. see Vol. III, VI for further entries

BARONE, E. & GIANNONE, A. & SCOZZAFAVA, R. [1980] *On some aspects of the theory and applications of finitely additive probability measures* (J 2821) Pubbl Ist Mat App Univ Stud Roma 16*43-53
- ◇ C20 H05 ◇ REV MR 82f:28001 Zbl 466 #28002
- ID 80631

BARONE, F. & GRASSINI, R. [1983] *Logico algebraic approach to Lagrangian systems* (J 2736) Int J Theor Phys 22*829-836
- ◇ B80 ◇ REV MR 84k:81016 Zbl 522 #70022 • ID 36787

BARONE, F. see Vol. II for further entries

BARRICELLI, N.A. [1982] *B-mathematics as a language* (X 3805) Blindern Theoretic Res Team: Oslo 20pp
- ◇ B80 ◇ REV MR 84k:03047b • ID 34984

BARRICELLI, N.A. & COLBAN, E.A. & HANSEN, B.B. [1983] *Arithmetical applications of B-mathematics* (X 3805) Blindern Theoretic Res Team: Oslo 60pp
- ◇ B35 ◇ REV MR 84m:03018 • ID 35723

BARRICELLI, N.A. [1983] *B-mathematics* (X 3805) Blindern Theoretic Res Team: Oslo 109pp
- ◇ B05 ◇ REV MR 84k:03047a • ID 34985

BARRICELLI, N.A. [1983] see AMLIEN, J.

BARRICELLI, N.A. see Vol. VI for further entries

BARROS DE, C.M. [1962] *Sur l'existence de fonctions recurrentes* (J 0068) Z Math Logik Grundlagen Math 8*117-123
- ◇ B28 E20 ◇ REV MR 27 #5683 • ID 00815

BARROS DE, C.M. see Vol. III, V for further entries

BARTHELEMY, J.-P. [1974] *Sur la refutabilite* (J 0306) Cah Topol & Geom Differ 15*21-46
- ◇ B20 F50 G30 ◇ REV MR 50 #1852 Zbl 327 #18008
- ID 04140

BARTHELEMY, J.-P. see Vol. V for further entries

BARTNICK, J. [1980] *Predicate logic without bound variables* (X 2865) Lang: Frankfurt vi+84pp
- ◇ B98 ◇ REV MR 82k:03010 • ID 70695

BARWISE, J. & EKLOF, P.C. [1969] *Lefschetz's principle* (J 0032) J Algeb 13*554-570
- ◇ B15 C60 C75 C85 ◇ REV MR 41 #5207 Zbl 194.517
- ID 00822

BARWISE, J. [1972] *The Hanf number of second order logic* (J 0036) J Symb Logic 37*588-594
- ◇ B15 C55 C85 E30 ◇ REV MR 58 #21603 Zbl 281 #02020 • ID 00839

BARWISE, J. [1975] *Admissible sets and structures. An approach to definability theory* (X 0811) Springer: Heidelberg & New York xiii+394pp
- ◇ B98 C40 C70 C98 D60 D98 E30 E98 ◇ REV MR 54 #12519 Zbl 316 #02047 JSL 43.139 • ID 60316

BARWISE, J. [1977] *An introduction to first-order logic* (C 1523) Handb of Math Logic 5-46
- ◇ B10 C07 C98 ◇ REV MR 58 #10395 JSL 49.968
- ID 24196

BARWISE, J. (ED.) [1977] *Handbook of mathematical logic* (X 0809) North Holland: Amsterdam xi+1165pp
• TRANSL [1982] (X 2027) Nauka: Moskva
⋄ B98 C98 D98 E98 F98 H98 ⋄ REV MR 56#15351 MR 84g:03004 MR 84j:03006 Zbl 443#03001 JSL 49.968 JSL 49.971 JSL 49.975 JSL 49.980 • REM 3rd ed 1982. Transl. in 4 parts. Russian suppl. by Mints,G.E. & Orevkov,V.P • ID 70117

BARWISE, J. & KEISLER, H.J. & KUNEN, K. (EDS.) [1980] *The Kleene Symposium* (S 3303) Stud Logic Found Math 101∗xx+425pp
⋄ B97 ⋄ REV MR 81j:03009 Zbl 436#00007 • ID 55837

BARWISE, J. see Vol. II, III, IV, V, VI for further entries

BARZIN, M. [1935] *Sur les demonstrations de non-contradiction des axiomes* (P 0639) Congr Nat des Sci (2);1935 Bruxelles 156-159
⋄ A05 B30 F99 ⋄ ID 00914

BARZIN, M. see Vol. V, VI for further entries

BASARAB, S.A. [1983] *Roth's theorem: non-standard aspects (Romanian)* (J 0197) Stud Cercet Mat Acad Romana 35∗105-113
⋄ H10 H15 ⋄ REV MR 85b:11116 Zbl 534#10050 • ID 38348

BASARAB, S.A. see Vol. III for further entries

BASSON, A.H. & O'CONNOR, D.J. [1953] *Introduction to symbolic logic* (X 0927) Univ Tutorial Pr: Slough 169pp
• LAST ED [1960] (X 0824) Free Press: New York viii+175pp
⋄ B98 ⋄ REV MR 22#3672 Zbl 52.8 Zbl 89.6 JSL 20.84 JSL 28.169 • ID 22460

BASU, M.S. & CHOUDHURY, A.K. [1961] *On detection of group invariance or total symmetry of a boolean function* (J 0274) Indian J Phys 36∗31-42
• REPR [1961] (J 0275) Proc Indian Assoc Cultivation Sci 45∗31-42
⋄ B05 ⋄ REV MR 25#35 JSL 36.694 • ID 02721

BATENS, D. [1980] *A completeness-proof method for extensions of the implicational fragment of the propositional calculus* (J 0047) Notre Dame J Formal Log 21∗509-517
⋄ B20 ⋄ REV MR 82b:03030 Zbl 419#03005 • ID 53912

BATENS, D. see Vol. II, V for further entries

BATLE, N. & CUFI, J. [1973] *Metamathematical justification of the substitution criteria of N.Bourbaki (Spanish)* (P 0725) Jorn Mat Luso-Espanol (1);1972 Lisbon 21-40
⋄ B15 ⋄ REV MR 50#111 Zbl 279#02053 • ID 17147

BATLE, N. see Vol. V, VI for further entries

BATOG, T. [1971] *Is there a contradiction in the theory of types?* (J 0286) Int Logic Rev 4∗284-287
⋄ A05 B15 ⋄ REV Zbl 333#02011 • ID 60344

BATOG, T. [1976] *On substitution for functorial variables* (J 2718) Fct Approximatio, Comment Math, Poznan 4∗141-142
⋄ B05 ⋄ REV MR 56#15358 Zbl 352#02011 • ID 50010

BATOG, T. see Vol. V for further entries

BAUDISCH, A. & SEESE, D.G. & TUSCHIK, H.-P. & WEESE, M. [1980] *Decidability and generalized quantifiers* (X 0911) Akademie Verlag: Berlin xii+235pp
⋄ B25 B98 C10 C60 C80 C98 ⋄ REV MR 82i:03048 Zbl 442#03011 JSL 47.907 • ID 56368

BAUDISCH, A. see Vol. III, IV, V for further entries

BAUER, F.L. [1949] *Zur Algebraik des Logikkalkuels* (J 0175) Methodos 1∗288-292
⋄ A05 B05 ⋄ REV MR 11.636 Zbl 37.294 JSL 16.62 • ID 00918

BAUER, F.L. [1960] *The formula-controlled logical computer "Stanislaus"* (J 0214) Math of Comp 14∗64-67
⋄ B35 ⋄ REV MR 24#B1730 Zbl 95.247 • ID 47982

BAUER, F.L. [1977] *Angstl's mechanism for checking wellformedness of parenthesis-free formulae* (J 0214) Math of Comp 31∗318-320
⋄ B05 ⋄ REV MR 54#7189 • ID 24982

BAUM, J.D. [1972] *An arithmetic method in symbolic logic* (J 0148) Math Gaz 56∗91-95
⋄ B05 ⋄ REV Zbl 234#02006 • ID 27792

BAUM, J.D. [1975] *The tangled tale of Sheffer's stroke* (J 3115) Gaz Austral Math Soc 2∗57-61
⋄ A10 B05 ⋄ REV Zbl 432#03001 • ID 53959

BAUSCH, A.F. [1965] *Modus ponens under hypotheses* (J 0036) J Symb Logic 30∗26
⋄ B05 B20 ⋄ REV MR 33#3886 Zbl 146.8 • ID 00885

BAXANDALL, P.R. & BROWN, W.S. & ROSE, G.S.C. & WATSON, F.R. [1978] *Proof in mathematics ("If", "then" and "perhaps"). A collection of material illustrating the nature and variety of the idea of proof in mathematics* (X 3357) Inst Educ Univ Keele: Keele v+130pp
⋄ B98 ⋄ REV MR 80g:00001 Zbl 426#03001 • ID 53597

BAXTER, L.D. [1978] *The undecidability of the third order dyadic unification problem* (J 0194) Inform & Control 38∗170-178
⋄ B15 B40 D80 ⋄ REV MR 80m:03077 Zbl 387#03006 • ID 52222

BAXTER, ROBERT & SEARBY, D. [1975] *Modus ponens eliminability in a formalisation of the predicate calculus* (J 3068) Atti Accad Sci Bologna Fis Ser 13 2/2∗51-61
⋄ B10 ⋄ REV MR 55#5387 Zbl 348#02017 • ID 60374

BAYACHOROVA, B.D. & PANKOV, P.S. & YUGAI, S.A. [1982] *Numerical theorem proving by electronic computers and its applications in various branches of mathematics* (J 0040) Kibernetika, Akad Nauk Ukr SSR 1982/6∗111-116
• TRANSL [1982] (J 0021) Cybernetics 18∗840-848
⋄ B35 ⋄ REV MR 84i:68158 Zbl 506#68074 • ID 40185

BAYLIS, C.A. & BENNETT, A.A. [1935] *A calculus for propositional concepts* (J 0094) Mind 44∗152-167
⋄ B05 ⋄ REV Zbl 11.241 FdM 61.53 • ID 40783

BAYLIS, C.A. & BENNETT, A.A. [1939] *Formal logic: A modern introduction* (X 0819) Prentice Hall: Englewood Cliffs xviii+407pp
⋄ B98 ⋄ REV Zbl 25.4 JSL 4.94 FdM 65.20 • ID 00893

BAYOD, J.M. [1982] *Spherical completeness with infinitesimals* (J 0236) Rev Mat Hisp-Amer, Ser 4 42∗3-14
⋄ H10 ⋄ REV Zbl 537#12017 • ID 43853

BAYOD, J.M. [1984] *A spherical completion within the infinitesimal hull of an ultrametric space* (J 2823) Quaest Math, S Africa 7*241-249
- ◊ H10 ◊ REV MR 86b:54028 • ID 44488

BAZILEVSKIJ, YU.YA. [1963] *The theory of sequential logical functions* (C 0751) Th of Math Mach 1-45
- ◊ B05 ◊ REV MR 29 # 1756 • ID 17002

BAZILEVSKIJ, YU.YA. see Vol. IV for further entries

BEACH, J. [1970] *Introduction to logic* (X 0802) Allyn & Bacon: London x+170pp
- ◊ B98 ◊ REV Zbl 215.318 • ID 22462

BEALER, G. [1983] *Completeness in the theory of properties, relations, and propositions* (J 0036) J Symb Logic 48*415-426
- ◊ B15 ◊ REV MR 85e:03061 Zbl 519 # 03017 • ID 37535

BEALER, G. see Vol. II for further entries

BECCHIO, D. [1970] *Demonstration algebrique de l'independance des schemas d'axiomes du calcul propositionnel classique* (J 2313) C R Acad Sci, Paris, Ser A-B 270*A1205-A1208
- ◊ B05 ◊ REV MR 41 # 3238 Zbl 194.306 • ID 42986

BECCHIO, D. [1970] *Demonstration algebrique du theoreme de Wajsberg et du theoreme methodologique de Jaskowski* (J 2313) C R Acad Sci, Paris, Ser A-B 270*A1301-A1304
- ◊ B05 ◊ REV MR 41 # 5189 Zbl 194.307 • ID 00925

BECCHIO, D. see Vol. II for further entries

BECK, JON M. [1979] *Simplicial sets and the foundations of analysis* (P 2901) Appl Sheaves;1977 Durham 113-124
- ◊ B28 F65 ◊ REV MR 80m:03029 Zbl 438 # 03057 • ID 55969

BECK, JON M. [1980] *On the relationship between algebra and analysis* (J 0326) J Pure Appl Algebra 19*43-60
- ◊ H05 ◊ REV MR 82a:18005 Zbl 455 # 03026 • ID 54287

BECK, JON M. see Vol. VI for further entries

BECKER, J.A. & LIPSHITZ, L. [1976] *An application of logic to analysis* (J 0017) Canad J Math 28*83-91
- ◊ B80 C65 ◊ REV MR 55 # 8397 Zbl 319 # 02051 • ID 29684

BECKER, J.A. see Vol. III, IV, V for further entries

BECKER, O. [1951] *Einfuehrung in die Logistik, vorzueglich in den Modalkalkuel* (X 0825) Westkulturverlag : Meisenheim 92pp
- ◊ B98 ◊ REV MR 13.309 Zbl 45.1 JSL 17.59 • ID 00927

BECKER, O. see Vol. II for further entries

BECKMAN, F.S. [1981] *Mathematical foundations of programming* (X 0832) Addison-Wesley: Reading xviii+443pp
- ◊ B98 ◊ REV MR 83g:68004 Zbl 443.68021 • ID 39144

BECVAR, J. (ED.) [1975] *Mathematical foundations of computer science 1975* (S 3302) Lect Notes Comput Sci 32*x+476pp
- ◊ B35 B75 ◊ REV MR 52 # 2265 Zbl 308 # 00007 • ID 48618

BECVAR, J. see Vol. II, IV, VI for further entries

BEHMANN, H. [1922] *Beitraege zur Algebra der Logik, insbesondere zum Entscheidungsproblem* (J 0043) Math Ann 86*163-229
- ◊ B20 B25 D35 ◊ REV FdM 48.1119 • ID 00942

BEHMANN, H. [1923] *Algebra der Logik und Entscheidungsproblem* (J 0157) Jbuchber Dtsch Math-Ver 32*66-67,2.Abteilung
- ◊ B20 B25 D35 ◊ ID 00943

BEHMANN, H. [1927] *Mathematik und Logik* (X 0823) Teubner: Stuttgart 59pp
- ◊ A05 B98 ◊ REV FdM 53.41 • ID 16737

BEHMANN, H. [1950] *Das Aufloesungsproblem in der Klassenlogik* (J 0009) Arch Math Logik Grundlagenforsch 1*17-29
- ◊ A05 B20 B25 ◊ REV MR 14.122 Zbl 41.348 JSL 18.74 • ID 01559

BEHMANN, H. [1958] *Ein logischer Abakus* (J 0009) Arch Math Logik Grundlagenforsch 4*42-52
- ◊ A05 B05 ◊ REV MR 20 # 3065 Zbl 81.10 JSL 23.450 • ID 00950

BEHMANN, H. [1959] *Der Praedikatenkalkuel mit limitierten Variablen. Grundlegung einer natuerlichen exakten Logik* (J 0036) J Symb Logic 24*112-140
- ◊ B10 ◊ REV MR 22 # 2545 Zbl 95.242 JSL 40.583 • ID 00952

BEHMANN, H. [1961] *Das Vereinfachungsproblem fuer aussagenlogische Normalformen* (J 0009) Arch Math Logik Grundlagenforsch 5*65-89
- ◊ B05 ◊ REV MR 23 # A3077 Zbl 117.10 • ID 00951

BEHMANN, H. see Vol. II, III, IV, V, VI for further entries

BEHRENS, M.F. [1974] *A local inverse function theorem* (P 1083) Victoria Symp Nonstand Anal;1972 Victoria 34-36
- ◊ H05 ◊ REV MR 57 # 12209 Zbl 275 # 26025 • ID 26613

BEHRENS, M.F. [1974] *Analytic sets in $\mathscr{M}(D)$* (P 1083) Victoria Symp Nonstand Anal;1972 Victoria 5-22
- ◊ H05 ◊ REV MR 58 # 23612 Zbl 301 # 46043 • ID 26611

BEHRENS, M.F. [1974] *Boundary values for meromorphic functions defined in the open unit disk* (P 1083) Victoria Symp Nonstand Anal;1972 Victoria 23-33
- ◊ H05 ◊ REV MR 58 # 6249 Zbl 293 # 30033 • ID 26612

BEHRENS, M.F. [1984] *Interpolation and Gleason parts in L-domains* (J 0064) Trans Amer Math Soc 286*203-225
- ◊ H05 ◊ REV MR 86b:46079 • ID 45802

BEHRMANN, E. & CRAEMER, D. [1974] *An algorithm for well-formed formulas in infix-notation* (J 0248) Math Student 42*205-211
- ◊ B03 ◊ REV MR 53 # 104 Zbl 361 # 02017 • ID 16568

BELL, E.T. [1927] *Algebraic arithmetic* (X 0803) Amer Math Soc: Providence iv+180pp
- ◊ B28 ◊ REV FdM 53.111 • ID 22575

BELL, E.T. [1927] *Arithmetic of logic* (J 0064) Trans Amer Math Soc 29*597-611
- ◊ B05 G05 ◊ REV FdM 53.44 • ID 00987

BELL, J.L. & COLE, J.C. & PRIEST, G. & SLOMSON, A. (EDS.) [1973] *The Proceedings of the Bertrand Russell Memorial Logic Conference (Uldum, 4 - 16 August 1971)* (X 2504) Russell Mem Conf: Leeds vi+404pp
- ◊ B97 ◊ REV MR 49 # 4747 • ID 70214

BELL, J.L. & MACHOVER, M. [1977] *A course in mathematical logic* (X 0809) North Holland: Amsterdam xviii+599pp
- ◊ B98 ◊ REV MR 57 # 12155 Zbl 359 # 02001 JSL 45.378 • ID 28152

BELL, J.L. see Vol. II, III, V for further entries

BELLACICCO, A. & FREGUGLIA, P. [1979] *Caratterizzazione semantica di tavole oggetto-predicato soggette a procedure di clustering* (J 3522) Rend Circ Mat Palermo, Ser 2   28*37-43
  ⋄ B80 ⋄ REV MR 81d:68042 Zbl 427 # 03007 • ID 53694

BELLACICCO, A. see Vol. V for further entries

BELLENOT, S.F. [1974] *Nonstandard topological vector spaces* (P 1083) Victoria Symp Nonstand Anal;1972 Victoria 37-39
  ⋄ H05 ⋄ REV MR 58 # 2124 Zbl 274 # 46012 • ID 26614

BELLENOT, S.F. [1976] *On nonstandard hulls on convex spaces* (J 0017) Canad J Math   28*141-147
  ⋄ H05 ⋄ REV MR 53 # 11329 Zbl 308 # 46001 • ID 23104

BELLENOT, S.F. [1980] *Basic sequences in non-Schwartz Frechet spaces* (J 0064) Trans Amer Math Soc   258*199-216
  ⋄ H05 ⋄ REV MR 83b:46008 Zbl 426 # 46001 • ID 39002

BELLENOT, S.F. see Vol. IV, V for further entries

BELLIA, M. & DAMERI, E. & DEGANO, P. & LEVI, G. & MARTELLI, M. [1982] *Applicative communicating processes in first order logic* (P 3867) Int Symp Progr (5);1982 Turin 1-14
  ⋄ B80 ⋄ REV Zbl 494 # 68033 • ID 48973

BELLIN, G. [1979] *Herbrand's theorem for calculi of sequents LK and LJ* (P 2615) Scand Logic Symp (5);1979 Aalborg 285-300
  ⋄ B10 ⋄ REV MR 83a:03058 Zbl 433 # 03035 • ID 35071

BELLOMO, A. [1978] *Analisi di un'argomentazione del "Convivio" dantesco mediante il calcolo dei predicati* (J 1515) Archimede   30*211-213
  ⋄ A10 B10 ⋄ REV Zbl 408 # 03002 • ID 56241

BELLOT, P. [1985] *A new proof for Craig's theorem* (J 0036) J Symb Logic   50*395-396
  ⋄ B10 B40 ⋄ ID 42554

BELNAP JR., N.D. [1959] see ANDERSON, A.R.

BELNAP JR., N.D. & THOMASON, R.H. [1963] *A rule-completeness theorem* (J 0047) Notre Dame J Formal Log   4*39-43
  ⋄ B22 F07 F50 ⋄ REV MR 27 # 28 Zbl 118.13 • ID 01010

BELNAP JR., N.D. & DUNN, J.M. [1968] *The substitution interpretation of the quantifiers* (J 0097) Nous, Quart J Phil   2*177-185
  ⋄ A05 B10 ⋄ ID 03177

BELNAP JR., N.D. [1975] see ANDERSON, A.R.

BELNAP JR., N.D. see Vol. II, VI for further entries

BEL'TYUKOV, A.P. [1976] *Decidability of the universal theory of natural numbers with addition and divisibility (Russian)* (S 0228) Zap Nauch Sem Leningrad Otd Mat Inst Steklov   60*15-28,221
  • TRANSL [1980] (J 1531) J Sov Math   14*1436-1444
  ⋄ B25 B28 F30 ⋄ REV MR 58 # 27419 Zbl 345 # 02035 • ID 29789

BEL'TYUKOV, A.P. see Vol. IV for further entries

BELYAKIN, N.V. [1979] *Nonstandardly finite sets (Russian)* (C 2967) Metodol Probl Mat   48-54
  ⋄ E30 H05 H15 ⋄ REV MR 81f:03004 • ID 70865

BELYAKIN, N.V. [1983] *A means of modeling a classical second-order arithmetic (Russian)* (J 0003) Algebra i Logika   22*3-25
  • TRANSL [1983] (J 0069) Algeb and Log   22*1-18
  ⋄ B15 C62 D55 F35 ⋄ REV MR 85h:03046 Zbl 538 # 03040 • ID 41479

BELYAKIN, N.V. see Vol. III, IV, VI for further entries

BELYANIN, P.N. & BRUEVICH, N.G. & CHELISHCHEV, B.E. & GONSALES-SABATER, A. [1979] *Mathematical theory of assembly technology (Russian)* (J 0023) Dokl Akad Nauk SSSR   246*1310-1313
  • TRANSL [1979] (J 0470) Sov Phys, Dokl   24*423-425
  ⋄ B80 ⋄ REV Zbl 427 # 90045 • ID 53759

BEN-ARI, M. [1980] *A simplified proof that regular resolution is exponential* (J 0232) Inform Process Lett   10*96-98
  ⋄ B35 D15 F20 ⋄ REV MR 81g:68062 Zbl 438 # 03054 • ID 55966

BEN-ARI, M. [1980] *Comments on "Tautology testing with a generalized matrix reduction method" by W. Bibel* (J 1426) Theor Comput Sci   11*341
  ⋄ B35 ⋄ REV MR 81f:03012 • REM The article was published ibid. 8(1979)/1*31-44, MR80i # 03023 • ID 70905

BEN-ARI, M. see Vol. II for further entries

BENCIVENGA, E. [1981] *Semantic tableaux for a logic with identity* (J 0068) Z Math Logik Grundlagen Math   27*241-247
  ⋄ B10 ⋄ REV MR 84e:03015 Zbl 459 # 03016 • ID 54456

BENCIVENGA, E. & LAMBERT, K. & MEYER, R.K. [1982] *The ineliminability of E! in free quantification theory without identity* (J 0122) J Philos Logic   11*229-231
  ⋄ B10 ⋄ REV MR 83i:03021 Zbl 488 # 03006 • ID 35510

BENCIVENGA, E. [1983] *Compactness of a supervaluation language* (J 0036) J Symb Logic   48*384-386
  ⋄ B15 ⋄ REV MR 84k:03081 Zbl 521 # 03017 • ID 35018

BENCIVENGA, E. [1983] *Dropping a few worlds* (J 0079) Logique & Anal, NS   26*241-246
  ⋄ B05 ⋄ REV MR 85e:03006 • ID 40541

BENCIVENGA, E. [1984] *Il primo libro di logica. Introduzione ai metodi della logica contemporanea* (X 0905) Boringhieri: Torino 228pp
  ⋄ B98 ⋄ REV Zbl 557 # 03001 • ID 46185

BENCIVENGA, E. see Vol. II, V, VI for further entries

BENDA, W. & HORNUNG, G. & RAULEFS, P. & VOLLMANN, F. [1983] *Der META-Beweiser fuer die Zahlentheorie* (P 3858) Adequate Modeling of Syst;1982 Bad Honnef 142-153
  ⋄ B35 ⋄ REV MR 85h:03017 Zbl 499 # 68040 • ID 43395

BENDALL, K. [1978] *Natural deduction, separation, and the meaning of logical operators* (J 0122) J Philos Logic   7*245-276
  ⋄ A05 B10 ⋄ REV MR 80d:03009 Zbl 387 # 03002 • ID 52218

BENDALL, K. [1979] *Negation as a sign of negative judgment* (J 0047) Notre Dame J Formal Log   20*68-76
  ⋄ B10 ⋄ REV MR 83d:03014 Zbl 315 # 02015 • ID 52378

BENDALL, K. see Vol. II for further entries

BENEJAM, J.-P. [1969] *Application du theoreme de Herbrand a la presentation de theses teratologiques de calcul des predicats elementaires* (J 2313) C R Acad Sci, Paris, Ser A-B 268*A757-A760
  ◊ B10 ◊ REV MR 40#22 Zbl 176.273 JSL 40.238
  • ID 01026

BENEJAM, J.-P. see Vol. III, IV, V for further entries

BENEYTO, R. [1971] *Analytic labyrinths (Spanish)* (J 0162) Teorema (Valencia) 1971/4*19-30
  ◊ B10 F07 ◊ REV MR 47#4770 • ID 01031

BENEYTO, R. [1973] *Trees, logic and decision mechanisms (Spanish)* (J 0162) Teorema (Valencia) 3*289-313
  ◊ B10 ◊ REV MR 50#4259 • ID 03888

BENEYTO, R. [1975] *A natural aspect of natural deduction* (J 0162) Teorema (Valencia) 5*361-381
  ◊ B10 ◊ REV MR 56#98 • ID 70933

BENIOFF, P.A. [1970] *Some aspects of the relationship between mathematical logic and physics I* (J 0209) J Math Phys 11*2553-2569
  ◊ B30 ◊ REV MR 42#4094 Zbl 199.279 • REM Part II 1971 • ID 01566

BENIOFF, P.A. [1971] *Some aspects of the relationship between mathematical logic and physics II* (J 0209) J Math Phys 12*360-376
  ◊ B30 ◊ REV MR 44#1324 Zbl 209.584 • REM Part I 1970 • ID 01567

BENIOFF, P.A. see Vol. II, III, IV, V for further entries

BENIS-SINACEUR, H. [1985] *La theorie d'Artin et Schreier et l'analyse non-standard d'Abraham Robinson* (J 0267) Arch Hist Exact Sci 34*257-264
  ◊ H05 ◊ ID 49520

BENNETT, A.A. [1935] see BAYLIS, C.A.

BENNETT, A.A. [1939] see BAYLIS, C.A.

BENNETT, A.A. see Vol. V for further entries

BENNETT, D.W. [1973] *An elementary completeness proof for a system of natural deduction* (J 0047) Notre Dame J Formal Log 14*430-432
  ◊ B10 C07 F07 ◊ REV MR 47#8284 Zbl 258#02048
  • ID 01038

BENNETT, D.W. [1977] *A note on the completeness proof for natural deduction* (J 0047) Notre Dame J Formal Log 18*145-146
  ◊ B10 C07 F07 ◊ REV MR 56#8355 Zbl 338#02014
  • ID 21957

BENNETT, D.W. [1980] *Junctions* (J 0047) Notre Dame J Formal Log 21*111-118
  ◊ B05 G05 ◊ REV MR 82e:03017 Zbl 363#02011
  • ID 53175

BENNETT, J.H. [1969] see BALLANTYNE, M.

BENNETT, J.H. & GUARD, J.R. & OGLESBY, F.C. & SETTLE, L.G. [1969] *Semi-automated mathematics* (J 0037) ACM J 16*49-62
  • REPR [1983] (C 4659) Autom of Reasoning 2*203-216
  ◊ B35 B75 ◊ ID 47335

BENNETT, J.H. see Vol. II for further entries

BENNINGHOFEN, B. [1984] *Superinfinitesimals and the calculus of the generalized Rieman integral* (P 2153) Logic Colloq;1983 Aachen 1*9-52
  ◊ H05 ◊ REV MR 86g:26008 • ID 45425

BENOIT, E. & CALLOT, J.-L. [1981] *Chasse au canard I,IV* (J 0264) Collect Math (Barcelona) 32*37-74,115-119
  ◊ H05 ◊ REV MR 85g:58062d Zbl 529#34046 • REM Part II 1981 by Benoit,E. Part III 1981 by Callot,J.-L. • ID 46233

BENOIT, E. [1981] *Equations differentielles: relation entree-sortie (English summary)* (J 3364) C R Acad Sci, Paris, Ser 1 293*293-296
  ◊ H05 ◊ REV MR 83h:34036 Zbl 485#34031 • ID 39261

BENOIT, E. [1981] *Tunnels et entonnoirs (English summary)* (J 3364) C R Acad Sci, Paris, Ser 1 292*283-286
  ◊ H05 ◊ REV MR 82g:58074 Zbl 484#34040 • ID 80684

BENOIT, E. [1982] *Les canards de R^3* (J 3364) C R Acad Sci, Paris, Ser 1 294*483-488
  ◊ H05 ◊ REV MR 84e:58047 Zbl 489#34061 • ID 39580

BENOIT, E. [1983] *Canards et chaos dans R^3* (P 4223) Math Tool & Model:1981/82 Toulouse & Paris 3*335-340
  ◊ H05 ◊ ID 44913

BENOIT, E. [1983] *Systemes lents-rapides dans R^3 et leurs canards* (J 1620) Asterisque 109-110*159-191
  ◊ H05 ◊ REV MR 86d:58103 Zbl 529#34037 • ID 45501

BENTHEM JUTTING VAN, L.S. [1977] *Checking Landau's "Grundlagen" in the AUTOMATH system* (X 3152) Techn Hogeschool: Eindhoven v+121pp
  • LAST ED [1979] (X 1121) Math Centr: Amsterdam iv+120pp
  ◊ B35 ◊ REV MR 58#32124a MR 58#32124b Zbl 352#68105 • ID 50069

BENTHEM VAN, J.F.A.K. [1974] *Semantic tableaus* (J 3077) Nieuw Arch Wisk, Ser 3 22*44-59
  ◊ B10 C07 F07 ◊ REV MR 49#2282 Zbl 285#02048
  • ID 03889

BENTHEM VAN, J.F.A.K. see Vol. II, III, V, VI for further entries

BENZAKEN, C. [1966] *Algorithmes de dualisation d'une fonction booleenne* (J 0186) Rev Franc Trait Info Chiffres 9*119-128
  ◊ B35 ◊ REV MR 34#5577 Zbl 166.257 • ID 01256

BERBAT, N.E. [1978] see ABDUL-KARIM, M.A.H.

BERESNEV, V.L. [1979] *Algorithms for the minimization of polynomials of boolean variables (Russian)* (J 0052) Probl Kibern 36*225-246
  ◊ B35 ◊ REV MR 83k:65053 Zbl 447#94028 • ID 69176

BERG, J. [1972] *Bolzano's theory of an ideal language* (C 0558) Contemp Phil Scand 405-415
  ◊ A10 B03 ◊ ID 14903

BERG, J. [1977] *Bolzano's contribution to logic and philosophy of mathematics* (P 1075) Logic Colloq;1976 Oxford 147-171
  ◊ A05 A10 B03 ◊ REV MR 58#26985 Zbl 434#03003
  • ID 16618

BERG, J. see Vol. II, IV, VI for further entries

BERG VAN DEN, I. & DIENER, M. [1981] *Diverses applications du lemme de Robinson en analyse non standard (English summary)* (J 3364) C R Acad Sci, Paris, Ser 1 293*501-504
  ◊ H05 H10 ◊ REV MR 83b:03074 Zbl 519#03053
  • ID 35114

BERG VAN DEN, I. & DIENER, M. [1981] *Halos et galaxies: une extension du lemme de Robinson (English summary)* (**J** 3364) C R Acad Sci, Paris, Ser 1   293*385-388
  ⋄ H05 ⋄ REV MR 82m:03071 • ID 72383

BERG VAN DEN, I. [1983] *Un point de vue nonstandard sur les developpements en serie de Taylor* (**J** 1620) Asterisque 109-110*209-223
  ⋄ H05 ⋄ REV MR 86e:26027 Zbl 529 #41030 • ID 38484

BERG VAN DEN, I. [1983] *Un principe de permanence general* (**J** 1620) Asterisque 109-110*193-208
  ⋄ E70 H05 ⋄ REV MR 86e:03061 Zbl 548 #03039 • ID 43210

BERGE, C. & RAMACHANDRA RAO, A. [1977] *A combinatorial problem in logic* (**J** 0193) Discr Math 17*23-26
  ⋄ B05 ⋄ REV MR 58 #21768 Zbl 352 #05047 • ID 50050

BERGE, C. see Vol. V for further entries

BERGER, M. & SLOAN, A. [1983] *Explicit solutions of partial differential equations* (**C** 3884) Nonstandard Anal - Recent Develop 1-14
  ⋄ H05 ⋄ REV MR 85f:35041 Zbl 508 #35021 • ID 38153

BERGMAN, GEORGE M. [1978] *Terms and cyclic permutations* (**J** 0004) Algeb Universalis 8*129-136
  ⋄ B03 D40 ⋄ REV MR 56 #15451 Zbl 327 #02012 • ID 50464

BERGMAN, GEORGE M. see Vol. V for further entries

BERGMANN, E. & NOLL, H. [1977] *Mathematische Logik mit Informatik-Anwendungen* (**X** 0811) Springer: Heidelberg & New York xv+324pp
  ⋄ A05 A10 B98 ⋄ REV MR 58 #21353 Zbl 384 #03001 • ID 52043

BERGMANN, G. [1943] *Notes on identity* (**J** 0153) Phil of Sci (East Lansing) 10*163-166
  ⋄ A05 B20 ⋄ REV JSL 8.86 • ID 41450

BERGMANN, G. [1956] *Propositional functions* (**J** 0103) Analysis (Oxford) 17*43-48
  ⋄ A05 B05 ⋄ REV JSL 36.177 • ID 01062

BERGMANN, G. see Vol. II, V for further entries

BERGMANN, H. [1971] *Eine Faerbungstheorie fuer endliche Graphen* (**J** 0127) J Reine Angew Math 247*87-91
  ⋄ B80 ⋄ REV MR 43 #1884 Zbl 223 #05102 • ID 01064

BERGMANN, H. see Vol. V for further entries

BERGMANN, MERRIE & MOOR, J. & NELSON, JACK [1980] *The logic book* (**X** 0981) Random House: New York ix+459pp
  ⋄ B98 ⋄ REV JSL 47.915 • ID 47407

BERGMANN, MERRIE see Vol. II for further entries

BERGSTRA, J.A. & TUCKER, J.V. [1982] *Two theorems about the completeness of Hoare's logic* (**J** 0232) Inform Process Lett 15*143-149
  ⋄ B10 B75 ⋄ REV MR 84f:68008 Zbl 455 #68020 • ID 54305

BERGSTRA, J.A. see Vol. II, III, IV, VI for further entries

BERKA, K. [1979] *A reinterpretation of Aristotle's syllogistic* (**J** 2808) Organon 15*35-48
  ⋄ A10 B20 ⋄ REV MR 82c:03014 • ID 70965

BERKA, K. see Vol. II for further entries

BERKELEY, E.C. [1942] *Conditions affecting the application of symbolic logic* (**J** 0036) J Symb Logic 7*160-168
  ⋄ B80 ⋄ REV MR 4.125 • ID 01066

BERKELEY, E.C. [1952] *A summary of symbolic logic and its practical applications* (**X** 0851) Berkeley: New York 24pp
  ⋄ B05 ⋄ REV Zbl 49.147 JSL 18.68 • ID 41851

BERKELEY, E.C. [1952] *Symbolic logic. Twenty problems and solutions* (**X** 0851) Berkeley: New York ii+28pp
  ⋄ B98 ⋄ REV Zbl 48.244 JSL 20.287 • ID 01544

BERLINSKI, D. & GALLIN, D. [1969] *Quine's definition of logical truth* (**J** 0097) Nous, Quart J Phil 3*111-128
  ⋄ A05 B10 ⋄ ID 28260

BERMAN, G. & COLIGN, A.W. [1977] *A computer representation of boolean and pseudo-boolean functions* (**P** 3425) SE Conf Combin, Graph Th & Comput (8);1977 Baton Rouge 163-181
  ⋄ B35 ⋄ REV MR 58 #15636 Zbl 397 #94018 • ID 69181

BERNARDO DI, G. [1972] *Introduzione alla logica dei sistemi normativi* (**X** 0881) Il Mulino: Bologna 183pp
  ⋄ B98 ⋄ REV MR 50 #57 JSL 40.466 • ID 17148

BERNARDO DI, G. see Vol. II for further entries

BERNAYS, P. [1922] *Ueber Hilberts Gedanken zur Grundlegung der Arithmetik* (**J** 0157) Jbuchber Dtsch Math-Ver 31*10-19
  ⋄ A05 B28 ⋄ REV FdM 48.1132 FdM 48.50 • ID 01074

BERNAYS, P. [1923] *Erwiderung auf die Note von Herrn Alois Mueller: "Ueber Zahlen als Zeichen"* (**J** 0043) Math Ann 90*159-163
  • LAST ED [1929] (**J** 1380) Ann Philos & Philos Kritik 4*492-497
  ⋄ A05 B28 ⋄ REV FdM 49.39 FdM 49.686 • ID 42966

BERNAYS, P. [1926] *Axiomatische Untersuchungen des Aussagenkalkuels der "Principia Mathematica"* (**J** 0044) Math Z 25*305-320
  ⋄ A05 B05 ⋄ REV FdM 52.49 • ID 01075

BERNAYS, P. [1927] *Probleme der theoretischen Logik* (**J** 0185) Unterrichtsbl Math Nat 33*369-377
  ⋄ A05 B10 ⋄ REV FdM 53.42 • ID 01259

BERNAYS, P. & SCHOENFINKEL, M. [1928] *Zum Entscheidungsproblem der mathematischen Logik* (**J** 0043) Math Ann 99*342-372
  ⋄ B20 B25 D35 ⋄ REV FdM 54.56 • ID 01076

BERNAYS, P. [1932] *Methoden des Nachweises von Widerspruchsfreiheit und ihre Grenzen* (**P** 0653) Int Congr Math (II, 4);1932 Zuerich 2*342-343
  ⋄ A05 B30 F99 ⋄ REV FdM 58.70 • ID 01289

BERNAYS, P. [1934] see HILBERT, D.

BERNAYS, P. [1935] *Hilberts Untersuchungen ueber die Grundlagen der Arithmetik* (**C** 1162) Hilbert: Ges Abhandlungen 3*196-217
  ⋄ A05 B28 ⋄ ID 01288

BERNAYS, P. [1936] *Logical calculus* (**X** 1335) Princeton Univ IAS: Princeton 125pp
  ⋄ B10 ⋄ REV JSL 3.162 • ID 16740

BERNAYS, P. [1939] *Bemerkungen zur Grundlagenfrage* (**C** 0656) Phil Mathematique 83-87
  ⋄ A05 B30 ⋄ REV JSL 5.78 • ID 01084

BERNAYS, P. [1939] see HILBERT, D.

BERNAYS, P. [1950] *Mathematische Existenz und Widerspruchsfreiheit* (C 1627) Etud Phil Sci (Gonseth) 11-25
⋄ A05 B30 F99 ⋄ REV JSL 22.210 • ID 37320

BERNAYS, P. [1953] *Existence et non-contradiction en mathematiques. Avec une note de G.Bouligand* (J 0180) Rev Phil France & Etranger 143*85-87
⋄ A05 B30 F99 ⋄ REV Zbl 52.9 • ID 47892

BERNAYS, P. [1955] *Betrachtungen ueber das Vollstaendigkeitsaxiom und verwandte Axiome* (J 0044) Math Z 63*219-229
⋄ B28 C60 C65 ⋄ REV MR 17.447 Zbl 68.269 • ID 01090

BERNAYS, P. & FRAENKEL, A.A. [1958] *Axiomatic set theory* (X 0809) North Holland: Amsterdam viii+227pp
⋄ B98 E98 ⋄ REV MR 21 #4912 Zbl 82.263 JSL 24.224
• REM 2nd ed. 1968. With a historical introduction by A.A.Fraenkel • ID 16969

BERNAYS, P. [1958] *Remarques sur le probleme de la decision en logique elementaire* (P 0576) Raisonn en Math & Sci Exper;1955 Paris 39-43
⋄ B10 D35 ⋄ REV MR 21 #3330 Zbl 85.251 JSL 25.285
• ID 01541

BERNAYS, P. [1959] *Die Mannigfaltigkeiten der Direktiven fuer die Gestaltung geometrischer Axiomensysteme* (P 0651) Axiomatic Method;1957 Berkeley 1-15
⋄ A05 B30 ⋄ REV MR 23 #A1535 Zbl 89.165 JSL 35.310
• ID 01286

BERNAYS, P. [1965] *Betrachtungen zum Sequenzen-Kalkuel* (C 0749) Contrib Logic & Methodol (Bochenski) 1-44
⋄ B20 F05 F07 ⋄ REV MR 51 #10043 Zbl 192.29
• ID 16287

BERNAYS, P. [1969] *Remark concerning the formalization of recursive definition in second order logic* (P 1841) Fct Recurs & Appl;1967 Tihany 19-23
⋄ B15 D75 ⋄ ID 32552

BERNAYS, P. [1970] *The original Gentzen consistency proof for number theory* (P 0603) Intuitionism & Proof Th;1968 Buffalo 409-417
⋄ A10 B28 F30 F50 ⋄ REV MR 43 #1810 Zbl 204.309 JSL 40.95 • ID 01284

BERNAYS, P. see Vol. II, III, V, VI for further entries

BERNERT, J. [1980] *The strongly adequate matrices of the form of product of Lindenbaum's matrices* (J 0387) Bull Sect Logic, Pol Acad Sci 9*102-107
⋄ B22 ⋄ REV MR 82c:03037 Zbl 449 #03020 • ID 56687

BERNERT, J. [1982] *A note on existence of matrices strongly adequate for some positive logics* (J 0302) Rep Math Logic, Krakow & Katowice 14*3-7
⋄ B22 ⋄ REV MR 84e:03021 Zbl 492 #03009 • ID 34360

BERNERT, J. see Vol. II for further entries

BERNHARDT, K. [1976] *Semantics for finitary predicate calculi* (J 0063) Studia Logica 35*227-241
⋄ B20 C90 ⋄ REV MR 56 #5228 Zbl 344 #02017
• ID 60487

BERNHARDT, K. see Vol. II, III, V for further entries

BERNSHTEJN, S.N. [1917] *Essay of axiomatical foundation of probability theory (Russian)* (J 1019) Vest Ser Mekh Mat, Univ Khar'kov 15*209-274
⋄ B30 ⋄ ID 40458

BERNSTEIN, A.R. & ROBINSON, A. [1966] *Solution of an invariant subspace problem of K.T.Smith and P.R.Halmos* (J 0048) Pac J Math 16*421-431
⋄ H05 ⋄ REV MR 33 #1724 Zbl 141.129 JSL 34.292
• ID 01106

BERNSTEIN, A.R. [1967] *Invariant subspaces of polynomially compact operators on Banach space* (J 0048) Pac J Math 21*445-464
⋄ H05 ⋄ REV MR 36 #3150 Zbl 148.383 • ID 01108

BERNSTEIN, A.R. & WATTENBERG, F. [1969] *Nonstandard measure theory* (P 0649) Appl Model Th to Algeb, Anal & Probab;1967 Pasadena 171-185
⋄ H05 ⋄ REV MR 40 #287 Zbl 195.14 • ID 01282

BERNSTEIN, A.R. [1970] *A new kind of compactness for topological spaces* (J 0027) Fund Math 66*185-193
⋄ E75 H05 ⋄ REV MR 40 #4924 Zbl 198.554 • ID 01109

BERNSTEIN, A.R. [1972] *Invariant subspaces for certain commuting operators on Hilbert space* (J 0120) Ann of Math, Ser 2 95*253-260
⋄ H05 ⋄ REV MR 45 #4185 Zbl 246 #47009 • ID 26615

BERNSTEIN, A.R. [1972] *The spectral theorem - a non-standard approach* (J 0068) Z Math Logik Grundlagen Math 18*419-434
⋄ H05 ⋄ REV MR 47 #4048 Zbl 268 #02040 • ID 01110

BERNSTEIN, A.R. [1973] *Non-standard analysis* (C 0654) Stud in Model Th 35-58
⋄ C60 C65 H05 ⋄ REV MR 49 #2366 • ID 03898

BERNSTEIN, A.R. [1974] *A non-standard integration theory for unbounded functions* (J 0068) Z Math Logik Grundlagen Math 20*97-108
⋄ H05 ⋄ REV MR 50 #13419 Zbl 399 #03054 • ID 03899

BERNSTEIN, A.R. & LOEB, P.A. [1974] *A nonstandard integration theory for unbounded functions* (P 1083) Victoria Symp Nonstand Anal;1972 Victoria 40-49
⋄ H05 ⋄ REV MR 58 #11313 Zbl 281 #26017 • ID 21195

BERNSTEIN, A.R. & WATTENBERG, F. [1974] *Cardinality-dependent properties of topological spaces* (P 1083) Victoria Symp Nonstand Anal;1972 Victoria 50-59
⋄ H05 ⋄ REV MR 58 #30953 Zbl 272 #54039 • ID 26617

BERNSTEIN, B.A. [1924] *Complete sets of representations of two-elements algebras* (J 0015) Bull Amer Math Soc 30*24-30
⋄ B05 ⋄ REV FdM 50.630 • ID 01113

BERNSTEIN, B.A. [1926] *Sets of postulates for the logic of propositions* (J 0064) Trans Amer Math Soc 28*472-478
⋄ B05 ⋄ REV FdM 52.50 • ID 01115

BERNSTEIN, B.A. [1927] *The dual of a logical expression* (J 0015) Bull Amer Math Soc 33*309-311
⋄ B05 ⋄ REV FdM 53.44 • ID 01118

BERNSTEIN, B.A. [1929] *Irredundant sets of postulates for the logic of propositions* (J 0015) Bull Amer Math Soc 35*545-548
⋄ B05 ⋄ REV FdM 55.30 • ID 01119

BERNSTEIN, B.A. [1931] *Application of boolean algebras to proving consistency and independence of postulates* (J 0015) Bull Amer Math Soc 37*715-719
⋄ B05 G05 ⋄ REV Zbl 3.291 FdM 57.59 • ID 01121

BERNSTEIN, B.A. [1931] *Whitehead and Russell's theory of deduction as a mathematical science* (J 0015) Bull Amer Math Soc 37∗480-488
⋄ A05 B05 ⋄ REV Zbl 2.2 FdM 57.59 • ID 01120

BERNSTEIN, B.A. [1932] *Note on the condition that a boolean equation has a unique solution* (J 0100) Amer J Math 54∗417-418
⋄ B05 ⋄ REV Zbl 4.146 FdM 58.66 • ID 01122

BERNSTEIN, B.A. [1932] *On proposition ∗4.78 of "Principia Mathematica"* (J 0015) Bull Amer Math Soc 38∗388-391
⋄ B05 E20 ⋄ REV Zbl 5.146 FdM 58.64 • ID 01123

BERNSTEIN, B.A. [1932] *On unit-zero boolean representations of operations and relations* (J 0015) Bull Amer Math Soc 38∗707-712
⋄ B05 G05 ⋄ REV Zbl 6.4 FdM 58.67 • ID 01125

BERNSTEIN, B.A. [1932] *On Nicod's reduction in the number of primitives of logic* (J 0171) Proc Cambridge Phil Soc Math Phys 28∗427-432
⋄ A05 B05 ⋄ REV Zbl 6.97 FdM 58.70 • ID 01126

BERNSTEIN, B.A. [1932] *Relation of Whitehead's and Russell's theory of deduction to the boolean logic of propositions* (J 0015) Bull Amer Math Soc 38∗589-593
⋄ A10 B10 ⋄ REV Zbl 5.146 FdM 58.65 • ID 01124

BERNSTEIN, B.A. [1933] *On section A of "Principia Mathematica"* (J 0015) Bull Amer Math Soc 39∗788-792
⋄ A05 A10 B05 ⋄ REV Zbl 8.98 FdM 59.50 • ID 01129

BERNSTEIN, B.A. [1933] *Remarks on propositions ∗1.1 and ∗3.35 of "Principia Mathematica"* (J 0015) Bull Amer Math Soc 39∗111-114
⋄ A05 B05 ⋄ REV Zbl 6.242 FdM 59.50 • ID 01127

BERNSTEIN, B.A. [1936] *Postulates for boolean algebra involving the operation of complete disjunction* (J 0120) Ann of Math, Ser 2 37∗317-325
⋄ B30 G05 ⋄ REV Zbl 14.98 JSL 1.68 FdM 62.34
• ID 01281

BERNSTEIN, B.A. [1937] *Remark on Nicod's reduction of "Principia Mathematica"* (J 0036) J Symb Logic 2∗165-166
⋄ B10 ⋄ REV Zbl 18.1 JSL 3.50 FdM 63.821 • ID 01131

BERNSTEIN, B.A. & PARKER, W.L. [1955] *On uniquely solvable boolean equations* (S 0183) Publ Math Univ California 3∗1-29
⋄ B05 G05 ⋄ REV MR 16.895 JSL 22.96 • ID 25049

BERRETA PION, M.M. & CARVALHO DE, R.L. & LOPES PASSOS, E.P. [1982] *Interactive system to construct minimal model on the Herbrand universe* (C 3881) Prog in Cybern & Syst Res, Vol 11 337-342
⋄ B35 ⋄ REV MR 84f:00034 • ID 45713

BERRY, GEORGE D.W. [1941] *On Quine's axioms of quantification* (J 0036) J Symb Logic 6∗23-27
⋄ B10 ⋄ REV MR 2.209 Zbl 26.244 JSL 6.102 • ID 01138

BERRY, GEORGE D.W. see Vol. III for further entries

BERTI, S.N. [1969] *Applications de la theorie des relations dans la logique mathematique (Russian)* (J 0197) Stud Cercet Mat Acad Romana 21∗3-11
⋄ A05 B10 ⋄ REV MR 42#4376 Zbl 176.273 JSL 35.584
• ID 01542

BERTOSSI, L. & CHUAQUI, R.B. [1985] *Approximation to truth and theory of errors* (P 2160) Latin Amer Symp Math Log (6);1983 Caracas 13-31
⋄ C90 H05 ⋄ REV Zbl 569#03011 • ID 41802

BESHENKOV, S.A. [1984] *Formal consistency as a positive principle of non-standard arithmetic (Russian)* (J 4404) Vopr Mat Logiki & Pril 1984∗107-111
⋄ H05 ⋄ ID 46563

BESSONOV, A.V. [1979] *Mixed interpretation of logical systems (Russian)* (C 2967) Metodol Probl Mat 129-144
⋄ B10 ⋄ REV MR 83j:03018 • ID 35331

BESSONOV, A.V. see Vol. II, VI for further entries

BETH, E.W. [1937] *Une demonstration de la non-contradiction de la logique des types au point de vue fini* (J 1793) Nieuw Arch Wisk, Ser 2 19∗59-62
⋄ B15 F35 ⋄ REV FdM 63.822 • ID 41045

BETH, E.W. [1940] *Einfuehrung in die Philosophie der Mathematik (Dutch)* (X 2110) Dekker & van de Vegt: Nijmegen 269pp
⋄ A05 B98 ⋄ REV Zbl 25.293 JSL 8.144 FdM 66.23
FdM 68.20 • REM 2nd ed. 1942 • ID 43798

BETH, E.W. [1941] *Hauptstuecke aus der modernen formalen Logik. I (Dutch)* (J 0290) Euclides 18∗93-107
⋄ B10 ⋄ REV FdM 67.43 • REM Part II 1942 • ID 41300

BETH, E.W. [1942] *Summulae logicales. Supplement to formal logic (Dutch)* (X 0812) Wolters-Noordhoff : Groningen 55pp
⋄ A05 B98 ⋄ REV MR 7.185 Zbl 26.242 • ID 01151

BETH, E.W. [1950] *Decision problems of logic and mathematics* (C 0643) Philosophie 1946-48 3-18
⋄ B20 B25 C10 D35 ⋄ REV JSL 22.359 • ID 01156

BETH, E.W. [1950] *Les fondements logiques des mathematiques* (X 0834) Gauthier-Villars: Paris 222pp
• TRANSL [1963] (X 0844) Feltrinelli: Milano
xiii+335pp (Italian)
⋄ A05 B98 ⋄ REV MR 12.71 MR 26#6035 Zbl 40.290
Zbl 65.1 JSL 16.153 JSL 25.269 JSL 36.325 • REM 2nd rev. ed. 1955, xv+241pp • ID 01155

BETH, E.W. [1953] *On Padoa's method in the theory of definition* (J 0028) Indag Math 15∗330-339
⋄ B10 C40 ⋄ REV MR 15.385 Zbl 53.344 JSL 21.194
• ID 01159

BETH, E.W. [1955] *Die Stellung der Logik im Gebaeude der heutigen Wissenschaft* (J 0178) Stud Gen 8∗425-431
⋄ B98 ⋄ REV Zbl 66.247 JSL 23.66 • ID 01164

BETH, E.W. [1955] *Remarks on natural deduction* (J 0028) Indag Math 17∗322-325
⋄ A05 B10 F07 ⋄ REV MR 17.4 Zbl 67.251 JSL 22.360
• ID 01163

BETH, E.W. [1955] *Semantic entailment and formal derivability* (J 0182) Kon Nederl Akad Wetensch Afd Let Med N S 18/13∗309-342
• TRANSL [1978] (X 2698) Univ Valencia Dept Log Filos Cienc: Valencia xvi+54pp • REPR [1969] (C 0569) Phil of Math Oxford Readings 9-41
⋄ B10 C07 ⋄ REV MR 19.625 JSL 22.360 • ID 01269

BETH, E.W. & TARSKI, A. [1956] *Equilaterality as the only primitive notion of Euclidean geometry* (**J** 0028) Indag Math 18*462-467
⋄ B30 ⋄ REV MR 18.328 Zbl 72.155 JSL 33.289
• ID 01165

BETH, E.W. [1956] *L'existence en mathematiques* (**X** 0834) Gauthier-Villars: Paris 60pp
⋄ A05 B98 F55 ⋄ REV MR 19.625 Zbl 75.231 • ID 01166

BETH, E.W. [1958] *On machines which prove theorems* (**J** 0061) Simon Stevin 32*49-60
• REPR [1983] (**C** 4659) Autom of Reasoning 1*79-92
⋄ B35 ⋄ REV MR 20 # 5727 Zbl 84.250 JSL 34.659
• ID 01169

BETH, E.W. [1958] *On the completeness of the classical sentential logic* (**J** 0028) Indag Math 20*434-437
⋄ B05 ⋄ REV MR 20 # 5729 Zbl 86.6 • ID 01168

BETH, E.W. [1959] *Moderne Logik* (**J** 0290) Euclides 34*257-266
⋄ B10 ⋄ REV Zbl 85.243 • ID 47977

BETH, E.W. [1959] *The foundation of mathematics. A study in the philosophy of science* (**X** 0809) North Holland: Amsterdam xxvi + 741pp
⋄ A05 B98 ⋄ REV MR 22 # 9445 Zbl 85.241 JSL 27.73 JSL 33.618 • REM 2nd ed. 1965 • ID 01170

BETH, E.W. & LEBLANC, H. [1960] *A note on the intuitionist and the classical propositional calculus* (**J** 0079) Logique & Anal, NS 3*174-176
⋄ B05 F50 ⋄ REV JSL 25.351 • ID 01171

BETH, E.W. [1960] *Completeness results for formal systems* (**P** 0660) Int Congr Math (II, 8);1958 Edinburgh 281-288
⋄ B10 C07 ⋄ REV MR 23 # A42 Zbl 119.251 JSL 27.110
• ID 01534

BETH, E.W. [1962] *Formal methods. An introduction to symbolic logic and to the study of effective operations in arithmetic and logic* (**X** 0835) Reidel: Dordrecht xiv + 170pp
⋄ B28 B98 F30 F50 F98 ⋄ REV MR 28 # 3920 Zbl 105.245 JSL 30.235 • ID 01176

BETH, E.W. [1962] *Logique inferentielle et logique bivalente de l'implication* (**J** 0082) Bull Soc Math Belg 14*120-132
⋄ B20 ⋄ REV MR 25 # 10 Zbl 112.244 • ID 01175

BETH, E.W. [1962] *Umformung einer abgeschlossenen deduktiven oder semantischen Tafel in eine natuerliche Ableitung auf Grund der derivativen bzw. klassischen Implikationslogik* (**C** 0712) Logik & Logikkalkuel 49-55
⋄ B10 F07 ⋄ REV Zbl 126.9 • ID 04250

BETH, E.W. [1963] *La logique de quanteurs. Axiomatisations faibles pour l'arithmetique* (**J** 0082) Bull Soc Math Belg 15*69-80
⋄ B28 ⋄ REV MR 26 # 6050 Zbl 116.6 • ID 01178

BETH, E.W. [1963] *Observations concerning computing, deduction and heuristics* (**C** 0659) Computer Progr & Formal Syst 21-32
⋄ B35 B80 ⋄ REV MR 27 # 4401 Zbl 108.9 JSL 33.118
• ID 01533

BETH, E.W. [1965] *Semantische Begruendung der derivativen Implikationslogik* (**J** 0009) Arch Math Logik Grundlagenforsch 7*23-28
⋄ B20 ⋄ REV MR 35 # 5290 Zbl 166.2 • ID 01184

BETH, E.W. [1970] *Aspects of modern logic* (**X** 0835) Reidel: Dordrecht xi + 176pp
⋄ B98 ⋄ REV Zbl 209.6 • ID 25631

BETH, E.W. see Vol. II, III, V, VI for further entries

BEZHANISHVILI, M.N. [1984] *Model-theoretic interpretation of Aristotle's apodictic syllogistics (Russian) (English summary)* (**J** 0233) Soobshch Akad Nauk Gruz SSR 115*181-184
⋄ A10 B20 ⋄ REV Zbl 557 # 03005 • ID 44649

BHATTACHARJEE, P.R. & BHOWMIK, KANTI BHUSAN [1981] *A simpler method for minimization of switching functions* (**J** 0382) Int J Comput Math 10*77-89
⋄ B35 ⋄ REV MR 82m:94054 Zbl 468 # 94013 • ID 69195

BHOWMIK, BIMAL & BHOWMIK, KANTI BHUSAN [1973] *The P-table and its application to second and third-order simplification of boolean functions* (**J** 1045) Int J Control 18*641-655
⋄ B35 ⋄ REV MR 48 # 8112 Zbl 261 # 02006 • ID 30455

BHOWMIK, KANTI BHUSAN [1973] see BHOWMIK, BIMAL

BHOWMIK, KANTI BHUSAN [1981] see BHATTACHARJEE, P.R.

BIBEL, W. [1969] *Schnittelimination in einem Teilsystem der einfachen Typenlogik* (**J** 0009) Arch Math Logik Grundlagenforsch 12*159-178
⋄ B15 F05 F15 F35 ⋄ REV MR 42 # 37 • ID 01198

BIBEL, W. [1974] *An approach to a systematic theorem proving procedure in first-order logic (German summary)* (**J** 0373) Comp Arch Inform & Numerik 12*43-55
⋄ B35 ⋄ REV MR 54 # 4930 Zbl 273 # 68058 • ID 24105

BIBEL, W. [1975] *Die systematische Beweismethode und ihre Anwendungen* (**P** 1641) Kuenstl Intelligenzforschg BRD;1975 Bonn 24-35
⋄ B35 ⋄ REV Zbl 352 # 68106 • ID 50070

BIBEL, W. [1975] *Effizienzvergleiche von Beweisprozeduren* (**P** 3527) GI Jahrestag (4);1974 Berlin 153-160
⋄ B35 ⋄ REV MR 52 # 12436 Zbl 311 # 68055 • ID 33379

BIBEL, W. & SCHREIBER, J. [1975] *Proof search in a Gentzen-like system of first-order logic* (**P** 3525) Int Comput Symp;1975 Antibes 205-212
⋄ B35 F07 ⋄ REV MR 58 # 32125 Zbl 324 # 68053
• ID 60499

BIBEL, W. [1976] *Maschinelles Beweisen* (**S** 0780) Jbuch Ueberblick Math 1976*115-142
⋄ B35 ⋄ REV MR 53 # 9741 Zbl 336 # 68036 • ID 60501

BIBEL, W. [1977] *A syntactic connection between proof procedures and refutation procedures* (**P** 3411) Theor Comput Sci (3);1977 Darmstadt 215-225
⋄ B35 ⋄ REV MR 58 # 19401 Zbl 359 # 68107 • ID 80717

BIBEL, W. [1979] *Tautology testing with a generalized matrix reduction method* (**J** 1426) Theor Comput Sci 8*31-44
⋄ B35 D15 F05 ⋄ REV MR 80i:03023 Zbl 421 # 03011
• ID 53410

BIBEL, W. [1980] *A theoretical basis for the systematic proof method* (**P** 3210) Math Founds of Comput Sci (9);1980 Rydzyna 154-167
⋄ B35 ⋄ REV MR 83c:68110 Zbl 465 # 68046 • ID 54951

BIBEL, W. & KOWALSKI, R. (EDS.) [1980] *5th Conference on Automated Deduction* (X 0811) Springer: Heidelberg & New York vii+385pp
◊ B35 B97 ◊ REV MR 81i:68006 Zbl 428 # 00026
• ID 80425

BIBEL, W. [1981] *On matrices with connections* (J 0037) ACM J 28*633-645
◊ B35 ◊ REV MR 83k:68092 Zbl 468 # 68097 • ID 55127

BIBEL, W. [1982] *A comparative study of several proof procedures* (J 0503) Artif Intell 18*269-293
◊ B35 F07 ◊ REV MR 83f:68106 Zbl 505 # 68041
• ID 40212

BIBEL, W. [1982] *Automated theorem proving* (X 0900) Vieweg: Wiesbaden xiii+293pp
◊ B35 ◊ REV MR 84k:03048 Zbl 492 # 68067 • ID 34986

BIBEL, W. [1982] *Computationally improved versions of Herbrand's theorem* (P 3708) Herbrand Symp Logic Colloq;1981 Marseille 11-28
◊ B35 ◊ REV MR 85f:03003 Zbl 518 # 68052 • ID 38385

BIBEL, W. [1982] *Deduktionsverfahren* (P 4072) Kuenstl Intell;1982 Teisendorf 99-140
◊ B35 F07 ◊ REV MR 85d:68004 • ID 40242

BIBEL, W. & HOERNIG, K.M. [1982] *Improvements of a tautology-testing algorithm* (P 3840) Autom Deduct (6);1982 New York 326-341
◊ B35 ◊ REV MR 85g:68064 Zbl 485 # 68084 • ID 38505

BIBEL, W. & SIEKMANN, J. (EDS.) [1982] *Kuenstliche Intelligenz* (X 0811) Springer: Heidelberg & New York 383pp
◊ B35 ◊ REV MR 85d:68004 Zbl 541 # 68059 • ID 40227

BIBEL, W. [1984] *Automatische Inferenz* (J 0503) Artif Intell 22*145-167
◊ B35 ◊ ID 40223

BIBEL, W. [1985] *Towards a connection machine for logical inference* (P 4306) Future Generat Comput Syst
◊ B35 ◊ ID 40320

BIBEL, W. see Vol. II for further entries

BIBILO, P.N. & ENIN, S.V. [1978] *Nonessential nature af arguments and decomposition of boolean functions (Russian)* (J 0474) Avtom Vychis Tekh, Akad Nauk Latv SSR 1978/4*16-21
• TRANSL [1978] (J 2666) Autom Control Comput Sci 12*14-19
◊ B05 ◊ REV MR 80b:94037 Zbl 387 # 94040 • ID 69519

BIBILO, P.N. & ENIN, S.V. [1979] *Joint decomposition of a system of vector boolean functions (Russian)* (J 0474) Avtom Vychis Tekh, Akad Nauk Latv SSR 1979/1*16-22
• TRANSL [1979] (J 2666) Autom Control Comput Sci 13/1*14-20
◊ B05 ◊ REV MR 81g:94059 Zbl 414 # 94039 • ID 69520

BIBILO, P.N. & ENIN, S.V. [1980] *Decomposition of a boolean function with minimum number of significant arguments of the subfunctions (Russian)* (J 0977) Izv Akad Nauk SSSR, Tekh Kibern 1980/3*123-129
• TRANSL [1981] (J 0522) Engin Cybern 18/3*75-81
◊ B05 ◊ REV MR 83i:94025 Zbl 458 # 94070 • ID 69197

BIBILO, P.N. & ENIN, S.V. [1980] *Joint decomposition of a system of boolean functions (Russian)* (J 0977) Izv Akad Nauk SSSR, Tekh Kibern 1980/2*108-113,221
• TRANSL [1980] (J 0522) Engin Cybern 18/2*96-102
◊ B05 ◊ REV MR 82d:94068 Zbl 458 # 94071 • ID 69196

BIBILO, P.N. [1981] *On the probability of the existence of multiple decomposition of a completely defined boolean function (Russian) (English summary)* (J 0413) Izv Akad Nauk Belor SSR, Ser Fiz-Mat 1981/5*120-121
◊ B05 ◊ REV MR 83e:94080 Zbl 469 # 94018 • ID 69198

BIBILO, P.N. [1982] *Probability of nonessentialness of subsets of arguments of Boolean functions (Russian)* (J 0474) Avtom Vychis Tekh, Akad Nauk Latv SSR 1982/4*68-70
• TRANSL [1982] (J 2666) Autom Control Comput Sci 16/4*68-68
◊ B05 ◊ REV MR 84i:06013 Zbl 511 # 94011 • ID 40095

BIELA, A. [1974] *On Lindenbaum's extensions* (J 0387) Bull Sect Logic, Pol Acad Sci 3/3-4*2-8
◊ B22 ◊ REV MR 53 # 2619 • REM Part I. Part II 1975
• ID 21512

BIELA, A. [1974] *Structural incompleteness of the Quine's formalization of the classical predicate calculus* (J 0302) Rep Math Logic, Krakow & Katowice 3*3-7
◊ B10 ◊ REV MR 52 # 2812 Zbl 301 # 02013 • ID 17612

BIELA, A. & POGORZELSKI, W.A. [1974] *The power of the class of Lindenbaum-Asser extensions of consistent sets of formulas* (J 0302) Rep Math Logic, Krakow & Katowice 2*5-7
◊ B22 ◊ REV MR 50 # 6823 Zbl 286 # 02014 • ID 04152

BIELA, A. [1975] *On Lindenbaum's extensions II* (J 0387) Bull Sect Logic, Pol Acad Sci 4*65-71
◊ B22 ◊ REV MR 53 # 2620 • REM Part I 1974. Part III 1976
• ID 21513

BIELA, A. [1976] *On the so-called Tarski's property in the theory of Lindenbaum's oversystems* (J 0302) Rep Math Logic, Krakow & Katowice 7*3-20
◊ B22 ◊ REV MR 57 # 9468 Zbl 398 # 03016 • REM Part I. Part II 1981 • ID 21923

BIELA, A. [1976] *On Lindenbaum's extensions. III* (J 0387) Bull Sect Logic, Pol Acad Sci 5*68-79
◊ B22 ◊ REV MR 58 # 21415 • REM Part II 1975 • ID 27112

BIELA, A. [1981] *On the so-called Tarski's property in the theory of Lindenbaum's oversystem. II: Axiomatizable systems* (J 0302) Rep Math Logic, Krakow & Katowice 11*13-48
◊ B22 ◊ REV MR 82k:03032 Zbl 465 # 03021 • REM Part I 1976 • ID 54924

BIELA, A. & DZIOBIAK, W. [1982] *On two properties of structurally complete logics* (J 0387) Bull Sect Logic, Pol Acad Sci 11*154-160
◊ B22 ◊ REV MR 86c:03027b Zbl 514 # 03011 • ID 37405

BIELA, A. & STEPIEN, T. [1983] *Lindenbaum's extensions* (J 0302) Rep Math Logic, Krakow & Katowice 15*3-11
◊ B22 ◊ REV MR 85b:03045 Zbl 523 # 03008 • ID 37021

BIELA, A. & DZIOBIAK, W. [1983] *On two properties of structurally complete logics* (J 0302) Rep Math Logic, Krakow & Katowice 16*51-54
◊ B22 ◊ REV MR 86c:03027a Zbl 537 # 03013 • ID 43727

BIELA, A. see Vol. II, VI for further entries

BIELECKA-HOLDA, M. [1980] *Independent basis for the consequence determined by nondegenerated distributive lattices* (J 0387) Bull Sect Logic, Pol Acad Sci 9*141-144
  ⋄ B22 G10 G25 ⋄ REV MR 82c:03038 Zbl 451 # 03001 • ID 54016

BIELTZ, P. [1981] *The logic of elections and social decision (Romanian)* (S 1613) Probl Logic (Bucharest) 8*217-239
  ⋄ B80 ⋄ ID 48161

BIELTZ, P. see Vol. VI for further entries

BIERMANN, K.R. & MAU, J. [1958] *Ueberpruefung einer fruehen Anwendung der Kombinatorik in der Logik* (J 0036) J Symb Logic 23*129-132
  ⋄ A10 B20 ⋄ REV MR 21 # 4075 Zbl 85.2 • ID 01201

BILLARD, B. [1981] *Polynomial manipulation with APL* (J 0212) ACM Commun 24*457-465 • ERR/ADD ibid 25*213
  ⋄ B35 ⋄ REV MR 82f:68042 MR 83g:68054 Zbl 459 # 68012 • ID 80723

BILLERA, L.J. [1970] *Clutter decomposition and monotonic Boolean functions* (J 1377) Ann New York Acad Sci 175*41-48
  ⋄ B05 ⋄ REV MR 46 # 7038 Zbl 244 # 05005 • ID 80725

BINFORD, F. [1965] *Solution to the exercises in: "First course in mathematical logic"* (X 0841) Blaisdell: New York ix+173pp
  ⋄ B98 ⋄ REV Zbl 126.8 JSL 32.422 • REM The book was published by Hill,S. & Suppes,P. 1964 • ID 01271

BING, K. [1946] *On B. Germansky's systems for the foundations of natural numbers. I,II (Hebrew)* (J 0173) Riveon Lematemat 1*21-28,57-60
  ⋄ B28 ⋄ REV MR 8.126 MR 8.558 Zbl 60.23 JSL 13.225 • ID 45442

BING, K. [1956] *On simplifying truth-functional formulas* (J 0036) J Symb Logic 21*253-254
  ⋄ B05 ⋄ REV MR 18.632 Zbl 72.1 JSL 22.221 • ID 01208

BING, K. [1957] *On the axiom of order and succession* (J 0036) J Symb Logic 22*141-144
  ⋄ B28 ⋄ REV MR 20 # 6981 Zbl 86.8 JSL 23.362 • ID 01209

BING, K. [1968] *Natural deduction with few restrictions on variables* (J 0191) Inform Sci 1*381-402
  ⋄ B10 F07 ⋄ REV MR 41 # 6664 • ID 01210

BING, K. see Vol. III for further entries

BINKLEY, R. & CLARK, R. [1967] *A cancellation algorithm for elementary logic* (J 0105) Theoria (Lund) 33*79-97
  • ERR/ADD ibid 34*85
  • REPR [1983] (C 4659) Autom of Reasoning 2*27-47
  ⋄ B10 ⋄ REV MR 38 # 3128 Zbl 174.9 • ID 01211

BINKLEY, R. see Vol. II for further entries

BIRD, O. [1960] *The formalizing of the topics in mediaeval logic* (J 0047) Notre Dame J Formal Log 1*138-149
  ⋄ A05 A10 B20 ⋄ REV JSL 34.497 • ID 01212

BIRD, O. [1961] *Topic and consequence in Ockham's logic* (J 0047) Notre Dame J Formal Log 2*65-78
  ⋄ A05 A10 B20 ⋄ REV JSL 34.497 • ID 01213

BIRD, O. [1964] *Syllogistic and its extensions* (X 0819) Prentice Hall: Englewood Cliffs xii+116pp
  ⋄ B20 ⋄ REV JSL 34.309 • ID 22463

BIRKELAND, B. [1980] *A singular Sturm-Liouville problem treated by nonstandard analysis* (J 0132) Math Scand 47*275-294
  ⋄ H05 ⋄ REV MR 82i:34022 Zbl 449 # 34017 • ID 80728

BIRKHOFF, GARRETT [1971] *Mathematics and psychology I,II (Bulgarian)* (J 0477) Spis Bulgar Akad Nauk 14(47)*212-231,297-322
  ⋄ A05 B80 ⋄ REV MR 53 # 13 • ID 16543

BIRKHOFF, GARRETT see Vol. II, III, IV, V for further entries

BIRYUKOV, B.V. & SPIRKIN, A.G. (EDS.) [1978] *Cybernetics and logic. The growth of cybernetical ideas and the development of computing machinery in their mathematical-logical aspects (Russian) (English summary)* (X 2027) Nauka: Moskva 334pp
  ⋄ B97 ⋄ REV Zbl 477 # 00045 • ID 55596

BISHOP, P. & GREIF, I. & HALE, R.L.V. & HEWITT, C. & MATSON, T. & SMITH, BRIAN & STEIGER, R. [1974] *Behavioral semantics of nonrecursive control structures* (P 3013) Progr Symp;1974 Paris 385-407
  ⋄ B35 B40 ⋄ REV Zbl 316 # 68013 • ID 62501

BISWAS, N.N. [1971] *Minimization of boolean functions* (J 0187) IEEE Trans Comp C-20*925-929
  ⋄ B35 ⋄ REV MR 45 # 8510 Zbl 224 # 94051 • ID 01524

BISWAS, N.N. [1975] *Introduction to logic and switching theory* (X 0836) Gordon & Breach: New York xiii+354pp
  ⋄ B98 ⋄ REV Zbl 314 # 94021 • ID 60523

BITTINGER, M.L. [1970] *Logic and proof* (X 0832) Addison-Wesley: Reading 129pp
  ⋄ B98 ⋄ REV MR 41 # 3229 Zbl 217.6 • ID 01245

BITTINGER, M.L. [1982] *Logic, proof, and sets* (X 0832) Addison-Wesley: Reading viii+131pp
  ⋄ B98 ⋄ REV JSL 50.860 • REM 2nd ed. • ID 47360

BIUNDO, S. & ZBORAY, F. [1984] *Automated induction proofs using methods of program synthesis* (J 3932) Comput Artif Intell (Bratislava) 3*473-481
  ⋄ B35 ⋄ REV Zbl 552 # 68076 • ID 43452

BIZAM, G. & HERCZEG, J. [1971] *Play and logic (Hungarian)* (X 1466) Mueszaki: Budapest 339pp
  • TRANSL [1975] (X 0885) Mir: Moskva 359pp (Russian) [1976] (X 0928) Akad Kiado: Budapest 391 pp (German)
  ⋄ B98 ⋄ REV MR 52 # 13295 MR 54 # 2410 Zbl 347 # 00001 • ID 21757

BLACK, M. [1949] *Language and philosophy: studies in method* (X 0992) Cornell Univ Pr: Ithaca xiii+264pp
  ⋄ A05 B98 ⋄ REV JSL 15.210 • ID 25304

BLACK, M. [1964] *Critical thinking: an introduction to logic and scientific method* (X 0819) Prentice Hall: Englewood Cliffs xv+402pp
  ⋄ A05 B98 ⋄ ID 25087

BLACK, M. see Vol. II, III for further entries

BLACKHURST, J.H. [1970] *Syllogistic and non-syllogistic aspects of the comparative argument* (J 0047) Notre Dame J Formal Log 11*34-36
  ⋄ A05 B20 ⋄ REV MR 43 # 3082 Zbl 187.235 • ID 01249

BLAIR, H.A. & TRYBULEC, A. [1985] *Computer aided reasoning* (P 4571) Log of Progr;1985 Brooklyn 406-412
  ⋄ B35 ⋄ REV Zbl 568 # 68070 • ID 49181

BLAIR, H.A. see Vol. III, IV for further entries

BLANCHARD, P. & TARSKI, J. [1978] *Renormalizable interactions in two dimensions and sharp-time fields* (J 2649) Acta Phys Austriaca 49∗129-152
⋄ H10 ⋄ REV MR 58 # 19991 • ID 80738

BLANCHARD, P. [1982] see ALBEVERIO, S.

BLANCHE, R. [1955] *L'axiomatique* (X 0840) Pr Univ France: Paris 102pp
⋄ B98 ⋄ REV MR 26 # 1259 JSL 23.438 • ID 01273

BLANCHE, R. [1957] *Introduction a la logique contemporaine* (X 0850) Colin: Paris 208pp
⋄ A05 B98 F50 ⋄ REV Zbl 292 # 02001 JSL 24.71
• ID 01538

BLANCHE, R. [1957] *Sur la structuration du tableau de connectifs interpropositionnels binaires* (J 0036) J Symb Logic 22∗17-18
⋄ B05 ⋄ REV MR 21 # 5556 Zbl 78.6 JSL 24.228
• ID 01291

BLANCHE, R. [1962] *Axiomatics* (X 0866) Routledge & Kegan Paul: Henley on Thames 65pp
⋄ A10 B25 B30 ⋄ REV MR 26 # 1259 Zbl 115.4 JSL 30.105 • ID 22464

BLANCHE, R. [1969] *Sur le systeme des connecteurs interpropositionnels* (J 1030) Formalisation 10∗131-149
⋄ B05 ⋄ ID 15161

BLANCHE, R. see Vol. II for further entries

BLASS, A.R. & GUREVICH, Y. & KOZEN, D. [1984] *A 0-1 law for logic with a fixed-point operator* (1111) Preprints, Manuscr., Techn. Reports etc. CRL-TR-38-84
⋄ B10 B75 C80 D70 ⋄ REM Univ. of Michigan, Computing Research Lab. • ID 47716

BLASS, A.R. [1984] *There are not exactly five objects* (J 0036) J Symb Logic 49∗467-469 • ERR/ADD ibid 50∗781
⋄ B20 C07 ⋄ REV MR 85e:03021 • ID 40306

BLASS, A.R. see Vol. III, IV, V, VI for further entries

BLEDSOE, W.W. [1970] see ANDERSON, ROBERT M.

BLEDSOE, W.W. [1971] *Splitting and reduction heuristics in automatic theorem proving* (J 0503) Artif Intell 2∗55-78
• REPR [1983] (C 4659) Autom of Reasoning 2∗508-530
⋄ B35 ⋄ REV Zbl 221 # 68052 • ID 28136

BLEDSOE, W.W. [1977] see BALLANTYNE, A.M.

BLEDSOE, W.W. [1979] *A maximal method for set variables in automatic theorem-proving* (J 0508) Machine Intelligence 9∗53-100
⋄ B35 ⋄ REV MR 82g:68089 • ID 80740

BLEDSOE, W.W. & HINES, L.M. [1980] *Variable elemination and chaining in a resolution-based prover for inequalities* (P 3063) Autom Deduct (5);1980 Les Arcs 70-87
⋄ B35 ⋄ REV Zbl 438 # 68050 • ID 69209

BLEDSOE, W.W. & LOVELAND, D.W. (EDS.) [1984] *Automated theorem proving* (S 3313) Contemp Math 29∗ix+360pp
⋄ B35 ⋄ REV MR 85d:68005 Zbl 545 # 00023 • ID 38962

BLEDSOE, W.W. [1984] *Some automatic proofs in analysis* (P 3084) Autom Theor Prov After 25 Yea;1983 Denver 89-118
⋄ B35 ⋄ REV MR 85d:68005 • ID 45263

BLEDSOE, W.W. & KUNEN, K. & SHOSTAK, R.E. [1985] *Completeness results for inequality provers* (J 0503) Artif Intell 27∗255-288
⋄ B35 ⋄ ID 49004

BLEDSOE, W.W. see Vol. V for further entries

BLIKLE, A.J. [1966] *Formalisation of parenthesis-free languages* (J 0068) Z Math Logik Grundlagen Math 12∗177-186
⋄ B03 ⋄ REV MR 35 # 2730 Zbl 193.320 • ID 01306

BLIKLE, A.J. see Vol. IV, V for further entries

BLISS, K. & CHIEN, R. & STOHL, F. [1971] *R2 a natural language question-answering system* (P 4201) AFIPS Proc;1971 303-308
⋄ B35 ⋄ ID 47291

BLOOM, S.L. & BROWN, D.J. & SUSZKO, R. [1970] *A note on abstract logics (Russian summary)* (J 0014) Bull Acad Pol Sci, Ser Math Astron Phys 18∗109-110
⋄ B22 ⋄ REV MR 42 # 7487 Zbl 209.11 • ID 01322

BLOOM, S.L. & BROWN, D.J. & SUSZKO, R. [1970] *Some theorems on abstract logics (Russian)* (J 0003) Algebra i Logika 9∗274-280
• TRANSL [1970] (J 0069) Algeb and Log 9∗165-168
⋄ B22 ⋄ REV MR 44 # 43 Zbl 225 # 02042 • ID 01325

BLOOM, S.L. [1972] *A note on regular congruences* (J 0387) Bull Sect Logic, Pol Acad Sci 1/4∗5-11
⋄ B22 C05 ⋄ REV MR 58 # 27473 • ID 71094

BLOOM, S.L. [1972] *Model-consequence operations and quasi-complete theories* (J 0387) Bull Sect Logic, Pol Acad Sci 1/4∗12-14
⋄ B22 ⋄ REV MR 58 # 27448 • ID 71095

BLOOM, S.L. [1973] *A theorem on well-finite standard consequence operations* (J 0387) Bull Sect Logic, Pol Acad Sci 2∗159-165
⋄ B22 ⋄ REV MR 53 # 12925 • ID 23173

BLOOM, S.L. & BROWN, D.J. [1973] *Classical abstract logics* (J 0202) Diss Math (Warsaw) 102∗43-52
⋄ B22 G05 ⋄ REV MR 56 # 5284 Zbl 317 # 02072
• ID 29660

BLOOM, S.L. [1974] *On "generalized logics"* (J 0063) Studia Logica 33∗65-68,121
⋄ B22 ⋄ REV MR 50 # 4239 Zbl 317 # 02027 • ID 03920

BLOOM, S.L. [1975] *A representation theorem for the lattice of standard consequence operations* (J 0063) Studia Logica 34∗235-237
⋄ B22 ⋄ REV MR 53 # 125 Zbl 316 # 02017 • ID 16646

BLOOM, S.L. [1975] *Some theorems on structural consequence operations* (J 0063) Studia Logica 34∗1-9
⋄ B22 C05 C20 ⋄ REV MR 53 # 2675 Zbl 311 # 02016
• ID 21621

BLOOM, S.L. [1976] *Projective and inductive generation of abstract logics* (J 0063) Studia Logica 35∗249-255
⋄ B22 ⋄ REV MR 55 # 12514 Zbl 348 # 02020 • ID 28147

BLOOM, S.L. [1977] *A note on Ψ-consequences* (J 0302) Rep Math Logic, Krakow & Katowice 8∗3-9
⋄ B22 F50 ⋄ REV MR 58 # 218 Zbl 377 # 02027
• ID 28146

BLOOM, S.L. see Vol. II, III, IV, V, VI for further entries

BLUDOV, V.S. [1967] *Algorithmus des Vergleichs der Wurzelformen von Funktionen der Logikalgebra (Russisch)* (S 3418) Pribory & Sist Avtomat (Khar'kov) 6*38-42
⋄ B05 ⋄ REV Zbl 243 # 94034 • ID 60568

BLUM, A. [1970] *The missing premiss* (J 0047) Notre Dame J Formal Log 11*203-204
⋄ B05 ⋄ REV MR 43 # 3083 Zbl 167.8 JSL 36.689 • ID 01335

BLUM, A. [1973] *A correction in Copi's account of boolean normal forms* (J 0047) Notre Dame J Formal Log 14*288
⋄ B05 ⋄ REV MR 47 # 6422 Zbl 225 # 02001 • ID 28900

BLUM, A. [1974] *A note on natural deduction* (J 0047) Notre Dame J Formal Log 15*349-350
⋄ B05 ⋄ REV MR 58 # 10281 Zbl 275 # 02032 • ID 01337

BLUM, A. see Vol. II for further entries

BLUMBERG, A.E. [1967] *Modern logic* (C 0601) Encycl of Philos 5*12-34
⋄ A05 B98 ⋄ REV JSL 35.299 • ID 01345

BLUMBERG, A.E. [1976] *Logic: A first course* (X 4512) Knopf: New York xiv+462pp
⋄ B98 ⋄ REV JSL 44.281 • ID 44584

BLUMBERG, H. [1913] *A set of postulates for arithmetic and algebra* (P 1646) Int Congr Math (5);1912 Cambridge GB 2*461-465
⋄ B28 ⋄ REV FdM 44.79 • ID 29485

BLUMBERG, H. see Vol. V for further entries

BLUMENTHAL, L.M. [1959] *New metric postulates for elliptic n-space* (P 0651) Axiomatic Method;1957 Berkeley 127-145
⋄ B30 ⋄ REV MR 21 # 5942 Zbl 92.151 • ID 03924

BLYTH, J.W. [1957] *A modern introduction to logic* (X 0847) Houghton Mifflin: Boston xvi+426pp
⋄ B98 ⋄ ID 01518

BLYTH, J.W. & JACOBSON JR., J.H. [1963] *Class logic. A programmed text* (X 0863) Harcourt: New York & London xxi+392pp
⋄ B98 ⋄ REM Review in J0075 1968, 29*292-294 • ID 04257

BOBROV, A.E. & ZHURKIN, V.A. [1972] *Synthesis of many-valued logic circuits in the system: maximum, minimum, cycle and special one-place functions (Russian) (English summary)* (J 1572) Izv Vyssh Ucheb Zaved, Radiofizika (Moskva) 15*1771-1778
• TRANSL [1972] (J 4330) Radiophys & Quant Electr 15*1361-1367
⋄ B35 ⋄ REV MR 56 # 18131 • ID 45304

BOBROV, A.E. & KOTS, V.S. [1973] *Minimierung von Funktionen einer mehrwertigen Logik, die durch das Maximum, das Minimum und alle einstelligen Funktionen bestimmt sind (Russisch)* (J 0474) Avtom Vychis Tekh, Akad Nauk Latv SSR 1973/4*36-37
⋄ B35 ⋄ REV Zbl 258 # 02021 • ID 60584

BOBROV, A.E. see Vol. II for further entries

BOCHAROV, V.A. [1980] *Algebraic reconstructions of syllogistics (Russian)* (C 3038) Log-Metodol Issl 285-312
⋄ B20 ⋄ REV MR 84a:03012 • ID 35559

BOCHAROV, V.A. [1983] *Subject-predicate calculus free from existential import* (J 0063) Studia Logica 42*209-221
⋄ B10 ⋄ REV MR 85i:03028 Zbl 552 # 03006 • ID 42317

BOCHAROV, V.A. [1984] *Aristotle and traditional logic* (X 0898) Moskov Gos Univ: Moskva 134pp
⋄ A10 B20 ⋄ ID 44338

BOCHENSKI, I.M. [1938] *Nove lezioni di logica simbolica* (X 2109) Angelicum: Roma 183pp
⋄ A05 B98 ⋄ REV FdM 65.1100 • ID 43782

BOCHENSKI, I.M. [1949] *Precis de logique mathematique* (X 0849) Kroonder: Bussum 90pp
⋄ B98 ⋄ REV MR 13.811 JSL 15.199 • ID 01537

BOCHENSKI, I.M. & MENNE, A. [1954] *Grundriss der Logistik* (X 0846) Schoeningh: Paderborn 124pp
⋄ B98 ⋄ REV JSL 24.220 • REM 2nd ed. 1962, 3rd ed. 1965, 4th ed. 1973 • ID 33428

BOCHENSKI, I.M. [1956] *Formale Logik* (X 0826) Alber: Freiburg xv+640pp
• TRANSL [1961] (X 0845) Univ Notre Dame Pr: Notre Dame xxii+567pp [1970] (X 0848) Chelsea: New York xxii+567pp
⋄ A10 B98 ⋄ REV MR 22 # 10899 Zbl 70.7 Zbl 98.241 JSL 25.57 • REM Transl. title: A history of formal logic • ID 42191

BOCHENSKI, I.M. [1959] *A precis of mathematical logic* (X 0835) Reidel: Dordrecht x+100pp
• TRANSL [1982] (X 3560) Paraninfo: Madrid 120pp
⋄ B98 ⋄ REV MR 31 # 2123 Zbl 90.9 JSL 25.78 • REM 2nd ed. 1963 • ID 22545

BOCHVAR, D.A. [1940] *Ueber einen Aussagenkalkuel mit abzaehlbaren logischen Summen und Produkten (Russian summary)* (J 0142) Mat Sb, Akad Nauk SSSR, NS 7(49)*65-100
⋄ B05 C75 ⋄ REV MR 1.321 Zbl 23.99 JSL 5.119 FdM 66.29 • ID 01348

BOCHVAR, D.A. [1945] *Some logical theorems on the normal sets and predicates (Russian) (English summary)* (J 0142) Mat Sb, Akad Nauk SSSR, NS 16(58)*345-352
⋄ B15 E70 ⋄ REV MR 7.356 Zbl 61.8 JSL 12.27 • ID 01351

BOCHVAR, D.A. [1957] *On paradoxes and the extended calculus of predicates (Russian)* (J 0142) Mat Sb, Akad Nauk SSSR, NS 42(84)*3-10
⋄ A05 B15 ⋄ REV MR 19.830 Zbl 78.243 • ID 01350

BOCHVAR, D.A. [1960] *Antinomies based on sets of definitions of predicates, each individually consistent (Russian)* (J 0142) Mat Sb, Akad Nauk SSSR, NS 52(94)*641-646
⋄ B10 ⋄ REV MR 26 # 30 Zbl 114.246 • ID 01352

BOCHVAR, D.A. [1969] *Measures of kernels of reducibility axioms (Russian)* (J 0023) Dokl Akad Nauk SSSR 185*1211-1214
• TRANSL [1969] (J 0062) Sov Math, Dokl 10*473-476
⋄ B15 E70 ⋄ REV MR 41 # 3251 Zbl 193.307 • ID 01353

BOCHVAR, D.A. & FUKSON, V.I. [1973] *On logical approximation operators (Russian)* (S 0066) Tr Mat Inst Steklov 133*65-77,274
• TRANSL [1973] (S 0055) Proc Steklov Inst Math 133*63-76
⋄ B15 E30 E35 ⋄ REV MR 49 # 2376 Zbl 295 # 02032 • ID 03926

BOCHVAR, D.A. & FUKSON, V.I. [1974] *Boolean operations over Russell cores (Russian)* (C 2577) Issl Formaliz Yazyk & Neklass Log 131-134
⋄ B10 E30 G05 ⋄ REV MR 58 # 27499 • ID 71137

BOCHVAR, D.A. see Vol. II, V for further entries

BODDENBERG, E. [1975] *Logik. I* (X 1039) Diesterweg: Frankfurt a.M. viii+103pp
- ◇ B98 ◇ REV MR 51#40 Zbl 323#02001 • ID 15248

BOEHM, C. & DEZANI, M. [1972] *A CUCH-machine: The automatic treatment of bound variables* (J 0435) Int J Comput & Inf Sci 1*171-191
- ◇ B35 B40 ◇ REV Zbl 277#68026 • ID 60576

BOEHM, C. see Vol. IV, VI for further entries

BOEHM, K. [1911] *Axiome der Arithmetik* (X 1372) Winter: Heidelberg 11pp
- ◇ B28 ◇ REV JSL 20.307 FdM 42.111 • ID 25516

BOEHME, G. [1981] *Einstieg in die mathematische Logik* (X 3223) Hanser: Muenchen 208pp
- ◇ B98 ◇ REV MR 83f:03002 Zbl 469#03001 • ID 55129

BOERGER, E. [1974] Σ_3-completude de l'ensemble des types de reduction (J 0079) Logique & Anal, NS 17*89-94
- ◇ B20 D35 D55 ◇ REV MR 55#2539 Zbl 294#02018
- • ID 28158

BOERGER, E. [1974] *Beitrag zur Reduktion des Entscheidungsproblems auf Klassen von Hornformeln mit kurzen Alternationen* (J 0009) Arch Math Logik Grundlagenforsch 16*67-84
- ◇ B20 B25 D35 ◇ REV MR 49#10535 Zbl 277#02009
- • ID 01368

BOERGER, E. [1978] *Bemerkung zu Gurevich's Arbeit ueber das Entscheidungsproblem fuer Standardklassen* (J 0009) Arch Math Logik Grundlagenforsch 19*111-114
- ◇ B20 D10 D35 ◇ REV MR 80a:03054b Zbl 402#03019
- • ID 29169

BOERGER, E. [1978] *Ein einfacher Beweis fuer die Unentscheidbarkeit der klassischen Praedikatenlogik* (J 0160) Math-Phys Sem-ber, NS 25*290-299
- ◇ B10 D03 D10 D35 ◇ REV MR 80c:03001 Zbl 399#03010 • ID 52807

BOERGER, E. [1979] see AANDERAA, S.O.

BOERGER, E. [1981] see AANDERAA, S.O.

BOERGER, E. [1982] see AANDERAA, S.O.

BOERGER, E. & OBERSCHELP, W. & RICHTER, M.M. & SCHINZEL, B. & THOMAS, WOLFGANG (EDS.) [1984] *Computation and proof theory* (S 3301) Lect Notes Math 1104*viii+475pp
- ◇ B97 D97 F97 ◇ REV MR 85k:03002b Zbl 547#00036
- • REM Proc. Log. Coll., Aachen 1983, Vol.II. Vol.I 1984 by Mueller,G.H. • ID 40062

BOERGER, E. [1984] *Decision problems in predicate logic* (P 3710) Logic Colloq;1982 Firenze 263-301
- ◇ B20 B25 D35 ◇ REV MR 85m:03008 Zbl 556#03012
- • ID 40047

BOERGER, E. & HASENJAEGER, G. & ROEDDING, D. (EDS.) [1984] *Logic and machines: decision problems and complexity* (P 2342) Symp Rek Kombin;1983 Muenster 171*vi+456pp
- ◇ B35 D97 ◇ REV MR 85k:68004 Zbl 538#00005
- • ID 40072

BOERGER, E. see Vol. III, IV for further entries

BOERI, F. [1978] see ANDRE, C.

BOETTGER, H. [1985] *Automatisches Theorembeweisen mit Konfigurationen (English and Russian summaries)* (J 0129) Elektr Informationsverarbeitung & Kybern 21*523-546
- ◇ B35 ◇ ID 49760

BOFFA, M. [1977] *A reduction of the theory of types* (P 1695) Set Th & Hierarch Th (3);1976 Bierutowice 95-100
- ◇ B15 E35 E70 ◇ REV MR 58#5014 Zbl 372#02038
- • ID 51446

BOFFA, M. [1977] *Modeles cumulatifs de la theorie des types* (J 0056) Publ Dep Math, Lyon 14/2*9-12
- ◇ B15 C62 E30 ◇ REV MR 58#27449 Zbl 432#03026
- • ID 53983

BOFFA, M. [1977] *The consistency problem for NF* (J 0036) J Symb Logic 42*215-220
- ◇ B15 E35 E70 ◇ REV MR 58#21604 Zbl 377#02040
- • ID 26448

BOFFA, M. & DALEN VAN, D. & MCALOON, K. (EDS.) [1979] *Logic colloquium '78. Proceedings of the colloquium held in Mons, August 1978* (S 3303) Stud Logic Found Math 97*x+434pp
- ◇ B97 ◇ REV MR 81a:03001 Zbl 423#00001 • ID 53511

BOFFA, M. [1981] *La theorie des types et NF* (J 3824) Bull Soc Math Belg, Ser A 33*21-31
- ◇ B15 E70 ◇ REV MR 83m:03016 Zbl 494#03036
- • ID 35424

BOFFA, M. [1982] *Algebres de Boole atomiques et modeles de la theorie des types* (P 3774) Th d'Ensembl de Quine;1981 Louvain-la-Neuve 1-5
- ◇ B15 C62 E70 ◇ REV MR 84g:03085 Zbl 536#03036
- • ID 34190

BOFFA, M. [1983] *Arithmetic and theory of types* (P 1601) Easter Conf on Model Th (1);1983 Diedrichshagen 10-16
- ◇ B15 C62 E70 F35 ◇ REV MR 84i:03008 Zbl 528#03020 • ID 37638

BOFFA, M. [1984] *Arithmetic and the theory of types* (J 0036) J Symb Logic 49*621-624
- ◇ B15 F30 F35 ◇ REV MR 85i:03176 Zbl 566#03024
- • ID 42423

BOFFA, M. see Vol. III, IV, V for further entries

BOJDAKOVA, V.N. & GLADKIJ, A.V. [1978] *On a system of axioms for propositional calculus (Russian)* (C 3211) Mat Ling & Teor Algor 35-40
- ◇ B05 ◇ REV Zbl 394#03013 • ID 52490

BOKSHTEJN, M.F. & PETRI, N.V. & SHCHEGOL'KOV, E.A. (EDS.) [1975] *Current questions in mathematical logic and set theory (Russian)* (X 2802) Moskov Ped Inst: Moskva 310pp
- ◇ B97 E97 ◇ REV MR 53#2566 • ID 80470

BOKSHTEJN, M.F. see Vol. V for further entries

BOLCK, F. (ED.) [1979] *Begriffsschrift -- Jenaer Frege-Konferenz* (X 2211) Schiller Univ: Jena iii+548pp
- ◇ A10 B97 ◇ REV MR 81m:03002 • ID 70039

BOLL, M. [1942] *Elements de logique scientifique* (X 0856) Dunod: Paris vii+243pp
- ◇ A05 B98 ◇ REV JSL 14.266 • ID 22393

BOLLMAN, D.A. [1967] *Formal nonassociative number theory* (J 0047) Notre Dame J Formal Log 8*9-16
- ◇ B28 D35 ◇ REV MR 39#5642 Zbl 183.13 • ID 01419

BOLLMAN, D.A. & TAPIA, M. [1972] *On the recursive unsolvability of the provability of the deduction theorem in partial propositional calculi* (J 0047) Notre Dame J Formal Log 13*124-128
- ◊ B20 D35 ◊ REV MR 46#3288 Zbl 227#02005
- • ID 01420

BOLLMAN, D.A. & LAPLAZA, M.L. [1973] *A set-theoretic model for nonassociative number theory* (J 0047) Notre Dame J Formal Log 14*107-110
- ◊ B28 E75 F30 ◊ REV MR 47#6487 Zbl 226#02046
- • ID 01422

BOLLMAN, D.A. see Vol. IV for further entries

BOLZANO, B. [1963] *Grundlegung der Logik. Ausgewaehlte Paragraphen aus der Wissenschaftslehre, Band I und II* (X 1088) Meiner: Hamburg lxxi+380pp
- • TRANSL [1973] (X 0835) Reidel: Dordrecht 398pp (English)
- ◊ A05 A10 B96 ◊ REV MR 80d:01024 Zbl 287#01027 JSL 31.104 • REM Edited and with introduction by Kambartel,F. 1st edition 1837 • ID 45044

BOLZANO, B. [1972] *Bernard Bolzano-Gesamtausgabe: Einleitung. Band 2. Abteilung I* (X 1267) Frommann: Stuttgart 180pp
- ◊ B96 ◊ REV MR 58#21347a • REM Ed. by Berg,J. & Kambartel,F. & Louzil,J. & Rootselaar van,B. & Winter,E.
- • ID 80445

BOLZANO, B. see Vol. V for further entries

BONACINI, R. & MELONI, G.C. [1974] *Teoria non-standard delle probabilita* (J 0207) Ist Lombardo Accad Sci Rend, A (Milano) 108*811-818
- ◊ H05 ◊ REV MR 51#5297 Zbl 368#60002 • ID 17404

BONDARENKO, A.S. [1968] *Use of semantic tableaux to establish the feasibility and general validity of certain formulae of predicate logic (Russian)* (J 0402) Uch Zap Ped Inst, Kemerovo 19*130-139
- ◊ B10 ◊ REV MR 52#2813 • ID 17613

BONEVAC, D. [1984] *Skolem fragments* (J 0047) Notre Dame J Formal Log 25*227-232
- ◊ B20 ◊ REV MR 85e:03022 • ID 40591

BONEVAC, D. see Vol. II for further entries

BONOTTO, C. & BRESSAN, A. [1983] *On a synonymy relation for extensional first order theories I: A notion of synonymy. II: A sufficient criterion for non synonymy. Applications* (J 0144) Rend Sem Mat Univ Padova 69*63-76,70*13-19
- ◊ B20 ◊ REV MR 85b:03012 Zbl 525#03019 Zbl 538#03025 • REM Part III 1984 • ID 38257

BONOTTO, C. & BRESSAN, A. [1984] *On a synonymy relation for extensional first order theories. III: A necessary and sufficient condition for synonymy* (J 0144) Rend Sem Mat Univ Padova 71*1-13
- ◊ B20 ◊ REV Zbl 553#03019 • REM Parts I,II 1983
- • ID 43309

BONOTTO, C. see Vol. II for further entries

BOOLE, G. [1947] *The mathematical analysis of logic, being an essay toward a calculus of deductive reasoning* (X 1096) Blackwell: Oxford 82pp
- ◊ A05 A10 B98 G05 ◊ REM 1st ed. 1847 • ID 16747

BOOLE, G. [1952] *Studies in logic and probability* (X 1368) Watts: London 500pp
- ◊ A05 B98 ◊ REV Zbl 49.8 JSL 24.203 • ID 25635

BOOLE, G. [1958] *An investigation of the laws of thought, on which are founded the mathematical theories of logic and probabilities* (X 0813) Dover: New York xi+424pp
- ◊ A05 B98 G05 ◊ REV JSL 16.224 • REM 1st ed. 1854
- • ID 16748

BOOLE, G. [1969] *L'analyse mathematique de la logique* (J 1030) Formalisation 10*27-34
- ◊ A05 B05 ◊ ID 15172

BOOLOS, G. & JEFFREY, R.C. [1974] *Computability and logic* (X 0805) Cambridge Univ Pr: Cambridge, GB x+262pp
- ◊ B98 D98 ◊ REV MR 49#7120 MR 82d:03001 Zbl 298#02003 JSL 42.585 • ID 03933

BOOLOS, G. [1975] *On second-order logic* (J 0301) J Phil 72*509-527
- ◊ A05 B15 ◊ ID 30641

BOOLOS, G. [1979] *The unprovability of consistency. An essay in modal logic* (X 0805) Cambridge Univ Pr: Cambridge, GB viii+184pp
- ◊ B25 B28 B45 B98 F30 F98 ◊ REV MR 81c:03013 Zbl 409#03009 JSL 46.871 • ID 56312

BOOLOS, G. [1980] *ω-consistency and the diamond* (J 0063) Studia Logica 39*237-243
- ◊ B28 F30 ◊ REV MR 81m:03068 Zbl 464#03049
- • ID 71208

BOOLOS, G. [1984] *Trees and finite satisfiability: proof of a conjecture of Burgess* (J 0047) Notre Dame J Formal Log 25*193-197
- ◊ B10 C13 F07 ◊ REV MR 85f:03005 Zbl 561#03004
- • ID 40442

BOOLOS, G. see Vol. II, III, IV, V, VI for further entries

BOOTH, K.S. [1978] see ADLEMAN, L.M.

BOOTH, K.S. see Vol. IV for further entries

BOPP, F. [1979] *Ueber den Zustandsraum der Quantenphysik* (J 1944) Sitzb, Akad Wiss, Bayern, Math-Nat Kl 1979*81-94
- ◊ H10 ◊ REV MR 82b:81008 • ID 80764

BORETZ, B. [1969] *Meta-variations: studies in the foundations of musical thought (I)* (J 1023) Perspectives of New Music 8/1*1-74
- ◊ B80 ◊ REV JSL 40.577 • REM Part II 1970: Sketch of a musical system (meta-variations, part II) • ID 15106

BORETZ, B. [1970] *Sketch of a musical system (meta-variations, part II)* (J 1023) Perspectives of New Music 8/2*49-111
- ◊ B80 ◊ REV JSL 40.577 • REM Part I 1969 • ID 15105

BORETZ, B. [1970] *The construction of musical syntax I* (J 1023) Perspectives of New Music 9*23-42
- ◊ B80 ◊ REV JSL 40.577 • REM Part II 1971: Musical syntax (II) • ID 15104

BORETZ, B. [1971] *Musical syntax (II)* (J 1023) Perspectives of New Music 10*232-270
- ◊ B80 ◊ REV JSL 40.577 • REM Part I 1970: The construction of musical syntax I • ID 15102

BORGA, M. [1978] *Il problema dell'indipendenza dei simboli primitivi di una teoria assiomatica* (J 0335) Atti Accad Ligure Sci Lett (Genova) 34*284-288
  ⋄ B30 ⋄ REV MR 81c:03001   Zbl 443 # 03017 • ID 56423

BORGA, M. & FREGUGLIA, P. & PALLADINO, D. (EDS.) [1980] *Rassegna di matematica. Logica matematica. Matematica applicata. Didattica della matematica* (X 2682) Casa Ed Tilgher: Genova 100pp
  ⋄ B97 D97 ⋄ REV MR 82b:03005   Zbl 459 # 00007
  • ID 54440

BORGA, M. see Vol. II, VI for further entries

BORGERS, A. [1949] *The natural number (Dutch)* (J 0061) Simon Stevin 26*32-64,65-73
  ⋄ B28 ⋄ REV MR 10.669   JSL 15.66 • ID 01468

BORGERS, A. [1960] *La methode axiomatique et la logique symbolique* (J 1411) Math 20 Siecle 1*25-34
  ⋄ A10 B30 ⋄ REV MR 24 # A1814   JSL 27.224 JSL 34.524
  • ID 28700

BORGERS, A. [1963] *Les lois et les regles de la logique symbolique elementaire des propositions et des predicats* (J 0082) Bull Soc Math Belg 15*131-154
  ⋄ B10 ⋄ REV MR 27 # 31   Zbl 106.4 • ID 01470

BORGERS, A. [1966] *La logique des classes* (J 0082) Bull Soc Math Belg 18*273-305
  ⋄ B20 ⋄ REV MR 37 # 3905   Zbl 192.26 • ID 01472

BORGERS, A. see Vol. V, VI for further entries

BORISOV, YU.F. [1979] *Mechanical determinism and the structure of the number line (Russian)* (C 2967) Metodol Probl Mat 39-48
  ⋄ A05 B28 ⋄ REV MR 81m:03009 • ID 71244

BORISOV, YU.F. (ED.) [1979] *Methodological problems of mathematics (Russian)* (X 2642) Nauka: Novosibirsk 303pp
  ⋄ B97 ⋄ REV MR 81a:03002 • ID 70053

BORKOWSKI, L. [1957] *Recent investigations on the propositional calculus (Polish)* (J 0063) Studia Logica 5*27-41
  ⋄ B05 ⋄ REV MR 19.240 • ID 33075

BORKOWSKI, L. [1957] *Systems of the propositional and of the functional calculus based on one primitive term (Polish and Russian summaries)* (J 0063) Studia Logica 6*7-55
  ⋄ B10 ⋄ REV MR 21 # 1267   JSL 24.242 • ID 01475

BORKOWSKI, L. [1957] *The first modern monography of Aristotle's syllogistic (Polish)* (J 0063) Studia Logica 5*13-26
  ⋄ B20 ⋄ REV MR 19.518 • ID 33074

BORKOWSKI, L. & SLUPECKI, J. [1958] *A logical system based on rules and its application in teaching mathematical logic (Polish and Russian summaries)* (J 0063) Studia Logica 7*71-113
  ⋄ B10 ⋄ REV MR 21 # 1268   Zbl 198.12 • ID 01476

BORKOWSKI, L. [1958] *On proper quantifiers I (Polish and Russian summaries)* (J 0063) Studia Logica 8*65-130
  ⋄ B20 B25 ⋄ REV MR 21 # 4095   Zbl 198.13   JSL 32.262
  • REM Part II 1960 • ID 01482

BORKOWSKI, L. [1958] *Reduction of arithmetic to logic based on the theory of types without the axiom of infinity and the typical ambiguity of arithmetical constant (Polish and Russian summaries)* (J 0063) Studia Logica 8*283-297
  ⋄ B15 F35 ⋄ REV MR 21 # 4094 • ID 01479

BORKOWSKI, L. [1960] *On proper quantifiers II (Polish and Russian summaries)* (J 0063) Studia Logica 10*7-28
  • ERR/ADD ibid 15*272
  ⋄ B20 B25 ⋄ REV MR 24 # A681   Zbl 198.13   JSL 32.263
  • REM Part I 1958 • ID 01483

BORKOWSKI, L. [1961] *A didactical approach to the zero-one decision procedure of the expressions of the first order monadic predicate calculus (Polish) (Russian and English summaries)* (J 0063) Studia Logica 11*57-76 • ERR/ADD ibid 15*271-272
  ⋄ B20 B25 C10 ⋄ REV MR 23 # A3076   Zbl 121.253
  • ID 01484

BORKOWSKI, L. & SLUPECKI, J. [1963] *Elements of mathematical logic and set theory (Polish)* (X 1034) PWN: Warsaw 285pp
  • TRANSL [1967] (X 0869) Pergamon Pr: Oxford xii+349pp
  ⋄ A05 B98 E98 ⋄ REV MR 37 # 3904   Zbl 171.248 • REM 2nd ed.1967, xii+349pp; 3rd corr. ed.1969, 306pp (errata insert) • ID 19509

BORKOWSKI, L. [1968] *Some remarks about the notion of definition (Polish) (Russian and English summaries)* (J 0063) Studia Logica 23*59-70
  ⋄ A05 B10 C40 ⋄ REV MR 39 # 1310   Zbl 305 # 02010
  JSL 35.468 • ID 01485

BORKOWSKI, L. [1976] *Formale Logik. Logische Systeme. Einfuehrung in die Metalogik* (X 0911) Akademie Verlag: Berlin xiv+578pp
  ⋄ A05 B98 ⋄ REV Zbl 357 # 02001 • REM Translated from Polish • ID 50349

BORKOWSKI, L. see Vol. II, V for further entries

BORODYANSKIJ, B.M. [1974] *Nonstandard analysis and enlargements of Hilbert spaces (Russian)* (S 2801) Sbor Nauch Trud (Ped Inst, Moskva) 39*265-269
  ⋄ H05 ⋄ REV MR 58 # 10427 • ID 71246

BORODYANSKIJ, B.M. [1974] *Nonstandard analysis and group representations (Russian)* (S 2801) Sbor Nauch Trud (Ped Inst, Moskva) 39*270-282
  ⋄ H05 ⋄ REV MR 58 # 22380 • ID 80768

BORODYANSKIJ, B.M. [1974] *Outline of the basic concepts of nonstandard analysis (Russian)* (S 2801) Sbor Nauch Trud (Ped Inst, Moskva) 39*257-264
  ⋄ H05 ⋄ REV MR 58 # 10426 • ID 71247

BOSANQUET, B. [1911] *Logic. Vol. II* (X 3333) Unknown Publisher: See Remarks
  ⋄ B98 ⋄ REM 2nd ed., Oxford • ID 16751

BOSCH, J.E. [1985] *Sur les nombres natureles et les fondements de l'arithmetique et la geometrie* (C 4181) Math Log & Formal Syst (Costa da) 17-42
  ⋄ B28 F30 ⋄ ID 48160

BOSCH, J.E. see Vol. V for further entries

BOSCH, W. & KRAJKIEWICZ, P. [1970] *A categorical system of axioms for the complex numbers* (J 0497) Math Mag 43*67-70
  ⋄ B28 ⋄ REV Zbl 191.58 • ID 15249

BOSTOCK, D. [1974] *Logic and arithmetic I: Natural numbers* (X 0815) Clarendon Pr: Oxford x+219pp
  ⋄ A05 B28 ⋄ REV MR 55 # 5382   Zbl 273 # 02005
  JSL 47.708 • ID 25637

BOSTOCK, D. [1979] *Logic and arithmetic. Vol. 2: Rational and irrational numbers* (X 0815) Clarendon Pr: Oxford ix + 307pp
⋄ A05   B28   ⋄ REV   MR 82b:03033   Zbl 417 # 03002   JSL 47.708 • ID 53239

BOSTOCK, D. [1980] *A study of type-neutrality I,II (II: Relations)* (J 0122) J Philos Logic 9∗211-296,363-414
⋄ A05   B15   ⋄ REV   MR 82b:03029a   MR 82b:03029b • ID 71259

BOTH, N. [1967] *On the bases of complete induction. Simultaneous induction (Romanian)* (J 0197) Stud Cercet Mat Acad Romana 19∗1251-1258
⋄ B28   F30   ⋄ REV   MR 40 # 4077 • ID 01491

BOTH, N. [1969] *A syntactic characterization of the independence of a system of axioms (Romanian) (French summary)* (J 0197) Stud Cercet Mat Acad Romana 21∗371-376
⋄ B30   ⋄ REV   MR 44 # 57   Zbl 181.4 • ID 01492

BOTH, N. [1969] *Monotony in propositional calculus (Romanian) (French summary)* (J 0197) Stud Cercet Mat Acad Romana 21∗543-551
⋄ B05   ⋄ REV   MR 44 # 27   Zbl 195.11 • ID 01493

BOTH, N. [1969] *On the completeness of an axiom system (Romanian) (French summary)* (J 0197) Stud Cercet Mat Acad Romana 21∗13-16
⋄ B05   ⋄ REV   MR 44 # 26   Zbl 177.7 • ID 01494

BOTH, N. [1971] *Fonctions de succession* (J 0060) Rev Roumaine Math Pures Appl 16∗7-11
⋄ B28   ⋄ REV   MR 43 # 6091   Zbl 216.13 • ID 01495

BOTH, N. [1972] *La caracterisation algebrique des systemes et theories deductives (Romanian) (French summary)* (J 0197) Stud Cercet Mat Acad Romana 24∗679-681
⋄ B22   C05   G25   ⋄ REV   MR 50 # 6984   Zbl 245 # 02046 • ID 28866

BOTH, N. [1972] *Monotonicity in the predicate calculus (Romanian) (French summary)* (J 0197) Stud Cercet Mat Acad Romana 24∗985-988
⋄ B10   ⋄ REV   MR 50 # 6783   Zbl 247 # 02014 • ID 03939

BOTH, N. [1974] *Considerations algebriques sur la theorie de la demonstration* (J 0068) Z Math Logik Grundlagen Math 20∗529-536
⋄ B10   E20   G15   ⋄ REV   MR 51 # 10044   Zbl 307 # 02020 • ID 03941

BOTH, N. [1974] *L'independance (Roumain) (Resume francais)* (J 0197) Stud Cercet Mat Acad Romana 26∗491-494
⋄ B22   ⋄ REV   MR 50 # 9743   Zbl 292 # 08001 • ID 60662

BOTH, N. [1974] *Metric notions in propositional logic (Romanian) (French summary)* (J 0197) Stud Cercet Mat Acad Romana 26∗325-332
⋄ B05   ⋄ REV   MR 50 # 4234   Zbl 283 # 02014 • ID 03940

BOTH, N. [1978] *Measures and metrics in the two-valued logic* (J 0517) Mathematica (Cluj) 20(43)∗113-118
⋄ B05   G05   ⋄ REV   MR 80f:03070   Zbl 402 # 03052 • ID 54699

BOTH, N. [1981] *Some remarks concerning Boolean functions* (J 3450) Studia Univ Babes-Bolyai, Math (Cluj) 26∗56-60
⋄ B05   ⋄ REV   MR 84b:06015   Zbl 481 # 03040 • ID 36832

BOTH, N. see Vol. V for further entries

BOUDREAUX, J.C. [1979] *Defining general structures* (J 0047) Notre Dame J Formal Log 20∗465-488
⋄ B15   C85   ⋄ REV   MR 80i:03021   Zbl 368 # 02051 • ID 56096

BOUDREAUX, J.C. [1980] *Frames versus minimally restricted structures* (J 0047) Notre Dame J Formal Log 21∗251-262
⋄ B15   C85   ⋄ REV   MR 81h:03025   Zbl 368 # 02052 • ID 53790

BOULIGAND, G. [1937] *Structure des theories. Problemes infinis* (X 0859) Hermann: Paris 58pp
⋄ A05   B98   ⋄ REV   FdM 63.819 • ID 25518

BOULIGAND, G. see Vol. V for further entries

BOURBAKI, N. [1949] *Foundations of mathematics for the working mathematician* (J 0036) J Symb Logic 14∗1-8
⋄ A05   B98   ⋄ REV   MR 11.73   Zbl 34.1   JSL 14.258 • ID 01499

BOURBAKI, N. see Vol. V for further entries

BOUVERE DE, K.L. [1965] *Synonymous theories* (P 0614) Th Models;1963 Berkeley 402-406
• REPR [1968] (C 0684) Automation in Lang Transl & Theorem Prov 123-127
⋄ B30   C40   ⋄ REV   Zbl 221 # 02041 • ID 04261

BOUVERE DE, K.L. [1974] *Some remarks concerning logical and ontological theories* (C 1936) Log Th & Semant Anal (Kanger) 103-112
⋄ A05   B30   ⋄ REV   Zbl 295 # 02030 • ID 60671

BOUVERE DE, K.L. see Vol. III, VI for further entries

BOWEN, K.A. [1972] *A note on cut elimination and completeness in first order theories* (J 0068) Z Math Logik Grundlagen Math 18∗173-176
⋄ B10   C07   F05   ⋄ REV   MR 46 # 8809   Zbl 243 # 02020 • ID 01505

BOWEN, K.A. [1973] *Cut elimination in transfinite type theory* (J 0068) Z Math Logik Grundlagen Math 19∗141-162
⋄ B15   F05   F35   ⋄ REV   MR 48 # 1893   Zbl 302 # 02009 • ID 01507

BOWEN, K.A. [1974] *Systems of transfinite type theory based on intuitionistic and modal logics* (J 0068) Z Math Logik Grundlagen Math 20∗355-372
⋄ B15   F05   F35   F50   ⋄ REV   MR 53 # 102   Zbl 299 # 02010 • ID 03943

BOWEN, K.A. see Vol. II, III, V, VI for further entries

BOYER, R.S. & MOORE, J.S. [1979] *A computational logic* (X 0801) Academic Pr: New York xiv + 397pp
⋄ B98   D70   ⋄ REV   MR 81d:68127   Zbl 448 # 68020 • ID 56665

BOYER, R.S. & MOORE, J.S. [1984] *A mechanical proof of the Turing completeness of pure LISP* (P 3084) Autom Theor Prov After 25 Yea;1983 Denver 133-157
⋄ B35   D20   ⋄ REV   MR 85d:68005 • ID 45265

BOYER, R.S. & MOORE, J.S. [1984] *Proof checking the RSA public key encryption algorithm* (J 0005) Amer Math Mon 91∗181-189
⋄ B35   ⋄ REV   MR 85k:68024   Zbl 548 # 68089 • ID 43254

BOYER, R.S. & MOORE, J.S. [1984] *Proof-checking, theorem-proving, and program verification* (P 3084) Autom Theor Prov After 25 Yea;1983 Denver 119-132
⋄ B35   ⋄ REV   MR 85d:68005   Zbl 552 # 68077 • ID 45264

BOYER, R.S. see Vol. IV for further entries

BOZOYAN, SH.E. & TOROSYAN, B.E. [1979] *Investigation of functions of the algebra of logic in terms of the activity of a set of variables (Russian) (English and Armenian summaries)* (**J 0312**) Izv Akad Nauk Armyan SSR, Ser Mat 14*124–141,156
  ⋄ B05 ⋄ REV MR 81b:94063 Zbl 414 # 94038 • ID 80779

BRACE, J.W. & KNEECE, R.R. [1972] *Approximation of strictly singular and strictly cosingular operators using nonstandard analysis* (**J 0064**) Trans Amer Math Soc 168*483–496
  ⋄ H05 ⋄ REV MR 58 # 30490 Zbl 235 # 47015 • ID 26618

BRADLEY, M.C. [1971] *Copi's method of deduction again* (**J 0047**) Notre Dame J Formal Log 12*454–458 • ERR/ADD ibid 14*584
  ⋄ A05 B10 ⋄ REV MR 45 # 4940 MR 48 # 8193 Zbl 224 # 02009 • ID 01575

BRADY, B. & BUTLER, G.A. & LANKFORD, D.S. [1984] *Abelian group unification algorithms for elementary terms* (**P 3084**) Autom Theor Prov After 25 Yea;1983 Denver 193–199
  ⋄ B35 ⋄ REV MR 85d:68005 Zbl 555 # 68065 • ID 45267

BRADY, J.M. [1977] *Hints on proofs by recursion induction* (**J 1193**) Comput J (London) 20*353–355
  ⋄ B35 D20 ⋄ REV Zbl 364 # 68015 • ID 50984

BRADY, J.M. see Vol. IV for further entries

BRADY, R.T. [1976] *A computer program for determining matrix models of propositional calculi* (**J 0079**) Logique & Anal, NS 19*233–253
  ⋄ B35 ⋄ REV Zbl 348 # 02005 • ID 30638

BRADY, R.T. [1977] *Unspecified constants in predicate calculus and first-order theories* (**J 0079**) Logique & Anal, NS 20*229–243
  ⋄ B10 C07 ⋄ REV MR 57 # 5672 Zbl 368 # 02013 • ID 30639

BRADY, R.T. see Vol. II, III, V for further entries

BRAFFORT, P. & HIRSCHBERG, D. (EDS.) [1963] *Computer programming and formal systems* (**X 0809**) North Holland: Amsterdam vii+161pp
  ⋄ B97 D05 D97 ⋄ REV MR 26 # 3225 Zbl 108.134 • ID 23565

BRAFFORT, P. & SCHEEPEN VAN, F. (EDS.) [1968] *Automation in language translation and theorem proving* (**X 1714**) Commiss Europ Comm : Bruxelles xv+295pp
  ⋄ B35 ⋄ REV MR 38 # 3119 Zbl 188.307 • ID 32459

BRAITHWAITE, R.B. [1959] *Axiomatizing a scietific system by axioms in the form of identifications* (**P 0651**) Axiomatic Method;1957 Berkeley 429–442
  ⋄ B30 ⋄ REV MR 21 # 6320 Zbl 126.7 • ID 27716

BRAITHWAITE, R.B. see Vol. II for further entries

BRAND, D. [1975] *Proving theorems with the modification method* (**J 1428**) SIAM J Comp 4*412–430
  ⋄ B35 ⋄ REV MR 54 # 9187 Zbl 333 # 68059 • ID 60694

BRANDOM, R.B. [1979] *A binary Sheffer operator which does the work of quantifiers and sentential connectives* (**J 0047**) Notre Dame J Formal Log 20*262–264
  ⋄ B10 ⋄ REV MR 80e:03008 Zbl 349 # 02008 • ID 52623

BRANDOM, R.B. [1981] *Semantic paradox of material implication* (**J 0047**) Notre Dame J Formal Log 22*129–132
  ⋄ A05 B20 ⋄ REV MR 83i:03018 Zbl 438 # 03011 • ID 54067

BRANDOM, R.B. see Vol. II for further entries

BRANZEI, D. & MIRON, R. [1983] *The foundations of arithmetic and geometry (Romanian)* (**X 0871**) Acad Rep Soc Romania: Bucharest 247pp
  ⋄ B28 ⋄ REV MR 86e:51001 Zbl 536 # 51001 • ID 48907

BRANZEI, D. see Vol. V for further entries

BRAUN, E.L. [1963] *Digital computer design: logic, circuitry, and synthesis* (**X 0801**) Academic Pr: New York xiii+606pp
  ⋄ B70 B98 ⋄ REV MR 28 # 2655 • ID 23515

BREDESON, J.G. [1978] see AREVALO, Z.

BREITKOPF, A. & KUTSCHERA VON, F. [1971] *Einfuehrung in die moderne Logik* (**X 0826**) Alber: Freiburg 175pp
  ⋄ B98 ⋄ REV MR 83e:03002 Zbl 478 # 03001 • REM 4th edition 1979; 194pp • ID 22414

BREJTBART, YU.YA. & REITER, A. [1975] *A branch-and-bound algorithm to obtain an optimal evaluation tree for monotonic boolean functions* (**J 1431**) Acta Inf 4*311–319
  ⋄ B35 ⋄ REV MR 52 # 9669 Zbl 326 # 02009 • ID 60698

BREJTBART, YU.YA. [1978] *A family of Boolean functions with nonlinear formula size* (**P 4048**) Allerton Conf Commun, Control & Comput (16);1978 Monticello 636–644
  ⋄ B05 ⋄ ID 46268

BREJTBART, YU.YA. & GAL, S. [1978] *Analysis of algorithms for the evaluation of monotonic boolean functions* (**J 0187**) IEEE Trans Comp C-27*1083–1087
  ⋄ B35 ⋄ REV MR 58 # 25106 Zbl 391 # 94030 • ID 69251

BREJTBART, YU.YA. & VAIRAVAN, K. [1979] *The computational complexity of a class of minimization algorithms for switching functions* (**J 0187**) IEEE Trans Comp C-28*941–943
  ⋄ B35 ⋄ REV MR 80m:94085 Zbl 428 # 94016 • ID 69250

BREJTBART, YU.YA. see Vol. IV for further entries

BRENDER, D.M. [1957] *The logical procedures needed for finding the minimals of a boolean function on a digital computer* (**P 1675**) Summer Inst Symb Log;1957 Ithaca 210
  ⋄ B35 ⋄ REV Zbl 145.407 JSL 25.368 • ID 01589

BRENNAN, J.G. [1957] *A handbook of logic* (**X 0837**) Harper & Row: New York x+222pp
  ⋄ B98 ⋄ REV JSL 24.186 JSL 25.334 • ID 22465

BRESSAN, A. [1974] *On the usefulness of modal logic in axiomatizations of physics* (**P 1791**) Proc Bienn Meet Phil of Sci Ass;1972 East Lansing 285–303
  ⋄ B45 B80 ⋄ REV Zbl 322 # 02014 • ID 60704

BRESSAN, A. [1983] see BONOTTO, C.

BRESSAN, A. [1984] see BONOTTO, C.

BRESSAN, A. see Vol. II, V for further entries

BREUER, M.A. & EPSTEIN, G. [1973] *The smallest many-valued logic for treatment of complemented and uncomplemented error signal* (**P 2009**) Int Symp Multi-Val Log (3);1973 Toronto 29–37
  ⋄ B80 ⋄ REV MR 55 # 7593 • ID 42051

BRICKHILL, C.J. [1972] see ASH, C.J.

BRODSKIJ, I.N. [1970] *On certain variant of the calculus of rejected formulae of propositional logic (Russian)* (C 0668) Neklass Log 332-348
⋄ B05 ⋄ REV MR 47 # 6452 • ID 02049

BRODSKIJ, I.N. [1972] *Elementare Einfuehrung in die Logik. 2.ueberarb. Aufl (Russian)* (X 0938) Leningrad Univ: Leningrad 64pp
⋄ B98 ⋄ REV Zbl 259 # 02001 • ID 30337

BRODY, B.A. [1973] *Logic, theoretical and applied* (X 0819) Prentice Hall: Englewood Cliffs viii+280pp
⋄ B98 ⋄ REV JSL 42.319 • ID 44487

BRONSTEIN, D.J. [1942] *A correction to the sentential calculus of Tarski's introduction to logic* (J 0036) J Symb Logic 7*34
⋄ B05 ⋄ REV MR 3.289 Zbl 60.20 • ID 01614

BROOKS, R.B.S. [1961] see ALLEN, L.E.

BROUWER, L.E.J. [1955] *The effect of intuitionism on classical algebra of logic* (J 0215) Proc Irish Acad, Sect A 57*113-116
⋄ B05 F50 ⋄ REV MR 17.446 Zbl 66.11 JSL 24.204 • ID 02051

BROUWER, L.E.J. see Vol. V, VI for further entries

BROWN, CYNTHIA A. & GOLDBERG, ALLEN & PURDOM JR., P.W. [1982] *Average time analyses of simplified Davis-Putnam procedures* (J 0232) Inform Process Lett 15*72-75
• ERR/ADD ibid 16/4*213
⋄ B35 D15 ⋄ REV MR 84d:68038a MR 84d:86038b Zbl 529 # 68065 • ID 39699

BROWN, D.J. [1970] see BLOOM, S.L.

BROWN, D.J. & ROBINSON, A. [1972] *A limit theorem on the cores of large standard exchange economies* (J 0054) Proc Nat Acad Sci USA 69*1258-1260
⋄ H10 ⋄ REV Zbl 242 # 90007 • ID 26147

BROWN, D.J. & SUSZKO, R. [1973] *Abstract logics* (J 0202) Diss Math (Warsaw) 102*9-41
⋄ B22 ⋄ REV MR 56 # 5284 Zbl 317 # 02071 • ID 60724

BROWN, D.J. [1973] see BLOOM, S.L.

BROWN, D.J. & LOEB, P.A. [1976] *The values of nonstandard exchange economies* (J 0029) Israel J Math 25*71-86
⋄ H05 H10 ⋄ REV MR 57 # 5003 Zbl 366 # 90012 • ID 26074

BROWN, D.J. & KHAN, M.A. [1980] *An extension of the Brown-Robinson equivalence theorem* (J 2662) Appl Math Comp 6*167-175
⋄ H10 ⋄ REV MR 81d:90023 Zbl 429 # 90012 • ID 80796

BROWN, F.M. [1978] *Towards the automation of set theory and its logic* (J 0503) Artif Intell 10*281-316
⋄ B35 ⋄ REV MR 80d:68124 Zbl 395 # 68082 • ID 52604

BROWN, F.M. & TAERNLUND, S.-A. [1979] *Inductive reasoning on recursive equations* (J 0503) Artif Intell 12*207-229
⋄ B35 ⋄ REV MR 81e:68029 Zbl 426 # 68089 • ID 53683

BROWN, F.M. [1979] *On a convenient division of labor in the generation of prime implicants* (J 3366) Comp & Electr Engin 6*267-271
⋄ B35 ⋄ REV MR 80m:94087 Zbl 421 # 94031 • ID 69252

BROWN, F.M. [1980] *An investigation into the goals of research in automatic theorem proving as related to mathematical reasoning* (J 0503) Artif Intell 14*221-242
⋄ B35 ⋄ REV MR 81i:68129 Zbl 446 # 68081 • ID 69253

BROWN, F.M. see Vol. II for further entries

BROWN, J.M. [1976] *Bernays's non-circular proof of the non-independence of the fourth axiom of "Principia Mathematica"* (J 0103) Analysis (Oxford) 36*207-208
⋄ A05 B05 ⋄ REV Zbl 347 # 02006 • ID 60727

BROWN, P.L. & STUBERMAN, W.E. [1965] *Elementary modern logic* (X 0880) Ronald Press: New York vii+269pp
⋄ B98 ⋄ ID 22467

BROWN, W.S. [1978] see BAXANDALL, P.R.

BROWNE, S.S.S. [1964] *Fundamentals of deductive logic* (X 1243) Brown: Dubuque vii+197pp
⋄ B98 ⋄ ID 22469

BROY, M. [1984] *On the Herbrand Kleene universe for nondeterministic computations* (P 3658) Math Founds of Comput Sci (11);1984 Prague 214-222
⋄ B35 ⋄ ID 44826

BROY, M. see Vol. II, IV for further entries

BRUEVICH, N.G. [1979] see BELYANIN, P.N.

BRUIJN DE, N.G. [1976] *Defining reals without the use of rationals* (J 0028) Indag Math 38(2)*100-108
⋄ B28 ⋄ REV MR 54 # 4983 Zbl 324 # 10054 • ID 72190

BRUIJN DE, N.G. [1977] *Construction of the system of real numbers* (J 0358) Versl Gewone Vergad Afd Natuurkd 86*121-125
⋄ B28 ⋄ REV MR 57 # 96 Zbl 367 # 10047 • ID 72189

BRUIJN DE, N.G. [1978] *Lambda-calculus with namefree formulas involving symbols that represent reference transforming mappings* (J 0028) Indag Math 40*348-356
⋄ B35 B40 ⋄ REV MR 80b:03016 Zbl 393 # 03009 • ID 29191

BRUIJN DE, N.G. [1980] *A survey of the project AUTOMATH* (C 3050) Essays Combin Log, Lambda Calc & Formalism (Curry) 579-606
⋄ B35 B40 ⋄ REV MR 81m:03017 Zbl 469 # 03006 • ID 72186

BRUIJN DE, N.G. see Vol. II, III, V, VI for further entries

BRUMM, G.L. [1962] *The method of possibility diagrams for testing the validity of certain types of inferences, based on Jevons' logical alphabet* (J 0047) Notre Dame J Formal Log 3*209-233
⋄ B35 ⋄ ID 01661

BRUNER, F.G. [1943] *Mathematical logic with transfinite types* (1111) Preprints, Manuscr., Techn. Reports etc. v+68pp
⋄ B15 ⋄ REV JSL 9.72 • REM Printed and distributed by the author, Chicago • ID 16758

BRUSENTSOV, N.P. [1971] *A mathematical theory of syllogistics (Russian)* (S 0716) Vychisl Tekh Vopr Kibern (Univ Leningrad) 8*154-176
⋄ B20 ⋄ REV MR 50 # 12641 • ID 04125

BRUSENTSOV, N.P. [1972] *An improved parenthesis-free notation for formulas (Russian)* (J 0199) Zh Vychisl Mat i Mat Fiz 12*820-822
• TRANSL [1972] (J 1049) USSR Comput Math & Math Phys 12/3*330-333
⋄ B03 ⋄ REV MR 47 # 1573 Zbl 252 # 68008 • ID 71360

BRUSENTSOV, N.P. [1977] *Lewis Carrol diagrams and the Aristotelian syllogistics (Russian)* (S 0716) Vychisl Tekh Vopr Kibern (Univ Leningrad) 13*164-182
⋄ B20 ⋄ REV MR 58 # 21416 Zbl 434 # 03012 • ID 55716

BRUSENTSOV, N.P. [1982] *Complete system of categorical syllogisms of Aristotle (Russian)* (S 0716) Vychisl Tekh Vopr Kibern (Univ Leningrad) 19*3-16
⋄ B20 ⋄ REV Zbl 554 # 03019 • ID 44941

BRUSENTSOV, N.P. [1982] *Representation of syllogistics in predicate logic (Russian)* (S 0716) Vychisl Tekh Vopr Kibern (Univ Leningrad) 18*119-126,227
⋄ B20 ⋄ REV MR 84j:03026 • ID 34620

BRUSENTSOV, N.P. see Vol. II for further entries

BRYLL, G. & SLUPECKI, J. & WYBRANIEC-SKARDOWSKA, U. [1970] *A certain theory that is equivalent to Tarski's theory of deductive systems (Polish) (English summary)* (S 1454) Zesz Nauk Wyz Szk Ped Mat, Opole 10*61-67
⋄ B22 ⋄ REV MR 53 # 12926 • ID 23175

BRYLL, G. & SLUPECKI, J. & WYBRANIEC-SKARDOWSKA, U. [1971] *Theory of rejected propositions I (Polish and Russian summaries)* (J 0063) Studia Logica 29*75-123
⋄ A10 B22 ⋄ REV MR 46 # 5116 Zbl 253 # 02049 • REM Part II 1972 • ID 12488

BRYLL, G. & SLUPECKI, J. & WYBRANIEC-SKARDOWSKA, U. [1972] *The theory of rejected propositions II (Polish and Russian summaries)* (J 0063) Studia Logica 30*97-145
⋄ A05 B22 ⋄ REV MR 47 # 8263 Zbl 282 # 02021 • REM Part I 1971 • ID 12493

BRYLL, G. & SLUPECKI, J. [1975] *Contributions to the general theory of deductive systems (Polish) (English summary)* (S 1454) Zesz Nauk Wyz Szk Ped Mat, Opole 15*53-62
⋄ B22 ⋄ REV MR 54 # 4962 Zbl 319 # 02041 • ID 24737

BRYLL, G. see Vol. II, III, IV for further entries

BUCHBERGER, B. [1985] *Basic features and development of the critical-pair/completion procedure* (P 4244) Rewriting Techn & Appl (1);1985 Dijon 1-45
⋄ B35 ⋄ ID 49761

BUCHBERGER, B. see Vol. IV for further entries

BUCHHOLZ, W. [1975] *Ein ausgezeichnetes Modell fuer die intuitionistische Typenlogik* (J 0009) Arch Math Logik Grundlagenforsch 17*55-60
⋄ B15 F05 F35 F50 ⋄ REV MR 52 # 5368 Zbl 317 # 02026 • ID 01681

BUCHHOLZ, W. see Vol. V, VI for further entries

BUCK, R.C. [1963] *Mathematical induction and recursive definitions* (J 0005) Amer Math Mon 70*128-135
⋄ B28 D20 ⋄ REV Zbl 113.3 • ID 48047

BUCUR, I. [1980] *Special chapters of algebra (Romanian)* (X 0871) Acad Rep Soc Romania: Bucharest 335pp
• TRANSL [1984] (X 0835) Reidel: Dordrecht viii + 406pp
⋄ B80 C98 D98 ⋄ REV MR 84c:14001 MR 86f:14001 • ID 46245

BUECHI, J.R. [1953] *Investigation of the equivalence of the axiom of choice and Zorn's lemma from the viewpoint of the hierarchy of types* (J 0036) J Symb Logic 18*125-135
⋄ B15 E25 ⋄ REV MR 15.2 Zbl 50.247 JSL 19.285 • ID 01683

BUECHI, J.R. [1960] *Weak second-order arithmetic and finite automata* (J 0068) Z Math Logik Grundlagen Math 6*66-92
⋄ B15 B25 C85 D05 F35 ⋄ REV MR 23 # A2317 Zbl 103.247 JSL 28.100 • ID 01690

BUECHI, J.R. [1962] *Turing machines and the "Entscheidungsproblem"* (J 0043) Math Ann 148*201-213
⋄ B20 D10 D35 ⋄ REV MR 29 # 5719 Zbl 118.16 • ID 01691

BUECHI, J.R. [1965] *Decision methods in the theory of ordinals* (J 0015) Bull Amer Math Soc 71*767-770
⋄ B15 B25 C85 D05 E10 ⋄ REV MR 32 # 7413 • ID 01697

BUECHI, J.R. [1965] *Transfinite automata recursions and weak second order theory of ordinals* (P 0623) Int Congr Log, Meth & Phil of Sci (2,Proc);1964 Jerusalem 3-23
⋄ B15 B25 C85 D05 E10 ⋄ REV MR 35 # 1480 • ID 01694

BUECHI, J.R. & LANDWEBER, L.H. [1969] *Definability in the monadic second-order theory of successor* (J 0036) J Symb Logic 34*166-170
⋄ B15 B25 C40 C85 D05 F35 ⋄ REV MR 42 # 4387 Zbl 209.22 • ID 01700

BUECHI, J.R. & SIEFKES, D. [1973] *Axiomatization of the monadic second order theory of ω_1* (C 3046) Decid Theories II - Monad 2nd Ord Th Count Ordinals 129-217
⋄ B15 B25 C85 E10 ⋄ REV Zbl 298 # 02050 • ID 71381

BUECHI, J.R. & SIEFKES, D. [1973] *The monadic second order theory of all countable ordinals* (C 3046) Decid Theories II Monad 2nd Ord Th Count Ordinals vi + 217pp
⋄ B15 B25 C85 E10 ⋄ REV MR 49 # 10534 Zbl 298 # 02050 • ID 01708

BUECHI, J.R. [1973] *The monadic second order theory of ω_1* (C 3046) Decid Theories II - Monad 2nd Ord Th Count Ordinals 1-127
⋄ B15 B25 C85 E10 ⋄ REV MR 57 # 16033 • ID 71377

BUECHI, J.R. [1977] *Using determinancy of games to eliminate quantifiers* (P 2588) FCT'77 Fund of Comput Th;1977 Poznan 367-378
⋄ B15 C10 C85 E60 ⋄ REV MR 58 # 16244 Zbl 367 # 02005 • ID 51165

BUECHI, J.R. & ZAIONTZ, C. [1983] *Deterministic automata and the monadic theory of ordinals $< \omega_2$* (J 0068) Z Math Logik Grundlagen Math 29*313-336
⋄ B15 B25 C85 D05 E10 ⋄ REV MR 85j:03008 Zbl 541 # 03004 • ID 40515

BUECHI, J.R. & SIEFKES, D. [1983] *The complete extensions of the monadic second order theory of countable ordinals* (J 0068) Z Math Logik Grundlagen Math 29*289-312
⋄ B15 C35 C85 D05 E10 ⋄ REV MR 85g:03058 Zbl 541 # 03003 • ID 33901

BUECHI, J.R. see Vol. III, IV, V, VI for further entries

BUFF, H.W. [1984] *ω-Konservativitaet der Nonstandardmengenlehre von Nelson bezueglich ZF+Kompaktheitssatz* (J 0068) Z Math Logik Grundlagen Math 30*133-144
⋄ E25 E35 E70 H05 ⋄ REV MR 86a:03059 Zbl 519 # 03042 • ID 43509

BUGAJSKI, S. & MOTYKA, Z. [1981] *Generalized Borel law and quantum probabilities* (J 2736) Int J Theor Phys 20*263-268
⋄ B30 G12 ⋄ REV MR 83f:81014 Zbl 487 # 60034
• ID 36710

BUGAJSKI, S. see Vol. II, III for further entries

BUKHARAEVA, Z.K. & GOLUNKOV, YU.V. [1978] *Complexity and running time of normal algorithms computing boolean functions (Russian)* (J 3937) Veroyat Met i Kibern (Kazan) 14*3-20
⋄ B05 B35 D03 D15 ⋄ REV MR 81j:68050 Zbl 411 # 03032 • ID 52886

BULL, R. [1984] *The classical propositional calculus of arguments* (J 0068) Z Math Logik Grundlagen Math 30*45-86
⋄ B22 ⋄ REV MR 86a:03008 Zbl 536 # 03039 • ID 37119

BULL, R.A. [1972] *Mathematical logic* (J 0329) Math Chron (Auckland) 2*17-27
⋄ B98 ⋄ REV MR 48 # 3689 Zbl 238 # 02001 • ID 27895

BULL, R.A. see Vol. II for further entries

BULLERS, W.I. & NOF, S.Y. & WHINSTON, A.B. [1980] *Artificial intelligence in manufacturing planning and control* (J 2669) AIIE Trans 12*351-363
⋄ B35 B75 ⋄ REV MR 82g:90055 • ID 80811

BULLOCK, A.M. & SCHNEIDER, H.H. [1972] *A calculus for finitely satisfiable formulas with identity* (J 0009) Arch Math Logik Grundlagenforsch 15*158-163
⋄ B10 C07 C13 ⋄ REV MR 48 # 8219 Zbl 262 # 02013
• ID 01746

BULLOCK, A.M. & SCHNEIDER, H.H. [1973] *On generating the finitely satisfiable formulas* (J 0047) Notre Dame J Formal Log 14*373-376
⋄ B10 C07 C13 D25 ⋄ REV MR 47 # 8280 Zbl 236 # 02014 • ID 01748

BULYGIN, E. [1971] see ALCHOURRON, C.E.

BULYGIN, E. see Vol. II for further entries

BUNDER, M.W. [1971] *Predicate calculus without free variables* (J 0079) Logique & Anal, NS 14*725-728
⋄ B20 ⋄ REV MR 47 # 6430 Zbl 238 # 02012 • ID 01760

BUNDER, M.W. [1979] *Alternative forms of propositional calculus for a given deduction theorem* (J 0047) Notre Dame J Formal Log 20*613-619
⋄ B22 ⋄ REV MR 80m:03031 Zbl 349 # 02012 • ID 56105

BUNDER, M.W. [1980] *Modus ponens free (implicational) logics* (P 3006) Brazil Conf Math Log (3);1979 Recife 23-34
⋄ B20 ⋄ REV MR 82f:03022 Zbl 446 # 03009 • ID 56538

BUNDER, M.W. [1980] *The consistency of a higher order predicate calculus and set theory based on combinatory logic* (P 2958) Latin Amer Symp Math Log (4);1978 Santiago 73-82
⋄ B15 B40 F05 ⋄ REV MR 81h:03106 Zbl 427 # 03012
• ID 53699

BUNDER, M.W. [1981] *Predicate calculus and naive set theory in pure combinatory logic* (J 0009) Arch Math Logik Grundlagenforsch 21*169-177
⋄ B10 B40 E70 ⋄ REV MR 84f:03012 Zbl 472 # 03011
• ID 55277

BUNDER, M.W. [1982] *Deduction theorems for weak implicational logics* (J 0063) Studia Logica 41*95-108
⋄ B22 ⋄ REV MR 86a:03009 Zbl 536 # 03001 • ID 37090

BUNDER, M.W. [1982] *Some results in Aczel-Feferman logic and set theory* (J 0068) Z Math Logik Grundlagen Math 28*269-276
⋄ B15 E70 ⋄ REV MR 83h:03079 Zbl 496 # 03034
• ID 36080

BUNDER, M.W. [1983] *A one axiom set theory based on higher order predicate calculus* (J 0009) Arch Math Logik Grundlagenforsch 23*99-107
⋄ B15 B40 E70 ⋄ REV Zbl 537 # 03010 • ID 31631

BUNDER, M.W. [1983] *Predicate calculus of arbitrarily high finite order* (J 0009) Arch Math Logik Grundlagenforsch 23*1-10
⋄ B15 ⋄ REV MR 85a:03015 Zbl 528 # 03005 • ID 34766

BUNDER, M.W. see Vol. II, V, VI for further entries

BUNDY, A. & WELHAM, B. [1981] *Using meta-level inference for selective application of multiple rewrite rule sets in algebraic manipulation* (J 0503) Artif Intell 16*189-211
⋄ B35 ⋄ REV MR 82h:68059 Zbl 438 # 68041 • ID 80814

BUNDY, A. [1983] *The computer modelling of mathematical reasoning* (X 0801) Academic Pr: New York xiv + 322pp
⋄ B35 ⋄ REV Zbl 541 # 68067 • ID 46350

BUNDY, A. see Vol. IV, VI for further entries

BUNYATOV, M.R. [1967] *Dirac spaces (Russian)* (S 0223) Nauch Trud NS, Politekh Inst Tashkent 43*24-38
⋄ C20 C60 H05 H10 ⋄ REV MR 43 # 32 • ID 02032

BURALI-FORTI, C. [1894] *Logica matematica* (X 1283) Hoepli: Milano vi + 158pp
⋄ B98 ⋄ REV FdM 25.115 • ID 22446

BURALI-FORTI, C. [1913] *Sur les lois generales de l'algorithme des symboles de fonction et d'operation* (P 1646) Int Congr Math (5);1912 Cambridge GB 2*480-491
⋄ B10 ⋄ REV FdM 44.75 • ID 29488

BURALI-FORTI, C. see Vol. V for further entries

BURCKHARDT, J.J. [1938] *Zur Neubegruendung der Mengenlehre* (J 0157) Jbuchber Dtsch Math-Ver 48*146-165
⋄ B28 E70 ⋄ REV MR 1.132 Zbl 19.201 JSL 3.165 FdM 64.33 • ID 41071

BURCKHARDT, J.J. see Vol. V for further entries

BURGINA, E.S. [1984] *A graduated semantic system for set theory (Russian)* (J 0087) Mat Zametki (Akad Nauk SSSR) 35*855-868
• TRANSL [1984] (J 1044) Math Notes, Acad Sci USSR 35*448-456
⋄ B35 E70 ⋄ REV MR 85k:03042 • ID 45486

BURGINA, E.S. see Vol. IV for further entries

BURKAMP, W. [1932] *Logik* (X 1308) Mittler: Herford vi + 175pp
⋄ B98 ⋄ REV FdM 58.54 • ID 25089

BURKS, A.W. & WARREN, D.W. & WRIGHT, J.B. [1954] *An analysis of a logical machine using parenthesis-free notation* (J 0235) Math Tables Other Aids Comp 8*53-57
⋄ B03 D05 ⋄ REV MR 15.833 Zbl 58.4 JSL 20.70
• ID 33359

BURKS, A.W. [1955] *Propositional statements* (J 0153) Phil of Sci (East Lansing) 22*175-193
⋄ B05 ⋄ REV MR 17.226 • ID 01778

BURKS, A.W. & WRIGHT, J.B. [1962] *Sequence generators, graphs and formal languages* (J 0194) Inform & Control 5*204-212
⋄ B20 D05 ⋄ REV MR 26#4902 Zbl 109.353 JSL 29.210
• ID 01780

BURKS, A.W. & WRIGHT, J.B. [1962] *Sequence generators and digital computers* (P 0613) Rec Fct Th;1961 New York 139-199
⋄ B20 D05 ⋄ REV Zbl 145.243 JSL 29.210 • ID 02718

BURKS, A.W. [1977] *Chance, cause, reason. An inquiry into the nature of scientific evidence* (X 0862) Univ Chicago Pr: Chicago xvi+694pp
⋄ A05 B98 ⋄ REV MR 58#10268 Zbl 421#03002 JSL 45.373 • ID 53402

BURKS, A.W. see Vol. IV for further entries

BURLACU, E. [1969] *On the inverses of boolean matrices (Romanian) (Russian and French summaries)* (J 4714) Bul Sti Tehn Inst Politeh Timisoara, NS 14(28)*15-20
⋄ B05 ⋄ REV MR 41#1608 Zbl 236#94032 • ID 02040

BUROSCH, G. & KUDRYAVTSEV, V.B. [1972] *Das Problem der Vollstaendigkeit fuer boolesche Funktionen ueber zwei Dualmengen* (J 0114) Math Nachr 54*105-125
⋄ B05 ⋄ REV MR 47#6436 Zbl 255#02010 • ID 07607

BUROSCH, G. see Vol. II for further entries

BURSKY, P. & SLAGLE, J.R. [1968] *Experiments with a multipurpose, theorem-proving heuristic program* (J 0037) ACM J 15*85-99
⋄ B35 ⋄ REV Zbl 159.212 JSL 35.596 • ID 01795

BUTH, M. [1976] *Logische Analyse eines mathematischen Lehrsatzes* (X 1039) Diesterweg: Frankfurt a.M. 49pp
⋄ A05 B10 ⋄ REV Zbl 333#02001 • ID 60797

BUTLER, G.A. [1984] see BRADY, B.

BUTLER, G.A. see Vol. IV for further entries

BUTOV, A.A. [1978] *A method of functional division of a completely defined boolean function (Russian)* (J 0011) Avtom Telemekh 1978/9*121-125
• TRANSL [1978] (J 0010) Autom & Remote Control 39*1361-1365
⋄ B05 ⋄ REV MR 80m:94088 Zbl 417#94019 • ID 69288

BUTRICK, R. [1981] *Deduction and analysis* (X 1946) Univ Pr Amer: Lanham x+109pp
⋄ B05 ⋄ REV Zbl 523#03006 • REM Revised ed. • ID 37019

BUTRICK, R. see Vol. III, V, VI for further entries

BUTTON, R.W. [1976] *A non-standard characterization of perfect mappings* (J 0017) Canad J Math 28*1277-1279
⋄ H05 ⋄ REV MR 54#13891 Zbl 361#54032 • ID 80841

BUTTON, R.W. [1976] *Monads for regular and normal spaces* (J 0047) Notre Dame J Formal Log 17*449-456
⋄ H05 ⋄ REV MR 56#3823 Zbl 292#02045 • ID 18103

BUTTON, R.W. [1977] *When do *continuous extensions exist?* (J 0047) Notre Dame J Formal Log 18*406-408
⋄ H05 H20 ⋄ REV MR 58#21586 Zbl 314#02065 • ID 24240

BUTTON, R.W. [1978] *A note on the Q-topology* (J 0047) Notre Dame J Formal Log 19*679-686
⋄ H05 H20 ⋄ REV MR 80a:54096 Zbl 332#02059 • ID 52166

BUTTON, R.W. [1979] *When do two topologies have the same monads?* (J 0254) Gen Topology Appl 10*7-11
⋄ C20 H05 ⋄ REV MR 81g:54070 Zbl 408#54033 • ID 80839

BUTTS, R.E. & HINTIKKA, K.J.J. (EDS.) [1977] *Proceedings of the fifth international congress of logic, methodology and philosophy of science. I,II,III,IV* (S 3308) Univ Western Ontario Ser in Philos of Sci 9*x+406pp,10*x+427pp,11*x+321pp,12*x+336pp
⋄ A05 A10 B97 ⋄ REV MR 56#15352 Zbl 363#00005 • ID 50826

BUZURKHANOV, V. [1981] see ABDULLAEVA, M.

BYERLY, H.C. [1973] *A primer of logic* (X 0837) Harper & Row: New York xiii+560pp
⋄ B98 ⋄ REV MR 53#5238 • ID 22870

CAFERRA, R. [1982] *Proof by matrix reduction as plan + validation* (P 3840) Autom Deduct (6);1982 New York 309-325
⋄ B35 ⋄ REV MR 85h:03018 Zbl 481#68085 • ID 43369

CAI, JINGQIU [1981] *Negative hyper ordered resolution principle (Chinese) (English summary)* (J 3923) Xiamen Daxue Xuebao, Ziran Kexue 20*297-304
⋄ B35 ⋄ REV Zbl 484#03005 • ID 36859

CAI, JINGQIU see Vol. V for further entries

CAICEDO, X. [1978] *A formal system for the non-theorems of the propositional calculus* (J 0047) Notre Dame J Formal Log 19*147-151
⋄ B05 ⋄ REV MR 58#129 Zbl 351#02009 • ID 27098

CAICEDO, X. see Vol. III, IV, V for further entries

CAIN, R.G. & HONG, SEJUNE & OSTAPKO, D.L. [1974] *Mini: a heuristic approach for logic minimization* (J 0284) IBM J Res Dev 18*443-458
⋄ B35 G05 ⋄ REV MR 54#12384 Zbl 289#02009 • ID 29979

CAIN, R.G. see Vol. II for further entries

CALABRESE, P. [1966] *The Menger algebras of 2-place functions in the 2-valued logic* (J 0047) Notre Dame J Formal Log 7*333-340
⋄ B05 G15 ⋄ REV MR 38#1998 Zbl 192.35 • ID 01812

CALL, R.L. [1972] *The Goedel-Herbrand theorems* (J 0047) Notre Dame J Formal Log 13*131-134
⋄ B10 C07 ⋄ REV MR 46#5097 Zbl 227#02014 • ID 71509

CALL, R.L. [1984] *Constructing sequent rules for generalized propositional logics* (J 0047) Notre Dame J Formal Log 25*171-178
⋄ B05 F07 ⋄ REV MR 85e:03020 Zbl 551#03036 • ID 40582

CALL, R.L. see Vol. II, III for further entries

CALLOT, J.-L. & DIENER, F. & DIENER, M. [1978] *Le probleme de la "chasse au canard" (English summary)* (J 2313) C R Acad Sci, Paris, Ser A-B 286*A1059
⋄ H05 ⋄ REV MR 80b:58080 Zbl 419#34039 • ID 53389

CALLOT, J.-L. [1981] see BENOIT, E.

CALLOT, J.-L. & SARI, T. [1983] *Stroboscopie infinitesimale et moyennisation dans les systemes d'equations differentielles a solution rapidement oscillantes* (P 4223) Math Tool & Model:1981/82 Toulouse & Paris 3*345-356
⋄ H05 ⋄ ID 44916

CALUDE, C. [1978] *On the category of recursive languages* (J 0517) Mathematica (Cluj) 19(24)*29-32
⋄ B03 D45 F40 G30 ⋄ REV MR 80b:03056 Zbl 384 # 03029 • ID 52071

CALUDE, C. & PAUN, G. [1981] *Global syntax and semantics for recursively enumerable languages* (J 2095) Fund Inform, Ann Soc Math Pol, Ser 4 4*245-254
⋄ B03 D25 ⋄ REV MR 83h:68133 Zbl 473 # 68068
• ID 55399

CALUDE, C. see Vol. II, IV, V, VI for further entries

CAMION, P. [1962] *Traitement de l'information par l'algebre de Boole* (P 1192) Symb Lang in Data Processing;1962 Roma 675-683
⋄ B05 G05 ⋄ ID 29446

CAMPBELL, P.J. [1977] *An answer to Armstrong's question about incompleteness in Copi* (J 0047) Notre Dame J Formal Log 18*262-264
⋄ A05 B05 ⋄ REV MR 55 # 12439 Zbl 332 # 02052
• ID 60817

CAMPBELL, P.J. see Vol. III, IV, V for further entries

CANGELOSI, V.E. [1967] *Compound statements and mathematical logic* (X 1164) Merrill: Columbus xii + 114pp
⋄ B98 ⋄ ID 22470

CANNONITO, F.B. [1962] *The Goedel incompleteness theorem and intelligent machines* (P 1197) AFIPS Spring Jt Computer Conf (21);1962 San Francisco 21*71-77
⋄ B35 D10 D25 F30 ⋄ REV JSL 36.693 • ID 21294

CANNONITO, F.B. see Vol. IV, VI for further entries

CANTINI, A. [1979] *I fondamenti della matematica da Dedekind a Tarski* (X 4086) Loescher: Torino 260pp
⋄ A05 B30 ⋄ ID 44055

CANTINI, A. see Vol. II, III, IV, V, VI for further entries

CANTY, J.T. [1963] *Completeness of Copi's method of deduction* (J 0047) Notre Dame J Formal Log 4*142-144
⋄ B05 C07 ⋄ REV MR 29 # 3361 Zbl 154.4 JSL 30.366
• ID 01832

CANTY, J.T. see Vol. II, III, V for further entries

CAPALDI, N. [1966] *Introduction to deductive logic* (X 1310) Monarch Pr: New York 166pp
⋄ B98 ⋄ ID 22471

CAPINSKI, M. [1977] *Ultraproducts of higher-order models and non-standard analysis* (S 2350) Zesz Nauk, Prace Mat, Uniw Krakow 18*71-79
⋄ H05 ⋄ REV MR 56 # 116 Zbl 353 # 02034 • ID 50117

CARBONI, A. [1982] *Analisi non-standard e topos* (J 1208) Rend Ist Mat Univ Trieste 14*1-16
⋄ H05 H10 ⋄ REV MR 85k:18005 Zbl 536 # 03050
• ID 37130

CARBONI, A. see Vol. V for further entries

CARDWELL, C.E. [1979] *Arguments and inference. An introduction to symbolic logic* (X 1164) Merrill: Columbus xi + 414pp
⋄ B98 ⋄ REV MR 83f:03003 Zbl 532 # 03002 • ID 34832

CARILLO GALLEGO, D. & MIRA ROS, J.M. [1975] *Ueber verschiedene Definitionen der "natuerlichen Zahl" (Spanisch)* (J 0299) Gac Mat, Ser 1a (Madrid) 27*92-99
⋄ B28 ⋄ REV MR 52 # 13047 Zbl 315 # 02001 • ID 29795

CARNAP, R. [1929] *Abriss der Logistik, mit besonderer Beruecksichtigung der Relationstheorie und ihrer Anwendungen* (X 0902) Springer: Wien vi + 114pp
⋄ A05 B98 ⋄ REV FdM 55.30 • ID 16764

CARNAP, R. [1935] *Ein Gueltigkeitskriterium fuer die Saetze der klassischen Mathematik* (J 0124) Monatsh Math-Phys 42*163-190
⋄ B30 F07 ⋄ REV Zbl 12.145 FdM 61.970 • ID 03962

CARNAP, R. [1936] see BACHMANN, F.

CARNAP, R. [1939] *Foundations of logic and mathematics* (X 0862) Univ Chicago Pr: Chicago viii + 71pp
⋄ A05 B98 ⋄ REV Zbl 23.97 JSL 4.117 FdM 65.1099
• ID 02724

CARNAP, R. [1943] *Formalization of logic* (X 0858) Harvard Univ Pr: Cambridge xviii + 159pp
⋄ A05 B98 ⋄ REV MR 4.209 Zbl 61.7 JSL 8.81
• ID 02228

CARNAP, R. [1954] *Einfuehrung in die symbolische Logik mit besonderer Beruecksichtigung ihrer Anwendungen* (X 0902) Springer: Wien x + 209pp
• TRANSL [1959] (X 0813) Dover: New York xiv + 241pp
⋄ A05 B30 B98 ⋄ REV MR 16.208 MR 21 # 2578
Zbl 56.6 Zbl 83.1 JSL 20.274 JSL 31.287 • ID 01852

CARNAP, R. [1964] *The logicist foundations of mathematics* (C 1105) Phil of Math. Sel Readings 31-41
⋄ B15 ⋄ ID 38052

CARNAP, R. & JEFFREY, R.C. (EDS.) [1971] *Studies in inductive logic and probability I* (X 0926) Univ Calif Pr: Berkeley vi + 264pp
⋄ B30 B48 ⋄ REV MR 48 # 49 Zbl 246 # 02024 • ID 25461

CARNAP, R. [1973] *Grundlagen der Logik und Mathematik* (X 0917) Nymphenburger: Muenchen 106pp
⋄ A05 B98 ⋄ REV MR 52 # 13308 Zbl 333 # 02002
• ID 21772

CARNAP, R. [1977] *Two essays on entropy.* (X 0926) Univ Calif Pr: Berkeley xxii + 115pp
⋄ A05 B30 ⋄ REV MR 57 # 4929 Zbl 399 # 03002 • REM Ed. with an introduction by A.Shimony • ID 52799

CARNAP, R. see Vol. II for further entries

CARNES, R.D. & WILCOX, W.C. [1968] *An infixed, punctuation-free notation* (J 0047) Notre Dame J Formal Log 9*171-178
⋄ B03 G05 ⋄ REV MR 38 # 6892 Zbl 239 # 02004
• ID 01857

CARNES, R.D. [1969] *A reduction procedure for Sheffer stroke formulas* (J 0047) Notre Dame J Formal Log 10*331-335
⋄ B05 ⋄ REV MR 40 # 18 Zbl 202.5 • ID 01859

CARNEY, J.D. & SCHEER, R.K. [1964] *Fundamentals of logic* (X 0843) Macmillan : New York & London xv + 474pp
⋄ B98 ⋄ ID 22474

CARNEY, J.D. [1970] *Introduction to symbolic logic* (**X** 0819) Prentice Hall: Englewood Cliffs ix+252pp
⋄ B98 ⋄ ID 22473

CARNIELLI, W.A. [1985] *An algorithm for axiomatizing and theorem proving in finite many-valued logics* (**J** 0079) Logique & Anal, NS 28∗363-368
⋄ B35 ⋄ ID 49653

CARNIELLI, W.A. see Vol. II, V for further entries

CARRERAS MATAS, J.M. [1967] *Class of formulas of minimum content of an axiomatic system with respect to a given formula (Spanish)* (**P** 0606) Reunion Mat Espanoles (7);1966 Valladolid 68-70
⋄ B10 ⋄ REV MR 41#5187 • ID 01868

CARRUCCIO, E. [1963] *Equazioni logiche nel calcolo delle proposzioni* (**J** 4408) Boll Unione Mat Ital, III Ser 18∗44-56
⋄ B05 ⋄ REV MR 27#30 Zbl 122.8 • ID 48083

CARRUCCIO, E. see Vol. VI for further entries

CARSON, D.F. & ROBINSON, G.A. & WOS, L. [1964] *The unit preference strategy in theorem proving* (**P** 1188) AFIPS Fall Jt Computer Conf (26);1964 San Francisco 26∗615-623
• REPR [1983] (**C** 4659) Autom of Reasoning 1∗387-396
⋄ B35 ⋄ REV Zbl 135.183 JSL 32.117 • ID 21986

CARSON, D.F. & ROBINSON, G.A. & WOS, L. [1965] *Efficiency and completeness of the set of support strategy in theorem proving* (**J** 0037) ACM J 12∗536-541
• REPR [1983] (**C** 4659) Autom of Reasoning 1∗484-492
⋄ B35 ⋄ REV MR 33#3488 Zbl 135.184 JSL 32.117
• ID 11284

CARSON, D.F. & ROBINSON, G.A. & WOS, L. [1966] *Automatic generation of proofs in the language of mathematics* (**P** 0573) Inform Processing (3);1965 New York 2∗325-326
⋄ B35 ⋄ ID 29438

CARSON, D.F. & ROBINSON, G.A. & SHALLA, L. & WOS, L. [1967] *The concept of demodulation in theorem proving* (**J** 0037) ACM J 14∗698-709
• REPR [1983] (**C** 4659) Autom of Reasoning 2∗66-84
⋄ B35 ⋄ REV Zbl 157.25 • ID 46546

CARTER, W.C. & RETTIG, A.S. [1953] *Analytic minimization methods I. Conjunctive forms* (**J** 0093) J Comp Syst 1∗179-195
⋄ B35 ⋄ REV MR 15.91 Zbl 53.2 JSL 19.232 • ID 01875

CARTIER, P. [1982] *Perturbations singulieres des equations differentielles ordinaires et analyse non-standard* (**S** 1567) Semin Bourbaki Exp.580∗21-44
• TRANSL [1984] (**J** 0067) Usp Mat Nauk 39/2(236)∗57-76
⋄ H10 ⋄ REV MR 84i:34072 MR 85j:34114 Zbl 508#03004 • REM Soc Math France: Paris; Asterisque 92-93 • ID 36931

CARTIER, P. see Vol. V, VI for further entries

CARTON, M. [1978] *La logique informationnelle recursive* (**S** 3216) Mem Publ Soc Sci Arts Lett Hainaut 89∗1-36
⋄ B35 B75 ⋄ REV Zbl 438#68039 • ID 55985

CARTWRIGHT, ROBERT & MCCARTHY, J. [1979] *Recursive programs as functions in a first order theory* (**P** 3479) Math Stud of Inform Process;1978 Kyoto 576-629
⋄ B10 B75 D20 ⋄ REV Zbl 407#68042 • ID 56231

CARTWRIGHT, ROBERT see Vol. VI for further entries

CARVALHO DE, R.L. [1982] see BERRETA PION, M.M.

CARVALHO DE, R.L. see Vol. II for further entries

CARVALLO, M. [1966] *Equations de Boole* (**J** 0186) Rev Franc Trait Info Chiffres 9∗109-118
⋄ B05 ⋄ REV MR 34#8884 Zbl 202.319 • ID 01877

CARVALLO, M. [1967] *Sur la resolution des equations de Post* (**J** 2313) C R Acad Sci, Paris, Ser A-B 265∗A601-A602
⋄ B05 ⋄ REV MR 37#36 Zbl 169.304 • ID 01878

CARVALLO, M. [1973] *Fonctions transitives generalisees* (**J** 2313) C R Acad Sci, Paris, Ser A-B 277∗A1023
⋄ B05 ⋄ REV MR 48#8314 Zbl 276#02043 • ID 29044

CARVALLO, M. [1973] *Tautologies a operateur non specifie* (**J** 2313) C R Acad Sci, Paris, Ser A-B 276∗A829-A830
⋄ B05 ⋄ REV MR 48#8329 Zbl 385#06019 • ID 01882

CARVALLO, M. [1977] *Periodicite de fonctions logiques recurrentes (English summary)* (**J** 2313) C R Acad Sci, Paris, Ser A-B 284∗A13-A16
⋄ B05 ⋄ REV MR 55#2373 Zbl 347#94025 • ID 60841

CARVALLO, M. [1985] *Sur la minimisation d'une expression representant une famille de fonctions booleennes (English summary)* (**J** 3441) RAIRO Inform Theor 19∗331-336
⋄ B35 ⋄ ID 49590

CARVALLO, M. see Vol. II, V for further entries

CASARI, E. [1960] *Lineamenti di logica matematica* (**X** 0844) Feltrinelli: Milano 324pp
⋄ B98 ⋄ REV Zbl 115.5 JSL 27.76 • ID 22395

CASARI, E. [1981] *Positively omitting types* (**C** 3515) Ital Studies in Phil of Sci 3-11
⋄ B10 C07 ⋄ REV MR 82i:03017 Zbl 449#03008
• ID 56675

CASARI, E. see Vol. II, III, V for further entries

CASE, J. & SMITH, C.H. [1978] *Anomaly hierarchies of mechanized inductive inference* (**P** 1740) ACM Symp Th of Comput (10);1978 San Diego 314-319
⋄ B35 D20 ⋄ REV MR 80d:68047 • ID 80855

CASE, J. & SMITH, C.H. [1983] *Comparison of identification criteria for machine inductive inference* (**J** 1426) Theor Comput Sci 25∗193-220
⋄ B35 B48 D20 ⋄ REV MR 84j:68029 Zbl 524#03025
• ID 37600

CASE, J. see Vol. IV for further entries

CASEY, H. & CLARK, M. [1970] *Logic: a practical approach* (**X** 1340) Regnery: Chicago 314pp
⋄ B98 ⋄ ID 22476

CASSEN, C.E. [1971] *Russell's distinction between the primary and secondary occurrence of definite descriptions* (**J** 0094) Mind 80∗620-622
⋄ A05 A10 B15 ⋄ REV MR 57#12168 • ID 71568

CASTANEDA, H.-N. [1976] *Ontology and grammar. I: Russell's paradox and the general theory of properties in natural language* (**J** 0105) Theoria (Lund) 42∗44-92
⋄ A05 B15 ⋄ REV MR 58#27208 Zbl 411#03007
• ID 52862

CASTANEDA, H.-N. see Vol. II for further entries

CAVADIA, I.C. [1971] *An algorithm for the formal calculus of boolean expressions* (J 3954) Rev Franc Inf & Rech Operat 5/B-3*65-86
⋄ B05 ⋄ REV Zbl 226 # 02004 • REM Part I. Part II 1973 • ID 48885

CAVADIA, I.C. [1973] *An algorithm for the formal calculus of boolean expressions. II* (J 0205) Rev Franc Autom, Inf & Rech Operat 7/B-3*63-84
⋄ B05 G05 ⋄ REV Zbl 267 # 68023 • REM Part I 1971 • ID 29908

CAVAILLES, J. [1938] *Methode axiomatique et formalisme. Essai sur le probleme du fondement des mathematiques* (X 0859) Hermann: Paris 196pp
⋄ A05 B30 ⋄ REV Zbl 21.289 Zbl 549 # 03001 JSL 4.32 FdM 64.26 FdM 64.930 • ID 25524

CAVAILLES, J. see Vol. V for further entries

CAVINESS, B.F. [1970] *On canonical forms and simplification* (J 0037) ACM J 17*385-396
⋄ B35 D35 ⋄ REV MR 43 # 7104 Zbl 193.313 • ID 03971

CAVINESS, B.F. see Vol. IV, VI for further entries

CAZANESCU, V.E. [1971] *Remarques sur les theories deductives* (C 0640) Log, Autom, Inform 23-29
⋄ B30 ⋄ REV MR 49 # 2254 Zbl 231 # 02073 • ID 03973

CAZANESCU, V.E. see Vol. IV for further entries

CEGIELSKI, P. [1981] *La theorie elementaire de la multiplication est consequence d'un nombre fini d'axiomes de $I\Sigma_0$* (J 3364) C R Acad Sci, Paris, Ser 1 293*351-352
⋄ B28 ⋄ REV MR 82m:03067 Zbl 474 # 03007 • ID 55411

CEGIELSKI, P. [1984] *La theorie elementaire de la divisibilite est finiment axiomatisable (English summary)* (J 3364) C R Acad Sci, Paris, Ser 1 299*367-369
⋄ B25 B28 F30 ⋄ REV MR 85j:03042 • ID 46773

CEGIELSKI, P. see Vol. III, IV, VI for further entries

CERNY, E. [1976] *Comments on "Equational logic"* (J 0187) IEEE Trans Comp C-25*102-103
⋄ B05 ⋄ REV Zbl 328 # 02009 • ID 60891

CERNY, E. [1977] *Unique and identity solutions of boolean equations* (J 2701) Digit Processes 3*331-337
⋄ B05 ⋄ REV MR 57 # 11939 Zbl 389 # 94013 • ID 69304

CERNY, E. see Vol. II for further entries

CERRUTI, U. [1983] *"Nonstandard" concepts in fuzzy topology* (C 3582) Adv Fuzzy Sets, Possibility Th & Appl 49-57
⋄ E72 E75 H10 ⋄ REV MR 85i:54002 • ID 44370

CERRUTI, U. see Vol. V for further entries

CERUTTI, E. & DAVIS, P.J. [1969] *Formac meets Pappus: Some observations on elementary analytic geometry by computer* (J 0005) Amer Math Mon 76*895-905
⋄ B35 ⋄ REV MR 41 # 1259 Zbl 184.242 • ID 01917

CHADADZE, O.S. & ROGAVA, M.G. [1974] *Sequential variants of predicate calculus with equality (Russian) (Georgian and English summaries)* (J 0233) Soobshch Akad Nauk Gruz SSR 76*537-539
⋄ B10 ⋄ REV MR 52 # 7843 Zbl 304 # 02013 • ID 18360

CHADWICK, J.J.M. & CROSS, R.W. [1972] *A characterization of pre-near-standardness in locally convex linear topological spaces* (J 0016) Bull Austral Math Soc 6*107-115
⋄ H05 ⋄ REV MR 45 # 5699 Zbl 233 # 46001 • ID 01920

CHADWICK, J.J.M. & WICKSTEAD, A.W. [1977] *A quotient of ultrapowers of Banach spaces and semi-Fredholm operators* (J 0161) Bull London Math Soc 9*321-325
⋄ C20 C65 H05 ⋄ REV MR 57 # 1080 Zbl 388 # 46028 • ID 80867

CHADWICK, J.J.M. [1977] *Standard biorthogonal systems in the enlargement of a normed space* (J 2823) Quaest Math, S Africa 2*433-438
⋄ H05 ⋄ REV MR 58 # 2178 Zbl 383 # 46013 • ID 80865

CHAKRAPANI, N. & RANGASWAMY, S.V. & TIKEKAR, V. [1979] *Development of a structured program for conversion to prenex normal form* (J 2755) J Indian Inst Sci 61/5*123-133
⋄ B35 ⋄ REV Zbl 419 # 68085 • ID 53398

CHAKRAPANI, N. & RANGASWAMY, S.V. & TIKEKAR, V. [1980] *A computational algorithm for the verification of tautologies in propositional calculus* (J 2755) J Indian Inst Sci 62/3*71-81
⋄ B35 ⋄ REV MR 81i:68131 Zbl 446 # 03007 • ID 56536

CHAMPEAUX DE, D. [1981] *Other directions for automatic theorem proving* (P 3642) Colloq Math Log in Computer Sci;1978 Salgotarjan 259-273
⋄ B35 ⋄ REV MR 83g:68007 • ID 46179

CHAN, S.P. & PICHAI, V. [1976] *A minimal reduction of cyclic prime implicant tables by a graph theoretic method* (P 2975) Asilomar Conf Circ, Syst & Comput (9);1975 Pacific Grove 31-35
⋄ B35 ⋄ REV MR 58 # 33574 Zbl 435 # 94036 • ID 69771

CHANDRA, A.K. & MARKOWSKY, G. [1978] *On the number of prime implicants* (J 0193) Discr Math 24*7-11
⋄ B05 ⋄ REV MR 82m:94055 Zbl 392 # 03038 • ID 52405

CHANDRA, A.K. & STOCKMEYER, L.J. & VISHKIN, U. [1984] *Constant depth reducibility* (J 1428) SIAM J Comp 13*423-439
⋄ B05 ⋄ REV MR 85g:68019 Zbl 538 # 68038 • ID 41501

CHANDRA, A.K. see Vol. II, IV for further entries

CHANG, C.C. & KEISLER, H.J. [1962] *An improved prenex normal form* (J 0036) J Symb Logic 27*317-326
⋄ B10 ⋄ REV MR 33 # 5463 Zbl 115.6 JSL 33.479 • ID 01958

CHANG, C.C. [1974] see ADDISON, J.W.

CHANG, C.C. see Vol. II, III, IV, V for further entries

CHANG, C.L. & LEE, R.C.T. & SLAGLE, J.R. [1969] *Completeness theorems for semantic resolution in consequence finding* (P 4250) Int Joint Conf Artif Intell (1);1969 Washington 281-285
⋄ B35 ⋄ ID 46539

CHANG, C.L. & LEE, R.C.T. & SLAGLE, J.R. [1970] *A new algorithm for generating prime implicants* (J 0187) IEEE Trans Comp C-19*304-310
⋄ B35 ⋄ REV MR 56 # 15213 Zbl 197.146 • ID 47295

CHANG, C.L. [1970] *Renamable paramodulation for automatic theorem proving with equality* (J 0503) Artif Intell 1*247-256
⋄ B35 ⋄ REV MR 46 # 10251 Zbl 207.22 • ID 80881

CHANG, C.L. [1970] *The unit and the input proof in theorem proving* (J 0037) ACM J 17*698-707
• REPR [1983] (C 4659) Autom of Reasoning 2*331-341
⋄ B35 ⋄ REV MR 44 # 7794 Zbl 212.342 • ID 47294

CHANG, C.L. & SLAGLE, J.R. [1971] *Completeness of linear refutation for theories with equality* (**J** 0037) ACM J 18*126-136
- ◊ B35 ◊ REV MR 46#4804   Zbl 239#68016 • ID 60906

CHANG, C.L. & LEE, R.C.T. [1973] *Symbolic logic and mechanical theorem proving* (**X** 0801) Academic Pr: New York xiii+331pp
- • TRANSL [1983] (**X** 2027) Nauka: Moskva 358pp
- ◊ B35 B98 ◊ REV MR 55#13894   MR 84k:03003   Zbl 263#68046   Zbl 518#03004 • ID 03980

CHANG, C.L. & SLAGLE, J.R. [1979] *Using rewriting rules for connection graphs to prove theorems* (**J** 0503) Artif Intell 12*159-178
- ◊ B35 ◊ REV MR 80j:68077   Zbl 429#68078 • ID 80882

CHANG, C.L. see Vol. II for further entries

CHANG, D.M.Y. & MOTT JR., T.H. [1963] *Computing irredundant normal forms from abbreviated presence functions* (**P** 0572) Inform Processing (2);1962 Muenchen 753-754
- ◊ B35 ◊ REV Zbl 142.129   JSL 33.633 • ID 27615

CHANG, D.M.Y. & MOTT JR., T.H. [1965] *Computing irredundant normal forms from abbreviated presence functions* (**J** 4305) IEEE Trans Electr Comp EC-14*335-342
- ◊ B35 ◊ REV Zbl 192.84   JSL 32.541 • ID 01966

CHANKVETADZE, O.E. [1975] *On some questions of the theory Ω_4 (Russian) (English summary)* (**J** 3112) Annot Dokl, Inst Prikl Mat, Tbilisi 10*19-21
- ◊ B10 ◊ REV MR 56#11743a   Zbl 402#03022 • ID 54669

CHANKVETADZE, O.E. [1975] *On some questions of the theory Ω_5 (Russian) (English summary)* (**J** 3112) Annot Dokl, Inst Prikl Mat, Tbilisi 10*23-25
- ◊ B10 ◊ REV MR 56#11743b   Zbl 402#03023 • ID 54670

CHANKVETADZE, O.E. [1976] *On the axiomatics of the Ω-calculus (Russian) (Georgian and English summaries)* (**J** 0233) Soobshch Akad Nauk Gruz SSR 81*37-39
- ◊ B10 ◊ REV MR 53#12862   Zbl 334#02006 • ID 23125

CHANKVETADZE, O.E. [1977] *Comparatively simple proofs of some criteria of N. Bourbaki (Russian) (Georgian and English summaries)* (**S** 2043) Issl Mat Log & Teor Algor (Tbilisi) 2*35-38
- ◊ B10 ◊ REV MR 57#2870 • ID 71638

CHANKVETADZE, O.E. [1978] *Some fundamental questions concerning the theory Ω_6 (Russian) (Georgian and English Summary)* (**S** 2043) Issl Mat Log & Teor Algor (Tbilisi) 1978*45-48
- ◊ B10 ◊ REV MR 80i:03019 • ID 71637

CHANKVETADZE, O.E. [1980] *On some criteria of N. Bourbaki's quantified theory and the corresponding criteria of the Ω-calculus (Russian) (English and Georgian summaries)* (**J** 0233) Soobshch Akad Nauk Gruz SSR 100*541-544
- ◊ B10 ◊ REV MR 84c:03025   MR 84i:03032   Zbl 469#03016 • ID 55144

CHANKVETADZE, O.E. [1981] *Invariance of homogeneity with respect to homogeneous substitutions in N.Bourbaki's τ-calculus (Russian) (Georgian and English summaries)* (**J** 0233) Soobshch Akad Nauk Gruz SSR 102*305-307
- ◊ B10 ◊ REV MR 83e:03023   Zbl 487#03023 • ID 35212

CHANKVETADZE, O.E. [1982] *The invariance of uniformity with respect to uniform substitutions in N. Bourbaki's τ-calculus (Russian) (English and Georgian summaries)* (**J** 0233) Soobshch Akad Nauk Gruz SSR 105*25-27
- ◊ B10 ◊ REV MR 83j:03019   Zbl 497#03005 • ID 35332

CHANKVETADZE, O.E. [1982] *The law of implication self-distributivity and the criterion for deduction in Bourbaki's τ-calculus (Russian) (English and Georgian summaries)* (**J** 0954) Tr Inst Prikl Mat, Tbilisi 11*120-122
- ◊ B10 ◊ REV MR 85h:03011 • ID 43285

CHANKVETADZE, O.E. [1982] *The rule of multiplication of implication and its application in τ-calculus (Russian) (English and Georgian summaries)* (**J** 0233) Soobshch Akad Nauk Gruz SSR 107*29-31
- ◊ B10 ◊ REV MR 84i:03032   Zbl 505#03006 • ID 34516

CHANKVETADZE, O.E. see Vol. II, V for further entries

CHAPIN JR., E.W. [1971] *Gentzen-like systems for partial propositional calculi I,II* (**J** 0047) Notre Dame J Formal Log 12*75-80,179-182
- ◊ B20 ◊ REV MR 45#1722   MR 45#4941   Zbl 177.7   Zbl 214.7   JSL 40.469 • ID 01988

CHAPIN JR., E.W. [1971] *On axioms and their corresponding deduction rules: A survey* (**J** 0079) Logique & Anal, NS 14*707-714
- ◊ B22 ◊ REV MR 55#5429   Zbl 234#02003 • ID 60911

CHAPIN JR., E.W. [1971] *The strong decidability of cut-logics I: Partial propositional calculi. II: Generalizations* (**J** 0047) Notre Dame J Formal Log 12*322-328,429-434
- ◊ B20 B25 ◊ REV MR 45#6571   MR 45#6572   Zbl 188.28   Zbl 214.18 • ID 01987

CHAPIN JR., E.W. [1972] *Measures of centrality and complexity for partial propositional calculi* (**J** 0009) Arch Math Logik Grundlagenforsch 15*7-18
- ◊ B20 ◊ REV MR 51#5257   Zbl 248#02018 • ID 01989

CHAPIN JR., E.W. & WEBB, S.M. [1973] *A non-standard proof in the theory of integration* (**J** 0047) Notre Dame J Formal Log 14*125-128
- ◊ H05 ◊ REV MR 47#8787   Zbl 232#02041 • ID 29510

CHAPIN JR., E.W. [1974] *Translations and structures for partial propositional calculi* (**J** 0063) Studia Logica 33*35-57
- ◊ B20 ◊ REV MR 51#7850   Zbl 293#02008 • ID 17993

CHAPIN JR., E.W. see Vol. V for further entries

CHAPMAN, F.M. & HENLE, P. [1933] *The fundamentals of logic* (**X** 1347) Scribner: New York xiii+384pp
- ◊ B98 ◊ REV FdM 59.65 • ID 22578

CHARRETTON, C. & RICHARD, D. [1973] *Elements d'une theorie non standard des groupes topologiques* (**J** 0056) Publ Dep Math, Lyon 9/2*1-29
- ◊ C60 H05 ◊ REV MR 50#2385   Zbl 281#22004 • ID 03982

CHARRETTON, C. & RICHARD, D. [1974] *Preuves non standards de resultats classiques des groupes topologiques et quelques resultats non standard* (**J** 2313) C R Acad Sci, Paris, Ser A-B 278*A55-A58
- ◊ H05 ◊ REV MR 50#13356   Zbl 274#22001 • ID 29033

CHARRETTON, C. & RICHARD, D. [1975] *Theoreme d'Ascoli et application aux groupes topologiques localement compacts en analyse non standard* (**J** 0056) Publ Dep Math, Lyon 12/2∗47-55
⋄ H05 ⋄ REV MR 52#9161 Zbl 318#54051 • ID 17994

CHARRETTON, C. see Vol. III, V for further entries

CHAUVIN, A. [1949] *Generalisation du theoreme de Goedel* (**J** 0109) C R Acad Sci, Paris 228∗1179-1180
⋄ B15 F30 ⋄ REV MR 10.668 Zbl 34.293 JSL 14.193 • ID 01996

CHAUVIN, A. see Vol. II, III, IV, V, VI for further entries

CHAUVINEAU, J. [1957] *La logique moderne* (**X** 0840) Pr Univ France: Paris 128pp
⋄ B98 ⋄ REV MR 19.1 JSL 24.70 • ID 22396

CHAVCHANIDZE, V.V. [1964] *Fundamental relations of the analytic theory of propositional algebra (Russian)* (**J** 0233) Soobshch Akad Nauk Gruz SSR 33∗27-34
⋄ B05 ⋄ REV MR 28#2961 • ID 02213

CHAVCHANIDZE, V.V. see Vol. II for further entries

CHEIFETZ, P.M. [1970] see AVENOSO, F.J.

CHELISHCHEV, B.E. [1979] see BELYANIN, P.N.

CHELLAS, B.F. [1969] *The logical form of imperatives* (**X** 1355) Stanford Univ Pr: Stanford 121pp
• LAST ED (**X** 2045) Perry Lane Pr: Stanford v+115pp
⋄ A05 B98 ⋄ ID 25267

CHELLAS, B.F. [1975] *Quantity and quantification* (**J** 0154) Synthese 31∗487-491
⋄ A05 B15 ⋄ REV Zbl 319#02009 • ID 31664

CHELLAS, B.F. [1980] *Modal logic. An introduction* (**X** 0805) Cambridge Univ Pr: Cambridge, GB xii+295pp
⋄ B45 B98 ⋄ REV MR 81i:03019 Zbl 431#03009 JSL 46.670 • ID 53915

CHELLAS, B.F. see Vol. II for further entries

CHEN, GUANGYI [1985] *Toeplitz generalized summation on the field of hyperreal numbers (Chinese) (English summary)* (**J** 2985) Dongbei Gongxueyuan Xuebao 3∗33-38
⋄ H05 ⋄ ID 49521

CHEN, JIYUAN [1984] *The satisfiability problem for simple boolean expressions belongs to P (Chinese) (English summary)* (**J** 2521) Beijing Shifan Daxue Xuebao, Ziran Kexue 1984/1∗14-21
⋄ B05 B25 D15 ⋄ REV MR 85j:03059 • ID 44191

CHENADEC LE, P. [1984] *Canonical forms in finitely presented algebras* (**P** 2633) Autom Deduct (7);1984 Napa 142-165
⋄ B35 C05 ⋄ REV Zbl 547#03028 • ID 43216

CHENG, CHUNGYING [1968] *On explanation of number progression* (**J** 0047) Notre Dame J Formal Log 9∗329-334
⋄ A05 B28 ⋄ REV Zbl 177.6 • ID 02070

CHENIQUE, F. [1974] *Comprendre la logique moderne. Tome I: Classes, propositions et predicats* (**X** 0856) Dunod: Paris xx+294+6pp
⋄ B98 E98 ⋄ REV MR 53#10539 Zbl 276#02001 • REM Vol. II 1974 • ID 23053

CHENIQUE, F. [1974] *Comprendre la logique moderne. Tome II: Logiques non classiques, relations et structures* (**X** 0856) Dunod: Paris xx+(295-480)+6pp
⋄ B98 ⋄ REV MR 53#10540 Zbl 276#02002 • REM Vol. I 1974 • ID 23054

CHEPURNOV, B.A. & KOGALOVSKIJ, S.R. [1976] *Some criteria of heredity and locality for formulas of higher order (Russian)* (**C** 3271) Issl Teor Mnozh & Neklass Logik 127-156
⋄ B15 C52 C85 ⋄ REV MR 58#21423 Zbl 412#03018 • ID 52949

CHEPURNOV, B.A. see Vol. III for further entries

CHEREDNICHENKO, N.D. & KOROTKOVA, M.A. & POPOV, S.V. [1977] *A program for finding proofs (Russian) (English summary)* (**X** 2512) Akad Nauk SSSR Inst Mat: Moskva 54∗36pp
⋄ B35 ⋄ REV MR 58#32127 • ID 81916

CHERLIN, G.L. & HIRSCHFELD, J. [1972] *Ultrafilters and ultraproducts in non-standard analysis* (**P** 0732) Contrib to Non-Standard Anal;1970 Oberwolfach 261-279
⋄ C20 E05 H05 ⋄ REV MR 58#5191 Zbl 262#02055 JSL 40.634 • ID 14684

CHERLIN, G.L. see Vol. III, IV, V for further entries

CHERNYAEV, V.G. [1968] *Ueber einige Arten der Dekomposition unvollstaendig definierter boolescher Funktionen (Russisch)* (**C** 3263) Sint Diskr Avtom Upravl Ustr 82-97
⋄ B05 ⋄ REV Zbl 275#94028 • ID 60886

CHERNYAVSKIJ, A.L. [1967] *Computer simulation of the process of solving complex logical problems (heuristic programming) (Russian)* (**J** 0011) Avtom Telemekh 1967/1∗166-187
• TRANSL [1967] (**J** 0010) Autom & Remote Control 28∗145-167
⋄ B35 ⋄ REV Zbl 173.194 JSL 33.303 • ID 01916

CHERNYSHEV, YU.O. [1977] *Optimization methods for combinatorial devices (Russian)* (**X** 2643) Sovet Radio: Moskva 158pp
⋄ B35 ⋄ REV MR 58#9306 Zbl 384#94022 • ID 69302

CHEVALLARD, Y. [1971] *Etude de compacite sur les calculs propositionnels incomplets* (**J** 2313) C R Acad Sci, Paris, Ser A-B 272∗A1033-A1034
⋄ B20 E25 ⋄ REV MR 43#4661 Zbl 212.12 • ID 02082

CHIEN, R. [1971] see BLISS, K.

CHIHARA, C.S. [1984] *A simple type theory without Platonic domains* (**J** 0122) J Philos Logic 13∗249-283
⋄ B15 ⋄ REV MR 86b:03015 Zbl 543#03002 • ID 40896

CHIHARA, C.S. see Vol. IV, V, VI for further entries

CHIKAWA, K. [1967] *On equivalences of laws in elementary protothetics. I* (**J** 0081) Proc Japan Acad 43∗743-747
⋄ B15 ⋄ REV MR 36#4960 Zbl 197.3 • REM Part II 1968 • ID 16325

CHIKAWA, K. [1968] *On equivalences of laws in elementary protothetics. II* (**J** 0081) Proc Japan Acad 44∗56-59
⋄ B15 ⋄ REV MR 37#2576 Zbl 197.3 • REM Part I 1967 • ID 16326

CHIKHACHEV, S.A. [1984] *An example of a theory without weakly (Σ, Σ)-atomic models (Russian)* (**J** 0003) Algebra i Logika 23∗336-340
• TRANSL [1984] (**J** 0069) Algeb and Log 23∗233-237
⋄ B20 C50 ⋄ REV MR 86h:03049 • ID 42695

CHIKHACHEV, S.A. see Vol. III for further entries

CHILAUSKY, R. & JACOBSEN, B. & MICHALSKI, R.S. [1976] *An application of variable-valued logic to inductive learning of plant disease diagnostic rules* (P 2011) Int Symp Multi-Val Log (6);1976 Logan 233-240
⋄ B80 ⋄ ID 35893

CHIMEV, K.N. [1964] *Some functions of P_2 (Bulgarian)(Russian summary)* (J 0224) God Vissh Tekh Ucheb Zaved Mat, Sofiya 1/2*27-35
⋄ B20 ⋄ REV MR 37 # 1226 Zbl 311 # 02019 • ID 02185

CHIMEV, K.N. [1972] *Separable pairs and strongly essential variables of functions of three, four and five arguments (Bulgarian)(Russian and French summaries)* (J 0224) God Vissh Tekh Ucheb Zaved Mat, Sofiya 8/3*31-39
⋄ B05 ⋄ REV MR 54 # 4917 Zbl 314 # 02014 • ID 24093

CHIMEV, K.N. & KUTIRKOV, G.A. [1972] *The minimal number of separable pairs of functions of the algebra of logic that depend essentially on not more than fifteen variables (Bulgarian)(Russian and French summaries)* (J 0224) God Vissh Tekh Ucheb Zaved Mat, Sofiya 8/4*155-162
⋄ B05 G15 ⋄ REV MR 54 # 55 Zbl 314 # 02015 • ID 24544

CHIMEV, K.N. [1976] *Separable pairs of functions of six arguments (Bulgarian) (Russian and English summaries)* (J 3171) God Vissh Ucheb Zaved, Prilozhna Mat, Sofiya 12/2*143-156
⋄ B05 ⋄ REV MR 80a:03072 Zbl 417 # 05060 • ID 71717

CHIMEV, K.N. [1977] *On strongly essential variables of functions of six arguments (Bulgarian)(Russian and English summaries)* (J 3171) God Vissh Ucheb Zaved, Prilozhna Mat, Sofiya 13/3*31-42
⋄ B05 ⋄ REV MR 81b:94076 Zbl 435 # 05055 • ID 80904

CHIMEV, K.N. [1981] *On some properties of functions* (P 2552) Conf Finite Algeb & Multi-Val Log;1979 Szeged 97-110
⋄ B05 E20 ⋄ REV MR 83j:08003 Zbl 481 # 04003 • ID 36837

CHIMEV, K.N. [1984] see ASLANSKI, M.

CHIMEV, K.N. see Vol. II, IV, V for further entries

CHIRKOV, M.K. & EGLITIS, L.V. [1975] *Generalized partial functions of the logic algebra and their minimization (Russian)* (S 0716) Vychisl Tekh Vopr Kibern (Univ Leningrad) 12*114-124
⋄ B35 ⋄ REV MR 57 # 11941 Zbl 448 # 94005 • ID 69321

CHIRKOV, M.K. & EGLITIS, L.V. [1977] *The minimization of systems of generalized partial functions of the logic algebra (Russian)* (S 0716) Vychisl Tekh Vopr Kibern (Univ Leningrad) 13*113-124
⋄ B35 ⋄ REV Zbl 434 # 94024 • ID 69318

CHIRKOV, M.K. [1978] *Simple implicators of a system of generalized partial functions (Russian)* (S 0716) Vychisl Tekh Vopr Kibern (Univ Leningrad) 15*96-103
⋄ B05 ⋄ REV MR 80g:94083 Zbl 531 # 94021 • ID 80908

CHIRKOV, M.K. see Vol. II, IV for further entries

CHISHOLM, R.M. & SYMONDS, B.K. [1957] *Inference by complementary elimination* (J 0036) J Symb Logic 22*233-236
⋄ B05 ⋄ REV Zbl 79.7 • ID 02088

CHISHOLM, R.M. see Vol. II for further entries

CHLEBUS, B.S. [1980] *Decidability and definability results concerning well-orderings and some extensions of first order logic* (J 0068) Z Math Logik Grundlagen Math 26*529-536
⋄ B15 B25 C40 C65 C80 D35 E07 ⋄ REV MR 82b:03072 Zbl 445 # 03018 • ID 56482

CHLEBUS, B.S. see Vol. II, III, IV for further entries

CHOMSKY, N. [1957] *Syntactic structures* (X 0873) Mouton: Paris 116pp
⋄ B03 ⋄ REV JSL 31.245 • ID 03590

CHOMSKY, N. [1965] *Aspects of the theory of syntax* (X 0865) MIT Pr: Cambridge, MA x+251pp
⋄ B98 ⋄ REV JSL 32.385 JSL 35.167 • ID 02429

CHOMSKY, N. [1968] *Language and mind* (X 0863) Harcourt: New York & London vii+88pp
⋄ A05 B65 B98 ⋄ ID 25312

CHOMSKY, N. [1972] *Studies on semantics in generative grammars* (X 0873) Mouton: Paris 207pp
⋄ B65 B98 D05 ⋄ ID 25314

CHOMSKY, N. see Vol. IV for further entries

CHONG, C.T. [1978] *Non-standard analysis* (J 1735) Bull South East Asian Soc 2*17-25
⋄ H05 ⋄ REV MR 83d:03070 Zbl 433 # 03043 • ID 31059

CHONG, C.T. & WICKS, M.J. (EDS.) [1983] *Southeast Asian conference on logic* (P 3669) SE Asian Conf on Log;1981 Singapore xiv+210pp
⋄ B97 ⋄ REV MR 84i:03004 Zbl 532 # 00005 • ID 34490

CHONG, C.T. see Vol. IV, V for further entries

CHOU, SHANGCHING [1984] *Proving elementary geometry theorems using Wu's algorithm* (P 3084) Autom Theor Prov After 25 Yea;1983 Denver 243-286
⋄ B35 ⋄ REV MR 85d:68005 • ID 45266

CHOUDHURY, A.K. [1961] see BASU, M.S.

CHOUDHURY, A.K. & DAS, S.R. [1966] *Computing irredundant normal forms from abbreviated presence functions* (J 4305) IEEE Trans Electr Comp EC-15*387
⋄ B35 ⋄ ID 02106

CHOW, C.K. [1961] *Boolean functions realizable with single threshold devices* (J 4711) IRE Proc 49*370-371
⋄ B35 ⋄ REV MR 23 # B1362 • ID 20832

CHRISTEN, C. [1976] *Spektralproblem und Komplexitaetstheorie* (P 3196) Kompl von Entscheid Probl;1973/74 Zuerich 102-126
⋄ B10 C13 D10 D15 D25 D35 ⋄ REV MR 57 # 18232 Zbl 391 # 03021 • ID 52348

CHRISTIAN, C.C. [1965] *Die Interpretation logischer Formen* (J 0238) Sitzb Oesterr Akad Wiss, Math-Nat Kl, Abt 2 174*201-233
⋄ B10 ⋄ REV MR 36 # 2480 Zbl 158.249 • ID 02223

CHRISTIAN, C.C. [1972] *Konsistenzkriterien fuer formale Theorien und ihre Anwendung auf Zahlen- und Mengentheorie* (J 0455) Phil Naturalis 13*405-442
⋄ B28 E30 F30 ⋄ REV MR 56 # 5263 • ID 71771

CHRISTIAN, C.C. [1976] *Die Bedeutung der Mengentheorie als Grundlagenwissenschaft* (J 0455) Phil Naturalis 16*238-271
⋄ A05 B98 E10 E30 ⋄ REV MR 56 # 5292 • ID 71770

CHRISTIAN, C.C. [1976] *Peano-Systeme* (J 0045) Monatsh Math 82∗81-116
- ◇ B28 F30 F35 ◇ REV MR 57 # 88 Zbl 344 # 02038
- • ID 60952

CHRISTIAN, C.C. [1983] *Der Beitrag Goedels fuer die Rechtfertigung der Leibnizschen Idee von den Infinitesimalien* (J 0238) Sitzb Oesterr Akad Wiss, Math-Nat Kl, Abt 2 192∗25-44
- ◇ A05 A10 H05 ◇ REV MR 85m:03006 MR 86a:01033
- • ID 48148

CHRISTIAN, C.C. see Vol. II, IV, V, VI for further entries

CHRISTIAN, R.R. [1958] *Introduction to logic and sets* (X 0943) Ginn: Boston 69pp
- • LAST ED [1965] (X 0841) Blaisdell: New York xii+116pp
- ◇ B98 E98 ◇ REV Zbl 156.4 JSL 33.631 • ID 22478

CHU, J.T. [1961] *A generalization of a theorem of Quine for simplifying truth functions* (J 0072) IRE Trans Electr Comp EC-10∗165-168
- ◇ B05 ◇ REV MR 25 # 3811 • ID 02115

CHU, J.T. see Vol. V for further entries

CHUAQUI, R.B. [1976] *Introduccion a la metamatematica y sus aplicaciones* (J 3685) Publ Inst Mat Univ Catolica Chile 83pp
- ◇ B98 ◇ ID 33786

CHUAQUI, R.B. [1977] *A model theoretical definition of probability* (P 1704) Int Congr Log, Meth & Phil of Sci (5);1975 London ON 6∗7-8
- ◇ B30 ◇ ID 33957

CHUAQUI, R.B. & MALITZ, J. [1983] *Preorderings compatible with probability measures* (J 0064) Trans Amer Math Soc 279∗811-824
- ◇ E75 H10 ◇ REV MR 85d:60010 Zbl 527 # 60001
- • ID 38558

CHUAQUI, R.B. [1985] see BERTOSSI, L.

CHUAQUI, R.B. see Vol. II, III, V for further entries

CHUBARYAN, A.A. & TSEJTIN, G.S. [1972] *Certain estimates of the length of logical deductions in classical propositional calculus (Russian) (Armenian summary)* (J 0346) Dokl Akad Nauk Armyan SSR 55∗10-12
- ◇ B05 F07 F20 ◇ REV MR 50 # 12639 Zbl 246 # 02011
- • ID 17157

CHUBARYAN, A.A. [1975] *A certain normal form, and a complexity characteristics of deduction in the classical propositional calculus (Russian) (Armenian and English summaries)* (J 0312) Izv Akad Nauk Armyan SSR, Ser Mat 10∗398-409,479
- ◇ B05 ◇ REV MR 53 # 12863 Zbl 364 # 02005 • ID 23126

CHUBARYAN, A.A. & TSEJTIN, G.S. [1975] *On some estimates of the length of derivation in classical propositional calculus (Russian)* (S 0422) Tr Vychisl Tsentra Akad Nauk Armyan SSR & Univ Erevan 8∗57-64
- ◇ B05 ◇ REV MR 57 # 9469 Zbl 398 # 03004 • ID 52732

CHUBARYAN, A.A. & NGUEN VAN TIN' [1975] *Some estimates of the complexity characteristics of deductions in classical propositional calculus (Russian)* (J 2892) Molodoj Nauch Rabotnik, Erevan 2(22)∗47-54
- ◇ B05 F20 ◇ REV MR 58 # 130 Zbl 364 # 02005
- • ID 76827

CHUBARYAN, A.A. [1977] *The complexity of deductions in formal arithmetic and predicate calculus (Russian)* (J 0346) Dokl Akad Nauk Armyan SSR 64∗193-196
- ◇ B10 F20 F30 ◇ REV MR 57 # 4613 Zbl 383 # 03039
- • ID 52021

CHUBARYAN, A.A. [1981] *On the complexity of conclusions in different systems of propositional calculus (Russian)* (C 3838) Priklad Mat, Vyp 1 81-89
- ◇ B22 ◇ REV Zbl 489 # 03023 • ID 37208

CHUBARYAN, A.A. [1982] *Complexity characteristics of inferences in systems of predicate calculus and formal arithmetic (Russian) (Armenian summary)* (S 0422) Tr Vychisl Tsentra Akad Nauk Armyan SSR & Univ Erevan 10∗124-139
- ◇ B10 F20 F30 ◇ REV MR 85e:03138 • ID 40752

CHUBARYAN, A.A. see Vol. VI for further entries

CHUDACEK, J. [1977] *Topological problems in alternative set theory* (P 1695) Set Th & Hierarch Th (3);1976 Bierutowice 119-133
- ◇ E70 E75 H05 H20 ◇ REV MR 58 # 16289 Zbl 389 # 03022 • ID 52311

CHUDNOVSKY, D.V. [1969] *Nonstandard analysis and homeomorphisms of B-spaces (Russian)* (J 0023) Dokl Akad Nauk SSSR 185∗772-774
- • TRANSL [1969] (J 0062) Sov Math, Dokl 10∗436-439
- ◇ H05 ◇ REV MR 39 # 5344 Zbl 189.124 • ID 02576

CHUDNOVSKY, D.V. see Vol. II, III, V for further entries

CHUKHROV, I.P. [1981] *Estimates of the number of minimal disjunctive normal forms for a zonal function I (Russian)* (J 0071) Met Diskr Analiz (Novosibirsk) 36∗74-92,95
- ◇ B05 ◇ REV MR 84h:03136 Zbl 483 # 94035 • ID 34322

CHUKHROV, I.P. [1982] *On the number of irredundant disjunctive normal forms (Russian)* (J 0023) Dokl Akad Nauk SSSR 262∗1329-1332
- • TRANSL [1982] (J 0062) Sov Math, Dokl 25∗254-257
- ◇ B05 ◇ REV MR 83d:03013 Zbl 538 # 03006 • ID 35168

CHUKHROV, I.P. [1984] *The number of minimal disjunctive normal forms (Russian)* (J 0023) Dokl Akad Nauk SSSR 276∗1335-1339
- • TRANSL [1984] (J 0062) Sov Math, Dokl 29∗714-718
- ◇ B05 ◇ REV MR 85j:94021 • ID 45491

CHURCH, A. [1928] *On the law of excluded middle* (J 0015) Bull Amer Math Soc 34∗75-78
- ◇ A05 B05 F50 ◇ REV FdM 54.53 • ID 02122

CHURCH, A. [1932] *A set of postulates for the foundation of logic* (J 0120) Ann of Math, Ser 2 33∗346-366
- ◇ B10 B40 D20 ◇ REV Zbl 4.145 JSL 23.23 FdM 58.997
- • REM Part I. Part II 1933 • ID 02123

CHURCH, A. [1933] *A set of postulates for the foundation of logic (second paper)* (J 0120) Ann of Math, Ser 2 34∗839-864
- ◇ B10 B40 D20 ◇ REV Zbl 8.289 JSL 24.94 FdM 59.52
- • REM Part I 1932 • ID 02124

CHURCH, A. [1934] *The Richard paradox* (J 0005) Amer Math Mon 41∗356-361
- ◇ B10 ◇ REV Zbl 9.146 FdM 60.25 • ID 02125

CHURCH, A. [1936] *A bibliography of symbolic logic* (J 0036) J Symb Logic 1∗121-218 • ERR/ADD ibid 3∗178-192
- ◇ A10 B98 ◇ REV Zbl 16.97 FdM 62.1046 • ID 02129

CHURCH, A. [1936] *A note on the "Entscheidungsproblem"*
(J 0036) J Symb Logic 1*40-41 • ERR/ADD ibid 1*101-102
⋄ B20 D20 D35 ⋄ REV Zbl 14.385 JSL 1.74
FdM 62.1058 • ID 02132

CHURCH, A. [1936] *Mathematical logic (mimeographed notes)*
(X 0857) Princeton Univ Pr: Princeton iii+113pp
⋄ B40 B98 D20 ⋄ REV JSL 2.39 FdM 62.1048 • ID 23466

CHURCH, A. [1939] *Schroeder's anticipation of the simple theory of types* (J 0748) Erkenntnis (Leipzig) 9*149-152
⋄ A10 B15 ⋄ REV JSL 5.71 • ID 31037

CHURCH, A. [1940] *A formulation of the simple theory of types*
(J 0036) J Symb Logic 5*56-68
⋄ B15 ⋄ REV MR 1.321 Zbl 23.289 JSL 5.114
FdM 66.1192 • ID 02140

CHURCH, A. [1940] *Elementary topics in mathematical logic*
(X 0915) Galois Inst Math Art: Brooklyn 102pp
⋄ B98 ⋄ REV Zbl 27.148 JSL 7.91 • ID 02460

CHURCH, A. [1944] *Introduction to mathematical logic. Part I*
(X 0857) Princeton Univ Pr: Princeton vi+118pp
• TRANSL [1960] (X 1656) Izdat Inostr Lit: Moskva
⋄ B98 ⋄ REV MR 18.631 MR 6.29 Zbl 60.20 JSL 10.19
JSL 22.286 JSL 23.362 • REM 2nd ed. 1956; x+376pp
• ID 02181

CHURCH, A. [1948] *Conditioned disjunction as a primitive connective for the propositional calculus* (J 0050) Port Math 7*87-90
⋄ B05 ⋄ REV MR 10.421 Zbl 34.291 JSL 14.197
• ID 02142

CHURCH, A. [1951] *The weak theory of implication* (C 0621)
Kontrolliertes Denken (Britzelmayr) 22-37
⋄ B20 ⋄ REV JSL 18.326 • ID 02143

CHURCH, A. [1952] *Brief bibliography of formal logic* (J 0251)
Proc Amer Acad Arts Sci 80*155-172
⋄ A10 B98 ⋄ REV Zbl 48.1 JSL 18.178 • ID 02458

CHURCH, A. & QUINE, W.V.O. [1952] *Some theorems on definability and decidability* (J 0036) J Symb Logic 17*179-187
⋄ B15 D35 ⋄ REV MR 14.233 Zbl 47.9 JSL 18.269
• ID 02144

CHURCH, A. [1972] *Axioms for functional calculi of higher order*
(C 1698) Logic & Art (Goodman) 197-213
⋄ B15 ⋄ REV MR 58#5015 JSL 23.22 • ID 31044

CHURCH, A. [1974] *Russellian simple type theory* (J 1027) Proc & Addr Amer Phil Ass 47*21-33
⋄ A10 B15 ⋄ ID 31047

CHURCH, A. [1976] *Comparison of Russell's resolution of the semantical antinomies with that of Tarski* (J 0036) J Symb Logic 41*747-760
⋄ A05 A10 B15 ⋄ REV MR 56#81 Zbl 383#03005
• ID 14576

CHURCH, A. see Vol. II, III, IV, V, VI for further entries

CHURCHMAN, C.W. [1940] *Elements of logic and formal science*
(X 1297) Lippincott: Philadelphia iv+337pp
⋄ B98 ⋄ REV JSL 6.169 • ID 22479

CHURCHMAN, C.W. see Vol. II for further entries

CHWISTEK, L.B. [1912] *The principle of contradiction in the light of recent investigations of Bertrand Russell (Russian)* (J 0300)
Rozpr Akad Krakow Histor-Filoz Ser 2 30*270-334
⋄ A05 B15 F35 ⋄ ID 04231

CHWISTEK, L.B. [1922] *Grundlagen der reinen Typentheorie (Polish)* (J 1125) Przeglad Filoz 25*359-391
⋄ A05 B15 E70 ⋄ REV FdM 48.216 • ID 41659

CHWISTEK, L.B. [1924] *The theory of constructive types (principles of logic and mathematics)* (J 0283) Ann Soc Pol Math 2*9-48
⋄ A05 B15 F65 ⋄ REV FdM 49.139 • ID 02670

CHWISTEK, L.B. [1926] *Ueber die Hypothesen der Mengenlehre*
(J 0044) Math Z 25*439-473
⋄ A05 B30 E30 ⋄ REV FdM 52.191 • ID 02159

CHWISTEK, L.B. [1929] *Neue Grundlagen der Logik und Mathematik* (J 0044) Math Z 30*704-724
⋄ A05 B30 F99 ⋄ REV FdM 55.626 • ID 02160

CHWISTEK, L.B. [1932] *Neue Grundlagen der Logik und Mathematik. Zweite Mitteilung* (J 0044) Math Z 34*527-534
⋄ A05 B30 E70 ⋄ REV Zbl 4.1 FdM 58.67 • ID 02161

CHWISTEK, L.B. [1937] *Foundations of general theory of classes (Polish)* (J 0282) Sprawozd Towarz Nauk, Lwow 17*247-249
⋄ B15 E70 ⋄ REV JSL 5.165 • ID 02671

CHWISTEK, L.B. & HETPER, W. [1938] *New foundation of formal metamathematics* (J 0036) J Symb Logic 3*1-36
⋄ B30 F35 ⋄ REV Zbl 18.337 JSL 3.120 FdM 64.29
• ID 02162

CHWISTEK, L.B. see Vol. II, V, VI for further entries

CHYTIL, M.K. & HAJEK, P. & HAVEL, I.M. [1966] *GUHA - the method of systematical hypotheses searching (Czech) (English summary)* (J 0156) Kybernetika (Prague) 2*31-47
• TRANSL [1966] (J 0373) Comp Arch Inform & Numerik 1*293-308
⋄ B35 B75 ⋄ REV MR 34#2468 Zbl 168.261
Zbl 312#68051 Zbl 316#68049 • ID 22310

CHYTIL, M.K. [1974] *The decomposition calculus and its semantics* (J 0063) Studia Logica 33*277-281
⋄ B20 ⋄ REV MR 51#2854 Zbl 296#02005 • ID 17391

CHYTIL, M.K. [1977] *Semantique de formules logiques en forme d'equivalence n-aire (demi-modeles)* (J 0047) Notre Dame J Formal Log 18*421-435
⋄ B20 ⋄ REV MR 58#16116 Zbl 236#02011
Zbl 348#02014 • ID 23637

CHYTIL, M.K. & HAJEK, P. & HAVRANEK, T. [1983] *The GUHA method (Czech)* (X 1226) Academia: Prague 315pp
⋄ B35 ⋄ REV MR 85m:68031 • ID 48186

CIPPO, C.P. & VENINI, E. [1979] *A neuro-model of the propositional calculus* (S 2990) Rend Semin Mat Brescia 3*102-118
⋄ B05 ⋄ REV MR 81b:68032 • ID 80930

CIRULIS, J. [1975] *Lectures on mathematical logic and set theory. Part I: Mathematical logic (Russian)* (X 0895) Latv Valsts (Gos) Univ : Riga 143pp
⋄ B98 ⋄ REV MR 58#27169a • REM Part II 1975
• ID 71815

CIRULIS, J. [1975] *Logic with inclusion (Russian)* (J 0068) Z Math Logik Grundlagen Math 21*247-266
◇ B15 ◇ REV MR 51 # 7814 Zbl 308 # 02014 • REM Part II 1978 • ID 17314

CIRULIS, J. [1978] *Logic with inclusion. II (Russian)* (J 0068) Z Math Logik Grundlagen Math 24*225-236
◇ B15 ◇ REV MR 58 # 5016 Zbl 394 # 03015 • REM Part I 1975. Part III 1979. • ID 52492

CIRULIS, J. [1978] *Superinductive classes (Russian)* (J 0337) Mat Ezheg, Akad Nauk Latv SSR 22*177-191,283-284
◇ B15 E20 E25 ◇ REV MR 80b:04006 Zbl 406 # 03060 • ID 56143

CIRULIS, J. [1979] *Logic with inclusion. III (Russian)* (J 0068) Z Math Logik Grundlagen Math 25*163-172
◇ B15 ◇ REV MR 80g:03011 Zbl 465 # 03001 • REM Part II 1978 • ID 54904

CIRULIS, J. [1979] *Theories with inclusion (Russian)* (J 0031) Izv Vyssh Ucheb Zaved, Mat (Kazan) 1979/5(204)*83-86
• TRANSL [1979] (J 3449) Sov Math 23/5*87-91
◇ A05 B10 ◇ REV MR 81m:03046 Zbl 438 # 03016 • ID 55458

CIRULIS, J. [1981] *On the EA-fragment of classical propositional logic* (J 0387) Bull Sect Logic, Pol Acad Sci 10*158-161
◇ B20 ◇ REV MR 84a:03013 Zbl 499 # 03001 • ID 35560

CIRULIS, J. [1981] *On the EK-fragments of positive and classical propositional logics* (J 0387) Bull Sect Logic, Pol Acad Sci 10*100-103
◇ B20 ◇ REV MR 83c:03009 Zbl 474 # 03004 • ID 55408

CIRULIS, J. [1982] *Axioms of protothetics with a strengthened capacity rule (Russian)* (J 0337) Mat Ezheg, Akad Nauk Latv SSR 26*264-270,285
◇ B15 B60 ◇ REV MR 84a:03028 Zbl 495 # 03007 • ID 35570

CIRULIS, J. see Vol. II, V for further entries

CLARK, M. [1970] see CASEY, H.

CLARK, M. see Vol. II for further entries

CLARK, R. & WELSH JR., P.J. [1962] *Introduction to logic* (X 0864) Van Nostrand: New York xii + 268pp
◇ B98 ◇ REV JSL 33.479 • ID 02437

CLARK, R. [1967] see BINKLEY, R.

CLARK, R. see Vol. II for further entries

CLARKE, E.M. [1985] *Using temporal logic for automatic verification of finite state systems* (P 4621) Log & Models of Concurrent Syst;1984 La Colle-sur-Loup 3-26
◇ B80 ◇ ID 49379

CLARKE, E.M. see Vol. II for further entries

CLAY, R.E. [1968] *The consistency of Lesniewski's mereology relative to the real number system* (J 0036) J Symb Logic 33*251-257
◇ B28 E35 E70 F25 ◇ REV MR 39 # 1321 Zbl 231 # 02030 • ID 02255

CLAY, R.E. see Vol. II, V for further entries

CLEAVE, J.P. [1971] *Cauchy, convergence and continuity* (J 0013) Brit J Phil Sci 22*27-37
◇ A10 H05 ◇ REV MR 44 # 3853 Zbl 242 # 01011 • ID 02270

CLEAVE, J.P. see Vol. II, III, IV, VI for further entries

CLEAVER, F.L. [1965] see ANDERSON, D.E.

CLIFFORD, W.K. [1879] *Lectures and essays. 2 Vols* (X 1253) Crowell Collier & Macmillan: New York 443pp
◇ B05 ◇ REM 3rd ed. 1901 • ID 25525

CLIFFORD, W.K. [1882] *Mathematical papers* (X 1253) Crowell Collier & Macmillan: New York 658pp
◇ B96 ◇ ID 25475

CLIMESCU, A. [1971] *Un calcul des applications (Romanian summary)* (J 0198) Bul Inst Politeh Iasi NS 17(21)/3-4*1-5
◇ B05 G25 ◇ REV MR 46 # 3404 Zbl 333 # 02009 • ID 02277

CLIMESCU, A. see Vol. V for further entries

CLOTE, P. & McALOON, K. [1983] *Two further combinatorial theorems equivalent to the 1-consistency of Peano arithmetic* (J 0036) J Symb Logic 48*1090-1104
◇ B28 C62 F30 ◇ REV MR 85e:03087 Zbl 545 # 03033 • ID 40361

CLOTE, P. see Vol. III, IV, V, VI for further entries

COATES, C.L. & LEWIS II, P.M. [1967] *Threshold logic* (X 0827) Wiley & Sons: New York xii + 483pp
◇ B70 B98 ◇ ID 19520

COBHAM, A. [1956] *Reduction to a symmetric predicate* (J 0036) J Symb Logic 21*56-59
◇ B10 D35 ◇ REV MR 17.1173 Zbl 71.10 JSL 22.297 • ID 02278

COBHAM, A. & FRIDSHAL, R. & NORTH, J.H. [1961] *An application of linear programming to the minimization of boolean functions* (P 0624) Switch Circ Th & Log Design (1,2);1960 Chicago;1961 Detroit 3-9
◇ B35 ◇ REV JSL 30.247 • ID 02280

COBHAM, A. see Vol. III, IV, VI for further entries

COBURN, B. & MILLER, DAVID [1977] *Two comments on Lemmon's beginning logic* (J 0047) Notre Dame J Formal Log 18*607-610
◇ B05 ◇ REV MR 58 # 5007 Zbl 322 # 02008 • ID 23699

COCCHIARELLA, N.B. [1968] *Some remarks on second order logic with existence attributes* (J 0097) Nous, Quart J Phil 2*165-175
◇ A05 B15 ◇ ID 32271

COCCHIARELLA, N.B. [1969] *A second order logic of existence* (J 0036) J Symb Logic 34*57-69
◇ A05 B15 ◇ REV MR 41 # 8215 Zbl 209.9 • ID 02288

COCCHIARELLA, N.B. [1969] *A substitution free axiom set for second order logic* (J 0047) Notre Dame J Formal Log 10*18-30
◇ B15 ◇ REV MR 41 # 3252 Zbl 179.10 • ID 02287

COCCHIARELLA, N.B. [1969] *Existence entailing attributes, modes of copulation, and modes of being in second order logic* (J 0097) Nous, Quart J Phil 3*33-48
◇ A05 B15 ◇ ID 32272

COCCHIARELLA, N.B. [1972] *Properties as individuals in formal ontology* (J 0097) Nous, Quart J Phil 6*165-187
◇ A05 B15 ◇ REV MR 58 # 27296 • ID 32273

COCCHIARELLA, N.B. [1973] *Whither Russell's paradox of predication?* (C 0699) Logic & Ontology 133-158
◇ A05 B15 ◇ ID 03604

COCCHIARELLA, N.B. [1974] *A new formulation of predicative second order logic* (J 0079) Logique & Anal, NS 17*61-87
⋄ A05 B15 F65 ⋄ REV MR 50 # 12643 Zbl 343 # 02010
• ID 04010

COCCHIARELLA, N.B. [1975] *A second order logic of variable-binding operators* (J 0302) Rep Math Logic, Krakow & Katowice 5*3-18
⋄ A05 B15 C80 ⋄ REV MR 57 # 5675 Zbl 348 # 02019
• ID 21896

COCCHIARELLA, N.B. [1975] *Second order theories of predication: old and new foundations* (J 0097) Nous, Quart J Phil 9*33-53
⋄ A05 A10 B15 E70 ⋄ REV MR 57 # 9473 • ID 32278

COCCHIARELLA, N.B. [1976] *A note on the definition of identity in Quine's new foundations* (J 0068) Z Math Logik Grundlagen Math 22*195-197
⋄ A05 B15 E35 E70 ⋄ REV MR 58 # 21615 Zbl 343 # 02049 • ID 18458

COCCHIARELLA, N.B. [1979] *The theory of homogeneous simple types as a second-order logic* (J 0047) Notre Dame J Formal Log 20*505-524
⋄ B15 E70 ⋄ REV MR 80k:03058 Zbl 314 # 02027
• ID 56097

COCCHIARELLA, N.B. [1980] *Nominalism and conceptualism as predicative second-order theories of predication* (J 0047) Notre Dame J Formal Log 21*481-500
⋄ A05 B15 C85 ⋄ REV MR 81j:03026 Zbl 416 # 03013
• ID 53911

COCCHIARELLA, N.B. [1985] *Two λ-extensions of the theory of homogeneous simple types as a second-order logic* (J 0047) Notre Dame J Formal Log 26*377-407
⋄ B15 B40 ⋄ ID 47532

COCCHIARELLA, N.B. see Vol. II, V for further entries

COERS, H. [1985] *Die Unabhaengigkeit des archimedischen Axioms vom starken Intervallschachtelungsaxiom* (J 2790) Math Sem-ber 32*50-60
⋄ B28 ⋄ ID 46290

COHEN, JACK K. & WATTS, D.E. [1980] *Computer-implemented set theory* (J 0005) Amer Math Mon 87*557-560
⋄ B35 E20 ⋄ REV Zbl 457 # 04001 • ID 54391

COHEN, JACQUES & RUBIN, A.L. [1979] *An interactive system for proving theorems in the predicate calculus* (P 3452) Symb & Algeb Comput;1979 Marseille
⋄ B35 ⋄ ID 47297

COHEN, JACQUES see Vol. IV for further entries

COHEN, L.J. [1970] *The implications of induction* (X 0816) Methuen: London & New York vii + 248pp
⋄ A05 B48 B98 ⋄ REV Zbl 298 # 02020 • ID 25404

COHEN, L.J. & LOS, J. & PFEIFFER, H. & PODEWSKI, K.-P. (EDS.) [1982] *Logic, methodology and philosophy of science VI* (S 3303) Stud Logic Found Math 104*856pp
⋄ B97 D98 ⋄ REV MR 83k:03004 Zbl 489 # 00005
• ID 36588

COHEN, L.J. see Vol. II for further entries

COHEN, M.R. & NAGEL, E. [1934] *An introduction to logic and scientific method* (X 0863) Harcourt: New York & London xii + 467pp
⋄ B98 ⋄ REV JSL 11.100 FdM 60.845 • ID 25096

COHEN, R.S. & DENEV, J.D. [1978] *Estimations of the number of subfunctions for almost all S-tuple boolean functions* (P 3397) Mat & Mat Obrazov (7);1978 Sl"nchev Bryag 289-297
⋄ B05 ⋄ REV Zbl 389 # 94014 • ID 69444

COHEN, R.S. & WARTOFSKY, M.W. (EDS.) [1983] *Language, logic and method* (S 3311) Boston St Philos Sci 31*464pp
⋄ B97 ⋄ REV Zbl 493 # 00003 • ID 38427

COHEN, R.S. see Vol. IV for further entries

COHEN, Y. [1982] *Logical polygraph* (J 0079) Logique & Anal, NS 25*425-433
⋄ B80 ⋄ REV MR 84j:03012 • ID 34606

COLBAN, E.A. [1983] see BARRICELLI, N.A.

COLE, J.C. [1973] see BELL, J.L.

COLE, J.C. see Vol. III, V for further entries

COLE, R. [1968] *Definitional boolean calculi* (J 0047) Notre Dame J Formal Log 9*343-350
⋄ B05 ⋄ REV MR 39 # 2598 Zbl 165.14 • ID 02318

COLE, R. see Vol. II for further entries

COLETTI, G. & REGOLI, G. & VINCENTI, R. [1979] *Algebraic and topological structure of nonstandard models of the reals* (J 2038) Rend Sem Mat, Torino 37/2*75-82 • ERR/ADD ibid 37/3*165
⋄ H05 H20 ⋄ REV MR 81k:03065a MR 81k:03065b Zbl 497 # 03052 • ID 71886

COLIGN, A.W. [1977] see BERMAN, G.

COLLINS, G.E. & HALPERN, J.D. [1970] *On the interpretability of arithmetic in set theory* (J 0047) Notre Dame J Formal Log 11*477-483
⋄ B28 D35 E30 F25 F30 ⋄ REV MR 45 # 4970 Zbl 212.17 • ID 02332

COLLINS, G.E. see Vol. III, IV, V for further entries

COLOMBETTI, M. & PAGELLO, E. [1976] *Programs, computations and temporal features* (P 1401) Math Founds of Comput Sci (5);1976 Gdansk 237-243
⋄ B35 B75 ⋄ REV Zbl 338 # 68067 • ID 61063

COMBES, M. [1971] *Fondements des mathematiques* (X 0840) Pr Univ France: Paris 104pp
⋄ A05 A10 B98 E98 ⋄ REV MR 44 # 1537 Zbl 259 # 02003 • ID 02334

COMER, S.D. [1985] *The elementary theory of interval real numbers* (J 0068) Z Math Logik Grundlagen Math 31*89-95
⋄ B25 B28 C65 ⋄ REV MR 86h:03069 Zbl 547 # 03007 Zbl 565 # 03008 • ID 42290

COMER, S.D. see Vol. III, IV, V for further entries

COMERFORD, J.D. [1980] *Affine and general linear equivalences of boolean functions* (J 0194) Inform & Control 45*156-169
⋄ B05 ⋄ REV MR 82d:94069 Zbl 461 # 94012 • ID 69346

COMMENTZ-WALTER, B. [1978] *Size-depth tradeoff in boolean formulas* (P 1872) Automata, Lang & Progr (5);1978 Udine 125-141
⋄ B35 ⋄ REV Zbl 383 # 94037 • ID 52041

COMMENTZ-WALTER, B. [1979] *Size-depth tradeoff in monotone boolean formulae* (J 1431) Acta Inf 12*227-243
⋄ B05 D15 ⋄ REV MR 80m:68038 Zbl 395 # 94036
• ID 53164

COMMENTZ-WALTER, B. & SATTLER, J. [1980] *Size-depth tradeoff in nonmonotone boolean formulae* (**J** 1431) Acta Inf 14*257-269
⋄ B05 ⋄ REV MR 82g:94026 Zbl 431 # 68053 • ID 80954

CONANT, L.L. [1896] *The number concept* (**X** 1253) Crowell Collier & Macmillan: New York vii+218pp
⋄ A10 B28 ⋄ REV FdM 27.46 • ID 25478

CONNES, A. [1970] *Determination de modeles minimaux en analyse non standard et application* (**J** 2313) C R Acad Sci, Paris, Ser A-B 271*A969-A971
⋄ H05 H15 ⋄ REV MR 44 # 3854 Zbl 255 # 02059 • ID 02352

CONNES, A. [1970] *Ultrapuissances et applications dans le cadre de l'analyse non standard* (**C** 1525) Semin Init Analyse (9-10) Paris 1969/71 9/8*25pp
⋄ C20 H05 ⋄ REV MR 44 # 3855 Zbl 216.12 • ID 26249

CONSTABLE, R.L. & ZLATIN, D.R. [1984] *The type theory of PL/CV3* (**J** 3120) ACM Trans Program Lang & Syst 6*94-117
⋄ B15 B75 ⋄ REV Zbl 522 # 68020 • ID 37056

CONSTABLE, R.L. & MENDLER, N.P. [1985] *Recursive definitions in type theory* (**P** 4571) Log of Progr;1985 Brooklyn 61-78
⋄ B15 F35 ⋄ ID 49190

CONSTABLE, R.L. see Vol. IV, VI for further entries

CONSTANTINESCU, M. [1965] *Das Logische und Historische in der Interpretation der aristotelischen Syllogistik* (**J** 0147) An Univ Bucuresti, Acta Logica 7-8*193-200
⋄ A10 B20 ⋄ REV Zbl 231 # 02005 • ID 61073

CONWAY, J.H. [1976] *On numbers and games* (**X** 0801) Academic Pr: New York ix+238pp
• TRANSL [1983] (**X** 0900) Vieweg: Wiesbaden vii+205pp
⋄ B28 E10 E70 E75 H15 H20 ⋄ REV MR 56 # 8365 MR 84m:03086 Zbl 334 # 00004 Zbl 491 # 00024
• ID 71909

CONWAY, J.H. see Vol. IV, V for further entries

COOK, S.A. & RECKHOW, R.A. [1974] *On the length of proofs in the propositional calculus: preliminary version* (**P** 1464) ACM Symp Th of Comput (6);1974 Seattle 135-148
⋄ B05 D15 F20 ⋄ REV MR 54 # 7215 Zbl 375 # 02004 • ID 25005

COOK, S.A. [1975] *Feasibly constructive proofs and the propositional calculus* (**P** 1618) ACM Symp Th of Comput (7);1975 Albuquerque 83-97
⋄ B05 F20 ⋄ REV MR 58 # 19341 Zbl 357 # 68061 • ID 28181

COOK, S.A. & RECKHOW, R.A. [1979] *The relative efficiency of propositional proof systems* (**J** 0036) J Symb Logic 44*36-50
⋄ B05 D15 F20 ⋄ REV MR 80e:03007 Zbl 408 # 03044 • ID 56282

COOK, S.A. see Vol. IV, V for further entries

COOLEY, J.C. [1942] *A primer of formal logic* (**X** 0843) Macmillan : New York & London xi+378pp
⋄ B98 ⋄ REV MR 4.125 Zbl 60.20 JSL 8.80 • ID 02359

COOLEY, J.E. [1975] *Theories of types and ordered pairs* (**J** 0047) Notre Dame J Formal Log 16*418-420
⋄ B15 E20 ⋄ REV MR 52 # 98 Zbl 254 # 02014 • ID 02360

COOPER, D.C. [1966] *Theorem-proving in computers* (**P** 2323) Non-Numer Comput Adv Progr;1963 Oxford 155-182
⋄ B35 ⋄ REV Zbl 265 # 68041 • ID 29853

COOPER, D.C. [1969] *Program scheme equivalences and second-order logic* (**J** 0508) Machine Intelligence 4*3-15
⋄ B15 D05 ⋄ REV MR 40 # 6806 Zbl 221 # 68015
• ID 28130

COOPER, D.C. [1972] *Theorem proving in arithmetic without multiplication* (**J** 0508) Machine Intelligence 7*91-99
⋄ B25 B35 F30 ⋄ REV Zbl 258 # 68046 • ID 61080

COOPER, J.L.B. [1969] *k-fold preordered sets* (**P** 0649) Appl Model Th to Algeb, Anal & Probab;1967 Pasadena 300-307
⋄ H05 ⋄ REV MR 54 # 180 Zbl 211.15 • ID 26623

COOPER, W.S. [1974] *Set theory and syntactic description* (**X** 0873) Mouton: Paris 52pp
⋄ B98 E75 ⋄ REV MR 53 # 2685 Zbl 279 # 68051
• ID 21633

COOPER, W.S. [1978] *Foundations of logico-linguistics: A unified theory of information, language, and logic* (**X** 0835) Reidel: Dordrecht 2*xvi+249pp
⋄ A05 B98 ⋄ REV Zbl 406 # 03001 • ID 28182

COPELAND SR., A.H. [1941] *Postulates for the theory of probability* (**J** 0100) Amer J Math 63*741-762
⋄ B30 ⋄ REV MR 3.167 Zbl 26.136 FdM 67.44 • ID 41304

COPI, I.M. [1950] *The inconsistency or redundancy of principia mathematica* (**J** 0075) Phil Phenom Research 11*190-199
⋄ A05 B15 ⋄ REV MR 12.664 Zbl 40.294 JSL 16.154
• ID 02370

COPI, I.M. [1953] *Introduction to logic* (**X** 0843) Macmillan : New York & London xvi+472pp
⋄ A10 B98 ⋄ REV Zbl 481 # 03001 Zbl 57.5 JSL 19.147 JSL 29.92 JSL 35.166 • REM 6th edition 1982; xiv+604pp
• ID 02371

COPI, I.M. [1954] *Symbolic logic* (**X** 0843) Macmillan : New York & London xiii+355pp
⋄ B98 ⋄ REV MR 17.223 MR 50 # 1824 MR 80c:03002 Zbl 281 # 02012 JSL 19.282 JSL 39.177 • ID 02372

COPI, I.M. [1956] *Another variant of natural deduction* (**J** 0036) J Symb Logic 21*52-55
⋄ B10 ⋄ REV MR 17.1038 Zbl 74.247 JSL 22.298
• ID 02373

COPI, I.M. & GOULD, JAMES A. (EDS.) [1967] *Contemporary readings in logical theory* (**X** 0843) Macmillan : New York & London viii+344pp
⋄ A05 A10 B97 ⋄ REV MR 35 # 14 • ID 25224

COPI, I.M. & GOULD, JAMES A. [1967] *Deontic logic, introduction* (**C** 0721) Contemp Readings in Log Th 301pp
⋄ B45 B98 ⋄ ID 04030

COPI, I.M. [1967] *The theory of types (Introduction)* (**C** 0721) Contemp Readings in Log Th 133-134
⋄ B15 ⋄ ID 04027

COPI, I.M. [1971] *The theory of logical types* (**X** 0866) Routledge & Kegan Paul: Henley on Thames 129pp
⋄ B15 ⋄ REV JSL 39.174 • ID 02454

COPI, I.M. also published under the name COPILOWISH, I.M.

COPI, I.M. see Vol. II, V for further entries

COPPOTELLI, F. [1968] *A first order type theory for the theory of sets* (J 0047) Notre Dame J Formal Log 9∗367-370
⋄ B15 E30 E70 ⋄ REV MR 39 # 2627 Zbl 181.303
• ID 02387

COPPOTELLI, F. [1977] *On two first order type theories for the theory of sets* (J 0047) Notre Dame J Formal Log 18∗147-150
⋄ B15 E30 E70 ⋄ REV MR 58 # 5215 Zbl 283 # 02048
• ID 21956

COQUAND, T. & HUET, G. [1985] *Constructions: a higher order proof system for mechanizing mathematics* (P 4601) EUROCAL;1985 Linz 1∗151-184
⋄ B35 ⋄ ID 49771

COQUAND, T. see Vol. III for further entries

CORCORAN, J. & HERRING, J.M. [1971] *Notes on a semantic analysis of variable binding term operators* (J 0079) Logique & Anal, NS 14∗644-657
⋄ A05 A10 B10 C80 ⋄ REV MR 46 # 6989 Zbl 239 # 02007 • ID 27917

CORCORAN, J. [1972] *Completeness of an ancient logic* (J 0036) J Symb Logic 37∗696-702
⋄ A10 B20 ⋄ REV MR 47 # 6435 Zbl 261 # 02004
• ID 02391

CORCORAN, J. [1972] *Strange arguments* (J 0047) Notre Dame J Formal Log 13∗206-210
⋄ B05 ⋄ REV MR 45 # 6567 Zbl 234 # 02007 • ID 02390

CORCORAN, J. & HATCHER, W.S. & HERRING, J.M. [1972] *Variable binding term operators* (J 0068) Z Math Logik Grundlagen Math 18∗177-182
⋄ B10 C80 ⋄ REV MR 46 # 5098 Zbl 257 # 02013
• ID 02392

CORCORAN, J. [1972] *Weak and strong completeness in sentential logics* (J 0079) Logique & Anal, NS 15∗429-434
⋄ B22 ⋄ REV MR 49 # 2245 Zbl 267 # 02007 • ID 04042

CORCORAN, J. [1973] *A mathematical model of Aristotle's syllogistic* (J 1615) Arch Geschichte Phil 55∗191-219
⋄ A05 A10 B20 ⋄ REV MR 56 # 2776 • ID 28174

CORCORAN, J. [1974] *Aristotle's natural deduction system* (P 1657) Ancient Log & Modern Interpr;1972 Buffalo 85-131
⋄ A05 A10 B20 ⋄ REV MR 58 # 16077 Zbl 286 # 02002
• ID 61097

CORCORAN, J. & FRANK, W. & MALONEY, M.J. [1974] *String theory* (J 0036) J Symb Logic 39∗625-637
⋄ A10 B03 ⋄ REV MR 53 # 2622 Zbl 298 # 02011
• ID 04043

CORCORAN, J. & ZIEWACZ, S. [1979] *Identity logics* (J 0047) Notre Dame J Formal Log 20∗777-784
⋄ B20 ⋄ REV MR 80h:03017 Zbl 368 # 02001 • ID 56169

CORCORAN, J. [1980] *Categoricity* (J 2028) Hist & Phil Log 1∗187-207
⋄ A05 B15 C35 C85 ⋄ REV MR 82j:03034 Zbl 504 # 03014 • ID 71934

CORCORAN, J. see Vol. II, III, VI for further entries

CORNIDES, T. [1974] *Ordinale Deontik* (X 0902) Springer: Wien x+210pp
⋄ B45 B98 ⋄ REV MR 57 # 15962 JSL 44.121 • ID 71948

CORNMAN, J.W. [1966] *Metaphysics, reference and language* (X 0875) Yale Univ Pr: New Haven xxi+288pp
⋄ A05 B98 ⋄ ID 25316

CORNMAN, J.W. see Vol. III for further entries

CORRADA, M. [1980] see ALCANTARA DE, L.P.

CORRADA, M. see Vol. III, V for further entries

COSTA, E.J.F. [1982] *Automatic program transformation viewed as theorem proving* (P 3867) Int Symp Progr (5);1982 Turin 37-46
⋄ B35 ⋄ REV Zbl 505 # 68040 • ID 48975

COSTA DA, N.C.A. [1964] see ARRUDA, A.I.

COSTA DA, N.C.A. [1966] see ARRUDA, A.I.

COSTA DA, N.C.A. [1973] *On the concept of transformation in the predicate calculus (Portuguese)* (C 1467) Notas Inst Mat & Estatica Sao Paulo 2∗51-57
⋄ B10 ⋄ REV MR 52 # 13315 Zbl 338 # 02006 • ID 21778

COSTA DA, N.C.A. & MORTENSEN, C. [1983] *Notes on the theory of variable binding term operators* (J 2028) Hist & Phil Log 4∗63-72
⋄ B10 ⋄ REV MR 85b:03013 Zbl 513 # 03014 • ID 37221

COSTA DA, N.C.A. see Vol. II, III, V, VI for further entries

COURVOISIER, M. & LAGASSE, J. & RICHARD, J. [1976] *Logique combinatoire. 3e ed* (X 0856) Dunod: Paris vi+73pp
⋄ B40 B98 ⋄ REV MR 57 # 15759 Zbl 376 # 02045
• ID 51687

COUTURAT, L. [1904] *L'algebre de la logique* (X 1324) Open Court: LaSalle 100pp
• TRANSL [1914] (X 1324) Open Court: LaSalle xiv+98pp (English) • REPR [1980] (X 0872) Blanchard: Paris 100pp
⋄ B98 G05 ⋄ REV MR 81j:01014 Zbl 426 # 03064 FdM 45.120 • REM 2nd ed. 1908. Reprint of the 2nd ed.
• ID 22581

COUTURAT, L. [1905] *Les principes des mathematiques. Avec un appendice sur la philosophie des mathematiques de Kant* (X 0892) Olms: Hildesheim viii+311pp
• LAST ED [1980] (X 0872) Blanchard: Paris viii+311pp
⋄ A05 A10 B30 ⋄ REV MR 81k:00006 Zbl 421 # 03001 FdM 36.81 • ID 25526

COUTURAT, L. see Vol. V for further entries

COUTURE, J. [1983] *Les classes dans les Principia mathematica sont-elles des expressions incompletes? (English and German summaries)* (J 0076) Dialectica 37∗249-267
⋄ A05 B15 E30 ⋄ REV MR 85d:03004 Zbl 568 # 03003
• ID 40942

COWEN, R.H. [1970] *A new proof of the compactness theorem for propositional logic* (J 0047) Notre Dame J Formal Log 11∗79-80
⋄ B05 ⋄ REV MR 43 # 4624 Zbl 167.12 • ID 02498

COWEN, R.H. [1975] *A characterization of logical consequences in quantification theory* (J 0047) Notre Dame J Formal Log 16∗375-377
⋄ B10 ⋄ REV MR 51 # 12502 Zbl 258 # 02016 • ID 02501

COWEN, R.H. [1977] *Generalizing Koenig's infinity lemma* (J 0047) Notre Dame J Formal Log 18∗243-247
⋄ B05 E05 E07 ⋄ REV MR 58 # 5306 Zbl 314 # 02071
• ID 21970

COWEN, R.H. [1982] *Solving algebraic problems in propositional logic by tableau* (**J 0009**) Arch Math Logik Grundlagenforsch 22∗187-190
◇ B05 ◇ REV MR 83k:03013 Zbl 487 # 03004 • ID 36157

COWEN, R.H. see Vol. III, V for further entries

COX, P.T. [1977] *A graphical proof procedure for first-order logic* (**P 3238**) Conf Theoret Comput Sci;1977 Waterloo ON 230-238
◇ B35 ◇ REV MR 58 # 25155 Zbl 429 # 68077 • ID 53897

COX, P.T. & PIETRZYKOWSKI, T. [1980] *A complete, nonredundant algorithm for reversed Skolemization* (**P 3063**) Autom Deduct (5);1980 Les Arcs 374-385
◇ B35 ◇ REV MR 82c:03017 Zbl 465 # 68045 • ID 54950

COX, P.T. & PIETRZYKOWSKI, T. [1981] *Deduction plans: A basis for intelligent backtracking* (**J 3191**) IEEE Trans Pattern Anal & Mach Intell 3∗52-65
◇ B35 ◇ REV Zbl 454 # 68114 • ID 54259

COX, P.T. & PIETRZYKOWSKI, T. [1984] *A complete, nonredundant algorithm for reversed Skolemization* (**J 1426**) Theor Comput Sci 28∗239-261
◇ B35 ◇ REV MR 85j:68097 Zbl 566 # 03005 • ID 45179

COY, W. [1975] *Drei Komplexitaetsmasse zweistufiger Normalformen boolescher Funktionen* (**P 3527**) GI Jahrestag (4);1974 Berlin 161-169
◇ B05 ◇ REV MR 55 # 5313 Zbl 307 # 94037 • ID 80986

COY, W. see Vol. II, IV for further entries

COZART, D. & MOORE JR., L.C. [1974] *The nonstandard hull of a Riesz space* (**J 0025**) Duke Math J 41∗263-276
◇ C60 H05 ◇ REV MR 50 # 10747 Zbl 293 # 46012
• ID 02503

CRABBE, M. [1975] *Types ambigus (English summary)* (**J 2313**) C R Acad Sci, Paris, Ser A-B 280∗A1-A2
◇ B15 E35 E70 ◇ REV MR 50 # 9583 Zbl 306 # 02016
• ID 04051

CRABBE, M. [1978] *Ambiguity and stratification* (**J 0027**) Fund Math 101∗11-17
◇ B15 E35 E70 F25 ◇ REV MR 80h:03073
Zbl 404 # 03041 • ID 29204

CRABBE, M. [1978] *Ramification et predicativite* (**J 0079**) Logique & Anal, NS 21∗399-419
◇ B15 ◇ REV MR 81e:03006 Zbl 449 # 03009 • ID 56676

CRABBE, M. [1983] *On the reduction of type theory* (**J 0068**) Z Math Logik Grundlagen Math 29∗235-237
◇ B15 ◇ REV MR 85g:03020 Zbl 554 # 03026 • ID 42211

CRABBE, M. [1983] *The axiom of choice in type theory with ambiguity axioms* (**P 1601**) Easter Conf on Model Th (1);1983 Diedrichshagen 28-38
◇ B15 E25 E70 ◇ REV MR 84i:03008 Zbl 554 # 03027
• ID 44943

CRABBE, M. see Vol. V, VI for further entries

CRAEMER, D. [1974] see BEHRMANN, E.

CRAIG, W. & QUINE, W.V.O. [1952] *On reduction to a symmetric relation* (**J 0036**) J Symb Logic 17∗188
◇ B10 C40 ◇ REV MR 14.233 Zbl 47.9 JSL 18.269
• ID 02505

CRAIG, W. [1957] *Analysis of first-order implications* (**P 1675**) Summer Inst Symb Log;1957 Ithaca 175-180
◇ B10 F05 ◇ REV JSL 38.519 • ID 02509

CRAIG, W. [1957] *Linear reasoning. A new form of the Herbrand-Gentzen theorem* (**J 0036**) J Symb Logic 22∗250-268
◇ B10 C07 C40 F05 ◇ REV MR 21 # 3317 Zbl 81.244 JSL 24.243 • ID 02510

CRAIG, W. [1957] *Three uses of the Herbrand-Gentzen theorem in relating model theory and proof theory* (**J 0036**) J Symb Logic 22∗269-285
◇ B10 C07 C40 F05 ◇ REV MR 21 # 3318 Zbl 79.245 JSL 24.243 • ID 02508

CRAIG, W. & VAUGHT, R.L. [1958] *Finite axiomatizability using additional predicates* (**J 0036**) J Symb Logic 23∗289-308
◇ B10 C52 ◇ REV MR 21 # 4909 Zbl 85.246 JSL 36.334
• ID 02511

CRAIG, W. [1960] *Bases for first-order theories and subtheories* (**J 0036**) J Symb Logic 25∗97-142
◇ B10 D25 F05 ◇ REV MR 24 # A1812 Zbl 108.3 JSL 37.616 • ID 02513

CRAIG, W. [1965] *Satisfaction for n-th order languages defined in n-th order languages* (**J 0036**) J Symb Logic 30∗13-25
◇ B15 C40 ◇ REV MR 33 # 3883 Zbl 199.3 • ID 02514

CRAIG, W. [1967] *Modus ponens and derivations from Horn formulas* (**J 0068**) Z Math Logik Grundlagen Math 13∗33-54
◇ B20 F05 F50 ◇ REV MR 35 # 22 Zbl 165.10 • ID 02516

CRAIG, W. [1974] see ADDISON, J.W.

CRAIG, W. see Vol. III, IV, V, VI for further entries

CRESSWELL, M.J. [1964] *Propositional arithmetic* (**J 0079**) Logique & Anal, NS 7∗185-189
◇ B05 D03 ◇ REV MR 31 # 2129 Zbl 121.10 • ID 02531

CRESSWELL, M.J. [1966] *Functions of propositions* (**J 0036**) J Symb Logic 31∗545-560
◇ B05 ◇ REV MR 35 # 1454 Zbl 158.245 • ID 02535

CRESSWELL, M.J. & HUGHES, G.E. [1968] *An introduction to modal logic* (**X 0816**) Methuen: London & New York xii+388pp
• TRANSL [1978] (**X 1174**) Gruyter: Berlin x+340pp [1973] (**X 1781**) Tecnos: Madrid 316pp [1973] (**X 3615**) Il Saggiatore: Milano xxxvii+436pp [1981] (**X 3711**) Koseisha Koseikaku: Tokyo xii+440pp
◇ B45 B98 ◇ REV MR 55 # 12472 Zbl 205.5 JSL 36.328
• ID 19031

CRESSWELL, M.J. [1972] *Second-order intensional logic* (**J 0068**) Z Math Logik Grundlagen Math 18∗297-320
◇ B15 B45 ◇ REV MR 46 # 5114 Zbl 246 # 02018
• ID 02551

CRESSWELL, M.J. [1973] *Logics and languages* (**X 0816**) Methuen: London & New York xi+273pp
• TRANSL [1979] (**X 1174**) Gruyter: Berlin xi+431pp (German)
◇ A05 B98 ◇ REV MR 55 # 2497 Zbl 408 # 03001 JSL 42.425 • REM See also 1975 • ID 02552

CRESSWELL, M.J. [1975] *Note on the use of sequences in "Logics and languages"* (**J 0047**) Notre Dame J Formal Log 16∗445-448
◇ B03 E75 ◇ REV MR 55 # 2498 Zbl 283 # 02028 • REM See 1973 • ID 02553

CRESSWELL, M.J. & GOLDBLATT, R.I. &
MEYERHOFF CRESSWELL, M. & SEGERBERG, K. [1982]
Symbolic logic. Proceedings of the 1981 Annual Conference of the Australian Association of Symbolic Logic held in Wellington, New Zealand, from 2-5 July 1981 (J 0063) Studia Logica 41*93-307
  ◇ B97 ◇ REV Zbl 529 # 00015 • ID 38476

CRESSWELL, M.J. see Vol. II, III, IV for further entries

CRISMA, L. & HOLZER, S. [1983] *Starconcepts* (J 3768) Boll Unione Mat Ital, VI Ser, D 2*175-191
  ◇ H05 ◇ REV Zbl 561 # 03035 • ID 44482

CROSS, R.W. [1972] see CHADWICK, J.J.M.

CROSSLEY, J.N. & DUMMETT, M. (EDS.) [1965] *Formal systems and recursive functions* (X 0809) North Holland: Amsterdam 320pp
  ◇ B97 D20 D97 ◇ REV Zbl 126.2 • ID 31683

CROSSLEY, J.N. [1966] *Some theorems in logic* (J 0248) Math Student 34*125-129
  ◇ B10 C07 ◇ ID 31222

CROSSLEY, J.N. (ED.) [1967] *Sets, models and recursion theory* (X 0809) North Holland: Amsterdam 331pp
  ◇ B97 C97 D97 ◇ REV MR 36 # 24 Zbl 158.2 • ID 31684

CROSSLEY, J.N. [1972] see ASH, C.J.

CROSSLEY, J.N. (ED.) [1975] *Algebra and logic* (X 0811) Springer: Heidelberg & New York viii + 307pp
  ◇ B97 C97 G97 ◇ REV MR 51 # 2851 Zbl 293 # 00006 • ID 31725

CROSSLEY, J.N. [1975] *What is mathematical logic?* (J 2053) Math Medley 1*3-6
  ◇ B98 F30 ◇ ID 31689

CROSSLEY, J.N. [1976] *Introduction to mathematical logic: a new area of mathematics with ancient roots* (J 2054) Math Scientist (Melbourne) 1*3-13
  ◇ B10 ◇ REV Zbl 336 # 02001 • ID 31691

CROSSLEY, J.N. see Vol. II, III, IV, V, VI for further entries

CSIRMAZ, L. [1980] *Structure of program runs of nonstandard time* (J 0380) Acta Cybern (Szeged) 4*325-331
  ◇ H05 ◇ REV MR 83c:03015 Zbl 441 # 68026 • ID 35125

CSIRMAZ, L. [1981] *Nonstandard runs and program verification* (J 0387) Bull Sect Logic, Pol Acad Sci 10*68-74
  ◇ B75 H10 ◇ REV Zbl 494 # 68015 • ID 36655

CSIRMAZ, L. [1982] *Nonstandard program runnings and correctness of programs* (J 0396) Mat Lapok 30*81-125
  ◇ B75 H10 ◇ ID 45210

CSIRMAZ, L. see Vol. III, IV, V, VI for further entries

CUDA, K. [1977] *The relation between ε-δ procedures and the infinitely small in nonstandard methods* (P 1695) Set Th & Hierarch Th (3);1976 Bierutowice 143-152
  ◇ H05 ◇ REV MR 58 # 21587 Zbl 372 # 02033 • ID 31033

CUDA, K. [1980] *An elimination of infinitely small quantities and infinitely large numbers (within the framework of AST)* (J 0140) Comm Math Univ Carolinae (Prague) 21*433-445
  ◇ E70 H05 H15 ◇ REV MR 83i:03104 Zbl 445 # 03042 • ID 56506

CUDA, K. & KUSSOVA, B. [1983] *Monads in basic equivalences* (J 0140) Comm Math Univ Carolinae (Prague) 24*437-452
  ◇ E70 H05 ◇ REV MR 85h:03059 Zbl 531 # 03032 • ID 37684

CUDA, K. [1984] *Translation of nonstandard definitions to standard ones* (J 0140) Comm Math Univ Carolinae (Prague) 25*615-634
  ◇ H05 ◇ REV Zbl 574 # 03051 • ID 44908

CUDA, K. see Vol. III, V, VI for further entries

CUESTA DUTARI, N. [1953] *Ordinal deductive models (Spanish)* (J 0236) Rev Mat Hisp-Amer, Ser 4 13*211-223
  ◇ B20 E07 ◇ REV MR 15.690 Zbl 51.289 • ID 02588

CUESTA DUTARI, N. [1954] *Deductive structures (Spanish)* (J 0236) Rev Mat Hisp-Amer, Ser 4 14*104-117
  ◇ B22 ◇ REV MR 16.987 Zbl 56.247 • ID 02589

CUESTA DUTARI, N. [1962] *Implication structures* (J 0296) Acta Salamantica Cienc 6*22pp
  ◇ B20 ◇ REV MR 28 # 3921 • ID 04217

CUESTA DUTARI, N. see Vol. V for further entries

CUFI, J. [1973] see BATLE, N.

CULBERTSON, J.T. [1958] *Mathematics and logic for digital devices* (X 0864) Van Nostrand: New York x + 224pp
  ◇ B98 G05 ◇ REV MR 19.1200 Zbl 80.335 JSL 23.366 • ID 23519

CULIK, K. [1969] *Problems in the theory of minimization of boolean expressions (Czech) (English summary)* (J 0086) Cas Pestovani Mat, Ceskoslov Akad Ved 94*340-353
  ◇ B35 ◇ REV MR 41 # 5190 Zbl 183.296 • ID 02596

CULIK, K. see Vol. IV, V for further entries

CUNNINGHAM, R.J. & DICK, A.J.J. [1985] *Rewrite systems on a lattice of types* (J 1431) Acta Inf 22*149-169
  ◇ B35 ◇ REV Zbl 575 # 68043 • ID 48875

CURRY, H.B. [1936] *A mathematical treatment of the rules of the syllogism* (J 0094) Mind 45*209-216
  ◇ B20 ◇ REV Zbl 13.289 JSL 1.114 FdM 62.32 • ID 33833

CURRY, H.B. [1937] *On the use of dots as brackets in logical expressions* (J 0036) J Symb Logic 2*26-28
  ◇ B03 ◇ REV Zbl 16.337 JSL 2.90 FdM 63.25 • ID 02607

CURRY, H.B. [1941] *A formalization of recursive arithmetic* (J 0100) Amer J Math 63*263-282
  • TRANSL [1970] (C 3626) Rekursiv Mat Analiz 3*437-461 (Russian)
  ◇ B28 D20 F30 ◇ REV MR 2.340 Zbl 25.5 JSL 7.42 • ID 02612

CURRY, H.B. [1950] *A theory of formal deducibility* (X 0845) Univ Notre Dame Pr: Notre Dame ix + 126pp
  ◇ B10 F50 F98 ◇ REV MR 11.487 Zbl 41.348 JSL 16.56 JSL 34.113 • REM 2nd ed. 1957; 3rd ed 1966, xi + 129pp • ID 04222

CURRY, H.B. [1952] *Leçons de logique algébrique* (X 0834) Gauthier-Villars: Paris 163pp
  ◇ B98 G15 G98 ◇ REV MR 13.613 Zbl 48.2 JSL 19.146 • ID 02741

CURRY, H.B. [1952] *On the definition of negation by a fixed proposition in inferential calculus* (J 0036) J Symb Logic 17*98-104
  ◇ B20 ◇ REV MR 14.122 Zbl 47.251 JSL 18.266 • ID 02622

CURRY, H.B. [1952] *On the definition of substitution, replacement and allied notions in an abstract formal system* (J 0252) Rev Philos Louvain 50*251-269
◇ B03 ◇ REV JSL 21.375 • ID 33827

CURRY, H.B. [1952] *The inferential theory of negation* (P 0593) Int Congr Math (II, 6);1950 Cambridge MA 1*722
◇ A05 B20 F50 ◇ ID 28059

CURRY, H.B. [1952] *The permutability of rules in the classical inferential calculus* (J 0036) J Symb Logic 17*245-248
◇ B10 ◇ REV MR 14.527 Zbl 48.2 JSL 20.66 • ID 02623

CURRY, H.B. [1952] *The system LD* (J 0036) J Symb Logic 17*35-42
◇ B10 F50 ◇ REV MR 13.811 Zbl 48.2 JSL 18.266 • ID 02620

CURRY, H.B. [1954] *Generalization of the deduction theorem* (P 0575) Int Congr Math (II, 7);1954 Amsterdam 2*399-400
◇ B05 ◇ ID 29464

CURRY, H.B. [1954] *Remarks on the definition and nature of mathematics* (J 0076) Dialectica 8*228-233
• TRANSL [1967] (C 2141) Filos Matematica 153-159 (Italian) • REPR [1964] (C 1105) Phil of Math. Sel Readings 152-156
◇ A05 B98 ◇ REV JSL 22.85 JSL 34.108 JSL 5.26 • ID 37345

CURRY, H.B. [1957] *Combinatory logic* (P 1675) Summer Inst Symb Log;1957 Ithaca 90-99
• REPR [1968] (C 0552) Phil Contemp - Chroniques 295-307
◇ B40 B98 ◇ REV Zbl 158.247 • ID 29324

CURRY, H.B. [1958] *Calculuses and formal systems* (J 0076) Dialectica 12*249-273
◇ A05 B22 ◇ REV MR 21 # 4092 Zbl 92.250 JSL 25.350 • ID 02626

CURRY, H.B. & FEYS, R. [1958] *Combinatory logic. Volume I* (X 0809) North Holland: Amsterdam xvi + 417pp
• TRANSL [1967] (X 1781) Tecnos: Madrid
◇ B40 B98 ◇ REV MR 20 # 817 MR 39 # 5368 Zbl 81.241 JSL 32.267 • REM With two sections by Craig,W.. 2nd ed. 1968. For vol.II 1972 see Curry,H.B. & Hindley,J.R. & Seldin,J.P. • ID 02627

CURRY, H.B. [1958] *On definitions in formal systems* (J 0079) Logique & Anal, NS 1*105-114
◇ B22 ◇ REV JSL 25.89 • ID 02629

CURRY, H.B. [1960] *The inferential approach to logical calculus I* (J 0079) Logique & Anal, NS 3*119-136
◇ B10 ◇ REM Part II 1961 • ID 33828

CURRY, H.B. [1961] *The inferential approach to logical calculus II* (J 0079) Logique & Anal, NS 4*5-22
◇ B10 ◇ REM Part I 1960 • ID 33829

CURRY, H.B. [1963] *Foundations of mathematical logic* (X 0822) McGraw-Hill: New York xii + 408pp
• TRANSL [1969] (X 0885) Mir: Moskva 567pp • LAST ED [1977] (X 0813) Dover: New York viii + 408pp
◇ B98 ◇ REV MR 26 # 6036 Zbl 163.242 Zbl 172.8 JSL 38.149 • ID 02634

CURRY, H.B. [1967] *Combinatory logic* (C 0601) Encycl of Philos 4*504-509
◇ A05 B40 B98 ◇ REV JSL 35.299 • ID 02637

CURRY, H.B. [1968] *A deduction theorem for inferential predicate calculus* (P 0608) Logic Colloq;1966 Hannover 91-108
◇ B10 ◇ REV MR 39 # 1289 Zbl 186.4 • ID 04227

CURRY, H.B. [1968] *The equivalence of two definitons of elementary formal system* (J 0020) Compos Math 20*13-20
• REPR [1968] (C 0727) Logic Found of Math (Heyting) 13-20
◇ B22 ◇ REV MR 39 # 35 Zbl 167.10 • ID 02640

CURRY, H.B. [1968] *The purposes of logical formalizations* (J 0079) Logique & Anal, NS 11*358-366
◇ B10 ◇ REV MR 38 # 4273 Zbl 165.302 • ID 02639

CURRY, H.B. see Vol. II, IV, VI for further entries

CURTIS, H.A. [1976] *Simplified decomposition of boolean functions* (J 0187) IEEE Trans Comp C-25*1033-1044
◇ B05 ◇ REV MR 57 # 9316 Zbl 352 # 94028 • ID 50076

CURTIS, H.A. [1979] *Short-cut method of deriving nearly optimal arrays of NAND trees* (J 0187) IEEE Trans Comp C-28*521-528
◇ B35 ◇ REV MR 80k:94051 Zbl 409 # 94032 • ID 69383

CUTLAND, N.J. [1982] *Infinitesimal methods in measure theory, probability theory and stochastic analysis* (J 2679) Bull Inst Math Appl (Southend oS) 18*52-57
◇ H10 ◇ REV MR 84j:03139e Zbl 494 # 03052 • ID 34733

CUTLAND, N.J. [1982] *On the existence of solutions to stochastic differential equations on Loeb spaces* (J 0982) Z Wahrscheinltheor & Verw Geb 60*335-357
◇ H10 ◇ REV MR 83h:60059 Zbl 544 # 60056 • ID 39265

CUTLAND, N.J. [1982] *Optimal controls for partially observed stochastic systems using nonstandard analysis* (P 3869) Stoch Diff Syts (2);1982 Bad Honnef 276-284
◇ H10 ◇ REV Zbl 509 # 93066 • ID 36804

CUTLAND, N.J. [1983] *Internal controls and relaxed controls* (J 3172) J London Math Soc, Ser 2 27*130-140
◇ H10 ◇ REV MR 84i:49019 Zbl 495 # 49002 • ID 40122

CUTLAND, N.J. [1983] *Nonstandard measure theory and its applications* (J 0161) Bull London Math Soc 15*529-589
◇ H10 H98 ◇ REV MR 85b:28001 Zbl 529 # 28009 • ID 38483

CUTLAND, N.J. [1983] *Optimal controls for partially observed stochastic systems: an infinitesimal approach* (J 3928) Stochastics 8*239-257
◇ H10 ◇ REV MR 84d:93065 Zbl 503 # 93073 • ID 36682

CUTLAND, N.J. [1984] *A question of Borel hyperdeterminacy* (J 0068) Z Math Logik Grundlagen Math 30*313-316
◇ E60 H05 ◇ REV MR 86f:03077 Zbl 526 # 90103 • ID 42232

CUTLAND, N.J. [1985] *Simplified existence for solutions to stochastic differential equations* (J 3928) Stochastics 14*319-325
◇ H10 ◇ REV Zbl 566 # 60059 • ID 48746

CUTLAND, N.J. see Vol. II, III, IV, V for further entries

CUTLER, R.B. & MUROGA, S. [1980] *Useless prime implicants of incompletely specified multiple-output switching functions* (J 0435) Int J Comput & Inf Sci 9*337-350
◇ B35 ◇ REV MR 81m:94031 Zbl 443 # 94034 • ID 69384

CVETKOVIC, D.M. & PEVAC, I. [1982] *Algorithms for transforming first order formulas in their natural form* (S 1003) Publ Elektroteh, Ser Mat Fiz, Beograde 735-762*155-160
⋄ B10 ⋄ REV Zbl 543 # 68075 • ID 40979

CVETKOVIC, D.M. [1983] *Discussing graph theory with a computer II. Theorems suggested by the computer* (J 0400) Publ Inst Math, NS (Belgrade) 33(47)*29-33
⋄ B35 ⋄ REV MR 85e:05151 Zbl 522 # 05068 • ID 39856

CVETKOVIC, D.M. & PEVAC, I. [1983] *Discussing graph theory with a computer III. Man-machine theorem proving* (J 0400) Publ Inst Math, NS (Belgrade) 34(48)*37-47
⋄ B35 ⋄ REV Zbl 568 # 05053 • REM Part II 1983. Part IV 1984 • ID 44661

CVETKOVIC, D.M. [1984] *Discussing graph theory with a computer. IV. Knowledge organization and examples of theorem proving* (P 4446) Graph Th;1983 Novi Sad 43-68
⋄ B35 ⋄ REV MR 85g:05004 Zbl 533 # 05058 • REM Part III 1983 by Cvetkovic,D.M. & Pevac,I. • ID 45836

CVETKOVIC, D.M. & PEVAC, I. [1984] *Some heuristic in automatic theorem proving* (J 0400) Publ Inst Math, NS (Belgrade) 35(49)*167-171
⋄ B35 ⋄ REV MR 86e:68091 • ID 48910

CZAJSNER, J. [1976] *Finitely axiomatizable sets on the basis of a system of the propositional calculus* (J 2718) Fct Approximatio, Comment Math, Poznan 4*149-157
⋄ B05 ⋄ REV MR 56 # 15399 Zbl 363 # 02053 • ID 50880

CZAJSNER, J. [1978] *Axiomatizable sets on the basis of a system of the propositional calculus* (J 2718) Fct Approximatio, Comment Math, Poznan 6*135-136
⋄ B05 ⋄ REV MR 80i:03018 Zbl 387 # 03004 • ID 52220

CZAJSNER, J. see Vol. V for further entries

CZELAKOWSKI, J. [1974] *Logics based on partial boolean σ-algebras. I* (J 0063) Studia Logica 33*371-396
⋄ B22 G05 ⋄ REV MR 50 # 15622 Zbl 331 # 02038 • REM Part II 1975 • ID 31654

CZELAKOWSKI, J. [1979] *A characterization of Matr(C)* (J 0387) Bull Sect Logic, Pol Acad Sci 8*83-86
⋄ B22 C05 ⋄ REV MR 80g:03024 Zbl 409 # 03016 • ID 31660

CZELAKOWSKI, J. [1979] *Large matrices which induce finite consequence operations* (J 0387) Bull Sect Logic, Pol Acad Sci 8*79-82
⋄ B22 C20 ⋄ REV MR 80g:03023 • ID 31659

CZELAKOWSKI, J. [1980] *Model-theoretic methods in methodology of propositional calculi* (X 2733) Acad Sci Inst Phi Soc: Wroclaw 75pp
⋄ B22 C05 G99 ⋄ REV MR 82c:03046 Zbl 445 # 03008 • ID 56472

CZELAKOWSKI, J. [1980] *Reduced products of logical matrices* (J 0063) Studia Logica 39*19-43
⋄ B22 C05 C20 ⋄ REV MR 81h:03060 Zbl 445 # 03009 • ID 56473

CZELAKOWSKI, J. [1981] *Equivalential logics. I,II* (J 0063) Studia Logica 40*227-236,355-372
⋄ B22 ⋄ REV MR 81h:03059 MR 84k:03072 MR 84m:03041 Zbl 476 # 03032 Zbl 492 # 03008 • ID 55544

CZELAKOWSKI, J. [1982] *Logical matrices and the amalgamation property* (J 0063) Studia Logica 41*329-341
⋄ B22 C40 C52 C90 ⋄ REV MR 85i:03091 Zbl 549 # 03014 • ID 42743

CZELAKOWSKI, J. [1983] *Algebraic aspects of deduction theorems* (J 0387) Bull Sect Logic, Pol Acad Sci 12*111-116
⋄ B22 ⋄ REV Zbl 554 # 03018 • ID 44940

CZELAKOWSKI, J. [1983] *Matrices, primitive satisfaction and finitely based logics* (J 0063) Studia Logica 42*89-104
⋄ B22 ⋄ ID 42322

CZELAKOWSKI, J. [1984] *Filter distributive logics* (J 0063) Studia Logica 43*353-378
⋄ B22 ⋄ ID 42385

CZELAKOWSKI, J. [1984] *Remarks on finitely based logics* (P 2153) Logic Colloq;1983 Aachen 1*147-168
⋄ B22 C90 ⋄ ID 41764

CZELAKOWSKI, J. [1985] *Sentential logics and Maehara interpolation property* (J 0063) Studia Logica 44*265-284
⋄ B22 C05 C40 C52 ⋄ ID 47502

CZELAKOWSKI, J. see Vol. II, III for further entries

CZERMAK, J. [1977] *A remark on Gentzen's calculus of sequents* (J 0047) Notre Dame J Formal Log 18*471-474
⋄ B20 F07 ⋄ REV MR 58 # 21500 Zbl 314 # 02026 • ID 24249

CZERMAK, J. see Vol. II, V for further entries

CZOGALA, E. & PEDRYCZ, W. [1981] *Methods of multivalued logics and fuzzy set theory in computer-aided decision making in diagnostic and therapeutic processes* (J 2815) Pol Tow Cybern, Kwart Nauk, Wroclaw 4*61-67
⋄ B80 ⋄ REV MR 82i:92004 • ID 81001

CZOGALA, E. see Vol. II, V for further entries

DABIJA, V. [1980] *On a method for automated minimization of boolean functions based on if-then-else trees* (J 3130) Bul Inst Politeh Bucuresti, Ser Electroteh 42/3*105-112
⋄ B35 ⋄ REV MR 82g:94027 Zbl 461 # 94013 • ID 69019

DABIJA, V. [1981] *On an IF-THEN-ELSE technique for automated theorem proving in propositional calculus* (J 3130) Bul Inst Politeh Bucuresti, Ser Electroteh 43/2*95-100
⋄ B35 ⋄ REV MR 83f:03011 Zbl 469 # 68087 • ID 55190

DACUNHA-CASTELLE, D. & KRIVINE, J.-L. [1975] *Sous-espaces de L^1 (English summary)* (J 2313) C R Acad Sci, Paris, Ser A-B 280*A645-A648
⋄ H05 ⋄ REV MR 51 # 1365 Zbl 295 # 46046 • ID 17495

DACUNHA-CASTELLE, D. see Vol. III, V for further entries

DAI, ZONGDUO & WAN, ZHEXIAN [1980] *Conditions for a Boolean function to be independent of some variable (Chinese)* (J 2771) Kexue Tongbao 25*135-137
⋄ B05 ⋄ ID 44165

DALE, A.J. [1983] *The non-independence of axioms in a propositional calculus formulated in terms of axiom schemata* (J 0079) Logique & Anal, NS 26*91-98
⋄ B05 ⋄ REV MR 85d:03015 Zbl 529 # 03001 • ID 37645

DALE, A.J. see Vol. II for further entries

DALE, T. [1979] *A natural deduction system for "if then"* (J 0079) Logique & Anal, NS 22*339-345
⋄ B20 ⋄ REV MR 81h:03026 • ID 72097

DALEN VAN, D. [1958] see BAR-HILLEL, Y.

DALEN VAN, D. [1972] *Logik und formale Theorien (Niederlaendisch)* (J 0290) Euclides 48*384-394
⋄ B98 ⋄ REV MR 54#12468 Zbl 264#02002 • ID 27492

DALEN VAN, D. [1979] see BOFFA, M.

DALEN VAN, D. [1980] *Logic and structure* (X 0811) Springer: Heidelberg & New York viii+172pp
⋄ B98 C98 ⋄ REV MR 81f:03001 MR 84k:03002 Zbl 434#03001 • REM 2nd ed. 1983; X+207pp • ID 55705

DALEN VAN, D. & LASCAR, D. & SMILEY, T.J. (EDS.) [1982] *Logic colloquium '80. Eur. Summer Meet., Prague 1980* (S 3303) Stud Logic Found Math 108*x+342pp
⋄ B97 D97 ⋄ REV MR 83i:03003 Zbl 489#00006 • ID 36589

DALEN VAN, D. [1982] see TROELSTRA, A.S.

DALEN VAN, D. see Vol. II, III, IV, V, VI for further entries

DALEY, R.P. [1977] *On the inference of optimal descriptions* (J 1426) Theor Comput Sci 4*301-319
⋄ B35 D15 ⋄ REV MR 56#8351 Zbl 375#02041 • ID 31707

DALEY, R.P. see Vol. II, IV for further entries

DALLA CHIARA SCABIA, M.L. [1976] *A general approach to non-distributive logics* (J 0063) Studia Logica 35*139-162
⋄ B22 ⋄ REV MR 55#10249 Zbl 333#02024 • ID 31064

DALLA CHIARA SCABIA, M.L. & TORALDO DI FRANCIA, G.G. [1976] *The logical dividing line between deterministic and indeterministic theories* (J 0063) Studia Logica 35*1-5
⋄ A05 B30 ⋄ REV MR 55#12448 Zbl 327#02007 • ID 31067

DALLA CHIARA SCABIA, M.L. & TORALDO DI FRANCIA, G.G. [1979] *Formal analysis of physical theories* (P 3234) Probl Founds of Physics;1977 Varenna 134-201
⋄ A05 B30 B51 ⋄ REV MR 82m:81003 Zbl 446#00020 • ID 56529

DALLA CHIARA SCABIA, M.L. (ED.) [1985] *Present state of the problem of the foundations of mathematics* (J 0154) Synthese 62*123-315
⋄ B98 ⋄ REV MR 86d:03003 • ID 45510

DALLA CHIARA SCABIA, M.L. see Vol. II, III, V, VI for further entries

DAMERI, E. [1982] see BELLIA, M.

DAMYANOV, B.P. & KHRISTOV, KH.YA. [1981] *Changing of the independent variable in the theory of asymptotic functions* (P 2267) Obobsh Funk & Primen Mat Fiz;1980 Moskva 535-551
⋄ H05 ⋄ REV MR 84e:46038 Zbl 525#46025 • ID 39585

DAMYANOV, B.P. see Vol. III for further entries

DANCY, R.M. [1975] *Sense and contradiction: A study in Aristotle* (X 0835) Reidel: Dordrecht xii+184pp
⋄ A10 B20 ⋄ REV Zbl 436#01001 • ID 55840

DANIELL, P.J. [1918] *Independence proofs and the theory of implication* (J 0320) Monist 28*451-453
⋄ A05 B20 ⋄ ID 04173

DANIELL, P.J. see Vol. V for further entries

DANIELS, C.B. & FREEMAN, J.B. [1977] *Classical second-order intensional logic with maximal propositions* (J 0122) J Philos Logic 6*1-31
⋄ B15 C85 C90 ⋄ REV MR 58#10310 Zbl 353#02005 • ID 30696

DANIELS, C.B. & FREEMAN, J.B. [1978] *A logic of generalized quantification* (J 0302) Rep Math Logic, Krakow & Katowice 10*9-42
⋄ B15 C80 C90 ⋄ REV MR 81m:03022 Zbl 446#03008 • ID 56537

DANIELS, C.B. see Vol. II for further entries

DANIELSSON, P.E. & PLAVSIC, V.M. [1979] *Sequential evaluation of boolean functions* (J 0187) IEEE Trans Comp C-28*879-887
⋄ B05 ⋄ REV Zbl 428#94015 • ID 69777

DANQUAH, J. [1976] *The circularity of the proof of the non-independence of the fourth axiom of "Principia Mathematica"* (J 0103) Analysis (Oxford) 36*110-111
⋄ A05 B05 ⋄ REV Zbl 347#02005 • ID 61233

DANTSIN, E.YA. [1979] *Parameters defining the time of tautology recognition by the splitting method (Russian)* (S 2582) Semiotika & Inf, Akad Nauk SSSR 12*8-17
⋄ B35 D15 ⋄ REV MR 81i:68130 • ID 81024

DANTSIN, E.YA. [1981] *Two systems for proving tautologies based on the splitting method (Russian) (English summary)* (S 0228) Zap Nauch Sem Leningrad Otd Mat Inst Steklov 105*24-44,198-199
• TRANSL [1983] (J 1531) J Sov Math 22*1293-1305
⋄ B35 D15 F20 ⋄ REV MR 83i:68140 Zbl 476#03020 Zbl 509#03004 • ID 55532

DANTSIN, E.YA. see Vol. IV, VI for further entries

DANTZIG VAN, D. [1949] *Significs and its relation to semiotics* (P 0682) Int Congr Philos (10);1948 Amsterdam 2*176-189
⋄ B03 ⋄ ID 37350

DANTZIG VAN, D. see Vol. II, V, VI for further entries

DAPUETO, C. [1982] *Sull' aritmetizzazione della sintassi del primo ordine (French summary)* (J 0549) Riv Mat Univ Parma, Ser 4 8*361-371
⋄ B28 F30 F40 ⋄ REV MR 85e:03146 Zbl 524#03009 • ID 45889

DAPUETO, C. see Vol. III, V for further entries

DARDZHANIYA, G.K. [1979] *Polynomial complexities of deduction of some logical calculi (Russian) (English summary)* (J 0288) Vest Ser Mat Mekh, Univ Moskva 1979/3*10-18,84
• TRANSL [1979] (J 0510) Moscow Univ Math Bull 34/3*8-17
⋄ B35 F20 ⋄ REV MR 81a:03055 Zbl 413#03032 • ID 53026

DARDZHANIYA, G.K. see Vol. II, VI for further entries

DARLINGTON, J.L. [1966] *Machine methods for proving logical arguments expressed in English* (P 0573) Inform Processing (3);1965 New York 2*530-531
⋄ B35 ⋄ ID 29442

DARLINGTON, J.L. [1968] *Automatic theorem-proving with equality substitutions and mathematical induction* (J 0508) Machine Intelligence 3*113-127
⋄ B35 ⋄ REV Zbl 214.26 • ID 47300

DARLINGTON, J.L. [1968] *Some theorem-proving strategies based on the resolution principle* (**J** 0508) Machine Intelligence 2∗51-71
⋄ B35 ⋄ REV Zbl 213.182 • ID 47301

DARLINGTON, J.L. [1969] *Theorem proving and information retrieval* (**J** 0508) Machine Intelligence 4∗173-181
⋄ B35 ⋄ REV Zbl 213.430 • ID 47302

DARLINGTON, J.L. [1969] *Theorem provers as question answerers* (**P** 4250) Int Joint Conf Artif Intell (1);1969 Washington 317
⋄ B35 ⋄ ID 47303

DARLINGTON, J.L. [1971] *Deductive question answering and plan formation in second order logic* (**C** 1566) Semin IRIA Log & Automates 1971 55-73
⋄ B35 ⋄ REV Zbl 268 # 68040 • ID 61234

DARLINGTON, J.L. [1976] *Automatic synthesis of SNOBOL programs* (**P** 3143) Comput Orient Learn Process;1974 Bonas 443-453
⋄ B35 ⋄ REV Zbl 376 # 68014 • ID 51708

DAS, S.R. [1966] see CHOUDHURY, A.K.

DASSOW, J. & WOLTER, U. [1985] *Remarks on equality languages* (**J** 3289) Wiss Z Tech Hochsch Madgeburg 29∗122-126
⋄ A05 B20 D05 ⋄ ID 49482

DASSOW, J. see Vol. II, IV for further entries

DAVAL, R. & GUILBAUD, G.T. [1945] *Le raisonnement mathematique* (**X** 0840) Pr Univ France: Paris 152pp
⋄ A05 B98 ⋄ REV MR 8.4 • ID 25653

DAVIDSON, D. & HINTIKKA, K.J.J. (EDS.) [1969] *Words and objections: Essays on the work of W.V.Quine* (**X** 0835) Reidel: Dordrecht vii + 366pp
⋄ A05 B97 ⋄ REV Zbl 188.311 • ID 25229

DAVIDSON, D. see Vol. II for further entries

DAVIES, M.K. [1980] *A note on substitutional quantification* (**J** 0097) Nous, Quart J Phil 14∗619-622
⋄ B10 ⋄ REV MR 82c:03012 • ID 72148

DAVIES, M.K. see Vol. II for further entries

DAVIO, M. & DESCHAMPS, J.-P. & THAYSE, A. [1978] *Optimization of multivalued decision algorithms* (**P** 2014) Int Symp Multi-Val Log (8);1978 Rosemont 171-178
⋄ B35 ⋄ REV MR 81b:94073 • ID 82968

DAVIO, M. see Vol. II for further entries

DAVIS, CHARLES C. [1975] *An investigation concerning the Hilbert-Sierpinski logical form of the axiom of choice* (**J** 0047) Notre Dame J Formal Log 16∗145-184 • ERR/ADD ibid 16∗608
⋄ B15 E25 ⋄ REV MR 51 # 10092 MR 51 # 153 Zbl 258 # 02062 • ID 02815

DAVIS, CHARLES C. see Vol. II, IV, V for further entries

DAVIS, L. [1979] *An alternate formulation of Kripke's theory of truth* (**J** 0122) J Philos Logic 8∗289-296
⋄ A05 B15 ⋄ REV MR 80j:03020 Zbl 413 # 03004 • REM Correction by Hazen,A. ibid. 10∗309-311 • ID 52998

DAVIS, MARTIN D. [1957] *A program for Presburger's algorithm* (**P** 1675) Summer Inst Symb Log;1957 Ithaca 215-223
• REPR [1983] (**C** 4659) Autom of Reasoning 1∗41-48
⋄ B35 ⋄ REV JSL 31.138 • ID 02825

DAVIS, MARTIN D. [1959] *Lecture notes on mathematical logic* (**X** 0888) New York Univ Inst Math Sci: New York 91pp
⋄ B98 ⋄ REV JSL 35.167 • ID 17204

DAVIS, MARTIN D. & PUTNAM, H. [1960] *A computing procedure for quantification theory* (**J** 0037) ACM J 7∗201-215
• REPR [1983] (**C** 4659) Autom of Reasoning 1∗125-139
⋄ B35 ⋄ REV MR 24 # B492 JSL 31.125 • ID 02823

DAVIS, MARTIN D. & LOGEMANN, G. & LOVELAND, D.W. [1962] *A machine program for theorem-proving* (**J** 0212) ACM Commun 5∗394-397
• REPR [1983] (**C** 4659) Autom of Reasoning 1∗267-270
⋄ B35 ⋄ REV MR 26 # 7175 JSL 32.118 • ID 04179

DAVIS, MARTIN D. [1963] *Eliminating the irrelevant from mechanical proofs* (**P** 0719) Exper Arith, High Speed Comput & Math;1962 Chicago 15-30
• REPR [1983] (**C** 4659) Autom of Reasoning 1∗315-330
⋄ B35 ⋄ REV MR 30 # 735 Zbl 131.12 JSL 32.118 • ID 04185

DAVIS, MARTIN D. [1974] *Computability* (**X** 1214) New York Univ Cour Inst Math: New York 248pp
⋄ B98 D98 ⋄ REV MR 50 # 77 Zbl 281 # 02038 • ID 21268

DAVIS, MARTIN D. [1977] *Applied nonstandard analysis* (**X** 0827) Wiley & Sons: New York xii + 181pp
• TRANSL [1980] (**X** 0885) Mir: Moskva 238pp
⋄ H05 H98 ⋄ REV MR 58 # 21590 Zbl 359 # 02060 JSL 43.383 • ID 50607

DAVIS, MARTIN D. & SCHWARTZ, J.T. [1979] *Metamathematical extensibility for theorem verifiers and proof-checkers* (**J** 2687) Comp Math Appl 5∗217-230
⋄ B35 ⋄ REV MR 81b:68113 Zbl 418 # 68079 • ID 69427

DAVIS, MARTIN D. see Vol. IV, VI for further entries

DAVIS, P.J. [1969] see CERUTTI, E.

DAVYDOV, G.V. & MASLOV, S.YU. & MINTS, G.E. & OREVKOV, V.P. & SHANIN, N.A. & SLISENKO, A.O. [1965] *An algorithm of computer searching for natural logical proof in propositional calculus (Russian)* (**X** 2027) Nauka: Moskva 39pp
• TRANSL [1983] (**C** 4659) Autom of Reasoning 1∗424-483
⋄ B35 ⋄ REV MR 33 # 5405 Zbl 198.36 • ID 33020

DAVYDOV, G.V. [1967] *A method of establishing deducibility in the classical predicate calculus (Russian)* (**S** 0228) Zap Nauch Sem Leningrad Otd Mat Inst Steklov 4∗8-17
• TRANSL [1967] (**J** 0521) Semin Math, Inst Steklov 4∗1-4
⋄ B10 B35 ⋄ REV MR 39 # 69 Zbl 165.317 • ID 02842

DAVYDOV, G.V. [1967] *The correction of unprovable formulas (Russian)* (**S** 0228) Zap Nauch Sem Leningrad Otd Mat Inst Steklov 4∗18-29
• TRANSL [1967] (**J** 0521) Semin Math, Inst Steklov 4∗5-8
⋄ B10 B35 ⋄ REV MR 39 # 70 Zbl 187.284 • ID 02841

DAVYDOV, G.V. [1968] *Some remarks on the search for a deduction in the predicate calculus (Russian)* (**S** 0228) Zap Nauch Sem Leningrad Otd Mat Inst Steklov 8∗8-20
• ERR/ADD ibid 16∗185-187
• TRANSL [1968] (**J** 0521) Semin Math, Inst Steklov 8∗1-6
⋄ B35 ⋄ REV MR 43 # 3089 Zbl 172.298 • ID 02843

DAVYDOV, G.V. & MASLOV, S.YU. & MINTS, G.E. & OREVKOV, V.P. & SLISENKO, A.O. [1969] *A computer algorithm for establishing deducibility based on inverse method (Russian)* (S 0228) Zap Nauch Sem Leningrad Otd Mat Inst Steklov 16*8-19 • ERR/ADD ibid 20*292-294
• TRANSL [1969] (J 0521) Semin Math, Inst Steklov 16*1-6
[1983] (C 4659) Autom of Reasoning 2*531-541
⋄ B35 ⋄ REV MR 41 #3278 Zbl 198.36 • ID 02844

DAVYDOV, G.V. [1971] *Synthesis of the resolution method with the inverse method (Russian) (English summary)* (S 0228) Zap Nauch Sem Leningrad Otd Mat Inst Steklov 20*24-35,282
• TRANSL [1973] (J 1531) J Sov Math 1*12-18
⋄ B35 ⋄ REV MR 46 #3261 Zbl 232 #90040 • ID 02847

DAVYDOV, G.V. & SUVOROV, P.YU. [1974] *Reduction of propositional tautology to graph coloring in three colors (Russian) (English summary)* (S 0228) Zap Nauch Sem Leningrad Otd Mat Inst Steklov 40*10-13,155
• TRANSL [1977] (J 1531) J Sov Math 8*251-254
⋄ B05 ⋄ REV MR 51 #43 Zbl 357 #02008 • ID 15243

DEDEKIND, R. [1888] *Was sind und was sollen die Zahlen?* (X 0900) Vieweg: Wiesbaden xv+58pp
⋄ B28 D20 E75 ⋄ REV FdM 20.49 • REM 10th ed. 1965. For an English transl. see 1948 • ID 35601

DEDEKIND, R. [1932] *Gesammelte mathematische Werke. Vol.III* (X 0900) Vieweg: Wiesbaden 508pp
⋄ B28 B96 E75 E96 ⋄ REV Zbl 4.337 FdM 58.42
• ID 38623

DEDEKIND, R. [1948] *Essays on the theory of numbers. I. Continuity and irrational numbers. II. The nature and meaning of numbers* (X 0813) Dover: New York 115pp
⋄ B28 E75 ⋄ REM Translations of papers of Dedekind,R.
• ID 23344

DEDEKIND, R. [1967] *Letter to Keferstein* (C 0675) From Frege to Goedel 99-103
⋄ B28 E75 ⋄ ID 14914

DEGANO, P. & SIROVICH, F. [1979] *Inductive generalization and proofs of function properties* (J 2293) Comp Linguist & Comp Lang 13*101-130
⋄ B35 B75 D20 D80 ⋄ REV MR 81f:68111
Zbl 449 #68054 • ID 56755

DEGANO, P. [1982] see BELLIA, M.

DEGEN, J.W. [1983] *Systeme der kumulativen Logik* (X 3111) Philosophia: Muenchen 265pp
⋄ B15 ⋄ REV Zbl 537 #03005 • ID 43720

DEGTYAREV, A.I. & KAPITONOVA, YU.V. & LYALETSKIJ, A.V. [1973] *Use of heuristic procedures in programs for the search for proofs of theorems (a survey) (Russian) (English summary)* (J 0040) Kibernetika, Akad Nauk Ukr SSR 1973/4*76-89
⋄ B35 ⋄ REV MR 48 #10220 Zbl 271 #68061 • ID 06914

DEGTYAREV, A.I. [1982] *On the forms of deduction in calculi with equality and the paramodulation rule (Russian)* (C 3832) Avtom Issl Mat 14-26
⋄ B35 ⋄ REV MR 85d:03023 Zbl 528 #03006 • ID 37625

DEGTYAREV, A.I. see Vol. VI for further entries

DEJON, B. & HENRICI, P. (EDS.) [1969] *Constructive aspects of the fundamental theorem of algebra: proceedings of a symposium, IBM* (X 0827) Wiley & Sons: New York 337pp
⋄ B28 C97 F97 ⋄ REV Zbl 175.1 • ID 23438

DELGADO, V.M. [1957] *La interpretacion formalista de la silogistica de Aristoteles* (J 1813) Estudios 13*167-176
⋄ A10 B20 ⋄ ID 32295

DELGADO, V.M. [1957] *Un nuevo modo de presentar la formalizacion de la ciencia* (J 1813) Estudios 13*39-64
⋄ B30 ⋄ ID 32294

DELGADO, V.M. [1968] *Los lenguajes formalizados de la logica moderna* (P 1815) Congr Int Ling & Filol Roman (11);1965 Madrid 327-347
⋄ B10 ⋄ ID 32303

DELGADO, V.M. [1972] *Lecciones de logica I. Introduccion general. Logica de proposiciones* (X 3333) Unknown Publisher: See Remarks 251pp
⋄ B05 B10 ⋄ REM Part II 1974 • ID 32305

DELGADO, V.M. [1974] *Lecciones de logica II. Logica de la cuantificacion. Algebra de clases. Otros temas. Historia de la logica deductiva. Logica inductiva* (X 3333) Unknown Publisher: See Remarks 254pp
⋄ A10 B10 ⋄ REM Part I 1972 • ID 32306

DELIYANNI, A. & KOWALSKI, R. [1979] *Logic and semantic networks* (J 0212) ACM Commun 22*184-192
⋄ B35 ⋄ REV Zbl 394 #68063 • ID 69440

DELL'ORCO, P. [1973] *A syntactic approach to automatic theorem proving* (P 3361) APL Congr '73;1973 Copenhagen 75-82
⋄ B35 ⋄ REV Zbl 272 #68061 • ID 61285

DELOBEL, C. & FAGIN, R. & PARKER JR., D.S. & SAGIV, Y. [1981] *An equivalence between relational database dependencies and a fragment of propositional logic* (J 0037) ACM J 28*435-453
⋄ B20 ⋄ REV MR 82h:68025 Zbl 462 #68082 • ID 54536

DELON, F. & LASCAR, D. & PARIGOT, M. & SABBAGH, G. (EDS.) [1984] *Logique, Octobre 1983, Paris. Compte rendu de la table ronde de logique des 15 et 16 octobre 1983, Paris* (S 3521) Mem Soc Math Fr 16*iii+103pp
⋄ B97 C97 ⋄ REV MR 86f:03003 Zbl 547 #00009
• ID 43258

DELON, F. & LASCAR, D. & LOUVEAU, A. & SABBAGH, G. (EDS.) [1985] *Seminaire general de logique 1982-83* (X 4643) Univ Paris VII, UER Math: Paris 186pp
⋄ B97 C98 D97 ⋄ REV MR 86f:03005 Zbl 567 #00003
• ID 47516

DELON, F. see Vol. III, IV, VI for further entries

DELONG, H. [1970] *A profile of mathematical logic* (X 0832) Addison-Wesley: Reading xiii+304pp
⋄ A05 A10 B98 ⋄ REV MR 41 #3230 Zbl 248 #02001 JSL 40.101 • ID 02907

DEMETROVICS, J. [1975] *The structural investigation of two-valued logic (Hungarian) (Russian summary)* (J 1458) Alkalmaz Mat Lapok 1*405-424
⋄ B05 ⋄ REV MR 57 #9532 Zbl 366 #02011 • ID 51094

DEMETROVICS, J. see Vol. II for further entries

DEMUTH, O. [1967] *Necessary and sufficient conditions for Riemann integrability of constructive functions (Russian)* (J 0023) Dokl Akad Nauk SSSR 176*757-758
• TRANSL [1967] (J 0062) Sov Math, Dokl 8*1176-1177
⋄ F60 H05 ⋄ REV MR 37 #65 Zbl 224 #02020 • ID 02915

DEMUTH, O. [1968] *The connection between Riemann and Lebesgue integrability of constructive functions (Russian)* (S 0228) Zap Nauch Sem Leningrad Otd Mat Inst Steklov 8*29-31
  ⋄ F60 H05 ⋄ REV MR 39 # 377   Zbl 172.299 • ID 28592

DEMUTH, O. [1968] *The Lebesgue integral and the concept of measureability of functions in constructive analysis (Russian)* (S 0228) Zap Nauch Sem Leningrad Otd Mat Inst Steklov 8*21-28
  • TRANSL [1968] (J 0521) Semin Math, Inst Steklov 8*7-10
  ⋄ F60 H05 ⋄ REV MR 39 # 1605   Zbl 224 # 02024
  • ID 16949

DEMUTH, O. see Vol. IV, V, VI for further entries

DENENBERG, L. & LEWIS, H.R. [1984] *The complexity of the satisfiability problem for Krom formulas* (J 1426) Theor Comput Sci 30*319-341
  ⋄ B20 B25 D15 ⋄ ID 43373

DENENBERG, L. see Vol. III, IV for further entries

DENEV, J.D. [1971] *Evaluation du nombre des subfonctions des presque toutes fonctions de Boole (Bulgare)(Resume francais)* (J 0255) God Fak Mat & Mekh, Univ Sofiya 66*353-362
  ⋄ B05 ⋄ REV MR 57 # 5443   Zbl 327 # 02009 • ID 61318

DENEV, J.D. [1972] *The complexity of the realization of almost all functions of the algebra of logic by the method of cascades (Russian)* (J 0087) Mat Zametki (Akad Nauk SSSR) 12*769-780
  • TRANSL [1972] (J 1044) Math Notes, Acad Sci USSR 12*897-903
  ⋄ B05 ⋄ REV MR 47 # 10147   Zbl 296 # 94019 • ID 61316

DENEV, J.D. [1978] see COHEN, R.S.

DENEV, J.D. & TONCHEV, V.D. [1979] *On the number of equivalence classes of boolean functions under a transformation group* (J 0137) C R Acad Bulgar Sci 32*1609-1610
  ⋄ B05 ⋄ REV MR 81g:94058   Zbl 427 # 94025 • ID 69443

DENISE, T.C. [1973] *The two logics: traditional and modern* (J 0047) Notre Dame J Formal Log 14*510-518
  ⋄ A05 B20 ⋄ REV MR 49 # 2248   Zbl 265 # 02002
  • ID 04080

DENNING, P.J. & DENNIS, J.B. & QUALITZ, J.E. [1978] *Machines, languages, and computation* (X 0819) Prentice Hall: Englewood Cliffs xxii + 601pp
  ⋄ B98 D05 D10 D98 ⋄ REV Zbl 492 # 68003 JSL 45.630
  • ID 37736

DENNIS, J.B. [1978] see DENNING, P.J.

DERSHOWITZ, N. [1982] *Orderings for term-rewriting systems* (J 1426) Theor Comput Sci 17*279-301
  ⋄ B35 ⋄ REV MR 83e:68137   Zbl 525 # 68054 • ID 40137

DERSHOWITZ, N. & HSIANG, JIEH [1983] *Rewrite methods for clausal and non-clausal theorem proving* (P 3851) Automata, Lang & Progr (10);1983 Barcelona 331-346
  ⋄ B35 ⋄ REV MR 85g:68065   Zbl 523 # 68080 • ID 37001

DERSHOWITZ, N. see Vol. IV for further entries

DESCHAMPS, J.-P. [1972] *Parametric solutions of boolean equations* (J 0193) Discr Math 3*333-342
  ⋄ B05 ⋄ REV Zbl 253 # 06011 • ID 61332

DESCHAMPS, J.-P. [1975] *Fermetures i-geneatrices - application aux fonctions booleennes permutantes* (J 0193) Discr Math 13*321-339
  ⋄ B05 G15 ⋄ REV MR 53 # 223   Zbl 324 # 02007
  • ID 16670

DESCHAMPS, J.-P. [1978] see DAVIO, M.

DESCHAMPS, J.-P. & LAPSCHER, F. [1979] *Presentation et optimisation d'un programme de recherche des decompositions disjointes d'une fonction booleenne. Extension aux fonctions booleennes generales* (J 0060) Rev Roumaine Math Pures Appl 24*893-911
  ⋄ B35 ⋄ REV MR 81a:94057   Zbl 421 # 94027 • ID 53459

DESCHAMPS, J.-P. see Vol. II, IV for further entries

DESTOUCHES, J.-L. & FEVRIER, P. & JOFFRE, D. (EDS.) [1953] *Les methodes formelles en axiomatiques* (S 1802) Colloq Int CNRS 36*197pp
  ⋄ B30 ⋄ ID 48622

DESTOUCHES, J.-L. & FEVRIER, P. (EDS.) [1954] *Applications scientifiques de la logique mathematique. Actes du 2eme colloque international de logique mathematique* (X 0834) Gauthier-Villars: Paris 176pp
  ⋄ B80 ⋄ ID 48623

DESTOUCHES, J.-L. [1954] *La logique et les theories physiques* (P 0646) Appl Sci de Log Math;1952 Paris 119-128
  ⋄ B30 ⋄ REV MR 16.437   Zbl 58.245 • ID 27972

DESTOUCHES, J.-L. (ED.) [1967] *Logic & foundations of science. E.W.Beth memorial colloquium* (X 0835) Reidel: Dordrecht viii + 140pp
  ⋄ B97 ⋄ REV MR 39 # 30   Zbl 157.325 • ID 48626

DESTOUCHES, J.-L. see Vol. II, IV, VI for further entries

DESTOUCHES-FEVRIER, P. [1945] *Rapport entre le calcul des problemes et le calcul des propositions* (J 0109) C R Acad Sci, Paris 220*484-486
  ⋄ A05 B05 F50 ⋄ REV MR 7.185   Zbl 61.471 JSL 12.133
  • ID 02962

DESTOUCHES-FEVRIER, P. also published under the name FEVRIER, P.

DESTOUCHES-FEVRIER, P. see Vol. II, VI for further entries

DETERING, L. & REUSCH, B. [1979] *On the generation of prime implicants* (J 2095) Fund Inform, Ann Soc Math Pol, Ser 4 2*167-186
  ⋄ B05 ⋄ REV MR 81d:94041   Zbl 445 # 94050 • ID 56524

DETLEFSEN, M. [1979] *On interpreting Goedel's second theorem* (J 0122) J Philos Logic 8*297-313
  ⋄ A05 B28 ⋄ REV MR 80h:03008   Zbl 405 # 03003
  • ID 54878

DETLEFSEN, M. see Vol. VI for further entries

DETLOVS, V.K. [1970] *Elements of mathematical logic and set theory (Latvian)* (X 0895) Latv Valsts (Gos) Univ : Riga 292pp
  ⋄ B98 E98 ⋄ REV MR 47 # 8230 • ID 17027

DETLOVS, V.K. [1974] *Mathematical logic (Latvian)* (X 2230) Zinatne: Riga 279pp
  ⋄ B98 ⋄ REV MR 55 # 2465 • ID 72316

DETLOVS, V.K. see Vol. IV for further entries

DEUSSEN, P. [1963] *Bericht ueber ein Programm zur Uebersetzung boolescher Ausdruecke in disjunktive Normalform* (P 1612) Schaltkreis & -werk Th (2);1961 Saarbruecken 66-81
  ◊ B35 ◊ REV Zbl 115.125 • ID 27837

DEUSSEN, P. see Vol. IV for further entries

DEUTSCH, M. [1975] *Zur Benutzung der Verkettung als Basis fuer die Arithmetik* (J 0068) Z Math Logik Grundlagen Math 21*145-158
  ◊ B28 F30 ◊ REV MR 51 # 12498 Zbl 357 # 02036
  • ID 04086

DEUTSCH, M. [1975] *Zur Theorie der spektralen Darstellung von Praedikaten durch Ausdruecke der Praedikatenlogik 1.Stufe* (J 0009) Arch Math Logik Grundlagenforsch 17*9-16
  ◊ B10 C13 D25 D35 ◊ REV MR 54 # 2436
  Zbl 337 # 02025 • ID 02972

DEUTSCH, M. [1977] *Eine mengentheoretische Grundlegung der Theorie der Berechenbarkeit II* (J 0160) Math-Phys Sem-ber, NS 24*56-70
  ◊ B15 D20 D35 E47 ◊ REV MR 58 # 27403b
  Zbl 365 # 02029 • REM Part I 1976 • ID 28193

DEUTSCH, M. [1977] *Zum Begriff der Wortmischung als Basis fuer die Arithmetik* (J 0068) Z Math Logik Grundlagen Math 23*241-264
  ◊ B28 D05 D75 ◊ REV MR 56 # 15375 Zbl 402 # 03042
  • ID 26483

DEUTSCH, M. [1981] *Zur Reduktionstheorie des Entscheidungsproblems* (J 0068) Z Math Logik Grundlagen Math 27*113-117
  ◊ B20 D35 ◊ REV MR 82f:03037 Zbl 465 # 03004
  • ID 54907

DEUTSCH, M. [1984] *Reductions for the satisfiability with a simple interpretation of the predicate variable* (P 2342) Symp Rek Kombin;1983 Muenster 285-311
  ◊ B20 D35 ◊ REV Zbl 547 # 03004 • ID 43185

DEUTSCH, M. see Vol. III, IV, V, VI for further entries

DEVIDE, V. [1955] *Ein Axiomensystem fuer die natuerlichen Zahlen* (J 0008) Arch Math (Basel) 6*408-412
  ◊ B28 ◊ REV MR 17.448 Zbl 67.248 JSL 22.357
  • ID 02973

DEVIDE, V. [1959] *Elementare Aufzaehlung der minimalen erzeugenden Operationssysteme der Aussagenlogik* (J 0068) Z Math Logik Grundlagen Math 5*265-279
  ◊ B05 ◊ REV MR 23 # A1512 Zbl 92.4 • ID 02794

DEVIDE, V. [1964] *Mathematical logic. Part I (The classical logic of propositions) (Croatian) (English summary)* (X 4380) Posebna Izd Mat Inst: Belgrade 228pp
  ◊ B05 B98 ◊ REV MR 29 # 5711 Zbl 115.4 JSL 35.326
  • ID 43409

DEVIDE, V. see Vol. IV, V for further entries

DEVITO, C.L. [1983] *Compactlike operators on locally convex spaces* (J 3172) J London Math Soc, Ser 2 28*543-550
  ◊ H05 ◊ REV MR 85e:47026 Zbl 547 # 47012 • ID 39933

DEWDNEY, M. & WALSH, T.R.S. & WEHLAU, D.L. [1983] *Average-time testing of satisfiability algorithms* (P 4015) SE Conf Combin, Graph Th & Comput (14);1983 Boca Raton 39*305-325
  ◊ B35 ◊ REV MR 85a:68068 • ID 38851

DEXTER, G.E. [1943] *The calculus of non-contradiction* (J 0100) Amer J Math 65*171-178
  ◊ A05 B05 ◊ REV MR 4.126 Zbl 60.20 JSL 8.54
  • ID 02990

DEZANI, M. [1972] see BOEHM, C.

DEZANI, M. also published under the name DEZANI-CIANCAGLINI, M.

DEZANI, M. see Vol. VI for further entries

DICK, A.J.J. [1985] see CUNNINGHAM, R.J.

DICKOFF, J.W. [1961] see ALLEN, L.E.

DICKOFF, J.W. & JAMES, P.A. [1965] *Symbolic logic and language: a programmed text* (X 0822) McGraw-Hill: New York 390pp
  ◊ B98 ◊ ID 22481

DICKOFF, J.W. & JAMES, P.A. [1969] *Principles of symbolic logic for pattern recognition* (J 1377) Ann New York Acad Sci 161*402-415
  ◊ B80 ◊ REV Zbl 388 # 03008 • ID 52263

DIDUKH, I.I. [1966] see ANUFRIEV, F.V.

DIENER, F. [1978] see CALLOT, J.-L.

DIENER, F. [1979] *Famille d'equations a cycle limite unique (English summary)* (J 2313) C R Acad Sci, Paris, Ser A-B 289*A571-A574
  ◊ H05 ◊ REV MR 80m:34025 Zbl 422 # 34034 • ID 81095

DIENER, M. [1978] see CALLOT, J.-L.

DIENER, M. [1981] see BERG VAN DEN, I.

DIENER, M. [1983] *Canard et bifurcations* (P 4223) Math Tool & Model:1981/82 Toulouse & Paris 3*289-313
  ◊ H05 ◊ ID 44922

DIENER, M. [1984] *The canard unchained or how fast/slow dynamical systems bifurcate* (J 2789) Math Intell 6*38-49
  ◊ H05 ◊ REV MR 85j:34120 Zbl 552 # 34055 • ID 45290

DIENES, P. [1938] *Logic of algebra* (X 0859) Hermann: Paris 76pp
  ◊ A05 B28 F55 G05 ◊ REV Zbl 20.98 JSL 4.100
  FdM 64.924 • ID 03008

DIENES, P. see Vol. II, V, VI for further entries

DIEUDONNE, J. [1939] *Les methodes axiomatiques modernes et les fondements des mathematiques* (J 0370) Rev Sci (Paris) 77*224-232
  ◊ A05 B30 ◊ REV JSL 4.163 FdM 65.1097 • ID 17035

DILGER, W. & JANSON, A. [1983] *Unifikationsgraphen fuer intelligentes Backtracking in Deduktionssystemen* (P 3087) Germ Worksh on Artif Intell (7);1983 Dassel 189-196
  ◊ B35 ◊ REV MR 85g:68007 Zbl 539 # 68084 • ID 44027

DILGER, W. & MUELLER, J. [1984] *An associative processor for theorem proving* (P 4076) IFAC Symp Artif Intell;1983 Leningrad 489-497
  ◊ B35 ◊ REV Zbl 561 # 68061 • ID 47493

DILLER, J. [1978] *Klassische Praedikatenlogik* (X 3176) Fernuniv Hagen: Hagen 573pp
  ◊ B98 ◊ REV MR 80c:03003 • ID 72386

DILLER, J. see Vol. III, IV, VI for further entries

DING, DECHENG & MO, SHAOKUI [1984] *Simplification of the axioms of recursive arithmetic (Chinese)* (J 3187) Shuxue Niankan, Xi A 5*169-176
  ◇ B28 F30 ◇ REV MR 86a:03065 Zbl 566 # 03029 • REM English summary in J4719 5*262 • ID 44972

DING, DECHENG see Vol. IV for further entries

DINKINES, F. [1964] *Elementary concepts of modern mathematics. Part 1. Elementary theory of sets. Part 2. Introduction to mathematical logic* (X 1228) Appleton-Century-Crofts: New York x+457pp
  ◇ B98 E98 ◇ REV Zbl 128.10 JSL 32.422 • ID 22482

DNEPROVSKAYA, N.V. [1969] *Generalization of Kolmogorov's criterion in a linear normed space over an enlarged field of real numbers (Russian)* (J 3558) Uch Zap Ped Inst, Kaliningrad 69*141-148
  ◇ H05 ◇ REV MR 55 # 6083 • ID 81103

DNEPROVSKAYA, N.V. [1969] *The Weierstrass theorems in the nonstandard space of continuous functions (Russian)* (J 3558) Uch Zap Ped Inst, Kaliningrad 69*51-56
  ◇ H05 ◇ REV MR 55 # 6082 • ID 81104

DODGE, C.W. [1969] *Sets, logic and numbers* (X 1337) Prindle Weber Schmidt: Boston xiii+346pp
  ◇ B98 E98 ◇ ID 22633

DODGSON, C.L. [1887] *The game of logic* (X 1253) Crowell Collier & Macmillan: New York 96pp
  ◇ B98 ◇ ID 22449

DODGSON, C.L. [1895] *Symbolic logic. Part I. Elementary* (X 1253) Crowell Collier & Macmillan: New York xxxi+188pp
  • LAST ED [1977] (X 3555) Potter: New York xxv+496pp
  • REPR [1955] (X 0851) Berkeley: New York xxxi+203pp
  ◇ B20 B98 ◇ REV MR 19.1 MR 80j:03002 JSL 22.309 JSL 25.264 JSL 29.135 • REM Last ed. contains part II: advanced, which was never published previously • ID 02237

DOEHMANN, K. [1967] *Der Gruppencharakter der Transformationen der dyadischen Aussagenverknuepfungen* (J 0079) Logique & Anal, NS 10*218-228
  ◇ B05 G25 ◇ REV MR 35 # 5285 Zbl 153.4 JSL 33.304 • ID 03065

DOEMOELKI, B. & GERGELY, T. (EDS.) [1981] *Mathematical logic in computer science (Colloquium held in Salgotarjan, Hungary, September 10-15, 1978)* (S 3312) Coll Math Soc Janos Bolyai 26*758pp
  ◇ B97 ◇ REV MR 83g:68007 Zbl 476 # 00024 • ID 55510

DOEPP, K. [1972] *Bemerkung zu Henkins Beweis fuer die Nichtstandard-Vollstaendigkeit der Typentheorie* (J 0047) Notre Dame J Formal Log 13*561-562
  ◇ B15 C62 C85 ◇ REV MR 50 # 12700 Zbl 232 # 02013 • ID 04094

DOEPP, K. [1972] *Ueber die Kennzeichnung von Junktoren durch die Gestalt der herleitbaren Ausdruecke* (J 0009) Arch Math Logik Grundlagenforsch 15*3-6
  ◇ B05 ◇ REV MR 51 # 10018 Zbl 248 # 02017 • ID 03071

DOEPP, K. [1984] *Filterkonvergenz in der Nichtstandard-Analysis* (J 0068) Z Math Logik Grundlagen Math 30*21-44
  ◇ H05 ◇ REV MR 85h:03074 Zbl 538 # 03054 • ID 41487

DOEPP, K. [1984] *Filterkonvergenz in der Nichtstandard-Analysis bei nichtelementaren Funktionen* (J 0068) Z Math Logik Grundlagen Math 30*353-384
  ◇ H05 ◇ REV MR 86e:03062 Zbl 555 # 03034 • ID 44328

DOEPP, K. see Vol. IV for further entries

DOLBY, G.R. [1982] *The role of statistics in the methodology of the life sciences* (J 3930) Biometrics 38*1069-1083
  ◇ B80 ◇ REV MR 84e:92003 Zbl 513 # 62105 • ID 37812

DOMOTOR, Z. [1978] *Axiomatization of Jeffrey utilities* (J 0154) Synthese 39*165-210
  ◇ H10 ◇ REV MR 80c:90003 Zbl 453 # 90005 • ID 54209

DOMOTOR, Z. see Vol. II for further entries

DOPP, J. [1962] *Logiques construites par une methode de deduction naturelle* (X 0834) Gauthier-Villars: Paris 191pp
  ◇ B98 F05 F07 F50 F98 ◇ REV MR 26 # 1243 Zbl 143.8 JSL 34.502 • ID 03085

DOPP, J. [1967] *Notions de logique formelle* (X 1313) Nauwelaerts: Louvain 304pp
  • LAST ED [1980] (X 3734) Cabay: Louvain-la-Neuve 304pp
  ◇ B98 ◇ REV MR 15.384 MR 83h:03001 Zbl 524 # 03001 • ID 22397

DOPP, J. see Vol. VI for further entries

DORN, G. & WEINGARTNER, P. (EDS.) [1985] *Foundations of logic and linguistics. Problems and their solutions. Papers from the seventh international congress on logic, methodology and philosophy of science* (X 1332) Plenum Publ: New York xi+715pp
  ◇ B97 ◇ REV MR 86h:03003 • ID 48657

DOSHITA, S. & ISHIBASHI, T. & YAMASAKI, S. [1980] *Unit resolution for a subclass of the Ackermann class* (J 2794) Mem Fac Engin, Kyoto Univ 42*63-75
  ◇ B35 D15 ◇ REV MR 81j:03065 • ID 80151

DOSHITA, S. & YAMASAKI, S. [1983] *The satisfiability problem for a class consisting of Horn sentences and some non-Horn sentences in propositional logic* (J 0194) Inform & Control 59*1-12 • ERR/ADD ibid 60*174
  ◇ B20 B25 D15 ◇ REV Zbl 564 # 03010 • ID 44209

DOSHITA, S. & HIRATA, M. & YAMASAKI, S. & YOSHIDA, M. [1984] *A new combination of input and unit deductions for Horn sentences* (J 0232) Inform Process Lett 18*209-213
  ◇ B35 ◇ REV MR 86c:68072 Zbl 563 # 68072 • ID 44207

DOTTERER, R.H. [1941] *A generalization of the antilogism* (J 0036) J Symb Logic 6*90-95
  ◇ A05 B20 ◇ REV MR 3.131 Zbl 26.245 JSL 7.38 • ID 03093

DOTTERER, R.H. [1943] *A supplementary note on the rules of the antilogism* (J 0036) J Symb Logic 8*24
  ◇ B20 ◇ REV MR 4.183 Zbl 60.19 • ID 03094

DOWLING, W.F. & GALLIER, J.H. [1984] *Linear-time algorithms for testing the satisfiability of propositional Horn formuae* (J 2551) J Log Progr 1/3*267-284
  ◇ B35 ◇ REV MR 86g:68144 • ID 44685

DOWNING, P. [1975] *Conditionals, impossibilities and material implications* (J 0103) Analysis (Oxford) 35*84-91
  ◇ A05 B05 ◇ REV Zbl 355 # 02003 • ID 50201

DRABBE, J. [1971] *Sur la definissabilite equationelle* (J 2313) C R Acad Sci, Paris, Ser A-B 273*A589
  ◇ B05 ◇ REV MR 45 # 6573 Zbl 222 # 02004 • ID 03107

DRABBE, J. see Vol. II, III, IV, V, VI for further entries

DRAGALIN, A.G. [1972] *On the use of classical calculi for establishing constructive truth (Russian) (English summary)* (J 0288) Vest Ser Mat Mekh, Univ Moskva 27/2*25-29 • TRANSL [1972] (J 0510) Moscow Univ Math Bull 27/1-2*76-79
- ◊ B10 F30 F50 ◊ REV MR 47#1581 Zbl 237#02007 • ID 03610

DRAGALIN, A.G. [1980] *Higher-order predicate logic in the form of calculus realization (Russian)* (C 2583) Aktual Probl Log & Metodol Nauki 236-252
- ◊ B15 F05 F35 F50 ◊ REV Zbl 533#03034 • ID 36552

DRAGALIN, A.G. see Vol. III, IV, V, VI for further entries

DRAGAN, I. [1968] *Un algorithme lexicographique pour la resolution des programmes polynomiaux en variables entieres* (J 3954) Rev Franc Inf & Rech Operat 2/14*81-89
- ◊ B35 ◊ REV MR 39#6622 Zbl 179.247 • ID 28704

DRAKE, F.R. [1985] *How recent work in mathematical logic relates to the foundations of mathematics* (J 2789) Math Intell 7*27-35
- ◊ A05 B98 E98 ◊ ID 47704

DRAKE, F.R. see Vol. II, IV, V for further entries

DRANGE, T. [1966] *Type crossings: sentential meaninglessness in the border area of linguistics and philosophy* (X 0873) Mouton: Paris 218pp
- ◊ A05 B15 B65 ◊ ID 25318

DREBEN, B. [1957] *Relation of m-valued quantificational logic to 2-valued quantificational logic* (P 1675) Summer Inst Symb Log;1957 Ithaca 303-304
- ◊ B10 B50 ◊ REV JSL 30.375 • ID 29366

DREBEN, B. [1957] *Systematic treatment of the decision problem* (P 1675) Summer Inst Symb Log;1957 Ithaca 363
- ◊ B20 B25 D35 ◊ ID 03611

DREBEN, B. & KAHR, A.S. & WANG, HAO [1962] *Classification of AEA formulas by letter atoms* (J 0015) Bull Amer Math Soc 68*528-532
- ◊ B20 B25 D35 ◊ REV MR 30#22 Zbl 112.5 JSL 29.101 • ID 03127

DREBEN, B. [1962] *Solvable Suranyi subclasses: an introduction to the Herbrand theory* (P 0698) Harvard Symp Digit Comput & Appl;1961 Cambridge MA 32-47
- ◊ B20 B25 ◊ REV MR 36#1318 JSL 30.390 • ID 03612

DREBEN, B. & WANG, HAO [1964] *A refutation procedure and its model-theoretic justification* (X 0858) Harvard Univ Pr: Cambridge
- ◊ B35 ◊ ID 47306

DREBEN, B. & GOLDFARB, W.D. [1971] *Note J* (C 0745) Herbrand: Log Writings 201-202
- ◊ B10 F05 F07 ◊ REV JSL 40.94 • REM Footnote to English translation of Herbrand 1930 • ID 17043

DREBEN, B. & GOLDFARB, W.D. [1971] *Note N* (C 0745) Herbrand: Log Writings 265-271
- ◊ A10 B10 F30 F35 ◊ REV JSL 40.94 • REM Footnote to English transl. of Herbrand 1931: Sur le probleme fondamental de la logique math. • ID 17045

DREBEN, B. see Vol. III, VI for further entries

DRESDEN, A. [1953] *Complete independence* (J 0287) Scripta Math 19*205-206
- ◊ B10 ◊ REV MR 15.279 • ID 03613

DRESDEN, A. see Vol. VI for further entries

DRESS, F. [1969] *Logique mathematique et analyse non-standard* (S 2348) Semin Th Nombres Bordeaux 13*12pp
- ◊ H05 H98 ◊ REV Zbl 299#02067 • ID 61409

DRIES VAN DEN, L. & SCHMIDT, K. [1984] *Bounds in the theory of polynomial rings over fields. A nonstandard approach* (J 0305) Invent Math 76*77-91
- ◊ C60 H10 H20 ◊ REV MR 85i:12016 Zbl 539#13011 • ID 41246

DRIES VAN DEN, L. see Vol. III, IV, V, VI for further entries

DUBARLE, H.D. [1957] *Initiation a la logique* (X 0834) Gauthier-Villars: Paris 91pp
- ◊ B98 ◊ REV MR 20#1621 Zbl 77.11 JSL 23.30 • ID 03139

DUBIKAJTIS, L. [1964] *Decompositions of theories* (J 0014) Bull Acad Pol Sci, Ser Math Astron Phys 12*597-601
- ◊ B10 ◊ REV MR 31#3315 Zbl 168.248 • ID 03142

DUBIKAJTIS, L. [1964] *Operations on deductive systems* (J 0014) Bull Acad Pol Sci, Ser Math Astron Phys 12*593-596
- ◊ B22 ◊ REV MR 31#3314 Zbl 168.248 • ID 03143

DUBIKAJTIS, L. see Vol. II for further entries

DUBISLAV, W. [1928] *Zur kalkuelmaessigen Characterisierung der Definitionen* (J 1380) Ann Philos & Philos Kritik 7*136-145
- ◊ A05 B10 ◊ REV FdM 54.52 • ID 28657

DUCASSE, C.J. [1939] *Symbols, signs and signals* (J 0036) J Symb Logic 4*41-52
- ◊ B03 ◊ REV JSL 5.79 FdM 65.20 • ID 03153

DUDEK, W. [1977] *Minimization of weakly defined multiargument logic functions by means of the reduction of quasi-implicants* (J 0141) Arch Autom & Telemech 22*407-420
- ◊ B35 ◊ REV MR 57#11943 Zbl 369#94017 • ID 51361

DUDEK, W. [1979] *Minimization of multiargument weakly-defined logic functions with multiargument bindings* (J 0141) Arch Autom & Telemech 24*123-134
- ◊ B35 ◊ REV MR 81a:94059 Zbl 406#94015 • ID 69462

DUERR, K. [1954] *Lehrbuch der Logistik* (X 0804) Birkhaeuser: Basel vii+181pp
- ◊ B98 ◊ REV MR 16.986 Zbl 56.5 JSL 21.88 • ID 03162

DULAC, M.-H. [1971] *Decidabilite et operations entre theories* (J 2313) C R Acad Sci, Paris, Ser A-B 273*A1113-A1114
- ◊ B10 B25 D35 ◊ REV MR 45#3204 Zbl 224#02038 • ID 03165

DUMITRIU, A. [1967] *La negation des quantificateurs* (J 0147) An Univ Bucuresti, Acta Logica 10*139-154
- ◊ B10 ◊ REV MR 38#1996 Zbl 248#02019 • ID 61422

DUMITRIU, A. [1971] *The antinomy of the theory of types* (J 0286) Int Logic Rev 3*51-54
- ◊ A05 B15 ◊ REV MR 58#10288 JSL 37.194 • ID 03621

DUMITRIU, A. [1974] *The antinomy of the theory of types and the solution of logico-mathematical paradoxes* (J 0286) Int Logic Rev 9*83-102
- ◊ A05 B15 ◊ REV Zbl 334#02005 • ID 61424

DUMITRIU, A. see Vol. II, VI for further entries

DUMMETT, M. [1965] see CROSSLEY, J.N.

DUMMETT, M. see Vol. II, VI for further entries

DUNHAM, B. & FRIDSHAL, R. [1959] *The problem of simplifying logical expressions* (J 0036) J Symb Logic 24*17-19
 ◇ B05 ◇ REV MR 22 # 668 JSL 25.300 • ID 03171

DUNHAM, B. & FRIDSHAL, R. & SWARD, G.L. [1960] *A non-heuristic program for proving elementary logical theorems* (P 0696) Inform Processing (1);1959 Paris 282-285
 • REPR [1983] (C 4659) Autom of Reasoning 1*93-98
 ◇ B35 ◇ REV MR 27 # 5387 Zbl 115.352 JSL 32.266
 • ID 03626

DUNHAM, B. & FRIDSHAL, R. & NORTH, J.H. [1962] *Exploratory mathematics by machine* (P 0695) Rect Devel in Inform & Decis Processes;1961 West Lafayette 149-160
 • REPR [1983] (C 4659) Autom of Reasoning 1*276-287
 ◇ B35 ◇ REV JSL 32.266 • ID 03629

DUNHAM, B. & NORTH, J.H. [1963] *Theorem testing by computer* (P 0674) Symp Math Th of Automata;1962 New York 173-177
 • REPR [1983] (C 4659) Autom of Reasoning 1*271-275
 ◇ B35 ◇ REV MR 30 # 1627 Zbl 116.346 JSL 32.266
 • ID 03173

DUNHAM, B. & WANG, HAO [1976] *Towards feasible solutions of the tautology problem* (J 0007) Ann Math Logic 10*117-154
 ◇ B35 D15 ◇ REV MR 54 # 14464 Zbl 349 # 02006
 • ID 23648

DUNHAM, B. see Vol. IV for further entries

DUNN, J.M. [1968] see BELNAP JR., N.D.

DUNN, J.M. see Vol. II, V for further entries

DURNEV, V.G. [1978] *Elements of set theory and mathematical logic (Russian)* (X 2766) Yaroslav Gos Univ: Yaroslavl' 116pp
 ◇ B98 E98 ◇ REV MR 80g:03001 • ID 72511

DURNEV, V.G. see Vol. III, IV for further entries

DUTHIE, G.D. [1974] *Logic of terms* (J 2811) Phil Quart (St Andrews) 24*37-51
 ◇ B20 ◇ REV MR 58 # 10286 • ID 72518

DUTHIE, G.D. see Vol. II for further entries

DUTHIE, W. [1938] *Boolean functions of bounded variation* (J 0025) Duke Math J 4*600-606
 ◇ B05 ◇ REV Zbl 19.392 JSL 3.164 FdM 64.31 • ID 03188

DUTHIE, W. see Vol. V for further entries

DUTTON, J.D. (ED.) [1966] *Logics: an introduction with exercises* (X 1248) Chandler: San Francisco 251pp
 ◇ B98 ◇ ID 22483

DYRDA, K. & PRUCNAL, T. [1980] *On finitely based consequence determined by a distributive lattice* (J 0387) Bull Sect Logic, Pol Acad Sci 9*60-66
 ◇ B22 G25 ◇ REV MR 81m:03037 Zbl 435 # 03006
 • ID 55767

DYRE, J.C. [1982] *Nonstandard characterizations of ideals in $C(X)$* (J 0132) Math Scand 50*44-54
 ◇ H10 ◇ REV MR 83i:54043 Zbl 468 # 46016 • ID 39398

DYWAN, Z. [1977] *On a certain condition of the finite structural axiomatization of the classical propositional calculus* (J 0302) Rep Math Logic, Krakow & Katowice 9*23-26
 ◇ B05 ◇ REV MR 58 # 27289 Zbl 391 # 03010 • ID 52337

DYWAN, Z. [1978] *Decidability of structural completeness for strongly finite propositional calculi* (J 0387) Bull Sect Logic, Pol Acad Sci 7*129-132
 ◇ B22 ◇ REV MR 81b:03033 Zbl 408 # 03018 • ID 56257

DYWAN, Z. [1978] *Dual counterparts of strongly finite consequences* (J 0387) Bull Sect Logic, Pol Acad Sci 7*75
 ◇ B22 ◇ REV MR 58 # 16136 Zbl 406 # 03049 • ID 56132

DYWAN, Z. [1979] *Finite structural axiomatization of every finite-valued propositional calculus* (J 0387) Bull Sect Logic, Pol Acad Sci 8*61-67
 ◇ B22 ◇ REV MR 80h:03042 Zbl 419 # 03016 • ID 53359

DYWAN, Z. & STEPIEN, T. [1980] *Every two-valued propositional calculus has the interpolation property* (J 0387) Bull Sect Logic, Pol Acad Sci 9*152-153
 ◇ B20 C40 ◇ REV MR 82b:03026 Zbl 453 # 03028
 • ID 54158

DYWAN, Z. [1980] *Finite structural axiomatization of every finite-valued propositional calculus* (J 0063) Studia Logica 39*1-4
 ◇ B22 ◇ REV MR 81h:03061 Zbl 446 # 03023 • ID 56552

DYWAN, Z. [1980] *Quasi-strongly finite sentential calculi* (J 0387) Bull Sect Logic, Pol Acad Sci 9*154-158
 ◇ B22 ◇ REV MR 82b:03064 Zbl 455 # 03009 • ID 54270

DYWAN, Z. [1981] *Finite unaxiomatizability of propositional cacluli with one variable* (J 0302) Rep Math Logic, Krakow & Katowice 12*3-7
 ◇ B22 ◇ REV MR 82i:03013 Zbl 463 # 03004 • ID 54544

DYWAN, Z. [1984] *An interpretation of Aristotle's syllogistic and a certain fragment of set theory in propositional calculi* (J 0387) Bull Sect Logic, Pol Acad Sci 13*85-91
 ◇ B20 E20 ◇ REV Zbl 552 # 03008 • ID 43341

DYWAN, Z. [1984] *An interpretation of a certain fragment of arithmetic in some propositional calculus* (J 0387) Bull Sect Logic, Pol Acad Sci 13*99-105
 ◇ B20 F30 ◇ REV Zbl 565 # 03007 • ID 45756

DYWAN, Z. [1985] *A new variant of the Goedel-Malcev theorem for the classical propositional calculus* (J 0387) Bull Sect Logic, Pol Acad Sci 14*8-14
 ◇ B05 ◇ ID 46501

DYWAN, Z. see Vol. II for further entries

DZEGELENOK, I.I. & MEDETOV, M.M. [1978] *Determination of the levels of complexity of a linear classifier (Russian)* (J 0199) Zh Vychisl Mat i Mat Fiz 18*1579-1588
 • TRANSL [1978] (J 1049) USSR Comput Math & Math Phys 18/6*219-229
 ◇ B05 ◇ REV MR 80k:68071 Zbl 398 # 68019 • ID 69472

DZHAVADOV, R.M. [1981] *On ε-completeness of sets of functions in the algebra of logic (Russian)* (J 0023) Dokl Akad Nauk SSSR 256*1042-1045
 • TRANSL [1981] (J 0062) Sov Math, Dokl 23*147-151
 ◇ B05 ◇ REV MR 83c:03055 Zbl 475 # 03003 • ID 55457

DZHAVADOV, R.M. [1982] *On the complexity of approximate assignment of functions in the algebra of logic (Russian)* (J 0023) Dokl Akad Nauk SSSR 265*24-27
• TRANSL [1982] (J 0062) Sov Math, Dokl 26*16-19
◊ B05 ◊ REV MR 83k:94046 Zbl 524 # 94023 • ID 38239

DZHIDZHYAN, R.Z. [1977] *Extended syllogistics (Russian)* (X 3559) Erevan Univ: Erevan 208pp
◊ B20 ◊ REV MR 81e:03003 • ID 72538

DZIK, W. [1975] *On structural completeness of some nonclassical predicate calculi* (J 0302) Rep Math Logic, Krakow & Katowice 5*19-26
◊ B22 F50 ◊ REV MR 58 # 16147 Zbl 343 # 02038
• ID 21897

DZIK, W. [1981] *On the content of lattices of logics I. The representation theorem for lattices of logics* (J 0302) Rep Math Logic, Krakow & Katowice 13*17-27
◊ B22 ◊ REV MR 83f:03070 Zbl 526 # 03041 • REM Part II 1982 • ID 35313

DZIK, W. [1981] *The existence of Lindenbaum's extensions is equivalent to the axiom of choice* (J 0302) Rep Math Logic, Krakow & Katowice 13*29-31
◊ B22 E25 ◊ REV MR 83f:04005 Zbl 492 # 03010
• ID 35320

DZIK, W. & TOKARZ, M. [1984] *Invariant matrix consequences* (J 0302) Rep Math Logic, Krakow & Katowice 18*37-43
◊ B22 ◊ ID 45523

DZIK, W. see Vol. II for further entries

DZIOBIAK, W. [1979] *An example concerning the lattice of the structural consequence operations* (J 0387) Bull Sect Logic, Pol Acad Sci 8*48-53
◊ B22 ◊ REV MR 80h:03043 Zbl 414 # 03019 • ID 53065

DZIOBIAK, W. [1979] *On strongly finite consequence operations* (J 0387) Bull Sect Logic, Pol Acad Sci 8*87-94
◊ B22 ◊ REV MR 80g:03026 Zbl 415 # 03021 • ID 53123

DZIOBIAK, W. & SACHWANOWICZ, W. [1979] *On two notions concerning the structural sentential calculi* (J 0387) Bull Sect Logic, Pol Acad Sci 8*54-60
◊ B22 ◊ REV MR 80g:03025 Zbl 419 # 03015 • ID 53358

DZIOBIAK, W. [1980] *An example of strongly finite consequence operation with $2^{\aleph_0}$ standard strengthenings* (J 0063) Studia Logica 39*375-379
◊ B22 ◊ REV MR 83a:03025 Zbl 463 # 03012 • ID 54552

DZIOBIAK, W. [1981] *The lattice of strengthenings of a strongly finite consequence operation* (J 0063) Studia Logica 40*177-193
◊ B22 ◊ REV MR 83m:03038 Zbl 479 # 03017 • ID 55679

DZIOBIAK, W. [1982] see BIELA, A.

DZIOBIAK, W. [1983] see BIELA, A.

DZIOBIAK, W. see Vol. II, III for further entries

EARLE, J.N.F. [1973] *Logic* (X 0843) Macmillan : New York & London ix + 131pp
◊ B98 ◊ ID 22484

EARMAN, J. [1975] *Infinities, infinitesimals, and indivisibles: The Leibnizian labyrinth* (J 3274) Stud Leibnitiana 7*236-251
◊ A05 A10 H05 ◊ REV MR 58 # 26874 Zbl 431 # 03041
• ID 53947

EATON, R.M. [1931] *General logic: an introductory survey* (X 1347) Scribner: New York xii + 630pp
◊ B98 ◊ ID 22582

EBBINGHAUS, H.-D. [1974] *Zur mengentheoretischen Begruendung der natuerlichen Zahlen* (J 0487) Math Unterricht 20*27-42
◊ B28 E75 ◊ ID 28198

EBBINGHAUS, H.-D. & FLUM, J. & THOMAS, WOLFGANG [1978] *Einfuehrung in die mathematische Logik* (X 0890) Wiss Buchges: Darmstadt ix + 288pp
• TRANSL [1984] (X 0811) Springer: Heidelberg & New York ix + 216pp
◊ B98 C95 C98 ◊ REV MR 81h:03001 Zbl 399 # 03001 Zbl 556 # 03001 • ID 28201

EBBINGHAUS, H.-D. see Vol. II, III, IV, V, VI for further entries

EBERLE, R.A. [1969] *Denotationless terms and predicates expressive of positive qualities* (J 0105) Theoria (Lund) 35*104-123
◊ B10 ◊ REV MR 42 # 31 Zbl 198.320 • ID 03222

EBERLE, R.A. see Vol. II, V for further entries

ECKHAUS, W. [1983] *Relaxation oscillations including a standard chase on French ducks* (C 3080) Asymptotic Anal, Vol 2 449-494
◊ H10 ◊ REV MR 85b:34066 Zbl 509 # 34037 • ID 39373

ECSEDI-TOTH, P. & MORICZ, F. & VARGA, A. [1977] *A note on symmetric boolean functions* (J 0380) Acta Cybern (Szeged) 3*321-326
◊ B05 ◊ REV MR 58 # 4663 Zbl 389 # 94016 • ID 69964

ECSEDI-TOTH, P. & VARGA, A. [1977] *An effective theorem proving algorithm* (J 0380) Acta Cybern (Szeged) 3*249-260
◊ B35 ◊ REV MR 57 # 11202 Zbl 375 # 68037 • ID 51641

ECSEDI-TOTH, P. & MORICZ, F. & VARGA, A. [1980] *A heuristic method for speeding up manual optimization of boolean functions* (J 3101) Acta Tech Acad Sci Hung 90*247-258
◊ B35 ◊ REV Zbl 456 # 94027 • ID 69982

ECSEDI-TOTH, P. & TURI, L. [1982] *On enumerability of interpolants* (P 3758) Algeb Conf (2);1981 Novi Sad 71-78
◊ B10 C40 ◊ REV MR 84b:03021 Zbl 536 # 03002
• ID 34907

ECSEDI-TOTH, P. see Vol. III for further entries

EDEL'MAN, S.L. [1975] *Mathematical logic (Russian)* (X 3407) "Vysshaya Shkola": Moskva 178pp
◊ B98 ◊ ID 33333

EDELSTEIN, R. [1975] *An interpolation lemma for the pure implicational calculus* (J 0036) J Symb Logic 40*443-444
◊ B20 ◊ REV MR 52 # 2810 Zbl 327 # 02010 • ID 14819

EDELSTEIN, R. see Vol. II for further entries

EDER, E. [1985] *Properties of substitutions and unifications* (J 4609) J Symb Comput 1*31-46
◊ B35 ◊ ID 49171

EDWARDS, R.E. [1979] *A formal background to mathematics. Ia, Ib: Logic, sets and numbers* (X 0811) Springer: Heidelberg & New York xxxiv + 933pp
◊ B98 E98 ◊ REV MR 80h:03001a MR 80h:03001b Zbl 413 # 03001 • REM Vol.II 1980 • ID 52995

EDWARDS, R.E. [1980] *A formal background to mathematics IIa,IIb: A critical approach to elementary analysis* (X 0811) Springer: Heidelberg & New York xlvii+606pp,vi+607-1170pp
⋄ B28 ⋄ REV MR 83i:00002 MR 83i:00003 Zbl 445 # 26001 • REM Vol.I 1979 • ID 39271

EFIMOV, E.I. [1982] *Intellectual problem solvers (Russian)* (X 2027) Nauka: Moskva 320pp
⋄ B35 ⋄ REV Zbl 506 # 68076 • ID 37830

EFIMOV, E.I. [1984] *Automatic construction of axioms of the applied propositional calculus (Russian)* (S 2582) Semiotika & Inf, Akad Nauk SSSR 23*60-73
⋄ B35 ⋄ REV Zbl 561 # 68057 • ID 46341

EFIMOV, E.I. see Vol. II for further entries

EFSTATHIOU, J. & MAMDANI, E.H. [1984] *An analysis of formal logics are inference mechanisms in expert systems* (J 1741) Int J Man-Mach Stud 21*213-227
⋄ B35 ⋄ REV Zbl 565 # 68090 • ID 48675

EFSTATHIOU, J. see Vol. II for further entries

EGIAZARYAN, EH.V. [1979] *Quantitative Charakteristiken von Systemen von Gleichungen mit partiellen booleschen Funktionen (Russisch)* (J 0346) Dokl Akad Nauk Armyan SSR 68*74-78
⋄ B05 ⋄ REV MR 81d:94032 Zbl 461 # 94014 • ID 69479

EGIAZARYAN, EH.V. [1980] *Estimates related to the number of solutions of systems of Boolean equations (Russian)* (S 2874) Vopr Kibern, Akad Nauk SSSR 64*124-130
⋄ B05 ⋄ REV Zbl 498 # 94024 • ID 38537

EGLITIS, L.V. [1975] see CHIRKOV, M.K.

EGLITIS, L.V. [1977] see CHIRKOV, M.K.

EGLITIS, L.V. see Vol. IV for further entries

EHDEL'MAN, G.S. [1974] *Certain radicals of metaideals (Russian)* (J 3536) Uch Zap Univ, Ivanovo 130*44-50
⋄ B10 C07 C60 ⋄ REV MR 58 # 16265 • ID 72569

EHDEL'MAN, G.S. see Vol. III for further entries

EHLERS, F. [1968] *Logic by way of set theory* (X 0818) Holt Rinehart & Winston: New York xi+386pp
⋄ B98 E98 ⋄ ID 22634

EHRENFEUCHT, A. & ORLOWSKA, E. [1967] *Mechanical proof procedure for propositional calculus* (J 0014) Bull Acad Pol Sci, Ser Math Astron Phys 15*25-30
⋄ B35 ⋄ REV MR 35 # 4084 Zbl 153.320 • ID 03249

EHRENFEUCHT, A. & PAWLAK, Z. [1967] *Some remarks on the bracket free notation* (J 0014) Bull Acad Pol Sci, Ser Math Astron Phys 15*105-106
⋄ B03 ⋄ REV MR 38 # 5580 Zbl 203.6 • ID 03247

EHRENFEUCHT, A. see Vol. II, III, IV, V, VI for further entries

EHRICH, H.-D. [1973] *Minimale und m-minimale Variablenmengen fuer partielle Boole'sche Funktionen* (J 1431) Acta Inf 2*172-179
⋄ B05 ⋄ REV MR 48 # 5732 Zbl 243 # 02008 • ID 61481

EICHHORN, H. [1983] *Conceptual and conventional definitions in the mathematical sciences* (J 0455) Phil Naturalis 20*147-159
⋄ A05 B98 ⋄ REV MR 84k:00018 • ID 39171

EIFRIG, B. [1972] *Ein nicht-standard Beweis fuer die Existenz eines starken Liftings in $\mathscr{L}_\infty(0,1]$* (P 0732) Contrib to Non-Standard Anal;1970 Oberwolfach 81-83
⋄ E05 E50 E75 H05 ⋄ REV MR 58 # 28420 Zbl 248 # 46036 • ID 17196

EIFRIG, B. [1972] *Ein Nicht-Standard-Beweis fuer die Existenz eines liftings* (J 0008) Arch Math (Basel) 23*425-427
⋄ H05 ⋄ REV MR 47 # 5593 Zbl 245 # 28008 • ID 28480

EIFRIG, B. [1973] *Zur Existenz eines Spielwertes bei Spielen auf kompakten Raeumen mit stetiger Auszahlungsfunktion* (J 0008) Arch Math (Basel) 24*671-672
⋄ H05 H10 ⋄ REV MR 49 # 4582 Zbl 275 # 90044 • ID 04109

EIFRIG, B. [1976] *Ein Nicht-Standard-Beweis fur die Existenz eines Liftings* (P 2965) Measure Th;1975 Oberwolfach 133-135
⋄ H05 ⋄ REV MR 56 # 1060 Zbl 333 # 46033 • ID 81152

EKLOF, P.C. [1969] see BARWISE, J.

EKLOF, P.C. see Vol. III, V for further entries

ELGOT, C.C. & RABIN, M.O. [1966] *Decidability and undecidability of extensions of second (first) order theory of (generalized) successor* (J 0036) J Symb Logic 31*169-181
⋄ B15 B25 C85 D05 D35 F30 F35 ⋄ REV Zbl 144.245 • ID 03291

ELGOT, C.C. see Vol. III, IV, VI for further entries

ELLERMAN, D.P. & ROTA, G.-C. [1978] *A measure theoretic approach to logical quantification* (J 0144) Rend Sem Mat Univ Padova 59*227-246
⋄ B10 C90 G25 ⋄ REV MR 81c:03027 Zbl 449 # 03028 • ID 56695

ELLERMAN, D.P. see Vol. III for further entries

ELLETT JR., F.S. & ERICSON, D.P. [1983] *The logic of causal methods in social science* (J 0154) Synthese 57*67-82
⋄ B80 ⋄ REV MR 84k:03042 Zbl 514 # 92029 • ID 34978

ELLIS, D. [1953] *Remarks on boolean functions* (J 0090) J Math Soc Japan 5*345-350
⋄ B05 ⋄ REV MR 16.788 Zbl 53.215 • REM Part I. Part II 1956 • ID 03336

ELLIS, D. [1956] *Remarks on boolean functions II* (J 0090) J Math Soc Japan 8*363-368
⋄ B05 ⋄ REV MR 19.380 Zbl 72.262 • REM Part I 1953 • ID 03434

ELLIS, D. see Vol. V for further entries

ELLIS, J.W. [1961] *Another very independent axiom system* (J 0005) Amer Math Mon 68*992
⋄ B30 ⋄ ID 04289

ELLIS, J.W. see Vol. V for further entries

EMCH, G. & JAUCH, J.M. [1965] *Structures logiques et mathematiques en physique quantique* (J 0076) Dialectica 19*259-279
⋄ B30 ⋄ ID 48783

EMDE, H. & REYERSBACH, W. & STROMBACH, W. [1972] *Mathematische Logik. Ihre Grundprobleme in Theorie und Anwendung* (X 0995) Beck'sche Verlagsbuchh: Muenchen 227pp
⋄ B98 G05 G10 ⋄ REV MR 48 # 1869 Zbl 238 # 02002 • ID 13231

EMDEN VAN, M.H. & KOWALSKI, R. [1976] *The semantics of predicate logic as a programming language* (J 0037) ACM J 23*733-742
⋄ B10 B75 ⋄ REV MR 56 # 13747 Zbl 339 # 68004
• ID 61527

EMDEN VAN, M.H. [1979] *Relational programming illustrated by a program for the game of mastermind* (J 2293) Comp Linguist & Comp Lang 13*131-150
⋄ B35 ⋄ REV Zbl 435 # 68067 • ID 55825

EMDEN VAN, M.H. [1982] see APT, K.R.

EMDEN VAN, M.H. see Vol. IV, V for further entries

EMEL'YANOV, A.M. [1981] *A method of analysis of human managerial work by frames and special modal logic (Russian)* (J 0977) Izv Akad Nauk SSSR, Tekh Kibern 1981/4*94-102
• TRANSL [1981] (J 0522) Engin Cybern 19/4*68-75
⋄ B80 ⋄ REV MR 84b:92083 Zbl 508 # 90055 • ID 38159

EMMONS, D.W. [1984] *Existence of Lindahl equilibria in measure theoretic economies without ordered preferences* (J 2750) J Econ Th 34*342-359
⋄ H10 ⋄ REV MR 86d:90018 Zbl 552 # 90012 • ID 48898

ENDERTON, H.B. [1970] *Finite partially-ordered quantifiers* (J 0068) Z Math Logik Grundlagen Math 16*393-397
⋄ B15 C80 C85 ⋄ REV MR 44 # 1546 Zbl 193.294
• ID 03351

ENDERTON, H.B. [1972] *A mathematical introduction to logic* (X 0801) Academic Pr: New York xiii+295pp
⋄ B98 C98 F30 ⋄ REV MR 49 # 2239 Zbl 298 # 02002 JSL 38.340 • ID 03355

ENDERTON, H.B. see Vol. III, IV, V, VI for further entries

ENDO, H. [1974] *Quantification, games and existence* (J 0260) Ann Jap Ass Phil Sci 4*231-234
⋄ B10 ⋄ REV Zbl 282 # 02003 • ID 29077

ENGELER, E. [1970] *Geometry and language* (J 0076) Dialectica 24*77-85
⋄ B30 C75 ⋄ REV MR 44 # 5216 Zbl 256 # 02008
• ID 03368

ENGELER, E. [1983] *Metamathematik der Elementarmathematik* (X 0811) Springer: Heidelberg & New York VII*132pp
⋄ B98 ⋄ REV MR 85d:03024 Zbl 515 # 03001 • ID 37835

ENGELER, E. see Vol. II, III, IV, V, VI for further entries

ENGLEBRETSEN, G. [1980] *Noncategorical syllogisms in the Analytics* (J 0047) Notre Dame J Formal Log 21*602-608
⋄ A05 A10 B20 ⋄ REV MR 81k:03014 Zbl 419 # 03003
• ID 53907

ENGLEBRETSEN, G. [1980] *On propositional form* (J 0047) Notre Dame J Formal Log 21*101-110
⋄ A05 A10 B05 ⋄ REV MR 81a:03003 Zbl 363 # 02003
• ID 53170

ENGLEBRETSEN, G. see Vol. II for further entries

ENIN, S.V. [1978] see BIBILO, P.N.

ENIN, S.V. [1979] see BIBILO, P.N.

ENIN, S.V. [1980] see BIBILO, P.N.

ENNIS, R.H. [1969] *Ordinary logic* (X 0819) Prentice Hall: Englewood Cliffs vi+151pp
⋄ B98 ⋄ ID 22485

ENRIQUES, F. [1911] *Sui numeri non archimedei e su alcune loro interpretazioni* (J 3967) Boll Mathesis Soc Ital Mat 3*87-105
⋄ B28 E10 E20 ⋄ REV FdM 42.92 • ID 37967

EPSTEIN, G. [1973] see BREUER, M.A.

EPSTEIN, G. & HORN, A. [1974] *Finite limitations on a propositional calculus for affirmation and negation* (J 0477) Spis Bulgar Akad Nauk 3*43-44
⋄ B20 G10 ⋄ REV MR 53 # 128 • ID 16649

EPSTEIN, G. see Vol. II, VI for further entries

ERICKSON, R.W. & MUSSER, D.R. [1980] *The AFFIRM theorem prover: Proof forests and management of large proofs* (P 3063) Autom Deduct (5);1980 Les Arcs 220-231
⋄ B35 ⋄ REV Zbl 438 # 68046 • ID 69523

ERICSON, D.P. [1983] see ELLETT JR., F.S.

ERMOLAEVA, N.M. [1973] *Logics akin to Hao Wang's calculus (Russian)* (J 0338) Nauch-Tekh Inf, Ser 2, Akad Nauk SSSR 1973/8*34-37
• TRANSL [1973] (J 2667) Autom Doc Math Linguist 7/3*28-33
⋄ B20 ⋄ REV MR 49 # 10524 Zbl 273 # 02014 • ID 28575

ERMOLAEVA, N.M. see Vol. II, IV for further entries

ERNE, M. [1981] *Isomorphismen und Identifikationen in der Ordnungstheorie* (J 2790) Math Sem-ber 28*74-91
⋄ B30 E07 ⋄ REV MR 82g:06007 Zbl 468 # 00020
• ID 81183

ERNE, M. see Vol. V for further entries

ERNI, W. & LAPSIEN, R. [1981] *On the time and tape complexity of weak unification* (J 0232) Inform Process Lett 12*146-150
⋄ B35 D15 ⋄ REV MR 82f:68047 Zbl 469 # 68056
• ID 81184

ERNST, C.J. [1982] *An approach to management expert systems using fuzzy logic* (P 4051) Fuzzy Set & Possibility Th;1980 Acapulco 196-203
⋄ B80 ⋄ REV MR 84b:03004 • ID 46324

ERNST, G.W. [1971] *The utility of independent subgoals in theorem proving* (J 0194) Inform & Control 18*237-252
⋄ B35 ⋄ REV MR 43 # 3119 Zbl 218 # 68019 • ID 03510

ERNST, G.W. [1976] *A definition-driven theorem prover* (J 0187) IEEE Trans Comp C-25*317-322
⋄ B35 ⋄ REV Zbl 326 # 68061 • ID 61580

ERSHOV, YU.L. & PALYUTIN, E.A. & TAJTSLIN, M.A. [1973] *Mathematical logic (Russian)* (X 0913) Novosibirsk Gos Univ: Novosibirsk 159pp
⋄ B98 C98 ⋄ REV MR 57 # 9448 • ID 32036

ERSHOV, YU.L. & KARGAPOLOV, M.I. & MERZLYAKOV, YU.I. & SHIRSHOV, A.I. & SMIRNOV, D.M. (EDS.) [1973] *Selected questions of algebra and logic (Russian,English)* (X 2642) Nauka: Novosibirsk 339pp
⋄ B97 ⋄ REV MR 48 # 16 Zbl 271 # 00005 • ID 80486

ERSHOV, YU.L. (ED.) [1979] *Fifth All-Union conference on mathematical logic. Dedicated to Academician A. I. Mal'tsev. Abstracts of papers (Russian)* (X 2652) Akad Nauk Sibirsk Otd Inst Mat: Novosibirsk 173pp
⋄ B97 ⋄ REV Zbl 476 # 03001 • ID 55513

ERSHOV, YU.L. & PALYUTIN, E.A. [1979] *Mathematical logic (Russian)* (X 2027) Nauka: Moskva 320pp
- TRANSL [1984] (X 0885) Mir: Moskva 303pp (English)
- ◇ B98 ◇ REV MR 81f:03002 MR 86a:03001 • ID 72749

ERSHOV, YU.L. [1979] *Some questions of the application of formalized languages for the investigation of philosophical problems (Russian)* (C 2967) Metodol Probl Mat 83-89
- ◇ A05 B80 ◇ REV MR 83m:03005 • ID 35417

ERSHOV, YU.L. & LAVROV, I.A. & PAVILENIS, R.I. & PETROV, V.V. & SMIRNOV, V.A. [1984] *Logic, the foundations of mathematics and linguistics (Russian) (English summary)* (J 2871) Vopr Fil, Moskva 1984/1*45-58
- ◇ B97 ◇ REV MR 85k:03004 • ID 45202

ERSHOV, YU.L. see Vol. II, III, IV, V, VI for further entries

ESSLER, W.K. [1966] *Einfuehrung in die Logik* (X 1292) Kroener: Stuttgart 239pp
- ◇ B98 ◇ REV JSL 45.381 JSL 45.382 • ID 22398

ESSLER, W.K. [1970] *Ein nichtkonstruktiver Beweis des ersten ε-Theorems* (J 0047) Notre Dame J Formal Log 11*369-371
- ◇ B10 ◇ REV MR 47 # 6431 Zbl 177.11 • ID 03559

ESSLER, W.K. & MARTINEZ CRUZADO, R.F. [1983] *Grundzuege der Logik I: Das logische Schliessen* (X 2870) Klostermann: Frankfurt xiv+306pp
- ◇ B10 ◇ REV Zbl 532 # 03001 • REM 3rd revised ed.
- • ID 38270

ESSLER, W.K. see Vol. II for further entries

ESTERLE, J. [1980] *Homomorphismes discontinus des algebres de Banach commutatives separables* (J 0343) Studia Math, Pol Akad Nauk 66*119-141
- ◇ H05 ◇ REV MR 81m:46067 Zbl 353 # 46045 • ID 50130

ESTERLE, J. see Vol. III, V for further entries

EVANGELIST, M. [1982] *Non-standard propositional logics and their application to complexity theory* (J 0047) Notre Dame J Formal Log 23*384-392
- ◇ B80 D15 F20 ◇ REV MR 83k:03014 Zbl 464 # 03028
- • ID 54618

EVANS, H.P. & KLEENE, S.C. [1939] *A postulational basis for probability* (J 0005) Amer Math Mon 46*141-148
- ◇ B30 B48 ◇ REV Zbl 21.145 JSL 4.120 FdM 65.549
- • ID 30933

EVANS, M.G. [1976] *A truth-functional analysis of Aristotelian logic* (J 0286) Int Logic Rev 14*198-215
- ◇ A10 B20 ◇ REV Zbl 356 # 02009 • ID 50281

EVENDEN, J. [1962] *A lattice-diagram for the propositional calculus* (J 0148) Math Gaz 46*119-122
- ◇ B05 ◇ REV Zbl 111.8 • ID 48043

EVENDEN, J. [1976] *A Begriffsschrift for sentential logic* (J 0079) Logique & Anal, NS 19*413-425
- ◇ B03 ◇ REV MR 57 # 5671 Zbl 348 # 02016 • ID 61617

EVENDEN, J. see Vol. II for further entries

EVES, H.W. [1965] *An introduction to the foundations and fundamental concepts of mathematics* (X 0818) Holt Rinehart & Winston: New York xv+398pp
- ◇ B98 ◇ REV Zbl 125.277 • ID 23350

EXNER, R.M. & ROSSKOPF, M.F. [1959] *Logic in elementary mathematics* (X 0822) McGraw-Hill: New York 274pp
- ◇ B98 ◇ REV JSL 39.179 • ID 22192

FACENDA AGUIRRE, J.A. [1984] *(HM)-spaces and measurable cardinals* (J 0018) Canad Math Bull 27*53-57
- ◇ E55 E75 H05 ◇ REV MR 85i:46005 Zbl 531 # 46002
- • ID 44369

FACENDA AGUIRRE, J.A. see Vol. III for further entries

FACIONE, P.A. & SCHERER, D. [1978] *Logic and logical thinking: a modular approach* (X 0822) McGraw-Hill: New York xii+495pp
- ◇ A05 B98 ◇ REV JSL 46.672 • ID 44754

FACIONE, P.A. see Vol. II for further entries

FADINI, A. [1972] *Algoritmo per la construzione de una funzione booleana composta mediante un'assegnata famiglia di funzioni booleana (English summary)* (J 0099) Ricerca, Riv Mat Pure & Appl 23/2*23-30
- ◇ B05 ◇ REV MR 47 # 4792 Zbl 261 # 94042 • ID 03640

FADINI, A. see Vol. II, V for further entries

FAGES, F. [1984] *Associative-commutative unification* (P 2633) Autom Deduct (7);1984 Napa 194-208
- ◇ B35 ◇ REV Zbl 547 # 03012 • ID 43197

FAGIN, R. [1981] see DELOBEL, C.

FAGIN, R. see Vol. III, IV, V for further entries

FAL'K, V.N. & MOROKHOVEK, YU.E. [1975] *A generalized non-parenthesis-free language (Russian)* (S 2850) Tr Ehnerg Inst Moskva 216*5-11
- ◇ B03 ◇ REV MR 57 # 2888 • ID 72789

FAL'K, V.N. see Vol. IV for further entries

FALMAGNE, J.-C. [1975] *A set of independent axioms for positive Holder systems* (J 0153) Phil of Sci (East Lansing) 42*137-151
- ◇ B30 ◇ REV MR 58 # 199 • ID 72793

FALMAGNE, J.-C. see Vol. II, V for further entries

FANG, J. [1978] *The illogical in the logical* (J 0286) Int Logic Rev 17-18*111-120
- ◇ A05 B05 ◇ REV MR 81a:03004 • ID 72795

FARAT, V.M. & KOZHEVNIKOVA, G.P. [1979] *Algorithms for translation of algebraic expressions into an improved parenthesis-free notation and an analysis of their effectiveness(Russian)* (C 3018) Vopr Anal Vychisl Slozhnosti Algor 23-34,47
- ◇ B03 D15 ◇ REV MR 82g:68044 • ID 81930

FARINAS DEL CERRO, L. [1982] *A simple deduction method for modal logic* (J 0232) Inform Process Lett 14*49-51
- ◇ B35 B45 ◇ REV MR 83k:03019 Zbl 515 # 03009
- • ID 36159

FARINAS DEL CERRO, L. [1982] *Logique modal et processus communicants* (P 3842) AFCET Math de l'Inf;1982 Paris 279-286
- ◇ B80 ◇ REV Zbl 509 # 03010 • ID 36522

FARINAS DEL CERRO, L. [1985] *Resolution modal logics* (P 4621) Log & Models of Concurrent Syst;1984 La Colle-sur-Loup 27-55
- REPR [1985] (J 0079) Logique & Anal, NS 28*153-172
- ◇ B35 ◇ ID 49345

FARINAS DEL CERRO, L. see Vol. II, III for further entries

FARIS, J.A. [1962] *Truth-functional logic* (X 0824) Free Press: New York vi+122pp
- ◇ B05 ◇ REV MR 26 # 6037 JSL 31.108 • ID 03666

FARIS, J.A. [1964] *Quantification theory* (X 0866) Routledge & Kegan Paul: Henley on Thames 147pp
⋄ B10 ⋄ REV MR 28 #4986 JSL 31.108 • REM 2nd edition 1969; 154pp • ID 03667

FARKAS, E.J. & SZABO, M.E. [1984] *A star-finite relational semantics for parallel programs* (P 2153) Logic Colloq;1983 Aachen 2*129-142
⋄ B75 H10 ⋄ ID 39591

FARKAS, E.J. & SZABO, M.E. [1984] *On the plausibility of nonstandard proofs in analysis* (J 0076) Dialectica 38*297-310
⋄ H05 ⋄ REV Zbl 556 #03048 • ID 39588

FARRELL, R.J. [1975] *A note on the truth-table for p ⊃ q* (J 0047) Notre Dame J Formal Log 16*301-304
⋄ B05 ⋄ REV Zbl 236 #02002 • ID 61633

FARRELL, R.J. see Vol. II for further entries

FARRUKH, M.O. [1975] *Application of nonstandard analysis to quantum mechanics* (J 0209) J Math Phys 16*177-200
⋄ H10 ⋄ REV MR 52 #9897 Zbl 296 #47020 • ID 26626

FARTOS MARTINEZ, M. [1975] *On correspondences between monadic quantified formulas and products of nonquantified formulas (Spanish)* (J 0162) Teorema (Valencia) 5*515-519
⋄ B15 ⋄ REV MR 55 #5388 • ID 72800

FARTOS MARTINEZ, M. [1976] *A system of rules for propositional calculus (Spanish)* (J 0299) Gac Mat, Ser 1a (Madrid) 28/3-4*43-47
⋄ A05 B05 ⋄ REV MR 54 #9969 Zbl 354 #02016 • ID 25595

FAUST, D.H. [1982] *The Boolean algebra of formulas of first-order logic* (J 0007) Ann Math Logic 23*27-53
⋄ B10 G05 ⋄ REV MR 84f:03007 Zbl 501 #03005 • ID 34426

FEDINA, A.M. [1970] *The completeness of systems of logical inference (propositional calculus) (Russian)* (C 0668) Neklass Log 80-90
⋄ B05 ⋄ REV MR 47 #6455 • ID 03673

FEDINA, A.M. [1970] *The syllogistics of classes (Russian)* (C 0668) Neklass Log 65-79
⋄ B20 ⋄ REV MR 47 #6454 • ID 03672

FEDYURKO, V.V. [1966] see ANUFRIEV, F.V.

FEFERMAN, S. [1957] *Some recent work of Ehrenfeucht and Fraisse* (P 1675) Summer Inst Symb Log;1957 Ithaca 201-209
⋄ B10 C07 C65 E10 ⋄ REV JSL 32.282 • ID 03683

FEFERMAN, S. [1964] *The number systems. Foundations of algebra and analysis* (X 0832) Addison-Wesley: Reading 418pp
• TRANSL [1971] (X 2027) Nauka: Moskva 440pp
⋄ B28 E98 ⋄ REV MR 29 #49 Zbl 117.258 JSL 38.151
• ID 03694

FEFERMAN, S. & KREISEL, G. [1966] *Persistent and invariant formulas relative to theories of higher order* (J 0015) Bull Amer Math Soc 72*480-485
⋄ B15 C40 C52 C85 ⋄ REV MR 33 #1229 Zbl 234 #02038 JSL 37.764 • ID 03697

FEFERMAN, S. [1977] *Theories of finite type related to mathematical practice* (C 1523) Handb of Math Logic 913-971
⋄ B15 D55 D65 F10 F15 F35 F98 ⋄ REV MR 58 #10343 JSL 49.980 • ID 27330

FEFERMAN, S. see Vol. II, III, IV, V, VI for further entries

FEHER, M. [1972] *Is there an antinomy in the theory of types* (J 0286) Int Logic Rev 5*126-128
⋄ B15 ⋄ REV MR 50 #45 Zbl 344 #02007 • ID 04235

FELSCHER, W. & SCHMIDT, J. [1958] *Natuerliche Zahlen, Ordnung, Nachfolge* (J 0009) Arch Math Logik Grundlagenforsch 4*81-94
⋄ B28 ⋄ REV MR 21 #5577 Zbl 84.247 JSL 27.91
• ID 03730

FELSCHER, W. [1969] *On the algebra of quantifiers (Russian summary)* (J 0014) Bull Acad Pol Sci, Ser Math Astron Phys 17*327-332
⋄ B10 C05 C07 G15 ⋄ REV MR 43 #7313 • ID 03737

FELSCHER, W. [1971] *An algebraic approach to first order logic* (P 0669) Conv Teor Modelli & Geom;1969/70 Roma 133-148
⋄ B10 C05 C07 C20 C90 G15 ⋄ REV MR 43 #1822 Zbl 209.12 • ID 03738

FELSCHER, W. [1978] *Naive Mengen und abstrakte Zahlen I,II (II: Algebraische und reelle Zahlen)* (X 0876) Bibl Inst: Mannheim 260pp,222pp
⋄ A10 B28 E98 ⋄ REV MR 80m:04001 MR 80m:12001 Zbl 377 #04001 • REM Part III 1979 • ID 51762

FELSCHER, W. [1979] *Naive Mengen und abstrakte Zahlen III* (X 0876) Bibl Inst: Mannheim 270pp
⋄ B28 E98 ⋄ REV MR 80m:04002 Zbl 409 #04002 • REM Parts I,II 1978 • ID 72847

FELSCHER, W. & SCHULTE-MOENTING, J. [1984] *Algebraic and deductive consequence operations* (P 3088) Univer Alg & Link Log, Alg, Combin, Comp Sci;1983 Darmstadt 41-66
⋄ B22 ⋄ REV Zbl 574 #03050 • ID 46641

FELSCHER, W. see Vol. III, V, VI for further entries

FENSKE, C. [1967] *Beweisprogramme fuer die Praedikatenlogik und der Vollstaendigkeitssatz von Beth* (X 0907) Westdeutscher Verlag: Wiesbaden 69pp
⋄ B35 ⋄ REV MR 36 #52 Zbl 244 #68042 • ID 17922

FENSTAD, J.E. [1959] *Notes on the application of formal methods in the soft sciences* (J 0310) Inquiry (Oslo) 2*34-64
⋄ B80 ⋄ ID 33211

FENSTAD, J.E. [1967] *A note on "standard" versus "non-standard" topology* (J 0028) Indag Math 29*378-380
⋄ E75 H05 ⋄ REV MR 36 #6281 Zbl 204.5 JSL 35.344
• ID 03742

FENSTAD, J.E. [1970] *Non-standard models for arithmetic and analysis* (P 0724) Skand Mat Kongr (15);1968 Oslo 30-47
⋄ H05 H15 ⋄ REV MR 41 #1523 Zbl 191.298 • ID 17153

FENSTAD, J.E. (ED.) [1971] *Proceedings of the Second Scandinavian Logic Symposium* (X 0809) North Holland: Amsterdam vii+405pp
⋄ B97 ⋄ REV MR 47 #6425 Zbl 219 #02001 • ID 70239

FENSTAD, J.E. & NYBERG, A.M. [1971] *Standard and non-standard methods in uniform topology* (P 0638) Logic Colloq;1969 Manchester 353-359
⋄ E75 H05 ⋄ REV MR 43 #2661 Zbl 215.519 • ID 03744

FENSTAD, J.E. [1979] *On the metaphysics of the real line* (C 1706) Essays Honour J. Hintikka 91-99
⋄ A05 H05 ⋄ REV Zbl 432#03041 • ID 53998

FENSTAD, J.E. [1979] see ALBEVERIO, S.

FENSTAD, J.E. [1980] *Nonstandard methods in stochastic analysis and mathematical physics* (J 0157) Jbuchber Dtsch Math-Ver 82*167-180
⋄ H10 ⋄ REV MR 81m:03075 Zbl 441#03028 • ID 56080

FENSTAD, J.E. [1984] see ALBEVERIO, S.

FENSTAD, J.E. [1985] *Is nonstandard analysis relevant for the philosophy of mathematics?* (J 0154) Synthese 62*289-301
⋄ A05 H05 ⋄ REV MR 86f:03010 • ID 39335

FENSTAD, J.E. see Vol. II, III, IV, V, VI for further entries

FERENCZI, M. [1979] *On the foundations of the axiomatic theory of probability in symbolic logic* (J 0286) Int Logic Rev 19-20*5-18
⋄ A05 B30 ⋄ REV MR 81j:03037 Zbl 439#03002 • ID 55994

FERENCZI, M. see Vol. II, III for further entries

FERRARI, P.L. [1981] *Certe estensioni di teorie del I ordine mediante operatori logici di selezione (English summary)* (J 0549) Riv Mat Univ Parma, Ser 4 7*181-186
⋄ B10 ⋄ REV MR 84d:03016 Zbl 492#03017 • ID 34056

FERRARI, P.L. see Vol. IV for further entries

FERRAZ DE ARAGON, D. [1979] *Finitary translation of mathematical theories* (P 2615) Scand Logic Symp (5);1979 Aalborg 271-283
⋄ B15 E70 ⋄ REV MR 82h:03059 Zbl 468#03033 • ID 55098

FERRO, A. & OMODEO, E.G. [1978] *An efficient validity test for formulae in extensional two-level syllogistic* (J 0319) Matematiche (Sem Mat Catania) 33*130-137
⋄ B20 B25 B35 ⋄ REV MR 83i:03025 Zbl 448#68021 • ID 56666

FERRO, A. see Vol. III, IV, V for further entries

FERRO, R. [1971] *Le formule positive e negative, un teorema debole di sostituzione (English summary)* (J 0330) Atti Ist Veneto, Fis Mat Nat 130*401-423
⋄ B20 ⋄ REV MR 48#5815 Zbl 333#02010 • ID 17165

FERRO, R. [1973] *Un'osservazione sulle funzioni booleane simmetriche (English summary)* (J 0088) Ann Univ Ferrara, NS, Sez 7 18*55-57
⋄ B05 ⋄ REV MR 49#168 Zbl 278#06008 • ID 04238

FERRO, R. [1978] *Interpolation theorem for $L^{2+}_{k,k}$* (J 0036) J Symb Logic 43*535-549
⋄ B15 C40 C75 C85 ⋄ REV MR 81a:03028 Zbl 397#03019 • ID 29281

FERRO, R. [1984] *Una nota sulla nozione di molto maggiore (English summary)* (J 0144) Rend Sem Mat Univ Padova 71*223-227
⋄ H05 ⋄ REV MR 86b:03087 Zbl 567#06003 • ID 44775

FERRO, R. see Vol. III, V for further entries

FEVRIER, P. [1953] see DESTOUCHES, J.-L.

FEVRIER, P. [1954] see DESTOUCHES, J.-L.

FEVRIER, P. [1959] *Logical structure of physical theories* (P 0651) Axiomatic Method;1957 Berkeley 376-389
⋄ B30 ⋄ ID 27713

FEVRIER, P. also published under the name DESTOUCHES-FEVRIER, P.

FEVRIER, P. see Vol. V, VI for further entries

FEYS, R. [1937] *Les logiques nouvelles des modalites* (J 1720) Rev Neoscolast Philos, Ser 2 40*517-553
⋄ B20 B48 ⋄ REV FdM 63.826 • REM See also 1938 • ID 41048

FEYS, R. [1939] *Principes de logistique, premier volume* (X 0879) Inst Sup Philos: Louvain 129pp
⋄ A05 B98 ⋄ REV JSL 5.38 • ID 03757

FEYS, R. [1944] *Logistic. Formal logic I. General survey. Logic of propositions and classes (Dutch)* (X 2110) Dekker & van de Vegt: Nijmegen 340pp
⋄ B98 ⋄ REV MR 7.185 Zbl 60.20 JSL 10.100 • ID 21325

FEYS, R. [1946] *Les methodes recentes de deduction naturelle* (J 0252) Rev Philos Louvain 44*370-400
⋄ B10 ⋄ REV JSL 12.95 • ID 03759

FEYS, R. [1947] *Note complementaire sur les methodes du deduction naturelle* (J 0252) Rev Philos Louvain 45*60-72
⋄ B10 ⋄ REV JSL 13.114 • ID 03760

FEYS, R. [1949] *A simple notation for relations* (J 0175) Methodos 1*79-93
⋄ B03 E07 ⋄ REV MR 11.1 Zbl 37.293 JSL 15.71 • ID 03761

FEYS, R. [1958] see CURRY, H.B.

FEYS, R. & MOTTE, M.-T. [1959] *Logique juridique, systemes juridiques* (J 0079) Logique & Anal, NS 2*143-147
⋄ B80 ⋄ REV JSL 33.141 • ID 43164

FEYS, R. [1961] *Logique formalisee et raissonement juridique* (C 0622) Essays Found of Math (Fraenkel) 312-321
⋄ B80 ⋄ REV MR 28#4985 Zbl 156.8 JSL 33.141 • ID 43162

FEYS, R. & FITCH, F.B. [1969] *Dictionary of symbols of mathematical logic* (X 0809) North Holland: Amsterdam xiv+171pp
• TRANSL [1980] (X 3560) Paraninfo: Madrid 189pp
⋄ B98 ⋄ REV MR 40#7082 Zbl 179.9 Zbl 489#03001 • ID 03766

FEYS, R. see Vol. II, VI for further entries

FIBY, R. & SOKOL, J. & SUDOLSKY, M. [1979] *Efficient resolution theorem proving in the propositional logic* (J 2293) Comp Linguist & Comp Lang 13*151-168
⋄ B35 ⋄ REV MR 81j:68112 Zbl 429#68076 • ID 53896

FIEDLER, H. [1965] *Zur Stufenreduktion von Kalkuelen* (J 0009) Arch Math Logik Grundlagenforsch 8*63
⋄ B15 ⋄ REV MR 33#3884 • ID 03772

FIESCHI, M. & JOUBERT, M. [1979] *Application d'une methode logique a l'aide a la decision en medecine. La methode de Davis et Putnam en logique propositionnelle* (P 3589) Reconn des Formes & Intell Artif;1979 Toulouse III*50-56
⋄ B80 ⋄ REV MR 81m:92009 • ID 81219

FIKES, R.E. & NILSSON, N.J. [1971] *STRIPS: A new approach to the application of theorem proving to problem solving* (P 4251) Int Joint Conf Artif Intell (2);1971 London 189-208
⋄ B35 ⋄ REV Zbl 234#68036 • ID 47325

FILIPOIU, A. [1978] *Many-sorted polynomials and algebras of polynomials (Romanian) (English summary)* (J 3130) Bul Inst Politeh Bucuresti, Ser Electroteh 40/2*7-15
⋄ B10 C05 ⋄ REV MR 80e:08003 Zbl 401#08001
• ID 54631

FILIPOIU, A. see Vol. II for further entries

FINE, T.L. [1973] *Theories of probability: An examination of foundations* (X 0801) Academic Pr: New York xii+263pp
⋄ A05 B30 ⋄ REV MR 55#6505 Zbl 275#60006
• ID 25410

FINE, T.L. see Vol. II, IV for further entries

FINKELSTEIN, D. [1982] *Quantum sets and Clifford algebras* (J 2736) Int J Theor Phys 21*489-503
⋄ B30 ⋄ REV MR 83m:81010 Zbl 492#03025 • ID 38097

FINKELSTEIN, D. see Vol. II, V for further entries

FINN, V.K. & SKVORTSOV, D.P. [1981] *A remark about an extension of the language of the many-sorted predicate logic (Russian)* (J 0338) Nauch-Tekh Inf, Ser 2, Akad Nauk SSSR 8*25-26
• TRANSL [1981] (J 2667) Autom Doc Math Linguist 4*89-91
⋄ B10 ⋄ REV Zbl 483#03007 • ID 38083

FINN, V.K. [1983] *Machine-oriented formalization of plausible arguments in the style of F.Bacon and J.S.Mill (Russian)* (S 2582) Semiotika & Inf, Akad Nauk SSSR 20*35-101
⋄ B35 ⋄ REV MR 85a:03018 Zbl 521#03012 • ID 34769

FINN, V.K. see Vol. II for further entries

FINSLER, P. [1933] *Die Existenz der Zahlenreihe und des Kontinuums* (J 2022) Comm Math Helvetici 5*88-94
⋄ A05 B28 E20 E50 ⋄ REV Zbl 5.339 FdM 59.883
• ID 03795

FINSLER, P. see Vol. IV, V, VI for further entries

FIORENTINI, M. & MARRUCCELLI, A. [1970] *Complementi di matematiche moderne; Logica matematicea, teorie degli insiemi, strutture algebriche* (X 0909) Cedam: Padova 150pp
⋄ B98 E98 ⋄ REV MR 44#1531 • ID 17901

FISCHER, MICHAEL J. & MEYER, A.R. & PATERSON, M.S. [1975] *Lower bounds on the size of boolean formulars: Preliminary report* (P 1618) ACM Symp Th of Comput (7);1975 Albuquerque 37-44
⋄ B05 ⋄ REV MR 55#11704 Zbl 381#94028 • ID 69553

FISCHER, MICHAEL J. & MEYER, A.R. & PATERSON, M.S. [1982] $\Omega(n \log n)$ *lower bounds on length of Boolean formulas* (J 1428) SIAM J Comp 11*416-427
⋄ B05 D15 ⋄ REV MR 83j:03065 Zbl 488#94036
• ID 35363

FISCHER, MICHAEL J. see Vol. II, III, IV, VI for further entries

FISCHER, R. & UHLIG, D. [1982] *Fehlerkorrigierende Realisierungen zu Booleschen Funktionen und Automatenfunktionen mit linearer Kompliziertheit* (S 2829) Rostocker Math Kolloq 19*91-100
⋄ B05 D05 ⋄ REV MR 84c:94026 Zbl 484#94041
• ID 42341

FISHBURN, P.C. [1983] *A generalization of comparative probability on finite sets* (J 0035) J Math Psychol 27*298-310
⋄ B80 ⋄ REV MR 85k:90007 Zbl 531#60004 • ID 45837

FISHBURN, P.C. see Vol. II, III, V for further entries

FISHER, A. [1982] *Formal number theory and computability* (X 0894) Oxford Univ Pr: Oxford xiii+190pp
⋄ B98 D20 F30 F98 ⋄ REV MR 85g:03001 Zbl 504#03002 • ID 36967

FISHER, D. [1982] *Extending functions to infinitesimals of finite order* (J 0005) Amer Math Mon 89*443-449
⋄ H05 ⋄ REV MR 84f:03058 Zbl 521#26010 • ID 34477

FISK, M. [1964] *A modern formal logic* (X 0819) Prentice Hall: Englewood Cliffs xi+116pp
⋄ B98 ⋄ REV JSL 30.87 • ID 22489

FISK, M. [1965] *The logic of either-or* (J 0047) Notre Dame J Formal Log 6*39-50
⋄ B20 ⋄ ID 04318

FITCH, F.B. [1938] *The consistency of the ramified Principia* (J 0036) J Symb Logic 3*140-149
⋄ B15 F25 ⋄ REV Zbl 20.97 JSL 4.97 FdM 64.925
• ID 04321

FITCH, F.B. [1941] *Closure and Quine's *101* (J 0036) J Symb Logic 6*18-22
⋄ B10 ⋄ REV MR 2.209 Zbl 26.244 JSL 6.102 • ID 04324

FITCH, F.B. [1944] *A minimum calculus for logic* (J 0036) J Symb Logic 9*89-94
⋄ B20 B40 ⋄ REV MR 7.45 Zbl 60.23 JSL 11.127
• ID 04326

FITCH, F.B. [1944] *Representation of calculi* (J 0036) J Symb Logic 9*57-62
⋄ B20 B40 ⋄ REV MR 6.197 Zbl 60.23 JSL 11.28
• ID 04328

FITCH, F.B. [1948] *An extension of basic logic* (J 0036) J Symb Logic 13*95-106
⋄ B20 B40 F35 ⋄ REV MR 9.559 Zbl 30.193 JSL 14.68
• ID 04329

FITCH, F.B. [1949] *A further consistent extension of basic logic* (J 0036) J Symb Logic 14*209-218
⋄ B20 B40 ⋄ REV MR 12.2 Zbl 38.5 JSL 15.219
• ID 04334

FITCH, F.B. [1949] *On natural numbers, integers, and rationals* (J 0036) J Symb Logic 14*81-84
⋄ B28 ⋄ REV MR 11.2 Zbl 34.9 JSL 14.258 • ID 04331

FITCH, F.B. [1949] *The Heine-Borel theorem in extended basic logic* (J 0036) J Symb Logic 14*9-15
⋄ B28 B40 E70 F35 ⋄ REV MR 10.669 Zbl 35.8 JSL 15.137 • ID 04333

FITCH, F.B. [1950] *A demonstrably consistent mathematics. Part I* (J 0036) J Symb Logic 15*17-24
⋄ B30 B40 E70 ⋄ REV MR 12.2 Zbl 39.245 JSL 16.268
• REM Part II 1951 • ID 17113

FITCH, F.B. [1951] *A demonstrably consistent mathematics. Part II* (J 0036) J Symb Logic 16*121-124
⋄ B30 B40 E70 ⋄ REV MR 13.4 Zbl 45.151 JSL 16.268
• REM Part I 1950 • ID 17114

FITCH, F.B. [1952] *Symbolic logic. An introduction* (X 0879) Inst Sup Philos: Louvain x+238pp
⋄ B40 B98 ⋄ REV MR 15.592 Zbl 49.5 JSL 17.266
• ID 04335

FITCH, F.B. [1969] see FEYS, R.

FITCH, F.B. [1974] *Elements of combinatory logic* (**X** 0875) Yale Univ Pr: New Haven viii+162pp
◇ B40 B98 ◇ REV MR 54#2429 JSL 41.789 • ID 24019

FITCH, F.B. [1984] *Correction to a definition of negation* (**J** 0036) J Symb Logic 49*47-50
◇ B10 ◇ REV MR 85h:03013 • ID 42442

FITCH, F.B. see Vol. II, IV, V, VI for further entries

FITTING, M. [1969] *Intuitionistic logic, model theory, and forcing* (**X** 0809) North Holland: Amsterdam 191pp
◇ B98 C90 C98 E25 E35 E45 E50 F50 F98 ◇ REV MR 41#6666 Zbl 188.320 JSL 36.166 • ID 04349

FITTING, M. [1971] *A tableau proof method admitting the empty domain* (**J** 0047) Notre Dame J Formal Log 12*219-224
◇ B20 ◇ REV MR 45#4946 Zbl 177.11 • ID 04352

FITTING, M. see Vol. II, III, IV, V, VI for further entries

FITTLER, R. [1984] *Some nonstandard quantum electrodynamics* (**J** 2146) Helv Phys Acta 57*579-609
◇ H10 ◇ REV MR 86d:81103 • ID 40128

FITTLER, R. see Vol. III for further entries

FITZPATRICK, P.J. [1973] *An extension of Venn diagrams* (**J** 0047) Notre Dame J Formal Log 14*77-86
◇ B05 ◇ REV MR 48#1872 Zbl 247#02013 • ID 29498

FLAGG, R.C. [1978] *On the independence of the Bigos-Kalmar axioms for sentential calculus* (**J** 0047) Notre Dame J Formal Log 19*285-288
◇ B05 ◇ REV MR 57#15947 Zbl 366#02012 • ID 51656

FLAGG, R.C. see Vol. II, V, VI for further entries

FLEISCHER, I. [1967] *Infinitesimals* (**J** 0311) Nordisk Mat Tidskr 15*151-155
◇ H05 ◇ REV MR 37#4215 Zbl 166.312 • ID 04376

FLEISCHER, I. see Vol. III, V for further entries

FLETCHER, P. & LINDGREN, W.F. [1972] *Transitive quasi-uniformities* (**J** 0034) J Math Anal & Appl 39*397-405
◇ E75 H05 ◇ REV MR 47#5828 Zbl 233#54014 • ID 17123

FLETCHER, P. see Vol. V for further entries

FLORES, A. [1976] *Sulla rappresentazione dei numeri razionali mediante terne di numeri naturali* (**J** 1515) Archimede 28*129-140
◇ B28 ◇ REV Zbl 367#08001 • ID 51201

FLOYD, R.W. & KING, J. [1970] *An interpretation-oriented theorem prover over integers* (**P** 0641) ACM Symp Th of Comput (2);1970 Northhampton 169-179
◇ B35 ◇ ID 47340

FLOYD, R.W. see Vol. IV for further entries

FLOYD, W.F. & WOODGER, J.-H. [1935] *A simple method of testing truth-functions* (**J** 0103) Analysis (Oxford) 3*92-96
◇ B05 ◇ REV JSL 2.59 • ID 41332

FLUM, J. [1971] *Eine Formulierung des Herbrandschen Satzes ohne Skolemfunktion* (**J** 0009) Arch Math Logik Grundlagenforsch 14*3-9
◇ B10 ◇ REV MR 44#2571 Zbl 223#02018 • ID 04391

FLUM, J. [1971] *Ganzgeschlossene und praedikatengeschlossene Logiken I,II* (**J** 0009) Arch Math Logik Grundlagenforsch 14*24-37,99-107
◇ B10 C85 C95 ◇ REV MR 44#38 MR 51#12506 Zbl 219#02007 Zbl 232#02012 • ID 04388

FLUM, J. [1974] *On Horn theories* (**J** 0044) Math Z 138*205-212
◇ B20 C20 C80 ◇ REV MR 51#7853 Zbl 275#02046 • ID 15020

FLUM, J. [1976] *Non-standard analysis (Spanish)* (**P** 1619) Coloq Log Simb;1975 Madrid 67-79
◇ H05 H98 ◇ REV MR 57#12211 Zbl 357#02051 • ID 50399

FLUM, J. [1978] see EBBINGHAUS, H.-D.

FLUM, J. see Vol. III, IV, V for further entries

FOGELIS, E. [1950] *On finite proofs of arithmetical theorems* (**J** 0324) Izv Akad Nauk Latv SSR 6*81-86
◇ B28 F30 ◇ REV MR 13.199 • ID 17125

FOKKINGA, M.M. [1981] *On the notion of strong typing* (**P** 3074) Algor Lang;1981 Amsterdam 305-320
◇ B15 ◇ REV MR 83g:68020 • ID 39116

FOLLESDAL, D. [1968] *Interpretation of quantifiers* (**P** 0627) Int Congr Log, Meth & Phil of Sci (3,Proc);1967 Amsterdam 271-281
◇ A05 B10 C90 ◇ REV MR 40#2525 Zbl 182.4 • ID 04443

FOLLESDAL, D. see Vol. II, III for further entries

FOMINA, N.I. & KALINICHENKO, L.A. [1979] *A method of syntactic analysis of constructions of a predicate logic language in data base control systems (Russian)* (**S** 2850) Tr Ehnerg Inst Moskva 412*56-60,157
◇ B03 ◇ REV MR 82g:68021 Zbl 435#68077 • ID 81760

FORDER, H.G. & KALMAN, J.A. [1962] *Implication in equational logic* (**J** 0148) Math Gaz 46*122-126
◇ B20 ◇ REV Zbl 116.4 JSL 36.162 • ID 04447

FORSYTHE, K. & MATWIN, S. [1984] *Implementation strategies for plan-based deduction* (**P** 2633) Autom Deduct (7);1984 Napa 426-444
◇ B35 ◇ ID 44446

FORTENBACHER, A. [1985] *An algebraic approach to unification under associativity and commutativity* (**P** 4244) Rewriting Techn & Appl (1);1985 Dijon 381-397
◇ B35 ◇ ID 49781

FORTUNE, S. & LEIVANT, D. & O'DONNELL, M.J. [1983] *The expressiveness of simple and second-order type structures* (**J** 0037) ACM J 30*151-185
◇ B15 ◇ REV MR 84j:03029 Zbl 519#68046 • ID 34622

FORTUNE, S. see Vol. IV for further entries

FOSTER, C.C. & TENNEY, R.L. [1976] *Non-transitive dominance* (**J** 0497) Math Mag 49*115-120
◇ B15 ◇ REV MR 54#101 • ID 79282

FOULIS, D.J. [1969] *Fundamental concepts of mathematics* (**X** 1337) Prindle Weber Schmidt: Boston 212pp
◇ B98 ◇ REV Zbl 222#00001 • ID 26299

FOULIS, D.J. see Vol. II for further entries

FOXLEY, E. [1961] *Testing the independence of a system of axioms, using a logical computer* (J 0171) Proc Cambridge Phil Soc Math Phys 57*443-448
⋄ B35 ⋄ REV MR 23 # A44 Zbl 103.247 • ID 04472

FOXLEY, E. [1962] *The determination of all Sheffer functions in 3-valued logic, using a logical computer* (J 0047) Notre Dame J Formal Log 3*41-50
⋄ B35 B50 ⋄ REV MR 27 # 3508 Zbl 112.5 JSL 27.681 • ID 04474

FOXLEY, E. [1964] *Determination of the set of all four-variable formulae corresponding to universal decision elements using a logical computer* (J 0068) Z Math Logik Grundlagen Math 10*302-314
⋄ B35 ⋄ REV MR 29 # 5408 JSL 35.160 • ID 04475

FOXLEY, E. [1967] *The construction of scale-of-two mechanisms from universal decision elements* (J 0068) Z Math Logik Grundlagen Math 13*281-287
⋄ B35 ⋄ REV MR 35 # 6488 Zbl 183.19 JSL 35.160 • ID 04476

FRAASSEN VAN, B.C. [1970] *On the extension of Beth's semantics of physical theories* (J 0153) Phil of Sci (East Lansing) 37*325-339
⋄ B30 ⋄ REV MR 42 # 2926 • ID 48030

FRAASSEN VAN, B.C. [1971] *Formal semantics and logic* (X 0843) Macmillan : New York & London xi+225pp
⋄ B98 ⋄ REV Zbl 253 # 02002 JSL 45.376 • ID 61762

FRAASSEN VAN, B.C. see Vol. II, III for further entries

FRAENKEL, A.A. [1925] *Die neueren Ideen zur Grundlegung der Analysis und Mengenlehre* (J 0157) Jbuchber Dtsch Math-Ver 33*97-103
⋄ A05 B28 B30 E30 E75 ⋄ REV FdM 50.146 • ID 04486

FRAENKEL, A.A. [1949] *The relation of equality in deduction systems* (P 0682) Int Congr Philos (10);1948 Amsterdam 752-755
⋄ B20 ⋄ REV MR 10.423 Zbl 31.104 JSL 14.130 • ID 17066

FRAENKEL, A.A. [1951] *On the crisis of the principle of the excluded middle* (J 0287) Scripta Math 17*5-16
⋄ A05 B20 F99 ⋄ REV Zbl 54.4 • ID 47908

FRAENKEL, A.A. [1958] see BERNAYS, P.

FRAENKEL, A.A. [1958] see BAR-HILLEL, Y.

FRAENKEL, A.A. [1959] *Mengenlehre und Logik* (X 1258) Duncker & Humblot: Berlin 110pp
• TRANSL [1966] (X 0832) Addison-Wesley: Reading 102pp (English)
⋄ B98 E98 ⋄ REV MR 22 # 1513 MR 34 # 24 Zbl 139.5 Zbl 86.244 JSL 34.112 • ID 28712

FRAENKEL, A.A. see Vol. V, VI for further entries

FRAISSE, R. [1950] *Sur les types de polyrelation et sur une hypothese d'origine logistique* (J 0109) C R Acad Sci, Paris 230*1557-1559
⋄ B10 C07 E07 E10 ⋄ REV MR 12.14 Zbl 40.164 • ID 04510

FRAISSE, R. [1955] *La construction des γ-operateurs et leur application au calcul logique du premier ordre* (J 0109) C R Acad Sci, Paris 240*2191-2193
⋄ B10 C07 E07 ⋄ REV MR 16.1006 Zbl 64.288 JSL 32.280 • ID 04516

FRAISSE, R. [1956] *Application des γ-operateurs au calcul logique du premier echelon* (J 0068) Z Math Logik Grundlagen Math 2*76-92
⋄ B10 C07 C65 ⋄ REV MR 19.829 Zbl 126.17 JSL 32.280 • ID 04518

FRAISSE, R. [1958] *Sur une extension de la polyrelation et des parentes tirant son origine du calcul logique du $k^{\grave{e}me}$-echelon* (P 0576) Raisonn en Math & Sci Exper;1955 Paris 45-50
⋄ B15 C07 C85 E07 ⋄ REV MR 21 # 4110 Zbl 126.18 JSL 25.285 • ID 04520

FRAISSE, R. [1967] *Cours de logique mathematique. Tome I. Relation, formule logique, compacite, completude* (X 0834) Gauthier-Villars: Paris xii+187pp
• TRANSL [1973] (X 0835) Reidel: Dordrecht xvi+186pp
⋄ B98 C98 ⋄ REV MR 37 # 3902 Zbl 247 # 02001 Zbl 247 # 02002 JSL 35.580 • REM Tome II 1972 • ID 04529

FRAISSE, R. [1970] *Aspects du theoreme de completude selon Herbrand* (P 0625) Symp Autom Demonst;1968 Versailles 73-86
⋄ B10 C07 ⋄ REV MR 42 # 5793 Zbl 205.305 • ID 04531

FRAISSE, R. [1972] *Cours de logique mathematique. Tome 2: Theorie des modeles* (X 0834) Gauthier-Villars: Paris xiv+177pp
• TRANSL [1974] (X 0835) Reidel: Dordrecht xix+191pp
⋄ B98 C98 ⋄ REV MR 49 # 10514 Zbl 247 # 02003 • REM Tome I 1967. Tome III 1975. • ID 04535

FRAISSE, R. [1972] *Reflexions sur la completude selon Herbrand* (J 0286) Int Logic Rev 5*86-98
⋄ B10 C07 ⋄ REV MR 51 # 125 Zbl 341 # 02042 JSL 40.238 • ID 04294

FRAISSE, R. [1974] *Isomorphisme local et equivalence associes a un ordinal; utilite en calcul des formules infinies a quanteurs finis* (P 0610) Tarski Symp;1971 Berkeley 241-254
⋄ B15 C07 C75 C85 ⋄ REV MR 50 # 9591 Zbl 327 # 02015 • ID 04537

FRAISSE, R. [1977] *Deux relations denombrables, logiquement equivalentes pour le second ordre, sont isomorphes* (J 0056) Publ Dep Math, Lyon 14/2*41-62
⋄ B15 C15 C85 E07 E45 ⋄ REV MR 58 # 27297 Zbl 397 # 03008 • ID 52674

FRAISSE, R. [1978] *Les axiomatiques sont-elles un jeu?* (J 0076) Dialectica 32*229-244
• REPR [1982] (P 3753) Penser Math;1981 Paris 39-57
⋄ A05 B30 ⋄ REV MR 83j:03008 MR 84a:03004 Zbl 402 # 03007 • ID 54654

FRAISSE, R. [1985] *Deux relations denombrables, logiquement equivalentes pour le second ordre, sont isomorphes (modulo un axiome de constructibilite)* (C 4181) Math Log & Formal Syst (Costa da) 161-182
⋄ B15 C15 C85 E45 ⋄ ID 48204

FRAISSE, R. see Vol. II, III, IV, V for further entries

FRANCIS, C.E. [1981] *Applications of nonstandard analysis to relativistic quantum mechanics. I* (J 2760) J Phys A Math & Gen 14*2539-2551
⋄ H10 ⋄ REV MR 83b:81013 Zbl 469 # 46027 • ID 39004

FRANCO, J. & PAULL, M. [1983] *Probabilistic analysis of the Davis-Putnam procedure for solving the satisfiability problem* (J 2702) Discr Appl Math 5*77-87
⋄ B35 D15 ⋄ REV MR 84e:68038 Zbl 497 # 68021 • ID 39281

FRANEK, M. [1972] *Ueber eine Konstruktion reeller Zahlen. I* (J 0128) Acta Math Univ Comenianae (Bratislava) 26*57-81
- ◇ B28 ◇ REV MR 48 # 468 Zbl 256 # 04002 • ID 61782

FRANEK, M. see Vol. III for further entries

FRANK, W. [1974] see CORCORAN, J.

FRANK, W. [1976] *A note on the adequacy of translations* (J 0047) Notre Dame J Formal Log 17*249-250
- ◇ B22 ◇ REV MR 55 # 12500 Zbl 258 # 02013 • ID 18146

FRAYNE, T.E. & MOREL, A.C. & SCOTT, D.S. [1962] *Reduced direct products* (J 0027) Fund Math 51*195-228 • ERR/ADD ibid 53*117
- ◇ B20 C20 C55 ◇ REV MR 26 # 28 MR 27 # 4751 Zbl 108.5 JSL 31.506 • ID 04557

FREEMAN, J.B. [1977] *A caution on propositional identity* (J 0103) Analysis (Oxford) 37*149-151
- ◇ A05 B05 ◇ REV Zbl 367 # 02004 • ID 51164

FREEMAN, J.B. [1977] see DANIELS, C.B.

FREEMAN, J.B. [1978] see DANIELS, C.B.

FREEMAN, J.B. see Vol. II for further entries

FREGE, F.L.G. [1879] *Begriffsschrift, eine der arithmetischen nachgebildete Formelsprache des reinen Denkens* (X 4059) Nebert: Halle x+88pp
- • TRANSL [1952] (C 1990) Transl Phil Writings Frege 1-20
- ◇ A05 B03 B05 B10 ◇ REV JSL 32.240 FdM 11.48
- • REM In Chapter I of the transl. some passages are omitted
- • ID 21331

FREGE, F.L.G. [1879] *Ueber Anwendungen der Begriffsschrift* (J 0363) Sitzb Jena Ges Med Natwiss 1879*89-93
- ◇ A05 A10 B80 ◇ REV JSL 32.240 • ID 17073

FREGE, F.L.G. [1882] *Ueber den Zweck der Begriffsschrift* (J 0363) Sitzb Jena Ges Med Natwiss 1882*97-106
- ◇ A05 A10 B30 ◇ REV JSL 32.240 • ID 17074

FREGE, F.L.G. [1882] *Ueber die wissenschaftliche Berechtigung einer Begriffsschrift* (J 0362) Z Phil & Philos Kritik 81*48-56
- ◇ A05 A10 B30 ◇ REV JSL 32.240 JSL 33.281 • ID 17075

FREGE, F.L.G. [1884] *Die Grundlagen der Arithmetik, eine logisch mathematische Untersuchung ueber den Begriff der Zahl* (X 4060) Koebner: Breslau xi+119pp
- • TRANSL [1950] (X 1096) Blackwell: Oxford
- ◇ A05 B28 E75 ◇ REV MR 11.487 JSL 16.67 • REM 2nd revised ed. 1953 • ID 21332

FREGE, F.L.G. [1885] *Ueber formale Theorien der Arithmetik* (J 0363) Sitzb Jena Ges Med Natwiss 1885*94-104
- ◇ A05 B28 ◇ REV JSL 46.175 • ID 28630

FREGE, F.L.G. [1893] *Grundgesetze der Arithmetik, Begriffsschriftlich abgeleitet. Vol.1* (X 4061) Pohle: Jena xxxii+254pp
- • TRANSL [1952] (C 1990) Transl Phil Writings Frege 137-158
- ◇ A05 B28 E30 E75 ◇ REV JSL 46.870 FdM 25.101
- • REM Transl. is selection of vol I • ID 21334

FREGE, F.L.G. [1895] *Kritische Beleuchtung einiger Punkte in E.Schroeder's Vorlesungen ueber die Algebra der Logik* (J 0361) Arch Systemat Phil 1*433-456
- • TRANSL [1952] (C 1990) Transl Phil Writings Frege 86-106
- ◇ A05 B20 ◇ REV JSL 18.92 JSL 33.282 JSL 46.870
- • ID 17077

FREGE, F.L.G. [1895] *Le nombre entier* (J 0145) Rev Metaph Morale 3*73-78
- ◇ A05 B28 ◇ REV FdM 27.49 • ID 04575

FREGE, F.L.G. [1896] *Ueber die Begriffsschrift des Herrn Peano und meine eigene* (J 0360) Ber Koenigl Ges Wiss Leipzig Math Kl 48*361-378
- ◇ A05 A10 B03 ◇ REV FdM 27.45 • ID 17078

FREGE, F.L.G. [1903] *Grundgesetze der Arithmetik, Begriffsschriftlich abgeleitet. Vol.2* (X 4061) Pohle: Jena xv+265pp
- ◇ A05 B28 E30 E75 ◇ REV JSL 46.870 FdM 34.71
- • ID 21336

FREGE, F.L.G. [1903] *Ueber die Grundlagen der Geometrie* (J 0157) Jbuchber Dtsch Math-Ver 12*319-324,368-375
- • TRANSL [1960] (J 0101) Phil Rev 69*3-17 (English)
- ◇ A05 B30 ◇ REV MR 45 # 6566 JSL 36.155 JSL 46.175 FdM 34.525 • REM Different from the articles in 1906 with the same title • ID 04578

FREGE, F.L.G. [1904] *Was ist eine Funktion ?* (X 1231) Barth: Leipzig 656-666
- • TRANSL [1952] (C 1990) Transl Phil Writings Frege 107-116 (English)
- ◇ A05 B28 E20 ◇ REV FdM 35.977 • REM Festschrift: Gewidmet Ludwig Boltzmann zu seinem 60. Geburtstag, 20 Februar 1904. • ID 21337

FREGE, F.L.G. [1906] *Ueber die Grundlagen der Geometrie* (J 0157) Jbuchber Dtsch Math-Ver 15*293-309,377-403,423-430
- ◇ A05 B30 ◇ REV JSL 46.175 FdM 37.485 • REM Different from the articles in 1903 with the same title
- • ID 04580

FREGE, F.L.G. [1908] *Die Unmoeglichkeit der Thomaeschen formalen Arithmetik aufs neue nachgewiesen* (J 0157) Jbuchber Dtsch Math-Ver 17*52-56
- ◇ A05 B28 ◇ REV JSL 46.175 FdM 39.230 • ID 04582

FREGE, F.L.G. [1915] *The fundamental laws of arithmetic* (J 0320) Monist 25*481-494
- ◇ A05 B28 E30 E75 ◇ ID 04583

FREGE, F.L.G. [1916] *The fundamental laws of arithmetic: Psychological logic* (J 0320) Monist 26*182-199
- ◇ A05 B28 ◇ ID 04584

FREGE, F.L.G. [1917] *Class, function, concept, relation* (J 0320) Monist 27*114-127
- ◇ A05 B28 E30 ◇ ID 04585

FREGE, F.L.G. [1952] *Translations from the philosophical writings of Gottlob Frege* (X 1237) Blackwood: Edinburgh x+244pp
- • LAST ED [1980] (X 3561) Rowman & Littlefield: Totowa x+228pp
- ◇ A05 A10 B96 E30 E96 ◇ REV MR 13.899 MR 82h:03011 Zbl 48.1 • REM Ed. by Black,M. & Geach,P.T. • ID 25773

FREGE, F.L.G. [1962] *Funktion, Begriff, Bedeutung. Fuenf logische Studien* (X 0903) Vandenhoeck & Ruprecht: Goettingen 104pp
- ◇ A05 B15 B28 B30 B96 E30 ◇ REV MR 53 # 7710 Zbl 355 # 02004 • ID 22970

FREGE, F.L.G. [1964] *Begriffsschrift und andere Aufsaetze* (X 0892) Olms: Hildesheim xvi+124pp
- ◇ A05 A10 B96 ◇ REV MR 29 # 2155 • ID 22344

FREGE, F.L.G. [1964] *The basic laws of arithmetic. Exposition of the system* (**X** 0926) Univ Calif Pr: Berkeley lxii + 144pp
 ⋄ A05 A10 B28 ⋄ REV MR 31 # 1173   MR 83m:01058
 Zbl 155.336   Zbl 543 # 03001   JSL 31.671   JSL 46.870 • REM
 2nd ed. 1982, lxiii + 144pp • ID 20892

FREGE, F.L.G. [1964] *The concept of number* (**C** 1105) Phil of Math. Sel Readings 85-112
 ⋄ B28 ⋄ ID 38057

FREGE, F.L.G. [1966] *Logische Untersuchungen* (**X** 0903) Vandenhoeck & Ruprecht: Goettingen 142pp
 ⋄ A05 B96 ⋄ REV Zbl 139.5 • ID 25662

FREGE, F.L.G. [1967] *Kleine Schriften* (**X** 0892) Olms: Hildesheim viii + 434pp
 ⋄ A05 B28 B30 B96 ⋄ REV MR 38 # 979   Zbl 193.439
 • REM Herausgegeben von Angelelli, I. • ID 24820

FREGE, F.L.G. [1969] *Nachgelassene Schriften und wissenschaftlicher Briefwechsel. Band I: Nachgelassene Schriften* (**X** 1088) Meiner: Hamburg xlv + 322pp
 ⋄ A05 A10 B96 ⋄ REV MR 42 # 1640   Zbl 185.5
 Zbl 553 # 01026 • REM 2nd. revised ed. 1983. xli + 388pp
 • ID 21219

FREGE, F.L.G. [1971] *On the foundations of geometry and formal theories of arithmetic* (**X** 0875) Yale Univ Pr: New Haven xlii + 163pp
 ⋄ B28 B30 ⋄ REV MR 45 # 6566   Zbl 236 # 02003
 JSL 36.155 • ID 27849

FREGUGLIA, P. [1975] *Sull'evoluzione algebrica dei principi logici classici* (**J** 1515) Archimede 27*169-175
 ⋄ A10 B05 ⋄ REV MR 55 # 2467   Zbl 325 # 02003
 • ID 30429

FREGUGLIA, P. [1978] *Su una interpretazione algebrico-categoriale del sillogismo* (**J** 1515) Archimede 30*32-48
 ⋄ A05 B20 ⋄ REV MR 81d:03066   Zbl 385 # 03008
 • ID 52116

FREGUGLIA, P. [1979] see BELLACICCO, A.

FREGUGLIA, P. [1980] see BORGA, M.

FREGUGLIA, P. see Vol. V, VI for further entries

FREJVALD, R.V. [1966] *Functional completeness for not everywhere defined functions of the algebra of logic (Russian)* (**J** 0071) Met Diskr Analiz (Novosibirsk) 8*55-68
 ⋄ B05 G15 ⋄ REV MR 35 # 1488   Zbl 199.6   JSL 35.592
 • ID 04597

FREJVALD, R.V. see Vol. II, IV for further entries

FREUDENTHAL, H. [1951] *On the foundations of geometry to be sought in intuition and abstraction (Latin)* (**J** 0359) Gregorianum (Vatican) 32*425-433
 ⋄ A05 B30 ⋄ REV MR 14.5 • ID 17079

FREUDENTHAL, H. [1955] *The concepts axiom and axiomatics in mathematics and physics (Dutch)* (**J** 0061) Simon Stevin 30*156-175
 ⋄ A05 B30 ⋄ REV MR 17.120   Zbl 67.248 • ID 28216

FREUDENTHAL, H. [1958] *Logique mathematique appliquee* (**X** 0834) Gauthier-Villars: Paris 58pp
 ⋄ A05 B05 B45 B70 ⋄ REV MR 20 # 5737   Zbl 84.5
 JSL 24.256 • ID 04610

FREUDENTHAL, H. [1959] *Logic as method and subject (Dutch)* (**J** 0290) Euclides 35*241-255
 ⋄ A05 B30 ⋄ REV MR 23 # A39 • ID 17081

FREUDENTHAL, H. [1960] *Lincos: Design of a language for cosmic intercourse. Part I* (**X** 0809) North Holland: Amsterdam v + 224pp
 ⋄ B80 ⋄ REV MR 22 # 9378   Zbl 95.5 • ID 25320

FREUDENTHAL, H. [1961] *Exact logic (Dutch)* (**X** 1408) Bohn: Amsterdam vii + 117pp
 • TRANSL [1966] (**X** 0838) Amer Elsevier: New York
 vi + 105pp (English)   [1969] (**X** 2027) Nauka: Moskva
 135pp (Russian)   [1975] (**X** 0814) Oldenbourg: Muenchen
 106pp (German)
 ⋄ A05 B98 ⋄ REV MR 25 # 3807   MR 57 # 9449
 Zbl 97.244   JSL 33.603 • ID 28713

FREUDENTHAL, H. see Vol. II, IV, VI for further entries

FREUND, H. & SORGER, P. [1974] *Aussagenlogik und Beweisverfahren* (**X** 1079) Teubner: Leipzig 136pp
 ⋄ B05 ⋄ REV MR 52 # 13297   Zbl 325 # 02001 • ID 21761

FREUND, H. & SORGER, P. [1976] *Logik, Mengen, Relationen. Praxis des mathematischen Beweisens* (**X** 0823) Teubner: Stuttgart 191pp
 ⋄ B05 ⋄ REV Zbl 328 # 02001 • ID 61814

FREY, G. [1983] *Nonstandard arithmetic and application to height functions* (**P** 4230) Groupe Etude Anal Ultrametrique (9);1982 Marseille 3/J8*2pp
 ⋄ C60 H10 H15 H20 ⋄ REV MR 85m:11035
 Zbl 513 # 12021 • ID 37806

FREY, G. see Vol. II, III for further entries

FREYD, P. [1972] *Aspects of topoi* (**J** 0016) Bull Austral Math Soc 7*1-76 • ERR/ADD ibid 7*467-480
 ⋄ B28 E70 E98 G30 ⋄ REV MR 53 # 576
 MR 54 # 7571   Zbl 252 # 18001 • ID 04618

FREYD, P. see Vol. III, V for further entries

FREYTAG-LOERINGHOFF BARON VON, B. [1951] *Logik I. Das System der reinen Logik und ihr Verhaeltnis zur Logistik* (**X** 0808) Kohlhammer: Stuttgart 222pp
 ⋄ A05 B98 ⋄ REV MR 22 # 2537   MR 50 # 12554
 Zbl 286 # 02005 • REM 5. ueberarbeitete Auflage 1972
 • ID 61820

FRIBOURG, L. [1984] *A narrowing procedure for theories with constructors* (**P** 2633) Autom Deduct (7);1984 Napa 259-281
 ⋄ B35 ⋄ REV Zbl 546 # 68076 • ID 43562

FRIBOURG, L. [1984] *Oriented equational clauses as a programming language* (**J** 2551) J Log Progr 1/2*165-177
 ⋄ B35 ⋄ REV MR 86b:68036   Zbl 574 # 68024 • ID 44194

FRIBOURG, L. [1985] *A superposition oriented theorem prover* (**J** 1426) Theor Comput Sci 35*129-164
 ⋄ B35 ⋄ REV Zbl 569 # 68075 • ID 49248

FRIDMAN, EH.I. & PENZIN, YU.G. [1972] *The elementary and the universal theory of the ordered group of integers with maximal subgroups (Russian)* (**C** 3549) Algebra, Vyp 1 (Irkutsk) 80-86
 ⋄ B20 B25 D35 ⋄ REV MR 56 # 11776 • ID 77209

FRIDMAN, EH.I. see Vol. III, IV, VI for further entries

FRIDMAN, G.SH. & KAREV, G.P. & TRESKOV, S.A. [1965] *Estimate of the complexity of a function in the algebra of logic (Russian)* (J 0023) Dokl Akad Nauk SSSR 165*745-747
• TRANSL [1965] (J 0062) Sov Math, Dokl 6*1493-1495
⋄ B05 G05 ⋄ REV MR 33 # 5459 Zbl 156.256 JSL 35.348
• ID 04624

FRIDMAN, G.SH. see Vol. IV for further entries

FRIDSHAL, R. [1957] *The Quine algorithm* (P 1675) Summer Inst Symb Log;1957 Ithaca 211-212
⋄ B35 ⋄ REV JSL 27.103 • ID 29348

FRIDSHAL, R. [1959] see DUNHAM, B.

FRIDSHAL, R. [1960] see DUNHAM, B.

FRIDSHAL, R. [1961] see COBHAM, A.

FRIDSHAL, R. [1962] see DUNHAM, B.

FRIEDMAN, H.M. [1973] *The consistency of classical set theory relative to a set theory with intuitionistic logic* (J 0036) J Symb Logic 38*315-319
⋄ B15 E30 E35 F50 ⋄ REV MR 50 # 68 Zbl 278 # 02045
• ID 04650

FRIEDMAN, H.M. [1975] *One hundred and two problems in mathematical logic* (J 0036) J Symb Logic 40*113-129
⋄ B98 C98 D98 E98 F98 ⋄ REV MR 51 # 5254 Zbl 318 # 02002 • ID 04296

FRIEDMAN, H.M. [1978] *Classically and intuitionistically provably recursive functions* (P 1864) Higher Set Th;1977 Oberwolfach 21-27
⋄ B15 D20 E70 F30 F50 ⋄ REV MR 80b:03093 Zbl 396 # 03045 • ID 31762

FRIEDMAN, H.M. [1980] *A strong conservative extension of Peano arithmetic* (P 2058) Kleene Symp;1978 Madison 113-122
⋄ B28 F30 F35 ⋄ REV MR 82d:03096 Zbl 471 # 03045
• ID 55240

FRIEDMAN, H.M. see Vol. III, IV, V, VI for further entries

FRIEDMAN, JOEL I. [1966] *The decision method for real algebra: Is it practical?* (1111) Preprints, Manuscr., Techn. Reports etc.
⋄ B35 ⋄ REM Rand Research Memorandum 4436-pr, January 1966 • ID 28208

FRIEDMAN, JOEL I. see Vol. II, V for further entries

FRIEDMAN, JOYCE [1963] *A computer program for a solvable case of the decision problem* (J 0037) ACM J 10*348-356
• REPR [1983] (C 4659) Autom of Reasoning 1*355-363
⋄ B35 ⋄ REV MR 27 # 2140 Zbl 118.332 JSL 29.101
• ID 04666

FRIEDMAN, JOYCE [1963] *A semi-decision procedure for the functional calculus* (J 0037) ACM J 10*1-24
• REPR [1983] (C 4659) Autom of Reasoning 1*331-354
⋄ B25 B35 ⋄ REV MR 26 # 4892 Zbl 144.244 JSL 29.101
• ID 04667

FRIEDMAN, JOYCE [1966] *Computer realization of a decision procedure in logic* (P 0573) Inform Processing (3);1965 New York 2*327-328
⋄ B25 B35 ⋄ REV Zbl 173.17 • ID 29439

FRIEDMAN, JOYCE see Vol. IV, VI for further entries

FRIEDMAN, W.H. [1980] *Calculemus* (J 0047) Notre Dame J Formal Log 21*166-174
⋄ B20 ⋄ REV MR 81e:03007 Zbl 299 # 02009 • ID 53173

FROENING, R. [1974] *Maschinelles Beweisen in Struktur-Theorien* (S 3180) Inform Ber, Inst Inf, Univ Bonn 2*79pp
⋄ B15 B35 ⋄ REV Zbl 353 # 68094 • ID 50137

FROIDEVAUX, C. [1983] *La fonction logique ε de Hilbert a travers les "Grundlagen der Mathematik"* (J 0392) Math Sci Hum 84*65-82
⋄ B10 ⋄ REV MR 85i:03001 Zbl 541 # 03001 • ID 41367

FROLOV, A.B. [1980] *Extension of the notion of resolvent deduction and the resolution principle (Russian)* (S 2850) Tr Ehnerg Inst Moskva 485*10-15,95
⋄ B35 ⋄ REV MR 84f:03010 • ID 34428

FROLOV, A.B. see Vol. V for further entries

FUENTES MIRAS, J.R. [1943] *Das Entscheidungsproblem im Aussagenkalkuel (Spanisch)* (J 4341) Euclides Madrid 3*669-672
⋄ B05 ⋄ REV Zbl 60.21 • ID 47949

FUENTES MIRAS, J.R. see Vol. III, VI for further entries

FUERER, M. [1981] *Alternation and the Ackermann case of the decision problem* (J 3370) Enseign Math, Ser 2 27*137-162
• REPR [1982] (P 3482) Logic & Algor (Specker);1980 Zuerich 161-186
⋄ B20 B25 D15 ⋄ REV MR 83d:03015a Zbl 479 # 03005 Zbl 502 # 03025 • ID 35169

FUERER, M. [1984] *The computational complexity of the unconstrained limited domino problem (with implications for logical decision problems)* (P 2342) Symp Rek Kombin;1983 Muenster 312-319
⋄ B20 D15 D80 ⋄ REV Zbl 549 # 68037 • ID 43160

FUERER, M. see Vol. III, IV, V for further entries

FUJIKAWA, Y. [1972] *The construction of an implicational normal form* (J 0260) Ann Jap Ass Phil Sci 4*151-156
⋄ B05 ⋄ REV MR 51 # 5258 Zbl 242 # 02007 • ID 17468

FUJIMURA, T. [1972] *System of modern logic (Japanese)* (C 1832) Katoba no Tetsugaku 148-217
⋄ B05 • ID 32336

FUJIMURA, T. see Vol. II for further entries

FUJITA, H. & NARUSHIMA, H. & NOJIMA, S. & ODAKA, A. [1976] *A new normal form of boolean functions* (J 0193) Discr Math 14*269-271
⋄ B05 ⋄ REV MR 53 # 15576 Zbl 323 # 02014 • ID 30408

FUJIWARA, T. [1971] *On the construction of the least universal Horn class containing a given class* (J 0351) Osaka J Math 8*425-436
⋄ B20 C05 C52 ⋄ REV MR 47 # 108 Zbl 264 # 08003
• ID 17104

FUJIWARA, T. [1973] *Note on generalized atomic sets of formulas* (J 0081) Proc Japan Acad 49*443-448
⋄ B20 C05 C40 ⋄ REV MR 49 # 2337 Zbl 276 # 02033
• ID 04714

FUJIWARA, T. [1975] *Universal sentences preserved under certain extensions* (J 0081) Proc Japan Acad 51*29-33
⋄ B20 C40 ⋄ REV MR 52 # 13367 Zbl 331 # 02030
• ID 21821

FUJIWARA, T. see Vol. III for further entries

FUKSON, V.I. [1973] see BOCHVAR, D.A.

FUKSON, V.I. [1974] see BOCHVAR, D.A.

FUKSON, V.I. see Vol. III, V for further entries

FULTON, J.A. [1974] *Unary predicates* (J 0047) Notre Dame J Formal Log 15*635-638
  ◇ B10 ◇ REV MR 50#9526 Zbl 245#02018 • ID 04718

FULTON, J.A. see Vol. II for further entries

FUNAHASHI, S. & NAGATA, S. & OSHIBA, T. [1979] *A method for obtaining proof figures of valid formulas in the first order predicate calculus* (J 0523) Bull Nagoya Inst Tech 31*117-126
  ◇ B10 F07 ◇ REV MR 81j:03093 • ID 77023

FUNAHASHI, S. & MIYAWAKI, M. & NAGATA, S. & OSHIBA, T. [1981] *A program for checking the validities of formulas in the first-order predicate calculus (Japanese) (English summary)* (J 0523) Bull Nagoya Inst Tech 32*91-96
  ◇ B35 ◇ REV MR 83f:68107 • ID 40214

FUNAHASHI, S. & NAGATA, S. & OSHIBA, T. [1984] *A procedure for checking validities of formulas by using adjoint formulas (Japanese) (English summary)* (J 0523) Bull Nagoya Inst Tech 36*109-112
  ◇ B35 ◇ ID 49351

FURMANOWSKI, T. [1982] *Adjoint interpretations of sentential calculi* (J 0063) Studia Logica 41*359-374
  ◇ B22 F25 G30 ◇ REV MR 85i:03195 Zbl 561#03013 • ID 42298

FURMANOWSKI, T. see Vol. II, III for further entries

GABBAY, D.M. [1973] *Applications of Scott's notion of consequence to the study of general binary intensional connectives and entailment* (J 0122) J Philos Logic 2*340-351
  ◇ B22 B46 ◇ REV MR 54#7202 Zbl 272#02034 • ID 30381

GABBAY, D.M. & JONGH DE, D.H.J. [1974] *A sequence of decidable finitely axiomatizable intermediate logics with the disjunction property* (J 0036) J Symb Logic 39*67-78
  ◇ B22 B25 B55 ◇ REV MR 51#10038 Zbl 289#02032 • ID 17555

GABBAY, D.M. [1978] *What is a classical connective?* (J 0068) Z Math Logik Grundlagen Math 24*37-44
  ◇ B22 F07 ◇ REV MR 57#9500 Zbl 374#02016 • ID 32163

GABBAY, D.M. & GUENTHNER, F. (EDS.) [1983] *Handbook of philosophical logic Vol. I: Elements of classical logic* (S 3307) Synth Libr 164*xi+493pp
  ◇ B98 C98 ◇ REV Zbl 538#03001 • ID 41457

GABBAY, D.M. & GUENTHNER, F. (EDS.) [1984] *Handbook of philosophical logic Vol II. Extensions of classical logic* (S 3307) Synth Libr 165*x+776pp
  ◇ B98 C90 C98 ◇ REV Zbl 572#03003 • REM Part I 1983 • ID 41831

GABBAY, D.M. see Vol. II, III, IV, VI for further entries

GABRIEL, G. [1978] *Implizite Definitionen -- eine Verwechslungsgeschichte* (J 2657) Ann of Sci 35*419-423
  ◇ A10 B28 ◇ REV MR 80g:01018 Zbl 385#01018 • ID 81294

GABRIELIAN, A. [1981] *Pure grammars and pure languages* (J 0382) Int J Comput Math 9*3-16
  ◇ B05 D03 D05 D10 ◇ REV MR 82f:68077 Zbl 454#68097 • ID 54258

GABRIELIAN, A. see Vol. IV for further entries

GACS, P. & LOVASZ, L. [1977] *Some remarks on generalized spectra* (J 0068) Z Math Logik Grundlagen Math 23*547-554
  ◇ B15 C13 D15 ◇ REV MR 58#10398 Zbl 398#03025 • ID 52753

GACS, P. see Vol. II, IV for further entries

GADSHIEV, M.M. [1971] *The maximal length of the abbreviated disjunctive normal form for boolean functions of five and six variables (Russian)* (J 0071) Met Diskr Analiz (Novosibirsk) 18*3-24
  ◇ B05 ◇ REV MR 45#3269 Zbl 235#02014 • ID 04740

GAGNON, L.S. [1976] *Nor logic: A system of natural deduction* (J 0047) Notre Dame J Formal Log 17*293-294
  ◇ B05 ◇ REV MR 53#10545 Zbl 305#02019 • ID 18151

GAGNON, L.S. see Vol. II for further entries

GAIFMAN, H. & SNIR, M. [1982] *Probabilities over rich languages, testing and randomness* (J 0036) J Symb Logic 47*495-548
  ◇ B28 ◇ REV MR 84b:03037 Zbl 501#60006 • ID 35629

GAIFMAN, H. see Vol. II, III, IV, V, VI for further entries

GAINES, B.R. [1979] *Logical foundations for database systems* (J 1741) Int J Man-Mach Stud 11*481-500
  ◇ B80 ◇ REV MR 80g:68023 Zbl 404#68098 • ID 81296

GAINES, B.R. & MAMDANI, E.H. (EDS.) [1981] *Fuzzy reasoning and its applications* (X 0801) Academic Pr: New York xviii+381pp
  ◇ B52 B98 ◇ REV MR 83d:03001 Zbl 488#03001 • ID 34823

GAINES, B.R. see Vol. II, IV, V for further entries

GAITANIS, N. & HALATSIS, C. [1978] *Irredundant normal forms and minimal dependence sets of a boolean function* (J 0187) IEEE Trans Comp C-27*1064-1068
  ◇ B05 ◇ REV MR 80a:94055 Zbl 388#94020 • ID 52288

GAL, S. [1978] see BREJTBART, YU.YA.

GALDA, K. & PASSOS, E.P. [1974] *Application of quantifier elimination to mechanical theorem proving* (P 3103) Adv Cybern Syst;1972 Oxford 1*335-348
  ◇ B35 C10 ◇ REV Zbl 332#68064 • ID 61899

GALIL, Z. [1977] *On resolution with clauses of bounded size* (J 1428) SIAM J Comp 6*444-459
  ◇ B35 D15 ◇ REV MR 56#13813 Zbl 368#68085 • ID 51297

GALIL, Z. [1977] *On the complexity of regular resolution and the Davis-Putnam procedure* (J 1426) Theor Comput Sci 4*23-46
  ◇ B35 D15 ◇ REV MR 56#4269 Zbl 385#68048 • ID 52174

GALIL, Z. see Vol. IV for further entries

GALLIE, R.D. [1975] *Substitutionalism and substitutional quantification* (J 0103) Analysis (Oxford) 35*97-101
  ◇ B10 ◇ REV Zbl 345#02006 • ID 29768

GALLIER, J.H. [1984] see DOWLING, W.F.

GALLIER, J.H. see Vol. III, IV for further entries

GALLIN, D. [1969] see BERLINSKI, D.

GALLIN, D. see Vol. II, III, V for further entries

GALVIN, F. [1967] *Reduced products, Horn sentences, and decision problems* (J 0015) Bull Amer Math Soc 73*59-64
⋄ B20 B25 C20 C30 C40 D35 ⋄ REV MR 35 # 39 Zbl 155.349 JSL 33.477 • ID 04765

GALVIN, F. [1970] *Horn sentences* (J 0007) Ann Math Logic 1*389-422
⋄ B20 C20 C30 C40 D35 ⋄ REV MR 48 # 3729 Zbl 206.278 JSL 38.651 • ID 04766

GALVIN, F. see Vol. III, IV, V for further entries

GANDY, R.O. [1954] *On the possibility of proving the consistency of the simple theory of types* (P 0575) Int Congr Math (II, 7);1954 Amsterdam 2*400-401
⋄ B15 ⋄ ID 29465

GANDY, R.O. [1956] *On the axiom of extensionality I* (J 0036) J Symb Logic 21*36-48
⋄ B15 E30 ⋄ REV MR 17.817 Zbl 73.8 JSL 29.142 • REM Part II 1959 • ID 21137

GANDY, R.O. [1974] *Set-theoretic functions for elementary syntax* (P 0693) Axiomatic Set Th;1967 Los Angeles 2*103-126
⋄ B03 D20 D65 E47 ⋄ REV MR 51 # 12524 Zbl 323 # 02067 • ID 17242

GANDY, R.O. & HYLAND, J.M.E. (EDS.) [1977] *Logic Colloquium 76. Proceedings of a conference held in Oxford in July 1976* (S 3303) Stud Logic Found Math 87*x+612pp
⋄ B97 C97 D97 ⋄ REV MR 57 # 53 Zbl 409 # 00002 • ID 16612

GANDY, R.O. [1977] *The simple theory of types* (P 1075) Logic Colloq;1976 Oxford 173-181
⋄ A10 B15 ⋄ REV MR 58 # 5017 Zbl 436 # 03005 • ID 16619

GANDY, R.O. see Vol. III, IV, V, VI for further entries

GAO, HENGSHAN [1983] *Comments on "The interpolation theorem for the propositional calculus $P(\kappa)$ when κ is a strongly inaccessible cardinal" by Luo,Libo (Chinese)* (J 3742) Shuxue Yanjiu yu Pinglun 3/4*133-134
⋄ B05 C40 C75 E55 ⋄ REV MR 86a:03007 • REM Comments to Luo,Libo 1980 • ID 45159

GAO, HENGSHAN see Vol. II, III, IV, V, VI for further entries

GARDIES, J.-L. [1968] *Les deux tetraedres des liaisons logiques interpropositionnelles bivalentes* (J 0063) Studia Logica 23*157-161
⋄ B05 ⋄ ID 33128

GARDIES, J.-L. [1976] *Modalites et normes* (J 1834) Arch Rechts- und Sozialphil 62*465-474
⋄ B80 ⋄ ID 32340

GARDIES, J.-L. see Vol. II for further entries

GARLAND, S.J. [1974] *Second-order cardinal characterizability* (P 0693) Axiomatic Set Th;1967 Los Angeles 2*127-146
⋄ B15 C40 C55 C85 D55 E10 E47 E55 ⋄ REV MR 54 # 4982 Zbl 319 # 02065 • ID 24144

GARLAND, S.J. see Vol. III, IV, V for further entries

GAROCHE, F. & LEONARD, M. [1984] *On a class of Boolean functions with matroid property* (J 0193) Discr Math 49*323-325
⋄ B05 ⋄ REV MR 85f:06023 Zbl 539 # 05022 • ID 39897

GASHKOV, S.B. [1978] *Ueber die Tiefe der boolschen Funktionen (Russisch)* (J 0052) Probl Kibern 34*265-268
⋄ B05 ⋄ REV MR 80k:94040 Zbl 415 # 94019 • ID 69614

GASHKOV, S.B. [1980] *Complexity of realization of boolean functions by schemes made up of functional elements and by formulas in bases whose elements realize continuous functions (Russian)* (J 0052) Probl Kibern 37*57-118,239
⋄ B35 ⋄ REV MR 82f:94025 Zbl 482 # 94035 • ID 81311

GASTEV, YU.A. [1959] *Ueber die Konstruktion der Analysis auf Grund einer axiomatisierten Geometrie der Geraden I (Russisch)* (S 0208) Uch Zap, Ped Inst, Moskva 3*46-57
⋄ B28 ⋄ REV Zbl 216.293 • REM Part II 1963 • ID 26264

GASTEV, YU.A. [1963] *Ueber die Konstruktion der Analysis auf Grund einer axiomatisierten Geometrie der Geraden II (Russisch)* (C 1654) Probl Logiki 137-143
⋄ B28 ⋄ REV Zbl 257 # 02045 • REM Part I 1959 • ID 29000

GASTEV, YU.A. [1970] *The expressible and deductive possibilities of logico-arithmetic calculi on the basis of the theory of types (Russian)* (C 1530) Issl Log Sist (Yanovskaya) 78-85
⋄ B15 F35 ⋄ REV MR 50 # 12644 Zbl 209.8 • ID 17919

GAUTHIER, Y. [1978] *Methodes et concepts de la logique formelle* (X 0893) Pr Univ Montreal: Montreal 238pp
⋄ A05 B98 ⋄ REV MR 80e:03001 Zbl 383 # 03004 • ID 51986

GAUTHIER, Y. see Vol. II, VI for further entries

GAVRILOV, G.P. & KUDRYAVTSEV, V.B. & YABLONSKIJ, S.V. [1966] *Boolesche Funktionen und Postsche Klassen (Russisch)* (X 2027) Nauka: Moskva 119pp
• TRANSL [1970] (X 0911) Akademie Verlag: Berlin ix+84pp
⋄ B05 B20 G05 G10 ⋄ REV MR 35 # 6489 MR 42 # 2868 Zbl 171.277 • ID 28531

GAVRILOV, G.P. & SAPOZHENKO, A.A. [1977] *Collection of problems in discrete mathematics (Russian)* (X 2027) Nauka: Moskva 368pp
⋄ B05 ⋄ REV MR 57 # 9371 Zbl 452 # 00010 • ID 54062

GAVRILOV, G.P. see Vol. II, V for further entries

GAVRILOV, M.A. [1963] *The present state of the theory of relay circuits (Russian)* (C 4103) Struktur Teor Relej Ustrojstv 5-73
⋄ B05 B70 ⋄ REV JSL 34.511 • ID 43318

GAZALE, M.J. [1957] *Irredundant disjunctive and conjunctive forms of a boolean function* (J 0284) IBM J Res Dev 1*171-176
⋄ B35 ⋄ REV JSL 30.106 • ID 04921

GAZALE, M.J. see Vol. II for further entries

GEACH, P.T. & WRIGHT VON, G.H. [1952] *On an extended logic of relations* (J 0990) Soc Sci Fennicae Comment Phys-Math 16/1*37pp
⋄ B20 B25 ⋄ REV Zbl 49.149 JSL 22.72 • ID 16851

GEACH, P.T. [1970] *A program for syntax* (J 0154) Synthese 22*3-17
⋄ B03 ⋄ ID 37360

GEACH, P.T. [1981] *Second-order quantification in Frege (Spanish)* (J 0162) Teorema (Valencia) 11*167-177
⋄ A10 B15 ⋄ REV MR 84a:03011 • ID 35558

GEACH, P.T. see Vol. II, V for further entries

GEERTSEMA, J.C. [1983] *Recent views in the foundational controversy in statistics* (J 3936) S Afrik Statist J 17*121-146
⋄ B30 ⋄ REV MR 85c:62003 Zbl 523 # 62003 • ID 36998

GEISER, J.R. [1968] *Absolute properties in higher order structures* (1111) Preprints, Manuscr., Techn. Reports etc.
⋄ B15 ⋄ REM Notes Dartmouth College • ID 21346

GEISER, J.R. [1968] *Nonstandard logic* (J 0036) J Symb Logic 33*236-250
⋄ C20 H10 H15 ⋄ REV MR 42 # 55 Zbl 157.17
• ID 04827

GEISER, J.R. [1970] *Nonstandard analysis* (J 0068) Z Math Logik Grundlagen Math 16*297-318
⋄ H05 ⋄ REV MR 43 # 6071 Zbl 174.16 • ID 04830

GEISER, J.R. [1974] *A formalization of Essenin-Volpin's proof theoretical studies by means of nonstandard analysis* (J 0036) J Symb Logic 39*81-87
⋄ C75 F50 H10 H15 ⋄ REV MR 55 # 5406 Zbl 296 # 02018 • ID 04831

GEISER, J.R. see Vol. III, IV, VI for further entries

GELDER VAN, A. [1984] *A satisfiability tester for non-clausal propositional calculus* (P 2633) Autom Deduct (7);1984 Napa 101-112
⋄ B35 ⋄ REV Zbl 546 # 68074 • ID 43560

GELERNTER, H.L. [1957] *Theorem proving by machine* (P 1675) Summer Inst Symb Log;1957 Ithaca 305-308
⋄ B35 ⋄ REV JSL 32.522 • ID 29367

GELERNTER, H.L. [1959] *A note on syntactic symmetry and the manipulation of formal systems by machine* (J 0194) Inform & Control 2*80-89
⋄ B35 ⋄ REV MR 21 # 3946 Zbl 92.134 JSL 31.514
• ID 04838

GELERNTER, H.L. [1960] *Realization of a geometry theorem proving machine* (P 0696) Inform Processing (1);1959 Paris 273-282
• REPR [1983] (C 4659) Autom of Reasoning 1*99-124
⋄ B35 ⋄ REV MR 26 # 4915 Zbl 114.69 JSL 32.522
• ID 04839

GELERNTER, H.L. & HANSEN, J.R. & LOVELAND, D.W. [1963] *Empirical explorations of the geometry-theorem proving machine* (C 4307) Comput & Thought 153-167
• REPR [1983] (C 4659) Autom of Reasoning 1*140-150
⋄ B35 ⋄ ID 47326

GELERNTER, H.L. [1965] *Machine-generated problem-solving graphs* (P 0797) Fonds des Math, Machines Math & Appl;1962 Tihany 283-307
• REPR [1983] (C 4659) Autom of Reasoning 1*288-314
⋄ B35 ⋄ REV MR 33 # 3485 Zbl 158.162 • ID 27606

GEL'FOND, M.G. [1972] *Relationship between the classical and constructive developments of mathematical analysis (Russian) (English summary)* (S 0228) Zap Nauch Sem Leningrad Otd Mat Inst Steklov 32*5-11,153
• TRANSL [1976] (J 1531) J Sov Math 6*347-352
⋄ B28 F35 F60 ⋄ REV MR 49 # 8827 Zbl 354 # 02028
• ID 04841

GEL'FOND, M.G. see Vol. VI for further entries

GENENZ, J. [1964] *Reduktionstheorie nach der Methode von Kahr-Moore-Wang* (X 3333) Unknown Publisher: See Remarks 88pp
⋄ B20 D35 D98 ⋄ REM Muenster • ID 21347

GENRICH, H.J. & LAUTENBACH, K. [1979] *The analysis of distributed systems by means of predicate/transition-nets* (P 3257) Semant of Concurr Comput;1979 Evian 123-146
⋄ B10 ⋄ REV MR 81f:68033 Zbl 407 # 68066 • ID 56235

GENRICH, H.J. see Vol. II, IV for further entries

GENSLER, H.J. [1973] *A simplified decision procedure for categorical syllogisms* (J 0047) Notre Dame J Formal Log 14*457-466
⋄ B20 B25 ⋄ REV MR 52 # 2814 Zbl 265 # 02009
• ID 18156

GENSLER, H.J. see Vol. II for further entries

GENTZEN, G. [1932] *Ueber die Existenz unabhaengiger Axiomensysteme zu unendlichen Satzsystemen* (J 0043) Math Ann 107*329-350
⋄ B30 ⋄ REV Zbl 5.338 FdM 58.63 • ID 04849

GENTZEN, G. [1935] *Untersuchungen ueber das logische Schliessen I,II* (J 0044) Math Z 39*176-210,405-431
• TRANSL [1955] (X 0840) Pr Univ France: Paris xi+170pp (French) [1964] (J 0325) Amer Phil Quart 1*288-306, 2*204-218 (English) [1969] (C 1409) Gentzen: Collected Papers 69-131 (English) [1978] (J 2751) J Fac Lib Art Yamaguchi Univ 12*1-20, 13*21-40 (Japanese)
• REPR [1969] (X 0890) Wiss Buchges: Darmstadt ii+62pp
⋄ B10 B25 F05 F07 F30 F50 ⋄ REV MR 17.3 MR 50 # 4228 MR 82m:01080 Zbl 10.145 Zbl 10.146 JSL 1.209 JSL 22.350 JSL 35.144 FdM 60.20 • REM 1969 transl. contains additional notes • ID 24221

GENTZEN, G. [1936] *Die Widerspruchsfreiheit der Stufenlogik* (J 0044) Math Z 41*357-366
• TRANSL [1969] (C 1409) Gentzen: Collected Papers 214-222
⋄ B15 F05 F35 ⋄ REV Zbl 15.193 JSL 1.119 FdM 62.43
• ID 04851

GENTZEN, G. [1936] *Die Widerspruchsfreiheit der reinen Zahlentheorie* (J 0043) Math Ann 112*493-565
• TRANSL [1969] (C 1409) Gentzen: Collected Papers 132-170 • LAST ED [1967] (X 0890) Wiss Buchges: Darmstadt 73pp
⋄ B28 F05 F07 F30 ⋄ REV MR 36 # 4973 Zbl 14.38 Zbl 169.308 JSL 1.75 FdM 62.44 • REM For additional sections IV and V see 1974 • ID 22402

GENTZEN, G. [1954] *Zusammenfassung von mehreren vollstaendigen Induktionen zu einer einzigen* (J 0009) Arch Math Logik Grundlagenforsch 2*1-3
⋄ B28 F30 ⋄ REV MR 18.272 Zbl 55.5 JSL 38.157
• ID 04853

GENTZEN, G. [1969] *The collected papers of Gerhard Gentzen* (X 0809) North Holland: Amsterdam xii+338pp
⋄ B96 F05 F07 F30 F50 F96 ⋄ REV MR 41 # 6660 Zbl 209.300 • REM Ed. by Szabo,M.E. • ID 29483

GENTZEN, G. [1974] *Ueber das Verhaeltnis zwischen intuitionistischer und klassischer Arithmetik* (J 0009) Arch Math Logik Grundlagenforsch 16*119-132
• TRANSL [1969] (C 1409) Gentzen: Collected Papers 53-67
⋄ B30 F30 F50 ⋄ REV MR 52 # 52 Zbl 286 # 02035
• REM Originally to appear in J0043; was withdrawn
• ID 04855

GENTZEN, G. see Vol. VI for further entries

GEORGE, F.H. [1977] *Precision, language and logic* (X 0869) Pergamon Pr: Oxford x+216pp
◇ A05 B98 ◇ REV Zbl 404 # 03005 • ID 54792

GEORGESCU, G. [1971] *Praedikatenkalkuel I* (J 0378) Gaz Mat (Bucharest) Ser A 76*361-366
◇ B98 ◇ REV Zbl 222 # 02006 • ID 26301

GEORGESCU, G. [1973] *The theory of categories and mathematical logic (Romanian)* (J 0378) Gaz Mat (Bucharest) Ser A 78*121-125
◇ B98 ◇ REV MR 52 # 10415 Zbl 256 # 18001 • ID 21712

GEORGESCU, G. [1977] *Information storage and retrieval systems based on the θ-valued logic (Romanian) (English summary)* (J 0197) Stud Cercet Mat Acad Romana 29*113-116
◇ B75 B80 ◇ REV MR 58 # 25178 Zbl 362 # 68132 • ID 50820

GEORGESCU, G. see Vol. II, III, V for further entries

GEORGIEVA, N.V. [1965] *A problem in propositional calculus (Bulgarian) (Russian and English summaries)* (J 0224) God Vissh Tekh Ucheb Zaved Mat, Sofiya 2/1*11-20
◇ B05 ◇ REV MR 36 # 1289 Zbl 311 # 02014 • ID 04882

GEORGIEVA, N.V. [1971] *A logical system which has $\equiv$ and $\vee$ as primitive connectives (Russian) (Polish and English summaries)* (J 0063) Studia Logica 28*71-76
◇ B20 ◇ REV MR 46 # 1547 Zbl 243 # 02007 • ID 04885

GEORGIEVA, N.V. [1971] *Independence of system of axioms and rules of inference of propositional calculus (Russian) (Polish and English summaries)* (J 0063) Studia Logica 28*65-70
◇ B05 ◇ REV MR 47 # 10 Zbl 243 # 02006 • ID 04883

GEORGIEVA, N.V. [1971] *Independence of the axiom and rules of inference of one system of the extended propositional calculus* (J 0047) Notre Dame J Formal Log 12*214-218
◇ B05 ◇ REV MR 45 # 24 Zbl 197.272 • ID 04884

GEORGIEVA, N.V. see Vol. II, III, IV, VI for further entries

GERGELY, T. [1973] see ANDREKA, H.

GERGELY, T. [1974] see ANDREKA, H.

GERGELY, T. [1975] see ANDREKA, H.

GERGELY, T. & VERSHININ, K.P. [1978] *Model theoretical investigation of theorem proving methods* (J 0047) Notre Dame J Formal Log 19*523-542
◇ B35 C07 ◇ REV MR 80b:03015 Zbl 333 # 68060 • ID 52253

GERGELY, T. & VERSHININ, K.P. [1981] *Concept sensitive formal language for task specification* (P 3642) Colloq Math Log in Computer Sci;1978 Salgotarjan 429-470
◇ B35 ◇ REV MR 83g:68007 Zbl 482 # 68090 • ID 37708

GERGELY, T. [1981] see DOEMOELKI, B.

GERGELY, T. see Vol. IV for further entries

GERICKE, H. [1952] *Algebraische Betrachtungen zu den Aristotelischen Syllogismen* (J 0008) Arch Math (Basel) 3*421-433
◇ A05 B20 ◇ REV MR 14.935 Zbl 49.146 JSL 22.308 • ID 04901

GERICKE, H. see Vol. III, V for further entries

GERMANO, G. [1970] *Metamathematische Begriffe in Standardtheorien* (J 0009) Arch Math Logik Grundlagenforsch 13*22-38
◇ A05 B30 F30 ◇ REV MR 44 # 70 Zbl 221 # 02037 • ID 04906

GERMANO, G. see Vol. III, IV, VI for further entries

GERMANSKY, B. [1961] *The induction axiom and the axiom of choice* (J 0068) Z Math Logik Grundlagen Math 7*219-223
◇ B28 E25 ◇ REV MR 27 # 2436 Zbl 101.251 JSL 27.237 • ID 04911

GHELFO, S. & OMODEO, E.G. [1985] *Towards practical implementations of syllogistic* (P 4601) EUROCAL;1985 Linz 2*40-49
◇ B35 ◇ ID 49675

GHITA, A. [1981] *A class of fields of hyperreal numbers* (J 0060) Rev Roumaine Math Pures Appl 26*571-579
◇ H05 ◇ REV MR 82m:03073 Zbl 489 # 12011 • ID 73350

GIANNONE, A. [1978] *An introduction to nonstandard methods via simply additive measures* (J 3523) Period Mat, Ser 5 54*41-58
◇ H05 ◇ REV MR 80d:03064 • ID 73353

GIANNONE, A. [1980] see BARONE, E.

GIANNONE, A. [1982] *Probabilita non σ-additive e analisi non-standard (English summary)* (J 3741) Rend Mat Appl, Ser 7 2*47-58
◇ C20 H10 ◇ REV MR 83i:03105 Zbl 492 # 60002 • ID 34858

GIBBINS, P. [1979] *Material implication, the sufficiency condition, and conditional proof* (J 0103) Analysis (Oxford) 39*21-24
◇ A05 B05 ◇ REV Zbl 406 # 03010 • ID 56093

GIBBINS, P. see Vol. II for further entries

GILBERT, M.A. [1976] *A heuristic procedure for natural deduction derivations using reductio ad absurdum* (J 0047) Notre Dame J Formal Log 17*638-639
◇ B35 F07 ◇ REV Zbl 305 # 02020 • ID 62011

GILES, R. [1976] *A pragmatic approach to the formalization of empirical theories* (P 1804) Form Meth in Methodol of Emp Sci;1974 Warsaw 113-135
◇ B30 ◇ REV MR 58 # 27305 Zbl 361 # 02021 • ID 50656

GILES, R. see Vol. II, III, IV, V for further entries

GILEZAN, K. [1971] *Une generalisation du theoreme de Loewenheim sur les equations de Boole* (J 0400) Publ Inst Math, NS (Belgrade) 11(25)*57-59
◇ B05 ◇ REV MR 49 # 2482 Zbl 237 # 02021 • ID 06087

GILEZAN, K. see Vol. II for further entries

GILL, A. [1976] *Applied algebra for the computer sciences* (X 0819) Prentice Hall: Englewood Cliffs xv+432pp
◇ B35 ◇ REV Zbl 346 # 68002 • ID 62016

GILL, A. see Vol. IV for further entries

GILLIES, D.A. [1982] *Frege, Dedekind and Peano on the foundations of arithmetic* (X 1994) Roy van Gorcum: Assen ix+103pp
◇ A10 B28 ◇ REV MR 84i:01060 Zbl 514 # 03004 • ID 36979

GILMORE, P.C. [1957] *The monadic theory of types in the lower predicate calculus* (P 1675) Summer Inst Symb Log;1957 Ithaca 309-312
  ⋄ B15 ⋄ REV JSL 37.766 • ID 04943

GILMORE, P.C. [1958] *An addition to logic of many-sorted theories* (J 0020) Compos Math 13∗277-281
  ⋄ B15 ⋄ REV MR 21#4910 JSL 32.521 • ID 04942

GILMORE, P.C. [1960] *A program for the production from axioms, of proofs for theorems derivable within the first order predicate calculus* (P 0696) Inform Processing (1);1959 Paris 265-273
  ⋄ B35 ⋄ REV MR 26#4914 Zbl 118.332 JSL 31.124
  • ID 21990

GILMORE, P.C. [1960] *A proof method for quantification theory: Its justification and realization* (J 0284) IBM J Res Dev 4∗28-35
  • REPR [1983] (C 4659) Autom of Reasoning 1∗151-161
  ⋄ B35 ⋄ REV MR 23#B3113 Zbl 97.3 JSL 31.124
  • ID 43088

GILMORE, P.C. [1970] *An examination of the geometry theorem machine* (J 0503) Artif Intell 1∗171-187
  ⋄ B35 ⋄ REV Zbl 205.316 • ID 47327

GILMORE, P.C. see Vol. II, III, V, VI for further entries

GINDIKIN, S.G. & MUCHNIK, A.A. [1962] *The completeness of a system made up of nonreliable elements realizing a function of algebraic logic (Russian)* (J 0023) Dokl Akad Nauk SSSR 144∗1007-1010
  • TRANSL [1962] (J 0470) Sov Phys, Dokl 7∗477-479
  ⋄ B05 ⋄ REV MR 26#2347 Zbl 121.15 • ID 48074

GINDIKIN, S.G. [1985] *Algebraic logic* (X 0811) Springer: Heidelberg & New York xviii+356pp
  ⋄ B98 ⋄ REM Transl. from the Russian • ID 49125

GINSBERG, A. [1981] *Quantum theory and the identity of indiscernibles revisited* (J 0153) Phil of Sci (East Lansing) 48∗487-491
  ⋄ A05 B30 ⋄ REV MR 83e:03109 • ID 35254

GIORELLO, G. [1972] *Strutture non-standard della teoria dei numeri reali* (J 1515) Archimede 24∗152-163
  ⋄ H05 ⋄ REV Zbl 241#02020 • ID 28824

GIORELLO, G. [1973] *Una rappresentazione non-standard delle distribuzioni functoriali* (J 0012) Boll Unione Mat Ital, IV Ser 7∗109-113
  ⋄ H05 ⋄ REV MR 47#3986 • ID 04997

GIORELLO, G. [1973] *Una rappresentazione nonstandard delle distribuzioni temperate e la trasformazione di Fourier* (J 0012) Boll Unione Mat Ital, IV Ser 7∗156-167
  ⋄ H05 ⋄ REV MR 47#3986 Zbl 261#46042 • ID 81344

GIOVANNETTI, E. [1983] *Une caracterisation des termes types dans un langage applicatif (English summary)* (J 3364) C R Acad Sci, Paris, Ser 1 296∗97-99
  ⋄ B10 B40 ⋄ REV MR 84b:03029 Zbl 567#68050
  • ID 35624

GIOVANNETTI, E. see Vol. VI for further entries

GIRARD, J.-Y. [1984] *The Ω-rule* (P 4313) Int Congr Math (II,14);1983 Warsaw 1∗307-321
  ⋄ B15 C62 C75 F35 ⋄ REV Zbl 569#03023 • ID 48565

GIRARD, J.-Y. see Vol. II, III, IV, V, VI for further entries

GIRLING, B. & MORING, H. [1973] *Logic and logic design* (X 3119) Intertext Books: Aylesbury vii+328pp
  ⋄ B70 B98 ⋄ REV Zbl 298#94033 • ID 62036

GIVANT, S. [1978] *Universal Horn classes categorical or free in power* (J 0007) Ann Math Logic 15∗1-53
  ⋄ B20 C05 C35 ⋄ REV MR 80c:03032 Zbl 401#03009
  • ID 29155

GIVANT, S. [1979] *A representation theorem for universal Horn classes categorical in power* (J 0007) Ann Math Logic 17∗91-116
  ⋄ B20 C05 C35 ⋄ REV MR 81b:03038 Zbl 436#03020
  • ID 55860

GIVANT, S. see Vol. III for further entries

GIVONE, D.D. [1968] see ALLEN, C.M.

GIVONE, D.D. [1973] see ALLEN, C.M.

GIVONE, D.D. see Vol. II for further entries

GIZOWA, H. [1965] *From Zygmunt Kramsztyk's work in the field of applied logic in medicine (Polish) (English and Russian summaries)* (S 0458) Zesz Nauk, Prace Log, Uniw Krakow 1∗75-84
  ⋄ B80 ⋄ ID 16503

GLADKIJ, A.V. [1974] *Lecture notes on mathematical logic and set theory (Russian)* (X 1434) Kalinin Gos Univ: Kalinin 163pp
  ⋄ B98 E98 ⋄ REV MR 52#13298 • ID 21763

GLADKIJ, A.V. [1977] *The language of mathematical logic* (X 1434) Kalinin Gos Univ: Kalinin
  ⋄ B98 ⋄ ID 33958

GLADKIJ, A.V. [1978] see BOJDAKOVA, V.N.

GLADKIJ, A.V. (ED.) [1983] *Mathematical logic, mathematical linguistics and theory of algorithms (Russian)* (X 1434) Kalinin Gos Univ: Kalinin 116pp
  ⋄ B97 D97 ⋄ REV MR 84k:03005 • ID 34944

GLADKIJ, A.V. see Vol. IV, V for further entries

GLADKIJ, M. [1981] *Mathematical logic and mathematical linguistics (Russian)* (X 1434) Kalinin Gos Univ: Kalinin 172pp
  ⋄ B98 ⋄ REV MR 83h:03004 • ID 45984

GLADSTONE, M.D. [1965] *Some ways of constructing a propositional calculus of any required degree of unsolvability* (J 0064) Trans Amer Math Soc 118∗192-210
  ⋄ B20 D25 D35 ⋄ REV MR 31#26 Zbl 168.12 JSL 34.505 • ID 05010

GLADSTONE, M.D. [1968] *A single-axiom implicational calculus of given unsolvability* (J 0068) Z Math Logik Grundlagen Math 14∗193-204
  ⋄ B20 D25 D30 D35 ⋄ REV MR 37#6180 Zbl 164.318
  • ID 05013

GLADSTONE, M.D. [1970] *On the number of variables in the axioms* (J 0047) Notre Dame J Formal Log 11∗1-15
  ⋄ B05 ⋄ REV MR 44#6440 Zbl 188.10 JSL 37.755
  • ID 05014

GLADSTONE, M.D. [1979] *The decidability of one-variable propositional calculi* (J 0047) Notre Dame J Formal Log 20∗438-450
  ⋄ B20 B25 ⋄ REV MR 80f:03016 Zbl 351#02010
  • ID 52622

GLADSTONE, M.D. see Vol. III, IV for further entries

GLAEWE, W. [1973] see BAGINSKI, M.

GLAGOLEV, V.V. [1967] *Certain estimates of disjunctive normal form of functions of the algebra of logic (Russian)* (J 0052) Probl Kibern 19*75-94
⋄ B05 ⋄ REV MR 37 # 1227   Zbl 251 # 94035 • ID 05017

GLAGOLEV, V.V. [1973] *Ueber die Laenge einer verkuerzten disjunktiven Normalform fuer boolesche Funktionen der Dimension 1 (Russisch)* (J 0071) Met Diskr Analiz (Novosibirsk) 22*29-33
⋄ B05 ⋄ REV MR 49 # 2202   Zbl 273 # 94042 • ID 62040

GLAZOWSKA, K. [1958] *The structure of valid n-term syllogisms (Polish) (Russian and English summaries)* (J 0063) Studia Logica 8*249-257
⋄ B20 ⋄ REV MR 21 # 6323 • ID 05029

GLAZUNOV, N.M. & KALUZHNIN, L.A. & SUSHCHANSKIJ, V.I. [1980] *A programming system for solution of combinatorial problems of contemporary algebra (Russian)* (P 3877) Rab Sov Sist & Met Anal Vych EhVM & Prim Teor Fiz;1979 Dubna 23-35
⋄ B35 ⋄ REV Zbl 525 # 05003 • ID 37483

GLEASON, G.G. [1974] *Normal and skew systems* (J 0047) Notre Dame J Formal Log 15*379-401
⋄ B05 F07 ⋄ REV MR 50 # 4235   Zbl 281 # 02017 • ID 23612

GLEBSKIJ, YU.V. & KOGAN, D.I. & LIOGON'KIJ, M.I. & TALANOV, V.A. [1969] *Range and degree of realizability of formulas in the restricted predicate calculus (Russian)* (J 0040) Kibernetika, Akad Nauk Ukr SSR 1969/2*17-27
• TRANSL [1969] (J 0021) Cybernetics 5*142-154
⋄ B10 C07 C13 ⋄ REV MR 46 # 42   Zbl 209.308   JSL 50.1073 • ID 05032

GLEBSKIJ, YU.V. see Vol. III, IV for further entries

GLIVENKO, V.I. [1936] *The crisis in the foundations of mathematics in the present state of their development (Russian)* (C 4066) Sb Stat po Filos Mat 69-83
⋄ B28 ⋄ ID 40772

GLIVENKO, V.I. see Vol. V, VI for further entries

GLUBRECHT, J.-M. [1982] *Ein Vollstaendigkeitsbeweis fuer schnittfreie Kalkuele mit der Maximalisierungsmethode von Henkin* (J 0009) Arch Math Logik Grundlagenforsch 22*159-166
⋄ B10 C07 F05 ⋄ REV MR 84b:03022   Zbl 496 # 03002 • ID 35619

GLUBRECHT, J.-M. & OBERSCHELP, A. & TODT, G. [1983] *Klassenlogik* (X 0876) Bibl Inst: Mannheim 467pp
⋄ B10 B98 E30 ⋄ REV MR 85j:03039   Zbl 514 # 03001 • ID 36976

GLUSHKOV, V.M. [1971] *Einfuehrung in die technische Kybernetik Band 1, 2* (X 1553) Dokumentation Saur: Muenchen 126pp
⋄ B98 ⋄ REV Zbl 243 # 94001 • ID 62053

GLUSHKOV, V.M. & KAPITONOVA, YU.V. & LETICHEVS'KIJ, O.A. & MALEVANYI, N.P. & VERSHININ, K.P. [1972] *On the construction of a practical formal language for the transcription of mathematical theories (Russian) (English summary)* (J 0040) Kibernetika, Akad Nauk Ukr SSR 1972/5*19-28
⋄ B35 ⋄ REV MR 47 # 6149   Zbl 257 # 68060 • ID 05049

GLUSHKOV, V.M. see Vol. IV, VI for further entries

GOAD, C.A. [1980] *Proofs as descriptions of computation* (P 3063) Autom Deduct (5);1980 Les Arcs 39-52
⋄ B35 F07 F50 ⋄ REV Zbl 438 # 68052 • ID 55988

GOAD, C.A. see Vol. II, III, VI for further entries

GODDARD, L. [1960] *The exclusive "or"* (J 0103) Analysis (Oxford) 20*97-106
⋄ B20 ⋄ ID 31086

GODDARD, L. [1966] *Predicates, relations and categories* (J 0273) Australasian J Phil 44*139-171
⋄ A05 B10 B50 ⋄ ID 15234

GODDARD, L. see Vol. II, IV, VI for further entries

GODLEVSKIJ, A.B. & LETICHEVS'KIJ, O.A. (EDS.) [1974] *Automata and algorithm theory, and mathematical logic (Russian)* (X 2522) Akad Nauk Inst Kibernet: Kiev 98pp
⋄ B35 B97 D05 D20 D97 ⋄ REV MR 55 # 1829 • ID 80465

GODLEVSKIJ, A.B. see Vol. IV for further entries

GOE, G. [1964] *Three axiom negation-alternation formulation of the truth-functional calculus* (J 0047) Notre Dame J Formal Log 5*129-132
⋄ B05 ⋄ REV MR 30 # 1935   Zbl 154.253   JSL 33.606 • ID 05065

GOE, G. [1966] *A reconstruction of formal logic* (J 0047) Notre Dame J Formal Log 7*129-157,158
⋄ A05 B10 ⋄ REV MR 35 # 2723   Zbl 154.254   JSL 33.137 • ID 05066

GOE, G. [1970] *Reconstructing formal logic: further developments and considerations* (J 0047) Notre Dame J Formal Log 11*37-75
⋄ A05 B10 ⋄ REV MR 44 # 2566   Zbl 167.10 • ID 05068

GOE, G. [1983] *Lezioni di logica* (X 3777) Angeli: Milano xiii + 515pp
⋄ B98 ⋄ REV MR 84f:03002   Zbl 509 # 03002   JSL 50.860 • ID 34422

GOE, G. see Vol. II, V for further entries

GOEDEL, K. [1929] *Ein Spezialfall des Entscheidungsproblems der theoretischen Logik* (J 1124) Ergebn Math Kolloquium 2*27-28
⋄ B20 B25 ⋄ REV FdM 57.1321 • ID 22123

GOEDEL, K. [1930] *Die Vollstaendigkeit der Axiome des logischen Funktionenkalkuels* (J 0124) Monatsh Math-Phys 37*349-360
• TRANSL [1967] (C 0675) From Frege to Goedel 583-591
⋄ B10 C07 ⋄ REV FdM 56.46 • ID 20884

GOEDEL, K. [1931] *Diskussion zur Grundlegung der Mathematik* (J 0748) Erkenntnis (Leipzig) 2*147-148,149-151
⋄ A05 B30 ⋄ ID 05069

GOEDEL, K. [1931] *Eine Eigenschaft der Realisierung des Aussagenkalkuels* (J 1124) Ergebn Math Kolloquium 3*20-21
⋄ B05 ⋄ REV FdM 57.1319 • ID 22125

GOEDEL, K. [1931] *Ueber formal unentscheidbare Saetze der "Principia Mathematica" und verwandter Systeme I* (J 0124) Monatsh Math-Phys 38*173-198
• TRANSL [1965] (C 0718) The Undecidable 5-38 [1967] (C 0675) From Frege to Goedel 596-616 [1962] (X 1323) Oliver & Boyd: Edinburgh vii+72pp
⋄ B28 B30 D20 D35 F25 F30 F35 ⋄ REV MR 27#1373 Zbl 124.4 Zbl 2.1 FdM 57.54 • ID 15052

GOEDEL, K. [1931] *Ueber Unabhaengigkeitsbeweise im Aussagenkalkuel* (J 1124) Ergebn Math Kolloquium 4*9-10
⋄ B05 ⋄ REV Zbl 7.193 FdM 59.865 • ID 22122

GOEDEL, K. [1933] *Zum Entscheidungsproblem des logischen Funktionenkalkuels* (J 0124) Monatsh Math-Phys 40*433-443
⋄ B10 B20 B25 C13 D35 ⋄ REV Zbl 8.289 FdM 59.865 • ID 20885

GOEDEL, K. [1934] *Ueber die Laenge von Beweisen* (J 1124) Ergebn Math Kolloquium 7*23-24
• TRANSL [1965] (C 0718) The Undecidable 82-83
⋄ B10 F20 F30 F35 ⋄ REV Zbl 14.241 JSL 1.116 JSL 31.484 FdM 62.43 • ID 35971

GOEDEL, K. [1938] *The consistency of the axiom of choice and of the generalized continuum-hypothesis* (J 0054) Proc Nat Acad Sci USA 24*556-557
⋄ B30 E25 E35 E45 E50 ⋄ REV Zbl 20.297 JSL 5.116 FdM 64.35 • ID 05071

GOEDEL, K. [1944] *Russell's mathematical logic* (C 4106) Phil of Russell 123-153
• TRANSL [1967] (C 2141) Filos Matematica 81-112 [1969] (J 1030) Formalisation 10*84-107
⋄ A10 B98 ⋄ REV JSL 11.75 JSL 34.313 • ID 15163

GOEDEL, K. [1947] *What is Cantor's continuum problem?* (J 0005) Amer Math Mon 54*515-525 • ERR/ADD ibid 55*151
• TRANSL [1967] (C 2141) Filos Matematica 113-136
• REPR [1964] (C 1105) Phil of Math. Sel Readings 258-273
⋄ A05 B98 E50 E98 ⋄ REV MR 9.403 Zbl 38.30 JSL 13.116 JSL 34.313 • ID 05073

GOEDEL, K. see Vol. II, IV, V, VI for further entries

GOESSEL, M. [1978] *A generalized principle of automata superposition relative to a pair of boolean functions (Russian) (English summary)* (J 0040) Kibernetika, Akad Nauk Ukr SSR 1978/3*33-37
• TRANSL [1978] (J 0021) Cybernetics 14*352-357
⋄ B05 D05 ⋄ REV MR 80m:68049 Zbl 393#94041 • ID 69601

GOETLIND, E. [1946] *On some equivalence propositions in two valued logic (Norwegian)* (J 4510) Norsk Mat Tidsskr 28*71-75
⋄ B05 ⋄ REV MR 10.277 JSL 13.51 • ID 28460

GOETLIND, E. [1947] *An axiom system for the calculus of propositions (Swedish)* (J 4510) Norsk Mat Tidsskr 29*1-4
⋄ B05 ⋄ REV MR 9.1 Zbl 30.193 JSL 13.52 • ID 24910

GOETLIND, E. [1952] *A note on Chwistek and Hetper's foundation of formal metamathematics* (P 0098) Skand Mat Kongr (11);1949 Trondheim 268-270
⋄ B30 F35 ⋄ REV MR 15.846 Zbl 48.247 JSL 19.140 • ID 21134

GOETLIND, E. see Vol. II for further entries

GOGUADZE, D.F. [1970] *The finite axiomatizability of a finite fragment of type theory (Russian) (Georgian and English summaries)* (J 0233) Soobshch Akad Nauk Gruz SSR 60*541-544
⋄ B15 F35 ⋄ REV MR 45#53 Zbl 216.10 • ID 05093

GOGUEN, J.A. & MESEGUER, J. [1985] *Completeness of many-sorted equational logic* (J 1447) Houston J Math 11*307-334
⋄ B20 C05 C07 ⋄ REV Zbl 498#03018 • ID 36903

GOGUEN, J.A. see Vol. II, IV, V for further entries

GOLDBERG, ADELE [1974] *Design of a computer-tutor for elementary mathematical logic* (P 1691) Inform Processing (6);1974 Stockholm 884-888
⋄ B80 ⋄ REV Zbl 296#68095 • ID 62084

GOLDBERG, ADELE & SUPPES, P. [1976] *Computer-assisted instruction in elementary logic at the university level* (J 2703) Educ Stud Math 6*447-474
⋄ B80 ⋄ REV Zbl 326#02001 • ID 62083

GOLDBERG, ALLEN [1982] see BROWN, CYNTHIA A.

GOLDBLATT, R.I. [1979] *Topoi. The categorial analysis of logic* (S 3303) Stud Logic Found Math 98*xv+486pp
• TRANSL [1983] (X 0885) Mir: Moskva 488pp
⋄ B98 C98 E98 F35 F50 F98 G30 ⋄ REV MR 81a:03063 Zbl 434#03050 JSL 47.445 • REM 2nd rev. ed. 1984; xvi+552pp • ID 55754

GOLDBLATT, R.I. [1982] see CRESSWELL, M.J.

GOLDBLATT, R.I. see Vol. II, III, V, VI for further entries

GOLDFARB, W.D. [1971] see DREBEN, B.

GOLDFARB, W.D. & LEWIS, H.R. [1973] *The decision problem for formulas with a small number of atomic subformulas* (J 0036) J Symb Logic 38*471-480
⋄ B20 D35 ⋄ REV MR 49#2328 Zbl 276#02029 • ID 08111

GOLDFARB, W.D. & LEWIS, H.R. [1975] *Skolem reduction classes* (J 0036) J Symb Logic 40*62-68
⋄ B20 D35 ⋄ REV MR 58#193 Zbl 306#02043 • ID 05117

GOLDFARB, W.D. [1981] *On the Goedel class with identity* (J 0036) J Symb Logic 46*354-364
⋄ B20 B25 D35 ⋄ REV MR 82g:03079 Zbl 472#03009 • ID 55275

GOLDFARB, W.D. [1981] *The undecidability of the second-order unification problem* (J 1426) Theor Comput Sci 13*225-230
⋄ B15 D35 F20 ⋄ REV MR 82c:03067 Zbl 457#03006 • ID 54331

GOLDFARB, W.D. & GUREVICH, Y. & SHELAH, S. [1984] *A decidable subclass of the minimal Goedel class with identity* (J 0036) J Symb Logic 49*1253-1261
⋄ B20 B25 ⋄ REV MR 86g:03015b Zbl 576#03009 • ID 38765

GOLDFARB, W.D. [1984] *The unsolvability of the Goedel class with identity* (J 0036) J Symb Logic 49*1237-1252
⋄ B20 D35 ⋄ REV MR 86g:03015a • ID 42450

GOLDFARB, W.D. [1984] *The Goedel class with identity is unsolvable* (J 0589) Bull Amer Math Soc (NS) 10*113-115
⋄ B20 D35 ⋄ REV MR 85f:03006 Zbl 534#03005 • ID 36555

GOLDFARB, W.D. see Vol. III, VI for further entries

GOL'DSHTEJN, B.G. [1983] *Decomposition of logics (Russian)* (J 0030) Izv Akad Nauk Uzb SSR, Ser Fiz-Mat 1983/6∗10–12
⋄ C95 H05 ⋄ REV MR 85f:81007 Zbl 565 # 03034
• ID 45227

GOL'DSHTEJN, M.SH. [1976] *Ueber die Darstellung abstrakter Logiken (Russisch)* (J 0024) Dokl Akad Nauk Uzb SSR 1976/5∗7–8
⋄ B22 G12 ⋄ REV Zbl 411 # 03060 • ID 52914

GOLL, P. [1973] see BAGINSKI, M.

GOLUNKOV, YU.V. [1978] see BUKHARAEVA, Z.K.

GOLUNKOV, YU.V. see Vol. II, IV for further entries

GOMEZ CALDERON, J. [1979] *Logica simbolica* (X 2686) CECSA: Mexico City 147pp
⋄ B98 ⋄ REV MR 82i:03001 Zbl 521 # 03001 • ID 73482

GONSALES-SABATER, A. [1979] see BELYANIN, P.N.

GONSETH, F. [1933] *Sur l'axiomatique de la theorie des ensembles et sur la logique des relations* (J 2022) Comm Math Helvetici 5∗108–136
⋄ A05 B20 E07 E30 ⋄ REV Zbl 5.339 FdM 59.883
• ID 05128

GONSETH, F. [1937] *Qu'est-ce que la logique* (X 0859) Hermann: Paris 91pp
⋄ B98 ⋄ REV JSL 3.165 FdM 63.836 • ID 25110

GONSETH, F. (ED.) [1941] *Les entretiens de Zuerich sur les fondements et la methode des sciences mathematiques: Exposes et discussions* (X 2220) Leemen: Zuerich 209pp
⋄ B97 E97 F97 ⋄ ID 48625

GONSETH, F. see Vol. VI for further entries

GONSHOR, H. [1971] *The ring of finite elements in a non-standard model of the reals* (J 3172) J London Math Soc, Ser 2 3∗493–500
⋄ H05 ⋄ REV MR 43 # 7418 Zbl 216.12 • ID 05140

GONSHOR, H. [1973] *Projective covers as subquotients of enlargements* (J 0029) Israel J Math 14∗257–261
⋄ H05 H20 ⋄ REV MR 50 # 1193 Zbl 259 # 02044
• ID 05142

GONSHOR, H. [1974] *Enlargements contain various kinds of completions* (P 1083) Victoria Symp Nonstand Anal;1972 Victoria 60–70
⋄ H05 H20 ⋄ REV MR 57 # 12208 Zbl 276 # 54045
• ID 26628

GONSHOR, H. [1979] *An application of nonstandard analysis to category theory* (J 0027) Fund Math 104∗75–83
⋄ H05 ⋄ REV MR 80k:54090 Zbl 424 # 18005 • ID 81367

GONSHOR, H. [1985] *Remarks on the Dedekind completion of a nonstandard model of the reals* (J 0048) Pac J Math 118∗117–132
⋄ H05 ⋄ REV Zbl 534 # 03034 • ID 44910

GONSHOR, H. see Vol. IV, V for further entries

GONZALEZ, M.O. & MANCILL, J.D. [1950] *On the system of natural numbers* (J 0005) Amer Math Mon 57∗104–112
⋄ B28 ⋄ REV JSL 15.138 • ID 05143

GOODMAN, NELSON & QUINE, W.V.O. [1940] *Elimination of extra-logical postulates* (J 0036) J Symb Logic 5∗104–109
⋄ A05 B10 ⋄ REV MR 2.65 Zbl 24.1 JSL 6.37
FdM 66.1194 • ID 16374

GOODMAN, NELSON [1949] *An improvement in the theory of simplicity* (J 0036) J Symb Logic 14∗228–229
⋄ A05 B10 ⋄ REV MR 12.233 Zbl 41.148 JSL 15.219
• ID 05154

GOODMAN, NELSON [1949] *The logical simplicity of predicates* (J 0036) J Symb Logic 14∗32–41
⋄ A05 B10 ⋄ REV MR 10.668 Zbl 41.148 JSL 15.219
• ID 05155

GOODMAN, NELSON see Vol. V, VI for further entries

GOODSTEIN, R.L. [1945] *Function theory in an axiom-free equation calculus* (J 1910) Proc London Math Soc, Ser 2 48∗401–434
⋄ B28 F30 F60 ⋄ REV MR 8.245 Zbl 60.23 JSL 11.24
• ID 05166

GOODSTEIN, R.L. [1951] *Constructive formalism: Essays on the foundations of mathematics* (X 0886) Leicester Univ Pr: Leicester 91pp
⋄ A05 B98 F30 F60 F98 ⋄ REV MR 14.123 Zbl 45.150
JSL 18.258 • ID 25669

GOODSTEIN, R.L. [1955] *On non-constructive theorems of analysis and the decision problem* (J 0132) Math Scand 3∗261–263
⋄ B25 B28 F30 F60 ⋄ REV MR 17.816 Zbl 67.248
• ID 05177

GOODSTEIN, R.L. [1957] *Mathematical logic* (X 0886) Leicester Univ Pr: Leicester viii + 104pp
⋄ B98 ⋄ REV MR 19.1 Zbl 77.12 JSL 28.98 • ID 22132

GOODSTEIN, R.L. [1957] *Recursive number theory. A development of recursive arithmetic in a logic-free equation calculus* (X 0809) North Holland: Amsterdam xii + 190pp
⋄ B28 F30 F50 F60 F98 ⋄ REV MR 21 # 1272
Zbl 77.14 JSL 23.227 • ID 05180

GOODSTEIN, R.L. [1958] *Models of propositional calculi in recursive arithmetic* (J 0132) Math Scand 6∗293–296
⋄ B05 B50 F30 F50 ⋄ REV MR 21 # 5570 Zbl 88.10
JSL 28.291 • ID 05182

GOODSTEIN, R.L. [1963] *A decidable fragment of recursive arithmetic* (J 0068) Z Math Logik Grundlagen Math 9∗199–201
⋄ B20 B25 F30 ⋄ REV MR 27 # 2403 Zbl 114.248
JSL 33.618 • ID 05189

GOODSTEIN, R.L. & LEE, R.D. [1966] *A decidable class of equations in recursive arithmetic* (J 0068) Z Math Logik Grundlagen Math 12∗235–239
⋄ B20 B25 F30 ⋄ REV MR 33 # 5491 Zbl 148.246
JSL 33.618 • ID 05190

GOODSTEIN, R.L. [1972] *A new proof of completeness* (J 0047) Notre Dame J Formal Log 13∗563–564
⋄ B05 C07 ⋄ REV MR 48 # 52 Zbl 225 # 02010 • ID 05195

GOODSTEIN, R.L. [1972] *The fundamental formula in the algebra of sets* (J 0148) Math Gaz 56∗199–207
⋄ B05 E20 G05 ⋄ REV MR 58 # 5217 Zbl 252 # 04001
• ID 27766

GOODSTEIN, R.L. [1974] *Satisfiable in a larger domain* (J 0047) Notre Dame J Formal Log 15∗598–600
⋄ B10 C07 ⋄ REV MR 51 # 47 Zbl 271 # 02013 • ID 05197

GOODSTEIN, R.L. [1976] *Arithmetic without sets* (J 0148) Math Gaz 60∗165–170
⋄ B28 ⋄ REV MR 58 # 10263 • ID 73514

GOODSTEIN, R.L. see Vol. II, III, IV, V, VI for further entries

GOODYEAR, P. [1984] *Double enlargements of topological spaces* (J 0068) Z Math Logik Grundlagen Math 30*389-392
 ⋄ H05 ⋄ REV MR 86a:54058 • ID 42271

GOOSENS, W.K. [1979] *Alternative axiomatizations of elementary probability theory* (J 0047) Notre Dame J Formal Log 20*227-239
 ⋄ A05 B30 ⋄ REV MR 81b:60002 Zbl 314#02058 • ID 52382

GORDON, E.I. & LYUBETSKIJ, V.A. [1981] *Some applications of nonstandard analysis in the theory of Boolean-valued measures (Russian)* (J 0023) Dokl Akad Nauk SSSR 256*1037-1041
 • TRANSL [1981] (J 0062) Sov Math, Dokl 23*142-146
 ⋄ H10 ⋄ REV MR 83e:03112 Zbl 485#03038 • ID 35257

GORDON, E.I. see Vol. III, IV, V for further entries

GORELIK, E.S. [1973] *The complexity of the realization of elementary conjunctions and disjunctions in the base $\{x/y\}$ (Russian)* (J 0052) Probl Kibern 26*27-36,327
 ⋄ B05 ⋄ REV MR 49#4674 Zbl 265#94025 • ID 05200

GOSTEV, YU.G. [1981] *Atomary languages and grammars. On the theory of data structure families (Russian)* (J 0040) Kibernetika, Akad Nauk Ukr SSR 1981/2*20-25
 • TRANSL [1981] (J 0021) Cybernetics 17*176-181
 ⋄ B05 D05 ⋄ REV MR 82k:68044 Zbl 469#68030 • ID 81380

GOTO, E. & TAKAHASI, H. [1963] *Some theorems useful in threshold logic for enumerating boolean functions* (P 0572) Inform Processing (2);1962 Muenchen 747-752
 ⋄ B05 ⋄ REV Zbl 143.395 • ID 27614

GOTO, M. & KAO, S. & NINOMIYA, T. [1977] *Determination of many-valued truth tables for undefined operators in axioms by a computer and their applications* (P 2013) Int Symp Multi-Val Log (7);1977 Charlotte 20-28
 ⋄ B35 ⋄ REV MR 57#15953 • ID 73549

GOTO, M. & KAO, S. & NINOMIYA, T. [1978] *Determination of the fittest number of truth-values and canonical forms of logical functions for a many-valued axiom set by a computer* (P 2014) Int Symp Multi-Val Log (8);1978 Rosemont 195-201
 ⋄ B35 ⋄ REV MR 81e:03018 • ID 73547

GOTO, M. & KAO, S. & NINOMIYA, T. [1979] *Axiomatization of Kleene's three-valued logic by a computer* (P 3003) Int Symp Multi-Val Log (9);1979 Bath 241-247
 ⋄ B35 ⋄ REV MR 80m:03052 • ID 73548

GOTO, M. & KAO, S. & NINOMIYA, T. [1981] *Axiomatization of Bochvar's three-valued logic by a computer* (P 3705) Int Symp Multi-Val Log (11);1981 Oklahoma City & Norman 146-151
 ⋄ B35 ⋄ REV MR 83m:94033 Zbl 541#03014 • ID 41378

GOTO, M. see Vol. II for further entries

GOTTLIEB, D. & MCCARTHY, T. [1979] *Substitutional quantification and set theory* (J 0122) J Philos Logic 8*315-331
 ⋄ A05 B20 ⋄ REV MR 80m:03012 Zbl 408#03006 • ID 56245

GOTTLOB, G. & LEITSCH, A. [1985] *Fast subsumption algorithms* (P 4646) Atti Incontri Log Mat (2);1983/84 Siena 64-77
 ⋄ B35 ⋄ ID 49783

GOTTSCHALK, W.H. [1953] *The theory of quaternality* (J 0036) J Symb Logic 18*193-196
 ⋄ B10 ⋄ REV MR 15.494 Zbl 53.341 JSL 19.229
 • ID 05210

GOTTSCHALK, W.H. see Vol. V for further entries

GOTTWALD, S. [1971] *Eine Konstruktion der reellen Zahlen "Unendlich"* (J 1209) Wiss Z Univ Leipzig, Math-Nat Reihe 20*359-364
 ⋄ B28 ⋄ REV MR 47#404 Zbl 209.362 • ID 81381

GOTTWALD, S. see Vol. II, V for further entries

GOULD, JAMES A. [1967] see COPI, I.M.

GOULD, JAMES A. see Vol. II for further entries

GOZE, M. & LUTZ, R. [1981] *Nonstandard analysis. A pratical guide with applications* (S 3301) Lect Notes Math 881*xiv+261pp
 ⋄ B98 H05 ⋄ REV MR 83i:03103 Zbl 506#03021 • ID 34857

GOZE, M. [1983] *Etude locale des courbes algebriques planes* (J 1620) Asterisque 109-110*245-259
 ⋄ H10 ⋄ REV MR 85m:14048 Zbl 534#14015 • ID 38349

GRABOWSKI, M. [1981] *Full weak second-order logic versus algorithmic logic* (P 3642) Colloq Math Log in Computer Sci;1978 Salgotarjan 471-483
 ⋄ B15 ⋄ REV MR 83g:68007 Zbl 541#03002 • ID 41368

GRABOWSKI, M. see Vol. III, IV for further entries

GRACIA, J.J.E. [1975] *Propositions as premises of syllogisms in medieval logic* (J 0047) Notre Dame J Formal Log 16*545-547
 ⋄ A10 B20 ⋄ REV MR 52#10364 Zbl 311#02004 • ID 21671

GRADSHTEJN, I.S. [1972] *Direct and inverse theorems (Russian)* (X 2027) Nauka: Moskva 128pp
 • TRANSL [1963] (X 0869) Pergamon Pr: Oxford xviii+173pp
 ⋄ B98 ⋄ REV MR 53#85 Zbl 102.247 • REM 5th edition
 • ID 16549

GRAINGER, A.D. [1975] *Invariant subspaces of compact operators on topological vector spaces* (J 0048) Pac J Math 56*477-493
 ⋄ C60 C65 H05 ⋄ REV MR 51#11133 Zbl 325#47004 • ID 17507

GRAINGER, A.D. [1979] *Finite points of filters in infinite-dimensional vector spaces* (J 0027) Fund Math 104*47-67
 ⋄ H05 ⋄ REV MR 80k:46008 Zbl 422#46061 • ID 81387

GRAINGER, A.D. [1980] *On the nonstandard duality theory of locally convex spaces* (J 0017) Canad J Math 32*460-479
 ⋄ H05 ⋄ REV MR 81g:46003 Zbl 476#46003 • ID 81386

GRANDY, R.E. [1977] *Advanced logic for applications* (S 3307) Synth Libr 110*xi+167pp
 ⋄ B98 ⋄ REV MR 57#9450 Zbl 381#03003 JSL 47.714 • ID 51871

GRANDY, R.E. see Vol. II, III, IV, VI for further entries

GRANT, J. [1975] *Inconsistent and incomplete logics* (J 0497) Math Mag 48*154-159
 ⋄ B22 B53 ⋄ REV MR 52#13311 Zbl 309#02010 • ID 21774

GRANT, J. see Vol. II, III for further entries

GRASS, W. & HALLER, R. (EDS.) [1980] *Language, logic, and philosophy* (X 2728) Hoelder-Pichler-Tempsky: Wien 617pp
- ◇ A05 B65 B97 ◇ REV MR 82e:03006 • ID 70020

GRASSINI, R. [1983] see BARONE, F.

GRASSMANN, G. [1966] *Die Formenlehre oder Mathematik. 1.die Groessenlehre. 2.die Begriffslehre der Logik. 3.die Bindelehre oder Kombinationslehre. 4.die Zahlenlehre oder Arithmetik. 5.die Aussenlehre oder Ausdehnungslehre (reprinted)* (X 0892) Olms: Hildesheim 229pp
- ◇ B98 ◇ ID 22622

GRAYSON, R.J. [1984] *Heyting-valued semantics* (P 3710) Logic Colloq;1982 Firenze 181-208
- ◇ B15 C90 F35 F50 G30 ◇ REV MR 86f:03099 Zbl 574#03048 • ID 41840

GRAYSON, R.J. see Vol. III, V, VI for further entries

GREBENSHCHIKOV, V.N. [1961] *The coalition of systems of equations of Boolean algebra and their solution (Russian)* (J 0023) Dokl Akad Nauk SSSR 141*1317-1319
- • TRANSL [1961] (J 0470) Sov Phys, Dokl 6*1040-1041
- ◇ B05 ◇ REV MR 25#33 Zbl 116.16 • ID 21006

GREEN, C.C. & RAPHAEL, B. [1967] *The use of theorem-proving techniques in question-answering systems* (P 1384) ACM Nat Conf (22);1967 169-181
- ◇ B35 ◇ ID 47332

GREEN, C.C. [1969] *Application of theorem proving to problem solving* (P 4250) Int Joint Conf Artif Intell (1);1969 Washington 219-239
- ◇ B35 ◇ ID 47331

GREEN, C.C. [1969] *Theorem-proving by resolution as a basis for question-answering systems* (J 0508) Machine Intelligence 4*183-205
- ◇ B35 ◇ REV Zbl 239#68015 • ID 47330

GREENBAUM, S. & NAGASAKA, A. & O'RORKE, P. & PLAISTED, D.A. [1982] *Comparison of natural deduction and locking resolution implementations* (P 3840) Autom Deduct (6);1982 New York 159-171
- ◇ B35 ◇ REV MR 85g:68063 • ID 43931

GREENE, C. & TAKEUTI, G. [1977] *On the decomposition of boolean polynomials* (J 2332) J Fac Sci, Univ Tokyo, Sect 1 A 24*23-28
- ◇ B05 ◇ REV MR 56#5384 • ID 39563

GREENSTEIN, C.H. [1978] *Dictionary of logical terms and symbols* (X 0864) Van Nostrand: New York xiii+188pp
- ◇ B03 ◇ REV MR 80b:03010 Zbl 486#03001 • ID 73613

GREENWOOD, P. & HERSH, R. [1975] *Stochastic differential and quasi-standard random variables* (P 1481) Probab Meth in Diff Equat;1974 Victoria 35-62
- ◇ H05 ◇ REV MR 54#3845 Zbl 312#60027 • ID 24676

GREGG, J.R. [1970] *Axiomatic quasi-natural deduction* (J 0047) Notre Dame J Formal Log 11*221-228
- ◇ B10 F07 ◇ REV MR 44#28 Zbl 198.317 • ID 05317

GREGG, J.R. [1971] *Two modes of deductive inference* (J 0047) Notre Dame J Formal Log 12*169-178
- ◇ B10 ◇ REV MR 44#2570 Zbl 198.318 • ID 05318

GREGORIO DI, S. [1982] see APOLLONI, B.

GREIF, I. [1974] see BISHOP, P.

GRELL, B. [1974] *Un simple systeme de logique fonde sur regles* (J 0302) Rep Math Logic, Krakow & Katowice 2*9-24
- ◇ B10 ◇ REV MR 50#1834 Zbl 286#02019 • ID 05331

GRENIEWSKI, H. [1950] *Certain notions of the theory of numbers as applied to the propositional calculus* (J 0086) Cas Pestovani Mat, Ceskoslov Akad Ved 74*132-136
- ◇ B05 ◇ REV MR 13.198 Zbl 40.294 JSL 33.304 • ID 05332

GRENIEWSKI, H. [1950] *Functors of the propositional calculus* (J 0283) Ann Soc Pol Math 22*78-86
- ◇ B05 ◇ REV MR 13.198 Zbl 54.6 JSL 33.304 • ID 28525

GRENIEWSKI, H. [1950] *Groups and fields definable in the propositional calculus* (J 0459) C R Soc Sci Lett Varsovie Cl 3 43*53-68
- ◇ B05 C60 ◇ REV MR 14.834 JSL 33.304 • ID 22000

GRENIEWSKI, H. [1951] *Arithmetics of natural numbers as part of the bi-valued propositional calculus* (S 0019) Colloq Math (Warsaw) 2*291-297
- ◇ B05 F30 ◇ REV MR 14.345 Zbl 45.295 JSL 33.305 • ID 05333

GRENIEWSKI, H. [1953] *An attempt at "rejuvenation" of the square of opposition (Polish, Russian) (English summary)* (J 0063) Studia Logica 1*276-286,287-297 • ERR/ADD ibid 1*301
- ◇ A05 B80 ◇ REV MR 16.987 JSL 20.83 • ID 05334

GRENIEWSKI, H. [1955] *Elements of formal logic (Polish)* (X 1034) PWN: Warsaw 492pp
- ◇ B98 ◇ REV Zbl 68.10 JSL 21.188 • ID 42617

GRENIEWSKI, H. see Vol. II for further entries

GRIEDER, A. [1980] *On the logic of relations* (J 0076) Dialectica 34*167-182
- ◇ B20 G15 ◇ REV MR 82f:03062 • ID 73621

GRIEDER, A. see Vol. V for further entries

GRIGOR'EV, D.YU. & OREVKOV, V.P. (EDS.) [1981] *Theoretical applications of methods of mathematical logic. III. Work collection (Russian)* (S 0228) Zap Nauch Sem Leningrad Otd Mat Inst Steklov 105*200pp
- ◇ B80 ◇ REV MR 82g:03007 Zbl 453#00010 • ID 54130

GRIGOR'EV, D.YU. see Vol. III, IV, V for further entries

GRIGORYAN, A.K. & SHIPILINA, L.B. [1979] *Effective algorithm for minimization of circuit complexity of underdetermined automata specified by nonbranching graphs (Russian)* (J 0011) Avtom Telemekh 1979/10*95-104
- • TRANSL [1979] (J 0010) Autom & Remote Control 40*1479-1487
- ◇ B35 ◇ REV MR 81d:94047 Zbl 433#68052 • ID 69623

GRIGOR'YAN, YU.G. [1967] *The variational problem of functions of logical algebra and a method for its computer realization (Russian)* (J 0040) Kibernetika, Akad Nauk Ukr SSR 1967/1*26-30
- • TRANSL [1967] (J 0021) Cybernetics 3/1*21-24
- ◇ B35 ◇ REV MR 44#5181 Zbl 148.253 • ID 05341

GRISHIN, V.N. [1976] *Reduction of comprehension axioms of a given depth to comprehension axioms of smaller depth (Russian)* (C 3271) Issl Teor Mnozh & Neklass Logik 174-180
- ◇ B15 E70 ◇ REV MR 58#16291 Zbl 406#03066 • ID 56149

GRISHIN, V.N. see Vol. II, V, VI for further entries

GRIZE, J.-B. [1955] *L'implication et la negation vues au travers des methodes de Gentzen et de Fitch* (J 0076) Dialectica 9*363-381
  ⋄ B05 ⋄ REV MR 17.1171 • ID 05367

GRIZE, J.-B. [1969] *Logique moderne. fasc.I: Logique des propositions et des predicats, deduction naturelle* (X 0834) Gauthier-Villars: Paris ii+90pp
  ⋄ B98 ⋄ REV MR 42#36 Zbl 194.306 • REM Part II 1971
  • ID 05369

GRIZE, J.-B. [1971] *Logique moderne. fasc.II: Logique des propositions et des predicats, tables de verite et axiomatisation* (X 0834) Gauthier-Villars: Paris iii+79pp
  ⋄ B98 ⋄ REV MR 48#3690 Zbl 264#02015 • REM Part I 1969. Part III 1973 • ID 05370

GRIZE, J.-B. see Vol. VI for further entries

GRONAU, H.-D.O.F. [1977] *Erzeugung dualer Vektoren durch gewisse abgeschlossene Mengen boolescher Funktionen* (S 2829) Rostocker Math Kolloq 3*45-56
  ⋄ B05 ⋄ REV MR 58#33547b Zbl 411#03058 • ID 52912

GRONAU, H.-D.O.F. [1981] *Sperner type theorems and complexity of minimal disjunctive normal forms of monotone Boolean functions* (J 0049) Period Math Hung 12*267-282
  ⋄ B05 ⋄ REV MR 83i:03096 Zbl 458#05003 • ID 35543

GROSJEAN, P.V. [1972] *Theorie algebrique du syllogisme categorique* (J 0079) Logique & Anal, NS 15*547-568
  ⋄ B20 ⋄ REV Zbl 275#02010 • ID 29692

GROSJEAN, P.V. see Vol. II for further entries

GROSS, M.W. & NORTHROP, F.S.C. [1953] *Alfred Whitehead: An anthology* (X 0805) Cambridge Univ Pr: Cambridge, GB 928pp
  ⋄ A05 B15 ⋄ ID 25789

GRUENBERG, T. [1983] *A tableau system of proof for predicate-functor logic with identity* (J 0036) J Symb Logic 48*1140-1144
  ⋄ B10 F07 ⋄ REV MR 86d:03011 Zbl 539#03003
  • ID 43736

GRUENDER, D. [1981] see AGAZZI, E.

GRUZINTSEV, G.A. [1927] *The concept of relation and the axiomatical definition of number (Russian) (French summary)* (J 4110) Zap Inst Nar Prosv, Dnepropetrovsk 1*25-43
  ⋄ B28 F99 ⋄ REV FdM 53.182 • ID 40886

GRUZINTSEV, G.A. see Vol. V for further entries

GRZEGORCZYK, A. & JASKOWSKI, S. & LOS, J. & MAZUR, S. & MOSTOWSKI, ANDRZEJ & RASIOWA, H. & SIKORSKI, R. [1955] *The present state of investigations on the foundations of mathematics (Polish)* (J 0051) Commentat Math, Ann Soc Math Pol, Ser 1 1*13-55
  • TRANSL [1954] (J 0067) Usp Mat Nauk 9/3*3-38 (Russian) [1954] (P 1924) Polnisch Math Kongr;1953 Warsaw 11-44 (German) [1955] (J 0202) Diss Math (Warsaw) 9*48pp (English)
  ⋄ A10 B98 E98 ⋄ REV MR 16.552 Zbl 57.243 Zbl 59.12 JSL 21.372 • ID 31447

GRZEGORCZYK, A. [1955] *The systems of Lesniewski in relation to contemporary logical research (Polish and Russian summaries)* (J 0063) Studia Logica 3*77-97 • ERR/ADD ibid 3*6
  ⋄ B15 ⋄ REV MR 17.1171 JSL 27.117 • ID 05394

GRZEGORCZYK, A. [1961] *An outline of mathematical logic (Polish)* (X 1034) PWN: Warsaw 477pp
  • TRANSL [1974] (S 3307) Synth Libr 70*x+596pp
  ⋄ B98 ⋄ REV MR 27#1347 MR 83f:03007 Zbl 132.245 JSL 48.220 • REM 5th ed. 1981, 510pp • ID 20967

GRZEGORCZYK, A. [1961] *Axiomatizability of geometry without points* (P 0711) Concept & Role of Model in Math & Sci;1960 Utrecht 104-111
  ⋄ B30 ⋄ REV MR 25#4394 JSL 37.201 • ID 21293

GRZEGORCZYK, A. [1961] *Logika popularna* (X 1034) PWN: Warsaw 131pp
  ⋄ B98 ⋄ REV MR 24#A675 Zbl 134.6 • ID 24867

GRZEGORCZYK, A. [1962] *On the validation of the sets of axioms in mathematical theories (Polish) (Russian and English summaries)* (J 0063) Studia Logica 13*197-202
  ⋄ A05 B30 ⋄ REV MR 26#4907 JSL 30.387 • ID 21263

GRZEGORCZYK, A. [1964] *A note on the theory of propositional types* (J 0027) Fund Math 54*27-29
  ⋄ B15 ⋄ REV MR 28#3925 Zbl 127.7 JSL 31.502
  • ID 05410

GRZEGORCZYK, A. [1971] *Outline of theoretical arithmetic (Polish)* (X 1034) PWN: Warsaw 314pp
  ⋄ B28 C62 F30 F35 F98 ⋄ REV MR 53#12856 MR 85e:03002 Zbl 251#02002 • REM 2nd ed. 1983
  • ID 23119

GRZEGORCZYK, A. [1972] *An approach to logical calculus (Polish and Russian summaries)* (J 0063) Studia Logica 30*33-43
  ⋄ B22 F50 ⋄ REV MR 47#4768 Zbl 286#02029
  • ID 05418

GRZEGORCZYK, A. see Vol. II, III, IV, V, VI for further entries

GRZEGOREK, E. [1976] *On axial maps of direct products, II* (S 0019) Colloq Math (Warsaw) 34*145-164
  ⋄ B20 E20 ⋄ REV MR 53#7790 Zbl 362#04009 • REM Part I 1975 by Ehrenfeucht,A. & Grzegorek,E. • ID 32138

GRZEGOREK, E. see Vol. V for further entries

GUARD, J.R. [1969] see BENNETT, J.H.

GUBA, W. [1977] *Ein maximaler lokaler Algorithmus fuer Klassen unverkuerzbarer Ueberdeckungen* (S 2829) Rostocker Math Kolloq 3*57-68
  ⋄ B35 ⋄ REV MR 58#5100 Zbl 408#05002 • ID 73688

GUCCIONE, S. & TORTORA, R. [1980] *A general deduction theorem* (J 2840) Stochastica, Univ Politec Barcelona 4*189-199
  ⋄ B22 ⋄ REV MR 83j:03016 Zbl 462#03006 • ID 54513

GUCCIONE, S. see Vol. II, IV for further entries

GUDDER, S.P. [1984] *An extension of classical measure theory* (J 0163) SIAM Review 26*71-89
  ⋄ B30 ⋄ REV MR 85d:81014 Zbl 559#28003 • ID 39095

GUDDER, S.P. see Vol. II, V for further entries

GUENTHNER, F. [1983] see GABBAY, D.M.

GUENTHNER, F. [1984] see GABBAY, D.M.

GUENTHNER, F. see Vol. II, III for further entries

GUESSARIAN, I. [1983] *Survey on classes of interpretations and some of their applications* (J 1456) SIGACT News 15,No.3*45-71
  ⋄ B20 ⋄ REV Zbl 524#68017 • ID 38235

GUESSARIAN, I. see Vol. IV, V for further entries

GUILBAUD, G.T. [1945] see DAVAL, R.

GUILLAUME, M. [1958] *Les tableaux semantiques du calcul des predicats restreint* (S 1567) Semin Bourbaki Exp.153*13pp
⋄ B20 F07 F98 ⋄ REV MR 21 # 4902 Zbl 84.8 • ID 42813

GUILLAUME, M. [1960] *Certains aspects syntaxiques d'une notion de modele: Relativisation d'une fonction logique de choix* (J 0154) Synthese 12*236-248
• REPR [1961] (P 0711) Concept & Role of Model in Math & Sci;1960 Utrecht 112-124
⋄ B10 ⋄ REV MR 24 # A1208 JSL 40.502 • ID 05438

GUILLAUME, M. [1960] *Sur une propriete remarquable du systeme de Bourbaki* (J 0109) C R Acad Sci, Paris 250*1776-1777
⋄ B10 ⋄ REV MR 22 # 12040 Zbl 149.16 • ID 05439

GUILLAUME, M. see Vol. II, V, VI for further entries

GUITEL, G. [1954] *Sur une representation symbolique du processus logique d'une demonstration* (P 0646) Appl Sci de Log Math;1952 Paris 25-28
⋄ B05 ⋄ REV Zbl 58.4 • ID 47914

GUMB, R.D. [1978] *Metaphor theory* (J 0302) Rep Math Logic, Krakow & Katowice 10*51-60
⋄ B10 ⋄ REV MR 81e:03005 Zbl 453 # 03007 • ID 54137

GUMB, R.D. see Vol. II, III, V for further entries

GUMIN, H. & HERMES, H. [1956] *Die Soundness des Praedikatenkalkuels auf der Basis der Quineschen Regel* (J 1114) Arch Phil 5*388-397
• REPR [1956] (J 0009) Arch Math Logik Grundlagenforsch 2*68-77
⋄ B10 ⋄ REV MR 17.1173 Zbl 71.9 JSL 30.386 • ID 21261

GUMIN, H. see Vol. IV for further entries

GUPTA, H.N. [1968] *On the rule of existential specification in systems of natural deduction* (J 0094) Mind 77*96-103
⋄ A05 B10 F07 ⋄ ID 14682

GUPTA, H.N. see Vol. III, VI for further entries

GUREVICH, I.B. & ZHURAVLEV, YU.I. [1974] *Minimierung boolescher Funktionen und effektive Erkennungsalgorithmen (Russisch)* (J 0040) Kibernetika, Akad Nauk Ukr SSR 1974/3*16-20
⋄ B35 ⋄ REV Zbl 286 # 94039 • ID 62231

GUREVICH, I.B. [1974] *The noncomputability within a class of local algorithms of certain predicates connected with the minimization of Boolean functions (Russian)* (J 0040) Kibernetika, Akad Nauk Ukr SSR 1974/2*24-30
• TRANSL [1974] (J 0021) Cybernetics 10*213-218
⋄ B35 D20 ⋄ REV MR 53 # 7628 Zbl 286 # 02038
• ID 62232

GUREVICH, R. [1984] *Decidability of the equational theory of positive numbers with raising to a power (Russian)* (J 0092) Sib Mat Zh 25/2(144)*216-219
⋄ B25 B28 C65 ⋄ REV MR 86f:03023 Zbl 555 # 03004
• ID 45002

GUREVICH, R. [1985] *Equational theory of positive numbers with exponentiation* (J 0053) Proc Amer Math Soc 94*135-141
⋄ B25 B28 C65 ⋄ REV Zbl 572 # 03014 • ID 44633

GUREVICH, R. see Vol. III, V for further entries

GUREVICH, Y. [1965] *Existential interpretation (Russian)* (J 0003) Algebra i Logika 4/4*71-85
⋄ B20 B25 C60 D35 F25 ⋄ REV MR 34 # 46
Zbl 294 # 02023 • REM Part II 1982 • ID 05452

GUREVICH, Y. [1966] *Effective recognition of satisfiability of formulae of the restricted predicate calculus (Russian)* (J 0003) Algebra i Logika 5/2*25-55
⋄ B20 B25 D35 ⋄ REV MR 35 # 4096 Zbl 198.324
• ID 05455

GUREVICH, Y. [1966] *On the decision problem for pure restricted predicate logic (Russian)* (J 0023) Dokl Akad Nauk SSSR 166*1032-1034
• TRANSL [1966] (J 0062) Sov Math, Dokl 7*217-219
⋄ B20 C13 D35 ⋄ REV MR 34 # 47 Zbl 158.252
• ID 05458

GUREVICH, Y. [1966] *The decision problem for the restricted predicate calculus (Russian)* (J 0023) Dokl Akad Nauk SSSR 168*510-511
• TRANSL [1966] (J 0062) Sov Math, Dokl 7*669-670
⋄ B20 C13 D35 ⋄ REV MR 34 # 60 Zbl 163.253
• ID 05456

GUREVICH, Y. [1970] *Minsky machines and the case $\forall\exists\forall$ & $\exists^\infty$ of the decision problem (Russian)* (J 0340) Mat Zap (Univ Sverdlovsk) 7/3*77-83
⋄ B20 D05 D35 ⋄ REV MR 44 # 2614 Zbl 317 # 02052
• ID 22243

GUREVICH, Y. & TURASHVILI, T.V. [1973] *A strengthening of a certain result of J. Suranyi (Russian) (Georgian and English summaries)* (J 0233) Soobshch Akad Nauk Gruz SSR 70*289-292
⋄ B20 D35 ⋄ REV MR 54 # 10003 Zbl 296 # 02024
• ID 25843

GUREVICH, Y. [1973] *Formulas with a single $\forall$ (Russian)* (C 0733) Izbr Vopr Algeb & Log (Mal'tsev) 97-110
⋄ B20 B25 C13 ⋄ REV MR 49 # 7115 Zbl 298 # 02049
• ID 32346

GUREVICH, Y. [1976] *Semi-conservative reduction* (J 0009) Arch Math Logik Grundlagenforsch 18*23-25
⋄ B20 D25 D35 ⋄ REV MR 57 # 2875 Zbl 351 # 02011
• ID 23714

GUREVICH, Y. [1976] *The decision problem for standard classes* (J 0036) J Symb Logic 41*460-464
⋄ B20 B25 D35 ⋄ REV MR 53 # 10572 Zbl 339 # 02045
• ID 14777

GUREVICH, Y. [1977] *Monadic theory of order and topology I* (J 0029) Israel J Math 27*299-319
⋄ B15 C65 C85 E07 E50 E75 ⋄ REV MR 56 # 5259
Zbl 359 # 02061 • REM Part II 1979 • ID 50608

GUREVICH, Y. [1979] *Monadic theory of order and topology II* (J 0029) Israel J Math 34*45-71
⋄ B15 C65 C85 E07 E45 E50 ⋄ REV MR 81f:03049
Zbl 428 # 03034 • REM Part I 1977 • ID 53793

GUREVICH, Y. [1982] *Existential interpretation II* (J 0009) Arch Math Logik Grundlagenforsch 22*103-120
⋄ B20 D35 F25 ⋄ REV MR 84b:03059 Zbl 493 # 03020
• REM Part I 1965 • ID 33757

GUREVICH, Y. [1982] see AANDERAA, S.O.

GUREVICH, Y. & SHELAH, S. [1983] *Interpreting second-order logic in the monadic theory of order* (J 0036) J Symb Logic 48*816-828
⋄ B15 C65 C85 D35 E07 E50 F25 ⋄ REV MR 85f:03007 Zbl 559 # 03008 • ID 33765

GUREVICH, Y. & SHELAH, S. [1983] *Rabin's uniformization problem* (J 0036) J Symb Logic 48*1105-1119
⋄ B15 C40 C65 C85 E40 ⋄ REV MR 85g:03055 Zbl 537 # 03006 • ID 33768

GUREVICH, Y. & SHELAH, S. [1983] *Random models and the Goedel case of the decision problem* (J 0036) J Symb Logic 48*1120-1124
⋄ B20 B25 C13 D35 ⋄ REV MR 85d:03019 Zbl 534 # 03006 • ID 33766

GUREVICH, Y. & MAGIDOR, M. & SHELAH, S. [1983] *The monadic theory of ω_2* (J 0036) J Symb Logic 48*387-398
⋄ B15 B25 C65 C85 D35 E10 E35 ⋄ REV MR 84i:03076 Zbl 549 # 03010 • ID 33764

GUREVICH, Y. [1984] see GOLDFARB, W.D.

GUREVICH, Y. [1984] see BLASS, A.R.

GUREVICH, Y. see Vol. II, III, IV, V, VI for further entries

GURMUND, L. [1955] *The problem of correct symbolism as related to some problems of social psychology* (X 1136) Elanders Bokt: Goteborg 176pp
⋄ B80 ⋄ ID 22100

GUSEV, L.A. [1963] see AJZERMAN, M.A.

GUSEV, L.A. see Vol. IV, V for further entries

GUTIERREZ-NOVOA, L. [1979] *Non-Euclidean real numbers* (J 3495) Boll Unione Mat Ital, V Ser, B 16*390-404
⋄ B28 C65 ⋄ REV MR 80j:51020 Zbl 452 # 12011 • ID 54116

GUTTENPLAN, S.D. & TAMNY, M. [1972] *Logic: a comprehensive introduction* (X 0837) Harper & Row: New York 384pp • LAST ED [1978] (X 2671) Basic Books: New York xiv+401pp
⋄ A05 B98 ⋄ REV Zbl 457 # 03002 JSL 45.383 • ID 22490

GUZ, W. [1980] *A non-symmetric transition probability in quantum mechanics* (J 1546) Rep Math Phys (Warsaw) 17*385-400
⋄ B30 ⋄ REV MR 83b:81019 Zbl 487 # 03037 • ID 38115

GUZ, W. see Vol. II for further entries

HAAS, G. [1984] *Konstruktive Einfuehrung in die formale Logik* (X 0876) Bibl Inst: Mannheim 268pp
⋄ B98 F98 ⋄ REV MR 86g:03001 Zbl 562 # 03001 • ID 44527

HACKER, E.A. [1967] *Number system for the immediate inferences and the syllogism in aristotelian logic* (J 0047) Notre Dame J Formal Log 8*318-320
⋄ B20 ⋄ REV Zbl 183.7 • ID 05467

HACKER, E.A. & PARRY, W.T. [1967] *Pure numerical boolean syllogisms* (J 0047) Notre Dame J Formal Log 8*321-324
⋄ A05 B20 ⋄ REV Zbl 183.7 • ID 05468

HACKER, E.A. [1975] *The octagon of opposition* (J 0047) Notre Dame J Formal Log 16*352-353
⋄ A10 B20 ⋄ REV Zbl 254 # 02003 • ID 62260

HACKING, I. [1972] *A concise introduction to logic* (X 0981) Random House: New York viii+339pp
⋄ B98 ⋄ REV JSL 38.341 • ID 15009

HACKING, I. [1977] *Do-it-yourself semantics for classical sequent calculi including ramified type theory* (P 1704) Int Congr Log, Meth & Phil of Sci (5);1975 London ON 1*371-390
⋄ B15 ⋄ REV MR 58 # 16128 Zbl 373 # 02014 JSL 47.689 • ID 51478

HACKING, I. see Vol. II for further entries

HACKSTAFF, L.H. [1966] *Systems of formal logic* (X 0835) Reidel: Dordrecht xi+354pp
⋄ B98 ⋄ REV Zbl 205.3 • ID 22549

HADAMARD, J. [1954] *La geometrie non euclidienne et les definitions axiomatiques* (J 0423) Pensee NS 58*74-81
⋄ B30 ⋄ REV Zbl 58.141 JSL 23.30 • ID 05475

HADAMARD, J. [1954] *La Geometrie non-euclidienne et les definitions axiomatiques* (J 0001) Acta Math Acad Sci Hung 5*95-104
⋄ A05 B30 ⋄ REV MR 16.1045 Zbl 58.141 JSL 22.296 JSL 35.349 • ID 05476

HADAMARD, J. [1954] *Sur l'impossibilite de demontrer la compatibilite des axiomes de l'arithmetique* (J 0423) Pensee NS 58*82
⋄ A05 B28 F99 ⋄ REV JSL 23.30 • REM Extract from Hadamard: Natures des mathematiques, principes fondamentaux • ID 42157

HADAMARD, J. see Vol. IV, V, VI for further entries

HADDAD, L. [1974] *Introduction a l'analyse nonstandard* (X 3175) Univ Liban Fac Sci: Hadath-Beyrouth 44pp
⋄ H05 ⋄ REV Zbl 356 # 46025 • ID 50328

HADDAD, L. [1978] *Comments on nonstandard topology* (J 1934) Ann Sci Univ Clermont Math 16*1-25
⋄ H05 ⋄ REV MR 80f:54049 Zbl 411 # 54050 • ID 81449

HADDAD, L. see Vol. III, V for further entries

HADGOPOULOS, D.J. [1979] *The principle of the division into four figures in traditional logic* (J 0047) Notre Dame J Formal Log 20*92-94
⋄ A05 A10 B10 ⋄ REV MR 80f:03003 Zbl 321 # 02001 • ID 32212

HAERLEN, H. [1930] *Die logische Grundlage eines mathematischen Beweisverfahrens* (P 0796) Congr Math Pays Slaves (1);1929 Warsaw 102-105
⋄ A05 B10 E30 ⋄ REV FdM 56.825 • ID 05487

HAERLEN, H. [1930] *Ueber Axiomensysteme als Satzfunktionen* (P 0741) Int Congr Math (II, 3);1928 Bologna 3*389-391
⋄ B05 ⋄ REV FdM 56.824 • ID 05486

HAERLEN, H. see Vol. III, V for further entries

HAERTIG, K. [1953] *Axiomatische Probleme in der klassischen Syllogistik* (P 0626) Ber Math-Tagung Berlin;1953 Berlin 19-20
⋄ B20 ⋄ REV Zbl 53.197 JSL 22.96 • ID 27701

HAERTIG, K. [1956] *Explizite Definitionen einiger Eigenschaften von Zeichenreihen* (J 0068) Z Math Logik Grundlagen Math 2*177-203
⋄ B03 ⋄ REV MR 19.933 Zbl 75.232 JSL 22.357 • ID 05489

HAERTIG, K. [1957] *Ein Spezialfall der Substitution als Grundbeziehung der elementaren Semiotik* (J 0068) Z Math Logik Grundlagen Math 3*151-156
⋄ B03 ⋄ REV MR 20 # 4478 Zbl 102.7 JSL 23.217
• ID 05490

HAERTIG, K. [1959] *Einstellige Funktionen als Grundbegriffe der elementaren Zahlentheorie* (J 0068) Z Math Logik Grundlagen Math 5*209-215
⋄ B28 ⋄ REV MR 31 # 32 JSL 27.91 • ID 05492

HAERTIG, K. [1960] *Zur Axiomatisierung der Nicht-Identitaeten des Aussagenkalkuels* (J 0068) Z Math Logik Grundlagen Math 6*240-247
⋄ B05 ⋄ REV MR 23 # A793 Zbl 148.245 JSL 27.367
• ID 05493

HAERTIG, K. see Vol. III, V for further entries

HAEUSSLER, A.F. [1976] *Polynomial beschraenkte nichtdeterministische Turingmaschinen und die Vollstaendigkeit des aussagelogischen Erfuellungsproblems* (P 3196) Kompl von Entscheid Probl;1973/74 Zuerich 20-35
⋄ B05 D10 D15 ⋄ REV Zbl 383 # 03025 • ID 52007

HAFNER, I. [1978] *The deduction theorem (Slovenian) (English summary)* (J 2310) Obz Mat Fiz, Ljubljana 25*34-40
⋄ B10 ⋄ REV MR 80m:03003 Zbl 366 # 02004 • ID 51087

HAFNER, I. [1979] *On proof length in the equivalential calculus* (S 3416) Prepr Ser Dep Math Univ Ljubljana 17*71-90
• REPR [1980] (J 3519) Glas Mat, Ser 3 (Zagreb) 15(35)*233-242
⋄ B20 F20 ⋄ REV MR 82d:03094 Zbl 426 # 03060 Zbl 453 # 03061 • ID 53656

HAFNER, I. [1981] *First-order language with equality* (J 2310) Obz Mat Fiz, Ljubljana 28*24-32
⋄ B10 ⋄ REV MR 82b:03001 Zbl 442 # 03010 • ID 56367

HAFNER, I. see Vol. IV, V, VI for further entries

HAGIWARA, Y. [1972] *A system for recognition of the implication relation between formulas in the logic of predicates* (J 0116) Electr & Comm Japan 55*140-147
⋄ B20 B35 ⋄ REV MR 56 # 10187 • ID 81452

HAILPERIN, T. [1953] *Quantification theory and empty individual domains* (J 0036) J Symb Logic 18*197-200
⋄ B20 ⋄ REV MR 15.277 Zbl 52.9 JSL 20.284 • ID 05503

HAILPERIN, T. [1954] *Remarks on identity and description in first-order axiom systems* (J 0036) J Symb Logic 19*14-20
⋄ B10 ⋄ REV MR 15.845 Zbl 55.4 JSL 20.81 • ID 05504

HAILPERIN, T. [1957] *A theory of restricted quantification I,II* (J 0036) J Symb Logic 22*19-35,113-129
⋄ B20 ⋄ REV MR 19.626 MR 21 # 5557 Zbl 81.245 JSL 25.175 • ID 16853

HAILPERIN, T. [1960] *Corrections to a theory of restricted quantification* (J 0036) J Symb Logic 25*54-56
⋄ B20 ⋄ REV MR 24 # A35 Zbl 106.4 • ID 05507

HAILPERIN, T. [1961] *A complete set of axioms for logical formulas invalid in some finite domain* (J 0068) Z Math Logik Grundlagen Math 7*84-96
⋄ B20 C13 ⋄ REV MR 26 # 3588 Zbl 111.8 JSL 27.108
• ID 05508

HAILPERIN, T. [1965] *An incorrect theorem* (J 0036) J Symb Logic 30*27
⋄ B10 ⋄ REV MR 33 # 3905 Zbl 137.248 JSL 31.128
• ID 05509

HAILPERIN, T. [1969] *A form of Herbrand's theorem* (J 0068) Z Math Logik Grundlagen Math 15*107-120
⋄ B10 F05 F07 ⋄ REV MR 40 # 23 Zbl 157.335
• ID 05608

HAILPERIN, T. see Vol. II, III, V for further entries

HAJEK, O. [1960] *Three principles of induction in mathematics (Czech)* (J 1527) Pokroky Mat Fyz Astron (Prague) 5*385-394
⋄ B28 E20 ⋄ REV MR 23 # A3663 Zbl 124.248 • ID 48087

HAJEK, P. [1966] see CHYTIL, M.K.

HAJEK, P. [1970] *Logische Kategorien* (J 0009) Arch Math Logik Grundlagenforsch 13*168-193
⋄ B10 C40 E40 F25 G30 ⋄ REV MR 43 # 7386 Zbl 226 # 02043 • ID 05521

HAJEK, P. [1973] *Automatic listing of important observational statements. I,II* (J 0156) Kybernetika (Prague) 9*187-205,251-271
⋄ B35 C13 C80 C90 ⋄ REV MR 53 # 5288 Zbl 289 # 68046 Zbl 289 # 68047 • REM Part III 1974
• ID 22921

HAJEK, P. [1973] *Some logical problems of automated research* (P 1448) Math Founds of Comput Sci (2);1973 Strbske Pleso 85-93
⋄ B35 ⋄ REV MR 58 # 19416 • ID 81456

HAJEK, P. [1974] *Automatic listing of important observational statements. III* (J 0156) Kybernetika (Prague) 10*95-124
⋄ B35 C13 C80 C90 ⋄ REV MR 54 # 4960 Zbl 289 # 68047 • REM Parts I,II 1973 • ID 24128

HAJEK, P. & HAVRANEK, T. [1977] *On generation of inductive hypotheses* (J 1741) Int J Man-Mach Stud 9*415-438
⋄ B35 C90 ⋄ REV Zbl 372 # 68026 • ID 31130

HAJEK, P. & HAVRANEK, T. [1978] *Mechanizing hypothesis formation. Mathematical foundations for a general theory* (X 0811) Springer: Heidelberg & New York xv+396pp
• TRANSL [1984] (X 2027) Nauka: Moskva 278pp
⋄ A05 B35 C13 C80 C90 D80 ⋄ REV MR 82f:03017 MR 86e:03022 Zbl 371 # 02002 • ID 31131

HAJEK, P. & HAVRANEK, T. [1978] *The GUHA method - its aims and techniques (twenty-four questions and answers)* (J 1741) Int J Man-Mach Stud 10*3-22
⋄ A05 B35 ⋄ REV Zbl 404 # 68092 • ID 54867

HAJEK, P. [1981] *On interpretability in theories containing arithmetics II* (J 0140) Comm Math Univ Carolinae (Prague) 22*667-688
⋄ B28 B45 F25 F30 ⋄ REV MR 83j:03094 Zbl 262 # 02049 Zbl 487 # 03032 • REM Part I 1972 by Hajek,P. & Hajkova,M. • ID 35379

HAJEK, P. [1983] *Arithmetical interpretations of dynamic logic* (J 0036) J Symb Logic 48*704-713
⋄ B28 B45 ⋄ REV MR 84j:03038 Zbl 546 # 03012
• ID 34630

HAJEK, P. [1983] see CHYTIL, M.K.

HAJEK, P. see Vol. II, III, IV, V, VI for further entries

HAKAMIES, A. (ED.) [1977] *Logik, Mathematik und Philosophie des Transzendenten* (X 0846) Schoeningh: Paderborn 180pp
⋄ A05 B97 ⋄ REV Zbl 396 # 00001 • ID 52608

HAKEN, A. [1985] *The intractability of resolution* (J 1426) Theor Comput Sci 39∗297-308
⋄ B35 D15 D35 ⋄ ID 49032

HALATSIS, C. [1978] see GAITANIS, N.

HALBERSTADT, W.H. [1960] *Introduction to modern logic: an elementary textbook of symbolic logic* (X 0837) Harper & Row: New York 221pp
⋄ B98 ⋄ REV JSL 29.43 • ID 22492

HALE, R.L.V. [1974] see BISHOP, P.

HALKOWSKA, K. [1967] *A note on the system of propositional calculus with primitive rule of extensionality (Polish) (Russian and English summaries)* (J 0063) Studia Logica 20∗145-150
⋄ B05 ⋄ REV MR 34 # 7334 Zbl 294 # 02003 • ID 62293

HALKOWSKA, K. [1973] *Conditional definitions and the idea of meaningful expression with conditionally defined terms (Polish) (English summary)* (S 1454) Zesz Nauk Wyz Szk Ped Mat, Opole 13∗5-56
⋄ B10 B20 ⋄ REV MR 54 # 7199 Zbl 314 # 02016 • ID 24990

HALKOWSKA, K. & PIROG-RZEPECKA, K. [1975] *On proofs in the theories containig conditional definition (Polish) (English summary)* (S 1454) Zesz Nauk Wyz Szk Ped Mat, Opole 15∗63-76
⋄ B20 C07 ⋄ REV MR 54 # 7200 Zbl 315 # 02014 • ID 29801

HALKOWSKA, K. & PIROG-RZEPECKA, K. & SLUPECKI, J. [1976] *Mathematical logic (Polish)* (X 1034) PWN: Warsaw 289pp
⋄ B98 C07 C98 E98 ⋄ REV MR 57 # 15933 Zbl 393 # 03001 • ID 52421

HALKOWSKA, K. & PIROG-RZEPECKA, K. & SLUPECKI, J. [1978] *Logic and set theory (Polish)* (X 1034) PWN: Warsaw 309pp
⋄ B98 E98 ⋄ REV MR 80a:03001 Zbl 404 # 03002 • ID 54789

HALKOWSKA, K. see Vol. II, III for further entries

HALLDEN, S. [1948] *Certain problems connected with the definition of identity and of a definite description given in "Principia Mathematica"* (J 0103) Analysis (Oxford) 9∗29-33
⋄ B10 ⋄ REV Zbl 33.243 JSL 14.136 • ID 05585

HALLDEN, S. see Vol. II, III for further entries

HALLER, R. [1980] see GRASS, W.

HALLERBERG, A.E. [1974] *Logic in mathematics: An elementary approach* (X 1274) Hafner: New York vi + 90pp
⋄ B98 ⋄ REV Zbl 293 # 02001 • ID 62297

HALLERBERG, A.E. [1974] *Mathematical proof: An elementary approach* (X 1274) Hafner: New York viii + 104pp
⋄ B98 ⋄ REV Zbl 293 # 02002 • ID 62298

HALMOS, P.R. [1966] *Invariant subspaces of polynomially compact operators* (J 0048) Pac J Math 16∗433-437
⋄ H05 ⋄ REV MR 33 # 1725 Zbl 141.229 • ID 05606

HALMOS, P.R. see Vol. V, VI for further entries

HALPERN, J.D. [1970] see COLLINS, G.E.

HALPERN, J.D. see Vol. III, V for further entries

HAMBLIN, C.L. [1962] *Translation to and from Polish notation* (J 1193) Comput J (London) 5∗210-213
⋄ B03 ⋄ REV Zbl 113.115 JSL 35.349 • ID 21975

HAMBLIN, C.L. [1967] *Elementary formal logic* (X 0816) Methuen: London & New York 182pp
⋄ B98 ⋄ ID 22493

HAMBLIN, C.L. [1973] *A felicitous fragment of the predicate calculus* (J 0047) Notre Dame J Formal Log 14∗433-447
⋄ B20 B25 ⋄ REV MR 51 # 115 Zbl 265 # 02008 • ID 17398

HAMBLIN, C.L. [1973] *Language types and logical theorems* (J 0194) Inform & Control 22∗183-187
⋄ B05 D05 ⋄ REV MR 48 # 3304 Zbl 256 # 68037 • ID 62308

HAMBLIN, C.L. see Vol. II for further entries

HAMBOURGER, R. [1977] *A difficulty with the Frege-Russell definition of number* (J 0301) J Phil 74∗409-414
⋄ A05 B28 ⋄ REV MR 56 # 8315 • ID 73805

HAMILTON, A.G. [1978] *Logic for mathematicians* (X 0805) Cambridge Univ Pr: Cambridge, GB viii + 224pp
• TRANSL [1981] (X 3560) Paraninfo: Madrid 243pp
⋄ B98 ⋄ REV MR 80c:03005 MR 83f:03004 Zbl 383 # 03003 Zbl 491 # 03001 JSL 45.379 • ID 28266

HAMILTON, A.G. [1982] *Numbers, sets and axioms* (X 0805) Cambridge Univ Pr: Cambridge, GB ix + 255pp
⋄ B28 ⋄ REV MR 84d:04001 Zbl 497 # 04001 JSL 49.1421 • ID 34110

HAMILTON, A.G. see Vol. III, IV for further entries

HAMILTON, J. [1973] see ADLER, A.

HAMILTON, N.T. & LANDIN, J. [1961] *Set theory: the structure of arithmetic* (X 0802) Allyn & Bacon: London xi + 264pp
⋄ B28 E75 E98 ⋄ REV Zbl 127.279 • ID 22640

HAN, BYUNG HO [1973] *Proof without employment of higher order predicate calculus of Maehara's ε-theorem* (J 0350) Sci Rep Tokyo Kyoiku Daigaku Sect A 12∗66-79
⋄ B15 ⋄ REV MR 54 # 4933 Zbl 288 # 02010 • ID 24106

HANAZAWA, M. [1979] *An interpretation of Skolem's paradox in the predicate calculus with ε-symbol* (J 1472) Sci Rep Saitama Univ, Ser A 9/2∗11-13
⋄ B10 C07 F05 F25 ⋄ REV MR 81h:03110 Zbl 457 # 03004 • ID 54329

HANAZAWA, M. [1980] *An extension of the notion of relativization to Hilbert's ε-symbol* (J 0068) Z Math Logik Grundlagen Math 26∗491-496
⋄ B10 ⋄ REV MR 81j:03023 Zbl 445 # 03011 • ID 56475

HANAZAWA, M. see Vol. III, V, VI for further entries

HANDSCHEL, G. [1979] *Eine graphische Veranschaulichung der logischen Operationen* (J 0160) Math-Phys Sem-ber, NS 26∗114-124
⋄ B10 G15 ⋄ REV MR 81j:03001 • ID 73818

HANF, W.P. & MYERS, D.L. [1983] *Boolean sentence algebras: isomorphism constructions* (J 0036) J Symb Logic 48∗329-338
⋄ B10 C52 ⋄ REV MR 84m:03096 Zbl 511 # 03005 • ID 35790

HANF, W.P. see Vol. III, IV, V, VI for further entries

HANNA, S.C. & SABER, J.C. [1971] *Sets and logic* (X 1290) Irwin: Homewood xi + 274pp
⋄ B98 E98 ⋄ ID 22641

HANSEL, C.E.M. & HENRY, DESMOND PAUL & MAYS, W. [1951] *Note on the exhibition of logical machines at the joint session, July 1950* (**J** 0094) Mind 60∗262-264
- ◇ B35 ◇ REV JSL 17.77 • ID 05638

HANSEN, B.B. [1983] see BARRICELLI, N.A.

HANSEN, J.C. [1975] *Some applications of a general theory of digraph measures* (**P** 1805) Int Symp Multi-Val Log (5,Proc);1975 Bloomington 262-276
- ◇ B80 ◇ REV MR 58 # 5343 • ID 35826

HANSEN, J.C. see Vol. II for further entries

HANSEN, J.R. [1963] see GELERNTER, H.L.

HANSEN, R.T. & SWANSON, L.G. [1980] *Placing the pigeonhole principle within the defining axioms of the integers* (**P** 3540) West-Coast Conf Combin, Graph Th & Comput;1979 Arcata 183-186
- ◇ B28 ◇ REV MR 83f:04001 Zbl 457 # 04008 • ID 35316

HANSSON, B. [1970] *Transitivity and topological structure of the preference space* (**P** 0785) Scand Logic Symp (1);1968 Aabo 3-18
- ◇ B30 ◇ REV MR 51 # 2595 Zbl 323 # 02046 • ID 30421

HANSSON, B. see Vol. II, III for further entries

HARARY, F. [1955] *Note on an enumeration theorem of Davis and Slepian* (**J** 0133) Michigan Math J 3∗149-153
- ◇ B05 ◇ REV MR 18.633 Zbl 74.250 • ID 05651

HARARY, F. [1961] *A very independent axiom system* (**J** 0005) Amer Math Mon 68∗159-162
- ◇ B30 E07 ◇ REV MR 32 # 5516 Zbl 133.243 JSL 39.604 • ID 05653

HARARY, F. [1963] *A measure of axiomatic independence* (**J** 0094) Mind 72∗143-144
- ◇ B30 ◇ REV JSL 39.604 • ID 44275

HARARY, F. see Vol. II, III, V for further entries

HARBECK, G. [1963] *Einfuehrung in die formale Logik* (**X** 0900) Vieweg: Wiesbaden vi+114pp
- ◇ B98 ◇ REV MR 29 # 1134 Zbl 249 # 02001 JSL 31.287 • ID 05654

HAREL, D. [1979] *Characterizing second order logic with first order quantifiers* (**J** 0068) Z Math Logik Grundlagen Math 25∗419-422
- ◇ B15 C80 C85 ◇ REV MR 80k:03034 Zbl 432 # 03007 • ID 53965

HAREL, D. see Vol. II, III, IV for further entries

HARMS, S. & KLIX, W.-D. [1977] *Ein Algorithmus zur automatischen Loesung konstruktiver Problemstellungen* (**S** 0410) Math Beitr Univ Halle-Wittenberg 6∗55-69
- ◇ B35 F65 ◇ REV MR 58 # 5101 • ID 73840

HARPER, L.H. [1975] *A note on some classes of boolean functions* (**J** 0548) Stud Appl Math 54∗161-164
- ◇ B05 D15 ◇ REV MR 56 # 5068 Zbl 307 # 02007 • ID 62346

HARPER, W.L. [1981] *A sketch of some recent developments in the theory of conditionals* (**C** 4140) Ifs 3-38
- ◇ B48 B98 ◇ REV MR 83a:03003 JSL 49.1411 • ID 47401

HARPER, W.L. see Vol. II for further entries

HARRINGTON, L.A. & MORLEY, M.D. & SCEDROV, A. & SIMPSON, S.G. [1985] *Harvey Friedman's research on the foundations of mathematics* (**X** 0809) North Holland: Amsterdam xvi+408pp
- ◇ B97 C97 D97 E97 F97 ◇ ID 49810

HARRINGTON, L.A. see Vol. III, IV, V, VI for further entries

HARRIS, J.H. [1971] *Ordinal theory in a conservative extension of predicate calculus* (**J** 0047) Notre Dame J Formal Log 12∗423-428
- ◇ B10 E10 E30 ◇ REV MR 45 # 6615 Zbl 188.23 • ID 05670

HARRIS, J.H. [1982] *What's so logical about the "logical" axioms?* (**J** 0063) Studia Logica 41∗159-171
- ◇ B10 ◇ REV MR 86a:03003 Zbl 566 # 03002 • ID 42301

HARRIS, J.H. see Vol. V for further entries

HARRISON, M.A. [1963] see ARNOLD, R.F.

HARRISON, M.A. [1963] *The number of transitivity sets of Boolean functions* (**J** 0514) SIAM Journ 11∗806-828
- ◇ B05 ◇ REV MR 28 # 1055 Zbl 119.12 JSL 35.160 • ID 43385

HARRISON, M.A. see Vol. II, IV for further entries

HARROP, R. [1954] *An investigation of the propositional calculus used in a particular system of logic* (**J** 0171) Proc Cambridge Phil Soc Math Phys 50∗495-512
- ◇ B20 B25 ◇ REV MR 16.661 Zbl 56.8 JSL 23.65 • ID 05676

HARROP, R. [1958] *On the existence of finite models and decision procedures for propositional calculi* (**J** 0171) Proc Cambridge Phil Soc Math Phys 54∗1-13
- ◇ B22 B25 ◇ REV MR 20 # 6 Zbl 80.8 JSL 25.180 • ID 05678

HARROP, R. [1959] *The finite model property and subsystems of classical propositional calculus* (**J** 0068) Z Math Logik Grundlagen Math 5∗29-32
- ◇ B22 ◇ REV MR 21 # 4093 Zbl 88.247 JSL 25.181 • ID 05679

HARROP, R. [1964] *A relativization procedure for propositional calculi, with an application to a generalized form of Post's theorem* (**J** 3240) Proc London Math Soc, Ser 3 14∗595-617
- ◇ B22 D35 ◇ REV MR 30 # 16 Zbl 158.251 JSL 32.125 • ID 05682

HARROP, R. [1965] *Some generalizations and applications of a relativization procedure for propositional calculi* (**P** 0688) Logic Colloq;1963 Oxford 12-41
- ◇ B22 B25 D35 ◇ REV MR 35 # 2737 Zbl 158.252 JSL 32.125 • ID 05684

HARROP, R. [1965] *Some structure results for propositional calculi* (**J** 0036) J Symb Logic 30∗271-292
- ◇ B22 ◇ REV MR 33 # 2523 Zbl 188.314 JSL 32.537 • ID 05683

HARROP, R. [1968] *Some forms of models of propositional calculi* (**P** 0608) Logic Colloq;1966 Hannover 163-174
- ◇ B22 ◇ REV MR 38 # 4279 Zbl 188.314 JSL 40.251 • ID 05685

HARROP, R. [1971] *On the equivalence for nonderivability testing of finite Smiley models and finite modified Smiley models* (**J** 0068) Z Math Logik Grundlagen Math 17∗137-143
- ◇ B22 ◇ REV MR 44 # 55 Zbl 188.313 JSL 40.251 • ID 05686

HARROP, R. [1973] *On simple, weak and strong models of propositional calculi* (J 0171) Proc Cambridge Phil Soc Math Phys 74*1-9
◇ B22 ◇ REV MR 50 # 1831   Zbl 267 # 02018 • ID 05687

HARROP, R. [1976] *Some results concerning finite model separability of propositional calculi* (J 0063) Studia Logica 35*179-189
◇ B22 B25 ◇ REV MR 55 # 7729   Zbl 361 # 02019
• ID 32206

HARROP, R. see Vol. IV, V, VI for further entries

HARTHONG, J. [1980] *Le moire (English summary)* (J 2313) C R Acad Sci, Paris, Ser A-B 290*A877-A879
◇ H10 ◇ REV MR 81g:78005 • ID 81498

HARTHONG, J. [1981] *La propagation des ondes (English summary)* (J 3364) C R Acad Sci, Paris, Ser 1 292*425-428
◇ H10 ◇ REV MR 82d:81010   Zbl 467 # 35039 • ID 81497

HARTHONG, J. [1981] *Le moire* (J 2650) Adv Appl Math 2*24-75
◇ H10 ◇ REV MR 82i:78003 • ID 81496

HARTHONG, J. [1983] *Elements pour une theorie du continu* (J 1620) Asterisque 109-110*235-244
◇ H05 ◇ REV MR 85k:03045   Zbl 543 # 03038 • ID 40940

HARTHONG, J. see Vol. II for further entries

HARTMANIS, J. [1976] *On effective speed-up and long proofs of trivial theorems in formal theories (French summary)* (J 4698) Rev Franc Autom, Inf & Rech Operat, Ser Rouge Inf Th 10/R-1*29-38
◇ B25 B35 D15 D20 F20 ◇ REV MR 54 # 6550
Zbl 399 # 03042 • ID 24160

HARTMANIS, J. see Vol. IV, VI for further entries

HASENJAEGER, G. [1950] *Ueber eine Art von Unvollstaendigkeit des Praedikatenkalkuels der ersten Stufe* (J 0036) J Symb Logic 15*273-276
◇ B10 ◇ REV MR 12.578   Zbl 41.149   JSL 16.146
• ID 05720

HASENJAEGER, G. [1952] *Topologische Untersuchungen zur Semantik und Syntax eines erweiterten Praedikatenkalkuels* (J 0009) Arch Math Logik Grundlagenforsch 1*99-129
◇ B10 C07 G05 ◇ REV MR 15.668   Zbl 49.6 • ID 05722

HASENJAEGER, G. [1953] *Eine Bemerkung zu Henkin's Beweis fuer die Vollstaendigkeit des Praedikatenkalkuels der ersten Stufe* (J 0036) J Symb Logic 18*42-48
◇ B10 C07 C57 D55 ◇ REV MR 14.1052   Zbl 51.5
JSL 31.268 • ID 05723

HASENJAEGER, G. [1955] *On definability and derivability* (P 1589) Math Interpr of Formal Systs;1954 Amsterdam 15-25
◇ B10 B15 C07 D55 ◇ REV MR 17.699   Zbl 68.246
JSL 24.171 • ID 27719

HASENJAEGER, G. [1958] *Ueber eine Interpretation der Praedikatenkalkuele hoeherer Stufe* (J 0009) Arch Math Logik Grundlagenforsch 4*71-80
◇ B15 ◇ REV MR 21 # 4907   Zbl 84.247 • ID 05726

HASENJAEGER, G. [1958] *Zur Axiomatisierung der k-zahlig allgemeingueltigen Ausdruecke des Stufenkalkuels* (J 0068) Z Math Logik Grundlagen Math 4*175-177
◇ B15 ◇ REV MR 21 # 4906   Zbl 117.11   JSL 25.176
• ID 05725

HASENJAEGER, G. [1960] *Unabhaengigkeitsbeweise in Mengenlehre und Stufenlogik durch Modelle* (J 0157) Jbuchber Dtsch Math-Ver 63*141-162
◇ B15 E25 E35 ◇ REV MR 23 # A3082   Zbl 222 # 02072
• ID 05727

HASENJAEGER, G. [1961] see SCHOLZ, H.

HASENJAEGER, G. [1962] *Einfuehrung in die Grundbegriffe und Probleme der modernen Logik* (X 0826) Alber: Freiburg 202pp
• TRANSL [1968] (X 4056) Labor: Barcelona 184pp [1972] (X 0835) Reidel: Dordrecht 180pp
◇ A05 A10 B98 ◇ REV MR 29 # 4667   Zbl 122.243
JSL 40.627 • ID 05729

HASENJAEGER, G. [1967] *On Loewenheim-Skolem-type insufficiencies of second order logic* (P 0691) Sets, Models & Recursion Th;1965 Leicester 173-182
◇ B15 C55 C85 ◇ REV MR 36 # 3637   Zbl 175.267
• ID 05731

HASENJAEGER, G. [1977] *Von der Syllogistik zur Mengentheorie* (P 1986) Dt Kongr Philos (11);1975 Goettingen 85-93
◇ A10 B20 E30 ◇ ID 47066

HASENJAEGER, G. [1978] *Praedikatenvariablen in der Zahlentheorie* (J 0076) Dialectica 32*209-220
◇ B28 ◇ REV MR 82g:03097   Zbl 402 # 03018 • ID 54665

HASENJAEGER, G. [1984] see BOERGER, E.

HASENJAEGER, G. see Vol. IV, V, VI for further entries

HASHIMOTO, N. [1978] *Writing a textbook on proof theory using nonstandard analysis (Japanese)* (P 4109) B-Val Anal & Nonstand Anal;1978 Kyoto 135-149
◇ H05 ◇ ID 47671

HASSE, M. [1966] *Grundbegriffe der Mengenlehre und Logik* (X 1079) Teubner: Leipzig 86pp
◇ B98 E98 ◇ REV MR 35 # 6560   Zbl 168.245 • ID 22643

HATCHER, W.S. [1964] *La notion d'equivalence entre systemes formels* (J 3370) Enseign Math, Ser 2 10*314-315
◇ B22 F25 ◇ ID 32200

HATCHER, W.S. [1966] *Logical truth and logical implication* (J 0036) J Symb Logic 31*561
◇ A05 B20 ◇ REV MR 34 # 7339   Zbl 166.2 • ID 05746

HATCHER, W.S. [1968] *Foundations of mathematics* (X 0810) Saunders: Philadelphia xiii+327pp
◇ B98 E30 E70 E98 G30 ◇ REV MR 38 # 5610
Zbl 191.282   JSL 51.467 • ID 22644

HATCHER, W.S. [1972] *Foundations as a branch of mathematics* (J 0122) J Philos Logic 1*349-358
◇ A05 B30 ◇ REV MR 55 # 2472   Zbl 266 # 02006
• ID 32201

HATCHER, W.S. [1972] see CORCORAN, J.

HATCHER, W.S. [1982] *Clone embeddings and the hyperreals* (P 3808) Rect Trends in Math;1982 Reinhardsbrunn 164-173
◇ H05 ◇ REV MR 84m:03103   Zbl 497 # 03051 • ID 35797

HATCHER, W.S. [1982] *The logical foundations of mathematics* (X 0869) Pergamon Pr: Oxford x+320pp
◇ B98 ◇ REV MR 84g:03003   Zbl 504 # 03001   JSL 51.467
• ID 34117

HATCHER, W.S. & LAFLAMME, C. [1983] *On the order structure of the hyperreal line* (J 0068) Z Math Logik Grundlagen Math 29∗197-202
◊ H05 ◊ REV MR 85d:03130 Zbl 537 # 03050 • ID 41116

HATCHER, W.S. [1985] *Elementary extension and the hyperreal numbers* (C 4181) Math Log & Formal Syst (Costa da) 205-219
◊ C65 H05 ◊ ID 48208

HATCHER, W.S. see Vol. III, IV, V, VI for further entries

HAUCK, J. & HERRE, H. & POSEGGA, M. [1972] *Zur Metatheorie formaler Systeme* (C 1533) Quantoren, Modal, Paradox 107-122
◊ B10 C07 F30 ◊ REV Zbl 256 # 02006 • ID 62383

HAUCK, J. see Vol. III, IV, VI for further entries

HAUSCHILD, K. [1968] *Metatheoretische Eigenschaften gewisser Klassen von elementaren Theorien* (J 0068) Z Math Logik Grundlagen Math 14∗205-244
◊ B28 C15 C35 E30 E70 ◊ REV MR 38 # 3146 Zbl 185.12 • ID 05770

HAUSCHILD, K. [1969] *Equivalence in respect to special classes of sentences* (J 0014) Bull Acad Pol Sci, Ser Math Astron Phys 17∗609-610
◊ B20 ◊ REV MR 42 # 4383 Zbl 193.302 • ID 05773

HAUSCHILD, K. [1971] *Nichtaxiomatisierbarkeit von Satzmengen durch Ausdruecke spezieller Gestalt* (J 0027) Fund Math 72∗245-253
◊ B28 C20 C40 C52 F30 ◊ REV MR 48 # 88 Zbl 215.48 JSL 38.161 • ID 05779

HAUSCHILD, K. & HERRE, H. & RAUTENBERG, W. [1972] *Entscheidbarkeit der monadischen Theorie 2. Stufe der n-separierten Graphen (Russian, English and French summaries)* (J 0115) Wiss Z Humboldt-Univ Berlin, Math-Nat Reihe 21∗507-511
◊ B15 B25 ◊ REV MR 50 # 79 Zbl 264 # 02046 • ID 05782

HAUSCHILD, K. [1972] *Ueber die Universalitaet axiomatisierbarer Fragmente der Zahlentheorie* (J 0068) Z Math Logik Grundlagen Math 18∗255-259
◊ B28 ◊ REV MR 46 # 8831 Zbl 249 # 02023 • ID 05793

HAUSCHILD, K. [1975] *Zur Uebertragbarkeit von Entscheidbarkeitsresultaten elementarer Theorien (Russian, English and French summaries)* (J 0115) Wiss Z Humboldt-Univ Berlin, Math-Nat Reihe 24∗780-783
◊ B15 B25 C65 C85 ◊ REV MR 58 # 5151 Zbl 341 # 02039 • ID 29757

HAUSCHILD, K. see Vol. II, III, IV, V, VI for further entries

HAUSNER, M. [1972] *On a non-standard construction of Haar measure* (J 0155) Commun Pure Appl Math 25∗403-405
◊ E75 H05 ◊ REV MR 46 # 2010 Zbl 239 # 28016 • ID 05802

HAVEL, I.M. [1966] see CHYTIL, M.K.

HAVEL, I.M. & STEPANKOVA, O. [1976] *A logical theory of robot problem solving* (J 0503) Artif Intell 7∗129-161
◊ B35 ◊ REV MR 54 # 1741 Zbl 328 # 68081 • ID 31136

HAVEL, I.M. & STEPANKOVA, O. [1977] *Incidental and state-dependent phenomena in robot problem solving* (J 0156) Kybernetika (Prague) 13∗421-438
◊ B35 ◊ REV MR 57 # 8212 Zbl 366 # 68062 • ID 31137

HAVEL, I.M. see Vol. IV for further entries

HAVRANEK, T. [1974] *An application of logical-probabilistic expressions to the realization of stochastic automata* (J 0156) Kybernetika (Prague) 10∗241-257
◊ B70 B80 D05 ◊ REV MR 49 # 12214 Zbl 283 # 94020 • ID 31143

HAVRANEK, T. [1977] see HAJEK, P.

HAVRANEK, T. & VOSAHLO, J. [1978] *A GUHA procedure with correlational quantifiers* (J 1741) Int J Man-Mach Stud 10∗67-74
◊ B35 C80 ◊ REV MR 80g:68118b Zbl 404 # 68095 • ID 54868

HAVRANEK, T. [1978] see HAJEK, P.

HAVRANEK, T. [1983] see CHYTIL, M.K.

HAVRANEK, T. see Vol. II, III, IV for further entries

HAWRANEK, J. & ZYGMUNT, J. [1980] *A theorem on the degree of complexity of some sentential logics* (J 0387) Bull Sect Logic, Pol Acad Sci 9∗67-70
◊ B22 ◊ REV MR 81h:03062 Zbl 446 # 03021 • ID 56550

HAWRANEK, J. [1980] *Some examples concerning uniformity and complexity of sentential logics* (J 0387) Bull Sect Logic, Pol Acad Sci 9∗71-72
◊ B22 ◊ REV MR 81h:03063 Zbl 446 # 03022 • ID 56551

HAWRANEK, J. & ZYGMUNT, J. [1981] *Another proof of Wojtylak's theorem* (J 0387) Bull Sect Logic, Pol Acad Sci 10∗80-82
◊ B22 ◊ REV Zbl 463 # 03022 • ID 54562

HAWRANEK, J. & ZYGMUNT, J. [1981] *On the degree of complexity of sentential logics. A couple of examples* (J 0063) Studia Logica 40∗141-153
◊ B22 ◊ REV MR 83i:03050 Zbl 484 # 03011 • REM Part I. Part II 1984 • ID 35516

HAWRANEK, J. & ZYGMUNT, J. [1983] *On normal extensions of sentential logics (Polish) (English summary)* (J 0481) Acta Univ Wroclaw 605(34)(Logika 10)∗17-29
◊ B22 ◊ REV Zbl 557 # 03006 • ID 46187

HAWRANEK, J. & ZYGMUNT, J. [1984] *On the degree of complexity of sentential logics. II. An example of the logic with semi-negation* (J 0063) Studia Logica 43∗405-414
◊ B22 ◊ REV Zbl 569 # 03012 • REM Part I 1981 • ID 42389

HAWRANEK, J. see Vol. II for further entries

HAY, L. [1978] *Convex subsets of 2^n and bounded truth-table reducibility* (J 0193) Discr Math 21∗31-46
◊ B05 D30 ◊ REV MR 81c:03031 Zbl 377 # 02036 • ID 30704

HAY, L. see Vol. II, III, IV, V for further entries

HAYASAKA, S. [1965] *On the axiom of mathematical induction (Japanese) (English summary)* (1111) Preprints, Manuscr., Techn. Reports etc. 2∗75-83
◊ B28 ◊ REV MR 53 # 12919 • REM Miyagi Technical College • ID 23169

HAYASHI, S. [1983] *Extracting Lisp programs from constructive proofs: a formal theory of constructive mathematics based on Lisp* (J 0390) Publ Res Inst Math Sci (Kyoto) 19∗169-191
◊ B35 F35 F50 ◊ REV MR 85f:68066 Zbl 513 # 68022 • ID 37813

HAYASHI, S. see Vol. V, VI for further entries

HAYES, J.P. [1976] *Enumeration of fanout-free boolean functions* (J 0037) ACM J 23*700-709
- ◊ B05 ◊ REV MR 55 # 12331 Zbl 352 # 94027 • ID 50075

HAYES, P.J. & KOWALSKI, R. [1969] *Semantic trees in automatic theorem-prooving* (J 0508) Machine Intelligence 4*87-101
- • REPR [1983] (C 4659) Autom of Reasoning 2*217-232
- ◊ B35 ◊ REV MR 41 # 1537 • ID 16940

HE, YANXIANG & HU, JIUQING & LI, WEIHUA [1981] *A way of mechanization of programs correctness proofs (Chinese) (English summary)* (J 2879) Wuhan Daxue Xuebao, Ziran Kexue 2/1*22-30
- ◊ B35 B75 ◊ REV MR 83i:68027 • ID 39309

HECHLER, S.H. [1972] *On monads in saturated enlargements* (J 0029) Israel J Math 12*49-50
- ◊ H05 ◊ REV MR 46 # 3290 Zbl 252 # 02055 • ID 05840

HECHLER, S.H. see Vol. III, V for further entries

HEDTSTUECK, U. [1984] *On the argument complexity of multiply transitive Boolean functions* (P 2342) Symp Rek Kombin;1983 Muenster 390-396
- ◊ B05 ◊ REV MR 85k:68004 Zbl 545 # 68029 • ID 45419

HEGENBERG, L. [1973] *Logica. O calculo sentential* (X 1279) Herder: Freiburg xvii+177pp
- ◊ B05 ◊ REV JSL 45.632 • ID 44720

HEGENBERG, L. [1973] *Logica. O calculo de predicados* (X 1279) Herder: Freiburg xi+226pp
- ◊ B10 ◊ REV JSL 45.632 • ID 44723

HEGENBERG, L. [1975] *Logica: Simbolizacao e deducao* (X 4328) Ed Pedag & Univ: Sao Paulo xiv+219pp
- ◊ B98 ◊ REV JSL 44.126 • ID 44580

HEIJENOORT VAN, J. (ED.) [1967] *From Frege to Goedel: a source book in mathematical logic, 1879-1931* (X 0858) Harvard Univ Pr: Cambridge x+660pp
- ◊ A10 B97 F97 ◊ REV MR 35 # 15 • ID 25903

HEIJENOORT VAN, J. see Vol. II, III, VI for further entries

HEILWEIL, M.F. & HOERNES, G.E. [1966] *Introduction a l'algebre de Boole et aux dispositifs logiques* (X 0856) Dunod: Paris viii+304pp
- • TRANSL [1972] (X 0814) Oldenbourg: Muenchen 291pp (2nd ed)
- ◊ B05 G05 ◊ REV Zbl 147.253 Zbl 285 # 94010 • REM Transl. from English • ID 62587

HEINRICH, S. [1980] *The isomorphic problem of envelopes* (S 3230) Prepr, Inst Math, Pol Acad Sci 213*17pp
- ◊ C20 C65 H05 ◊ REV MR 84h:46025 Zbl 448 # 46020 • ID 56660

HEINRICH, S. [1980] *Ultraproducts in Banach space theory* (J 0127) J Reine Angew Math 313*72-104
- ◊ C20 C65 C98 H05 ◊ REV MR 82b:46013 Zbl 412 # 46017 • ID 81516

HEINRICH, S. [1981] *Ultraproducts of L_1-predual spaces* (J 0027) Fund Math 113*221-234
- ◊ C20 C65 H05 ◊ REV MR 83m:46020 Zbl 427 # 46009 Zbl 472 # 46022 • ID 53749

HEINRICH, S. & MANKIEWICZ, P. [1982] *Applications of ultrapowers to the uniform and Lipschitz classification of Banach spaces* (J 0343) Studia Math, Pol Akad Nauk 73*225-251
- ◊ C20 C65 H05 ◊ REV MR 84h:46026 Zbl 448 # 46021 Zbl 506 # 46008 • ID 37826

HEINRICH, S. [1982] *The isomorphic problem of envelopes* (J 0343) Studia Math, Pol Akad Nauk 73*41-49
- ◊ C20 C65 H05 ◊ REV MR 84h:46025 Zbl 519 # 46015 • ID 46460

HEINRICH, S. [1983] *Ultrapowers of locally convex spaces and applications I* (S 3414) Prepr, Akad Wiss DDR, Inst Math 10/83*58pp
- ◊ C20 C65 H10 ◊ REV Zbl 522 # 46002 • REM Part II 1983 • ID 37053

HEINRICH, S. [1983] *Ultrapowers of locally convex spaces and applications II* (S 3414) Prepr, Akad Wiss DDR, Inst Math 11/83*37pp
- • REPR [1985] (J 0114) Math Nachr 121*211-229
- ◊ C20 C65 H10 ◊ REV Zbl 522 # 46003 • REM Part I 1983 • ID 37054

HEINRICH, S. see Vol. III for further entries

HEINTZ, JOHN [1968] *Identity, quantification and predicables* (J 0079) Logique & Anal, NS 11*390-402
- ◊ A05 B10 ◊ REV Zbl 184.10 • ID 05873

HELBIG, H. [1977] *A new method for deductive answer finding in a question-answering system* (P 1694) Inform Processing (7);1977 Toronto 389-393
- ◊ B35 ◊ REV MR 58 # 8561 Zbl 363 # 68126 • ID 50921

HELMAN, GLEN [1983] *An interpretation of classical proofs* (J 0122) J Philos Logic 12*39-71
- ◊ A05 B10 ◊ REV MR 85e:03011 Zbl 513 # 03007 • ID 37215

HELMAN, GLEN see Vol. VI for further entries

HELMER, O. [1935] *Axiomatischer Aufbau der Geometrie in formalisierter Darstellung* (J 1131) Schr Berlin Math Sem 2*175-201
- ◊ B30 ◊ REV FdM 61.598 • ID 21184

HELMER, O. see Vol. II, III, VI for further entries

HELMS, L.L. & LOEB, P.A. [1979] *Applications of nonstandard analysis to spin models* (J 0034) J Math Anal & Appl 69*341-352
- ◊ H10 ◊ REV MR 80d:82038 Zbl 426 # 60095 • ID 53674

HELMS, L.L. & LOEB, P.A. [1982] *A nonstandard proof of the martingale convergence theorem* (J 0308) Rocky Mountain J Math 12*165-170
- ◊ H05 H10 ◊ REV MR 83g:60054 Zbl 479 # 60049 • ID 39168

HELMS, L.L. [1983] *Hyperfinite spin models* (C 3884) Nonstandard Anal - Recent Develop 15-26
- ◊ H10 ◊ REV MR 84h:60171 Zbl 507 # 60097 • ID 37718

HENDRY, H.E. [1975] *Another system of natural deduction* (J 0047) Notre Dame J Formal Log 16*491-495
- ◊ B10 F07 ◊ REV MR 53 # 5248 Zbl 254 # 02013 • ID 18186

HENDRY, H.E. see Vol. II, III, VI for further entries

HENKIN, L. [1949] *Fragments of propositional calculus* (J 0036) J Symb Logic 14*42-48
⋄ B20 ⋄ REV MR 11.487 Zbl 34.7 JSL 14.197 • ID 05893

HENKIN, L. [1949] *The completeness of the first-order functional calculus* (J 0036) J Symb Logic 14*159-166
• REPR [1969] (C 0569) Phil of Math Oxford Readings 42-50
⋄ B10 C07 ⋄ REV MR 11.487 Zbl 34.6 JSL 15.68 • ID 05892

HENKIN, L. [1950] *An algebraic characterization of quantifiers* (J 0027) Fund Math 37*63-74
⋄ B20 ⋄ REV MR 12.662 JSL 16.290 • ID 05895

HENKIN, L. [1950] *Completeness in the theory of types* (J 0036) J Symb Logic 15*81-91
• REPR [1969] (C 0569) Phil of Math Oxford Readings 51-63
⋄ A05 B15 C85 F35 ⋄ REV MR 12.70 Zbl 39.8 JSL 16.72 • ID 05894

HENKIN, L. [1953] *Banishing the rule of substitution for functional variables* (J 0036) J Symb Logic 18*201-208
⋄ B10 B15 ⋄ REV MR 15.277 Zbl 53.200 JSL 20.179 • ID 05896

HENKIN, L. [1954] *A generalization of the notion of ω-consistency* (J 0036) J Symb Logic 19*183-196
⋄ B10 C07 F30 ⋄ REV MR 16.103 Zbl 56.11 JSL 23.40 • ID 05900

HENKIN, L. [1955] *Boolean representation through propositional calculus* (J 0027) Fund Math 41*89-96
⋄ B05 ⋄ REV MR 16.103 Zbl 56.9 JSL 38.521 • REM A remark was published by Los,J. ibid 44*82-83 • ID 05901

HENKIN, L. & MONTAGUE, R. [1956] *On the definition of "formal deduction"* (J 0036) J Symb Logic 21*129-136
⋄ B22 ⋄ REV MR 17.1173 Zbl 73.7 • ID 05906

HENKIN, L. [1957] *A generalization of the concept of ω-completeness* (J 0036) J Symb Logic 22*1-14
⋄ B10 C07 F30 ⋄ REV MR 20#1626 Zbl 81.12 JSL 24.172 • ID 05910

HENKIN, L. & SUPPES, P. & TARSKI, A. (EDS.) [1959] *The axiomatic method, with special reference to geometry and physics* (X 0809) North Holland: Amsterdam xi+488pp
⋄ B30 B97 C65 D35 ⋄ REV Zbl 88.244 • ID 22387

HENKIN, L. [1960] *On mathematical induction* (J 0005) Amer Math Mon 67*323-338
⋄ A05 B28 ⋄ REV MR 22#10913 JSL 27.92 • ID 05916

HENKIN, L. & SMITH, N.W. & VARINEAU, V.J. & WALSH, M.J. [1962] *Retracing elementary mathematics* (X 0843) Macmillan : New York & London xviii+418pp
⋄ B28 E98 ⋄ REV Zbl 121.52 JSL 29.209 • ID 05921

HENKIN, L. [1963] *A theory of propositional types* (J 0027) Fund Math 52*323-344 • ERR/ADD ibid 53/1*119
⋄ B15 ⋄ REV MR 27#3497 Zbl 127.6 JSL 30.385 • ID 21069

HENKIN, L. [1967] *Logical systems containing only a finite number of symbols* (X 0893) Pr Univ Montreal: Montreal 48pp
⋄ B20 ⋄ REV MR 39#1313 Zbl 164.307 • ID 26576

HENKIN, L. [1968] *Relativization with respect to formulas and its use in proofs of independence* (J 0020) Compos Math 20*88-106
• REPR [1968] (C 0727) Logic Found of Math (Heyting) 88-106
⋄ B10 ⋄ REV MR 38#3126 Zbl 155.23 JSL 40.499 • ID 05926

HENKIN, L. [1970] *Extending boolean operations* (J 0048) Pac J Math 32*723-752
⋄ B05 ⋄ REV MR 41#3345 Zbl 218#06004 • ID 05927

HENKIN, L. & MONK, J.D. & TARSKI, A. [1971] *Cylindric algebras. Vol.I* (S 3303) Stud Logic Found Math vi+508pp
⋄ B10 G15 ⋄ REV MR 47#3171 Zbl 214.13 • REM 2nd ed. 1985. Part II 1985 • ID 26577

HENKIN, L. [1974] see ADDISON, J.W.

HENKIN, L. & JOJA, A. & MOISIL, G.C. & SUPPES, P. (EDS.) [1975] *Logic, methodology and philosophy of science, IV. Proceedings of the fourth international congress for logic, methodology and philosophy of science* (X 0809) North Holland: Amsterdam x+981pp
⋄ B97 ⋄ REV MR 55#7657 • ID 80463

HENKIN, L. [1977] *The logic of equality* (J 0005) Amer Math Mon 84*597-612
⋄ B20 C05 ⋄ REV MR 57#12345 Zbl 376#02017 • ID 27268

HENKIN, L. & MONK, J.D. & TARSKI, A. [1985] *Cylindric algebras. Part II* (S 3303) Stud Logic Found Math ix+302pp
⋄ B10 G15 ⋄ REM Part I 1971 • ID 44588

HENKIN, L. see Vol. III, IV, V, VI for further entries

HENLE, J.M. & KLEINBERG, E.M. [1979] *Infinitesimal calculus* (X 0865) MIT Pr: Cambridge, MA ix+135pp
⋄ H05 ⋄ REV MR 82b:26026 Zbl 439#26010 • ID 81522

HENLE, J.M. [1984] *Tangent planes with infinitesimals* (J 0005) Amer Math Mon 91*433-435
⋄ H05 ⋄ REV MR 86f:26020 • ID 44330

HENLE, J.M. see Vol. III, V for further entries

HENLE, P. [1932] *The independence of the postulates of logic* (J 0015) Bull Amer Math Soc 38*409-414
⋄ B10 ⋄ REV Zbl 5.146 FdM 58.64 • ID 05935

HENLE, P. [1933] see CHAPMAN, F.M.

HENLE, P. & SMITH, HENRY BRADFORD [1935] *A note on the validity of aristotelian logic* (J 0153) Phil of Sci (East Lansing) 2*111-114
⋄ A05 A10 B20 ⋄ REV FdM 62.1050 • ID 40878

HENLE, P. & KALLEN, H.M. & LANGER, S.K. (EDS.) [1951] *Structure, method, and meaning. Essays in honor of Henry M. Scheffer* (X 4563) Liberal Arts Pr: New York xvi+306pp
⋄ B98 ⋄ REV Zbl 44.1 • ID 48573

HENLE, P. see Vol. II for further entries

HENLE, R. [1935] *A definition of abstract systems* (J 0094) Mind 44*341-346
⋄ B22 ⋄ REV FdM 61.50 • ID 40774

HENLEY, E.J. & OGUNBIYI, E.I. [1981] *Irredundant forms and prime implicants of a function with multistate variables* (J 3192) IEEE Trans on Reliab R-30*39-42
⋄ B05 ⋄ REV Zbl 456#90027 • ID 54325

HENNESSY, M. [1980] *A proof system for the first-order relational calculus* (J 0119) J Comp Syst Sci 20*96-110
⋄ B35 C07 ⋄ REV MR 81h:03023 Zbl 431 # 68084
• ID 53954

HENNESSY, M. see Vol. VI for further entries

HENRICI, P. [1969] see DEJON, B.

HENRY, DESMOND PAUL [1951] see HANSEL, C.E.M.

HENRY, DESMOND PAUL [1961] *The truncation of truth-functional calculation* (J 0047) Notre Dame J Formal Log 2*193-205
⋄ B05 ⋄ REV MR 25 # 3812 Zbl 116.5 JSL 39.174
• ID 05941

HENRY, DESMOND PAUL see Vol. III for further entries

HENSCHEN, L.J. [1976] *Semantic resolution for Horn sets* (J 0187) IEEE Trans Comp C-25*816-822
⋄ B20 B35 ⋄ REV MR 56 # 1841 Zbl 331 # 68054
• ID 62470

HENSCHEN, L.J. [1979] *Theorem proving by covering expressions* (J 0037) ACM J 26*385-400
⋄ B35 ⋄ REV MR 80e:68231 Zbl 403 # 68079 • ID 54782

HENSCHEN, L.J. & NAQVI, S.A. [1982] *Representing infinite sequences of resolvents in recursive first-order Horn databases* (P 3840) Autom Deduct (6);1982 New York 342-359
⋄ B35 ⋄ REV MR 85e:68014 Zbl 481 # 68088 • ID 38461

HENSCHEN, L.J. & WOS, L. [1983] *Automated theorem proving 1965-1970* (C 4659) Autom of Reasoning 2*1-26
⋄ B35 ⋄ REV Zbl 567 # 03002 • ID 48957

HENSCHEN, L.J. & SMITH, BRIAN & VEROFF, R. & WINKER, S.K. & WOS, L. [1983] *Questions concerning possible shortest single axioms for the equivalential calculus: an application of automated theorem proving to infinite domains* (J 0047) Notre Dame J Formal Log 24*205-223
⋄ B35 ⋄ REV MR 84g:03018 Zbl 488 # 03008 • ID 34133

HENSON, C.W. [1972] *On the nonstandard representation of measures* (J 0064) Trans Amer Math Soc 172*437-446
⋄ H05 ⋄ REV MR 47 # 3631 Zbl 255 # 28006 • ID 05953

HENSON, C.W. [1972] *The nonstandard hulls of a uniform space* (J 0048) Pac J Math 43*115-137
⋄ C65 H05 ⋄ REV MR 47 # 2559 Zbl 245 # 54046
• ID 05952

HENSON, C.W. & MOORE JR., L.C. [1972] *The nonstandard theory of topological vector spaces* (J 0064) Trans Amer Math Soc 172*405-435 • ERR/ADD ibid 184*509
⋄ C65 H05 ⋄ REV MR 46 # 7836 Zbl 254 # 46001
• ID 22286

HENSON, C.W. & MOORE JR., L.C. [1973] *Invariance of the nonstandard hulls of locally convex spaces* (J 0025) Duke Math J 40*193-206
⋄ C65 H05 ⋄ REV MR 51 # 8775 Zbl 256 # 46001
• ID 05956

HENSON, C.W. & MOORE JR., L.C. [1974] *Invariance of the nonstandard hulls of a uniform space* (P 1083) Victoria Symp Nonstand Anal;1972 Victoria 85-98
⋄ C65 H05 ⋄ REV MR 58 # 2784 Zbl 272 # 54040
• ID 26633

HENSON, C.W. & MOORE JR., L.C. [1974] *Nonstandard hulls of the classical Banach spaces* (J 0025) Duke Math J 41*277-284
⋄ C65 H05 ⋄ REV MR 51 # 3867 Zbl 298 # 46028
• ID 05962

HENSON, C.W. & MOORE JR., L.C. [1974] *Semi-reflexivity of the nonstandard hulls of a locally convex space* (P 1083) Victoria Symp Nonstand Anal;1972 Victoria 71-84
⋄ C65 H05 ⋄ REV MR 57 # 12159 Zbl 276 # 46006
• ID 26632

HENSON, C.W. & MOORE JR., L.C. [1974] *Subspaces of the nonstandard hull of a normed space* (J 0064) Trans Amer Math Soc 197*131-143
⋄ C65 H05 ⋄ REV MR 51 # 1351 Zbl 307 # 46008
• ID 05960

HENSON, C.W. [1974] *The isomorphism property in nonstandard analysis and its use in the theory of Banach spaces* (J 0036) J Symb Logic 39*717-731
⋄ C65 H05 ⋄ REV MR 50 # 12713 Zbl 306 # 02055
• ID 05959

HENSON, C.W. [1975] *The monad system of the finest compatible uniform structure* (J 0053) Proc Amer Math Soc 51*163-170
⋄ C65 H05 ⋄ REV MR 51 # 6779 Zbl 307 # 54023
• ID 17439

HENSON, C.W. [1975] *When do two Banach spaces have isometrically isomorphic nonstandard hulls ?* (J 0029) Israel J Math 22*57-67
⋄ C65 H05 ⋄ REV MR 52 # 6386 Zbl 314 # 46023
• ID 05963

HENSON, C.W. [1976] *Nonstandard hulls of Banach spaces* (J 0029) Israel J Math 25*108-144
⋄ C65 H05 ⋄ REV MR 57 # 1089 Zbl 348 # 46014
• ID 26078

HENSON, C.W. [1976] *Ultraproducts of Banach spaces* (C 4232) Altgeld Book
⋄ C20 C65 H05 ⋄ ID 46653

HENSON, C.W. & JOCKUSCH JR., C.G. & RUBEL, L.A. & TAKEUTI, G. [1977] *First-order topology* (J 0202) Diss Math (Warsaw) 143*40pp
⋄ B25 C65 C75 D35 H05 ⋄ REV MR 55 # 5434 Zbl 399 # 03019 • ID 30718

HENSON, C.W. [1979] *Analytic sets, Baire sets and the standard part map* (J 0017) Canad J Math 31*663-672
⋄ H05 ⋄ REV MR 80i:28019 Zbl 373 # 54035 • ID 81527

HENSON, C.W. [1979] *Unbounded Loeb measures* (J 0053) Proc Amer Math Soc 74*143-150
⋄ H05 H10 ⋄ REV MR 80b:28011 Zbl 397 # 28001
• ID 52718

HENSON, C.W. & WATTENBERG, F. [1981] *Egoroff's theorem and the distribution of standard points in a nonstandard model* (J 0053) Proc Amer Math Soc 81*455-461
⋄ H05 ⋄ REV MR 83c:03059 Zbl 481 # 03044 • ID 35152

HENSON, C.W. & MOORE JR., L.C. [1983] *Nonstandard analysis and the theory of Banach spaces* (C 3884) Nonstandard Anal - Recent Develop 27-112
⋄ C65 H05 ⋄ REV MR 85f:46033 Zbl 511 # 46070
• ID 38543

HENSON, C.W. & RUBEL, L.A. [1984] *Some applications of Nevanlinna theory to mathematical logic:identities of exponential functions* (J 0064) Trans Amer Math Soc 282∗1-32 • ERR/ADD ibid 294∗381
  ◊ B20 C65 F30 ◊ REV MR 85h:03015 Zbl 533 # 03015
  • ID 36539

HENSON, C.W. & KAUFMANN, M. & KEISLER, H.J. [1984] *The strength of nonstandard methods in arithmetic* (J 0036) J Symb Logic 49∗1039-1058
  ◊ C62 E30 F30 F35 H05 H15 ◊ REV MR 86h:03115
  • ID 39860

HENSON, C.W. see Vol. III, IV, V for further entries

HERBRAND, J. [1928] *Sur la theorie de la demonstration* (J 0109) C R Acad Sci, Paris 186∗1274-1276
  • TRANSL [1971] (C 0745) Herbrand: Log Writings 29-32
  • REPR [1968] (C 2486) Herbrand: Ecrits Logiques 21-23
  ◊ A05 B10 F07 ◊ REV JSL 36.523 JSL 40.94 FdM 54.52
  • ID 05972

HERBRAND, J. [1929] *Sur le probleme fondamental des mathematiques* (J 0109) C R Acad Sci, Paris 189∗554-556
  • TRANSL [1971] (C 0745) Herbrand: Log Writings 41-43
  ◊ A05 B10 F99 ◊ REV JSL 36.523 JSL 40.94 FdM 55.32
  • ID 05968

HERBRAND, J. [1929] *Sur quelques proprietes des propositions vraies et leurs applications* (J 0109) C R Acad Sci, Paris 189∗1076-1078
  • TRANSL [1971] (C 0745) Herbrand: Log Writings 38-40
  ◊ B10 F30 ◊ REV JSL 36.523 JSL 40.94 FdM 55.32
  • ID 05967

HERBRAND, J. [1930] *Les bases de la logique Hilbertienne* (J 0145) Rev Metaph Morale 37∗243-255
  • TRANSL [1971] (C 0745) Herbrand: Log Writings 203-214
  ◊ A05 B10 ◊ REV JSL 36.523 JSL 40.94 FdM 56.46
  • ID 05969

HERBRAND, J. [1930] *Recherches sur la theorie de la demonstration* (J 0459) C R Soc Sci Lett Varsovie Cl 3 33∗128pp
  • TRANSL [1967] (C 0675) From Frege to Goedel 525-581
  [1971] (C 0745) Herbrand: Log Writings 46-188,272-276
  • REPR [1968] (C 2486) Herbrand: Ecrits Logiques 35-153
  [1931] (J 1092) Ann Univ Paris 6∗186-189
  ◊ B10 B25 F05 F07 F25 F30 ◊ REV JSL 36.523 JSL 40.94 FdM 56.824 • ID 20866

HERBRAND, J. [1931] *Sur le probleme fondamental de la logique mathematique* (J 0459) C R Soc Sci Lett Varsovie Cl 3 24∗12-56
  • TRANSL [1971] (C 0745) Herbrand: Log Writings 215-271
  ◊ A05 B25 B30 F30 F99 ◊ REV Zbl 3.290 JSL 36.523 FdM 57.1320 • ID 16812

HERBRAND, J. [1968] *Ecrits logiques* (X 0840) Pr Univ France: Paris iii+244pp
  • TRANSL [1971] (X 0858) Harvard Univ Pr: Cambridge viii+312pp
  ◊ A05 A10 B96 F96 ◊ REV MR 37 # 24 MR 51 # 2844 Zbl 194.303 • REM Edited by Heijenoort van,J. • ID 24825

HERBRAND, J. see Vol. VI for further entries

HERCZEG, J. [1971] see BIZAM, G.

HERINGER, H.J. [1972] *Formale Logik und Grammatik* (X 0877) Niemeyer: Tuebingen vi+104pp
  ◊ B65 B98 D05 ◊ REV MR 52 # 13299 Zbl 337 # 68003
  • ID 21764

HERMES, H. & SCHOLZ, H. [1936] *Ein neuer Vollstaendigkeitsbeweis fuer das reduzierte Fregesche Axiomensystem des Aussagenkalkuels* (J 0426) Dt Math 1∗733-772
  • REPR [1937] (J 0956) Forsch Logik Grundl exakt Wiss 1∗40pp
  ◊ B05 ◊ REV Zbl 16.1 JSL 2.94 FdM 62.1058 • ID 05998

HERMES, H. [1937] *Ein Axiomensystem fuer die Syntax des (klassischen) Logikkalkuels* (P 0756) Congr Int Phil (9);1937 Paris VI∗43-45
  ◊ B10 ◊ REV FdM 63.836 • ID 32167

HERMES, H. [1938] *Eine Axiomatisierung der allgemeinen Mechanik* (J 0956) Forsch Logik Grundl exakt Wiss 3∗48pp
  ◊ B30 ◊ REV Zbl 19.98 JSL 3.119 FdM 64.32 • ID 32169

HERMES, H. [1938] *Semiotik. Eine Theorie der Zeichengestalten als Grundlage fuer Untersuchungen von formalisierten Sprachen* (J 0956) Forsch Logik Grundl exakt Wiss 5∗22pp
  ◊ B03 ◊ REV Zbl 20.97 JSL 4.87 FdM 64.27 • ID 16986

HERMES, H. [1951] *Zum Begriff der Axiomatisierbarkeit* (J 0114) Math Nachr 4∗343-347
  ◊ A05 B30 D25 ◊ REV MR 12.578 Zbl 42.7 JSL 22.83
  • ID 06000

HERMES, H. [1951] *Zur Theorie der aussagenlogischen Matrizen* (J 0044) Math Z 53∗414-418
  ◊ B22 ◊ REV MR 12.663 Zbl 42.7 JSL 16.275 • ID 06003

HERMES, H. [1952] *Maschinen zur Entscheidung von mathematischen Problemen* (J 0160) Math-Phys Sem-ber, NS 2∗179-189
  ◊ B35 ◊ REV MR 13.784 Zbl 46.7 JSL 22.376 • ID 06006

HERMES, H. & SCHOLZ, H. [1952] *Mathematische Logik* (X 1079) Teubner: Leipzig 82pp
  ◊ B98 ◊ REV MR 16.435 Zbl 47.248 JSL 19.278
  • ID 06004

HERMES, H. [1952] *Ueber den Begriff der Grenze in der Mathematik* (J 0178) Stud Gen 5∗585-591
  ◊ B30 ◊ REV MR 15.92 Zbl 47.285 • ID 32172

HERMES, H. [1953] *Sur le concept d'axiomatisabilite* (P 0644) Meth Form en Axiom;1950 Paris 23-25 (discussion p.26)
  ◊ B30 ◊ REV MR 14.1051 Zbl 50.3 JSL 22.83 • ID 42140

HERMES, H. [1956] see GUMIN, H.

HERMES, H. [1957] *Einfuehrung in die mathematische Logik. Klassische Praedikatenlogik* (X 0910) Aschendorffsche Verlagsbuchh: Muenster v+176pp
  • TRANSL [1973] (X 0811) Springer: Heidelberg & New York xi+242pp (English) • LAST ED [1969] (X 0823) Teubner: Stuttgart 204pp
  ◊ B10 B98 ◊ REV MR 28 # 2035 MR 40 # 1256 MR 49 # 10518 MR 55 # 68 Zbl 115.5 JSL 30.355 JSL 38.647 • ID 21083

HERMES, H. [1959] *Zur Axiomatisierung der Mechanik* (P 0651) Axiomatic Method;1957 Berkeley 282-290
  ◊ B30 ◊ REV MR 22 # 337 Zbl 87.388 • ID 14848

HERMES, H. [1965] *Eine Termlogik mit Auswahloperator*
(X 0811) Springer: Heidelberg & New York iv+42pp
• TRANSL [1970] (S 3301) Lect Notes Math
6*iv+55pp (English)
⋄ B10 B60 ⋄ REV MR 33 # 3906 MR 41 # 8207
Zbl 166.254 JSL 34.679 JSL 35.440 • ID 06014

HERMES, H. [1966] *Zum Folgerungsbegriff* (J 0178) Stud Gen
19*140-145
⋄ A05 B22 ⋄ REV Zbl 192.25 • ID 16285

HERMES, H. [1967] *Die Rolle der Logik beim Aufbau
naturwissenschaftlicher Theorien* (J 2093) Arbgem Forsch
Nordrhein-Westfalen Nat, Ing, Ges-Wiss 168*32pp
⋄ B30 ⋄ REV Zbl 246 # 02007 • ID 06016

HERMES, H. [1968] *Methodik der Mathematik und Logik*
(C 4678) Enzykl Geisteswiss Arbeitsmethoden 3-43
⋄ A05 A10 B98 ⋄ REV MR 38 # 3117 Zbl 185.6
• ID 28565

HERMES, H. [1968] *Praedikatenlogik und Theorie der rekursiven
Funktionen* (C 0552) Phil Contemp - Chroniques 254-265
⋄ B10 D20 ⋄ ID 14956

HERMES, H. [1971] *A simplified proof for the unsolvability of the
decision problem in the case* ∀∃∀ (P 0638) Logic Colloq;1969
Manchester 307-310
⋄ B20 D35 ⋄ REV MR 43 # 1840 Zbl 221 # 02032
• ID 06021

HERMES, H. see Vol. II, III, IV, V, VI for further entries

HERNANDEZ S., J.L. [1967] *Teoremas de reversibilidad isomorfica
de operaciones logicas multivariables* (J 0367) Scientia
(Valparaiso) 132*52-56
⋄ B05 ⋄ REV MR 38 # 26 • ID 48132

HEROLD, A. [1983] *Universal unification and a class of
equational theories* (P 3858) Adequate Modeling of Syst;1982
Bad Honnef 177-190
⋄ B35 C05 ⋄ REV MR 85i:03036 Zbl 499 # 68039
• ID 44073

HERRE, H. [1972] see HAUSCHILD, K.

HERRE, H. [1972] see HAUCK, J.

HERRE, H. see Vol. III, IV, V, VI for further entries

HERRERA MIRANDA, J. [1978] *Les theories convexes de Horn
(English summary)* (J 2313) C R Acad Sci, Paris, Ser A-B
287*A593-A594
⋄ B20 C05 C52 G30 ⋄ REV MR 83k:03038
Zbl 407 # 03038 • ID 36172

HERRERA MIRANDA, J. see Vol. III for further entries

HERRING, J.M. [1971] see CORCORAN, J.

HERRING, J.M. [1972] see CORCORAN, J.

HERRING, J.M. [1976] *Equivalence of several notions of theory
completeness in a free logic* (J 0302) Rep Math Logic, Krakow
& Katowice 6*87-91
⋄ B20 C35 ⋄ REV MR 57 # 9517 Zbl 384 # 03005
• ID 21916

HERRMANN, E. & WOLTER, H. [1980] *Untersuchungen zu
schwachen Logiken der zweiten Stufe* (J 0068) Z Math Logik
Grundlagen Math 26*59-68
⋄ B15 C15 C85 ⋄ REV MR 81k:03013 Zbl 445 # 03019
• ID 56483

HERRMANN, E. see Vol. III, IV for further entries

HERRMANN, R.A. [1975] *Nonstandard topological extensions*
(J 0016) Bull Austral Math Soc 13*269-290 • ERR/ADD ibid
13*472
⋄ H05 ⋄ REV MR 53 # 9192a MR 53 # 9192b
Zbl 309 # 54038 Zbl 309 # 54039 • ID 23046

HERRMANN, R.A. [1977] *A nonstandard generalization for
perfect maps* (J 0068) Z Math Logik Grundlagen Math
23*223-236
⋄ H05 H20 ⋄ REV MR 56 # 5282 Zbl 399 # 03055
• ID 26481

HERRMANN, R.A. [1977] *Nonstandard quasi-Hausdorff, Urysohn
and regular-closed extensions* (J 0406) Bull Inst Math, Acad
Sin (Taipei) 5*13-25
⋄ H05 ⋄ REV MR 57 # 7571 Zbl 362 # 02051 • ID 50773

HERRMANN, R.A. [1978] *Perfect maps and remoteness* (J 0195)
Bull Calcutta Math Soc 70*413-419
⋄ H05 ⋄ REV MR 81j:54082 Zbl 463 # 54049 • ID 81542

HERRMANN, R.A. [1978] *The nonstandard theory of
semi-uniform spaces* (J 0068) Z Math Logik Grundlagen
Math 24*237-256
⋄ C65 H05 H20 ⋄ REV MR 58 # 12992 Zbl 468 # 54037
• ID 55117

HERRMANN, R.A. [1979] *A nonstandard approach to S-closed
spaces* (S 2848) Topology Proc 3*123-138
⋄ H05 ⋄ REV MR 82i:54084 Zbl 413 # 54050 • ID 81541

HERRMANN, R.A. [1979] *Convergence spaces and nonstandard
compactifications* (J 2128) C R Math Acad Sci, Soc Roy
Canada 1*187-190
⋄ H05 ⋄ REV MR 80g:54058 Zbl 416 # 54001 • ID 81543

HERRMANN, R.A. [1979] *Nonstandard implication algebras*
(J 0042) Mat Vesn, Drust Mat Fiz Astron Serb
3(16)(31)*403-411
⋄ G10 G25 H05 H20 ⋄ REV MR 82i:03076
Zbl 456 # 06002 • ID 54314

HERRMANN, R.A. [1979] *Point monads and p-closed spaces*
(J 0047) Notre Dame J Formal Log 20*395-400
⋄ H05 H20 ⋄ REV MR 83c:54075 Zbl 368 # 02054
• ID 52659

HERRMANN, R.A. [1980] *A nonstandard approach to
pseudotopological compactifications* (J 0068) Z Math Logik
Grundlagen Math 26*361-384
⋄ H05 ⋄ REV MR 82b:03113 Zbl 489 # 54020 • ID 74081

HERRMANN, R.A. [1981] *Rigorous infinitesimal modelling*
(J 0352) Math Jap 26*461-465
⋄ H10 ⋄ REV MR 83j:00027 Zbl 475 # 03039 • ID 55493

HERRMANN, R.A. [1985] *Supernear functions* (J 0352) Math
Jap 30*169-185
⋄ H05 ⋄ ID 47701

HERSH, R. [1975] see GREENWOOD, P.

HERSH, R. see Vol. IV, V for further entries

HERTZ, P. [1922] *Ueber Axiomensysteme fuer beliebige
Satzsysteme I* (J 0043) Math Ann 87*246-269
⋄ A05 B30 ⋄ REV FdM 48.1117 • REM Part II 1923
• ID 28771

HERTZ, P. [1923] *Ueber Axiomensysteme fuer beliebige
Satzsysteme II* (J 0043) Math Ann 89*76-102
⋄ A05 B30 ⋄ REV FdM 49.683 • REM Part I 1922
• ID 28772

HERTZ, P. [1929] *Ueber Axiomensysteme von Satzsystemen* (J 0157) Jbuchber Dtsch Math-Ver 38*45-46,2.Abt.
⋄ A05 B30 ⋄ REV FdM 55.39 • ID 06039

HERTZ, P. [1929] *Ueber Axiomensysteme fuer beliebige Satzsysteme* (J 0043) Math Ann 101*457-514
⋄ A05 B30 ⋄ REV FdM 55.627 • ID 06040

HERTZ, P. [1931] *Vom Wesen des Logischen, insbesondere der Bedeutung des modus barbara* (J 0748) Erkenntnis (Leipzig) 2*369-392
⋄ A05 B20 ⋄ REV Zbl 4.145 FdM 57.1314 • ID 40767

HETPER, W. [1934] *Semantische Arithmetik* (J 0459) C R Soc Sci Lett Varsovie Cl 3 27*9-26
⋄ B28 F30 ⋄ REV Zbl 11.1 FdM 61.977 • ID 40820

HETPER, W. [1937] *Grundlagen der Semantik (Polish)* (J 4710) Pol Tow Mat, Wiad Mat 43*57-86
⋄ B03 ⋄ REV FdM 63.827 • ID 41050

HETPER, W. [1937] *Problem of completeness of the system of elementary semantics (Polish)* (S 0281) Arch Towarz Nauk Lwow, Sect 3 17*249
⋄ B05 E70 ⋄ REV JSL 5.75 • ID 41432

HETPER, W. [1938] *Le calcul des proposition etabli sans axiomes (Polish)* (S 0281) Arch Towarz Nauk Lwow, Sect 3 10*234-240
⋄ B05 ⋄ REV Zbl 21.386 JSL 5.37 FdM 64.925 • ID 33683

HETPER, W. [1938] *Le role des schemas independants dans le systeme de la semantique elementaire (Polish) (French summary)* (S 0281) Arch Towarz Nauk Lwow, Sect 3 9*253-264
⋄ B30 ⋄ REV Zbl 19.145 JSL 4.33 FdM 64.929 • ID 41419

HETPER, W. [1938] see CHWISTEK, L.B.

HETPER, W. [1938] *Relations ancestrales dans le systeme de la semantique (Polish) (French summary)* (S 0281) Arch Towarz Nauk Lwow, Sect 3 9*265-281
⋄ B30 E07 ⋄ REV Zbl 19.145 JSL 4.34 FdM 64.929 • ID 41420

HEWITT, C. [1969] *PLANNER: A language for proving theorems in robots* (P 4250) Int Joint Conf Artif Intell (1);1969 Washington 295-301
⋄ B35 ⋄ ID 47338

HEWITT, C. [1974] see BISHOP, P.

HEYTING, A. [1958] *Intuitionism in mathematics* (C 0742) Phil Mid-Century 101-115
• TRANSL [1967] (C 2141) Filos Matematica 249-267
⋄ A05 B98 F50 F55 F98 ⋄ REV JSL 34.313 JSL 39.609 JSL 40.472 • ID 06062

HEYTING, A. [1959] *Axioms for intuitionistic plane affine geometry* (P 0651) Axiomatic Method;1957 Berkeley 160-173
⋄ B30 F55 ⋄ REV MR 22#10911 Zbl 92.250 • ID 27711

HEYTING, A. [1961] *Axiomatic method and intuitionism* (C 0622) Essays Found of Math (Fraenkel) 237-247
⋄ B30 F55 ⋄ REV MR 29#21 JSL 36.522 • ID 06065

HEYTING, A. see Vol. III, IV, VI for further entries

HIGHT, S.L. [1973] *Complex disjunctive decomposition of incompletely specified boolean functions* (J 0187) IEEE Trans Comp C-22*103-110
⋄ B05 ⋄ REV MR 58#15714 Zbl 255#02005 • ID 28953

HIGUCHI, T. & KAMEYAMA, M. [1982] *Construction of a processor exclusively for picture processing using many-valued logic (Japanese)* (P 4081) Many-Val Log & Appl;1982 Kyoto 191-206
⋄ B80 ⋄ ID 47633

HILBERT, D. [1899] *Grundlagen der Geometrie* (X 0823) Teubner: Stuttgart 92pp
• TRANSL [1899] (X 1324) Open Court: LaSalle ix+226pp
⋄ A05 B30 F99 ⋄ REV MR 43#1019 FdM 30.424 • REM 10th ed. 1968;vii+271pp • ID 23454

HILBERT, D. [1905] *Ueber die Grundlagen der Logik und der Arithmetik* (P 1091) Int Congr Math (3);1904 Heidelberg 174-185
• TRANSL [1905] (J 0152) Enseign Math 7*89-103 [1905] (J 0320) Monist 15*338-352 [1967] (C 0675) From Frege to Goedel 130-138
⋄ A05 B28 ⋄ REV FdM 36.84 • ID 16825

HILBERT, D. [1918] *Axiomatisches Denken* (J 0043) Math Ann 78*405-415
• TRANSL [1918] (J 0152) Enseign Math 20*122-136 • REPR [1935] (C 1162) Hilbert: Ges Abhandlungen 3*146-156
⋄ A05 B30 ⋄ REV Zbl 13.56 FdM 46.62 FdM 46.64 • ID 22081

HILBERT, D. [1923] *Die logischen Grundlagen der Mathematik* (J 0043) Math Ann 88*151-165
• REPR [1935] (C 1162) Hilbert: Ges Abhandlungen 3*178-191
⋄ A05 B30 ⋄ ID 22077

HILBERT, D. [1926] *Ueber das Unendliche* (J 0043) Math Ann 95*161-190
• TRANSL [1964] (C 1105) Phil of Math. Sel Readings 134-151 (English) [1967] (C 0675) From Frege to Goedel 367-392 (English) [1967] (C 2141) Filos Matematica 161-183 (Spanish) • REPR [1927] (J 0157) Jbuchber Dtsch Math-Ver 36*201-215
⋄ A05 B28 D20 E10 E30 ⋄ REV FdM 53.41 • REM Reprint is a shortened version • ID 45196

HILBERT, D. [1928] *Die Grundlagen der Mathematik* (J 0107) Abh Math Sem Univ Hamburg 6*65-85
• TRANSL [1967] (C 0675) From Frege to Goedel 464-479
⋄ A05 B30 E50 ⋄ REV FdM 54.55 FdM 54.56 • ID 06083

HILBERT, D. & ACKERMANN, W. [1928] *Grundzuege der theoretischen Logik* (X 0811) Springer: Heidelberg & New York viii+120pp
• TRANSL [1950] (X 0848) Chelsea: New York xii+172pp (English) [1950] (X 1876) Kexue Chubanshe: Beijing
⋄ B25 B98 D35 ⋄ REV MR 50#4230 Zbl 239#02001 JSL 15.59 JSL 16.52 JSL 25.158 JSL 3.83 FdM 54.55
• REM 4th ed. 1959;viii+188pp. • ID 00107

HILBERT, D. [1929] *Probleme der Grundlegung der Mathematik* (P 0741) Int Congr Math (II, 3);1928 Bologna 1*135-141
• REPR [1930] (J 0043) Math Ann 102*1-9
⋄ A05 B30 ⋄ REV FdM 55.31 FdM 55.626 • ID 20829

HILBERT, D. [1930] *Die Grundlegung der elementaren Zahlenlehre* (J 0043) Math Ann 104*485-494
• REPR [1935] (C 1162) Hilbert: Ges Abhandlungen 3*192-195
⋄ A05 B28 ⋄ REV Zbl 1.260 Zbl 13.56 FdM 57.54
• ID 22075

HILBERT, D. [1930] *Naturerkennen und Logik* (J 1859) Naturwissenschaften 18*959-963
⋄ A05 B30 ⋄ REV Zbl 1.49 FdM 56.50 • ID 42992

HILBERT, D. [1931] *Beweis des "tertium non datur"* (J 1109) Nachr Akad Wiss Goettingen, Math-Phys Kl 1931*120-125
⋄ A05 B28 F50 ⋄ REV Zbl 3.49 FdM 57.55 • ID 21186

HILBERT, D. & BERNAYS, P. [1934] *Grundlagen der Mathematik I* (X 0811) Springer: Heidelberg & New York xii+471pp
• TRANSL [1979] (X 2027) Nauka: Moskva 558pp
⋄ A05 B98 F05 F30 F98 ⋄ REV MR 81c:03002 Zbl 191.284 Zbl 478#03002 JSL 35.321 FdM 60.17 • REM 2nd edition 1968; xv+473pp. Part II 1939 • ID 01098

HILBERT, D. & BERNAYS, P. [1939] *Grundlagen der Mathematik II* (X 0811) Springer: Heidelberg & New York xii+498pp
• TRANSL [1982] (X 2027) Nauka: Moskva 556pp
⋄ A05 B98 F05 F15 F30 F40 F98 ⋄ REV MR 42#7477 Zbl 20.193 Zbl 211.9 Zbl 518#03001 JSL 5.16 FdM 65.21 • REM 2nd edition 1970; xiv+561pp. Part I 1934 • ID 01082

HILBERT, D. see Vol. IV, V, VI for further entries

HILL, S. & SUPPES, P. [1964] *First course in mathematical logic* (X 0841) Blaisdell: New York ix+274pp
⋄ B98 ⋄ REV Zbl 126.8 JSL 32.421 • REM See 1965 by Binford,F. • ID 22542

HILPINEN, R. & NIINILUOTO, I. & SAARINEN, E. (EDS.) [1977] *Foundations of probability and statistics. I,II,III* (J 0154) Synthese 36*1-176,179-281,397-498
⋄ A05 B30 ⋄ REV MR 56#11739 MR 56#9751 MR 56#9752 • ID 80458

HILPINEN, R. see Vol. II for further entries

HILTON, A.M. [1963] *Logic, computing machines, and automation* (X 1354) Spartan Books : Sutton xxi+427pp
⋄ A05 B98 D05 ⋄ REV MR 28#741 Zbl 109.101 JSL 38.341 • ID 23532

HINDLEY, J.R. & LERCHER, B. & SELDIN, J.P. [1972] *Introduction to combinatory logic* (X 0805) Cambridge Univ Pr: Cambridge, GB 179pp
• TRANSL [1975] (X 0905) Boringhieri: Torino 153pp (Italian)
⋄ B40 B98 F98 ⋄ REV MR 49#25 Zbl 269#02005 JSL 38.518 • ID 23471

HINDLEY, J.R. see Vol. IV, VI for further entries

HINES, L.M. [1980] see BLEDSOE, W.W.

HINNION, R. [1979] *Modele constructible de la theorie des ensembles de Zermelo dans la theorie des types* (J 3133) Bull Soc Math Belg, Ser B 31*3-11
⋄ B15 C62 E30 E35 E45 ⋄ REV MR 82e:03048 Zbl 439#03032 • ID 56023

HINNION, R. see Vol. III, V for further entries

HINTIKKA, K.J.J. [1953] *A new approach to sentential logic* (J 0990) Soc Sci Fennicae Comment Phys-Math 17*14pp
⋄ B05 ⋄ REV MR 22#9446 Zbl 52.10 JSL 22.361
• ID 28443

HINTIKKA, K.J.J. [1953] *Distributive normal forms in the calculus of predicates* (J 0096) Acta Philos Fenn 6*71pp
⋄ B10 C07 ⋄ REV MR 16.1079 Zbl 50.246 JSL 20.75
• ID 06111

HINTIKKA, K.J.J. [1954] *An application of logic to algebra* (J 0132) Math Scand 2*243-246
⋄ B80 G10 ⋄ REV MR 17.449 Zbl 57.23 JSL 22.216
• ID 28444

HINTIKKA, K.J.J. [1955] *Form and content in quantification theory* (J 0096) Acta Philos Fenn 8*7-55
⋄ A05 B10 ⋄ REV MR 16.1079 Zbl 67.1 JSL 22.361
• ID 06113

HINTIKKA, K.J.J. [1955] *Notes on quantification theory* (J 0990) Soc Sci Fennicae Comment Phys-Math 17*13pp
⋄ B10 ⋄ REV MR 22#4633 Zbl 67.250 JSL 22.361
• ID 28445

HINTIKKA, K.J.J. [1955] *Reductions in the theory of types* (J 0096) Acta Philos Fenn 8*57-115
⋄ B15 C85 ⋄ REV MR 17.119 Zbl 67.2 JSL 31.660
• ID 06112

HINTIKKA, K.J.J. [1956] *Identity, variables and impredicative definitions* (J 0036) J Symb Logic 21*225-245
⋄ A05 B20 F65 ⋄ REV MR 18.455 Zbl 71.11 JSL 32.258
• ID 06114

HINTIKKA, K.J.J. [1964] *Distributive normal forms and deductive interpolation* (J 0068) Z Math Logik Grundlagen Math 10*185-191
⋄ B10 C07 C40 ⋄ REV MR 34#5641 JSL 31.267
• ID 06122

HINTIKKA, K.J.J. [1965] *Distributive normal forms in first-order logic* (P 0688) Logic Colloq;1963 Oxford 48-91
• TRANSL [1980] (C 4675) Hintikka: Logiko-Epist Issled 105-157
⋄ B10 C07 ⋄ REV MR 35#2726 JSL 31.267 • ID 06123

HINTIKKA, K.J.J. [1969] see DAVIDSON, D.

HINTIKKA, K.J.J. & NIINILUOTO, I. [1973] *On the surface semantics of quantificational proof procedures* (J 0963) Ajatus (Helsinki) 35*197-215
⋄ B35 C07 F07 ⋄ REV Zbl 291#02005 • ID 32405

HINTIKKA, K.J.J. [1973] *Surface semantics: definition and its motivation* (P 0783) Truth, Syntax & Modal;1970 Philadelphia 128-147
⋄ A05 B10 B45 ⋄ REV MR 53#7717 Zbl 261#02007 JSL 42.315 • ID 44480

HINTIKKA, K.J.J. [1974] *Quantifiers vs. quantification theory* (J 0076) Dialectica 27*329-358
• REPR [1979] (C 4695) Game-Th Semantics 49-79
⋄ A05 B10 C80 ⋄ REV Zbl 362#02008 JSL 51.240
• ID 50730

HINTIKKA, K.J.J. [1977] see BUTTS, R.E.

HINTIKKA, K.J.J. [1981] see AGAZZI, E.

HINTIKKA, K.J.J. [1981] *Standard vs. nonstandard logic: higher order, modal, and first-order logics* (C 2617) Modern Log Survey 283-296
⋄ B10 B15 B45 ⋄ REV MR 82f:03002 Zbl 464#03001
• ID 42772

HINTIKKA, K.J.J. [1982] *A dialogical model of teaching* (J 0154) Synthese 51*39-59
⋄ B80 ⋄ REV MR 84h:03057 • ID 34255

HINTIKKA, K.J.J. & HINTIKKA, M.P. [1982] *Sherlock Holmes confronts modern logic: toward a theory of information-seeking through questioning* (P 3754) Argumentation;1978 Groningen 55-76
⋄ A05 H05 ⋄ REV MR 84c:03052 • ID 34936

HINTIKKA, K.J.J. see Vol. II, III, V for further entries

HINTIKKA, M.P. [1982] see HINTIKKA, K.J.J.

HINTIKKA, M.P. see Vol. II for further entries

HIRANO, J. [1937] *Zum Zerlegungssatz im erweiterten einstelligen Praedikatenkalkuel* (J 0428) Proc Phys-Math Soc Japan 19*395-412
⋄ B20 ⋄ REV Zbl 16.337 JSL 3.91 FdM 63.825 • ID 06133

HIRANO, J. see Vol. V for further entries

HIRANO, T. [1934] *Die kontradiktorische Logik* (J 1124) Ergebn Math Kolloquium 7*6-7
⋄ B20 ⋄ REV Zbl 14.3 FdM 62.41 • ID 40862

HIRATA, M. [1984] see DOSHITA, S.

HIRSCHBERG, D. [1963] see BRAFFORT, P.

HIRSCHFELD, J. & MACHOVER, M. [1969] *Lectures on non-standard analysis* (X 0811) Springer: Heidelberg & New York 79pp
⋄ B98 H05 ⋄ REV MR 40 # 2531 Zbl 182.559 • ID 08459

HIRSCHFELD, J. [1972] see CHERLIN, G.L.

HIRSCHFELD, J. [1976] *Non standard analysis and the compactification of groups* (J 0029) Israel J Math 25*145-153
⋄ C60 H05 H20 ⋄ REV MR 58 # 22374 Zbl 348 # 22002 • ID 26079

HIRSCHFELD, J. see Vol. III, IV, V, VI for further entries

HIRSCHFELDER, J.J. [1974] *Nonstandard analysis in a nutshell* (C 1217) Value Distr Th Compl Anal & Rel Topics Diff Geom 13-27
⋄ H05 ⋄ REV MR 50 # 1875 Zbl 302 # 02020 • ID 21266

HIRSCHHORN, E. [1958] *Simplification of a class of boolean functions* (J 0037) ACM J 5*67-75
⋄ B05 ⋄ REV MR 20 # 452 Zbl 86.10 JSL 23.236 • ID 06143

HIRST, K.E. & RHODES, F. [1971] *Conceptual models in mathematics* (X 0959) Allen & Unwin: London xii+182pp
⋄ B30 ⋄ REV MR 50 # 12497 • ID 24219

HISCHER, Horst & LUCHT, L. [1976] *Zum Verstaendnis des Induktionsaxioms* (J 0160) Math-Phys Sem-ber, NS 23*228-236
⋄ B28 F30 ⋄ REV MR 54 # 9964 • ID 25590

HIZ, H. [1946] *Remarques sur le degre de completude* (J 0109) C R Acad Sci, Paris 223*973-974
⋄ B22 F50 ⋄ REV MR 8.245 Zbl 60.21 JSL 12.57 • ID 06149

HIZ, H. [1957] *Complete sentential calculus admitting extensions* (P 1675) Summer Inst Symb Log;1957 Ithaca 260-262
⋄ B22 ⋄ ID 06150

HIZ, H. [1957] *Inferential equivalence and natural deduction* (J 0036) J Symb Logic 22*237-240
⋄ A05 B05 ⋄ REV MR 20 # 3064 Zbl 81.242 JSL 35.325 • ID 06151

HIZ, H. [1959] *Extendible sentential calculus* (J 0036) J Symb Logic 24*193-202
⋄ B22 ⋄ REV Zbl 96.242 JSL 25.299 • ID 06153

HIZ, H. [1973] *A completeness proof for C-calculus* (J 0047) Notre Dame J Formal Log 14*253-258 • ERR/ADD ibid 17*640
⋄ B20 ⋄ REV MR 47 # 8235 MR 53 # 12864 Zbl 232 # 02010 • ID 06157

HIZ, H. see Vol. IV for further entries

HOAK, D. [1975] *An all-rule axiomatization of the propositional calculus and the equivalence of some well-known axiomatizations* (J 0243) Pi Mu Epsilon J 6*59-68
⋄ B05 ⋄ REV MR 51 # 7813 Zbl 335 # 02005 • ID 17313

HOARE, C.A.R. [1984] see ATIYAH, M.

HOARE, C.A.R. [1985] *A couple of novelties in the propositional calculus* (J 0068) Z Math Logik Grundlagen Math 31*173-178
⋄ B22 ⋄ REV Zbl 547 # 03003 • ID 42295

HOARE, C.A.R. see Vol. IV for further entries

HOCKNEY, R. [1962] *An intersection algorithm giving all irredundant normal forms from a prime implicant list* (J 0072) IRE Trans Electr Comp EC-11*289-290
⋄ B05 B35 ⋄ REV Zbl 123.334 • ID 06166

HODES, L. & SPECKER, E. [1968] *Lengths of formulas and elimination of quantifiers I* (P 0608) Logic Colloq;1966 Hannover 175-188
• TRANSL [1973] (J 3079) Kiber Sb Perevodov, NS 10*99-113
⋄ B35 C10 C40 ⋄ REV MR 39 # 32 MR 48 # 5714 Zbl 191.285 JSL 40.503 • ID 24939

HODES, L. [1970] *The logical complexity of geometric properties on the plane* (J 0037) ACM J 17*339-347
⋄ B35 C40 ⋄ REV MR 45 # 8527 Zbl 198.35 • ID 06168

HODGES, W. (ED.) [1972] *Conference in Mathematical Logic -- London '70* (S 3301) Lect Notes Math 255*viii+351pp
⋄ B97 C97 ⋄ REV MR 48 # 5804 Zbl 227 # 02011 • ID 70226

HODGES, W. [1977] *Logic* (X 0868) Penguin Books: Harmondsworth & New York 331pp
⋄ B98 ⋄ REV JSL 45.382 • ID 30709

HODGES, W. [1983] *Elementary predicate logic* (C 4085) Handb Philos Log 1*1-131
⋄ B10 C98 ⋄ REV Zbl 538 # 03001 • ID 39725

HODGES, W. see Vol. III, IV, V, VI for further entries

HOEEGH-KROHN, R. [1979] see ALBEVERIO, S.

HOEEGH-KROHN, R. [1982] see ALBEVERIO, S.

HOEEGH-KROHN, R. [1984] see ALBEVERIO, S.

HOELDER, O. [1900] *Anschauung und Denken in der Geometrie (Antrittsvorlesung)* (X 0823) Teubner: Stuttgart 75pp
⋄ B30 ⋄ REV FdM 31.467 FdM 31.67 • ID 25542

HOELDER, O. [1914] *Die Arithmetik in strenger Begruendung* (X 1079) Teubner: Leipzig 74pp
• LAST ED [1929] (X 0811) Springer: Heidelberg & New York 74pp
⋄ B28 ⋄ ID 25543

HOELDER, O. [1924] *Die mathematische Methode. Logisch-erkenntnistheoretische Untersuchungen im Gebiet der Mathematik, Mechanik und Physik* (X 0811) Springer: Heidelberg & New York x+563pp
⋄ A05 B30 ⋄ REV FdM 50.22 • ID 25544

HOELDER, O. see Vol. V for further entries

HOENEN, P. [1947] *Recherches de logique formelle. La structure du systeme des syllogismes et des sorites - la logique des notions "au moins" et "tout au plus"* (X 2245) Gregorian Univ Pr: Vatican City vii+384pp
⋄ B20 ⋄ REV JSL 19.302 • ID 42410

HOERNES, G.E. [1966] see HEILWEIL, M.F.

HOERNIG, K.M. [1982] see BIBEL, W.

HOERNIG, K.M. see Vol. IV for further entries

HOFF-HANSEN, E. [1943] *A mathematical Interpretation of the classical propositional calculus (Norwegian)* (J 4510) Norsk Mat Tidsskr 25*6-12
⋄ B05 ⋄ REV MR 8.125 Zbl 27.385 JSL 13.169 • ID 28461

HOFFMANN, GEERD-RUEDIGER & VEENKER, G. [1971] *The unit-clause proof procedure with equality (German summary)* (J 0373) Comp Arch Inform & Numerik 7*91-105
⋄ B35 ⋄ REV MR 44 # 5227 Zbl 219 # 68049 • ID 06186

HOFSTADTER, D.R. [1979] *Goedel, Escher, Bach: an eternal golden braid* (X 2671) Basic Books: New York xxi+777pp
⋄ A05 B98 D99 ⋄ REV MR 80j:03009 Zbl 457 # 03001 JSL 48.864 • ID 74228

HOLZER, S. [1983] see CRISMA, L.

HONG, SEJUNE & OSTAPKO, D.L. [1974] *Generating test examples for heuristic boolean minimization* (J 0284) IBM J Res Dev 18*459-464
⋄ B35 ⋄ REV MR 55 # 14432 Zbl 289 # 02010 • ID 29980

HONG, SEJUNE [1974] see CAIN, R.G.

HONG, SEJUNE see Vol. II for further entries

HOOK, J.L. [1985] *A note on interpretations of many-sorted theories* (J 0036) J Symb Logic 50*372-374
⋄ B10 C07 F25 ⋄ REV Zbl 571 # 03026 • ID 41804

HOOKER, C.A. [1972] *Definite descriptions* (J 0095) Philos Stud 23*365-375
⋄ A05 B10 ⋄ REV MR 55 # 7722 • ID 74248

HOOKER, C.A. & LEACH, J.J. & MCCLENNEN, E. (EDS.) [1978] *Foundations and applications of decision theory. Papers resulting from a workshop* (S 3308) Univ Western Ontario Ser in Philos of Sci 13*xxiii+208pp
⋄ B80 ⋄ REV MR 81a:90002b Zbl 398 # 90003 • ID 52793

HOOKER, C.A. see Vol. II for further entries

HOOVER, D.N. & PERKINS, E. [1983] *Nonstandard construction of the stochastic integral and applications to stochastic differential equations I,II* (J 0064) Trans Amer Math Soc 275*1-36,37-58
⋄ H05 ⋄ REV MR 85d:60111 Zbl 533 # 60063 • ID 33149

HOOVER, D.N. & KEISLER, H.J. [1984] *Adapted probability distributions* (J 0064) Trans Amer Math Soc 286*159-201
⋄ C50 C65 H10 ⋄ REV Zbl 548 # 60019 • ID 43234

HOOVER, D.N. see Vol. II, III, IV for further entries

HORECKY, J. [1982] *COLING 82. Proceedings of the Ninth International Conference on Computational Linguistics, Prague, July 5-10, 1982* (X 0809) North Holland: Amsterdam 432pp
⋄ B97 ⋄ REV Zbl 529 # 68041 • ID 38490

HORN, A. [1974] see EPSTEIN, G.

HORN, A. see Vol. II, III, V, VI for further entries

HORN VAN, C.E. [1916] *An axiom in symbolic logic* (J 0171) Proc Cambridge Phil Soc Math Phys 19*22-31
⋄ B05 ⋄ ID 06234

HORNUNG, G. [1983] see BENDA, W.

HOROVITZ, J. [1967] *La logique et le droit* (J 0079) Logique & Anal, NS 10*43-56
⋄ B80 ⋄ ID 06235

HOROVITZ, J. [1972] *Law and logic. A critical account of legal argument* (X 0902) Springer: Wien xvi+214pp
⋄ B80 ⋄ REV JSL 39.619 • ID 06236

HORT, C. & OSSWALD, H. [1984] *On nonstandard models in higher order logic* (J 0036) J Symb Logic 49*204-219
⋄ B15 C55 C85 E55 H05 H20 ⋄ REV MR 85i:03122 • ID 44206

HORTALA-GONZALEZ, M.T. & RODRIGUEZ ARTALEJO, M. [1985] *Hoare's logic for nondeterministic regular programs: a nonstandard completeness theorem* (P 4628) Automata, Lang & Progr (12);1985 Nafplion 270-280
⋄ B75 H10 ⋄ ID 49497

HOSKINS, R.F. [1982] *Infinitesimals, nonstandard analysis and generalised functions* (J 2679) Bull Inst Math Appl (Southend oS) 18*49-51
⋄ H05 ⋄ REV MR 84j:03139d Zbl 514 # 26010 • ID 34732

HOSKINS, R.F. [1982] *Superreals and superfunctions* (J 2734) Int J Math Educ Sci Technol 13*67-75
⋄ H05 ⋄ REV MR 83d:26012 Zbl 499 # 26010 • ID 38300

HOSOI, T. [1966] *Algebraic proof of the separation theorem on classical propositional calculus* (J 0081) Proc Japan Acad 42*67-69
⋄ B20 ⋄ REV MR 34 # 2436 Zbl 154.253 • ID 37164

HOSOI, T. [1966] *The separation theorem on the classical system* (J 0434) J Fac Sci Univ Tokyo, Sect 1 12*223-230
⋄ B10 F50 ⋄ REV MR 33 # 5484 Zbl 146.246 JSL 33.128 • ID 06242

HOSOI, T. see Vol. II for further entries

HOSSAIN, M.F. [1983] *On Beth trees in mathematical logic (Bengali summary)* (J 2525) Chittagong Univ Stud Part II Sci 7*113-118
⋄ B10 ⋄ REV MR 85g:03017 • ID 43489

HOTOMSKI, P. [1975] *A certain method of finding the values of formulas of the propositional calculus (Russian)* (J 0042) Mat Vesn, Drust Mat Fiz Astron Serb 12(27)*341-345
⋄ B05 ⋄ REV MR 53 # 5247 Zbl 328 # 02008 JSL 47.440 • ID 22877

HOTOMSKI, P. [1982] *An induction law in proofs by contradiction with an application to automatic theorem proving (Russian)* (J 0400) Publ Inst Math, NS (Belgrade) 31(45)*51-63
⋄ B35 F30 ⋄ REV MR 84k:03049 Zbl 521 # 03007 • ID 37063

HOTOMSKI, P. [1982] *Some characteristics of tables of Boolean functions, and the construction of expressions in Sheffer operators (Russian)* (**P 3758**) Algeb Conf (2);1981 Novi Sad 83-86
⬦ B05 B70 ⬦ REV MR 84b:94036 • ID 38981

HOTOMSKI, P. [1983] *A way of incorporating an induction rule in an automatic theorem-proving procedure with resolution (Russian)* (**J 0400**) Publ Inst Math, NS (Belgrade) 33(47)*89-95
⬦ B35 ⬦ REV MR 85a:03019 Zbl 515#68066 • ID 34770

HOTOMSKI, P. [1985] *An automatic theorem-proving system with resolution, induction and symmetry (Russian)* (**P 4661**) Algeb & Log;1984 Zagreb 55-61
⬦ B35 ⬦ ID 49008

HOUSE, R.W. & RADO, T. [1965] *A generalization of Nelson's algorithm for obtaining prime implicants* (**J 0036**) J Symb Logic 30*8-12
⬦ B35 ⬦ REV MR 33#1223 Zbl 163.257 JSL 32.265 • ID 06257

HOWSON, C. [1976] *The development of logical probability* (**C 2103**) Develop of Log Probab (Lakatos) 277-298
⬦ A05 B30 B48 ⬦ REV Zbl 346#02004 • ID 62614

HOWSON, C. see Vol. II for further entries

HRBACEK, K. [1978] *Axiomatic foundations for nonstandard analysis* (**J 0027**) Fund Math 98*1-19
⬦ E35 E70 H05 H20 ⬦ REV MR 84b:03084 Zbl 373#02039 • ID 29206

HRBACEK, K. [1979] *Nonstandard set theory* (**J 0005**) Amer Math Mon 86*659-677
⬦ E70 H05 H20 ⬦ REV MR 81c:03055 Zbl 444#03038 • ID 74287

HRBACEK, K. see Vol. III, IV, V for further entries

HSIANG, JIEH [1983] see DERSHOWITZ, N.

HSIANG, JIEH & SRIVAS, M. [1985] *PROLOG-based inductive theorem proving* (**P 4672**) Found of Softw Tech & Th Comput Sci (5);1985 New Delhi 129-149
⬦ B35 ⬦ ID 49785

HSIANG, JIEH [1985] *Two results in term rewriting theorem proving* (**P 4244**) Rewriting Techn & Appl (1);1985 Dijon 301-324
⬦ B35 ⬦ ID 49702

HU, JIUQING [1981] see HE, YANXIANG

HU, SHIGENG [1980] *The general form of separation axiom (Chinese) (English summary)* (**J 2754**) Huazhong Gongxueyuan Xuebao 8/4*1-6
⬦ H05 ⬦ REV MR 83h:54029 • ID 39263

HU, SHIHUA [1957] *Mathematical logic, its fundamental properties and scientific significance (Chinese)* (**J 4452**) Zhexue Yanjiu 1*1-45
⬦ B98 ⬦ ID 48510

HU, SHIHUA [1964] *Classic predicate calculus (Chinese)* (**J 0420**) Shuxue Jinzhan 7/4*349-396
⬦ B10 ⬦ ID 48512

HU, SHIHUA & LU, ZHONGWAN [1981] *Foundation of mathematical logic I (Chinese)* (**X 1876**) Kexue Chubanshe: Beijing
⬦ B98 ⬦ REM Vol.II 1982 • ID 48514

HU, SHIHUA & LU, ZHONGWAN [1982] *Foundation of mathematical logic II (Chinese)* (**X 1876**) Kexue Chubanshe: Beijing
⬦ B98 ⬦ REM Vol.I 1981 • ID 48515

HU, SHIHUA see Vol. II, IV, VI for further entries

HU, SZETIEN [1965] *Threshold logic* (**X 0926**) Univ Calif Pr: Berkeley xiv+338pp
⬦ B70 B98 ⬦ REV MR 36#2435 JSL 40.250 • ID 22214

HUANG, CHENGGUI [1979] *Two-phase calculus (Chinese) (English summary)* (**J 2754**) Huazhong Gongxueyuan Xuebao 7/4*8,17-32
⬦ H05 H10 ⬦ REV MR 82f:03063 • ID 74292

HUANG, CHENGGUI & SHI, ZUIJIAN [1980] *The problem of singularity of the δ-function (Chinese English summary)* (**J 2754**) Huazhong Gongxueyuan Xuebao 8/2*9-18 • TRANSL [1980] (**J 2684**) J Huazhong Inst Tech (Engl Ed)
⬦ H05 ⬦ REV MR 83j:46051 • ID 40015

HUANG, QIEYUAN [1985] *Type structures (Chinese)* (**J 0418**) Shuxue Xuebao 1/2*161-176
⬦ B15 ⬦ ID 48685

HUBIEN, H. [1977] *A new basis for classical propositional calculus* (**J 0079**) Logique & Anal, NS 20*225-228
⬦ B05 ⬦ REV MR 57#2871 Zbl 393#03003 • ID 52423

HUBIEN, H. see Vol. II for further entries

HUET, G. [1973] *The undecidability of unification on third order logic* (**J 0194**) Inform & Control 22*257-267
⬦ B15 F35 ⬦ REV MR 54#2442 Zbl 257#02038 • ID 06291

HUET, G. [1985] see COQUAND, T.

HUET, G. see Vol. III, IV, VI for further entries

HUGHES, C.E. [1976] *A reduction class containing formulas with one monadic predicate and one binary function symbol* (**J 0036**) J Symb Logic 41*45-49
⬦ B20 D35 ⬦ REV MR 54#4955 Zbl 332#02051 • ID 14752

HUGHES, C.E. [1976] *Two variable implicational calculi of prescribed many-one degrees of unsolvability* (**J 0036**) J Symb Logic 41*39-44
⬦ B20 D25 ⬦ REV MR 53#12911 Zbl 339#02044 • ID 14753

HUGHES, C.E. see Vol. II, IV for further entries

HUGHES, G.E. & LONDEY, D.G. [1965] *The elements of formal logic* (**X 0816**) Methuen: London & New York xii+398pp
⬦ B98 ⬦ ID 22495

HUGHES, G.E. [1968] see CRESSWELL, M.J.

HUGHES, G.E. see Vol. II for further entries

HUGLY, P. & SAYWARD, C.W. [1976] *Prior on propositional identity* (**J 0103**) Analysis (Oxford) 36*182-184
⬦ A05 B05 ⬦ REV Zbl 361#02006 • ID 50641

HUGLY, P. & SAYWARD, C.W. [1979] *A semantical account of the vicious circle principle* (**J 0047**) Notre Dame J Formal Log 20*595-598
⬦ A05 B20 ⬦ REV MR 80f:03018 Zbl 363#02028 • ID 56135

HUGLY, P. [1980] *Reflections on an extensionality theorem* (**J 0047**) Notre Dame J Formal Log 21*45-50
⬦ B10 ⬦ REV MR 81a:03006 Zbl 363#02012 • ID 53177

HUGLY, P. & SAYWARD, C.W. [1982] *Indenumerability and substitutional quantification* (J 0047) Notre Dame J Formal Log 23*358-366
⋄ A05 B10 C07 ⋄ REV MR 83k:03016 Zbl 452 # 03004 • ID 54068

HUGLY, P. see Vol. II, III for further entries

HULLOT, J.-M. [1980] *Canonical forms and unification* (P 3063) Autom Deduct (5);1980 Les Arcs 318-334
⋄ B35 ⋄ REV MR 82m:68067 Zbl 441 # 68108 • ID 81610

HUNT, E.B. & QUINLAND, J.R. [1968] *A formal deductive problem-solving system* (J 0037) ACM J 15*625-646
⋄ B35 ⋄ REV Zbl 246 # 68015 • ID 46592

HUNTER, G. [1971] *Metalogic: an introduction to the metatheory of standard firstorder logic* (X 0926) Univ Calif Pr: Berkeley xiii + 288pp
⋄ B10 B98 ⋄ REV MR 56 # 5186 Zbl 284 # 02001 • ID 22356

HUNTINGTON, E.V. [1902] *A complete set of postulates for the theory of absolute continuous magnitude* (J 0064) Trans Amer Math Soc 3*264-279
⋄ B28 ⋄ REV FdM 33.297 • ID 06327

HUNTINGTON, E.V. [1904] *Sets of independent postulates for the algebra of logic* (J 0064) Trans Amer Math Soc 5*288-309
⋄ B05 B30 G05 ⋄ REV FdM 35.87 • ID 06328

HUNTINGTON, E.V. [1905] *The continuum as a type of order: an exposition of the modern theory* (J 0120) Ann of Math, Ser 2 6*151-184,7*15-43
• TRANSL [1907] (X 0834) Gauthier-Villars: Paris x + 126pp (Esperanto) • LAST ED [1955] (X 0813) Dover: New York vii + 82pp • REPR [1905] (X 0858) Harvard Univ Pr: Cambridge 63pp
⋄ B28 E07 E10 E75 ⋄ REV FdM 36.98 FdM 46.1451
• REM The last edition has the title: The continuum, and other types of serial orders • ID 37876

HUNTINGTON, E.V. [1913] *A set of postulates for abstract geometry, expressed in terms of the simple relation of inclusion* (P 1646) Int Congr Math (5);1912 Cambridge GB 2*466-470
⋄ B30 ⋄ REV FdM 44.544 FdM 44.82 • ID 29486

HUNTINGTON, E.V. [1917] *Complete existential theory of the postulates for serial order* (J 0015) Bull Amer Math Soc 23*276-280
⋄ B30 C13 C35 E07 ⋄ REV FdM 46.308 • ID 41774

HUNTINGTON, E.V. [1917] *Complete existential theory of the postulates for well ordered sets* (J 0015) Bull Amer Math Soc 23*280-282
⋄ B30 C13 C35 E07 ⋄ REV FdM 46.308 • ID 41776

HUNTINGTON, E.V. [1934] *Independent postulates for the "informal" part of "Principia Mathematica"* (J 0015) Bull Amer Math Soc 40*127-136
⋄ A05 B30 ⋄ REV Zbl 9.3 FdM 60.23 • ID 06333

HUNTINGTON, E.V. [1934] *Independent postulates for an "informal Principia System with equality"* (J 0015) Bull Amer Math Soc 40*137-143
⋄ A05 B10 ⋄ REV Zbl 9.3 FdM 60.24 • ID 06334

HUNTINGTON, E.V. [1934] *The postulational method in mathematics* (J 0005) Amer Math Mon 41*84-92
⋄ A05 B30 ⋄ REV Zbl 9.3 FdM 60.23 • ID 06332

HUNTINGTON, E.V. [1935] *The inter-deducibility of the new Hilbert-Bernays theory and "Principia Mathematica"* (J 0120) Ann of Math, Ser 2 36*313-324
⋄ B30 E30 ⋄ REV Zbl 12.1 FdM 61.53 • ID 06337

HUNTINGTON, E.V. [1937] *Postulates for assertion, conjunction, negation, and equality* (J 0251) Proc Amer Acad Arts Sci 72*1-44
⋄ A10 B20 ⋄ REV Zbl 17.145 JSL 2.91 FdM 63.821 • ID 06341

HUNTINGTON, E.V. [1939] *Note on a recent set of postulates for the calculus of propositions* (J 0036) J Symb Logic 4*10-14
⋄ B05 ⋄ REV Zbl 20.194 JSL 4.90 FdM 65.27 • ID 06342

HUNTINGTON, E.V. see Vol. II, VI for further entries

HURD, A.E. & ROBINSON, A. [1968] *On flexural wave propagation on nonhomogeneous elastic plates* (J 0374) SIAM J Appl Math 16*1081-1089
⋄ H10 ⋄ REV Zbl 188.583 • ID 26141

HURD, A.E. [1971] *Local conditions for equivalence of compact dynamical systems* (J 0100) Amer J Math 93*742-752
⋄ H10 ⋄ REV MR 46 # 5750 Zbl 227 # 54039 • ID 06343

HURD, A.E. [1971] *Nonstandard analysis of dynamical systems. I: Limit motions, stability* (J 0064) Trans Amer Math Soc 160*1-26
⋄ H10 ⋄ REV MR 43 # 6546 Zbl 251 # 34032 • ID 26636

HURD, A.E. [1974] *Near periods and Bohr compactifications* (P 1083) Victoria Symp Nonstand Anal;1972 Victoria 106-112
⋄ H05 ⋄ REV MR 58 # 12212 Zbl 278 # 43014 • ID 26638

HURD, A.E. [1974] *Nonstandard dynamical systems* (P 1083) Victoria Symp Nonstand Anal;1972 Victoria xi
⋄ H10 ⋄ ID 26637

HURD, A.E. [1981] *Nonstandard analysis and lattice statistical mechanics: A variational principle* (J 0064) Trans Amer Math Soc 263*89-110
⋄ H10 ⋄ REV MR 84a:82033 Zbl 457 # 03065 • ID 54390

HURD, A.E. (ED.) [1983] *Nonstandard analysis - recent developments* (S 3301) Lect Notes Math 983*iv + 213pp
⋄ H05 ⋄ REV MR 84c:03005 Zbl 497 # 00011 • REM Incl. papers pres. at the 2nd Vict. Symp. on nonstandard analysis at the Univ. of Victoria,B.C.,'80 • ID 35672

HURD, A.E. & LOEB, P.A. [1985] *An introduction to nonstandard real analysis* (X 0801) Academic Pr: New York xii + 232pp
⋄ H05 ⋄ ID 48930

HURD, A.E. see Vol. V for further entries

HURDLE JR., B.G.N. [1977] see ARCHIE, L.C.

HURST, S.L. [1968] *An extension of binary minimisation techniques to ternary equations* (J 1193) Comput J (London) 11*277-286
⋄ B35 ⋄ REV Zbl 169.17 • ID 41935

HURST, S.L. [1984] *The relationship between the self-dualized classification of Boolean functions and spectral coefficient classification* (J 0379) Int J Electron 56*801-808
⋄ B05 ⋄ REV MR 86d:94038 • ID 44282

HUSAIN, M. [1982] *"If...then" and Aristotle's syllogism* (J 0286) Int Logic Rev 25*30-35
⋄ A10 B20 ⋄ REV MR 85j:03010 • ID 45826

HUSSMANN, H. [1985] *Unification in conditional-equational theories* (P 4601) EUROCAL;1985 Linz 2∗543-553
⋄ B35 ⋄ ID 49751

HUSSON, L. [1973] *L'infrastructure du raisonnement juridique* (J 0079) Logique & Anal, NS 16∗3-20
⋄ B80 ⋄ REV MR 48 # 5826 • ID 47900

HUTCHINS, G.M. & MOORE, G.W. [1981] *A Hintikka possible worlds model for certainty levels in medical decision making* (J 0154) Synthese 48∗87-119
⋄ B80 ⋄ REV MR 83d:03021 Zbl 469 # 92004 • ID 55192

HUZINO, S. [1959] *On the existence of Sheffer stroke class in the sequential machines* (J 0106) Mem Fac Sci, Kyushu Univ, Ser A 13∗53-68
⋄ B05 D05 ⋄ REV MR 21 # 4917 Zbl 114.331 • ID 06355

HUZINO, S. see Vol. IV for further entries

HYLAND, J.M.E. [1977] see GANDY, R.O.

HYLAND, J.M.E. see Vol. III, IV, V, VI for further entries

HYLTON, P. [1980] *Russell's substitutional theory* (J 0154) Synthese 45∗1-31
⋄ A05 A10 B10 ⋄ REV MR 82k:03006 Zbl 517 # 01019 • ID 74355

HYON, JONGRAK [1982] *A graph-theoretic method in the algebra of logic I (Korean) (English summary)* (J 2602) Su-Hak Kwa Mul-li, People's Rep Korea 1982/4∗10-13
⋄ B05 ⋄ REV MR 84g:94031 • ID 39454

IANNACCI, R. [1976] *Un teorema relativo alla estensione di Henkin del calcolo predicativo (English summary)* (J 2311) Rend Mat, Ser 6 9∗337-346
⋄ B10 ⋄ REV MR 54 # 2432 Zbl 343 # 02008 • ID 24022

IBRAHIM, R.L.R. [1981] *A computer system for axiomatic investigation* (J 1741) Int J Man-Mach Stud 15∗179-200
⋄ B35 ⋄ REV MR 82j:68074 Zbl 459 # 68052 • ID 81631

IBSEN, K. [1971] *Equivalent propositions (Danish) (English summary)* (J 0311) Nordisk Mat Tidskr 19∗77-80,108
⋄ B05 ⋄ REV MR 48 # 53 Zbl 222 # 02005 • ID 06367

IBUKI, K. & NAEMURA, K. & NOZAKI, A. [1963] *General theory of complete sets of logical functions* (J 0116) Electr & Comm Japan 46/7∗55-65
⋄ B05 B70 ⋄ REV JSL 37.416 • ID 06368

IDZIAK, P.M. [1983] *A finite base for the consequence operation determined by the ring Z_p in the language without constants* (J 0387) Bull Sect Logic, Pol Acad Sci 12∗76-82
⋄ B22 ⋄ REV MR 84i:03053 Zbl 535 # 03033 • ID 34532

IGARASHI, Y. [1980] *The size of arrays for a prime implicant generating algorithm* (J 1193) Comput J (London) 23∗73-77
⋄ B35 ⋄ REV MR 80m:68031 Zbl 427 # 68060 • ID 53758

IGARASHI, Y. see Vol. IV for further entries

IMAI, Y. & ISEKI, K. [1965] *On axiom systems of propositional calculi. I* (J 0081) Proc Japan Acad 41∗436-439
⋄ B20 ⋄ REV MR 32 # 2313 Zbl 223 # 02007 JSL 34.122 • REM Part II 1965 by Arai,Y. • ID 06426

IMAI, Y. & ISEKI, K. [1966] *On axiom systems of propositional calculi. XIV* (J 0081) Proc Japan Acad 42∗19-22
⋄ B20 ⋄ REV MR 33 # 3902 Zbl 156.248 • REM Part XIII 1965 by Tanaka,S. Part XV 1966 by Iseki,K. • ID 06380

IMAI, Y. see Vol. VI for further entries

IMIELINSKI, T. [1979] *Functional dependencies and boolean forms. Deeper examination of connection* (J 1927) Prace Inst Podstaw Inf, Pol Akad Nauk 381∗19pp
⋄ B05 ⋄ REV Zbl 417 # 03033 • ID 53269

INGLETON, A.W. [1982] *An introduction to nonstandard analysis* (J 2679) Bull Inst Math Appl (Southend oS) 18∗34-37
⋄ H05 ⋄ REV MR 84j:03139a Zbl 514 # 26009 • ID 34729

INOUE, K. & NAKAMURA, A. [1975] *On the expressive power of logical metalanguages I^n and I_+* (S 4568) Hiroshima Univ Fac Engin, Mem 6/1∗1-8
⋄ B15 F35 ⋄ REV MR 53 # 2041 • ID 48761

INOUE, K. see Vol. IV for further entries

IRVIN, A.A. [1970] *Connective implication (Russian)* (C 1530) Issl Log Sist (Yanovskaya) 142-165
⋄ B20 ⋄ REV MR 48 # 3697 Zbl 217.7 • ID 06401

ISAACS, G.L. [1968] *Real numbers: a development of the real numbers in an axiomatic set theory* (X 0822) McGraw-Hill: New York viii+112pp
⋄ B28 E75 ⋄ REV MR 37 # 6138 Zbl 159.343 • ID 23369

ISAACS, R. [1978] *An example of a first-order language* (J 0377) Bol Mat (Bogota) 12∗55-74
⋄ B10 ⋄ REV Zbl 414 # 03005 • ID 53051

ISEKI, K. [1965] *Algebraic formulations of propositional calculi* (J 0081) Proc Japan Acad 41∗803-807
⋄ B20 G05 G25 ⋄ REV MR 33 # 5457a Zbl 143.6 JSL 33.625 • ID 06433

ISEKI, K. [1965] see IMAI, Y.

ISEKI, K. [1965] *On axiom systems of propositional calculi. IV* (J 0081) Proc Japan Acad 41∗575-577
⋄ B05 ⋄ REV MR 33 # 5453 JSL 34.122 • REM Part III 1965 by Arai,Y. Part V 1965 by Iseki,K. & Tanaka,S. • ID 06427

ISEKI, K. [1965] see ARAI, Y.

ISEKI, K. & TANAKA, S. [1965] *On axiom systems of propositional calculi. V,X* (J 0081) Proc Japan Acad 41∗661-662,801-802
⋄ B05 ⋄ REV MR 33 # 3898 MR 33 # 5454 Zbl 156.248 JSL 34.122 • REM Part IV 1965 by Iseki,K. Parts VI,IX 1965 by Tanaka,S. Part XI 1965 by Tanaka,K. • ID 42973

ISEKI, K. [1966] *A characterization of the NB-system* (J 0081) Proc Japan Acad 42∗871-872
⋄ B20 ⋄ REV MR 35 # 1441 Zbl 192.27 • ID 06440

ISEKI, K. [1966] *Algebraic formulations of propositional calculi with variable forming functors* (J 0081) Proc Japan Acad 42∗1058-1059
⋄ B20 ⋄ REV MR 35 # 4085 Zbl 202.294 • ID 06441

ISEKI, K. [1966] *An algebraic formulation of $K - N$ propositional calculus* (J 0081) Proc Japan Acad 42∗1164-1167
⋄ B05 ⋄ REV MR 35 # 2719 Zbl 166.249 • REM Part II 1967 by Tanaka,S. • ID 06442

ISEKI, K. [1966] see IMAI, Y.

ISEKI, K. [1966] *On axiom systems of propositional calculi. XV,XXI,XXIII,XXIV* (J 0081) Proc Japan Acad 42∗217-220,42∗875-877,42∗878-879,42∗441-442
⋄ B05 ⋄ REV MR 33 # 3903 MR 34 # 2432a Zbl 156.248 Zbl 156.249 JSL 38.521 • REM Part XIV 1966 by Imai,Y. & Iseki,K. Parts XVI,XX,XXII 1966, Part XXV 1967 by Tanaka,S. • ID 06436

ISEKI, K. [1967] *Algebraic formulation of propositional calculi with general detachment rule* (J 0081) Proc Japan Acad 43*31-34
⋄ B20 G25 ⋄ REV MR 35 # 2721 Zbl 174.8 JSL 35.465
• ID 06444

ISEKI, K. [1967] *Axiom systems of aristotle traditional logic* (J 0081) Proc Japan Acad 43*125-128
⋄ B20 ⋄ REV Zbl 164.306 • REM Part I. Part II 1967 by Tanaka,S. • ID 20904

ISEKI, K. [1968] *Symbolic logic (propositional logic) (Japanese)* (X 3552) Iwanami Shoten: Tokyo 303pp
⋄ B05 ⋄ REV Zbl 194.306 JSL 35.580 • REM Part II 1973
• ID 41937

ISEKI, K. [1973] *Symbolic logic. II: Predicate logic (Japanese)* (X 2327) Maki Shoten: Tokyo 1-3,305-568
⋄ B10 ⋄ REV Zbl 265 # 02001 • REM Part I 1968 • ID 29813

ISEKI, K. see Vol. II, V, VI for further entries

ISEMINGER, G. [1968] *An introduction to deductive logic* (X 1228) Appleton-Century-Crofts: New York vii + 184pp
⋄ B98 ⋄ ID 22497

ISEMINGER, G. see Vol. II for further entries

ISHIBASHI, T. [1980] see DOSHITA, S.

ISHIKAWA, S. [1977] *The weak law of large numbers by counting probability* (J 2770) Keio Math Sem Rep (Yokohama) 2*27-29
⋄ H05 ⋄ REV MR 56 # 16741 Zbl 389 # 60018 • ID 81670

ISHIMOTO, A. [1977] *A propositional fragment of Lesniewski's ontology* (J 0063) Studia Logica 36*285-299
⋄ A05 B20 ⋄ REV MR 58 # 16123 Zbl 391 # 03004
• ID 52331

ISHIMOTO, A. & KOBAYASHI, M. [1982] *A propositional fragment of Lesniewski's ontology and its formulation by the tableau method* (J 0063) Studia Logica 41*181-195
⋄ B20 ⋄ REV MR 86a:03025 • ID 42302

ISHIMOTO, A. see Vol. II for further entries

ISSEL, W. [1969] *Semantische Untersuchungen ueber Quantoren I* (J 0068) Z Math Logik Grundlagen Math 15*353-358
⋄ B10 C55 C80 ⋄ REV MR 41 # 6675 Zbl 188.16 • REM Parts II,III 1970 • ID 06459

ISSEL, W. see Vol. III for further entries

ITO, J. & ONO, K. [1966] *On a characteristic feature of the positive logics* (J 0111) Nagoya Math J 28*193-196
⋄ B20 ⋄ REV MR 34 # 7342 Zbl 163.7 • ID 90071

ITO, J. [1970] *On reducibility of provability in the primitive logic* [*LO*] (J 0111) Nagoya Math J 37*137-144
⋄ B20 ⋄ REV MR 41 # 6662 Zbl 205.9 • ID 06463

ITO, MAKOTO [1934] *Einige Anwendungen der Theorie des Entscheidungsproblems zur Axiomatik I* (J 0261) Tohoku Math J 37*222-235
⋄ B30 C75 F25 ⋄ REV Zbl 7.385 FdM 61.51 • REM Part II 1935 • ID 32185

ITO, MAKOTO [1935] *Einige Anwendungen der Theorie des Entscheidungsproblems zur Axiomatik. II* (J 0261) Tohoku Math J 40*241-251
⋄ B30 F25 ⋄ REV Zbl 11.97 FdM 61.51 • REM Part I 1934
• ID 28786

ITO, MAKOTO [1956] *General solution of boolean equation with m-variables* (J 4442) Kyushu Daigaku Kagaku Syuho 28/4*246-248
⋄ B05 G05 ⋄ ID 32190

ITO, MAKOTO [1959] *On boolean equation with many unknown elements and generalized Poretzky's formula* (J 0203) Rev Ser A Mat Fis, Univ Tucuman 12*107-112
⋄ B05 ⋄ REV Zbl 87.262 • ID 28624

ITO, MAKOTO see Vol. II, III, IV for further entries

IVANOV, P.M. [1977] *On an approach to the simplification of boolean functions (Russian)* (C 3354) Algeb & Teor Chisel ('77) 82-87
⋄ B35 ⋄ REV MR 58 # 33550 Zbl 392 # 94020 • ID 52418

IVANOV, V.A. [1973] *The noise immunity of Boolean functions (Russian)* (J 0040) Kibernetika, Akad Nauk Ukr SSR 1973/5*19-23
• TRANSL [1973] (J 0021) Cybernetics 9*737-743
⋄ B05 ⋄ REV MR 51 # 9994 Zbl 275 # 94027 • ID 62706

IWAMARU, Y. & NAGATA, M. & NAKANISHI, M. & NISHIMURA, T. [1974] *Implementation of Gentzen-type formal system representing properties of functions* (J 0407) Comm Math Univ St Pauli (Tokyo) 23*45-66
⋄ B35 ⋄ REV MR 52 # 102 Zbl 341 # 68058 • ID 18203

IWANUS, B. [1969] *An extension of the traditional logic containing the elementary ontology and the algebra of classes* (J 0063) Studia Logica 25*97-139
⋄ B20 G05 ⋄ REV Zbl 261 # 02009 • ID 30456

IWANUS, B. [1969] *Remarks about syllogistic with negative terms (Polish and Russian summaries)* (J 0063) Studia Logica 24*131-141
⋄ B20 ⋄ REV MR 40 # 1257 Zbl 243 # 02005 • ID 06473

IWANUS, B. see Vol. II, III, V for further entries

IZUMI, Y. [1954] *Sur les formes normales* (J 0261) Tohoku Math J 6*26-29
⋄ B10 ⋄ REV MR 16.783 Zbl 56.11 JSL 22.296 • ID 06478

IZUMI, Y. & WADA, T. [1955] *Sur la notion de la perfection* (J 0261) Tohoku Math J 7*132-135
⋄ A05 B05 ⋄ REV MR 17.701 Zbl 66.10 • ID 06479

IZUMI, Y. see Vol. V, VI for further entries

JACKSON, C.L. [1967] see ANKERLIN, R.A.

JACOBS, W. [1980] *The existential presuppositions of Aristotle's logic* (J 0095) Philos Stud 37*419-428
⋄ A05 B20 ⋄ REV MR 81k:03001 • ID 74446

JACOBSEN, B. [1976] see CHILAUSKY, R.

JACOBSON, A. & KLEMKE, E.D. & ZABECH, F. (EDS.) [1974] *Readings in semantics* (X 1285) Univ Ill Pr: Urbana v + 853pp
⋄ A05 B98 ⋄ ID 48637

JACOBSON JR., J.H. [1963] see BLYTH, J.W.

JACUNOV, A.I. & ROMANKEVICH, A.M. [1974] *On a representation method of boolean functions (Russian)* (J 0474) Avtom Vychis Tekh, Akad Nauk Latv SSR 1974/3*30-35
⋄ B05 ⋄ REV MR 50 # 4137 Zbl 296 # 02004 • ID 64890

JAIN, R.C. [1980] *On representation of boolean functions in simple form* (J 3435) Pure Appl Math Sci 11*65-70
⋄ B05 ⋄ REV MR 81e:94038 Zbl 431 # 94057 • ID 53956

JAMES, P.A. [1961] see ALLEN, L.E.

JAMES, P.A. [1965] see DICKOFF, J.W.

JAMES, P.A. [1969] see DICKOFF, J.W.

JANKOWSKI, A.W. [1985] *Galois structures* (J 0063) Studia Logica 44*109-124
⋄ B22 C95 ⋄ ID 47505

JANKOWSKI, A.W. see Vol. II, III, V, VI for further entries

JANSOHN, H.-S. & LANDWEHR, R. & WRIGHTSON, G. [1982] *An interactive proof system for higher order logic* (C 3881) Prog in Cybern & Syst Res, Vol 11 343-347
⋄ B15 B35 F35 ⋄ REV MR 84f:00034 Zbl 484 # 68072
• ID 36620

JANSON, A. [1983] see DILGER, W.

JANSSEN, G. [1972] *Restricted ultraproducts of finite von Neumann algebras* (P 0732) Contrib to Non-Standard Anal;1970 Oberwolfach 101-114
⋄ C20 C60 H05 ⋄ REV MR 58 # 2335 Zbl 248 # 46049
• ID 06540

JANSSEN, T.M.V. [1980] *Logical investigations on PTQ arising from programming requirements* (J 0154) Synthese 44*361-390
⋄ B80 ⋄ REV MR 83f:03027 Zbl 457 # 03011 • ID 54336

JANZ, A. [1980] *Eine neue Variante der Nichtstandart-Analysis und einige ihrer Anwendungen in der allgemeinen Topologie* (S 3382) Sem-ber, Humboldt-Univ Berlin, Sekt Math 30*v+124pp
⋄ H05 ⋄ REV MR 82m:03072 Zbl 468 # 54038 • ID 74471

JANZ, A. [1980] *Konstruktion von Kompaktifizierungen topologischer Raeume mit Hilfe von Nichtstandard-Modellen* (J 0115) Wiss Z Humboldt-Univ Berlin, Math-Nat Reihe 29*421-427
⋄ H05 H20 ⋄ REV MR 82h:54034 Zbl 457 # 54036
• ID 54399

JANZ, A. [1980] *Zwei Nichtstandard-Metrisationstheoreme* (J 0115) Wiss Z Humboldt-Univ Berlin, Math-Nat Reihe 29*429-431
⋄ H05 H20 ⋄ REV MR 82h:54046 Zbl 457 # 54037
• ID 54400

JARDINE, C.J. & JARDINE, N. [1971] *The matching of parts of things (Polish and Russian summaries)* (J 0063) Studia Logica 27*123-132
⋄ A05 B80 E70 ⋄ REV MR 46 # 3127 Zbl 264 # 02013
• ID 27501

JARDINE, N. [1971] see JARDINE, C.J.

JASKOWSKI, S. [1934] *On the rules of suppositions in formal logic* (J 0063) Studia Logica 1*5-32
• TRANSL [1967] (C 0615) Polish Logic 1920-39 232-258
⋄ A05 B20 ⋄ REV Zbl 11.97 FdM 60.846 • REM The source is an older series of Studia Logica • ID 06557

JASKOWSKI, S. [1948] *Sur les variables propositionelles dependantes* (J 0451) Studia Soc Sci Torunensis Sect A 1*17-21
⋄ B05 ⋄ REV MR 10.2 Zbl 41.351 JSL 14.65 • ID 06548

JASKOWSKI, S. [1948] *Trois contributions au calcul des propositions bivalent* (J 0451) Studia Soc Sci Torunensis Sect A 1*1-15
⋄ B05 ⋄ REV MR 10.2 JSL 13.164 • ID 06547

JASKOWSKI, S. [1955] see GRZEGORCZYK, A.

JASKOWSKI, S. [1963] *Ueber Tautologien, in welchen keine Variable mehr als zweimal vorkommt* (J 0068) Z Math Logik Grundlagen Math 9*219-228
⋄ B05 ⋄ REV MR 27 # 2397 Zbl 115.5 • ID 06554

JASKOWSKI, S. [1966] *On formulas in which no individual variable occurs more than twice* (J 0036) J Symb Logic 31*1-6
⋄ B05 B20 ⋄ REV MR 33 # 5458 Zbl 192.37 • ID 06555

JASKOWSKI, S. [1969] *On the interpretations of aristotelian categorical propositions in the predicate calculus* (J 0063) Studia Logica 24*161-174
⋄ A10 B20 ⋄ REV MR 40 # 5414 Zbl 246 # 02012 JSL 17.268 • ID 06558

JASKOWSKI, S. [1975] *About certain groups of classes of sets and their application to the definitions of numbers* (J 0063) Studia Logica 34*133-144
⋄ B28 ⋄ REV MR 51 # 7878 Zbl 313 # 02030 • ID 29626

JASKOWSKI, S. [1975] *Three contributions to the two-valued propositional calculus* (J 0063) Studia Logica 34*121-132
⋄ B05 ⋄ REV MR 51 # 7846 Zbl 313 # 02005 • ID 18209

JASKOWSKI, S. see Vol. II, IV, V, VI for further entries

JAUCH, J.M. [1965] see EMCH, G.

JAUCH, J.M. see Vol. II for further entries

JEANROND, H. [1979] *A unique termination theorem for a theory with generalized commutative axioms* (P 1873) Automata, Lang & Progr (6);1979 Graz 71*316-330
⋄ B35 ⋄ REV Zbl 421 # 68085 • ID 53458

JEANROND, H. see Vol. IV for further entries

JEFFREY, R.C. [1965] *The logic of decision* (X 0822) McGraw-Hill: New York xiv+201pp
• LAST ED [1983] (X 0862) Univ Chicago Pr: Chicago xiv+231pp
⋄ B48 B98 ⋄ REV MR 38 # 1770 MR 85g:62009 Zbl 167.471 • ID 25416

JEFFREY, R.C. [1967] *Formal logic: Its scope and limits* (X 0822) McGraw-Hill: New York xii+238pp
⋄ B98 ⋄ REV JSL 38.646 JSL 49.1408 • REM 2nd ed. 1981; xvi+198pp • ID 47400

JEFFREY, R.C. [1971] see CARNAP, R.

JEFFREY, R.C. [1974] see BOOLOS, G.

JEFFREY, R.C. see Vol. II for further entries

JEFFREYS, H. [1931] *Scientific inference* (X 0805) Cambridge Univ Pr: Cambridge, GB vi+247pp
⋄ A05 B98 ⋄ REV JSL 29.194 • ID 25417

JEFFREYS, H. see Vol. II for further entries

JENSEN, A. [1970] *The possible influence of non-standard analysis on elementary mathematics* (P 0785) Scand Logic Symp (1);1968 Aabo 61-68
⋄ H05 ⋄ REV MR 51 # 137 Zbl 324 # 02042 • ID 15230

JENSEN, A. [1972] *A computer oriented version of "non-standard analysis"* (P 0732) Contrib to Non-Standard Anal;1970 Oberwolfach 281-289
⋄ H05 ⋄ REV MR 58 # 5192 Zbl 255 # 02061 • ID 06591

JENSEN, A. see Vol. III, V for further entries

JENSEN, D.C. & PIETRZYKOWSKI, T. [1977] *Mechanizing ω-order type theory through unification* (J 1426) Theor Comput Sci 3∗123-171
 ◇ B35 ◇ REV MR 57 # 59 Zbl 361 # 02020 • ID 50655

JENSEN, D.C. see Vol. III, VI for further entries

JENSEN, F.V. & MAYOH, B.H. & MOELLER, K.K. (EDS.) [1979] *Proceedings from 5th Scandinavian Logic Symposium, Aalborg, 17-19, January 1979* (X 2646) Aalborg Univ Pr: Aalborg| vii + 361pp
 ◇ B97 ◇ REV MR 81m:03006 Zbl 419 # 00002 • ID 53341

JENSEN, F.V. see Vol. III, IV, V for further entries

JERVELL, H.R. [1973] *Skolem and Herbrand theorems in first order logic* (S 1626) Oslo Preprint Ser 6∗77pp
 ◇ B10 C07 F05 ◇ ID 33265

JERVELL, H.R. see Vol. II, III, IV, V, VI for further entries

JEVONS, W.S. [1880] *Studies in deductive logic: a manual for students* (X 1253) Crowell Collier & Macmillan: New York xxviii + 304pp
 ◇ B98 ◇ ID 22566

JHA, M.N. [1978] *Generalised metric via ultra-real numbers* (J 0437) Prog Math (Allahabad) 12∗21-26
 ◇ H10 ◇ REV MR 81f:03078 Zbl 439 # 03050 • ID 56041

JHA, S.N. [1977] *A non-standard treatment of some aspects of two variables function theory* (J 0437) Prog Math (Allahabad) 11∗49-57
 ◇ H05 ◇ REV MR 56 # 8776 Zbl 368 # 26005 • ID 81718

JI, ZHERUI [1981] *On the axiomatics of hyperreal number I (Chinese) (English summary)* (J 3942) Zhongshan Daxue Xuebao, Ziran Kexue 2∗106-112
 ◇ H05 ◇ REV Zbl 515 # 03042 • ID 37850

JIRKU, P. [1981] *Logical and linguistic aspects of computer-based inference processes* (J 2817) Prague Bull Math Linguist 35∗41-53
 ◇ B35 ◇ REV MR 82h:68124 • ID 81719

JIRKU, P. see Vol. IV for further entries

JOCKUSCH JR., C.G. & SIMPSON, S.G. [1976] *A degree theoretic definition of the ramified analytical hierarchy* (J 0007) Ann Math Logic 10∗1-32
 ◇ D30 D55 E40 E45 H05 ◇ REV MR 58 # 10370 Zbl 333 # 02039 • ID 18212

JOCKUSCH JR., C.G. [1977] see HENSON, C.W.

JOCKUSCH JR., C.G. see Vol. III, IV, V, VI for further entries

JOERGENSEN, J. [1937] *Outline of the recent development of the theory of deduction (Danish)* (X 1494) Munksgaard: Copenhagen 117pp
 ◇ B10 ◇ REV Zbl 17.337 JSL 3.43 • ID 41366

JOFFRE, D. [1953] see DESTOUCHES, J.-L.

JOHANSSON, I. [1953] *Sur le concept de "le" (ou de "ce qui") dans le calcul affirmatif et dans les calculs intuitionnistes* (P 0644) Meth Form en Axiom;1950 Paris 65-72
 ◇ B20 F50 ◇ REV MR 15.2 Zbl 53.341 JSL 23.346 • ID 06644

JOHANSSON, I. [1954] *Symboles logiques dans l'enseignement des theories deductives* (P 0646) Appl Sci de Log Math;1952 Paris 21-24
 ◇ B03 ◇ REV Zbl 58.4 • ID 27954

JOHANSSON, I. see Vol. VI for further entries

JOHNSON, D.R. & THOMASON, R.H. [1969] *Predicate calculus with free quantifier variables* (J 0036) J Symb Logic 34∗1-7
 ◇ B15 C80 ◇ REV MR 39 # 5328 Zbl 188.314 • ID 06648

JOHNSON, D.R. & MATTSON, D.A. [1975] *Some applications of non-standard analysis to proximity spaces* (S 0019) Colloq Math (Warsaw) 34∗17-24
 ◇ H05 ◇ REV MR 53 # 9154 Zbl 287 # 54052 • ID 81723

JOHNSON, F.A. [1979] *Copi's method of deduction* (J 0047) Notre Dame J Formal Log 20∗295-300
 ◇ B05 ◇ REV MR 80g:03010 Zbl 315 # 02013 • ID 52619

JOHNSON, F.A. see Vol. II for further entries

JOHNSON, J.S. [1973] *Axiom systems for first order logic with finitely many variables* (J 0036) J Symb Logic 38∗576-578
 ◇ B20 C05 G15 ◇ REV MR 49 # 8847 Zbl 286 # 02017 • ID 06681

JOHNSON, L.E. [1975] *A perspective on propositions* (J 0286) Int Logic Rev 12∗243-255
 ◇ B05 ◇ REV Zbl 323 # 02011 • ID 30406

JOHNSON WU, K. [1980] *On a tableau rule for identity* (J 0047) Notre Dame J Formal Log 21∗175-178
 ◇ B10 ◇ REV MR 80m:03027 Zbl 365 # 02017 • ID 53198

JOHNSON WU, K. see Vol. II for further entries

JOHNSTONE JR., H.W. [1954] *Elementary deductive logic* (X 1253) Crowell Collier & Macmillan: New York viii + 241pp
 ◇ B98 ◇ REV Zbl 59.14 JSL 20.165 • ID 22499

JOHNSTONE JR., H.W. [1962] see ANDERSON, J.M.

JOHNSTONE JR., H.W. & PRICE, ROBERT [1964] *Axioms for the implicational calculus with one variable* (J 0105) Theoria (Lund) 30∗1-4
 ◇ B20 ◇ REV JSL 35.584 • ID 06685

JOHNSTONE JR., H.W. [1966] *A definition of conjunction in the pure implicational calculus with one variable* (J 0079) Logique & Anal, NS 9∗310-312
 ◇ B20 ◇ REV MR 35 # 1442 Zbl 154.4 JSL 35.584 • ID 06687

JOHNSTONE JR., H.W. [1982] *The syllogism on the negative-entailing interpretation of affirmative propositions* (J 0079) Logique & Anal, NS 25∗203-205
 ◇ B20 ◇ REV MR 84c:03026 • ID 35686

JOHNSTONE JR., H.W. see Vol. III for further entries

JOJA, A. [1975] see HENKIN, L.

JOJA, A. see Vol. II for further entries

JONES, C.K.R.T. & KELEMEN, P.J. [1981] *The ϱ-calculus* (J 0068) Z Math Logik Grundlagen Math 27∗97-110
 ◇ H05 ◇ REV MR 82k:03105 Zbl 469 # 03049 • ID 55176

JONES, JOHN [1985] *Formalisations of many valued propositional calculi with variable functors* (J 0068) Z Math Logik Grundlagen Math 31∗403-422
 ◇ B22 ◇ REV Zbl 571 # 03009 • ID 47547

JONES, JOHN see Vol. II for further entries

JONGELING, T.B. [1985] *On an axiomatization of evolutionary theory* (J 1885) J Theor Biol 117∗529-543
 ◇ A05 B30 ◇ ID 49327

JONGH DE, D.H.J. [1974] see GABBAY, D.M.

JONGH DE, D.H.J. see Vol. II, III, IV, V, VI for further entries

JORDAN, I.B. & MOUFTAH, H.T. [1975] *A design technique for an integrable ternary arithmetic unit* (P 1805) Int Symp Multi-Val Log (5,Proc);1975 Bloomington 359-372
⋄ B20 B70 ⋄ REV MR 58 # 9867 • ID 35835

JORDAN, I.B. see Vol. II for further entries

JORGENSEN, J. [1931] *A treatise of formal logic: its evolution and main branches, with its relations to mathematics and philosophy. I,II,III* (X 0894) Oxford Univ Pr: Oxford xvi + 266pp,vi + 273pp,vi + 321pp
• LAST ED (X 1343) Russell: New York
⋄ A05 A10 B98 ⋄ REV FdM 57.50 • ID 22584

JORGENSEN, P. & STRUNZ, H. [1976] *Anwendung graphentheoretischer Verfahren in der Entscheidungstabellentechnik* (J 1633) Angew Inf 18∗65-73
⋄ B35 ⋄ REV Zbl 317 # 68042 • ID 65561

JOUANNAUD, J.-P. & KIRCHNER, C. & KIRCHNER, H. [1983] *Incremental construction of unification algorithms in equational theories* (P 3851) Automata, Lang & Progr (10);1983 Barcelona 361-373
⋄ B35 ⋄ REV Zbl 516 # 68067 • ID 38594

JOUANNAUD, J.-P. see Vol. II, IV for further entries

JOUBERT, M. [1979] see FIESCHI, M.

JOYNER JR., W.M. [1973] *Automatic theorem-proving and the decision problem* (P 3062) IEEE Symp Switch & Automata Th (14);1973 Iowa City 159-166
⋄ B25 B35 ⋄ REV MR 55 # 6989 • ID 81740

JOYNER JR., W.M. [1976] *Resolution strategies as decision procedures* (J 0037) ACM J 23∗398-417
⋄ B25 B35 ⋄ REV MR 53 # 15007 Zbl 335 # 68062
• ID 23231

JOZSA, R. [1984] *Sheaf models and massless fields* (J 2736) Int J Theor Phys 23∗67-97
⋄ B30 C90 F50 G30 ⋄ REV MR 85m:81076 • ID 45334

JUHASZ, I. & MACHOVER, M. [1969] *A note on non-standard topology* (J 0028) Indag Math 31∗482-484
⋄ H05 ⋄ REV MR 40 # 6488 Zbl 201.245 JSL 35.487
• ID 06761

JUHASZ, I. [1972] *Non-standard notes on the hyperspace* (P 0732) Contrib to Non-Standard Anal;1970 Oberwolfach 171-177
⋄ E75 H05 ⋄ REV MR 58 # 2786 Zbl 247 # 54008
• ID 06764

JUHASZ, I. see Vol. V for further entries

JUHOS VON, B. [1954] *Elemente der neuen Logik* (X 1026) Humboldt: Frankfurt & Wien 256pp
⋄ B98 ⋄ ID 22410

JUHOS VON, B. see Vol. II for further entries

JUSSILA, E. [1975] *Functional logic. Part I. Sentences* (X 3297) Erkki Juhani Jussila: Tampere 52pp
⋄ B05 ⋄ REV Zbl 334 # 02004 • ID 62821

JUSTEN, K. [1981] *A note on regular resolution* (J 0373) Comp Arch Inform & Numerik 26∗87-89
⋄ B35 F20 ⋄ REV MR 82k:03012 Zbl 431 # 03036
• ID 55967

JUZA, M. [1982] *About the sixth Hilbert's problem* (J 0022) Cheskoslov Mat Zh 32(107)∗1-52
⋄ B30 ⋄ REV MR 84f:03009 Zbl 498 # 70003 • ID 34427

JUZA, M. see Vol. V for further entries

KAAZ, M.A. [1977] *Elemente der mathematischen Logik fuer den Gebrauch in Physik und Technik* (X 0814) Oldenbourg: Muenchen 243pp
⋄ B98 ⋄ REV MR 58 # 16076 Zbl 361 # 02022 • ID 50657

KAAZ, M.A. see Vol. V for further entries

KABULOV, A.V. & LOSEV, G.F. [1978] *On local simplification algorithms for disjunctive normal forms of boolean functions (Russian)* (J 0199) Zh Vychisl Mat i Mat Fiz 18∗728-734
• TRANSL [1978] (J 1049) USSR Comput Math & Math Phys 18/3∗201-207
⋄ B35 ⋄ REV MR 80b:94038 Zbl 395 # 03041 • ID 52592

KABULOV, A.V. [1982] see BAJKHUMANOV, A.A.

KABULOV, A.V. see Vol. II, V for further entries

KABZINSKI, J.K. [1973] *On problems of definability of propositional connectives* (J 0387) Bull Sect Logic, Pol Acad Sci 2∗127-130
⋄ A10 B05 ⋄ REV MR 55 # 7730 • ID 74590

KABZINSKI, J.K. [1975] *Some problems connected with equivalential formalization of classical sentential calculus (Polish and English)* (S 1454) Zesz Nauk Wyz Szk Ped Mat, Opole 15∗29-36
⋄ B20 ⋄ REV MR 55 # 5384 Zbl 316 # 02016 • ID 62826

KABZINSKI, J.K. [1981] *On equivalential fragment of implicative extensional consequence* (J 0387) Bull Sect Logic, Pol Acad Sci 10∗135-139
⋄ B22 G25 ⋄ REV MR 83b:03010 Zbl 472 # 03021
• ID 55287

KABZINSKI, J.K. see Vol. II, VI for further entries

KAC, M. & ULAM, S.M. [1968] *Mathematics and logic: Retrospect and prospects* (X 1334) Praeger: New York ix + 170pp
• REPR [1979] (X 0868) Penguin Books: Harmondsworth & New York 204pp
⋄ A05 A10 B98 ⋄ REV MR 38 # 964 MR 82a:00007 Zbl 205.289 Zbl 479 # 00011 JSL 36.677 • REM Rep. abridged • ID 25688

KACZMARZ, S. [1932] *Axioms for arithmetic* (J 0039) J London Math Soc 7∗179-182
⋄ B28 ⋄ REV Zbl 5.146 JSL 20.307 FdM 58.92 • ID 06787

KAESBAUER, M. & KUTSCHERA VON, F. (EDS.) [1962] *Logik und Logikkalkuel* (X 0826) Alber: Freiburg 248pp
⋄ B98 ⋄ REV Zbl 119.242 • ID 22441

KAESBAUER, M. [1962] *Logisches System mit Praedikatquantoren* (C 0712) Logik & Logikkalkuel 237-249
⋄ B15 ⋄ REV Zbl 126.10 • ID 06790

KAHANE, H. [1969] *Logic and philosophy. A modern introduction* (X 0821) Wadsworth Publ: Belmont xv + 450pp
⋄ B98 ⋄ REV JSL 39.613 • ID 06792

KAHR, A.S. [1962] see DREBEN, B.

KAHR, A.S. & MOORE, E.F. & WANG, HAO [1962] *Entscheidungsproblem reduced to the ∀∃∀ case* (J 0054) Proc Nat Acad Sci USA 48∗365-377
⋄ B20 D35 ⋄ REV MR 30 # 21 Zbl 102.8 JSL 27.225
• ID 19074

KAHR, A.S. see Vol. IV for further entries

KAISER, KLAUS [1972] see ARMBRUST, M.

KAISER, KLAUS [1974] *Direct limits in quasi-universal model classes* (J 3172) J London Math Soc, Ser 2 9*585-588
⋄ B15 C30 C52 C85 ⋄ REV MR 52 # 77 Zbl 309 # 02054
• ID 06804

KAISER, KLAUS [1975] *Various remarks on quasi-universal model classes* (J 0009) Arch Math Logik Grundlagenforsch 17*91-95
⋄ B15 C05 C52 C85 ⋄ REV MR 52 # 5407 Zbl 331 # 02031 • ID 06805

KAISER, KLAUS [1976] *Quasi-axiomatic classes* (J 0009) Arch Math Logik Grundlagenforsch 17*129-134
⋄ B15 C20 C52 C75 C85 ⋄ REV MR 54 # 4968 Zbl 331 # 02032 • ID 06806

KAISER, KLAUS see Vol. III, IV for further entries

KAKUDA, Y. [1980] *Non-standard analysis in boolean-valued models* (P 2611) Symp Founds of Math;
⋄ E40 H05 ⋄ ID 43408

KAKUDA, Y. [1984] *Nonstandard analysis without nonstandard models* (P 3668) Log & Founds of Math;1983 Kyoto 10-11
⋄ H05 ⋄ ID 42929

KAKUDA, Y. see Vol. III, V for further entries

KALICKI, J. [1950] *Note on truth tables* (J 0036) J Symb Logic 15*174-181,IV
⋄ B05 ⋄ REV MR 12.663 Zbl 39.6 JSL 16.65 • ID 06812

KALICKI, J. [1950] *On the structure of bracket-free formulae* (J 4510) Norsk Mat Tidsskr 32*33-39
⋄ B03 ⋄ REV Zbl 41.150 JSL 16.151 • ID 31567

KALICKI, J. [1952] *A test for the equality of truth-tables* (J 0036) J Symb Logic 17*161-163
⋄ B05 ⋄ REV MR 14.344 Zbl 49.147 JSL 18.268
• ID 06813

KALICKI, J. [1954] *An undecidable problem in the algebra of truth-tables* (J 0036) J Symb Logic 19*172-176
⋄ B05 D35 ⋄ REV MR 16.324 Zbl 58.246 JSL 20.283
• ID 06815

KALICKI, J. see Vol. II, III for further entries

KALINICHENKO, L.A. [1979] see FOMINA, N.I.

KALISH, D. & MONTAGUE, R. [1957] *Remarks on descriptions and natural deduction I,II* (J 0009) Arch Math Logik Grundlagenforsch 3*50-64,65-73
⋄ B10 F07 ⋄ REV MR 19.1032 MR 19.724 Zbl 81.245 JSL 23.449 • ID 24907

KALISH, D. & MONTAGUE, R. [1964] *Logic. Techniques of formal reasoning* (X 0863) Harcourt: New York & London x+350pp
⋄ B98 ⋄ REV JSL 34.641 • ID 06824

KALISH, D. & MONTAGUE, R. [1965] *On Tarski's formalisation of predicate logic with identity* (J 0009) Arch Math Logik Grundlagenforsch 7*81-101
⋄ B10 ⋄ REV MR 34 # 2438 Zbl 166.1 • ID 06826

KALLEN, H.M. [1951] see HENLE, P.

KALLICK, B. [1969] *A decision procedure based on the resolution method* (P 0594) Inform Processing (4);1968 Edinburgh 1*269-275
⋄ B25 B35 ⋄ REV Zbl 213.19 JSL 38.656 • ID 22190

KALMAN, J.A. [1962] see FORDER, H.G.

KALMAN, J.A. [1976] *Computer studies of $T_{\to} - W - I$* (J 1893) Relevance Logic Newslett 1*181-188
⋄ B35 B46 ⋄ REV Zbl 346 # 02003 • ID 62846

KALMAN, J.A. [1978] *A shortest single axiom for the classical equivalential calculus* (J 0047) Notre Dame J Formal Log 19*141-144
⋄ B20 ⋄ REV MR 58 # 10282 Zbl 368 # 02009 • ID 27094

KALMAN, J.A. [1983] *Condensed detachment as a rule of inference* (J 0063) Studia Logica 42*443-451
⋄ B35 ⋄ REV Zbl 568 # 03010 • ID 42333

KALMAN, J.A. see Vol. II, V for further entries

KALMAR, L. [1928] *Eine Bemerkung zur Entscheidungstheorie* (J 0460) Acta Univ Szeged, Sect Mat 4*248-252
⋄ B20 B25 F30 ⋄ REV FdM 55.32 • ID 06835

KALMAR, L. [1932] *Zum Entscheidungsproblem der mathematischen Logik* (P 0653) Int Congr Math (II, 4);1932 Zuerich 2*337-338
⋄ B20 B25 D35 ⋄ REV FdM 58.70 • ID 06837

KALMAR, L. [1933] *Ueber die Erfuellbarkeit derjenigen Zaehlausdruecke, welche in der Normalform zwei benachbarte Allzeichen enthalten* (J 0043) Math Ann 108*466-484
⋄ B20 B25 ⋄ REV Zbl 6.385 FdM 59.864 • ID 06838

KALMAR, L. [1934] *Ueber die Axiomatisierbarkeit des Aussagenkalkuels* (J 0460) Acta Univ Szeged, Sect Mat 7*222-243
⋄ B05 ⋄ REV Zbl 14.194 FdM 61.972 • ID 06840

KALMAR, L. [1937] *Zur Reduktion des Entscheidungsproblems* (J 4510) Norsk Mat Tidsskr 19*121-130
⋄ B20 D35 ⋄ REV Zbl 17.337 JSL 3.46 FdM 63.823
• REM A note by Skolem ibid 19*130-133 (Norwegian)
• ID 06842

KALMAR, L. [1937] *Zurueckfuehrung des Entscheidungsproblems auf den Fall von Formeln mit einer einzigen, binaeren Funktionsvariablen* (J 0020) Compos Math 4*137-144
⋄ B20 D35 ⋄ REV Zbl 15.338 JSL 2.48 FdM 62.1058
• ID 06841

KALMAR, L. [1939] *On the reduction of the decision problem I: Ackermann prefix, a single binary predicate* (J 0036) J Symb Logic 4*1-9
⋄ B20 D35 ⋄ REV Zbl 20.195 JSL 4.127 FdM 65.27
• REM Part II 1947 by Kalmar,L. & Suranyi,J. • ID 06843

KALMAR, L. [1940] *On the possibility of definition by recursion* (J 0460) Acta Univ Szeged, Sect Mat 9*227-232
⋄ B28 ⋄ REV MR 1.132 Zbl 22.194 JSL 5.70 FdM 66.32
• ID 06844

KALMAR, L. & SURANYI, J. [1947] *On the reduction of the decision problem II: Goedel prefix, a single binary predicate* (J 0036) J Symb Logic 12*65-73
⋄ B20 D35 ⋄ REV MR 11.303 Zbl 29.99 JSL 13.48 • REM Part I 1939. Part III 1950 • ID 06846

KALMAR, L. [1950] *Contribution to the reduction theory of the decision problem I. Prefix $(x_1) (x_2) (Ex_3)...(Ex_{n-1}) (x_n)$, a single binary predicate* (J 0001) Acta Math Acad Sci Hung 1*64-73
⋄ B20 D35 ⋄ REV MR 12.661 Zbl 39.246 JSL 17.73
• REM Part II 1950 by Suranyi,J. • ID 06854

KALMAR, L. & SURANYI, J. [1950] *On the reduction of the decision problem III: Pepis prefix, a single binary predicate* (J 0036) J Symb Logic 15*161-173
⋄ B20 D35 ⋄ REV MR 12.661 Zbl 41.352 JSL 16.215
• REM Part II 1947 • ID 06851

KALMAR, L. [1950] *Ueber die Cantorsche Theorie der reellen Zahlen* (J 0057) Publ Math (Univ Debrecen) 1*150-159
⋄ B28 ⋄ REV MR 12.15 Zbl 37.34 • ID 06853

KALMAR, L. [1951] *Contributions to the reduction theory of the decision problem. III Prefix $(x_1)(Ex_2)...(Ex_{n-2})(x_{n-1})(x_n)$, a single binary predicate* (J 0001) Acta Math Acad Sci Hung 2*19-38
⋄ B10 D35 ⋄ REV MR 13.715 Zbl 44.3 JSL 18.264 • REM Part II 1950 by Suranyi,J. Part IV 1951 • ID 42204

KALMAR, L. [1956] *A direct proof of the unsolvability of the decision problem by means of a general recursive algorithm (Hungarian)* (J 0462) Mat Fiz Oszt Koezlem, Acad Sci Hung 6*1-25
⋄ B20 D35 ⋄ REV MR 20#4 Zbl 75.5 JSL 24.173
• ID 16972

KALMAR, L. [1956] *Ein direkter Beweis fuer die allgemein-rekursive Unloesbarkeit des Entscheidungsproblems des Praedikatenkalkuels der ersten Stufe mit Identitaet* (J 0068) Z Math Logik Grundlagen Math 2*1-14
⋄ B10 D35 ⋄ REV MR 18.369 Zbl 75.5 JSL 27.86
• ID 06861

KALMAR, L. (ED.) [1965] *Colloque sur les fondements des mathematiques, les machines mathematiques, et leurs applications* (X 0928) Akad Kiado: Budapest 320pp
⋄ B97 D05 D10 D97 ⋄ REV MR 32#5493 Zbl 148.1
• ID 48630

KALMAR, L. see Vol. III, IV, V, VI for further entries

KALUZHNIN, L.A. [1980] see GLAZUNOV, N.M.

KALUZHNIN, L.A. see Vol. II, IV, V for further entries

KAMAE, T. [1982] *A simple proof of the ergodic theorem using nonstandard analysis* (J 0029) Israel J Math 42*284-290
⋄ H10 ⋄ REV MR 84i:28019 Zbl 499#28011 • ID 38302

KAMAE, T. see Vol. IV for further entries

KAMENOV, K. [1978] *Optimization of the special functional $\Theta(f)$ in factorization of boolean functions into factors (Russian) (English and Lithuanian summaries)* (C 3409) Chisl Met Optim & Primen 55-66
⋄ B05 ⋄ REV MR 80g:90076 Zbl 415#94017 • ID 69652

KAMENSKIJ, M.I. & PETROVA, L.P. & SADOVSKIJ, B.N. [1982] *Mathematical logic. Methodical instructions (Russian)* (X 0898) Moskov Gos Univ: Moskva 63pp
⋄ B05 ⋄ REV Zbl 554#03006 • ID 44934

KAMEYAMA, M. [1982] see HIGUCHI, T.

KAMINSKI, S. [1958] *On the number of concluding syllogistic moods (Polish) (Russian and English summaries)* (J 0063) Studia Logica 8*165-176
⋄ B20 ⋄ REV MR 21#6322 • ID 06879

KAMINSKI, S. [1961] *Traditional theory of immediate inference as a fragment of two-valued propositional calculus (Polish) (Russian and English summaries)* (J 0063) Studia Logica 11*7-21
⋄ B20 ⋄ REV MR 23#A3079 Zbl 121.253 • ID 06880

KAMINSKI, S. [1965] *Rules for syllogisms with the consideration of schemata with negated subject terms (Polish) (Russian and English summaries)* (J 0063) Studia Logica 16*45-52
⋄ B20 ⋄ REV MR 32#2315 Zbl 292#02011 JSL 32.544
• ID 62851

KANGER, S. [1955] *A note on partial postulate sets for propositional logic* (J 0105) Theoria (Lund) 21*99-104
⋄ B05 ⋄ REV MR 19.3 Zbl 66.256 JSL 22.330 • ID 06884

KANGER, S. [1957] *Provability in logic* (X 1163) Almqvist & Wiksell: Stockholm 49pp
⋄ B10 F05 F07 F98 ⋄ REV MR 19.239 Zbl 77.12
JSL 23.37 • ID 22373

KANGER, S. [1963] *A simplified proof method for elementary logic* (C 0659) Computer Progr & Formal Syst 87-94
• REPR [1983] (C 4659) Autom of Reasoning 1*364-371
⋄ B10 F07 ⋄ REV MR 27#1350 JSL 32.119 • ID 06888

KANGER, S. [1970] *Equational calculi and automatic demonstration* (C 0735) Logic & Value (Dahlquist) 220-226
⋄ B25 B35 C05 ⋄ REV MR 53#12897 • ID 23148

KANGER, S. [1972] *Law and logic* (J 0105) Theoria (Lund) 38*105-132
⋄ A05 B80 ⋄ REV MR 58#16151 Zbl 265#02020
• ID 29818

KANGER, S. see Vol. II, III for further entries

KANOVEJ, V.G. [1983] *Some problems of descriptive set theory and definability in the theory of types (Russian)* (C 3807) Issl Neklass Log & Formal Sist 21-81
⋄ B15 E15 ⋄ REV MR 86b:03060 • ID 44359

KANOVEJ, V.G. see Vol. III, IV, V, VI for further entries

KANOVICH, M.I. [1971] *On complexity of Boolean function minimzation (Russian)* (J 0023) Dokl Akad Nauk SSSR 198*35-38
• TRANSL [1971] (J 0062) Sov Math, Dokl 12*720-724
⋄ B35 D15 ⋄ REV MR 43#7334 Zbl 235#02026
• ID 06900

KANOVICH, M.I. [1974] *A theorem on speeding up in formal systems (Russian)* (C 2319) Slozh Vychisl & Algor 186-189
⋄ B35 D15 ⋄ REV Zbl 307#68034 • ID 62867

KANOVICH, M.I. [1975] *A hierachical semantic system with set variables (Russian)* (J 0023) Dokl Akad Nauk SSSR 221*1256-1259
• TRANSL [1975] (J 0062) Sov Math, Dokl 16*504-509
⋄ B20 B65 C40 D05 F50 ⋄ REV MR 52#5385
Zbl 325#02030 • ID 18218

KANOVICH, M.I. [1984] *On the independence of invariant propositions (Russian)* (J 0023) Dokl Akad Nauk SSSR 276*27-31
• TRANSL [1984] (J 0062) Sov Math, Dokl 29*425-429
⋄ B30 D20 • ID 45178

KANOVICH, M.I. [1985] *Efficient logical algorithms for analysis and syntheis of dependencies (Russian)* (J 0023) Dokl Akad Nauk SSSR 285*1301-1305
⋄ B35 • ID 49033

KANOVICH, M.I. see Vol. III, IV, VI for further entries

KANTARIYA, G.V. [1972] *Eine optimale Auswahl mit Vertraeglichkeit alternativer Hypothesen (Russian) (Georgian and English summary)* (J 0233) Soobshch Akad Nauk Gruz SSR 68*553-555
⋄ B35 ⋄ REV MR 50#12342 Zbl 247#68039 • ID 62887

KAO, S. [1977] see GOTO, M.

KAO, S. [1978] see GOTO, M.

KAO, S. [1979] see GOTO, M.

KAO, S. [1981] see GOTO, M.

KAO, S. see Vol. II for further entries

KAPETANOVIC, M. [1983] *Some simple decidability proofs* (J 0042) Mat Vesn, Drust Mat Fiz Astron Serb 7(20)(35)*27-29
⋄ B20 B25 ⋄ REV MR 85f:03008 Zbl 553 # 03004 • ID 40445

KAPETANOVIC, M. see Vol. II for further entries

KAPHENGST, H. [1985] *Zum Aufbau einer mehrsortigen elementaren Logik* (J 0068) Z Math Logik Grundlagen Math 31*39-56
⋄ B10 ⋄ ID 42285

KAPHENGST, H. see Vol. IV for further entries

KAPITONOVA, YU.V. [1972] see GLUSHKOV, V.M.

KAPITONOVA, YU.V. [1972] see ANUFRIEV, F.V.

KAPITONOVA, YU.V. [1973] see DEGTYAREV, A.I.

KAPITONOVA, YU.V. see Vol. VI for further entries

KAPLAN, DAVID & MONTAGUE, R. [1965] *Foundations of higher-order logic* (P 0623) Int Congr Log, Meth & Phil of Sci (2,Proc);1964 Jerusalem 101-111
⋄ B15 ⋄ REV MR 35 # 4093 Zbl 156.23 • ID 06918

KAPLAN, DAVID see Vol. II for further entries

KAPUR, D. & KRISHNAMURTHY, B. [1984] *A natural proof sytem based on rewriting techniques* (P 2633) Autom Deduct (7);1984 Napa 53-64
⋄ B35 F07 ⋄ REV Zbl 546 # 68075 • ID 43561

KAPUR, D. see Vol. IV for further entries

KARAKHANYAN, L.M. & SAPOZHENKO, A.A. [1979] *Estimates for parameters of DNFs of not everywhere defined (partial) functions of the algebra of logic (Russian)* (S 3911) Komb-Algeb Met Prikl Mat (Gor'kij) 1979*48-56,121
⋄ B05 ⋄ ID 46418

KARAKHANYAN, L.M. see Vol. II for further entries

KARAPETYAN, B.K. & POGOSYAN, EH.M. [1977] *Inductors and their connection with the method of empirical prediction (Russian)* (S 0507) Vychisl Sist (Akad Nauk SSSR Novosibirsk) 69*102-112
⋄ B35 D70 ⋄ REV Zbl 449 # 68055 • ID 56756

KARATUEV, V.G. & MATROSOV, V.M. & NOVIKOV, M.A. & SUMENKOV, E.A. & VASIL'EV, S.N. & YADYKIN, S.A. [1979] *Machine deduction of theorems on dynamic properties with vector functions of Lyapunov (Russian)* (J 0040) Kibernetika, Akad Nauk Ukr SSR 1979/2*27-36
• TRANSL [1979] (J 0021) Cybernetics 15*184-194
⋄ B35 ⋄ REV MR 80m:68082 Zbl 444 # 68080 • ID 69698

KAREV, G.P. [1965] see FRIDMAN, G.SH.

KARGAPOLOV, M.I. [1973] see ERSHOV, YU.L.

KARGAPOLOV, M.I. see Vol. III, IV for further entries

KARPOV, V.G. & MOSHCHENSKIJ, V.A. [1977] *Mathematical logic and discrete mathematics (Russian)* (X 1574) Vyssheyshaya Shkola: Minsk 254pp
⋄ B98 E98 ⋄ REV MR 57 # 51 Zbl 371 # 02001 • ID 51362

KARPOVA, N.A. [1980] *Complexity of realization of functions of the algebra of logic in certain infinite bases (Russian)* (J 0071) Met Diskr Analiz (Novosibirsk) 35*9-14,104
⋄ B05 ⋄ REV MR 83h:94026 Zbl 473 # 68036 • ID 39258

KARWOWSKI, W. [1984] see ALBEVERIO, S.

KASAHARA, S. [1973] *A characterization of nonstandard real fields* (J 1508) Math Sem Notes, Kobe Univ 1*vi+8pp
• REPR [1974] (J 0081) Proc Japan Acad 50*53-56
⋄ C20 C60 H05 ⋄ REV MR 51 # 12804 Zbl 287 # 02039 • ID 30535

KASAHARA, S. [1974] *A remark on nonstandard real fields* (J 1508) Math Sem Notes, Kobe Univ 2/3*4pp
⋄ H05 ⋄ REV MR 53 # 8028 Zbl 306 # 02054 • ID 62900

KASAHARA, S. [1975] *A remark on simplifications of boolean polynomials* (J 1508) Math Sem Notes, Kobe Univ 3/2*xxi+7pp
⋄ B05 ⋄ REV MR 53 # 10497 Zbl 329 # 06009 • ID 62902

KASAHARA, S. see Vol. V for further entries

KASHER, A. (ED.) [1976] *Language in focus: foundations, methods and systems. Essays in memory of Yehoshua Bar-Hillel* (X 0835) Reidel: Dordrecht xxviii+679pp
⋄ B97 ⋄ REV Zbl 367 # 00005 • ID 51156

KASHER, A. see Vol. II, III for further entries

KASSLER, M. [1963] *A sketch of the use of formalized languages for the assertion of music* (J 1023) Perspectives of New Music 1*83-94
⋄ B80 ⋄ REV JSL 40.576 • ID 15108

KASSLER, M. [1967] *Towards a theory that is the twelve-note-class system* (J 1023) Perspectives of New Music 5*1-80
⋄ B80 ⋄ REV JSL 40.576 • ID 15107

KASSLER, M. [1976] *The decidability of languages that assert music* (J 1023) Perspectives of New Music 14*249-251
⋄ B25 B80 ⋄ ID 32363

KASYMOV, N.KH. [1981] see ABDULLAEVA, M.

KATERINOCHKINA, N.N. [1982] *Search for a maximal upper zero for discrete monotone functions (Russian)* (S 2829) Rostocker Math Kolloq 21*5-10
⋄ B05 ⋄ REV MR 85g:94020 Zbl 539 # 03006 • ID 43737

KATERINOCHKINA, N.N. see Vol. II, IV for further entries

KATZ, JOSE & THURBER, J.K. [1974] *Applications of fractional powers of delta functions* (P 1083) Victoria Symp Nonstand Anal;1972 Victoria 272-302
⋄ H10 ⋄ REV MR 58 # 4005 • ID 27650

KATZ, R. [1964] *Axiomatic analysis: an introduction to logic and the real number system* (X 1004) Heath: Lexington xiv+336pp
⋄ B28 ⋄ REV Zbl 146.245 • ID 23374

KAUFMANN, F. [1978] *The infinite in mathematics. Logico-mathmatical writings* (X 0835) Reidel: Dordrecht xvii+235pp
⋄ A05 A10 B30 ⋄ REV MR 58 # 16066 Zbl 402 # 03001
• REM With an introduction by Nagel,E. Transl. from several German articles • ID 54648

KAUFMANN, F. see Vol. V for further entries

KAUFMANN, M. [1984] see HENSON, C.W.

KAUFMANN, M. see Vol. III, IV, V for further entries

KAVUN, N.I. [1947] *A development of the theory of real numbers by the method of A.N. Kolmogorov (Russian)* (**J** 0067) Usp Mat Nauk 2/5*199-229
  ⋄ B28 ⋄ REV MR 10.669 Zbl 36.165 • ID 06977

KAWAI, T. [1979] *An application of nonstandard analysis to characters of groups of continuous functions* (**J** 2826) Rep Fac Sci, Kagoshima Univ, Math Phys Chem 12*43-45
  ⋄ C65 H10 ⋄ REV MR 81i:22002 Zbl 499 # 22003
  • ID 38297

KAWAI, T. [1979] *An axiom system for nonstandard set theory* (**J** 2826) Rep Fac Sci, Kagoshima Univ, Math Phys Chem 12*37-42
  ⋄ E70 H05 H20 ⋄ REV MR 81d:03056 Zbl 442 # 03043
  • ID 56400

KAWAI, T. [1981] *Axiom systems of nonstandard set theory* (**P** 3201) Logic Symposia;1979/80 Hakone 57-65
  ⋄ E70 H05 H20 ⋄ REV MR 81d:03056 MR 84j:03140 Zbl 478 # 03031 • ID 55646

KAWAI, T. [1981] *Nonstandardization of Feferman's set theory and a conservation theorem (Japanese)* (**P** 4153) B-Val Anal & Nonstand Anal;1981 Kyoto 83-92
  ⋄ E35 E70 H05 H20 ⋄ ID 47766

KAWAI, T. [1983] *Nonstandard analysis by axiomatic method* (**P** 3669) SE Asian Conf on Log;1981 Singapore 55-76
  ⋄ H05 ⋄ REV MR 85e:03161 Zbl 542 # 03046 • ID 40796

KAYUNOV, O.N. [1983] *Nonstandard solutions of equations of mathematical physics (Russian)* (**C** 3798) Mat Log, Mat Ling & Teor Algor 52-57
  ⋄ H10 ⋄ REV MR 86d:35010 • ID 45017

KAZAKOV, V.D. [1961] *On higher level minimal expressions for boolean functions* (**P** 1607) Hungar Math Congr (2);1960 Budapest V27-V31
  ⋄ B05 ⋄ ID 29294

KEANE, O. [1975] *Abstract Horn theories* (**C** 0772) Model Th & Topoi 15-50
  ⋄ B20 G30 ⋄ REV MR 52 # 2871 JSL 46.158 JSL 46.158
  • ID 17656

KEARNS, J.T. [1966] *Quantifiers and universal validity* (**J** 0079) Logique & Anal, NS 9*298-309
  ⋄ A05 B10 ⋄ REV MR 35 # 6523 Zbl 192.27 JSL 33.137
  • ID 06979

KEARNS, J.T. [1969] *Deductive logic: a programmed introduction* (**X** 1228) Appleton-Century-Crofts: New York
  ⋄ B98 ⋄ ID 22500

KEARNS, J.T. see Vol. II, III, VI for further entries

KECSKIC, J.D. [1968] *An axiomatization of the propositional caculus and the completeness theorem* (**J** 0042) Mat Vesn, Drust Mat Fiz Astron Serb 5(20)*361-366
  ⋄ B05 ⋄ REV MR 38 # 3122 Zbl 165.503 • ID 48131

KEENE, G.B. [1964] *First-order functional calculus* (**X** 0866) Routledge & Kegan Paul: Henley on Thames vi + 82pp
  ⋄ B98 ⋄ REV JSL 36.167 • ID 06993

KEENE, G.B. [1969] *The relational syllogism. A systematic approach to relational logic* (**X** 1194) Univ Exeter: Exeter iv + 35pp
  ⋄ B20 ⋄ REV JSL 35.448 • ID 21974

KEENE, G.B. see Vol. II, V for further entries

KEISLER, H.J. [1962] see CHANG, C.C.

KEISLER, H.J. [1965] *Reduced products and Horn classes* (**J** 0064) Trans Amer Math Soc 117*307-328
  ⋄ B20 C20 C52 C55 ⋄ REV MR 30 # 1047 Zbl 199.11 JSL 31.507 • ID 07008

KEISLER, H.J. & WALKOE JR., W.J. [1973] *The diversity of quantifier prefixes* (**J** 0036) J Symb Logic 38*79-85
  ⋄ B10 C07 C13 D55 ⋄ REV MR 51 # 12472 Zbl 259 # 02007 • ID 07033

KEISLER, H.J. [1974] *Monotone complete fields* (**P** 1083) Victoria Symp Nonstand Anal;1972 Victoria 113-115
  ⋄ C60 C65 H05 ⋄ REV MR 58 # 21585 Zbl 289 # 02040
  • ID 30004

KEISLER, H.J. [1976] *Elementary calculus* (**X** 1337) Prindle Weber Schmidt: Boston xviii + 880pp
  ⋄ H05 H98 ⋄ REV Zbl 325 # 26001 JSL 46.673 • ID 44758

KEISLER, H.J. [1976] *Foundations of infinitesimal calculus* (**X** 1337) Prindle Weber Schmidt: Boston x + 214pp
  ⋄ C20 H05 H98 ⋄ REV Zbl 333 # 26001 JSL 46.643
  • ID 33913

KEISLER, H.J. [1980] see BARWISE, J.

KEISLER, H.J. [1984] see HOOVER, D.N.

KEISLER, H.J. [1984] *An infinitesimal approach to stochastic analysis* (**S** 2450) Mem Amer Math Soc, NS 48/297*x + 184pp
  ⋄ H05 ⋄ REV MR 86c:60086 Zbl 529 # 60062 • ID 33914

KEISLER, H.J. [1984] see HENSON, C.W.

KEISLER, H.J. see Vol. II, III, V for further entries

KELEMEN, P.J. & ROBINSON, A. [1972] *The nonstandard λ : $\varphi_2^4(x)$: model. I: The technique of nonstandard analysis in theoretical physics. II: The standard model from a nonstandard point of view* (**J** 0209) J Math Phys 13*1870-1874,1875-1878
  ⋄ H05 H10 ⋄ REV MR 47 # 8015 MR 47 # 8016 Zbl 247 # 02051 Zbl 247 # 02052 • ID 62934

KELEMEN, P.J. [1974] *Quantum mechanics, quantum field theory, hyper-quantum mechanics* (**P** 1083) Victoria Symp Nonstand Anal;1972 Victoria 116-121
  ⋄ H10 ⋄ REV MR 58 # 4003 • ID 26642

KELEMEN, P.J. [1981] see JONES, C.K.R.T.

KELLER, J.P. [1978] *The intermittent server* (**J** 0056) Publ Dep Math, Lyon 15/1*83-95
  ⋄ H10 ⋄ REV MR 81h:60118 Zbl 418 # 60091 • ID 81819

KELLER, J.P. see Vol. III for further entries

KELLNER, W.G. & MULLIN, A.A. [1958] *A residue test for boolean functions* (**J** 0466) Trans Illinois State Acad Sci 51/3-4*14-19
  ⋄ B05 ⋄ REV JSL 25.185 • ID 28762

KEMENY, J.G. [1948] *Models of logical systems* (**J** 0036) J Symb Logic 13*16-30
  ⋄ B22 ⋄ REV MR 9.487 Zbl 35.4 JSL 13.154 • ID 07048

KEMENY, J.G. [1956] *A new approach to semantics - part I,II*
(J 0036) J Symb Logic 21*1-27,149-161
⋄ A05 B15 ⋄ REV MR 18.270 Zbl 73.246 JSL 35.310
• ID 20814

KEMENY, J.G. [1956] *Semantics as a branch of logic* (C 0647)
Encycl Britannica 20*313,313A-313D
⋄ B98 ⋄ REV JSL 23.23 • ID 16857

KEMENY, J.G. see Vol. II, III, VI for further entries

KEMPSKI VON, J. [1956] *Relationen- und praedikatenlogische Untersuchungen zur Syllogistik* (J 1114) Arch Phil 5*407-419
• REPR [1956] (J 0009) Arch Math Logik Grundlagenforsch 2*87-99
⋄ B20 ⋄ REV Zbl 71.9 JSL 24.79 • ID 07056

KENELLY, J.W. [1967] *Informal logic* (X 0802) Allyn & Bacon: London viii+134pp
⋄ B98 ⋄ REV Zbl 162.11 • ID 22502

KENNEDY, H.C. [1972] *The origins of modern axiomatics: Pasch to Peano* (J 0005) Amer Math Mon 79*133-136
⋄ A10 B30 ⋄ REV MR 46#7 Zbl 227#50003 • ID 07057

KENT, C.F. [1975] *Independence versus logical independence in the countable case* (P 0775) Logic Colloq;1973 Bristol 399-408
⋄ B10 ⋄ REV MR 52#2815 Zbl 331#02029 • ID 17614

KENT, C.F. see Vol. III, IV, VI for further entries

KENTON, S.A. [1979] *Mathematical foundations of constitutional law* (J 0497) Math Mag 52*223-227
⋄ A05 B80 ⋄ REV Zbl 418#03017 • ID 53302

KERNTOPF, P. [1975] *Iterative compositions of boolean functions* (P 3300) Asilomar Conf Circ, Syst & Comput (8);1974 Pacific Grove 78-81
⋄ B05 ⋄ REV Zbl 352#94026 • ID 50074

KERNTOPF, P. see Vol. II for further entries

KETONEN, J. & WEYHRAUCH, R.W. [1984] *A decidable fragment of predicate calculus* (J 1426) Theor Comput Sci 32*297-307
⋄ B20 B25 ⋄ REV MR 86g:03018 • ID 44002

KETONEN, J. [1984] *EKL-a mathematically oriented proof checker* (P 2633) Autom Deduct (7);1984 Napa 65-79
⋄ B35 ⋄ REV MR 85k:68003 • ID 44450

KETONEN, J. see Vol. III, V, VI for further entries

KETONEN, O. [1941] *On the completeness of the predicate calculus (Finnish)* (J 0963) Ajatus (Helsinki) 10*77-92
⋄ B10 ⋄ REV JSL 7.126 • ID 41448

KETONEN, O. [1944] *Untersuchungen zum Praedikatenkalkuel* (J 0990) Soc Sci Fennicae Comment Phys-Math 23*71pp
⋄ B10 ⋄ REV MR 8.125 JSL 10.127 • ID 07076

KEYNES, J.N. [1884] *Studies and exercises in formal logic, including a generalisation of logical processes in their application to complex inferences* (X 1253) Crowell Collier & Macmillan: New York xii+414pp
⋄ B98 ⋄ REV JSL 29.135 • ID 22551

KEYSER, C.J. [1902] *Concerning the axiom of infinity and mathematical induction* (J 0015) Bull Amer Math Soc 9*424-434
⋄ A05 B28 E30 ⋄ REV FdM 34.71 • ID 07078

KFOURY, A.J. [1982] *Some connections between iterative programs, recursive programs, and first-order logic* (P 3738) Log of Progr;1981 Yorktown Heights 150-166
⋄ B10 B75 ⋄ REV MR 83i:68028 Zbl 481#68018 • ID 38455

KFOURY, A.J. see Vol. II, IV for further entries

KHACHATRYAN, M.A. [1982] *Complexity of derivation of certain propositional formulas (Russian) (Armenian summary)* (S 0422) Tr Vychisl Tsentra Akad Nauk Armyan SSR & Univ Erevan 10*103-123
⋄ B05 ⋄ REV MR 85f:03059 • ID 40802

KHACHATRYAN, M.A. see Vol. VI for further entries

KHAN, M.A. [1974] *Approximately convex average sums of unbounded sets* (J 0053) Proc Amer Math Soc 43*181-185
⋄ H05 ⋄ REV MR 49#3691 • ID 07079

KHAN, M.A. [1976] *Oligopoly in markets with a continuum of traders: an asymptotic interpretation* (J 2750) J Econ Th 12*273-297
⋄ H10 ⋄ REV MR 53#15305 Zbl 345#90007 • ID 81837

KHAN, M.A. [1977] *Some remarks on sets with unbounded non-convexities* (J 3217) Metroeconomica 29*149-158
⋄ H10 ⋄ REV MR 80i:52004 Zbl 415#03056 • ID 53158

KHAN, M.A. [1980] see BROWN, D.J.

KHARIN, N.N. [1963] *Mathematical logic and the theory of sets (Russian)* (X 1649) Rusvozizdat: Leningrad 192pp
⋄ A05 B98 ⋄ REV MR 30#10 • ID 42867

KHASIN, L.S. [1969] *Complexity bounds for the realization of monotonic symmetrical functions by means of formulas in the basis* ∨, ∧, ¬ *(Russian)* (J 0023) Dokl Akad Nauk SSSR 189*752-755
• TRANSL [1969] (J 0470) Sov Phys, Dokl 14*1149-1151
⋄ B05 D20 ⋄ REV MR 41#6621 Zbl 206.290 • ID 07080

KHASIN, L.S. [1969] *The realization of monotone symmetric functions by formulae in the base* ∨, ∧, ¬ *(Russian)* (J 0052) Probl Kibern 21*253-257
• TRANSL [1969] (J 0471) Syst Th Res 21*254-259
⋄ B05 B50 ⋄ REV MR 44#1575 Zbl 203.490 • ID 05733

KHASIN, L.S. [1970] *The use of negation for the realization of the monotone symmetric functions of the algebra of logic by formulae in the basis* ∨, ∧, ¬ *(Russian)* (J 0071) Met Diskr Analiz (Novosibirsk) 17*45-55
⋄ B05 ⋄ REV MR 44#7798 Zbl 206.290 • ID 05734

KHINCHIN, A.YA. [1923] *Das Stetigkeitsaxiom des Linearkontinuums als Induktionsprinzip betrachtet* (J 0027) Fund Math 4*164-166
⋄ B28 E75 ⋄ REV FdM 49.149 • ID 43534

KHINCHIN, A.YA. [1929] *The role and the character of induction in mathematics (Russian)* (S 4104) Sbor Rab Mat Razdela Komm Akad 1*5-7
⋄ B28 ⋄ ID 43543

KHINCHIN, A.YA. [1949] *The simplest linear continuum (Russian)* (J 0067) Usp Mat Nauk 4/2*180-197
⋄ B28 E07 ⋄ REV MR 11.2 • ID 06096

KHINCHIN, A.YA. see Vol. VI for further entries

KHOMICH, V.I. [1970] *On the complexity of realization of propositional formulae (Russian)* (J 0023) Dokl Akad Nauk SSSR 195∗1050-1051
- TRANSL [1970] (J 0062) Sov Math, Dokl 11∗1637-1639
  ◇ B35 F50 ◇ REV MR 44 # 1576 Zbl 218 # 02028
- ID 06201

KHOMICH, V.I. [1975] *Weakly and strongly nonrealizable propositional formulae (Russian)* (J 0068) Z Math Logik Grundlagen Math 21∗267-288
  ◇ B20 D15 D20 F50 ◇ REV MR 51 # 10048 Zbl 309 # 02021 • ID 17549

KHOMICH, V.I. [1979] *Separability of superintuitionistic propositional calculi (Russian)* (S 0554) Issl Teor Algor & Mat Logik (Moskva) 3∗98-114
  ◇ B20 B55 ◇ REV MR 82d:03039 Zbl 422 # 03007 Zbl 464 # 03025 • ID 53468

KHOMICH, V.I. see Vol. II, IV, VI for further entries

KHRAPCHENKO, V.M. [1976] *The complexity of the realization of symmetric functions of the algebra of logic by finitely based formulas (Russian)* (J 0052) Probl Kibern 31∗231-234,238
  ◇ B05 ◇ REV MR 54 # 14881 Zbl 414 # 94044 • ID 81600

KHRAPCHENKO, V.M. [1978] *The relation between the complexity and depth of formulas (Russian)* (J 0071) Met Diskr Analiz (Novosibirsk) 32∗76-94,97
  ◇ B05 ◇ REV MR 81d:94035 Zbl 456 # 94024 • ID 54326

KHRAPCHENKO, V.M. see Vol. IV for further entries

KHRISTOV, KH.YA. & TODOROV, T.D. [1973] *Asymptotic numbers - a generalization of the notion of number* (J 0344) Izv Bulgar Akad Nauk Fiz Inst 24∗221-231
  ◇ H05 H20 ◇ REV MR 49 # 10657 Zbl 357 # 10032
- ID 17160

KHRISTOV, KH.YA. [1974] *Eine neue Art von verallgemeinerten Funktionen - die asymptotischen Funktionen* (J 3408) Nova Acta Akad Leopoldina, NF (Halle) 39∗181-197
  ◇ H05 ◇ REV Zbl 456 # 46037 • ID 54317

KHRISTOV, KH.YA. & TODOROV, T.D. [1976] *Asymptotic numbers - algebraic operations with them* (J 2547) Serdica, Bulgar Math Publ 2∗87-102
  ◇ H05 ◇ REV MR 54 # 5702 Zbl 357 # 10033 • ID 50430

KHRISTOV, KH.YA. [1981] see DAMYANOV, B.P.

KHRISTOV, KH.YA. see Vol. III for further entries

KHROMOJ, YA.V. [1978] *A collection of exercises and problems in mathematical logic (Ukrainian)* (X 2645) Vishcha Shkola: Kiev 159pp
  ◇ B98 ◇ REV MR 58 # 16075 • ID 74291

KIELKOPF, C.F. [1972] *Premisses are not axioms* (J 0047) Notre Dame J Formal Log 13∗129-130
  ◇ B05 ◇ REV Zbl 227 # 02004 • ID 27353

KIELKOPF, C.F. see Vol. II for further entries

KIESOW, H. [1958] *Einige Bemerkungen zur Grundlegung der Wahrscheinlichkeitstheorie in elementaren Sprachen* (J 0009) Arch Math Logik Grundlagenforsch 4∗124-127
  ◇ B30 ◇ REV MR 21 # 6603 Zbl 89.131 • ID 33308

KILGORE, W.J. [1968] *An introductory logic* (X 0818) Holt Rinehart & Winston: New York xiv+352pp
  ◇ B98 ◇ ID 22503

KILMISTER, C.W. [1967] *Language, logic and mathematics* (X 1165) Hodder & Stroughton: London v+124pp
  ◇ A05 B98 D35 ◇ REV MR 37 # 1223 Zbl 162.11
- ID 22338

KIM, K.H. [1982] *Boolean matrix theory and applications* (S 3310) Lect Notes Pure Appl Math 425pp
  ◇ B05 ◇ REV MR 84a:15001 Zbl 495 # 15003 • ID 38516

KING, J. [1970] see FLOYD, R.W.

KINOKUNIYA, Y. [1973] *Relativities between sets and measurements* (S 0445) Mem Muroran Inst Tech 8∗29-41
  ◇ B30 E50 E75 ◇ REV MR 54 # 5414 • ID 81856

KINOKUNIYA, Y. see Vol. V for further entries

KIPNIS, M.M. [1967] *A property of propositional formulas (Russian)* (J 0023) Dokl Akad Nauk SSSR 174∗277-278
- TRANSL [1967] (J 0062) Sov Math, Dokl 8∗620-622
  ◇ B05 ◇ REV MR 35 # 6547 Zbl 183.7 JSL 33.606
- ID 19098

KIPNIS, M.M. see Vol. IV, VI for further entries

KIRCHNER, C. [1983] see JOUANNAUD, J.-P.

KIRCHNER, H. [1983] see JOUANNAUD, J.-P.

KIREMITDJIAN, G. [1975] *Verification interactive de demonstrations mathematiques* (X 2854) Univ Paris XI UER Math: Paris 94pp
  ◇ B35 ◇ REV MR 58 # 32126 • ID 81857

KIRIN, V.G. [1975] *On free and independent terms of a formula (Serbo-Croatian summary)* (J 3519) Glas Mat, Ser 3 (Zagreb) 10(30)∗207-217
  ◇ B10 ◇ REV MR 54 # 2449 Zbl 319 # 02011 • ID 24037

KIRIN, V.G. [1978] *A semantic characterization of terms free for some variable in a formula* (J 0063) Studia Logica 37∗337-340
  ◇ B10 ◇ REV MR 81h:03024 Zbl 406 # 03011 • ID 56094

KIRIN, V.G. [1978] *Prolegomena to mathematics. Elements of mathematical logic (Serbo-Croatian)* (X 3290) Skolska Knjiga: Zagreb 31pp
  ◇ B10 ◇ REV Zbl 458 # 03005 • ID 54411

KIRIN, V.G. [1980] *On some applications of a tautology* (J 3519) Glas Mat, Ser 3 (Zagreb) 15(35)∗243-247
  ◇ B80 ◇ REV MR 82b:03028 Zbl 453 # 03008 • ID 54138

KIRIN, V.G. [1982] *On decomposition of formulas of a first-order language* (J 1008) Demonstr Math (Warsaw) 15∗647-658
  ◇ B10 ◇ REV MR 84e:03016 Zbl 527 # 03040 • ID 34354

KIRIN, V.G. [1983] *A note on the deduction theorem* (J 3944) Rad Jugosl Akad Znan Umjet, Mat Znan 2∗97-109
  ◇ B10 ◇ REV Zbl 528 # 03003 • ID 37623

KIRIN, V.G. [1984] *On sentential tautologies within first order languages (Serb.-Croat. summary)* (J 3519) Glas Mat, Ser 3 (Zagreb) 19(39)∗203-216
  ◇ B10 ◇ REV Zbl 561 # 03003 • ID 46527

KIRIN, V.G. see Vol. II, V for further entries

KIRK, R.B. [1975] *The Haar integral via non-standard analysis* (J 0048) Pac J Math 58∗517-527
  ◇ E75 H05 ◇ REV MR 51 # 13197 Zbl 309 # 43002
- ID 17233

KIRK, R.B. see Vol. V for further entries

KIRK, R.E. [1981] *A complete semantics for implicational logics* (J 0068) Z Math Logik Grundlagen Math 27*381-383
⋄ B20 ⋄ REV MR 83a:03021 Zbl 498 # 03016 • ID 33202

KIRK, R.E. [1982] *A result on propositional logics having the disjunction property* (J 0047) Notre Dame J Formal Log 23*71-74
⋄ B22 ⋄ REV MR 83e:03036 Zbl 452 # 03027 • ID 55080

KIRK, R.E. see Vol. II, III, VI for further entries

KIRKERUD, B. [1982] *Completeness of Hoare-calculi revisited* (J 0130) BIT 22*402-418
⋄ B10 ⋄ REV MR 84g:68013 Zbl 506 # 68030 • ID 39425

KIRSANOV, G.M. & TSEJTLIN, G.E. & YUSHCHENKO, E.L. [1979] *ANALYST: A program package for proving identities (theorems) in axiomatic algorithmic-algebra systems (Russian)* (J 0040) Kibernetika, Akad Nauk Ukr SSR 1979/4*28-33
• TRANSL [1979] (J 0021) Cybernetics 15*467-473
⋄ B35 ⋄ REV Zbl 454 # 68115 • ID 69657

KIRSCH, A. [1970] *Die Einfuehrung der natuerlichen Zahlen als Operatoren* (J 0160) Math-Phys Sem-ber, NS 17*57-67
⋄ B28 ⋄ REV MR 41 # 5184 Zbl 199.285 • ID 48068

KIRSCHENMANN, P. [1972] *Concepts of randomness* (J 0122) J Philos Logic 1*395-414
⋄ B30 ⋄ REV MR 54 # 14027 Zbl 268 # 60001 • ID 81859

KIRUTA, A.YA. [1979] *The theory of expected utility for intransitive preferences (Russian)* (S 2874) Vopr Kibern, Akad Nauk SSSR 58*117-126
⋄ B30 ⋄ REV MR 81j:90017 • ID 81860

KISELEV, A.A. & KOGALOVSKIJ, S.R. [1979] *Some remarks on higher order monadic languages (Russian)* (P 2554) All-Union Symp Th Log Infer;1974 Moskva 274-283
⋄ B15 C85 ⋄ REV MR 84k:03033 • ID 32589

KISELEV, A.A. see Vol. IV, V for further entries

KITAHASHI, T. & OGAWA, H. & TANAKA, K. [1977] *Formal deduction based on a directed AND/OR graph and its completeness (Japanese)* (J 0979) Denshi Tsushin Gakkai Ronbunshi, Sect A-D 60-D/11*913-920
• TRANSL [1977] (J 0464) Syst-Comp-Controls 8/6*9-16
⋄ B35 ⋄ REV MR 80g:68120 • ID 82387

KITAHASHI, T. & OGAWA, H. & TANAKA, K. [1977] *Minimum deduction process by D graph deduction (Japanese)* (J 0979) Denshi Tsushin Gakkai Ronbunshi, Sect A-D 60-D/11*905-912
• TRANSL [1977] (J 0464) Syst-Comp-Controls 8/6*1-8
⋄ B35 ⋄ REV MR 80g:68119 • ID 82388

KITAHASHI, T. see Vol. II for further entries

KLAUA, D. [1961] *Konstruktion ganzer, rationaler und reeller Ordinalzahlen und die diskontinuierliche Struktur der transfiniten reellen Zahlenraeume* (X 0911) Akademie Verlag: Berlin 141pp
⋄ B28 E10 ⋄ REV MR 25 # 5002 Zbl 206.12 • ID 21070

KLAUA, D. see Vol. II, III, V, VI for further entries

KLAUS, G. [1964] *Moderne Logik. Abriss der formalen Logik* (X 0806) Dt Verlag Wiss: Berlin xii+452pp
⋄ B98 ⋄ ID 14564

KLEENE, S.C. [1935] *A theory of positive integers in formal logic I,II* (J 0100) Amer J Math 57*153-173,219-244
⋄ B28 B40 ⋄ REV Zbl 11.2 FdM 61.55 FdM 61.56
• ID 28412

KLEENE, S.C. [1939] see EVANS, H.P.

KLEENE, S.C. [1945] *On the interpretation of intuitionistic number theory* (J 0036) J Symb Logic 10*109-124
⋄ B28 F50 ⋄ REV MR 7.406 JSL 12.91 • ID 07169

KLEENE, S.C. [1952] *Introduction to metamathematics* (X 0809) North Holland: Amsterdam x+550pp
• TRANSL [1957] (X 1656) Izdat Inostr Lit: Moskva 526pp (Russian) [1984] (X 1876) Kexue Chubanshe: Beijing xii+234pp,x+235-688pp (Chinese) [1974] (X 1781) Tecnos: Madrid (Spanish)
⋄ A05 B98 D98 F30 F50 F98 ⋄ REV MR 14.525 Zbl 47.7 JSL 19.215 JSL 25.280 JSL 33.290 JSL 35.350 JSL 38.333 • REM Co-publisher: Wolters-Noordhoff; 8th revised ed. 1980. Chinese transl. in 2 parts • ID 07173

KLEENE, S.C. [1960] *Mathematical logic: constructive and non-constructive operations* (P 0660) Int Congr Math (II, 8);1958 Edinburgh 137-153 • ERR/ADD [1963] (J 0064) Trans Amer Math Soc 108*142
⋄ B98 D98 ⋄ REV MR 22 # 5569 Zbl 126.19 JSL 27.78
• ID 07186

KLEENE, S.C. [1967] *Mathematical logic* (X 0827) Wiley & Sons: New York xiii+398pp
• TRANSL [1967] (X 3636) Tokyo Tosho: Tokyo 200pp+266pp [1971] (X 0850) Colin: Paris [1973] (X 0885) Mir: Moskva 480pp
⋄ B98 D20 D25 D55 D98 F30 F98 ⋄ REV MR 36 # 25 Zbl 149.243 JSL 35.438 • ID 45895

KLEENE, S.C. [1969] *The new logic* (J 1843) Amer Sci 57*333-347
⋄ B98 ⋄ ID 32371

KLEENE, S.C. see Vol. II, IV, VI for further entries

KLEIN-BARMEN, F. [1933] *Ueber gekoppelte Axiomensysteme in der Theorie der abstrakten Verknuepfungen* (J 0044) Math Z 37*39-60
⋄ B30 ⋄ REV FdM 59.923 • ID 07204

KLEIN-BARMEN, F. see Vol. III, V for further entries

KLEINBERG, E.M. [1979] see HENLE, J.M.

KLEINBERG, E.M. see Vol. III, IV, V, VI for further entries

KLEINE BUENING, H. [1981] *Some undecidable theories with monadic predicates and without equality* (J 0009) Arch Math Logik Grundlagenforsch 21*137-148
⋄ B20 D35 ⋄ REV MR 83k:03052 Zbl 475 # 03022
• ID 55476

KLEINE BUENING, H. see Vol. III, IV for further entries

KLEINKNECHT, R. [1981] *Eliminability, noncreativity and explicit definability* (J 0079) Logique & Anal, NS 24*223-229
⋄ B10 ⋄ REV MR 84j:03025 Zbl 504 # 03015 • ID 34619

KLEINKNECHT, R. see Vol. III, V for further entries

KLEMKE, E.D. [1974] see JACOBSON, A.

KLETTE, R. & LINDNER, R. [1979] *Zweidimensional arbeitende Vektormaschinen und ihr Leistungsvermoegen bei der Loesung von Entscheidungsproblemen der Aussagenlogik* (J 0129) Elektr Informationsverarbeitung & Kybern 15*37-46
⋄ B35 D05 D15 ⋄ REV MR 80g:68081 Zbl 414 # 68024
• ID 53095

KLETTE, R. see Vol. IV for further entries

KLIMOVSKY, G. [1956] *Tres enunciados equivalentes al teorema de Zorn* (S 0473) Contrib Cient, Ser Mat 2/1*29pp
⋄ B05 E25 G05 ⋄ REV MR 20 # 1628 Zbl 74.12 JSL 36.681 • ID 07233

KLIMOVSKY, G. see Vol. III, V for further entries

KLIX, W.-D. [1977] see HARMS, S.

KLOETZER, G. & RAUTENBERG, W. [1972] *Im Grenzbereich Algebra, Logik, Maschinen. 10 Jahre Forschungsarbeit der Nowosibirsker Schule A.I. Malcev's (eine Studie)* (J 1670) Mitt Math Ges DDR 1972/1-2*43-102
⋄ A10 B98 C05 D05 D98 ⋄ REV Zbl 243 # 02001 • ID 30031

KLUG, U. [1951] *Juristische Logik* (X 0811) Springer: Heidelberg & New York viii+160pp
⋄ B80 ⋄ REV JSL 17.274 JSL 24.87 • ID 25274

KLUG, U. [1962] *Bemerkungen zur logischen Analyse einiger rechtstheoretischer Begriffe und Behauptungen* (C 0712) Logik & Logikkalkuel 115-125
⋄ B80 ⋄ ID 07243

KNEEBONE, G.T. [1963] *Mathematical logic and the foundations of mathematics: an introduction* (X 0864) Van Nostrand: New York xiv+435pp
⋄ A05 A10 B98 ⋄ REV MR 27 # 26 Zbl 166.247 • ID 19088

KNEEBONE, G.T. [1971] *Mathematical logic in relation to ordinary mathematics* (J 0178) Stud Gen 24*946-959
⋄ A05 B98 ⋄ REV Zbl 245 # 02005 • ID 28859

KNEEBONE, G.T. see Vol. V for further entries

KNEECE, R.R. [1972] see BRACE, J.W.

K"NEV, P. [1972] *An algorithm for the determination of the absolute minimum form of weakly determined functions* (J 3121) Izv Bulgar Akad Nauk Tekhn Kibern 14*5-14
⋄ B35 ⋄ REV Zbl 258 # 94031 • ID 62822

KNOPP, K. [1939] *Darstellung der reellen Zahlen durch Grenzprozesse* (X 0823) Teubner: Stuttgart 30pp
⋄ B28 ⋄ REV Zbl 21.213 JSL 5.154 FdM 65.192 • ID 23385

KNOPP, K. see Vol. V for further entries

KOBAYASHI, M. [1982] see ISHIMOTO, A.

KOBRZYNSKI, Z. [1937] *La theorie des determinants logiques* (J 0459) C R Soc Sci Lett Varsovie Cl 3 30*75-82
⋄ A05 B05 ⋄ REV Zbl 17.290 JSL 3.52 FdM 63.825 • ID 07266

KOCHEN, S. [1957] *Completeness of algebraic systems in higher order calculi* (P 1675) Summer Inst Symb Log;1957 Ithaca 370-376
⋄ B15 C35 C60 C85 ⋄ REV JSL 27.97 • ID 07272

KOCHEN, S. see Vol. II, III, V for further entries

KOCHKAREV, B.S. [1963] *The improvement of E. N. Hilbert's estimates for the number of monotone functions of the algebra of logic (Russian)* (P 4037) Itog Nauch Konf Kazan Univ;1962 Kazan 106-108
⋄ B05 ⋄ REV MR 32 # 5562 • ID 42948

KOCHKAREV, B.S. [1965] *Estimation of the complexity of formulas for monotone functions of the algebra of logic in the class of disjunctive normal forms (d.n.f.) (Russian)* (J 0968) Uch Zap Univ, Kazan 125/6*49-57
⋄ B05 ⋄ REV MR 37 # 28 Zbl 256 # 94046 • ID 07277

KOCK, A. & MIKKELSEN, C.J. [1974] *Topos-theoretic factorization of non-standard extensions* (P 1083) Victoria Symp Nonstand Anal;1972 Victoria 122-143
⋄ G30 H05 ⋄ REV MR 58 # 841 Zbl 276 # 18001 • REM A preprint appeared 1972 by Aarhus University • ID 29047

KOCK, A. see Vol. III, V, VI for further entries

KODERA, H. [1980] *Remark on classical logic and intuitionistic logic* (J 2678) Bull Aichi Univ Educ Nat Sci 29*31-37
⋄ B20 F05 F50 ⋄ REV MR 81m:03015 • ID 74921

KOEGST, M. [1973] *On the problem of the decomposition of Boolean functions* (P 3272) Strukt & Metodol Rek Soz Sist Avtom Ustr;1973 Budapest 29-31
⋄ B05 ⋄ REV MR 50 # 16067 Zbl 334 # 94013 • ID 63029

KOEGST, M. & OBERST, E. [1979] *Zur graphischen Darstellung Boolescher Funktionen* (J 1633) Angew Inf 21*206-209
⋄ B05 ⋄ REV Zbl 397 # 94017 • ID 69748

KOEN, R.I. & MANEV, K.N. [1978] *Algebraic properties of the universal functions of the algebra of logic (Russian)* (J 0137) C R Acad Bulgar Sci 31*1237-1240
⋄ B05 ⋄ REV MR 81c:03053 Zbl 418 # 94017 • ID 53338

KOEN, R.I. [1979] *Subfunctions and invariant classes of functions of the algebra of logic (Russian)* (J 0137) C R Acad Bulgar Sci 32*1605-1608
⋄ B05 ⋄ REV MR 81j:03095 Zbl 427 # 03053 • ID 53740

KOENIG, E.C. [1979] *Establishing valid arguments by computer and storing their meanings: A premiss of form a of Aristotelian logic* (J 1429) Kybernetes 8*299-303
⋄ B35 ⋄ REV Zbl 413 # 68108 • ID 53045

KOENIG, J. [1914] *Neue Grundlagen der Logik, Arithmetik und Mengenlehre* (X 2636) Veit: Leipzig viii+259pp
⋄ A05 B30 E10 E98 ⋄ REV FdM 45.124 • ID 21392

KOENIG, J. see Vol. V for further entries

KOGALOVSKIJ, S.R. [1966] *On higher order logic (Russian)* (J 0023) Dokl Akad Nauk SSSR 171*1272-1274
• TRANSL [1966] (J 0062) Sov Math, Dokl 7*1642-1645
⋄ B15 C40 C85 ⋄ REV MR 35 # 35 Zbl 204.1 • ID 07292

KOGALOVSKIJ, S.R. [1966] *On the semantics of the theory of types (Russian)* (J 0031) Izv Vyssh Ucheb Zaved, Mat (Kazan) 1966/1*89-98
• TRANSL [1970] (J 0225) Amer Math Soc, Transl, Ser 2 94*93-104
⋄ B15 C40 C85 ⋄ REV MR 34 # 2442 Zbl 202.301 JSL 37.193 • ID 24967

KOGALOVSKIJ, S.R. [1968] *Some remarks on higher order logic (Russian)* (J 0023) Dokl Akad Nauk SSSR 178*1007-1009
• TRANSL [1968] (J 0062) Sov Math, Dokl 9*227-229
⋄ B15 C40 C85 F35 ⋄ REV MR 39 # 43 Zbl 182.321 • ID 07299

KOGALOVSKIJ, S.R. [1970] *Some reduction theorems for higher order logic (Russian)* (J 0023) Dokl Akad Nauk SSSR 190*519-522
• TRANSL [1970] (J 0062) Sov Math, Dokl 11*138-142
⋄ B15 F35 ⋄ REV MR 41 # 8216 Zbl 212.312 • ID 07300

KOGALOVSKIJ, S.R. & RORER, M.A. [1973] *On the question of the definability of the concept of definability (Russian)* (J 0226) Uch Zap Ped Inst, Ivanovo 125*46-72
⋄ B15 C40 C85 ⋄ REV MR 58 # 10289 • ID 74934

KOGALOVSKIJ, S.R. [1974] *Certain criteria for localness for higher order formulas (Russian)* (J 3536) Uch Zap Univ, Ivanovo 130*3-18
⋄ B15 C85 ⋄ REV MR 57 # 5730 • ID 74930

KOGALOVSKIJ, S.R. [1974] *Reductions in second-order logic (Russian)* (C 2578) Filos & Logika 363-397
⋄ B15 C85 ⋄ REV MR 58 # 27298 • ID 74929

KOGALOVSKIJ, S.R. [1976] see CHEPURNOV, B.A.

KOGALOVSKIJ, S.R. [1979] see KISELEV, A.A.

KOGALOVSKIJ, S.R. see Vol. III, IV, V for further entries

KOGAN, A.YU. & ZHURAVLEV, YU.I. [1985] *Realization of Boolean functions with a small number of zeros by disjunctive normal forms and related problems (Russian)* (J 0023) Dokl Akad Nauk SSSR 285*795-799
⋄ B05 B75 ⋄ ID 48994

KOGAN, D.I. [1969] see GLEBSKIJ, YU.V.

KOGAN, D.I. see Vol. IV for further entries

KOHOUT, L.J. [1976] *Application of multi-valued logics to the study of human movement control and of movement disorders* (P 2011) Int Symp Multi-Val Log (6);1976 Logan 224-232
⋄ B80 ⋄ REV MR 58 # 20563 • ID 81899

KOHOUT, L.J. see Vol. II, IV, V for further entries

KOKOSZYNSKA, M. & KUBINSKI, T. & SLUPECKI, J. [1956] *The application of logistic concepts to the explication of some concepts on natural sciences (Polish) (Russian and English summaries)* (J 0063) Studia Logica 4*155-211
⋄ B80 ⋄ REV MR 18.457 Zbl 74.247 • ID 33072

KOKOSZYNSKA, M. see Vol. II for further entries

KOLKER, R.J. [1966] *Computing boolean functions with finite state machines* (P 0573) Inform Processing (3);1965 New York 2*523
⋄ B35 ⋄ ID 29441

KOLMOGOROV, A.N. [1946] *On the foundations of the theory of real numbers (Russian)* (J 0067) Usp Mat Nauk 1/1(11)*217-219
• TRANSL [1951] (J 0086) Cas Pestovani Mat, Ceskoslov Akad Ved 76*155-157 (Czech)
⋄ B28 ⋄ REV MR 14.255 • ID 07317

KOLMOGOROV, A.N. [1969] *Logical basis of information theory and probability theory (Russian)* (J 2320) Probl Peredachi Inf, Akad Nauk SSSR 5/3*3-7
• TRANSL [1969] (J 3419) Probl Inf Transmiss 5/3*1-4
⋄ B30 D15 D20 D80 ⋄ REV MR 47 # 3105 Zbl 265 # 94010 • ID 63036

KOLMOGOROV, A.N. see Vol. III, IV, V, VI for further entries

KOLODZIEJ, R. [1977] *Algorithms for the validity of formulas of predicate calculus through hypergraphs (Polish) (English and Russian summaries)* (S 4733) Prace Inst Mat, Politech Wroclaw, Ser Stud Mater 13*9-18
⋄ B35 ⋄ REV MR 58 # 21559 Zbl 378 # 05051 • ID 74947

KOLODZIEJ, R. see Vol. VI for further entries

KOLPAKOV, V.I. [1969] *Estimate of the number of covers of the n-dimensional cube (Russian)* (J 0071) Met Diskr Analiz (Novosibirsk) 14*16-17
⋄ B05 ⋄ REV MR 46 # 79 JSL 37.627 • ID 07319

KOLPAKOV, V.I. [1970] *The correspondence between monotone functions and a set of irredundant tests for tables (Russian)* (J 0071) Met Diskr Analiz (Novosibirsk) 16*44-50
⋄ B05 ⋄ REV MR 44 # 6407 • ID 07320

KOLPAKOV, V.I. see Vol. II for further entries

KOMKOV, V. & WAID, C.C. [1973] *Asymptotic behavior of non-linear inhomogeneous equations via non-standard analysis I: Second order equations* (J 1405) Ann Pol Math 28*67-87
⋄ H05 ⋄ REV MR 48 # 4412 Zbl 252 # 34038 • REM Part II 1974 • ID 26647

KOMKOV, V. [1974] *Asymptotic behavior of non-linear, inhomogeneous differential equations via non-standard analysis II: Some applications to higher order equation* (J 1405) Ann Pol Math 30*205-218
⋄ H05 ⋄ REV MR 49 # 9347 Zbl 277 # 34035 • REM Part I 1973 by Komkov,V. & Waid,C.C. Part III 1980 • ID 07321

KOMKOV, V. [1979] *A note on a formal manipulation of divergent series and integrals* (J 1008) Demonstr Math (Warsaw) 12*421-428
⋄ H05 ⋄ REV MR 81c:40003 Zbl 418 # 40002 • ID 81907

KOMKOV, V. [1980] *Asymptotic behavior of nonlinear differential equations via nonstandard analysis III: Boundedness and monotone behavior of the equation* $(a(t)\varphi(x)x')' + c(t)f(x) = q(t)$ (J 1405) Ann Pol Math 38*101-108
⋄ H05 ⋄ REV MR 82d:34044 Zbl 468 # 34038 • REM Part II 1974 • ID 81906

KOMKOV, V. & MCLAUGHLIN, T.G. [1984] *Local analysis of nonstandard C^∞ functions of predistributional type* (J 1405) Ann Pol Math 44*15-38
⋄ H05 ⋄ REV Zbl 558 # 46041 • ID 44284

KOMORI, Y. & ONO, H. [1982] *Logics without the contradiction rule* (P 3833) Symp Semigroups (6);1982 Kyoto 78-79
⋄ B20 B53 ⋄ REV Zbl 501 # 03007 • ID 37775

KOMORI, Y. & ONO, H. [1985] *Logics without the contraction rule* (J 0036) J Symb Logic 50*169-201
⋄ B20 ⋄ ID 42534

KOMORI, Y. see Vol. II, III for further entries

KONDAKOV, N.I. [1978] *Woerterbuch der Logik* (X 2600) DEB - Verlag: Berlin 554pp
⋄ A05 A10 B98 ⋄ REV Zbl 382 # 03001 • ID 51925

KONDO, K. [1980] *Hering's opponent colour scheme as tristimulus projective - geometrical correlation and a boolean logical analysis of colour naming* (J 0996) RAAG Res Notes 105*36pp
⋄ B80 ⋄ REV Zbl 435 # 92002 • ID 55834

KONDO, M. [1944] *Une methode operationelle dans la theorie des nombres naturels* (J 0081) Proc Japan Acad 20*564-568
⋄ B28 ⋄ REV MR 7.355 Zbl 61.10 • ID 07325

KONDO, M. [1956] *Sur les nombres reels et nommables* (J 0109) C R Acad Sci, Paris 242*1945-1948
⋄ B28 D55 E15 F65 ⋄ REV MR 17.933 Zbl 70.278 JSL 22.299 • ID 07331

KONDO, M. [1958] *Sur les ensembles nommables et le fondement de l'analyse mathematique I* (J 2307) Japan J Math 28*1-116
⋄ B28 E15 ⋄ REV MR 22 # 4638 Zbl 93.10 • ID 37182

KONDO, M. & MURATA, H. [1965] *Problem-solving machines I,II,III* (J 0081) Proc Japan Acad 41*254-259,299-303,355-359
⋄ B35 ⋄ REV MR 32 # 4884a MR 32 # 4884b Zbl 139.331 Zbl 173.453 JSL 34.132 • ID 07342

KONDO, M. & MURATA, H. [1970] *Problem-solving machines for Euclidean goemetry* (J 0191) Inform Sci 2*395-429
⋄ B35 ⋄ ID 37184

KONDO, M. see Vol. II, IV, V, VI for further entries

KOPIEKI, R. & SARALSKI, B. & WALIGORA, G. [1975] *The research of Jaskowski on decidability theory of first order sentences I* (J 0063) Studia Logica 34*201-214
⋄ A10 B20 B25 D35 ⋄ REV MR 54 # 86 Zbl 315 # 02044 • ID 23964

KORCIK, A. [1958] *Verification of Aristotle's theory of syllogism by means of Gergonne's method (Polish) (English summary)* (J 4467) Rocz Filoz 5*5-15
⋄ B20 ⋄ REV JSL 24.81 • ID 42173

KORDOS, M. & SZCZERBA, L.W. [1969] *On the ΠΣ-axiom systems of hyperbolic and some related geometries (Russian summary)* (J 0014) Bull Acad Pol Sci, Ser Math Astron Phys 17*175-180
⋄ B30 ⋄ REV MR 39 # 6746 Zbl 185.13 • ID 07366

KORFHAGE, R. [1966] *Logic and algorithms with applications to the computer and information sciences* (X 0827) Wiley & Sons: New York xii+194pp
⋄ B98 D98 ⋄ REV MR 35 # 5279 Zbl 148.7 JSL 36.344 • ID 07376

KORNEV, YU.N. [1967] *Prinzip der Erhoehung des Unbestimmtheitsgrades boolescher Funktionen. Algorithmus zur Bestimmung der kuerzesten disjunktiven Normalformen (Russian)* (J 0040) Kibernetika, Akad Nauk Ukr SSR 1967/4*9-16
⋄ B35 ⋄ REV Zbl 248 # 94044 • ID 63058

KORNEV, YU.N. [1967] *Reduced disjunctive normal forms of partially defined boolean functions (Russian)* (J 0040) Kibernetika, Akad Nauk Ukr SSR 1967/1*10-15
• TRANSL [1967] (J 0021) Cybernetics 3/1*8-12
⋄ B05 ⋄ REV MR 43 # 126 Zbl 153.12 • ID 19131

KOROBKOV, V.K. [1956] *Realization of symmetric functions in the class of π-circuits (Russian)* (J 0023) Dokl Akad Nauk SSSR 109*260-263
⋄ B05 ⋄ REV MR 18.372 Zbl 72.249 • ID 43580

KOROBKOV, V.K. & REZNIK, T.L. [1962] *Certain algorithms for the computation of monotonic functions in the algebra of logic (Russian)* (J 0023) Dokl Akad Nauk SSSR 147*1022-1025
• TRANSL [1962] (J 0062) Sov Math, Dokl 3*1763-1767
⋄ B05 ⋄ REV MR 32 # 1142 Zbl 145.9 • ID 07377

KOROBKOV, V.K. [1963] *Estimation of the number of monotonic functions of the algebra of logic and of the complexity of the algorithm for finding the resolvent set for an arbitrary monotonic function of the algebra of logic (Russian)* (J 0023) Dokl Akad Nauk SSSR 150*744-747
• TRANSL [1963] (J 0062) Sov Math, Dokl 4*753-756
⋄ B05 G10 ⋄ REV MR 32 # 1143 Zbl 178.333 • ID 07379

KOROBKOV, V.K. [1965] *Monotone functions of the algebra of logic (Russian)* (J 0052) Probl Kibern 13*5-28
⋄ B05 ⋄ REV MR 34 # 8885 Zbl 281 # 94023 • ID 63059

KOROTKOVA, M.A. [1977] see CHEREDNICHENKO, N.D.

KORSELT, A. [1914] *Allgemeinste vollstaendige Induktion* (J 3975) Arch Math & Phys 22*280-281
⋄ B28 E07 ⋄ REV FdM 45.128 • ID 38011

KORSELT, A. see Vol. V for further entries

KORSHUNOV, A.D. [1969] *Comparison of the complexity of the largest and shortest disjunctive normal forms and a lower estimate of the number of irredundant disjunctive normal forms for almost all Boolean functions (Russian)* (J 0040) Kibernetika, Akad Nauk Ukr SSR 1969/4*1-11
• TRANSL [1969] (J 0021) Cybernetics 5*357-369
⋄ B05 D20 ⋄ REV MR 46 # 3215 Zbl 195.29 • ID 16309

KORSHUNOV, A.D. [1969] *The upper complexity bound of the shortest disjunctive normal forms of almost all boolean functions (Russian)* (J 0040) Kibernetika, Akad Nauk Ukr SSR 1969/6*1-8
• TRANSL [1969] (J 0021) Cybernetics 5*705-715
⋄ B05 ⋄ REV MR 46 # 3216 Zbl 221 # 94063 • ID 07382

KORSHUNOV, A.D. [1977] *Solution of Dedekind's problem on the number of monotonic Boolean functions (Russian)* (J 0023) Dokl Akad Nauk SSSR 233*543-546
• TRANSL [1977] (J 0062) Sov Math, Dokl 18*442-445
⋄ B05 ⋄ REV MR 58 # 33598 Zbl 366 # 06020 • ID 81917

KORSHUNOV, A.D. [1981] *Complexity of shortest disjunctive normal forms of random Boolean functions (Russian)* (J 0071) Met Diskr Analiz (Novosibirsk) 37*9-41,85
⋄ B05 ⋄ REV MR 84e:94022 MR 85c:08002 • ID 44836

KORSHUNOV, A.D. [1983] *The maximal length of dead-end disjunctive normal forms for almost all Boolean functions (Russian)* (J 0052) Probl Kibern 40*255-260
⋄ B05 ⋄ REV MR 85e:06018 Zbl 523 # 94029 • ID 39926

KORSHUNOV, A.D. see Vol. IV for further entries

KOSCIUK, S.A. [1983] *Stochastic solutions to partial differential equations* (C 3884) Nonstandard Anal - Recent Develop 113-119
⋄ H05 ⋄ REV MR 85j:60106 Zbl 521 # 60073 • ID 45791

KOSOVSKIJ, N.K. [1978] *The complexity of the solvability of boolean functional equations (Russian)* (S 0716) Vychisl Tekh Vopr Kibern (Univ Leningrad) 15*104-111
⋄ B05 D15 G05 ⋄ REV MR 80d:68062 Zbl 534 # 94007 • ID 90060

KOSOVSKIJ, N.K. [1981] *Elements of mathematical logic and its applications to the theory of subrecursive algorithms (Russian)* (X 0938) Leningrad Univ: Leningrad 192pp
⋄ B98 D20 D98 F30 F60 F98 ⋄ REV MR 83m:03013 Zbl 479 # 03001 • ID 55663

KOSOVSKIJ, N.K. see Vol. III, IV, VI for further entries

KOSPANOV, EH.SH. [1971] *The product of the shortest disjunctive normal forms (Russian)* (J 0071) Met Diskr Analiz (Novosibirsk) 18*35-40
⋄ B05 ⋄ REV MR 45 # 4895 Zbl 235 # 02015 • ID 27817

KOSSEJ, I.P. [1982] *Four fast algorithms in the method of resolutions (Russian)* (J 0977) Izv Akad Nauk SSSR, Tekh Kibern 1982/5*217-221
• TRANSL [1982] (J 0522) Engin Cybern 20/5*112-117
⋄ B35 ⋄ REV MR 85e:68090 Zbl 525 # 68056 • ID 37491

KOSSOWSKI, P. [1970] *On the axiomatic treatment of the theory of models IV: Independence of some axiom system of model theory (Polish summary)* (S 0458) Zesz Nauk, Prace Log, Uniw Krakow 5∗15-24
⋄ B22 C07 ⋄ REV MR 45 # 1748 • REM Part III 1970 by Surma,S.J. • ID 07392

KOSTYAKOV, V.M. [1972] see ANUFRIEV, F.V.

KOSTYRKO, V.F. [1964] *The reduction class $\forall\exists^n\forall$ (Russian)* (J 0003) Algebra i Logika 3/5-6∗45-55
⋄ B20 D35 ⋄ REV MR 31 # 27 Zbl 178.323 • ID 19124

KOSTYRKO, V.F. [1966] *On the AEA reduction class (Russian) (English summary)* (J 0040) Kibernetika, Akad Nauk Ukr SSR 1966/1∗17-22
⋄ B20 D35 ⋄ REV MR 34 # 4136 Zbl 145.242 • ID 19123

KOSTYRKO, V.F. [1971] *The reduction class $\forall x \forall y \exists z F(x,y,z) \wedge \forall^m \mathfrak{A}(F)$ (Russian)(English summary)* (J 0040) Kibernetika, Akad Nauk Ukr SSR 1971/5∗1-3
⋄ B20 C13 D35 ⋄ REV MR 46 # 1550 Zbl 231 # 02055 • ID 19122

KOSTYRKO, V.F. see Vol. III, IV for further entries

KOSTYUKOVICH, A.I. [1978] *On recognizing the equivalence of boolean functions with respect to the inversion of variables (Russian) (English summary)* (J 0413) Izv Akad Nauk Belor SSR, Ser Fiz-Mat 1978/1∗106-109,143
⋄ B35 ⋄ REV MR 58 # 4672 Zbl 393 # 94042 • ID 69660

KOSTYUKOVICH, A.I. [1978] *On recognizing the membership of boolean functions to one class and one type (Russian)* (J 0413) Izv Akad Nauk Belor SSR, Ser Fiz-Mat 1978/2∗125
⋄ B35 ⋄ REV Zbl 384 # 06017 • ID 69661

KOTARBINSKA, J. [1957] *The concept of sign (Polish) (Russian and English summaries)* (J 0063) Studia Logica 6∗57-143
⋄ A05 B03 ⋄ REV MR 21 # 1262 • ID 33079

KOTARBINSKI, T. [1929] *Elements of epistemology, formal logic, and methodology of the sciences (Polish)* (X 3333) Unknown Publisher: See Remarks 483pp
• LAST ED [1961] (X 2885) Zakl Narod Wyd Pol Ak: Wroclaw 648pp
⋄ A05 B98 ⋄ REM Lwow • ID 37375

KOTAS, J. & PIECZKOWSKI, A. [1970] *Allgemeine logische und mathematische Theorien* (J 0068) Z Math Logik Grundlagen Math 16∗353-376
⋄ B22 ⋄ REV MR 43 # 7291 Zbl 274 # 02024 • ID 07402

KOTAS, J. [1971] *Logical systems with implications (Polish and Russian summaries)* (J 0063) Studia Logica 28∗101-117
⋄ B22 G25 ⋄ REV MR 46 # 43 Zbl 317 # 02070 • ID 07406

KOTAS, J. (ED.) [1975] *Stanislaw Jaskowski's achievements in mathematical logic* (J 0063) Studia Logica 34∗107-214
⋄ A10 B97 ⋄ REV MR 51 # 7800 • ID 70194

KOTAS, J. see Vol. II, III, VI for further entries

KOTS, V.S. [1973] see BOBROV, A.E.

KOTS, V.S. see Vol. II for further entries

KOWALSKI, R. [1969] *Search strategies for theorem-proving* (J 0508) Machine Intelligence 5∗181-201
⋄ B35 ⋄ REV MR 45 # 6521 Zbl 218 # 68018 • ID 47342

KOWALSKI, R. [1969] see HAYES, P.J.

KOWALSKI, R. & KUEHNER, D. [1970] *Linear resolution with selection function* (1111) Preprints, Manuscr., Techn. Reports etc. 34∗
• REPR [1983] (C 4659) Autom of Reasoning 1∗542-577
⋄ B35 ⋄ REM Edinburgh Univ, Research Report 34
• ID 48959

KOWALSKI, R. [1970] *The case for using equality axioms in automatic demonstration* (P 4474) Autom Demonstr;1968 Versailles 112-127
• REPR [1983] (C 4659) Autom of Reasoning 2∗377-398
⋄ B35 ⋄ REV MR 43 # 1469 Zbl 226 # 68045 • ID 47341

KOWALSKI, R. [1976] see EMDEN VAN, M.H.

KOWALSKI, R. [1979] see DELIYANNI, A.

KOWALSKI, R. [1979] *Logic for problem solving* (X 0809) North Holland: Amsterdam xi+287pp
⋄ B35 ⋄ REV MR 81j:68003 Zbl 436 # 68002 JSL 47.477
• ID 81926

KOWALSKI, R. [1980] see BIBEL, W.

KOZEN, D. [1984] see BLASS, A.R.

KOZEN, D. see Vol. II, III, IV, VI for further entries

KOZHEVNIKOVA, G.P. [1979] see FARAT, V.M.

KRAJKIEWICZ, P. [1970] see BOSCH, W.

KRAKOVSKAYA, O.S. & MAJSTROVA, T.D. [1964] *On a method of minimization of normal forms of boolean functions (Russian)* (C 0570) Form Log & Metodol Nauk 269-300
⋄ B35 ⋄ REV Zbl 166.272 • ID 14846

KRAKOWSKI, I. [1980] *The four color problem reconsidered* (J 0095) Philos Stud 38∗91-96
⋄ A05 B35 ⋄ REV MR 82a:03007 • ID 75050

KRAMOSIL, I. [1972] *A method for random sampling of well-formed formulas (A method for random sampling of formulas of an elementary theory and statistical estimation of their deducibility equipped by a program. I)* (J 0156) Kybernetika (Prague) 8∗133-148
⋄ B35 ⋄ REV MR 48 # 8156a Zbl 242 # 02014 • ID 26338

KRAMOSIL, I. [1973] *A method for statistical testing of an at random sampled formula* (J 0156) Kybernetika (Prague) 9∗162-173
⋄ B35 ⋄ REV MR 49 # 8156b Zbl 275 # 02018 • ID 29694

KRAMOSIL, I. [1973] *Statistical estimation of deducibility in a random sequence of formulas* (P 0762) Inform Th, Stat Decis Fcts & Random Proc (6);1971 Prague 449-463
⋄ A05 B35 ⋄ REV MR 58 # 27429 Zbl 299 # 02014
• ID 63116

KRAMOSIL, I. [1973] *Statistical estimation of deducibility in formalized theories* (P 3000) Conf Probab Th (4);1971 Brasov 281-298
⋄ B35 ⋄ REV MR 55 # 5402 Zbl 282 # 62035 • ID 75055

KRAMOSIL, I. [1979] *A note on computational complexity of a statistical deducibility testing procedure* (P 2059) Math Founds of Comput Sci (8);1979 Olomouc 337-345
⋄ B35 D15 ⋄ REV MR 81e:68113 Zbl 417 # 68079
• ID 53285

KRAMOSIL, I. [1979] *Statistical approach to proof theory* (J 0156) Kybernetika (Prague) Suppl.15∗97pp
⋄ B35 ⋄ REV MR 81d:68128 Zbl 445 # 68068 • ID 56523

KRAMOSIL, I. [1980] *Computational complexity of a statistical verification procedure for propositional calculus* (P 3897) Inform Th (3);1980 Liblice 123-130
  ⋄ B35 D15 ⋄ REV Zbl 529 # 68067 • ID 38493

KRAMOSIL, I. [1980] *Gentzen-like random axiomatic systems* (P 3056) Inform Th, Stat Decis Fcts & Random Proc (7);1974 Prague A*345-352pp
  ⋄ B35 ⋄ REV MR 58 # 10294 Zbl 412 # 60002 • ID 75052

KRAMOSIL, I. [1980] *Statistical testing procedure for lengths of formalized proofs* (J 0156) Kybernetika (Prague) 16*209-224
  ⋄ B35 F20 ⋄ REV MR 83b:68097 Zbl 444 # 03007
  • ID 56455

KRAMOSIL, I. [1982] *Three semantical interpretations of a statistical theoremhood testing procedure* (J 0156) Kybernetika (Prague) 18*440-446
  ⋄ B35 ⋄ REV MR 84c:03030 Zbl 524 # 03007 • ID 35688

KRAMOSIL, I. [1983] *Statistical verification procedures for propositional calculus* (J 3932) Comput Artif Intell (Bratislava) 2*235-258
  ⋄ B35 ⋄ REV MR 85h:68030 Zbl 529 # 68066 • ID 38492

KRAMOSIL, I. & SINDELAR, J. [1985] *Computational complexity of probabilistic searching algorithms over Herbrand universes* (J 3932) Comput Artif Intell (Bratislava) 4*97-108
  ⋄ B35 ⋄ REV Zbl 561 # 68062 • ID 47494

KRAMOSIL, I. see Vol. II, III, IV for further entries

KRASZEWSKI, Z. [1962] *Modification of a decision method of theorems of 1-place functional calculus (Polish)* (J 0063) Studia Logica 13*237-247
  ⋄ B20 ⋄ ID 33112

KRASZEWSKI, Z. see Vol. II, V for further entries

KRAVTSOV, S.S. [1972] *Certain topological properties of the functions of the algebra of logic (Russian)* (J 0071) Met Diskr Analiz (Novosibirsk) 21*10-25
  ⋄ B05 ⋄ REV MR 49 # 8871 Zbl 273 # 94038 • ID 63126

KRAWCZYK, A. & MAREK, W. [1977] *On the rules of proof generated by hierarchies* (P 1695) Set Th & Hierarch Th (3);1976 Bierutowice 227-239
  ⋄ B15 C55 C85 E45 ⋄ REV MR 58 # 21621 Zbl 424 # 03026 • ID 31404

KRAWCZYK, A. see Vol. III, V for further entries

KREISEL, G. [1950] *Note on arithmetic models for consistent formulae of the predicate calculus I* (J 0027) Fund Math 37*265-285
  ⋄ B10 C57 D45 D55 F30 ⋄ REV MR 12.790 JSL 18.180
  • REM Part II 1953 • ID 07481

KREISEL, G. [1953] *A variant to Hilbert's theory of the foundations of arithmetic* (J 0013) Brit J Phil Sci 4*107-129,357
  ⋄ A05 B28 D30 D55 ⋄ REV MR 15.670 JSL 22.304
  • ID 07487

KREISEL, G. [1953] *Note on arithmetic models for consistent formulae of the predicate calculus II* (P 0645) Int Congr Philos (11);1953 Bruxelles 14*39-49
  ⋄ B10 C57 D45 D55 F30 ⋄ REV MR 15.668 Zbl 53.200 JSL 21.403 • REM Part I 1951 • ID 20815

KREISEL, G. [1954] *Applications of mathematical logic to various branches of mathematics* (P 0646) Appl Sci de Log Math;1952 Paris 37-49
  ⋄ B98 C65 ⋄ REV MR 16.782 Zbl 57.245 JSL 24.236
  • ID 16973

KREISEL, G. [1954] *Remark on complete interpretations by models* (J 0009) Arch Math Logik Grundlagenforsch 2*4-9
  • REPR [1954] (J 1114) Arch Phil 5*84-89
  ⋄ B15 F25 F30 ⋄ REV MR 17.119 Zbl 55.6 JSL 36.169
  • ID 22027

KREISEL, G. & TAIT, W.W. [1961] *Finite definability of number-theoretic functions and parametric completeness of equational calculi* (J 0068) Z Math Logik Grundlagen Math 7*28-38
  ⋄ B20 D20 ⋄ REV MR 24 # A1822 Zbl 116.5 JSL 32.270
  • ID 07524

KREISEL, G. [1965] *Mathematical logic* (C 1602) Lect on Modern Math 3*95-195
  • TRANSL [1965] (C 4652) Lect on Modern Math (Japanese) 128-291
  ⋄ B98 F10 F35 F50 F98 ⋄ REV MR 31 # 2124
  Zbl 147.247 JSL 32.419 • ID 27577

KREISEL, G. [1966] see FEFERMAN, S.

KREISEL, G. & KRIVINE, J.-L. [1967] *Elements de logique mathematique. Theorie des modeles* (X 0856) Dunod: Paris viii + 213pp
  • TRANSL [1972] (X 0811) Springer: Heidelberg & New York xv + 274pp [1967] (X 0809) North Holland: Amsterdam xi + 222pp (English)
  ⋄ B98 C98 ⋄ REV MR 34 # 7331 MR 36 # 2463
  MR 50 # 4231 Zbl 146.7 Zbl 238 # 02003 JSL 34.112
  • ID 22411

KREISEL, G. [1967] *Informal rigour and completeness proofs* (P 2268) Int Colloq Philos of Sci;1965 London 1*138-186
  • TRANSL [1978] (C 4649) Paradiso di Cantor 59-93 • REPR [1969] (C 0569) Phil of Math Oxford Readings 78-94
  ⋄ A05 B30 C07 E30 E50 F50 ⋄ ID 07533

KREISEL, G. [1969] *Axiomatisations of nonstandard analysis that are conservative extensions of formal systems for classical standard analysis* (P 0649) Appl Model Th to Algeb, Anal & Probab;1967 Pasadena 93-106
  ⋄ B30 F35 H05 ⋄ REV MR 39 # 50 Zbl 188.322
  • ID 26648

KREISEL, G. [1970] *Hilbert's programme and the search for automatic proof procedures* (P 0625) Symp Autom Demonst;1968 Versailles 128-146
  ⋄ A05 B35 F99 ⋄ REV MR 44 # 1571 Zbl 206.277
  • ID 48003

KREISEL, G. [1973] *Bertrand Arthur William Russell, Earl Russell 1872-1970* (J 0480) Roy Soc Bibl Mem Fellows 582-620
  ⋄ A10 B15 E99 ⋄ ID 48845

KREISEL, G. [1974] *A notion of mechanistic theory* (J 0154) Synthese 29*11-26
  • REPR [1976] (C 2953) Log & Probab in Quant Mech 3-18
  ⋄ A05 B30 D10 ⋄ REV Zbl 307 # 02028 Zbl 335 # 02031
  • ID 63130

KREISEL, G. [1975] *Was hat die Logik in den letzten 25 Jahren fur die Mathematik geleistet?* (J 2688) Conceptus (Wien) 9*40-45
  ⋄ A05 B98 C75 D40 ⋄ REV MR 58 # 21356 • ID 75092

KREISEL, G. [1977] *From foundations to science: Justifying and unwinding proofs* (**P** 3269) Set Th Found Math (Kurepa);1977 Beograd 63-72
⋄ A05 B35 F07 ⋄ REV MR 58 # 16100   Zbl 414 # 03033
• ID 53079

KREISEL, G. [1981] *Aussagenquantoren: Ein Beitrag der mathematischen Logik zur Erkenntnistheorie* (**C** 3202) Logik, Ethik & Sprache (Freundlich) 105-117
⋄ A05 B80 ⋄ REV Zbl 472 # 03002 • ID 55268

KREISEL, G. [1985] *Mathematical logic: tool and object lesson for science* (**J** 0154) Synthese 62*139-151
⋄ A05 B98 ⋄ ID 45516

KREISEL, G. see Vol. II, III, IV, V, VI for further entries

KREISER, L. (ED.) [1973] *Schriften zur Logik. Written by Frege, G.* (**X** 0911) Akademie Verlag: Berlin 310pp
⋄ A10 B96 ⋄ REM Aus dem Nachlass • ID 48656

KREJNOVICH, V.YA. & OSWALD, U. [1982] *A decision method for the universal theorems of Quine's new foundations* (**J** 0068) Z Math Logik Grundlagen Math 28*181-187
⋄ B20 B25 E70 ⋄ REV MR 84b:03071   Zbl 515 # 03030
• ID 35653

KREJNOVICH, V.YA. see Vol. VI for further entries

KREMPA, J. & MAZBIC-KULMA, B. [1977] *Elements of logic, set theory and algebra (Elementy logiki, teorii mnogosci i algebry) (Polish)* (**X** 2880) Wydawn Nauk Techn: Warsaw 170pp
⋄ B98 E98 ⋄ REV MR 58 # 4968   Zbl 383 # 03001
• ID 51983

KREMPA, J. see Vol. IV for further entries

KRETZMANN, N. [1965] *Elements of formal logic* (**X** 1238) Bobbs-Merril: Indianapolis xi + 243pp
⋄ B98 ⋄ ID 22505

KREYCHE, R.J. [1954] *Logic for undergraduates* (**X** 1257) Dryden Pr: Hinsdale 308pp
• LAST ED [1970] (**X** 0818) Holt Rinehart & Winston: New York xi + 224pp
⋄ B98 ⋄ ID 22506

KRIEGEL, K. & WAACK, S. [1985] *Lower bounds for Boolean formulae of depth 3 and the topology of the n-cube* (**P** 4647) FCT'85 Fund of Comput Th;1985 Cottbus 227-234
⋄ B05 ⋄ ID 46643

KRIEGEL, K. see Vol. IV for further entries

KRIPKE, S.A. & POUR-EL, M.B. [1967] *Deduction-preserving "recursive isomorphisms" between theories* (**J** 0015) Bull Amer Math Soc 73*145-148
⋄ B30 D25 D35 F30 ⋄ REV MR 35 # 6548 Zbl 174.20
• ID 31215

KRIPKE, S.A. & POUR-EL, M.B. [1967] *Deduction-preserving "recursive isomorphisms" between theories* (**J** 0027) Fund Math 61*141-163
⋄ B30 D25 D35 F30 ⋄ REV MR 40 # 5447 Zbl 174.20
• ID 31216

KRIPKE, S.A. see Vol. II, III, IV, V, VI for further entries

KRISHNAMURTHY, B. [1984] see KAPUR, D.

KRISHNAMURTHY, B. [1985] *Short proofs for tricky formulas* (**J** 1431) Acta Inf 22*253-275
⋄ B35 ⋄ REV Zbl 552 # 03009 • ID 48223

KRIVINE, J.-L. [1967] see KREISEL, G.

KRIVINE, J.-L. [1975] see DACUNHA-CASTELLE, D.

KRIVINE, J.-L. see Vol. II, III, V for further entries

KRIVTSOV, V.N. [1984] *A type of formal negationfree systems (Russian)* (**J** 0288) Vest Ser Mat Mekh, Univ Moskva 1984/2*27-31
• TRANSL [1984] (**J** 0510) Moscow Univ Math Bull 39/2*37-41
⋄ B20 ⋄ REV MR 85j:03033 • ID 45005

KRIVTSOV, V.N. [1984] *Deductive potentialities of the negationless predicate calculus (Russian)* (**J** 0288) Vest Ser Mat Mekh, Univ Moskva 1984/4*3-5
• TRANSL [1984] (**J** 0510) Moscow Univ Math Bull 39/4*1-4
⋄ B20 ⋄ REV MR 86d:03015 • ID 44183

KRIVTSOV, V.N. see Vol. VI for further entries

KRIZEK, P. [1978] *On rewriting sentences into formulas* (**J** 0063) Studia Logica 37*115-128
⋄ A05 B03 ⋄ REV MR 80a:03034   Zbl 406 # 03012
• ID 56095

KRNIC, L. [1965] *Types of bases of algebra of logic (Russian) (Serbo-Croatian summary)* (**J** 0371) Glas Mat-Fiz Astron, Ser 2 (Zagreb) 20*23-32
⋄ B05 ⋄ REV MR 33 # 2524   JSL 34.121 • ID 31410

KRNIC, L. [1978] *On generating systems and bases for F_2 (Serbo-Croatian) (English summary)* (**J** 0042) Mat Vesn, Drust Mat Fiz Astron Serb 2(15)(30)*363-367
⋄ B05 G05 ⋄ REV MR 81f:03072   Zbl 423 # 03058
• ID 53570

KRNIC, L. see Vol. II for further entries

KROEGER, F. [1985] *On termporal program verification rules (French summary)* (**J** 3441) RAIRO Inform Theor 19*261-280
⋄ B35 ⋄ REV Zbl 563 # 68007 • ID 49194

KROEGER, F. see Vol. II, VI for further entries

KROM, MELVEN R. [1963] *Separation principles in the hierarchy theory of pure first-order logic* (**J** 0036) J Symb Logic 28*222-236
⋄ B10 C40 C52 D55 ⋄ REV MR 31 # 2133 Zbl 137.9 JSL 31.503 • ID 07571

KROM, MELVEN R. [1964] *A decision procedure for a class of formulas of first order predicate calculus* (**J** 0048) Pac J Math 14*1305-1319
⋄ B20 B25 ⋄ REV MR 31 # 3316 Zbl 171.270 • ID 07570

KROM, MELVEN R. [1967] *The decision problem for segregated formulas in first-order logic* (**J** 0132) Math Scand 21*233-240
⋄ B20 B25 D35 ⋄ REV MR 39 # 1286 Zbl 169.310
• ID 07574

KROM, MELVEN R. [1967] *The decision problem for a class of first-order formulas in which all disjunctions are binary* (**J** 0068) Z Math Logik Grundlagen Math 13*15-20
⋄ B20 ⋄ REV MR 35 # 24 Zbl 162.316 • ID 45437

KROM, MELVEN R. [1968] *Some interpolation theorems for first-order formulas in which all disjunctions are binary* (**J** 0079) Logique & Anal, NS 11*403-412
⋄ B20 C40 ⋄ REV MR 39 # 2603 • ID 07575

KROM, MELVEN R. [1970] *The decision problem for formulas in prenex conjunctive normal form with binary disjunctions* (J 0036) J Symb Logic 35*210-216
♦ B20 D03 D35 ♦ REV MR 43#55 Zbl 207.13
• ID 07576

KROM, MELVEN R. [1977] *Complete rings of sets and sentential logic* (J 0063) Studia Logica 36*173-175
♦ B05 E20 G05 ♦ REV MR 58#10283 Zbl 357#02007
• ID 50355

KROM, MELVEN R. see Vol. III, IV, V for further entries

KRONECKER, L. [1887] *Ueber den Zahlbegriff* (J 0127) J Reine Angew Math 101*337-355
• REPR [1899] (C 4394) Kroneckers Werke 3*1,249-274
♦ B28 F55 F65 ♦ REV FdM 19.63 • ID 37377

KRUGER, A.N. & MANICAS, P.T. [1968] *Essentials of logic* (X 0864) Van Nostrand: New York xi+483PP
♦ B98 ♦ ID 22520

KRUPA, A. & ZAWISZA, B. [1984] *Applications of ultrapowers in analysis of unbounded selfadjoint operators* (J 3417) Bull Pol Acad Sci, Math 32*581-588
♦ C20 C65 H10 ♦ REV MR 86h:47029 • ID 44797

KRUPKO, N.A. [1977] *On one automatic system of logical conclusion (Russian)* (J 0085) Vest Ser Mat Mekh Astron, Univ Leningrad 1977/4*152-153
♦ B35 ♦ REV Zbl 383#03041 • ID 52023

KRYNICKI, M. [1979] *On the expressive power of the language using the Henkin quantifier* (P 1705) Scand Logic Symp (4);1976 Jyvaeskylae 259-265
♦ B15 B25 C80 C85 ♦ REV MR 82c:03056 Zbl 411#03027 • ID 52881

KRYNICKI, M. & LACHLAN, A.H. [1979] *On the semantics of the Henkin quantifier* (J 0036) J Symb Logic 44*184-200
♦ B15 B25 C40 C55 C80 C85 ♦ REV MR 80h:03054 Zbl 418#03026 • ID 53311

KRYNICKI, M. see Vol. III, IV, VI for further entries

KUBINSKI, T. [1956] see KOKOSZYNSKA, M.

KUBINSKI, T. [1961] *On the scope of Lindenbaum's theorem on complete supersystems (Polish) (Russian and English summaries)* (J 0063) Studia Logica 12*83-98
♦ B22 ♦ REV MR 27#3551 Zbl 121.254 • ID 07593

KUBINSKI, T. [1963] *A proof of consistency of Borkowski's logical system containing Peano's arithmetic (Polish and Russian summaries)* (J 0063) Studia Logica 14*197-225
♦ B15 F25 F30 ♦ REV MR 32#3990 Zbl 292#02029
• ID 07595

KUBINSKI, T. see Vol. II, IV, V, VI for further entries

KUDIELKA, V. [1963] *Programmierung von Minimisierungsverfahren fuer zweistufige Logik* (P 1612) Schaltkreis & -werk Th (2);1961 Saarbruecken 49-65
♦ B35 ♦ REV Zbl 115.125 • ID 27836

KUDIELKA, V. & OLIVA, P. [1966] *Complete sets of functions of two and three binary variables* (J 4305) IEEE Trans Electr Comp EC-15*930-931
♦ B05 ♦ REV Zbl 154.180 JSL 37.417 • ID 07601

KUDRYAVTSEV, V.B. [1966] see GAVRILOV, G.P.

KUDRYAVTSEV, V.B. [1972] see BUROSCH, G.

KUDRYAVTSEV, V.B. see Vol. II, IV, V for further entries

KUEHNER, D. [1970] see KOWALSKI, R.

KUEHNRICH, M. [1981] *Zur Praedikatentheorie der ersten und zweiten Stufe* (S 3414) Prepr, Akad Wiss DDR, Inst Math 11/81*24pp
♦ B15 ♦ REV Zbl 453#03029 • ID 54159

KUEHNRICH, M. [1983] *Eine aequivalente Formalisierung der Logik von Feferman und Aczel* (J 0068) Z Math Logik Grundlagen Math 29*565-568
♦ B15 E70 ♦ REV MR 85d:03054 Zbl 534#03027
• ID 36566

KUEHNRICH, M. [1983] *On the Hermes term logic* (J 0302) Rep Math Logic, Krakow & Katowice 16*3-16
♦ B10 C07 C20 ♦ REV MR 86f:03051 Zbl 406#03053
• ID 36540

KUEHNRICH, M. see Vol. III, V for further entries

KUGLER, L.D. [1969] *A nonstandard approach to linear functions* (J 0133) Michigan Math J 16*157-160
♦ H05 ♦ REV MR 39#7314 Zbl 181.159 • ID 33463

KUGLER, L.D. [1969] *Nonstandard almost periodic functions on a group* (J 0053) Proc Amer Math Soc 22*527-533
♦ H05 ♦ REV MR 39#4603 Zbl 181.150 JSL 35.349
• ID 07625

KUGLER, L.D. [1969] *Nonstandard analysis of almost periodic functions* (P 0649) Appl Model Th to Algeb, Anal & Probab;1967 Pasadena 150-166
♦ H05 ♦ REV MR 39#711 Zbl 203.297 • ID 26649

KUHN, STEVEN T. [1983] *An axiomatization of predicate functor logic* (J 0047) Notre Dame J Formal Log 24*233-241
♦ B10 ♦ REV MR 84e:03038 Zbl 508#03031 • ID 34377

KUHN, STEVEN T. see Vol. II for further entries

KUIPERS, T.A.F. [1976] *A two-dimensional continuum of a priori probability distributions on constituents* (P 1804) Form Meth in Methodol of Emp Sci;1974 Warsaw 82-92
♦ A05 B30 ♦ REV MR 58#27430 Zbl 364#60008
• ID 50981

KUIPERS, T.A.F. see Vol. II for further entries

KUL'YANOV, E.G. [1975] *An algorithm for finding the maximal upper zero of an arbitrary monotonic function of the algebra of logic (Russian)* (J 0199) Zh Vychisl Mat i Mat Fiz 15*1080-1082
• TRANSL [1975] (J 1049) USSR Comput Math & Math Phys 15/4*267-269
♦ B35 ♦ REV Zbl 318#94034 • ID 63189

KUNEN, K. [1980] see BARWISE, J.

KUNEN, K. [1985] see BLEDSOE, W.W.

KUNEN, K. see Vol. III, IV, V for further entries

KUPPERMAN, J. & McGRADE, A.S. [1966] *Fundamentals of logic* (X 0878) Doubleday: London 272pp
♦ B98 ♦ ID 22508

KURATOWSKI, K. & TARSKI, A. [1931] *Les operations logiques et les ensembles projectifs* (J 0027) Fund Math 17*240-248
• TRANSL [1956] (C 1159) Tarski: Logic, Semantics, Metamathematics 143-151
♦ B10 D55 E15 ♦ REV Zbl 3.105 JSL 34.99 FdM 57.92
• ID 07656

KURATOWSKI, K. see Vol. III, IV, V for further entries

KUREPA, D. [1950] *Demonstration du principe de l'induction totale* (J 0109) C R Acad Sci, Paris 230*703-705
- ⋄ B28 E20 ⋄ REV MR 11.412 Zbl 35.8 • ID 28728

KUREPA, D. see Vol. V for further entries

KUR'EROV, YU.N. [1979] *Normal form of mutual-absorption tactics (Russian)* (J 0040) Kibernetika, Akad Nauk Ukr SSR 1979/6*138-141
- • TRANSL [1979] (J 0021) Cybernetics 15*930-935
- ⋄ B35 F05 ⋄ REV MR 81m:68077 Zbl 508 # 68059
- • ID 81977

KUR'EROV, YU.N. [1981] *Equiprobable canonical calculi (Russian)* (S 2582) Semiotika & Inf, Akad Nauk SSSR 17*90-97
- ⋄ B35 D03 ⋄ REV MR 84m:03060 Zbl 484 # 03018
- • ID 35760

KUR'EROV, YU.N. see Vol. IV for further entries

KURMIT, A.A. [1958] *Independence of a system of axioms for the propositional calculus (Russian) (Latvian summary)* (J 0483) Uch Zap, Univ Riga 20/3*21-25
- ⋄ B05 ⋄ REV MR 21 # 1933 • ID 07679

KURMIT, A.A. see Vol. IV for further entries

KURODA, S. [1958] *An investigation on the logical structure of mathematics. I. A logical system* (J 0107) Abh Math Sem Univ Hamburg 22*242-266
- ⋄ B15 B28 E70 F07 ⋄ REV MR 23 # A48 Zbl 81.247
- • REM Part II 1959 • ID 07681

KURODA, S. [1958] *An investigation on the logical structure of mathematics. III. Fundamental deductions. IV. Compendium for deductions* (J 0111) Nagoya Math J 13*21-52,123-133
- ⋄ B15 B28 E70 F07 ⋄ REV MR 23 # A50 MR 23 # A51 Zbl 87.245 • REM Part II 1959. Part V 1958 • ID 07683

KURODA, S. [1958] *An investigation on the logical structure of mathematics XII. The principle of extensionality and of choice* (J 0081) Proc Japan Acad 34*400-403
- ⋄ B15 B28 E25 E70 ⋄ REV MR 23 # A57 Zbl 92.4
- • REM Corrections in part VI(1959). Part X 1959. Part XIII 1960 • ID 07691

KURODA, S. [1959] *An investigation on the logical structure of mathematics. II. Transformation of the proof* (J 0107) Abh Math Sem Univ Hamburg 23*201-227
- ⋄ B28 B30 F05 ⋄ REV MR 23 # A49 Zbl 87.245 • REM Parts I,III,IV 1958 • ID 25622

KURODA, S. [1959] *An investigation on the logical structure of mathematics. VI. Consistent V-systems T(V). VII. Set-theoretical contradictions* (J 0111) Nagoya Math J 14*95-107,109-127 • ERR/ADD to part XII
- ⋄ B28 B30 E70 F30 ⋄ REV MR 23 # A53 MR 23 # A54 Zbl 94.7 • REM Part V 1958. Part VIII 1959 • ID 25455

KURODA, S. [1959] *An investigation on the logical structure of mathematics. VIII. Consistency of the natural-number theory $T_1(N)$* (J 0111) Nagoya Math J 14*129-158
- ⋄ B28 B30 F30 ⋄ REV MR 23 # A55a Zbl 94.7 • REM Parts VI,VII,XI,X 1959 • ID 25434

KURODA, S. [1959] *An investigation on the logical struture of mathematics. IX. Deductions in the natural-number theory $T_1(N)$. X. Concepts and sets* (J 1770) Osaka Math J 11*7-42,213-248
- ⋄ B15 B28 B30 F30 ⋄ REV MR 23 # A55b MR 23 # A56 Zbl 87.245 Zbl 92.4 • REM Part VIII 1959
- • ID 15552

KURODA, S. [1960] *An investigation on the logical structure of mathematics. XIII. A method of programming of proofs in mathematics for electronic computing machines* (J 0111) Nagoya Math J 16*195-203
- ⋄ B35 ⋄ REV Zbl 90.10 • REM Part XII 1958 • ID 90106

KURODA, S. see Vol. V, VI for further entries

KUSSOVA, B. [1983] see CUDA, K.

KUSSOVA, B. see Vol. V for further entries

KUTATELADZE, S.S. [1985] *Nonstandard analysis of tangent cones (Russian)* (J 0023) Dokl Akad Nauk SSSR 284*525-527
- • TRANSL [1985] (J 0062) Sov Math, Dokl 32*437-439
- ⋄ H05 ⋄ ID 49119

KUTATELADZE, S.S. see Vol. III, V for further entries

KUTIRKOV, G.A. [1972] see CHIMEV, K.N.

KUTIRKOV, G.A. [1981] *One approach to the expression of functions of the algebra of logic* (J 2889) Zesz Nauk, Mat, Politech Lodz 13*53-58
- ⋄ B05 ⋄ REV MR 82j:03079 • ID 75211

KUTIRKOV, G.A. see Vol. II for further entries

KUTSCHERA VON, F. [1962] see KAESBAUER, M.

KUTSCHERA VON, F. [1962] *Zum Deduktionsbegriff der klassischen Praedikatenlogik erster Stufe* (C 0712) Logik & Logikkalkuel 211-236
- ⋄ B10 ⋄ REV Zbl 121.253 • ID 07708

KUTSCHERA VON, F. [1967] *Elementare Logik* (X 0902) Springer: Wien viii+392pp
- ⋄ B98 ⋄ REV MR 37 # 3903 Zbl 105.3 • ID 07711

KUTSCHERA VON, F. [1971] see BREITKOPF, A.

KUTSCHERA VON, F. see Vol. II, VI for further entries

KUZICHEV, A.S. [1970] *Solution of certain problems of mathematical logic by means of Venn diagrams (Russian)* (C 1530) Issl Log Sist (Yanovskaya) 282-331
- ⋄ B20 ⋄ REV MR 47 # 8236 Zbl 217.7 • ID 28035

KUZICHEV, A.S. [1977] *A system of λ-conversion with logical operators and an equality operator (Russian)* (J 0023) Dokl Akad Nauk SSSR 236*796-799
- • TRANSL [1977] (J 0062) Sov Math, Dokl 18*1268-1272
- ⋄ B10 B40 F05 ⋄ REV MR 57 # 15983 Zbl 395 # 03009
- • ID 52560

KUZICHEV, A.S. see Vol. V, VI for further entries

KUZ'MIN, V.A. [1965] *Realization of functions of the algebra of logic by means of automata, normal algorithms and Turing machines (Russian)* (J 0052) Probl Kibern 13*75-96
- ⋄ B05 D03 D05 D10 ⋄ REV MR 33 # 7212 Zbl 263 # 02020 • ID 29866

KUZNETSOV, A.V. [1963] *Undecidability of the general problems of completeness solvability and equivalence for propositional calculi (Russian)* (J 0003) Algebra i Logika 2/4*47-66
• TRANSL [1966] (J 0225) Amer Math Soc, Transl, Ser 2 59*56-72
⋄ B22 D35 ⋄ REV MR 28#3935 Zbl 166.263 JSL 37.756
• ID 19144

KUZNETSOV, A.V. (ED.) [1976] *Fourth All-Union Conference on Mathematical Logic (Russian)* (X 2741) Shtiintsa: Kishinev 170pp
⋄ B97 ⋄ REV MR 56#15246 • ID 80457

KUZNETSOV, A.V. & RATSA, M.F. [1979] *A criterion for functional completeness in classical first-order predicate logic (Russian)* (J 0023) Dokl Akad Nauk SSSR 249*540-544
• TRANSL [1979] (J 0062) Sov Math, Dokl 20*1305-1309
⋄ B10 ⋄ REV MR 80m:03026 Zbl 443#03007 • ID 75246

KUZNETSOV, A.V. [1979] *Means for detection of nondeducibility and inexpressibility (Russian)* (P 2554) All-Union Symp Th Log Infer;1974 Moskva 5-33
⋄ B20 ⋄ REV MR 84k:03036 • ID 34973

KUZNETSOV, A.V. see Vol. II, III, IV, VI for further entries

KUZNETSOV, S.E. [1983] *A lower bound for the length of the shortest d.n.f. of almost all Boolean functions (Russian)* (J 3937) Veroyat Met i Kibern (Kazan) 19*44-47
⋄ B05 ⋄ REV MR 85b:94016 Zbl 524#94032 • ID 39421

KUZNETSOV, S.E. [1983] *Complexity of realization of a sequence of Boolean functions by formulas of depth 3 in the basis $\{\vee, \&, \neg\}$ (Russian)* (J 3937) Veroyat Met i Kibern (Kazan) 19*40-43
⋄ B05 ⋄ REV MR 84i:00006 Zbl 524#94031 • ID 44883

KUZNETSOV, S.E. see Vol. IV for further entries

KYBURG JR., H.E. [1978] *Subjective probability: Criticisms, reflections, and problems* (J 0122) J Philos Logic 7*157-180
⋄ A05 B30 ⋄ REV MR 58#24430 Zbl 374#02005
• ID 51529

KYBURG JR., H.E. [1984] *Theory and measurement* (X 0805) Cambridge Univ Pr: Cambridge, GB viii+273pp
⋄ B30 ⋄ ID 45059

KYBURG JR., H.E. see Vol. II for further entries

L'ABBE, M. [1951] *On the independence of Henkin's Axiom for fragments of the propositional calculus* (J 0036) J Symb Logic 16*43-45
⋄ B20 ⋄ REV MR 12.662 Zbl 43.8 JSL 16.228 • ID 07729

L'ABBE, M. [1953] *Systems of transfinite types involving λ-conversion* (J 0036) J Symb Logic 18*209-224
⋄ B15 B40 ⋄ REV MR 15.593 Zbl 53.203 JSL 23.361
• ID 07730

LABORDE, J.-M. [1978] *Un theoreme d'algebre de Boole et le theoreme d'Herbrand (English summary)* (J 2313) C R Acad Sci, Paris, Ser A-B 286*A439-A441
⋄ B10 G05 G10 ⋄ REV MR 58#16278 Zbl 383#06002
• ID 69670

LABORDE, J.-M. [1980] *A propos d'un probleme d'algebre de Boole* (S 3358) Ann Discrete Math 8*127-128
⋄ B05 ⋄ REV MR 81k:05066 Zbl 442#05052 • ID 56401

LABORDE, J.-M. [1980] *Sur le cardinal maximum de la base complete d'une fonction Booleenne, en fonction du nombre de conjonctions de l'une de ses formes normales* (J 0193) Discr Math 32*209-212
⋄ B05 ⋄ REV MR 81m:94035 Zbl 455#05011 • ID 81997

LACHLAN, A.H. [1979] see KRYNICKI, M.

LACHLAN, A.H. see Vol. II, III, IV, V, VI for further entries

LACOMBE, D. & LAUDET, M. & NOLIN, L. & SCHUETZENBERGER, M.-P. (EDS.) [1970] *Symposium on automatic demonstration* (S 3301) Lect Notes Math 125*v+310pp
⋄ B35 B97 F97 ⋄ REV MR 42#2716 Zbl 196.21
• ID 48620

LACOMBE, D. see Vol. IV, VI for further entries

LADEGAILLERIE, Y. [1974] *Une propriete des consensus de monomes et une application* (J 2313) C R Acad Sci, Paris, Ser A-B 278*A879-A880
⋄ B05 ⋄ REV MR 49#7196 Zbl 286#06004 • ID 07773

LADNER, R.E. [1977] *Application of model theoretic games to discrete linear orders and finite automata* (J 0194) Inform & Control 33*281-303
• TRANSL [1980] (J 3079) Kiber Sb Perevodov, NS 17*164-191
⋄ B15 B25 C07 C13 C65 C85 D05 E07 E60 ⋄ REV MR 58#10387 Zbl 387#68037 • ID 52252

LADNER, R.E. see Vol. II, III, IV for further entries

LADRIERE, J. [1957] *Les limitations internes des formalismes. Etude sur la signification du theoreme de Goedel et des theoremes apparentes dans la theorie des fondements des mathematiques* (X 0834) Gauthier-Villars: Paris xv+715pp
⋄ A05 B98 F30 F98 ⋄ REV MR 20#2 Zbl 78.242 JSL 25.270 • ID 07777

LADRIERE, J. see Vol. III, IV, VI for further entries

LAEUCHLI, H. [1968] *A decision procedure for the weak second order theory of linear order* (P 0608) Logic Colloq;1966 Hannover 189-197
⋄ B15 B25 C65 C85 ⋄ REV MR 39#5343 • ID 07787

LAEUCHLI, H. see Vol. II, III, V, VI for further entries

LAFLAMME, C. [1983] see HATCHER, W.S.

LAGASSE, J. [1976] see COURVOISIER, M.

LAKATOS, I. [1978] *Cauchy and the continuum: the significance of non-standard analysis for the history and philosophy of mathematics* (J 2789) Math Intell 1*151-161
⋄ H05 ⋄ REV MR 80j:01029 Zbl 398#01009 • ID 52727

LAKATOS, I. [1978] *Mathematics, science and epistemology. Philosophical papers. Vol. 2* (X 0805) Cambridge Univ Pr: Cambridge, GB x+285pp
• TRANSL [1982] (X 0900) Vieweg: Wiesbaden x+279pp
⋄ A05 B30 ⋄ REV MR 81e:01030 MR 83j:01085 Zbl 373#02003 Zbl 476#03003 • ID 51467

LAKATOS, I. see Vol. II, VI for further entries

LAKE, J. [1975] *Comparing type theory and set theory* (J 0068) Z Math Logik Grundlagen Math 21*355-356
⋄ B15 E30 E70 ⋄ REV MR 51#12525 Zbl 323#02021
• ID 07802

LAKE, J. see Vol. III, V for further entries

LAMBALGEN VAN, M. [1983] *Quantum set theory* (J 3150) Delft Prog Rep, Ser F 8*17-24
  ◊ E70 G12 H10 ◊ REV MR 85e:03126 Zbl 537#03035 • ID 40731

LAMBEK, J. [1958] *The mathematics of sentence structures* (J 0005) Amer Math Mon 65*154-170
  ◊ B30 B65 F07 ◊ REV MR 21#4904 Zbl 80.7 JSL 33.627 • ID 07815

LAMBEK, J. & SCOTT, P.J. [1984] *Aspects of higher order categorical logic* (P 2180) Math Appl Categ Th;1983 Denver 145-174
  ◊ B15 F35 F50 G30 ◊ REV MR 85i:03196 Zbl 549#03058 • ID 44302

LAMBEK, J. see Vol. III, IV, V, VI for further entries

LAMBERT, K. & SCHARLE, T.W. [1967] *A translation theorem for two systems of free logic* (J 0079) Logique & Anal, NS 10*328-341
  ◊ B20 B60 F25 ◊ REV MR 37#1229 Zbl 164.309 • ID 07823

LAMBERT, K. [1969] *Logical truth and microphysics* (C 1134) Log Way of Doing Things 93-117
  ◊ B30 ◊ REV Zbl 184.281 • ID 22098

LAMBERT, K. [1972] *Notes on free description theory: some philosophical issues and consequences* (J 0122) J Philos Logic 1*184-191
  ◊ A05 B20 ◊ REV MR 53#12865 Zbl 247#02015 • ID 23127

LAMBERT, K. [1982] see BENCIVENGA, E.

LAMBERT, K. see Vol. II for further entries

LAMBERT JR., W.M. [1977] *Un tipo de solidez para un sistema de deduccion natural* (J 1680) Cienc Tecnol, Costa Rica 1*35-44
  ◊ B10 F07 ◊ REV MR 81j:03024 • ID 31153

LAMBERT JR., W.M. see Vol. IV for further entries

LAMBROS, C.H. [1979] *A generalized theorem concerning a restricted rule of substitution in the field of propositional calculi* (J 0047) Notre Dame J Formal Log 20*760-764
  ◊ B22 ◊ REV MR 82m:03011 Zbl 321#02007 • ID 63304

LAMBROS, C.H. [1979] *A shortened proof of Sobocinski's theorem concerning a restricted rule of substitution in the field of propositional calculi* (J 0047) Notre Dame J Formal Log 20*112-114
  ◊ B22 ◊ REV MR 83b:03011 Zbl 299#02007 • ID 52380

LAMBROS, C.H. see Vol. II for further entries

LANDAU, E. [1930] *Grundlagen der Analysis* (X 0911) Akademie Verlag: Berlin ixv+134pp
  • TRANSL [1951] (X 0848) Chelsea: New York xiv+134pp
  ◊ B28 E75 E98 ◊ REV MR 12.397 Zbl 42.278 JSL 11.126 JSL 16.223 FdM 56.191 • ID 23388

LANDIN, J. [1961] see HAMILTON, N.T.

LANDIS, E.H. & RICHARDSON, R.P. [1916] *Fundamental conceptions of modern mathematics. Variables and quantifiers, with a discussion of the general conception of the functional relation* (X 1324) Open Court: LaSalle xxi+216pp
  ◊ A05 B30 ◊ REV FdM 46.312 FdM 46.318 • ID 25578

LANDMAN, F. & VELTMAN, F. (EDS.) [1984] *Varieties of formal semantics* (X 4217) Foris: Dordrecht xii+425pp
  ◊ B97 ◊ REV Zbl 556#00006 • REM Proceedings of the fourth Amsterdam colloquium 1982 • ID 46159

LANDWEBER, L.H. [1969] see BUECHI, J.R.

LANDWEBER, L.H. see Vol. IV for further entries

LANDWEHR, R. [1982] see JANSOHN, H.-S.

LANGER, S.K. [1929] *A set of postulates for the logical structure of music* (J 0320) Monist 39*561-570
  ◊ B80 ◊ ID 07847

LANGER, S.K. [1937] *An introduction to symbolic logic* (X 0847) Houghton Mifflin: Boston 363pp
  • LAST ED [1953] (X 0813) Dover: New York 367pp
  ◊ B98 ◊ REV MR 14.1051 Zbl 51.5 JSL 3.83 FdM 63.819 • REM 2nd ed.1953 • ID 07848

LANGER, S.K. [1951] see HENLE, P.

LANGFORD, C.H. [1926] *Analytic completeness of postulate sets* (J 1910) Proc London Math Soc, Ser 2 25*115-142
  ◊ B20 C35 C65 E07 ◊ REV FdM 52.49 • ID 07849

LANGFORD, C.H. [1926] *Some theorems on deducibility* (J 0120) Ann of Math, Ser 2 28*16-40
  ◊ B30 C10 C35 C65 ◊ REV FdM 52.48 • REM Part I. Part II 1926 • ID 07850

LANGFORD, C.H. [1926] *Theorems on deducibility (second paper)* (J 0120) Ann of Math, Ser 2 28*459-471
  ◊ B30 C10 C35 C65 ◊ REV FdM 53.43 • REM Part II. Part I 1926 • ID 07851

LANGFORD, C.H. [1927] *An analysis of some general propositions* (J 0015) Bull Amer Math Soc 33*666-672
  ◊ B10 ◊ REV FdM 53.44 • ID 07854

LANGFORD, C.H. [1927] *On inductive relations* (J 0015) Bull Amer Math Soc 33*599-607
  ◊ A05 B28 ◊ REV FdM 53.43 • ID 07853

LANGFORD, C.H. & LEWIS, C.I. [1932] *Symbolic logic* (X 1228) Appleton-Century-Crofts: New York xi+506pp
  • LAST ED [1959] (X 0813) Dover: New York ix+518pp
  ◊ B25 B98 C10 C98 ◊ REV MR 21#4091 Zbl 87.8 FdM 58.56 • ID 22592

LANGFORD, C.H. [1939] *A theorem on deducibility for second-order functions* (J 0036) J Symb Logic 4*77-79
  ◊ B15 C35 C65 C85 ◊ REV Zbl 21.98 JSL 5.30 FdM 65.28 • ID 07855

LANGFORD, C.H. see Vol. III, V for further entries

LANGLEY, P. & LARSON, P. & SILAS, S. & WERTZ, S.K. [1983] *A proof of CN_qN_p from C_{pq} by the rule of detachment in Jeffrey's system 5.6* (J 0286) Int Logic Rev 27*37-40
  ◊ B05 ◊ REV MR 86b:03014 • ID 45588

LANKFORD, D.S. [1984] see BRADY, B.

LANKFORD, D.S. see Vol. IV for further entries

LANTSCHOOT VAN, E. [1962] *New concepts in mathematical logic* (J 4529) Rev A Tijdschrift 4*119-127
  ◊ B05 ◊ REV Zbl 109.100 • ID 48038

LAPLAZA, M.L. [1973] see BOLLMAN, D.A.

LAPLAZA, M.L. see Vol. IV, V for further entries

LAPSCHER, F. [1979] see DESCHAMPS, J.-P.

LAPSCHER, F. see Vol. V for further entries

LAPSIEN, R. [1981] see ERNI, W.

LAROUCHE, L. [1968] *Examination of the axiomatic foundations of a theory of change I* (J 0047) Notre Dame J Formal Log 9*371-384
  ◊ B30 ◊ REV MR 40 # 7088   Zbl 177.6 • REM Parts II,III 1969 • ID 07862

LAROUCHE, L. [1969] *Examination of the axiomatic foundations of a theory of change II,III* (J 0047) Notre Dame J Formal Log 10*277-284,385-409
  ◊ B30 ◊ REV MR 40 # 7088   Zbl 179.312   Zbl 179.315 • REM Part I 1968. Part IV 1971 • ID 07863

LAROUCHE, L. [1971] *Examination of the axiomatic foundations of a theory of change IV* (J 0047) Notre Dame J Formal Log 12*378-380
  ◊ A05   B30 ◊ REV MR 45 # 8502   Zbl 218 # 02006 • REM Part III 1969. Part V 1972 • ID 75350

LAROUCHE, L. [1972] *Examination of the axiomatic foundations of a theory of change V* (J 0047) Notre Dame J Formal Log 13*53-68
  ◊ B30 ◊ REV MR 46 # 7002   Zbl 229 # 02007 • REM Part IV 1971 • ID 19193

LARSON, P. [1983] see LANGLEY, P.

LASCAR, D. [1982] see DALEN VAN, D.

LASCAR, D. [1984] see DELON, F.

LASCAR, D. [1985] see DELON, F.

LASCAR, D. see Vol. III, IV, V, VI for further entries

LASKI, J. [1975] *An algorithm for generating orthogonal expansions of logical functions* (J 0014) Bull Acad Pol Sci, Ser Math Astron Phys 23*209-213
  ◊ B35 ◊ REV MR 52 # 13001   Zbl 314 # 94028 • ID 63317

LAUDET, M. [1970] see LACOMBE, D.

LAUGWITZ, D. & SCHMIEDEN, C. [1958] *Eine Erweiterung der Infinitesimalrechnung* (J 0044) Math Z 69*1-39
  ◊ B28   C65   E20   H05 ◊ REV MR 20 # 2404   Zbl 82.42 • ID 48699

LAUGWITZ, D. [1973] *A new theory of contact angles* (P 3042) Conv Algeb Commut & Conv Geom;1971/72 Roma 325-336
  ◊ B30   H05 ◊ REV MR 49 # 3798   Zbl 271 # 50006 • ID 82032

LAUGWITZ, D. [1978] *Infinitesimalkalkuel: Kontinuum und Zahlen - eine elementare Einfuehrung in die Nichtstandard-Analysis* (X 0876) Bibl Inst: Mannheim 187pp
  ◊ H05 ◊ REV MR 80d:26018   Zbl 412 # 26010   JSL 48.217 • ID 52979

LAUGWITZ, D. [1979] *A nonstandard approach to distributions and operational calculus* (P 3548) Generalized Fcts & Operat Calc;1975 Varna 156-162
  ◊ H05 ◊ REV MR 81d:46043   Zbl 428 # 44001 • ID 82028

LAUGWITZ, D. [1981] *Verallgemeinerte Grenzwerte beschraenkter Zahlfolgen* (J 2790) Math Sem-ber 28*30-38
  ◊ H05 ◊ REV MR 82i:46015   Zbl 478 # 40009 • ID 82029

LAUGWITZ, D. [1983] *Ω-calculus as a generalization of field extension; an alternative approach to nonstandard analysis* (C 3884) Nonstandard Anal - Recent Develop 120-133
  ◊ H05 ◊ REV MR 84h:26024   Zbl 511 # 03031 • ID 36916

LAUGWITZ, D. [1983] *Nichtstandard-Mathematik, begruendet durch eine Verallgemeinerung der Koerpererweiterung* (J 2699) Expo Math 1*307-333
  ◊ A10   H05 ◊ REV Zbl 538 # 26010 • ID 41496

LAUGWITZ, D. [1984] *Infinitesimals in physics (an introduction to the application of nonstandard methods)* (C 4187) Math Struct, Comp Math, Math Modell 233-243
  ◊ H10 ◊ ID 46425

LAUTENBACH, K. [1979] see GENRICH, H.J.

LAUTENBACH, K. see Vol. IV for further entries

LAVIT, C. [1976] *Une generalisation de la notion de fonction Booleenne symetrique* (J 0089) Calcolo 12*391-404
  ◊ B05 ◊ REV MR 57 # 15760   Zbl 341 # 06007 • ID 63325

LAVROV, I.A. [1970] *Logic and algorithms (Russian)* (X 0913) Novosibirsk Gos Univ: Novosibirsk 173pp
  ◊ B98   D98 ◊ REV MR 48 # 8188 • ID 07889

LAVROV, I.A. & MAKSIMOVA, L.L. [1970] *Problems in logic (Russian)* (X 0913) Novosibirsk Gos Univ: Novosibirsk 112pp
  ◊ B98   E98 ◊ REV MR 48 # 8189 • ID 07888

LAVROV, I.A. & MAKSIMOVA, L.L. [1975] *Problems in set theory, mathematical logic and the theory of algorithms (Russian)* (X 2027) Nauka: Moskva 240pp
  ◊ B98   D98   E98 ◊ REV MR 52 # 13300   Zbl 307 # 02001 • REM 2nd ed. 1984; 224pp • ID 21765

LAVROV, I.A. [1984] see ERSHOV, YU.L.

LAVROV, I.A. see Vol. III, IV, VI for further entries

LAWLER, E.L. [1964] *An approach to multilevel boolean minimization* (J 0037) ACM J 11*283-295
  ◊ B35 ◊ REV MR 29 # 3014   Zbl 126.328   JSL 33.630 • ID 07890

LAWLER, E.L. see Vol. IV for further entries

LAZEROWITZ, M. [1948] see AMBROSE, A.

LAZEROWITZ, M. [1961] see AMBROSE, A.

LAZEROWITZ, M. see Vol. II for further entries

LEACH, J.J. [1978] see HOOKER, C.A.

LEACH, J.J. see Vol. II for further entries

LEBESGUE, H. [1941] *Les controverses sur la theorie des ensembles et la question des fondements* (P 0652) Entretiens Zuerich Fond & Method Sci Math;1938 Zuerich 109-124
  ◊ A05   B30   E30 ◊ REV MR 2.339   Zbl 61.7   JSL 7.35 • ID 37379

LEBESGUE, H. see Vol. IV, V, VI for further entries

LEBLANC, H. [1955] *An introduction to deductive logic* (X 0827) Wiley & Sons: New York xii + 244pp
  ◊ B98 ◊ REV MR 16.661   Zbl 64.5   JSL 23.210 • ID 07909

LEBLANC, H. [1956] *Two probability concepts* (J 0301) J Phil 53*679-688
  ◊ B30 ◊ ID 07910

LEBLANC, H. [1960] see BETH, E.W.

LEBLANC, H. [1960] *On requirements for conditional probability functions* (J 0036) J Symb Logic 25*238-242
 ◊ B05 B30 B48 ◊ REV MR 33 # 750 Zbl 243 # 02010
 • ID 07913

LEBLANC, H. [1961] *An extension of the equivalence calculus* (J 0068) Z Math Logik Grundlagen Math 7*104-105
 ◊ B20 ◊ REV MR 25 # 2959 Zbl 117.10 JSL 27.248
 • ID 07918

LEBLANC, H. [1961] *The algebra of logic and the theory of deduction* (J 0301) J Phil 58*553-558
 ◊ B05 ◊ REV JSL 37.755 • ID 07915

LEBLANC, H. [1962] *Boolean algebra and the propositional calculus* (J 0094) Mind 71*383-386
 ◊ B05 G05 ◊ REV JSL 37.755 • ID 07919

LEBLANC, H. [1963] *Proof routines for the propositional calculus* (J 0047) Notre Dame J Formal Log 4*81-104
 ◊ B35 ◊ REV MR 29 # 2167 • ID 07923

LEBLANC, H. [1965] *Marginalia on Gentzen's Sequenzen-Kalkule* (C 0749) Contrib Logic & Methodol (Bochenski) 73-83
 ◊ B10 F07 F50 ◊ REV MR 48 # 3711 Zbl 156.6
 • ID 75405

LEBLANC, H. [1966] *Techniques of deductive inference* (X 0819) Prentice Hall: Englewood Cliffs vii+216pp
 ◊ B98 ◊ ID 22510

LEBLANC, H. [1966] *Two separation theorems for natural deduction* (J 0047) Notre Dame J Formal Log 7*159-180
 ◊ B10 F07 F50 ◊ REV MR 36 # 6263 Zbl 173.3
 • ID 07926

LEBLANC, H. [1966] *Two shortcomings of natural deduction* (J 0301) J Phil 63*29-37
 ◊ B10 ◊ REV MR 36 # 1290 • ID 24778

LEBLANC, H. & THOMASON, R.H. [1967] *All or none: a novel choice of primitives for elementary logic* (J 0036) J Symb Logic 32*345-351
 ◊ B10 ◊ REV MR 36 # 27 JSL 34.124 • ID 07927

LEBLANC, H. [1968] *A simplified account of validity and implication for quantificational logic* (J 0036) J Symb Logic 33*231-235
 ◊ B10 ◊ REV MR 41 # 3240 Zbl 185.10 JSL 35.466
 • ID 07933

LEBLANC, H. & THOMASON, R.H. [1968] *Completeness theorem for some presupposition-free logics* (J 0027) Fund Math 62*125-164
 ◊ B20 ◊ REV MR 38 # 983 Zbl 162.312 JSL 37.424
 • ID 07929

LEBLANC, H. [1968] *Syntactically free, semantically bound (a note on variables)* (J 0047) Notre Dame J Formal Log 9*167-170
 ◊ B10 ◊ REV MR 41 # 3236 • ID 07931

LEBLANC, H. & MEYER, R.K. [1969] *Open formulas and the empty domain* (J 0009) Arch Math Logik Grundlagenforsch 12*78-84
 ◊ B20 ◊ REV MR 40 # 1258 Zbl 181.298 • ID 07934

LEBLANC, H. [1969] *Three generalizations of a theorem of Beth's* (J 0079) Logique & Anal, NS 12*205-220
 ◊ B15 C07 ◊ REV MR 44 # 3832 Zbl 209.302 • ID 07936

LEBLANC, H. & MEYER, R.K. [1970] *On prefacing $(\forall X)A \supset A(Y/X)$ with $(\forall Y)$ – A free quantification theory without identity* (J 0068) Z Math Logik Grundlagen Math 16*447-462
 ◊ B20 ◊ REV MR 43 # 7293 Zbl 225 # 02021 • ID 19184

LEBLANC, H. & MEYER, R.K. [1970] *Truth-value semantics for the theory of types* (P 0559) Phil Probl in Logic;1968 Irvine 77-101
 ◊ B15 ◊ REV MR 45 # 3176 Zbl 245 # 02045 • ID 07937

LEBLANC, H. [1971] *Truth-value semantics for a logic of existence* (J 0047) Notre Dame J Formal Log 12*153-168
 ◊ B20 ◊ REV MR 45 # 1725 Zbl 239 # 02006 • ID 07939

LEBLANC, H. & WISDOM, W.A. [1972] *Deductive logic* (X 0802) Allyn & Bacon: London 368pp
 ◊ B98 ◊ REV JSL 40.628 • ID 22358

LEBLANC, H. & SNYDER, D.P. [1972] *Duals of Smullyan trees* (J 0047) Notre Dame J Formal Log 13*387-393 • ERR/ADD ibid 15*648
 ◊ B10 F07 ◊ REV MR 46 # 8816 MR 50 # 1853 Zbl 197.273 • ID 19182

LEBLANC, H. & MEYER, R.K. [1972] *Matters of separation* (J 0047) Notre Dame J Formal Log 13*229-236
 ◊ B10 F07 F50 ◊ REV MR 45 # 6575 Zbl 234 # 02008
 • ID 07940

LEBLANC, H. & WEAVER, G.E. [1973] *Truth-functionality and the ramified theory of types* (P 0783) Truth, Syntax & Modal;1970 Philadelphia 148-167
 ◊ B15 ◊ REV MR 54 # 2416 Zbl 273 # 02010 JSL 42.313
 • ID 24612

LEBLANC, H. [1975] *That "Principia Mathematica", first edition, has a predicative interpretation after all* (J 0122) J Philos Logic 4*67-70
 ◊ B15 ◊ REV MR 57 # 12167 Zbl 304 # 02007 • ID 63339

LEBLANC, H. [1976] *Truth-value semantics* (X 0809) North Holland: Amsterdam xii+319pp
 ◊ B98 ◊ REV MR 56 # 15360 Zbl 348 # 02001 JSL 43.376
 • ID 63340

LEBLANC, H. & PAULOS, J.A. & WEAVER, G.E. [1977] *Rules of deduction and truth-tables* (J 0302) Rep Math Logic, Krakow & Katowice 8*71-79
 ◊ B05 ◊ REV MR 58 # 21422 Zbl 378 # 02005 • ID 24191

LEBLANC, H. [1979] *Generalization in first order logic* (J 0047) Notre Dame J Formal Log 20*835-857
 ◊ B10 ◊ REV MR 81d:03011 Zbl 368 # 02012 • ID 56183

LEBLANC, H. [1984] *A new semantics for first-order logic, multivalent and mostly intensional* (J 3781) Topoi 3*55-62
 ◊ B10 C30 ◊ REV MR 86f:03038 • ID 44672

LEBLANC, H. see Vol. II, III, VI for further entries

LEBLANC, L. [1960] *Dualite pour les egalites Booleennes* (J 0109) C R Acad Sci, Paris 250*3552-3553
 ◊ B05 ◊ REV MR 22 # 6746 Zbl 101.271 JSL 29.207
 • ID 07945

LEBLANC, L. [1962] *Duality for boolean equalities* (J 0053) Proc Amer Math Soc 13*74-79
 ◊ B05 ◊ REV MR 25 # 5013 Zbl 101.271 JSL 29.53
 • ID 07948

LEBLANC, L. see Vol. III for further entries

LEE, H.N. [1961] *Symbolic logic* (X 0981) Random House: New York ix+356pp
⋄ B98 ⋄ REV JSL 25.262 • ID 22511

LEE, J.M. [1973] *The form of reductio ad absurdum* (J 0047) Notre Dame J Formal Log 14*381-386
⋄ A05 B05 ⋄ REV Zbl 236 # 02010 • ID 63346

LEE, R.C.T. [1969] see CHANG, C.L.

LEE, R.C.T. [1970] see CHANG, C.L.

LEE, R.C.T. [1972] *An algorithm to generate prime implicants and its application to the selection problem* (J 0191) Inform Sci 4*251-259
⋄ B35 ⋄ REV MR 46 # 1464 Zbl 251 # 94037 • ID 63349

LEE, R.C.T. [1972] *Fuzzy logic and the resolution principle* (J 0037) ACM J 19*109-119
⋄ B35 B52 ⋄ REV MR 52 # 4736 Zbl 245 # 02020 • ID 28860

LEE, R.C.T. [1973] see CHANG, C.L.

LEE, R.C.T. see Vol. II for further entries

LEE, R.D. [1966] see GOODSTEIN, R.L.

LEE, R.D. see Vol. VI for further entries

LEE, S.C. [1976] *Vector boolean algebra and calculus* (J 0187) IEEE Trans Comp C-25*865-874
⋄ B80 G05 ⋄ REV Zbl 335 # 94020 • ID 63350

LEE, S.C. see Vol. II for further entries

LEGENHAUSEN, G. [1985] *New semantics for the lower predicate calculus* (J 0079) Logique & Anal, NS 28*317-339
⋄ B10 ⋄ ID 49704

LEHMANN, A. [1973] *Two sets of perfect syllogisms* (J 0047) Notre Dame J Formal Log 14*425-429
⋄ B20 ⋄ REV Zbl 232 # 02011 • ID 29006

LEHMANN, G. [1985] *Modell- und rekursionstheoretische Grundlagen psychologischer Theorienbildung* (X 0811) Springer: Heidelberg & New York xxii+297pp
⋄ B10 C98 D80 D99 ⋄ ID 42737

LEHMANN, S.K. [1976] *An interpretation of "finite" modal first-order languages in classical second-order languages* (J 0036) J Symb Logic 41*337-340
⋄ B15 B45 ⋄ REV MR 53 # 10551 Zbl 341 # 02015 • ID 14765

LEHMANN, S.K. see Vol. II for further entries

LEHMER, D.H. [1963] *Some high-speed logic* (P 0719) Exper Arith, High Speed Comput & Math;1962 Chicago 141-145
⋄ B35 ⋄ REV MR 28 # 1732 Zbl 122.48 • ID 46514

LEISENRING, A.C. [1968] *An abstract property of formalized languages which contain Hilbert's ε-symbol* (J 0068) Z Math Logik Grundlagen Math 14*81-92
⋄ B22 ⋄ REV MR 36 # 6262 Zbl 157.14 • ID 07982

LEISENRING, A.C. [1969] *Mathematical logic and Hilbert's ε-symbol* (X 0836) Gordon & Breach: New York 142pp
⋄ B10 E25 F05 F98 ⋄ REV MR 43 # 1807 Zbl 188.315 • ID 07983

LEITSCH, A. [1985] see GOTTLOB, G.

LEITSCH, A. see Vol. IV for further entries

LEIVANT, D. [1983] see FORTUNE, S.

LEIVANT, D. see Vol. II, IV, VI for further entries

LEJEWSKI, C. [1958] *On implicational definitions (Polish and Russian summaries)* (J 0063) Studia Logica 8*189-211
⋄ B05 B20 ⋄ REV MR 21 # 7154 JSL 24.246 • ID 07985

LEJEWSKI, C. [1961] *On prosleptic syllogisms* (J 0047) Notre Dame J Formal Log 2*158-176
⋄ A05 A10 B20 ⋄ ID 07990

LEJEWSKI, C. [1964] *Aristotle's syllogistic and its extensions* (C 0567) Form & Strategy in Sci (Woodger) 203-232
⋄ A10 B20 ⋄ ID 14876

LEJEWSKI, C. [1968] *A propositional calculus in which three mutually undefinable functors are used as primitive terms (Polish and Russian summaries)* (J 0063) Studia Logica 22*17-50
⋄ B22 ⋄ REV MR 38 # 3123 Zbl 311 # 02013 • ID 07993

LEJEWSKI, C. see Vol. II, V for further entries

LEKOMTSEV, YU.K. [1975] *The algebraic approach to the syntax of colors in painting (Russian)* (S 0393) Uch Zap Univ, Tartu 365*193-205
⋄ B80 ⋄ REV MR 53 # 10692 • ID 23095

LEMMON, E.J. [1961] *Quantifier rules and natural deduction* (J 0094) Mind 70*235-238
⋄ B10 ⋄ REV MR 23 # A3664 JSL 31.127 • ID 08002

LEMMON, E.J. [1965] *A further note on natural deduction* (J 0094) Mind 74*594-597
⋄ A05 B10 ⋄ ID 14681

LEMMON, E.J. [1965] *Beginning logic* (X 1035) Nelson: Walton on Thames x+225pp
⋄ B98 ⋄ REV Zbl 158.244 JSL 40.287 • ID 15171

LEMMON, E.J. [1977] *An introduction to modal logic* (X 1096) Blackwell: Oxford x+94pp
⋄ A10 B45 B98 ⋄ REV MR 57 # 15931 Zbl 388 # 03006 JSL 44.653 • ID 52261

LEMMON, E.J. [1978] *Beginning logic* (X 2725) Hackett Publ: Indianapolis x+225pp
⋄ B98 ⋄ REV MR 58 # 107 JSL 46.421 • ID 75449

LEMMON, E.J. see Vol. II, V for further entries

LENAT, D.B. [1984] *Automated theory formation in mathematics* (P 3084) Autom Theor Prov After 25 Yea;1983 Denver 287-314
⋄ B35 ⋄ REV MR 85d:68005 Zbl 563 # 68073 • ID 45268

LENSTRA JR., H.W. [1968] *A definition of the system of natural numbers, equivalent to that of Peano* (J 0028) Indag Math 30*390-392
⋄ B28 ⋄ REV MR 38 # 3217 Zbl 169.9 JSL 35.474 • ID 08007

LENSTRA JR., H.W. see Vol. V for further entries

LENZ, H. [1958] *Zur Axiomatik der Zahlen* (J 0001) Acta Math Acad Sci Hung 9*33-44
⋄ B28 ⋄ REV MR 20 # 2284 Zbl 85.245 JSL 20.375 JSL 25.352 • ID 08008

LEONARD, H.S. [1967] *Principles of reasoning: an introduction to logic, methodology, and the theory of signs* (X 0813) Dover: New York xviii+620pp
⋄ A05 B98 ⋄ REV JSL 23.435 • ID 22513

LEONARD, H.S. see Vol. II, V for further entries

LEONARD, M. [1984] see GAROCHE, F.

LEONE, M. [1985] *On temporal invariance of menu systems* (**P** 2999) Proc Conf Databasis (Calzone);1985 Heidelberg 1-12
⋄ B05   B75   C90 ⋄ ID 49912

LEONG, Y. [1980] *Reductio ad absurdum (Proof by contradiction)* (**J** 2053) Math Medley 8∗12-16
⋄ A10   B05 ⋄ REV Zbl 473 # 03002 • ID 55333

LEONTOVICH, A.M. & VIKTOROVA, I.I. [1978] *On the number of tests for the determination of significant variables of boolean functions (Russian)* (**J** 2320) Probl Peredachi Inf, Akad Nauk SSSR 14/4∗85-97
• TRANSL [1978] (**J** 3419) Probl Inf Transmiss 14/4∗289-306
⋄ B05 ⋄ REV MR 80k:94055   Zbl 423 # 94046 • ID 69990

LERCHER, B. [1972] see HINDLEY, J.R.

LERCHER, B. see Vol. VI for further entries

LERMAN, M. & SCHMERL, J.H. & SOARE, R.I. (EDS.) [1981] *Logic year 1979-80, The University of Connecticut, USA* (**S** 3301) Lect Notes Math 859∗viii+326pp
⋄ B97   C97 ⋄ REV MR 82e:03007   Zbl 456 # 00006
• ID 54311

LERMAN, M. see Vol. III, IV, V for further entries

LEROY, S.F. [1980] *Entry and equilibrium under adjustment costs* (**J** 2750) J Econ Th 23∗348-360
⋄ H10 ⋄ REV Zbl 452 # 90011 • ID 54127

LESCANNE, P. [1984] *Term rewriting systems and algebra* (**P** 2633) Autom Deduct (7);1984 Napa 166-174
⋄ B35 ⋄ REV Zbl 546 # 68079 • ID 44451

LESISZ, W. & POGORZELSKI, W.A. [1976] *A simplified definition of the notion of similarity between formulas of the first-order predicate calculus* (**J** 0302) Rep Math Logic, Krakow & Katowice 7∗63-69
⋄ B10 ⋄ REV MR 57 # 9472   Zbl 358 # 02011 • ID 21927

LESISZ, W. [1984] *On propositional calculus with a variable functor* (**J** 0302) Rep Math Logic, Krakow & Katowice 17∗19-25
⋄ B22 ⋄ REV MR 86c:03006   Zbl 559 # 03005 • ID 44044

LESNIEWSKI, S. [1932] *Ueber die Definitionen in der sogenannten Theorie der Deduktion* (**J** 0459) C R Soc Sci Lett Varsovie Cl 3   24∗289-309
• TRANSL [1967] (**C** 0615) Polish Logic 1920-39 170-187 (Polish)
⋄ A05   B10 ⋄ REV Zbl 6.97   JSL 35.442   FdM 58.998
• ID 08033

LESNIEWSKI, S. [1967] *The collected works of Stanislaw Lesniewski* (**X** 0845) Univ Notre Dame Pr: Notre Dame
⋄ B96   E96 ⋄ ID 28535

LESNIEWSKI, S. see Vol. V for further entries

LETICHEVS'KIJ, O.A. [1966] see ANUFRIEV, F.V.

LETICHEVS'KIJ, O.A. [1972] see GLUSHKOV, V.M.

LETICHEVS'KIJ, O.A. [1974] see GODLEVSKIJ, A.B.

LETICHEVS'KIJ, O.A. see Vol. IV for further entries

LEVEILLE, J.P. & ROMAN, P. [1976] *Group representations in certain lattices of propositions* (**J** 2736) Int J Theor Phys 14∗73-90
⋄ B30 ⋄ REV MR 53 # 10669   Zbl 362 # 20028 • ID 50812

LEVI, B. [1940] *La nocion de "dominio deductivo" como elemento de orientacion en las cuestiones de fundamentos de las teorias matematicas* (**J** 0491) Publ Inst Mat Univ Nac Litoral (Rosario RA) 2∗179-208
⋄ A05   B30 ⋄ REV MR 2.339   Zbl 60.19   JSL 7.44
• ID 08040

LEVI, B. see Vol. V for further entries

LEVI, G. [1982] see BELLIA, M.

LEVI, I. [1979] *Serious possibility* (**C** 1706) Essays Honour J. Hintikka 219-236
⋄ A05   B30 ⋄ REV Zbl 429 # 03004 • ID 53835

LEVI, I. see Vol. II for further entries

LEVIN, A.A. [1969] *The relative complexity of the reduced disjunctive normal form (Russian)* (**J** 0071) Met Diskr Analiz (Novosibirsk) 15∗25-34
⋄ B05 ⋄ REV MR 44 # 5185   Zbl 251 # 94036 • ID 08045

LEVIN, A.A. [1974] *Complexity of the disjunctive normal form of a sum of terminal disjunctive normal forms with respect to the disjunctive normal form of a sum of minimal disjunctive normal forms (Russian)* (**J** 0071) Met Diskr Analiz (Novosibirsk) 24∗50-68,96
⋄ B05 ⋄ REV MR 54 # 9993   Zbl 298 # 94051 • ID 25762

LEVIN, A.A. [1981] *The comparative complexity of disjunctive normal forms (Russian)* (**J** 0071) Met Diskr Analiz (Novosibirsk) 36∗23-38,93
⋄ B05 ⋄ REV MR 85d:03123   Zbl 483 # 94034 • ID 36814

LEVIN, A.A. [1983] *Gluing of Boolean functions and its application in estimates of the comparative complexity of DNF (Russian)* (**J** 0071) Met Diskr Analiz (Novosibirsk) 40∗54-71,101
⋄ B05 ⋄ ID 44842

LEVIN, A.G. [1978] *A method for the solution of difference logic equations (Russian)* (**J** 1516) Vest Ser Fiz Mat Mekh, Univ Minsk 1∗28-31
⋄ B35 ⋄ REV MR 58 # 9855 • ID 82059

LEVIN, A.G. see Vol. IV for further entries

LEVIN, A.M. [1975] *The axiom of choice in classical analysis (Russian) (English summary)* (**J** 0288) Vest Ser Mat Mekh, Univ Moskva 30/4∗59-65
• TRANSL [1975] (**J** 0510) Moscow Univ Math Bull 30/3-4∗106-111
⋄ B30   E15   E25   E35   E75   F35 ⋄ REV MR 54 # 7254   Zbl 316 # 02070 • ID 25810

LEVIN, A.M. [1979] *On an interesting axiomatic theory (Russian)* (**C** 2581) Issl Neklass Log & Teor Mnozh 137-142
⋄ B28   F35 ⋄ REV MR 81j:03089   Zbl 422 # 03032
• ID 53493

LEVIN, A.M. see Vol. V, VI for further entries

LEVIN, N.P. [1949] *Computational logic* (**J** 0036) J Symb Logic 14∗167-172
⋄ B05   G05 ⋄ REV MR 11.151   Zbl 35.5   JSL 15.69
• ID 08048

LEVIN, V.A. & PEREL'ROIZEN, E.Z. [1980] *Logical methods in the field of computer vision (Russian)* (**S** 0507) Vychisl Sist (Akad Nauk SSSR Novosibirsk) 82∗95-109,112
⋄ B80 ⋄ REV MR 82m:68131 • ID 82064

LEVIN, V.A. [1980] *Qualitative analysis of the problem of combinatorial optimization by the method of logical determinants (Russian)* (J 0474) Avtom Vychis Tekh, Akad Nauk Latv SSR 1980/6*49-55,97
- TRANSL [1980] (J 2666) Autom Control Comput Sci 14*40-45
  ⋄ B80 ⋄ REV MR 82i:68025 Zbl 479 # 68047 • ID 82061

LEVIN, V.A. see Vol. II, IV for further entries

LEVITZ, H. [1974] *Non-standard analysis: an exposition* (J 3370) Enseign Math, Ser 2 20*9-32
  ⋄ H05 ⋄ REV MR 51 # 837 Zbl 286 # 26006 • ID 17490

LEVITZ, H. & LEVITZ, K. [1979] *Logic and Boolean algebra* (X 4259) Barron's Educ: Woodbury viii + 132pp
  ⋄ B05 ⋄ REV JSL 46.420 • ID 44744

LEVITZ, H. [1982] *Calculation of an order type: an application of nonstandard methods* (J 0068) Z Math Logik Grundlagen Math 28*219-228
  ⋄ H05 ⋄ REV MR 83m:04004 Zbl 522 # 03057 • ID 35479

LEVITZ, H. see Vol. III, IV, V, VI for further entries

LEVITZ, K. [1979] see LEVITZ, H.

LEVY, A. [1958] see BAR-HILLEL, Y.

LEVY, A. see Vol. III, IV, V, VI for further entries

LEVY, G. [1971] *Operations booleennes generalisees* (J 2313) C R Acad Sci, Paris, Ser A-B 272*A1471-A1473
  ⋄ B05 ⋄ REV MR 43 # 7309 Zbl 231 # 02075 • ID 08090

LEVY, G. see Vol. II for further entries

LEVY, M. & RAYNAUD, Y. [1984] *Ultrapuissances des espaces $L^p(L^q)$ (English summary)* (J 3364) C R Acad Sci, Paris, Ser 1 299*81-84
  ⋄ C20 C65 H10 H20 ⋄ REV MR 85i:46044 Zbl 569 # 46008 • ID 44350

LEVY, M. see Vol. III for further entries

LEWANDOWSKI, A. [1963] *A modification of Hilbert's axiom of order (Polish) (Russian and English summaries)* (J 0051) Commentat Math, Ann Soc Math Pol, Ser 1 8*93-97
  ⋄ B28 ⋄ REV MR 32 # 4009 Zbl 138.417 • ID 08096

LEWANDOWSKI, A. & SUSZKO, R. [1968] *A note concerning the theory of descriptions (Polish and Russian summaries)* (J 0063) Studia Logica 22*51-56
  ⋄ B10 ⋄ REV MR 38 # 4295 Zbl 313 # 02007 • ID 08097

LEWIS, A.A. [1985] *Hyperfinite von Neumann games* (J 3914) Math Soc Sci 9*189-194
  ⋄ H05 ⋄ ID 47762

LEWIS, A.A. [1985] *Loeb-measurable solutions to *finite games* (J 3914) Math Soc Sci 9*197-247
  ⋄ H05 ⋄ REV Zbl 571 # 90108 • ID 48280

LEWIS, A.A. see Vol. IV, V for further entries

LEWIS, C.I. [1918] *A survey of symbolic logic* (X 0926) Univ Calif Pr: Berkeley vi + 406pp
- REPR [1960] (X 0813) Dover: New York
  ⋄ B45 B46 B98 ⋄ REV JSL 16.225 FdM 47.870
  • ID 22590

LEWIS, C.I. [1932] see LANGFORD, C.H.

LEWIS, C.I. [1960] *A survey of symbolic logic (Reprinted edition)* (X 0813) Dover: New York x + 327pp
  ⋄ B45 B98 ⋄ ID 25502

LEWIS, C.I. see Vol. II for further entries

LEWIS, H.R. [1973] see AANDERAA, S.O.

LEWIS, H.R. [1973] see GOLDFARB, W.D.

LEWIS, H.R. [1974] see AANDERAA, S.O.

LEWIS, H.R. [1975] see GOLDFARB, W.D.

LEWIS, H.R. [1976] *Krom formulas with one dyadic predicate letter* (J 0036) J Symb Logic 41*341-362
  ⋄ B20 D05 D35 ⋄ REV MR 53 # 10573 Zbl 381 # 03012
  • ID 14766

LEWIS, H.R. [1978] *Complexity of solvable cases of the decision problem for the predicate calculus* (P 3578) IEEE Symp Found of Comput Sci (19);1978 Ann Arbor 35-47
  ⋄ B20 B25 D15 ⋄ REV MR 80e:03041 • ID 75527

LEWIS, H.R. [1978] *Renaming a set of clauses as a Horn set* (J 0037) ACM J 25*134-135
  ⋄ B20 D15 ⋄ REV MR 57 # 8151 Zbl 365 # 68082
  • ID 82072

LEWIS, H.R. [1978] *Satisfiability problems for propositional calculi* (P 4048) Allerton Conf Commun, Control & Comput (16);1978 Monticello 513-520
  ⋄ B05 ⋄ REV MR 84b:94002 Zbl 428 # 03035 • ID 46283

LEWIS, H.R. [1979] *Satisfiability problems for propositional calculi* (J 0041) Math Syst Theory 13*45-53
  ⋄ B05 D15 ⋄ REV MR 80m:03025 Zbl 428 # 03035
  • ID 53794

LEWIS, H.R. [1979] *Unsolvable classes of quantificational formulas* (X 0832) Addison-Wesley: Reading xvii + 198pp
  ⋄ B20 B25 D35 D98 ⋄ REV MR 81i:03069
  Zbl 423 # 03003 JSL 47.221 • ID 53515

LEWIS, H.R. [1980] *Complexity results for classes of quantificational formulas* (J 0119) J Comp Syst Sci 21*317-353
- TRANSL [1983] (J 3079) Kiber Sb Perevodov, NS 20*64-106
  ⋄ B20 B25 D10 D15 ⋄ REV MR 82m:03052
  Zbl 471 # 03034 • ID 36546

LEWIS, H.R. [1982] see AANDERAA, S.O.

LEWIS, H.R. [1984] see DENENBERG, L.

LEWIS, H.R. see Vol. III, IV for further entries

LEWIS II, P.M. [1967] see COATES, C.L.

LEWIS II, P.M. see Vol. IV for further entries

LI, BANGHE [1978] *Nonstandard analysis and multiplication of distributions (Chinese)* (J 1024) Zhongguo Kexue 21*561-585
  ⋄ H05 ⋄ REV MR 80e:46028 Zbl 392 # 46026 • ID 82073

LI, BANGHE [1979] *Differential and integral calculus on a non-Archimedean field (Chinese) (English summary)* (J 0418) Shuxue Xuebao 22/1*14-27
  ⋄ H05 H10 ⋄ REV MR 80c:26016 Zbl 397 # 26006
  • ID 48677

LI, BANGHE [1980] *Integral mean value theorems on the field of formal power series (Chinese)* (J 4572) Shuxue Jikan, Sichuan Shang Shuxuehui 1/1*45-56
⋄ H05 H10 ⋄ ID 48681

LI, BANGHE [1980] *The differential and integral calculus on the field of formal power series (Chinese)* (J 2771) Kexue Tongbao 25*24-26
⋄ H05 H10 ⋄ REM A special issue on Math., Phys and Chem. • ID 48680

LI, BANGHE & LI, YACHING [1981] *On multiplication of distributions (Chinese) (English summary)* (J 3750) Jilin Daxue, Ziran Kexue Xuebao 1981/1*13-30
⋄ H10 ⋄ ID 48678

LI, BANGHE & LI, YACHING [1985] *Nonstandard analysis and multiplication of distributions in any dimension (Chinese)* (J 3766) Zhongguo Kexue, Xi A 28*716-726
⋄ H10 ⋄ ID 48679

LI, DAFA [1981] *A method for determining the unsatisfiability of a set of ground clauses (Chinese) (English summary)* (J 2754) Huazhong Gongxueyuan Xuebao 3*67-70
⋄ B35 ⋄ REV MR 84d:03021 • ID 34061

LI, DAFA [1982] *Some results on unit proofs (Chinese) (English summary)* (J 2754) Huazhong Gongxueyuan Xuebao 10/2*7-12
⋄ B35 ⋄ REV MR 83j:68114 • ID 39380

LI, DAFA [1983] *Comparing the efficiency of input and unit resolution (Chinese) (English summary)* (J 3218) J Huazhong Univ Sci Tech (Engl Ed) 11/6*55-58
⋄ B35 ⋄ ID 45269

LI, DAFA [1983] *Proof of the conjecture of Henschen and Wos (Chinese) (English summary)* (J 2157) Qinghua Daxue Xuebao 23*83-88
⋄ B35 ⋄ REV MR 85i:03037 • ID 44078

LI, WEIHUA [1981] see HE, YANXIANG

LI, WEIHUA [1981] *Three fast algorithms for theorem proving (Chinese) (English summary)* (J 2879) Wuhan Daxue Xuebao, Ziran Kexue 1981/1*22-30
⋄ B35 ⋄ REV MR 83m:03022 • ID 35429

LI, YACHING [1980] *Some results on products of distributions (Chinese)* (J 2771) Kexue Tongbao 25*16-20
⋄ H10 ⋄ REM A special issue on Math., Phys. and Chem.
• ID 48682

LI, YACHING [1981] see LI, BANGHE

LI, YACHING [1985] see LI, BANGHE

LIBER, A.E. [1966] *Binaere boolesche Algebra und ihre Anwendungen (Russisch)* (X 0958) Saratov Univ: Saratov 80pp
⋄ B05 B70 G05 ⋄ REV Zbl 299 # 02001 • ID 63440

LIEBER, L.R. [1947] *Mits, wits, and logic* (X 0942) Norton: New York 240pp
⋄ B60 B98 ⋄ REV Zbl 223 # 02031 JSL 13.55 • ID 22514

LIEBER, L.R. see Vol. V for further entries

LIEBERHERR, K.J. & SPECKER, E. [1979] *Complexity of partial satisfaction* (P 3535) IEEE Symp Founds of Comput Sci (20);1979 San Juan 132-139
⋄ B05 D15 ⋄ REV MR 83a:68036 • ID 38843

LIEBERHERR, K.J. & SPECKER, E. [1981] *Complexity of partial satisfaction* (J 0037) ACM J 28*411-421
⋄ B05 D15 ⋄ REV MR 83a:03033 Zbl 456 # 68078
• ID 69677

LIEBERHERR, K.J. & VAVASIS, S.A. [1982] *Analysis of polynomial approximation algorithms for constant expressions* (P 3862) Theor Comput Sci (6);1983 Dortmund 187-197
⋄ B05 D15 ⋄ REV MR 84b:68003 Zbl 495 # 68028
• ID 46285

LIEDL, R. [1970] *Harmonische Analysis bei Aussagenkalkuelen* (J 0009) Arch Math Logik Grundlagenforsch 13*158-167
⋄ B05 ⋄ REV MR 46 # 3297 Zbl 217.7 • ID 08136

LIFSCHITZ, V. [1967] *Deductive general validity and reduction classes (Russian)* (S 0228) Zap Nauch Sem Leningrad Otd Mat Inst Steklov 4*69-77
⋄ B20 D35 ⋄ REV MR 38 # 5604 Zbl 165.19 • ID 08138

LIFSCHITZ, V. [1980] *Semantical completeness theorems in logic and algebra* (J 0053) Proc Amer Math Soc 79*89-96
⋄ B10 F07 ⋄ REV MR 81g:03073 Zbl 453 # 03060
• ID 54190

LIFSCHITZ, V. [1985] *Calculable natural numbers* (C 3659) Intens Math 173-190
⋄ B28 F30 F50 ⋄ ID 44843

LIFSCHITZ, V. see Vol. III, IV, VI for further entries

LIGHTSTONE, A.H. & ROBINSON, A. [1957] *Syntactical transforms* (J 0064) Trans Amer Math Soc 86*220-245
• REPR [1979] (C 4594) Sel Pap Robinson 1*120-145
⋄ B10 C52 C60 ⋄ REV MR 19.935 Zbl 208.10 JSL 24.244 • ID 08149

LIGHTSTONE, A.H. [1964] *The axiomatic method: an introduction to mathematical logic* (X 0819) Prentice Hall: Englewood Cliffs x + 246pp
⋄ B98 ⋄ REV MR 29 # 1133 Zbl 129.256 JSL 31.106
• ID 22360

LIGHTSTONE, A.H. [1966] *Symbolic logic and the real number system: An introduction to the foundations of number systems* (X 0837) Harper & Row: New York ix + 225pp
⋄ B28 ⋄ REV Zbl 146.245 • ID 22361

LIGHTSTONE, A.H. [1968] *Group theory and the principle of duality* (J 0018) Canad Math Bull 11*43-50
⋄ B10 ⋄ REV MR 37 # 5081 Zbl 174.10 • ID 08151

LIGHTSTONE, A.H. [1969] *The notion of "consequence" in the predicate calculus* (J 0497) Math Mag 42*57-60
⋄ B10 ⋄ REV MR 39 # 33 Zbl 182.4 • ID 08152

LIGHTSTONE, A.H. [1972] *Infinitesimals* (J 0005) Amer Math Mon 79*242-251
⋄ H05 ⋄ REV MR 46 # 49 Zbl 248 # 02066 • ID 08153

LIGHTSTONE, A.H. [1973] *Infinitesimals and integration* (J 0497) Math Mag 46*20-30
⋄ H05 ⋄ REV MR 47 # 4790 Zbl 249 # 26018 • ID 08154

LIGHTSTONE, A.H. & WONG, KAM [1975] *Dirac delta functions via nonstandard analysis* (J 0018) Canad Math Bull 18*759-762
⋄ H05 ⋄ REV MR 53 # 14118 Zbl 329 # 26020 • ID 82076

LIGHTSTONE, A.H. & ROBINSON, A. [1975] *Nonarchimedean fields and asymptotic expansions* (X 0809) North Holland: Amsterdam x + 204pp
⋄ H05 ⋄ REV MR 54 # 2457 Zbl 303 # 26013 JSL 46.163
• ID 24041

LIGHTSTONE, A.H. [1978] *Mathematical logic. An introduction to model theory. Edited by H. B. Enderton* (X 1332) Plenum Publ: New York xiii+338pp
⋄ B98 C98 H05 ⋄ REV MR 80i:03002 Zbl 382 # 03002
• ID 51926

LIGHTSTONE, A.H. see Vol. III for further entries

LIGMANOWSKI, M. & SORDYL, E. [1967] *Eine Methode zur bestimmung der Primimplikanten einer Menge logischer Funktionen (Polish) (English and Russian summaries)* (J 1665) Appl Mathematicae, Pol Akad Nauk 114*79-86
⋄ B35 ⋄ REV Zbl 257 # 02009 • ID 28991

LILLO DE, N.J. [1979] *Models of an extension of the theory ORD* (J 0047) Notre Dame J Formal Log 20*729-734
⋄ B28 C40 C65 E10 E30 ⋄ REV MR 81f:03043 Zbl 349 # 02009 • ID 56184

LILLO DE, N.J. see Vol. III, IV, V for further entries

LIN, JINHUO & MENG, ZHANGRONG [1965] *To compute Boolean expression with best order (Chinese)* (P 4564) Math Logic;1963 Xi-An 123-124
⋄ B35 ⋄ ID 49360

LIN, PEIKEE [1985] *Unconditional bases and fixed points of nonexpansive mappings* (J 0048) Pac J Math 116*69-76
⋄ C20 C65 H05 ⋄ REV MR 86c:47075 Zbl 566 # 47038
• ID 48745

LINDENBAUM, A. & TARSKI, A. [1927] *Sur l'independance des notions primitives dans les systemes mathematiques* (J 0283) Ann Soc Pol Math 5*111-113
⋄ B30 C40 E75 ⋄ ID 30918

LINDENBAUM, A. [1930] *Remarques sur une question de la methode axiomatique* (J 0027) Fund Math 15*313-321
⋄ A05 B30 ⋄ REV FdM 56.488 • ID 08160

LINDENBAUM, A. & TARSKI, A. [1935] *Ueber die Beschraenktheit der Ausdrucksmittel deduktiver Theorien* (J 1124) Ergebn Math Kolloquium 7*15-22
• TRANSL [1956] (C 1159) Tarski: Logic, Semantics, Metamathematics 384-392
⋄ B30 C07 C35 C65 ⋄ REV JSL 1.115 FdM 62.39
• ID 30921

LINDENBAUM, A. [1936] *Sur la simplicite formelle des notions* (P 0632) Congr Int Phil des Sci;1935 Paris 7*29-38
⋄ B30 ⋄ REV JSL 2.55 FdM 62.1050 • ID 38723

LINDENBAUM, A. see Vol. V, VI for further entries

LINDGREN, W.F. [1972] see FLETCHER, P.

LINDNER, R. [1979] see KLETTE, R.

LINDNER, R. see Vol. II, IV, V, VI for further entries

LINDSTROEM, P. [1973] *A note on weak second order logic with variables for elementarily definable relations* (P 0710) Russell Mem Logic Conf;1971 Uldum 221-233
⋄ B15 C55 C62 ⋄ REV MR 50 # 4238 • ID 08177

LINDSTROEM, P. [1974] *On characterizing elementary logic* (C 1936) Log Th & Semant Anal (Kanger) 129-146
⋄ B10 C95 ⋄ REV Zbl 289 # 02001 • ID 29976

LINDSTROEM, P. see Vol. III, VI for further entries

LINDSTROEM, T.L. [1980] *Hyperfinite stochastic integration. I: The nonstandard theory. II: Comparison with the standard-theory. III: Hyperfinite representations of standard martingales* (J 0132) Math Scand 46*265-292,293-314,315-331 • ERR/ADD ibid 46*332-333
⋄ H10 ⋄ REV MR 83a:60091 Zbl 427 # 60057 Zbl 427 # 60058 Zbl 427 # 60059 • ID 55820

LINDSTROEM, T.L. [1982] *A Loeb-measure approach to theorems by Prohorov, Sazonov and Gross* (J 0064) Trans Amer Math Soc 269*521-534
⋄ B50 H05 ⋄ REV MR 83m:60008 Zbl 484 # 60007
• ID 36617

LINDSTROEM, T.L. [1984] see ALBEVERIO, S.

LINES ESCARDO, E. [1962] *Ueber die Struktur der Menge der natuerlichen Zahlen (Spanisch)* (J 0264) Collect Math (Barcelona) 14*287-300
⋄ B28 E20 ⋄ REV MR 29 # 4693 Zbl 121.256 • ID 48076

LINIAL, N. & TARSI, M. [1985] *Deciding hypergraph 2-colourability by H-resolution* (J 1426) Theor Comput Sci 38*343-347
⋄ B35 ⋄ ID 49196

LIOGON'KIJ, M.I. [1969] *A certain property of the equivalence axiom in first order predicate calculus (Russian)* (J 0499) Uch Zap Univ, Gor'kij 105*17-21
⋄ B10 ⋄ REV MR 45 # 8498 • ID 08180

LIOGON'KIJ, M.I. [1969] *On the conditional satisfiability ratio of logical formulae (Russian)* (J 0087) Mat Zametki (Akad Nauk SSSR) 6*651-662
• TRANSL [1969] (J 1044) Math Notes, Acad Sci USSR 6*856-861
⋄ B10 ⋄ REV MR 42 # 49 Zbl 212.24 • ID 08181

LIOGON'KIJ, M.I. [1969] see GLEBSKIJ, YU.V.

LIOGON'KIJ, M.I. [1970] *On the question of quantitative characteristics of logical formulae (Russian) (English summary)* (J 0040) Kibernetika, Akad Nauk Ukr SSR 1970/3*16-22
• TRANSL [1970] (J 0021) Cybernetics 6*205-212
⋄ B20 B25 C13 ⋄ REV MR 47 # 1597 Zbl 229 # 02040
• ID 08182

LIOGON'KIJ, M.I. see Vol. IV for further entries

LIPCZYNSKA, M. & WOLTER, H. [1969] *Elements of logic. An exposition for law* (X 1016) Uniw Bieruta : Wroclaw 353pp
⋄ B80 ⋄ ID 14285

LIPSHITZ, L. [1976] see BECKER, J.A.

LIPSHITZ, L. see Vol. III, IV for further entries

LIS, Z. [1960] *Logical consequence, semantic and formal* (J 0063) Studia Logica 10*39-60
⋄ B10 C07 ⋄ REV MR 24 # A1813 Zbl 121.252 • ID 08189

LIS, Z. [1968] *An algebraic approach to traditional logic (Polish) (English and Russian summaries)* (J 0063) Studia Logica 22*99-122
⋄ B10 ⋄ REV MR 38 # 1997 Zbl 306 # 02015 • ID 63461

LISOVIK, L.P. [1984] *Monadic second-order theories of two successor functions with an additional predicate (Russian)* (J 0270) Dokl Akad Nauk Ukr SSR, Ser A 8*80-82
⋄ B15 B25 C10 C85 D05 ⋄ REV MR 86c:03008
• ID 48409

LISOVIK, L.P. see Vol. III, IV for further entries

LIST, G. [1973] see BAGINSKI, M.

LIU, CHUCHANG [1980] *The mechanical algorithm for solving Dedekind's problem (Chinese) (English summary)* (**J 2879**) Wuhan Daxue Xuebao, Ziran Kexue 1980/3∗30–38
⋄ B35 ⋄ REV MR 82j:03010 • ID 75589

LIU, SHICHAO [1980] *A proof-theoretic approach to nonstandard analysis with emphasis on distinguishing between constructive and nonconstructive results* (**P 2058**) Kleene Symp;1978 Madison 391–414
⋄ E70 F99 H05 H20 ⋄ REV MR 81m:03076 Zbl 464 # 03060 • ID 75591

LIU, SHICHAO [1984] *A proof-theoretic approach to nonstandard analysis (continued)* (**P 2153**) Logic Colloq;1983 Aachen 1∗281–296
⋄ E70 F50 H05 ⋄ ID 45423

LIU, SHICHAO see Vol. II, IV, V, VI for further entries

LIU, XUHUA [1982] *Delection strategy in the resolution principle (Chinese) (English summary)* (**J 3750**) Jilin Daxue, Ziran Kexue Xuebao 1982/2∗97–106
⋄ B35 ⋄ REV MR 83m:03023 • ID 34886

LIU, XUHUA & WANG, XIANGHAO [1982] *Generalized resolution (Chinese) (English summary)* (**J 3793**) Jisuanjii Xuebao 5∗81–92
⋄ B35 ⋄ REV MR 84m:03019 • ID 35724

LIU, XUHUA & WANG, XIANGHAO [1983] *Factoring problem in resolution (Chinese)* (**J 3766**) Zhongguo Kexue, Xi A 26∗753–760
⋄ B35 ⋄ REV MR 85a:03020 Zbl 512 # 68076 • ID 34771

LIU, XUHUA [1985] see AN, ZHI

LIU, XUHUA [1985] *The input semicancellation resolution principle on Horn sets (Chinese)* (**J 2771**) Kexue Tongbao 30∗1201–1202
⋄ B35 ⋄ ID 49332

LIU, XUHUA see Vol. II for further entries

LIU, ZUNQUAN & QIN, CHAOBIN [1981] *Mechanical deduction of formulas of differential equations. I (Chinese)* (**J 1024**) Zhongguo Kexue 24∗313–323
⋄ B35 ⋄ REV MR 84m:65096a • REM Part II 1981 • ID 48992

LIU, ZUNQUAN & QIN, CHAOBIN [1981] *Mechanical deduction of formulas of differential equations. II (Chinese)* (**J 2771**) Kexue Tongbao 26∗257–258
• TRANSL [1981] (**J 3769**) Sci Bull, Foreign Lang Ed 26∗779–780
⋄ B35 ⋄ REV MR 84m:65096b Zbl 493 # 34031 • REM Part I 1981 • ID 45995

LIVOVSCHI, L. [1971] *Representation of boolean functions in arbitrary bases* (**P 1560**) Tr Mezhdurn Semin Priklad Aspekt Teor Avtom;1971 Varna 1∗179–184
⋄ B05 ⋄ REV Zbl 255 # 94020 • ID 63474

LOBOVIKOV, V.O. [1982] *Some applications of modal logic to ethical categories (Russian)* (**C 3743**) Probl Log & Metodol Nauk 256–268
⋄ B80 ⋄ REV MR 84j:03041 Zbl 555 # 03002 • ID 34633

LOBOVIKOV, V.O. [1984] *A non-logical interpretation of the mathematical apparatus of the classical first order predicate logic (Russian)* (**C 4403**) Logika, Pozn, Otrazh 33–58
⋄ B10 ⋄ ID 46599

LOCKS, M.O. [1976] *Reversal in boolean minimalization* (**J 0079**) Logique & Anal, NS 19∗285–297
⋄ B35 ⋄ REV MR 57 # 5465 Zbl 364 # 02006 • ID 50936

LOCKS, M.O. [1978] *Logical and probability analysis of systems* (**J 0047**) Notre Dame J Formal Log 19∗123–136
⋄ A05 A10 B05 ⋄ REV MR 58 # 10439 Zbl 376 # 02044 • ID 51686

LOCKS, M.O. [1978] *Minimalization of boolean polynomials, truth functions, and lattices* (**J 0047**) Notre Dame J Formal Log 19∗264–270
⋄ B35 G10 ⋄ REV MR 58 # 4675 Zbl 398 # 03050 • ID 52778

LOEB, M.H. [1953] *Concatenation as basis for a complete system arithmetic* (**J 0036**) J Symb Logic 18∗1–6
⋄ B03 ⋄ REV MR 14.938 Zbl 52.250 JSL 35.150 • ID 08224

LOEB, M.H. [1968] *Die Vollstaendigkeit der verzweigten Typenlogik mit unendlicher Terminduktion* (**J 0009**) Arch Math Logik Grundlagenforsch 11∗68–72
⋄ B15 F35 ⋄ REV MR 38 # 42 Zbl 182.323 • ID 08229

LOEB, M.H. [1972] *A reduction theorem for predicate logic* (**J 0036**) J Symb Logic 37∗352–354
⋄ B10 ⋄ REV MR 47 # 6432 Zbl 347 # 02030 • ID 08236

LOEB, M.H. [1976] *Embedding first order predicate logic in fragments of intuitionistic logic* (**J 0036**) J Symb Logic 41∗705–718
⋄ B10 D35 F50 ⋄ REV MR 56 # 79 Zbl 358 # 02012 • ID 14746

LOEB, M.H. see Vol. II, III, IV, VI for further entries

LOEB, P.A. [1971] *A nonstandard representation of measurable spaces and L_∞* (**J 0015**) Bull Amer Math Soc 77∗540–544
⋄ H05 ⋄ REV MR 43 # 2488 • ID 27620

LOEB, P.A. [1972] *A non-standard representation of measurable spaces, L_∞, and L_∞^** (**P 0732**) Contrib to Non-Standard Anal;1970 Oberwolfach 65–80
⋄ E75 H05 ⋄ REV MR 58 # 2215 Zbl 246 # 26017 • ID 19227

LOEB, P.A. [1973] *A combinatorial analog of Lyapunov's theorem for infinitesimal generated atomic vector measures* (**J 0053**) Proc Amer Math Soc 39∗585–586
⋄ H05 ⋄ REV Zbl 271 # 28007 • ID 27623

LOEB, P.A. [1974] see BERNSTEIN, A.R.

LOEB, P.A. [1974] *A nonstandard representation of Borel measures and σ-finite measures* (**P 1083**) Victoria Symp Nonstand Anal;1972 Victoria 144–152
⋄ H05 ⋄ REV MR 57 # 16537 Zbl 285 # 28002 • ID 27621

LOEB, P.A. [1974] *A note on continuity for Robinson's predistribution* (**P 1083**) Victoria Symp Nonstand Anal;1972 Victoria 153–154
⋄ H05 ⋄ REV MR 58 # 7074 Zbl 287 # 26017 • ID 27622

LOEB, P.A. [1975] *Conversion from nonstandard to standard measure spaces and applications in probability theory* (**J 0064**) Trans Amer Math Soc 211∗113–122
⋄ H05 ⋄ REV MR 52 # 10980 Zbl 312 # 28004 JSL 50.243 • ID 18262

LOEB, P.A. [1976] *Applications of nonstandard analysis to ideal boundaries in potential theory* (J 0029) Israel J Math 25*154-187
◊ H05 ◊ REV MR 56 # 15961 Zbl 346 # 31007 • ID 26080

LOEB, P.A. [1976] see BROWN, D.J.

LOEB, P.A. [1979] *An introduction to nonstandard analysis and hyperfinite probability theory* (C 2987) Probab Anal & Rel Topics, Vol 2 105-142
◊ H05 ◊ REV MR 80m:60005 Zbl 441 # 03027 • ID 82100

LOEB, P.A. [1979] see HELMS, L.L.

LOEB, P.A. [1979] *Weak limits of measures and the standard part map* (J 0053) Proc Amer Math Soc 77*128-135
◊ H05 ◊ REV MR 80i:28020 Zbl 394 # 03061 • ID 56218

LOEB, P.A. [1982] see HELMS, L.L.

LOEB, P.A. [1984] *A functional approach to nonstandard measure theory* (P 4227) Modern Anal Probab;1982 New Haven 251-261
◊ H05 ◊ REV MR 86b:28026 Zbl 533 # 28008 • ID 45256

LOEB, P.A. [1984] *Measure spaces in nonstandard models underlying standard stochastic processes* (P 4313) Int Congr Math (II,14);1983 Warsaw 1*323-335
◊ H10 ◊ ID 48567

LOEB, P.A. [1985] *A nonstandard functional approach to Fubini's theorem* (J 0053) Proc Amer Math Soc 93*343-346
◊ C60 C65 H05 H20 ◊ REV Zbl 566 # 28004 • ID 44489

LOEB, P.A. [1985] see HURD, A.E.

LOESCHAU, G. [1973] see BAGINSKI, M.

LOEWENHEIM, L. [1908] *Ueber das Aufloesungsproblem im logischen Klassenkalkuel* (J 0366) Sitzber Berlin Math Ges 7*89-94
◊ B20 B25 G05 ◊ REV FdM 39.89 • ID 08241

LOEWENHEIM, L. [1940] *Einkleidung der Mathematik in Schroederschen Relativkalkuel* (J 0036) J Symb Logic 5*1-15
◊ B30 E30 ◊ REV MR 1.321 Zbl 24.2 JSL 5.127 FdM 66.30 • ID 08247

LOEWENHEIM, L. see Vol. III, V, VI for further entries

LOEWNER, K. [1939] *Grundzuege einer Inhaltslehre im Hilbertschen Raume* (J 0120) Ann of Math, Ser 2 40*816-833
◊ H05 ◊ REV MR 1.48 FdM 65.1174 • ID 41168

LOGAN, G.J. [1976] *Note on the Frattini-Neumann-Schmidt intersection theorem* (J 0329) Math Chron (Auckland) 5*65-69
◊ B22 E20 ◊ REV MR 55 # 5501 Zbl 352 # 08010 • ID 50056

LOGEMANN, G. [1962] see DAVIS, MARTIN D.

LOKTIONOV, V.I. [1970] *A certain theorem on logical semantics (Russian)* (C 0668) Neklass Log 262-275
◊ B22 ◊ REV MR 49 # 2249 • ID 08250

LOLLI, G. [1978] *Alcune applicazioni della compattezza* (J 0319) Matematiche (Sem Mat Catania) 33*321-332
◊ C20 C60 H05 ◊ REV MR 84i:03115 Zbl 469 # 03050 • ID 55177

LOLLI, G. [1978] *Lezioni di logica matematica* (X 0905) Boringhieri: Torino 203pp
◊ B98 ◊ ID 31832

LOLLI, G. & LONGO, G. & MARCJA, A. (EDS.) [1984] *Logic colloquium '82. Proceedings of the colloquium held in Florence, 23-28 August, 1982* (S 3303) Stud Logic Found Math 112*viii+358pp
◊ B97 C97 D97 F97 ◊ REV MR 85g:03006 Zbl 538 # 00003 • ID 41493

LOLLI, G. see Vol. III, IV, V, VI for further entries

LONDEY, D. [1970] *On the strong completeness of some equivalential systems* (J 3672) Philosophia Arhusiensis 4*11-16
◊ B20 ◊ ID 33658

LONDEY, D.G. [1965] see HUGHES, G.E.

LONDON, F. [1925] *Ueber die Irreversibilitaet deduktiver Schlussweisen* (J 0157) Jbuchber Dtsch Math-Ver 34*84-86
◊ A05 B10 F07 ◊ REV FdM 51.47 • ID 08256

LONGINI COHEN, E. [1977] see ANDREWS, P.B.

LONGINI COHEN, E. [1982] see ANDREWS, P.B.

LONGINI COHEN, E. [1984] see ANDREWS, P.B.

LONGO, G. [1974] *I problemi di decisione e la loro complessita* (J 3436) Quad, Ist Appl Calcolo, Ser 3 8*87-108
◊ B25 B35 D15 F20 ◊ REV Zbl 427 # 03010 • ID 53697

LONGO, G. [1984] see LOLLI, G.

LONGO, G. see Vol. III, IV, V, VI for further entries

LOOMIS JR., H.H. & WYMAN JR., R.H. [1965] *On complete sets of logic primitives* (J 4305) IEEE Trans Electr Comp EC-14*173-174
◊ B05 ◊ REV Zbl 132.249 JSL 35.160 • ID 19225

LOPARIC, A. [1977] *Une etude semantique de quelques calculs propositionnels (English summary)* (J 2313) C R Acad Sci, Paris, Ser A-B 284*A835-A838
◊ B05 B25 ◊ REV MR 55 # 5399 Zbl 378 # 02010 • ID 27299

LOPARIC, A. see Vol. II for further entries

LOPES DOS SANTOS, L.H. [1980] *Constructive completeness proofs for positive propositional calculi* (P 3006) Brazil Conf Math Log (3);1979 Recife 199-209
◊ B20 ◊ REV MR 83j:03021 Zbl 459 # 03005 • ID 35334

LOPES DOS SANTOS, L.H. see Vol. II for further entries

LOPES PASSOS, E.P. [1982] see BERRETA PION, M.M.

LOPEZ-ESCOBAR, E.G.K. [1967] *A complete, infinitary axiomatization of weak second-order logic* (J 0027) Fund Math 61*93-103
◊ B15 C75 C85 F35 ◊ REV MR 36 # 2482 Zbl 174.13 JSL 35.467 • ID 08262

LOPEZ-ESCOBAR, E.G.K. [1972] *Constructions and negationless logic (Polish and Russian summaries)* (J 0063) Studia Logica 30*7-22
◊ B20 E70 F50 ◊ REV MR 48 # 5827 Zbl 287 # 02015 • ID 08267

LOPEZ-ESCOBAR, E.G.K. & VELDMAN, W. [1975] *Intuitionistic completeness of a restricted second-order logic* (P 1440) ⊢ ISILC Proof Th Symp (Schuette);1974 Kiel 500*198-232
◊ B15 C90 F50 ◊ REV MR 54 # 67 Zbl 334 # 02015 • ID 24555

LOPEZ-ESCOBAR, E.G.K. see Vol. II, III, V, VI for further entries

LORENS, C.S. [1964] *Invertible boolean functions* (J 4305) IEEE Trans Electr Comp EC-13∗529-541
⋄ B05 ⋄ REV MR 30#5888   JSL 36.347 • ID 08274

LORENZEN, P. [1948] *Einfuehrung in die Logik* (X 0908) Univ Math Inst: Bonn i+34pp
⋄ B98 ⋄ REV MR 10.1   Zbl 30.2 • ID 28576

LORENZEN, P. [1954] *Zur Begruendung der zweiwertigen Aussagenlogik* (J 1114) Arch Phil 5∗109-112
• REPR [1954] (J 0009) Arch Math Logik Grundlagenforsch 2∗29-32
⋄ A05 B05 ⋄ REV MR 17.223   Zbl 56.7   JSL 24.175
• ID 33309

LORENZEN, P. [1955] *Einfuehrung in die operative Logik und Mathematik* (X 0811) Springer: Heidelberg & New York vii+298pp
⋄ A05 B98 F50 F65 F98 ⋄ REV MR 17.223   Zbl 66.248   Zbl 68.8   JSL 22.289   JSL 35.330 • REM 2nd ed. 1969
• ID 08290

LORENZEN, P. [1956] *Zur Interpretation der Syllogistik* (J 0009) Arch Math Logik Grundlagenforsch 2∗100-103
⋄ A05 B20 ⋄ REV MR 19.521   Zbl 70.245   JSL 23.229
• ID 08292

LORENZEN, P. [1957] *Ueber die Syllogismen als Relationenmultiplikationen* (J 0009) Arch Math Logik Grundlagenforsch 3∗112-116
⋄ A05 B20 ⋄ REV MR 20#6349   Zbl 86.8   JSL 24.80
• ID 08293

LORENZEN, P. [1962] *Metamathematik* (X 0876) Bibl Inst: Mannheim 173pp
• TRANSL [1967] (X 0834) Gauthier-Villars: Paris 162pp (French) [1971] (X 1781) Tecnos: Madrid (Spanish)
⋄ A05 B98 C60 D20 D35 D98 F98 ⋄ REV MR 28#3932   Zbl 105.246   JSL 31.106 • ID 08303

LORENZEN, P. [1965] *Die klassische Analysis als eine konstruktive Theorie* (P 3696) Logic Colloq;1964 Bristol 18∗81-94
⋄ B28 F65 ⋄ REV MR 34#2458   Zbl 158.5 • ID 08304

LORENZEN, P. [1970] *Logica formal* (X 1781) Tecnos: Madrid
⋄ B98 ⋄ ID 33353

LORENZEN, P. [1974] *Justification of the deductive method (Russian)* (C 2578) Filos & Logika 78-83
⋄ A05 B30 ⋄ REV MR 58#27367 • ID 75660

LORENZEN, P. [1976] *Die Vollstaendigkeit einer unverzweigten Variante des "analytischen" Entscheidungsverfahrens der klassischen Logik* (J 0009) Arch Math Logik Grundlagenforsch 18∗19-22
⋄ B05 ⋄ REV MR 57#2872   Zbl 344#02009 • ID 24316

LORENZEN, P. [1978] *Eine konstruktive Deutung des Dualismus in der Wahrscheinlichkeitstheorie* (J 0989) Z Allg Wissth 9∗256-275
⋄ A05 B30 ⋄ REV MR 80i:03012 • ID 75670

LORENZEN, P. see Vol. II, IV, V, VI for further entries

LOS, J. [1946] *Une preuve d'axiomatisation de la logique traditionelle* (J 0501) Ann Univ Lublin, Sect Math 1∗211-228
⋄ B20 ⋄ REV MR 10.93   JSL 13.166 • ID 08312

LOS, J. [1949] *On logical matrices (Polish)* (S 0502) Prace Wroclaw Tow Nauk, Ser B 19∗42pp
⋄ B22 B50 ⋄ REV MR 19.724   Zbl 45.296   JSL 16.59
• ID 19296

LOS, J. [1950] *Sur la notion independance dans la metamathematique (Polish summary)* (J 0086) Cas Pestovani Mat, Ceskoslov Akad Ved 74∗138-140
⋄ B30 ⋄ REV Zbl 39.245   JSL 20.88 • ID 19295

LOS, J. [1951] *An algebraic proof of completeness for the two-valued propositional calculus* (S 0019) Colloq Math (Warsaw) 2∗236-240
⋄ B05 G05 ⋄ REV MR 14.345   Zbl 45.295 • ID 08316

LOS, J. [1955] *On the axiomatic treatment of probability* (S 0019) Colloq Math (Warsaw) 3∗125-137
⋄ B30 ⋄ REV Zbl 64.127 • ID 08327

LOS, J. [1955] *The algebraic treatment of the methodology of elementary deductive system(Polish and Russian summaries)* (J 0063) Studia Logica 2∗151-212,328
⋄ B22 C07 ⋄ REV MR 18.785   Zbl 67.251   JSL 21.193
• ID 08332

LOS, J. [1955] see GRZEGORCZYK, A.

LOS, J. & MOSTOWSKI, ANDRZEJ & RASIOWA, H. [1956] *A proof of Herbrand's theorem* (J 3941) J Math Pures Appl, Ser 9 35∗19-24 • ERR/ADD ibid 40∗129-134
⋄ B10 C07 F05 F50 ⋄ REV MR 17.699   MR 28#22   Zbl 199.7   Zbl 73.7   JSL 36.168 • ID 08333

LOS, J. [1957] *Remarks on Henkin's paper: boolean representation through propositional calculus* (J 0027) Fund Math 44∗82-83
⋄ B05 ⋄ REV MR 19.724   Zbl 87.251   JSL 38.521 • REM The article was published ibid 41∗89-96 • ID 08337

LOS, J. & SUSZKO, R. [1958] *Remarks on sentential logics* (J 0028) Indag Math 20∗177-183
⋄ B22 ⋄ REV MR 20#5125   Zbl 92.248   JSL 40.603
• ID 08338

LOS, J. [1960] *Fields of events and their definition in the axiomatic treatment of probability (Polish) (Russian and English summaries)* (J 0063) Studia Logica 9∗95-132
⋄ B30 ⋄ REV MR 27#3005   Zbl 201.491 • ID 33089

LOS, J. [1963] *Remarks on foundations of probability. Semantical interpretation of the probability of formulas* (P 0677) Int Congr Math (II, 9,Proc);1962 Djursholm 225-229
⋄ B30 ⋄ REV MR 28#3919   Zbl 137.112 • ID 08348

LOS, J. [1963] *Semantic representation of the probability of formulas in formalized theories (Polish summary)* (J 0063) Studia Logica 14∗183-196
• REPR [1977] (C 3174) Twenty-five Years Log Meth Poland 327-340
⋄ A05 B10 B48 B60 ⋄ REV MR 32#3087   Zbl 292#02008 • ID 31415

LOS, J. [1982] see COHEN, L.J.

LOS, J. see Vol. II, III, V for further entries

LOSEV, G.F. [1973] *Best local algorithm of index 1 for constructing a sum of minimal disjunctive normal forms of a boolean function (Russian)* (J 0023) Dokl Akad Nauk SSSR 212∗816-817
• TRANSL [1973] (J 0062) Sov Math, Dokl 14∗1481-1483
⋄ B35 ⋄ REV MR 57#19130   Zbl 325#02006 • ID 30433

LOSEV, G.F. [1974] *The best local algorithm for constructing the sum of minimal-essential disjunctive normal forms of a boolean function, using neighbourhoods of minimal order (Russian)* (J 0199) Zh Vychisl Mat i Mat Fiz 14*470-478
- TRANSL [1974] (J 1049) USSR Comput Math & Math Phys 14/2*193-201
◇ B35 ◇ REV MR 49#6678 Zbl 286#02010 • ID 63517

LOSEV, G.F. [1978] see KABULOV, A.V.

LOSEV, G.F. [1982] *Local algorithms for information calculation and the minimum covering problem (Russian)* (J 0023) Dokl Akad Nauk SSSR 267*544-548
- TRANSL [1982] (J 0062) Sov Math, Dokl 26*649-653
◇ B35 ◇ REV MR 85d:94018 Zbl 546#94023 • ID 39106

LOSEV, G.F. see Vol. IV for further entries

LOURENCO, M. (ED.) [1979] *Goedel's theorem and the continuum hypothesis (Portuguese)* (X 2719) Fundacao Calouste Gulbenkian: Lisbon xcv+900pp
◇ B97 E50 F30 F97 ◇ REV MR 82b:03003 Zbl 486#03005 • REM Transl from several English articles • ID 70030

LOUVEAU, A. [1985] see DELON, F.

LOUVEAU, A. see Vol. IV, V for further entries

LOVASZ, L. [1977] see GACS, P.

LOVASZ, L. [1980] *Efficient algorithms: an approach by formal logic* (C 3494) Stud on Math Progr. Math Meth Oper Res, Vol 1 1*119-126
◇ B35 C13 D15 ◇ REV MR 82e:68042 Zbl 427#68050 • ID 53755

LOVASZ, L. see Vol. III, V for further entries

LOVELAND, D.W. [1962] see DAVIS, MARTIN D.

LOVELAND, D.W. [1963] see GELERNTER, H.L.

LOVELAND, D.W. [1968] *Mechanical theorem-proving by model elimination* (J 0037) ACM J 15*236-251
- REPR [1983] (C 4659) Autom of Reasoning 2*117-134
◇ B35 ◇ REV Zbl 162.28 • ID 31161

LOVELAND, D.W. [1969] *Theorem-provers combining model elimination and resolution* (J 0508) Machine Intelligence 4*73-86
- REPR [1983] (C 4659) Autom of Reasoning 2*249-263
◇ B35 ◇ REV MR 40#5169 Zbl 257#68083 • ID 31162

LOVELAND, D.W. [1970] *A linear format for resolution* (P 0625) Symp Autom Demonst;1968 Versailles 147-162
- REPR [1983] (C 4659) Autom of Reasoning 2*399-416
◇ B35 ◇ REV MR 43#4299 Zbl 202.15 • ID 08358

LOVELAND, D.W. [1972] *A unifying view of some linear Herbrand procedures* (J 0037) ACM J 19*366-384
◇ B35 ◇ REV MR 47#1336 Zbl 243#68012 • ID 08359

LOVELAND, D.W. & STICKEL, M.E. [1976] *A hole in goal trees: some quidance from resolution theory* (J 0187) IEEE Trans Comp C-25*335-341
◇ B35 ◇ REV Zbl 324#68055 • ID 31163

LOVELAND, D.W. [1978] *Automated theorem proving: a logical Basis* (X 0809) North Holland: Amsterdam xiii+405pp
◇ B35 ◇ REV MR 57#14640 Zbl 364#68082 JSL 45.629 • ID 31165

LOVELAND, D.W. & REDDY, C.R. [1981] *Deleting repeated goals in the problem reduction format* (J 0037) ACM J 28*646-661
◇ B35 ◇ REV MR 83k:68093 Zbl 468#68096 • ID 40068

LOVELAND, D.W. (ED.) [1982] *6th conference on automated deduction* (S 3302) Lect Notes Comput Sci 138*vii+389pp
◇ B35 ◇ REV MR 85e:68002 Zbl 476#00025 • ID 39972

LOVELAND, D.W. [1984] see BLEDSOE, W.W.

LOVELAND, D.W. see Vol. III, IV, VI for further entries

LOZHKIN, S.A. [1981] *The relation between depth and complexity of equivalent formulas and the depth of monotone functions of the algebra of logic (Russian)* (J 0052) Probl Kibern 38*269-271,272
◇ B05 G05 ◇ REV MR 84h:03137 Zbl 531#03005 • ID 34323

LOZHKIN, S.A. [1983] *Depth of functions of the algebra of logic in certain bases (Russian)* (J 3955) Ann Univ Budapest, Sect Comp 4*113-125
◇ B05 ◇ REV MR 86b:03079 Zbl 549#94044 • ID 45850

LU, RUQIAN [1982] *A resolution theorem on Horn sets (Chinese)* (J 3766) Zhongguo Kexue, Xi A 25*310-317
◇ B35 ◇ REV MR 84c:03031 Zbl 488#68054 • ID 34934

LU, RUQIAN [1982] *Semantic mappings on Herbrand base (Chinese)* (J 2771) Kexue Tongbao 27*67-69
- TRANSL [1982] (J 3769) Sci Bull, Foreign Lang Ed 10*1042-1045
◇ B35 ◇ REV MR 85d:68046 Zbl 497#03010 • ID 38107

LU, RUQIAN [1982] *The principle of strong ordered input resolution (Chinese)* (J 3766) Zhongguo Kexue, Xi A 25*430-439
◇ B35 ◇ REV MR 83m:68162 Zbl 488#68055 • ID 40373

LU, YANGCI & LU, ZHONGWAN & TANG, ZHISONG & WAN, ZHEXIAN [1960] *Mathematical logic and its application (Chinese)* (J 2771) Kexue Tongbao 2*50-52
◇ B98 ◇ ID 48538

LU, ZHONGWAN [1960] see LU, YANGCI

LU, ZHONGWAN [1981] see HU, SHIHUA

LU, ZHONGWAN [1981] *Mathematical logic and mechanical proofs (Chinese)* (X 1876) Kexue Chubanshe: Beijing
◇ B35 ◇ ID 48518

LU, ZHONGWAN [1982] see HU, SHIHUA

LUCAS, J.R. [1980] *Truth, probability and set theory* (P 2958) Latin Amer Symp Math Log (4);1978 Santiago 209-218
◇ A05 B30 ◇ REV MR 81i:03008 Zbl 428#03004 • ID 53764

LUCAS, T. [1973] *L'analyse non-standard* (J 1917) Rev Questions Sci 144*477-491
◇ H05 ◇ REV MR 56#5187 Zbl 265#26016 • ID 75702

LUCAS, T. see Vol. II, III, V, VI for further entries

LUCHINS, A.S. & LUCHINS, E.H. [1965] *Logical foundation of mathematics for behavioral scientists* (X 0818) Holt Rinehart & Winston: New York xii+436pp
◇ B98 ◇ ID 22516

LUCHINS, E.H. [1965] see LUCHINS, A.S.

LUCHT, L. [1976] see HISCHER, HORST

LUCKHAM, D.C. [1967] *The resolution principle in theorem-proving* (J 0508) Machine Intelligence 1*47-61
⋄ B35 ⋄ ID 46532

LUCKHAM, D.C. [1968] *Some tree-paring strategies for theorem-proving* (J 0508) Machine Intelligence 3*95-112
⋄ B35 ⋄ REV Zbl 197.148 • ID 46534

LUCKHAM, D.C. [1969] see ALLEN, JOHN

LUCKHAM, D.C. [1970] *Refinement theorems in resolution theory* (P 0625) Symp Autom Demonst;1968 Versailles 163-190
• REPR [1983] (C 4659) Autom of Reasoning 2*435-465
⋄ B35 ⋄ REV MR 43#4300 Zbl 216.241 • ID 08374

LUCKHAM, D.C. & NILSSON, N.J. [1971] *Extracting information from resolution proof trees* (J 0503) Artif Intell 2*27-54
⋄ B35 ⋄ REV MR 46#7003 Zbl 221#68051 • ID 08375

LUCKHAM, D.C. see Vol. IV, VI for further entries

LUCKHARDT, H. [1969] *Kodifikation und Aussagenlogik* (J 0009) Arch Math Logik Grundlagenforsch 12*18-38
⋄ B05 B25 F50 ⋄ REV MR 41#3239 Zbl 182.6
• ID 08380

LUCKHARDT, H. [1969] *Skolem-Normalformen (English summary)* (J 0504) Manuscr Math 1*241-257
⋄ B15 ⋄ REV MR 41#37 Zbl 188.16 • ID 08379

LUCKHARDT, H. [1984] *Obere Komplexitaetsschranken fuer TAUT- Entscheidungen* (P 3621) Frege Konferenz (2);1984 Schwerin 331-337
⋄ B05 D15 ⋄ REV MR 86d:03036 • ID 45393

LUCKHARDT, H. see Vol. II, III, IV, VI for further entries

LUDWIG, G. & NEUMANN, H. [1981] *Connections between different approaches to the foundations of quantum mechanics* (P 3185) Interpr & Found of Quantum Th;1979 Marburg 133-143
⋄ B30 ⋄ REV MR 84j:81006 Zbl 495#03043 • ID 36891

LUE, QICI [1980] *Concurrent relations and nonstandard models (Chinese) (English summary)* (J 2684) J Huazhong Inst Tech (Engl Ed) 8/2*19-22
⋄ H10 ⋄ ID 46954

LUE, QICI see Vol. II, III for further entries

LUEKOE, G. [1977] *Truth functions and problems in graph colouring* (J 0380) Acta Cybern (Szeged) 3*327-333
⋄ B80 ⋄ REV MR 58#10565 Zbl 404#05027 • ID 54850

LUKASIEWICZ, J. [1921] *Two-valued logic (Polish)* (J 1125) Przeglad Filoz 23*189-205
⋄ A05 B05 B50 ⋄ REV FdM 48.1125 • ID 28632

LUKASIEWICZ, J. [1924] *Demonstration de la compatibilite des axiomes de la theorie de la deduction* (J 0283) Ann Soc Pol Math 3*149
⋄ A05 B10 ⋄ REV FdM 51.55 • ID 08389

LUKASIEWICZ, J. [1929] *Elements of mathematical logic (Polish)* (X 1034) PWN: Warsaw 99pp
• TRANSL [1963] (X 0869) Pergamon Pr: Oxford xi+124pp
⋄ B98 ⋄ REV MR 22#6692 Zbl 126.7 JSL 30.237 JSL 31.284 • REM 2nd ed. 1958 • ID 20964

LUKASIEWICZ, J. & TARSKI, A. [1930] *Untersuchungen ueber den Aussagenkalkuel* (J 0459) C R Soc Sci Lett Varsovie Cl 3 23*30-50
• TRANSL [1956] (C 1159) Tarski: Logic, Semantics, Metamathematics 38-59
⋄ A05 B05 ⋄ REV JSL 33.130 JSL 34.99 FdM 57.1319
• ID 08390

LUKASIEWICZ, J. [1932] *Ein Vollstaendigkeitsbeweis des zweiwertigen Aussagenkalkuels* (J 0459) C R Soc Sci Lett Varsovie Cl 3 24*153-183
⋄ A05 B05 ⋄ REV Zbl 4.385 FdM 58.998 • ID 08393

LUKASIEWICZ, J. [1937] *In defense of logistic (Polish)* (S 4530) Stud Gnieseusia 15*22pp
• TRANSL [1970] (C 1786) Lukasiewicz: Select Works 236-249 [1961] (C 0609) Lukasiewicz: Log & Philos Papers 210-219
⋄ A05 B98 ⋄ REV JSL 3.43 JSL 33.132 • ID 41703

LUKASIEWICZ, J. [1939] *The equivalential calculus (Polish)* (J 0490) Collecteana Logica 1*145-169
• TRANSL [1967] (C 0615) Polish Logic 1920-39 1920-39 88-115
⋄ A05 B20 ⋄ REV Zbl 22.289 JSL 33.133 JSL 35.442 FdM 65.22 • ID 16523

LUKASIEWICZ, J. [1941] *Die Logik und das Grundlagenproblem* (P 0652) Entreteins Zuerich Fond & Method Sci Math;1938 Zuerich 82-100
• TRANSL [1970] (C 1786) Lukasiewicz: Select Works 278-294
⋄ A05 B98 ⋄ REV MR 2.338 Zbl 61.8 JSL 7.35
• ID 16528

LUKASIEWICZ, J. [1948] *The shortest axiom of the implicational calculus of proposition* (J 0215) Proc Irish Acad, Sect A 52*25-33
⋄ B20 ⋄ REV MR 10.93 Zbl 29.98 JSL 13.164 • ID 08396

LUKASIEWICZ, J. [1950] *Concerning an axiom system of the implicational propositional calculus (Polish)* (J 0283) Ann Soc Pol Math Suppl 22*27-92
⋄ B20 ⋄ REV JSL 20.173 • ID 42454

LUKASIEWICZ, J. [1953] *Comment on K.J.Cohen's remark* (J 0028) Indag Math 15*113
⋄ B05 F50 ⋄ REV MR 14.1053 Zbl 51.6 JSL 19.217
• ID 25951

LUKASIEWICZ, J. [1953] *Sur la formalisation des theories mathematiques* (P 0644) Meth Form en Axiom;1950 Paris 11-19
⋄ B30 ⋄ REV MR 15.1 Zbl 50.247 JSL 22.214 • ID 08400

LUKASIEWICZ, J. [1961] *Logistic and philosophy (Polish)* (C 0609) Lukasiewicz: Log & Philos Papers 195-209
⋄ A05 B98 ⋄ REV JSL 33.132 • ID 45092

LUKASIEWICZ, J. [1961] *Remarks on Nicod's axiom and on "generalizing deduction" (Polish)* (C 0609) Lukasiewicz: Log & Philos Papers 164-177
⋄ B05 ⋄ REV JSL 30.376 • ID 44967

LUKASIEWICZ, J. see Vol. II, VI for further entries

LUPANOV, O.B. [1960] *On the complexity of the realization by formulas of the function of logical algebra (Russian)* (J 0052) Probl Kibern 3*61-80
- TRANSL [1960] (J 1195) Probl Cybernet 3*782-811 [1963] (J 0449) Probl Kybern 3*68-90 (German)
  ◊ B05 ◊ REV MR 23 # A3681 Zbl 126.326 JSL 36.547
  • ID 08413

LUPANOV, O.B. [1961] *Implementing the algebra of logic functions in terms of bounded depth formulas in the basis of &, ∨, ¬ (Russian)* (J 0023) Dokl Akad Nauk SSSR 136*1041-1042
- TRANSL [1961] (J 0470) Sov Phys, Dokl 6*107-108
  ◊ B05 ◊ REV Zbl 104.241 • ID 45279

LUPANOV, O.B. [1961] *On the realization of functions of logical algebra by formulae of finite classes (formulae of limited depth) in the basis &, ∨, ¬ (Russian)* (J 0052) Probl Kibern 6*5-14
- TRANSL [1965] (J 1195) Probl Cybernet 6*1-14
  ◊ B05 ◊ REV JSL 36.571 • ID 45280

LUPANOV, O.B. see Vol. IV for further entries

LUSK, E.L. & OVERBEEK, R.A. [1980] *Data structures and control architecture for implementation of theorem-proving programs* (P 3063) Autom Deduct (5);1980 Les Arcs 232-249
  ◊ B35 ◊ REV Zbl 438 # 68048 • ID 69752

LUSK, E.L. & WINKER, S.K. & WOS, L. [1981] *Semigroups, antiautomorphisms, and involutions: a computer solution to an open problem. I* (J 0214) Math of Comp 37*533-545
  ◊ B35 ◊ REV MR 82k:20103 Zbl 517.20041 • ID 83153

LUSK, E.L. & OVERBEEK, R.A. [1982] *Experiments with resolution-based theorem-proving algorithms* (J 2687) Comp Math Appl 8*141-152
  ◊ B35 ◊ REV MR 83k:68094 Zbl 487 # 68080 • ID 40070

LUSK, E.L. & MCCUNE, W. & OVERBEEK, R.A. [1982] *Logic machine architecture: inference mechanisms* (P 3840) Autom Deduct (6);1982 New York 70-84
  ◊ B35 ◊ REV MR 85e:68002 • ID 45188

LUSZCZEWSKA-ROMAHNOWA, S. [1953] *An analysis and generalization of Venn's diagrammatic decision procedure (Polish, Russian) (English summary)* (J 0063) Studia Logica 1*185-213,214-246 • ERR/ADD ibid 1*301
  ◊ B20 ◊ REV MR 16.892 Zbl 59.16 JSL 21.193 • ID 28467

LUSZCZEWSKA-ROMAHNOWA, S. see Vol. II, V for further entries

LUTZ, R. [1981] see GOZE, M.

LUTZ, R. & SARI, T. [1981] *Sur le comportement asymptotique des solutions dans un probleme aux limites non lineaires (English summary)* (J 3364) C R Acad Sci, Paris, Ser 1 292*925-928
  ◊ H05 ◊ REV MR 82f:34015 Zbl 464 # 34039 • ID 82124

LUTZ, R. & SARI, T. [1982] *Applications of nonstandard analysis to boundary value problems in singular perturbation theory* (P 4018) Th & Appl Sing Perturbations;1981 Oberwolfach 113-135
  ◊ H10 ◊ REV MR 84a:34057 Zbl 518 # 34049 • ID 38852

LUTZ, R. [1983] *L'intrusion de l'analyse nonstandard dans l'etude des perturbations singulieres* (J 1620) Asterisque 109-110*101-139
  ◊ H05 ◊ REV MR 86d:58105 Zbl 529 # 58025 • ID 45502

LUXEMBURG, W.A.J. [1962] *Nonstandard analysis. Lectures on A. Robinson's theory of infinitesimals and infinitely large numbers* (X 1246) Caltech: Pasadena
  ◊ C20 H05 H98 ◊ ID 21419

LUXEMBURG, W.A.J. [1962] *Two applications of the method of construction by ultrapowers to analysis* (J 0015) Bull Amer Math Soc 68*416-419
  ◊ C20 E05 E75 H05 ◊ REV MR 25 # 3837 Zbl 109.8
  • ID 08426

LUXEMBURG, W.A.J. [1967] *A new approach to the theory of monads* (1111) Preprints, Manuscr., Techn. Reports etc.
  ◊ H05 ◊ REM Technical Report No.1, Office of Naval Research • ID 21420

LUXEMBURG, W.A.J. [1969] *A general theory of monads* (P 0649) Appl Model Th to Algeb, Anal & Probab;1967 Pasadena 18-86
  ◊ B15 C20 H05 ◊ REV MR 39 # 6244 Zbl 207.524 JSL 36.541 • ID 21147

LUXEMBURG, W.A.J. [1969] *Reduced powers of the real number system and equivalents of the Hahn-Banach extension theorem* (P 0649) Appl Model Th to Algeb, Anal & Probab;1967 Pasadena 123-137
  ◊ C20 E25 E75 H05 ◊ REV MR 38 # 5616 Zbl 181.401
  • ID 24816

LUXEMBURG, W.A.J. & TAYLOR, R.F. [1970] *Almost commuting matrices are near commuting matrices* (J 0028) Indag Math 32*96-98
  ◊ C60 H05 ◊ REV MR 41 # 3502 Zbl 195.50 • ID 08429

LUXEMBURG, W.A.J. [1972] *A nonstandard analysis approach to Fourier analysis* (P 0732) Contrib to Non-Standard Anal;1970 Oberwolfach 15-39
  ◊ H05 ◊ REV Zbl 249 # 42001 • ID 08432

LUXEMBURG, W.A.J. [1972] *A remark on the Cantor-Lebesque lemma* (P 0732) Contrib to Non-Standard Anal;1970 Oberwolfach 41-46
  ◊ H05 ◊ REV MR 57 # 10341 Zbl 247 # 26013 • ID 08431

LUXEMBURG, W.A.J. & ROBINSON, A. (EDS.) [1972] *Contributions to non-standard analysis* (X 0809) North Holland: Amsterdam vii+289pp
  ◊ B97 H05 H97 ◊ REV MR 57 # 16517 Zbl 236 # 00005
  • ID 25074

LUXEMBURG, W.A.J. [1972] *On some concurrent binary relations occuring in analysis* (P 0732) Contrib to Non-Standard Anal;1970 Oberwolfach 85-100
  ◊ H05 ◊ REV MR 58 # 2355 Zbl 248 # 26018 • ID 08433

LUXEMBURG, W.A.J. [1973] *Non-standard analysis* (X 1246) Caltech: Pasadena v+151pp
  ◊ H05 ◊ REV MR 58 # 10428 • ID 75725

LUXEMBURG, W.A.J. [1974] *On a theorem of helly and a theorem about liftings* (P 1083) Victoria Symp Nonstand Anal;1972 Victoria xiii
  ◊ H05 ◊ ID 27627

LUXEMBURG, W.A.J. [1975] *Nichtstandard-Zahlsysteme und die Begruendung des Leibnizschen Infinitesimalkalkuels* (S 0780) Jbuch Ueberblick Math 1975*31-44
  ◊ H05 ◊ REV MR 58 # 10430 Zbl 312 # 26017 • ID 75724

LUXEMBURG, W.A.J. [1977] *Nichtstandard-Zahlsysteme und die Begruendung des Leibnizschen Infinitesimalkalkuels (Slowakisch)* (J 1527) Pokroky Mat Fyz Astron (Prague) 22*316-325
  ◇ H05 ◇ REV MR 58 # 10431 Zbl 376 # 26014 • ID 51705

LUXEMBURG, W.A.J. [1977] *Non-standard analysis* (P 1704) Int Congr Log, Meth & Phil of Sci (5);1975 London ON 1*107-119
  ◇ A10 H05 ◇ REV MR 57 # 12212 Zbl 387 # 03032
  • ID 52248

LUXEMBURG, W.A.J. see Vol. III, V for further entries

LYALETSKIJ, A.V. [1973] see DEGTYAREV, A.I.

LYALETSKIJ, A.V. & MALASHONOK, A.I. [1975] *A calculus of k-disjuncts with the rule of a latent clash-resolution (Russian)* (C 2962) Mat Voprosy Teor Intell Mashin 3-33
  ◇ B35 F07 ◇ REV MR 58 # 21501 Zbl 421 # 03045
  • ID 53444

LYALETSKIJ, A.V. [1975] *A k-disjunct calculus (Russian)* (C 2962) Mat Voprosy Teor Intell Mashin 34-48
  ◇ B35 F07 ◇ REV MR 58 # 21502 Zbl 421 # 03046
  • ID 53445

LYALETSKIJ, A.V. [1981] *A variant of Herbrand's theorem for formulas in prenex form (Russian)* (J 0040) Kibernetika, Akad Nauk Ukr SSR 1981/1*112-116,152
  • TRANSL [1981] (J 0021) Cybernetics 17*125-129
  ◇ B35 F05 ◇ REV MR 84a:03072 Zbl 462 # 68074
  • ID 35594

LYALETSKIJ, A.V. [1982] *Testing the admissibility of substitutions (Russian)* (C 3832) Avtom Issl Mat 40-46
  ◇ B35 ◇ REV MR 85a:68163 Zbl 535 # 03006 • ID 38311

LYANTSE, V.E. [1983] *Can nonstandard analysis be ignored? (Jordan form of an operator in an infinite-dimensional space) (Russian)* (C 4685) Gen Th Bound Value Probl 108-112
  ◇ H05 ◇ REV MR 85m:47011 • ID 45504

LYNCH, E.P. [1980] *Applied symbolic logic* (X 0820) Intersci Publ: New York xi+260pp
  ◇ B98 ◇ REV Zbl 528 # 94018 • ID 36772

LYNDON, R.C. [1951] *Identities in two-valued calculi* (J 0064) Trans Amer Math Soc 71*457-465
  ◇ B05 B20 C05 ◇ REV MR 13.422 Zbl 44.2 JSL 18.69
  • ID 08436

LYNDON, R.C. [1959] *Existential Horn sentences* (J 0053) Proc Amer Math Soc 10*994-998
  ◇ B20 C30 C40 C52 ◇ REV MR 22 # 6712 Zbl 114.249 JSL 30.253 • ID 08439

LYNDON, R.C. see Vol. III, IV, V for further entries

LYNGHOLM, C. & YOURGRAU, W. [1960] *A double-iteration property of boolean functions* (J 0047) Notre Dame J Formal Log 1*111-114
  ◇ B05 ◇ REV MR 25 # 5014 JSL 25.299 • ID 18072

LYSENKO, E.V. & MOKLYAK, N.G. & POPOV, V.A. & SKIBENKO, I.T. [1973] *Ueber einen Algorithmus der Suche von Arten unbestimmter boolescher Funktionen (Russian)* (S 3418) Pribory & Sist Avtomat (Khar'kov) 26*158-165
  ◇ B35 ◇ REV Zbl 296 # 02003 • ID 63541

LYUBCHENKO, G.G. [1958] *Methods for determining the identical truth or falsity of the calculus of assertion formulae of bivalent logic (Ukrainian) (Russian and English summaries)* (J 0270) Dokl Akad Nauk Ukr SSR, Ser A 1958*1153-1156
  ◇ B35 ◇ REV MR 21 # 647 Zbl 81.243 • ID 08449

LYUBCHENKO, G.G. [1960] *Representing boolean functions by formulae (Ukrainian) (Russian and English summaries)* (J 0270) Dokl Akad Nauk Ukr SSR, Ser A 1960*1011-1015
  ◇ B05 ◇ REV MR 23 # A825 Zbl 126.8 • ID 21009

LYUBCHENKO, G.G. [1961] *Binary codes for logical machines (Ukrainian) (Russian and English summaries)* (J 0270) Dokl Akad Nauk Ukr SSR, Ser A 1961*604-607
  ◇ B80 ◇ REV MR 24 # A1217 Zbl 146.248 • ID 21008

LYUBCHENKO, G.G. [1961] *On finding formulas of minimum length for functions of algebraic logic (Ukrainian) (Russian summary)* (J 4406) Zb Prats Obchis Mat Tekhn 3*3-12
  ◇ B05 ◇ REV MR 35 # 18 Zbl 129.100 • ID 45436

LYUBETSKIJ, V.A. [1981] see GORDON, E.I.

LYUBETSKIJ, V.A. [1985] *Some algebraic questions of nonstandard analysis (Russian)* (J 0023) Dokl Akad Nauk SSSR 280*38-41
  • TRANSL [1985] (J 0062) Sov Math, Dokl 31*30-34
  ◇ C60 C90 H05 ◇ REV MR 86e:03063 • ID 45424

LYUBETSKIJ, V.A. see Vol. III, IV, V for further entries

LYUSTERNIK, L.A. [1925] *To the question of the foundation of analysis and the founding of geometry without set theory (Russian)* (J 4122) Vest Komm Akad 13*214-222
  ◇ B30 ◇ ID 43716

LYUSTERNIK, L.A. [1952] see ABRAMOV, A.A.

MA, HUAXIAO [1982] *A simplified algorithm for finding Boolean differences and its applications in engineering (Chinese) (English summary)* (J 3959) Chengdu Keji Daxue Xuebao 1982/2*79-94
  ◇ B80 ◇ REV MR 84g:94026 Zbl 489 # 94036 • ID 39427

MA, XIWEN [1982] *A relational approach in semantics (Chinese) (English summary)* (J 3793) Jisuanjii Xuebao 5*1-10
  ◇ B10 ◇ REV MR 84e:68013 • ID 39270

MA, XIWEN see Vol. II for further entries

MAC LANE, S. [1939] *Symbolic logic* (J 0005) Amer Math Mon 46*289-294
  ◇ B10 ◇ REV FdM 65.20 • ID 43752

MAC LANE, S. see Vol. V, VI for further entries

MACALUSO, A.T. & TORTORICI, M. [1977] *Un processo inferenziale per la risolutione, in calcolo automatico, di equazioni e sistemi di equazioni logiche* (J 0104) Atti Accad Sci Lett Arti Palermo, Ser 4/I 36*181-233
  ◇ B35 ◇ REV Zbl 478 # 03003 • ID 55618

MACCOLL, H. [1896] *The calculus of equivalent statements* (J 0077) Proc London Math Soc 28*156-183,555-579
  ◇ B20 ◇ REV FdM 28.64 • ID 08452

MACCOLL, H. [1906] *Symbolic logic and its applications* (X 1296) Longman: Harlow & New York xi+141pp
  ◇ B05 B98 ◇ REV FdM 37.66 • ID 22595

MACDONALD, A.L. [1976] *Sturm-Liouville theory via nonstandard analysis* (J 0452) Indiana Univ Math J 25*531-540
  ◇ H05 ◇ REV MR 54 # 3078 Zbl 364 # 34009 • ID 24068

MACDOWELL, R. & SPECKER, E. [1961] *Modelle der Arithmetik* (P 0633) Infinitist Meth;1959 Warsaw 257-263
⋄ B28 C20 C55 C62 ⋄ REV MR 27#2425 Zbl 126.11 JSL 38.651 • ID 19285

MACE, C.A. [1933] *The principles of logic: an introductory survey* (X 1296) Longman: Harlow & New York xii+388pp
⋄ B98 ⋄ REV FdM 59.860 • ID 22596

MACHOVER, M. [1969] see JUHASZ, I.

MACHOVER, M. [1969] see HIRSCHFELD, J.

MACHOVER, M. [1977] see BELL, J.L.

MACHOVER, M. see Vol. III, IV, V for further entries

MACINTYRE, A. & PACHOLSKI, L. & PARIS, J.B. (EDS.) [1978] *Logic colloquium '77. Proceedings of the colloquium held in Wroclaw, august 1977* (S 3303) Stud Logic Found Math 96*x+311pp
⋄ B97 C97 ⋄ REV MR 80a:03003 Zbl 426#00004 • ID 53596

MACINTYRE, A. see Vol. III, IV, V, VI for further entries

MACK, J.M. [1975] see BARNES, D.W.

MACKENZIE, J. [1985] *A pragmatic requirement for classically valid arguments* (J 0079) Logique & Anal, NS 28*75-78
⋄ B10 ⋄ REV Zbl 572#03005 • ID 49218

MADUCH, M. [1970] *Bases of classical consequences* (S 1454) Zesz Nauk Wyz Szk Ped Mat, Opole 10*37-41
⋄ B22 ⋄ REV MR 57#5678 • ID 75795

MADUCH, M. [1973] *Consequence operation defined by cartesian product of matrices (Polish) (English summary)* (S 1454) Zesz Nauk Wyz Szk Ped Mat, Opole 13*159-162
⋄ B22 ⋄ REV MR 57#5680 Zbl 314#02023 • ID 63576

MADUCH, M. [1973] *Lukasiewicz's rules of rejection (Polish) (English summary)* (S 1454) Zesz Nauk Wyz Szk Ped Mat, Opole 13*115-121
⋄ B05 ⋄ REV MR 54#4918 Zbl 314#02021 • ID 24094

MADUCH, M. see Vol. II for further entries

MAEHARA, S. [1954] *Gentzen's theorem on an extended predicate calculus* (J 0081) Proc Japan Acad 30*923-926
⋄ B15 F05 ⋄ REV MR 18.271 Zbl 57.247 JSL 27.109 • ID 08513

MAEHARA, S. [1955] *The predicate calculus with ε-symbol* (J 0090) J Math Soc Japan 7*323-344
⋄ B10 ⋄ REV MR 18.271 Zbl 67.250 JSL 27.109 • ID 08514

MAEHARA, S. [1957] *Equality axiom on Hilbert's ε-symbol* (J 0434) J Fac Sci Univ Tokyo, Sect 1 7*419-435
⋄ B10 ⋄ REV MR 18.711 Zbl 86.8 JSL 27.109 • ID 08517

MAEHARA, S. [1957] *Remark on Skolem's theorem concerning the impossibility of characterization of the natural number sequence* (J 0081) Proc Japan Acad 33*588-590
⋄ B28 C62 F30 ⋄ REV MR 20#6364 Zbl 79.8 JSL 31.659 • ID 08516

MAEHARA, S. [1957] *Ueber die rekursive Einfuehrung der Funktionen in der reinen Zahlentheorie* (J 0081) Proc Japan Acad 33*111-113
⋄ B28 ⋄ REV MR 20#6363 Zbl 78.245 JSL 27.90 • ID 08515

MAEHARA, S. & NISHIMURA, T. & SEKI, S. [1960] *Non-constructive proofs of a metamathematical theorem concerning the consistency of analysis and its extension* (J 0260) Ann Jap Ass Phil Sci 1*269-288
⋄ B15 F05 F35 ⋄ REV MR 23#A794 JSL 32.283 • ID 08521

MAEHARA, S. [1960] *On the interpolation theorem of Craig (Japanese)* (J 0091) Sugaku 12*235-237
⋄ B10 C40 F05 ⋄ REV MR 29#3356 Zbl 123.246 • ID 08520

MAEHARA, S. [1962] *Cut-elimination theorem concerning a formal system for ramified theory of types which admits quantification on types* (J 0260) Ann Jap Ass Phil Sci 2*55-64
⋄ B15 F05 F30 F35 F65 ⋄ REV MR 27#3504 Zbl 109.8 JSL 35.325 • ID 08526

MAEHARA, S. [1966] *Introduction to mathematical logic (Japanese)* (X 1168) Kyoritsu Shuppan: Tokyo i+3+214+5pp
⋄ B98 ⋄ REV MR 35#2716 • ID 22167

MAEHARA, S. [1969] *A system of simple type theory with type variables* (J 0260) Ann Jap Ass Phil Sci 3*131-137
⋄ B15 ⋄ REV MR 40#4096 Zbl 172.294 JSL 39.604 • ID 08527

MAEHARA, S. & TAKEUTI, G. [1971] *Two interpolation theorems for a Π_1^1 predicate calculus* (J 0036) J Symb Logic 36*262-270
⋄ B15 C40 F35 ⋄ REV MR 46#6991 Zbl 278#02013 • ID 08528

MAEHARA, S. see Vol. III, IV, VI for further entries

MAGAJNA, B. [1983] *Infinitesimals (Slovenian) (English summary)* (J 2310) Obz Mat Fiz, Ljubljana 30*33-41
⋄ C20 H05 ⋄ REV MR 84i:03116 Zbl 512#26010 • ID 34592

MAGARI, R. [1964] *Sui sistemi di assiomi "minimali" per una data teoria* (J 4408) Boll Unione Mat Ital, III Ser 19*423-435
⋄ B30 ⋄ REV MR 31#56 Zbl 145.6 • ID 08530

MAGARI, R. [1966] *Calcoli generali e spazi V_α* (J 0319) Matematiche (Sem Mat Catania) 21*83-108
⋄ B22 B60 ⋄ REV MR 33#7242 Zbl 145.245 • REM Part I. Part II 1966 • ID 49174

MAGARI, R. [1966] *Calcoli generali con piu di un operatore deduzione* (J 0319) Matematiche (Sem Mat Catania) 21*135-149
⋄ B22 ⋄ REV MR 33#7243 Zbl 145.246 • REM Part II. Parts I,III 1966 • ID 49175

MAGARI, R. see Vol. II, VI for further entries

MAGGIOLO-SCHETTINI, A. & WINKOWSKI, J. [1978] *An algebraic characterization of derivability relations* (J 1929) Prace Centr Oblicz Pol Akad Nauk 307*24pp
⋄ B22 ⋄ REV Zbl 383#68011 • ID 69685

MAGGIOLO-SCHETTINI, A. see Vol. IV for further entries

MAGIDOR, M. & SHELAH, S. & STAVI, J. [1983] *On the standard part of nonstandard models of set theory* (J 0036) J Symb Logic 48*33-38
⋄ C62 E45 H05 ⋄ REV MR 84m:03058 Zbl 522#03060 • ID 35758

MAGIDOR, M. [1983] see GUREVICH, Y.

MAGIDOR, M. see Vol. III, IV, V for further entries

MAGNARADZE, D.G. & PKHAKADZE, SH.S. (EDS.) [1975] *Studies in mathematical logic and the theory of algorithms (Russian)* (X 1052) Tbilisi Univ: Tbilisi 152pp
⋄ B97 D97 ⋄ REV MR 53 # 2614 • REM Part I. Part II 1977
• ID 21507

MAGNARADZE, L.G. & PKHAKADZE, SH.S. (EDS.) [1977] *Studies in mathematical logic and the theory of algorithms. No. II (Russian)* (X 1052) Tbilisi Univ: Tbilisi 39pp
⋄ B97 D97 ⋄ REV MR 56 # 2779 • REM Part I 1975. Part III 1978 • ID 70125

MAGNARADZE, L.G. & PKHAKADZE, SH.S. (EDS.) [1978] *Studies in mathematical logic and the theory of algorithms (Russian)* (X 1052) Tbilisi Univ: Tbilisi 54pp
⋄ A05 B97 ⋄ REV MR 80c:03006 • ID 70062

MAHE, L. [1973] *Topos infinitesimal* (J 2313) C R Acad Sci, Paris, Ser A-B 277*A497-A500
⋄ G30 H05 ⋄ REV MR 49 # 8986 Zbl 266 # 18010
• ID 08550

MAHE, L. see Vol. IV, V, VI for further entries

MAINZER, K. [1976] *Das Begruendungsproblem des mathematischen Kontinuums in der neuzeitlichen Entwicklung der Grundlagenforschung* (J 0455) Phil Naturalis 16*125-137
⋄ A10 B28 ⋄ REV MR 57 # 52 • ID 75823

MAINZER, K. see Vol. VI for further entries

MAJSTROVA, T.D. [1964] see KRAKOVSKAYA, O.S.

MAKANIN, G.S. [1964] *A new solvable case of the decision problem for first-order predicate calculus (Russian)* (C 0570) Form Log & Metodol Nauk 125-153
⋄ B03 B25 ⋄ REV Zbl 148.8 • ID 14845

MAKANIN, G.S. [1977] *The problem of the solvability of equations in a free semigroup (Russian)* (J 0142) Mat Sb, Akad Nauk SSSR, NS 103(145)*147-236,319
• TRANSL [1977] (J 0349) Math of USSR, Sbor 32*129-198
⋄ B03 B25 D40 ⋄ REV MR 57 # 9874 Zbl 371 # 20047
• ID 51395

MAKANIN, G.S. [1977] *The problem of solvability of equations in a free semigroup (Russian)* (J 0023) Dokl Akad Nauk SSSR 233/2*287-290
• TRANSL [1977] (J 0062) Sov Math, Dokl 18*330-334
⋄ B03 B25 D40 ⋄ REV MR 58 # 5997 Zbl 379 # 20046
• ID 51863

MAKANIN, G.S. [1980] *Equations in a free semigroup (Russian)* (P 1959) Int Congr Math (II,13);1978 Helsinki 1*263-268
• TRANSL [1981] (J 0225) Amer Math Soc, Transl, Ser 2 117*1-6
⋄ B03 B25 D40 ⋄ REV MR 82b:20081 Zbl 436 # 20039
• ID 82143

MAKANIN, G.S. see Vol. III, IV for further entries

MAKEEV, A.I. [1974] *On the Dedekind construction of the field of real numbers (Russian)* (J 3536) Uch Zap Univ, Ivanovo 130*19-24
⋄ B28 ⋄ REV MR 57 # 99 • ID 75827

MAKINSON, D. [1969] *Remarks on the concept of distribution in traditional logic* (J 0097) Nous, Quart J Phil 3*103-108
⋄ A05 A10 B20 ⋄ REV JSL 40.608 • ID 14473

MAKINSON, D. [1973] *Topics in modern logic* (X 0816) Methuen: London & New York viii+107pp
⋄ B98 ⋄ REV MR 58 # 27172 • ID 22363

MAKINSON, D. & SEGERBERG, K. [1974] *Post completeness and ultrafilters* (J 0068) Z Math Logik Grundlagen Math 20*385-388
⋄ B22 B45 G20 ⋄ REV MR 52 # 2825 Zbl 298 # 02014
• ID 08573

MAKINSON, D. [1976] *A characterization of structural completeness of a structural consequence operation* (J 0302) Rep Math Logic, Krakow & Katowice 6*99-102
⋄ B22 ⋄ REV MR 57 # 5726 Zbl 358 # 02063 • ID 21918

MAKINSON, D. [1979] *Qu'est-ce la completude structurale?* (J 0056) Publ Dep Math, Lyon 16/3-4*65-66
⋄ B22 ⋄ REV MR 82c:03039 Zbl 474 # 03010 • ID 55414

MAKINSON, D. see Vol. II for further entries

MAKOWIECKA, H. [1965] *On a primitive notion of one-dimensional geometries over some fields* (J 0014) Bull Acad Pol Sci, Ser Math Astron Phys 13*43-47
⋄ B30 C60 ⋄ REV MR 30 # 4752 Zbl 132.408 • ID 08587

MAKOWIECKA, H. [1975] *An elementary geometry in connection with decomposition of a plane* (J 0014) Bull Acad Pol Sci, Ser Math Astron Phys 23*665-674
⋄ B30 C60 ⋄ REV MR 52 # 5410 Zbl 312 # 50003
• ID 08589

MAKOWIECKA, H. [1975] *On primitive notions in n-dimensional elementary geometries* (J 0014) Bull Acad Pol Sci, Ser Math Astron Phys 23*675-682
⋄ B30 C60 ⋄ REV MR 52 # 5411 Zbl 312 # 50004
• ID 08588

MAKOWIECKA, H. [1975] *Quartenary relations in weak euclidean geometries (Russian summary)* (J 0014) Bull Acad Pol Sci, Ser Math Astron Phys 23*683-692
⋄ B30 C60 ⋄ REV MR 52 # 5412 Zbl 312 # 50005
• ID 18272

MAKOWIECKA, H. [1975] *The norm relation in weak euclidean geometry with the axiom on two circles (Russian summary)* (J 0014) Bull Acad Pol Sci, Ser Math Astron Phys 23*1181-1187
⋄ B30 C60 ⋄ REV MR 52 # 13363 Zbl 325 # 50002
• ID 21816

MAKOWIECKA, H. [1975] *The theory of bi-proportionality as a geometry* (J 0014) Bull Acad Pol Sci, Ser Math Astron Phys 23*657-664
⋄ B30 C60 ⋄ REV MR 52 # 5409 Zbl 312 # 50006
• ID 08590

MAKOWIECKA, H. [1976] *A general property of ternary relations in elementary geometry (Russian summary)* (J 0014) Bull Acad Pol Sci, Ser Math Astron Phys 24*163-169
⋄ B30 C60 ⋄ REV MR 54 # 2448 Zbl 325 # 50003
• ID 24642

MAKOWIECKA, H. [1977] *On minimal systems of primitives in elementary euclidean geometry* (J 0014) Bull Acad Pol Sci, Ser Math Astron Phys 25*269-277
⋄ B30 ⋄ REV MR 57 # 3970 Zbl 356 # 50002 • ID 50330

MAKSIMOVA, L.L. [1970] see LAVROV, I.A.

MAKSIMOVA, L.L. [1973] *Implication lattices (Russian)* (J 0003) Algebra i Logika 12*445-467,492-493
- TRANSL [1973] (J 0069) Algeb and Log 12*249-261
- ◊ B20 G10 ◊ REV MR 52 # 10373 Zbl 301 # 02065
- • ID 21675

MAKSIMOVA, L.L. [1975] see LAVROV, I.A.

MAKSIMOVA, L.L. [1976] *The principle of separation of variables in propositional logics (Russian)* (J 0003) Algebra i Logika 15*168-184,245
- TRANSL [1976] (J 0069) Algeb and Log 15*105-114
- ◊ B22 ◊ REV MR 58 # 21417 Zbl 354 # 02021 • ID 26037

MAKSIMOVA, L.L. [1977] *Craig's interpolation theorem and amalgamable varieties (Russian)* (J 0023) Dokl Akad Nauk SSSR 237*1281-1284
- TRANSL [1977] (J 0062) Sov Math, Dokl 18*1550-1553
- ◊ B22 B55 C05 C40 C90 G10 ◊ REV MR 57 # 5698 Zbl 393 # 03013 • ID 52433

MAKSIMOVA, L.L. see Vol. II, III for further entries

MALASHONOK, A.I. [1972] see ANUFRIEV, F.V.

MALASHONOK, A.I. [1975] see LYALETSKIJ, A.V.

MALATESTA, M. [1976] *Logistica I: Introduzione. La logica degli enunciati* (X 1229) Ateneo: Napoli 170pp
- ◊ B98 ◊ REM Part II 1978 • ID 35977

MALATESTA, M. [1978] *Logistica II: Le tautologie. L'interdefinibilita dei funtori* (X 1229) Ateneo: Napoli 175pp
- ◊ B05 ◊ REM Part I 1976 • ID 35978

MALEVANYI, N.P. [1972] see GLUSHKOV, V.M.

MALIAUKIENE, L. [1985] *Elimination of the induction axiom in the multiplicative arithmetic with restricted difference (Russian) (English and Lithuanian summaries)* (J 3939) Mat Logika Primen (Akad Nauk Litov SSR) 1985/4*9-15,135
- ◊ B28 F30 ◊ ID 49123

MALIAUKIENE, L. see Vol. VI for further entries

MALINA, S. [1981] *Notes on a method of introducing real numbers (Slovak) (English and Russian summaries)* (J 0128) Acta Math Univ Comenianae (Bratislava) 38*153-168
- ◊ B28 ◊ REV MR 83e:04010 Zbl 502 # 26003 • ID 35266

MALINOWSKI, G. & PRZELECKI, M. & SZANIAWSKI, K. & WOJCICKI, R. (EDS.) [1976] *Formal methods in the methodology of empirical sciences* (X 0835) Reidel: Dordrecht 457pp
- ◊ B30 ◊ REV MR 58 # 21365 Zbl 344 # 00007 • ID 70083

MALINOWSKI, G. [1978] *A proof of Ryszard Wojcicki's conjecture* (J 0387) Bull Sect Logic, Pol Acad Sci 7*20-25
- ◊ B22 ◊ REV MR 58 # 5167 Zbl 414 # 03017 • ID 53063

MALINOWSKI, G. [1979] *Topics in the theory of strengthenings of sentential calculi* (X 2733) Acad Sci Inst Phi Soc: Wroclaw 112pp
- ◊ B22 G20 ◊ REV MR 81f:03032 Zbl 423 # 03028
- • ID 53540

MALINOWSKI, G. [1981] *A characterization of strengthenings of a two-valued nonuniform sentential calculus* (J 0302) Rep Math Logic, Krakow & Katowice 12*17-33
- ◊ B22 ◊ REV MR 82g:03046 Zbl 502.03013 • ID 75906

MALINOWSKI, G. [1983] *Pseudo-referential matrix semantics for propositional logics* (J 0387) Bull Sect Logic, Pol Acad Sci 12*90-98
- ◊ B22 ◊ REV Zbl 542 # 03013 • ID 43680

MALINOWSKI, G. [1984] *Matrix referentiality* (P 3621) Frege Konferenz (2);1984 Schwerin 148-161
- ◊ B22 ◊ REV Zbl 551 # 03012 • ID 43894

MALINOWSKI, G. [1985] *Notes on sentential logic with identity* (J 0079) Logique & Anal, NS 28*341-352
- ◊ B10 B60 ◊ ID 49715

MALINOWSKI, G. [1985] *The problem of degrees of maximality (a survey) (Polish summary)* (J 3815) Acta Univ Lodz Folia Philos 3*37-57
- ◊ B22 ◊ ID 49713

MALINOWSKI, G. see Vol. II, III for further entries

MALITZ, J. [1979] *Introduction to mathematical logic. Set theory, computable functions, model theory* (X 0811) Springer: Heidelberg & New York xii+198pp
- ◊ B98 C98 D98 ◊ REV MR 81h:03002 Zbl 407 # 03001 JSL 49.672 • ID 56167

MALITZ, J. & RUBIN, M. [1980] *Compact fragments of higher order logic* (P 2958) Latin Amer Symp Math Log (4);1978 Santiago 219-238
- ◊ B15 C55 C80 C85 ◊ REV MR 82k:03055 Zbl 434 # 03013 • ID 55717

MALITZ, J. [1983] see CHUAQUI, R.B.

MALITZ, J. see Vol. III, IV, V for further entries

MALONEY, M.J. [1974] see CORCORAN, J.

MAL'TSEV, A.I. [1936] *Untersuchungen aus dem Gebiete der mathematischen Logik (Russian summary)* (J 0142) Mat Sb, Akad Nauk SSSR, NS 1(43)*323-336
- TRANSL [1971] (C 2621) Mal'tsev: Metamath of Algeb Syst 1-14
- ◊ B10 C07 ◊ REV Zbl 14.385 JSL 2.84 FdM 62.42
- • ID 08602

MAL'TSEV, A.I. [1959] *Model correspondences (Russian)* (J 0216) Izv Akad Nauk SSSR, Ser Mat 23*313-336
- TRANSL [1971] (C 2621) Mal'tsev: Metamath of Algeb Syst 66-94
- ◊ B15 C05 C07 C52 C60 C85 ◊ REV MR 22 # 10909 Zbl 100.12 JSL 34.299 JSL 35.299 • ID 19260

MAL'TSEV, A.I. [1976] *Iterative Post algebras (Russian)* (X 0913) Novosibirsk Gos Univ: Novosibirsk 100pp
- ◊ B20 ◊ REV MR 57 # 9535 • ID 75891

MAL'TSEV, A.I. see Vol. II, III, IV, V, VI for further entries

MAL'TSEV, I.A. [1967] *A function of the algebra of logic (Russian)* (J 0071) Met Diskr Analiz (Novosibirsk) 10*61-68
- ◊ B05 ◊ REV MR 37 # 29 • ID 08613

MAL'TSEV, I.A. see Vol. II, IV for further entries

MAMATOV, YU.A. [1976] *Estimation of the number of functions of the algebra of logic realized by the plane d.n.f (Russian)* (J 0199) Zh Vychisl Mat i Mat Fiz 16*170-188
- TRANSL [1976] (J 1049) USSR Comput Math & Math Phys 16/1*161-179
- ◊ B05 B70 ◊ REV MR 55 # 2383 Zbl 326 # 94025
- • ID 63641

MAMATOV, YU.A. [1979] *Concerning a principle for obtaining lower bounds on the complexity of formulars (Russian)* (J 0023) Dokl Akad Nauk SSSR 245*782-784
- TRANSL [1979] (J 0062) Sov Math, Dokl 20*339-342
- ◊ B05 ◊ REV MR 83e:03108 Zbl 427 # 68049 • ID 53754

MAMATOV, YU.A. [1979] *On a principle for obtaining high (exponential for some parameter values) lower bounds for the complexity of disjunctive normal forms (Russian)* (J 0023) Dokl Akad Nauk SSSR 245*1054-1057
• TRANSL [1979] (J 0062) Sov Math, Dokl 20*399-401
⋄ B05 ⋄ REV MR 83h:68062 Zbl 418 # 68047 • ID 69708

MAMATOV, YU.A. [1980] *A method for the synthesis of planar disjunctive normal forms for logic functions (Russian)* (J 0199) Zh Vychisl Mat i Mat Fiz 20*1075-1077,1088
• TRANSL [1980] (J 1049) USSR Comput Math & Math Phys 20/4*281-283
⋄ B05 ⋄ REV MR 81k:94053 Zbl 441 # 94013 • ID 82158

MAMDANI, E.H. [1981] see GAINES, B.R.

MAMDANI, E.H. [1984] see EFSTATHIOU, J.

MAMDANI, E.H. see Vol. II for further entries

MANARA, C.F. [1965] *Un teorema di Beppo Levi riguardante la logica formale* (J 0313) Period Mat (Bologna) 43*177-182
⋄ B05 ⋄ REV MR 32 # 2330 Zbl 192.25 • ID 08639

MANASTER, A.B. [1975] *Completeness, compactness, and undecidability: an introduction to mathematical logic* (X 0819) Prentice Hall: Englewood Cliffs vi+154pp
⋄ B98 C07 D35 D98 ⋄ REV MR 53 # 12857 Zbl 306 # 02001 JSL 42.320 • ID 23120

MANASTER, A.B. see Vol. III, IV, V for further entries

MANCILL, J.D. [1950] see GONZALEZ, M.O.

MANDERS, K.L. [1979] *The theory of all substructures of a structure: Characterisation and decision problems* (J 0036) J Symb Logic 44*583-598
⋄ B20 B25 C57 C60 ⋄ REV MR 81i:03047 Zbl 429 # 03007 • ID 53838

MANDERS, K.L. [1982] *On the space-time ontology of physical theories* (J 0153) Phil of Sci (East Lansing) 49*575-590
⋄ B30 ⋄ REV MR 84d:03020 • ID 34060

MANDERS, K.L. see Vol. II, III, IV, V for further entries

MANESSE, G. [1975] *Resolution d'une equation booleenne au moyen de l'arbre de developpement relatif a ses inconnues (English summary)* (J 2313) C R Acad Sci, Paris, Ser A-B 281*A1067-A1069
⋄ B05 ⋄ REV MR 56 # 11595 Zbl 325 # 02041 • ID 30447

MANEV, K.N. [1978] see KOEN, R.I.

MANGANI, P. [1966] *Sur certe algebre connesse con sistemi di logica elementare dotati dell' operator tau di Hilbert* (J 0319) Matematiche (Sem Mat Catania) 21*66-82
⋄ B10 G25 ⋄ REV MR 33 # 5488 Zbl 145.6 • ID 08654

MANGANI, P. see Vol. II, III for further entries

MANGIONE, C. [1964] *Elementi di logica matematica* (X 0905) Boringhieri: Torino 127pp
⋄ B98 ⋄ REV Zbl 144.243 • ID 22418

MANGIONE, C. see Vol. II, III, VI for further entries

MANICAS, P.T. [1968] see KRUGER, A.N.

MANIN, YU.I. [1977] *A course in mathematical logic* (X 0811) Springer: Heidelberg & New York xiii+286pp
⋄ B98 C07 D98 E35 E50 F30 G12 ⋄ REV MR 56 # 15345 Zbl 383 # 03002 • REM Translated from Russian • ID 51984

MANIN, YU.I. [1979] *Provable and unprovable (Russian)* (X 2643) Sovet Radio: Moskva 168pp
• TRANSL [1981] (X 0885) Mir: Moskva 265pp (Spanish)
⋄ A05 B98 E35 E50 F30 F98 G12 ⋄ REV Zbl 403 # 03002 • ID 54716

MANIN, YU.I. see Vol. IV, V, VI for further entries

MANKIEWICZ, P. [1982] see HEINRICH, S.

MANKIEWICZ, P. see Vol. III for further entries

MANNA, Z. & WALDINGER, R. [1981] *Deductive synthesis of the unification algorithm* (J 3298) Sci Comput Programming 1*5-48
⋄ B35 ⋄ REV MR 83f:68009 Zbl 472 # 68054 • ID 40204

MANNA, Z. & WALDINGER, R. [1985] *Deduction with relation matching* (P 4672) Found of Softw Tech & Th Comput Sci (5);1985 New Delhi 212-224
⋄ B35 ⋄ ID 49767

MANNA, Z. [1985] see ABADI, M.

MANNA, Z. & WALDINGER, R. [1985] *Special relations in automated deduction* (P 4628) Automata, Lang & Progr (12);1985 Nafplion 413-423
⋄ B35 ⋄ ID 49502

MANNA, Z. see Vol. II, IV for further entries

MANNILA, H. [1985] see BACK, R.J.R.

MANNILA, H. see Vol. III, V for further entries

MANSFIELD, R. [1984] *A complete axiomatization of computer arithmetic* (J 0214) Math of Comp 42*623-635
⋄ B28 ⋄ REV MR 85k:65040 Zbl 575 # 68053 • ID 45216

MANSFIELD, R. see Vol. III, IV, V for further entries

MARACCHIA, S. [1974] *Matematica antica e moderna in un caso di logica proposizionale* (J 1515) Archimede 26*28-35
⋄ A10 B05 ⋄ REV MR 54 # 12470 • ID 75975

MARAVALL CASESNOVES, D. [1981] *Attempt at a theory of the infinitely large and the infinitely small, certain and random, and the laws of large numbers (Spanish) (English summary)* (J 0234) Rev Acad Cienc Exact Fis Nat Madrid 75*555-577
⋄ H10 ⋄ REV MR 84h:03142 • ID 34328

MARC-WOGAU, K. [1950] *Modern logic. An elementary text-book (Swedish)* (X 4499) Ehlins: Stockholm 211pp
⋄ B98 ⋄ REV JSL 17.288 • ID 41850

MARCHENKOV, S.S. [1975] *Elementary Skolem functions (Russian)* (J 0087) Mat Zametki (Akad Nauk SSSR) 17*133-141
• TRANSL [1975] (J 1044) Math Notes, Acad Sci USSR 17*79-83
⋄ B10 ⋄ REV MR 51 # 92 Zbl 317 # 02040 • ID 29649

MARCHENKOV, S.S. [1982] *Undecidability of the positive ∀∃-theory of a free semigroup (Russian)* (J 0092) Sib Mat Zh 23/1*196-198,223
⋄ B20 D10 D35 ⋄ REV MR 83e:03067 • ID 35236

MARCHENKOV, S.S. [1984] *Existence of finite bases in closed classes of Boolean functions (Russian)* (J 0003) Algebra i Logika 23*88-99
• TRANSL [1984] (J 0069) Algeb and Log 23*66-74
⋄ B05 ⋄ REV MR 86d:06019 Zbl 555 # 03029 • ID 46127

MARCHENKOV, S.S. see Vol. II, IV for further entries

MARCJA, A. & TULIPANI, S. [1974] *Questioni di teoria dei modelli per linguaggi universali positivi (English summary)* (J 0149) Atti Accad Naz Lincei Fis Mat Nat, Ser 8 56∗915-923
- ⋄ B20 C07 C10 ⋄ REV MR 52#10409 Zbl 317#02055
- REM Part I. Part II 1976 by Tulipani,S. • ID 21706

MARCJA, A. [1984] see LOLLI, G.

MARCJA, A. see Vol. II, III for further entries

MAREK, I. [1984] *Consequence operations defined by partial matrices* (J 0302) Rep Math Logic, Krakow & Katowice 17∗47-55
- ⋄ B22 ⋄ REV MR 85m:03022 Zbl 562#03013 • ID 44048

MAREK, J.C. [1981] *Ueber weder notwendige noch hinreichende Bedingungen. Zur logischen Analyse einer gebraeuchlichen philosophischen Redewendung* (C 3202) Logik, Ethik & Sprache (Freundlich) 154-165
- ⋄ A05 B05 ⋄ REV Zbl 479#03003 • ID 55665

MAREK, W. & ONYSZKIEWICZ, J. [1972] *Elements of logic and foundations of mathematics in problems (Polish)* (X 1034) PWN: Warsaw 278pp
- TRANSL [1982] (X 0835) Reidel: Dordrecht viii+276pp
- ⋄ B98 C98 D98 E98 ⋄ REV MR 84h:03001 Zbl 288#02001 Zbl 574#03001 • ID 34216

MAREK, W. [1973] *Consistance d'une hypothese de Fraisse sur la definissabilite dans un langage du second ordre* (J 2313) C R Acad Sci, Paris, Ser A-B 276∗A1147-A1150
- ⋄ B15 C65 E10 E35 E45 ⋄ REV MR 52#2879a Zbl 259#02048 • ID 17662

MAREK, W. [1973] *Sur la consistance d'une hypothese de Fraisse sur la definissabilite dans un langage du second ordre* (J 2313) C R Acad Sci, Paris, Ser A-B 276∗A1169-A1172
- ⋄ B15 C65 E10 E35 E45 ⋄ REV MR 52#2879b Zbl 257#02050 • ID 17663

MAREK, W. [1977] see KRAWCZYK, A.

MAREK, W. see Vol. II, III, IV, V, VI for further entries

MARGARIS, A. [1961] *Successor axioms for the integers* (J 0005) Amer Math Mon 68∗441-444
- ⋄ B28 ⋄ REV MR 23#A1534 Zbl 109.241 • ID 31892

MARGARIS, A. [1967] *First order mathematical logic* (X 0841) Blaisdell: New York 211pp
- ⋄ B98 ⋄ REV MR 36#26 Zbl 156.247 JSL 37.616
- ID 08762

MARGARIS, A. see Vol. II for further entries

MARGENAU, H. [1945] *On the frequency theory of probability* (J 0075) Phil Phenom Research 6∗11-25
- ⋄ A05 B30 ⋄ REV MR 7.189 JSL 11.29 • ID 08763

MARINI, D. & MIGLIOLI, P.A. & ORNAGHI, M. [1975] *First order logic as a tool to solve and classify problems* (P 0784) GI Jahrestag (5);1975 Dortmund 669-679
- ⋄ B10 B25 F50 ⋄ REV MR 53#5282 Zbl 329#68079
- ID 22915

MARINO SARMIENTO, R. [1972] *Konsistenz und Inkonsistenz in der Mathematik (aus einem deduktiven Axiomensystem)* (J 0377) Bol Mat (Bogota) 5/2-6∗99-103
- ⋄ B30 ⋄ REV Zbl 266#02002 • ID 63712

MARINO SARMIENTO, R. [1974] *Topology in logic: the compactness theorem (Spanish)* (J 0377) Bol Mat (Bogota) 8∗89-94
- ⋄ B05 ⋄ REV MR 54#56 Zbl 335#02006 • ID 23940

MARKOV, A.A. [1957] *On the inversion complexity of a system of functions (Russian)* (J 0023) Dokl Akad Nauk SSSR 116∗917-919
- TRANSL [1958] (J 0037) ACM J 5∗331-334
- ⋄ B05 ⋄ REV MR 20#3773 Zbl 79.246 JSL 30.378
- ID 19375

MARKOV, A.A. [1963] *On the inversion complexity of a system of boolean functions (Russian)* (J 0023) Dokl Akad Nauk SSSR 150∗477-479
- TRANSL [1963] (J 0062) Sov Math, Dokl 4∗694-696
- ⋄ B05 D55 ⋄ REV MR 27#1369 Zbl 171.279 • ID 08773

MARKOV, A.A. [1964] *Normal algorithms which compute Boolean functions (Russian)* (J 0023) Dokl Akad Nauk SSSR 157∗262-264
- TRANSL [1964] (J 0062) Sov Math, Dokl 5∗922-924
- ⋄ B05 D03 ⋄ REV MR 30#3841 Zbl 127.10 • ID 08774

MARKOV, A.A. [1967] *Normal algorithms connected with computation of boolean functions (Russian)* (J 0216) Izv Akad Nauk SSSR, Ser Mat 31∗161-208
- TRANSL [1967] (J 0448) Math of USSR, Izv 1∗151-194
- ⋄ B05 D03 ⋄ REV MR 35#1474 Zbl 164.318 • ID 08776

MARKOV, A.A. [1984] *Elements of mathematical logic (Russian)* (X 0898) Moskov Gos Univ: Moskva 80pp
- ⋄ B98 ⋄ REV MR 85h:03001 • ID 43264

MARKOV, A.A. see Vol. II, III, IV, V, VI for further entries

MARKOVA, V.P. [1977] *Calculation of the Fourier-spectrum of a system of boolean functions that are given by a disjunctive normal form (Russian)* (S 0507) Vychisl Sist (Akad Nauk SSSR Novosibirsk) 69∗59-69,155
- ⋄ B05 ⋄ REV MR 58#33564 Zbl 418#94018 • ID 82181

MARKOVA, V.P. see Vol. II for further entries

MARKOVIC, Z. [1981] *On axiomatizability and preservation in Kripke models* (J 1943) Recueil Travaux Inst Math (Beograd) 30(44)∗111-112
- ⋄ B20 ⋄ REV MR 84a:03014 Zbl 497#03028 • ID 35561

MARKOVIC, Z. [1983] *Some preservation results for classical and intuitionistic satisfiability in Kripke models* (J 0047) Notre Dame J Formal Log 24∗395-398
- ⋄ B20 C25 C90 F50 ⋄ REV MR 85a:03016 Zbl 487#03015 • ID 37386

MARKOVIC, Z. see Vol. III, VI for further entries

MARKOWSKY, G. [1978] see CHANDRA, A.K.

MARKOWSKY, G. see Vol. V for further entries

MARKUSZ, Z. [1983] *On first order many-sorted logic* (J 2845) Tanulmanyok 151∗85pp
- ⋄ B10 C07 ⋄ REV MR 86e:03009 • ID 44560

MARKWALD, W. [1971] *Praedikatenlogik mit partiell definierten Funktionen I* (J 0009) Arch Math Logik Grundlagenforsch 14∗10-23
- ⋄ B10 B60 ⋄ REV MR 44#2567 Zbl 215.46 • REM Part II 1974 • ID 08782

MARKWALD, W. [1972] *Einfuehrung in die formale Logik und Metamathematik* (X 0918) Klett: Stuttgart 168pp
- ⋄ B98 ⋄ REV MR 49#7112 Zbl 304#02001 • ID 08783

MARKWALD, W. [1974] *Praedikatenlogik mit partiell definierten Funktionen. II* (J 0009) Arch Math Logik Grundlagenforsch 16∗15-22
⋄ B10   B60 ⋄ REV MR 51 # 10024   Zbl 289 # 02012 • REM Part I 1971 • ID 08784

MARKWALD, W. see Vol. III, IV, VI for further entries

MARLOW, A.R. [1974] *Physical axiomatics* (J 0209) J Math Phys 15∗2236-2241
⋄ B30 ⋄ REV MR 52 # 4833   Zbl 352 # 02027 • ID 50026

MARLOW, A.R. see Vol. II for further entries

MARRUCCELLI, A. [1970] see FIORENTINI, M.

MARSDEN, E.L. [1974] *Reducible implicative models* (J 0246) J Nat Sci Math 14∗23-34
⋄ B20   B45   C05   G25 ⋄ REV MR 56 # 2819 Zbl 334 # 02028 • ID 63732

MARTELLI, A. & MONTANARI, U. [1982] *An efficient unification algorithm* (J 3120) ACM Trans Program Lang & Syst 4∗258-282
⋄ B35 ⋄ REV Zbl 478 # 68093 • ID 55659

MARTELLI, M. [1982] see BELLIA, M.

MARTEM'JANOV, YU.S. [1981] *Controlled derivation of sentences (Russian) (English summary)* (S 0393) Uch Zap Univ, Tartu 594∗71-88
⋄ B35 ⋄ ID 47226

MARTIN, J.N. [1975] *A many-valued semantics for category mistakes* (J 0154) Synthese 31∗63-83
⋄ A05   B80 ⋄ REV Zbl 316 # 02008 • ID 63749

MARTIN, J.N. see Vol. II, VI for further entries

MARTIN, R.M. [1943] *A homogeneous system for formal logic* (J 0036) J Symb Logic 8∗1-23
⋄ A05   B28   G05 ⋄ REV MR 4.182   Zbl 60.22   JSL 8.54
• ID 08814

MARTIN, R.M. [1950] *On virtual classes and real numbers* (J 0036) J Symb Logic 15∗131-134
⋄ B28   E70 ⋄ REV MR 12.234   Zbl 41.147   JSL 16.64
• ID 08817

MARTIN, R.M. [1970] *On the proto-theory of musical structure* (J 1023) Perspectives of New Music 9∗68-73
⋄ B80 ⋄ REV JSL 40.577 • ID 15103

MARTIN, R.M. [1975] see ANDERSON, A.R.

MARTIN, R.M. see Vol. II, IV for further entries

MARTIN-LOEF, P. [1970] *On the notion of randomness* (P 0603) Intuitionism & Proof Th;1968 Buffalo 73-78
• TRANSL [1974] (C 2319) Slozh Vychisl & Algor 364-369
⋄ B30   D55   D80   E15 ⋄ REV MR 43 # 1237   Zbl 203.299 JSL 40.450 • ID 20933

MARTIN-LOEF, P. [1971] *Complexity oscillations in infinite binary sequences* (J 0982) Z Wahrscheinltheor & Verw Geb 19∗225-230
⋄ B30 ⋄ REV MR 56 # 9609   Zbl 212.231 • ID 82186

MARTIN-LOEF, P. [1973] *Hauptsatz for intuitionistic simple type theory* (P 0793) Int Congr Log, Meth & Phil of Sci (4,Proc);1971 Bucharest 279-290
⋄ B15   F05   F35   F50 ⋄ REV MR 56 # 8332 • ID 76088

MARTIN-LOEF, P. see Vol. III, IV, VI for further entries

MARTINEZ CRUZADO, R.F. [1983] see ESSLER, W.K.

MART'YANOV, V.I. [1984] *Invariant transformations of formulas (Russian)* (J 0087) Mat Zametki (Akad Nauk SSSR) 36∗571-582
• TRANSL [1984] (J 1044) Math Notes, Acad Sci USSR 36∗782-788
⋄ B35 ⋄ REV MR 86e:03010 • ID 44357

MART'YANOV, V.I. see Vol. III, IV, VI for further entries

MASLOV, S.YU. [1964] *An inverse method of establishing deducibilities in the classical predicate calculus (Russian)* (J 0023) Dokl Akad Nauk SSSR 159∗17-20
• TRANSL [1964] (J 0062) Sov Math, Dokl 5∗1420-1424
⋄ B35 ⋄ REV MR 30 # 3005   Zbl 146.246 • ID 24957

MASLOV, S.YU. [1965] see DAVYDOV, G.V.

MASLOV, S.YU. & MINTS, G.E. & OREVKOV, V.P. [1965] *Unsolvability in the constructive predicate calculus of certain classes of formulas containing only monadic predicate variables (Russian)* (J 0023) Dokl Akad Nauk SSSR 163∗295-297
• TRANSL [1965] (J 0062) Sov Math, Dokl 6∗918-920
⋄ B20   D35   F50 ⋄ REV MR 33 # 52   Zbl 173.10 JSL 35.143 • ID 19335

MASLOV, S.YU. [1966] *Application of the inverse method for establishing deducibility to the theory of decidable fragments in the classical predicate calculus (Russian)* (J 0023) Dokl Akad Nauk SSSR 171∗1282-1285
• TRANSL [1966] (J 0062) Sov Math, Dokl 7∗1653-1657
⋄ B20   B25 ⋄ REV MR 35 # 25   Zbl 162.21 • ID 19334

MASLOV, S.YU. [1967] *An inverse method for establishing deducibility of nonprenex formulas of predicate calculus (Russian)* (J 0023) Dokl Akad Nauk SSSR 172∗22-25
• TRANSL [1967] (J 0062) Sov Math, Dokl 8∗16-19 [1983] (C 4659) Autom of Reasoning 2∗48-54
⋄ B35 ⋄ REV MR 35 # 19   Zbl 165.318 • ID 19332

MASLOV, S.YU. [1968] *An inverse method of establishing deducibility for logical calculi (Russian)* (S 0066) Tr Mat Inst Steklov 98∗26-87 • ERR/ADD ibid 121∗167
• TRANSL [1968] (S 0055) Proc Steklov Inst Math 98∗25-95
⋄ B35 ⋄ REV MR 40 # 5416   Zbl 172.298 • ID 19330

MASLOV, S.YU. [1969] see DAVYDOV, G.V.

MASLOV, S.YU. [1969] *The connection between tactics of the inverse method and the resolution method (Russian)* (S 0228) Zap Nauch Sem Leningrad Otd Mat Inst Steklov 16∗137-146
• TRANSL [1969] (J 0521) Semin Math, Inst Steklov 16∗69-73 [1983] (C 4659) Autom of Reasoning 2∗264-272
⋄ B35 ⋄ REV MR 42 # 1645   Zbl 199.315 • ID 08846

MASLOV, S.YU. [1971] *Extension of the inverse method to calculus with an equality (Russian) (English summary)* (S 0228) Zap Nauch Sem Leningrad Otd Mat Inst Steklov 20∗80-96,285
⋄ B35 ⋄ REV MR 45 # 6579   Zbl 227 # 68043 • ID 08847

MASLOV, S.YU. & MINTS, G.E. & OREVKOV, V.P. [1971] *Mechanical proof-search and the theory of logical deduction in the USSR* (J 2076) Rev Int Philos 25∗575-584
• REPR [1983] (C 4659) Autom of Reasoning 1∗29-40
⋄ B35 ⋄ REV MR 58 # 32128 • ID 82191

MASLOV, S.YU. [1971] *Proof-search strategies for methods of the resolution type* (J 0508) Machine Intelligence 6∗77-90
⋄ B35 ⋄ REV Zbl 281 # 68042 • ID 46542

MASLOV, S.YU. & OREVKOV, V.P. [1972] *Decidable classes that reduce to a single quantifier class (Russian)* (S 0066) Tr Mat Inst Steklov 121*57-66,165
- TRANSL [1972] (S 0055) Proc Steklov Inst Math 121*61-72
- ◊ B20 B25 ◊ REV MR 50 # 12678   Zbl 301 # 02041
- ID 08848

MASLOV, S.YU. [1972] *Deduction search in calculi of general type (Russian) (English summary)* (S 0228) Zap Nauch Sem Leningrad Otd Mat Inst Steklov 32*59-65,156
- TRANSL [1976] (J 1531) J Sov Math 6*395-400
- ◊ B35 D03 F07 ◊ REV MR 49 # 8845   Zbl 344 # 02026
- ID 08850

MASLOV, S.YU. [1972] *Search for deduction as a model of a heuristic process (Russian) (English summary)* (J 0040) Kibernetika, Akad Nauk Ukr SSR 1972/5*74-78
- ◊ B35 ◊ REV MR 47 # 6151 • ID 08851

MASLOV, S.YU. [1972] *The inverse method, and tactics for establishing deducibility for a calculus with functional symbols (Russian)* (S 0066) Tr Mat Inst Steklov 121*14-56,165
- TRANSL [1972] (S 0055) Proc Steklov Inst Math 121*11-60
- ◊ B35 ◊ REV MR 51 # 5269   Zbl 305 # 68065 • ID 17454

MASLOV, S.YU. [1975] *The theory of deduction search, and some of its applications (Russian)* (J 0040) Kibernetika, Akad Nauk Ukr SSR 1975/4*134-144
- TRANSL [1975] (J 0021) Cybernetics 11*659-669
- ◊ B35 ◊ REV MR 58 # 19402   Zbl 314 # 68034 • ID 82189

MASLOV, S.YU. & MINTS, G.E. (EDS.) [1975] *Theoretical applications of the methods of mathematical logic I (Russian)* (X 2641) Nauka: Leningrad 180pp
- ◊ B97 ◊ REV MR 51 # 7801 • ID 70193

MASLOV, S.YU. & NORGELA, S.A. [1977] *Herbrand strategies and the "greater deducibility" relation (Russian) (English summary)* (S 0228) Zap Nauch Sem Leningrad Otd Mat Inst Steklov 68*51-61,144
- TRANSL [1981] (J 1531) J Sov Math 15*28-33
- ◊ B25 B35 F05 ◊ REV MR 58 # 21504   Zbl 358 # 02030
- ID 32672

MASLOV, S.YU. [1979] *Theory of inference search and questions of the psychology of creation (Russian)* (S 2582) Semiotika & Inf, Akad Nauk SSSR 13*17-46
- ◊ A05 B35 ◊ REV MR 82e:03025   Zbl 443 # 68086
- ID 76108

MASLOV, S.YU. [1983] *Deductive systems and their economic applications (Russian)* (X 2593) Nauch Sov Probl Kompl Kibern Akad Nauk SSSR: Moskva 56pp
- ◊ B80 ◊ REV MR 85g:93010   Zbl 516 # 90004 • REM Preprint • ID 38595

MASLOV, S.YU. see Vol. II, IV, VI for further entries

MASON, I. [1985] *The metatheory of the classical propositional calculus is not axiomatizable* (J 0036) J Symb Logic 50*451-457
- ◊ B05 ◊ ID 42558

MASSEY, G.J. [1963] *Note on Copi's system* (J 0047) Notre Dame J Formal Log 4*140-141
- ◊ B10 ◊ REV MR 29 # 3360   JSL 30.366 • ID 08853

MASSEY, G.J. [1970] *Understanding symbolic logic* (X 0837) Harper & Row: New York 428pp
- ◊ A05 B98 ◊ REV MR 43 # 1808   Zbl 237 # 02001 JSL 36.678 • ID 08860

MASSEY, G.J. see Vol. II, V for further entries

MATEI, S. [1969] *Formulae for solving some boolean equations (Romanian) (Russian and English summaries)* (J 4714) Bul Sti Tehn Inst Politeh Timisoara, NS 14(28)*37-40
- ◊ B05 ◊ REV MR 41 # 1611   Zbl 235 # 94035 • ID 08868

MATERNA, P. [1978] *Theory of types and data description* (J 0156) Kybernetika (Prague) 14*313-327
- ◊ B15 ◊ REV MR 80b:68119   Zbl 402 # 68060 • ID 54712

MATERNA, P. see Vol. II, III for further entries

MATES, B. [1965] *Elementary logic* (X 0894) Oxford Univ Pr: Oxford x+227pp
- TRANSL [1969] (X 0903) Vandenhoeck & Ruprecht: Goettingen 269pp • LAST ED [1972] (X 0815) Clarendon Pr: Oxford viii+227pp
- ◊ A10 B98 ◊ REV MR 40 # 15   MR 48 # 45   Zbl 146.246 Zbl 375 # 02001 JSL 31.483 • REM 2nd ed. 1978, 305pp
- ID 08872

MATHIAS, A.R.D. & ROGERS JR., H. (EDS.) [1973] *Cambridge summer school in mathematical logic (Held in Cambridge, England, August 1 - 21, 1971)* (S 3301) Lect Notes Math 337*ix+660pp
- ◊ B97 ◊ REV MR 48 # 47   Zbl 261 # 00006 • ID 70236

MATHIAS, A.R.D. see Vol. III, IV, V for further entries

MATIYASEVICH, YU.V. [1970] *Solution of the tenth problem of Hilbert* (J 0396) Mat Lapok 21*83-87
- ◊ B28 D25 D35 ◊ REV MR 46 # 3447   Zbl 223 # 02041
- ID 82198

MATIYASEVICH, YU.V. [1972] *The application of the methods of the theory of logical derivation to graph theory (Russian)* (J 0087) Mat Zametki (Akad Nauk SSSR) 12*781-790
- TRANSL [1972] (J 1044) Math Notes, Acad Sci USSR 12*904-910
- ◊ B80 ◊ REV MR 48 # 148   Zbl 247 # 05115 • ID 08886

MATIYASEVICH, YU.V. [1974] *A proof-scheme in discrete mathematics (Russian) (English summary)* (S 0228) Zap Nauch Sem Leningrad Otd Mat Inst Steklov 40*94-100,158
- TRANSL [1977] (J 1531) J Sov Math 8*312-316
- ◊ B80 ◊ REV MR 51 # 78   Zbl 359 # 68106 • ID 15215

MATIYASEVICH, YU.V. [1975] *Metamathematical approach to proving theorems in discrete mathematics (Russian) (English summary)* (S 0228) Zap Nauch Sem Leningrad Otd Mat Inst Steklov 49*31-50,177
- TRANSL [1978] (J 1531) J Sov Math 10*517-533
- ◊ B80 ◊ REV MR 51 # 12503   Zbl 325 # 68048 • ID 17290

MATIYASEVICH, YU.V. [1977] *Some purely mathematical results inspired by mathematical logic* (P 1704) Int Congr Log, Meth & Phil of Sci (5);1975 London ON 1*121-127
- ◊ B30 D25 D35 D80 ◊ REV MR 58 # 5508 Zbl 377 # 02001 • ID 51718

MATIYASEVICH, YU.V. & SLISENKO, A.O. (EDS.) [1979] *Studies in constructive mathematics and mathematical logic VIII (Russian)* (S 0228) Zap Nauch Sem Leningrad Otd Mat Inst Steklov 88*1-252
- TRANSL [1982] (J 1531) J Sov Math 20*2263-2390
- ◊ B97 F97 ◊ REV MR 80j:03005 • ID 70057

MATIYASEVICH, YU.V. see Vol. IV, VI for further entries

MATROSOV, V.M. [1979] see KARATUEV, V.G.

MATROSOV, V.M. & VASIL'EV, S.N. [1981] *Comparison principles in the dynamics of systems with distributed parameters (Russian)* (C 2351) Ustojchivost' Dvizheniya 198-217,303-304
⋄ B10 ⋄ REV MR 83k:93041 • ID 40322

MATSON, T. [1974] see BISHOP, P.

MATSUMOTO, K. [1970] *Mathematical logic (Japanese)* (X 1168) Kyoritsu Shuppan: Tokyo 192PP
⋄ B98 ⋄ ID 90096

MATSUMOTO, K. see Vol. II, III, IV, VI for further entries

MATTREY, R.F. [1973] see ALLEN, C.M.

MATTSON, D.A. [1975] see JOHNSON, D.R.

MATULIS, V.A. [1962] *Two variantes of the classical predicate calculus without structural inference rules (Russian)* (J 0023) Dokl Akad Nauk SSSR 147*1029-1031
• TRANSL [1962] (J 0062) Sov Math, Dokl 3*1770-1772
⋄ B10 ⋄ REV MR 26 # 1241 • ID 08900

MATULIS, V.A. [1963] *Variants of the classical predicate calculus with a single deduction tree (Russian)* (J 0023) Dokl Akad Nauk SSSR 148*768-770
• TRANSL [1963] (J 0062) Sov Math, Dokl 4*176-179
⋄ B10 F07 ⋄ REV MR 26 # 4894 Zbl 166.249 • ID 08901

MATWIN, S. & PIETRZYKOWSKI, T. [1982] *Exponential improvement of efficient backtracking: a strategy for plan-based deduction* (P 3840) Autom Deduct (6);1982 New York 223-239
⋄ B35 ⋄ REV MR 85i:68041 • ID 44329

MATWIN, S. [1984] see FORSYTHE, K.

MAU, J. [1958] see BIERMANN, K.R.

MAUTNER, F.I. [1946] *An extension of Klein's Erlanger program: logic as invariant-theory* (J 0100) Amer J Math 68*345-384
⋄ A05 B10 ⋄ REV MR 8.3 Zbl 60.20 JSL 11.134
• ID 08910

MAYEGA, J. [1974] *Statistical decidability of theorems* (J 1550) Creation Math 7*3-10
⋄ B35 ⋄ REV Zbl 301 # 68085 • ID 63797

MAYOH, B.H. [1970] *The relation between an object and its name: notation systems and their fixed point theorems* (P 0785) Scand Logic Symp (1);1968 Aabo 77-95
⋄ A05 B03 D20 D35 D45 F30 F60 ⋄ REV MR 52 # 2850 Zbl 318 # 02048 • ID 17641

MAYOH, B.H. [1974] *Extracting information from logical proofs* (C 1936) Log Th & Semant Anal (Kanger) 61-72
⋄ B35 F07 ⋄ REV Zbl 297 # 68073 • ID 63802

MAYOH, B.H. [1979] see JENSEN, F.V.

MAYOH, B.H. see Vol. III, IV, VI for further entries

MAYS, W. & PRINZ, D.G. [1950] *A relay machine for the demonstration of symbolic logic* (J 0512) Nature 165*197-198
⋄ B35 ⋄ REV Zbl 35.78 JSL 15.138 • ID 08927

MAYS, W. [1951] see HANSEL, C.E.M.

MAYS, W. see Vol. IV for further entries

MAZBIC-KULMA, B. [1977] see KREMPA, J.

MAZUR, S. [1955] see GRZEGORCZYK, A.

MAZUR, S. see Vol. V, VI for further entries

MAZURKIEWICZ, A.W. [1969] *On the simple form of the deduction systems* (J 0485) Algorytmy, Pol Akad Nauk 5*45-50
⋄ B22 ⋄ REV MR 41 # 49 Zbl 185.24 • ID 08930

MAZURKIEWICZ, A.W. see Vol. IV for further entries

MAZZANTI, G. & MIROLLI, M. [1981] *Loeb operators and interior operators* (J 0144) Rend Sem Mat Univ Padova 65*77-84
⋄ H05 ⋄ REV MR 83f:03065 Zbl 484 # 03035 • ID 35311

MAZZANTI, S. & SCHMITT, B.V. [1981] *Solutions periodiques symetriques de l'equation de Duffing sans dissipation* (J 0415) J Differ Equations 42*199-214
⋄ H10 ⋄ REV MR 83b:34050 Zbl 439 # 34033 • ID 39001

MAZZANTI, S. see Vol. IV for further entries

MCALOON, K. [1979] see BOFFA, M.

MCALOON, K. [1983] see CLOTE, P.

MCALOON, K. see Vol. III, IV, V, VI for further entries

MCCALLUM, D.B. & SMITH, J.B. [1951] *Mechanized reasoning - logical computers and their design* (J 0513) Electronic Engin 23*126-133
⋄ B35 ⋄ REV MR 13.389 JSL 17.77 • ID 08951

MCCARTHY, J. [1962] *Computer programs for checking mathematical proofs* (P 0613) Rec Fct Th;1961 New York 219-227
⋄ B35 ⋄ REV Zbl 143.400 JSL 32.523 • ID 08956

MCCARTHY, J. [1979] see CARTWRIGHT, ROBERT

MCCARTHY, J. see Vol. II, IV for further entries

MCCARTHY, T. [1979] see GOTTLIEB, D.

MCCARTHY, T. see Vol. III, V for further entries

MCCAWLEY, J.D. [1981] *Everything that linguists have always wanted to know about logic, but were ashamed to ask* (X 1096) Blackwell: Oxford xv+508pp
⋄ B98 ⋄ REV JSL 49.1407 • ID 47399

MCCAWLEY, J.D. see Vol. II, IV for further entries

MCCLENNEN, E. [1978] see HOOKER, C.A.

MCCLUSKEY JR., E.J. [1956] *Detection of group invariance or total symmetry of a boolean function* (J 0432) Bell Syst Tech J 35*1445-1453
⋄ B35 ⋄ REV MR 18.624 JSL 23.236 • ID 08962

MCCLUSKEY JR., E.J. [1956] *Minimization of boolean functions* (J 0432) Bell Syst Tech J 35*1417-1444
⋄ B35 ⋄ REV MR 18.624 JSL 23.235 • ID 08963

MCCLUSKEY JR., E.J. [1961] *An essay on prime implicant tables* (J 0514) SIAM Journ 9*604-631
⋄ B35 ⋄ REV JSL 29.52 • ID 08964

MCCLUSKEY JR., E.J. [1961] *Minimal sums for boolean functions having many unspecified fundamental products* (P 0624) Switch Circ Th & Log Design (1,2);1960 Chicago;1961 Detroit 10-17
• REPR [1962] (J 0244) Trans Amer Inst Elect Engin 81*387-392
⋄ B35 ⋄ REV JSL 32.263 • ID 19318

MCCLUSKEY JR., E.J. & PYNE, I.B. [1962] *The reduction of redundancy in solving prime implicant tables* (J 0072) IRE Trans Electr Comp EC-11*473-482
⋄ B35 ⋄ REV MR 30 # 739 JSL 32.540 • ID 16372

MCCLUSKEY JR., E.J. & SCHORR, H. [1963] *Essential multiple-output prime implicants* (P 0674) Symp Math Th of Automata;1962 New York 437-457
◊ B35 ◊ REV MR 30#3004 Zbl 118.330 JSL 35.483
• ID 08966

MCCOLL, W.F. [1976] *The depth of boolean functions* (P 1870) Automata, Lang & Progr (3);1976 Edinburgh 307-321
◊ B05 ◊ REV Zbl 398#94034 • ID 69712

MCCOLL, W.F. & PATERSON, M.S. [1977] *The depth of all boolean functions* (J 1428) SIAM J Comp 6*373-380
• TRANSL [1980] (J 3079) Kiber Sb Perevodov, NS 17*154-163 (Russian)
◊ B05 ◊ REV Zbl 361#94051 • ID 69711

MCCOLL, W.F. [1978] *Complexity hierarchies for boolean functions* (J 1431) Acta Inf 11*71-77
◊ B05 B70 ◊ REV MR 80b:94041 Zbl 382#94032
• ID 69713

MCCOLL, W.F. [1978] *The circuit depth of symmetric boolean functions* (J 0119) J Comp Syst Sci 17*108-115
◊ B05 ◊ REV MR 80a:94061 Zbl 389#94017 • ID 69714

MCCORD, M.C. [1972] *Non-standard analysis and homology* (J 0027) Fund Math 74*21-28
◊ H05 ◊ REV MR 45#9316 Zbl 233#55005 • ID 08973

MCCUNE, W. [1982] see LUSK, E.L.

MCCUNE, W. & SMITH, BRIAN & VEROFF, R. & WOS, L. [1984] *The linked inference principle. II. The user's viewpoint* (P 2633) Autom Deduct (7);1984 Napa 316-332
◊ B35 ◊ ID 44463

MCDONALD, J.L. & SUPPES, P. [1984] *Student use of an interactive theorem prover* (P 3084) Autom Theor Prov After 25 Yea;1983 Denver 315-360
◊ B35 ◊ REV MR 85d:68005 Zbl 553#68052 • ID 43442

MCGEE, V. [1985] *A counterexample to modus ponens* (J 0301) J Phil 82*462-471
◊ B05 ◊ ID 48962

MCGEE, V. see Vol. II for further entries

MCGILL, V.J. [1939] *Concerning the laws of contradiction and excluded middle* (J 0153) Phil of Sci (East Lansing) 6*196-211
◊ B05 ◊ ID 38639

MCGRADE, A.S. [1966] see KUPPERMAN, J.

MCKEAN, J.M. [1972] see ASENJO, F.G.

MCKEE, T.A. [1980] *Monadic characterizations in nonstandard topology* (J 0068) Z Math Logik Grundlagen Math 26*395-397
◊ C10 H05 H20 ◊ REV MR 81m:03074 Zbl 449#54046
• ID 76216

MCKEE, T.A. [1984] *Logical aspects of combinatorial duality* (J 0018) Canad Math Bull 27*251-256
◊ B30 ◊ REV MR 85e:03026 Zbl 514#05058 • ID 36818

MCKEE, T.A. see Vol. II, III for further entries

MCKELLAR, A.C. & SHEN, V.Y. & WEINER, P. [1969] *A fast algorithm for testing switching functions for disjunctive decompositions* (P 1129) Princeton Conf Inform Sci & Syst (3);1969 Princeton 480-483
◊ B35 B70 ◊ REV Zbl 293#94019 • ID 65251

MCKENZIE, R. & MYCIELSKI, J. & THOMPSON, D. [1971] *On boolean functions and connected sets* (J 0041) Math Syst Theory 5*259-270
◊ B20 ◊ REV MR 54#11876 Zbl 221#02047 • ID 09001

MCKENZIE, R. [1975] *On spectra, and the negative solution of the decision problem for identities having a finite nontrivial model* (J 0036) J Symb Logic 40*186-196
◊ B20 C05 C13 D35 ◊ REV MR 51#12499
Zbl 316#02052 • ID 09009

MCKENZIE, R. see Vol. III, IV, V, VI for further entries

MCKINSEY, J.C.C. [1936] *On boolean functions of many variables* (J 0064) Trans Amer Math Soc 40*343-362
◊ B05 G05 ◊ REV Zbl 15.387 JSL 2.58 FdM 62.1055
• ID 09010

MCKINSEY, J.C.C. [1936] *On the generation of the functions Cpq and Np of Lukasiewicz and Tarski by means of a single binary operation* (J 0015) Bull Amer Math Soc 42*849-851
◊ B05 ◊ REV Zbl 16.2 JSL 2.59 FdM 62.1053 • ID 09011

MCKINSEY, J.C.C. [1936] *On the independence of Hilbert and Ackermann's postulates for the calculus of propositional functions* (J 0100) Amer J Math 58*336-344
◊ B05 ◊ REV Zbl 14.98 JSL 1.64 FdM 62.40 • ID 09013

MCKINSEY, J.C.C. [1936] *Reducible boolean functions* (J 0015) Bull Amer Math Soc 42*263-267
◊ B05 ◊ REV Zbl 14.194 JSL 1.69 FdM 62.34 • ID 09014

MCKINSEY, J.C.C. [1937] *A condition that a first boolean function vanish wherever a second does not* (J 0015) Bull Amer Math Soc 43*694-696
◊ B05 G05 ◊ REV Zbl 17.387 JSL 3.47 FdM 63.873
• ID 09015

MCKINSEY, J.C.C. [1943] *The decision problem for some classes of sentences without quantifiers* (J 0036) J Symb Logic 8*61-76
◊ B20 B25 C05 C60 G10 ◊ REV MR 5.85 JSL 9.30
• ID 09021

MCKINSEY, J.C.C. & SUPPES, P. [1955] *On the notion of invariance in classical mechanics* (J 0013) Brit J Phil Sci 5*290-302
◊ B30 ◊ ID 41178

MCKINSEY, J.C.C. see Vol. II, III, V, VI for further entries

MCLAUGHLIN, T.G. [1961] *A muted variation on a theme of Mendelson* (J 0068) Z Math Logik Grundlagen Math 7*57-60
◊ B10 ◊ REV MR 24#A684 Zbl 115.6 JSL 38.660
• ID 09037

MCLAUGHLIN, T.G. [1984] see KOMKOV, V.

MCLAUGHLIN, T.G. see Vol. III, IV, V for further entries

MCMINN, T.J. [1974] *A formal number-termed number system based on recursion* (J 0308) Rocky Mountain J Math 4*649-672
◊ B28 F30 ◊ REV MR 50#7084 Zbl 293#02032
• ID 09065

MCNAUGHTON, R. [1954] *A non-standard truth definition* (J 0053) Proc Amer Math Soc 5*505-509
◊ B28 C62 E30 H10 H20 ◊ REV MR 15.925 Zbl 56.12
• ID 09068

MCNAUGHTON, R. [1957] *On the measure of normal formulas* (J 0048) Pac J Math 7*969-982
◊ B05   B70 ◊ REV MR 19.240   Zbl 219 # 02003 • ID 09070

MCNAUGHTON, R. [1961] *Unate truth functions* (J 0072) IRE Trans Electr Comp EC-10*1-6
◊ B05   B70 ◊ REV MR 22 # 9462   JSL 32.263 • ID 09072

MCNAUGHTON, R. [1965] *Undefinability of addition from one unary operator* (J 0064) Trans Amer Math Soc 117*329-337
◊ B28   C10 ◊ REV MR 31 # 1191   Zbl 143.250   JSL 31.270 • ID 09073

MCNAUGHTON, R. see Vol. II, IV, V, VI for further entries

MCNULTY, G.F. [1977] *Fragments of first order logic, I: Universal Horn logic* (J 0036) J Symb Logic 42*221-237
◊ B20 ◊ REV MR 58 # 16255   Zbl 381 # 03011 • ID 26449

MCNULTY, G.F. see Vol. III, IV, V for further entries

MCROBBIE, M.A. & MEYER, R.K. & THISTLEWAITE, P.B. [1980] *A mechanized decision procedure for non-classical logics: The program KRIPKE* (J 0387) Bull Sect Logic, Pol Acad Sci 9*189-192
◊ B25   B35 ◊ REV MR 82b:03035   Zbl 453 # 03002 • ID 54132

MCROBBIE, M.A. see Vol. II for further entries

MEDETOV, M.M. [1978] see DZEGELENOK, I.I.

MEDVEDEV, F.A. [1978] *Cantor's theory of real numbers (Russian)* (S 0450) Istor-Mat Issl, Akad Nauk SSSR 23*56-70,357
◊ A10   B28   E30 ◊ REV MR 82h:01035a   Zbl 418 # 01003 • ID 82222

MEDVEDEV, F.A. see Vol. V for further entries

MEDVEDEV, S.S. [1978] *Simplification of boolean functions on L-symmetrical matrices (Russian)* (J 0474) Avtom Vychis Tekh, Akad Nauk Latv SSR 1978/1*6-11,91
• TRANSL [1978] (J 2666) Autom Control Comput Sci 12/1*6-11
◊ B05 ◊ REV MR 58 # 20861   Zbl 374 # 94018 • ID 51577

MEILER, M. [1974] see BAER, G.

MEISTERS, G.H. & MONK, J.D. [1973] *Construction of the reals via ultrapowers* (J 0308) Rocky Mountain J Math 3*141-158
◊ C20   H05 ◊ REV MR 50 # 292   Zbl 259 # 12104 • ID 09087

MEISTERS, G.H. see Vol. III, V for further entries

MEJLBO, L.C. [1981] *On nonstandard analysis* (J 3075) Normat 29*7-19
◊ H05 ◊ REV MR 82b:03002   Zbl 456 # 26010 • ID 76268

MELLIAR-SMITH, P.M. & SCHWARTZ, R.L. & SHOSTAK, R.E. [1982] *STP: a mechanized logic for specification and verification* (P 3840) Autom Deduct (6);1982 New York 32-49
◊ B35 ◊ REV MR 85e:68062   Zbl 481 # 68084 • ID 40037

MELLIAR-SMITH, P.M. see Vol. II for further entries

MELONI, G.C. [1971] *Il reticolo dei filtri dal punto di vista dell'analisi non standard* (J 0207) Ist Lombardo Accad Sci Rend, A (Milano) 105*756-768
◊ H05 ◊ REV MR 46 # 50   Zbl 238 # 02052 • ID 09091

MELONI, G.C. [1973] *Analisi non standard di anelli e corpi topologici (English summary)* (J 0207) Ist Lombardo Accad Sci Rend, A (Milano) 107*503-510
◊ C60   H05   H20 ◊ REV MR 52 # 5414   Zbl 282 # 54028 • ID 18293

MELONI, G.C. [1974] see BONACINI, R.

MELONI, G.C. [1975] *Introduzione semantica alla logica matematica* (X 2739) Ist Edit Intern: Milano ii+37pp.
◊ B10 ◊ REV MR 56 # 5188 • ID 76276

MELONI, G.C. see Vol. II, III, V, VI for further entries

MELTER, R.A. & RUDEANU, S. [1984] *Alternative definitions of Boolean functions and relations* (J 0008) Arch Math (Basel) 43*16-20
◊ B05 ◊ REV MR 86g:06023   Zbl 535 # 06009 • ID 46966

MELTER, R.A. see Vol. III for further entries

MELTZER, B. [1966] *Theorem-proving for computers: Some results on resolution and renaming* (J 1193) Comput J (London) 8*341-343
• REPR [1983] (C 4659) Autom of Reasoning 1*493-496
◊ B35 ◊ REV MR 32 # 8537   Zbl 158.260 • ID 46550

MELTZER, B. [1968] *A new look at mathematics and its mechanization* (J 0508) Machine Intelligence 3*63-70
◊ B35 ◊ REV Zbl 214.25 • ID 46553

MELTZER, B. [1968] *Some notes on resolution strategies* (J 0508) Machine Intelligence 3*71-75
◊ B35 ◊ REV Zbl 214.25 • ID 46556

MELTZER, B. [1968] *Some recent developments in complete strategies for theorem-proving by computer* (J 0068) Z Math Logik Grundlagen Math 14*377-382
◊ B35 ◊ REV MR 39 # 6691   Zbl 182.25 • ID 09093

MELTZER, B. [1969] *Power amplification for theorem provers* (J 0508) Machine Intelligence 5*165-179
◊ B35 ◊ REV MR 44 # 2381   Zbl 219 # 68051 • ID 46597

MELTZER, B. [1969] *The use of symbolic logic in proving mathematical theorems by means of a digital computer* (C 0705) Found of Math (Goedel) 39-45
◊ B35 ◊ REV MR 41 # 3279   Zbl 184.29   JSL 40.505 • ID 09094

MELTZER, B. [1971] *Prolegomena to a theory of efficiency of proof procedures* (P 4216) Artif Intell & Heuristic Progr;1970 Menaggio 15-33
◊ B35 ◊ REV MR 56 # 17252 • ID 46568

MELTZER, B. see Vol. II for further entries

MELZI, G. [1976] *I supporti fisici dell'inferenza formale. Vol. I* (X 1364) Vita e Pensiero: Milano xv+243pp
◊ B10   B75   D05 ◊ REV MR 55 # 75   Zbl 376 # 94023 • ID 76279

MENDELSON, E. [1964] *Introduction to mathematical logic* (X 0864) Van Nostrand: New York x+300pp
• TRANSL [1971] (X 2027) Nauka: Moskva 320pp
◊ B98 ◊ REV MR 29 # 2158   MR 50 # 1825
MR 80d:03001   Zbl 192.19   Zbl 498 # 03001   JSL 34.110
• REM 2nd ed. 1979; viii+328pp. 3rd ed. 1984 • ID 09105

MENDELSON, E. [1973] *Number systems of the foundations of analysis* (X 0801) Academic Pr: New York xii+358pp
◊ B28 ◊ REV MR 50 # 10162   Zbl 268 # 26001 • ID 09108

MENDELSON, E. see Vol. II, III, IV, V, VI for further entries

MENDES DES SANTOS, S. & MILLAN, M.R. [1974] *Automatic proofs for theorems on predicate calculus* (P 3103) Adv Cybern Syst;1972 Oxford 1*349-361
◇ B35 ◇ REV Zbl 332 # 68062 • ID 63864

MENDES SILVA, A.J. [1970] *Geometrisierung der Aussagenlogik (Portugiesisch)* (J 0084) Gaz Mat (Lisboa) 31/117-120*63-66
◇ B05 ◇ REV Zbl 261 # 02005 • ID 30454

MENDLER, N.P. [1985] see CONSTABLE, R.L.

MENG, ZHANGRONG [1965] see LIN, JINHUO

MENGER, K. [1928] *Bemerkungen zu Grundlagenfragen I* (J 0157) Jbuchber Dtsch Math-Ver 37*213-226
◇ A05 B30 E30 F99 ◇ REV FdM 54.95 • REM Part II 1928 • ID 39109

MENGER, K. [1928] *Bemerkungen zu Grundlagenfragen II* (J 0157) Jbuchber Dtsch Math-Ver 37*298-302
◇ A05 B30 E30 ◇ REV FdM 54.95 • REM Parts I,III 1928 • ID 39110

MENGER, K. [1928] *Bemerkungen zu Grundlagenfragen, IV. Axiomatik der endlichen Mengen und der elementargeometrischen Verknuepfungsbeziehungen* (J 0157) Jbuchber Dtsch Math-Ver 37*309-325
◇ B30 E30 ◇ REV FdM 54.95 • REM Part III 1928 • ID 31950

MENGER, K. [1931] *Eine elementare Bemerkung ueber die Struktur logischer Formeln* (J 1124) Ergebn Math Kolloquium 3*22-23
◇ B03 ◇ REV FdM 57.1320 • ID 31951

MENGER, K. [1938] *A new foundation of non-euclidean, affine, real projective and euclidean geometry* (J 0054) Proc Nat Acad Sci USA 24*486-490
◇ B30 ◇ REV Zbl 20.158 FdM 64.1259 • ID 09118

MENGER, K. [1959] *An axiomatic theory of functions and fluents* (P 0651) Axiomatic Method;1957 Berkeley 454-473
◇ A05 B30 ◇ REV MR 22 # 2540 Zbl 87.268 • ID 21011

MENGER, K. [1962] *Function algebra and propositional calculus* (P 0578) Self Organizing Systs;1962 Chicago 525-532
◇ B05 G05 ◇ REV JSL 31.272 • ID 21993

MENGER, K. & SCHULTZ, MARTIN H. [1963] *Postulates for the substitutive algebra of the 2-place functors in the 2-valued calculus of propositions* (J 0047) Notre Dame J Formal Log 4*188-192
◇ B05 G25 ◇ REV MR 29 # 4669 Zbl 281 # 02013 JSL 31.272 • ID 09122

MENGER, K. see Vol. II, IV, V, VI for further entries

MENNE, A. [1951] see ANGSTL, H.

MENNE, A. [1954] see BOCHENSKI, I.M.

MENNE, A. [1962] *Some results of investigation of the syllogism and their philosophical consequences* (C 2389) Logico-Philos Studies 55-63
◇ A05 A10 B20 ◇ REV JSL 30.363 • ID 44963

MENNE, A. [1966] *Einfuehrung in die Logik* (X 1987) Francke: Muenchen
◇ B98 ◇ ID 33438

MENNE, A. see Vol. II for further entries

MEO, A.R. [1962] *On the minimal third order expression of a boolean function* (P 0616) Switch Circ Th & Log Design (3);1962 Chicago 5-24
◇ B35 ◇ REV JSL 32.540 • ID 09129

MEREDITH, C.A. [1953] *A single axiom of positive logic* (J 0093) J Comp Syst 1*169-170
◇ B20 F50 ◇ REV MR 15.1 Zbl 53.197 JSL 19.144 • ID 09134

MEREDITH, C.A. [1953] *Single axioms for the system (C,N), (C,O) and (A,N) of the two-valued propositional calculus* (J 0093) J Comp Syst 1*155-164
◇ B05 ◇ REV MR 15.1 Zbl 53.2 JSL 19.143 • ID 09133

MEREDITH, C.A. [1953] *The figures and moods of the n-term aristotelian syllogism* (J 0505) Dominican Studies 6*42-47
◇ A10 B20 ◇ ID 31891

MEREDITH, C.A. & PRIOR, A.N. [1963] *Notes on the axiomatics of the propositional calculus* (J 0047) Notre Dame J Formal Log 4*171-187
◇ B05 ◇ REV MR 30 # 1033 Zbl 146.8 JSL 33.306 • ID 09136

MEREDITH, C.A. [1966] *Postulates for implicational calculi* (J 0036) J Symb Logic 31*7-9
◇ B20 ◇ REV MR 33 # 3892 Zbl 146.247 JSL 33.306 • ID 09142

MEREDITH, C.A. & PRIOR, A.N. [1968] *Equational logic* (J 0047) Notre Dame J Formal Log 9*212-226 • ERR/ADD ibid 10*452
◇ B20 F50 G05 ◇ REV MR 40 # 24a Zbl 175.259 • ID 09143

MEREDITH, C.A. & PRIOR, A.N. [1969] *Equational postulates for the Sheffer stroke* (J 0047) Notre Dame J Formal Log 10*266-270
◇ B05 G05 ◇ REV MR 39 # 6731 Zbl 184.8 • ID 09146

MEREDITH, C.A. [1969] *Terminal functors permissible with syllogistic* (J 0047) Notre Dame J Formal Log 10*309-312
◇ A05 B20 ◇ REV MR 40 # 21 Zbl 179.313 • ID 09145

MEREDITH, C.A. see Vol. II for further entries

MEREDITH, D. [1973] *On a property of certain propositional formulae* (J 0047) Notre Dame J Formal Log 14*103-106
◇ B05 ◇ REV MR 48 # 1873 Zbl 212.11 • ID 09148

MEREDITH, D. [1978] *Positive logic and λ-constants* (J 0063) Studia Logica 37*269-285
◇ B20 B25 B40 ◇ REV MR 80c:03019 Zbl 393 # 03008 • ID 31932

MEREDITH, D. [1979] *Axiomatics for implication* (J 0047) Notre Dame J Formal Log 20*89-91
◇ B20 F50 ◇ REV MR 80e:03022 Zbl 314 # 02025 • ID 31933

MEREDITH, D. [1980] *A positive logic proof procedure* (C 3050) Essays Combin Log, Lambda Calc & Formalism (Curry) 503-510
◇ B20 B35 B40 F50 ◇ REV MR 82a:03014 Zbl 469 # 03006 • ID 76287

MEREDITH, D. see Vol. II, III, VI for further entries

MERLIER, T. [1975] *Applications du calcul propositionnel aux structures algebriques ordonnees* (C 3259) Semin Algeb Non Commutative Paris 1974/75 12*8pp
◇ B05 ◇ REV MR 52 # 13567 Zbl 373 # 06010 • ID 51516

MERTENS, A. [1973] see BAGINSKI, M.

MERZLYAKOV, YU.I. [1973] see ERSHOV, YU.L.

MERZLYAKOV, YU.I. see Vol. III for further entries

MESAROVIC, M.D. [1968] *On some metamathematical results as properties of general systems* (J 0041) Math Syst Theory 2*357-361
  ⋄ B22 ⋄ REV MR 39 # 2597 Zbl 198.322 • ID 09157

MESEGUER, J. [1985] see GOGUEN, J.A.

MESEGUER, J. see Vol. IV, V for further entries

MESERVE, B.E. [1955] *Decision methods for elementary algebra* (J 0005) Amer Math Mon 62*1-8
  ⋄ B20 ⋄ REV MR 16.555 Zbl 64.8 JSL 22.295 • ID 09162

METAKIDES, G. (ED.) [1982] *Patras logic symposium* (S 3303) Stud Logic Found Math 109*ix+391pp
  ⋄ B97 C97 ⋄ REV MR 84b:03008 Zbl 504 # 00001
  • ID 35609

METAKIDES, G. see Vol. III, IV, V, VI for further entries

METZLER, W. [1970] *Eine Einfuehrung der positiven rationalen und positiven reellen Zahlen auf Grund von Messvorgaengen* (J 0160) Math-Phys Sem-ber, NS 17*68-87
  ⋄ B28 ⋄ REV MR 41 # 5185 Zbl 199.286 • ID 48069

MEULEN TER, A.G.B. (ED.) [1983] *Studies in modeltheoretic semantics* (X 4217) Foris: Dordrecht x+206pp
  ⋄ B97 C97 ⋄ REV Zbl 563 # 00003 • ID 47464

MEYER, A.R. [1975] see FISCHER, MICHAEL J.

MEYER, A.R. [1982] see FISCHER, MICHAEL J.

MEYER, A.R. & MITCHELL, J.C. [1985] *Second-order logical relations* (P 4571) Log of Progr;1985 Brooklyn 225-236
  ⋄ B15 B22 ⋄ REV Zbl 565 # 68029 • ID 48667

MEYER, A.R. see Vol. II, III, IV, VI for further entries

MEYER, R.K. [1969] see LEBLANC, H.

MEYER, R.K. [1970] see LEBLANC, H.

MEYER, R.K. [1972] see LEBLANC, H.

MEYER, R.K. [1973] *On conserving positive logics* (J 0047) Notre Dame J Formal Log 14*224-236
  ⋄ B20 ⋄ REV MR 48 # 5813 Zbl 225 # 02019 JSL 42.311
  • ID 09199

MEYER, R.K. [1980] see MCROBBIE, M.A.

MEYER, R.K. [1982] see BENCIVENGA, E.

MEYER, R.K. see Vol. II, IV, V, VI for further entries

MEYERHOFF CRESSWELL, M. [1982] see CRESSWELL, M.J.

MEYERS, L.F. [1973] *Simultaneous versus successive quantification* (J 0047) Notre Dame J Formal Log 14*247-249
  ⋄ B10 ⋄ REV MR 47 # 6423 Zbl 251 # 02017 • ID 28906

MEZHLUMBEKOVA, V.F. [1975] *Cut-elimination in a system of negationless arithmetic (Russian) (English summary)* (J 0288) Vest Ser Mat Mekh, Univ Moskva 30/6*10-16
  • TRANSL [1975] (J 0510) Moscow Univ Math Bull 30/5-6*61-66
  ⋄ B20 F05 F30 F50 ⋄ REV MR 53 # 7735 Zbl 322 # 02031 • ID 63913

MEZHLUMBEKOVA, V.F. [1975] *On systems of the negationless calculus of predicates (Russian) (English summary)* (J 0288) Vest Ser Mat Mekh, Univ Moskva 30/5*3-9
  • TRANSL [1975] (J 0510) Moscow Univ Math Bull 30/5-6*1-6
  ⋄ B20 F50 ⋄ REV MR 52 # 13337 Zbl 325 # 02016
  • ID 21799

MEZHLUMBEKOVA, V.F. see Vol. VI for further entries

MICHALOS, A.C. [1969] *Principles of logic* (X 0819) Prentice Hall: Englewood Cliffs xiii+433pp
  ⋄ B98 ⋄ ID 22523

MICHALOS, A.C. see Vol. II for further entries

MICHALSKI, R.S. [1976] see CHILAUSKY, R.

MICHALSKI, R.S. see Vol. II for further entries

MIGLIOLI, P.A. [1975] see MARINI, D.

MIGLIOLI, P.A. see Vol. II, III, IV, VI for further entries

MIHAILESCU, E.G. [1937] *Recherches sur un sous-systeme du calcul de propositions* (J 0269) Ann Sci Univ Jassy Sec 1 23*106-124
  ⋄ B20 ⋄ REV Zbl 16.1 JSL 2.51 FdM 63.24 • ID 16534

MIHAILESCU, E.G. [1937] *Recherches sur la negation et l'equivalence dans le calcul des propositions* (J 0269) Ann Sci Univ Jassy Sec 1 23*360-408
  ⋄ B20 ⋄ REV Zbl 16.337 JSL 2.173 FdM 63.24 • ID 09215

MIHAILESCU, E.G. [1937] *Sur le principe de contradiction* (J 0525) C R Acad Sci Roumanie 1*284-286
  ⋄ B20 ⋄ REV FdM 63.822 • ID 41044

MIHAILESCU, E.G. [1938] *Recherches sur les formes normales* (J 0269) Ann Sci Univ Jassy Sec 1 24*1-81
  ⋄ B05 ⋄ REV Zbl 20.99 JSL 4.91 FdM 64.29 • ID 09217

MIHAILESCU, E.G. [1938] *Recherches sur l'equivalence et la reciprocite dans le calcul des propositions* (J 0269) Ann Sci Univ Jassy Sec 1 24*116-153
  ⋄ B05 ⋄ REV Zbl 18.1 JSL 3.55 FdM 64.28 • ID 41384

MIHAILESCU, E.G. [1938] *Sur certains sous-systemes de la logique positive classique* (J 0525) C R Acad Sci Roumanie 2*119-120
  ⋄ B20 ⋄ REV Zbl 18.193 FdM 64.28 • ID 41066

MIHAILESCU, E.G. [1938] *Sur le calcul des propositions* (J 0494) Bull Math Soc Sci Roumanie 40*241-244
  ⋄ B05 ⋄ REV Zbl 20.98 FdM 64.925 • ID 41080

MIHAILESCU, E.G. [1939] *Recherches sur l'equivalence, la negation et la reciprocite dans le calcul des propositions* (J 0517) Mathematica (Cluj) 15*81-118
  ⋄ B20 ⋄ REV Zbl 20.337 JSL 5.125 FdM 65.1103
  • ID 09218

MIHAILESCU, E.G. [1939] *Recherches sur les formes normales par rapport a l'equivalence et la disjonction, dans le calcul des propositions* (J 0269) Ann Sci Univ Jassy Sec 1 25*73-152
  ⋄ B20 ⋄ REV Zbl 20.99 JSL 4.91 • ID 09219

MIHAILESCU, E.G. [1951] *Researches on sub-systems of the propositional calculus (Roumanian)* (J 0197) Stud Cercet Mat Acad Romana 2*1-44
  ⋄ B20 ⋄ REV MR 16.554 JSL 17.277 • ID 41849

MIHAILESCU, E.G. [1956] *Formes normales dans le calcul des propositions bivalentes (Romanian) (Russian and French summaries)* (J 0518) Bul Sti Mat-Fiz, Acad Romina 8*297-327
   ◇ B05 ◇ REV MR 18.866 Zbl 75.231 JSL 34.122
   • ID 41297

MIHAILESCU, E.G. [1956] *Formes normales dans l'ensemble S(C) (Romanian) (Russian and French summaries)* (J 0518) Bul Sti Mat-Fiz, Acad Romina 8*329-361
   ◇ B05 ◇ REV MR 18.865 Zbl 75.231 JSL 35.329
   • ID 09221

MIHAILESCU, E.G. [1956] *Formes normales dans l'ensemble S(D) du calcul bivalent des propositions (Romanian) (Russian and French summaries)* (J 0230) An Univ Iasi, NS, Sect Ia 2*15-25,25-27
   ◇ B05 ◇ REV MR 20 # 4480 Zbl 75.4 • ID 47963

MIHAILESCU, E.G. [1958] *Recherches sur quelques systemes du calcul des propositions* (J 0147) An Univ Bucuresti, Acta Logica 1*173-185
   ◇ B20 ◇ REV MR 21 # 2580 Zbl 84.247 JSL 34.122
   • ID 09222

MIHAILESCU, E.G. [1959] *Formules normales des foncteurs de la logique classique (Romanian) (Russian and French summaries)* (J 0197) Stud Cercet Mat Acad Romana 10*117-144
   ◇ B05 ◇ REV MR 22 # 1509 Zbl 102.7 JSL 35.465
   • ID 19439

MIHAILESCU, E.G. [1959] *Les formes normales dans le calcul bivalent des propositions* (J 0147) An Univ Bucuresti, Acta Logica 2*201-227
   ◇ B05 ◇ REV MR 22 # 4632 Zbl 92.248 JSL 36.172
   • ID 09223

MIHAILESCU, E.G. [1959] *Sur quelques theorems de la logique classique* (J 0060) Rev Roumaine Math Pures Appl 4*233-248
   ◇ B05 ◇ REV MR 22 # 3677 Zbl 99.6 JSL 36.173
   • ID 09224

MIHAILESCU, E.G. [1960] *Sur quelques theorems dans les calculus des propositions bivalentes* (J 0147) An Univ Bucuresti, Acta Logica 3*105-115
   ◇ B05 ◇ REV MR 25 # 9 Zbl 249 # 02006 JSL 36.173
   • ID 09225

MIHAILESCU, E.G. [1962] *Sur les proprietes de l'implication par rapport a l'equivalence et la disjonction* (J 0147) An Univ Bucuresti, Acta Logica 5*119-134
   ◇ B20 ◇ REV MR 27 # 3495 Zbl 118.12 JSL 35.584
   • ID 09228

MIHAILESCU, E.G. [1962] *Systeme d'equivalence, conjonction et reciprocite (Romanian) (Russian and French summaries)* (J 0197) Stud Cercet Mat Acad Romana 13*445-458
   ◇ B20 ◇ REV MR 26 # 2355 Zbl 108.2 • ID 42873

MIHAILESCU, E.G. [1962] *The properties of the logical difference operator with respect to recirpocity and to disjunction (Romanian) (Russian and French summaries)* (J 0237) Stud Univ Cluj, Ser Math Phys Chem 7/1*9-18
   ◇ B05 ◇ REV MR 31 # 28 Zbl 121.253 • ID 09227

MIHAILESCU, E.G. [1963] *Generalization of some normal forms* (J 0060) Rev Roumaine Math Pures Appl 8*101-115
   ◇ B05 ◇ REV MR 27 # 1352 Zbl 168.14 JSL 35.329
   • ID 09229

MIHAILESCU, E.G. [1965] *Properties of the Nicode functor in connection with eqivalence and disjunction* (J 0147) An Univ Bucuresti, Acta Logica 7-8*205-208
   ◇ B05 ◇ REV MR 33 # 24 Zbl 231 # 02014 • ID 09232

MIHAILESCU, E.G. [1965] *Properties of the Sheffer functor in connection with reciprocity and conjunction* (J 0147) An Univ Bucuresti, Acta Logica 7-8*209-212
   ◇ B05 ◇ REV MR 33 # 25 Zbl 231 # 00215 • ID 09233

MIHAILESCU, E.G. [1965] *Sur les formes normales par rapport a l'eqivalence, la reciprocite et la conjonction* (J 0147) An Univ Bucuresti, Acta Logica 7-8*141-150
   ◇ B05 ◇ REV MR 33 # 22 Zbl 231 # 02012 • ID 09231

MIHAILESCU, E.G. [1965] *Sur quelques proprietes de la conjonction par rapport a la reciprocite* (J 0147) An Univ Bucuresti, Acta Logica 7-8*179-187
   ◇ B05 ◇ REV MR 33 # 23 Zbl 231 # 02013 • ID 09230

MIHAILESCU, E.G. [1966] *The properties of the converse implication in respect to the equivalence and disjunction* (J 0147) An Univ Bucuresti, Acta Logica 9*165-186
   ◇ B20 ◇ REV MR 34 # 27 Zbl 248 # 02015 • ID 63927

MIHAILESCU, E.G. [1969] *L'ordre d'incompletitude pour le systeme d'equivalence la negation et la reciprocite* (J 0047) Notre Dame J Formal Log 10*425-452
   ◇ B20 ◇ REV MR 41 # 6663 Zbl 188.9 • ID 09236

MIHAILESCU, E.G. [1969] *Les proprietes du foncteur Sheffer par rapport a l'equivalence et la disjonction* (J 0060) Rev Roumaine Math Pures Appl 14*1129-1160
   ◇ B05 ◇ REV MR 40 # 5411 Zbl 193.286 • ID 09235

MIHAILESCU, E.G. [1969] *Mathematical logic. Elements of propositional and predicate calculus (Romanian)* (X 0871) Acad Rep Soc Romania: Bucharest 334pp
   ◇ B98 ◇ REV MR 41 # 5191 Zbl 199.297 • ID 21042

MIHAILESCU, E.G. [1970] *Normal forms of the Nicod functors (Romanian)* (S 1613) Probl Logic (Bucharest) 2*79-111
   ◇ B05 ◇ REV Zbl 229 # 02013 • ID 47815

MIHAILESCU, E.G. [1973] *Les proprietes du foncteur Nicod par rapport a la reciprocite et conjonction I* (J 0047) Notre Dame J Formal Log 14*527-535
   ◇ B05 ◇ REV MR 49 # 8808 Zbl 266 # 02008 • REM Part II 1974 • ID 19437

MIHAILESCU, E.G. [1974] *Les proprietes du foncteur Nicod par rapport a la reciprocite et conjonction II* (J 0047) Notre Dame J Formal Log 15*85-96
   ◇ B05 ◇ REV MR 49 # 8808 Zbl 273 # 02007 • REM Part I 1973 • ID 19438

MIHAILESCU, E.G. see Vol. II, III, IV for further entries

MIHALJINEC, M. (ED.) [1977] *Colloque international de logique, Clermont-Ferrand, 18-25 juillet 1975* (S 1802) Colloq Int CNRS 249*224pp.
   ◇ B97 ◇ REV Zbl 426 # 00003 • ID 53595

MIHALJINEC, M. see Vol. V, VI for further entries

MIJAJLOVIC, Z. [1974] *On decidability of one class of boolean formulas* (J 0042) Mat Vesn, Drust Mat Fiz Astron Serb 11(26)*48-54
   ◇ B05 B25 ◇ REV MR 50 # 80 Zbl 286 # 02065 • ID 09242

MIJAJLOVIC, Z. [1977] *Some remarks on boolean terms - model theoretic approach* (J 0400) Publ Inst Math, NS (Belgrade) 21(35)*135-140
⋄ B05 C50 G05 ⋄ REV MR 57#2909 Zbl 362#02055 • ID 50777

MIJAJLOVIC, Z. see Vol. III, V for further entries

MIJOULE, R. [1976] *Theorie des types et modeles de la theorie des ensembles (English summary)* (J 2313) C R Acad Sci, Paris, Ser A-B 283*A733-A735
⋄ B15 C62 G30 ⋄ REV MR 54#10010 Zbl 355#02050 • ID 25854

MIJOULE, R. see Vol. IV, VI for further entries

MIKELADZE, L.L. [1980] *A system for drawing relevant conclusions from a large volume of data (Russian)* (J 2694) Cybern & Syst 30-34
⋄ B35 ⋄ REV MR 83e:03025 • ID 35214

MIKELADZE, Z.N. [1979] *A class of logical concepts (Russian)* (P 2554) All-Union Symp Th Log Infer;1974 Moskva 287-299
⋄ B15 ⋄ REV MR 84k:03034 • ID 34972

MIKELADZE, Z.N. see Vol. II for further entries

MIKHEEVA, L. & SALUM, H. [1969] *Construction of abbreviated disjunctive normal forms of boolean functions by the method of masks (Russian)* (J 0080) Izv Akad Nauk Ehston SSR, Fiz, Mat 18*458-460
⋄ B05 ⋄ REV MR 43#6099 Zbl 191.313 • ID 09240

MIKKELSEN, C.J. [1974] see KOCK, A.

MIKKELSEN, C.J. see Vol. V, VI for further entries

MIKOLAJEWICZ, B. [1983] *Some classes of models for Peano arithmetic and for some related theories (Polish) (English summary)* (J 0481) Acta Univ Wroclaw 605(34)(Logika 10)*31-69
⋄ B28 F30 ⋄ REV Zbl 549#03022 • ID 43117

MIKOLAJEWICZ, B. see Vol. VI for further entries

MIKUSINSKI, J. & MIKUSINSKI, P. [1976] *Natuerliche Zahlen und ihre Axiomatik (Polnisch)* (J 0519) Wiad Mat, Ann Soc Math Pol, Ser 2 19*137-140
⋄ B28 ⋄ REV MR 55#10264 Zbl 327#02001 • ID 63937

MIKUSINSKI, J. see Vol. V for further entries

MIKUSINSKI, P. [1976] see MIKUSINSKI, J.

MIKUSINSKI, P. see Vol. V for further entries

MILEV, L.K. [1974] *Die Konstruktion rationaler Zahlen in $\mathbf{N} \times \{1,2\} \times \mathbf{N}$ (Bulgarian) (Russian and French summaries)* (J 3171) God Vissh Ucheb Zaved, Prilozhna Mat, Sofiya 10/1*111-121
⋄ B28 ⋄ REV Zbl 341#02045 • ID 29759

MILEV, L.K. see Vol. V for further entries

MILICI, C. [1974] *A remark on axiom-system for the classical two-valued $\{\rightarrow, \neg\}$ propositional logic (Serbo-Croatian summary)* (J 3519) Glas Mat, Ser 3 (Zagreb) 9(29)*3-5
⋄ B05 ⋄ REV MR 50#48 Zbl 286#02013 • ID 19434

MILICI, C. [1974] *On the propositional calculus defined with $\{\rightarrow, \&, \neg\}$* (J 0304) An Univ Timisoara, Sti Mat 12*119-124
⋄ B05 ⋄ REV MR 57#2873 Zbl 443#03006 • ID 56412

MILICI, C. [1975] *Note on the C-calculus* (J 0047) Notre Dame J Formal Log 16*548
⋄ B20 ⋄ REV MR 52#7841 Zbl 299#02008 • ID 18297

MILICI, C. & RACHIN, N. [1980] *On the structures of formulas of propositional calculus defined with $(\vee, \neg)$ (Romanian) (English summary)* (J 4715) Bul Sti Tehn Inst Politeh Timisoara, Ser Mat-Fiz 25(39)*43-44
⋄ B05 ⋄ REV MR 83j:03017 • ID 35330

MILICI, C. see Vol. V for further entries

MILLAN, M.R. [1974] see MENDES DES SANTOS, S.

MILLER, D.MICHAEL & MUZIO, J.C. [1974] *Two-place decomposition of binary functions* (P 3242) Manitoba Conf Num Math (3);1973 Winnipeg 293-306
⋄ B05 ⋄ REV MR 50#4135 Zbl 319#94022 • ID 63945

MILLER, D.MICHAEL see Vol. II for further entries

MILLER, DALE A. [1982] see ANDREWS, P.B.

MILLER, DALE A. [1984] see ANDREWS, P.B.

MILLER, DALE A. [1984] *Expansion tree proofs and their conversion to natural deduction proofs* (P 2633) Autom Deduct (7);1984 Napa 375-393
⋄ B35 F07 ⋄ REV Zbl 547#03031 • ID 43218

MILLER, DAVID [1974] *On the comparison of false theories by their bases* (J 0013) Brit J Phil Sci 25*178-188
⋄ A05 B30 ⋄ REV Zbl 377#02008 • ID 09251

MILLER, DAVID [1977] see COBURN, B.

MILLER, DAVID [1982] *A geometry of logic* (P 3845) Conf Math Service of Man (2,Proc)(Feriet);1982 Las Palmas 505-509
⋄ B05 ⋄ REV Zbl 511#03027 • ID 36912

MILLER, DAVID see Vol. II for further entries

MILLER, H.I. & ZIVALJEVIC, B. [1984] *Remarks on the zero-one law (Russian summary)* (J 1522) Math Slovaca 34*375-384
⋄ H05 ⋄ REV MR 86e:60025 • ID 48919

MILLER, H.I. see Vol. V for further entries

MILLER, JAMES WILKIN [1938] *The structure of Aristotelian logic* (X 1326) Kegan Paul Trench Trubner: London 97pp
⋄ A10 B20 ⋄ REV JSL 4.121 • ID 42550

MILLER, JAMES WILKIN [1955] *Exercises in introductory symbolic logic* (X 0828) Edwards Brothers: Ann Arbor ix+59pp
⋄ B98 ⋄ REV JSL 22.310 • ID 09255

MILLER, JAMES WILKIN [1958] *Logic workbook* (X 0894) Oxford Univ Pr: Oxford vii+88pp
⋄ B98 ⋄ REV Zbl 99.5 JSL 23.213 • ID 09256

MILLER, JAMES WILKIN see Vol. II for further entries

MILNE, R. [1978] *Transforming predicate transformers* (P 2929) Form Descr of Progr Concepts (1);1977 St.Andrews 31-65
⋄ B10 ⋄ REV MR 80e:68018 Zbl 374#68013 • ID 51570

MILNER, R. [1976] *Program semantics and mechanized proof* (C 3517) Found of Comput Sci, Vol 2, Part 2 3-44
⋄ B35 ⋄ REV MR 58#19303 Zbl 437#68006 • ID 69724

MILNER, R. [1979] *ICF: A way of doing proofs with a machine* (P 2059) Math Founds of Comput Sci (8);1979 Olomouc 146-159
⋄ B35 B40 ⋄ REV MR 81g:68124 Zbl 423#68049 • ID 53582

MILNER, R. [1984] *The use of machines to assist in rigorous proof. Mathematical logic and programming languages (with discussion)* (J 0354) Phil Trans Roy Soc London, Ser A 312*411-422
⋄ B35 ⋄ ID 44452

MILNER, R. see Vol. II, IV, VI for further entries

MINICOZZI, E. & REITER, R. [1972] *A note on linear resolution in consequence-finding* (J 0503) Artif Intell 3*175-180
⋄ B35 ⋄ REV MR 50#9071 Zbl 252#68049 • ID 09300

MINICOZZI, E. see Vol. II, IV for further entries

MINKER, J. & ZANON, G. [1982] *An extension to linear resolution with selection function* (J 0232) Inform Process Lett 14*191-194
⋄ B35 ⋄ REV MR 83i:68141 Zbl 477#68100 • ID 69725

MINKER, J. [1982] *On closed databases and the closed world assumption* (P 3840) Autom Deduct (6);1982 New York 292-308
⋄ B35 ⋄ REV MR 85e:68017 Zbl 481#68087 • ID 39990

MINKER, J. see Vol. II for further entries

MINOGUE, B.P. [1977] *Numbers, properties, and Frege* (J 0095) Philos Stud 31*423-427
⋄ A05 B28 ⋄ REV MR 58#27247 • ID 76431

MINTS, G.E. [1965] see DAVYDOV, G.V.

MINTS, G.E. [1965] see MASLOV, S.YU.

MINTS, G.E. [1966] *Herbrand's theorem for the predicate calculus with equality and functional symbols (Russian)* (J 0023) Dokl Akad Nauk SSSR 169*273-275
• TRANSL [1966] (J 0062) Sov Math, Dokl 7*911-914
⋄ B10 F05 ⋄ REV MR 34#7341 Zbl 156.250 JSL 35.325
• ID 09276

MINTS, G.E. [1967] *Herbrand's theorem (Russian)* (C 2542) Mat Teor Log Vyvoda 311-350
⋄ B10 F05 F07 ⋄ REV MR 36#4958 Zbl 156.250
• ID 37138

MINTS, G.E. [1967] *Variation in the deduction search tactics in sequential calculi (Russian)* (S 0228) Zap Nauch Sem Leningrad Otd Mat Inst Steklov 4*134-151
• TRANSL [1967] (J 0521) Semin Math, Inst Steklov 4*52-59
⋄ B35 F07 F50 ⋄ REV MR 41#1505 Zbl 166.249
• ID 19424

MINTS, G.E. [1968] *Admissible and deductive rules (Russian)* (S 0228) Zap Nauch Sem Leningrad Otd Mat Inst Steklov 8*189-191
• TRANSL [1968] (J 0521) Semin Math, Inst Steklov 8*90-91
⋄ B10 F07 F50 ⋄ REV MR 43#4626 Zbl 207.4
• ID 19422

MINTS, G.E. [1969] see DAVYDOV, G.V.

MINTS, G.E. (ED.) [1970] *Mathematical logic (Russian)* (X 2673) Bibl Akad Nauk SSSR: Leningrad 174pp
⋄ A10 B98 ⋄ REV MR 53#12855 • ID 70167

MINTS, G.E. [1971] see MASLOV, S.YU.

MINTS, G.E. [1971] *Quantifier-free and one-quantifier systems (Russian) (English summary)* (S 0228) Zap Nauch Sem Leningrad Otd Mat Inst Steklov 20*115-133,285
• TRANSL [1973] (J 1531) J Sov Math 1*71-84
⋄ B20 D20 F30 ⋄ REV MR 44#6486 Zbl 222#02022
• ID 09286

MINTS, G.E. [1974] *Functional form. Herbrand's theorem for non-prenex formulas (Russian)* (C 4137) Kleene: Mat Logika 447-450
⋄ B10 F05 ⋄ ID 37140

MINTS, G.E. [1974] *Gentzen's formal systems (Russian)* (C 2546) Ehntsikl Kibern 1*222-224
⋄ B10 F05 F07 ⋄ ID 37142

MINTS, G.E. [1975] see MASLOV, S.YU.

MINTS, G.E. & TYUGU, E.KH. [1982] *The completeness of structural synthesis rules (Russian)* (J 0023) Dokl Akad Nauk SSSR 263*291-295
• TRANSL [1982] (J 0062) Sov Math, Dokl 25*343-346
⋄ B20 B75 D20 F50 ⋄ REV MR 84m:03013 Zbl 517#68058 • ID 35718

MINTS, G.E. see Vol. II, III, IV, V, VI for further entries

MIRA ROS, J.M. [1975] see CARILLO GALLEGO, D.

MIROLLI, M. [1981] see MAZZANTI, G.

MIROLLI, M. see Vol. II, III, VI for further entries

MIRON, R. [1983] see BRANZEI, D.

MISERCQUE, D. [1980] *Sur le treillis distributif des $\forall_1$-formules fermees de l'arithmetique de Peano (English summary)* (J 2313) C R Acad Sci, Paris, Ser A-B 290*A571-A573
⋄ B20 F30 G10 ⋄ REV MR 81m:03069 Zbl 424#03031
• ID 76441

MISERCQUE, D. see Vol. III, IV, VI for further entries

MISES VON, R. [1928] *Wahrscheinlichkeit, Statistik und Wahrheit* (X 0811) Springer: Heidelberg & New York vii+189pp
• TRANSL [1981] (X 0813) Dover: New York xii+244pp (English)
⋄ B30 ⋄ REV MR 83j:60001 FdM 54.520 • REM 2nd edition 1951: vii+189p. The transl. is a revised version
• ID 33703

MITANI, S. [1978] *Completion theory performed on the basis of LJ* (J 0352) Math Jap 23*461-490
⋄ B35 ⋄ REV MR 80b:68112 Zbl 427#54001 • ID 82271

MITANI, S. [1983] *Dimension theory constructed on the basis of LJ* (J 0352) Math Jap 28*595-599
⋄ B35 F55 ⋄ REV MR 85h:03019 Zbl 538#03010
• ID 41462

MITCHELL, D. [1962] *An introduction to logic* (X 0939) Hutchinson: London 192pp
⋄ B98 ⋄ ID 25144

MITCHELL, D. [1970] *An introduction to logic* (X 0878) Doubleday: London ix+227pp
⋄ B98 ⋄ REV MR 50#12629 JSL 38.345 • ID 09320

MITCHELL, J.C. [1985] see MEYER, A.R.

MITEW, W. & POSTHOFF, C. [1975] *Ein Verfahren zur Loesung boolescher Gleichungen* (J 2803) Nachrichtentech Elektr 25*A,222-224
⋄ B35 ⋄ REV MR 54#12475 • ID 77471

MIURA, S. [1966] *A remark on the intersection of two logics* (J 0111) Nagoya Math J 26*167-171
⋄ B20 ⋄ REV MR 33#3927 Zbl 202.295 JSL 34.504
• ID 09333

MIURA, S. & ONO, K. [1967] *On pairs of very-close formal systems* (J 0081) Proc Japan Acad 43∗175-177
⋄ B22 ⋄ REV MR 36#2466 Zbl 166.254 • ID 10128

MIURA, S. see Vol. II, III for further entries

MIYAWAKI, M. [1981] see FUNAHASHI, S.

MIZUTANI, C. [1977] *Monadic second order logic with an added quantifier Q* (J 2606) Tsukuba J Math 1∗45-76
⋄ B15 B25 C40 C55 C80 ⋄ REV MR 58#10291 Zbl 414#03023 • ID 53069

MIZUTANI, C. see Vol. VI for further entries

MLCEK, J. [1979] *Valuations of structures* (J 0140) Comm Math Univ Carolinae (Prague) 20∗681-695
⋄ E70 H05 ⋄ REV MR 81f:03063 Zbl 449#03052 • ID 56719

MLCEK, J. see Vol. III, V, VI for further entries

MLEZIVA, M. [1958] *Theory of propositions (Czech)* (J 4445) Moderni Logica 7∗29-75
⋄ B05 ⋄ ID 48782

MLEZIVA, M. [1959] *Die Unabhaengigkeit des Axiomensystems des Aussagenkalkuels von Hermes und Scholz* (J 0086) Cas Pestovani Mat, Ceskoslov Akad Ved 84∗454-460
⋄ B05 ⋄ REV MR 22#6695 JSL 35.488 • ID 19414

MLEZIVA, M. [1966] *Ueber das Enthaltensein des klassischen Aussagenkalkuels in den nicht-klassischen Aussagenkalkuelen (Czech summary)* (J 2830) Rozpr, Mat Prirod, Cheskoslov Akad Ved 76/13∗71pp
⋄ B22 ⋄ REV MR 37#6160 • ID 42955

MLEZIVA, M. see Vol. II, III for further entries

MO, SHAOKUI [1950] *The deduction theorems and two new logical systems* (J 0175) Methodos 2∗56-75
⋄ B20 B45 ⋄ REV MR 12.662 JSL 17.153 • ID 12028

MO, SHAOKUI [1952] *A note on the theory of quantification* (J 0036) J Symb Logic 17∗243-244
⋄ B10 ⋄ REV MR 14.440 Zbl 48.246 JSL 18.179 • ID 12029

MO, SHAOKUI [1955] *Some axiom systems for propositional calculus (Chinese) (English summary)* (J 0418) Shuxue Xuebao 5∗117-135
⋄ B05 ⋄ REV MR 17.225 JSL 25.182 • ID 09349

MO, SHAOKUI [1957] *On the rules of logical systems (Chinese)* (J 2804) Nanjing Daxue Xuebao, Ziran Kexue 1957/3∗77-85
⋄ B22 ⋄ ID 47121

MO, SHAOKUI [1963] *Notes on mathematical logic (Chinese)* (J 0418) Shuxue Xuebao 13∗485-507
• TRANSL [1963] (J 0419) Chinese Math Acta 4∗527-552
⋄ B98 ⋄ REV MR 29#1143 • ID 19399

MO, SHAOKUI [1964] *Enumeration quantifiers and predicate calculus (Chinese)* (J 0418) Shuxue Xuebao 14∗218-230
• TRANSL [1964] (J 0419) Chinese Math Acta 5∗239-253 (English)
⋄ B10 C07 C80 ⋄ REV Zbl 192.28 • ID 47141

MO, SHAOKUI [1965] *Introduction to mathematical logic (Chinese)* (X 0740) Kexue Jishu Chubanshe: Shanghai
⋄ B98 ⋄ ID 47171

MO, SHAOKUI [1979] *Study on mathematical logic (Chinese) (English summary)* (J 2754) Huazhong Gongxueyuan Xuebao 7/1∗2-3,28-45
• TRANSL [1979] (J 2684) J Huazhong Inst Tech (Engl Ed) 1/1∗19-30
⋄ B10 ⋄ REV MR 83i:03016ab • ID 35493

MO, SHAOKUI [1980] *Introduction to mathematical logic (Chinese)* (X 4589) Shanghai Jiaoyu Chubanshe: Shanghai 172pp
⋄ B98 ⋄ ID 48906

MO, SHAOKUI [1980] *On the nature of higher order functions and operators (Chinese) (English summary)* (J 2804) Nanjing Daxue Xuebao, Ziran Kexue 1980/1∗1-10
⋄ B15 ⋄ REV MR 82i:03016 Zbl 432#03008 • ID 53966

MO, SHAOKUI & SHEN, BAIYING & XU, YONGSHEN [1984] *Mathematical logic (Chinese)* (X 4579) Gnodeng Jiaoyu Chubanshe: Beijing 250pp
⋄ B98 ⋄ ID 48527

MO, SHAOKUI [1984] see DING, DECHENG

MO, SHAOKUI see Vol. II, III, IV, V, VI for further entries

MOELLER, K.K. [1979] see JENSEN, F.V.

MOGILEVSKIJ, G.L. & OSTROUKHOV, D.A. [1978] *On a mechanical propositional calculus using Smullyan's analytical tableaux (Russian)* (J 0040) Kibernetika, Akad Nauk Ukr SSR 1978/4∗43-46
• TRANSL [1978] (J 0021) Cybernetics 14∗526-529
⋄ B35 ⋄ REV MR 80b:68113 Zbl 416#68085 • ID 53233

MOISIL, G.C. [1937] *Le principe d'identite et le principe du syllogisme* (J 0525) C R Acad Sci Roumanie 1∗276-278
⋄ B20 ⋄ REV FdM 63.822 • ID 41046

MOISIL, G.C. [1939] *Recherches sur le syllogisme* (J 0269) Ann Sci Univ Jassy Sec 1 25∗341-384
⋄ B20 ⋄ REV Zbl 21.290 JSL 4.167 FdM 65.29 • ID 41706

MOISIL, G.C. [1955] *Sur le fonctionnement des schemas a boutons reels (Romanian) (Russian and French summaries)* (J 0518) Bul Sti Mat-Fiz, Acad Romina 7∗33-49
⋄ B05 B70 ⋄ REV JSL 37.414 • ID 09366

MOISIL, G.C. [1958] *Sur la logique positive (Russian and English summaries)* (J 0147) An Univ Bucuresti, Acta Logica 1∗149-171
⋄ B20 ⋄ REV MR 21#2579 Zbl 92.247 • ID 09367

MOISIL, G.C. [1965] *Sur la logique strictement positive (Romanian and Russian summaries)* (J 0230) An Univ Iasi, NS, Sect Ia IIB∗15-24
⋄ B20 ⋄ REV MR 34#4109 Zbl 192.25 • ID 24969

MOISIL, G.C. [1965] *Sur les lois distributives dans le calcul des propositions* (J 0147) An Univ Bucuresti, Acta Logica 7-8∗125-130
⋄ B20 ⋄ REV MR 33#29 Zbl 238#02009 • ID 27897

MOISIL, G.C. [1966] *Remarques sur les modeles de la logique elementaire* (J 0070) Bull Soc Sci Math Roumanie, NS 10∗27-29
⋄ B20 ⋄ REV MR 40#5423 Zbl 183.9 • ID 09374

MOISIL, G.C. [1968] *Classical logic of propositions of higher type (Roumanian)* (J 0197) Stud Cercet Mat Acad Romana 20∗238-258
⋄ B15 ⋄ REV MR 41#1516 Zbl 195.298 • ID 09376

MOISIL, G.C. [1971] *La logique elementaire* (C 0640) Log, Autom, Inform 31-39
⋄ B20 ⋄ REV MR 49#2246 Zbl 235#02013 • ID 09379

MOISIL, G.C. [1972] *La logique formelle et son probleme actuel* (C 3094) Moisil: Essais Logiques Non Chrysippiennes 11-31
⋄ B10 ⋄ REV JSL 13.160 • REM Appeared already 1939 • ID 41707

MOISIL, G.C. [1975] see HENKIN, L.

MOISIL, G.C. see Vol. II, IV, V, VI for further entries

MOKLYAK, N.G. [1973] see LYSENKO, E.V.

MOKLYAK, N.G. see Vol. II for further entries

MOLER, N. & SUPPES, P. [1968] *Quantifier-free axioms for constructive plane geometry* (J 0020) Compos Math 20*143-152
• REPR [1968] (C 0727) Logic Found of Math (Heyting) 143-152
⋄ B30 ⋄ REV MR 37#4699 Zbl 183.249 JSL 40.499
• ID 09383

MOLLERUP, J. [1907] *Die Definition des Mengenbegriffs* (J 0043) Math Ann 64*190-194
⋄ A05 B28 E30 ⋄ REV FdM 38.95 • ID 09387

MOLLERUP, J. [1907] *Sur la theorie des ensembles et le concept de nombre* (C 4534) Kjoebenhavn Overs 127-149
⋄ B28 E20 ⋄ REV FdM 38.95 • ID 37914

MOLLERUP, J. see Vol. V for further entries

MONK, J.D. [1965] *Substitutionless predicate logic with identity* (J 0009) Arch Math Logik Grundlagenforsch 7*102-121
⋄ B20 ⋄ REV MR 34#4111 Zbl 158.246 • ID 09398

MONK, J.D. [1971] see HENKIN, L.

MONK, J.D. [1971] *Provability with finitely many variables* (J 0053) Proc Amer Math Soc 27*353-358
⋄ B20 ⋄ REV MR 43#1811 Zbl 216.10 • ID 09406

MONK, J.D. [1973] see MEISTERS, G.H.

MONK, J.D. [1976] *Mathematical logic* (X 0811) Springer: Heidelberg & New York x+531pp
⋄ B98 ⋄ REV MR 57#5656 Zbl 354#02002 JSL 44.283
• ID 30765

MONK, J.D. [1985] see HENKIN, L.

MONK, J.D. see Vol. III, V for further entries

MONTAGUE, R. [1956] see HENKIN, L.

MONTAGUE, R. & TARSKI, A. [1957] *Independent recursive axiomatizability* (P 1675) Summer Inst Symb Log;1957 Ithaca 270
⋄ B30 D25 ⋄ ID 29361

MONTAGUE, R. [1957] *Non-finite axiomatizability* (P 1675) Summer Inst Symb Log;1957 Ithaca 256-259
⋄ B30 F25 F30 ⋄ REV Zbl 148.245 • ID 29356

MONTAGUE, R. [1957] see KALISH, D.

MONTAGUE, R. [1961] *Semantical closure and non-finite axiomatizability I* (P 0633) Infinitist Meth;1959 Warsaw 45-69
⋄ B15 E30 F25 F30 ⋄ REV MR 27#38 Zbl 116.7
JSL 29.59 • ID 09424

MONTAGUE, R. [1964] see KALISH, D.

MONTAGUE, R. [1965] see KAPLAN, DAVID

MONTAGUE, R. [1965] see KALISH, D.

MONTAGUE, R. [1965] *Reductions of higher-order logic* (P 0614) Th Models;1963 Berkeley 251-264
⋄ B15 ⋄ REV MR 34#4114 Zbl 202.301 • ID 09430

MONTAGUE, R. [1965] *Set theory and higher-order logic* (P 0688) Logic Colloq;1963 Oxford 131-148
⋄ B15 C62 E30 ⋄ REV MR 36#3638 Zbl 129.259
JSL 40.459 • ID 24776

MONTAGUE, R. see Vol. II, III, IV, V, VI for further entries

MONTALI, T. [1971] *Pseudolimiti e formule SH* (J 1526) Riv Mat Univ Parma, Ser 2 12*245-258
⋄ B20 C30 G30 ⋄ REV MR 52#39 Zbl 293#02036
• ID 17210

MONTANARI, U. [1982] see MARTELLI, A.

MONTEIRO, A. [1960] *Matrices de Morgan characteristiques pour le calcul propositionnel classique* (J 0110) Anais Acad Bras Cienc 32*1-7
⋄ B05 G05 ⋄ REV MR 28#18 MR 50#6832 Zbl 94.6
JSL 28.174 • ID 09436

MONTEIRO, A. [1965] *Generalisation d'un theoreme de R.Sikorski sur les algebres de Boole* (J 0247) Bull Sci Math, Ser 2 89*65-74
• REPR [1974] (S 0889) Notas Logica Mat 10*ii+10pp
⋄ B05 ⋄ REV MR 32#4054 Zbl 133.272 • ID 09442

MONTEIRO, A. [1974] *Matrices de Morgan caracteristiques pour le calcul propositionnel classique* (S 0889) Notas Logica Mat 6*7pp
⋄ B05 ⋄ REV Zbl 292#02010 • ID 64009

MONTEIRO, A. see Vol. II, V, VI for further entries

MONTELLA, E. [1974] *Il connettivo: "ne...ne..."* (J 1515) Archimede 26*127-131
⋄ B05 ⋄ REV MR 54#9970 • ID 25596

MONTEVERDI, D. [1973] *SH a piu sorte di variabili* (J 3254) Riv Mat Univ Parma, Ser 3 2*67-77
⋄ B20 C30 G30 ⋄ REV MR 52#13372 Zbl 341#02043
• ID 21823

MONTEVERDI, D. see Vol. III for further entries

MONTGOMERY, H. & ROUTLEY, R. [1968] *On systems containing Aristotle's thesis* (J 0036) J Symb Logic 33*82-96
⋄ B20 ⋄ REV MR 38#3129 • ID 16935

MONTGOMERY, H. see Vol. II, III for further entries

MOOR, J. [1980] see BERGMANN, MERRIE

MOORE, E.F. [1962] see KAHR, A.S.

MOORE, E.F. see Vol. IV for further entries

MOORE, G.W. [1981] see HUTCHINS, G.M.

MOORE, J.S. [1979] see BOYER, R.S.

MOORE, J.S. [1979] *A mechanical proof of the termination of Takeuchi's function* (J 0232) Inform Process Lett 9*176-181
⋄ B35 D20 ⋄ REV MR 81f:68113 Zbl 411#68079
• ID 52930

MOORE, J.S. [1984] see BOYER, R.S.

MOORE, J.S. see Vol. IV for further entries

MOORE, R.L. [1935] *A set of axioms for plane analysis situs* (J 0027) Fund Math 25*13-28
⋄ B30 ⋄ REV Zbl 11.275 FdM 61.638 • ID 09466

MOORE, R.L. see Vol. V for further entries

MOORE, S.M. [1980] *Nonstandard analysis and generalized functions* (J 0307) Rev Colomb Mat 14*73-94
⋄ H05 ⋄ REV MR 82a:46041 Zbl 453 # 46029 • ID 82291

MOORE, S.M. [1980] *Stochastic fields from stochastic mechanics* (J 0209) J Math Phys 21*2102-2110
⋄ H10 ⋄ REV MR 82f:82015 Zbl 438 # 60089 • ID 82290

MOORE, S.M. [1982] *Nonstandard analysis applied to path integrals and generalized functions (Italian and Russian summaries)* (J 2775) Nuovo Cimento B, Ser 2 70/2*277-290
⋄ C65 H10 H20 ⋄ REV MR 84a:81011 • ID 38856

MOORE JR., L.C. [1972] see HENSON, C.W.

MOORE JR., L.C. [1973] see HENSON, C.W.

MOORE JR., L.C. [1974] see HENSON, C.W.

MOORE JR., L.C. [1974] see COZART, D.

MOORE JR., L.C. [1976] *Hyperfinite extensions of bounded operators on a Hilbert space* (J 0064) Trans Amer Math Soc 218*285-295
⋄ H05 ⋄ REV MR 53 # 6343 Zbl 341 # 47018 • ID 22963

MOORE JR., L.C. [1980] *Hyperfinite-dimensional subspaces of the nonstandard hull of c_0* (J 0053) Proc Amer Math Soc 80*597-603
⋄ H05 ⋄ REV MR 81m:46037 Zbl 472 # 46017 • ID 82287

MOORE JR., L.C. [1981] *Approximately finite-dimensional Banach spaces* (J 2752) J Fct Anal 42*1-11
⋄ H05 ⋄ REV MR 82h:46035 Zbl 513.46055 • ID 82286

MOORE JR., L.C. [1983] see HENSON, C.W.

MOORE JR., L.C. see Vol. III, V for further entries

MOORE VAN DER, W.A. [1968] *A weak ramified type theory* (C 0684) Automation in Lang Transl & Theorem Prov 269-295
⋄ B15 ⋄ REV MR 39 # 1325 Zbl 197.3 • ID 24938

MORDUKHAJ-BOLTOVSKOJ, D. [1927] *Lobachevskij and fundamental logical problems in mathematics (Russian)* (J 2744) Izv Sev-Kavk Nauch Tsentra, Estestv (Rostov nD) 12*78-96
⋄ B28 ⋄ REV FdM 53.45 • ID 45560

MORDUKHAJ-BOLTOVSKOJ, D. [1938] *Sur les syllogismes en logique et les hypersyllogismes en metalogique* (J 4279) Bull Soc Phys Math Kazan, Ser 3 10*161-172
⋄ B20 ⋄ REV Zbl 21.98 JSL 5.120 • ID 45563

MORDUKHAJ-BOLTOVSKOJ, D. see Vol. V for further entries

MORE JR., T. [1971] *An interactive method for algebraic proofs* (P 2335) Canad Math Congr (25);1971 Thunder Bay 129-217
⋄ B35 ⋄ REV MR 57 # 18273 • ID 82296

MORE JR., T. see Vol. V for further entries

MOREL, A.C. [1962] see FRAYNE, T.E.

MOREL, A.C. also published under the name DAVIS, A.C.

MOREL, A.C. see Vol. III, V for further entries

MORGAN, C.G. [1971] *Hypothesis generation by machine* (J 0503) Artif Intell 2*179-187
⋄ B35 ⋄ REV MR 46 # 10256 Zbl 227 # 68045 • ID 31888

MORGAN, C.G. [1973] *On the algorithmic generation of hypotheses* (J 0367) Scientia (Valparaiso) 108*585-598
⋄ B35 ⋄ REV MR 56 # 5114 • ID 31885

MORGAN, C.G. [1973] *Proper definitions in Principia Mathematica* (J 0286) Int Logic Rev 7*80-85
⋄ B05 ⋄ ID 31886

MORGAN, C.G. [1973] *Sentential calculus for logical falsehoods* (J 0047) Notre Dame J Formal Log 14*347-353
⋄ B05 ⋄ REV MR 47 # 8238 Zbl 198.13 • ID 09483

MORGAN, C.G. [1973] *Truth, falsehood, and contigency in first-order predicate calculus* (J 0047) Notre Dame J Formal Log 14*536-542
⋄ A05 B10 ⋄ REV MR 48 # 8195 Zbl 226 # 02008 • ID 09482

MORGAN, C.G. [1976] *Methods for automated theorem proving in nonclassical logics* (J 0187) IEEE Trans Comp C-25*852-862
⋄ B35 F50 ⋄ REV MR 55 # 9619 Zbl 329 # 68074 • ID 27095

MORGAN, C.G. see Vol. II, III, IV, VI for further entries

MORGAN DE, A. [1966] *Logic* (C 0553) Morgan de: On Syllogism & other Log Writings 247-270
⋄ B20 ⋄ REV Zbl 168.1 JSL 41.546 • ID 14489

MORGAN DE, A. [1966] *On the syllogism I: On the structure of syllogism. II: On the symbols of logic, the theory of syllogism, and in particular the copula. III: And on logic in general. IV: And on the logic of relations. V: And on various points of the onymatic system* (C 0553) Morgan de: On Syllogism & other Log Writings 1-21,22-68,74-146,208-246,271-345
⋄ B20 ⋄ REV MR 37 # 6147 Zbl 168.1 JSL 41.546 • ID 15093

MORGAN DE, A. [1966] *On the syllogism and other logical writings* (X 0866) Routledge & Kegan Paul: Henley on Thames xxxi+355pp
⋄ B96 ⋄ REV MR 37 # 6147 • ID 25655

MORGAN DE, A. [1966] *Some suggestions in logical phraseology* (C 0553) Morgan de: On Syllogism & other Log Writings 69-73
⋄ B03 ⋄ REV Zbl 168.1 JSL 41.546 • ID 15095

MORGAN DE, A. [1966] *Syllabus of a proposed system of logic* (C 0553) Morgan de: On Syllogism & other Log Writings 147-207
⋄ B20 ⋄ REV Zbl 168.1 JSL 41.546 • ID 14488

MORGAN DE, A. see Vol. II for further entries

MORICZ, F. [1977] see ECSEDI-TOTH, P.

MORICZ, F. [1980] see ECSEDI-TOTH, P.

MORING, H. [1973] see GIRLING, B.

MORLEY, M.D. [1985] see HARRINGTON, L.A.

MORLEY, M.D. see Vol. III, IV, V for further entries

MOROKHOVEK, YU.E. [1975] see FAL'K, V.N.

MOROKHOVETS, M.K. & VERSHININ, K.P. [1983] *Strategies of the search for derivation of statements with restricted quantifiers (Russian)* (J 0040) Kibernetika, Akad Nauk Ukr SSR 1983/3*9-15
- TRANSL [1983] (J 0021) Cybernetics 19*298-308
- ◇ B35 ◇ REV MR 85c:68080 Zbl 562#68069 • ID 39096

MOROKHOVETS, M.K. [1985] *Procedures for inference search, and transitive relations (Russian) (English summary)* (J 0040) Kibernetika, Akad Nauk Ukr SSR 1985/5*111-115,136
- ◇ B35 ◇ ID 49037

MOROZ, B.Z. [1971] *Formal systems arising in analysis of physical theories (Russian)* (J 0023) Dokl Akad Nauk SSSR 198*1018-1020
- TRANSL [1971] (J 0062) Sov Math, Dokl 12*934-937
- ◇ B30 ◇ REV MR 43#8302 Zbl 235#02022 • ID 27832

MOROZ, B.Z. see Vol. II for further entries

MORREALE, E. [1967] *Partitioned list algorithms for prime implicant determination from canonical forms* (J 4305) IEEE Trans Electr Comp EC-16*611-620
- ◇ B35 ◇ REV Zbl 171.147 • ID 09501

MORRIS, J.B. [1969] *E-resolution: Extensions of resolution to include the equality relation* (P 4250) Int Joint Conf Artif Intell (1);1969 Washington 287-294
- ◇ B35 ◇ ID 46570

MORSCHER, E. [1977] *Betrachtungen zur Praedikatenlogik* (C 1849) Grundzuege & Verfahren der Rechtslog 2*160-188
- ◇ B80 ◇ ID 32392

MORSCHER, E. see Vol. II, IV for further entries

MORTENSEN, C. [1983] see COSTA DA, N.C.A.

MORTENSEN, C. see Vol. II, V, VI for further entries

MORTIMER, M. [1975] *On languages with two variables* (J 0068) Z Math Logik Grundlagen Math 21*135-140
- ◇ B20 ◇ REV MR 53#113 Zbl 343#02009 • ID 09507

MORTIMER, M. see Vol. II, III for further entries

MOSAHEB, G.-H. [1955] *Introduction to formal logic (Iranian)* (X 4502) Univ Tehran: Tehran 709pp
- ◇ B98 ◇ REV JSL 22.354 • ID 42153

MOSCHOVAKIS, J.R. [1969] *A note on k-axiomatizations of identity* (J 0009) Arch Math Logik Grundlagenforsch 12*76-77
- ◇ B10 ◇ REV MR 39#6739 Zbl 193.288 • ID 09511

MOSCHOVAKIS, J.R. see Vol. VI for further entries

MOSHCHENSKIJ, V.A. [1973] *Lectures on mathematical logic (Russian)* (X 1212) Belor Gos Univ: Minsk 159pp
- ◇ B98 D10 D20 ◇ REV MR 50#6775 Zbl 275#02001 • ID 21271

MOSHCHENSKIJ, V.A. [1974] *Abnehmende Folge von Aussagenkalkuelen mit kompliziert werdender Deduktion* (J 0413) Izv Akad Nauk Belor SSR, Ser Fiz-Mat 1974/6*128-129
- ◇ B22 ◇ ID 64048

MOSHCHENSKIJ, V.A. [1975] *Uniqueness of a certain normal form for propositional formulae (Russian)* (J 1516) Vest Ser Fiz Mat Mekh, Univ Minsk 3*73-75,95
- ◇ B05 ◇ REV MR 54#9971 • ID 25597

MOSHCHENSKIJ, V.A. [1977] see KARPOV, V.G.

MOSHCHENSKIJ, V.A. see Vol. II, IV for further entries

MOSS, J.M.B. & SCOTT, D.S. [1975] *Bibliography of logic books* (X 0894) Oxford Univ Pr: Oxford v+106pp
- ◇ A10 B98 ◇ ID 32398

MOSTERIN, J. [1976] *Logica de primer order* (X 1744) Ariel: Esplugas de Llobregat 140pp
- ◇ B98 ◇ ID 31171

MOSTERIN, J. see Vol. III, V for further entries

MOSTOWSKI, A.WLODZIMIERZ & PAWLAK, Z. [1970] *Logik fuer Ingenieure (Polnisch)* (X 1034) PWN: Warsaw 315pp
- ◇ B98 ◇ REV Zbl 213.11 • ID 27979

MOSTOWSKI, A.WLODZIMIERZ [1979] *A note concerning the complexity of a decision problem for positive formulas in SkS* (P 2952) CAAP'79 Arbres en Algeb & Progr (4);1979 Lille 173-180
- ◇ B15 B25 D15 ◇ REV MR 80m:03028 • ID 76601

MOSTOWSKI, A.WLODZIMIERZ [1981] *The complexity of automata and subtheories of monadic second order arithmetics* (P 3165) FCT'81 Fund of Comput Th;1981 Szeged 453-466
- ◇ B15 B25 D05 D15 F35 ◇ REV MR 83g:03036 Zbl 469#03025 • ID 55152

MOSTOWSKI, A.WLODZIMIERZ see Vol. III, IV for further entries

MOSTOWSKI, ANDRZEJ [1947] *On absolute properties of relations* (J 0036) J Symb Logic 12*33-42
- ◇ B15 C75 C85 E15 E47 ◇ REV MR 10.93 Zbl 29.100 JSL 13.46 • ID 09538

MOSTOWSKI, ANDRZEJ [1948] *Logique mathematique. Cours donne a l'universite (Polish)* (S 0257) Monograf Mat 18*viii+388pp
- ◇ B98 ◇ REV MR 10.229 Zbl 34.147 JSL 14.189 • ID 28463

MOSTOWSKI, ANDRZEJ [1949] *Sur l'interpretation geometrique et topologique des notions logiques* (P 0682) Int Congr Philos (10);1948 Amsterdam 767-769
- ◇ B10 C90 ◇ REV MR 10.423 Zbl 31.193 JSL 14.184 • ID 09545

MOSTOWSKI, ANDRZEJ [1951] *On the rules of proof in the pure functional calculus of the first order* (J 0036) J Symb Logic 16*107-111
- ◇ B10 ◇ REV MR 13.3 Zbl 43.9 JSL 16.272 • ID 09548

MOSTOWSKI, ANDRZEJ & RASIOWA, H. [1953] *A geometric interpretation of logical formulae (Polish) (English and Russian summaries)* (J 0063) Studia Logica 1*254-275,301
- ◇ B10 F50 G05 G10 G15 ◇ REV MR 19.378 Zbl 59.14 • ID 16376

MOSTOWSKI, ANDRZEJ [1955] see GRZEGORCZYK, A.

MOSTOWSKI, ANDRZEJ [1956] see LOS, J.

MOSTOWSKI, ANDRZEJ [1961] *Concerning the problem of axiomatizability of the field of real numbers in the weak second order logic* (C 0622) Essays Found of Math (Fraenkel) 269-286
- ◇ B15 B28 C85 ◇ REV MR 28#5003 JSL 32.130 • ID 09570

MOSTOWSKI, ANDRZEJ [1962] *On invariant, dual invariant and absolute formulas* (J 0202) Diss Math (Warsaw) 29*1-38
- ◇ B15 C40 C85 ◇ REV MR 29#5721 Zbl 106.5 • ID 22070

MOSTOWSKI, ANDRZEJ [1965] *Thirty years of foundational studies. Lectures on the development of mathematical logic and the studies of the foundations of mathematics in 1930-1964* (J 0096) Acta Philos Fenn 17*1-180
- REPR [1966] (X 1096) Blackwell: Oxford 180pp
- ◇ A10 B98 C98 D98 E98 F98 ◇ REV MR 33 # 18 MR 33 # 5445 Zbl 146.245 JSL 33.111 • ID 09578

MOSTOWSKI, ANDRZEJ [1979] *An excerpt from the book Mathematical logic* (J 0519) Wiad Mat, Ann Soc Math Pol, Ser 2 22*77-78
- ◇ B98 ◇ REV MR 82c:01043 • ID 82306

MOSTOWSKI, ANDRZEJ [1979] *Foundational studies. Selected works Vol. I,II* (X 0809) North Holland: Amsterdam xlvi+635pp,viii+605pp
- ◇ B96 C96 D96 E96 ◇ REV MR 81i:01018 Zbl 425 # 01021 • ID 82308

MOSTOWSKI, ANDRZEJ see Vol. II, III, IV, V, VI for further entries

MOTOHASHI, N. [1972] *A new theorem on definability in a positive second order logic with countable conjunctions and disjunctions* (J 0081) Proc Japan Acad 48*153-156
- ◇ B15 C40 C75 C85 ◇ REV MR 47 # 4761 Zbl 252 # 02006 • ID 09599

MOTOHASHI, N. [1973] *Model theory on a positive second order logic with countable conjunctions and disjunctions* (J 0090) J Math Soc Japan 25*27-42
- ◇ B15 C40 C75 C85 ◇ REV MR 47 # 6476 Zbl 244 # 02019 • ID 09602

MOTOHASHI, N. [1973] *Two theorems on mix-relativization* (J 0081) Proc Japan Acad 49*161-163
- ◇ B10 C40 C75 ◇ REV MR 49 # 2344 Zbl 289 # 02011 • ID 09600

MOTOHASHI, N. [1978] *An elimination theorem of uniqueness condition* (P 2607) IBM Symp Math Log & Comput Sci (3);
- ◇ B10 C40 ◇ ID 90092

MOTOHASHI, N. [1982] ε-*theorems and elimination theorems of uniqueness conditions* (P 3634) Patras Logic Symp;1980 Patras 373-387
- ◇ B10 C40 F05 ◇ REV MR 84h:03027 Zbl 535 # 03009 • ID 33418

MOTOHASHI, N. [1982] *An axiomatization theorem* (J 0090) J Math Soc Japan 34*551-560
- ◇ B10 C07 C40 C75 ◇ REV MR 83h:03018 Zbl 476 # 03022 • ID 55534

MOTOHASHI, N. [1982] *Elimination theorems of uniqueness conditions* (J 0068) Z Math Logik Grundlagen Math 28*511-524
- ◇ B10 ◇ REV MR 84b:03023 Zbl 528 # 03036 • ID 35620

MOTOHASHI, N. [1984] *A normal form theorem for first order formulas and its application to Gaifman's splitting theorem* (J 0036) J Symb Logic 49*1262-1267
- ◇ B10 C07 C62 ◇ ID 42487

MOTOHASHI, N. [1984] *Equality and Lyndon's interpolation theorem* (J 0036) J Symb Logic 49*123-128
- ◇ B20 C40 C75 ◇ REV MR 86f:03053 Zbl 574 # 03014 • ID 42488

MOTOHASHI, N. see Vol. II, III, V, VI for further entries

MOTT JR., T.H. [1960] *Determination of the irredundant normal forms of a truth function by iterated consensus of the prime implicants* (P 0696) Inform Processing (1);1959 Paris 422
- ◇ B35 ◇ REV MR 23 # A3093 • ID 27694

MOTT JR., T.H. [1960] *Determination of the irredundant normal forms of a truth function by iterated consensus of the prime implicants* (J 0072) IRE Trans Electr Comp EC-9*245-252
- ◇ B35 ◇ REV MR 23 # A3093 JSL 32.541 • ID 09603

MOTT JR., T.H. [1963] see CHANG, D.M.Y.

MOTT JR., T.H. [1965] see CHANG, D.M.Y.

MOTTE, M.-T. [1959] see FEYS, R.

MOTYKA, Z. [1981] see BUGAJSKI, S.

MOUFTAH, H.T. [1975] see JORDAN, I.B.

MOUFTAH, H.T. see Vol. II for further entries

MOURANT, J.A. [1963] *Formal logic: an introductory text book* (X 0843) Macmillan : New York & London 421pp
- ◇ B98 ◇ ID 22524

MOVSISYAN, YU.M. [1975] *Algebraic systems of the second level (Russian)* (S 0166) Mat Issl, Mold SSR 10/2*182-191,285-286
- ◇ B15 C05 ◇ REV MR 53 # 235 Zbl 347 # 08005 • ID 16671

MOZHAROV, R.V. [1965] *One method for testing the completeness of systems of logical functions (Russian)* (J 0011) Avtom Telemekh 26*1644-1645
- TRANSL [1965] (J 0010) Autom & Remote Control 26*1592-1593
- ◇ B05 ◇ REV MR 32 # 9158 Zbl 143.181 JSL 38.343 • ID 09606

MOZZOCHI, C.J. [1976] *Another approach to some recursion theorems of Landau* (J 1519) Kyungpook Math J (Taegu) 16*21-25
- ◇ B28 ◇ REV MR 54 # 9987 Zbl 349 # 04002 • ID 25754

MUCHNIK, A.A. & YANOV, YU.I. [1959] *Existence of k-valued closed classes without a finite basis (Russian)* (J 0023) Dokl Akad Nauk SSSR 127*44-46
- ◇ B30 B50 ◇ REV MR 21 # 7174 Zbl 100.10 JSL 27.247 • ID 00382

MUCHNIK, A.A. [1962] see GINDIKIN, S.G.

MUCHNIK, A.A. [1985] *Games on infinite trees and automata with dead ends. A new proof of decidability for the monadic theory with two successor functions (Russian)* (S 2582) Semiotika & Inf, Akad Nauk SSSR 24*16-40,142
- ◇ B15 B25 C85 D05 ◇ REV Zbl 576 # 03010 • ID 47256

MUCHNIK, A.A. see Vol. II, IV, VI for further entries

MUCHNIK, B.A. [1967] *A criterion for the comparability of bases in the realization of functions of the algebra of logic by formulae (Russian)* (J 0087) Mat Zametki (Akad Nauk SSSR) 1*515-524 • ERR/ADD ibid 3*760
- TRANSL [1967] (J 1044) Math Notes, Acad Sci USSR 1*341-346
- ◇ B05 ◇ REV MR 36 # 2464 Zbl 165.324 • ID 09614

MUCHNIK, B.A. [1968] *Letter to the editors (Russian)* (J 0087) Mat Zametki (Akad Nauk SSSR) 3*760
- ◇ B05 ◇ REV MR 37 # 3906 • ID 09615

MUELLER, D.W. [1969] *Nonstandard proofs of invariance principles in probability theory* (P 0649) Appl Model Th to Algeb, Anal & Probab;1967 Pasadena 186-194
⋄ H05 ⋄ REV MR 39 # 1002   Zbl 248 # 60009 • ID 27628

MUELLER, D.W. [1975] *Endliche Mengen und natuerliche Zahlen* (J 0160) Math-Phys Sem-ber, NS 22*42-45
⋄ A10 B28 E10 ⋄ REV MR 52 # 10425   Zbl 299 # 04001 • ID 21717

MUELLER, D.W. see Vol. IV, V for further entries

MUELLER, GERT H. [1957] *Zur operativen Begruendung von Logik und Mathematik* (J 0170) Ratio (Frankfurt) 1*77-86
• TRANSL [1957] (J 1022) Ratio (Oxford) 1*85-95 (English)
⋄ B30 B60 ⋄ ID 33928

MUELLER, GERT H. [1966] *Definitionen und Postulatensysteme* (J 0178) Stud Gen 19*145-159
⋄ B30 ⋄ REV Zbl 178.304 • ID 33931

MUELLER, GERT H. & OBERSCHELP, A. & POTTHOFF, K. (EDS.) [1975] *ISILC Logic Conference* (X 0811) Springer: Heidelberg & New York iv+651pp
⋄ B97 ⋄ REV MR 52 # 10365   Zbl 307 # 00007 • ID 70183

MUELLER, GERT H. [1980] *Mathematisierung* (C 4082) Handb Wiss Begriffe 401-405
⋄ B30 ⋄ ID 49861

MUELLER, GERT H. [1981] *Framing Mathematics* (J 2386) Epistemologia 4*253-285
⋄ A05 B28 B30 E30 ⋄ ID 49862

MUELLER, GERT H. & TAKEUTI, G. & TUGUE, T. (EDS.) [1981] *Logic symposia, Hakone 1979, 1980. Proceedings* (S 3301) Lect Notes Math 891*xi+394pp
⋄ B97 E97 ⋄ REV MR 83d:03003   Zbl 465 # 00005 • ID 54903

MUELLER, GERT H. & RICHTER, M.M. (EDS.) [1984] *Models and sets* (S 3301) Lect Notes Math 1103*viii+484pp
⋄ B97 C97 D97 H97 ⋄ REV MR 85k:03002a Zbl 547 # 00008 • REM Logic Colloquium;1983 Aachen, Vol.I. Vol.II 1984 by Boerger,E. • ID 41750

MUELLER, GERT H. see Vol. II, III, IV, V, VI for further entries

MUELLER, H.R. [1940] *Algebraischer Aussagenkalkuel* (J 0931) Anz Oesterr Akad Wiss, Math-Nat Kl 149*77-115
⋄ B05 ⋄ REV MR 3.130   Zbl 25.194   JSL 7.126   FdM 66.28 • ID 21092

MUELLER, I. [1979] *The completeness of Stoic propositional logic* (J 0047) Notre Dame J Formal Log 20*201-215
⋄ A10 B20 ⋄ REV MR 80d:03011   Zbl 314 # 02003 • ID 52368

MUELLER, J. [1984] see DILGER, W.

MUELLER, R.K. [1956] *A topological method for the determination of the minimal forms of a boolean function* (J 0072) IRE Trans Electr Comp EC-5*126-132
⋄ B05 ⋄ REV JSL 25.368 • ID 16978

MUKHOPADHYAY, A. [1971] *Complete sets of logic primitives* (C 1202) Rect Devel Switch Th 1-26
⋄ B05 ⋄ REV MR 43 # 4678 • ID 21283

MULLER, D.E. & PREPARATA, F.P. [1970] *Generation of near optimal boolean functions* (J 0119) J Comp Syst Sci 4*93-102
⋄ B05 ⋄ REV Zbl 191.312 • ID 10746

MULLER, D.E. & SCHUPP, P.E. [1981] *Context-free languages, groups, the theory of ends, second-order logic, tiling problems, cellular automata, and vector addition systems* (J 0589) Bull Amer Math Soc (NS) 4*331-334
⋄ B15 B25 D05 D80 ⋄ REV MR 82m:03051 Zbl 484 # 03019 • ID 76648

MULLER, D.E. & SCHUPP, P.E. [1985] *The theory of ends, pushdown automata, and second-order logic* (J 1426) Theor Comput Sci 37*51-75
⋄ B15 B25 D10 D80 ⋄ ID 47646

MULLER, D.E. see Vol. III, IV, V for further entries

MULLIN, A.A. [1958] see KELLNER, W.G.

MULLIN, A.A. see Vol. IV, V, VI for further entries

MULMULEY, K. [1984] *The mechanization of existence proofs of recursive predicates* (P 2633) Autom Deduct (7);1984 Napa 460-475
⋄ B35 ⋄ REV Zbl 546 # 68067 • ID 44454

MUNDICI, D. [1981] *A group-theoretical invariant for elementary equivalence and its role in representations of elementary classes* (J 0063) Studia Logica 40*253-267
⋄ B10 C07 C52 ⋄ REV MR 84e:03045   Zbl 482 # 03013 • ID 90124

MUNDICI, D. [1983] *A lower bound for the complexity of Craig's interpolants in sentential logic* (J 0009) Arch Math Logik Grundlagenforsch 23*27-36
⋄ B20 C40 D15 ⋄ REV MR 85c:03003   Zbl 511 # 03004 • ID 37383

MUNDICI, D. see Vol. II, III, IV, V, VI for further entries

MUNDY, B. [1983] *Relational theories of Euclidean space and Minkowski spacetime* (J 0153) Phil of Sci (East Lansing) 50*205-226
⋄ A05 B30 ⋄ REV MR 84m:03017 • ID 35722

MUNOZ, V. [1961] *De la axiomatica a los sistemas formales* (X 1407) Inst Mat J.Juan: Madrid 80pp
⋄ B30 B98 ⋄ REV MR 27 # 3494 • ID 28705

MURAKAMI, H. & NAKAGIRI, S. & YEH, CHEHCHIH [1977] *Asymptotic behavior of solutions of nonlinear functional equations via nonstandard analysis (Italian summary)* (J 0149) Atti Accad Naz Lincei Fis Mat Nat, Ser 8 62*749-754
⋄ H05 ⋄ REV MR 58 # 6796   Zbl 417 # 34113 • ID 82320

MURAKAMI, H. & NAKAGIRI, S. & YEH, CHEHCHIH [1983] *Asymptotic behavior of solutions of nonlinear differential equations with deviating arguments via nonstandard analysis* (J 0051) Commentat Math, Ann Soc Math Pol, Ser 1 41*203-208
⋄ H10 ⋄ REV MR 85k:34185   Zbl 551 # 34039 • ID 45322

MURATA, H. [1965] see KONDO, M.

MURATA, H. [1970] see KONDO, M.

MUROGA, S. [1980] see CUTLER, R.B.

MUROGA, S. see Vol. II for further entries

MURRAY, N.V. [1982] *Completely non-clausal theorem proving* (J 0503) Artif Intell 18*67-85
⋄ B35 ⋄ REV MR 84g:03020   Zbl 472 # 68053 • ID 55330

MURRAY, N.V. & ROSENTHAL, E. [1985] *Path resolution and semantic graphs* (P 4601) EUROCAL;1985 Linz 2*50-63
⋄ B35 ⋄ ID 49723

MURSKIJ, V.L. [1965] *The existence in three-valued logic of a closed class with finite basis, not having a finite complete system of identities (Russian)* (J 0023) Dokl Akad Nauk SSSR 163*815-818
- TRANSL [1965] (J 0062) Sov Math, Dokl 6*1020-1024
- ◊ B22 B50 ◊ REV MR 32 # 3998 Zbl 154.255 JSL 37.762
- • ID 09654

MURSKIJ, V.L. see Vol. III, IV for further entries

MUSES, C. [1977] *Applied hypernumbers. Computational concepts* (J 2662) Appl Math Comp 3*211-226
- ◊ B75 H05 ◊ REV MR 58 # 16813 Zbl 359 # 10050 • REM Part I. Part II 1978 • ID 48949

MUSES, C. [1978] *Hypernumbers. II. Further concepts and computational applications* (J 2662) Appl Math Comp 4*45-66
- ◊ A05 H05 ◊ REV MR 80a:03007 Zbl 377 # 10029 • REM Part I 1977 • ID 76677

MUSSER, D.R. [1980] see ERICKSON, R.W.

MUZIO, J.C. [1974] see MILLER, D.MICHAEL

MUZIO, J.C. [1976] *A complete classification of three-place functors in two valued logic* (J 0047) Notre Dame J Formal Log 17*429-437
- ◊ B05 ◊ REV MR 57 # 9470 Zbl 258 # 02010 • ID 18306

MUZIO, J.C. [1978] *A note concerning a sole sufficient operator* (J 0047) Notre Dame J Formal Log 19*419-420
- ◊ B20 ◊ REV MR 58 # 10303 Zbl 323 # 02017 • ID 51482

MUZIO, J.C. [1979] *Classes of universal decision elements using negative substitutions* (J 0047) Notre Dame J Formal Log 20*314-320
- ◊ B05 B70 ◊ REV MR 80h:03016 Zbl 314 # 02012
- • ID 52621

MUZIO, J.C. see Vol. II for further entries

MYCIELSKI, J. [1971] see MCKENZIE, R.

MYCIELSKI, J. [1981] *Finitistic real analysis* (S 2824) Real Anal Exchange 6*127-130
- ◊ B28 F65 ◊ REV Zbl 453 # 03064 • ID 54194

MYCIELSKI, J. [1983] *The meaning of the conjecture $P \ne NP$ for mathematical logic* (J 0005) Amer Math Mon 90*129-130
- ◊ B30 D15 ◊ REV MR 85f:03001 Zbl 513 # 03028
- • ID 37228

MYCIELSKI, J. [1985] *Sullivan's lamination of a planar region* (J 3370) Enseign Math, Ser 2 31*67-70
- ◊ H05 ◊ ID 48054

MYCIELSKI, J. see Vol. III, IV, V, VI for further entries

MYERS, D.L. [1983] see HANF, W.P.

MYERS, D.L. see Vol. III, IV, V for further entries

MYHILL, J.R. [1950] *A complete theory of natural, rational, and real numbers* (J 0036) J Symb Logic 15*185-196,IX
- ◊ B28 ◊ REV MR 12.579 Zbl 41.342 JSL 16.65 • ID 09700

MYHILL, J.R. [1950] *A reduction in the number of primitive ideas of arithmetic* (J 0036) J Symb Logic 15*130
- ◊ B28 ◊ REV MR 12.233 Zbl 38.6 JSL 16.74 • ID 09701

MYHILL, J.R. [1950] *A system which can define its own truth* (J 0027) Fund Math 37*190-192
- ◊ B30 D20 F25 F30 ◊ REV MR 13.97 Zbl 41.150 JSL 21.319 • ID 09699

MYHILL, J.R. [1951] *Report on some investigations concerning the consistency of the axiom of reducibility* (J 0036) J Symb Logic 16*35-42
- ◊ B15 F35 ◊ REV MR 12.791 Zbl 42.8 JSL 16.217
- • ID 09702

MYHILL, J.R. [1952] *A derivation of number theory from ancestral theory* (J 0036) J Symb Logic 17*192-197
- ◊ B28 ◊ REV MR 14.527 Zbl 47.14 JSL 18.77 • ID 09706

MYHILL, J.R. [1953] *On the interpretation of the sign " $\to$ "* (J 0036) J Symb Logic 18*60-62
- ◊ B20 ◊ REV MR 14.936 Zbl 50.246 JSL 20.178
- • ID 09708

MYHILL, J.R. [1971] *Embedding classical type theory in "intuitionistic" type theory* (P 0693) Axiomatic Set Th;1967 Los Angeles 1*267-270 • ERR/ADD ibid 2*185-188
- ◊ B15 F35 F50 ◊ REV MR 43 # 7298 Zbl 219 # 02043
- • ID 28094

MYHILL, J.R. [1974] *Embedding classical type theory in "intuitionistic" type theory, a correction* (P 0693) Axiomatic Set Th;1967 Los Angeles 2*185-188
- ◊ B15 F35 F50 ◊ REV MR 43 # 7298 Zbl 219 # 02043
- • REM This is not simply a correction of the 1971 paper, it is a completely new paper • ID 48354

MYHILL, J.R. see Vol. II, III, IV, V, VI for further entries

NABEBIN, A.A. [1977] *Expressibility in restricted second-order arithmetic (Russian)* (J 0092) Sib Mat Zh 18*830-837,957
- TRANSL [1977] (J 0475) Sib Math J 18*588-593
- ◊ B15 C40 D05 F35 ◊ REV MR 58 # 16250 Zbl 385 # 03047 • ID 52155

NABEBIN, A.A. see Vol. IV for further entries

NADIU, G.S. [1969] *The interpolation theorem in strict positive logic* (J 0070) Bull Soc Sci Math Roumanie, NS 13(61)*185-193
- ◊ B20 C40 ◊ REV MR 47 # 23 Zbl 204.310 • ID 09749

NADIU, G.S. see Vol. III, IV, VI for further entries

NAEMURA, K. [1963] see IBUKI, K.

NAESS, A. [1969] *Introduction to logic and scientific method* (X 1233) Bedminster Pr: Totowa
- ◊ B98 ◊ ID 22525

NAGASAKA, A. [1982] see GREENBAUM, S.

NAGATA, M. [1974] see IWAMARU, Y.

NAGATA, M. & NAKANISHI, M. & NISHIMURA, T. [1975] *TKP1.2 - the extension of TKP1 by adding some facilities* (J 0350) Sci Rep Tokyo Kyoiku Daigaku Sect A 19*82-89
- ◊ B35 F35 ◊ REV MR 52 # 16171 Zbl 335 # 68063
- • ID 21892

NAGATA, M. see Vol. II for further entries

NAGATA, S. [1979] see FUNAHASHI, S.

NAGATA, S. [1981] see FUNAHASHI, S.

NAGATA, S. [1984] see FUNAHASHI, S.

NAGATA, S. see Vol. II for further entries

NAGEL, E. [1934] see COHEN, M.R.

NAGEL, E. & NEWMAN, J.R. [1958] *Goedel's proof* (X 0924) New York Univ Pr: New York ix+118pp
- ◊ B28 D35 F30 F98 ◊ REV MR 20 # 1625 Zbl 86.246 JSL 24.222 • ID 09765

NAGEL, E. & NEWMAN, J.R. [1961] *Discussion: Putnam's review of Goedel's proof* (J 0153) Phil of Sci (East Lansing) 28*209-211
⋄ B28 F30 ⋄ REV MR 24#A5 • ID 09768

NAGEL, E. & SUPPES, P. & TARSKI, A. (EDS.) [1962] *Logic, methodology and philosophy of science* (X 1355) Stanford Univ Pr: Stanford ix+661pp
⋄ B97 ⋄ REV Zbl 128.241 • REM Reprinted 1965 • ID 32720

NAGEL, E. see Vol. II, IV, VI for further entries

NAGORNYJ, N.M. [1976] *Normal algorithms and first order languages (Russian)* (S 0554) Issl Teor Algor & Mat Logik (Moskva) 2*46-50,157
⋄ B10 D03 ⋄ REV MR 58#16205 • ID 76745

NAGORNYJ, N.M. see Vol. IV, VI for further entries

NAGUMO, M. [1977] *Quantities and real numbers* (J 0351) Osaka J Math 14*1-10
⋄ B28 ⋄ REV MR 57#9677 Zbl 379#02019 • ID 51852

NAKAGIRI, S. [1977] see MURAKAMI, H.

NAKAGIRI, S. [1983] see MURAKAMI, H.

NAKAMURA, A. [1970] *On a propositional calculus whose decision problem is recursively unsolvable* (J 0111) Nagoya Math J 38*145-152
⋄ B22 D35 ⋄ REV MR 41#3277 Zbl 202.13 • ID 09788

NAKAMURA, A. [1975] see INOUE, K.

NAKAMURA, A. & ONO, H. [1981] *Undecidability of extensions of the monadic first-order theory of successor and two-dimensional finite automata* (P 3201) Logic Symposia;1979/80 Hakone 155-174
⋄ B15 B25 D05 D35 ⋄ REV MR 83k:03054 Zbl 474#03023 • ID 55427

NAKAMURA, A. see Vol. II, III, IV, VI for further entries

NAKANISHI, M. [1974] see IWAMARU, Y.

NAKANISHI, M. [1975] see NAGATA, M.

NAKANISHI, M. see Vol. II, VI for further entries

NAQVI, S.A. [1982] see HENSCHEN, L.J.

NARENS, L. [1971] *A nonstandard proof of the Jordan curve theorem* (J 0048) Pac J Math 36*219-229
⋄ H05 ⋄ REV MR 43#2680 Zbl 212.554 • ID 09808

NARENS, L. [1972] *Topologies of closed subsets* (J 0064) Trans Amer Math Soc 174*55-76
⋄ H05 ⋄ REV MR 47#1007 Zbl 258#54006 • ID 09809

NARENS, L. [1974] *Field embeddings of generalized metric spaces* (P 1083) Victoria Symp Nonstand Anal;1972 Victoria 155-170
⋄ H05 ⋄ REV MR 58#31037 Zbl 276#54029 • ID 27629

NARENS, L. [1974] *Homeomorphism types of generalized metric spaces* (P 1083) Victoria Symp Nonstand Anal;1972 Victoria 171-179
⋄ H05 ⋄ REV MR 57#16061 Zbl 276#54030 • ID 76762

NARENS, L. [1974] *Measurement without Archimedean axioms* (J 0153) Phil of Sci (East Lansing) 41*374-393
⋄ H05 ⋄ REV MR 57#101 • ID 76763

NARENS, L. see Vol. II, III for further entries

NARUSHIMA, H. [1976] see FUJITA, H.

NARUSHIMA, H. see Vol. IV, V for further entries

NATORP, P. [1904] *Logik. Grundlegung und logischer Aufbau der Mathematik und mathematischen Naturwissenschaft* (X 1263) Elwert: Marburg 71pp
⋄ B98 ⋄ REV FdM 35.79 • ID 25563

NATORP, P. [1910] *Die logischen Grundlagen der exakten Wissenschaften* (X 0823) Teubner: Stuttgart xx+416pp
⋄ A05 B30 ⋄ REV FdM 41.81 • ID 25564

NAZAROV, A.G. [1970] *Certain properties of transfinite real numbers (Russian) (Armenian summary)* (J 0346) Dokl Akad Nauk Armyan SSR 51*273-282
⋄ B28 ⋄ REV MR 48#8239 Zbl 215.329 • ID 09826

NAZAROV, A.G. [1970] *Transfinite numbers, endowed with properties of real numbers (Russian) (Armenian summary)* (J 0346) Dokl Akad Nauk Armyan SSR 51*150-156
⋄ B28 ⋄ REV MR 48#8238 Zbl 215.329 • ID 09825

NAZAROV, A.G. [1971] *A transfinite number line (Russian) (Armenian summary)* (J 0346) Dokl Akad Nauk Armyan SSR 52*151-156
⋄ B28 ⋄ REV MR 48#1929 Zbl 222#04002 • ID 76774

NAZARYAN, G.A. [1972] *Certain estimates of the realization of Boolean functions in algorithmic languages (Russian) (Armenian summmary)* (J 0346) Dokl Akad Nauk Armyan SSR 55*129-133
⋄ B05 B75 ⋄ REV MR 48#12893 Zbl 255#68013 • ID 64166

NAZARYAN, G.A. [1975] *On the complexity of frequency computations and extensions of boolean functions (Russian) (Armenian summary)* (J 0346) Dokl Akad Nauk Armyan SSR 61*70-75
⋄ B05 B75 ⋄ REV MR 55#11712 Zbl 334#68030 • ID 64165

NAZARYAN, G.A. [1975] *On the realization of boolean functions in algorithmic languages (Russian)* (S 0422) Tr Vychisl Tsentra Akad Nauk Armyan SSR & Univ Erevan 8*36-56
⋄ B05 B75 D03 ⋄ REV MR 56#5243 Zbl 426#03063 • ID 53659

NAZARYAN, G.A. [1976] *Complexity classes of sets of boolean functions (Russian) (Armenian summary)* (J 0346) Dokl Akad Nauk Armyan SSR 63*257-263
⋄ B05 D15 ⋄ REV MR 58#10359 Zbl 356#94050 • ID 76769

NAZARYAN, G.A. [1976] *Some comparative complexity characteristics of computations of boolean functions (Russian) (English and Armenian summaries)* (J 0312) Izv Akad Nauk Armyan SSR, Ser Mat 11*355-365,368
⋄ B35 ⋄ REV MR 55#10263 Zbl 356#02030 • ID 50302

NAZARYAN, G.A. [1978] *Ueber eine Synthese von Algorithmen approximativ berechenbarer boolescher Funktionen (Russian)* (J 0346) Dokl Akad Nauk Armyan SSR 66*15-21
⋄ B35 D15 D20 ⋄ REV MR 80c:65040 Zbl 383#94036 • ID 52040

NAZARYAN, G.A. [1979] *Ueber disjunkte Zerlegungen von Mengen boolescher Funktionen (Russian)* (J 0346) Dokl Akad Nauk Armyan SSR 69*88-94
⋄ B05 D15 ⋄ REV Zbl 421#03027 • ID 53426

NAZARYAN, G.A. [1980] *The relations of some complexity characteristics of sets of boolean functions (Russian) (Armenian summary)* (J 0346) Dokl Akad Nauk Armyan SSR 71*217-220
- ◊ B05 D15 ◊ REV MR 82g:03073 Zbl 473 # 68035
- • ID 76768

NAZARYAN, G.A. [1982] *Realization of Boolean functions in algorithmic languages under constraints on the running time of the algorithms (Russian) (Armenian summary)* (S 0422) Tr Vychisl Tsentra Akad Nauk Armyan SSR & Univ Erevan 10*41-52
- ◊ B05 D15 ◊ REV MR 85a:03076 • ID 34748

NECHIPORUK, EH.I. [1964] *On the synthesis of logical nets in incomplete and degenerate basis (Russian)* (J 0023) Dokl Akad Nauk SSSR 155*299-301
- • TRANSL [1964] (J 0470) Sov Phys, Dokl 9*207-208
- ◊ B05 ◊ REV MR 29 # 1106 Zbl 168.261 JSL 36.548
- • ID 45284

NECHIPORUK, EH.I. [1966] *A boolean function (Russian)* (J 0023) Dokl Akad Nauk SSSR 169*765-766
- • TRANSL [1966] (J 0062) Sov Math, Dokl 7*999-1000
- ◊ B05 ◊ REV MR 36 # 1237 Zbl 161.9 • ID 19489

NECHIPORUK, EH.I. [1969] *On a boolean matrix (Russian)* (J 0052) Probl Kibern 21*237-240
- • TRANSL [1969] (J 0471) Syst Th Res 21*236-239
- ◊ B05 ◊ REV MR 43 # 7265 Zbl 214.261 • ID 09833

NECHIPORUK, EH.I. see Vol. II for further entries

NECULA, N.N. [1967] *A numerical procedure for determination of the prime implicants of a boolean function* (J 4305) IEEE Trans Electr Comp EC-16*687-689
- ◊ B35 ◊ REV Zbl 171.148 • ID 09834

NEDERPELT, R.P. [1977] *Presentation of natural deduction* (P 3269) Set Th Found Math (Kurepa);1977 Beograd 115-126
- ◊ B10 F07 ◊ REV MR 58 # 16177 Zbl 415 # 03008
- • ID 53110

NEDERPELT, R.P. [1980] *An approach to theorem proving on the basis of a typed lambda-calculus* (P 3063) Autom Deduct (5);1980 Les Arcs 182-194
- ◊ B35 B40 F60 ◊ REV Zbl 438 # 68053 • ID 55989

NEDZYNSKI, T.G. [1979] *Quantification, domains of discourse, and existence* (J 0047) Notre Dame J Formal Log 20*130-140
- ◊ A05 B10 ◊ REV MR 80f:03011 Zbl 323 # 02019
- • ID 52379

NEGRU, I.S. [1972] *Certain sublattices of the lattice of all propositional logics with the ordinary concept of formula (Russian)* (S 0166) Mat Issl, Mold SSR 7/4(26)*174-196,257
- ◊ B22 ◊ REV MR 51 # 00048 Zbl 275 # 02052 • ID 24787

NEGRU, I.S. [1973] *Certain cardinality properties of the lattice of primitive logics (Russian)* (S 0166) Mat Issl, Mold SSR 3(29)*161-166,184
- ◊ B20 ◊ REV MR 51 # 2855 Zbl 312 # 02047 • ID 17392

NEGRU, I.S. [1974] *On algebraic properties of the totality of propositional logics (Russian)* (J 0023) Dokl Akad Nauk SSSR 218*1276-1279
- • TRANSL [1974] (J 0062) Sov Math, Dokl 15*1491-1496
- ◊ B22 ◊ REV MR 52 # 2865 Zbl 312 # 02046 • ID 64178

NEGRU, I.S. see Vol. II for further entries

NEIDORF, R. [1967] *Deductive forms: An elementary logic* (X 0837) Harper & Row: New York xiii+407pp
- ◊ B98 ◊ REV Zbl 202.5 • ID 22526

NELSON, D. [1966] *Non-null implication* (J 0036) J Symb Logic 31*562-572
- ◊ B20 B28 F50 ◊ REV MR 35 # 26 Zbl 148.244 JSL 33.129 • ID 09846

NELSON, D. see Vol. II, VI for further entries

NELSON, EDWARD [1977] *Internal set theory: A new approach to nonstandard analysis* (J 0015) Bull Amer Math Soc 83*1165-1198
- ◊ E70 H05 ◊ REV MR 57 # 9544 Zbl 373 # 02040 JSL 48.1203 • ID 51504

NELSON, EVELYN [1977] *Classes defined by implications* (J 0004) Algeb Universalis 7*405-407
- ◊ B20 C05 C52 ◊ REV MR 56 # 218 Zbl 378 # 08004
- • ID 26606

NELSON, EVELYN see Vol. III, IV, V for further entries

NELSON, GEORGE C. [1973] *Nonconstructivity of models of the reals (Russian summary)* (J 0014) Bull Acad Pol Sci, Ser Math Astron Phys 21*1067-1071
- ◊ C62 C65 D55 H05 H15 ◊ REV MR 51 # 10066 Zbl 293 # 02042 • ID 17539

NELSON, GEORGE C. see Vol. III, IV, VI for further entries

NELSON, GREG [1984] *Combining satisfiability procedures by equality-sharing* (P 3084) Autom Theor Prov After 25 Yea;1983 Denver 201-211
- ◊ B25 B35 ◊ REV MR 85d:68005 Zbl 564 # 03011
- • ID 45270

NELSON, GREG see Vol. IV for further entries

NELSON, JACK [1980] see BERGMANN, MERRIE

NELSON, RAYMOND J. [1955] *Simplest normal truth functions* (J 0036) J Symb Logic 20*105-108,627-631
- ◊ B05 ◊ REV MR 17.224 Zbl 66.256 JSL 21.328
- • ID 20812

NELSON, RAYMOND J. [1955] *Weak simplest normal truth functions* (J 0036) J Symb Logic 20*232-234
- ◊ B05 ◊ REV MR 17.701 Zbl 67.249 JSL 21.330
- • ID 09855

NELSON, RAYMOND J. see Vol. IV for further entries

NEMESSZEGHY, E.A. & NEMESSZEGHY, E.Z. [1971] *Is $(p \supset q) = (\sim p \vee q)$ Df. a proper definition in the system of Principia Mathematica?* (J 0094) Mind 80*282-283
- ◊ B05 ◊ REV MR 56 # 11740 • ID 76802

NEMESSZEGHY, E.A. [1976] *Note on an independence proof of Johansson* (J 0047) Notre Dame J Formal Log 17*438
- ◊ B20 ◊ REV MR 56 # 2791 Zbl 331 # 02011 • ID 18309

NEMESSZEGHY, E.Z. [1971] see NEMESSZEGHY, E.A.

NEMETI, I. [1973] see ANDREKA, H.

NEMETI, I. [1974] see ANDREKA, H.

NEMETI, I. [1975] see ANDREKA, H.

NEMETI, I. [1982] see ANDREKA, H.

NEMETI, I. see Vol. II, III, V for further entries

NEPEJVODA, N.N. [1979] *Some remarks about constructive non-standard analysis (Russian)* (S 2579) Teor Mnozhestv & Topol (Izhevsk) 2*74-76
⋄ F99 H05 ⋄ REV Zbl 496 # 03042 • ID 36869

NEPEJVODA, N.N. see Vol. II, III, IV, V, VI for further entries

NERODE, A. & SHORE, R.A. [1980] *Second order logic and first order theories of reducibility orderings* (P 2058) Kleene Symp;1978 Madison 181-200
⋄ B10 B15 D30 D35 F35 F40 ⋄ REV MR 82g:03078 Zbl 465 # 03024 • ID 54927

NERODE, A. see Vol. III, IV, V, VI for further entries

NESGOVOROVA, G.P. [1974] *Ueberfuehrung von Formeln des Praedikatenkalkuels erster Ordnung in die Praefixform (Russisch)* (S 0507) Vychisl Sist (Akad Nauk SSSR Novosibirsk) 59*139-150
⋄ B10 ⋄ REV Zbl 312 # 02010 • ID 29093

NEUBRUNN, T. & RIECAN, B. & RIECANOVA, Z. [1984] *An elementary approach to some applications of nonstandard analysis* (P 4043) Winter School on Abstract Anal (11);1983 Zelezna Ruda 197-200
⋄ H10 ⋄ REV MR 85j:46138 Zbl 548 # 46061 • ID 45143

NEUBRUNN, T. & RIECANOVA, Z. [1984] *Elementary nonstandard approach to metric spaces* (J 0128) Acta Math Univ Comenianae (Bratislava) 44-45*301-312
⋄ H05 ⋄ REV MR 86b:54030 Zbl 565 # 54020 • ID 45426

NEUBRUNN, T. see Vol. II, V for further entries

NEUBUESER, J. [1970] *Investigations of groups on computers* (P 0690) Comput Prob in Abstr Algeb;1967 Oxford 1-19
⋄ B35 ⋄ REV MR 41 # 5480 Zbl 191.19 • ID 09891

NEUMANN, H. [1981] see LUDWIG, G.

NEUMANN, H. see Vol. II, III, IV for further entries

NEUMANN, O. [1984] *On the definition of ordered n-tuples* (J 0068) Z Math Logik Grundlagen Math 30*197-198
⋄ B15 E20 ⋄ REV MR 86d:03048 Zbl 559 # 03031 • ID 42222

NEUMANN VON, J. [1928] *Die Axiomatisierung der Mengenlehre* (J 0044) Math Z 27*669-752
⋄ B28 E30 E70 ⋄ REV FdM 54.88 • ID 09902

NEUMANN VON, J. [1961] *Collected works. Vol.1. Logic, theory of sets and quantum mechanics* (X 0869) Pergamon Pr: Oxford x+654pp
⋄ B96 E96 ⋄ REV MR 28 # 1100 Zbl 100.2 • ID 25726

NEUMANN VON, J. [1964] *The formalist foundations of mathematics* (C 1105) Phil of Math. Sel Readings 50-54
⋄ B30 ⋄ ID 38053

NEUMANN VON, J. see Vol. II, IV, V, VI for further entries

NEVINS, A.J. [1975] *Plane geometry theorem proving using forward chaining* (J 0503) Artif Intell 6*1-23
• TRANSL [1979] (J 3079) Kiber Sb Perevodov, NS 16*145-170 (Russian)
⋄ B35 ⋄ REV Zbl 301 # 68086 • ID 69737

NEWELL, A. & SHAW, J.C. & SIMON, H.A. [1957] *Empirical explorations of the logic theory machine: a case study in heuristic* (P 1178) West Joint Comput Conf;1957 Los Angeles 218-230
• REPR [1963] (C 4307) Comput & Thought 109-133
[1983] (C 4659) Autom of Reasoning 1*49-74
⋄ B35 ⋄ REV JSL 27.102 • ID 12334

NEWELL, A. & SHAW, J.C. & SIMON, H.A. [1960] *Report on general problem-solving program* (P 0696) Inform Processing (1);1959 Paris 256-264
⋄ B35 ⋄ REV MR 23 # B2139 Zbl 112.85 • ID 27690

NEWMAN, J.R. [1958] see NAGEL, E.

NEWMAN, J.R. [1961] see NAGEL, E.

NEWMAN, J.R. see Vol. IV, VI for further entries

NEWMAN, M.H.A. & TURING, A.M. [1942] *A formal theorem in Church's theory of types* (J 0036) J Symb Logic 7*28-33
⋄ B15 ⋄ REV MR 3.290 JSL 7.122 • ID 09931

NEWMAN, M.H.A. [1942] *On theories with a combinatorial definition of "equivalence"* (J 0120) Ann of Math, Ser 2 43*223-243
⋄ B10 B40 ⋄ REV MR 4.126 Zbl 60.125 JSL 7.123 • ID 09934

NEWMAN, M.H.A. [1943] *Stratified systems of logic* (J 0171) Proc Cambridge Phil Soc Math Phys 39*69-83
⋄ B15 ⋄ REV MR 4.182 Zbl 60.21 JSL 9.50 • ID 09935

NEWMAN, M.H.A. see Vol. V, VI for further entries

NGUEN VAN TIN' [1975] see CHUBARYAN, A.A.

NICKLES, T. (ED.) [1980] *Scientific discovery, logic, and rationality* (X 0835) Reidel: Dordrecht xii+385pp
⋄ A05 B97 ⋄ REV MR 82g:03006 • ID 70012

NICOD, J. [1916] *A reduction in the number of primitive propositions of logic* (J 0171) Proc Cambridge Phil Soc Math Phys 19*32-41
⋄ B05 ⋄ ID 09940

NICOD, J. [1924] *Geometry in the sensible world (French)* (X 1071) Alcan:Paris
• LAST ED [1962] (X 0840) Pr Univ France: Paris
⋄ A05 B80 F99 ⋄ REV MR 42 # 33 Zbl 242 # 00023 FdM 50.30 • REM For transl. see Nicod,J. 1930 • ID 25431

NICOD, J. [1930] *Geometry and induction, containing "Geometry in the sensible world", and "The logical problem of induction"* (X 0866) Routledge & Kegan Paul: Henley on Thames xx+245pp
• LAST ED [1970] (X 0926) Univ Calif Pr: Berkeley 245pp
⋄ A05 B30 F98 ⋄ REV MR 42 # 33 Zbl 242 # 00023
• REM Transl. of Nicod,J. 1924 (both) • ID 09941

NICOD, J. see Vol. II for further entries

NIDDITCH, P.H. [1957] *Introductory formal logic of mathematics* (X 0927) Univ Tutorial Pr: Slough vii+188pp
⋄ B98 ⋄ REV MR 19.723 Zbl 82.13 JSL 25.77 • ID 09944

NIDDITCH, P.H. [1962] *Propositional calculus* (X 0866) Routledge & Kegan Paul: Henley on Thames 383pp
⋄ B05 ⋄ REV MR 26 # 6042 JSL 30.357 • ID 09948

NIELAND, J.J.F. [1967] *Beth's tableau-method* (P 0683) Log & Founds of Sci (Beth);1964 Paris 19-38
⋄ B05 B25 B50 F07 ⋄ REV MR 39 # 1287 Zbl 203.9 • ID 09951

NIELAND, J.J.F. see Vol. II for further entries

NIETHAMMER, W. & VEENKER, G. [1969] *Maschinen und mathematische Beweise* (J 0160) Math-Phys Sem-ber, NS 16*170-188
⋄ B35 ⋄ REV MR 42 # 4335 Zbl 175.282 • ID 46580

NIGMATULLIN, R.G. [1976] *Computation of boolean functions on digital machines (Russian)* (J 2605) Programmirovanie 1976/6*26-34
  ⋄ TRANSL [1976] (J 2604) Progr Comput Software 2*432-439
  ⋄ B35 ⋄ REV Zbl 397 # 68060 • ID 69739

NIGMATULLIN, R.G. see Vol. IV for further entries

NIINILUOTO, I. [1973] *Mathematical logic (Finnish)* (X 3333) Unknown Publisher: See Remarks 94+58pp
  ⋄ B98 ⋄ ID 32404

NIINILUOTO, I. [1973] see HINTIKKA, K.J.J.

NIINILUOTO, I. [1977] see HILPINEN, R.

NIINILUOTO, I. see Vol. II, V for further entries

NILSSON, N.J. [1971] see LUCKHAM, D.C.

NILSSON, N.J. [1971] see FIKES, R.E.

NILSSON, N.J. [1979] *A production system for automatic deduction* (J 0508) Machine Intelligence 9*101-126
  ⋄ B35 ⋄ REV MR 81g:68125 • ID 82357

NILSSON, N.J. [1980] *Principles of artificial intelligence* (X 2846) Tioga Publ Co: Palo Alto xv+476pp
  ⋄ B35 ⋄ REV MR 81j:68106 Zbl 422 # 68039 • ID 69740

NINOMIYA, I. [1955] *On the number of types of symmetric boolean output matrices* (J 0534) Mem Fac Engin, Nagoya Univ 7*115-124
  ⋄ B05 B70 ⋄ REV JSL 25.185 • ID 09967

NINOMIYA, T. [1977] see GOTO, M.

NINOMIYA, T. [1978] see GOTO, M.

NINOMIYA, T. [1979] see GOTO, M.

NINOMIYA, T. [1981] see GOTO, M.

NINOMIYA, T. see Vol. II for further entries

NISHIMURA, T. [1960] see MAEHARA, S.

NISHIMURA, T. [1974] see IWAMARU, Y.

NISHIMURA, T. [1975] see NAGATA, M.

NISHIMURA, T. see Vol. II, III, IV, V, VI for further entries

NOBIS, K. & WOZNIAK, C. [1982] *Nonstandard analysis and balance equations in the theory of porous media (Russian summary)* (J 2677) Bull Acad Pol Sci, Ser Sci Tech 29*213-218
  ⋄ H10 ⋄ REV MR 83m:03078b Zbl 502 # 73004 • ID 35475

NOBIS, K. [1984] *On the application of nonstandard analysis in mechanics of porous media* (J 2677) Bull Acad Pol Sci, Ser Sci Tech 32*383-387
  ⋄ H10 ⋄ REV Zbl 551 # 73008 • ID 44926

NOBIS, K. & WIERZBICKI, E. & WOZNIAK, C. [1984] *On the physical interpretation of nonstandard methods in mechanics* (J 2677) Bull Acad Pol Sci, Ser Sci Tech 32*379-382
  ⋄ H10 ⋄ REV Zbl 551 # 73007 • ID 44925

NOF, S.Y. [1980] see BULLERS, W.I.

NOGUCHI, S. & TOGASHI, A. [1984] *A finite termination problem for term rewriting systems (Japanese) (English summary)* (J 0979) Denshi Tsushin Gakkai Ronbunshi, Sect A-D 67*753-760
  ⋄ TRANSL [1984] (J 0464) Syst-Comp-Controls 15/6*27-35
  ⋄ B35 ⋄ REV MR 86d:68042 • ID 48880

NOJIMA, S. [1976] see FUJITA, H.

NOJIMA, S. see Vol. IV for further entries

NOLIN, L. [1958] *Sur l'algebre des predicats* (P 0576) Raisonn en Math & Sci Exper;1955 Paris 33-37
  ⋄ B10 G15 ⋄ REV MR 21 # 3357 Zbl 85.244 JSL 24.235 • ID 09990

NOLIN, L. [1958] *Sur les proprietes communes aux calculs des propositions et a leurs modeles* (J 0068) Z Math Logik Grundlagen Math 4*101-107
  ⋄ B22 G25 ⋄ REV Zbl 108.1 • ID 09989

NOLIN, L. [1958] *Sur un systeme de "deduction naturelle"* (J 0109) C R Acad Sci, Paris 246*1128-1131
  ⋄ B10 ⋄ REV MR 20 # 2273 Zbl 83.2 • ID 09988

NOLIN, L. [1970] see LACOMBE, D.

NOLIN, L. [1971] *Types a deux variables (methode de Lindenbaum)* (C 1566) Semin IRIA Log & Automates 1971 151-167
  ⋄ B20 ⋄ REV Zbl 268 # 02021 • ID 64224

NOLIN, L. see Vol. IV, VI for further entries

NOLL, H. [1977] see BERGMANN, E.

NOLL, H. [1980] *A note on resolution: How to get rid of factoring without loosing completeness* (P 3063) Autom Deduct (5);1980 Les Arcs 250-263
  ⋄ B35 ⋄ REV MR 81k:68070 Zbl 438 # 68051 • ID 55987

NORGELA, S.A. [1976] *On approximating reduction classes of CPC by decidable classes (Russian) (English summary)* (S 0228) Zap Nauch Sem Leningrad Otd Mat Inst Steklov 60*103-108,224
  • TRANSL [1980] (J 1531) J Sov Math 14*1493-1496
  ⋄ B20 B25 D35 ⋄ REV MR 58 # 27422 Zbl 342 # 02033
  • ID 32671

NORGELA, S.A. [1977] see MASLOV, S.YU.

NORGELA, S.A. [1977] *On recursive nonseparability of the strategies of deduction-search in the classical predicate calculus (Russian)* (J 2574) Litov Mat Sb (Vil'nyus) 17/3*146-147
  ⋄ B35 D35 ⋄ ID 32680

NORGELA, S.A. [1978] *Herbrand strategies of deduction-search in predicate calculus I (Russian) (English and Lithuanian summaries)* (J 2574) Litov Mat Sb (Vil'nyus) 18/4*95-100,201
  • TRANSL [1978] (J 3283) Lith Math J 18*513-517
  ⋄ B25 B35 D35 F07 ⋄ REV MR 80e:03050 Zbl 394 # 03017 • REM Part II 1979 • ID 32674

NORGELA, S.A. [1979] *Herbrand's strategies of deduction-search in predicate calculus II (Russian) (Lithuanian summary)* (J 2574) Litov Mat Sb (Vil'nyus) 19/1*161-167,232
  • TRANSL [1979] (J 3283) Lith Math J 19*113-117
  ⋄ B35 ⋄ REV MR 81b:03013 Zbl 433 # 03007 • REM Part I 1978 • ID 32675

NORGELA, S.A. [1980] *On approximation of some classes of classical predicate calculus by decidable classes. I,II (Russian) (English and Lithuanian summaries)* (J 2574) Litov Mat Sb (Vil'nyus) 20/1*135-143,219,4*89-96,210
  ⋄ B10 B20 B25 ⋄ REV MR 81i:03016 MR 82i:03015 Zbl 432 # 03006 Zbl 455 # 03006 • ID 32676

NORGELA, S.A. [1982] *Some applications of the inverse method of proof search (Russian) (French and Lithuanian summaries)* (J 3939) Mat Logika Primen (Akad Nauk Litov SSR) 2∗41-47
⋄ B35 ⋄ REV MR 85g:03018 Zbl 516 # 03033 • ID 43490

NORGELA, S.A. see Vol. III, VI for further entries

NORMANN, D. [1983] *Characterizing the continuous functionals* (J 0036) J Symb Logic 48∗965-969
⋄ D65 H05 ⋄ REV MR 85d:03094 Zbl 536 # 03026 • ID 37108

NORMANN, D. see Vol. IV, V, VI for further entries

NORTH, J.H. [1957] *A machine evaluation of logical building blocks* (P 1675) Summer Inst Symb Log;1957 Ithaca 56-58
⋄ B35 ⋄ ID 29318

NORTH, J.H. [1961] see COBHAM, A.

NORTH, J.H. [1962] see DUNHAM, B.

NORTH, J.H. [1963] see DUNHAM, B.

NORTHROP, F.S.C. [1953] see GROSS, M.W.

NORWOOD, F.H. [1982] *Long proofs* (J 0005) Amer Math Mon 89∗110-112
⋄ B35 F20 ⋄ REV MR 82m:03002 Zbl 498 # 03002 • ID 76885

NOSSUM, R. [1985] *Automated theorem proving methods* (J 0130) BIT 25∗51-64
⋄ B35 ⋄ REV Zbl 564 # 03016 • ID 45081

NOTCUTT, B. [1935] *A set of independent postulates for propositional functions of one variable* (J 0120) Ann of Math, Ser 2 36∗670-678
⋄ A05 B05 ⋄ REV Zbl 14.97 • ID 09998

NOTTALE, L. & SCHNEIDER, J. [1984] *Fractals and nonstandard analysis* (J 0209) J Math Phys 25∗1296-1300
⋄ H10 ⋄ REV MR 85h:03075 • ID 43353

NOVAK, I.L. [1950] *A construction for models of consistent systems* (J 0027) Fund Math 37∗87-110
⋄ B15 C07 E70 ⋄ REV MR 12.791 JSL 16.273 • ID 09999

NOVAK, JOSEPH A. [1980] *Some recent work on the assertoric syllogistic* (J 0047) Notre Dame J Formal Log 21∗229-242
⋄ A10 B20 ⋄ REV MR 81h:03027 Zbl 351 # 02005 • ID 53767

NOVIKOV, M.A. [1979] see KARATUEV, V.G.

NOVIKOV, P.S. [1939] *On some existence theorems (Russian)* (J 0023) Dokl Akad Nauk SSSR 23∗438-440
⋄ B05 C75 ⋄ REV Zbl 21.290 JSL 5.69 FdM 65.1106 • ID 10016

NOVIKOV, P.S. [1943] *On the consistency of certain logical calculus (Russian summary)* (J 0142) Mat Sb, Akad Nauk SSSR, NS 12(54)∗231-261
⋄ B05 B28 C75 ⋄ REV MR 5.197 Zbl 60.21 JSL 11.129 • ID 19479

NOVIKOV, P.S. [1949] *On regularity classes (Russian)* (J 0023) Dokl Akad Nauk SSSR 64∗293-295
⋄ B10 B45 ⋄ REV MR 11.1 Zbl 38.150 JSL 14.255 • ID 10018

NOVIKOV, P.S. [1949] *On the axiom of complete induction (Russian)* (J 0023) Dokl Akad Nauk SSSR 64∗457-459
⋄ B25 B28 D35 F30 ⋄ REV MR 11.304 Zbl 38.151 JSL 14.256 • ID 19476

NOVIKOV, P.S. [1959] *Elements of mathematical logic (Russian)* (X 3709) Izdat Fiz-Mat Lit: Moskva 400pp
• TRANSL [1973] (X 0900) Vieweg: Wiesbaden 286pp (German) [1964] (X 1323) Oliver & Boyd: Edinburgh xi+296pp (English) • REPR [1973] (X 2027) Nauka: Moskva 400pp
⋄ B98 ⋄ REV MR 22 # 5565 MR 50 # 1827 Zbl 288 # 02002 Zbl 90.8 JSL 30.356 JSL 31.672 • ID 22364

NOVIKOV, P.S. [1979] *Selected works. Theory of sets and functions. Mathematical logic and algebra (Russian)* (X 2027) Nauka: Moskva 396pp
⋄ B96 C96 E96 ⋄ REV MR 80i:01017 Zbl 485 # 01027 • ID 38498

NOVIKOV, P.S. see Vol. III, IV, V, VI for further entries

NOVIKOV, YA.A. [1982] *The algebraic construction of implicative normal form (Russian)* (J 1516) Vest Ser Fiz Mat Mekh, Univ Minsk 1982∗57-60,80
⋄ B05 ⋄ REV MR 83e:03020 Zbl 512 # 03007 • ID 35209

NOVOSELOV, V.G. [1972] *Statistical estimation of the efficiency of methods for minimization of boolean functions (Russian)* (P 3117) All-Union Conf Autom Contr-Eng Cybern (4);1968 Tbilisi 96-104
⋄ B35 ⋄ REV MR 48 # 8120 Zbl 263 # 94015 • ID 64234

NOZAKI, A. [1963] see IBUKI, K.

NOZAKI, A. see Vol. II, IV for further entries

NURLYBAEV, A.N. [1979] *A boolean function (Russian)* (C 2065) Teor Nereg Kriv Raz Geom Post 62-69
⋄ B05 ⋄ REV MR 81f:03013 • ID 76901

NURLYBAEV, A.N. [1979] *On the construction of a reduced disjunctive normal form by the Nelson method (Russian)* (C 2065) Teor Nereg Kriv Raz Geom Post 70-74
⋄ B35 ⋄ REV MR 81f:03014 • ID 76900

NURLYBAEV, A.N. [1984] *The shortest disjunctive normal functions of the algebra of logic (Russian) (Kazakh summary)* (J 0403) Izv Akad Nauk Kazak SSR, Ser Fiz-Mat 1984/3∗65-67
⋄ B05 ⋄ REV MR 86b:94025 Zbl 538 # 94025 • ID 41506

NURLYBAEV, A.N. see Vol. II for further entries

NURMEEV, N.N. [1976] *Computation of boolean functions by Turing machines (Russian)* (J 3937) Veroyat Met i Kibern (Kazan) 12-13∗60-76
⋄ B05 D10 D15 ⋄ REV MR 58 # 27389 Zbl 397 # 68043 • ID 52721

NUTE, D.E. [1981] *Essential formal semantics* (X 0854) Littlefield Adams: Totowa xiii+186pp
⋄ B98 ⋄ REV MR 83e:03003 Zbl 494 # 03001 JSL 51.252 • ID 35269

NUTE, D.E. see Vol. II, III for further entries

NYBERG, A.M. [1971] see FENSTAD, J.E.

NYBERG, A.M. see Vol. III, IV, V for further entries

OBERSCHELP, A. [1958] *Ueber die Axiome produkt-abgeschlossener arithmetischer Klassen* (J 0009) Arch Math Logik Grundlagenforsch 4∗95-123
⋄ B20 B25 C30 C40 C52 ⋄ REV MR 21 # 6330 Zbl 108.6 JSL 32.532 • ID 10042

OBERSCHELP, A. [1962] *Eine Bemerkung ueber den Kalkuel des natuerlichen Schliessens* (J 0009) Arch Math Logik Grundlagenforsch 6*3-6
⋄ B10 ⋄ REV MR 26 # 3590   Zbl 112.5 • ID 10046

OBERSCHELP, A. [1962] *Untersuchungen zur mehrsortigen Quantorenlogik* (J 0043) Math Ann 145*297-333
⋄ B10 ⋄ REV MR 27 # 3534   Zbl 101.250   JSL 27.225 • ID 10047

OBERSCHELP, A. [1968] *Aufbau des Zahlensystems* (X 0903) Vandenhoeck & Ruprecht: Goettingen 184pp
⋄ B28 ⋄ REV MR 39 # 27   Zbl 175.256 • REM 3rd ed. 1976;196pp • ID 10051

OBERSCHELP, A. [1970] *Mengenlehre und Logik* (S 1747) Christiana Albertina (Kiel) 10*47-55
⋄ B98   E98 • ID 31198

OBERSCHELP, A. [1974] *Elementare Logik und Mengenlehre I* (X 0876) Bibl Inst: Mannheim 259pp
⋄ B98   E98 ⋄ REV MR 55 # 2466   Zbl 295 # 02001 • REM Part II 1978 • ID 31195

OBERSCHELP, A. [1975] see MUELLER, GERT H.

OBERSCHELP, A. [1976] *On Frege's numbers in set theory* (P 1748) Lang & Pensee Math;1976 Luxembourg 309-327
⋄ B28   E70 ⋄ REV Zbl 369 # 04001 • ID 31200

OBERSCHELP, A. [1977] *Ein Ansatz zur Beruecksichtigung von Sprechhandlungen in der Praedikatenlogik* (C 1749) Kasusth, Klassifik, Semant Interpr 263-283
⋄ B10   B65 ⋄ ID 47084

OBERSCHELP, A. [1978] *Elementare Logik und Mengenlehre II* (X 0876) Bibl Inst: Mannheim 229pp
⋄ B98   E98 ⋄ REV MR 58 # 108   Zbl 367 # 04001 • REM Part I 1974 • ID 31196

OBERSCHELP, A. [1983] see GLUBRECHT, J.-M.

OBERSCHELP, A. see Vol. III, IV, V for further entries

OBERSCHELP, W. [1959] *Ueber Einfachheitsprinzipien in der Wahrscheinlichkeitstheorie I* (J 0009) Arch Math Logik Grundlagenforsch 5*3-25
⋄ B30 ⋄ REV MR 23 # A675   Zbl 135.5 • ID 10054

OBERSCHELP, W. [1962] *Ueber die Begruendung wahrscheinlichkeitstheoretischer Axiome durch Wetten* (J 0009) Arch Math Logik Grundlagenforsch 6*35-51
⋄ B30 ⋄ REV MR 32 # 4713   Zbl 108.304 • ID 10055

OBERSCHELP, W. [1984] see BOERGER, E.

OBERSCHELP, W. see Vol. III, IV for further entries

OBERST, E. [1966] *Ueber eine Moeglichkeit zur Bestimmung der Primkonjunktionen einer kanonischen alternativen Normalform* (J 0129) Elektr Informationsverarbeitung & Kybern 2*159-163
⋄ B35 ⋄ REV MR 34 # 5584   Zbl 166.271   JSL 35.348 • ID 10057

OBERST, E. [1979] see KOEGST, M.

O'BRIAN, N.R. [1976] *Local properties of analytic functions and non-standard analysis* (P 2915) Compl Anal & Appl;1975 Trieste 89-106
⋄ H05 ⋄ REV MR 58 # 6320   Zbl 347 # 32003 • ID 82383

O'CONNOR, D.J. [1953] see BASSON, A.H.

ODAKA, A. [1976] see FUJITA, H.

O'DONNELL, M.J. [1983] see FORTUNE, S.

O'DONNELL, M.J. see Vol. IV, VI for further entries

OEFFENBERGER, G.N. [1969] *Zur Frage der Bestimmbarkeit des Wahrheitswertes der schlusskraeftigen syllogistischen Modi im Falle falscher Praemissenkonjunktionen (Polish summary)* (S 0458) Zesz Nauk, Prace Log, Uniw Krakow 4*7-12
⋄ B20 ⋄ REV MR 41 # 29 • ID 10065

OGAWA, H. [1977] see KITAHASHI, T.

OGLESBY, F.C. [1969] see BENNETT, J.H.

OGLESBY, F.C. see Vol. III, VI for further entries

OGUNBIYI, E.I. [1981] see HENLEY, E.J.

OHAMA, S. [1968] *On a formalism which makes any sequence of symbols well-formed* (J 0111) Nagoya Math J 32*1-4
⋄ B03 ⋄ REV MR 38 # 33   Zbl 165.12 • ID 10069

OHAMA, S. see Vol. II, VI for further entries

OHLBACH, H.-J. [1983] *Ein regelbasiertes Klauselgraph-Beweisverfahren* (P 3087) Germ Worksh on Artif Intell (7);1983 Dassel 216-224
⋄ B35 ⋄ REV MR 85g:68007   Zbl 539 # 68086 • ID 44041

OHLBACH, H.-J. & WRIGHTSON, G. [1984] *Solving a problem in relevance logic with an automated theorem prover* (P 2633) Autom Deduct (7);1984 Napa 496-508
⋄ B35   B46 ⋄ REV MR 86b:03024   Zbl 547 # 03013 • ID 43199

OIKKONEN, J. [1979] *Hierarchies of model-theoretic definability - an approach to second-order logics* (P 1705) Scand Logic Symp (4);1976 Jyvaeskylae 197-225
⋄ B15   C40   C70   C75   C80   C85 ⋄ REV MR 80i:03048   Zbl 399 # 03035 • ID 52831

OIKKONEN, J. see Vol. III, V for further entries

OKOL'NISHNIKOVA, E.A. [1979] *On the role of negations in the realization of monotone boolean functions by means of formulas in the basis (∨,&, ¬) (Russian)* (J 0071) Met Diskr Analiz (Novosibirsk) 33*68-76,110-111
⋄ B05 ⋄ REV MR 82e:94072   Zbl 449 # 03068 • ID 56735

OKOL'NISHNIKOVA, E.A. [1982] *The effect of negations on the complexity of realization of monotone Boolean functions by formulas of bounded depth (Russian)* (J 0071) Met Diskr Analiz (Novosibirsk) 38*74-80
⋄ B05 ⋄ REV MR 85c:94048   Zbl 516 # 94027 • ID 39103

OKOL'NISHNIKOVA, E.A. see Vol. II for further entries

OLIVA, P. [1966] see KUDIELKA, V.

OMELYANOVSKIJ, M. [1977] *Axiomatics and the search for the foundations of physics* (P 1704) Int Congr Log, Meth & Phil of Sci (5);1975 London ON 2*47-65
⋄ B30 ⋄ REV MR 58 # 10393 • ID 76972

OMODEO, E.G. [1978] see FERRO, A.

OMODEO, E.G. [1982] *The linked conjunct method for automatic deduction and related search techniques* (J 2687) Comp Math Appl 8*185-203
⋄ B35 ⋄ REV MR 83i:68142   Zbl 506 # 68075 • ID 39316

OMODEO, E.G. [1985] see GHELFO, S.

OMODEO, E.G. see Vol. II, III, V for further entries

OMORI, S. [1953] *Formalization of an intensional logic (Japanese)* (S 4293) Proc Dept Humanities 3*129-149
⋄ B20 ⋄ REV JSL 20.173 • ID 42457

ONICESCU, O. [1971] *Principes de logique et de philosophie mathematique* (X 0871) Acad Rep Soc Romania: Bucharest 229pp
⋄ A05 B98 ⋄ REV MR 54 # 12435 Zbl 235 # 02005 • ID 27815

ONICESCU, O. see Vol. V for further entries

ONO, H. [1967] *Normalizing procedure of sequences over ordinary logical constants* (J 0539) Sci Pap Coll Gen Educ, Univ Tokyo 17*25-34
⋄ B03 ⋄ REV MR 36 # 29 Zbl 209.303 • ID 10108

ONO, H. [1981] see NAKAMURA, A.

ONO, H. [1982] see KOMORI, Y.

ONO, H. [1985] see KOMORI, Y.

ONO, H. [1985] *Semantical analysis of predicate logics without the contraction rule* (J 0063) Studia Logica 44*187-196
⋄ B10 C90 F50 ⋄ ID 47510

ONO, H. see Vol. II, III, IV, V, VI for further entries

ONO, K. [1965] *A certain kind of formal theories* (J 0111) Nagoya Math J 25*59-86
⋄ B20 ⋄ REV MR 30 # 4672 Zbl 142.246 JSL 34.503 • ID 10121

ONO, K. [1966] *A formalism for primitive logic and mechanical proof-checking* (J 0111) Nagoya Math J 26*195-203
⋄ B35 ⋄ REV MR 33 # 3893 Zbl 192.88 JSL 34.504 • ID 10125

ONO, K. [1966] *A formalism for the classical sentence-logic* (J 0111) Nagoya Math J 28*69-71
⋄ B05 ⋄ REV MR 34 # 7348 Zbl 163.6 • ID 90068

ONO, K. [1966] *A lemma which distinguishes minimal logics from other logics* (J 0111) Nagoya Math J 28*197-201
⋄ B20 ⋄ REV MR 34 # 7343 Zbl 163.6 • ID 90072

ONO, K. [1966] *Formal system having just one primitive notion* (J 0111) Nagoya Math J 28*73-77
⋄ B20 ⋄ REV MR 34 # 7344 Zbl 163.7 • ID 90069

ONO, K. [1966] see ITO, J.

ONO, K. [1966] *On development of formal systems starting from primitive logic* (J 0111) Nagoya Math J 28*79-83
⋄ B20 ⋄ REV MR 34 # 7345 Zbl 163.7 • ID 10123

ONO, K. [1966] *On universal character on the primitive logic* (J 0111) Nagoya Math J 27*331-353
⋄ B20 ⋄ REV MR 34 # 31 Zbl 163.5 JSL 34.503 • ID 10124

ONO, K. [1966] *Reinforced logics* (J 0111) Nagoya Math J 28*15-25
⋄ B15 ⋄ REV MR 34 # 7347 Zbl 163.6 • ID 90066

ONO, K. [1967] see MIURA, S.

ONO, K. [1967] *Reduction of logics to the primitive logic* (J 0090) J Math Soc Japan 19*384-398
⋄ B20 ⋄ REV MR 36 # 3626 Zbl 166.254 • ID 10129

ONO, K. [1968] *A study on formal deductions in the primitive logic* (J 0111) Nagoya Math J 31*1-14
⋄ B20 ⋄ REV MR 36 # 6264 Zbl 162.13 • ID 90073

ONO, K. [1968] *On a class of truth-value evaluations of the primitive logic* (J 0111) Nagoya Math J 31*71-80
⋄ B20 ⋄ REV MR 36 # 6278 Zbl 162.13 • ID 90075

ONO, K. [1968] *On formal theories* (J 0111) Nagoya Math J 32*361-371
⋄ B20 ⋄ REV MR 38 # 2001 Zbl 165.12 • ID 10130

ONO, K. [1969] *On a class of set-theoretical interpretations of the primitive logic* (J 0090) J Math Soc Japan 21*313-329
⋄ B20 E75 ⋄ REV MR 39 # 3972 Zbl 184.282 • ID 10135

ONO, K. see Vol. II, V, VI for further entries

ONYSZKIEWICZ, J. [1972] see MAREK, W.

ONYSZKIEWICZ, J. see Vol. III, V for further entries

OPPEN, D.C. [1980] *Complexity, convexity and combinations of theories* (J 1426) Theor Comput Sci 12*291-302
⋄ B25 B35 D15 ⋄ REV MR 82a:03013 Zbl 437 # 03007 • ID 55875

OPPEN, D.C. see Vol. III, IV, VI for further entries

OREVKOV, V.P. [1965] see DAVYDOV, G.V.

OREVKOV, V.P. [1965] see MASLOV, S.YU.

OREVKOV, V.P. [1968] *Glivenko classes of sequents (Russian)* (S 0228) Zap Nauch Sem Leningrad Otd Mat Inst Steklov 8*196-201
• TRANSL [1968] (J 0521) Semin Math, Inst Steklov 8*95-97
⋄ B10 F50 ⋄ REV MR 42 # 5775 Zbl 198.318 • ID 10152

OREVKOV, V.P. [1968] *Glivenko's sequence classes (Russian)* (S 0066) Tr Mat Inst Steklov 98*131-154 • ERR/ADD ibid 121*167
• TRANSL [1968] (S 0055) Proc Steklov Inst Math 98*147-173
⋄ B10 B20 F50 ⋄ REV MR 39 # 2606 Zbl 172.290 • ID 19539

OREVKOV, V.P. (ED.) [1968] *The calculi of symbolic logic I (Russian)* (S 0066) Tr Mat Inst Steklov 98*5-202 • ERR/ADD ibid 121*167
• TRANSL [1968] (S 0055) Proc Steklov Inst Math 98*iv+229pp
⋄ B97 F97 ⋄ REV MR 43 # 4620 MR 49 # 18 • ID 37197

OREVKOV, V.P. [1968] *Two undecidable classes of formulas in classical predicate calculus (Russian)* (S 0228) Zap Nauch Sem Leningrad Otd Mat Inst Steklov 8*202-210 • ERR/ADD ibid 20*292-294
• TRANSL [1968] (J 0521) Semin Math, Inst Steklov 8*98-102
⋄ B20 D35 ⋄ REV MR 42 # 5776 Zbl 182.18 • ID 10153

OREVKOV, V.P. [1969] see DAVYDOV, G.V.

OREVKOV, V.P. [1969] *On nonlengthening applications of equality rules (Russian)* (S 0228) Zap Nauch Sem Leningrad Otd Mat Inst Steklov 16*152-156 • ERR/ADD ibid 20*292-294
• TRANSL [1969] (J 0521) Semin Math, Inst Steklov 16*77-79
⋄ B25 B35 F05 F07 ⋄ REV MR 41 # 6665 Zbl 197.273 • ID 10154

OREVKOV, V.P. [1971] see MASLOV, S.YU.

OREVKOV, V.P. [1971] *On biconjunctive reduction classes (Russian) (English summary)* (S 0228) Zap Nauch Sem Leningrad Otd Mat Inst Steklov 20∗170-174,287
• TRANSL [1973] (J 1531) J Sov Math 1∗106-109
◊ B20 D35 ◊ REV MR 45 # 4944 Zbl 222 # 02034
• ID 10157

OREVKOV, V.P. [1972] see MASLOV, S.YU.

OREVKOV, V.P. (ED.) [1972] *Logical and logico-mathematical calculi 2 (Russian)* (S 0066) Tr Mat Inst Steklov 121∗167pp
• TRANSL [1974] (S 0055) Proc Steklov Inst Math v + 183pp
◊ B97 ◊ REV MR 48 # 5810 MR 49 # 4746 • ID 70220

OREVKOV, V.P. [1981] see GRIGOR'EV, D.YU.

OREVKOV, V.P. see Vol. II, III, IV, VI for further entries

OREY, S. [1957] *Strongly standard formulas* (P 1675) Summer Inst Symb Log;1957 Ithaca 169-173
◊ B15 ◊ ID 29341

OREY, S. [1959] *Model theory for the higher order predicate calculus* (J 0064) Trans Amer Math Soc 92∗72-84
◊ B15 C85 ◊ REV MR 21 # 6325 Zbl 88.10 JSL 27.96
• ID 10165

OREY, S. [1964] *New foundations and the axiom of counting* (J 0025) Duke Math J 31∗655-660
◊ B15 E70 ◊ REV MR 32 # 38 Zbl 143.9 JSL 34.649
• ID 10167

OREY, S. see Vol. III, V, VI for further entries

ORLOWSKA, E. [1967] see EHRENFEUCHT, A.

ORLOWSKA, E. [1967] *Mechanical proof procedure for the n-valued propositional calculus (Russian summary)* (J 0014) Bull Acad Pol Sci, Ser Math Astron Phys 15∗537-541
◊ B35 ◊ REV MR 36 # 3654 Zbl 162.320 • ID 10170

ORLOWSKA, E. [1969] *Automatic theorem proving in a certain class of formulae of predicate calculus* (J 0014) Bull Acad Pol Sci, Ser Math Astron Phys 17∗117-119
◊ B35 ◊ REV MR 41 # 1447 Zbl 258 # 68047 • ID 64304

ORLOWSKA, E. [1969] *Mechanical theorem proving in a certain class of formulae of the predicate calculus (Polish) (Russian summary)* (J 0063) Studia Logica 25∗17-29
◊ B35 ◊ REV MR 44 # 1572 Zbl 257 # 68085 • ID 10171

ORLOWSKA, E. [1973] *Theorem-proving systems* (J 0202) Diss Math (Warsaw) 103∗51pp
◊ B35 ◊ REV MR 48 # 8208 Zbl 268 # 68041 • ID 10173

ORLOWSKA, E. [1975] *On the Jaskowski's method of suppositions* (J 0063) Studia Logica 34∗187-200
◊ B10 ◊ REV MR 52 # 5343 Zbl 325 # 02022 • ID 18318

ORLOWSKA, E. [1976] *Herbrand systems for proving theorems of predicate calculus (Polish) (English and Russian summaries)* (X 1034) PWN: Warsaw 26pp
◊ B35 ◊ REV MR 55 # 7746 Zbl 378 # 68043 • ID 51823

ORLOWSKA, E. [1978] *The resolution principle for ω^+-valued logic* (J 2095) Fund Inform, Ann Soc Math Pol, Ser 4 2∗1-15
◊ B35 ◊ REV MR 80b:03025 Zbl 433 # 03006 • ID 54012

ORLOWSKA, E. [1979] *A generalization of the resolution principle* (J 3293) Bull Acad Pol Sci, Ser Math 27∗227-234
◊ B35 ◊ REV MR 83j:03027 Zbl 444 # 03006 • ID 56454

ORLOWSKA, E. [1980] *Resolution systems and their applications. I,II* (J 2095) Fund Inform, Ann Soc Math Pol, Ser 4 3∗235-267,333-361
◊ B35 ◊ REV MR 82i:03021 Zbl 472 # 68052 Zbl 472 # 68057 • ID 55329

ORLOWSKA, E. see Vol. II, VI for further entries

ORNAGHI, M. [1975] see MARINI, D.

ORNAGHI, M. see Vol. II, III, IV, VI for further entries

O'RORKE, P. [1982] see GREENBAUM, S.

OSDOL VAN, D.H. [1972] *Truth with respect to an ultrafilter or how to make intuition rigorous* (J 0005) Amer Math Mon 79∗355-363
◊ C20 H05 ◊ REV MR 46 # 1545 Zbl 241 # 26004
• ID 22272

OSHIBA, T. [1979] see FUNAHASHI, S.

OSHIBA, T. [1981] *A method for obtaining proof figures of valid formulas in the first order predicate calculus* (J 0407) Comm Math Univ St Pauli (Tokyo) 30∗49-62
◊ B10 ◊ REV MR 82m:03066 Zbl 467 # 03005 • ID 55003

OSHIBA, T. [1981] see FUNAHASHI, S.

OSHIBA, T. [1984] see FUNAHASHI, S.

OSHIBA, T. see Vol. VI for further entries

OSSWALD, H. [1984] see HORT, C.

OSSWALD, H. see Vol. III, VI for further entries

OSTAPKO, D.L. [1974] see HONG, SEJUNE

OSTAPKO, D.L. [1974] see CAIN, R.G.

OSTAPKO, D.L. see Vol. II for further entries

OSTROUKHOV, D.A. [1978] see MOGILEVSKIJ, G.L.

OSTROUKHOV, D.A. see Vol. IV, VI for further entries

OSWALD, U. [1982] see KREJNOVICH, V.YA.

OSWALD, U. see Vol. III, V for further entries

OTEPANOV, V.I. & VERKHOZIN, O.M. [1972] *Problems in logic (Russian)* (X 1006) Irkutsk Gos Univ: Irkutsk 73pp
◊ B98 E98 ◊ REV MR 50 # 1826 • ID 13931

OVERBEEK, R.A. [1980] see LUSK, E.L.

OVERBEEK, R.A. [1982] see LUSK, E.L.

OVERBEEK, R.A. see Vol. IV for further entries

PABION, J.F. [1974] *L'axiomatisation de la syntaxe et le second theoreme de Goedel* (J 0056) Publ Dep Math, Lyon 11/4∗27-87
◊ B30 F30 ◊ REV MR 52 # 2862 Zbl 331 # 02017
• ID 17648

PABION, J.F. [1976] *Logique mathematique* (X 0859) Hermann: Paris xxxii + 263pp
◊ B98 ◊ REV MR 56 # 15346 Zbl 345 # 02001 JSL 44.282
• ID 64331

PABION, J.F. [1980] *TT_3I est equivalent a l'arithmetique du second ordre (English summary)* (J 2313) C R Acad Sci, Paris, Ser A-B 290∗A1117-A1118
◊ B15 F25 F35 ◊ REV MR 82f:03051 Zbl 437 # 03004
• ID 55872

PABION, J.F. & RICHARD, D. [1981] *Synonymy and re-interpretation for some sublanguages of Peano arithmetic* (P 2614) Open Days in Model Th & Set Th;1981 Jadwisin 231-236
⋄ B28   C62   F30 ⋄ ID 33728

PABION, J.F. see Vol. II, III, V for further entries

PACHOLSKI, L. & WEGLORZ, B. [1968] *Topologically compact structures and positive formulas* (S 0019) Colloq Math (Warsaw) 19∗37-42
⋄ B20   C40   C50 ⋄ REV MR 37 # 871   Zbl 184.13   JSL 40.88 • ID 10215

PACHOLSKI, L. [1978] see MACINTYRE, A.

PACHOLSKI, L. see Vol. III, V for further entries

PADOA, A. [1902] *Essai d'une theorie algebrique des nombres entiers, precede d'une introduction logique a une theorie deductive quelconque* (P 1484) Int Congr Math (2);1900 Paris 3∗309-365
• TRANSL [1967] (C 0675) From Frege to Goedel 119-123
⋄ A10   B30   C40 ⋄ REM Only a part is translated
• ID 14911

PADOA, A. [1912] *La logique deductive dans sa derniere phase de developpement* (X 0834) Gauthier-Villars: Paris 106pp
⋄ A10   B98 ⋄ REV FdM 43.98 • ID 22597

PADOA, A. [1930] *Proposizioni assiomatiche* (P 0741) Int Congr Math (II, 3);1928 Bologna 3∗381-387
⋄ B30 ⋄ REV FdM 56.822 • ID 27772

PADOA, A. [1936] *Les extensions successives de l'ensemble des nombres au point de vue deductif* (P 0632) Congr Int Phil des Sci;1935 Paris 7∗53-59
⋄ B28 ⋄ REV FdM 62.1069 • ID 38722

PADOA, A. see Vol. V for further entries

PAEPPINGHAUS, P. [1983] *Completeness properties of classical theories of finite type and the normal form theorem* (J 0202) Diss Math (Warsaw) 207,62pp
⋄ B15   F05   F35 ⋄ REV MR 85b:03100   Zbl 537 # 03038
• ID 40742

PAEPPINGHAUS, P. see Vol. II, IV, VI for further entries

PAGELLO, E. [1976] see COLOMBETTI, M.

PAGER, D. [1962] *An emendation of the axiom system of Hilbert and Ackermann for the restricted calculus of predicates* (J 0036) J Symb Logic 27∗131-138
⋄ B20 ⋄ REV MR 27 # 2396   Zbl 114.245   JSL 34.520
• ID 10241

PAGER, D. see Vol. IV for further entries

PAGLI, P. [1970] *Ammissibilita di connetivi in un calcolo generale (English summary)* (J 0088) Ann Univ Ferrara, NS, Sez 7 15∗13-33
⋄ B22 ⋄ REV MR 45 # 3181   Zbl 214.13 • ID 10244

PAGLI, P. see Vol. VI for further entries

PAHI, B. [1970] see APPLEBEE, R.C.

PAHI, B. [1971] *Full models and restricted extensions of propositional calculi* (J 0068) Z Math Logik Grundlagen Math 17∗5-10
⋄ B20 ⋄ REV MR 43 # 3084   Zbl 229 # 02014 • ID 10246

PAHI, B. [1971] *Restricted extensions of implicational calculi* (J 0068) Z Math Logik Grundlagen Math 17∗11-16
⋄ B20 ⋄ REV MR 43 # 3085   Zbl 229 # 02015 • ID 10245

PAHI, B. [1972] *A method for proving the nonexistence of finite characteristic models for implicational calculi* (J 0068) Z Math Logik Grundlagen Math 18∗169-172
⋄ B22 ⋄ REV MR 46 # 25   Zbl 264 # 02022 • ID 10248

PAHI, B. [1972] *A theorem on the interrelationship of axiom systems for implicational calculi* (J 0068) Z Math Logik Grundlagen Math 18∗165-167
⋄ B22 ⋄ REV MR 46 # 24   Zbl 264 # 02021 • ID 10249

PAHI, B. [1974] *On the non-existence of finite characteristic models for some classes of implicational calculi* (J 0068) Z Math Logik Grundlagen Math 20∗113-119
⋄ B22 ⋄ REV MR 50 # 9538   Zbl 317 # 02009 • ID 10251

PAHI, B. see Vol. II for further entries

PAILLET, J.L. [1973] *La methode des systemes logiquement inductifs* (J 0060) Rev Roumaine Math Pures Appl 18∗1393-1412
⋄ B10   C07   C30 ⋄ REV MR 55 # 5386   Zbl 273 # 02037
• ID 30495

PAILLET, J.L. [1975] *Une characterisation de l'equivalence des formules modulo une theorie (English summary)* (J 2313) C R Acad Sci, Paris, Ser A-B 281∗A745-A747
⋄ B10 ⋄ REV MR 53 # 118   Zbl 319 # 02044 • ID 16582

PAILLET, J.L. see Vol. III, VI for further entries

PALLADINO, D. [1980] see BORGA, M.

PALLADINO, D. see Vol. III, V for further entries

PALLASCHKE, D. [1975] *Eine algebraische Formulierung des Beobachtbarkeitsbegriffes in der Kontrolltheorie* (P 2980) Num Beh Variat & Steuerungsprobl;1974 Bonn 37-45
⋄ B80   G05 ⋄ REV MR 53 # 2439   Zbl 315 # 93005
• ID 82429

PALMA DE, R. [1971] *L'algebre binaire de Boole et ses applications a l'informatique* (X 0856) Dunod: Paris x+148pp
⋄ B05   B70   G05 ⋄ REV Zbl 271 # 94025 • ID 64367

PALTANEA, R. [1980] *Example of a grammar that generates the formulas of propositional calculus* (J 3071) Bul Univ Brasov, Ser C 22∗93-100
⋄ B05   D05 ⋄ REV MR 83e:03062 • ID 35233

PALYUTIN, E.A. [1971] *Boolean algebras with a categorical theory in a weak second order logic (Russian)* (J 0003) Algebra i Logika 10∗523-534
• TRANSL [1971] (J 0069) Algeb and Log 10∗325-331
⋄ B15   C35   C85   D35   G05 ⋄ REV MR 46 # 3302
Zbl 248 # 02055 • ID 64365

PALYUTIN, E.A. [1973] see ERSHOV, YU.L.

PALYUTIN, E.A. [1979] see ERSHOV, YU.L.

PALYUTIN, E.A. see Vol. III, IV, V for further entries

PANCZAKIEWICZ, M. [1968] *Set-theoretic foundations of the arithmetic of natural numbers (Polish) (English summary)* (S 0543) Zesz Nauk Wyz Ped Mat, Katowice 6∗61-71
⋄ B28 ⋄ REV MR 48 # 10810 • ID 10265

PANCZAKIEWICZ, M. [1970] *Mathematical logic (Polish)* (S 0544) Prace Mat Uniw Katowice 9∗269pp
⋄ B98 ⋄ REV MR 50 # 12631 • ID 19525

PANKAJAM, S. [1941] *On the formal structure of the propositional calculus I* (J 0441) J Indian Math Soc, NS 5*49–61
⋄ B05 ⋄ REV MR 3.130 Zbl 61.8 JSL 7.39 • REM Part II 1942 • ID 10268

PANKAJAM, S. [1942] *On the formal structure of the propositional calculus II* (J 0441) J Indian Math Soc, NS 6*51–62,102
⋄ B05 ⋄ REV MR 4.125 Zbl 61.8 JSL 8.84 • REM Part I 1941 • ID 10270

PANKOV, P.S. [1978] *A combined method for proving certain theorems of mathematical analysis via a computer (Russian) (English summary)* (J 0040) Kibernetika, Akad Nauk Ukr SSR 1978/3*119–125
• TRANSL [1978] (J 0021) Cybernetics 14*441–448
⋄ B35 ⋄ REV MR 80k:68084 Zbl 384#68077 • ID 82430

PANKOV, P.S. [1982] see BAYACHOROVA, B.D.

PAOLA DI, R.A. [1971] *The relational data file and the decision problem for classes of proper formulas* (P 1681) Symp Inform Storage & Retrieval;1971 College Park 95–105
⋄ B20 B75 D35 ⋄ ID 30683

PAOLA DI, R.A. [1973] *The solvability of the decision problem for classes of proper formulas and related results* (J 0037) ACM J 20*112–126
⋄ B20 B25 ⋄ REV MR 48#7698 Zbl 285#68008 • ID 30684

PAOLA DI, R.A. see Vol. III, IV, VI for further entries

PAP, A. [1954] *Propositions, sentences, and the semantic definition of truth* (J 0105) Theoria (Lund) 20*23–35
⋄ A05 B05 B10 ⋄ REV JSL 21.381 • ID 10272

PAPAKONSTANTINOU, G. [1979] *Minimization of modulo-2 sum of products* (J 0187) IEEE Trans Comp C-28*163–167
⋄ B35 ⋄ REV MR 80b:94043 Zbl 415#94016 • ID 69756

PAPIC, P. [1971] *Sur le continu de Souslin (Serbo-Croatian summary)* (J 3519) Glas Mat, Ser 3 (Zagreb) 6(26)*351–355
⋄ B28 E65 ⋄ REV MR 46#9950 Zbl 232#04001 • ID 27414

PAPP, B. [1972] *Procede pour determiner les formes normales minimales des fonctions booleennes, en utilisant les regles de minimisation de la fonction de Cout* (J 0380) Acta Cybern (Szeged) 1*241–250
⋄ B35 ⋄ REV MR 49#10618 Zbl 288#02008 • ID 10275

PARIGOT, M. [1984] see DELON, F.

PARIGOT, M. see Vol. III for further entries

PARIKH, R. [1969] *A nonstandard theory of topological groups* (P 0649) Appl Model Th to Algeb, Anal & Probab;1967 Pasadena 279–284
⋄ H05 ⋄ REV MR 39#1318 Zbl 188.21 • ID 10281

PARIKH, R. & PARNES, M. [1972] *Conditional probability can be defined for all pairs of sets of reals* (J 0345) Adv Math 9*313–315
⋄ H05 ⋄ REV MR 48#3085 Zbl 254#28009 • ID 10283

PARIKH, R. & PARNES, M. [1974] *Conditional probabilities and uniform sets* (P 1083) Victoria Symp Nonstand Anal;1972 Victoria 180–194
⋄ H05 ⋄ REV MR 58#2937 Zbl 279#60004 • ID 31223

PARIKH, R. (ED.) [1975] *Logic Colloquium* (X 0811) Springer: Heidelberg & New York iv+251pp
⋄ B97 ⋄ REV MR 51#5255 Zbl 298#00007 • ID 70197

PARIKH, R. see Vol. II, III, IV, V, VI for further entries

PARIS, J.B. [1978] see MACINTYRE, A.

PARIS, J.B. [1978] *Note on an induction axiom* (J 0036) J Symb Logic 43*113–117
⋄ B28 C62 F30 ⋄ REV MR 81e:03057 Zbl 399#03009 • ID 29247

PARIS, J.B. & WILKIE, A.J. [1985] *Counting problems in bounded arithmetic* (P 2160) Latin Amer Symp Math Log (6);1983 Caracas 317–340
⋄ B28 C62 F30 ⋄ REV Zbl 572#03034 • ID 41796

PARIS, J.B. see Vol. III, IV, V, VI for further entries

PARKER, F.D. [1964] *Boolean matrices and logic* (J 0497) Math Mag 37*33–38
⋄ B05 ⋄ REV Zbl 124.3 JSL 40.614 • ID 14472

PARKER, W.L. [1955] see BERNSTEIN, B.A.

PARKER-RHODES, A.F. [1981] *The theory of indistinguishables. A search for explanatory principles below the level of physics* (X 0835) Reidel: Dordrecht xvi+216pp
⋄ B30 ⋄ REV MR 83m:03020 Zbl 492#03001 • ID 35427

PARKER JR., D.S. [1981] see DELOBEL, C.

PARKIN, I.A. [1978] *On determining the prime implicants of a boolean function without recourse to its minterm form* (J 3116) Austral Comp J 10*153–156
⋄ B35 ⋄ REV MR 80g:94090 Zbl 392#94021 • ID 52419

PARKS, R.Z. & RESCHER, N. [1971] *Restricted inference* (J 0079) Logique & Anal, NS 14*675–683
⋄ B10 F07 ⋄ REV MR 46#17 Zbl 232#02022 • ID 11123

PARKS, R.Z. see Vol. II for further entries

PARNES, M. [1972] see PARIKH, R.

PARNES, M. [1974] see PARIKH, R.

PARNES, M. see Vol. IV for further entries

PARRY, W.T. [1954] *A new symbolism for the propositional calculus* (J 0036) J Symb Logic 19*161–168,VI
⋄ B05 ⋄ REV MR 16.103 Zbl 56.8 JSL 23.63 • ID 10300

PARRY, W.T. [1965] *Comments on a variant form of natural deduction* (J 0036) J Symb Logic 30*119–122
⋄ B10 F07 ⋄ REV Zbl 173.4 JSL 31.286 • ID 10301

PARRY, W.T. [1966] *Quantification of the predicate and many-sorted logic* (J 0075) Phil Phenom Research 26*342–360
⋄ B15 ⋄ REV JSL 40.606 • ID 14476

PARRY, W.T. [1967] see HACKER, E.A.

PARRY, W.T. see Vol. II for further entries

PARSONS, C. [1971] *A plea for substitutional quantification* (J 0301) J Phil 68*231–237
⋄ A05 B10 ⋄ ID 28349

PARSONS, C. [1983] *Mathematics in philosophy. Selected essays* (X 0992) Cornell Univ Pr: Ithaca 365pp
⋄ A05 B80 ⋄ REV MR 86a:01050 • ID 45008

PARSONS, C. see Vol. II, IV, V, VI for further entries

PASCH, M. [1882] *Vorlesungen ueber neuere Geometrie* (X 0823) Teubner: Stuttgart iv+202pp
⋄ B30 ⋄ REV FdM 14.498 • ID 23461

PASCH, M. [1909] *Grundlagen der Analysis* (X 0823) Teubner: Stuttgart v+140pp
⋄ B28 ⋄ ID 23413

PASINI, A. [1979] *Model theoretic approach to geometries* (J 2748) J Comb Inf Syst Sci (New Dehli) 4∗66-75
  ◇ B30 ◇ REV MR 80j:03051 Zbl 412 # 51003 • ID 52980

PASINI, A. see Vol. III for further entries

PASQUINELLI, A. [1957] *Introduzione alla logica simbolica* (X 1262) Einaudi: Torino x+120pp
  ◇ B98 ◇ REV Zbl 85.243 JSL 32.105 • ID 22424

PASSOS, E.P. [1974] see GALDA, K.

PASTEL, A.M. [1970] *Caracterisation geometrique des fonctions logiques permutantes* (J 2313) C R Acad Sci, Paris, Ser A-B 270∗A102-A105
  ◇ B05 ◇ REV MR 41 # 1613 Zbl 188.18 • ID 10319

PASTEL, A.M. [1971] *Sur les fonctions logiques permutantes* (J 2313) C R Acad Sci, Paris, Ser A-B 272∗A1154-A1156
  ◇ B20 ◇ REV MR 43 # 7379 Zbl 222 # 02067 • ID 26323

PASTRE, D. [1978] *Automatic theorem proving in set theory* (J 0503) Artif Intell 10∗1-27
  ◇ B35 ◇ REV MR 81e:68114 Zbl 374 # 68059 • ID 51575

PASZTOR-VARGA, K. [1972] *On some minimizing algorithms of boolean functions* (J 3101) Acta Tech Acad Sci Hung 73∗349-362
  ◇ B35 ◇ REV MR 48 # 13503 Zbl 315 # 94048 • ID 64404

PATERSON, M.S. [1975] see FISCHER, MICHAEL J.

PATERSON, M.S. [1977] see MCCOLL, W.F.

PATERSON, M.S. [1982] see FISCHER, MICHAEL J.

PATERSON, M.S. see Vol. II, IV for further entries

PATTEN, P.R. [1976] *A nonstandard model of the real numbers with applications to limits and continuity* (J 0243) Pi Mu Epsilon J 6∗272-280
  ◇ H05 ◇ REV MR 55 # 580 Zbl 346 # 26019 • ID 82444

PATTERSON, G.W. [1960] *What is a code ?* (J 0212) ACM Commun 3∗315-318
  ◇ B03 ◇ REV MR 25 # 1071 Zbl 87.326 JSL 30.385 • ID 10322

PATTON, T.E. [1963] *A system of quantificational deduction* (J 0047) Notre Dame J Formal Log 4∗105-112
  ◇ B10 ◇ REV MR 32 # 5501 Zbl 171.253 • ID 10323

PATTON, T.E. [1964] *A liberalized system of quantificational deduction* (J 0047) Notre Dame J Formal Log 5∗293-294
  ◇ B10 ◇ REV MR 34 # 7340 • ID 10325

PATTON, T.E. see Vol. IV for further entries

PATZIG, G. [1959] *Die Aristotelische Syllogistik. Logisch-Philosophische Untersuchungen ueber Buch "A" der "Ersten Analytiken"* (X 0903) Vandenhoeck & Ruprecht: Goettingen 207pp
  • TRANSL [1968] (X 0835) Reidel: Dordrecht 215pp
  ◇ A05 B20 ◇ REV Zbl 185.5 JSL 31.102 • REM 3rd ed. 1969 • ID 10327

PAUL, E. [1984] *A new interpretation of the resolution principle* (P 2633) Autom Deduct (7);1984 Napa 333-355
  ◇ B35 ◇ REV Zbl 547 # 03010 • ID 43195

PAUL, E. [1985] *On solving the equality problem in theories defined by Horn clauses* (P 4601) EUROCAL;1985 Linz 2∗363-377
  ◇ B20 B35 ◇ ID 49772

PAUL, E. see Vol. III for further entries

PAUL, W.J. [1977] *A 2,5n-lower bound on the combinational complexity of boolean functions* (J 1428) SIAM J Comp 6∗427-443
  • TRANSL [1979] (J 3079) Kiber Sb Perevodov, NS 16∗23-44 (Russian)
  ◇ B05 B70 ◇ REV MR 56 # 10138 Zbl 358 # 68081 • ID 69763

PAUL, W.J. see Vol. IV for further entries

PAULL, M. [1983] see FRANCO, J.

PAULOS, J.A. [1977] see LEBLANC, H.

PAULOS, J.A. [1981] *Probabilistic, truth-value, and standard semantics and the primacy of predicate logic* (J 0047) Notre Dame J Formal Log 22∗11-16
  ◇ A05 B10 B48 ◇ REV MR 82b:03051 Zbl 416 # 03012 • ID 56466

PAULOS, J.A. see Vol. II, III for further entries

PAULSON, L. [1983] *A higher-order implementation of rewriting* (J 3298) Sci Comput Programming 3∗119-149
  ◇ B35 ◇ REV MR 84m:68088 Zbl 551 # 68076 • ID 39530

PAUN, G. [1981] see CALUDE, C.

PAUN, G. see Vol. II, IV, VI for further entries

PAVILENIS, R.I. [1984] see ERSHOV, YU.L.

PAWLAK, Z. [1960] *New method of parenthesis-free notations of formulae* (J 0014) Bull Acad Pol Sci, Ser Math Astron Phys 8∗197-198
  ◇ B03 ◇ REV JSL 31.115 • ID 10334

PAWLAK, Z. [1967] see EHRENFEUCHT, A.

PAWLAK, Z. [1970] *Definitional approach to automatic demonstration* (P 0625) Symp Autom Demonst;1968 Versailles 191-193
  ◇ B35 ◇ REV MR 42 # 2886 Zbl 195.30 • ID 16300

PAWLAK, Z. [1970] see MOSTOWSKI, A.WLODZIMIERZ

PAWLAK, Z. see Vol. II, IV, V for further entries

PAWLOWSKY, V. & TRILLAS, E. [1983] *On order and morphisms related to a Sheffer stroke* (J 2840) Stochastica, Univ Politec Barcelona 7∗201-207
  ◇ B05 ◇ REV Zbl 566 # 06006 • ID 45421

PAWLOWSKY, V. see Vol. II for further entries

PEANO, G. [1889] *The principles of arithmetic, presented by a new method (Latin)* (X 4721) Bocca & Clausen: Torino xvi+20pp
  • TRANSL [1967] (C 0675) From Frege to Goedel 85-97 (English)
  ◇ B03 B28 E75 F30 ◇ REV FdM 21.51 • ID 19497

PEANO, G. [1891] *Formole di logica matematica* (J 0391) Riv Mat Univ Parma 1∗24-31,182-184
  ◇ B20 ◇ REV FdM 23.51 • ID 42995

PEANO, G. [1891] *Principii di logica matematica* (J 0391) Riv Mat Univ Parma 1∗1-10
  ◇ B20 ◇ REV FdM 23.51 • ID 42994

PEANO, G. [1894] *Notations de la logique mathematique. Introduction au formulaire de mathematique* (X 1478) Guadagnini: Torino 52pp
  ◇ A10 B03 ◇ REV FdM 25.103 • ID 19496

PEANO, G. [1895] *Review of : Frege : Grundgesetze der Arithmetik, begriffsschriftlich abgeleitet, Vol.1* (J 0391) Riv Mat Univ Parma 5*122-128
⋄ B28 ⋄ ID 19495

PEANO, G. [1897] *Formulaire de mathematiques Vol.2, chap.1: Logique mathematique* (X 4721) Bocca & Clausen: Torino 63pp
⋄ B20 ⋄ REV FdM 28.63 • REM Chap.2 of Vol.2 1898
• ID 33670

PEANO, G. [1897] *Studii di logica matematica* (J 1742) Atti Accad Sci Torino, Fis Mat Nat 32*565-583
⋄ A10 B20 B28 ⋄ REV JSL 45.177 FdM 28.340
• ID 10345

PEANO, G. [1898] *Formulaire de mathematiques Vol.2, chap.2: Arithmetique* (X 4721) Bocca & Clausen: Torino viii+59pp
⋄ B28 ⋄ REV FdM 29.47 • REM Chap.1 of Vol.2 1897. Vol.4 1903 • ID 33668

PEANO, G. [1900] *Formules de logique mathematique* (J 0391) Riv Mat Univ Parma 7*1-40
⋄ B03 B10 ⋄ REV FdM 31.69 • ID 20850

PEANO, G. [1902] *Les definitions mathematiques* (P 1484) Int Congr Math (2);1900 Paris III*279-288
⋄ B30 ⋄ REV FdM 33.83 • ID 38660

PEANO, G. [1903] *Formulaire mathematique Vol.4* (X 4721) Bocca & Clausen: Torino xvi+406pp
⋄ B28 ⋄ REV FdM 34.72 • REM Vol.2 1897. Vol.5 1908
• ID 33669

PEANO, G. [1908] *Formulario mathematice Vol.5* (X 4721) Bocca & Clausen: Torino xxxvi+463pp
⋄ B28 ⋄ REV FdM 39.84 • REM Turin. Vol.4 1903 • ID 33671

PEANO, G. [1913] *Review of : A.N.Whitehead and B.Russell : Principia Mathematica I,II* (J 1204) Boll Bibl Sci Mat 1913*47-53,75-81
⋄ A10 B28 ⋄ ID 28523

PEANO, G. [1957] *Opere scelte Vol. 1* (X 0860) Cremonese: Firenze viii+531pp
⋄ B96 ⋄ REV MR 19.827 Zbl 78.41 • ID 25709

PEANO, G. [1958] *Opere scelte Vol. 2* (X 0860) Cremonese: Firenze vi+518pp
⋄ B96 ⋄ REV MR 20#6339 Zbl 83.1 • ID 25710

PEANO, G. [1959] *Opere scelte Vol. 3* (X 0860) Cremonese: Firenze vii+470pp
⋄ B96 ⋄ REV MR 21#3302 Zbl 84.245 • ID 25711

PEANO, G. see Vol. V for further entries

PECORA, L.M. [1982] *A nonstandard infinite-dimensional vector space approach to Gaussian functional measures* (J 0209) J Math Phys 23*969-982
⋄ H05 ⋄ REV MR 84i:60054 Zbl 491#28015 • ID 40138

PEDRYCZ, W. [1981] see CZOGALA, E.

PEDRYCZ, W. see Vol. II, V for further entries

PEIRCE, C.S. [1880] *On the algebra of logic, chapter I.-Syllogistic. chapter II.-The logic of non-relative terms, chapter III. The logic of relatives* (J 0100) Amer J Math 3*15-57
⋄ A05 A10 B20 ⋄ REV JSL 34.494 FdM 12.41 • ID 10347

PEIRCE, C.S. (ED.) [1883] *Studies in logic, by members of the Johns Hopkins University* (X 0960) Little, Brown & Co: Boston & Toronto vi+203pp
⋄ B97 ⋄ ID 22556

PEIRCE, C.S. [1885] *On the algebra of logic: A contribution to the philosophy of notation* (J 0100) Amer J Math 7*180-202
⋄ A05 A10 B20 ⋄ REV FdM 17.44 • ID 10349

PEIRCE, C.S. [1976] *The new elements of mathematics. Vol.III parts 1 and 2. Mathematical miscellanea* (X 0873) Mouton: Paris xxxix+1153pp
⋄ A05 B96 G05 ⋄ REV MR 58#27155c MR 58#27155d Zbl 443#01023 JSL 47.705 • REM Edited by Eisele,C. • ID 47406

PEIRCE, C.S. see Vol. II for further entries

PEKLO, B.T. [1971] *Recht und Logik* (J 0147) An Univ Bucuresti, Acta Logica 14*59-80
⋄ B80 ⋄ REV MR 57#9485 Zbl 315#02035 • ID 29809

PEKLO, B.T. [1972] *Mancherlei ueber rechtslogische Fragen (juristisch-logisches quodlibet.)* (J 0079) Logique & Anal, NS 15*461-488
⋄ B80 ⋄ REV MR 49#2241 Zbl 278#02028 • ID 29064

PEKLO, B.T. [1973] *Logical inconsistencies* (J 0286) Int Logic Rev 1973*95-97
⋄ B05 ⋄ REV Zbl 344#02008 • ID 64423

PEKLO, B.T. see Vol. II for further entries

PENZIN, YU.G. [1972] see FRIDMAN, EH.I.

PENZIN, YU.G. see Vol. III, IV, VI for further entries

PENZOV, YU.E. [1968] *Elements of mathematical logic and set theory (Russian)* (X 0958) Saratov Univ: Saratov 143pp
⋄ B98 E98 ⋄ REV MR 53#86 • ID 16550

PENZOV, YU.E. see Vol. V for further entries

PEPIS, J. [1936] *Beitraege zur Reduktionstheorie des logischen Entscheidungsproblemes* (J 0460) Acta Univ Szeged, Sect Mat 8*7-41
⋄ B20 D35 ⋄ REV Zbl 14.98 JSL 2.84 FdM 62.1059
• ID 10360

PEPIS, J. [1937] *Ueber das Entscheidungsproblem des engeren logischen Funktionskalkuels (Polish) (German summary)* (S 0281) Arch Towarz Nauk Lwow, Sect 3 7/8*1-172
⋄ B20 B25 D35 ⋄ REV Zbl 19.97 JSL 4.93 FdM 63.823
• ID 19493

PEPIS, J. [1938] *Ein Verfahren der mathematischen Logik* (J 0036) J Symb Logic 3*61-76
⋄ B20 D35 ⋄ REV JSL 3.161 FdM 64.29 • ID 10361

PEPIS, J. [1938] *Untersuchungen ueber das Entscheidungsproblem der mathematischen Logik* (J 0027) Fund Math 30*257-348
⋄ B20 D35 ⋄ REV Zbl 18.385 JSL 3.160 FdM 64.29
• ID 10362

PEREL'ROIZEN, E.Z. [1980] see LEVIN, V.A.

PERKINS, E. [1981] *A global intrinsic characterization of Brownian local time* (J 2661) Ann Probab 9*800-817
⋄ H10 ⋄ REV MR 82k:60158 Zbl 469#60081 • ID 82462

PERKINS, E. [1982] *On the construction and distribution of a local martingale with a given absolute value* (J 0064) Trans Amer Math Soc 271*261-281
⋄ H10 ⋄ REV MR 83h:60044 Zbl 506#60043 • ID 37829

PERKINS, E. [1982] *Weak invariance principles for local time* (J 0982) Z Wahrscheinltheor & Verw Geb 60*437-451
⋄ H10 ⋄ REV MR 83h:60080 Zbl 465#60065 • ID 39266

PERKINS, E. [1983] see HOOVER, D.N.

PERKINS, E. [1983] *Stochastic processes and nonstandard analysis* (C 3884) Nonstandard Anal - Recent Develop 162-185
◇ H10 ◇ REV MR 85d:60104 Zbl 526 # 60028 • ID 38190

PERRIN, M.J. & ZALC, A. [1975] *Un theoreme de compacite en analyse. Application aux β_n-modeles et β_n-extensions (English summary)* (J 2313) C R Acad Sci, Paris, Ser A-B 280∗A177-A180
◇ B15 C62 ◇ REV MR 53 # 12933 Zbl 301 # 02056
• ID 23181

PERRIN, M.J. see Vol. III, V, VI for further entries

PERZANOWSKI, J. [1973] *A linguistic criterion of structural incompleteness* (J 0302) Rep Math Logic, Krakow & Katowice 1∗13-14
◇ B22 ◇ REV MR 49 # 8849 Zbl 278 # 02017 • ID 10388

PERZANOWSKI, J. see Vol. II, V, VI for further entries

PESCHEL, K. [1979] *Moeglichkeiten einer formallogischen Darstellung kommunikativer Situationen* (S 2877) Wiss Z Univ Leipzig, Ges-Sprachwiss Reihe 28∗325-331
◇ A05 B05 ◇ REV MR 81b:03030 Zbl 421 # 03008
• ID 53407

PESCHEL, K. see Vol. II for further entries

PETER, R. [1962] *Ueber die "kuerzeste" Form von Booleschen Funktionen (Russian summary)* (J 0462) Mat Fiz Oszt Koezlem, Acad Sci Hung 7∗79-93
◇ B05 ◇ REV MR 32 # 2324 Zbl 112.8 • ID 10413

PETER, R. [1969] *Ueber zweistufig definierte Sprachen* (P 1841) Fct Recurs & Appl;1967 Tihany 12-18
◇ B15 D20 ◇ ID 32551

PETER, R. [1973] *Mathematische Fassung der sogenannten "Entscheidungs-Tabellen" (Russian summary)* (J 0380) Acta Cybern (Szeged) 2∗89-108
◇ B05 ◇ REV MR 58 # 8454 Zbl 271 # 02026 • ID 29916

PETER, R. also published under the name POLITZER, R.

PETER, R. see Vol. III, IV, V, VI for further entries

PETERS, F.E. [1974] *Einfuehrung in mathematische Methoden der Informatik* (X 0876) Bibl Inst: Mannheim 342pp
◇ B70 B98 D80 D98 ◇ REV MR 53 # 2499 Zbl 319 # 94001 • ID 64464

PETERS, R. [1964] *Two remarks concerning Menger's and Schultz's postulates for the substitutive algebra of the 2-place functors in the 2-valued calculus of propositions* (J 0047) Notre Dame J Formal Log 5∗125-128
◇ B05 ◇ REV MR 31 # 33 Zbl 148.243 • ID 10428

PETERSON, G.E. [1976] *Theorem proving with lemmas* (J 0037) ACM J 23∗573-581
◇ B35 ◇ REV MR 56 # 1842 • ID 27096

PETERSON, G.E. [1983] *A technique for establishing completeness results in theorem proving with equality* (J 1428) SIAM J Comp 12∗82-100
◇ B35 ◇ REV MR 84b:68135 Zbl 522 # 68087 • ID 36784

PETERSON, J.G. [1976] *Shortest single axiom for the classical equivalential calculus* (J 0047) Notre Dame J Formal Log 17∗267-271
◇ B20 ◇ REV MR 56 # 15359 Zbl 313 # 02006 • ID 18333

PETERSON, J.G. [1977] *Single axioms for the classical equivalential calculus* (J 0329) Math Chron (Auckland) 6∗21-47
◇ B20 B30 ◇ REV MR 58 # 21418 Zbl 381 # 03014
• ID 51882

PETERSON, J.G. [1978] *An automatic theorem prover for substitution and detachment systems* (J 0047) Notre Dame J Formal Log 19∗119-122
◇ B35 ◇ REV MR 58 # 8552 Zbl 368 # 68086 • ID 27093

PETRI, N.V. [1969] *Algorithms connected with predicates and Boolean functions (Russian)* (J 0023) Dokl Akad Nauk SSSR 185∗37-99
• TRANSL [1969] (J 0062) Sov Math, Dokl 10∗294-297
◇ B35 D15 ◇ REV MR 40 # 4113 Zbl 193.315 • ID 10436

PETRI, N.V. [1969] *The complexity of algorithms and their operating time (Russian)* (J 0023) Dokl Akad Nauk SSSR 186∗30-31
• TRANSL [1969] (J 0062) Sov Math, Dokl 10∗547-549
◇ B35 D15 D20 ◇ REV MR 40 # 7115 Zbl 272 # 02053
• ID 10438

PETRI, N.V. [1975] see BOKSHTEJN, M.F.

PETRI, N.V. see Vol. IV, VI for further entries

PETRICK, S.R. [1960] *On the minimization of boolean functions* (P 0696) Inform Processing (1);1959 Paris 422-423
◇ B35 ◇ ID 27693

PETROSYAN, A.V. [1964] *Einige Eigenschaften von Funktionen der Logikalgebra (Russisch)* (S 0422) Tr Vychisl Tsentra Akad Nauk Armyan SSR & Univ Erevan 2∗38-50
◇ B05 ◇ ID 16249

PETROSYAN, A.V. [1982] *Some differential properties of Boolean functions (Russian) (English summary)* (J 2845) Tanulmanyok 135∗15-37
◇ B05 ◇ REV MR 84g:94028 • ID 45859

PETROV, V.V. [1984] see ERSHOV, YU.L.

PETROVA, L.P. [1982] see KAMENSKIJ, M.I.

PEVAC, I. [1982] see CVETKOVIC, D.M.

PEVAC, I. [1983] see CVETKOVIC, D.M.

PEVAC, I. [1984] see CVETKOVIC, D.M.

PFANZAGL, J. [1968] *Theory of measurement* (X 0827) Wiley & Sons: New York 235pp
◇ A05 B30 E75 ◇ REV MR 40 # 4180 Zbl 186.536
• ID 10447

PFEIFFER, H. [1982] see COHEN, L.J.

PFEIFFER, H. see Vol. II, V, VI for further entries

PFENNING, F. [1984] *Analytic and non-analytic proofs* (P 2633) Autom Deduct (7);1984 Napa 394-413
◇ B35 ◇ REV Zbl 547 # 03032 • ID 43231

PFENNING, F. [1984] see ANDREWS, P.B.

PHILLIPS, N.C.K. [1984] *Theorems ad infinitum* (J 2823) Quaest Math, S Africa 7∗295-298
◇ B10 ◇ REV MR 86d:03012 Zbl 561 # 03006 • ID 44358

PHILLIPS, N.C.K. see Vol. III, V for further entries

PHILLIPS, R.G. [1969] *Liouville's theorem* (J 0048) Pac J Math 28∗397-405
◇ C65 H05 ◇ REV MR 42 # 4715 • ID 32232

PHILLIPS, R.G. see Vol. III for further entries

PIAGET, J. [1952] *Essai sur les transformations des operations logiques. Les 256 operations ternaires de la logique bivalente des propositions* (X 0840) Pr Univ France: Paris xi+239pp
⋄ B05 ⋄ ID 22425

PICCOLI, A. [1976] *Anelli nel calcolo delle proposizioni* (J 0220) Atti Accad Sci Torino, Fis Mat Nat 110*295-297
⋄ B05 ⋄ REV MR 55 # 2564 Zbl 355 # 02008 • ID 50206

PICCOLI, A. [1976] *Gruppi del calcolo delle proposizioni* (J 0220) Atti Accad Sci Torino, Fis Mat Nat 110*257-260
⋄ B05 ⋄ REV MR 54 # 7183 Zbl 354 # 02017 • ID 24978

PICH, W. [1933] *Ueber Unabhaengigkeitsbeweise im Aussagenkalkuel* (J 0931) Anz Oesterr Akad Wiss, Math-Nat Kl 70*194-195
⋄ B05 ⋄ REV Zbl 7.385 JSL 2.46 FdM 59.876 • ID 10464

PICH, W. [1936] *Ueber Unabhaengigkeitsbeweise im Aussagenkalkuel* (J 0124) Monatsh Math-Phys 45*1-7
⋄ B05 ⋄ REV Zbl 15.193 JSL 2.46 FdM 62.1057 • ID 10465

PICHAI, V. [1976] see CHAN, S.P.

PICKERT, G. [1958] *Ebene Inzidenzgeometrie: Beispiele zur Axiomatik mit einer Einfuehrung in die formale Logik* (X 0932) Salle: Frankfurt 92pp
⋄ B30 ⋄ REV MR 20 # 5139 Zbl 80.139 • ID 10468

PICKERT, G. see Vol. V for further entries

PIECZKOWSKI, A. [1964] *On the equivalence of the calculus of dependent sentential variables and the cylindrical algebra without diagonal elements* (J 0014) Bull Acad Pol Sci, Ser Math Astron Phys 12*143-146
⋄ B20 G15 ⋄ REV MR 29 # 2176 • ID 10470

PIECZKOWSKI, A. [1968] *Remarks on J.-L. Gardies "Les deux tetraedres des liaisons logiques interpropositionnelles bivalentes" (Polish)* (J 0063) Studia Logica 23*163-164
⋄ B05 ⋄ REV MR 39 # 1283 • ID 33129

PIECZKOWSKI, A. [1968] *Undecidability of the homogeneous formulas of degree 3 of the predicate calculus (Polish and Russian summaries)* (J 0063) Studia Logica 22*7-16
⋄ B20 D35 ⋄ REV MR 38 # 4311 Zbl 315 # 02045 • ID 10474

PIECZKOWSKI, A. [1970] see KOTAS, J.

PIECZKOWSKI, A. [1974] *Ueber Theorien im erweiterten Sinne* (J 0063) Studia Logica 33*317-331
⋄ B10 ⋄ REV MR 51 # 122 Zbl 311 # 02056 • ID 15196

PIECZKOWSKI, A. see Vol. II, VI for further entries

PIETRZYKOWSKI, T. [1973] *A complete mechanization of second-order type theory* (J 0037) ACM J 20*333-364
⋄ B15 B35 F35 ⋄ REV MR 49 # 8443 Zbl 253 # 68021 • ID 10491

PIETRZYKOWSKI, T. [1977] see JENSEN, D.C.

PIETRZYKOWSKI, T. [1980] see COX, P.T.

PIETRZYKOWSKI, T. [1981] see COX, P.T.

PIETRZYKOWSKI, T. [1982] see MATWIN, S.

PIETRZYKOWSKI, T. [1984] see COX, P.T.

PINCUS, D. [1974] *The strength of the Hahn-Banach theorem* (P 1083) Victoria Symp Nonstand Anal;1972 Victoria 369*203-248
⋄ E25 E75 H05 ⋄ REV MR 57 # 16072 Zbl 279 # 02044 • ID 21224

PINCUS, D. see Vol. III, V for further entries

PINKAVA, V. [1971] *Logical models of sexual deviations* (J 1741) Int J Man-Mach Stud 3*351-374
⋄ B80 ⋄ REV Zbl 226 # 92015 • ID 41991

PINKAVA, V. see Vol. II for further entries

PINTER, C. [1963] *Formes minimales des fonctions booleennes* (J 0109) C R Acad Sci, Paris 256*3407-3409
⋄ B05 ⋄ REV MR 27 # 2427 Zbl 125.278 • ID 10503

PINTER, C. [1971] *On simplifying truth functions: a preliminary reduction of coreless formulas* (J 0187) IEEE Trans Comp C-20*938-941
⋄ B05 ⋄ REV MR 47 # 8237 Zbl 227 # 94026 • ID 10505

PINTER, C. see Vol. II, III, V, VI for further entries

PINUS, A.G. [1973] *A weak second order theory of fixed sets (Russian)* (C 1443) Algebra, Vyp 2 (Irkutsk) 154-160
⋄ B15 D35 ⋄ REV MR 54 # 4956 • ID 24124

PINUS, A.G. [1979] *A calculus of one-place predicates (Russian)* (J 0031) Izv Vyssh Ucheb Zaved, Mat (Kazan) 1979/1(200)*54-60
• TRANSL [1979] (J 3449) Sov Math 23/1*43-47
⋄ B20 B25 C55 C80 C85 ⋄ REV MR 80j:03053 Zbl 408 # 03008 • ID 53695

PINUS, A.G. see Vol. III, IV, V, VI for further entries

PIOCHI, B. [1983] *Logical matrices and non-structural consequence operators* (J 0063) Studia Logica 42*33-42
⋄ B22 ⋄ REV Zbl 537 # 03048 • ID 43783

PIOCHI, B. see Vol. II for further entries

PIROG-RZEPECKA, K. [1975] see HALKOWSKA, K.

PIROG-RZEPECKA, K. [1976] see HALKOWSKA, K.

PIROG-RZEPECKA, K. [1978] see HALKOWSKA, K.

PIROG-RZEPECKA, K. see Vol. II, V for further entries

PITRAT, J. [1966] *Realization of a program which chooses the theorems it proves* (P 0573) Inform Processing (3);1965 New York 2*324-325
⋄ B35 ⋄ REV Zbl 173.17 • ID 29437

PKHAKADZE, SH.S. [1975] see MAGNARADZE, D.G.

PKHAKADZE, SH.S. [1977] *Some questions of notation theory (Russian)* (X 1052) Tbilisi Univ: Tbilisi 195pp
⋄ B30 ⋄ REV MR 80e:03010 Zbl 381 # 03004 • ID 51872

PKHAKADZE, SH.S. [1977] see MAGNARADZE, L.G.

PKHAKADZE, SH.S. [1978] see MAGNARADZE, L.G.

PKHAKADZE, SH.S. [1979] *Some questions of notation theory (Russian)* (P 2539) Frege Konferenz (1);1979 Jena 319-340
⋄ B30 ⋄ REV MR 82j:03009 • ID 77269

PKHAKADZE, SH.S. [1982] *Some problems of the theory of abbreviating symbols (Russian) (English and Georgian summaries)* (J 0954) Tr Inst Prikl Mat, Tbilisi 11*42-50
⋄ B03 ⋄ REV MR 85f:03009 • ID 40446

PKHAKADZE, SH.S. see Vol. IV, V for further entries

PKHAKADZE, V.SH. [1982] *Substitutability of equivalences and equalities in the equality theory of Bourbaki (Russian) (English and Georgian summaries)* (J 0954) Tr Inst Prikl Mat, Tbilisi 11*33-41
⋄ B10 ⋄ REV MR 85b:03014 • ID 40549

PLA I CARRERA, J. [1980] *On Keisler's axiomatization of nonstandard analysis (Catalan) (English summary)* (P 3512) Jorn Mat Luso-Espanol (7);1980 St Feliu de Guixois 1*93-95
⋄ H05 ⋄ ID 44985

PLA I CARRERA, J. see Vol. II, III, IV, V, VI for further entries

PLAISTED, D.A. [1980] *Abstraction mappings in mechanical theorem proving* (P 3063) Autom Deduct (5);1980 Les Arcs 264-280
⋄ B35 ⋄ REV MR 81k:68071 Zbl 442 # 68096 • ID 56404

PLAISTED, D.A. [1980] *The application of multivariate polynomials to inference rules and partial tests for unsatisfiability* (J 1428) SIAM J Comp 9*698-705
⋄ B35 D15 F20 ⋄ REV MR 82g:68045 Zbl 448 # 68022 • ID 56667

PLAISTED, D.A. [1981] *Theorem proving with abstraction* (J 0503) Artif Intell 16*47-108
• TRANSL [1984] (J 3079) Kiber Sb Perevodov, NS 21*139-212
⋄ B35 ⋄ REV MR 82m:68134 Zbl 454 # 68113 • ID 69776

PLAISTED, D.A. [1982] *A simplified problem reduction format* (J 0503) Artif Intell 18*227-261
⋄ B35 ⋄ REV MR 83f:68108 Zbl 497 # 68058 • ID 36642

PLAISTED, D.A. [1982] see GREENBAUM, S.

PLAISTED, D.A. [1984] *Complete problems in the first-order predicate calculus* (J 0119) J Comp Syst Sci 29*8-35
⋄ B10 D15 ⋄ REV MR 86b:03045 • ID 44042

PLAISTED, D.A. [1984] *Using examples, case analysis, and dependency graphs in theorem proving* (P 2633) Autom Deduct (7);1984 Napa 356-374
⋄ B35 ⋄ REV Zbl 546 # 68073 • ID 44455

PLAISTED, D.A. see Vol. II, IV for further entries

PLASS, M. [1979] see ASPVALL, B.

PLAVSIC, V.M. [1979] see DANIELSSON, P.E.

PLISKO, V.E. [1978] *Some variants of the notion of realizability for predicate formulas (Russian)* (J 0216) Izv Akad Nauk SSSR, Ser Mat 42*637-653
• TRANSL [1978] (J 0448) Math of USSR, Izv 12*588-604
⋄ B10 F50 ⋄ REV MR 80b:03035 Zbl 384 # 03042
• ID 52084

PLISKO, V.E. see Vol. IV, VI for further entries

PLIUSKEVICIENE, A. [1971] *Specialization of the use of axioms for deduction search in axiomatic theories with equality (Russian) (English summary)* (S 0228) Zap Nauch Sem Leningrad Otd Mat Inst Steklov 20*175-185,287
• TRANSL [1973] (J 1531) J Sov Math 1*110-116
⋄ B35 ⋄ REV MR 45 # 4947 MR 48 # 5808 Zbl 231 # 02035 • ID 27741

PLIUSKEVICIENE, A. [1972] *Extension of the inverse method to axiomatic theories with equality (Russian) (English summary)* (S 0228) Zap Nauch Sem Leningrad Otd Mat Inst Steklov 32*108-115,157
⋄ B35 ⋄ REV MR 52 # 5376 Zbl 347 # 68049 • ID 18338

PLIUSKEVICIENE, A. [1985] *Generalized disjunction and existence properties for the logic of provability (Russian)* (J 3939) Mat Logika Primen (Akad Nauk Litov SSR) 1985/4*16-24
⋄ B35 B45 ⋄ ID 49019

PLIUSKEVICIENE, A. see Vol. VI for further entries

PLIUSKEVICIUS, R. [1968] *Kanger's version of the predicate calculus with symbols for not everywhere defined functions (Russian)* (S 0228) Zap Nauch Sem Leningrad Otd Mat Inst Steklov 8*211-224
• TRANSL [1968] (J 0521) Semin Math, Inst Steklov 8*103-109
⋄ B10 B60 ⋄ REV MR 45 # 1727 Zbl 175.273 • ID 10543

PLIUSKEVICIUS, R. see Vol. II, VI for further entries

PLOTKIN, G.D. [1969] *A note on inductive generalization* (J 0508) Machine Intelligence 5*153-163
⋄ B35 ⋄ REV MR 41 # 4861 Zbl 219 # 68045 • ID 46584

PLOTKIN, G.D. [1972] *Building-in equational theories* (J 0508) Machine Intelligence 7*73-90
⋄ B35 ⋄ REV Zbl 262 # 68036 • ID 64560

PLOTKIN, G.D. [1980] *λ-definability in the full type hierarchy* (C 3050) Essays Combin Log, Lambda Calc & Formalism (Curry) 363-373
⋄ B15 B40 ⋄ REV MR 82a:03016 Zbl 469 # 03006
• ID 77373

PLOTKIN, G.D. see Vol. IV, VI for further entries

PLOTKIN, J.M. & ROSENTHAL, J.W. [1982] *The expected complexity of analytic tableaux analysis in propositional calculus* (J 0047) Notre Dame J Formal Log 23*409-426
⋄ B05 B35 D15 F20 ⋄ REV MR 83k:03044 Zbl 464 # 03011 • ID 35398

PLOTKIN, J.M. see Vol. II, III, V for further entries

PLUHACEK, A. [1973] *Ueber einige Eigenschaften der Loesungen boolescher Gleichungen* (J 3289) Wiss Z Tech Hochsch Madgeburg 17*253-255
⋄ B05 ⋄ REV Zbl 298 # 94049 • ID 64561

PLYMEN, R.J. [1961] *A model of the arithmetic of alephs in the equation calculus* (J 0068) Z Math Logik Grundlagen Math 7*257-258
⋄ B20 E10 ⋄ REV MR 25 # 2958 Zbl 118.253 JSL 35.321
• ID 10567

PNUELI, A. [1985] *In transition from global to modular temporal reasoning about programs* (P 4621) Log & Models of Concurrent Syst;1984 La Colle-sur-Loup 123-144
⋄ B35 B45 B75 ⋄ ID 49386

PNUELI, A. [1985] *Linear and branching structures in the semantics and logics of reactive systems* (P 4628) Automata, Lang & Progr (12);1985 Nafplion 15-32
⋄ B35 B45 B75 ⋄ ID 49524

PNUELI, A. see Vol. II, IV for further entries

PODEWSKI, K.-P. [1982] see COHEN, L.J.

PODEWSKI, K.-P. see Vol. III, V for further entries

POEL VAN DER, W.L. [1969] *The foundation of the propositional calculus by means of one axiom (Dutch)* (X 1121) Math Centr: Amsterdam ZW-001*6pp
⋄ B05 ⋄ REV MR 42 # 34 • ID 21170

POEL VAN DER, W.L. see Vol. VI for further entries

POGORZELSKI, H.A. & RYAN, W.J. [1982] *Foundations of semilogical theory of numbers. Vol.1* (X 3782) Univ Maine: Orono 608pp
  ◊ B28 ◊ REV MR 84g:03019 Zbl 532#03003 • REM Vol.2 1985 • ID 34134

POGORZELSKI, H.A. & RYAN, W.J. [1985] *Foundations of semilogical theory of numbers II* (X 3782) Univ Maine: Orono iii+743pp
  ◊ B28 ◊ REM Vol.I 1982 • ID 48966

POGORZELSKI, H.A. see Vol. IV, VI for further entries

POGORZELSKI, W.A. & SLUPECKI, J. [1960] *Basic properties of deductive systems based on nonclassical logics I,II (Polish) (Russian and English summaries)* (J 0063) Studia Logica 9*163-176,10*77-95
  ◊ B22 F50 ◊ REV MR 26#6039 Zbl 129.256 • ID 21001

POGORZELSKI, W.A. & SLUPECKI, J. [1961] *A variant of the proof of the completeness of the first order functional calculus (Polish and Russian summaries)* (J 0063) Studia Logica 12*125-134
  ◊ B10 C07 ◊ REV MR 26#6040 Zbl 156.7 JSL 36.688 • ID 31969

POGORZELSKI, W.A. & SLUPECKI, J. [1962] *A proof of the completeness of the classical propositional calculus on the ground of an axiomatic methodology (Polish)* (J 0481) Acta Univ Wroclaw 12*11-18
  ◊ B05 G05 ◊ REV MR 38#980 JSL 32.536 • ID 21240

POGORZELSKI, W.A. & SLUPECKI, J. [1962] *On mathematical proof (Polish)* (X 1166) PZW: Warsaw 128pp
  ◊ B98 ◊ REV JSL 31.284 • ID 31970

POGORZELSKI, W.A. [1962] *The adequacy of theory of deductive systems with respect to sentential calculi (Polish) (Russian and English summaries)* (J 0063) Studia Logica 13*103-131
  ◊ B22 ◊ REV MR 26#4888 Zbl 121.253 • ID 31968

POGORZELSKI, W.A. [1964] *A schema of deduction theorems for the propositional calculus (Polish) (Russian and English summaries)* (J 0063) Studia Logica 15*181-188
  ◊ B05 ◊ REV MR 32#5495 Zbl 292#02004 • ID 10588

POGORZELSKI, W.A. [1964] *A survey of deduction theorems for the propositional calculi (Polish) (Russian and English summaries)* (J 0063) Studia Logica 15*163-179
  ◊ B05 ◊ REV MR 32#5494 Zbl 292#02003 • ID 10589

POGORZELSKI, W.A. [1968] *On the scope of the classical deduction theorem* (J 0036) J Symb Logic 33*77-81
  ◊ A05 B10 ◊ REV MR 39#2600 Zbl 175.260 JSL 40.606 • ID 10590

POGORZELSKI, W.A. [1968] *Some remarks on the concept of completeness of the propositional calculus I (Polish) (Russian and English summaries)* (J 0063) Studia Logica 23*43-58
  ◊ B05 ◊ REV MR 39#1284 Zbl 311#02015 • REM Part II 1973 by Pogorzelski,W.A. & Prucnal,T. • ID 10591

POGORZELSKI, W.A. [1969] *Classical calculus of propositions. Outline of the theory (Polish)* (X 1034) PWN: Warsaw 203pp
  ◊ B05 ◊ ID 19513

POGORZELSKI, W.A. [1969] *Non-creativity and translability of definitions in propositional calculus (Polish)* (J 0481) Acta Univ Wroclaw 101*3-12
  ◊ B05 ◊ ID 31974

POGORZELSKI, W.A. [1969] *Two-valued propositional calculus and deduction theorem (Polish)* (J 0481) Acta Univ Wroclaw 101*13-17
  ◊ B05 ◊ ID 31973

POGORZELSKI, W.A. [1971] *Structural completeness of the propositional calculus (Russian summary)* (J 0014) Bull Acad Pol Sci, Ser Math Astron Phys 19*349-351
  ◊ B05 ◊ REV MR 46#1548 Zbl 214.7 JSL 40.604 • ID 10593

POGORZELSKI, W.A. & PRUCNAL, T. [1973] *Some remarks on the notion of completeness of the propositional calculus. II* (J 0302) Rep Math Logic, Krakow & Katowice 1*15-19
  ◊ B05 ◊ REV MR 52#7888 Zbl 283#02012 • REM Part I 1968 • ID 18340

POGORZELSKI, W.A. [1974] *Concerning the notion of completeness of invariant propositional calculi* (J 0063) Studia Logica 33*69-72,121
  ◊ B22 ◊ REV MR 50#83 Zbl 283#02013 • ID 10597

POGORZELSKI, W.A. & PRUCNAL, T. [1974] *Equivalence of the structural completeness theorem for propositional calculus and the boolean representation theorem* (J 0302) Rep Math Logic, Krakow & Katowice 3*37-39
  ◊ B05 G05 ◊ REV MR 50#9569 Zbl 301#02045 • ID 10594

POGORZELSKI, W.A. & PRUCNAL, T. [1974] *Introduction to mathematical logic. Part I. Elements of the algebra of propositional logic (Polish)* (X 1425) Univ Slaski: Katowice 36pp
  ◊ B98 G05 ◊ REV MR 52#10361 • ID 21667

POGORZELSKI, W.A. & PRUCNAL, T. [1974] *Structural completeness of the first-order predicate calculus* (J 0014) Bull Acad Pol Sci, Ser Math Astron Phys 22*215-217
  ◊ B10 ◊ REV MR 50#1835 Zbl 286#02018 • ID 10598

POGORZELSKI, W.A. [1974] see BIELA, A.

POGORZELSKI, W.A. [1975] *On the notion of the rule of inference and completeness of systems: some comments on H.C. Wasserman's remarks "Admissible rules, derivable rules, and extendible logistic systems"* (J 0302) Rep Math Logic, Krakow & Katowice 5*73-75
  ◊ B22 ◊ REV MR 53#12923 • REM The article was published in J0047 15(1974)*265-278 • ID 23171

POGORZELSKI, W.A. & PRUCNAL, T. [1975] *Structural completeness of the first-order predicate calculus* (J 0068) Z Math Logik Grundlagen Math 21*315-320
  ◊ B10 ◊ REV MR 52#5400 Zbl 312#02011 • ID 18342

POGORZELSKI, W.A. & PRUCNAL, T. [1975] *The substitution rule for predicate letters in the first-order predicate calculus* (J 0302) Rep Math Logic, Krakow & Katowice 5*77-90
  ◊ B10 ◊ REV MR 58#16125 Zbl 335#02007 • ID 21904

POGORZELSKI, W.A. [1976] see LESISZ, W.

POGORZELSKI, W.A. [1981] *Classical calculus of quantifiers. Outline of the theory (Polish)* (X 1034) PWN: Warsaw 1981*228pp
  ◊ B98 ◊ REV MR 83e:03004 • ID 35270

POGORZELSKI, W.A. [1981] *On Hilbert's operation on logical rules I* (J 0302) Rep Math Logic, Krakow & Katowice 12*35-50
  ◊ B22 ◊ REV MR 84e:03040 Zbl 465#03002 • REM Part II 1984 • ID 54905

POGORZELSKI, W.A. & WOJTYLAK, P. [1982] *Elements of the theory of completeness in propositional logic (Polish summary)* (S 0544) Prace Mat Uniw Katowice 512∗144pp
⋄ B22 ⋄ REV MR 84g:03001 Zbl 539 # 03002 • ID 34115

POGORZELSKI, W.A. [1984] *On Hilbert's operation on logical rules. II* (J 0302) Rep Math Logic, Krakow & Katowice 17∗3-11
⋄ B22 ⋄ REV MR 86e:03031 Zbl 556 # 03011 • REM Part I 1981 • ID 44053

POGORZELSKI, W.A. see Vol. II for further entries

POGOSYAN, EH.M. [1977] see KARAPETYAN, B.K.

POGOSYAN, EH.M. see Vol. II, IV for further entries

POHM, A. & REID, A. & SCHAUER, R. & STEWART, R. [1960] *Some applications of magnetic film parametrons as logical devices* (J 0072) IRE Trans Electr Comp EC-9∗315-320
⋄ B80 ⋄ ID 41877

POINCARE, H. [1902] *Du role de l'intuition et de la logique en mathematiques* (P 1484) Int Congr Math (2);1900 Paris 115-130
⋄ A05 B98 ⋄ ID 38663

POINCARE, H. [1903] *La science et l'hypothese* (X 2075) Flammarion: Paris
• TRANSL [1904] (X 1079) Teubner: Leipzig
⋄ A05 B30 ⋄ REV FdM 34.80 • REM Repr. 1906,1914,1974. English transl. 1905 • ID 38664

POINCARE, H. [1908] *Science et methode* (X 2075) Flammarion: Paris 311pp
• TRANSL [1914] (X 1079) Teubner: Leipzig 311pp (German) [1914] (X 1035) Nelson: Walton on Thames 288pp (English)
⋄ A05 B30 ⋄ REV JSL 18.327 • ID 38666

POINCARE, H. [1909] *La logique de l'infini* (J 0145) Rev Metaph Morale 17∗461-482
⋄ A05 B15 E30 ⋄ REV FdM 40.97 • ID 37938

POINCARE, H. see Vol. V for further entries

POIZAT, B. [1975] *Theoremes globaux (English summary)* (J 2313) C R Acad Sci, Paris, Ser A-B 280∗A845-A847
⋄ B20 C40 ⋄ REV MR 52 # 2863 Zbl 301 # 02047
• ID 17649

POIZAT, B. [1982] *Deux ou trois choses que je sais de L_n* (J 0036) J Symb Logic 47∗641-658
⋄ B20 C07 C13 C50 ⋄ REV MR 84b:03055 Zbl 507 # 03014 • ID 35642

POIZAT, B. see Vol. III, IV, V for further entries

POLJAK, S. & TURZIK, D. [1982] *A polynomial algorithm for constructing a large bipartite subgraph, with an application to a satisfiability problem* (J 0017) Canad J Math 34∗519-524
⋄ B05 D15 ⋄ REV MR 83j:05048 Zbl 471 # 68041
• ID 39883

POLLOCK, J.L. [1969] *Introduction to symbolic logic* (X 0818) Holt Rinehart & Winston: New York xii+241pp
⋄ B98 ⋄ REV JSL 40.101 • ID 10626

POLLOCK, J.L. [1970] *Zermelo-Fraenkel set theory and cumulative type theory* (J 0079) Logique & Anal, NS 13∗452-466
⋄ A05 B15 E30 ⋄ REV MR 44 # 6478 Zbl 232 # 02042
• ID 10627

POLLOCK, J.L. [1971] *Henkin style completeness proofs in theories lacking negation* (J 0047) Notre Dame J Formal Log 12∗509-511
⋄ B20 ⋄ REV MR 45 # 6576 Zbl 223 # 02051 • ID 10628

POLLOCK, J.L. see Vol. II, III for further entries

POLYA, G. [1940] *Sur les types des propositions composees* (J 0036) J Symb Logic 5∗98-103
⋄ B05 ⋄ REV MR 2.65 Zbl 24.1 JSL 5.168 FdM 66.28
• ID 10630

POLYA, G. [1941] *Heuristic reasoning and the theory of probability* (J 0005) Amer Math Mon 48∗450-465
⋄ B30 ⋄ REV MR 3.131 • ID 10631

POLYA, G. see Vol. II for further entries

PONASSE, D. [1966] *Structures et techniques de la logique mathematique* (J 1750) Atomes & Nucleus 237∗605-609
⋄ B05 ⋄ ID 31203

PONASSE, D. [1967] *Algebrisation du calcul propositionnel au moyen d'anneaux booleens universels* (J 0179) Ann Fac Sci Clermont 35∗33-36
⋄ B05 G05 ⋄ REV MR 44 # 3829 • ID 10637

PONASSE, D. [1967] *Logique mathematique. Elements de base: calcul propositionnel, calcul des predicats* (X 1172) Office Central Lib (OCDL): Paris 164pp
⋄ B98 ⋄ REV MR 39 # 31 Zbl 153.5 JSL 35.579 • ID 22069

PONASSE, D. [1973] *Mathematical logic* (X 0836) Gordon & Breach: New York x+126pp
⋄ B98 ⋄ REV MR 49 # 8805 Zbl 252 # 02001 JSL 41.790
• ID 10640

PONASSE, D. see Vol. III, V for further entries

POPESCU, E. [1969] *A construction of the set of rational numbers starting from the cartesian product $N \times N \times (N \setminus \{0\})$* (J 0937) Bul Inst Politeh Brasov Ser A 11∗17-23
⋄ B28 ⋄ REV MR 40 # 5578 • ID 10645

POPESCU, E. see Vol. V for further entries

POPOV, A.I. [1959] *Introduction to mathematical logic (Russian)* (X 0938) Leningrad Univ: Leningrad 108pp
⋄ B98 ⋄ REV MR 22 # 3674 Zbl 97.244 • ID 10646

POPOV, M.M. [1983] *Logical connectives as derivatives of the rules of inference (Russian)* (S 2582) Semiotika & Inf, Akad Nauk SSSR 21∗89-107
⋄ B05 ⋄ REV MR 85k:03006 Zbl 574 # 03003 • ID 44855

POPOV, S.V. [1974] *An equivalence relation and a complete system of schemes of equivalent transformations of deductions in propositional calculus (Russian)* (J 0052) Probl Kibern 29∗103-150,245
⋄ B05 F07 ⋄ REV MR 51 # 79 Zbl 309 # 02029 • ID 15197

POPOV, S.V. [1975] *The complexity of deductions in certain propositional calculi (Russian)* (J 0052) Probl Kibern 30∗57-146
⋄ B05 F20 ⋄ REV MR 52 # 5377 Zbl 414 # 03035
• ID 18344

POPOV, S.V. [1976] *On the complexity of derivations in classical propositional calculus (Russian)* (J 0023) Dokl Akad Nauk SSSR 228∗1041-1044
• TRANSL [1976] (J 0062) Sov Math, Dokl 17∗876-880
⋄ B05 F20 ⋄ REV MR 54 # 9972 Zbl 357 # 02009
• ID 25598

POPOV, S.V. [1977] see CHEREDNICHENKO, N.D.

POPOV, S.V. [1977] *Some complexity characteristics of derivations in propositional calculus (Russian) (English summary)* (S 2651) Prepr Inst Prikl Mat, Akad Nauk SSSR 116*63pp
⋄ B05 B70 F20 ⋄ REV MR 58 # 10284 • ID 77440

POPOV, S.V. & ZAKHAR'YASHCHEV, M.V. [1978] *Deduction procedure based on the syntactic tree method (Russian)* (S 2651) Prepr Inst Prikl Mat, Akad Nauk SSSR 111*51pp
⋄ B35 ⋄ REV MR 81b:03014 Zbl 416 # 68084 • ID 53232

POPOV, S.V. [1982] *Diagrams of natural deductions (Russian)* (J 0052) Probl Kibern 39*5-66
⋄ B20 ⋄ REV MR 84e:03013 Zbl 517 # 03002 • ID 34352

POPOV, S.V. see Vol. II, III, IV, VI for further entries

POPOV, V.A. [1973] see LYSENKO, E.V.

POPOV, V.A. see Vol. II for further entries

POPOV, V.M. [1982] *Solvability of the syllogistic with negative terms (Russian)* (C 3849) Modal & Relevant Log, Vyp 1 36-46
⋄ B20 ⋄ REV Zbl 526 # 03005 • ID 38165

POPOV, V.M. see Vol. II for further entries

POPOVIC, M.V. (ED.) [1980] *Actual problems of logic and methodology of science (Russian)* (X 2199) Naukova Dumka: Kiev 336pp
⋄ A05 B97 ⋄ REV MR 82b:03004 Zbl 505 # 00006 • ID 38202

POPOVICI, C.P. [1972] *Boolean expressions (Romanian)* (J 0378) Gaz Mat (Bucharest) Ser A 77*461-465
⋄ B05 ⋄ REV MR 52 # 2805 • ID 17609

POPOVICI, C.P. [1973] *Boolean functions (Romanian)* (J 0378) Gaz Mat (Bucharest) Ser A 78(A)*10-16
⋄ B05 ⋄ REV MR 52 # 2806 Zbl 317 # 06010 • ID 17610

POPOWICZ, Z. [1975] *Remarks on dual structures in a Dirac space (Russian summary)* (J 0014) Bull Acad Pol Sci, Ser Math Astron Phys 23*1119-1124
⋄ H05 ⋄ REV MR 52 # 14948 Zbl 318 # 46015 • ID 21887

POPPER, K.R. [1947] *Functional logic without axioms or primitive rules of inference* (J 0028) Indag Math 9*561-571
⋄ A05 B20 F50 ⋄ REV MR 9.321 JSL 13.173 • ID 10648

POPPER, K.R. [1948] *On the theory of deduction I: Derivation and its generalizations. II: The definitions of classical and intuitionistic negation* (J 0028) Indag Math 10*44-54,111-120
⋄ B20 ⋄ REV MR 9.486 MR 9.847 Zbl 30.101 Zbl 30.3 JSL 14.62 • ID 10650

POPPER, K.R. see Vol. II for further entries

POPPLESTONE, R.J. [1967] *Beth tree methods in theorem proving* (J 0508) Machine Intelligence 1*31-46
⋄ B35 ⋄ ID 46589

PORTA, HORACIO [1963] *Sur un theoreme de Skolem* (J 0109) C R Acad Sci, Paris 256*5262-5264
⋄ B20 G25 ⋄ REV MR 27 # 3496 Zbl 192.35 • ID 10661

PORTE, J. [1957] *Un systeme de postulats pour le calcul des predicats* (J 0109) C R Acad Sci, Paris 245*817-819
⋄ B10 ⋄ REV MR 19.626 Zbl 78.5 JSL 23.41 • ID 10662

PORTE, J. [1958] *Deux systemes simples pour le calcul des propositions* (J 0364) Publ Sci Univ Alger Ser A Math 5*5-16
⋄ B05 ⋄ REV MR 21 # 1266 Zbl 105.4 JSL 24.247 • ID 24891

PORTE, J. [1958] *Schemas pour le calcul des propositions fonde sur la conjonction et la negation* (J 0036) J Symb Logic 23*421-431
⋄ B05 ⋄ REV MR 22 # 2542 Zbl 99.6 JSL 25.179 • ID 10665

PORTE, J. [1960] *Un systeme pour le calcul des propositions classiques ou la regle de detachement n'est pas valable* (J 0109) C R Acad Sci, Paris 251*188-189
⋄ B05 ⋄ REV MR 22 # 12041 Zbl 95.6 • ID 10666

PORTE, J. [1962] *Quelques extensions du theoreme de deduction* (J 0188) Rev Union Mat Argentina 20*259-266
⋄ B05 ⋄ REV MR 26 # 3576 Zbl 124.247 • ID 10667

PORTE, J. [1962] *Un systeme logistique tres faible pour le calcul propositionnel classique* (J 0109) C R Acad Sci, Paris 254*2500-2502
⋄ B05 ⋄ REV MR 24 # A2529 Zbl 116.4 JSL 27.248 • ID 10668

PORTE, J. [1965] *Recherches sur la theorie generale des systemes formels et sur les systemes connectifs* (X 0834) Gauthier-Villars: Paris vi+146pp
⋄ B22 ⋄ REV MR 32 # 36 Zbl 239 # 02025 • ID 23480

PORTE, J. [1972] *La logique mathematique et le calcul mecanique* (S 0889) Notas Logica Mat 8*105pp
⋄ B98 D98 ⋄ REV MR 51 # 5253 Zbl 279 # 02001 JSL 24.70 • ID 17471

PORTE, J. [1980] *Simplifying the axioms of the predicate calculus* (J 0047) Notre Dame J Formal Log 21*346-350
⋄ B10 ⋄ REV MR 81f:03016 Zbl 394 # 03014 • ID 53768

PORTE, J. see Vol. II, IV, VI for further entries

PORTER, A. & VASWANI, P.K.T. [1959] *The optimization of logical goal-seeking procedures* (J 0940) J Electron Control Ser 1 6*168-185
⋄ B35 ⋄ REV JSL 35.597 • ID 10669

POSEGGA, M. [1972] see HAUCK, J.

POSPESEL, H. [1971] *Arguments: deductive logic exercises* (X 0819) Prentice Hall: Englewood Cliffs ix+208pp
⋄ B98 ⋄ ID 22527

POSPESEL, H. [1974] *Introduction to logic. Propositional logic* (X 0819) Prentice Hall: Englewood Cliffs xii+211pp
⋄ B05 B98 ⋄ REV JSL 43.383 • ID 44546

POSPESEL, H. [1976] *Introduction to logic. Predicate logic* (X 0819) Prentice Hall: Englewood Cliffs xiii+205pp
⋄ B10 B98 ⋄ REV JSL 43.383 • ID 44547

POSPESEL, H. see Vol. II for further entries

POST, E.L. [1921] *Introduction to a general theory of elementary propositions* (J 0100) Amer J Math 43*163-185
• REPR [1967] (C 0675) From Frege to Goedel 265-283
⋄ B05 B20 B50 ⋄ REV FdM 48.1122 • ID 10681

POST, E.L. [1941] *The two-valued iterative systems of mathematical logic* (X 0857) Princeton Univ Pr: Princeton 122pp
⋄ B05 B20 ⋄ REV MR 2.337 JSL 6.114 • ID 10683

POST, E.L. see Vol. IV for further entries

POSTHOFF, C. [1975] see MITEW, W.

POSTHOFF, C. [1978] *Die Loesung und Aufloesung binaerer Gleichungen mit Hilfe des booleschen Differentialkalkuels* (J 0129) Elektr Informationsverarbeitung & Kybern 14*53-80
⋄ B05 ⋄ REV MR 58 # 20869 Zbl 383 # 06003 • ID 69782

POSTLEY, J.A. [1955] *A method for the evaluation of a system of boolean algebraic equations* (J 0235) Math Tables Other Aids Comp 9*5-8
⋄ B05 ⋄ REV MR 18.107 Zbl 64.244 JSL 21.335
• ID 10689

POTTHOFF, K. [1972] *Ordnungseigenschaften von Nichtstandardmodellen* (C 0648) Th of Sets and Topology (Hausdorff) 403-426
⋄ C20 C62 H05 H15 ⋄ REV MR 50 # 12712 Zbl 269 # 02024 • ID 10697

POTTHOFF, K. [1975] see MUELLER, GERT H.

POTTHOFF, K. [1983] *Quelques applications des methodes non-standard a la theorie des groupes* (J 4149) Publ Inst Rech Math Avancee 222/S-07*36-39
⋄ C60 H05 H20 ⋄ ID 40113

POTTHOFF, K. see Vol. III, V for further entries

POTTS, T.C. [1978] *Fregean grammar: A formal outline* (J 0063) Studia Logica 37*7-26
⋄ A10 B03 ⋄ REV MR 80a:03012 Zbl 413 # 03020
• ID 53014

POTTS, T.C. see Vol. VI for further entries

POUR-EL, M.B. [1967] see KRIPKE, S.A.

POUR-EL, M.B. [1968] *Effectively extensible theories* (J 0036) J Symb Logic 33*56-68
⋄ B30 D25 ⋄ REV MR 38 # 3148 Zbl 179.19 • ID 10705

POUR-EL, M.B. see Vol. III, IV, VI for further entries

POUZET, M. [1979] *Chaines de theories universelles* (J 0027) Fund Math 103*133-149
⋄ B20 C52 ⋄ REV MR 81j:03025 Zbl 365 # 02037
• ID 51038

POUZET, M. see Vol. III, V for further entries

POVAROV, G.N. [1954] *On functional separability of Boolean functions (Russian)* (J 0023) Dokl Akad Nauk SSSR 94*801-803
⋄ B05 ⋄ REV MR 16.107 Zbl 56.9 JSL 21.102 • ID 19582

POVAROV, G.N. [1955] *To the study of symmetric Boolean functions from the point of view of the theory of relay-contact circuits (Russian)* (J 0023) Dokl Akad Nauk SSSR 104*183-185
⋄ B05 B70 ⋄ REV MR 18.860 Zbl 68.109 JSL 22.99
• ID 45646

POVAROV, G.N. [1959] *On symmetries of boolean functions (Russian)* (P 0607) All-Union Math Conf (3);1956 Moskva 4*88
⋄ B05 ⋄ REV JSL 22.99 • ID 31282

POVAROV, G.N. [1980] *Simple methods of comparing boolean functions (Russian)* (J 0977) Izv Akad Nauk SSSR, Tekh Kibern 1980/6*116-118
• TRANSL [1980] (J 0522) Engin Cybern 18/6*94-97
⋄ B05 ⋄ REV MR 83b:94055 Zbl 478 # 94028 • ID 69785

PRADE, H. [1985] *A computational approach to approximate and plausible reasoning with applications to expert systems* (J 3191) IEEE Trans Pattern Anal & Mach Intell 7*260-283
⋄ B35 ⋄ REV Zbl 565 # 68089 • ID 48674

PRADE, H. see Vol. II, V for further entries

PRASAD, S.N. [1970] *An extension of Bolzano-Weierstrass theorem to the field of ultra real products* (J 0545) Proc Indian Acad Sci, Sect A A40*273-274
⋄ H05 ⋄ REV MR 45 # 6983 Zbl 233 # 26002 • ID 21276

PRATHER, R.E. [1960] *Computational aids for determining the minimal form of a truth function* (J 0037) ACM J 7*299-310
⋄ B35 ⋄ REV MR 22 # 12060 Zbl 94.9 JSL 33.630
• ID 10722

PRATHER, R.E. see Vol. IV for further entries

PRATT, V.R. [1975] *The effect of basis on size of boolean expressions* (P 1513) IEEE Symp Founds of Comput Sci (16);1975 Berkeley 119-121
• TRANSL [1980] (J 3079) Kiber Sb Perevodov, NS 17*114-123
⋄ B05 ⋄ REV MR 55 # 2391 Zbl 469 # 94017 • ID 69788

PRATT, V.R. see Vol. II, III, IV for further entries

PRAWITZ, D. & PRAWITZ, H. & VOGHERA, N. [1960] *A mechanical proof procedure and its realization in an electronic computer* (J 0037) ACM J 7*102-129
• REPR [1983] (C 4659) Autom of Reasoning 1*202-228
⋄ B35 ⋄ REV MR 23 # B1084 Zbl 213.24 JSL 31.126
• ID 10726

PRAWITZ, D. [1960] *An improved proof procedure* (J 0105) Theoria (Lund) 26*102-139
• REPR [1983] (C 4659) Autom of Reasoning 1*162-201
⋄ B35 ⋄ REV MR 22 # 5567 Zbl 99.8 JSL 31.126
• ID 10725

PRAWITZ, D. [1965] *Natural deduction. A proof-theoretical study* (X 1163) Almqvist & Wiksell: Stockholm 113pp
⋄ B15 B98 F05 F07 F50 F98 ⋄ REV MR 33 # 1227 Zbl 173.2 JSL 32.255 • ID 22164

PRAWITZ, D. [1967] *A note on existential instantiation* (J 0036) J Symb Logic 32*81-82
⋄ B10 ⋄ REV MR 35 # 2724 Zbl 155.338 • ID 10730

PRAWITZ, D. [1968] *Hauptsatz for higher order logic* (J 0036) J Symb Logic 33*452-457
⋄ B15 F05 F35 ⋄ REV MR 39 # 44 Zbl 164.310 JSL 39.607 • ID 10733

PRAWITZ, D. [1969] *Advances and problems in mechanical proof procedures* (J 0508) Machine Intelligence 4*59-71
• REPR [1983] (C 4659) Autom of Reasoning 2*283-297
⋄ B35 ⋄ REV MR 43 # 5760 Zbl 257 # 68082 • ID 10734

PRAWITZ, D. [1970] *A proof procedure with matrix reduction* (P 0625) Symp Autom Demonst;1968 Versailles 207-214
⋄ B35 ⋄ REV MR 46 # 8484 Zbl 212.30 • ID 10736

PRAWITZ, D. [1975] *Comments on Gentzen-type procedures and the classical notion of truth* (P 1440) ⊢ ISILC Proof Th Symp (Schuette);1974 Kiel 500*290-319
⋄ B10 F05 F07 F30 ⋄ REV MR 55 # 12482 Zbl 342 # 02022 • ID 27280

PRAWITZ, D. [1982] *Beweise und die Bedeutung und Vollstaendigkeit der logischen Konstanten* (J 2688) Conceptus (Wien) 16*31-44
⋄ A05 B20 F07 ⋄ REV MR 84c:03021 • ID 34933

PRAWITZ, D. [1985] *Remarks on some approaches to the concept of logical consequence* (J 0154) Synthese 62*153-171
⋄ A05 B22 F07 ⋄ ID 45636

PRAWITZ, D. see Vol. II, V, VI for further entries

PRAWITZ, H. [1960] see PRAWITZ, D.

PREPARATA, F.P. [1970] see MULLER, D.E.

PREPARATA, F.P. [1978] see ADLEMAN, L.M.

PREPARATA, F.P. see Vol. II for further entries

PRESBURGER, M. [1930] *Ueber die Vollstaendigkeit eines gewissen Systems der Arithmetik ganzer Zahlen, in welchem die Addition als einzige Operation hervortritt* (P 0796) Congr Math Pays Slaves (1);1929 Warsaw 92-101,395
⋄ B25 B28 C10 C35 F30 ⋄ REV FdM 56.825 • ID 10749

PRESIC, M.D. [1980] *On the embedding of propositional models* (J 0400) Publ Inst Math, NS (Belgrade) 28(42)*151-158
⋄ B05 ⋄ REV MR 84h:03023 Zbl 491 # 03014 • ID 34237

PRESIC, M.D. see Vol. III, V for further entries

PRESIC, S.B. [1971] *Une methode de resolution des equations dont toutes les solutions appartiennent a un ensemble fini donne* (J 2313) C R Acad Sci, Paris, Ser A-B 272*A654-A657
⋄ B05 ⋄ REV MR 43 # 3124 Zbl 226 # 05004 • ID 48016

PRESIC, S.B. [1972] *Ein Satz ueber reproduktive Loesungen* (J 0400) Publ Inst Math, NS (Belgrade) 14(28)*133-136
⋄ B05 ⋄ REV MR 52 # 5509 Zbl 272 # 08002 • ID 64668

PRESIC, S.B. [1972] *Tautologies. I,II (Serbian)* (J 2349) Mat Struchno-Metod Chasopis 1/2-3*29-41,1/4*37-43
⋄ B05 ⋄ REV Zbl 267 # 02005 • ID 29878

PRESIC, S.B. [1975] *Equational reformulation of formal theories* (J 0400) Publ Inst Math, NS (Belgrade) 19(33)*131-138
⋄ B10 ⋄ REV MR 54 # 4921 Zbl 335 # 08004 • ID 24097

PRESIC, S.B. see Vol. II, III, IV for further entries

PRESTEL, A. & SZCZERBA, L.W. [1979] *Nonaxiomatizability of real general affine geometry* (J 0027) Fund Math 104*193-202
⋄ B30 ⋄ REV MR 81a:03013 Zbl 497 # 03009 • ID 77534

PRESTEL, A. see Vol. III, IV, V for further entries

PREVIALE, F. [1966] *Il "problema del prefisso" per i calcoli logici con piu tipi di variabili* (J 2038) Rend Sem Mat, Torino 26*111-136
⋄ B10 ⋄ REV MR 37 # 46 Zbl 204.5 • ID 10751

PREVIALE, F. [1968] *Introduzione a una teoria generale della rappresentazione per le teorie assiomatiche* (J 0220) Atti Accad Sci Torino, Fis Mat Nat 6*42pp
⋄ B30 ⋄ REV MR 39 # 5331 Zbl 258 # 02001 • ID 10753

PREVIALE, F. [1969] *Rappresentabilita ed equipollenza di teorie assiomatiche I* (J 0315) Ann Sc Norm Sup Pisa Fis Mat, Ser 3 23*635-655
⋄ B15 ⋄ REV MR 41 # 3253 Zbl 188.319 • REM Part II 1970 • ID 10754

PREVIALE, F. [1970] *Rappresentabilita ed equipollenza di teorie assiomatiche II. Un 'applicazione alla geometria "senza punti"* (J 0315) Ann Sc Norm Sup Pisa Fis Mat, Ser 3 24*165-200
⋄ B15 ⋄ REV MR 42 # 5144 Zbl 198.18 • REM Part I 1969 • ID 10755

PREVIALE, F. [1973] *Diversificazione delle specie di individuo nella logica del I ordine (English summary)* (J 0012) Boll Unione Mat Ital, IV Ser 7*443-452
⋄ B10 ⋄ REV MR 52 # 2816 Zbl 268 # 02007 • ID 17615

PREVIALE, F. see Vol. II, III, VI for further entries

PRICE, ROBERT [1961] *The stroke function in natural deduction* (J 0068) Z Math Logik Grundlagen Math 7*117-123
⋄ B05 F07 ⋄ REV MR 28 # 2049 Zbl 107.6 JSL 38.149 • ID 10756

PRICE, ROBERT [1964] see JOHNSTONE JR., H.W.

PRIDA, J.F. [1982] *A non-standard study of the theory of relative recursivity (Spanish)* (J 0264) Collect Math (Barcelona) 33*201-214
⋄ D20 D30 H05 ⋄ REV MR 85b:03066 Zbl 536 # 03023 • ID 37105

PRIDA, J.F. see Vol. VI for further entries

PRIEST, G. [1973] see BELL, J.L.

PRIEST, G. see Vol. II, III, V, VI for further entries

PRIJATELJ, N. [1960] *Introduction to mathematical logic (Slovenian)* (X 2799) Mladinska Knjiga: Ljubljana 150pp
⋄ B98 ⋄ REV MR 40 # 16 Zbl 123.245 JSL 33.480 • REM 2nd ed. 1969 • ID 43180

PRIJATELJ, N. [1982] *Foundations of mathematical logic I (Slovenian)* (J 2310) Obz Mat Fiz, Ljubljana 188pp
⋄ B98 ⋄ REV MR 83f:03005 Zbl 486 # 03002 • ID 35289

PRIJATELJ, N. see Vol. III, V for further entries

PRIMAS, H. [1977] *Theory reduction and non-boolean theories* (J 2756) J Math Biol 4*281-301
⋄ B30 G12 ⋄ REV MR 57 # 15993 Zbl 357 # 92006 • ID 50452

PRIMENKO, EH.A. [1977] *On the number of types of invertible boolean functions (Russian)* (J 0474) Avtom Vychis Tekh, Akad Nauk Latv SSR 1977/6*12-14
⋄ B05 ⋄ REV MR 57 # 9817 Zbl 382 # 94030 • ID 69792

PRIMENKO, EH.A. see Vol. II for further entries

PRINZ, D.G. [1950] see MAYS, W.

PRIOR, A.N. [1955] *Formal logic* (X 0815) Clarendon Pr: Oxford ix+329pp
⋄ A05 A10 B98 E30 ⋄ REV MR 17.569 Zbl 124.2 Zbl 67.249 JSL 27.218 • REM 2nd ed.; 1962; 341pp • ID 22325

PRIOR, A.N. [1958] *Pierce's axioms for propositional calculus* (J 0036) J Symb Logic 23*135-136
⋄ B05 ⋄ REV MR 21 # 3320 Zbl 84.6 JSL 25.87 • ID 10780

PRIOR, A.N. [1961] *Some axiom-pairs for material and strict implication* (J 0068) Z Math Logik Grundlagen Math 7*61-65
⋄ B05 B45 ⋄ REV MR 25 # 11 Zbl 116.4 JSL 37.184 • ID 10782

PRIOR, A.N. [1963] see MEREDITH, C.A.

PRIOR, A.N. [1964] *Two additions to positive implications* (J 0036) J Symb Logic 29*31-32
⋄ B20 B45 ⋄ REV MR 33 # 42 • ID 10787

PRIOR, A.N. [1968] see MEREDITH, C.A.

PRIOR, A.N. [1969] see MEREDITH, C.A.

PRIOR, A.N. [1969] *Propositional calculus in implication and non-equivalence* (J 0047) Notre Dame J Formal Log 10∗271-272
◊ B20 ◊ REV MR 39 # 6733 Zbl 188.313 • ID 10802

PRIOR, A.N. see Vol. II, V for further entries

PRISCO DI, C.A. (ED.) [1985] *Methods in mathematical logic. Proceedings of the 6th Latin American Symposium on Mathematical Logic held in Caracas, Venezuela, Aug. 1-6, 1983* (S 3301) Lect Notes Math 1130∗vii+407pp
◊ B97 C97 ◊ REV MR 86d:03002 Zbl 556 # 00007
• ID 41792

PRISCO DI, C.A. see Vol. III, V for further entries

PRUCNAL, T. [1973] see POGORZELSKI, W.A.

PRUCNAL, T. & WRONSKI, A. [1974] *An algebraic characterization of the notion of structural completeness* (J 0387) Bull Sect Logic, Pol Acad Sci 3/1∗30-33
◊ B22 ◊ REV MR 53 # 7759 • ID 23001

PRUCNAL, T. [1974] see POGORZELSKI, W.A.

PRUCNAL, T. [1975] see POGORZELSKI, W.A.

PRUCNAL, T. [1980] see DYRDA, K.

PRUCNAL, T. [1985] *On finitely e-based consequence determined by Wronski's matrix* (J 0387) Bull Sect Logic, Pol Acad Sci 14∗15-20
◊ B22 ◊ REV Zbl 575 # 03008 • ID 48820

PRUCNAL, T. see Vol. II, V for further entries

PRULLAGE, M.M. [1976] *A theory of restricted variables without existence assumptions* (J 0047) Notre Dame J Formal Log 17∗589-612
◊ B20 ◊ REV MR 56 # 5208 Zbl 315 # 02017 • ID 21946

PRZELECKI, M. & WOJCICKI, R. [1969] *The problem of analyticity* (J 0154) Synthese 19∗374-399
• REPR [1977] (C 3174) Twenty-five Years Log Meth Poland 589-614
◊ A05 B15 ◊ REV Zbl 182.323 Zbl 372 # 02008 • ID 51416

PRZELECKI, M. [1974] *A set theoretic versus a model thoretic approach to the logical structure of physical theories* (J 0063) Studia Logica 33∗91-112
◊ B30 ◊ REV MR 54 # 7195 Zbl 361 # 02009 • ID 24986

PRZELECKI, M. [1976] see MALINOWSKI, G.

PRZELECKI, M. see Vol. III for further entries

PSHENICHNIKOVA, S.V. [1964] *On an algorithm for the automatic proof of certain theorems in analysis (Russian) (Azerbaijani summary)* (J 0135) Izv Akad Nauk Azerb SSR, Ser Fiz-Tekh Mat 1964/4∗65-71
◊ B35 ◊ REV MR 30 # 19 Zbl 127.12 • ID 10817

PTAK, V. [1977] *Nondiscrete mathematical induction* (P 2937) Gen Topol & Rel to Mod Anal & Algeb (4);1976 Prague 166-178
◊ B28 E75 ◊ REV MR 58 # 7237 Zbl 367 # 46007
• ID 51205

PUCCIANO, G. [1913] *I principii dell' ordinamento naturale e della continuita* (J 0336) Giorn Mat Battaglini 51∗221-239
◊ B28 ◊ REV FdM 44.89 • ID 37984

PUCCIANO, G. [1914] *Il continuo lineare aperto ed omogeneo e la geometria archimedea della retta* (J 0336) Giorn Mat Battaglini 52∗19-52
◊ B28 E75 ◊ REV FdM 45.130 • ID 38014

PUDLAK, P. [1975] *Polynomially complete problems in the logic of automated discovery* (P 0454) Math Founds of Comput Sci (4);1975 Marianske Lazne 32∗358-361
◊ B35 D15 ◊ REV MR 52 # 12404 Zbl 318 # 68055
• ID 21740

PUDLAK, P. [1975] *The observational predicate calculus and complexity of computations* (J 0140) Comm Math Univ Carolinae (Prague) 16∗395-398
◊ B10 D15 ◊ REV MR 52 # 7202 Zbl 311 # 02021
• ID 29600

PUDLAK, P. [1977] *Generalized quantifiers and semisets* (P 1639) Set Th & Hierarch Th (1);1974 Karpacz 109-116
◊ B10 C80 E70 ◊ REV MR 58 # 5212 Zbl 411 # 03043
• ID 32425

PUDLAK, P. & SPRINGSTEEL, F. [1979] *Complexity in mechanized hypothesis formation* (J 1426) Theor Comput Sci 8∗203-225
◊ B35 D15 ◊ REV MR 81b:68054 Zbl 404 # 68097
• ID 54869

PUDLAK, P. [1983] *A definition of exponentiation by a bounded arithmetical formula* (J 0140) Comm Math Univ Carolinae (Prague) 24∗667-671
◊ B28 C62 F30 ◊ REV MR 85e:03142 Zbl 533 # 03032
• ID 36550

PUDLAK, P. [1984] *Bounds for Hodes-Specker theorem* (P 2342) Symp Rek Kombin;1983 Muenster 421-445
◊ B05 D15 ◊ REV Zbl 551 # 03022 • ID 43900

PUDLAK, P. see Vol. III, V, VI for further entries

PULATOV, A.K. [1977] *On the influence of zero chains on the complexity of realization of boolean functions by contact schemes (Russian)* (J 0071) Met Diskr Analiz (Novosibirsk) 30∗30-37
◊ B05 ◊ REV MR 80m:94099 Zbl 407 # 94018 • ID 69794

PUPYREV, E.I. [1976] *Application of probability axioms to problems in the analysis of boolean functions (Russian)* (J 0011) Avtom Telemekh 1976/7∗132-137
• TRANSL [1976] (J 0010) Autom & Remote Control 37∗1089-1094
◊ B05 ◊ REV MR 55 # 10156 Zbl 405 # 94023 • ID 69797

PUPYREV, E.I. [1977] *Identification of redundant subformulae in boolean function formulae (Russian. Loose English translation)* (J 2819) Probl Contr Inf Th, Akad Nauk SSSR & Acad Sci Hung 6∗243-247
◊ B05 ◊ REV MR 57 # 19137 Zbl 367 # 94045 • ID 51213

PURDOM JR., P.W. [1982] see BROWN, CYNTHIA A.

PURITZ, C.W. [1972] *Almost perpendicular vectors* (P 0732) Contrib to Non-Standard Anal;1970 Oberwolfach 115-121
◊ H05 ◊ REV MR 58 # 12306 Zbl 257 # 02046 • ID 10822

PURITZ, C.W. [1976] *Quasimonad spaces: a nonstandard approach to convergence* (J 3240) Proc London Math Soc, Ser 3 32∗230-250
◊ H05 ◊ REV MR 53 # 1570 Zbl 319 # 54033 • ID 16692

PURITZ, C.W. see Vol. III, V for further entries

PURTILL, R.L. [1969] *Doing logic by computer* (J 0047) Notre Dame J Formal Log 10∗150-162
◊ B35 ◊ REV MR 39 # 2368 Zbl 175.9 • ID 10823

PURTILL, R.L. [1971] *Logic for philosophers* (X 0837) Harper & Row: New York xxii+419pp
◊ B98 ◊ REV JSL 39.614 • ID 10825

PURTILL, R.L. [1972] *Logical thinking* (X 0837) Harper & Row: New York xv+157pp
⋄ B98 ⋄ ID 22528

PURTILL, R.L. [1979] *Logic. Argument, refutation, and proof* (X 0837) Harper & Row: New York xix+412pp
⋄ B98 ⋄ REV Zbl 473#03001 • ID 55332

PURTILL, R.L. see Vol. II for further entries

PUTNAM, H. [1960] see DAVIS, MARTIN D.

PUTNAM, H. [1979] *Philosophical papers. Vol. 2* (X 0805) Cambridge Univ Pr: Cambridge, GB xvii+457pp
⋄ A05 B96 ⋄ REV MR 82c:01071b Zbl 485#01026
• ID 82554

PUTNAM, H. [1980] *Models and reality* (J 0036) J Symb Logic 45*464-482
• TRANSL [1982] (J 2688) Conceptus (Wien) 16*9-30
⋄ A05 B98 C99 ⋄ REV MR 81h:03016 Zbl 443#03003
• ID 36156

PUTNAM, H. [1980] *Philosophical papers. Vol. 1* (X 0805) Cambridge Univ Pr: Cambridge, GB xiv+364pp
⋄ A05 B96 ⋄ REV MR 82c:01071a Zbl 426#00022
• ID 82555

PUTNAM, H. see Vol. II, III, IV, V, VI for further entries

PYNE, I.B. [1962] see MCCLUSKEY JR., E.J.

QIN, CHAOBIN [1981] see LIU, ZUNQUAN

QUAISSER, E. [1973] *Winkelmetrik in affin-metrischen Ebenen* (J 0068) Z Math Logik Grundlagen Math 19*17-32
⋄ B30 ⋄ REV MR 48#2887 Zbl 252#50014 • ID 10844

QUALITZ, J.E. [1978] see DENNING, P.J.

QUINE, W.V.O. [1932] *A note on Nicod's postulate* (J 0094) Mind 41*345-350
⋄ A10 B05 ⋄ REV FdM 58.64 • ID 30778

QUINE, W.V.O. [1934] *A method of generating part of arithmetic without use of intuitive logic* (J 0015) Bull Amer Math Soc 40*753-761
⋄ B28 ⋄ REV Zbl 10.147 FdM 60.853 • ID 10854

QUINE, W.V.O. [1934] *A system of logistic* (X 0858) Harvard Univ Pr: Cambridge x+204pp
⋄ A05 B15 B98 ⋄ REV FdM 60.845 • ID 19580

QUINE, W.V.O. [1934] *Ontological remarks on the propositional calculus* (J 0094) Mind 43*472-476
⋄ A05 B05 ⋄ REV Zbl 10.146 FdM 60.847 • ID 30779

QUINE, W.V.O. [1936] *Definition of substitution* (J 0015) Bull Amer Math Soc 42*561-569
⋄ B03 ⋄ REV Zbl 15.50 JSL 1.116 FdM 62.41 • ID 10857

QUINE, W.V.O. [1936] *On the axiom of reducibility* (J 0094) Mind 45*498-500
⋄ B15 ⋄ REV JSL 2.60 FdM 62.1055 • ID 30780

QUINE, W.V.O. [1937] *Logic based on inclusion and abstraction* (J 0036) J Symb Logic 2*145-152
⋄ B15 E30 ⋄ REV Zbl 18.194 JSL 3.53 FdM 63.821
• ID 10863

QUINE, W.V.O. [1937] *On derivability* (J 0036) J Symb Logic 2*113-119
⋄ B10 ⋄ REV Zbl 18.2 JSL 3.53 FdM 63.822 • ID 10861

QUINE, W.V.O. [1938] *Completeness of the propositional calculus* (J 0036) J Symb Logic 3*37-40
⋄ B05 ⋄ REV Zbl 18.338 JSL 3.118 FdM 64.27 • ID 10866

QUINE, W.V.O. [1938] *On the theory of types* (J 0036) J Symb Logic 3*125-139
⋄ B15 E70 ⋄ REV Zbl 20.338 JSL 4.34 FdM 64.924
• ID 10865

QUINE, W.V.O. [1940] see GOODMAN, NELSON

QUINE, W.V.O. [1940] *Mathematical logic* (X 0942) Norton: New York xiii+348pp
• LAST ED [1951] (X 0858) Harvard Univ Pr: Cambridge xii+346pp
⋄ A05 B98 E70 ⋄ REV MR 13.613 MR 2.65 MR 25#6 Zbl 44.247 JSL 12.56 JSL 17.149 JSL 5.163 FdM 66.27
• REM 3rd ed. 1951, repr. 1981 • ID 10867

QUINE, W.V.O. [1941] *Elementary logic* (X 0943) Ginn: Boston vi+170pp
• LAST ED [1966] (X 0837) Harper & Row: New York x+129pp
⋄ B98 ⋄ REV MR 3.129 MR 34#4100 JSL 35.166 JSL 6.99 • REM 2nd ed. 1965 • ID 10869

QUINE, W.V.O. [1944] *O sentido da nova logica* (X 0949) St Martin's Pr: New York 252pp
⋄ B98 ⋄ REV MR 7.45 Zbl 60.20 JSL 12.16 • ID 22427

QUINE, W.V.O. [1945] *On the logic of quantification* (J 0036) J Symb Logic 10*1-12
⋄ B10 B25 ⋄ REV MR 7.45 Zbl 60.22 JSL 12.17
• ID 10872

QUINE, W.V.O. [1946] *A short course in logic* (X 0944) Harvard Coop Soc: Cambridge iv+130pp
⋄ B98 ⋄ REV JSL 12.60 • ID 10876

QUINE, W.V.O. [1946] *Concatenation as a basis for arithmetic* (J 0036) J Symb Logic 11*105-114
⋄ B28 ⋄ REV MR 8.307 JSL 13.219 • ID 10874

QUINE, W.V.O. [1948] *Theory of deduction. Parts I-IV* (X 0944) Harvard Coop Soc: Cambridge 156pp
⋄ B98 ⋄ REV Zbl 41.148 JSL 14.190 • ID 10880

QUINE, W.V.O. [1950] *Methods of logic* (X 0818) Holt Rinehart & Winston: New York xx+264pp
• LAST ED [1974] (X 0866) Routledge & Kegan Paul: Henley on Thames viii+280pp
⋄ B98 ⋄ REV MR 12.233 Zbl 38.148 JSL 15.203 JSL 24.219 • ID 10883

QUINE, W.V.O. [1950] *On natural deduction* (J 0036) J Symb Logic 15*93-102,IV
⋄ B10 ⋄ REV MR 12.70 Zbl 41.350 JSL 17.76 • ID 10882

QUINE, W.V.O. [1951] *The ordered pair in number theory* (C 0585) Struct, Meth & Meaning (Sheffer) 84-87
⋄ B28 ⋄ REV Zbl 54.7 JSL 16.289 • ID 10886

QUINE, W.V.O. [1952] see CRAIG, W.

QUINE, W.V.O. [1952] see CHURCH, A.

QUINE, W.V.O. [1952] *The problem of simplifying truth functions* (J 0005) Amer Math Mon 59*521-531
⋄ B05 ⋄ REV MR 14.440 Zbl 48.245 JSL 18.280
• ID 10888

QUINE, W.V.O. [1953] *Two theorems about truth-functions* (J 0250) Bol Soc Mat Mexicana 10∗64-70
⋄ B05 B45 ⋄ REV MR 15.277 Zbl 53.198 JSL 19.142
• ID 10892

QUINE, W.V.O. [1954] *Quantification and the empty domain* (J 0036) J Symb Logic 19∗177-179
⋄ B10 B60 ⋄ REV MR 16.324 Zbl 56.10 JSL 20.284
• ID 10895

QUINE, W.V.O. [1955] *A proof procedure for quantification theory* (J 0036) J Symb Logic 20∗141-149
⋄ B10 ⋄ REV MR 17.224 Zbl 68.14 JSL 31.657 • ID 10898

QUINE, W.V.O. [1955] *A way to simplify truth functions* (J 0005) Amer Math Mon 62∗627-631
⋄ B05 ⋄ REV MR 17.814 Zbl 68.242 JSL 21.328
• ID 10899

QUINE, W.V.O. [1955] *On Frege's way out* (J 0094) Mind 64∗145-159
⋄ B15 E30 ⋄ REV MR 18.455 Zbl 64.5 • ID 10896

QUINE, W.V.O. [1959] *On cores and prime implicants of truth functions* (J 0005) Amer Math Mon 66∗755-760
• REPR [1966] (C 0587) Quine: Sel Logic Papers 164-171
⋄ B05 ⋄ REV MR 21 # 7155 Zbl 201.322 JSL 35.329
• ID 21099

QUINE, W.V.O. [1961] *A basis for number theory in finite classes* (J 0015) Bull Amer Math Soc 67∗391-392
⋄ B28 E75 ⋄ REV MR 23 # A3678 Zbl 100.11 • ID 10906

QUINE, W.V.O. [1963] *Set theory and its logic* (X 0858) Harvard Univ Pr: Cambridge xv+359pp
• TRANSL [1973] (X 0900) Vieweg: Wiesbaden xv+263pp [1978] (X 2866) Ullstein: Berlin (German)
⋄ A10 B98 E30 E98 ⋄ REV MR 43 # 37 Zbl 122.246 JSL 37.768 • REM 2nd rev. ed. 1969 • ID 23295

QUINE, W.V.O. [1966] *On boolean functions* (C 0587) Quine: Sel Logic Papers 176-180
⋄ B05 ⋄ ID 32228

QUINE, W.V.O. [1966] *Selected logic papers* (X 0981) Random House: New York x+250pp
⋄ A05 B96 ⋄ REV MR 34 # 7333 • ID 25163

QUINE, W.V.O. [1981] *Predicate functors revisited* (J 0036) J Symb Logic 46∗649-652
⋄ B10 B40 ⋄ REV MR 83e:03028 Zbl 472 # 03007
• ID 55273

QUINE, W.V.O. see Vol. II, III, IV, V, VI for further entries

QUINLAND, J.R. [1968] see HUNT, E.B.

RABIN, M.O. [1961] *Non-standard models and independence of the induction axiom* (C 0622) Essays Found of Math (Fraenkel) 287-299
⋄ B28 C62 F30 H15 ⋄ REV MR 28 # 4999 Zbl 143.10 JSL 38.159 • ID 10934

RABIN, M.O. [1966] see ELGOT, C.C.

RABIN, M.O. [1968] *Decidability of second-order theories and automata on infinite trees* (J 0015) Bull Amer Math Soc 74∗1025-1029
⋄ B15 B25 C85 D05 ⋄ REV MR 38 # 44 Zbl 313 # 02029
• ID 29625

RABIN, M.O. [1969] *Decidability of second order theories and automata on infinite trees* (J 0064) Trans Amer Math Soc 141∗1-35
• TRANSL [1971] (J 3079) Kiber Sb Perevodov, NS 8∗72-116
⋄ B15 B25 C85 D05 ⋄ REV MR 40 # 30 Zbl 221 # 02031 JSL 37.618 • ID 10942

RABIN, M.O. [1970] *Weakly definable relations and special automata* (P 1072) Math Log & Founds of Set Th;1968 Jerusalem 1-23
⋄ B15 B25 C40 C85 D05 ⋄ REV MR 43 # 3121 Zbl 214.22 JSL 40.622 • ID 22233

RABIN, M.O. [1971] *Decidability and definability in second-order theories* (P 0743) Int Congr Math (II,11,Proc);1970 Nice 1∗239-244
⋄ B15 B25 C40 C85 D05 ⋄ REV MR 54 # 12512 Zbl 226 # 02041 JSL 40.623 • ID 14690

RABIN, M.O. [1972] *Automata on infinite objects and Church's problem* (X 0803) Amer Math Soc: Providence iii+22pp
⋄ B15 B25 C85 D05 ⋄ REV MR 48 # 75 Zbl 315 # 02037 JSL 40.623 • ID 10943

RABIN, M.O. see Vol. III, IV, V, VI for further entries

RACHIN, N. [1980] see MILICI, C.

RADEV, S.R. [1984] *Propositional logics of formal languages* (J 2095) Fund Inform, Ann Soc Math Pol, Ser 4 7∗447-482
⋄ B05 ⋄ ID 46295

RADEV, S.R. see Vol. II, III for further entries

RADINOVIC, S. [1984] *A hereditary property of HM-spaces* (J 0482) Publ Math Univ Belgrade 35(49)∗153-156
⋄ H10 ⋄ ID 44780

RADO, T. [1962] *Comments on the presence function of gazale* (J 0284) IBM J Res Dev 6∗268-269
⋄ B35 ⋄ REV JSL 30.106 • ID 10963

RADO, T. [1965] see HOUSE, R.W.

RADO, T. see Vol. IV, V for further entries

RADOJEVIC, P. [1973] *Les tautologies dans les regles des deduction (Serbian)* (J 2349) Mat Struchno-Metod Chasopis 2/3∗48-57
⋄ B05 ⋄ REV Zbl 267 # 02006 • ID 29879

RADU, E. [1978] *L'oeuvre de Grigore C. Moisil en logique mathematique I,II,III* (J 0060) Rev Roumaine Math Pures Appl 23∗463-477,605-610,1077-1092
⋄ A05 A10 B96 ⋄ REV MR 58 # 4925 MR 81h:01013 Zbl 375 # 02002 Zbl 375 # 02003 Zbl 389 # 03004 • REM Part III includes a bibliography • ID 51580

RADU, E. see Vol. II, VI for further entries

RAGGIO, A.R. [1964] *Direct consistency proof of Gentzen's system of natural deduction* (J 0047) Notre Dame J Formal Log 5∗27-30
⋄ B10 F07 ⋄ REV MR 31 # 29 Zbl 148.6 • ID 10969

RAGGIO, A.R. [1974] *A simple proof of Herbrand's theorem* (J 0047) Notre Dame J Formal Log 15∗487-488
⋄ B10 F05 F07 ⋄ REV MR 51 # 5270 Zbl 272 # 02017
• ID 10972

RAGGIO, A.R. [1977] *Semi-formal Beth tableaux* (P 1076) Latin Amer Symp Math Log (3);1976 Campinas 239-241
⋄ B10 F07 ⋄ REV MR 57 # 15949 Zbl 385 # 03050
• ID 16607

RAGGIO, A.R. see Vol. II, VI for further entries

RAMACHANDRA RAO, A. [1977] see BERGE, C.

RAMSEY, F.P. [1926] *The foundations of mathematics* (J 1910) Proc London Math Soc, Ser 2 25*338-384
- ⋄ A05 B98 ⋄ REV FdM 52.46 • ID 28767

RAMSEY, F.P. [1930] *On a problem of formal logic* (J 1910) Proc London Math Soc, Ser 2 30*264-286
- ⋄ B10 B25 E05 E20 E75 ⋄ ID 28781

RAMSEY, F.P. [1931] *The foundations of mathematics and other logical essays* (X 0866) Routledge & Kegan Paul: Henley on Thames xviii+292pp
- • LAST ED [1960] (X 0854) Littlefield Adams: Totowa xviii+292pp
- ⋄ A05 B96 ⋄ REV MR 22#10912 Zbl 2.5 JSL 15.157 JSL 35.312 FdM 31.47 • ID 24865

RAMSEY, F.P. [1978] *Foundations* (X 2696) Humanities Pr: Atlantic Highlands viii+287pp
- • TRANSL [1980] (X 1267) Frommann: Stuttgart 237pp
- ⋄ A05 A10 B96 ⋄ REV MR 57#15930 Zbl 436#01019
- • REM Ed. by Mellor,D.H. • ID 82569

RANDOLPH, J.F. [1965] *Cross-examining propositional calculus and set operations* (J 0005) Amer Math Mon 72*117-127
- ⋄ B05 G05 ⋄ REV Zbl 131.6 • ID 30833

RANDOLPH, J.F. see Vol. V for further entries

RANGASWAMY, S.V. [1979] see CHAKRAPANI, N.

RANGASWAMY, S.V. [1980] see CHAKRAPANI, N.

RANTALA, V. [1975] *Urn models: A new kind of non-standard model for first-order logic* (J 0122) J Philos Logic 4*455-474
- ⋄ A05 B10 C07 C90 ⋄ REV MR 57#12166 Zbl 324#02043 • ID 64743

RANTALA, V. [1978] *Correspondence and non-standard models: A case study* (J 0096) Acta Philos Fenn 30*366-378
- ⋄ A05 H10 ⋄ REV MR 82i:00024 Zbl 417#03003
- • ID 53240

RANTALA, V. see Vol. II, III, V for further entries

RAPHAEL, B. [1967] see GREEN, C.C.

RAS, Z. [1971] *Deductive systems of computing machines* (J 0014) Bull Acad Pol Sci, Ser Math Astron Phys 19*517-524
- ⋄ B05 D05 ⋄ REV MR 49#4333 Zbl 227#02018
- • ID 27358

RAS, Z. [1971] *On a relationship between the propositional calculus and a grammar (Russian summary)* (J 0014) Bull Acad Pol Sci, Ser Math Astron Phys 19*635-637
- ⋄ B05 D05 ⋄ REV MR 47#1584 Zbl 227#02019
- • ID 10995

RAS, Z. [1972] *On a relationship between certain grammars and enumerable first order predicate calculi (Russian summary)* (J 0014) Bull Acad Pol Sci, Ser Math Astron Phys 20*95-99
- ⋄ B20 D05 ⋄ REV MR 47#1585 Zbl 237#02009
- • ID 10996

RAS, Z. see Vol. IV for further entries

RASHID, S. [1978] see ANDERSON, ROBERT M.

RASHID, S. [1978] *Existence of equilibrium in infinite economies with production* (J 1400) Econometrica 46*1155-1164
- ⋄ H10 ⋄ REV MR 80a:90026 Zbl 434#90022 • ID 55758

RASHID, S. [1985] *Nonstandard analysis and infinite economies: The Cournot-Nash solution* (P 4200) Adv in Equilibrum Th;1984 Indianapolis 244*173-184
- ⋄ H10 ⋄ REV Zbl 562#90006 • ID 47444

RASIOWA, H. [1947] *Axiomatisation d'un systeme partiel de la theorie de la deduction* (J 0459) C R Soc Sci Lett Varsovie Cl 3 40*22-27
- ⋄ B20 ⋄ REV MR 11.303 Zbl 39.6 JSL 15.139 • ID 41613

RASIOWA, H. [1949] *Sur un certain systeme d'axiomes du calcul des propositions* (J 4510) Norsk Mat Tidsskr 31*1-3
- ⋄ B05 ⋄ REV MR 11.303 Zbl 40.146 JSL 14.197
- • ID 10997

RASIOWA, H. [1953] see MOSTOWSKI, ANDRZEJ

RASIOWA, H. & SIKORSKI, R. [1953] *Algebraic treatment of the notion of satisfiability* (J 0027) Fund Math 40*62-95
- ⋄ B10 C90 G05 ⋄ REV MR 15.668 Zbl 53.2 JSL 20.78
- • ID 11006

RASIOWA, H. & SIKORSKI, R. [1953] *On satisfiability and decidability in non-classical functional calculi* (J 0014) Bull Acad Pol Sci, Ser Math Astron Phys 1*229-231
- ⋄ B10 B25 B45 C90 F50 ⋄ REV MR 15.668 Zbl 51.245
- • ID 11004

RASIOWA, H. [1955] *A proof of ε-theorems* (J 0014) Bull Acad Pol Sci, Ser Math Astron Phys 3*299-302
- ⋄ B10 ⋄ REV MR 19.112 Zbl 66.10 JSL 33.286 • ID 11016

RASIOWA, H. [1955] *Algebraic models of axiomatic theories* (J 0027) Fund Math 41*291-310
- ⋄ B30 C90 F50 G05 G10 G25 ⋄ REV MR 19.111 Zbl 65.4 JSL 33.285 • ID 11010

RASIOWA, H. [1955] *On a fragment of the implicative propositional calculus (Polish) (Russian and English summaries)* (J 0063) Studia Logica 3*208-226
- ⋄ B20 G25 ⋄ REV MR 17.226 Zbl 68.10 JSL 22.330
- • ID 24899

RASIOWA, H. [1955] see GRZEGORCZYK, A.

RASIOWA, H. [1956] see LOS, J.

RASIOWA, H. [1956] *On the ε-theorems* (J 0027) Fund Math 43*156-165
- ⋄ B10 G05 ⋄ REV MR 18.711 Zbl 74.10 JSL 33.286
- • ID 19577

RASIOWA, H. [1957] *Sur la methode algebrique dans la methodologie des systemes deductifs elementaires* (J 0070) Bull Soc Sci Math Roumanie, NS 1*223-231
- ⋄ B10 C07 G05 G10 ⋄ REV MR 21#3329 • ID 11017

RASIOWA, H. & SIKORSKI, R. [1958] *On the isomorphism of Lindenbaum algebras with fields of sets* (S 0019) Colloq Math (Warsaw) 5*143-158
- ⋄ B10 G05 ⋄ REV MR 20#6353 Zbl 86.6 JSL 27.98
- • ID 11020

RASIOWA, H. & SIKORSKI, R. [1960] *On the Gentzen theorem* (J 0027) Fund Math 48*57-69
- ⋄ B10 C07 F05 ⋄ REV MR 22#1510 Zbl 99.6 • ID 11024

RASIOWA, H. & SIKORSKI, R. [1963] *The mathematics of metamathematics* (S 0257) Monograf Mat 41*522pp
- • TRANSL [1972] (X 2027) Nauka: Moskva 591pp (Russian)
- ⋄ B98 C98 F50 F98 G05 G10 G98 ⋄ REV MR 29#1149 MR 50#4232 Zbl 122.243 JSL 32.274
- • REM 3rd ed 1979 • ID 11028

RASIOWA, H. [1964] *A generalization of a formalized theory of fields of sets on nonclassical logics* (J 0202) Diss Math (Warsaw) 42*30pp
⋄ B20 E70 G10 G25 ⋄ REV MR 29 # 3358 Zbl 265 # 02035 • ID 25051

RASIOWA, H. [1965] *On non-classical calculi of classes* (P 0797) Fonds des Math, Machines Math & Appl;1962 Tihany 53-55
⋄ B20 B50 E70 G10 G25 ⋄ REV MR 33 # 3894 Zbl 265 # 02036 • ID 29826

RASIOWA, H. [1971] *Introduction to modern mathematics (Polish)* (X 1034) PWN: Warsaw 302pp
• TRANSL [1973] (X 0809) North Holland: Amsterdam xi + 339pp (English)
⋄ B30 B98 C05 E98 ⋄ REV MR 48 # 8190b Zbl 204.308 JSL 43.153 • ID 22289

RASIOWA, H. [1972] *Introduction to set theory and mathematical logic (Bulgarian)* (X 1925) Nauka i Izkustwo: Sofia 343pp
⋄ B98 C05 E98 ⋄ ID 31450

RASIOWA, H. [1974] *An algebraic approach to non-classical logics* (X 0809) North Holland: Amsterdam xv + 403pp
⋄ B98 F50 G05 G10 G20 G25 G98 ⋄ REV MR 56 # 5285 Zbl 299 # 02069 JSL 42.432 • ID 31451

RASIOWA, H. see Vol. II, III, IV, V, VI for further entries

RASKOVIC, M. [1979] *On existence of expansion of a complex function* (J 0400) Publ Inst Math, NS (Belgrade) 26(40)*269-271
⋄ H05 ⋄ REV MR 82d:30051 Zbl 435 # 03050 • ID 55810

RASKOVIC, M. [1985] *An application of nonstandard analysis to functional equations* (J 0400) Publ Inst Math, NS (Belgrade) 37(51)*23-24
⋄ H05 ⋄ ID 49710

RASKOVIC, M. see Vol. III, V for further entries

RATSA, M.F. [1979] see KUZNETSOV, A.V.

RATSA, M.F. see Vol. II, VI for further entries

RAULEFS, P. [1983] see BENDA, W.

RAULEFS, P. see Vol. VI for further entries

RAUTENBERG, W. [1960] see ASSER, G.

RAUTENBERG, W. [1962] *Ueber metatheoretische Eigenschaften einiger geometrischer Theorien* (J 0068) Z Math Logik Grundlagen Math 8*5-41
⋄ B30 C65 D35 ⋄ REV MR 26 # 4883 Zbl 112.247 • ID 11041

RAUTENBERG, W. [1966] *Ueber Hilberts Schnittpunktsaetze* (J 0068) Z Math Logik Grundlagen Math 12*57-59
⋄ B30 C65 ⋄ REV MR 34 # 3402 Zbl 168.250 • ID 11045

RAUTENBERG, W. [1968] *Unterscheidbarkeit endlicher geordneter Mengen mit gegebener Anzahl von Quantoren* (J 0068) Z Math Logik Grundlagen Math 14*267-272
⋄ B20 C07 C13 C65 D35 ⋄ REV MR 37 # 2595 Zbl 169.7 • ID 11048

RAUTENBERG, W. [1969] *Euklidische und Minkowskische Orthogonalitaetsrelationen* (J 0027) Fund Math 64*189-196
⋄ B30 ⋄ REV MR 40 # 825 Zbl 176.173 • ID 33686

RAUTENBERG, W. [1972] see HAUSCHILD, K.

RAUTENBERG, W. [1972] see KLOETZER, G.

RAUTENBERG, W. [1972] *Zur praktischen und theoretischen Wirksamkeit der mathematischen Logik* (C 1533) Quantoren, Modal, Paradox 95-106
⋄ B98 ⋄ ID 39979

RAUTENBERG, W. [1975] *Eine Synthese der axiomatischen und der kardinalen Definition der natuerlichen Zahlen* (J 0160) Math-Phys Sem-ber, NS 22*225-239
⋄ B28 ⋄ REV MR 52 # 13597 Zbl 314 # 02001 • ID 21875

RAUTENBERG, W. [1979] *Klassische und nichtklassische Aussagenlogik* (X 0900) Vieweg: Wiesbaden xi + 361pp
⋄ B45 B50 B98 C90 ⋄ REV MR 81i:03002 Zbl 424 # 03007 • ID 77698

RAUTENBERG, W. [1979] *Reelle Zahlen in elementarer Darstellung* (X 0918) Klett: Stuttgart 150pp
⋄ B28 ⋄ REV Zbl 425 # 00001 • ID 33551

RAUTENBERG, W. [1981] *2-element matrices* (J 0063) Studia Logica 40*315-353
⋄ B20 B22 ⋄ REV MR 84m:03035 Zbl 493 # 03006 • ID 35739

RAUTENBERG, W. [1985] *Consequence relations of 2-element algebras* (P 4180) Int Congr Log, Meth & Phil of Sci (7,Pap);1983 Salzburg 3-22
⋄ B05 B22 ⋄ ID 48112

RAUTENBERG, W. see Vol. II, III, IV, VI for further entries

RAYNAUD, Y. [1984] see LEVY, M.

RAYNAUD, Y. see Vol. III for further entries

RAZBOROV, A.A. [1985] *Lower bounds on the monotone complexity of some Boolean functions (Russian)* (J 0023) Dokl Akad Nauk SSSR 281*798-801
• TRANSL [1985] (J 0062) Sov Math, Dokl 31*354-357
⋄ B05 ⋄ ID 45127

RECKHOW, R.A. [1974] see COOK, S.A.

RECKHOW, R.A. [1979] see COOK, S.A.

RECKHOW, R.A. see Vol. IV for further entries

REDDY, C.R. [1981] see LOVELAND, D.W.

REDDY, C.R. see Vol. III, IV, VI for further entries

REEB, G. & STERN, J. [1977] *Seance debat sur l'analyse nonstandard* (J 3169) Gaz Math, Soc Math Fr 8*8-18
⋄ H05 ⋄ REV Zbl 359 # 02059 • ID 50606

REEB, G. [1979] *Equations differentielles et analyse non classique (d'apres J.-L. Callot (Oran))* (P 1984) Colloq on Diff Geom (4);1978 Santiago de Compostela 240-245
⋄ H05 ⋄ REV MR 83f:58032 Zbl 468 # 34031 • ID 40260

REEB, G. [1982] *Analyse non standard et theorie du moire* (P 3730) Actual Math;1981 Luxembourg 189-195
⋄ H10 ⋄ REV MR 84a:58063 Zbl 511 # 03030 • ID 36915

REGOLI, G. [1979] see COLETTI, G.

REICHBACH, J. [1953] *Ueber den auf Alternative und Negation aufgebauten Aussagenkalkuel (Polish and Russian summaries)* (J 0063) Studia Logica 1*13-18 • ERR/ADD ibid 1*299
⋄ B05 ⋄ REV MR 17.118 Zbl 59.14 JSL 21.87 • ID 11059

REICHBACH, J. [1955] *Completeness of the functional calculus of first order (Polish, Russian) (English summary)* (J 0063) Studia Logica 2*213-228,229-250 • ERR/ADD ibid 2*329
⋄ B10 C07 ⋄ REV MR 17.446 Zbl 67.250 JSL 21.194 • ID 33706

REICHBACH, J. [1961] *A note on my paper: on characterizations of the first-order functional calculus* (J 0047) Notre Dame J Formal Log 2*251-252
⋄ B10 ⋄ REV MR 25#3815c Zbl 242#02010 • ID 11063

REICHBACH, J. [1961] *On characterizations of the first-order functional calculus* (J 0047) Notre Dame J Formal Log 2*1-15
⋄ B10 ⋄ REV MR 25#3815b Zbl 119.251 • ID 11062

REICHBACH, J. [1961] *On theses of the first-order functional calculus* (J 0068) Z Math Logik Grundlagen Math 7*175-184
⋄ B10 ⋄ REV MR 25#3815a Zbl 243#02009 • ID 11064

REICHBACH, J. [1962] *On generalization of the satisfiability definition and proof rules with remarks to my paper: On theses of the first-order functional calculus* (J 0068) Z Math Logik Grundlagen Math 8*267-276
⋄ B10 ⋄ REV MR 26#3591 Zbl 112.6 • ID 11066

REICHBACH, J. [1962] *On the connection of the first-order functional calculus with many-valued propositional calculi* (J 0047) Notre Dame J Formal Log 3*102-107
⋄ B10 B50 ⋄ REV MR 26#4897 Zbl 242#02011 • ID 11065

REICHBACH, J. [1963] *About connection of the first-order functional calculus with many valued propositional calculi* (J 0068) Z Math Logik Grundlagen Math 9*117-124
⋄ B10 B50 ⋄ REV MR 28#1124 Zbl 163.243 • ID 11067

REICHBACH, J. [1963] *Some characterizations of theses of the first-order functional calculus* (J 0043) Math Ann 152*195-207
⋄ B10 ⋄ REV MR 29#25 Zbl 119.251 • ID 24874

REICHBACH, J. [1964] *A note about connection of the first-order functional calculus with many valued propositional calculi* (J 0047) Notre Dame J Formal Log 5*158-160
⋄ B10 B50 ⋄ REV MR 31#2138 Zbl 163.4 • ID 11068

REICHBACH, J. [1965] *On the connection of the first-order functional calculus with $\aleph_0$ propositional calculus* (J 0047) Notre Dame J Formal Log 6*73-80
⋄ B10 B50 ⋄ REV MR 34#7353 Zbl 163.4 • ID 11070

REICHBACH, J. [1967] *On generalizations of the satisfiability definition and Gentzen-Jaskowski's sequent proof rules* (X 3333) Unknown Publisher: See Remarks 21pp
⋄ B10 F07 ⋄ REV Zbl 189.8 • REM Tel-Aviv • ID 19564

REICHBACH, J. [1968] *A note on theses of the first-order functional calculus* (J 0047) Notre Dame J Formal Log 9*335-339
⋄ B10 ⋄ REV MR 39#2604 Zbl 179.10 • ID 11072

REICHBACH, J. [1969] *Propositional calculi and completeness theorem* (J 0537) Yokohama Math J 17*1-16
⋄ B05 C07 ⋄ REV MR 40#1259 Zbl 184.279 • ID 11073

REICHBACH, J. [1969] *Some examples of different methods of formal proofs with generalizations of the satisfiability definition* (J 0047) Notre Dame J Formal Log 10*214-224
⋄ B10 ⋄ REV MR 39#1292 Zbl 175.266 • ID 11074

REICHBACH, J. [1970] *On statistical tests in generalizations of Herbrand's theorems and asymptotic probabilistic models (main theorems of mathematics)* (J 1550) Creation Math 1*6-22
⋄ B10 C40 C90 ⋄ REV Zbl 237#02018 • ID 27880

REICHBACH, J. [1971] *Some methods of formal proofs. III* (J 0047) Notre Dame J Formal Log 12*479-482
⋄ B10 B15 ⋄ REV MR 46#19 Zbl 212.16 • REM Part II 1976 • ID 11075

REICHBACH, J. [1975] *Generalized models for classical and intuitionistic predicate calculi* (J 0537) Yokohama Math J 23*5-30
⋄ B10 B25 C90 F50 ⋄ REV MR 55#2556 Zbl 343#02040 • ID 32003

REICHBACH, J. see Vol. II, III, IV, VI for further entries

REICHENBACH, H. [1932] *Axiomatik der Wahrscheinlichkeitsrechnung* (J 0044) Math Z 34*568-619
⋄ B30 ⋄ REV Zbl 3.354 FdM 58.1151 • ID 11079

REICHENBACH, H. [1932] *Wahrscheinlichkeitslogik* (J 0277) Sitzb Preuss Akad Wiss Phys Math Kl 1932*476-488
⋄ A05 B30 B48 ⋄ REV Zbl 6.67 JSL 2.54 FdM 58.59 • ID 11080

REICHENBACH, H. [1933] *Die logischen Grundlagen des Wahrscheinlichkeitsbegriffs* (J 0748) Erkenntnis (Leipzig) 3*401-425
⋄ A05 B30 ⋄ REV Zbl 7.252 FdM 59.61 • ID 11081

REICHENBACH, H. [1939] *Introduction a la logistique* (X 0859) Hermann: Paris 68pp
⋄ B98 ⋄ REV Zbl 21.290 JSL 12.86 FdM 65.1106 • ID 22598

REICHENBACH, H. [1939] *Ueber die semantische und die Objekt-Auffassung von Wahrscheinlichkeitsausdruecken* (J 0748) Erkenntnis (Leipzig) 8*50-68
⋄ B30 B48 ⋄ REV Zbl 21.385 FdM 65.1107 • ID 43787

REICHENBACH, H. [1941] *Note on probability implication* (J 0015) Bull Amer Math Soc 47*265-267
⋄ B30 ⋄ REV MR 2.227 Zbl 21.385 FdM 67.44 • ID 41305

REICHENBACH, H. [1947] *Elements of symbolic logic* (X 0843) Macmillan : New York & London xiii+444pp
• LAST ED [1980] (X 0813) Dover: New York iv+444pp
⋄ A05 B98 ⋄ REV MR 8.556 Zbl 34.3 Zbl 486#03003 JSL 14.50 • REM 2nd ed. 1980 • ID 11084

REICHENBACH, H. [1949] *The theory of probability* (X 0926) Univ Calif Pr: Berkeley xvi+492pp
⋄ A05 B98 ⋄ REV MR 11.152 Zbl 38.286 JSL 16.48 • ID 25436

REICHENBACH, H. [1952] *The syllogism revised* (J 0153) Phil of Sci (East Lansing) 19*1-16
⋄ B20 ⋄ REV JSL 25.81 • ID 42532

REICHENBACH, H. see Vol. II for further entries

REID, A. [1960] see POHM, A.

REISCHER, C. & SIMOVICI, D.A. [1984] *Graph functions of Boolean functions* (J 0187) IEEE Trans Comp C-33*97-99
⋄ B05 ⋄ REV MR 86c:94040 Zbl 528#94028 • ID 44164

REISCHER, C. see Vol. II, IV for further entries

REITER, A. [1975] see BREJTBART, YU.YA.

REITER, R. [1970] *The predicate elimination strategy in theorem-proving* (P 0641) ACM Symp Th of Comput (2);1970 Northhampton 180-183
⋄ B35 ⋄ ID 46605

REITER, R. [1971] *Two results on ordering for resolution with merging and linear format* (J 0037) ACM J 18*630–646
⋄ B35 ⋄ REV Zbl 238 # 68030 • ID 46609

REITER, R. [1972] see MINICOZZI, E.

REITER, R. [1974] *On self-modifying programs* (J 0191) Inform Sci 7*157–169
⋄ B35 B75 ⋄ REV MR 48 # 12887 Zbl 286 # 68002 • ID 64792

REITER, R. [1976] *A semantically guided deductive system for automatic theorem proving* (J 0187) IEEE Trans Comp C-25*328–334
⋄ B35 ⋄ REV Zbl 324 # 68052 • ID 64793

REITER, R. [1980] *A logic for default reasoning* (J 0503) Artif Intell 13*81–132
⋄ B35 ⋄ REV MR 83e:68138 Zbl 435 # 68069 • ID 55826

REITER, R. see Vol. V, VI for further entries

RENARDEL DE LAVALETTE, G.R. [1981] *The interpolation theorem in fragments of logics* (J 0028) Indag Math 43*71–86
⋄ B20 B55 C40 F50 ⋄ REV MR 82d:03019 Zbl 471 # 03012 • ID 55207

RENARDEL DE LAVALETTE, G.R. [1984] *Descriptions in mathematical logic* (J 0063) Studia Logica 43*281–294
⋄ B10 F50 ⋄ ID 46669

RENARDEL DE LAVALETTE, G.R. see Vol. VI for further entries

RENNIE, M.K. [1971] *Completeness in the logic of predicate modifiers* (J 0079) Logique & Anal, NS 14*627–643
⋄ B10 ⋄ REV MR 48 # 5844 Zbl 239 # 02010 • ID 11103

RENNIE, M.K. [1974] *Some uses of type theory in the analysis of language* (X 0904) ANU RSSS Philos: Canberra viii+166pp
⋄ A05 B15 B65 ⋄ REV MR 52 # 45 Zbl 295 # 02004 • ID 17214

RENNIE, M.K. see Vol. II, III, IV for further entries

RENYI, A. [1970] *Foundations of probability* (X 1167) Holden-Day: San Francisco 366pp
⋄ B98 ⋄ REV MR 41 # 9314 • ID 25437

RESCHER, N. [1964] *Introduction to logic* (X 0949) St Martin's Pr: New York xv+360pp
⋄ A05 B98 ⋄ REV JSL 35.579 • ID 22009

RESCHER, N. [1969] *Topics in philosophical logic* (S 3307) Synth Libr xiv+347pp
⋄ A05 B45 B98 C90 ⋄ REV Zbl 175.264 • ID 25168

RESCHER, N. [1971] see PARKS, R.Z.

RESCHER, N. & URQUHART, A.I.F. [1971] *Temporal logic* (X 0902) Springer: Wien xviii+273pp
⋄ A05 B45 B98 ⋄ REV MR 49 # 2267 Zbl 229 # 02027 JSL 40.252 • ID 11120

RESCHER, N. [1975] *A theory of possibility* (X 1331) Univ Pittsburgh Pr: Pittsburgh xvi+255pp
⋄ A05 B98 ⋄ REV MR 55 # 12455 JSL 43.158 • ID 77745

RESCHER, N. see Vol. II, III, V, VI for further entries

RESNIK, M.D. [1966] *A note on natural deduction* (J 0047) Notre Dame J Formal Log 7*206–208
⋄ B10 ⋄ REV MR 35 # 2725 Zbl 154.4 • ID 11125

RESNIK, M.D. [1968] *Professor Goddard and the simple theory of types* (J 0094) Mind 77*565–568
⋄ B15 ⋄ ID 32430

RESNIK, M.D. [1969] *A set theoretic approach to the simple theory of types* (J 0105) Theoria (Lund) 35*239–258
⋄ B15 E70 ⋄ REV MR 41 # 6683 Zbl 231 # 02017 • ID 11127

RESNIK, M.D. [1970] *Elementary logic* (X 0822) McGraw-Hill: New York 457pp
⋄ B98 ⋄ REV MR 42 # 7478 Zbl 235 # 02002 • ID 11128

RESNIK, M.D. see Vol. II, VI for further entries

RETTIG, A.S. [1953] see CARTER, W.C.

REUSCH, B. [1979] see DETERING, L.

REUSCH, B. see Vol. IV for further entries

REVZIN, I.I. [1957] *Some questions of the formalization of syntax I,II (Russian)* (J 4093) Byull Ob'ed Probl Mashin Perevodov 1*5–36,3*20–29
⋄ B03 ⋄ ID 45658

REYERSBACH, W. [1972] see EMDE, H.

REYNOLDS, J.C. [1969] *A generalized resolution principle based upon context-free grammars* (P 0594) Inform Processing (4);1968 Edinburgh 2*1405–1411
⋄ B35 D05 ⋄ REV Zbl 211.315 • ID 46610

REYNOLDS, J.C. [1969] *Transformational systems and the algebraic structure of atomic formulas* (J 0508) Machine Intelligence 5*135–152
⋄ B35 G10 ⋄ REV MR 41 # 4863 Zbl 219 # 68044 • ID 14515

REYNOLDS, J.C. see Vol. IV, VI for further entries

REZNIK, T.L. [1962] see KOROBKOV, V.K.

REZNIKOFF, I. [1963] *Chaines de formules* (J 0109) C R Acad Sci, Paris 256*5021–5023
⋄ B05 C40 C90 F50 ⋄ REV MR 27 # 27 Zbl 143.7 • ID 11135

REZNIKOFF, I. [1965] *Tout ensemble de formules de la logique classique est equivalent a un ensemble independant* (J 0109) C R Acad Sci, Paris 260*2385–2388
⋄ B05 C40 ⋄ REV MR 31 # 2131 Zbl 143.7 • ID 11136

REZNIKOFF, I. see Vol. III, VI for further entries

REZUS, A. [1982] *On a theorem of Tarski* (J 3757) Libertas Math 2*63–97
⋄ B05 B22 ⋄ REV MR 84b:03019 Zbl 497 # 03004 • ID 34906

REZUS, A. [1983] *Abstract AUTOMATH* (S 1605) Math Centr Tracts 160*vi+188pp
⋄ B35 ⋄ REV MR 84j:03030 Zbl 507 # 03001 • ID 34623

REZUS, A. see Vol. VI for further entries

RHODES, F. [1971] see HIRST, K.E.

RIBEIRO, H. [1969] *Elementary languages and mathematical structures (Portuguese)* (J 0084) Gaz Mat (Lisboa) 30/113–116*1–8
⋄ B10 C07 ⋄ REV MR 46 # 4798 Zbl 248 # 02004 • ID 64815

RIBEIRO, H. see Vol. III, V for further entries

RICHARD, D. [1973] see CHARRETTON, C.

RICHARD, D. [1974] see CHARRETTON, C.

RICHARD, D. [1975] see CHARRETTON, C.

RICHARD, D. [1976] *Mesure de Haar en analyse non-standard* (J 0056) Publ Dep Math, Lyon 13/1∗15-21
- ⋄ H05 ⋄ REV MR 55 # 628   Zbl 363 # 28014 • ID 82604

RICHARD, D. [1981] see PABION, J.F.

RICHARD, D. [1984] *Les relations arithmetiques sur les entiers primaires sont definissables au premier ordre par successeur et coprimarite (English summary)* (J 3364) C R Acad Sci, Paris, Ser 1 299∗795-798
- ⋄ B28 F30 ⋄ REV MR 85m:03040 FdM 61.1028
- • ID 39634

RICHARD, D. [1984] *The arithmetics as theories of two orders (English and French summaries)* (P 2167) Orders: Descr & Roles;1982 L'Arbresle 287-311
- ⋄ B28 C62 D35 F30 ⋄ REV MR 85k:06001 MR 86h:03102   Zbl 555 # 03026 • ID 41779

RICHARD, D. [1985] *All arithmetical sets of powers of primes are first-order definable in terms of the successor function and the coprimeness predicate* (J 0193) Discr Math 53∗221-247
- ⋄ B28 F30 ⋄ REV MR 86h:03103   Zbl 562 # 03006
- • ID 45095

RICHARD, D. [1985] *Definissabilite de l'arithmetique par successeur, coprimarite et puissance* (J 3364) C R Acad Sci, Paris, Ser 1 300∗415-418
- ⋄ B28 F30 ⋄ ID 46289

RICHARD, D. see Vol. III, IV, V, VI for further entries

RICHARD, J. [1976] see COURVOISIER, M.

RICHARD, J. see Vol. V for further entries

RICHARDSON, R.P. [1916] see LANDIS, E.H.

RICHTER, H. [1952] *Zur Grundlegung der Wahrscheinlichkeitstheorie I,II,III* (J 0043) Math Ann 125∗129-139,223-234,335-343
- ⋄ B30 ⋄ REV MR 14.484 MR 15.634 Zbl 48.359 Zbl 51.102 Zbl 53.405 • REM Part IV 1953 • ID 33687

RICHTER, M.M. [1975] *Automatisches Beweisen und Gleichheitslogik* (P 1641) Kuenstl Intelligenzforschg BRD;1975 Bonn 15-23
- ⋄ B35 ⋄ REV Zbl 371 # 02008 • ID 51369

RICHTER, M.M. [1976] *Mathematische Logik I* (S 1642) Schr Inf Angew Math, Ber (Aachen) 31∗iii+101pp
- ⋄ B98 ⋄ ID 33426

RICHTER, M.M. [1976] *Ueber die unendlich kleinen Groessen in der Analysis* (J 0160) Math-Phys Sem-ber, NS 23∗59-74
- ⋄ H05 ⋄ REV MR 53 # 8361   Zbl 318 # 26004 • ID 23044

RICHTER, M.M. [1978] *Logikkalkuele* (X 0823) Teubner: Stuttgart 232pp
- ⋄ B35 B98 F05 G10 ⋄ REV MR 80g:03002 Zbl 381 # 03002 • ID 51870

RICHTER, M.M. [1982] *Ideale Punkte, Monaden und Nichtstandard-Methoden* (X 0900) Vieweg: Wiesbaden vii+264pp
- ⋄ C20 C60 E70 H05 H98 ⋄ REV MR 84g:03107 Zbl 487 # 03040 • ID 34209

RICHTER, M.M. & SZABO, M.E. [1983] *Towards a nonstandard analysis of programs* (C 3884) Nonstandard Anal - Recent Develop 186-203
- ⋄ B75 H10 ⋄ REV MR 84e:68015 Zbl 507 # 03030
- • ID 37244

RICHTER, M.M. [1983] *Variationen des Endlichkeitsbegriffes und Perspektiven fuer moegliche Anwendungen* (C 4414) Math Syst in Econ 457-463
- ⋄ E75 H10 ⋄ REV MR 84j:90005 Zbl 527 # 04005
- • ID 44846

RICHTER, M.M. [1984] see BOERGER, E.

RICHTER, M.M. [1984] see MUELLER, GERT H.

RICHTER, M.M. [1984] *Some aspects of nonstandard methods in general algebra* (P 3088) Univer Alg & Link Log, Alg, Combin, Comp Sci;1983 Darmstadt 78-84
- ⋄ C05 C30 H05 ⋄ REV Zbl 551 # 03038 • ID 43919

RICHTER, M.M. & SZABO, M.E. [1985] *Nonstandard computation theory* (C 4213) Algeb, Combin & Log in Comput Sci
- ⋄ D20 D75 H10 ⋄ ID 39589

RICHTER, M.M. see Vol. III, V, VI for further entries

RICKEY, V.F. [1971] *On weak and strong validity of rules for the propositional calculus* (J 0047) Notre Dame J Formal Log 12∗115-116
- ⋄ B05 ⋄ REV MR 44 # 2563   Zbl 197.272 • ID 11174

RICKEY, V.F. [1972] *Axiomatic inscriptional syntax. I. general syntax* (J 0047) Notre Dame J Formal Log 13∗1-33
- ⋄ B03 ⋄ REV MR 48 # 1880   Zbl 228 # 02019 • REM Part II 1973 • ID 11175

RICKEY, V.F. [1973] *Axiomatic inscriptional syntax; part II: The syntax of protothetic* (J 0047) Notre Dame J Formal Log 14∗1-52
- ⋄ B03 ⋄ REV MR 50 # 4257   Zbl 226 # 02027 • REM Part I 1972 • ID 11176

RICKEY, V.F. [1974] *The one variable implicational calculus* (J 0047) Notre Dame J Formal Log 15∗478-480
- ⋄ B20 ⋄ REV MR 51 # 2856   Zbl 261 # 02012 • ID 11177

RICKEY, V.F. [1975] *Creative definitions in propositional calculi* (J 0047) Notre Dame J Formal Log 16∗273-294
- ⋄ B05 C40 ⋄ REV MR 52 # 38   Zbl 232 # 02008 • ID 11178

RICKEY, V.F. [1975] *On creative definitions in the Principia Mathematica* (J 0079) Logique & Anal, NS 18∗175-182
- ⋄ B15 C40 ⋄ REV MR 53 # 7721   Zbl 316 # 02003
- • ID 22975

RICKEY, V.F. [1978] *On creative definitions in first order functional calculi* (J 0047) Notre Dame J Formal Log 19∗307-309
- ⋄ B10 C40 ⋄ REV MR 58 # 5012   Zbl 368 # 02011
- • ID 51658

RICKEY, V.F. see Vol. II, V for further entries

RIDDER, J. [1946] *Ueber den Aussagen- und den engeren Praedikatenkalkuel* (J 0028) Indag Math 8∗701-712
- ⋄ B10 ⋄ REV MR 8.306   JSL 12.135   JSL 13.174 • REM Part I. Part II 1947 • ID 11179

RIDDER, J. [1947] *Ueber den Aussagen- und den engeren Praedikatenkalkuel. II* (J 0028) Indag Math 9∗9-15
- ⋄ B20 ⋄ REV MR 8.306   MR 9.221   Zbl 29.196   JSL 12.135
- • REM Part I 1946 • ID 11180

RIDDER, J. [1948] *Logic of propositions* (J 0154) Synthese 6*496-502
◇ B05 ◇ REV MR 10.499 JSL 15.231 • ID 11181

RIDDER, J. see Vol. II, VI for further entries

RIEBER, C.H. [1918] *Footnotes to formal logic* (X 0826) Alber: Freiburg 177pp
◇ A05 B98 ◇ REV FdM 46.73 • ID 25580

RIECAN, B. [1984] see NEUBRUNN, T.

RIECAN, B. see Vol. V for further entries

RIECANOVA, Z. [1984] see NEUBRUNN, T.

RIEGER, L. [1955] *On a fundamental theorem of mathematical logic (Czech) (Russian and English summaries)* (J 0086) Cas Pestovani Mat, Ceskoslov Akad Ved 80*217-231
◇ B10 C07 G05 G25 ◇ REV MR 19.378 Zbl 68.242 JSL 35.489 • ID 19556

RIEGER, L. [1964] *Zu den Strukturen der klassischen Praedikatenlogik* (J 0068) Z Math Logik Grundlagen Math 10*121-138
◇ B10 C07 G05 ◇ REV MR 29 # 2168 Zbl 124.247 JSL 35.440 • ID 11193

RIEGER, L. [1967] *Algebraic methods of mathematical logic* (X 0801) Academic Pr: New York 210pp
◇ A05 B98 G05 G15 ◇ REV Zbl 218 # 02001 JSL 35.440 • ID 11194

RIEGER, L. see Vol. III, IV, V, VI for further entries

RINE, D.C. [1975] *Associative and multi-valued logic for possible improvements in some X-ray image processing* (P 1805) Int Symp Multi-Val Log (5,Proc);1975 Bloomington 146-161
◇ B80 ◇ ID 35820

RINE, D.C. see Vol. II, IV for further entries

RINON, S. & YOELI, M. [1964] *Application of ternary algebra to the study of static hazards* (J 0037) ACM J 11*84-97
◇ B20 ◇ REV Zbl 137.339 • ID 41903

RISER, J. [1967] *A Gentzen-type calculus for single-operator proposititonal logic* (J 0036) J Symb Logic 32*75-80
◇ B05 ◇ REV MR 36 # 2467 Zbl 153.6 JSL 33.129 • ID 11203

RISER, J. [1979] *A simplification procedure for alternational normal schemata* (J 0047) Notre Dame J Formal Log 20*765-767
◇ B05 ◇ REV MR 80k:03014 Zbl 349 # 02007 • ID 56182

RISTEA, T.G. [1968] *On propositional, truth and boolean functions* (J 0047) Notre Dame J Formal Log 9*160-166
◇ B05 ◇ REV MR 38 # 4274 Zbl 253 # 02007 • ID 11204

RISTEA, T.G. [1971] *La theorie des systemes deductifs chez A. Tarski (Romanian) (French summary)* (S 1613) Probl Logic (Bucharest) 3*307-322
◇ B22 ◇ REV Zbl 215.324 • ID 28022

RISTEA, T.G. see Vol. V for further entries

RITCHIE, R.W. [1965] *A rudimentary definition of addition* (J 0036) J Symb Logic 30*350-354
◇ A05 B28 ◇ REV MR 33 # 2541 Zbl 244 # 02013 JSL 35.475 • ID 11206

RITCHIE, R.W. see Vol. IV, VI for further entries

ROBBIN, J.W. [1969] *Mathematical logic: a first course* (X 0867) Benjamin: Reading xii + 212pp
◇ B98 ◇ REV MR 40 # 4078 Zbl 186.3 • ID 19552

ROBBIN, J.W. see Vol. III for further entries

ROBERT, A. [1984] *L'analyse non standard* (X 0997) Queen's Univ: Kingston 119pp
◇ H05 ◇ REV MR 85m:03003 Zbl 538 # 03053 • ID 41486

ROBERT, A. [1984] *Une approche naive de l'analyse non-standard* (J 0076) Dialectica 38*287-296
◇ H05 ◇ REV Zbl 559 # 03043 • ID 44521

ROBERT, A. [1985] *Analyse non standard* (X 4457) Pr Polytechn: Lausanne xvii + 119pp
◇ B98 H05 ◇ REV Zbl 565 # 03033 • ID 48149

ROBERT, P. [1965] *Methodes fonctorielles sur l'axiomatique des systemes de generateurs, des rangs, etc* (J 0109) C R Acad Sci, Paris 260*4291-4294,5983-5986
◇ B30 E20 G10 G30 ◇ REV MR 31 # 2301 MR 31 # 2303 • ID 26664

ROBERT, P. [1967] *Sur l'axiomatique des systemes generateurs, des rangs, etc* (J 2313) C R Acad Sci, Paris, Ser A-B 265*A649-A651
◇ B30 ◇ REV MR 36 # 5051 Zbl 237 # 08013 • ID 26665

ROBERTS, C. [1974] *Dale on material implication* (J 0103) Analysis (Oxford) 35*27-28
◇ B05 ◇ REV Zbl 336 # 02007 • ID 64849

ROBERTS, D.D. [1964] *The existential graphs and natural deduction* (C 0707) Stud Phil of C.S. Pierce 109-121
◇ A05 B10 ◇ REV JSL 35.320 • ID 11216

ROBERTS, F.S. [1984] *Applications of the theory of meaningfulness to order and matching experiments* (P 4275) Trends in Math Psychol;1983 Bruxelles 20*283-292
◇ B80 ◇ REV Zbl 546 # 92016 • ID 45593

ROBERTS, F.S. see Vol. III, V for further entries

ROBINSON, A. [1954] *L'application de la logique formelle aux mathematiques* (P 0646) Appl Sci de Log Math;1952 Paris 51-64
◇ B30 C60 ◇ REV MR 16.782 Zbl 57.7 JSL 23.218 • ID 11226

ROBINSON, A. [1957] *Proving a theorem (as done by man, logician or machine)* (P 1675) Summer Inst Symb Log;1957 Ithaca 350-352
• REPR [1983] (C 4659) Autom of Reasoning 1*74-78
◇ A05 B35 ◇ REV JSL 32.522 • ID 29376

ROBINSON, A. [1957] see LIGHTSTONE, A.H.

ROBINSON, A. [1958] *Outline of an introduction to mathematical logic I,II,III* (J 0018) Canad Math Bull 1*41-54,113-127,193-208
◇ B98 C07 G05 ◇ REV MR 20 # 5123 MR 20 # 5124 MR 21 # 6321 Zbl 80.5 Zbl 82.15 Zbl 84.4 • REM Part IV 1959 • ID 25442

ROBINSON, A. [1959] *Outline of an introduction to mathematical logic IV* (J 0018) Canad Math Bull 2*33-42
◇ B10 B98 C07 C98 G05 ◇ REV MR 21 # 6321 Zbl 84.4 • REM Parts I-III 1958 • ID 19551

ROBINSON, A. [1960] *On the mechanization of the theory of equations* (J 0493) Bull Res Counc Israel Sect F 9*47-70
◇ B35 ◇ REV MR 26 # 4910 Zbl 213.24 JSL 33.118 • ID 11243

ROBINSON, A. [1961] *Non-standard analysis* (**J** 0028) Indag Math 23*432-440
- REPR [1979] (**C** 4594) Sel Pap Robinson 2*3-11
- ◇ H05 H15 ◇ REV MR 26#33 Zbl 102.7 JSL 34.292
- ID 11251

ROBINSON, A. [1961] *On the d-calculus for linear differential equations with constant coefficients* (**J** 0148) Math Gaz 45*202-206
- ◇ H05 ◇ ID 26127

ROBINSON, A. [1962] *Complex function theory over non-archimedean fields* (1111) Preprints, Manuscr., Techn. Reports etc.
- ◇ H05 ◇ REM Technical-Scientific Note No.30 U.S.A.F. Contract No.61 (052)-187, Jerusalem • ID 27635

ROBINSON, A. [1963] *Introduction to model theory and to the metamathematics of algebra* (**X** 0809) North Holland: Amsterdam ix+284pp
- TRANSL [1967] (**X** 2027) Nauka: Moskva 376pp
- ◇ B98 C60 C98 H15 ◇ REV MR 27#3533 MR 36#3642 JSL 25.172 • REM 2nd rev. ed. 1974 • ID 28804

ROBINSON, A. [1963] *On languages which are based on non-standard arithmetic* (**J** 0111) Nagoya Math J 22*83-117
- REPR [1979] (**C** 4594) Sel Pap Robinson 2*12-46
- ◇ C75 H05 H15 ◇ REV MR 27#3532 Zbl 166.261 JSL 34.516 • ID 11254

ROBINSON, A. [1964] *On generalized limits and linear functionals* (**J** 0048) Pac J Math 14*269-283
- REPR [1979] (**C** 4594) Sel Pap Robinson 2*47-61
- ◇ C20 H05 ◇ REV MR 29#1534 Zbl 121.95 JSL 34.292
- ID 11258

ROBINSON, A. [1965] *On the theory of normal families* (**J** 0096) Acta Philos Fenn 18*159-184
- ◇ H05 ◇ REV MR 32#5845 Zbl 148.306 • ID 26129

ROBINSON, A. [1965] *Topics in non-archimedean mathematics* (**P** 0614) Th Models;1963 Berkeley 285-298
- REPR [1979] (**C** 4594) Sel Pap Robinson 2*99-112
- ◇ H05 H15 H20 ◇ REV MR 33#5489 Zbl 156.250 JSL 34.292 • ID 11259

ROBINSON, A. [1966] *Non-standard analysis* (**X** 0809) North Holland: Amsterdam ix+293pp
- ◇ A10 B98 H05 H10 H15 H98 ◇ REV MR 34#5680 Zbl 151.8 JSL 35.292 • REM 2d edition 1968; 291pp
- ID 11263

ROBINSON, A. [1966] *On some applications of model theory to algebra and analysis* (**J** 0384) Rend Mat, Ser 5 25*562-592
- REPR [1979] (**C** 4594) Sel Pap Robinson 2*158-188
- ◇ C60 C98 H05 H98 ◇ REV MR 36#2489 Zbl 157.28
- ID 22170

ROBINSON, A. [1966] see BERNSTEIN, A.R.

ROBINSON, A. [1968] see HURD, A.E.

ROBINSON, A. & ZAKON, E. [1969] *A set-theoretical characterization of enlargements* (**P** 0649) Appl Model Th to Algeb, Anal & Probab;1967 Pasadena 109-122
- REPR [1979] (**C** 4594) Sel Pap Robinson 2*206-219
- ◇ C20 C62 E75 H05 H20 ◇ REV MR 39#1319 Zbl 193.303 • ID 11267

ROBINSON, A. [1969] *Compactification of groups and rings and non-standard analysis* (**J** 0036) J Symb Logic 34*576-588
- REPR [1979] (**C** 4594) Sel Pap Robinson 2*243-255
- ◇ C40 C60 H05 H20 ◇ REV MR 44#1765 Zbl 196.11
- ID 11269

ROBINSON, A. [1969] *Germs* (**P** 0649) Appl Model Th to Algeb, Anal & Probab;1967 Pasadena 138-149
- ◇ H05 ◇ REV MR 38#4723 Zbl 187.271 • ID 26143

ROBINSON, A. [1969] *Topics in nonstandard algebraic number theory* (**P** 0649) Appl Model Th to Algeb, Anal & Probab;1967 Pasadena 1-17
- REPR [1979] (**C** 4594) Sel Pap Robinson 2*189-205
- ◇ C60 H05 H15 H20 ◇ REV MR 42#3054 Zbl 182.327
- ID 21051

ROBINSON, A. [1972] see BROWN, D.J.

ROBINSON, A. [1972] see LUXEMBURG, W.A.J.

ROBINSON, A. [1972] see KELEMEN, P.J.

ROBINSON, A. [1973] *Function theory on some nonarchimedian fields* (**J** 0005) Amer Math Mon 80*87-109
- ◇ C60 C65 H05 ◇ REV MR 48#8464 Zbl 269#26020
- ID 11280

ROBINSON, A. [1973] *Metamathematical problems* (**J** 0036) J Symb Logic 38*500-516
- TRANSL [1973] (**J** 0396) Mat Lapok 24*1-17 • REPR [1979] (**C** 4594) Sel Pap Robinson 1*43-59
- ◇ A05 B30 C60 C98 H05 ◇ REV MR 49#2240 Zbl 289#02002 • ID 11276

ROBINSON, A. [1973] *Standard and nonstandard number systems* (**J** 3077) Nieuw Arch Wisk, Ser 3 21*115-133
- REPR [1979] (**C** 4594) Sel Pap Robinson 2*426-444
- ◇ C25 C60 C98 H05 H15 H20 H98 ◇ REV MR 55#7767 Zbl 267#02041 • ID 26149

ROBINSON, A. [1974] *Enlarged sheaves* (**P** 1083) Victoria Symp Nonstand Anal;1972 Victoria 369*249-260
- ◇ H05 ◇ REV MR 58#214 Zbl 277#02014 • ID 29528

ROBINSON, A. [1975] see LIGHTSTONE, A.H.

ROBINSON, A. & ROQUETTE, P. [1975] *On the finiteness theorem of Siegel and Mahler concerning diophantine equations* (**J** 0401) J Number Th 7*121-176
- ◇ C60 H05 H15 H20 ◇ REV MR 51#10222 Zbl 299#12107 • ID 17511

ROBINSON, A. [1979] *Selected papers of Abraham Robinson. Vol.II: Nonstandard analysis and philosophy* (**X** 0875) Yale Univ Pr: New Haven xlv+582pp
- LAST ED [1979] (**X** 0809) North Holland: Amsterdam xlv+582pp
- ◇ A05 B96 C96 H05 H10 H96 ◇ REV MR 80h:01039b Zbl 424#01031 JSL 47.203 • REM Ed. by Keisler,H.J. & Koerner,S. & Luxemburg,W.A.J. & Young,A.D. Vol.II 1979 • ID 53900

ROBINSON, A. also published under the name ROBINSOHN, A.

ROBINSON, A. see Vol. III, IV, V for further entries.

ROBINSON, G.A. [1964] see CARSON, D.F.

ROBINSON, G.A. [1965] see CARSON, D.F.

ROBINSON, G.A. [1966] see CARSON, D.F.

ROBINSON, G.A. [1967] see CARSON, D.F.

ROBINSON, G.A. & WOS, L. [1969] *Paramodulation and theorem-proving in first order theories with equality* (J 0508) Machine Intelligence 4*135-150
- REPR [1983] (C 4659) Autom of Reasoning 2*298-316
- ◇ B35 ◇ REV MR 43 # 1473 Zbl 219 # 68047 • ID 46613

ROBINSON, G.A. & WOS, L. [1970] *Axiom systems in automatic theorem proving* (P 0625) Symp Autom Demonst;1968 Versailles 215-236
- ◇ B35 ◇ REV MR 42 # 2889 • ID 46614

ROBINSON, G.A. & WOS, L. [1970] *Paramodulation and set of support* (P 0625) Symp Autom Demonst;1968 Versailles 276-310
- ◇ B35 ◇ REV MR 42 # 7512 Zbl 208.23 • ID 14300

ROBINSON, G.A. & WOS, L. [1973] *Maximal models and refutation completeness: semidecision procedures in automatic theorem proving* (P 0678) Word Probl: Decis & Burnside Probl in Group Th;1969 Irvine 609-639
- REPR [1983] (C 4659) Autom of Reasoning 2*578-608
- ◇ B35 ◇ REV MR 53 # 7154 Zbl 283 # 68058 JSL 41.786
- • ID 22968

ROBINSON, JOHN ALAN [1963] *Theorem-proving on the computer* (J 0037) ACM J 10*163-174
- REPR [1983] (C 4659) Autom of Reasoning 1*372-386
- ◇ B35 ◇ REV MR 26 # 7178 Zbl 109.356 JSL 31.514
- • ID 11285

ROBINSON, JOHN ALAN [1964] *On automatic deduction* (J 0952) Rice Univ Studies 50*69-89
- ◇ B35 ◇ REV JSL 31.516 • ID 11286

ROBINSON, JOHN ALAN [1965] *A machine-oriented logic based on the resolution principle* (J 0037) ACM J 12*23-41
- REPR [1983] (C 4659) Autom of Reasoning 1*397-415
- ◇ B35 ◇ REV MR 30 # 732 Zbl 139.123 JSL 31.515
- • ID 11287

ROBINSON, JOHN ALAN [1965] *Automatic deduction with hyper-resolution* (J 0382) Int J Comput Math 1*227-234
- REPR [1983] (C 4659) Autom of Reasoning 1*416-423
- ◇ B35 ◇ REV MR 32 # 5522 Zbl 158.260 JSL 39.189
- • ID 11288

ROBINSON, JOHN ALAN [1967] *A review of automatic theorem-proving* (P 0737) Math Aspects Comput Sci;1966 New York 1-18
- ◇ B35 ◇ REV MR 39 # 2540 Zbl 174.292 JSL 39.190
- • ID 11289

ROBINSON, JOHN ALAN [1967] *Heuristic and complete processes in the mech. of theorem proving* (P 1390) Syst & Comput Sci;1965 London ON 116-124
- ◇ B35 ◇ REV MR 38 # 3092 • ID 46617

ROBINSON, JOHN ALAN [1968] *The generalized resolution principle* (J 0508) Machine Intelligence 3*77-93
- REPR [1983] (C 4659) Autom of Reasoning 2*135-151
- ◇ B35 ◇ REV Zbl 195.311 • ID 46615

ROBINSON, JOHN ALAN [1969] *A note on mechanizing higher order logic* (J 0508) Machine Intelligence 5*123-133
- ◇ B15 B35 ◇ REV MR 41 # 4864 Zbl 221 # 68053
- • ID 46618

ROBINSON, JOHN ALAN [1969] *Mechanizing higher-order logic* (J 0508) Machine Intelligence 4*151-170
- ◇ B35 B40 F35 ◇ REV MR 46 # 6992 Zbl 228 # 68025
- • ID 11291

ROBINSON, JOHN ALAN [1969] *New directions in mechanical theorem proving* (P 0594) Inform Processing (4);1968 Edinburgh 1*63-69
- REPR [1983] (C 4659) Autom of Reasoning 2*152-158
- ◇ B35 ◇ REV MR 42 # 4049 Zbl 213.24 • ID 11290

ROBINSON, JOHN ALAN [1970] *An overview of mechanical theorem proving* (P 4248) Th Approaches Non-Num Probl Solving 2-20
- ◇ B35 ◇ REV Zbl 195.30 • ID 46530

ROBINSON, JOHN ALAN [1970] *Computational logic: The unification computation* (J 0508) Machine Intelligence 5
- ◇ B35 ◇ ID 46619

ROBINSON, JOHN ALAN [1970] *The present state of mechanical theorem proving* (P 4248) Th Approaches Non-Num Probl Solving
- ◇ B35 ◇ ID 46616

ROBINSON, JOHN ALAN [1971] *Building deduction machines. Art. int. and heuristic programming* (J 0508) Machine Intelligence 6
- ◇ B35 ◇ ID 46529

ROBINSON, JOHN ALAN [1979] *Logic: form and function. The mechanization of deductive reasoning* (X 1261) Edinburgh Univ Pr: Edinburgh vi+312pp
- ◇ A10 B10 B35 B98 ◇ REV MR 84d:03022 Zbl 428 # 03009 JSL 51.227 • ID 53769

ROBINSON, R.M. [1957] *Restricted set-theoretical definitions in arithmetic* (P 1675) Summer Inst Symb Log;1957 Ithaca 139-140
- ◇ B28 F30 F35 ◇ REV MR 20 # 3 Zbl 112.7 JSL 31.659
- REM Summary. See 1958 • ID 11315

ROBINSON, R.M. [1958] *Restricted set-theoretical definitions in arithmetic* (J 0053) Proc Amer Math Soc 9*238-242
- ◇ B28 F30 F35 ◇ REV MR 20 # 3 Zbl 112.7 JSL 31.659
- REM For a summary see 1957 • ID 48434

ROBINSON, R.M. see Vol. III, IV, V, VI for further entries

ROBINSON, T.T. [1968] *Independence of two nice sets of axioms for the propositional calculus* (J 0036) J Symb Logic 33*265-270
- ◇ B05 ◇ REV MR 39 # 3961 Zbl 162.11 JSL 35.139
- • ID 11328

ROBINSON, T.T. see Vol. VI for further entries

ROBISON, G.B. [1969] *An introduction to mathematical logic* (X 0819) Prentice Hall: Englewood Cliffs xii+212pp
- ◇ B98 ◇ REV JSL 36.679 • ID 11329

ROBITASHVILI, N.G. [1965] *On the problem of reducing a pp-formula of the pure first-order functional calculus to some special case of the decision problem (Georgian) (Russian summary)* (J 0233) Soobshch Akad Nauk Gruz SSR 39*269-276
- ◇ B20 B25 ◇ REV MR 32 # 5502 • ID 11330

ROBITASHVILI, N.G. [1971] *A combination of the inverse method and the method of resolution (Russian) (Georgian and English summaries)* (J 0233) Soobshch Akad Nauk Gruz SSR 64*269-272
- ◇ B35 F07 ◇ REV MR 45 # 4945 Zbl 251 # 68048
- • ID 11331

ROBITASHVILI, N.G. see Vol. II for further entries

RODEK, I. [1976] *Normal forms as another method for testing whether rational expressions of the sentential calculus are tautologies* (S 2890) Zesz Nauk, Mat Fiz, Politech Slask (Gliwice) 26*297-303
  ◇ B05 ◇ REV MR 58 # 16117 • ID 77839

RODENHAUSEN, H. [1982] *A characterization of nonstandard liftings of measurable functions and stochastic processes* (J 0029) Israel J Math 43*1-22
  ◇ H10 ◇ REV MR 85h:28008   Zbl 504 # 60003 • ID 38419

RODRIGUEZ ARTALEJO, M. [1981] *Eine syntaktisch-algebraische Methode zur Konstruktion von Modellen* (J 0068) Z Math Logik Grundlagen Math 27*59-71
  ◇ B10   C07   C30   C57 ◇ REV   MR 82e:03027   Zbl 481 # 03016 • ID 77840

RODRIGUEZ ARTALEJO, M. [1985] see HORTALA-GONZALEZ, M.T.

RODRIGUEZ ARTALEJO, M. see Vol. III for further entries

ROEDDING, D. & SCHWICHTENBERG, H. [1972] *Bemerkungen zum Spektralproblem* (J 0068) Z Math Logik Grundlagen Math 18*1-12
  ◇ B15   C13   D10   D15 ◇ REV   MR 46 # 5128   Zbl 242 # 02049 • ID 11343

ROEDDING, D. [1984] see BOERGER, E.

ROEDDING, D. see Vol. III, IV for further entries

ROEDDING, W. [1968] *Eine Art von Gleichgewicht zahlentheoretischer und mengentheoretischer Axiomensysteme* (J 0009) Arch Math Logik Grundlagenforsch 11*17-31
  ◇ B30   E30   F30 ◇ REV   MR 37 # 2602   Zbl 169.12
  • ID 11344

ROEDDING, W. see Vol. IV for further entries

ROEPER, P. [1980] *Intervals and tenses* (J 0122) J Philos Logic 9*451-470
  ◇ A05   B30   B45 ◇ REV   MR 83i:03035   Zbl 451 # 03007
  • ID 54022

ROEPER, P. [1985] *Generalisation of first-order logic to nonatomic domains* (J 0036) J Symb Logic 50*815-838
  ◇ B10 ◇ ID 47383

ROESSLER, K. [1934] *Beweis der Widerspruchsfreiheit des Funktionenkalkuels der mathematischen Logik (French summary)* (J 0945) Riv Filos 18*8pp
  ◇ B10   F25 ◇ REV   Zbl 11.3   FdM 60.850 • ID 19547

ROGAVA, M.G. [1967] *On sequential modifications of applied predicate calculi (Russian)* (S 0228) Zap Nauch Sem Leningrad Otd Mat Inst Steklov 4*189-200
  • TRANSL [1967] (J 0521) Semin Math, Inst Steklov 4*77-81
  ◇ B10 ◇ REV MR 39 # 1293   Zbl 208.9 • ID 37196

ROGAVA, M.G. [1969] *A sequential version of the calculus of equivalences (Russian)* (J 0954) Tr Inst Prikl Mat, Tbilisi 2*9-11
  ◇ B20 ◇ REV MR 43 # 1812   Zbl 223 # 02011 • ID 11347

ROGAVA, M.G. [1969] *A specialization of certain logical rules in applied predicate calculi (Russian) (Georgian and English summaries)* (J 0233) Soobshch Akad Nauk Gruz SSR 56*273-276
  ◇ B20 ◇ REV MR 43 # 1809   Zbl 218 # 02012 • ID 11346

ROGAVA, M.G. [1972] *Permutability of the applications of rules in applied predicate calculi (Russian) (Georgian and English summaries)* (J 0233) Soobshch Akad Nauk Gruz SSR 65*273-276
  ◇ B10 ◇ REV MR 45 # 6577   Zbl 229 # 02016 • ID 11348

ROGAVA, M.G. [1972] *Sequential variants of applied predicate calculi without structural deductive rules (Russian)* (S 0066) Tr Mat Inst Steklov 121*136-164,167
  • TRANSL [1972] (S 0055) Proc Steklov Inst Math 121*151-181
  ◇ B10   F07 ◇ REV   MR 54 # 12495   Zbl 286 # 02037
  • ID 64879

ROGAVA, M.G. [1974] see CHADADZE, O.S.

ROGAVA, M.G. [1981] *On the sequential version of the propositional fragment of the first-order calculus of terms (Russian) (Georgian and English summaries)* (J 0233) Soobshch Akad Nauk Gruz SSR 101*545-548
  ◇ B20 ◇ REV MR 82m:03065   Zbl 457 # 03003 • ID 54328

ROGAVA, M.G. see Vol. II, III, VI for further entries

ROGERS, L. [1982] *Infinitesimal, continuity and meaning* (J 2679) Bull Inst Math Appl (Southend oS) 18*38-42
  ◇ H05 ◇ REV MR 84j:03139b   Zbl 516 # 26010 • ID 34730

ROGERS, L. see Vol. VI for further entries

ROGERS, R. [1971] *Mathematical logic and formalized theories. A survey of basic concepts and results* (X 0809) North Holland: Amsterdam xi+235pp
  ◇ B98   C98 ◇ REV   MR 47 # 9   Zbl 245 # 02001 • ID 11365

ROGERS JR., H. [1956] *Certain logical reduction and decision problems* (J 0120) Ann of Math, Ser 2   64*264-284
  ◇ B20   D35 ◇ REV   MR 18.271   Zbl 74.14   JSL 22.217
  • ID 11355

ROGERS JR., H. [1973] see MATHIAS, A.R.D.

ROGERS JR., H. see Vol. III, IV, VI for further entries

ROHLEDER, H. [1956] *Zur Umformung logischer Ausdruecke mit Hilfe programmgesteuerter Rechenanlagen* (J 0068) Z Math Logik Grundlagen Math 2*57-58
  ◇ B35 ◇ REV MR 18.457   Zbl 74.15 • ID 11371

ROHLEDER, H. [1959] *Ein Verfahren zum Aufstellen optimaler Normalformen bei gegebenen Primimplikanden* (J 0068) Z Math Logik Grundlagen Math 5*334-339
  ◇ B35 ◇ REV MR 23 # A3665   Zbl 93.14 • ID 11374

ROHLEDER, H. [1961] *Zum Ausmultiplizieren der Klammern beim Verfahren von Nelson* (J 1046) Z Angew Math Mech 41*77-78
  ◇ B35   D05 ◇ REV   Zbl 103.248 • ID 20926

ROHLEDER, H. [1963] *Ein Verfahren zum Aufsuchen minimaler Ausdruecke* (J 0068) Z Math Logik Grundlagen Math 9*125-140
  ◇ B35 ◇ REV MR 26 # 4891   Zbl 143.22 • ID 11379

ROHLEDER, H. [1963] *Eine Halbordnung im Aussagekalkuel* (J 0068) Z Math Logik Grundlagen Math 9*21-52
  ◇ B05 ◇ REV MR 26 # 3582   Zbl 143.22 • ID 11378

ROHLEDER, H. [1965] *Eine Bemerkung zur Zeilendominanz beim Aufsuchen optimaler Normalformen* (J 0068) Z Math Logik Grundlagen Math 11*273-276
  ◇ B35 ◇ REV MR 32 # 3961   Zbl 171.278 • ID 11380

ROHLEDER, H. [1967] see BAER, G.

ROHLEDER, H. see Vol. II for further entries

ROHRBACH, H. [1951] *Das Axiomensystem von Erhard Schmidt fuer die Menge der natuerlichen Zahlen* (J 0114) Math Nachr 4*315-321
  ⋄ B28 ⋄ REV MR 14.714 Zbl 42.8 JSL 20.68 • ID 11382

ROHRBACH, H. [1966] *Was sind und was sollen Zahlen ?* (X 1272) Gutenberg: Mainz 28pp
  ⋄ B28 ⋄ ID 25716

ROITMAN, J. [1982] *Non-isomorphic hyper-real fields from non-isomorphic ultrapowers* (J 0044) Math Z 181*93-96
  ⋄ C20 H05 H20 ⋄ REV MR 84a:54030 Zbl 474 # 54003 • ID 55446

ROITMAN, J. see Vol. V for further entries

ROMAN, P. [1976] see LEVEILLE, J.P.

ROMANKEVICH, A.M. [1974] see JACUNOV, A.I.

ROMANKEVICH, A.M. see Vol. II for further entries

ROMANOV, A.M. [1978] *The shortest disjunctive normal forms for products of boolean functions (Russian)* (J 0071) Met Diskr Analiz (Novosibirsk) 32*52-65,96
  ⋄ B35 ⋄ REV MR 81b:94071 Zbl 456 # 94020 • ID 82641

ROMANSKI, R. [1980] *Algorithms for functional decomposition of boolean function (Bulgarian) (Russian and English summaries)* (J 3360) Avtom Izchisl Tekh, Sofiya 14/6*36-42
  ⋄ B35 ⋄ REV Zbl 451 # 94012 • ID 69876

RONCHI DELLA ROCCA, S. & VENNERI, B.M. [1984] *Principal type schemes for an extended type theory* (J 1426) Theor Comput Sci 28*151-169
  ⋄ B15 ⋄ REV MR 86d:03017 Zbl 535 # 03007 • ID 38312

RONCHI DELLA ROCCA, S. see Vol. IV, VI for further entries

ROOTSELAAR VAN, B. & STAAL, J.F. (EDS.) [1968] *Logic, methodology and philosophy of science III* (S 3303) Stud Logic Found Math xii+553pp
  ⋄ B98 ⋄ REV MR 40 # 14 Zbl 177.293 • ID 33328

ROOTSELAAR VAN, B. see Vol. II, IV, VI for further entries

ROQUETTE, P. [1975] see ROBINSON, A.

ROQUETTE, P. see Vol. III, IV for further entries

RORER, M.A. [1973] see KOGALOVSKIJ, S.R.

ROSE, A. [1949] *A reduction in the number of the axioms of the propositional calculus* (J 4510) Norsk Mat Tidsskr 31*113-115
  ⋄ B05 ⋄ REV MR 11.303 Zbl 40.146 JSL 15.139 • ID 16886

ROSE, A. [1950] *Completeness of Lukasiewicz-Tarski propositional calculi* (J 0043) Math Ann 122*296-298
  ⋄ B05 ⋄ REV MR 12.662 Zbl 41.149 JSL 16.228 • ID 11402

ROSE, A. [1951] *A formalization of the c-o-propositional calculus* (J 0171) Proc Cambridge Phil Soc Math Phys 47*635-636
  ⋄ B05 ⋄ REV MR 13.309 Zbl 43.7 JSL 17.66 • ID 11413

ROSE, A. [1951] *Strong completeness of fragments of the propositional calculus* (J 0036) J Symb Logic 16*204
  ⋄ B20 ⋄ REV MR 13.309 Zbl 43.8 JSL 17.147 • ID 11412

ROSE, A. [1951] *The degree of completeness of some Lukasiewicz-Tarski propositional calculi* (J 0039) J London Math Soc 26*47-49 • ERR/ADD ibid 44*587-591
  ⋄ B20 B50 ⋄ REV MR 12.662 MR 39 # 5321 Zbl 44.251 JSL 17.147 JSL 39.350 • ID 11403

ROSE, A. [1951] *The degrees of completeness of a partial system of the 2-valued propositional calculus* (J 0044) Math Z 54*181-183
  ⋄ B20 ⋄ REV MR 13.614 Zbl 45.296 JSL 17.147 • ID 11411

ROSE, A. [1952] *An extension of computational logic* (J 0036) J Symb Logic 17*32-34
  ⋄ B35 ⋄ REV MR 13.811 Zbl 48.244 JSL 17.204 • ID 11418

ROSE, A. [1952] *Sur un ensemble de fonctions primitives pour le calcul des predicats du premier ordre lequel constitue son propre dual* (J 0109) C R Acad Sci, Paris 234*1830-1831
  ⋄ B10 ⋄ REV MR 14.3 Zbl 48.245 JSL 18.343 • ID 11416

ROSE, A. [1953] *Conditioned disjunction as a primitive connective for the erweiterter Aussagenkalkuel* (J 0036) J Symb Logic 18*63-65
  ⋄ B05 ⋄ REV MR 14.936 Zbl 53.199 JSL 18.344 • ID 11425

ROSE, A. [1954] *A formalisation of the 2-valued propositional calculus with self-dual primitives* (J 0043) Math Ann 127*255-257
  ⋄ B05 ⋄ REV MR 15.771 Zbl 55.5 JSL 19.295 • ID 11433

ROSE, A. [1955] *A single axiom for a partial system of the propositional calculus* (J 0068) Z Math Logik Grundlagen Math 1*196-197
  ⋄ B20 ⋄ REV MR 17.932 Zbl 68.242 JSL 24.176 • ID 11435

ROSE, A. [1964] *Fragments du calcul propositionnel bivalent a foncteurs variables* (J 0109) C R Acad Sci, Paris 258*1363-1365
  ⋄ B20 ⋄ REV MR 28 # 2038 Zbl 178.305 • ID 11470

ROSE, A. [1965] *Two non-henkinian fragments of the 2-valued propositional calculus with variable functors* (J 0068) Z Math Logik Grundlagen Math 11*45-55
  ⋄ B20 ⋄ REV MR 31 # 30 • ID 11474

ROSE, A. [1966] *Sur quelques resultats touchant la formalisation des calculs propositionnels a foncteurs variables* (J 2313) C R Acad Sci, Paris, Ser A-B 262*A1085-A1087
  ⋄ B05 B20 ⋄ REV MR 34 # 1158 Zbl 214.9 • ID 11479

ROSE, A. [1969] *Calcul implicatif de non-contradiction* (J 2313) C R Acad Sci, Paris, Ser A-B 269*A289-A290
  ⋄ B20 ⋄ REV MR 40 # 5412 Zbl 184.7 • ID 11486

ROSE, A. [1969] *Single generators for Henkinian fragments of the 2-valued propositional calculus* (J 0068) Z Math Logik Grundlagen Math 15*85-92
  ⋄ B20 ⋄ REV MR 40 # 19 Zbl 182.314 • ID 11487

ROSE, A. [1971] *Tautologies sans constants* (J 2313) C R Acad Sci, Paris, Ser A-B 272*A1617-A1619
  ⋄ B05 ⋄ REV MR 43 # 7292 Zbl 215.318 • ID 11493

ROSE, A. [1976] *Formalisations with non-standard degrees of completeness* (J 0068) Z Math Logik Grundlagen Math 22*177-186
  ⋄ B20 ⋄ REV MR 57 # 2874 Zbl 341 # 02011 • ID 18478

ROSE, A. [1977] *Simplified formalizations of fragments of the propositional calculus* (J 0047) Notre Dame J Formal Log 18*255-261
⋄ B20 ⋄ REV MR 56#8321   Zbl 323#02015 • ID 23617

ROSE, A. [1978] *A note on formalisation by the method of description of truth-tables* (J 0068) Z Math Logik Grundlagen Math 24*109-112
⋄ B05 ⋄ REV MR 80a:03010   Zbl 393#03004 • ID 52424

ROSE, A. [1978] *Sur certaines tautologies sans constantes qui possedent des variables de trois sortes (English summary)* (J 2313) C R Acad Sci, Paris, Ser A-B 286*A847-A848
⋄ B05 ⋄ REV MR 81e:03004   Zbl 394#03011 • ID 52488

ROSE, A. [1978] *Sur une extension du concept d'une tautologie sans constantes (English summary)* (J 2313) C R Acad Sci, Paris, Ser A-B 286*A803-A805
⋄ B05 ⋄ REV MR 58#16118   Zbl 393#03005 • ID 52425

ROSE, A. [1982] *A generalisation of the concept of functional completeness and applications to modus ponens* (J 0068) Z Math Logik Grundlagen Math 28*317-322
⋄ B05 ⋄ REV MR 84j:03023   Zbl 499#03007 • ID 34617

ROSE, A. see Vol. II, IV, V, VI for further entries

ROSE, G.S.C. [1978] see BAXANDALL, P.R.

ROSE, H.E. & SHEPHERDSON, J.C. (EDS.) [1975] *Proceedings of the Logic Colloquium (Bristol, July, 1973)* (X 0809) North Holland: Amsterdam viii+513pp
⋄ B97 ⋄ REV MR 51#10015 • ID 70192

ROSE, H.E. see Vol. IV, VI for further entries

ROSE, L.E. [1968] *Aristotle's syllogistic* (X 1359) Thomas: Springfield vii+149pp
⋄ A10   B20 ⋄ REV JSL 36.670 • ID 43498

ROSENBAUM, I. [1950] *Introduction to mathematical logic and its applications* (X 1306) Univ Miami Pr: Austin iii+98pp
⋄ B98 ⋄ REV Zbl 41.347   JSL 16.156 • ID 22534

ROSENBAUM, I. [1952] *A logistic proof of a theorem related to Landau's theorem 4* (P 0593) Int Congr Math (II, 6);1950 Cambridge MA 1*730-731
⋄ B28 ⋄ ID 28064

ROSENBERG, I.G. [1964] *Detection et identification des fonctions Booleennes symetriques generalisees* (J 0060) Rev Roumaine Math Pures Appl 9*465-473
⋄ B05 ⋄ REV MR 32#9159   Zbl 166.257 • ID 26681

ROSENBERG, I.G. [1966] *Ein anschauliches Modell fuer die Minimisierung boolescher Funktionen* (J 0060) Rev Roumaine Math Pures Appl 11*189-193
⋄ B35 ⋄ REV MR 34#8888   Zbl 178.338 • ID 26682

ROSENBERG, I.G. see Vol. II, III, V for further entries

ROSENBERG, S. [1972] *A note on propositional calculus* (J 0047) Notre Dame J Formal Log 13*506-510
⋄ B05 ⋄ REV MR 47#6428   Zbl 242#02006 • ID 11522

ROSENBERG, S. see Vol. II for further entries

ROSENBERG, Y. [1975] *In defense of Copi* (J 0047) Notre Dame J Formal Log 16*607
⋄ B05 ⋄ REV MR 51#10014   Zbl 316#02014 • ID 18364

ROSENBLOOM, P.C. [1950] *The elements of mathematical logic* (X 0813) Dover: New York 6+iv+214pp
⋄ B98 ⋄ REV MR 12.789   JSL 18.277 • ID 11524

ROSENBLOOM, P.C. see Vol. IV, VI for further entries

ROSENTHAL, E. [1985] see MURRAY, N.V.

ROSENTHAL, J.W. [1982] see PLOTKIN, J.M.

ROSENTHAL, J.W. see Vol. III, IV, V for further entries

ROSSER, J.B. [1935] *A mathematical logic without variables II* (J 0025) Duke Math J 1*328-355
⋄ A05   B10 ⋄ REV Zbl 12.386   FdM 61.56 • REM Part I 1935 • ID 16819

ROSSER, J.B. [1939] *An informal exposition of proofs of Goedel's theorems and Church's theorem* (J 0036) J Symb Logic 4*53-60
⋄ A05   B20   D35   F30 ⋄ REV Zbl 22.292   JSL 4.165   FdM 65.27 • ID 11547

ROSSER, J.B. & WANG, HAO [1950] *Non-standard models for formal logics* (J 0036) J Symb Logic 15*113-129
⋄ B15   C62   E30   E70   F25   H15   H20 ⋄ REV MR 12.384   Zbl 37.295   JSL 16.145 • ID 11558

ROSSER, J.B. [1953] *Logic for mathematicians* (X 0822) McGraw-Hill: New York xiv+530pp
• LAST ED [1978] (X 0848) Chelsea: New York xv+574pp
⋄ B15   B98   E70   E98 ⋄ REV MR 14.935   MR 80b:03001   Zbl 404#03001   Zbl 68.7   JSL 18.326   JSL 45.631 • REM 2nd ed. 1978 • ID 11565

ROSSER, J.B. [1955] *Deux esquisses de logique* (X 0834) Gauthier-Villars: Paris 69pp
⋄ B10   B40   D20   E30 ⋄ REV MR 16.661   Zbl 64.7 • ID 23483

ROSSER, J.B. see Vol. II, III, IV, V, VI for further entries

ROSSKOPF, M.F. [1959] see EXNER, R.M.

ROTA, G.-C. [1978] see ELLERMAN, D.P.

ROTA, G.-C. see Vol. IV, VI for further entries

ROURE, M.-L. [1967] *Elements de logique contemporaine* (X 0840) Pr Univ France: Paris 127pp
⋄ B98 ⋄ ID 22428

ROUSSEAU, C. [1978] *Topos theory and complex analysis* (J 0326) J Pure Appl Algebra 10*299-313
⋄ F50   G30   H05 ⋄ REV MR 57#413   Zbl 378#02028 • ID 51810

ROUSSEAU, C. [1979] *Topos theory and complex analysis* (P 2901) Appl Sheaves;1977 Durham 623-659
⋄ F50   F55   G30   H05 ⋄ REV MR 81d:32014   Zbl 433#32003 • ID 82662

ROUSSEAU, C. see Vol. V for further entries

ROUSSEAU, G. [1966] *A decidable class of number theoretic equations* (J 0039) J London Math Soc 41*737-741
⋄ B20   B25   F30 ⋄ REV MR 34#1182   Zbl 166.5 • ID 11593

ROUSSEAU, G. [1968] *Note on the decidability of a certain class of number theoretic equations* (J 0039) J London Math Soc 43*385-386
⋄ B20   B25   F30 ⋄ REV MR 37#1244   Zbl 162.317 • ID 11598

ROUSSEAU, G. see Vol. II, V, VI for further entries

ROUTLEDGE, N.A. [1956] *Logic on electronic computers: a practical method for reducing expressions to conjunctive normal form* (J 0171) Proc Cambridge Phil Soc Math Phys 52*161-173
- ◇ B35 ◇ REV MR 18.422 Zbl 74.121 JSL 24.255
- • ID 11606

ROUTLEDGE, N.A. see Vol. IV, VI for further entries

ROUTLEY, R. [1965] *What numbers are* (J 0079) Logique & Anal, NS 8*196-208
- ◇ A05 B28 E10 ◇ ID 30839

ROUTLEY, R. [1968] see MONTGOMERY, H.

ROUTLEY, R. [1969] *A simple natural deduction system* (J 0079) Logique & Anal, NS 12*129-152
- ◇ B10 ◇ REV MR 44 #2568 Zbl 207.4 • ID 11611

ROUTLEY, R. [1975] *Universal semantics?* (J 0122) J Philos Logic 4*327-356
- ◇ A05 B15 ◇ REV MR 58 #27345 Zbl 333 #02007
- • ID 64938

ROUTLEY, R. see Vol. II, III, IV, V, VI for further entries

ROWICKI, A. [1965] *On minimal programming of formulas in Lukasiewicz's notation* (J 0014) Bull Acad Pol Sci, Ser Math Astron Phys 13*669-674
- ◇ B35 ◇ REV MR 34 #5585 Zbl 133.255 • ID 30846

ROWICKI, A. see Vol. IV for further entries

ROYCE, J. [1951] *An extension of the algebra of logic* (C 4505) Royce's Log Ess 293-309
- ◇ B05 ◇ REV JSL 17.144 • ID 41845

ROYSE, JAMES R. [1969] *Mathematical induction in ramified type theory* (J 0068) Z Math Logik Grundlagen Math 15*7-10
- ◇ B15 F35 F65 ◇ REV MR 38 #5613 Zbl 184.7
- • ID 11634

ROYSE, JOSIAH [1951] *The relation of the principles of logic to the foundations of geometry* (C 4505) Royce's Log Ess 379-441
- ◇ B05 ◇ REV JSL 17.145 • ID 41847

ROZENBLAT, B.V. [1979] *Positive theories of free inverse semigroups (Russian)* (J 0092) Sib Mat Zh 20*1282-1293,1408
- • TRANSL [1979] (J 0475) Sib Math J 20*910-918
- ◇ B20 C05 D35 D40 ◇ REV MR 80m:03079 Zbl 431 #20045 • ID 53949

ROZENBLAT, B.V. & VAZHENIN, YU.M. [1983] *On positive theories of free algebraic systems (Russian)* (J 0031) Izv Vyssh Ucheb Zaved, Mat (Kazan) 1983/3*70-73
- • TRANSL [1983] (J 3449) Sov Math 27/3*88-91
- ◇ B20 B25 C05 ◇ REV MR 84j:03028 Zbl 526 #03010
- • ID 34621

ROZENBLAT, B.V. see Vol. III, IV for further entries

ROZENFEL'D, T.K. & SILAEV, V.N. [1979] *Boolean equations and decomposition of boolean functions (Russian)* (J 0977) Izv Akad Nauk SSSR, Tekh Kibern 1979/1*119-126
- • TRANSL [1979] (J 0522) Engin Cybern 17/1*85-92
- ◇ B05 ◇ REV MR 81a:94064 Zbl 435 #94029 • ID 69878

ROZONOEHR, L.I. [1963] see AJZERMAN, M.A.

ROZONOEHR, L.I. see Vol. II, IV for further entries

RUBCHINSKIJ, A.A. & VINOGRADSKAYA, T.M. [1980] *Logical forms of choice functions (Russian)* (J 0023) Dokl Akad Nauk SSSR 254*1362-1366
- • TRANSL [1980] (J 0470) Sov Phys, Dokl 25*812-814
- ◇ B10 E25 ◇ REV MR 82f:04005 Zbl 478 #90005
- • ID 55660

RUBEL, L.A. [1977] see HENSON, C.W.

RUBEL, L.A. [1983] *Conformal inequivalence of annuli and the first-order theory of subgroups of PSL(2,R)* (J 0053) Proc Amer Math Soc 88*679-683
- ◇ B10 ◇ REV MR 85a:30018 Zbl 512 #03008 • ID 36504

RUBEL, L.A. [1984] see HENSON, C.W.

RUBEL, L.A. see Vol. III, IV, V for further entries

RUBIN, A.L. [1979] see COHEN, JACQUES

RUBIN, A.L. see Vol. V for further entries

RUBIN, H. [1969] *A new approach to the foundations of probability* (C 0705) Found of Math (Goedel) 46-50
- ◇ B30 ◇ REV MR 39 #7639 Zbl 218 #60003 • ID 14482

RUBIN, H. see Vol. II, V for further entries

RUBIN, M. [1980] see MALITZ, J.

RUBIN, M. see Vol. III, IV, V for further entries

RUBINSTEIN, A. [1984] *The single profile analogues to multiprofile theorems: mathematical logic's approach* (J 3217) Metroeconomica 25*719-730
- ◇ B80 ◇ REV MR 86g:90009 • ID 44413

RUBIO DE FRANCIA, J.L. [1981] *Contribuciones al analisis funcional no-standard* (J 0236) Rev Mat Hisp-Amer, Ser 4 41*31-44
- ◇ H05 ◇ REV Zbl 513 #46056 • ID 37811

RUBY, L. [1960] *Logic: An introduction* (X 1297) Lippincott: Philadelphia xviii+522pp
- ◇ B98 ◇ ID 22535

RUDEANU, S. [1959] *Independent systems of axioms in lattice theory* (J 0070) Bull Soc Sci Math Roumanie, NS 3*475-488
- ◇ B30 G10 ◇ REV MR 25 #2982 Zbl 114.15 JSL 38.521
- • ID 11659

RUDEANU, S. [1965] *Remarks on Motinonyi Goto's papers on boolean equations* (J 0060) Rev Roumaine Math Pures Appl 10*311-317
- ◇ B05 ◇ REV MR 36 #3695 Zbl 135.33 • ID 30856

RUDEANU, S. [1966] *On solving boolean equations in the theory of graphs* (J 0060) Rev Roumaine Math Pures Appl 11*653-664
- ◇ B05 ◇ REV MR 34 #99 Zbl 161.210 • ID 11660

RUDEANU, S. [1970] *On reproductive solutions of boolean equations* (J 0400) Publ Inst Math, NS (Belgrade) 10(24)*71-78
- ◇ B05 ◇ REV MR 45 #6603 Zbl 209.12 • ID 11662

RUDEANU, S. [1972] *On boolean matrix equations* (J 0060) Rev Roumaine Math Pures Appl 17*1075-1090
- ◇ B05 ◇ REV MR 47 #8570 Zbl 259 #06011 • ID 30353

RUDEANU, S. [1974] *An algebraic approach to boolean equations* (J 0187) IEEE Trans Comp C-23*206-207
- ◇ B05 ◇ REV MR 51 #2803 Zbl 282 #02004 • ID 29078

RUDEANU, S. [1976] *Square roots and functional decomposition of boolean functions* (J 0187) IEEE Trans Comp C-25*528-532
  ◊ B05 ◊ REV MR 54#2352 Zbl 327#94045 • ID 64965

RUDEANU, S. [1983] *Linear Boolean equations and generalized minterms* (J 0193) Discr Math 43*241-248
  ◊ B05 ◊ REV MR 84e:06019 Zbl 525#06005 • ID 37484

RUDEANU, S. [1984] see MELTER, R.A.

RUDEANU, S. see Vol. II, V for further entries

RUDIN, M.E. & SHELAH, S. [1979] *Unordered types of ultrafilters* (S 2848) Topology Proc 3*199-204
  ◊ E05 E75 H05 ◊ REV MR 80k:04002 Zbl 431#03033 • ID 53939

RUDIN, M.E. also published under the name ESTILL, M.E.

RUDIN, M.E. see Vol. III, V for further entries

RUKHAYA, KH.M. [1977] *A formal theory $\mathcal{T}$ (Russian) (Georgian and English summaries)* (S 2043) Issl Mat Log & Teor Algor (Tbilisi) 2*7-28
  ◊ B10 E30 ◊ REV MR 57#5674a • ID 78000

RUKHAYA, KH.M. [1977] *A generalization of the theory $\mathcal{T}$ (Russian) (Georgian and English summaries)* (S 2043) Issl Mat Log & Teor Algor (Tbilisi) 2*29-33
  ◊ B20 E30 ◊ REV MR 57#5674b • ID 77999

RUKHAYA, KH.M. [1978] *A formal equality theory $\mathcal{T}_{eg}$ (Russian) (Georgian and English Summary)* (S 2043) Issl Mat Log & Teor Algor (Tbilisi) 1978*9-23
  ◊ B10 ◊ REV MR 80j:03017 • ID 77998

RUKHAYA, KH.M. [1979] *A variant of theory T extended by abbreviating symbols* (P 2539) Frege Konferenz (1);1979 Jena 365-381
  ◊ B10 ◊ REV MR 82m:03012 • ID 78002

RUKHAYA, KH.M. [1982] *Description of a derived formal mathematical theory $\mathcal{T}^+$ (Russian) (English and Georgian summaries)* (J 0954) Tr Inst Prikl Mat, Tbilisi 11*75-99
  ◊ B10 ◊ REV MR 85g:03019 • ID 43494

RUKHAYA, KH.M. see Vol. V for further entries

RUS, T. [1964] *Ueber ein formales System I,II (Rumaenisch)* (J 0197) Stud Cercet Mat Acad Romana 15*459-470,595-615
  ◊ B10 D20 ◊ REV MR 32#3991 Zbl 199.3 • ID 16208

RUS, T. [1967] *Some observations concerning the application of the electronic computers in order to solve nonarithmetical Problems* (J 0517) Mathematica (Cluj) 9*343-360
  ◊ B35 ◊ REV Zbl 155.21 • ID 30865

RUS, T. see Vol. IV for further entries

RUSINOWITCH, M. [1985] *Path of subterms ordering and recursive decomposition ordering revisited* (P 4244) Rewriting Techn & Appl (1);1985 Dijon 225-240
  ◊ B35 ◊ ID 49786

RUSSELL, B. [1897] *An essay on the foundations of geometry* (X 0805) Cambridge Univ Pr: Cambridge, GB 201pp
  • LAST ED [1956] (X 0813) Dover: New York 201pp
  ◊ A05 B30 ◊ REV MR 17.1121 Zbl 75.153 FdM 28.413 • ID 25495

RUSSELL, B. [1903] *The principles of mathematics. Vol. 1* (X 0805) Cambridge Univ Pr: Cambridge, GB xxix+534pp
  • LAST ED [1937] (X 0959) Allen & Unwin: London xxxix+534pp [1938] (X 0942) Norton: New York xxxix+534pp
  ◊ A05 B28 B30 E30 ◊ REV JSL 3.156 FdM 34.62 • REM Repr. 1964 • ID 25581

RUSSELL, B. [1905] *The existential import of propositions* (J 0094) Mind 14*398-401
  ◊ A05 B10 ◊ ID 11674

RUSSELL, B. [1906] *The theory of implication* (J 0100) Amer J Math 28*159-202
  ◊ A05 B20 ◊ REV FdM 37.60 • ID 11677

RUSSELL, B. [1908] *Mathematical logic as based on the theory of types* (J 0100) Amer J Math 30*222-262
  • REPR [1967] (C 0675) From Frege to Goedel 150-182 [1967] (C 0721) Contemp Readings in Log Th 135-153
  ◊ A05 B15 ◊ REV JSL 39.355 FdM 39.85 • REM C0721 partial reprint • ID 11679

RUSSELL, B. [1910] *La theorie des types logiques* (J 0145) Rev Metaph Morale 18*263-301
  • REPR [1969] (J 1030) Formalisation 10*53-83
  ◊ A05 B15 ◊ REV FdM 41.83 • ID 11680

RUSSELL, B. [1910] see WHITEHEAD, A.N.

RUSSELL, B. [1912] see WHITEHEAD, A.N.

RUSSELL, B. [1913] see WHITEHEAD, A.N.

RUSSELL, B. [1932] see WHITEHEAD, A.N.

RUSSELL, B. & WHITEHEAD, A.N. [1967] *Incomplete symbols: descriptions* (C 0675) From Frege to Goedel 216-223
  ◊ B30 ◊ REV JSL 40.472 • ID 14529

RUSSELL, B. see Vol. V for further entries

RUSSELL, L.J. [1964] *A layout for logical operations* (J 0273) Australasian J Phil 42*313-321
  ◊ B05 ◊ ID 30870

RUSSELL, L.J. [1966] *Note on a layout for logical operations* (J 0273) Australasian J Phil 44*89-91
  ◊ B05 ◊ ID 30869

RUTKOWSKI, A. [1978] *Elements of mathematical logic (Polish)* (X 2881) Wydawn Szkol Ped: Warsaw 200pp
  ◊ B98 ◊ REV MR 58#4969 • ID 78008

RUTKOWSKI, A. see Vol. III, IV, V for further entries

RUTZ, P. [1973] *Zweiwertige und mehrwertige Logik - ein Beitrag zur Geschichte und Einheit der Logik* (X 2036) Ehrenwirth: Muenchen 108pp
  ◊ A10 B98 ◊ REV MR 55#10235 Zbl 454#03002 • ID 32236

RUZZO, W.L. [1978] see ADLEMAN, L.M.

RUZZO, W.L. see Vol. IV for further entries

RVACHEV, V.L. [1973] *The concept of solution structure of a boundary value problem (Russian)* (J 2867) Vest Politekh Inst Khar'kov 72*3-9,68
  ◊ B80 ◊ REV MR 49#1809 • ID 82678

RVACHEV, V.L. & SLESARENKO, A.P. [1976] *Algebra of logic and integral transforms in boundary value problems (Russian)* (X 2199) Naukova Dumka: Kiev 287pp
  ◊ B80 G05 ◊ REV MR 57#18510 • ID 82680

RVACHEV, V.L. see Vol. II for further entries

RYABTSEV, I.I. [1976] *The local definition of a generalized function by means of nonstandard analysis (Russian)* (J 0031) Izv Vyssh Ucheb Zaved, Mat (Kazan) 1976/10(173)*70-73
• TRANSL [1976] (J 3449) Sov Math 20/10*56-58
⋄ H05 ⋄ REV MR 58 # 7075   Zbl 366 # 46036   • ID 82615

RYAN, R. [1975] *Basic digital electronics - understanding number systems, boolean algebra, & logic circuits* (X 0911) Akademie Verlag: Berlin 210pp
⋄ B05   B70   G05 ⋄ REV JSL 41.549   • ID 14493

RYAN, W.J. [1982] see POGORZELSKI, H.A.

RYAN, W.J. [1985] see POGORZELSKI, H.A.

RYAN, W.J. see Vol. IV, VI for further entries

RYBAK, JANET & RYBAK, JOHN [1976] *Venn diagrams extended: Map logic* (J 0047) Notre Dame J Formal Log 17*469-475
⋄ B20   E20 ⋄ REV MR 57 # 103   Zbl 286 # 02023
• ID 64981

RYBAK, JANET & RYBAK, JOHN [1984] *Mechanizing logic I: Map logic extended formally to relational arguments. II: Automated map logic method for relational arguments on paper and by computer* (J 0047) Notre Dame J Formal Log 25*250-264,265-282
⋄ B35 ⋄ REV MR 86b:03020ab   Zbl 554 # 03008
Zbl 554 # 03009   • ID 42572

RYBAK, JANET see Vol. V for further entries

RYBAK, JOHN [1976] see RYBAK, JANET

RYBAK, JOHN [1984] see RYBAK, JANET

RYBAK, JOHN see Vol. V for further entries

RYLL-NARDZEWSKI, C. [1952] *The role of the axiom of induction in elementary arithmetic* (J 0027) Fund Math 39*239-263
⋄ B28 ⋄ REV MR 14.938   Zbl 50.7   JSL 19.287   • ID 11707

RYLL-NARDZEWSKI, C. see Vol. III, IV, V, VI for further entries

SAARINEN, E. [1977] see HILPINEN, R.

SAARINEN, E. see Vol. II for further entries

SABBAGH, G. [1971] *Logique mathematique I. Generalites* (C 1495) Encycl Universalis 10*53-56
⋄ B98 ⋄ REV JSL 38.341   • REM Part II 1971 by Reznikoff,I.
• ID 28607

SABBAGH, G. [1984] see DELON, F.

SABBAGH, G. [1985] see DELON, F.

SABBAGH, G. see Vol. III, IV, V for further entries

SABER, J.C. [1971] see HANNA, S.C.

SACHWANOWICZ, W. [1979] see DZIOBIAK, W.

SADE, A. [1970] *Morphismes sur le systeme des operateurs propositionnels* (J 0114) Math Nachr 44*231-251
⋄ B05 ⋄ REV MR 45 # 1724   Zbl 204.309   • ID 11760

SADE, A. see Vol. II, V for further entries

SADOVSKIJ, B.N. [1982] see KAMENSKIJ, M.I.

SADOWSKI, W. [1961] *A proof of completeness of the two-valued propositional calculus (Polish) (Russian and English summaries)* (J 0063) Studia Logica 11*49-55
⋄ B05 ⋄ REV MR 23 # A3075   Zbl 126.259   JSL 31.274
• ID 11763

SADOWSKI, W. see Vol. II for further entries

SAENDIG, A.-M. [1974] *Integraltransformationen von Predistributionen (Italian summary)* (J 0012) Boll Unione Mat Ital, IV Ser 10*60-75
⋄ H05 ⋄ REV MR 51 # 11030   Zbl 311 # 46026   • ID 17508

SAENDIG, A.-M. [1974] *Lokale Werte und Grenzwerte von Predistributionen* (J 0411) Studia Sci Math Hung 9*133-148
⋄ H05 ⋄ REV MR 53 # 3691   Zbl 302 # 46027   • ID 21657

SAENDIG, A.-M. [1974] *Ueber die Regularisierung und Ordnung von Predistributionen* (J 0411) Studia Sci Math Hung 9*149-154
⋄ H05 ⋄ REV MR 53 # 3692   Zbl 302 # 46028   • ID 21658

SAGIV, Y. [1981] see DELOBEL, C.

SAIN, I. [1982] see ANDREKA, H.

SAIN, I. [1983] *Total correctness in nonstandard dynamic logic* (J 0387) Bull Sect Logic, Pol Acad Sci 12*64-70
⋄ B75   H10 ⋄ REV MR 85b:03043   Zbl 538 # 68017
• ID 40674

SAIN, I. see Vol. II, III, V for further entries

SAITO, M. [1976] *Ultraproducts and nonstandard analysis (Japanese)* (X 3636) Tokyo Tosho: Tokyo vi+160pp
⋄ C20   H05 ⋄ ID 33414

SAITO, M. [1977] *On the non-standard representation of linear mappings from a function space* (J 0407) Comm Math Univ St Pauli (Tokyo) 26*165-185
⋄ H05 ⋄ REV MR 80a:03091   Zbl 381 # 46042   • ID 78078

SAITO, M. [1978] *Introduction of a few set theories within nonstandard analysis (Japanese)* (P 4109) B-Val Anal & Nonstand Anal;1978 Kyoto 117-125
⋄ E70   H05 ⋄ ID 46816

SAITO, S. [1963] *Truth value assignment in predicate calculus of first order* (J 0047) Notre Dame J Formal Log 4*216-223
⋄ B10 ⋄ REV MR 29 # 3355   Zbl 124.246   JSL 31.268
• ID 11766

SAITO, S. [1969] *A theory of categorical syllogism* (J 0047) Notre Dame J Formal Log 10*327-330
⋄ B20 ⋄ REV MR 39 # 6734   Zbl 179.313   • ID 11769

SAITO, S. see Vol. II for further entries

SAKAI, H. [1974] *On necessary but not-sufficient conditions* (J 0009) Arch Math Logik Grundlagenforsch 16*143-146
⋄ B10   F30   F35   G05 ⋄ REV MR 51 # 2881
Zbl 291 # 02024   • ID 11772

SAKAI, H. see Vol. IV, VI for further entries

SALAMUCHA, J. [1958] *The proof "ex motu" for the existence of God: Logical analysis of St. Thomas' arguments* (J 4077) New Scholast 32*334-372
⋄ A05   B80 ⋄ REV JSL 34.647   • ID 45213

SALIMOV, F.I. [1979] *On the question of modeling Boolean random values by functions of the algebra of logic (Russian)* (J 3937) Veroyat Met i Kibern (Kazan) 15*68-89
⋄ A05   B05   G05 ⋄ REV MR 83j:03104   Zbl 432 # 60006
• ID 35386

SALLANTIN, J. [1972] *Systemes de propositions et informations* (J 2313) C R Acad Sci, Paris, Ser A-B 274*A986-A988
⋄ B05 ⋄ REV MR 46 # 3386   Zbl 252 # 94008   • ID 11777

SALLE, P. [1978] *Une extension de la theorie des types en λ-calcul* (P 1872) Automata, Lang & Progr (5);1978 Udine 398-410
- ◊ B15 B40 ◊ REV MR 80k:03019 Zbl 384 # 03008
- ID 52050

SALLE, P. [1980] *Une generalisation de la theorie des types en λ-calcul I,II* (J 3441) RAIRO Inform Theor 14*143-167,301-314
- ◊ B15 B40 ◊ REV MR 82f:03012 Zbl 461 # 03002
- ID 56539

SALLE, P. see Vol. VI for further entries

SALMON, W.C. [1963] *Logic* (X 0819) Prentice Hall: Englewood Cliffs xiv+114pp
- ◊ B98 ◊ REV JSL 29.89 JSL 42.107 • ID 11778

SALMON, W.C. see Vol. II for further entries

SALOMAA, A. [1963] *On essential variables of functions, especially in the algebra of logic* (J 3994) Ann Acad Sci Fennicae Ser A I 339*11pp
- REPR [1964] (J 0498) Ann Univ Turku, Ser A I 69*14pp
- ◊ B05 E20 G15 ◊ REV MR 30 # 24 Zbl 134.7 JSL 32.539
- ID 11786

SALOMAA, A. see Vol. II, IV, V for further entries

SALUM, H. [1969] see MIKHEEVA, L.

SAMBORSKIJ, S.N. [1985] *Limit trajectories of singular pertubed differential equations (Russian) (English summary)* (J 0270) Dokl Akad Nauk Ukr SSR, Ser A 1985/9*22-25,86
- ◊ H05 ◊ ID 49587

SAMET, H. [1978] *A canonical form algorithm for proving equivalence of conditional forms* (J 0232) Inform Process Lett 7*103-106
- ◊ B35 ◊ REV MR 58 # 5009 Zbl 368 # 68021 • ID 78102

SAMITOV, R.K. [1983] *On the complexity of linear deductions in resolution theory (Russian)* (J 3937) Veroyat Met i Kibern (Kazan) 19*106-112
- ◊ B35 ◊ REV MR 85b:03019 Zbl 535 # 03005 • ID 38310

SAMOKHVALOV, K.F. [1978] *On the axiomatic representation of empirical theories (Russian)* (S 0507) Vychisl Sist (Akad Nauk SSSR Novosibirsk) 76*15-25,154
- ◊ A05 B30 ◊ REV MR 80k:03012 Zbl 435 # 03002
- ID 55763

SAMOKHVALOV, K.F. [1981] *Ways of measuring the simplicity of empirical theories (Russian)* (S 0507) Vychisl Sist (Akad Nauk SSSR Novosibirsk) 91*68-75
- ◊ B30 ◊ REV MR 84k:03045 Zbl 525 # 03002 • ID 34982

SAMOKHVALOV, K.F. see Vol. II, VI for further entries

SAMPEI, Y. [1950] *On the orthogonal expansion of the boolean polynomial and its applications I* (J 0438) J Hokkaido Univ Educ Ser 1 11*114-125
- ◊ B05 ◊ REV MR 14.4 JSL 21.401 • REM Part II 1953
- ID 90100

SAMPEI, Y. [1950] *Some remarks concerning identity* (J 0438) J Hokkaido Univ Educ Ser 1 11*109-112
- ◊ B10 ◊ REV MR 13.521 JSL 21.402 • ID 90101

SAMPEI, Y. [1953] *On the orthogonal expansion of the boolean polynomial and its application II* (J 0407) Comm Math Univ St Pauli (Tokyo) 1*51-57
- ◊ B05 ◊ REV MR 15.190 Zbl 53.198 JSL 21.401 • REM Part I 1950 • ID 11806

SAMPEI, Y. see Vol. IV, V for further entries

SANCHEZ, E. [1979] *Inverses of fuzzy relations. Application to possibility distributions and medical diagnosis* (J 2720) Fuzzy Sets Syst 2*75-86
- ◊ B80 E07 E72 ◊ REV MR 80e:03070 Zbl 399 # 03040
- ID 52836

SANCHEZ, E. [1980] *Fuzzy logics with application to medical diagnosis* (P 4254) Joint Autom Control Conf;1980 San Francisco 4pp
- ◊ B80 ◊ ID 46073

SANCHEZ, E. & SOULA, G. [1982] *Soft deduction rules in medical diagnostic processes* (C 3786) Approx Reason in Decis Anal 77-88
- ◊ B80 ◊ REV MR 84i:03059 • ID 34538

SANCHEZ, E. see Vol. II, V for further entries

SANCHEZ-MAZAS, M. [1971] *Arithmetical calculus of propositions (Spanish)* (J 0162) Teorema (Valencia) 1971/3*63-92
- ◊ B20 ◊ REV MR 47 # 8195 Zbl 227 # 68042 • ID 11819

SANCHEZ-MAZAS, M. [1972] *Calcul arithmetique des propositions* (J 0286) Int Logic Rev 6*222-245
- ◊ B05 ◊ REV MR 50 # 4237 Zbl 333 # 68064 • ID 11820

SANCHEZ-MAZAS, M. [1982] *Un modele arithmetique de la syllogistique et ses extensions* (P 3622) Int Congr Log, Meth & Phil of Sci (6,Proc);1979 Hannover
- ◊ B20 ◊ ID 35807

SANCHEZ-MAZAS, M. see Vol. II for further entries

SANCHIS, L.E. [1961] *Nueva demonstracion de la completicidad functional del calculo proposicional bivalente* (J 0047) Notre Dame J Formal Log 2*33-40
- ◊ B05 ◊ REV MR 26 # 8 Zbl 116.4 JSL 27.471 • ID 11822

SANCHIS, L.E. [1965] *A predicative extension of elementary logic I* (J 0188) Rev Union Mat Argentina 22*123-138
- ◊ B15 F05 F65 ◊ REV MR 32 # 7398 Zbl 242 # 02009
- ID 11824

SANCHIS, L.E. [1971] *A generalization of the Gentzen Hauptsatz* (J 0047) Notre Dame J Formal Log 12*499-504
- ◊ B10 F05 ◊ REV MR 46 # 30 Zbl 188.11 • ID 11827

SANCHIS, L.E. [1971] *Cut elimination, consistency and completeness in classical logic* (J 0079) Logique & Anal, NS 14*715-723
- ◊ B10 F05 ◊ REV MR 47 # 4771 Zbl 242 # 02032
- ID 11826

SANCHIS, L.E. [1974] *Logical completeness of directed resolution* (J 0079) Logique & Anal, NS 17*335-341
- ◊ B35 ◊ REV MR 55 # 10251 Zbl 327 # 68080 • ID 65046

SANCHIS, L.E. see Vol. III, IV, VI for further entries

SANDEWALL, E. [1976] *Conversion of predicate-calculus axioms to corresponding deterministic programs* (J 0187) IEEE Trans Comp C-25*342-346
- ◊ B35 B75 ◊ REV Zbl 324 # 68011 • ID 65051

SANDFORD, D.M. [1980] *Using sophisticated models in resolution theorem proving* (S 3302) Lect Notes Comput Sci 90*xi+239pp
- ◊ B35 ◊ REV MR 82m:68135 Zbl 453 # 03014 • ID 54144

SANMARTIN ESPLUGUES, J. [1973] *Syllogistics, many-valued logic and model theory (Spanish)* (J 0162) Teorema (Valencia) 3*355-365
- ◊ A10 B20 B50 C90 ◊ REV MR 50 # 9522 • ID 78125

SANMARTIN ESPLUGUES, J. see Vol. III, V for further entries

SANTOS, J. [1957] see ARANGO, H.

SAPOZHENKO, A.A. [1967] *Estimate of the complexity of a function (Russian)* (J 0071) Met Diskr Analiz (Novosibirsk) 9∗67-72
⋄ B05 ⋄ REV Zbl 174.38 • ID 19643

SAPOZHENKO, A.A. [1972] *Ueber die Kompliziertheit der disjunktiven Normalformen, die sich mittels des Gradientenalgorithmus ergeben (Russisch)* (J 0071) Met Diskr Analiz (Novosibirsk) 21∗62-71
⋄ B05 ⋄ REV MR 49 # 12188 Zbl 273 # 94041 • ID 65062

SAPOZHENKO, A.A. [1977] see GAVRILOV, G.P.

SAPOZHENKO, A.A. [1979] see KARAKHANYAN, L.M.

SAPOZHENKO, A.A. [1980] *Estimation of the length and the number of dead-end disjunctive normal forms for almost all partial boolean functions (Russian)* (J 0087) Mat Zametki (Akad Nauk SSSR) 28∗279-300,320
• TRANSL [1980] (J 1044) Math Notes, Acad Sci USSR 28∗603-615
⋄ B05 ⋄ REV MR 82b:94044 Zbl 441 # 94014 • ID 69890

SAPOZHENKO, A.A. see Vol. II for further entries

SARALSKI, B. [1975] see KOPIEKI, R.

SARI, T. [1981] see LUTZ, R.

SARI, T. [1981] *Sur le comportement asymptotique des solutions dans un probleme aux limites semi-lineaire (English summary)* (J 3364) C R Acad Sci, Paris, Ser 1 292∗867-870
⋄ H05 ⋄ REV MR 82g:34083 Zbl 464 # 34038 • ID 82710

SARI, T. [1982] see LUTZ, R.

SARI, T. [1983] see CALLOT, J.-L.

SATO, M. [1983] *Theory of symbolic expressions I* (J 1426) Theor Comput Sci 22∗19-55
⋄ B03 ⋄ REV Zbl 535 # 68013 • REM Part II 1985 • ID 38341

SATO, M. see Vol. II, III, VI for further entries

SATTLER, J. [1980] see COMMENTZ-WALTER, B.

SAURIOL, P. [1976] *La structure tetrahexaedrique du systeme complet des propositions categoriques* (J 0488) Dialogue (Ottawa) 15∗479-501
⋄ A05 B20 ⋄ REV MR 54 # 4915 • ID 24091

SAYRE, K.M. [1964] *Syllogistic inference within the propositional calculus* (J 0047) Notre Dame J Formal Log 5∗238-240
⋄ A05 A10 B05 ⋄ REV Zbl 171.252 • ID 11873

SAYWARD, C.W. [1976] see HUGLY, P.

SAYWARD, C.W. [1979] see HUGLY, P.

SAYWARD, C.W. [1982] see HUGLY, P.

SAYWARD, C.W. see Vol. II, III for further entries

SCANLON, T.M. [1973] *The consistency of number theory via Herbrand's theorem* (J 0036) J Symb Logic 38∗29-58
⋄ B10 F05 F30 ⋄ REV MR 58 # 16191 Zbl 255 # 02027 • ID 11874

SCANLON, T.M. see Vol. VI for further entries

SCARPELLINI, B. [1963] *Eine Anwendung der unendlichwertigen Logik auf topologische Raeume* (J 0027) Fund Math 52∗129-150 • ERR/ADD ibid 53∗345
⋄ B80 C65 ⋄ REV MR 34 # 8378a Zbl 149.193 • ID 11875

SCARPELLINI, B. [1984] *Complete second order spectra* (J 0068) Z Math Logik Grundlagen Math 30∗509-524
⋄ B15 C13 ⋄ REV MR 86f:03059 Zbl 564 # 03007
• ID 39663

SCARPELLINI, B. [1984] *Second order spectra* (P 2342) Symp Rek Kombin;1983 Muenster 171∗380-389
⋄ B15 C13 ⋄ REV MR 85k:68004 Zbl 548 # 03004
• ID 41786

SCARPELLINI, B. [1985] *Lower bound results on lengths of second-order formulas* (J 0073) Ann Pure Appl Logic 29∗29-58
⋄ B15 C13 D10 F20 F35 ⋄ ID 47473

SCARPELLINI, B. see Vol. II, III, IV, V, VI for further entries

SCEDROV, A. [1985] *Extending Goedel's modal interpretation to type theory and set theory* (C 3659) Intens Math 81-119
⋄ B15 B45 E70 F35 F50 ⋄ REV Zbl 569 # 03025
• ID 44771

SCEDROV, A. [1985] see HARRINGTON, L.A.

SCEDROV, A. see Vol. II, III, IV, V, VI for further entries

SCHAEFER, T.J. [1978] *The complexity of satisfiability problems* (P 1740) ACM Symp Th of Comput (10);1978 San Diego 216-226
⋄ B05 D15 ⋄ REV MR 80d:68058 • ID 82718

SCHAGRIN, M.L. [1968] *The language of logic. A programmed text* (X 0981) Random House: New York vii + 247pp
⋄ B98 ⋄ REV JSL 39.612 • ID 12857

SCHARLE, T.W. [1962] *A diagram of the functors of the two-valued propositional calculus (Added note ibid 287-288)* (J 0047) Notre Dame J Formal Log 3∗243-255
⋄ B05 ⋄ REV MR 27 # 3501 MR 27 # 3502 Zbl 114.244 JSL 27.230 JSL 28.175 • ID 12858

SCHARLE, T.W. [1965] *Axiomatization of porpositional calulus with Sheffer functors* (J 0047) Notre Dame J Formal Log 6∗209-217
⋄ B05 ⋄ REV MR 33 # 2525 Zbl 144.244 • ID 12860

SCHARLE, T.W. [1966] *Single axiom schemata for D and S* (J 0047) Notre Dame J Formal Log 7∗344-348
⋄ B05 ⋄ REV MR 38 # 1999 Zbl 169.304 • ID 12861

SCHARLE, T.W. [1967] see LAMBERT, K.

SCHARLE, T.W. see Vol. II for further entries

SCHAUER, R. [1960] see POHM, A.

SCHEEPEN VAN, F. [1968] see BRAFFORT, P.

SCHEER, R.K. [1964] see CARNEY, J.D.

SCHEIBE, E. [1983] *Two types of successor relations between theories* (J 0989) Z Allg Wissth 14∗68-80
⋄ B30 ⋄ REV MR 84k:03046 • ID 34983

SCHEIBE, E. see Vol. II for further entries

SCHENK, G. [1974] *Die Logik* (J 0424) Wiss Z Univ Halle-Wittenberg, Math-Nat Reihe 23∗27-54
⋄ A10 B98 ⋄ REV MR 57 # 2859 • ID 78168

SCHEPERS, H. [1975] *Leibniz Disputation "de conditionibus". Ansaetze zu einer juristischen Aussagenlogik* (P 1998) Int Leibniz Kongr (2);1972 Hannover 4∗1-17
⋄ A10 B80 ⋄ REV MR 58 # 27186 Zbl 317 # 02001
• ID 29629

SCHERER, D. [1971] *The form of reductio ad absurdum* (J 0094) Mind 80∗247-252
⋄ A05 B05 ⋄ REV MR 56 # 11741 • ID 78170

SCHERER, D. [1978] see FACIONE, P.A.

SCHINZEL, B. [1984] see BOERGER, E.

SCHINZEL, B. see Vol. IV for further entries

SCHIPPER, E.W. & SCHUH, E. [1959] *A first course in modern logic* (X 0818) Holt Rinehart & Winston: New York xvii + 398pp
• REPR [1960] (X 0866) Routledge & Kegan Paul: Henley on Thames xviii + 398pp
⋄ B98 ⋄ REV JSL 24.220 • ID 22536

SCHIRN, M. (ED.) [1976] *Studien zu Frege. I,II,III* (X 1267) Frommann: Stuttgart 317pp,303pp,201pp
⋄ A05 A10 B97 ⋄ REV MR 57 # 9455 Zbl 352 # 02003 JSL 47.226 JSL 47.227 • ID 50002

SCHLINK, B. [1971] *On a principle of contradiction in normative logic and jurisprudence* (J 0472) Theory Decis 2∗35-48
⋄ B80 ⋄ REV MR 45 # 4958 Zbl 259 # 02023 • ID 30346

SCHMERL, J.H. [1981] see LERMAN, M.

SCHMERL, J.H. & SIMPSON, S.G. [1982] *On the role of Ramsey quantifiers in first order arithmetic* (J 0036) J Symb Logic 47∗423-435
⋄ B28 C10 C62 C80 ⋄ REV MR 83j:03062 Zbl 492 # 03015 JSL 50.1078 • ID 35360

SCHMERL, J.H. see Vol. III, IV, V for further entries

SCHMID, J. [1978] *∗-compactifications* (J 0050) Port Math 37∗191-201
⋄ H05 ⋄ REV MR 82j:54040 Zbl 488 # 54012 • ID 82724

SCHMID, J. see Vol. III for further entries

SCHMIDT, DAVID A. [1984] *A programming notation for tactical reasoning* (P 2633) Autom Deduct (7);1984 Napa 445-459
⋄ B35 ⋄ REV Zbl 546 # 68080 • ID 44457

SCHMIDT, GARFIELD C. [1981] *Nonreflexive, well founded sets and natural numbers* (J 2734) Int J Math Educ Sci Technol 12∗299-318
⋄ B28 E20 ⋄ REV MR 84b:03025 Zbl 455 # 04002 • ID 54289

SCHMIDT, GUNTHER & STROEHLEIN, T. [1976] *A boolean matrix iteration in timetable construction* (J 3200) Linear Algeb & Appl 15∗27-51
⋄ B05 ⋄ REV MR 56 # 10922 Zbl 358 # 94049 • ID 50546

SCHMIDT, GUNTHER see Vol. II, IV, V for further entries

SCHMIDT, H.A. [1938] *Ueber deduktive Theorien mit mehreren Sorten von Grunddingen* (J 0043) Math Ann 115∗485-506
⋄ B10 ⋄ REV Zbl 18.338 JSL 4.98 • ID 12883

SCHMIDT, H.A. [1950] *Mathematische Grundlagenforschung* (X 1079) Teubner: Leipzig 48pp
⋄ B98 ⋄ REV MR 13.4 JSL 17.198 • ID 22429

SCHMIDT, H.A. [1951] *Die Zulaessigkeit der Behandlung mehrsortiger Theorien mittels der ueblichen einsortigen Praedikatenlogik* (J 0043) Math Ann 123∗187-200
⋄ B10 ⋄ REV MR 13.614 Zbl 42.6 JSL 17.76 • ID 38680

SCHMIDT, H.A. [1960] *Mathematische Gesetze der Logik. I. Vorlesungen ueber Aussagenlogik* (X 0811) Springer: Heidelberg & New York xxiv + 555pp
⋄ B05 B98 F50 F98 ⋄ REV MR 29 # 1135 Zbl 253 # 08002 Zbl 88.246 • ID 12892

SCHMIDT, H.A. & SCHUETTE, K. & THIELE, H. (EDS.) [1968] *Contribtions to mathematical logic. Proc. of the logic colloq., Hannover 1966* (X 0809) North Holland: Amsterdam 298pp
⋄ B97 ⋄ REV MR 38 # 3120 Zbl 177.293 • ID 38681

SCHMIDT, H.A. see Vol. II, III, VI for further entries

SCHMIDT, J. [1953] *Einige grundlegende Begriffe und Saetze aus der Theorie der Huellenoperatoren* (P 0626) Ber Math-Tagung Berlin;1953 Berlin 21-48
⋄ B22 C05 E20 ⋄ REV MR 16.1083 Zbl 52.26 • ID 12901

SCHMIDT, J. [1958] see FELSCHER, W.

SCHMIDT, J. [1966] *Der baryzentrische Kalkuel als axiomatische Grundlage der affinen Geometrie* (J 0127) J Reine Angew Math 224∗44-57
⋄ B30 ⋄ REV MR 34 # 6599 • ID 32243

SCHMIDT, J. [1981] *Algebraic studies of first-order enlargements* (J 0047) Notre Dame J Formal Log 22∗315-343
⋄ B10 G15 G30 H20 ⋄ REV MR 83d:03071 Zbl 459 # 03028 • ID 54468

SCHMIDT, J. see Vol. II, III, V, VI for further entries

SCHMIDT, K. [1981] *Ein Rechenverfahren fuer die elementare Logik (English summary)* (J 0989) Z Allg Wissth 12∗110-115
⋄ B25 B35 ⋄ REV MR 83e:03024 • ID 35213

SCHMIDT, K. [1984] see DRIES VAN DEN, L.

SCHMIDT, K. see Vol. III, V for further entries

SCHMIEDEN, C. [1958] see LAUGWITZ, D.

SCHMITT, B.V. [1981] see MAZZANTI, S.

SCHMUCKER, K.J. [1984] *Fuzzy sets, natural language computations, and risk analysis, Foreword by Lotfi A. Zadeh* (X 3581) Comput Sci Press: Rockville xv + 192pp
⋄ B52 B98 ⋄ REV Zbl 552 # 90025 • ID 43453

SCHNEIDER, H.H. [1958] *Semantics of the predicate calculus with identity and the validity in the empty individual-domain* (J 0050) Port Math 17∗85-96
⋄ B10 ⋄ REV Zbl 84.247 JSL 30.385 • ID 12925

SCHNEIDER, H.H. [1961] *A syntactical characterization of the predicate calculus with identity and the validity in all individual-domains* (J 0050) Port Math 20∗105-117
⋄ B10 ⋄ REV MR 32 # 3992 Zbl 97.245 JSL 30.385 • ID 12926

SCHNEIDER, H.H. [1972] see BULLOCK, A.M.

SCHNEIDER, H.H. [1973] see BULLOCK, A.M.

SCHNEIDER, H.H. [1976] *A deduction system for the full first-order predicate logic* (J 0047) Notre Dame J Formal Log 17∗439-445
⋄ B10 ⋄ REV MR 56 # 111 Zbl 258 # 02017 • ID 18404

SCHNEIDER, H.H. [1980] *Substitutions for predicate variables and functional variables* (J 0047) Notre Dame J Formal Log 21∗33-44
⋄ B10 ⋄ REV MR 81b:03009 Zbl 305 # 02021 • ID 53176

SCHNEIDER, H.H. see Vol. V for further entries

SCHNEIDER, J. [1984] see NOTTALE, L.

SCHNORR, C.-P. [1971] *Zufaelligkeit und Wahrscheinlichkeit. Eine algorithmische Begruendung der Wahrscheinlichkeitstheorie* (X 0811) Springer: Heidelberg & New York iv+212pp
- ◊ B30 B75 D80 ◊ REV MR 54#2328 Zbl 232#60001
- • ID 23998

SCHNORR, C.-P. [1974] *On maximal merging of information in boolean computations* (P 1869) Automata, Lang & Progr (2);1974 Saarbruecken 294-300
- ◊ B35 ◊ REV MR 55#12570 Zbl 284#68032 • ID 82736

SCHNORR, C.-P. [1976] *The combinational complexity of equivalence* (J 1426) Theor Comput Sci 1*289-295
- • TRANSL [1979] (J 3079) Kiber Sb Perevodov, NS 16*74-81
- ◊ B05 D15 ◊ REV Zbl 333#68032 • ID 69903

SCHNORR, C.-P. [1977] *A survey of the theory of random sequences* (P 1704) Int Congr Log, Meth & Phil of Sci (5);1975 London ON 3*193-211
- ◊ B30 ◊ REV MR 58#24431 • ID 82733

SCHNORR, C.-P. [1977] *The network complexity and the breadth of Boolean functions* (P 1075) Logic Colloq;1976 Oxford 491-504
- ◊ B05 B70 D15 ◊ REV MR 58#8494 Zbl 418#68048
- • ID 16638

SCHNORR, C.-P. see Vol. III, IV, VI for further entries

SCHOCK, R. [1964] *Contribution to syntax, semantics, and the philosophy of science* (J 0047) Notre Dame J Formal Log 5*241-289
- ◊ A05 B10 ◊ REV MR 34#4105 JSL 37.423 • ID 12943

SCHOCK, R. [1964] *On the logic of variable binders* (J 0009) Arch Math Logik Grundlagenforsch 6*71-90
- ◊ B10 C80 ◊ REV MR 31#34 Zbl 121.11 • ID 12942

SCHOCK, R. [1967] *Logik* (X 1163) Almqvist & Wiksell: Stockholm 105pp
- ◊ B98 ◊ REV Zbl 169.296 JSL 34.642 • ID 21980

SCHOCK, R. see Vol. II, III, V for further entries

SCHOENFELD, W. [1981] *Gleichungen in der Algebra der binaeren Relationen* (X 2797) Minerva Publ: Muenchen 74pp
- ◊ B35 E07 ◊ REV MR 84e:68037 Zbl 521#03049 • REM Habil.-Schr.,Univ. Stuttgart 1979 • ID 37083

SCHOENFELD, W. [1982] *Upper bounds for proof-search in a sequent calculus for relational equations* (J 0068) Z Math Logik Grundlagen Math 28*239-246
- ◊ B35 F20 ◊ REV MR 84d:03072 Zbl 506#03001
- • ID 34105

SCHOENFELD, W. [1983] *Proof search for unprovable formulas* (P 3087) Germ Worksh on Artif Intell (7);1983 Dassel 207-215
- ◊ B35 F07 F20 ◊ REV MR 85g:68007 Zbl 548#68088
- • ID 43253

SCHOENFELD, W. see Vol. III, IV for further entries

SCHOENFINKEL, M. [1924] *Ueber die Bausteine der mathematischen Logik* (J 0043) Math Ann 92*305-316
- • TRANSL [1967] (C 0675) From Frege to Goedel 355-366
- ◊ A05 B10 B40 ◊ REV FdM 50.23 • ID 12955

SCHOENFINKEL, M. [1928] see BERNAYS, P.

SCHOENFLIES, A. [1911] *Ueber die Stellung der Definition in der Axiomatik* (J 0157) Jbuchber Dtsch Math-Ver 20*222-255
- ◊ A05 B30 ◊ REV FdM 42.75 • ID 12959

SCHOENFLIES, A. see Vol. V for further entries

SCHOLZ, H. [1930] *Die Axiomatik der Alten* (J 0168) Blaetter Deutsch Philos 4*259-278
- ◊ A10 B30 ◊ REV FdM 56.42 • ID 38685

SCHOLZ, H. [1933] see BACHMANN, F.

SCHOLZ, H. & SCHWEITZER, H. [1935] *Die sogenannten Definitionen durch Abstraktion. Eine Theorie der Definitionen durch Bildung von Gleichheitsverwandtschaften* (X 1282) Hirzel: Stuttgart v+108pp
- • REPR [1935] (X 4718) Meiner: Leipzig v+108pp
- ◊ B30 ◊ REV Zbl 13.241 FdM 61.971 • ID 22599

SCHOLZ, H. [1936] see HERMES, H.

SCHOLZ, H. [1939] *Was ist eine formalisierte Theorie?* (J 2074) Sem-ber, Muenster 13*13-53
- ◊ B30 ◊ REV FdM 65.20 • ID 43751

SCHOLZ, H. [1941] *Eine neue Gestalt der Grundlagenforschung* (J 0983) Forsch Fortschritte 17*382-384
- ◊ A05 A10 B30 ◊ REV Zbl 26.242 JSL 12.58 • ID 12963

SCHOLZ, H. [1941] *Metaphysik als strenge Wissenschaft* (X 2111) Staufenverlag: Koeln 188pp
- ◊ A05 B15 ◊ REV MR 3.291 Zbl 27.147 JSL 6.156 FdM 67.32 • ID 41298

SCHOLZ, H. [1948] *Vorlesungen ueber Grundzuege der mathematischen Logik I* (X 0910) Aschendorffsche Verlagsbuchh: Muenster xiv+276pp
- ◊ B98 ◊ REV MR 12.661 Zbl 31.103 JSL 15.200 JSL 19.115 • REM Vol.II 1949, 2nd revised edition 1950
- • ID 22431

SCHOLZ, H. [1949] *Vorlesungen ueber Grundzuege der mathematischen Logik II* (X 0910) Aschendorffsche Verlagsbuchh: Muenster x+311pp
- ◊ B98 ◊ REV Zbl 32.242 JSL 15.200 • REM Vol.I 1948, 2nd revised edition 1951 • ID 20807

SCHOLZ, H. [1952] *Ein ungeloestes Problem in der symbolischen Logik (problem 1)* (J 0036) J Symb Logic 17*160
- ◊ B10 C13 ◊ ID 33304

SCHOLZ, H. [1952] see HERMES, H.

SCHOLZ, H. [1953] *Der klassische und der moderne Begriff einer mathematischen Theorie* (J 0160) Math-Phys Sem-ber, NS 3*30-43
- ◊ A05 A10 B30 ◊ REV MR 14.714 Zbl 50.3 JSL 21.310
- • ID 12969

SCHOLZ, H. & HASENJAEGER, G. [1961] *Grundzuege der mathematischen Logik* (X 0811) Springer: Heidelberg & New York xvi+504pp
- ◊ B98 ◊ REV MR 23#A2306 Zbl 137.4 JSL 28.245
- • ID 22434

SCHOLZ, H. [1961] *Logik, Grammatik, Metaphysik* (C 4422) Mathesis Univers 399-436
- ◊ A05 B10 ◊ REV JSL 28.283 • ID 45695

SCHORR, H. [1963] see MCCLUSKEY JR., E.J.

SCHREIBER, A. [1975] *Theorie und Rechtfertigung. Untersuchungen zum Rechtfertigungsproblem axiomatischer Theorien in der Wissenschaftstheorie* (X 0900) Vieweg: Wiesbaden 204pp
  ◊ A05 B30 ◊ REV Zbl 356 # 02005 • ID 50277

SCHREIBER, J. [1975] see BIBEL, W.

SCHREIBER, P. [1965] *Untersuchungen ueber die Modelle der Typentheorie* (J 0068) Z Math Logik Grundlagen Math 11∗343-372
  ◊ B15 C85 ◊ REV MR 34 # 5654 Zbl 166.259 • ID 12976

SCHREIBER, P. [1968] *Lexikographische Ordnung als Grundbegriff der Semiotik* (J 0114) Math Nachr 38∗53-60
  ◊ B03 ◊ REV MR 38 # 4292 Zbl 164.318 • ID 12979

SCHREIBER, P. [1969] *Ueber die Quantorenbasen der klassischen Praedikatenlogik* (J 0068) Z Math Logik Grundlagen Math 15∗77-80
  ◊ B10 ◊ REV MR 39 # 1288 Zbl 175.261 • ID 12980

SCHREIBER, P. [1974] *Durch Schlussregeln definierte logische Kalkuele* (J 0309) Wiss Z Univ Greifswald, Math-Nat Reihe 23∗35-38
  ◊ B22 ◊ REV MR 51 # 5260 Zbl 361 # 02043 • ID 17473

SCHREIBER, P. [1977] *Grundlagen der Mathematik* (X 0806) Dt Verlag Wiss: Berlin 239pp
  ◊ B98 ◊ REV MR 57 # 15932 Zbl 362 # 02001 • REM 2nd ed. 1984 • ID 50723

SCHREIBER, P. [1979] *Die allgemeinste Form der formalisierten Sprachen* (P 2539) Frege Konferenz (1);1979 Jena 414-432
  ◊ B10 ◊ REV MR 83j:03048 • ID 35350

SCHREIBER, P. see Vol. III, IV, VI for further entries

SCHREIBER, R. [1962] *Logik des Rechts* (X 0811) Springer: Heidelberg & New York viii+100pp
  ◊ B80 ◊ REV JSL 31.286 • ID 25283

SCHREINER, H. [1982] *Information systems and artificial intelligence in law. Logical procedures for the application of technical intelligence in juridical decisions* (P 4041) Log, Inform, Law;1981 Firenze 1∗165-179
  ◊ B80 ◊ REV MR 85h:03010 • ID 43387

SCHRODER, F.W.K.E. [1890] *Vorlesungen ueber die Algebra der Logik (exakte Logik) Vol.1* (X 1079) Teubner: Leipzig xii+717pp
  • LAST ED [1966] (X 0848) Chelsea: New York 819pp
  ◊ A05 A10 B20 B98 G98 ◊ REV MR 33 # 1219 MR 33 # 1220 Zbl 188.307 JSL 40.609 FdM 22.73
  • ID 19702

SCHRODER, F.W.K.E. [1891] *Vorlesungen ueber die Algebra der Logik (exakte Logik) Vol.2* (X 1079) Teubner: Leipzig xiii+400pp
  ◊ A05 A10 B20 B98 G05 G15 G98 ◊ REV MR 33 # 1220 JSL 31.283 FdM 23.51 • ID 19703

SCHRODER, F.W.K.E. [1895] *Vorlesungen ueber die Algebra der Logik (exakte Logik) Vol.3 Algebra und Logik der Relative* (X 1079) Teubner: Leipzig viii+649pp
  • TRANSL [1966] (X 0848) Chelsea: New York vi+819pp
  ◊ A05 A10 B20 B98 G98 ◊ REV MR 33 # 1221 FdM 26.74 • ID 19704

SCHRODER, F.W.K.E. [1898] *Ueber zwei Definitionen der Endlichkeit und G.Cantor'sche Saetze* (J 0985) Nova Acta Acad Caes Leopoldina (Halle) 71∗303-362
  ◊ B30 E20 E25 ◊ REV FdM 29.49 • ID 12983

SCHRODER, F.W.K.E. [1909] *Abriss der Algebra der Logik 1: Elementarlehre* (X 0823) Teubner: Stuttgart v+50pp
  ◊ B20 G05 ◊ REV FdM 40.92 • REM Edited by Mueller, E.
  • ID 22558

SCHRODER, F.W.K.E. [1910] *Abriss der Algebra der Logik 2: Aussagentheorie, Funktionen, Gleichungen und Ungleichungen* (X 0823) Teubner: Stuttgart vi+51+159pp
  ◊ B20 G05 ◊ REV FdM 41.87 • REM Edited by Mueller, E.
  • ID 22559

SCHRODER, F.W.K.E. [1966] *Der Operationskreis des Logikkalkuels* (X 0823) Teubner: Stuttgart v+37pp
  ◊ A05 B20 ◊ REV FdM 9.38 • REM 1st ed. 1877 • ID 19705

SCHROEDER-HEISTER, P. [1984] *A natural extension of natural deduction* (J 0036) J Symb Logic 49∗1284-1300
  ◊ B22 F07 F50 ◊ REV Zbl 574 # 03045 • ID 42500

SCHROEDER-HEISTER, P. see Vol. VI for further entries

SCHROETER, K. [1941] *Ein allgemeiner Kalkuelbegriff* (J 0956) Forsch Logik Grundl exakt Wiss 6∗43pp
  ◊ B03 ◊ REV MR 6.29 Zbl 26.243 JSL 8.77 JSL 9.20
  • ID 16993

SCHROETER, K. [1943] *Axiomatisierung der Fregeschen Aussagenkalkuele* (S 4550) Forsch Log & Grundl Exakt Wiss, NS 26pp
  ◊ B05 ◊ REV JSL 9.69 • ID 19699

SCHROETER, K. [1952] *Deduktiv abgeschlossene Mengen ohne Basis* (J 0114) Math Nachr 7∗293-304
  ◊ B22 ◊ REV MR 14.234 Zbl 48.3 JSL 22.331 • ID 12987

SCHROETER, K. [1955] *Methoden zur Axiomatisierung beliebiger Aussagen- und Praedikatenkalkuele* (J 0068) Z Math Logik Grundlagen Math 1∗241-251
  ◊ B10 ◊ REV MR 17.1038 Zbl 66.256 JSL 35.140
  • ID 12990

SCHROETER, K. [1955] *Theorie des logischen Schliessens I* (J 0068) Z Math Logik Grundlagen Math 1∗37-86
  ◊ B10 F07 F50 ◊ REV MR 17.814 Zbl 64.7 • REM Part II 1958 • ID 12989

SCHROETER, K. [1956] *Die Unabhaengigkeit der elementaren praedikatenlogischen Schlussregeln* (J 0068) Z Math Logik Grundlagen Math 2∗218-227
  ◊ B10 ◊ REV MR 18.633 Zbl 74.10 JSL 37.194 • ID 12991

SCHROETER, K. [1956] *Ueber den Zusammenhang der in den Implikationsaxiomen vollstaendigen Axiomensysteme des zweiwertigen mit denen des intuitionistischen Aussagenkalkuel* (J 0068) Z Math Logik Grundlagen Math 2∗173-176
  ◊ B20 F50 ◊ REV MR 18.866 Zbl 75.6 JSL 35.583
  • ID 12993

SCHROETER, K. [1957] *Die Vollstaendigkeit der die Implikation enthaltenden zweiwertigen Aussagenkalkuele und Praedikatenkalkuele der ersten Stufe* (J 0068) Z Math Logik Grundlagen Math 3∗81-107
  ◊ B20 ◊ REV MR 20 # 2275 Zbl 80.6 • ID 12995

SCHROETER, K. [1958] see ASSER, G.

SCHROETER, K. [1958] *Theorie des logischen Schliessens II* (J 0068) Z Math Logik Grundlagen Math 4∗10-65
  ◊ B10 C07 F05 F07 ◊ REV MR 20 # 6979 Zbl 80.8 JSL 32.418 • REM Part I 1955 • ID 24841

SCHROETER, K. see Vol. III, VI for further entries

SCHUBERT, H. [1896] *Mathematical essays and recreations* (X 1324) Open Court: LaSalle 149pp
⋄ B98 ⋄ ID 25498

SCHUBERT, H. [1904] *Principes fondamentaux de l'arithmetique* (X 0834) Gauthier-Villars: Paris 63pp
⋄ B28 ⋄ ID 25584

SCHUERFELD, U. [1983] *New lower bounds on the formula size of Boolean functions* (J 1431) Acta Inf 19*183-194
⋄ B05 ⋄ REV MR 84g:06023 Zbl 492#94022 • ID 38538

SCHUETTE, K. [1933] *Ueber einen Teilbereich des Aussagenkalkuels* (J 0459) C R Soc Sci Lett Varsovie Cl 3 26*30-32
⋄ B05 ⋄ REV FdM 59.863 • ID 12996

SCHUETTE, K. [1933] *Untersuchungen zum Entscheidungsproblem der mathematischen Logik* (J 0043) Math Ann 109*572-603
⋄ B20 B25 D35 ⋄ REV Zbl 9.2 FdM 60.21 • ID 12997

SCHUETTE, K. [1934] *Ueber die Erfuellbarkeit einer Klasse von logischen Formeln* (J 0043) Math Ann 110*161-194
⋄ B20 B25 ⋄ REV Zbl 9.337 FdM 60.21 • ID 12998

SCHUETTE, K. [1950] *Schlussweisen-Kalkuele der Praedikatenlogik* (J 0043) Math Ann 122*47-65
⋄ B10 B25 F05 F07 F50 ⋄ REV MR 12.233 Zbl 36.148 JSL 16.155 • ID 12999

SCHUETTE, K. [1953] *Zur Widerspruchsfreiheit einer typenfreien Logik* (J 0043) Math Ann 125*394-400
⋄ B15 B40 F05 F65 ⋄ REV MR 15.386 Zbl 50.244 JSL 20.67 • ID 13003

SCHUETTE, K. [1956] *Ein System des verknuepfenden Schliessens* (J 1114) Arch Phil 5*375-387
• REPR [1956] (J 0009) Arch Math Logik Grundlagenforsch 2*55-67
⋄ B10 F07 F30 ⋄ REV MR 19.3 Zbl 71.8 JSL 22.297 • ID 19693

SCHUETTE, K. [1958] *Aussagenlogische Grundeigenschaften formaler Systeme* (J 0076) Dialectica 12*422-442
⋄ B05 F05 F07 ⋄ REV MR 21#3324 Zbl 90.9 JSL 27.230 • ID 13008

SCHUETTE, K. [1960] *Beweistheorie* (X 0811) Springer: Heidelberg & New York x+355pp
• TRANSL [1977] (X 0811) Springer: Heidelberg & New York xii+299pp
⋄ B98 F05 F15 F30 F35 F98 ⋄ REV MR 22#9438 MR 58#21497 Zbl 102.247 Zbl 367#02012 JSL 25.243 JSL 47.218 • REM 2nd rev. ed. Comment 1960 by Frey,G. • ID 13010

SCHUETTE, K. [1960] *Syntactical and semantical properties of simple type theory* (J 0036) J Symb Logic 25*305-326
⋄ B15 F05 F35 ⋄ REV MR 26#2354 Zbl 109.5 JSL 32.418 • ID 13011

SCHUETTE, K. [1961] *Ein formales System der klassischen Aussagenlogik mit einer einzigen Grundverknuepfung* (J 0009) Arch Math Logik Grundlagenforsch 5*113-118
⋄ B05 ⋄ REV MR 23#A46 Zbl 112.245 • ID 13009

SCHUETTE, K. [1962] *Lecture notes in mathematical logic. Vol.I* (X 1328) Penn State Math: University Park 105pp
⋄ B98 F15 ⋄ REM Vol.II 1963 • ID 22382

SCHUETTE, K. [1963] *Lecture notes in mathematical logic. Vol.II* (X 1328) Penn State Math: University Park 109pp
⋄ B98 F15 ⋄ REM Vol.I 1962 • ID 22383

SCHUETTE, K. [1965] *Eine Grenze fuer die Beweisbarkeit der transfiniten Induktion in der verzweigten Typenlogik* (J 0009) Arch Math Logik Grundlagenforsch 7*45-60
⋄ B15 F15 F35 F65 ⋄ REV MR 33#7240 Zbl 158.7 JSL 32.284 • ID 13015

SCHUETTE, K. [1968] see SCHMIDT, H.A.

SCHUETTE, K. [1968] *On simple type theory with extensionality* (P 0627) Int Congr Log, Meth & Phil of Sci (3,Proc);1967 Amsterdam 179-184
⋄ B15 ⋄ REV MR 40#4097 Zbl 228#02021 • ID 13018

SCHUETTE, K. [1978] *Ein Ansatz zum Entscheidungsverfahren fuer eine Formelklasse der Praedikatenlogik mit Identitaet* (J 0380) Acta Cybern (Szeged) 4*141-148
⋄ B10 B25 ⋄ REV MR 80b:03013 Zbl 406#03015 • ID 56098

SCHUETTE, K. see Vol. II, III, IV, V, VI for further entries

SCHUETZENBERGER, M.-P. [1970] see LACOMBE, D.

SCHUETZENBERGER, M.-P. see Vol. IV for further entries

SCHUH, E. [1959] see SCHIPPER, E.W.

SCHUH, E. see Vol. II for further entries

SCHULTE-MOENTING, J. [1984] see FELSCHER, W.

SCHULTE-MOENTING, J. see Vol. III, VI for further entries

SCHULTZ, MARTIN H. [1963] see MENGER, K.

SCHULTZ, P. [1975] *Semantic paradoxes* (J 2761) J Recreational Math 7*140-142
⋄ A05 B05 ⋄ REV MR 55#2477 Zbl 374#02009 • ID 51533

SCHULZE, B. [1969] *Einfuehrung einer Halbordnung im Aussagenkalkuel* (J 0068) Z Math Logik Grundlagen Math 15*25-35
⋄ B05 ⋄ REV MR 39#1285 Zbl 174.36 • ID 13037

SCHUMM, G.F. [1975] *A Henkin-style completeness proof for the pure implicational calculus* (J 0047) Notre Dame J Formal Log 16*402-404
⋄ B20 ⋄ REV MR 51#10073 Zbl 258#02014 • ID 13045

SCHUMM, G.F. see Vol. II, III, V for further entries

SCHUPP, P.E. [1981] see MULLER, D.E.

SCHUPP, P.E. [1985] see MULLER, D.E.

SCHUPP, P.E. see Vol. III, IV, V for further entries

SCHUSTER, P. [1976] *Probleme, die zum Erfuellungsproblem der Aussagenlogik polynomial aequivalent sind* (P 3196) Kompl von Entscheid Probl;1973/74 Zuerich 36-48
⋄ B05 D15 ⋄ REV MR 57#18232 Zbl 386#68048 • ID 69906

SCHWABHAEUSER, W. [1953] *Zur Definition des geordneten Paares im einstelligen Stufenkalkuel* (P 0626) Ber Math-Tagung Berlin;1953 Berlin 13-14
⋄ B15 E20 ⋄ ID 27700

SCHWABHAEUSER, W. [1956] *Ueber die Vollstaendigkeit der elementaren euklidischen Geometrie* (J 0068) Z Math Logik Grundlagen Math 2∗137-165
⋄ B30 C35 C65 ⋄ REV MR 18.863 Zbl 74.12 JSL 36.156
• ID 13051

SCHWABHAEUSER, W. [1959] *Entscheidbarkeit und Vollstaendigkeit der elementaren hyperbolischen Geometrie* (J 0068) Z Math Logik Grundlagen Math 5∗132-205
⋄ B25 B30 C35 C65 ⋄ REV MR 27 # 3549 Zbl 95.348 JSL 36.156 • ID 13052

SCHWABHAEUSER, W. [1965] *Metamathematical methods in foundations of geometry* (P 0623) Int Congr Log, Meth & Phil of Sci (2,Proc);1964 Jerusalem 152-165
⋄ B30 ⋄ REV MR 35 # 5317 Zbl 199.12 JSL 35.474
• ID 13057

SCHWABHAEUSER, W. [1967] *Zur Axiomatisierbarkeit von Theorien in der schwachen Logik der zweiten Stufe* (J 0009) Arch Math Logik Grundlagenforsch 10∗60-96
⋄ B15 C85 ⋄ REV MR 37 # 3915 Zbl 165.15 • ID 13058

SCHWABHAEUSER, W. [1979] *Non-finitizability of a weak second-order theory* (J 0027) Fund Math 103∗83-102
⋄ B15 C62 C65 C85 F35 ⋄ REV MR 81h:03079 Zbl 422 # 03010 • ID 53471

SCHWABHAEUSER, W. & SZMIELEW, W. & TARSKI, A. [1983] *Metamathematische Methoden in der Geometrie* (X 0811) Springer: Heidelberg & New York viii+482pp
⋄ B30 C65 C98 D35 ⋄ REV MR 85e:03004 Zbl 564 # 51001 • ID 40225

SCHWABHAEUSER, W. [1983] *Metamathematische Methoden in der Geometrie. Teil II: Metamathematische Betrachtungen* (C 4602) Metamath Meth in Geom 173-452
⋄ B30 C65 C98 ⋄ REV MR 85e:03004 Zbl 564 # 51001
• ID 41875

SCHWABHAEUSER, W. see Vol. III, V for further entries

SCHWANITZ, G. [1973] see BAGINSKI, M.

SCHWARTZ, DIETRICH [1975] *Ultraprodukte in der Theorie der logischen Auswahlfunktionen* (J 0068) Z Math Logik Grundlagen Math 21∗385-394
⋄ B10 C20 ⋄ REV MR 52 # 13375 Zbl 333 # 02043
• ID 13068

SCHWARTZ, DIETRICH [1977] *Sequenzenschliessen in der algebraischen Attributenlogik* (J 0068) Z Math Logik Grundlagen Math 23∗487-495
⋄ B10 G25 ⋄ REV MR 58 # 5013 Zbl 393 # 03021
• ID 52441

SCHWARTZ, DIETRICH see Vol. II, III, IV, V for further entries

SCHWARTZ, J.T. (ED.) [1967] *Mathematical aspects of computer science* (X 0803) Amer Math Soc: Providence vi+226pp
⋄ B35 B65 B75 B97 D10 D80 D97 ⋄ REV
MR 38 # 2975 Zbl 165.2 • ID 23572

SCHWARTZ, J.T. [1979] see DAVIS, MARTIN D.

SCHWARTZ, J.T. [1980] *Fast probabilistic algorithms for verification of polynomial identities* (J 0037) ACM J 27∗701-717
⋄ B35 ⋄ REV MR 82m:68078 Zbl 452 # 68050 • ID 82752

SCHWARTZ, J.T. see Vol. III, IV, V for further entries

SCHWARTZ, R.L. [1982] see MELLIAR-SMITH, P.M.

SCHWARTZ, R.L. see Vol. II for further entries

SCHWARTZ, T. [1982] *No minimally reasonable collective-choice process can be strategy-proof* (J 3914) Math Soc Sci 3∗57-72
⋄ B35 ⋄ REV MR 83m:90014 Zbl 489 # 90013 • ID 46688

SCHWARTZ, T. see Vol. II, IV for further entries

SCHWEITZER, H. [1935] see SCHOLZ, H.

SCHWEIZER, B. & SKLAR, A. [1977] *The axiomatic characterization of functions* (J 0068) Z Math Logik Grundlagen Math 23∗373-382
⋄ B30 E20 E30 ⋄ REV MR 58 # 10441 Zbl 442 # 03025
• ID 56382

SCHWEIZER, B. see Vol. IV for further entries

SCHWICHTENBERG, H. [1972] see ROEDDING, D.

SCHWICHTENBERG, H. [1983] *On Martin-Loef's theory of types* (P 3829) Atti Incontri Log Mat (1);1982 Siena 299-325
⋄ B15 F35 F50 ⋄ REV Zbl 537 # 03039 • ID 40090

SCHWICHTENBERG, H. see Vol. III, IV, V, VI for further entries

SCHWIND, C.B. [1978] *A formalism for the description of question answering systems* (C 2972) Natural Lang Commun with Computers 1-48
⋄ B65 B80 ⋄ REV MR 81i:68133 Zbl 378 # 68045
• ID 82755

SCHWIND, C.B. [1980] *Natural language analysis by theorem proving methods: disambiguating pronouns in natural language texts* (P 3580) Repr des Conn & Raison dans Sci Homme;1979 St Maximin 123-138
⋄ B35 B45 B65 ⋄ REV MR 82g:03045 • ID 78311

SCHWIND, C.B. see Vol. II for further entries

SCIENZA, G. [1979] *Elementary mathematics from an information-theoretic viewpoint: the real numbers* (J 2825) Rend Sem Fac Sci Univ Cagliari 49∗475-504
⋄ B28 F60 ⋄ REV MR 82f:03055 • ID 78312

SCIENZA, G. see Vol. V for further entries

SCOGNAMIGLIO, G. [1963] *Un metodo di calcolo dei prodotti delle matrici booleane elementari* (J 2328) Ann Pont Ist Sup Sci Lett Napoli 13∗413-429
⋄ B05 D03 ⋄ REV Zbl 265 # 06013 • ID 29833

SCOTT, D.S. [1956] *A symmetric primitive notion for euclidean geometry* (J 0028) Indag Math 18∗456-461
⋄ B30 ⋄ REV MR 18.328 Zbl 72.155 JSL 33.288
• ID 11900

SCOTT, D.S. & SUPPES, P. [1958] *Foundational aspects of theories of measurement* (J 0036) J Symb Logic 23∗113-128
⋄ B30 C13 C52 ⋄ REV MR 22 # 6716 Zbl 84.246
JSL 33.287 • ID 11906

SCOTT, D.S. [1959] *Dimension in elementary euclidean geometry* (P 0651) Axiomatic Method;1957 Berkeley 53-67
⋄ B30 ⋄ REV MR 21 # 4920 Zbl 92.137 JSL 34.514
• ID 11908

SCOTT, D.S. [1962] see FRAYNE, T.E.

SCOTT, D.S. [1967] *Existence and description in formal logic* (C 1032) Philosopher of the Century (Russell) 181-200
⋄ A05 B10 E30 ⋄ REV JSL 38.166 • ID 22187

SCOTT, D.S. [1969] *Boolean models and nonstandard analysis* (P 0649) Appl Model Th to Algeb, Anal & Probab;1967 Pasadena 87-92
⋄ E35 E40 E50 H05 ⋄ REV MR 38 # 4300 Zbl 187.271
• ID 24818

SCOTT, D.S. [1974] see ADDISON, J.W.

SCOTT, D.S. [1974] *Rules and derived rules* (C 1936) Log Th & Semant Anal (Kanger) 147-161
⋄ B22 ⋄ REV Zbl 296 # 02012 • ID 65167

SCOTT, D.S. [1975] see MOSS, J.M.B.

SCOTT, D.S. [1979] *A note on distributive normal forms* (C 1706) Essays Honour J. Hintikka 75-90
⋄ B10 C07 ⋄ REV Zbl 452 # 03005 • ID 54069

SCOTT, D.S. [1982] *Domains for denotational semantics* (P 3836) Automata, Lang & Progr (9);1982 Aarhus 577-613
⋄ B22 B40 ⋄ REV MR 83m:68029 Zbl 495 # 68025 • ID 38520

SCOTT, D.S. [1982] *Lectures on a mathematical theory of computation* (P 3906) Th Found of Progr Methodol;1981 Marktoberdorf 145-292
⋄ B40 B98 D75 ⋄ REV MR 85g:68043 Zbl 516 # 68064 • ID 38593

SCOTT, D.S. see Vol. II, III, IV, V, VI for further entries

SCOTT, P.J. [1984] see LAMBEK, J.

SCOTT, P.J. see Vol. V, VI for further entries

SCOZZAFAVA, R. [1980] see BARONE, E.

SCOZZAFAVA, R. see Vol. II for further entries

SEARBY, D. [1975] see BAXTER, ROBERT

SEARLES, H.L. [1948] *Logic and scientific methods. An introductory course* (X 0880) Ronald Press: New York xii+326pp
⋄ A05 B80 ⋄ REV JSL 13.144 • ID 41604

SECKENDORFF VON, V. [1937] *Beweis des Induktionsschlusses der natuerlichen Zahlen aus der Dedekindschen Definition endlicher Mengen* (J 0366) Sitzber Berlin Math Ges 36∗16-24
⋄ B28 E10 F30 ⋄ REV Zbl 17.156 FdM 63.30 • ID 41030

SECKENDORFF VON, V. [1940] *Beweis und Definition durch verallgemeinerte transfinite Induktion. Ihr Aufbau und ihre Stellung in einem logischen Kalkuel* (J 0366) Sitzber Berlin Math Ges 38-39∗1-8
⋄ B15 E20 ⋄ REV Zbl 25.5 FdM 66.1192 • ID 43819

SEDIVY, J. [1969] *Solution of simple logical problems by colouring graphs* (J 0156) Kybernetika (Prague) 5∗501-512
⋄ B05 ⋄ REV Zbl 187.263 JSL 38.150 • ID 11928

SEDIVY, J. see Vol. V for further entries

SEELAND, H. [1978] *Algorithmische Theorien und konstruktive Geometrie* (X 2727) Hochschul Verl: Stuttgart viii+72pp
⋄ B30 B75 ⋄ REV MR 82b:03034 Zbl 404 # 03054 • ID 54841

SEELY, R.A.G. [1982] *Locally Cartesian closed categories and type theory 1* (J 2128) C R Math Acad Sci, Soc Roy Canada 4∗271-275
⋄ B15 F35 F50 G30 ⋄ REV MR 83m:03076 Zbl 509 # 03034 • ID 34890

SEELY, R.A.G. see Vol. V, VI for further entries

SEESE, D.G. [1975] *Zur Entscheidbarkeit der monadischen Theorie 2. Stufe baumartiger Graphen (Russian, English and French summaries)* (J 0115) Wiss Z Humboldt-Univ Berlin, Math-Nat Reihe 24∗768-772
⋄ B15 B25 ⋄ REV MR 58 # 5157 Zbl 362 # 02037 • ID 50759

SEESE, D.G. [1977] *Second order logic, generalized quantifiers and decidability* (J 0014) Bull Acad Pol Sci, Ser Math Astron Phys 25∗725-732
⋄ B15 B25 C55 C65 C80 C85 D35 ⋄ REV MR 57 # 87 Zbl 383 # 03010 • ID 27132

SEESE, D.G. [1979] *Some graph-theoretical operations and decidability* (J 0114) Math Nachr 87∗15-21
⋄ B15 B25 C65 C85 ⋄ REV MR 81j:03028 Zbl 414 # 03007 • ID 53053

SEESE, D.G. [1980] see BAUDISCH, A.

SEESE, D.G. see Vol. III, IV, V for further entries

SEGERBERG, K. [1973] *Hallden's theorem on Post completeness* (C 1389) Modality, Morality, Probl of Sense & Nonsense (Hallden) 206-209
⋄ B22 ⋄ REV MR 56 # 11759 • ID 28639

SEGERBERG, K. [1974] see MAKINSON, D.

SEGERBERG, K. [1982] *Classical propositional operators. An exercise in the foundations of logic* (X 0815) Clarendon Pr: Oxford x+152pp
⋄ B05 B22 ⋄ REV MR 83i:03001 Zbl 491 # 03003 JSL 49.993 • ID 35481

SEGERBERG, K. [1982] see CRESSWELL, M.J.

SEGERBERG, K. [1983] *Arbitrary truth-value functions and natural deduction* (J 0068) Z Math Logik Grundlagen Math 29∗557-564
⋄ B05 B20 ⋄ REV MR 85a:03014 Zbl 549 # 03007 • ID 34765

SEGERBERG, K. see Vol. II, III, VI for further entries

SEIFFERT, H. [1973] *Einfuehrung in die Logik. Logische Propaedeutik und formale Logik* (X 0995) Beck'sche Verlagsbuchh: Muenchen 231pp
⋄ B98 ⋄ REV MR 52 # 13302 Zbl 346 # 02001 • ID 22435

SEILER, E. [1970] *An application of higher order predicate logic (Romanian)* (S 1613) Probl Logic (Bucharest) 2∗313-353
⋄ B15 ⋄ REV Zbl 229 # 02017 • ID 47821

SEILER, E. [1973] *On the consistency of the theory of limits in topological unions. I (Romanian)* (S 1613) Probl Logic (Bucharest) 5∗149-239
⋄ B15 E75 ⋄ REV MR 58 # 27478 • REM Part II 1975 • ID 78359

SEILER, E. [1975] *On the consistency of the theory of limits in topological unions. II (Romanian)* (S 1613) Probl Logic (Bucharest) 6∗197-247
⋄ B05 E75 ⋄ REV MR 57 # 16046 • REM Part I 1973 • ID 78361

SEILER, E. see Vol. V for further entries

SEKI, S. [1953] *On the weakened type-logic (note on metamathematics I)* (J 0407) Comm Math Univ St Pauli (Tokyo) 2∗29-40
⋄ B15 ⋄ REV MR 15.90 Zbl 53.343 • ID 24918

SEKI, S. [1960] see MAEHARA, S.

SEKI, S. [1960] *Programs for mathematical proofs by machines-present status of inferential analysis (Japanese)* (J 0091) Sugaku 12∗114-118
  ⋄ B35 ⋄ REV MR 26 # 4916 Zbl 114.248 • ID 11957

SEKI, S. see Vol. IV, V, VI for further entries

SEKIMOTO, T. [1972] *On the uniqueness of the shortest single axiom for the implicational calculus of propositions* (J 0081) Proc Japan Acad 48∗290-292
  ⋄ B20 ⋄ REV MR 51 # 10019 Zbl 268 # 02006 • ID 17566

SEKIMOTO, T. [1973] *Propositional and implicational propositional calculus (Japanese)* (J 0381) J Tsuda College (Tokyo) 5∗85-103
  ⋄ B05 B20 ⋄ REV MR 49 # 2247 • ID 11958

SELDIN, J.P. [1972] see HINDLEY, J.R.

SELDIN, J.P. [1975] *Arithmetic as a study of formal systems* (J 0047) Notre Dame J Formal Log 16∗449-464
  ⋄ A05 B28 ⋄ REV MR 54 # 7184 Zbl 236 # 02001
  • ID 18377

SELDIN, J.P. see Vol. VI for further entries

SEMENOV, A.L. [1977] *Presburgerness of predicates regular in two number systems (Russian)* (J 0092) Sib Mat Zh 18∗403-418,479
  • TRANSL [1977] (J 0475) Sib Math J 18∗289-300
  ⋄ B10 D20 F30 ⋄ REV MR 56 # 8349 Zbl 369 # 02023
  • ID 51323

SEMENOV, A.L. see Vol. III, IV, VI for further entries

SEN, M. [1983] *Minimization of Boolean functions of any number of variables using decimal labels* (J 0191) Inform Sci 30∗37-45
  ⋄ B35 ⋄ REV MR 85a:93083 Zbl 569 # 94023 • ID 38878

SENYUKOVA, A.G. [1973] *Methode der logischen Gleichungen in Anwendung auf das Problem der Vergleichung nichtorientierter Graphen, die durch Nachbarschaftmatrizen vorgegeben sind (Russian)* (J 2339) Vychisl Met & Progr, Univ Moskva 21∗211-214
  ⋄ B80 ⋄ REV Zbl 275 # 05127 • ID 29717

SEREBRYANNIKOV, O.F. [1972] *Heuristic principles and logical calculi* (X 2737) Israel Progr Sci Transl: Jerusalem III∗182pp
  ⋄ B98 ⋄ REV MR 58 # 16186 Zbl 302 # 02008 • REM Translated from Russian • ID 65215

SEREBRYANNIKOV, O.F. see Vol. II, VI for further entries

SERIKOV, YU.A. [1972] *An algebraic method of solving logic equations (Russian)* (J 0977) Izv Akad Nauk SSSR, Tekh Kibern 1972/2∗114-124
  • TRANSL [1972] (J 0522) Engin Cybern 10/2∗273-282
  ⋄ B05 ⋄ REV MR 55 # 12338 Zbl 265 # 06014 • ID 29834

SERRUS, C. [1945] *Traite de logique* (X 1312) Aubier-Montaigne: Paris 383pp
  ⋄ A05 B98 ⋄ REV JSL 12.57 • ID 22436

SESMAT, A. [1951] *Logique. vol.II: les raisonnements, la logistique* (X 0859) Hermann: Paris 361-776pp
  ⋄ A05 B98 ⋄ REM Vol.I 1950 • ID 22438

SETHI, I.K. [1980] *Fast sequential evaluation of monotonic boolean functions* (J 0191) Inform Sci 20∗101-113
  ⋄ B35 ⋄ REV MR 80m:94102 Zbl 451 # 94013 • ID 69912

SETLUR, R.V. [1970] *A method to determine the expressive power of a set of connectives* (J 0187) IEEE Trans Comp C-19∗1223-1225
  ⋄ B05 ⋄ REV MR 48 # 5820 Zbl 217.293 • ID 11990

SETLUR, R.V. [1970] *On the equivalence of strong and weak validity of rule schemes in the two-valued propositional calculus* (J 0047) Notre Dame J Formal Log 11∗249-253 • ERR/ADD ibid 15∗648
  ⋄ B05 ⋄ REV MR 44 # 2564 Zbl 167.7 • ID 19629

SETLUR, R.V. [1970] *The product of implication and counter-implication systems* (J 0047) Notre Dame J Formal Log 11∗241-248
  ⋄ B20 ⋄ REV MR 44 # 3830 Zbl 167.13 • ID 11989

SETLUR, R.V. see Vol. II for further entries

SETO, Y. [1968] *Proofs of some axioms by stroke function* (J 0081) Proc Japan Acad 44∗1024-1027
  ⋄ B05 ⋄ REV MR 38 # 4275 Zbl 182.314 • ID 48089

SETTE, A.-M. [1973] *On the propositional calculus P^1* (J 0352) Math Jap 18∗173-180
  ⋄ B05 ⋄ REV MR 51 # 10020 Zbl 238 # 02026 Zbl 289 # 02013 • ID 17565

SETTE, A.-M. & SETTE, J. [1978] *Functorialization of first-order language with finitely many predicates* (P 1800) Brazil Conf Math Log (1);1977 Campinas 293-303
  ⋄ B10 G30 ⋄ REV MR 80h:18003 Zbl 393 # 03006
  • ID 52426

SETTE, A.-M. see Vol. II, III, VI for further entries

SETTE, J. [1978] see SETTE, A.-M.

SETTLE, L.G. [1969] see BENNETT, J.H.

SHABANOV-KUSHNARENKO, YU.P. [1980] *Analytic methods for implicit determination of finite alphabetical operators (Russian)* (J 2668) Avtom Sist Upravl & Prib Avtom, Khar'kov 54∗96-102,III-IV
  ⋄ B03 ⋄ REV MR 81j:03039 • ID 78417

SHABANOV-KUSHNARENKO, YU.P. [1980] *Formulas of the algebra of finite predicates (Russian)* (J 2668) Avtom Sist Upravl & Prib Avtom, Khar'kov 53∗4-10,I
  ⋄ B20 ⋄ REV MR 82e:04002 • ID 78416

SHABANOV-KUSHNARENKO, YU.P. see Vol. III for further entries

SHAFAAT, A. [1967] *Principle of localization for a more general type of languages* (J 3240) Proc London Math Soc, Ser 3 17∗629-643
  ⋄ B10 C55 C75 C90 ⋄ REV MR 36 # 1308 Zbl 207.296
  • ID 12002

SHAFAAT, A. [1973] *On products of relational structures* (J 0001) Acta Math Acad Sci Hung 24∗13-19
  ⋄ B15 C30 ⋄ REV MR 52 # 80 Zbl 263 # 08003 • ID 18382

SHAFAAT, A. see Vol. III for further entries

SHALLA, L. [1967] see CARSON, D.F.

SHANIN, N.A. [1965] see DAVYDOV, G.V.

SHANIN, N.A. see Vol. II, III, IV, V, VI for further entries

SHAPIRO, S. [1977] *Incomplete translations of complete logics* (J 0047) Notre Dame J Formal Log 18∗248-250
  ⋄ B10 ⋄ REV MR 56 # 11745 Zbl 353 # 02028 • ID 21971

SHAPIRO, S. [1985] *Epistemic and intuitionistic arithmetic* (C 3659) Intens Math 11-46
⋄ B28 F30 F50 ⋄ REV Zbl 559#03036 • ID 44774

SHAPIRO, S. see Vol. II, III, IV, VI for further entries

SHAPIRO, S.I.M. [1984] *Solution of logical and game-theoretic problems. Logical-psychological studies (Russian)* (X 3775) Radio i Svyaz: Moskva 153pp
⋄ B80 ⋄ REV MR 86e:03007 Zbl 561#03001 • ID 45535

SHARONOV, V.I. & ZAMOV, N.K. [1968] *A certain algorithm for finding a full amplification of sequences in propositional calculus (Russian)* (J 0468) Uch Zap Ped Inst (Alma Ata) 128*67-70
⋄ B05 ⋄ REV MR 44#1573 Zbl 228#02008 • ID 14351

SHARONOV, V.I. & ZAMOV, N.K. [1969] *A certain class of strategies used in theorem proving by the resolution method (Russian)* (S 0228) Zap Nauch Sem Leningrad Otd Mat Inst Steklov 3(16)*54-64 • ERR/ADD ibid 20*292-294
• TRANSL [1969] (S 1490) Stud Constructive Math 3*
⋄ B35 ⋄ REV MR 42#2715 Zbl 206.291 • ID 14353

SHARONOV, V.I. & ZAMOV, N.K. [1969] *A class of strategies for the determination of provability by the resolution method (Russian)* (S 0228) Zap Nauch Sem Leningrad Otd Mat Inst Steklov 16*54-64
• TRANSL [1969] (J 0521) Semin Math, Inst Steklov 16*26-31
⋄ B35 ⋄ REV Zbl 231#68036 • ID 46548

SHARONOV, V.I. & ZAMOV, N.K. [1969] *The strengthening of formulae that are provable in propositional calculus (Russian)* (J 0468) Uch Zap Ped Inst (Alma Ata) 129/4*21-31
⋄ B05 ⋄ REV MR 50#49 Zbl 283#68056 • ID 14355

SHARONOV, V.I. & ZAMOV, N.K. [1970] *Amplifications of formulae of predicate calculus (Russian)* (J 0468) Uch Zap Ped Inst (Alma Ata) 130/3*54-59
⋄ B10 B20 D35 ⋄ REV MR 44#2573 Zbl 223#02010 • ID 14357

SHARONOV, V.I. & ZAMOV, N.K. [1973] *Anwendung von Isogrammen bei der Ermittlung des Schlusses (Russisch)* (S 2738) Issl Prikl Mat (Univ Kazan') 1*94-102
⋄ B35 ⋄ REV MR 57#11204 Zbl 352#68109 • ID 50071

SHAROV, C.I. [1984] *On some generalizations of Beth's semantics (Russian)* (J 4404) Vopr Mat Logiki & Pril 58-69
⋄ B10 ⋄ ID 46596

SHAROV, N.N. [1979] see AMBROSIMOV, A.S.

SHAW, J.C. [1957] see NEWELL, A.

SHAW, J.C. [1957] *Programming the logic theory machine* (P 1178) West Joint Comput Conf;1957 Los Angeles 230-240
⋄ B35 ⋄ REV JSL 27.103 • ID 12026

SHAW, J.C. [1960] see NEWELL, A.

SHCHEGOL'KOV, E.A. [1975] see BOKSHTEJN, M.F.

SHCHEGOL'KOV, E.A. see Vol. IV, V for further entries

SHCHEGOL'KOVA, G.M. [1970] *Certain theorems of the theory of quantifiers (Russian)* (C 0668) Neklass Log 91-106
⋄ B10 ⋄ REV MR 47#8258 • ID 11895

SHEFFER, H.M. [1913] *A set of five independent postulates for boolean algebras, with application to logical constants* (J 0064) Trans Amer Math Soc 14*481-488
⋄ B05 G05 ⋄ ID 12031

SHEJNBERGAS, I.M. [1970] *Simple bases and the number of functions in certain Post classes (Russian)* (J 0087) Mat Zametki (Akad Nauk SSSR) 8*105-114
• TRANSL [1970] (J 1044) Math Notes, Acad Sci USSR 8*528-533
⋄ B20 ⋄ REV MR 43#4640 Zbl 209.11 • ID 11948

SHEJNBERGAS, I.M. see Vol. II for further entries

SHEKHTMAN, V.B. [1982] *Undecidable propositional calculi (Russian)* (S 2874) Vopr Kibern, Akad Nauk SSSR 75*74-116
⋄ B22 B55 D35 ⋄ REV Zbl 499#03003 • ID 38116

SHEKHTMAN, V.B. see Vol. II, III, IV for further entries

SHELAH, S. [1973] *First order theory of permutation groups* (J 0029) Israel J Math 14*149-162 • ERR/ADD ibid 15*437-441
⋄ B15 C55 C60 C85 ⋄ REV MR 54#4972 Zbl 284#20003 • ID 19627

SHELAH, S. [1973] *There are just four second-order quantifiers* (J 0029) Israel J Math 15*282-300
⋄ B15 C80 C85 ⋄ REV MR 49#20 Zbl 273#02009 JSL 51.234 • ID 12052

SHELAH, S. [1977] *Decidability of a portion of the predicate calculus* (J 0029) Israel J Math 28*32-44
⋄ B20 B25 ⋄ REV MR 58#21562 Zbl 368#02014 • ID 27202

SHELAH, S. & STERN, J. [1978] *The Hanf number of the first order theory of Banach spaces* (J 0064) Trans Amer Math Soc 244*147-171
⋄ C55 C65 H10 ⋄ REV MR 80a:03047 Zbl 396#03047 • ID 31519

SHELAH, S. [1979] see RUDIN, M.E.

SHELAH, S. [1983] see GUREVICH, Y.

SHELAH, S. [1983] see MAGIDOR, M.

SHELAH, S. [1984] see GOLDFARB, W.D.

SHELAH, S. see Vol. II, III, IV, V, VI for further entries

SHEN, BAIYING [1984] see MO, SHAOKUI

SHEN, BAIYING see Vol. II, IV, VI for further entries

SHEN, V.Y. [1969] see MCKELLAR, A.C.

SHEN, YOUDING [1981] *The part of the first-order logical calculus (with identity) which is independent of quantifiers (Chinese)* (J 0418) Shuxue Xuebao 24*650-655
⋄ B20 ⋄ REV MR 83k:03018 Zbl 524#03008 • ID 36158

SHEN, YOUDING see Vol. II, V, VI for further entries

SHENG, C.L. [1965] *Detection of totally symmetric boolean functions* (J 4305) IEEE Trans Electr Comp EC-14*924-926
⋄ B35 ⋄ REV Zbl 135.34 JSL 36.694 • ID 12063

SHENG, C.L. [1969] *Threshold logic* (X 0801) Academic Pr: New York 206pp
⋄ B75 B98 ⋄ ID 48529

SHENG, C.L. see Vol. IV for further entries

SHEPHERDSON, J.C. [1956] *Note on a system of Myhill* (J 0036) J Symb Logic 21*261-264
⋄ B28 ⋄ REV MR 18.369 Zbl 71.10 • ID 12068

SHEPHERDSON, J.C. [1956] *On the interpretation of Aristotelian syllogistic* (J 0036) J Symb Logic 21*137-147
  ◊ A10 B20 ◊ REV MR 18.1 Zbl 72.247 JSL 22.381
  • ID 12069

SHEPHERDSON, J.C. [1975] see ROSE, H.E.

SHEPHERDSON, J.C. [1984] see ATIYAH, M.

SHEPHERDSON, J.C. see Vol. III, IV, V, VI for further entries

SHEPPARD, D.A. & VRANESIC, Z.G. [1974] *Fault detection of binary sequential machines using R-valued test machines* (J 0187) IEEE Trans Comp C-23*352-358
  ◊ B35 ◊ REV MR 52 # 13017 Zbl 288 # 94018 • ID 65255

SHESTAKOV, V.I. [1952] see ABRAMOV, A.A.

SHESTAKOV, V.I. see Vol. II for further entries

SHESTOPAL, G.A. [1961] *On the number of simple bases of Boolean functions (Russian)* (J 0023) Dokl Akad Nauk SSSR 140*314-317
  • TRANSL [1961] (J 0062) Sov Math, Dokl 2*1215-1219
  ◊ B05 ◊ REV MR 24 # A3107 JSL 31.501 • ID 19631

SHESTOPAL, G.A. [1966] *Simple basis in closed classes of functions of the algebra of logic (Russian)* (J 0023) Dokl Akad Nauk SSSR 168*1023-1026
  • TRANSL [1966] (J 0062) Sov Math, Dokl 7*792-795
  ◊ B20 ◊ REV MR 34 # 33 Zbl 163.245 • ID 11988

SHESTOPAL, G.A. see Vol. III, VI for further entries

SHI, ZUIJIAN [1980] see HUANG, CHENGGUI

SHIMADA, K. [1974] *A theorem prover for intuitionistic propositional logic* (J 0381) J Tsuda College (Tokyo) 6*39-44
  ◊ B35 F50 ◊ REV MR 58 # 10331 • ID 78527

SHIPILINA, L.B. [1979] see GRIGORYAN, A.K.

SHIRSHOV, A.I. [1973] see ERSHOV, YU.L.

SHIRSHOV, A.I. see Vol. IV for further entries

SHOENFIELD, J.R. [1954] *A relative consistency proof* (J 0036) J Symb Logic 19*21-28
  ◊ B30 E30 E35 E70 F25 ◊ REV MR 15.668 Zbl 55.4 JSL 22.367 • ID 12087

SHOENFIELD, J.R. [1967] *Mathematical logic* (X 0832) Addison-Wesley: Reading viii + 344pp
  • TRANSL [1975] (X 2027) Nauka: Moskva 527pp
  ◊ B98 C98 D98 E98 F98 ◊ REV MR 37 # 1224 MR 53 # 87 Zbl 155.11 JSL 40.234 • ID 22384

SHOENFIELD, J.R. see Vol. III, IV, V, VI for further entries

SHOESMITH, D.J. & SMILEY, T.J. [1978] *Multiple-conclusion logic* (X 0805) Cambridge Univ Pr: Cambridge, GB xiii + 396pp
  ◊ B22 F07 F98 ◊ REV MR 80k:03001 Zbl 381 # 03001 JSL 46.161 • ID 51869

SHOESMITH, D.J. see Vol. II for further entries

SHOLOMOV, L.A. [1966] *Complexity criteria for Boolean functions (Russian)* (J 0052) Probl Kibern 17*91-127
  ◊ B05 ◊ REV MR 36 # 3603 Zbl 192.87 • ID 16242

SHOLOMOV, L.A. [1970] *On calculating the complexity of boolean functions on Turing machines (Russian)* (J 0052) Probl Kibern 22*53-66
  • TRANSL [1972] (J 0471) Syst Th Res 22*51-56
  ◊ B05 ◊ REV Zbl 248 # 02036 • ID 65451

SHOLOMOV, L.A. [1975] *A sequence of complexly computable functions (Russian)* (J 0087) Mat Zametki (Akad Nauk SSSR) 17*957-966
  • TRANSL [1975] (J 1044) Math Notes, Acad Sci USSR 17*574-579
  ◊ B05 ◊ REV MR 52 # 13340 Zbl 327 # 94049 • ID 21801

SHOLOMOV, L.A. [1978] *Informationseigenschaften von Kompliziertheits-Funktionalen fuer Systeme nicht total definierter boolscher Funktionen (Russisch)* (J 0052) Probl Kibern 34*133-150
  ◊ B05 B70 ◊ REV MR 80b:94010 Zbl 415 # 94018 • ID 69915

SHOLOMOV, L.A. [1984] *Complexity of the composition representation of choice functions. I,II (Russian) (English summary)* (J 0040) Kibernetika, Akad Nauk Ukr SSR 1984/2*5-9,1984/4*10-14,22
  • TRANSL [1984] (J 0021) Cybernetics 20*172-179,477-484
  ◊ B05 ◊ REV MR 86f:04001 Zbl 568 # 05002 Zbl 575 # 90007 • ID 45428

SHOLOMOV, L.A. see Vol. V for further entries

SHORB, A.M. [1975] *Completely additive measure and integration* (J 0053) Proc Amer Math Soc 53*453-459
  ◊ H05 ◊ REV MR 52 # 3461 Zbl 336 # 02041 • ID 17691

SHORB, A.M. see Vol. III for further entries

SHORE, R.A. [1980] see NERODE, A.

SHORE, R.A. see Vol. III, IV, V, VI for further entries

SHOSTAK, R.E. [1976] *Refutation graphs* (J 0503) Artif Intell 7*51-64 • ERR/ADD ibid 7/3*283
  ◊ B35 ◊ REV MR 54 # 6584a MR 54 # 6584b Zbl 328 # 68080 • ID 24167

SHOSTAK, R.E. [1977] *On the role of unification in mechanical theorem proving* (J 1431) Acta Inf 7*319-323
  ◊ B35 ◊ REV MR 55 # 1869 Zbl 325 # 68049 • ID 65277

SHOSTAK, R.E. [1977] *On the SUP-INF method for proving Presburger formulas* (J 0037) ACM J 24*529-543
  ◊ B25 B35 F30 ◊ REV MR 58 # 13992 Zbl 423 # 68052
  • ID 69916

SHOSTAK, R.E. [1978] *An algorithm for reasoning about equality* (J 0212) ACM Commun 21*583-585
  ◊ B35 ◊ REV MR 58 # 4976 Zbl 378 # 68044 • ID 78572

SHOSTAK, R.E. [1979] *A practical decision procedure for arithmetic with function symbols* (J 0037) ACM J 26*351-360
  ◊ B25 B35 F30 ◊ REV MR 80f:68020 Zbl 496 # 03003
  • ID 82799

SHOSTAK, R.E. [1982] *Deciding combinations of theories* (P 3840) Autom Deduct (6);1982 New York 209-222
  ◊ B25 B35 ◊ REV MR 85h:68075 Zbl 481 # 68089
  • ID 38462

SHOSTAK, R.E. [1982] see MELLIAR-SMITH, P.M.

SHOSTAK, R.E. [1984] *Deciding combinations of theories* (J 0037) ACM J 31*1-12
  ◊ B35 ◊ ID 49024

SHOSTAK, R.E. (ED.) [1984] *7th international conference on automated deduction* (X 0811) Springer: Heidelberg & New York vi + 508pp
  ◊ B35 ◊ REV MR 85k:68003 Zbl 537 # 00025 • ID 47107

SHOSTAK, R.E. [1985] see BLEDSOE, W.W.

SHUBIN, M.A. & ZVONKIN, A.K. [1984] *Nonstandard analysis and singular perturbations of the ordinary differential equations (Russian)* (J 0067) Usp Mat Nauk 39/2*77-127
• TRANSL [1984] (J 1399) Russ Math Surv 39/2*69-131
⋄ H10 ⋄ REV MR 85j:34119 Zbl 549 # 34055 • ID 46087

SHUKLA, A. [1965] *A set of axioms for the propositional calculus with implication and converse nonimplication* (J 0047) Notre Dame J Formal Log 6*123-128
⋄ B20 ⋄ REV MR 34 # 7336 Zbl 145.241 JSL 31.664 • ID 12117

SHUKLA, A. [1966] *A set of axioms for the propositional calculus with implication and non-equivalence* (J 0047) Notre Dame J Formal Log 7*281-286
⋄ B20 ⋄ REV MR 35 # 5288 Zbl 156.6 • ID 12118

SHUKLA, A. [1969] *A note on independence* (J 0047) Notre Dame J Formal Log 10*410-412
⋄ B20 ⋄ REV MR 40 # 20 Zbl 209.8 • ID 12120

SHUKLA, A. see Vol. II, III for further entries

SHUKLA, R. & SINGH, R.K.P. [1950] *A note on Goetlind's axiom system for the calculus of propositions* (J 0248) Math Student 18*108-110
⋄ B05 ⋄ REV MR 14.345 Zbl 43.249 • ID 12347

SHUKLA, R. see Vol. II, V for further entries

SHURA-BURA, M.R. [1952] see ABRAMOV, A.A.

SHVARTS, G.F. [1979] *Some extensions of intuitionistic type theory (Russian) (English summary)* (J 0288) Vest Ser Mat Mekh, Univ Moskva 1979/3*31-34
• TRANSL [1979] (J 0510) Moscow Univ Math Bull 34/3*32-36
⋄ B15 F35 F50 ⋄ REV MR 80i:03068 Zbl 415 # 03049 • ID 56037

SHVARTS, G.F. see Vol. VI for further entries

SHVARTSMAN, M.I. [1979] *Generalized approach to minimization of boolean functions (Russian)* (J 0474) Avtom Vychis Tekh, Akad Nauk Latv SSR 1979/6*11-14
• TRANSL [1979] (J 2666) Autom Control Comput Sci 13/6*9-12
⋄ B35 ⋄ REV MR 82h:94030 Zbl 442 # 94019 • ID 69919

SIBAJIBAN [1970] *A remark on note on duality* (J 0047) Notre Dame J Formal Log 11*99-100
⋄ B05 ⋄ REV MR 43 # 3086 Zbl 193.287 • ID 12131

SIBERT JR., E.E. [1969] *A machine-oriented logic incorporating the equality relation* (J 0508) Machine Intelligence 4*103-133
⋄ B35 ⋄ REV MR 41 # 1538 Zbl 219 # 68048 • ID 12132

SIDORENKO, J.A. [1972] *Die logische Folgebeziehung und ihre Formalisierung* (C 1533) Quantoren, Modal, Paradox 213-224
⋄ A05 B22 ⋄ REV Zbl 254 # 02028 • ID 65289

SIEBER, K. [1985] *A partial correctness logic for procedures (in an ALGOL-like language)* (P 4571) Log of Progr;1985 Brooklyn 320-342
⋄ B35 B75 ⋄ REV Zbl 564 # 68009 • ID 49206

SIEFKES, D. [1968] *Recursion theory and the theorem of Ramsey in one-place second order successor arithmetic* (P 0608) Logic Colloq;1966 Hannover 237-254
⋄ B15 B25 C85 ⋄ REV MR 39 # 64 Zbl 186.11 JSL 40.504 • ID 21146

SIEFKES, D. [1970] *Decidable theories I. Buechi's monadic second order successor arithmetic* (S 3301) Lect Notes Math 120*xii+130pp
⋄ B15 B25 C85 D05 F35 F98 ⋄ REV MR 44 # 6488 Zbl 399 # 03011 • ID 12145

SIEFKES, D. [1970] *Decidable extensions of monadic second order successor arithmetic* (P 0577) Automatenth & Formale Sprachen;1969 Oberwolfach 441-472
⋄ B15 B25 C85 D05 F35 ⋄ REV MR 56 # 8354 Zbl 213.19 • ID 27984

SIEFKES, D. [1971] *Undecidable extensions of monadic second order successor arithmetic* (J 0068) Z Math Logik Grundlagen Math 17*385-394
⋄ B15 B25 C85 D35 F35 ⋄ REV MR 45 # 1763 Zbl 193.312 • ID 12146

SIEFKES, D. [1973] see BUECHI, J.R.

SIEFKES, D. [1978] *An axiom system for the weak monadic second order theory of two successors* (J 0029) Israel J Math 30*264-284
⋄ B15 B25 C85 D05 ⋄ REV MR 80a:03015 Zbl 397 # 03009 • ID 29136

SIEFKES, D. [1983] see BUECHI, J.R.

SIEFKES, D. see Vol. IV, VI for further entries

SIEKMANN, J. & STEPHAN, W. [1979] *Completeness and soundness of the connection graph proof procedure* (P 3108) AISB/GI Conf Artif Intel;1978 Hamburg 340-344
⋄ B35 ⋄ REV Zbl 399 # 68088 • ID 69933

SIEKMANN, J. & WRIGHTSON, G. [1980] *Paramodulated connection graphs* (J 1431) Acta Inf 13*67-86
⋄ B35 ⋄ REV MR 82b:03036 Zbl 407 # 05070 • ID 78610

SIEKMANN, J. [1982] see BIBEL, W.

SIEKMANN, J. & SZABO, P. [1982] *Universal unification and a classification of equational theories* (P 3840) Autom Deduct (6);1982 New York 369-389
• TRANSL [1984] (J 3079) Kiber Sb Perevodov, NS 21*213-234
⋄ B25 B35 C05 ⋄ REV MR 85i:08005 Zbl 494 # 68087 • ID 36659

SIEKMANN, J. & WRIGHTSON, G. (EDS.) [1983] *Automation of reasoning I,II* (X 0811) Springer: Heidelberg & New York xii+525pp,xii+637pp
⋄ B35 ⋄ REV MR 84f:03011 Zbl 567 # 03001 Zbl 567 # 03002 • ID 34429

SIEKMANN, J. [1984] *Universal unification* (P 2633) Autom Deduct (7);1984 Napa 1-42
⋄ B35 ⋄ REV MR 86g:03025 Zbl 547 # 03011 • ID 44374

SIEKMANN, J. see Vol. IV for further entries

SIEMENS JR., D.F. [1961] *An extension of "Fitch's rules"* (J 0068) Z Math Logik Grundlagen Math 7*199-204
⋄ B05 ⋄ REV MR 28 # 2050 Zbl 107.7 • ID 12148

SIEMENS JR., D.F. see Vol. II for further entries

SIGWART, C. [1895] *Logic. Vol.II: logical methods* (X 0959) Allen & Unwin: London viii+584pp
⋄ B98 ⋄ REM Vol.I 1904 • ID 22561

SIGWART, C. [1904] *Logik Vol.1* (X 1309) Mohr: Tuebingen
• TRANSL [1908] (X 4465) Obshch Pol'za & Provints: Leningrad
⋄ B98 ⋄ REM 3rd ed.; Vol.II 1895 • ID 43001

SIKORSKI, R. [1953] see RASIOWA, H.

SIKORSKI, R. [1955] see GRZEGORCZYK, A.

SIKORSKI, R. [1958] see RASIOWA, H.

SIKORSKI, R. [1958] *On Herbrand's theorem* (S 0019) Colloq Math (Warsaw) 6∗55-58
⋄ B10 F05 G05 ⋄ REV MR 20 # 6352 Zbl 88.10 JSL 35.587 • ID 12303

SIKORSKI, R. [1960] see RASIOWA, H.

SIKORSKI, R. [1961] *A topological characterization of open theories* (J 0014) Bull Acad Pol Sci, Ser Math Astron Phys 9∗259-260
⋄ B20 C07 G05 ⋄ REV MR 25 # 2961 Zbl 124.249 JSL 38.163 • ID 12306

SIKORSKI, R. [1962] *Applications of topology to foundations of mathematics* (P 1766) Gen Topol & Rel to Mod Anal & Algeb (1);1961 Prague 322-330
⋄ B30 E75 F50 ⋄ REV MR 26 # 3583 • ID 42875

SIKORSKI, R. [1962] *On open theories* (S 0019) Colloq Math (Warsaw) 9∗171-182
⋄ B20 C07 G05 ⋄ REV MR 27 # 1370 Zbl 124.249 JSL 38.163 • ID 12308

SIKORSKI, R. [1963] see RASIOWA, H.

SIKORSKI, R. see Vol. II, III, IV, V, VI for further entries

SILAEV, V.N. [1979] see ROZENFEL'D, T.K.

SILAS, S. [1983] see LANGLEY, P.

SILTERE, M.Y. [1969] *Mechanical deduction of arithmetical identities* (J 0141) Arch Autom & Telemech 6∗110-114
⋄ B35 ⋄ ID 46535

SILVER, B. [1982] *The application of homogenization to simultaneous equations* (P 3840) Autom Deduct (6);1982 New York 132-143
⋄ B35 ⋄ REV MR 85i:03038 Zbl 481 # 68041 • ID 44081

SILVER, C.L. [1980] *A simple strong-completeness proof for sentential logic* (J 0047) Notre Dame J Formal Log 21∗179-181
⋄ B05 ⋄ REV MR 81d:03010 Zbl 394 # 03012 • ID 53174

SILVESTRINI, D. [1981] *Alcune considerazioni sul metodo delle supervalutazioni e una semantica a supervalutazioni per il calcolo predicativo classico* (P 3092) Congr Naz Logica;1979 Montecatini Terme 335-350
⋄ B35 B60 ⋄ ID 48725

SIMANOV, A.L. & TAJMANOV, A.D. [1979] *On the problem of the logic of the foundation of scientific theories (Russian)* (C 2967) Metodol Probl Mat 145-155
⋄ B30 ⋄ REV MR 83j:03014 • ID 35329

SIMAUTI, T. [1956] *Proof of a special case of the fundamental conjecture of Takeuti's GLC* (J 0090) J Math Soc Japan 8∗135-144
⋄ B15 F05 F35 ⋄ REV MR 18.1 Zbl 75.234 JSL 24.64 • ID 12080

SIMAUTI, T. [1963] *Mechanization of mathematics* (J 0116) Electr & Comm Japan 46/11∗64-70
⋄ B35 ⋄ REV JSL 35.484 • ID 12323

SIMAUTI, T. see Vol. V, VI for further entries

SIMON, D. [1984] *A linear time algorithm for a subcase of second order instantiation* (P 2633) Autom Deduct (7);1984 Napa 209-223
⋄ B35 ⋄ REV Zbl 546 # 68081 • ID 44458

SIMON, H.A. [1957] see NEWELL, A.

SIMON, H.A. [1959] *Definable terms and primitives in axiom systems* (P 0651) Axiomatic Method;1957 Berkeley 443-453
⋄ B30 ⋄ REV MR 22 # 1516 Zbl 126.7 JSL 25.355 • ID 12335

SIMON, H.A. [1960] see NEWELL, A.

SIMON, H.A. see Vol. IV for further entries

SIMON, J. [1977] *Polynomially bounded quantification over higher types and a new hierarchy of the elementary sets* (P 1076) Latin Amer Symp Math Log (3);1976 Campinas 267-281
⋄ B15 D10 D15 D20 ⋄ REV MR 58 # 179 Zbl 393 # 03028 • ID 16610

SIMON, J. see Vol. IV for further entries

SIMONS, L. [1974] *Logic without tautologies* (J 0047) Notre Dame J Formal Log 15∗411-431
⋄ B20 ⋄ REV MR 51 # 2896 Zbl 226 # 02028 • ID 12338

SIMONS, L. [1978] *More logics without tautologies* (J 0047) Notre Dame J Formal Log 19∗543-557
⋄ B20 ⋄ REV MR 80a:03014 Zbl 272 # 02026 Zbl 422 # 03008 • ID 53469

SIMONS, L. see Vol. II, V for further entries

SIMOVICI, D.A. [1984] see REISCHER, C.

SIMOVICI, D.A. see Vol. II, IV for further entries

SIMPSON, S.G. [1976] see JOCKUSCH JR., C.G.

SIMPSON, S.G. [1982] see SCHMERL, J.H.

SIMPSON, S.G. [1985] see HARRINGTON, L.A.

SIMPSON, S.G. see Vol. III, IV, V, VI for further entries

SIMS, B. [1982] *"Ultra"-techniques in Banach space theory* (X 0997) Queen's Univ: Kingston iv+117pp
⋄ H05 ⋄ REV MR 86h:46032 • ID 44784

SINDELAR, J. [1985] see KRAMOSIL, I.

SINDELAR, J. see Vol. II, III for further entries

SINGER, M.F. [1976] *One parameter subgroups and nonstandard analysis* (J 0504) Manuscr Math 18∗1-13
⋄ H05 ⋄ REV MR 53 # 700 Zbl 328 # 22011 • ID 16691

SINGER, M.F. see Vol. III, IV for further entries

SINGH, R.K.P. [1950] see SHUKLA, R.

SINGH, S. [1967] *The natural number arithmetic in Goedel's axiomatic set theory* (J 0545) Proc Indian Acad Sci, Sect A 37∗221-228
⋄ B28 F30 ⋄ REV MR 41 # 6684 Zbl 241 # 02026 • ID 12349

SINGH, S. see Vol. V for further entries

SINGLETARY, W.E. [1964] *A complex of problems proposed by Post* (J 0015) Bull Amer Math Soc 70∗105-109 • ERR/ADD ibid 70∗826
⋄ B20 D03 D25 D30 D35 ⋄ REV MR 28 # 2040 MR 29 # 5714 Zbl 171.274 JSL 31.273 • ID 12350

SINGLETARY, W.E. [1967] *A note on finite axiomatization of partial propositional calculi* (J 0036) J Symb Logic 32*352-354
  ◊ B20 ◊ REV MR 36 # 51 Zbl 162.312 • ID 12354

SINGLETARY, W.E. [1968] *Results regarding the axiomatization of partial propositional calculi* (J 0047) Notre Dame J Formal Log 9*193-211
  ◊ B20 ◊ REV MR 39 # 4002 Zbl 195.298 JSL 36.172
  • ID 12355

SINGLETARY, W.E. [1974] *Many-one degrees associated with partial propositional calculi* (J 0047) Notre Dame J Formal Log 15*335-343
  ◊ B20 D25 ◊ REV MR 50 # 1862 Zbl 232 # 02009
  • ID 12356

SINGLETARY, W.E. see Vol. II, IV for further entries

SIOSON, F.M. [1965] *A characterization of tautologies in the propositional calculus with three variables (Spanish)* (J 4443) Gac Mat (Madrid) 17*68-74
  ◊ B05 ◊ REV MR 33 # 38 Zbl 156.7 • ID 12361

SIOSON, F.M. see Vol. II for further entries

SIROVICH, F. [1979] see DEGANO, P.

SKALA, H.L. [1966] *The irreducible generating sets of 2-place functions in the 2-valued logic* (J 0047) Notre Dame J Formal Log 7*341-343
  ◊ B05 ◊ REV MR 38 # 2000 Zbl 192.35 • ID 12377

SKALA, H.L. see Vol. V for further entries

SKIBENKO, I.T. [1973] see LYSENKO, E.V.

SKIBENKO, I.T. see Vol. II for further entries

SKLAR, A. [1977] see SCHWEIZER, B.

SKLAR, A. see Vol. IV for further entries

SKOLEM, T.A. [1919] *Untersuchungen ueber die Axiome des Klassenkalkuels und ueber Produktions- und Summationsprobleme, welche gewisse Klassen von Aussagen betreffen* (J 0974) Norsk Vid-Akad Oslo Mat-Natur Kl Skr 3*37pp
  • REPR [1970] (C 1098) Skolem: Select Works in Logic 67-101
  ◊ B20 B25 C10 D35 G05 ◊ ID 12383

SKOLEM, T.A. [1920] *Logisch-kombinatorische Untersuchungen ueber die Erfuellbarkeit oder Beweisbarkeit mathematischer Saetze nebst einem Theorem ueber dichte Mengen* (J 0974) Norsk Vid-Akad Oslo Mat-Natur Kl Skr 1920/4*1-36
  • TRANSL [1967] (C 0675) From Frege to Goedel 254-263
  • REPR [1970] (C 1098) Skolem: Select Works in Logic 103-136
  ◊ B10 B25 C07 C65 C75 ◊ REV FdM 48.1121
  • ID 12382

SKOLEM, T.A. [1923] *Begruendung der elementaren Arithmetik durch die rekurrierende Denkweise ohne Anwendung scheinbarer Veraenderlichen mit unendlichem Ausdehnungsbereich* (J 1145) Vidensk Selsk Kristiana Skrifter Ser 1 6*1-38
  • TRANSL [1967] (C 0675) From Frege to Goedel 303-333 (English) [1970] (C 1098) Skolem: Select Works in Logic 153-188
  ◊ A05 B28 D20 F30 ◊ REV Zbl 228 # 02001 • ID 38689

SKOLEM, T.A. [1923] *Einige Bemerkungen zur axiomatischen Begruendung der Mengenlehre* (P 1086) Skand Mat Kongr (5);1922 Helsinki 217-232
  • TRANSL [1967] (C 0675) From Frege to Goedel 290-301 (English) • REPR [1970] (C 1098) Skolem: Select Works in Logic 137-152
  ◊ A05 B30 C07 C62 E30 ◊ REV FdM 49.138 • ID 21221

SKOLEM, T.A. [1928] *Ueber die mathematische Logik* (J 4510) Norsk Mat Tidsskr 10*125-142
  • TRANSL [1967] (C 0675) From Frege to Goedel 508-524
  • REPR [1970] (C 1098) Skolem: Select Works in Logic 189-206
  ◊ A05 B25 B98 C07 C10 C35 ◊ REV FdM 54.58
  • ID 12385

SKOLEM, T.A. [1929] *Ueber einige Grundlagenfragen der Mathematik* (J 0974) Norsk Vid-Akad Oslo Mat-Natur Kl Skr 4*1-49
  • REPR [1970] (C 1098) Skolem: Select Works in Logic 227-273
  ◊ A05 A10 B30 C07 E30 ◊ REV FdM 55.31 • ID 12386

SKOLEM, T.A. [1931] *Ueber die symmetrisch allgemeinen Loesungen im identischen Kalkuel* (J 0974) Norsk Vid-Akad Oslo Mat-Natur Kl Skr 6*32pp
  • REPR [1970] (C 1098) Skolem: Select Works in Logic 307-336
  ◊ B05 ◊ REV Zbl 4.385 FdM 58.65 • ID 12390

SKOLEM, T.A. [1931] *Ueber einige Satzfunktionen in der Arithmetik* (J 0974) Norsk Vid-Akad Oslo Mat-Natur Kl Skr 7*28pp
  • REPR [1970] (C 1098) Skolem: Select Works in Logic 281-306
  ◊ B25 B28 C10 F30 ◊ REV Zbl 2.3 FdM 57.1320
  • ID 12388

SKOLEM, T.A. [1932] *Ueber die symmetrisch allgemeinen Loesungen im Klassenkalkuel* (J 0027) Fund Math 18*61-76
  ◊ B20 ◊ REV Zbl 4.385 FdM 58.66 • ID 12389

SKOLEM, T.A. [1933] *Ein kombinatorischer Satz mit Anwendung auf ein logisches Entscheidungsproblem* (J 0027) Fund Math 20*254-261
  • REPR [1970] (C 1098) Skolem: Select Works in Logic 337-344
  ◊ B20 B25 E05 ◊ REV Zbl 7.97 FdM 59.54 • ID 12391

SKOLEM, T.A. [1933] *Ueber die Unmoeglichkeit einer Charakterisierung der Zahlenreihe mittels eines endlichen Axiomensystems* (J 0975) Norsk Mat Forenings Skr 10*73-82
  • REPR [1970] (C 1098) Skolem: Select Works in Logic 345-354
  ◊ B28 C20 C62 F30 H15 ◊ REV Zbl 7.193 FdM 59.53
  • ID 12392

SKOLEM, T.A. [1934] *Ein Satz ueber Zaehlausdruecke* (J 0460) Acta Univ Szeged, Sect Mat 7*193-199
  • REPR [1970] (C 1098) Skolem: Select Works in Logic 367-373
  ◊ B28 ◊ REV Zbl 12.385 FdM 61.973 • ID 12394

SKOLEM, T.A. [1934] *Ueber die Nichtcharakterisierbarkeit der Zahlenreihe mittels endlich oder abzaehlbar unendlich vieler Aussagen mit ausschliesslich Zahlenvariablen* (J 0027) Fund Math 23*150-161
• REPR [1970] (C 1098) Skolem: Select Works in Logic 355-366
⋄ B28 C20 C62 F30 H15 ⋄ REV Zbl 10.49 FdM 60.25
• ID 12393

SKOLEM, T.A. [1935] *Ueber die Erfuellbarkeit gewisser Zaehlausdruecke* (J 0974) Norsk Vid-Akad Oslo Mat-Natur Kl Skr 6*14pp
⋄ B25 B28 ⋄ REV Zbl 13.97 FdM 61.973 • ID 12395

SKOLEM, T.A. [1936] *Ein Satz ueber die Erfuellbarkeit von einigen Zaehlausdruecken der Form*
$(x)(\exists y_1,...,y_n)K_1(x,y_1,...,y_n) \& (x_1,x_2,x_3)K_2(x_1,x_2,x_3)$
(J 0752) Avh Norske Vid-Akad Oslo I 1935/8*1-10
• REPR [1970] (C 1098) Skolem: Select Works in Logic 375-394
⋄ B10 B25 ⋄ REV Zbl 13.242 JSL 1.111 FdM 62.42
• ID 40864

SKOLEM, T.A. [1936] *Einige Reduktionen des Entscheidungsproblems* (J 0752) Avh Norske Vid-Akad Oslo I 1936/6*17pp
• REPR [1970] (C 1098) Skolem: Select Works in Logic 395-409
⋄ B20 D35 ⋄ REV Zbl 15.338 FdM 62.1059 • ID 40914

SKOLEM, T.A. [1936] *Selected chapters of mathematical logic (Norwegian) (German summary)* (X 4350) Michelsens Inst Videnskap, Griegs: Bergen 73pp
⋄ B96 ⋄ REV JSL 2.87 FdM 62.1048 • ID 41345

SKOLEM, T.A. [1936] *Ueber die Zurueckfuehrbarkeit einiger durch Rekursionen definierter Relationen auf "arithmetische"* (J 0460) Acta Univ Szeged, Sect Mat 8*73-88
• REPR [1970] (C 1098) Skolem: Select Works in Logic 425-440
⋄ B28 F30 ⋄ REV Zbl 16.194 JSL 2.85 FdM 63.25
• ID 12396

SKOLEM, T.A. [1937] *Eine Bemerkung zum Entscheidungsproblem* (P 1608) Int Congr Math (II, 5);1936 Oslo 2*268-270
⋄ B20 B25 D35 ⋄ REV JSL 3.57 FdM 63.31 • ID 27938

SKOLEM, T.A. [1943] *Some remarks on the preceding article of E.Hoff-Hansen (Norwegian)* (J 4510) Norsk Mat Tidsskr 25*13-16
⋄ B05 C05 G25 ⋄ REV MR 8.125 Zbl 27.385 JSL 13.169
• ID 12400

SKOLEM, T.A. [1952] *Consideraciones sobre los fundamentos de la matematica I* (J 0236) Rev Mat Hisp-Amer, Ser 4 12*169-200
⋄ B98 F30 F50 ⋄ REV MR 15.92 Zbl 52.8 JSL 21.373
• REM Part II 1953 • ID 21071

SKOLEM, T.A. [1952] *On the proofs of independence of the axioms of the classical sentential calculus* (J 0121) Kon Norske Vidensk Selsk Forh 24*20-25
• REPR [1970] (C 1098) Skolem: Select Works in Logic 529-534
⋄ B05 ⋄ REV MR 14.526 Zbl 47.13 JSL 18.67 • ID 12405

SKOLEM, T.A. [1953] *Consideraciones sobre los fundamentos de la matematica II* (J 0236) Rev Mat Hisp-Amer, Ser 4 13*149-174
⋄ B98 F30 F50 ⋄ REV MR 15.92 Zbl 52.8 JSL 21.373
• REM Part I 1952 • ID 21072

SKOLEM, T.A. [1953] *The logical background of arithmetic* (J 0082) Bull Soc Math Belg 6*23-34
• REPR [1970] (C 1098) Skolem: Select Works in Logic 541-552
⋄ A05 B28 ⋄ REV MR 16.553 Zbl 57.6 JSL 23.228
• ID 12411

SKOLEM, T.A. [1954] *Results in investigations in the foundations (Norwegian)* (P 0788) Skand Mat Kongr (12);1953 Lund 273-289
⋄ A05 B98 D03 F55 ⋄ REV MR 16.553 Zbl 56.245
• ID 21133

SKOLEM, T.A. [1955] *Peano's axioms and models of arithmetic* (P 1589) Math Interpr of Formal Systs;1954 Amsterdam 1-14
• REPR [1970] (C 1098) Skolem: Select Works in Logic 587-600
⋄ B28 C20 C62 F30 H15 ⋄ REV MR 17.699
Zbl 68.246 JSL 22.306 • ID 27718

SKOLEM, T.A. [1956] *A version of the proof of equivalence between complete induction and the uniqueness of primitive recursion* (J 0121) Kon Norske Vidensk Selsk Forh 29*10-15
• REPR [1970] (C 1098) Skolem: Select Works in Logic 606-606
⋄ B28 F30 ⋄ REV MR 18.2 Zbl 72.6 JSL 24.69 • ID 12416

SKOLEM, T.A. [1958] *Reduction of axiom systems with axiom schemes to systems with only simple axioms* (J 0076) Dialectica 12*443-450
• REPR [1970] (C 1098) Skolem: Select Works in Logic 645-652
⋄ B30 ⋄ REV MR 21 #3325 Zbl 93.12 JSL 27.232
• ID 12421

SKOLEM, T.A. [1958] *Une relativisation des notions mathematiques fondamentales* (P 0576) Raisonn en Math & Sci Exper;1955 Paris 13-18
• REPR [1970] (C 1098) Skolem: Select Works in Logic 633-638
⋄ A05 B30 ⋄ REV MR 21 #4895 Zbl 87.8 JSL 25.284
• ID 12423

SKOLEM, T.A. [1961] *Interpretation of mathematical theories in the first order predicate calculus* (C 0622) Essays Found of Math (Fraenkel) 218-225
• REPR [1970] (C 1098) Skolem: Select Works in Logic 673-680
⋄ B15 B30 ⋄ REV MR 29 #4671 Zbl 158.245 • ID 12427

SKOLEM, T.A. [1970] *Selected works in logic* (X 1554) Universitesforlaget: Oslo 732pp
⋄ B96 C96 ⋄ REV MR 44 #2562 Zbl 228 #02001 • REM Edited by Fenstad,J.E. • ID 29109

SKOLEM, T.A. see Vol. II, III, IV, V, VI for further entries

SKORUBSKIJ, V.I. [1980] *Arithmetical and logical foundations of digital machines. Textbook (Russian)* (X 3219) Tochnoj Mekhaniki Optiki: Leningrad 60pp
⋄ B70 B98 ⋄ REV Zbl 479 #94029 • ID 55704

SKVORTSOV, D.P. [1981] see FINN, V.K.

SKVORTSOV, D.P. [1983] *On some ways of construction logical languages with quantifiers over finite sequences (Russian)* (S 2582) Semiotika & Inf, Akad Nauk SSSR 20*102-126
⋄ B15 C80 ⋄ REV MR 84m:03011 Zbl 512 # 03027
• ID 40156

SKVORTSOV, D.P. [1985] *The question of "how many": definition of the notion of cardinality for finite sets in some arithmetic systems (Russian)* (S 2582) Semiotika & Inf, Akad Nauk SSSR 24*41-57,142
⋄ B28 E30 ⋄ ID 47594

SKVORTSOV, D.P. see Vol. II, III, VI for further entries

SKYRMS, B. [1966] *Choice and chance* (X 1201) Dickenson: Belmont viii+165pp
⋄ A05 B48 B98 ⋄ REV JSL 41.547 • ID 21285

SKYRMS, B. [1976] *Possible worlds physics and metaphysics* (J 0095) Philos Stud 30*323-332
⋄ B30 ⋄ ID 28383

SKYRMS, B. see Vol. II for further entries

SLAGHT, R.L. [1974] *A concise method for translating propositional formulae containing the standard truth-functional connectives into a Sheffer stroke equivalent; plus an extension of the method* (J 0047) Notre Dame J Formal Log 15*161-164
⋄ B05 ⋄ REV MR 49 # 2285 Zbl 232 # 02007 • ID 12449

SLAGHT, R.L. see Vol. II for further entries

SLAGLE, J.R. [1965] *Experiments with a deductive question-answering program* (J 0212) ACM Commun 8*792-798
⋄ B35 ⋄ REV JSL 35.596 • ID 12450

SLAGLE, J.R. [1966] *A multipurpose theorem proving heuristic program that learns* (P 0573) Inform Processing (3);1965 New York 2*323-324
⋄ B35 ⋄ REV Zbl 173.17 • ID 29436

SLAGLE, J.R. [1967] *Automatic theorem proving with renamable and semantic resolution* (J 0037) ACM J 14*687-697
• REPR [1983] (C 4659) Autom of Reasoning 2*55-65
⋄ B35 ⋄ REV MR 38 # 3093 Zbl 157.24 JSL 35.595
• ID 12451

SLAGLE, J.R. [1968] see BURSKY, P.

SLAGLE, J.R. [1969] see CHANG, C.L.

SLAGLE, J.R. [1970] see CHANG, C.L.

SLAGLE, J.R. [1970] *Interpolation theorems for resolution in lower predicate calculus* (J 0037) ACM J 17*535-542
⋄ B35 C40 ⋄ REV MR 50 # 6804 Zbl 198.36 • ID 12452

SLAGLE, J.R. [1971] see CHANG, C.L.

SLAGLE, J.R. [1979] see CHANG, C.L.

SLAGLE, J.R. see Vol. VI for further entries

SLATER, B.H. [1978] *A fragment of new propositional logic* (J 0286) Int Logic Rev 17-18*121-133
⋄ B20 ⋄ REV MR 81c:03011 • ID 78702

SLATER, B.H. [1979] *Aristotle's propositional logic* (J 0095) Philos Stud 36*35-49
⋄ A05 A10 B20 ⋄ REV MR 81b:03032 • ID 78703

SLATER, B.H. see Vol. II, V for further entries

SLATER, J.G. [1966] *The required correction to Copi's statement of UG* (J 0079) Logique & Anal, NS 9*267
⋄ B10 ⋄ REV MR 35 # 20 JSL 32.255 • ID 12455

SLEPIAN, D. [1953] *On the number of symmetry types of boolean functions of n variables* (J 0017) Canad J Math 5*185-193
⋄ B05 ⋄ REV MR 15.93 Zbl 51.248 JSL 20.70 • ID 12459

SLESARENKO, A.P. [1976] see RVACHEV, V.L.

SLISENKO, A.O. [1965] see DAVYDOV, G.V.

SLISENKO, A.O. [1969] see DAVYDOV, G.V.

SLISENKO, A.O. [1975] *Finite approach to the problem of optimizing theorem-proving algorithms (Russian) (English summary)* (S 0228) Zap Nauch Sem Leningrad Otd Mat Inst Steklov 49*123-130,178
• TRANSL [1978] (J 1531) J Sov Math 10*597-603
⋄ B10 B35 ⋄ REV MR 51 # 12489 Zbl 308 # 68076
• ID 54645

SLISENKO, A.O. [1979] see MATIYASEVICH, YU.V.

SLISENKO, A.O. [1984] *Linguistic considerations in devising effective algorithms* (P 4313) Int Congr Math (II,14);1983 Warsaw 1*347-357
⋄ B35 D20 ⋄ ID 48589

SLISENKO, A.O. see Vol. IV, VI for further entries

SLOAN, A. [1978] *A note on exponentials of distributions* (J 0048) Pac J Math 79*207-224
⋄ H05 ⋄ REV MR 80m:46035 Zbl 423 # 46028 • ID 82829

SLOAN, A. [1981] *The strong convergence of Schroedinger propagators* (J 0064) Trans Amer Math Soc 264*557-570
⋄ H10 ⋄ REV MR 84c:47045 Zbl 517 # 35009 • ID 39666

SLOAN, A. [1983] see BERGER, M.

SLOMSON, A. [1969] *An undecidable two sorted predicate calculus* (J 0036) J Symb Logic 34*21-23
⋄ B10 D35 ⋄ REV MR 39 # 2636 Zbl 187.277 • ID 12473

SLOMSON, A. [1973] see BELL, J.L.

SLOMSON, A. see Vol. III, V for further entries

SLUPECKI, J. [1946] *Les remarques sur la syllogistique d'Aristotle (Polish)* (J 0501) Ann Univ Lublin, Sect Math 1*187-191
⋄ A10 B20 ⋄ REV MR 10.93 JSL 13.166 • ID 41600

SLUPECKI, J. [1950] *On Aristotelian syllogistic* (J 4716) Stud Philos, Leopolis (Poznan) 4*275-300
⋄ A10 B20 ⋄ REV Zbl 45.294 JSL 17.210 • ID 12477

SLUPECKI, J. [1953] *Ueber die Regeln des Aussagenkalkuels (Polish and Russian summaries)* (J 0063) Studia Logica 1*19-43,299
⋄ B05 ⋄ REV MR 16.891 Zbl 59.14 JSL 21.87 • ID 12479

SLUPECKI, J. [1955] *A logical system without operators (Polish, Russian) (English summary)* (J 0063) Studia Logica 3*98-108,109-124 • ERR/ADD ibid 3*6
⋄ B15 ⋄ REV MR 17.1171 Zbl 68.11 • ID 24884

SLUPECKI, J. [1956] see KOKOSZYNSKA, M.

SLUPECKI, J. [1958] see BORKOWSKI, L.

SLUPECKI, J. [1958] *On some partial systems of the propositional calculus (Polish) (Russian and English summaries)* (J 0063) Studia Logica 8*177-187
⋄ B20 ⋄ REV MR 21 # 5558 Zbl 128.11 • ID 24893

SLUPECKI, J. [1959] *The Lukasiewicz function (Polish)* (J 0481) Acta Univ Wroclaw 3*33-40
⋄ B05 B20 ⋄ REV JSL 32.536 • ID 21242

SLUPECKI, J. [1960] see POGORZELSKI, W.A.

SLUPECKI, J. [1961] see POGORZELSKI, W.A.

SLUPECKI, J. [1962] see POGORZELSKI, W.A.

SLUPECKI, J. [1963] see BORKOWSKI, L.

SLUPECKI, J. [1970] see BRYLL, G.

SLUPECKI, J. [1971] see BRYLL, G.

SLUPECKI, J. [1972] *L-decidability and decidability* (J 0387) Bull Sect Logic, Pol Acad Sci 1/3∗38-43
 ⋄ B22 B25 ⋄ REV MR 51 # 116 • ID 15183

SLUPECKI, J. [1972] *Proof of the completeness of a fragmentary system of the propositional calculus (Polish) (English summary)* (S 0458) Zesz Nauk, Prace Log, Uniw Krakow 7∗23-29
 ⋄ B20 ⋄ REV MR 47 # 6429 • ID 12496

SLUPECKI, J. [1972] see BRYLL, G.

SLUPECKI, J. [1975] see BRYLL, G.

SLUPECKI, J. [1976] see HALKOWSKA, K.

SLUPECKI, J. [1978] see HALKOWSKA, K.

SLUPECKI, J. see Vol. II, III, V for further entries

SMARANDACHE, F.G. [1979] *Deducibility theorems in mathematical logic* (J 0304) An Univ Timisoara, Sti Mat 17∗163-168
 ⋄ B05 ⋄ REV MR 82a:03012 Zbl 446 # 03006 • ID 56535

SMART, J.J.C. [1949] *Whitehead and Russell's theory of types* (J 0103) Analysis (Oxford) 10∗93-96
 ⋄ B15 ⋄ REV JSL 15.218 • ID 31251

SMART, J.J.C. [1950] *The theory of types again* (J 0103) Analysis (Oxford) 11∗131-133
 ⋄ B15 ⋄ ID 31252

SMART, J.J.C. [1951] *The theory of types - a further note* (J 0103) Analysis (Oxford) 12∗24
 ⋄ B15 ⋄ ID 31253

SMART, J.J.C. [1953] *A note on categories* (J 0013) Brit J Phil Sci 4∗227-228
 ⋄ B15 ⋄ ID 31254

SMART, J.J.C. [1961] *Goedel's theorem, Church's theorem and mechanism* (J 0154) Synthese 13∗105-110
 ⋄ A05 B10 D35 F30 ⋄ REV Zbl 104.242 • ID 31260

SMETANICH, YA.S. [1960] *On the completeness of the propositional calculus with additional operations in one argument (Russian)* (J 0065) Tr Moskva Mat Obshch 9∗357-371
 ⋄ B22 ⋄ REV MR 24 # A680 Zbl 117.252 • ID 12504

SMETANICH, YA.S. [1961] *On propositional calculi with an additional operation (Russian)* (J 0023) Dokl Akad Nauk SSSR 139∗309-312
 • TRANSL [1961] (J 0062) Sov Math, Dokl 2∗937-939
 ⋄ B22 ⋄ REV MR 26 # 26 Zbl 122.8 • ID 12505

SMETANICH, YA.S. see Vol. VI for further entries

SMILEY, T.J. [1962] *Syllogism and quantification* (J 0036) J Symb Logic 27∗58-72
 ⋄ A05 A10 B20 ⋄ REV MR 27 # 1351 Zbl 121.10 JSL 40.606 • ID 12510

SMILEY, T.J. [1962] *The independence of connectives* (J 0036) J Symb Logic 27∗426-436
 ⋄ B22 ⋄ REV MR 30 # 3003 Zbl 139.6 JSL 40.250 • ID 12511

SMILEY, T.J. [1973] *What is a syllogism?* (J 0122) J Philos Logic 2∗136-154
 ⋄ A05 B20 ⋄ REV Zbl 259 # 02005 • ID 30339

SMILEY, T.J. [1978] see SHOESMITH, D.J.

SMILEY, T.J. [1982] see DALEN VAN, D.

SMILEY, T.J. see Vol. II for further entries

SMIRNOV, D.M. [1973] see ERSHOV, YU.L.

SMIRNOV, D.M. see Vol. III, V for further entries

SMIRNOV, S.V. [1970] *Chebyshev approximations and the theory of models (Russian)* (J 0003) Algebra i Logika 9∗73-79
 • TRANSL [1970] (J 0069) Algeb and Log 9∗45-49
 ⋄ H05 ⋄ REV Zbl 209.14 • ID 32585

SMIRNOV, S.V. [1978] *Approximations of mappings into a group (Russian)* (J 0003) Algebra i Logika 17∗86-101
 • TRANSL [1978] (J 0069) Algeb and Log 17∗63-74
 ⋄ H05 ⋄ REV MR 81h:41038 Zbl 395 # 03045 • ID 52596

SMIRNOV, V.A. [1963] *Bemerkungen zum System der Syllogistik und der allgemeinen Theorie der Deduktion (Russian)* (C 1654) Probl Logiki 64-83
 ⋄ B10 ⋄ REV Zbl 285 # 02015 • ID 29945

SMIRNOV, V.A. [1967] *Imbedding syllogistics into predicate calculus (Russian)* (C 0582) Log Semant & Modal Logika 254-258
 ⋄ B20 ⋄ REV MR 36 # 1288 Zbl 203.1 • ID 12518

SMIRNOV, V.A. [1972] *Formal derivation and logical calculi (Russian)* (X 2027) Nauka: Moskva 272pp
 ⋄ B98 C07 C40 C98 F05 F07 F98 ⋄ REV Zbl 269 # 02004 • ID 30359

SMIRNOV, V.A. [1973] *An absolute first order predicate calulus* (J 0387) Bull Sect Logic, Pol Acad Sci 2∗38-45
 ⋄ B10 F05 ⋄ REV MR 52 # 13339 • ID 21800

SMIRNOV, V.A. & TAVANETS, P.V. (EDS.) [1974] *Philosphy and logic (Russian)* (X 2027) Nauka: Moskva 479pp
 ⋄ A05 B97 ⋄ REV MR 58 # 21368 • ID 70080

SMIRNOV, V.A. (ED.) [1979] *Logical inference (Russian)* (X 2027) Nauka: Moskva 312pp
 ⋄ B97 ⋄ REV MR 84j:03007 • ID 34601

SMIRNOV, V.A. [1979] *Theory of quantification and $\mathfrak{E}$-calculi* (P 1705) Scand Logic Symp (4);1976 Jyvaeskylae 41-47
 ⋄ B10 F50 ⋄ REV MR 81c:03010 Zbl 431 # 03015 • ID 53921

SMIRNOV, V.A. [1980] *Adequate conversion of assertions of syllogistics in predicate calculus (Russian)* (C 2583) Aktual Probl Log & Metodol Nauki 230-236
 ⋄ B20 ⋄ REV Zbl 528 # 03004 • ID 37624

SMIRNOV, V.A. [1983] *Embedding the elementary ontology of Stanislaw Lesniewski into the monadic second-order calculus of predicates* (J 0063) Studia Logica 42∗197-207
 ⋄ B15 E70 ⋄ REV MR 85m:03019 • ID 42348

SMIRNOV, V.A. [1984] see ERSHOV, YU.L.

SMIRNOV, V.A. see Vol. II, III, VI for further entries

SMIRNOVA, E.D. & TAVANETS, P.V. [1967] *Semantics in logic (Russian)* (C 0582) Log Semant & Modal Logika 3-53
 • TRANSL [1972] (C 1533) Quantoren, Modal, Paradox 123-178
 ⋄ A05 B98 ⋄ REV MR 37 # 48 Zbl 204.1 • ID 12523

SMIRNOVA, E.D. [1974] *Consistency and eliminability in proof theory (Russian)* (C 2578) Filos & Logika 84-101
⋄ A05 B10 F99 ⋄ REV MR 58 # 27265 • ID 78744

SMIRNOVA, E.D. [1980] *Analysis of expressive possibilities of languages and theories (Russian)* (C 3038) Log-Metodol Issl 246-264
⋄ B30 ⋄ REV MR 83i:03017 • ID 35495

SMIRNOVA, I.M. [1963] see AJZERMAN, M.A.

SMIRNOVA, I.M. see Vol. IV, V for further entries

SMIRNOVA, O.S. [1973] *Some examples of nonstandard extensions (Russian)* (J 0226) Uch Zap Ped Inst, Ivanovo 123*91-98
⋄ H05 ⋄ REV MR 58 # 27462 • ID 78745

SMIRNOVA, O.S. see Vol. III for further entries

SMITH, BRIAN [1974] see BISHOP, P.

SMITH, BRIAN [1983] see HENSCHEN, L.J.

SMITH, BRIAN & WINKER, S.K. & WOS, L. [1984] *A new use of an automated reasoning assistant: Open questions in equivalential calculus and the study of infinite domains* (J 0503) Artif Intell 22*303-356
⋄ B35 ⋄ REV Zbl 553 # 68051 • ID 43440

SMITH, BRIAN [1984] see MCCUNE, W.

SMITH, C.H. [1978] see CASE, J.

SMITH, C.H. [1983] see CASE, J.

SMITH, C.H. see Vol. II, IV for further entries

SMITH, D.R. [1982] *Derived preconditions and their use in program synthesis* (P 3840) Autom Deduct (6);1982 New York 172-193
⋄ B35 ⋄ REV MR 85g:68066 Zbl 481 # 68083 • ID 43932

SMITH, G.C. [1982] *The Boole-De Morgan correspondence, 1842-1864* (X 0894) Oxford Univ Pr: Oxford vii+156pp
⋄ A10 B05 ⋄ REV MR 84f:01088 Zbl 505 # 01018 JSL 49.657 • ID 38204

SMITH, HENRY BRADFORD [1935] see HENLE, P.

SMITH, HENRY BRADFORD [1936] *The algebra of propositions* (J 0153) Phil of Sci (East Lansing) 3*551-578
⋄ B05 B45 ⋄ REV JSL 2.43 FdM 62.1052 • ID 41315

SMITH, HENRY BRADFORD see Vol. II for further entries

SMITH, J.B. [1951] see MCCALLUM, D.B.

SMITH, JAN [1984] *An interpretation of Martin-Loef's type theory in a type-free theory of propositions* (J 0036) J Symb Logic 49*730-753
⋄ B15 ⋄ REV MR 85j:03106 • ID 42502

SMITH, JAN see Vol. VI for further entries

SMITH, N.W. [1962] see HENKIN, L.

SMITH, ROBIN [1978] *The mathematical origins of Aristotle's syllogistic* (J 0267) Arch Hist Exact Sci 19*201-210
⋄ A10 B20 ⋄ REV MR 81f:01004 Zbl 403 # 01002 • ID 54714

SMITH, S.F. [1983] see ARNON, D.S.

SMOLIN, V.P. [1975] *Monadic algebras and the one-place predicate calculus* (J 2604) Progr Comput Software 1*455-462
⋄ B20 C07 G15 ⋄ REV Zbl 367 # 02032 • ID 51192

SMOLIN, V.P. see Vol. III for further entries

SMOLKA, G. [1983] *Completeness of the connection graph proof procedure for unit-refutable clause sets* (P 3858) Adequate Modeling of Syst;1982 Bad Honnef 191-204
⋄ B35 ⋄ REV MR 85k:03007 Zbl 499 # 68038 • ID 45208

SMULLYAN, A. [1962] *Fundamentals of logic* (X 0819) Prentice Hall: Englewood Cliffs 131pp
⋄ B98 ⋄ REV JSL 29.41 • ID 12537

SMULLYAN, A. see Vol. II for further entries

SMULLYAN, R.M. [1957] *Languages in which self reference is possible* (J 0036) J Symb Logic 22*55-67
• REPR [1969] (C 0569) Phil of Math Oxford Readings 64-77
⋄ A05 B10 B28 F30 F40 ⋄ REV MR 19.830 Zbl 218 # 02026 JSL 24.228 • ID 12538

SMULLYAN, R.M. [1961] *Theory of formal systems* (S 3513) Ann Math Stud xi+142pp
• TRANSL [1981] (X 2027) Nauka: Moskva 208pp
⋄ B98 D03 D05 D20 D25 D98 ⋄ REV MR 22 # 12042 MR 27 # 2409 MR 83h:03002 Zbl 529 # 03014 Zbl 97.245 JSL 30.88 • ID 21025

SMULLYAN, R.M. [1963] *A unifying principle in quantification theory* (J 0054) Proc Nat Acad Sci USA 49*828-832
⋄ B10 F05 F07 ⋄ REV MR 27 # 2410 Zbl 118.249 • ID 12547

SMULLYAN, R.M. [1963] *First order logic* (X 0811) Springer: Heidelberg & New York xii+158pp
⋄ B98 F07 F98 ⋄ REV MR 39 # 5311 Zbl 172.289 JSL 40.237 • REM 3rd edition 1971 • ID 28562

SMULLYAN, R.M. [1965] *A unifying principle in quantification theory* (P 0614) Th Models;1963 Berkeley 433-434
⋄ B10 F05 F07 ⋄ REV Zbl 199.7 • ID 16212

SMULLYAN, R.M. [1965] *Analytic natural deduction* (J 0036) J Symb Logic 30*123-139
⋄ B10 F05 F07 ⋄ REV MR 34 # 34 Zbl 149.4 • ID 12551

SMULLYAN, R.M. [1966] *Finite nest structures and propositional logic* (J 0036) J Symb Logic 31*322-324
⋄ B05 ⋄ REV MR 34 # 5650 Zbl 149.6 • ID 12553

SMULLYAN, R.M. [1966] *Trees and nest structures* (J 0036) J Symb Logic 31*303-321
⋄ B05 ⋄ REV MR 34 # 5649 Zbl 149.5 • ID 12552

SMULLYAN, R.M. [1968] *Analytic cut* (J 0036) J Symb Logic 33*560-564
⋄ B10 F05 F07 ⋄ REV MR 39 # 3969 Zbl 181.5 • ID 12555

SMULLYAN, R.M. [1968] *Uniform Gentzen system* (J 0036) J Symb Logic 33*549-559
⋄ B10 ⋄ REV MR 39 # 3968 Zbl 197.273 • ID 12556

SMULLYAN, R.M. [1970] *Abstract quantification theory* (P 0603) Intuitionism & Proof Th;1968 Buffalo 79-91
⋄ B10 C07 F07 ⋄ REV MR 42 # 2928 Zbl 206.272 • ID 12557

SMULLYAN, R.M. [1978] *What is the name of this book? The riddle of Dracula and other logical puzzles* (X 0819) Prentice Hall: Englewood Cliffs xi+241pp
⋄ B98 ⋄ REV Zbl 432 # 00028 • ID 53958

SMULLYAN, R.M. [1983] *Dame oder Tiger? Logische Denkspiele und eine mathematische Novelle ueber Goedels grosse Entdeckung* (**X** 1265) Fischer: Stuttgart 239pp
⋄ B98 F99 ⋄ REV Zbl 545 # 00007 • REM Transl. from English • ID 41200

SMULLYAN, R.M. see Vol. II, IV, V, VI for further entries

SMYTH, M.B. [1971] *A diagrammatic treatment of syllogistic* (**J** 0047) Notre Dame J Formal Log 12*483-488
⋄ A05 B20 ⋄ REV Zbl 224 # 02007 • ID 12558

SMYTH, M.B. [1974] *Involution as a basis for propositional calculi* (**J** 0047) Notre Dame J Formal Log 15*569-588
⋄ B05 F50 ⋄ REV MR 51 # 10021 Zbl 225 # 02018 • ID 12559

SMYTH, M.B. see Vol. IV for further entries

SNEED, J.D. [1971] *The logical structure of mathematical physics* (**S** 3307) Synth Libr xv+311pp
⋄ A05 B30 G12 ⋄ REV MR 81d:03008 Zbl 324 # 02001 Zbl 426 # 03002 • REM 2nd rev. ed. 1979. 35*xxv+320pp • ID 25445

SNEED, J.D. [1982] *The logical structure of Bayesian decision theory* (**P** 3759) Phil of Econ;1981 Muenchen 201-222
⋄ B30 ⋄ REV MR 84b:03026 • ID 34908

SNIR, M. [1979] *The covering problem of complete uniform hypergraphs* (**J** 0193) Discr Math 27*103-105
⋄ B05 ⋄ REV MR 83c:05102 Zbl 412 # 05031 • ID 39067

SNIR, M. [1982] see GAIFMAN, H.

SNIR, M. see Vol. IV for further entries

SNYDER, D.P. [1972] see LEBLANC, H.

SNYDER, D.P. see Vol. II for further entries

SOARE, R.I. [1981] see LERMAN, M.

SOARE, R.I. see Vol. III, IV, VI for further entries

SOBEL, J.H. [1974] *Principia mathematica description theory: the classical and an alternative notation* (**J** 0047) Notre Dame J Formal Log 15*63-72
⋄ B03 ⋄ REV MR 49 # 4748 Zbl 272 # 02021 • ID 12571

SOBEL, J.H. [1976] *Alternative notations for principia mathematica description theory: possible modifications* (**J** 0047) Notre Dame J Formal Log 17*476-478
⋄ B03 ⋄ REV MR 56 # 82 Zbl 283 # 02015 • ID 18395

SOBEL, J.H. [1979] *Sentential notations: Unique decomposition* (**J** 0047) Notre Dame J Formal Log 20*377-382
⋄ B03 ⋄ REV MR 80c:03016 Zbl 314 # 02024 • ID 52620

SOBOCINSKI, B. [1935] *Die Axiomatisierung der implikativkonjunktiven Deduktionstheorie (Polish)* (**J** 1125) Przeglad Filoz 38*85-95
⋄ B20 ⋄ REV FdM 62.1056 • ID 40910

SOBOCINSKI, B. [1939] *Axiomatisierung des "konjunktiv-negativen" Systems des Aussagenkalkuels (Polish)* (**J** 0490) Collecteana Logica 1*179-193
⋄ B05 ⋄ REV Zbl 23.97 FdM 65.24 • ID 43754

SOBOCINSKI, B. [1954] *Axiomatization of a conjunctive-negative calculus of propositions* (**J** 0093) J Comp Syst 1*229-242
⋄ B20 ⋄ REV MR 16.891 Zbl 58.7 JSL 20.303 • ID 12576

SOBOCINSKI, B. [1955] *Note on a problem of Paul Bernays* (**J** 0036) J Symb Logic 20*109-114
⋄ B05 ⋄ REV MR 17.4 Zbl 65.1 JSL 21.328 • ID 12577

SOBOCINSKI, B. [1956] *On well-constructed axiom systems* (**J** 1586) Rocz Pol Towarz Nauk Obczyznie 6*54-65
⋄ B10 ⋄ REV JSL 22.358 • ID 27077

SOBOCINSKI, B. [1961] *On the single axioms of protothetic II,III* (**J** 0047) Notre Dame J Formal Log 2*111-126,129-148
⋄ A05 B05 B15 ⋄ REV MR 27 # 2420 Zbl 125.278 JSL 30.245 • REM Part I 1960 • ID 21078

SOBOCINSKI, B. [1964] *On the propositional system A of Vuchkovich and its extension I,II* (**J** 0047) Notre Dame J Formal Log 5*141-153,223-237
⋄ B20 B60 ⋄ REV MR 36 # 4965 Zbl 133.254 JSL 31.118 • ID 21264

SOBOCINSKI, B. [1974] *A theorem concerning a restricted rule of substitution in the field of propositional calculi I & II* (**J** 0047) Notre Dame J Formal Log 15*465-476,589-597
⋄ B05 B22 ⋄ REV MR 50 # 6781 Zbl 287 # 02005 • ID 12619

SOBOCINSKI, B. [1978] *Awkward axiom - systems* (**J** 0047) Notre Dame J Formal Log 19*315-320
⋄ B22 ⋄ REV MR 58 # 132 Zbl 368 # 02047 • ID 51682

SOBOCINSKI, B. [1978] *Note about Lukasiewicz's theorem concerning the system of axioms of the implicational propositional calculus* (**J** 0047) Notre Dame J Formal Log 19*457-460
⋄ B20 ⋄ REV MR 58 # 10285 Zbl 368 # 02010 • ID 51657

SOBOCINSKI, B. see Vol. II, III, V for further entries

SOCHILINA, A.V. [1978] *Algorithm and program for establishing derivability resolving wide classes of formulas (Russian) (English summary)* (**J** 0040) Kibernetika, Akad Nauk Ukr SSR 1978/3*136-138
⋄ B35 ⋄ REV Zbl 382 # 03009 • ID 51933

SOCHOR, A. [1976] *The alternative set theory* (**P** 1476) Set Th & Hierarch Th (2) (Mostowski);1975 Bierutowice 259-271
⋄ B30 E70 E75 H20 ⋄ REV MR 57 # 2913 Zbl 344 # 02049 • ID 23811

SOCHOR, A. [1977] *Differential calculus in the alternative set theory* (**P** 1695) Set Th & Hierarch Th (3);1976 Bierutowice 273-284
⋄ E70 E75 H05 ⋄ REV MR 57 # 12221 Zbl 434 # 03037 • ID 31250

SOCHOR, A. [1979] *Some remarks to the connection between the alternative set theory and nonstandard methods* (**P** 2539) Frege Konferenz (1);1979 Jena 458-467
⋄ E70 H05 ⋄ REV MR 82d:03104 • ID 78823

SOCHOR, A. see Vol. III, V, VI for further entries

SOKOL, J. [1979] see FIBY, R.

SOLOMONOFF, R.J. [1978] *Complexity-based induction systems: Comparisons and convergence theorems* (**J** 2745) IEEE Trans Inf Theory IT-24*422-432
⋄ B30 D10 ⋄ REV MR 58 # 20912 Zbl 382 # 60003 • ID 51980

SOLON, B.Ya. [1981] *PC-degrees inside an e-degree of a hyperimmune retraceable set (Russian)* (**C** 3865) Algeb Sistemy (Ivanovo) 203-217
⋄ D25 D30 D50 H05 ⋄ REV MR 85g:03063 Zbl 549 # 03029 • ID 43887

SOLON, B.Ya. see Vol. IV for further entries

SOLS, I. [1975] *Bon ordre dans l'objet des nombres naturels d'un topos booleen (English summary)* (J 2313) C R Acad Sci, Paris, Ser A-B 281*A601-A603
  ◊ B28 E70 G30 ◊ REV MR 52 # 96   Zbl 346 # 18010
  • ID 18397

SOLS, I. see Vol. IV, V for further entries

SORDYL, E. [1967] see LIGMANOWSKI, M.

SORGER, P. [1974] see FREUND, H.

SORGER, P. [1976] see FREUND, H.

SOTIROV, V. [1973] *Is only one bracket sufficient? (Bulgarian)* (J 0477) Spis Bulgar Akad Nauk 16(49)*26-28
  ◊ B03 ◊ REV Zbl 264 # 02001 • ID 27491

SOTIROV, V.KH. [1970] *A certain set of axioms for the propositional calculus (Russian)* (J 0137) C R Acad Bulgar Sci 23*133-135
  ◊ B05 ◊ REV MR 42 # 35   Zbl 198.12 • ID 12653

SOTIROV, V.KH. see Vol. II for further entries

SOUDIEUX, C. [1960] *De l'infini arithmetique* (X 0978) Schulthess: Zuerich iii+115pp
  ◊ B28 F30 F98 ◊ REV MR 24 # A1823 Zbl 91.50
  • ID 12654

SOULA, G. [1982] see SANCHEZ, E.

SPASOWSKI, M. [1973] *Some connections between Cn, Cn^{-1} and dCn* (J 0387) Bull Sect Logic, Pol Acad Sci 2*46-50
  ◊ B22 ◊ REV MR 52 # 13313 • ID 21776

SPASOWSKI, M. [1974] *Some properties of the operation d* (J 0302) Rep Math Logic, Krakow & Katowice 3*53-56
  ◊ B22 ◊ REV MR 52 # 5342 Zbl 306 # 02012 • ID 18400

SPASOWSKI, M. [1983] *Some properties of dual counterparts of consequence operations (Polish) (English summary)* (J 0481) Acta Univ Wroclaw 10*71-116
  ◊ B22 ◊ REV Zbl 572 # 03007 • ID 49222

SPASOWSKI, M. see Vol. II for further entries

SPECKER, E. [1958] *Dualitaet* (J 0076) Dialectica 12*451-465
  ◊ B30 C07 E70 ◊ REV MR 21 # 5551 Zbl 91.7 JSL 27.231 • ID 12669

SPECKER, E. [1961] see MACDOWELL, R.

SPECKER, E. [1962] *Typical ambiguity* (P 0612) Int Congr Log, Meth & Phil of Sci (1,Proc);1960 Stanford 116-124
  ◊ B15 E70 ◊ REV MR 28 # 1129 Zbl 156.21 • ID 12670

SPECKER, E. [1968] see HODES, L.

SPECKER, E. [1979] see LIEBERHERR, K.J.

SPECKER, E. [1981] see LIEBERHERR, K.J.

SPECKER, E. see Vol. II, III, IV, V, VI for further entries

SPIRKIN, A.G. [1978] see BIRYUKOV, B.V.

SPIVAKOV, YU.L. [1974] *Algebraic extensions of the field of formal power series $F_D((t))$ (Russian) (Uzbek summary)* (J 0030) Izv Akad Nauk Uz SSR, Ser Fiz-Mat 18/6*15-21,78
  ◊ C20 C60 H05 ◊ REV MR 52 # 5639 Zbl 303 # 12102
  • ID 18401

SPRINGSTEEL, F. [1979] see PUDLAK, P.

SPRINGSTEEL, F. [1981] *Complexity of hypothesis formation problems* (J 1741) Int J Man-Mach Stud 15*319-332
  ◊ B35 D15 ◊ REV MR 82m:68091 • ID 82856

SPRINGSTEEL, F. see Vol. IV for further entries

SRIVAS, M. [1985] see HSIANG, JIEH

STAAL, J.F. [1960] *Formal structures in Indian logic* (J 0154) Synthese 12*279-286
  • REPR [1961] (P 0711) Concept & Role of Model in Math & Sci;1960 Utrecht 155-162
  ◊ A05 A10 B20 ◊ REV MR 24 # A1190 JSL 35.572
  • ID 19687

STAAL, J.F. [1968] see ROOTSELAAR VAN, B.

STABLER, E.R. [1952] *Applied logic and modern problems* (P 0593) Int Congr Math (II, 6);1950 Cambridge MA 1*732-733
  ◊ A05 B80 ◊ ID 28066

STABLER, E.R. [1953] *An introduction to mathematical thought* (X 0832) Addison-Wesley: Reading xviii+268pp
  ◊ A05 B98 ◊ REV MR 14.441 Zbl 50.4 JSL 20.288
  • ID 23429

STACHNIAK, Z. [1978] *Some notes on characteristic consequence operations* (J 0387) Bull Sect Logic, Pol Acad Sci 7*159-166
  ◊ B22 G25 ◊ REV MR 80j:03041 Zbl 413 # 03019
  • ID 53013

STACHNIAK, Z. [1985] *Note on structural logics* (J 0068) Z Math Logik Grundlagen Math 31*317-320
  ◊ B22 ◊ ID 47558

STACHNIAK, Z. see Vol. III, V for further entries

STAHL, G. [1956] *Introduccion a la logica simbolica* (X 1715) Edunsa Ed Univ: Santiago 206pp
  ◊ B98 ◊ REV JSL 32.519 • REM 2nd ed. 1962 • ID 32460

STAHL, G. [1956] *La suficiencia de la logica bivalente para la fisca de los cuantos* (J 0988) Rev Filos (Santiago) 3*18-27
  ◊ B05 G12 ◊ ID 13086

STAHL, G. [1957] *Les univers du discours et les calculs correspondants* (J 1092) Ann Univ Paris 4*530-539
  ◊ B10 ◊ REV JSL 24.240 • ID 32467

STAHL, G. [1958] *An opposite and an expanded System* (J 0068) Z Math Logik Grundlagen Math 4*244-247
  ◊ B05 ◊ REV MR 21 # 4905 Zbl 201.5 JSL 25.298
  • ID 13087

STAHL, G. [1958] *Aspectos formales de algunas paradojas semanticas* (J 0988) Rev Filos (Santiago) 5*31-41
  ◊ B15 ◊ REV JSL 34.140 • ID 32471

STAHL, G. [1958] *Enfoque moderno de la logica clasica* (X 1715) Edunsa Ed Univ: Santiago 187pp
  ◊ B05 B10 ◊ REV JSL 32.519 • ID 32461

STAHL, G. [1959] *Los universos del discurso y los sistemas correspondientes* (J 1717) An Univ Chile 116*50-55
  ◊ B10 ◊ REV JSL 33.605 • ID 32468

STAHL, G. [1963] *A paratheory of type theory* (J 0068) Z Math Logik Grundlagen Math 9*169-171
  ◊ A05 B15 ◊ REV MR 26 # 6049 Zbl 112.11 JSL 37.194
  • ID 13090

STAHL, G. [1964] *Elementos de la metalogica y metamatematica* (X 1715) Edunsa Ed Univ: Santiago 160pp
  ◊ B98 ◊ REV JSL 32.519 • ID 32462

STAHL, G. [1973] *A ZF System combined with type theory (Spanish) (English summary)* (**J** 0367) Scientia (Valparaiso) 144∗105-108
⋄ B15 E70 ⋄ REV MR 48 # 8233 • ID 13092

STAHL, G. [1973] *Elementos de metamatematica* (**X** 1715) Edunsa Ed Univ: Santiago
⋄ B98 ⋄ ID 32465

STAHL, G. see Vol. II, IV, V, VI for further entries

STANDLEY, G.B. [1954] *Ideographic computation in the propositional calculus* (**J** 0036) J Symb Logic 19∗169-171
⋄ B05 ⋄ REV MR 16.103 Zbl 56.8 JSL 23.63 • ID 13101

STANDLEY, G.B. [1962] *Two arithmetical techniques with numbered classes* (**J** 0036) J Symb Logic 27∗437-438
⋄ B20 ⋄ REV Zbl 119.12 JSL 30.376 • ID 13102

STANDLEY, G.B. [1971] *New methods in symbolic logic* (**X** 0847) Houghton Mifflin: Boston 217pp
⋄ B98 ⋄ REV JSL 39.178 • ID 13104

STANDLEY, G.B. see Vol. II for further entries

STANFORD, P.H. [1975] *Polish circles* (**J** 0068) Z Math Logik Grundlagen Math 21∗395-399
⋄ B03 ⋄ REV MR 53 # 2621 Zbl 339 # 02008 • ID 13105

STANFORD, P.H. see Vol. VI for further entries

STANLEY, R.L. [1953] *An extended procedure in quantification logic* (**J** 0036) J Symb Logic 18∗97-104
⋄ B10 ⋄ REV MR 15.1 Zbl 52.248 JSL 21.197 • ID 13108

STANLEY, R.L. [1955] *Simplified foundations for mathematical logic* (**J** 0036) J Symb Logic 20∗123-139
⋄ B10 ⋄ REV MR 17.118 Zbl 64.246 JSL 21.322 • ID 13109

STANLEY, R.L. [1965] *Natural deduction, inference and consistency* (**J** 0053) Proc Amer Math Soc 16∗367-374
⋄ B15 ⋄ REV MR 31 # 3306 Zbl 143.9 • ID 13111

STANLEY, R.L. see Vol. V for further entries

STAPLES, J. [1983] *Two-level expression representation for faster evaluation* (**P** 3908) Graph-Gram & Appl to Comput Sci (2);1982 Neunkirchen 392-404
⋄ B35 B40 D03 ⋄ REV Zbl 522 # 68036 • ID 37059

STAPLES, J. see Vol. VI for further entries

STARK, W.R. [1979] *Type hierarchies and type-level reduction* (**J** 2695) Cybernetica 22∗17-31
⋄ B15 ⋄ REV MR 81g:03009 Zbl 413 # 68094 • ID 53044

STARK, W.R. [1980] *Automatic model construction* (**J** 0191) Inform Sci 22∗99-106
⋄ B35 ⋄ REV MR 81m:68078 Zbl 455 # 03001 • ID 54262

STARK, W.R. see Vol. II, III, IV, V for further entries

STASZEK, W. [1971] *On proofs of rejection (Polish and Russian summaries)* (**J** 0063) Studia Logica 29∗17-25
⋄ B20 ⋄ REV MR 46 # 3274 Zbl 251 # 02019 • ID 13131

STASZEK, W. [1972] *A certain Interpretation of the theory of rejected propositions (Polish and Russian summaries)* (**J** 0063) Studia Logica 30∗147-152
⋄ B20 ⋄ REV MR 51 # 10022 Zbl 295 # 02008 • ID 17564

STASZEK, W. see Vol. III for further entries

STATMAN, R. [1977] *Complexity and derivations from quantifier-free Horn formulae, mechanical introduction of explicit definitions, and refinement of completeness theorems* (**P** 1075) Logic Colloq;1976 Oxford 7∗505-518
⋄ B35 F20 ⋄ REV MR 58 # 170 Zbl 441 # 03021 • ID 16639

STATMAN, R. [1978] *Bounds for proof-search and speed-up in the predicate calculus* (**J** 0007) Ann Math Logic 15∗225-287
⋄ B35 F05 F20 ⋄ REV MR 81c:03049 Zbl 411 # 03047 • ID 52901

STATMAN, R. [1980] *Solution to a problem of Chang and Lee* (**J** 0047) Notre Dame J Formal Log 21∗518-520
⋄ B35 ⋄ REV MR 83i:03026 Zbl 416 # 03049 • ID 53941

STATMAN, R. see Vol. IV, VI for further entries

STAVI, J. [1983] see MAGIDOR, M.

STAVI, J. see Vol. II, III, IV, V, VI for further entries

STEBBING, L.S. [1930] *A modern introduction to logic* (**X** 0816) Methuen: London & New York xviii+505pp
• LAST ED [1961] (**X** 0837) Harper & Row: New York xvii+525pp
⋄ B98 ⋄ REV MR 22 # 10900 FdM 56.829 • ID 22602

STEBBING, L.S. [1943] *A modern elementary logic* (**X** 0816) Methuen: London & New York viii+214pp
⋄ B98 ⋄ REV JSL 9.63 • ID 13139

STECK, M. [1941] *Unbekannte Briefe Frege's ueber die Grundlagen der Geometrie und Antwortbrief Hilbert's an Frege* (**J** 0417) Sitzb Akad Wiss, Heidelberg, Math-Nat Kl 2∗31pp
⋄ A10 B30 ⋄ REV MR 11.150 Zbl 26.242 • ID 14947

STEEN, S.W.P. [1972] *Mathematical logic, with special reference to the natural numbers* (**X** 0805) Cambridge Univ Pr: Cambridge, GB xvi+638pp
⋄ B98 F30 F98 ⋄ REV MR 57 # 9451 Zbl 275 # 02002 • ID 22386

STEGMUELLER, W. [1957] *Das Wahrheitsproblem und die Idee der Semantik. Eine Einfuehrung in die Theorien von A.Tarski und R.Carnap* (**X** 0902) Springer: Wien x+328pp
⋄ A05 B30 ⋄ REV MR 19.1031 Zbl 88.8 JSL 31.496
• REM 2nd ed. 1968; x+328pp • ID 13144

STEGMUELLER, W. [1970] *Probleme und Resultate der Wissenschaftstheorie und Analytischen Philosophie. Band II: Theorie und Erfahrung. 1. Halbband: Begriffsformen, Wissenschaftssprache, empirische Signifikanz und theoretische Begriffe* (**X** 0811) Springer: Heidelberg & New York xv+485pp
⋄ A05 B30 ⋄ REV Zbl 211.7 • REM Revised ed. 1974; xiv+485pp. Part I 1969. Part II,2 1973 • ID 65517

STEGMUELLER, W. see Vol. II, III, IV, V, VI for further entries

STEIGER, R. [1974] see BISHOP, P.

STEIN, M. [1981] *Eine Bemerkung zur Wohlordungseigenschaft der natuerlichen Zahlen* (**J** 0026) Elem Math 36∗134-137
⋄ B28 ⋄ REV MR 83f:04002 • ID 35317

STEIN, M. see Vol. VI for further entries

STEINAUER, P.H. [1979] *La logique au service du droit. Etude de logique contemporaine pour une meilleure communication de la pensee juridique* (**X** 2862) Univ Fribourg: Fribourg xii+198pp
⋄ B80 ⋄ REV JSL 48.1206 • ID 47424

STEINER, H.-G. [1969] *Representation de la syllogistique aristotelicienne sur la base de la theorie des relations* (J 0082) Bull Soc Math Belg 21∗271-284
⋄ A10 B20 ⋄ REV MR 41 # 1504 Zbl 196.8 • ID 13159

STEINER, H.-G. see Vol. IV, VI for further entries

STENGER, H.-J. [1984] *Algebraic characterisations of NTIME(F) and NTIME(F,A)* (J 3441) RAIRO Inform Theor 18∗365-385
⋄ B20 B35 D15 D20 ⋄ REV Zbl 547 # 68080 • ID 43303

STENIUS, E. [1951] *Modern logic (Swedish)* (J 3665) Finsk Tidskrift 150∗102-112
⋄ B98 ⋄ REV JSL 24.190 • ID 33917

STENIUS, E. see Vol. II, III, VI for further entries

STEPANKOVA, O. [1976] see HAVEL, I.M.

STEPANKOVA, O. [1977] see HAVEL, I.M.

STEPANKOVA, O. [1979] *Planning in uncertain environments through situation calculus* (P 3108) AISB/GI Conf Artif Intel;1978 Hamburg 330-339
⋄ B80 ⋄ REV Zbl 412 # 68084 • ID 52989

STEPANKOVA, O. [1979] *Uniform formulas and their proof in a situation theory (Russian and Czech summaries)* (J 3350) Acta Polytech Prague CVUT, Ser 4 1979/1∗35-49
⋄ B75 B80 ⋄ REV MR 83i:68143 • ID 33185

STEPANKOVA, O. [1981] *Normal form of proof of certain formulas of situation calculus* (P 3642) Colloq Math Log in Computer Sci;1978 Salgotarjan 725-738
⋄ B80 ⋄ REV MR 83g:68007 Zbl 488 # 68058 • ID 36748

STEPANKOVA, O. see Vol. V for further entries

STEPANOV, V.A. [1984] *Propositional logic of reflexive sentences (Russian)* (X 2235) VINITI: Moskva 6119-84
⋄ B05 ⋄ REM Part I. Part II 1985 • ID 46962

STEPANOV, V.A. [1985] *Propositional logic of reflexive sentences. II (Russian)* (X 2235) VINITI: Moskva 4276-85
⋄ B05 F30 ⋄ REM Part I 1984 • ID 46515

STEPHAN, W. [1979] see SIEKMANN, J.

STEPIEN, T. [1973] *A survey of minor Wajsberg's results concerning fragmentary systems of the classical propositional calculus* (J 0387) Bull Sect Logic, Pol Acad Sci 2∗103-106
⋄ B20 ⋄ REV MR 55 # 7731 • ID 78998

STEPIEN, T. [1980] see DYWAN, Z.

STEPIEN, T. [1981] *Craig-Goedel-Lindenbaum's property and Sobocinski-Tarski's property in propositional calculi* (J 0387) Bull Sect Logic, Pol Acad Sci 10∗116-121
⋄ B22 C40 ⋄ REV MR 83b:03030 Zbl 481 # 03007 • ID 33327

STEPIEN, T. [1983] see BIELA, A.

STEPIEN, T. [1983] *System $\bar{S}$* (J 0302) Rep Math Logic, Krakow & Katowice 15∗59-65
⋄ B22 ⋄ REV MR 85d:03064 Zbl 505 # 03008 • ID 38132

STEPIEN, T. [1984] *A sufficient and necessary condition for Tarski's property in Lindenbaum's extensions* (J 0068) Z Math Logik Grundlagen Math 30∗447-453
⋄ B22 ⋄ REV Zbl 563 # 03005 • ID 42274

STEPIEN, T. [1984] *The sufficient and necessary condition for Tarski's property in Lindenbaum's extensions* (J 0387) Bull Sect Logic, Pol Acad Sci 13∗222-225
⋄ B22 ⋄ REV Zbl 565 # 03006 • ID 48614

STEPIEN, T. [1985] *Logic based on atomic entailment* (J 0387) Bull Sect Logic, Pol Acad Sci 14∗65-71
⋄ B22 B60 ⋄ ID 48971

STEPIEN, T. [1985] *On number of Lindenbaum's oversystems of propositional and predicate calculi* (J 0068) Z Math Logik Grundlagen Math 31∗333-344
⋄ B22 ⋄ ID 47560

STEPIEN, T. see Vol. II for further entries

STERN, J. [1977] see REEB, G.

STERN, J. [1978] see SHELAH, S.

STERN, J. (ED.) [1982] *Proceedings of the Herbrand Symposium. Logic colloquium '81, held in Marseille, France, July 1981* (S 3303) Stud Logic Found Math 107∗xi + 384pp
⋄ B97 D97 ⋄ REV MR 85f:03003 Zbl 489 # 00007 • ID 36590

STERN, J. see Vol. III, IV, V for further entries

STERN, S.T. [1969] *A number theory for the seminaturals* (J 0068) Z Math Logik Grundlagen Math 15∗401-410
⋄ B28 ⋄ REV MR 42 # 3017 Zbl 198.24 • ID 33194

STERN, S.T. see Vol. III, V for further entries

STERNFELD, R. [1979] *Frege's achievements and literal scientific discourse* (J 2076) Rev Int Philos 33∗723-738
⋄ A05 A10 B30 ⋄ REV MR 82d:03006 • ID 79014

STEVENSON, L. [1975] *A formal theory of sortal quantification* (J 0047) Notre Dame J Formal Log 16∗185-207
⋄ A05 B10 ⋄ REV MR 51 # 2858 Zbl 298 # 02010 • ID 13188

STEVENSON, L. [1976] *Freges zwei Definitionen der Quantifikation* (C 1891) Studien zu Frege II∗103-124
⋄ A05 A10 B10 ⋄ REV MR 58 # 27268 • ID 79020

STEWART, R. [1960] see POHM, A.

STICKEL, M.E. [1976] see LOVELAND, D.W.

STICKEL, M.E. [1981] *A unification algorithm for associative-commutative functions* (J 0037) ACM J 28∗423-434
⋄ B35 ⋄ REV MR 82h:68123 Zbl 462 # 68075 • ID 82888

STICKEL, M.E. [1984] *A case study of theorem proving by the Knuth-Bendix method: discovering that $x^3 = x$ implies ring commutativity* (P 2633) Autom Deduct (7);1984 Napa 248-258
⋄ B35 ⋄ REV Zbl 546 # 68078 • ID 44459

STICKEL, M.E. see Vol. II for further entries

STIEFEL, B. [1982] *Ueber Iterierte einer Booleschen Funktion* (S 2829) Rostocker Math Kolloq 19∗119-128
⋄ B05 ⋄ REV MR 84c:94028 Zbl 484 # 94043 • ID 39655

STIHI, T. [1973] *A case of decidability in the first order predicate calculus (Romanian)* (S 1613) Probl Logic (Bucharest) 5∗257-265
⋄ B20 B25 ⋄ REV MR 58 # 195 • ID 79027

STIHI, T. [1973] *Une generalisation du carre logique* (J 0047) Notre Dame J Formal Log 14*215-223
  ◊ A10 B20 ◊ REV MR 49 # 8809 Zbl 251 # 02014
  • ID 13189

STIHI, T. see Vol. V for further entries

STILLWELL, J.C. [1972] see ASH, C.J.

STILLWELL, J.C. [1977] *Concise survey of mathematical logic* (J 3194) J Austral Math Soc, Ser A 24*139-161
  ◊ B98 C98 D10 D35 ◊ REV MR 57 # 5655 Zbl 393 # 03002 • ID 52422

STILLWELL, J.C. see Vol. III, IV for further entries

STOCKMEYER, L.J. [1984] see CHANDRA, A.K.

STOCKMEYER, L.J. see Vol. IV for further entries

STOHL, F. [1971] see BLISS, K.

STOICHITA, R. [1963] *La transcription du carre logique en calcul propositionnel* (J 0147) An Univ Bucuresti, Acta Logica 6*145-162
  ◊ B20 ◊ REV MR 30 # 1923 Zbl 231 # 02016 • ID 27427

STOICHITA, R. see Vol. II for further entries

STOJAKOVIC, M. [1972] *On the exponential properties of the implication* (J 0400) Publ Inst Math, NS (Belgrade) 13(27)*121-125
  ◊ B05 ◊ REV MR 48 # 8229 Zbl 253 # 02008 • ID 13196

STOJAKOVIC, M. [1975] *Inductive and Peano models (Serbo-Croatian) (English summary)* (J 2855) Zbor Rad, Prir-Mat Fak, Ser Mat (Novi Sad) 5*1-8
  ◊ B28 C62 ◊ REV MR 56 # 15416 • ID 79037

STOJAKOVIC, M. see Vol. II, V for further entries

STOLL, R.R. [1961] *Sets, logic and axiomatic theories* (X 0994) Freeman: San Francisco x+206pp
  ◊ B98 E98 ◊ REV MR 23 # A791 Zbl 106.235 Zbl 292 # 02002 JSL 25.278 • ID 19683

STOLL, R.R. [1963] *Set theory and logic* (X 0994) Freeman: San Francisco xiv+474pp
  ◊ B98 E98 ◊ REV MR 30 # 11 Zbl 112.244 JSL 29.40 • ID 13197

STOLYAR, A.A. [1970] *Introduction to elementary mathematical logic* (X 0865) MIT Pr: Cambridge, MA vii+209pp
  • TRANSL [1971] (X 1574) Vyssheyshaya Shkola: Minsk 224pp
  ◊ B98 ◊ REV Zbl 215.46 Zbl 234 # 02001 • ID 22539

STOLZ, O. [1885] *Vorlesungen ueber allgemeine Arithmetik. Vol. I* (X 0823) Teubner: Stuttgart 344pp
  ◊ B28 ◊ REV FdM 17.116 FdM 18.130 • REM Vol.II 1886
  • ID 23464

STOLZ, O. [1886] *Vorlesungen ueber allgemeine Arithmetik. Vol. II* (X 0823) Teubner: Stuttgart 326pp
  ◊ B28 ◊ REV FdM 18.130 • REM Vol.I 1885 • ID 23465

STOLZ, O. [1891] *Groessen und Zahlen* (X 0823) Teubner: Stuttgart 30pp
  ◊ B28 ◊ REV FdM 23.60 • ID 25499

STONE, A.L. [1965] *Extensive ultraproducts and Haar measures* (P 0614) Th Models;1963 Berkeley 419-423
  ◊ C20 C65 E75 H05 ◊ REV MR 34 # 1441 • ID 27645

STONE, A.L. [1969] *Nonstandard analysis in topological algebra* (P 0649) Appl Model Th to Algeb, Anal & Probab;1967 Pasadena 285-299
  ◊ H05 ◊ REV MR 38 # 3309 Zbl 186.72 • ID 27646

STONE, M.H. [1937] *Note on formal logic* (J 0100) Amer J Math 59*506-514
  ◊ B28 G05 G15 ◊ REV Zbl 17.145 JSL 2.174 FdM 63.25
  • ID 13210

STONE, M.H. see Vol. VI for further entries

STOPES-ROE, H.V. [1969] *An economy in the formation rules for quantification theory* (J 0047) Notre Dame J Formal Log 10*313-316
  ◊ B10 ◊ REV MR 39 # 6735 Zbl 167.7 • ID 13217

STORK, H.-G. [1979] *Remarks on the satisfiability problem of propositional logic* (J 3359) Appl Comp Sci, Ber Prakt Inf 13*31-43
  ◊ B05 D15 ◊ REV MR 81b:68058 Zbl 398 # 03024
  • ID 52752

STRAUSS, M. [1938] *Mathematics as logical syntax - a method to formalize the language of a physical theory* (J 0748) Erkenntnis (Leipzig) 7*147-153
  ◊ B30 ◊ REV JSL 4.25 FdM 64.931 • ID 41085

STRAUSS, P. [1967] *Some systems of natural deduction* (J 0047) Notre Dame J Formal Log 8*286-290
  ◊ B05 B10 ◊ REV MR 38 # 4278 Zbl 189.281 JSL 35.466
  • ID 13222

STRAUSS, P. see Vol. V, VI for further entries

STRAWSON, P.F. [1952] *Introduction to logical theory* (X 0816) Methuen: London & New York x+266pp
  ◊ A05 B98 ◊ REV Zbl 49.146 JSL 18.273 • ID 25196

STRAZDINJA, D.P. & STRAZDINS, I.E. [1973] *The possibilities of enlarging the Harvard classification of boolean functions (Russian)* (S 0764) Teor Konech Avtom & Prilozh (Riga) 2*24-30,79
  ◊ B05 ◊ REV MR 51 # 3018 • ID 17363

STRAZDINS, I.E. [1969] *Automorphism groups of an algebra of boolean functions and Post classes (Russian)* (C 1074) Tr Algeb Sem Riga 1969 253-266
  ◊ B20 ◊ REV MR 42 # 7492 • ID 20957

STRAZDINS, I.E. [1973] *Structure of the semigroup of endomorphisms of the algebra of two-place boolean functions. I (Russian) (Latvian and English summaries)* (J 0337) Mat Ezheg, Akad Nauk Latv SSR 13*102-114
  ◊ B05 ◊ REV MR 49 # 4749 Zbl 301 # 20063 • REM Part II 1974 • ID 13225

STRAZDINS, I.E. [1973] see STRAZDINJA, D.P.

STRAZDINS, I.E. [1974] *A linear selfdual group over GF(2), and its action on the algebra of boolean functions (Russian) (English summary)* (J 0040) Kibernetika, Akad Nauk Ukr SSR 1974/5*146-147
  ◊ B05 ◊ REV MR 52 # 13379 Zbl 299 # 20032 • ID 79055

STRAZDINS, I.E. [1974] *Structure of the semigroup of endomorphisms of the algebra of two-place boolean functions. II (Russian) (Latvian and English summaries)* (J 0337) Mat Ezheg, Akad Nauk Latv SSR 14*199-211
  ◊ B05 ◊ REV MR 51 # 2905 Zbl 332 # 20019 • REM Part I 1973 • ID 17347

STRAZDINS, I.E. [1975] *Affine classification of Boolean functions of five variables (Russian)* (**J** 0474) Avtom Vychis Tekh, Akad Nauk Latv SSR 1975/1*1-9
• TRANSL [1975] (**J** 2666) Autom Control Comput Sci 9/1*1-7
⋄ B20 ⋄ REV MR 57 # 9341 Zbl 294 # 94024 • ID 65556

STRAZDINS, I.E. [1976] *Self-complementary types of boolean functions (Russian)* (**J** 0337) Mat Ezheg, Akad Nauk Latv SSR 18*190-197,223
⋄ B05 ⋄ REV MR 57 # 9342 Zbl 429 # 94037 • ID 69937

STRAZDINS, I.E. [1981] *On fundamental transformation groups in the algebra of logic* (**P** 2552) Conf Finite Algeb & Multi-Val Log;1979 Szeged 669-691
⋄ B05 B50 ⋄ REV MR 84g:03105 Zbl 478 # 03005 • ID 55620

STRAZDINS, I.E. see Vol. II for further entries

STREET, R. & WALTERS, R.F.C. [1973] *The comprehensive factorization of a functor* (**J** 0015) Bull Amer Math Soc 79*936-941
⋄ B15 G30 ⋄ REV MR 49 # 10753 Zbl 274 # 18001 • ID 32482

STREET, R. [1974] *Elementary cosmoi I* (**C** 1721) Sydney Category Th Sem, Sydney 1972/73 134-180
⋄ B15 G30 ⋄ REV MR 50 # 7290 Zbl 325 # 18005 • ID 32478

STREET, R. [1978] *A survey of topos theory* (**X** 4435) Macquarie Univ: North Ryde 60pp
⋄ B15 G30 ⋄ ID 32486

STREET, R. [1978] *Cosmoi of internal categories* (1111) Preprints, Manuscr., Techn. Reports etc. iv+87pp
⋄ B15 G30 ⋄ REV Zbl 393 # 18009 • REM Sydney Metropolitan Category Seminar Report • ID 32484

STREET, R. & WALTERS, R.F.C. [1978] *Yoneda structures on 2-categories* (**J** 0032) J Algeb 50*350-379
⋄ B15 G30 ⋄ REV MR 57 # 3214 Zbl 401 # 18004 • ID 32014

STREET, R. see Vol. V for further entries

STROEHLEIN, T. [1976] see SCHMIDT, GUNTHER

STROEHLEIN, T. see Vol. V for further entries

STROMBACH, W. [1972] see EMDE, H.

STROYAN, K.D. [1972] *Additional remarks on the theory of monads* (**P** 0732) Contrib to Non-Standard Anal;1970 Oberwolfach 245-259
⋄ H05 ⋄ REV MR 58 # 31039 Zbl 247 # 54023 • ID 13235

STROYAN, K.D. [1972] *Uniform continuity and rates of growth of meromorphic functions* (**P** 0732) Contrib to Non-Standard Anal;1970 Oberwolfach 47-64
⋄ H05 ⋄ REV MR 58 # 28550 Zbl 248 # 30038 • ID 13236

STROYAN, K.D. [1973] *A characterization of the Mackey uniformity* $m(L^\infty, L^1)$ *for finite measures* (**J** 0048) Pac J Math 49*223-228
⋄ H05 ⋄ REV MR 49 # 1090 Zbl 271 # 46023 • ID 27649

STROYAN, K.D. [1974] *A nonstandard characterization of mixed topologies* (**P** 1083) Victoria Symp Nonstand Anal;1972 Victoria 261-271
⋄ H05 ⋄ REV MR 58 # 2160 Zbl 275 # 46020 • ID 27648

STROYAN, K.D. [1974] *Infinitesimal relations on the space of bounded holomorphic functions* (**P** 1083) Victoria Symp Nonstand Anal;1972 Victoria xvi
⋄ H05 ⋄ ID 27647

STROYAN, K.D. [1977] *Infinitesimal analysis of curves and surfaces* (**C** 1523) Handb of Math Logic 197-231
⋄ H05 ⋄ REV MR 58 # 10395 JSL 49.968 • ID 24201

STROYAN, K.D. [1978] *Infinitesimal calculus on locally convex spaces I: Fundamentals* (**J** 0064) Trans Amer Math Soc 240*363-383
⋄ H05 ⋄ REV MR 58 # 12351 Zbl 389 # 46032 • REM Part II 1983 • ID 82907

STROYAN, K.D. [1983] *Infinitesimal analysis of* l^∞ *in its Mackey topology* (**C** 3884) Nonstandard Anal - Recent Develop 204-213
⋄ H05 H10 ⋄ REV MR 84m:46013 Zbl 507 # 46066 • ID 37719

STROYAN, K.D. [1983] *Locally convex infinitesimal calculus II. Computations on Mackey* (l^∞) (**J** 2752) J Fct Anal 53*1-15
⋄ H05 H10 ⋄ REV MR 85d:46099 Zbl 533 # 46048 • REM Part I 1978 • ID 39087

STRUNZ, H. [1976] see JORGENSEN, P.

STRUVE, H. [1984] *Affine Ebenen mit Orthogonalitaetsrelation* (**J** 0068) Z Math Logik Grundlagen Math 30*223-231
⋄ B30 ⋄ REV MR 85k:51005 Zbl 546 # 51005 • ID 42225

STUBERMAN, W.E. [1965] see BROWN, P.L.

STUERMANN, W.E. [1960] *Plotting boolean functions* (**J** 0005) Amer Math Mon 67*170-172
⋄ B05 ⋄ REV JSL 27.246 • ID 13238

STYAZHKIN, N.I. [1959] *Vereinfachung gewisser Algorithmen des klassischen Aussagenkalkuels durch P.S. Poretskij (Russian)* (**C** 1155) Log Issl (Moskva) 33-47
⋄ B05 ⋄ REV Zbl 106.4 • ID 48034

STYAZHKIN, N.I. [1974] *Logic with the elements of mathematical logic (Russian)* (**X** 1453) Moskov Gos Istor-Arkh Inst: Moskva 163pp
⋄ B98 ⋄ REV MR 53 # 10541 • ID 23055

SUBBOTIN, A.L. [1964] *Aristotelian syllogistics from an algebraic point of view (Russian)* (**C** 0570) Form Log & Metodol Nauk 65-79
⋄ B20 G05 ⋄ REV Zbl 125.278 • ID 14843

SUBBOTIN, A.L. see Vol. II for further entries

SUBBOTOVSKAYA, B.A. [1961] *Realizations of linear functions by formulas using* ∨,&,− *(Russian)* (**J** 0023) Dokl Akad Nauk SSSR 136*553-555
• TRANSL [1961] (**J** 0062) Sov Math, Dokl 2*110-112
⋄ B35 ⋄ REV MR 25 # 1088 Zbl 100.10 • ID 47986

SUDOLSKY, M. [1979] see FIBY, R.

SUGIHARA, T. [1960] *Modern logic (Japanese)* (**X** 1187) Sankibo: Tokyo 134pp
⋄ B98 ⋄ REV JSL 32.543 • ID 28570

SUGIHARA, T. see Vol. II for further entries

SULLIVAN, D.J. [1963] *Fundamentals of logic* (**X** 0822) McGraw-Hill: New York 288pp
⋄ B98 ⋄ ID 22540

SUMENKOV, E.A. [1979] see KARATUEV, V.G.

SUMMERS, G.J. [1968] *New puzzles in logical deduction* (X 0813) Dover: New York vi+121pp
  ◇ B98 ◇ REV Zbl 235#02001 • ID 27814

SUMMERSBEE, S. & WALTERS, A. [1962] *Programming the functions of formal logic* (J 0047) Notre Dame J Formal Log 3*133-141
  ◇ B35 ◇ REV MR 27#1374 Zbl 129.5 • REM Part I. Part II 1963 • ID 12722

SUMMERSBEE, S. & WALTERS, A. [1963] *Programming the functions of formal logic II (multi-valued logics)* (J 0047) Notre Dame J Formal Log 4*293-305
  ◇ B35 ◇ REV MR 29#4673 Zbl 128.44 JSL 35.596 • REM Part I 1962 • ID 12724

SUPPES, P. [1951] *A set of independent axioms for extensive quantities (Portuguese)* (J 0050) Port Math 10*163-172
  ◇ B28 ◇ REV MR 13.897 Zbl 44.171 • ID 12727

SUPPES, P. [1955] see MCKINSEY, J.C.C.

SUPPES, P. [1957] *Introduction to logic* (X 0864) Van Nostrand: New York xviii+312pp
  ◇ B98 ◇ REV MR 26#4875 Zbl 77.11 JSL 22.353 • ID 12728

SUPPES, P. [1958] see SCOTT, D.S.

SUPPES, P. [1959] see HENKIN, L.

SUPPES, P. [1962] see NAGEL, E.

SUPPES, P. & ZINNES, J.L. [1963] *Basic measurement theory* (C 0676) Handb of Math Psychology 1*1-76
  ◇ B30 ◇ REV JSL 36.322 • ID 12734

SUPPES, P. [1964] see HILL, S.

SUPPES, P. [1965] *Logics appropriate to empirical theories* (P 0614) Th Models;1963 Berkeley 364-375
  ◇ A05 B30 ◇ REV MR 34#7133 Zbl 156.20 JSL 35.129 • ID 12736

SUPPES, P. [1968] see MOLER, N.

SUPPES, P. [1969] *Studies in the methodology and foundations of science. Selected papers from 1951-1969* (X 0835) Reidel: Dordrecht 483pp
  ◇ A05 B96 ◇ REV MR 40#4049 Zbl 188.310 • ID 12737

SUPPES, P. [1974] *The axiomatic method in the empirical sciences* (P 0610) Tarski Symp;1971 Berkeley 465-480
  ◇ A05 A10 B30 ◇ REV MR 51#2829 Zbl 317#02069 • ID 12739

SUPPES, P. [1975] see HENKIN, L.

SUPPES, P. [1976] see GOLDBERG, ADELE

SUPPES, P. [1984] see MCDONALD, J.L.

SUPPES, P. [1984] *The next generation of interactive theorem provers* (P 2633) Autom Deduct (7);1984 Napa 303-315
  ◇ B35 ◇ REV Zbl 546#68072 • ID 43558

SUPPES, P. see Vol. II, V for further entries

SURANYI, J. [1943] *Zur Reduktion des Entscheidungsproblems des logischen Funktionskalkuels (Hungarian) (German summary)* (J 0461) Mat Fiz Lapok 50*51-74
  ◇ B20 D35 ◇ REV MR 9.129 JSL 9.22 • ID 19712

SURANYI, J. [1947] see KALMAR, L.

SURANYI, J. [1949] *Reduction of the decision problem to formulas containing a bounded number of quantifiers only* (P 0682) Int Congr Philos (10);1948 Amsterdam 759-762
  ◇ B20 D35 ◇ REV MR 11.303 Zbl 34.153 JSL 14.131 • ID 20805

SURANYI, J. [1950] *Contributions to the reduction theory of the decision problem II: Three universal, one existential quantifier* (J 0001) Acta Math Acad Sci Hung 1*261-271
  ◇ B20 D35 ◇ REV MR 13.715 Zbl 41.351 JSL 18.264 • REM Part I 1950 by Kalmar,L. Part III 1951 by Kalmar,L. • ID 12740

SURANYI, J. [1950] see KALMAR, L.

SURANYI, J. [1951] *Contributions to the reduction theory of the decision problem V: Ackermann prefix with three universal quantifiers* (J 0001) Acta Math Acad Sci Hung 2*325-335
  ◇ B20 D35 ◇ REV MR 14.344 Zbl 45.2 JSL 18.265 • REM Part IV 1951 by Kalmar,L. • ID 12741

SURANYI, J. [1955] *On the reduction theory of the decision problem of symbolic logic (Hungarian) (Russian and English summaries)* (J 0396) Mat Lapok 6*180-197
  ◇ B20 D35 ◇ REV MR 17.447 JSL 22.296 • ID 12742

SURANYI, J. [1959] *Reduktionstheorie des Entscheidungsproblems im Praedikatenkalkuel der ersten Stufe* (X 0928) Akad Kiado: Budapest 216pp
  • LAST ED (X 0806) Dt Verlag Wiss: Berlin 212pp
  ◇ B20 D35 ◇ REV MR 21#7156 Zbl 91.11 JSL 25.274 • ID 12743

SURANYI, J. [1971] *Reduction of the decision problem of the first order predicate calculus to reflexive and symmetrical binary predicates* (J 0049) Period Math Hung 1*97-106
  ◇ B20 D35 ◇ REV MR 44#2569 Zbl 242#02050 • ID 12744

SURANYI, J. see Vol. V for further entries

SURDU, A. [1971] *Aristotle's definition of the syllogism (Romanian)* (S 1613) Probl Logic (Bucharest) 4*305-325
  ◇ A10 B20 ◇ ID 47840

SURMA, S.J. [1965] *On the relation of formal inference and some related concepts (Polish) (English and Russian summaries)* (S 0458) Zesz Nauk, Prace Log, Uniw Krakow 1*37-55
  ◇ B22 ◇ REV MR 37#1228 • ID 12745

SURMA, S.J. [1966] *The concept of measurable magnitude in set-theoretical treatment (Polish) (English summary)* (S 0458) Zesz Nauk, Prace Log, Uniw Krakow 2*31-37
  ◇ B30 ◇ ID 16507

SURMA, S.J. [1966] *The termal completeness of sets of propositions (Polish) (English summary)* (S 0458) Zesz Nauk, Prace Log, Uniw Krakow 2*39-45
  ◇ B05 ◇ ID 16508

SURMA, S.J. [1967] *Indirect-deduction theorems (Polish) (Russian and English summaries)* (J 0063) Studia Logica 20*151-166
  ◇ B05 ◇ REV MR 35#1446 Zbl 307#02021 • ID 12747

SURMA, S.J. [1968] *A simplified axiom system for Tarski's consequence theory (Polish) (English summary)* (S 1454) Zesz Nauk Wyz Szk Ped Mat, Opole 4*77-86
  ◇ B22 ◇ ID 33419

SURMA, S.J. [1968] *Four studies in metamathematics (Polish) (Russian and English summaries)* (J 0063) Studia Logica 23*79-114
  ⋄ B22 C07 C35 E25 ⋄ REV MR 39 # 3965 Zbl 313 # 02028 • ID 12748

SURMA, S.J. [1968] *Investigations into the theory of deduction (Polish) (English summary)* (S 0458) Zesz Nauk, Prace Log, Uniw Krakow 3*56pp
  ⋄ B22 ⋄ ID 33817

SURMA, S.J. [1968] *Kinds of completeness of deductive theories (Polish) (English summary)* (S 0458) Zesz Nauk, Prace Log, Uniw Krakow 3*85-89
  ⋄ B22 ⋄ ID 33820

SURMA, S.J. [1968] *On metamathematical finite character properties* (S 0458) Zesz Nauk, Prace Log, Uniw Krakow 3*81-84
  ⋄ B22 ⋄ ID 33819

SURMA, S.J. [1968] *Theorems on deduction for descending implications (Polish) (English and Russian summaries)* (J 0063) Studia Logica 22*62-83
  ⋄ B20 ⋄ REV MR 38 # 4280 Zbl 313 # 02020 • ID 12749

SURMA, S.J. [1969] *On the axiomatic treatment of the theory of models I: Theory of models as an extension of Tarski's consequence theory (Polish summary)* (S 0458) Zesz Nauk, Prace Log, Uniw Krakow 4*39-45
  ⋄ B22 C07 ⋄ REV MR 41 # 1521 • REM Part II 1970 • ID 12752

SURMA, S.J. [1970] *On the axiomatic treatment of the theory of models II: Syntactical characterization of a fragment of the theory of models (Polish summary)* (S 0458) Zesz Nauk, Prace Log, Uniw Krakow 5*43-55
  ⋄ B22 C07 C75 ⋄ REV MR 45 # 1746 • REM Part I 1969. Part III 1970 • ID 79124

SURMA, S.J. [1970] *On the axiomatic treatment of the theory of models III: The function of model content (Polish summary)* (S 0458) Zesz Nauk, Prace Log, Uniw Krakow 5*57-63
  ⋄ B22 C07 ⋄ REV MR 45 # 1747 • REM Part II 1970. Part IV 1970 by Kossowski,P. • ID 12754

SURMA, S.J. [1971] *Method of natural deduction in the equivalential and equivalential-negational propositional calculus (Polish summary)* (S 0458) Zesz Nauk, Prace Log, Uniw Krakow 6*55-68
  ⋄ B20 ⋄ REV MR 45 # 3156 • ID 12756

SURMA, S.J. [1972] *A method of axiomatization of two-valued propositional connectives* (J 0387) Bull Sect Logic, Pol Acad Sci 1/4*32-37
  ⋄ B20 ⋄ REV MR 58 # 27293 • ID 79104

SURMA, S.J. [1972] *A survey of the results and methods of investigations of the equivalential propositional calculus (Polish summary)* (S 0458) Zesz Nauk, Prace Log, Uniw Krakow 7*51-75
  ⋄ B20 ⋄ REV MR 58 # 16119 • ID 16530

SURMA, S.J. [1972] *A uniform method of proof of the completeness theorem for the equivalential propositional calculus and for some of its extensions (Polish summary)* (S 0458) Zesz Nauk, Prace Log, Uniw Krakow 7*35-50
  ⋄ B20 ⋄ REV MR 54 # 9973b • ID 16529

SURMA, S.J. [1973] *A method of axiomatization of two-valued propositional connectives* (J 0302) Rep Math Logic, Krakow & Katowice 1*27-32
  ⋄ B20 ⋄ REV MR 49 # 10525 Zbl 279 # 02006 • ID 12759

SURMA, S.J. [1974] *An algorithm for axiomatizing every finite logic* (J 0302) Rep Math Logic, Krakow & Katowice 3*57-61
  ⋄ B22 ⋄ REV MR 51 # 123 Zbl 314 # 02057 • ID 15187

SURMA, S.J. [1974] *On axiomatization of finite logics* (P 1385) Int Symp Multi-Val Log (4);1974 Morgantown 315-321
  ⋄ B22 ⋄ ID 33559

SURMA, S.J. [1976] *An axiomatization of the Sheffer function (Polish) (English summary)* (J 0481) Acta Univ Wroclaw 5*51-54
  ⋄ B05 ⋄ ID 33825

SURMA, S.J. [1976] *Every reduct of two-element boolean algebra can be finitely axiomatized by modus ponens or substitution rule* (P 2011) Int Symp Multi-Val Log (6);1976 Logan 110-114
  ⋄ B20 ⋄ REV MR 58 # 16120 • ID 79108

SURMA, S.J. [1977] *A uniform method for axiomatization of finite logics* (C 2012) Comput Sci & Multi-Val Logic 107-143
  ⋄ B22 ⋄ REV MR 51 # 123 • ID 33560

SURMA, S.J. (ED.) [1977] *On Lesniewski's systems. Proceedings of XXII conference on the history of logic* (J 0063) Studia Logica 36*245-426
  ⋄ A10 B15 B97 ⋄ REV MR 57 # 15936 Zbl 372 # 00005 • ID 51405

SURMA, S.J. [1980] *On the axiomatic treatment of the Sheffer stroke and of the dual Sheffer stroke* (P 3673) Int Symp Multi-Val Log (10);1980 Evanston 279-281
  ⋄ B20 ⋄ REV MR 83m:94032 Zbl 552 # 03007 • ID 43337

SURMA, S.J. [1982] *On the axiomatic treatment of the Sheffer function* (J 3772) J Nigerian Math Soc 1*83-87
  ⋄ B05 ⋄ REV MR 84e:03014 Zbl 548 # 03003 • ID 34353

SURMA, S.J. see Vol. II, V, VI for further entries

SUSHCHANSKIJ, V.I. [1980] see GLAZUNOV, N.M.

SUSZKO, R. [1948] *Concerning logic without axioms (Russian)* (J 0479) Kwart Filoz 17*199-205
  ⋄ B20 ⋄ REV MR 10.421 JSL 15.66 • ID 28420

SUSZKO, R. [1950] *Canonic axiomatic systems* (J 4716) Stud Philos, Leopolis (Poznan) 4*301-330
  ⋄ B22 E30 ⋄ REV Zbl 45.294 JSL 17.211 • ID 38697

SUSZKO, R. [1958] see LOS, J.

SUSZKO, R. [1961] *Concerning the method of logical schemes, the notion of logical calculus and the role of consequence relations (Polish and Russian summaries)* (J 0063) Studia Logica 11*185-216
  ⋄ A05 B22 ⋄ REV MR 33 # 5450 Zbl 121.252 • ID 12769

SUSZKO, R. [1962] *A note concerning the binary quantifiers* (J 0105) Theoria (Lund) 28*269-276
  ⋄ B10 ⋄ REV JSL 36.689 • ID 12770

SUSZKO, R. [1965] *A note concerning the rules of inference for quantifiers* (J 0009) Arch Math Logik Grundlagenforsch 7*124-127
  ⋄ B10 ⋄ REV MR 34 # 5647 Zbl 166.255 • ID 12771

SUSZKO, R. [1968] see LEWANDOWSKI, A.

SUSZKO, R. [1970] see BLOOM, S.L.

SUSZKO, R. [1971] *Sentential variables versus arbitrary sentential constants (Polish summary)* (S 0458) Zesz Nauk, Prace Log, Uniw Krakow 6*85-88
⋄ A05 B05 ⋄ REV MR 50 # 66 • ID 12776

SUSZKO, R. [1972] *Equational logic and theories in sentential language* (J 0387) Bull Sect Logic, Pol Acad Sci 1/2*2-9
⋄ B05 C05 C07 ⋄ REV MR 50 # 12714 • ID 12781

SUSZKO, R. [1973] see BROWN, D.J.

SUSZKO, R. [1974] *Equational logic and theories in sentential languages* (S 0019) Colloq Math (Warsaw) 29*19-23
⋄ B05 C05 C07 ⋄ REV MR 49 # 39 Zbl 296 # 02027 • ID 12782

SUSZKO, R. & WEINFELD, T. [1979] *Filters and natural extensions of closure systems* (J 0387) Bull Sect Logic, Pol Acad Sci 8*130-132
⋄ B22 ⋄ REV MR 80m:03063 Zbl 419 # 03042 • ID 53385

SUSZKO, R. see Vol. II, III, V, VI for further entries

SUVOROV, P.YU. [1974] see DAVYDOV, G.V.

SUVOROV, P.YU. [1976] *On the recognition of the tautological nature of propositional formulas (Russian) (English summary)* (S 0228) Zap Nauch Sem Leningrad Otd Mat Inst Steklov 60*197-206,226-227
• TRANSL [1980] (J 1531) J Sov Math 14*1556-1562
⋄ B35 ⋄ REV MR 58 # 27294 Zbl 343 # 68041 • ID 65614

SUVOROV, P.YU. [1979] *Representation of proofs by colored graphs and the Hadwiger conjecture (Russian)* (S 0228) Zap Nauch Sem Leningrad Otd Mat Inst Steklov 88*209-217,246
• TRANSL [1982] (J 1531) J Sov Math 20*2376-2381
⋄ B10 F07 ⋄ REV MR 81g:05060 Zbl 494 # 05022
• ID 36649

SVENONIUS, L. [1972] *Translation and reduction* (J 0122) J Philos Logic 1*297-316
⋄ A05 B10 ⋄ REV MR 55 # 94 Zbl 259 # 02006 • ID 30340

SVENONIUS, L. [1973] *On the first-order logic of terms* (J 0036) J Symb Logic 38*177-188
⋄ B10 ⋄ REV MR 51 # 5261 Zbl 273 # 02012 • ID 12798

SVENONIUS, L. [1973] *Translation and reduction* (P 1556) Exact Philos: Probl, Tools & Goals;1971 Montreal 31-50
⋄ A05 B10 ⋄ REV Zbl 273 # 02013 • ID 30484

SVENONIUS, L. see Vol. III, IV, VI for further entries

SVOBODA, A. [1967] *Ordering of implicants* (J 4305) IEEE Trans Electr Comp EC-16*100-105
⋄ B05 ⋄ ID 12799

SVOBODA, A. [1975] *The concept of term exclusiveness and its effect on the theory of boolean functions* (J 0037) ACM J 22*425-440
⋄ B05 ⋄ REV MR 51 # 12434 Zbl 319 # 94021 • ID 65618

SWANSON, L.G. [1980] see HANSEN, R.T.

SWARD, G.L. [1960] see DUNHAM, B.

SWARD, G.L. see Vol. V, VI for further entries

SWIGGART, P. [1979] *Domain restrictions in standard deductive logic* (J 0047) Notre Dame J Formal Log 20*115-129
⋄ A05 B10 ⋄ REV MR 80f:03015 Zbl 322 # 02007
• ID 52377

SWIGGART, P. see Vol. V for further entries

SYMONDS, B.K. [1957] see CHISHOLM, R.M.

SZABO, M.E. [1983] see RICHTER, M.M.

SZABO, M.E. [1984] see FARKAS, E.J.

SZABO, M.E. [1985] see RICHTER, M.M.

SZABO, M.E. see Vol. II, VI for further entries

SZABO, P. [1978] *The undecidability of the D_A-unification problem* (X 3159) TH Karlsruhe Fak Informatik: Karlsruhe 19pp
⋄ B35 D35 ⋄ REV Zbl 401 # 68066 • ID 54642

SZABO, P. & UNVERICHT, E. [1978] *The unification problem for distributive terms* (X 3159) TH Karlsruhe Fak Informatik: Karlsruhe 27pp
⋄ B35 ⋄ REV Zbl 401 # 68067 • ID 54643

SZABO, P. [1982] see SIEKMANN, J.

SZABO, P. see Vol. IV for further entries

SZANIAWSKI, K. [1976] see MALINOWSKI, G.

SZANIAWSKI, K. see Vol. II for further entries

SZCZECH, W. [1973] *On one of Wajsberg's theorem* (J 0302) Rep Math Logic, Krakow & Katowice 1*33-37
⋄ B20 ⋄ REV MR 49 # 7114 Zbl 278 # 02012 • ID 12830

SZCZECH, W. see Vol. II, V for further entries

SZCZERBA, L.W. & TARSKI, A. [1965] *Metamathematical properties of some affine geometries* (P 0623) Int Congr Log, Meth & Phil of Sci (2,Proc);1964 Jerusalem 166-178
⋄ B25 B30 C52 C65 D35 ⋄ REV MR 35 # 843 Zbl 149.385 JSL 36.333 • ID 12831

SZCZERBA, L.W. [1969] see KORDOS, M.

SZCZERBA, L.W. [1970] *Independence of Pasch's axiom* (J 0014) Bull Acad Pol Sci, Ser Math Astron Phys 18*491-498
⋄ B30 ⋄ REV MR 42 # 8376 Zbl 201.532 • ID 39483

SZCZERBA, L.W. [1970] *On the euclidean geometry without the Pasch axiom* (J 0014) Bull Acad Pol Sci, Ser Math Astron Phys 18*659-666
⋄ B30 ⋄ REV MR 43 # 2596 Zbl 204.209 • ID 39485

SZCZERBA, L.W. [1970] *Semantic method of proving theorems (Russian summary)* (J 0014) Bull Acad Pol Sci, Ser Math Astron Phys 18*507-512
⋄ B10 ⋄ REV MR 43 # 22 Zbl 212.30 • ID 12834

SZCZERBA, L.W. [1971] *Undefinability of order in Pasch-free geometry* (J 0014) Bull Acad Pol Sci, Ser Math Astron Phys 19*315-317
⋄ B30 ⋄ REV MR 44 # 4611 Zbl 212.522 • ID 39496

SZCZERBA, L.W. [1972] *A paradoxical model of euclidean affine geometry* (J 0014) Bull Acad Pol Sci, Ser Math Astron Phys 20*845-851
⋄ B30 ⋄ REV MR 47 # 943 • ID 39507

SZCZERBA, L.W. [1972] *Automatic theorem proving* (J 0519) Wiad Mat, Ann Soc Math Pol, Ser 2 14*25-38
⋄ B35 ⋄ REV MR 56 # 13814 Zbl 253 # 68019 • ID 82922

SZCZERBA, L.W. [1977] *Interpretability of elementary theories* (P 1704) Int Congr Log, Meth & Phil of Sci (5);1975 London ON 1*129-145
⋄ B30 F25 ⋄ REV MR 57 # 16040 Zbl 375 # 02005
• ID 51583

SZCZERBA, L.W. [1979] *Interpretability and axiomatizability* (J 3293) Bull Acad Pol Sci, Ser Math 27*425-429
⋄ B30 F25 ⋄ REV MR 81f:03017 Zbl 445 # 03013
• ID 56477

SZCZERBA, L.W. & TARSKI, A. [1979] *Metamathematical discussion of some affine geometries* (J 0027) Fund Math 104*155-192
⋄ B25 B30 C35 C52 C65 D35 ⋄ REV MR 81a:03012 Zbl 497 # 03008 • ID 79177

SZCZERBA, L.W. [1979] see PRESTEL, A.

SZCZERBA, L.W. [1983] *Interpretations* (J 0387) Bull Sect Logic, Pol Acad Sci 12*208-213
⋄ B30 C07 F25 ⋄ ID 39510

SZCZERBA, L.W. see Vol. III, IV, VI for further entries

SZMIELEW, W. [1959] *Absolute calculus of segments and its metamathematical implications* (J 0014) Bull Acad Pol Sci, Ser Math Astron Phys 7*213-220
⋄ B30 ⋄ REV MR 21 # 7464 Zbl 95.152 JSL 27.238
• ID 12852

SZMIELEW, W. [1959] *Some metamathematical problems concerning elementary hyperbolic geometry* (P 0651) Axiomatic Method;1957 Berkeley 30-52
⋄ B30 ⋄ REV MR 21 # 7463 Zbl 92.386 JSL 27.237
• ID 12851

SZMIELEW, W. [1962] *New foundations of absolute geometry* (P 0612) Int Congr Log, Meth & Phil of Sci (1,Proc);1960 Stanford 168-175
⋄ B30 C65 ⋄ REV MR 28 # 1517 Zbl 135.207 JSL 37.201
• ID 12850

SZMIELEW, W. [1974] *The order and the semi-order of n-dimensional euclidean space in the axiomatic and model-theoretic aspects* (P 0777) Grundl Geom & Algeb Meth;1973 Potsdam 69-79
⋄ B30 C65 ⋄ REV MR 53 # 112 Zbl 291 # 50003
• ID 16578

SZMIELEW, W. [1983] see SCHWABHAEUSER, W.

SZMIELEW, W. see Vol. III, IV, V, VI for further entries

SZUKSZTA, W. [1965] *Die Definition benachbarter Konstituenten boolescher Funktionen mehrerer Veraenderlicher (Polish) (Russian and English summaries)* (S 1655) Zesz Nauk, Mat, Politech Gdansk 66*3-12
⋄ B05 ⋄ REV Zbl 257 # 94022 • ID 65635

SZYNDLER, J. [1970] *A note on a certain property of Hilbert's implicative sentential calculus (Polish) (English summary)* (S 1454) Zesz Nauk Wyz Szk Ped Mat, Opole 10*51-59
⋄ B20 ⋄ REV MR 53 # 10544 • ID 23059

TABAKOV, M. [1979] *A formal system without well-formed formulas* (J 0387) Bull Sect Logic, Pol Acad Sci 8*27-29
⋄ B20 ⋄ REV MR 80d:03010 Zbl 429 # 03006 • ID 53837

TABAKOV, M. see Vol. VI for further entries

TABATA, H. [1969] *Free structures and universal Horn sentences* (J 0352) Math Jap 14*101-104
⋄ B20 C05 C52 ⋄ REV MR 44 # 2596 Zbl 199.327
• ID 13247

TABATA, H. [1971] *A generalized free structure and several properties of universal Horn classes* (J 0352) Math Jap 16*91-102
⋄ B20 C05 C52 ⋄ REV MR 46 # 120 Zbl 309 # 08003
• ID 13248

TABATA, H. see Vol. III for further entries

TACON, D.G. [1972] *Weak compactness in normed linear spaces* (J 0038) J Austral Math Soc 14*9-16
⋄ H05 ⋄ REV MR 47 # 3954 Zbl 238 # 46011 • ID 13249

TACON, D.G. [1973] *Weak compactness in locally convex spaces* (J 0064) Trans Amer Math Soc 180*463-474
⋄ H05 ⋄ REV Zbl 274 # 46001 • ID 48154

TACON, D.G. [1979] *Two characterizations of power compact operators* (J 0053) Proc Amer Math Soc 73*356-360
⋄ H05 ⋄ REV MR 80d:47031 Zbl 418 # 47007 • ID 53333

TACON, D.G. [1980] *Nonstandard extensions of transformations between Banach spaces* (J 0064) Trans Amer Math Soc 260*147-158
⋄ H05 ⋄ REV MR 81e:47021 Zbl 452 # 47021 • ID 82929

TACON, D.G. [1983] *Generalized semi-Fredholm transformations* (J 3194) J Austral Math Soc, Ser A 34*60-70
⋄ H05 ⋄ REV MR 84d:47015 Zbl 531 # 47011 • ID 38569

TAERNLUND, S.-A. [1979] see BROWN, F.M.

TAERNLUND, S.-A. see Vol. IV for further entries

TAIT, W.W. [1961] see KREISEL, G.

TAIT, W.W. [1966] *A nonconstructive proof of Gentzen's Hauptsatz for second order predicate logic* (J 0015) Bull Amer Math Soc 72*980-983
⋄ B15 F05 F35 ⋄ REV MR 34 # 5655 Zbl 199.8 JSL 33.289 • ID 13283

TAIT, W.W. see Vol. III, IV, V, VI for further entries

TAJMANOV, A.D. [1966] *On formulae of Horn-type (Russian)* (J 0216) Izv Akad Nauk SSSR, Ser Mat 30*523-524
⋄ B20 D35 ⋄ REV MR 34 # 1160 Zbl 156.251 • ID 19765

TAJMANOV, A.D. [1967] *Systems with a solvable universal theory (Russian)* (J 0003) Algebra i Logika 6/5*33-43
⋄ B20 B25 C20 ⋄ REV MR 37 # 6173 Zbl 165.318
• ID 13274

TAJMANOV, A.D. [1979] see SIMANOV, A.L.

TAJMANOV, A.D. see Vol. III, IV, V for further entries

TAJTELBAUM, A. [1923] *On the primitive term of logistic (Polish)* (J 1125) Przeglad Filoz 26*68-89
• TRANSL [1956] (C 1159) Tarski: Logic, Semantics, Metamathematics 1-23
⋄ B15 B20 ⋄ REV JSL 34.99 • ID 30912

TAJTELBAUM, A. [1923] *Sur le terme primitif de la logistique* (J 0027) Fund Math 4*196-200
⋄ A05 B05 ⋄ ID 13393

TAJTELBAUM, A. [1924] *Sur les truth-functions au sens de MM. Russell et Whitehead* (J 0027) Fund Math 5*59-74
⋄ A05 B05 ⋄ REV JSL 34.99 • ID 13396

TAJTELBAUM, A. also published under the name TARSKI, A.

TAJTELBAUM, A. see Vol. V for further entries

TAJTSLIN, M.A. [1970] *Model theory (Russian)* (X 0913) Novosibirsk Gos Univ: Novosibirsk 214pp
⋄ B98 C98 ⋄ REV MR 49#2350 • ID 28675

TAJTSLIN, M.A. [1973] see ERSHOV, YU.L.

TAJTSLIN, M.A. [1979] *A description of algebraic systems in weak logic of order ω and in program logic (Russian)* (C 2065) Teor Nereg Kriv Raz Geom Post 91-98
⋄ B15 B75 C35 C85 ⋄ REV MR 82e:68031 • ID 82932

TAJTSLIN, M.A. see Vol. II, III, IV for further entries

TAKAGI, T. [1945] *Zur Axiomatik der ganzen und der reellen Zahlen* (J 0081) Proc Japan Acad 21*111-113
⋄ B28 ⋄ REV MR 11.151 Zbl 61.91 • ID 13287

TAKAGI, T. see Vol. II for further entries

TAKAHASHI, MOTO-O [1968] *On simple type theory (Japanese)* (J 0091) Sugaku 20*129-141
⋄ B15 F05 F35 ⋄ REV MR 40#2535 • ID 13295

TAKAHASHI, MOTO-O see Vol. II, III, IV, V, VI for further entries

TAKAHASHI, S. [1969] *Analysis in categories* (X 0997) Queen's Univ: Kingston 131pp
⋄ G30 H05 ⋄ REV MR 42#7747 Zbl 213.296 • ID 13301

TAKAHASHI, S. [1974] *Methodes logiques en geometrie diophantienne* (X 0893) Pr Univ Montreal: Montreal 48*178pp
⋄ B28 C98 D98 G30 H98 ⋄ REV MR 51#424 Zbl 325#02038 • ID 17488

TAKAHASHI, S. see Vol. III, IV for further entries

TAKAHASI, H. [1963] see GOTO, E.

TAKAMATSU, T. [1977] *Computer experiment on proof of chemical reaction* (J 1725) Bull Daito-Bunka Univ 15*103-110
⋄ B35 ⋄ ID 32494

TAKAMATSU, T. see Vol. II, VI for further entries

TAKEUCHI, Y. [1976] *Representation of nonstandard numbers by means of hermitian operators (Spanish) (English summary)* (J 0307) Rev Colomb Mat 10*125-140
⋄ H05 ⋄ REV MR 58#21588 Zbl 354#26010 • ID 79206

TAKEUCHI, Y. [1977] *Construction of nonstandard numbers from the rationals (Spanish)* (J 0377) Bol Mat (Bogota) 11*185-207
⋄ C20 H05 ⋄ REV MR 80h:03096 Zbl 494#03053 • ID 79205

TAKEUCHI, Y. [1978] *Representation of non-standard numbers by asymptotic behaviour of real functions (Spanish)* (J 0377) Bol Mat (Bogota) 12*17-37
⋄ H05 ⋄ REV MR 83c:03060 Zbl 435#03051 • ID 55811

TAKEUCHI, Y. [1983] *Nonstandard functions and distribution theory (Spanish) (English summary)* (J 0307) Rev Colomb Mat 17*117-151
⋄ H10 ⋄ REV Zbl 539#46051 • ID 45195

TAKEUCHI, Y. see Vol. V for further entries

TAKEUTI, G. [1953] *On a generalized logic calculus* (J 2307) Japan J Math 23*39-96 • ERR/ADD ibid 24*149-156
⋄ B15 F35 ⋄ REV MR 17.701 Zbl 53.202 JSL 22.351
• ID 37768

TAKEUTI, G. [1955] *On the fundamental conjecture of GLC I,II* (J 0090) J Math Soc Japan 7*249-275,394-408
⋄ B15 F05 F35 ⋄ REV MR 18.1 Zbl 86.7 JSL 24.62
• REM Parts III,IV 1956 • ID 28446

TAKEUTI, G. [1956] *A metamathematical theorem on functions* (J 0090) J Math Soc Japan 8*65-78
⋄ B15 F05 F10 F35 ⋄ REV Zbl 86.7 JSL 24.65
• ID 13308

TAKEUTI, G. [1956] *Construction of ramified real numbers* (J 0260) Ann Jap Ass Phil Sci 1*41-61
⋄ B15 F05 F35 F65 ⋄ REV MR 18.271 Zbl 75.233 JSL 22.352 • ID 13307

TAKEUTI, G. [1956] *On the fundamental conjecture of GLC III,IV* (J 0090) J Math Soc Japan 8*54-64,145-155
⋄ B15 F05 F35 ⋄ REV MR 18.1 Zbl 75.233 JSL 24.62
• REM Parts I,II 1955. Part V 1958 • ID 28447

TAKEUTI, G. [1957] *On the theory of ordinal numbers* (J 0090) J Math Soc Japan 9*93-113 • ERR/ADD ibid 12*127
⋄ B28 E10 E30 E35 E45 F05 ⋄ REV MR 19.237 Zbl 84.8 JSL 24.67 • REM Part I. Part II 1958 • ID 19761

TAKEUTI, G. [1958] *On the fundamental conjecture of GLC V* (J 0090) J Math Soc Japan 10*121-134
⋄ B15 F05 F35 ⋄ REV MR 20#6348 Zbl 75.233 JSL 24.62 • REM Parts III,IV 1956. Part VI 1961 • ID 28448

TAKEUTI, G. [1958] *Remark on the fundamental conjecture of GLC* (J 0090) J Math Soc Japan 10*44-45
⋄ B15 F05 F35 ⋄ REV MR 20#6347 Zbl 75.234
• ID 13312

TAKEUTI, G. [1960] *An example on the fundamental conjecture of GLC* (J 0090) J Math Soc Japan 12*238-242
⋄ B15 F05 F35 ⋄ REV MR 24#A3056 Zbl 75.234 JSL 28.173 • ID 13316

TAKEUTI, G. [1961] *On the fundamental conjecture of GLC VI* (J 0081) Proc Japan Acad 37*440-443
⋄ B15 F05 F35 ⋄ REV MR 25#1995 Zbl 106.237 JSL 29.147 • REM Part V 1958 • ID 13321

TAKEUTI, G. [1961] *On the inductive definition with quantifiers of second order* (J 0090) J Math Soc Japan 13*333-341
⋄ B15 F15 F35 ⋄ REV MR 26#1269 Zbl 109.6 JSL 29.147 • ID 13320

TAKEUTI, G. [1962] *Dirac space* (J 0081) Proc Japan Acad 38*414-418
⋄ H10 ⋄ REV MR 26#6751 Zbl 129.415 • ID 13326

TAKEUTI, G. [1968] *Formalization principle* (P 0627) Int Congr Log, Meth & Phil of Sci (3,Proc);1967 Amsterdam 105-118
⋄ B15 E45 E55 E70 ⋄ REV MR 40#5441 Zbl 182.328
• ID 13334

TAKEUTI, G. [1971] see MAEHARA, S.

TAKEUTI, G. [1977] see HENSON, C.W.

TAKEUTI, G. [1977] see GREENE, C.

TAKEUTI, G. [1978] *Two applications of logic to mathematics* (X 3552) Iwanami Shoten: Tokyo viii+137pp
• LAST ED [1978] (X 0857) Princeton Univ Pr: Princeton viii+137pp
⋄ B98 C90 E40 E75 F05 F30 F35 G12 ⋄ REV MR 58#21591 Zbl 393#03027 • ID 52447

TAKEUTI, G. [1981] *Logic and set theory* (C 2617) Modern Log Survey 167-171
⋄ B51 B98 E40 E70 F50 ⋄ REV MR 82f:03002 Zbl 464 # 03001 • ID 39565

TAKEUTI, G. [1981] see MUELLER, GERT H.

TAKEUTI, G. see Vol. II, III, IV, V, VI for further entries

TAL', A.A. [1963] see AJZERMAN, M.A.

TAL', A.A. see Vol. IV for further entries

TALANOV, V.A. [1969] see GLEBSKIJ, YU.V.

TALANOV, V.A. see Vol. III for further entries

TALL, D. [1979] *The calculus of Leibniz - an alternative modern approach* (J 2789) Math Intell 2*54-55
⋄ A05 H05 ⋄ REV MR 82f:26022 Zbl 499 # 26009 • ID 82943

TALL, D. [1980] *Looking at graphs through infinitesimal microscopes, windows and telescopes* (J 0148) Math Gaz 64*22-49
⋄ A10 H05 ⋄ REV MR 81g:03083 Zbl 499 # 26008 • ID 79226

TALL, D. [1982] *Elementary axioms and pictures for infinitesimal calculus* (J 2679) Bull Inst Math Appl (Southend oS) 18*43-48
⋄ H05 ⋄ REV MR 84j:03139c Zbl 519 # 26010 • ID 34731

TAMAKI, I. [1974] *Syllogistic and calculus of classes* (J 0079) Logique & Anal, NS 17*191-196
⋄ B20 ⋄ REV MR 51 # 2860 Zbl 294 # 02001 • ID 17393

TAMARI, D. [1954] *Some mutual applications of logic and mathematics* (P 0646) Appl Sci de Log Math;1952 Paris 89-90
⋄ B80 ⋄ REV MR 16.555 Zbl 56.245 • ID 13345

TAMARI, D. [1974] *Formulae for well formed formulae and their enumeration* (J 0038) J Austral Math Soc 17*154-162
⋄ B03 ⋄ REV MR 50 # 131 Zbl 286 # 08007 • ID 13348

TAMARI, D. see Vol. IV, V for further entries

TAMAS, G. (ED.) [1983] *Studien zur Logik* (X 0928) Akad Kiado: Budapest 345pp
⋄ B98 ⋄ REV MR 85j:01005a • ID 46783

TAMNY, M. [1972] see GUTTENPLAN, S.D.

TAMURA, S. [1976] *On NBN-systems* (J 2751) J Fac Lib Art Yamaguchi Univ 10*205-209
⋄ B20 ⋄ REV MR 84e:03018a • ID 34356

TAMURA, S. [1977] *The separation theorem of NBN-system* (J 2751) J Fac Lib Art Yamaguchi Univ 11*123-127
⋄ B20 ⋄ REV MR 84e:03018b • ID 34357

TAMURA, S. see Vol. II, III, IV for further entries

TANAKA, H. [1965] *On an arithmetical set of real numbers* (J 0014) Bull Acad Pol Sci, Ser Math Astron Phys 13*519-520
⋄ B15 ⋄ REV MR 33 # 3926 Zbl 166.5 • ID 13359

TANAKA, H. see Vol. II, III, IV, V, VI for further entries

TANAKA, K. [1965] *On axiom systems of propositional calculi. XI* (J 0081) Proc Japan Acad 41*898-900
⋄ B05 ⋄ REV MR 33 # 3899 Zbl 156.248 JSL 34.122 • REM Part X 1965 by Tanaka,S. & Iseki,K. Part XII 1965 by Arai,Y. • ID 13369

TANAKA, K. [1977] see KITAHASHI, T.

TANAKA, K. see Vol. II, IV, V for further entries

TANAKA, S. [1965] *On axiom systems of propositional calculi. VI,VIII,IX,XIII* (J 0081) Proc Japan Acad 41*122-124,795-797,798-800,904-907
⋄ B20 ⋄ REV MR 33 # 3896 MR 33 # 3897 MR 33 # 3901 MR 33 # 4555 Zbl 156.248 Zbl 156.249 JSL 34.122 • REM Parts V,X 1965 by Iseki,K. & Tanaka,S. Part VII 1965 by Arai,Y. & Iseki,K. Part XII 1965 by Arai,Y. Part XIV 1966 by Imai,Y. & Iseki,K. • ID 13372

TANAKA, S. [1965] see ISEKI, K.

TANAKA, S. [1966] *Axiom systems of B-algebra. VI* (J 0081) Proc Japan Acad 42*448-451
⋄ B05 G05 ⋄ REV MR 34 # 4110 Zbl 156.6 • REM Part V 1966 by Tanaka,K. • ID 13378

TANAKA, S. [1966] *On axiom systems of propositional calculi. XVI,XVIII,XX,XXII* (J 0081) Proc Japan Acad 42*221-224,355-357,361-363,873-874
⋄ B20 ⋄ REV MR 34 # 1159 MR 34 # 2431b,d MR 35 # 1445a Zbl 156.248 • REM Parts XV,XXI,XXI,XXIII 1966 by Iseki,K. Part XVII 1966 by Arai,Y. Part XIX 1966 by Tanaka,S. & Arai ,Y. • ID 13377

TANAKA, S. [1966] see ARAI, Y.

TANAKA, S. [1966] *On the propositional calculus with a variable functor $C\delta pC\delta Np\delta q$* (J 0081) Proc Japan Acad 42*1161-1163
⋄ B05 ⋄ REV MR 35 # 2722 Zbl 192.27 • ID 19758

TANAKA, S. [1966] *On variants of axiom systems of propositional calculus I,II* (J 0081) Proc Japan Acad 42*108-110,452-456
⋄ B20 ⋄ REV MR 34 # 2432c MR 34 # 2435 Zbl 192.27 • ID 13375

TANAKA, S. [1967] *Axiom systems of aristotle traditional logic. II* (J 0081) Proc Japan Acad 43*194-197
⋄ B20 ⋄ REV Zbl 164.306 • REM Part I 1967 by Iseki,K. Part III 1968 by Seto,Y. • ID 48447

TANAKA, S. [1967] *On axiom systems of propositional calculi. XXV* (J 0081) Proc Japan Acad 43*192-193
⋄ B20 ⋄ REV MR 35 # 6520 Zbl 156.249 • REM Part XXIV 1966 by Iseki,K. • ID 13379

TANAKA, S. [1968] *An algebraic formulation of $K-N$ propositional calculus. IV* (J 0081) Proc Japan Acad 44*786-787
⋄ B20 ⋄ REV MR 38 # 5586 • REM Part III 1966 • ID 48092

TANAKA, S. [1968] *Axiom systems of Aristotle traditional logic. IV* (J 0081) Proc Japan Acad 44*340-341
⋄ A10 B20 ⋄ REV MR 37 # 2573 Zbl 169.296 • REM Part III 1968 by Seto,Y. • ID 13380

TANAKA, S. [1968] *On proofs of some axioms with Sheffer functor 'D'* (J 0081) Proc Japan Acad 44*1018-1023
⋄ B05 ⋄ REV MR 38 # 4276 Zbl 182.314 • ID 48090

TANAKA, S. [1969] *On the proposition $C\delta CpqC\delta p\delta q$ with a variable functor* (J 0081) Proc Japan Acad 45*95-96
⋄ B20 ⋄ REV MR 39 # 3962 Zbl 177.7 • ID 13383

TANAKA, S. [1985] *A new axiom system of propositional calculus* (J 0352) Math Jap 30*971-979
⋄ B05 ⋄ ID 49026

TANAKA, S. see Vol. II, V for further entries

TANG, TONGGAO [1981] *A note on Herbrand's theorem (Chinese) (English summary)* (J 2754) Huazhong Gongxueyuan Xuebao 9/5*1-4
⋄ B10 F05 ⋄ REV MR 84d:03017 • ID 34057

TANG, TONGGAO see Vol. II, IV for further entries

TANG, ZHISONG [1960] see LU, YANGCI

TANG, ZHISONG see Vol. IV for further entries

TAPIA, M. [1972] see BOLLMAN, D.A.

TAPSCOTT, B.L. [1976] *Elementary applied symbolic logic* (X 0819) Prentice Hall: Englewood Cliffs xii+496pp
⋄ B98 ⋄ REV JSL 44.281 • ID 44589

TAPSCOTT, B.L. see Vol. II for further entries

TARASOV, V.V. [1975] *The completeness problem for systems of functions of the algebra of logic with unreliable realization (Russian)* (J 0142) Mat Sb, Akad Nauk SSSR, NS 98(140)*378-394,495
• TRANSL [1975] (J 0349) Math of USSR, Sbor 27*339-354
⋄ B05 ⋄ REV MR 55 # 12340 Zbl 322 # 94021 • ID 27726

TARASOV, V.V. see Vol. II for further entries

TARJAN, R.E. [1978] *Complexity of monotone networks for computing conjunctions* (S 3358) Ann Discrete Math 2*121-133
⋄ B05 B70 D15 ⋄ REV MR 81f:68058 Zbl 383 # 94038
• ID 69946

TARJAN, R.E. [1979] see ASPVALL, B.

TARJAN, R.E. see Vol. IV for further entries

TARSI, M. [1985] see LINIAL, N.

TARSKI, A. [1924] *Sur les ensembles finis* (J 0027) Fund Math 6*45-95
⋄ B28 E10 E20 E25 ⋄ REV FdM 50.135 • ID 13394

TARSKI, A. [1925] *Une remarque concernant les principes d'arithmetique theorique* (J 0283) Ann Soc Pol Math 3*150
⋄ B28 ⋄ REV FdM 51.55 • ID 30914

TARSKI, A. [1927] see LINDENBAUM, A.

TARSKI, A. [1929] *Les fondements de la geometrie des corps* (J 0283) Ann Soc Pol Math Appendix*29-33
• TRANSL [1956] (C 1159) Tarski: Logic, Semantics, Metamathematics 24-29
⋄ B30 G05 ⋄ REV JSL 34.99 • ID 30915

TARSKI, A. [1929] *Remarques sur les notions fondamentales de la methodologie des mathematiques* (J 0283) Ann Soc Pol Math 7*270-272
⋄ A05 B22 ⋄ REV FdM 55.38 • ID 13399

TARSKI, A. [1930] *Fundamentale Begriffe der Methodologie der deduktiven Wissenschaften I* (J 0124) Monatsh Math-Phys 37*361-404
• TRANSL [1956] (C 1159) Tarski: Logic, Semantics, Metamathematics 60-109
⋄ A05 B15 B22 ⋄ REV JSL 34.99 FdM 56.46 • ID 13404

TARSKI, A. [1930] *Ueber einige fundamentale Begriffe der Metamathematik* (J 0459) C R Soc Sci Lett Varsovie Cl 3 23*22-29
• TRANSL [1956] (C 1159) Tarski: Logic, Semantics, Metamathematics 30-37
⋄ A05 B22 ⋄ REV JSL 34.99 FdM 57.1318 • ID 13403

TARSKI, A. [1930] see LUKASIEWICZ, J.

TARSKI, A. [1930] *Untersuchungen ueber den Aussagenkalkuel* (J 1124) Ergebn Math Kolloquium 2*13-14
• REPR [1931] (J 0124) Monatsh Math-Phys 38*24-25
⋄ A05 B05 ⋄ REV FdM 57.1319 • ID 20991

TARSKI, A. [1931] see KURATOWSKI, K.

TARSKI, A. [1931] *Sur les ensembles definissables de nombres reels I* (J 0027) Fund Math 17*210-239
• TRANSL [1956] (C 1159) Tarski: Logic, Semantics, Metamathematics 110-142 (English)
⋄ B25 B28 C10 C40 C60 C65 D55 E15 E47 F35 ⋄
REV Zbl 75.4 JSL 34.99 FdM 57.60 • ID 16924

TARSKI, A. [1932] *Der Wahrheitsbegriff in den Sprachen der deduktiven Disziplinen* (J 0931) Anz Oesterr Akad Wiss, Math-Nat Kl 69*23-25
⋄ A05 B30 C07 C40 F30 ⋄ REV Zbl 4.1 FdM 58.997
• REM Summary. The full article appeared in 1933 • ID 13406

TARSKI, A. [1933] *Einige Betrachtungen ueber die Begriffe der ω-Widerspruchsfreiheit und der ω-Vollstaendigkeit* (J 0124) Monatsh Math-Phys 40*97-112
• TRANSL [1956] (C 1159) Tarski: Logic, Semantics, Metamathematics 279-295 (English)
⋄ B28 C62 F30 ⋄ REV MR 17.1171 Zbl 7.97 JSL 34.99
FdM 59.53 • ID 13407

TARSKI, A. [1933] *On the notion of truth in reference to formalized deductive sciences (Polish)* (J 0459) C R Soc Sci Lett Varsovie Cl 3 34*vii+116pp
• TRANSL [1935] (J 4716) Stud Philos, Leopolis (Poznan) 1*261-405 (German) [1956] (C 1159) Tarski: Logic, Semantics, Metamathematics 152-278 (English) [1961] (C 0769) L'Antinom Ment nel Pensiero Contemp 391-677 (Italian)
⋄ A05 B15 B30 C07 C40 F30 ⋄ REV Zbl 13.289
Zbl 4.1 Zbl 75.7 JSL 34.99 FdM 58.997 FdM 62.1051
• REM Polish summary in J1093, 1930/31, 210-211; German summary in J0931, 1932 • ID 28816

TARSKI, A. [1934] *Some methodological investigations on the definability of terms (Polish)* (J 1125) Przeglad Filoz 37*438-460
• TRANSL [1956] (C 1159) Tarski: Logic, Semantics, Metamathematics 296-319 [1935] (J 0748) Erkenntnis (Leipzig) 5*80-100
⋄ B10 C35 C40 ⋄ REV Zbl 12.1 JSL 34.99 FdM 61.50
• ID 30917

TARSKI, A. [1935] *Grundzuege des Systemkalkuels. Teil I* (J 0027) Fund Math 25*503-526
• TRANSL [1956] (C 1159) Tarski: Logic, Semantics, Metamathematics 342-364
⋄ B22 C07 ⋄ REV Zbl 12.385 JSL 34.100 FdM 62.38
• REM Part II 1936 • ID 13409

TARSKI, A. [1935] see LINDENBAUM, A.

TARSKI, A. [1935] *Ueber die Erweiterungen der unvollstaendigen Systeme des Aussagenkalkuels* (J 1124) Ergebn Math Kolloquium 7*51-57
• TRANSL [1956] (C 1159) Tarski: Logic, Semantics, Metamathematics 393-400
⋄ B05 B22 ⋄ REV Zbl 14.386 JSL 1.116 JSL 34.100
FdM 62.39 • ID 30922

TARSKI, A. [1936] *Grundzuege des Systemkalkuels. Teil II* (J 0027) Fund Math 26*283-301
• TRANSL [1956] (C 1159) Tarski: Logic, Semantics, Metamathematics 364-383
⋄ A05 B22 C07 C35 G15 ⋄ REV Zbl 14.387 JSL 1.71 FdM 62.38 • REM Part I 1935 • ID 13415

TARSKI, A. [1936] *On mathematical logic and the deductive method (Polish)* (X 1764) Ksiaznica Atlas: Warsaw 167pp
• TRANSL [1960] (X 0834) Gauthier-Villars: Paris xv+224pp [1969] (X 1056) Valention Bompiani: Milano
⋄ B98 ⋄ REV MR 40 # 5410 Zbl 171.252 JSL 30.236 FdM 62.39 • REM For futher transl. see also 1937 • ID 32714

TARSKI, A. [1936] *The establishment of scientific semantics (Polish)* (J 1125) Przeglad Filoz 39*50-57
• TRANSL [1936] (P 0632) Congr Int Phil des Sci;1935 Paris 3*1-8 (German) [1956] (C 1159) Tarski: Logic, Semantics, Metamathematics 401-408 (English)
⋄ A05 B10 B30 C07 ⋄ REV MR 17.1171 JSL 2.83 JSL 34.100 FdM 62.1065 • ID 13413

TARSKI, A. [1937] *Einfuehrung in die mathematische Logik und in die Methodologie der Mathematik* (X 0902) Springer: Wien x+166pp
• TRANSL [1941] (X 0894) Oxford Univ Pr: Oxford xviii+239pp (English) 3rd ed 1965 [1948] (X 2027) Nauka: Moskva 322pp (Russian) [1951] (X 1042) Espasa-Calpe: Madrid (Spanish) 2nd ed 1968 [1953] (X 0809) North Holland: Amsterdam (Dutch) 2nd ed 1964 [1956] (X 0929) Weizmann Science Pr Israel: Jerusalem (Hebrew) 2nd ed 1966 [1966] (X 1226) Academia: Prague (Czech) 2nd ed 1969 [1971] (X 1052) Tbilisi Univ: Tbilisi (Georgian) [1980] (X 4569) Commercial Pr: Beijing (Chinese) 2nd ed. • LAST ED [1970] (X 0903) Vandenhoeck & Ruprecht: Goettingen 252pp
⋄ A05 B30 B98 ⋄ REV MR 2.209 MR 40 # 5409 MR 55 # 7716 Zbl 18.1 Zbl 25.4 Zbl 352 # 02001 JSL 12.61 JSL 3.51 JSL 31.647 FdM 63.22 • REM For the Polish original and further transl. see 1936 • ID 19757

TARSKI, A. [1937] *Ideale in den Mengenkoerpern* (J 0283) Ann Soc Pol Math 15*186-189
⋄ B22 G05 ⋄ REV JSL 3.47 • ID 13418

TARSKI, A. [1938] *Der Aussagenkalkuel und die Topologie* (J 0027) Fund Math 31*103-134
• TRANSL [1956] (C 1159) Tarski: Logic, Semantics, Metamathematics 421-454
⋄ B05 F50 G05 G10 ⋄ REV Zbl 20.337 JSL 34.100 JSL 4.26 FdM 64.928 • ID 13420

TARSKI, A. [1938] *Ein Beitrag zur Axiomatik der abelschen Gruppen* (J 0027) Fund Math 30*253-256
⋄ B30 C60 ⋄ REV Zbl 19.52 FdM 64.53 • ID 30925

TARSKI, A. [1939] *On undecidable statements in enlarged systems of logic and the concept of truth* (J 0036) J Symb Logic 4*105-112
⋄ B10 F30 ⋄ REV MR 1.34 Zbl 21.385 JSL 5.115 FdM 65.20 • ID 13428

TARSKI, A. [1941] *On the calculus of relations* (J 0036) J Symb Logic 6*73-89
⋄ B20 E07 G15 ⋄ REV MR 3.130 Zbl 26.244 • ID 13430

TARSKI, A. [1944] *The semantic conception of truth and the foundations of semantics* (J 0075) Phil Phenom Research 4*341-376
• TRANSL [1955] (J 0290) Euclides 30*1-43 (Dutch) [1963] (X 1364) Vita e Pensiero: Milano 43pp
⋄ A05 B10 B30 C07 ⋄ REV MR 16.438 MR 6.31 Zbl 61.8 Zbl 64.4 JSL 9.68 • ID 13432

TARSKI, A. [1948] *A decision method for elementary algebra and geometry* (X 4282) Rand Cor: Santa Monica iii+60pp
• LAST ED [1951] (X 0926) Univ Calif Pr: Berkeley iii+63pp
⋄ B25 B30 C10 C35 C60 ⋄ REV MR 10.499 MR 13.423 Zbl 44.251 JSL 14.188 JSL 17.207 • ID 19756

TARSKI, A. [1948] *A problem concerning the notion of definability* (J 0036) J Symb Logic 13*107-111
⋄ B10 B15 C40 ⋄ REV MR 10.176 Zbl 29.242 JSL 13.172 • ID 13437

TARSKI, A. [1948] *Axiomatic and algebraic aspects of two theorems on sums of cardinals* (J 0027) Fund Math 35*79-104
⋄ B28 C15 E10 E25 G05 ⋄ REV MR 10.687 Zbl 31.289 JSL 14.257 • ID 13436

TARSKI, A. [1956] *A general theorem concerning primitive notions of Euclidean geometry* (J 0028) Indag Math 18*468-474
⋄ B30 ⋄ REV MR 18.328 Zbl 72.155 JSL 33.289
• ID 13449

TARSKI, A. [1956] see BETH, E.W.

TARSKI, A. [1956] *Logic, semantics, metamathematics. Papers from 1923 to 1938* (X 0815) Clarendon Pr: Oxford xiv+471pp
• LAST ED [1983] (X 2725) Hackett Publ: Indianapolis xxx+506pp
⋄ B96 C96 ⋄ REV MR 17.1171 MR 85e:01065 Zbl 75.7 JSL 34.99 • REM See also 1972 • ID 28568

TARSKI, A. [1957] see MONTAGUE, R.

TARSKI, A. [1959] see HENKIN, L.

TARSKI, A. [1959] *What is elementary geometry?* (P 0651) Axiomatic Method;1957 Berkeley 16-29
• TRANSL [1958] (J 3127) Bol Soc Mat Mexicana, Ser 2 3*41-51 • REPR [1969] (C 0569) Phil of Math Oxford Readings 164-175
⋄ B30 ⋄ REV MR 21 # 2575 MR 21 # 4919 Zbl 92.385 JSL 27.93 • ID 21118

TARSKI, A. [1962] see NAGEL, E.

TARSKI, A. [1965] *A simplified formalization of predicate logic with identity* (J 0009) Arch Math Logik Grundlagenforsch 7*61-79
⋄ B10 ⋄ REV MR 34 # 2437 Zbl 166.1 JSL 39.602
• ID 13462

TARSKI, A. [1965] see SZCZERBA, L.W.

TARSKI, A. [1967] *The completeness of elementary algebra and geometry* (X 0999) CNRS Inst B Pascal: Paris iv+50pp
⋄ B25 B30 C10 C35 C60 C65 ⋄ REV JSL 34.302
• ID 13463

TARSKI, A. [1971] see HENKIN, L.

TARSKI, A. [1972] *Logique, semantique, metamathematique 1923-1944. Tome I* (X 0850) Colin: Paris 276pp
⋄ B96 C96 ⋄ REV Zbl 259 # 02002 • REM See also 1956
• ID 30338

TARSKI, A. [1979] see SZCZERBA, L.W.

TARSKI, A. [1983] see SCHWABHAEUSER, W.

TARSKI, A. [1985] see HENKIN, L.

TARSKI, A. also published under the name TAJTELBAUM, A.

TARSKI, A. see Vol. II, III, IV, V, VI for further entries

TARSKI, J. [1978] see BLANCHARD, P.

TARSKI, J. [1978] *Short introduction to nonstandard analysis and its physical applications* (P 2955) Int Summer Inst Theor Phys (8);1976 Bielefeld 225-239
⋄ H10 ⋄ REV MR 80d:81047 • ID 82953

TARSKI, J. see Vol. II for further entries

TASEV, A.G. [1975] *Eine funktionale Abhaengigkeit boolescher Funktionen und ihre Anwendung in der Theorie der funktionalen Dekomposition (Bulgarian) (Russian and English summaries)* (J 3171) God Vissh Ucheb Zaved, Prilozhna Mat, Sofiya 10/3*61-69
⋄ B05 ⋄ REV Zbl 339 # 02007 • ID 65671

TASEV, A.G. [1981] *Boolean differences and their properties (Bulgarian) (English and Russian summaries)* (J 3360) Avtom Izchisl Tekh, Sofiya 15/1*34-40
⋄ B05 ⋄ REV MR 83g:94042 Zbl 465 # 94031 • ID 69947

TAUTS, A. [1964] *Solution of logical equations in the first-order one-place predicate calculus (Russian) (English summary)* (J 4701) Trudy Akad Nauk Ehston SSR, Fiz Astron (Tartu) 24*3-16
⋄ B20 B25 ⋄ REV MR 33 # 2526 • ID 13475

TAUTS, A. [1964] *Solution of logical equations by an iteration method in the first-order predicate calculus (Russian) (English summary)* (J 4701) Trudy Akad Nauk Ehston SSR, Fiz Astron (Tartu) 24*17-24
⋄ B35 ⋄ REV MR 33 # 2527 • ID 42960

TAUTS, A. [1970] *An analogue of Herbrand's theorem for second order predicate calculus (Russian) (Estonian and German summaries)* (S 3468) Tr Mat & Mekh (Tartu) 9(253)*55-65
⋄ B15 ⋄ REV MR 43 # 7308 Zbl 235 # 02017 • ID 13480

TAUTS, A. [1978] *Ambiguity of the axiom of normal functions (Russian) (Estonian and German summaries)* (S 3468) Tr Mat & Mekh (Tartu) 22(464)*13-27
⋄ B15 C75 E10 E55 E65 ⋄ REV MR 80b:03086 Zbl 461 # 03006 • ID 54483

TAUTS, A. [1985] *Extraction of a program from a derivation and its regularity I (Russian) (English summary)* (J 0040) Kibernetika, Akad Nauk Ukr SSR 1985/1*45-55,134
⋄ B35 ⋄ ID 49180

TAUTS, A. see Vol. II, III, IV, V, VI for further entries

TAVANETS, P.V. [1967] see SMIRNOVA, E.D.

TAVANETS, P.V. (ED.) [1970] *Studies in systems of logic (Russian)* (X 2027) Nauka: Moskva 334pp
⋄ B97 ⋄ REV MR 47 # 6426 • ID 70238

TAVANETS, P.V. (ED.) [1973] *Theorie der logischen Folgerungen (Russian)* (X 2027) Nauka: Moskva 272pp
⋄ B22 B97 ⋄ REV Zbl 259 # 00004 • ID 48636

TAVANETS, P.V. [1974] see SMIRNOV, V.A.

TAVANETS, P.V. see Vol. II for further entries

TAYLOR, J.S. [1912] *A set of five postulates for boolean algebras in terms of the operation "exception"* (S 0183) Publ Math Univ California 1*241-248
⋄ A05 B30 G05 ⋄ ID 13485

TAYLOR, J.S. [1917] *Complete existential theory of Bernstein's set of four postulates for boolean algebras* (J 0120) Ann of Math, Ser 2 19*64-69
⋄ B30 G05 ⋄ REV FdM 46.92 • ID 13486

TAYLOR, J.S. [1920] *Sheffer's set of five postulates for boolean algebras in terms of the operation "rejection" made completely independent* (J 0015) Bull Amer Math Soc 26*449-454
⋄ B05 G05 ⋄ REV FdM 47.42 • ID 13487

TAYLOR, J.S. see Vol. V for further entries

TAYLOR, R.F. [1969] *On some properties of bounded internal functions* (P 0649) Appl Model Th to Algeb, Anal & Probab;1967 Pasadena 167-170
⋄ H05 ⋄ REV MR 39 # 1320 Zbl 198.20 • ID 13493

TAYLOR, R.F. [1970] see LUXEMBURG, W.A.J.

TEMPLE, G. [1977] *Inference without axiom or paradoxes* (P 1075) Logic Colloq;1976 Oxford 221-233
⋄ A10 B20 ⋄ REV MR 58 # 21420 Zbl 428 # 03001 • ID 16622

TEMPLE, G. see Vol. V for further entries

TENNANT, N. [1978] *Natural logic* (X 1261) Edinburgh Univ Pr: Edinburgh ix+196pp
⋄ B98 ⋄ REV MR 81i:03004 Zbl 483 # 03001 JSL 48.215 • ID 79279

TENNANT, N. [1979] *La barre de Scheffer dans la logique des sequents et des syllogismes* (J 0079) Logique & Anal, NS 22*505-514
⋄ B05 ⋄ REV MR 82c:03013 Zbl 443 # 03016 • ID 56422

TENNANT, N. see Vol. II, III, VI for further entries

TENNEY, R.L. [1975] *Second-order Ehrenfeucht games and the decidability of the second-order theory of an equivalence relation* (J 0038) J Austral Math Soc 20*323-331
⋄ B15 B25 C85 ⋄ REV MR 52 # 2861 Zbl 319 # 02036 • ID 17647

TENNEY, R.L. [1976] see FOSTER, C.C.

TERENKOV, V.I. [1973] *The accuracy of algorithms for the computation of estimates for tables that are generated by monotone boolean functions (Russian)* (J 0199) Zh Vychisl Mat i Mat Fiz 13*1620-1625,1640
• TRANSL [1973] (J 1049) USSR Comput Math & Math Phys 13/6*315-323
⋄ B35 ⋄ REV MR 50 # 1575 Zbl 291 # 94021 • ID 65684

TERRELL, D.B. [1967] *Logic: A modern introduction to deductive reasoning* (X 0818) Holt Rinehart & Winston: New York xii+355pp
⋄ B98 ⋄ REV MR 35 # 2717 • ID 22306

TERZILER, M. [1983] *La representation Booleienne via le calcul propositionnel* (J 3980) Karadeniz Univ Math J (Trabzon) 5*245-252
⋄ B05 ⋄ REV MR 85f:06022 Zbl 534 # 03032 • ID 36567

TERZILER, M. see Vol. IV for further entries

THARP, L.H. [1973] *The characterization of monadic logic* (J 0036) J Symb Logic 38*481-488
⋄ B15 B20 C95 ⋄ REV MR 49 # 2250 Zbl 275 # 02015 • ID 13527

THARP, L.H. [1975] *Which logic is the right logic?* (J 0154) Synthese 31*1-21
⋄ A05 B10 ⋄ REV MR 57 # 5665 Zbl 331 # 02002
• ID 65685

THARP, L.H. see Vol. III, IV, V, VI for further entries

THATCHER, J.W. & WRIGHT, J.B. [1968] *Generalized finite automata theory with an application to a decision problem of second order logic* (J 0041) Math Syst Theory 2*57-81
⋄ B15 B25 D05 ⋄ REV MR 37 # 75 Zbl 157.22 JSL 37.619 • ID 28596

THATCHER, J.W. see Vol. IV, V for further entries

THAYSE, A. [1975] *La detection des aleas dans les circuits logiques au moyen du calcul differentiel Booleen* (J 2701) Digit Processes 1*141-169
⋄ B05 B70 ⋄ REV MR 56 # 15216 Zbl 429 # 94033
• ID 69952

THAYSE, A. [1978] see DAVIO, M.

THAYSE, A. [1980] *Boolean difference calculus* (J 2701) Digit Processes 6*175-184
⋄ B05 ⋄ REV MR 82c:94014 Zbl 443 # 94032 • ID 69950

THAYSE, A. [1981] *Boolean calculus of differences* (S 3302) Lect Notes Comput Sci 101*vii+144pp
⋄ B05 ⋄ REV MR 83e:94084 Zbl 452 # 94033 • ID 69953

THAYSE, A. [1981] *Universal algorithms for evaluating boolean functions* (J 2702) Discr Appl Math 3*53-65
⋄ B05 ⋄ REV MR 82d:94083 Zbl 469 # 94019 • ID 69951

THAYSE, A. see Vol. II for further entries

THIEL, C. [1972] *Auf dem Weg zur Vereinheitlichung der logischen Symbolik* (J 1515) Archimede 24*48-49
⋄ B03 ⋄ ID 30586

THIEL, C. [1978] *Duality lost? Transforming Gentzen derivations into winning strategies for dialogue games* (J 0162) Teorema (Valencia) 8*57-66
⋄ B10 F07 ⋄ REV MR 58 # 133 • ID 30592

THIEL, C. [1979] *Zur Bestimmung der Arithmetik* (C 2537) Spez Wissenschaftsth 1*29-34
⋄ B28 ⋄ ID 32568

THIEL, C. see Vol. V, VI for further entries

THIELE, H. [1956] *Eine Axiomatisierung der zweiwertigen Praedikatenkalkuele der ersten Stufe, welche die Implikation erhalten* (J 0068) Z Math Logik Grundlagen Math 2*93-106
⋄ B20 ⋄ REV MR 19.3 Zbl 74.248 JSL 32.260 • ID 13537

THIELE, H. [1957] *Vollstaendigkeit im Stufenkalkuel* (J 0068) Z Math Logik Grundlagen Math 3*211-224
⋄ B15 ⋄ REV MR 20 # 2277 Zbl 79.6 • ID 13538

THIELE, H. [1968] see SCHMIDT, H.A.

THIELE, H. [1976] *Ein graphentheoretisches Modell zur Beschreibung von Klassifizierungsprozessen bei nichtdisjunkten Systemen von Klassen* (P 2898) Algor Kompl, Lern-& Erkenn-Prozess;1976 Jena 93-107
⋄ B05 B75 ⋄ REV MR 56 # 10220 Zbl 426 # 68091
• ID 53684

THIELE, H. see Vol. II, IV for further entries

THIELE, R. [1979] *Mathematische Beweise* (X 1079) Teubner: Leipzig 172pp
⋄ B98 ⋄ REV MR 82k:03002 Zbl 405 # 00018 • REM 2nd ed. 1981; 176pp • ID 54872

THISTLEWAITE, P.B. [1980] see MCROBBIE, M.A.

THOM, P. [1979] *Aristotle's syllogistic* (J 0047) Notre Dame J Formal Log 20*751-759
⋄ A05 A10 B20 ⋄ REV MR 80h:03004 Zbl 254 # 02004 Zbl 421 # 03009 • ID 53408

THOM, P. [1981] *The syllogism* (X 3111) Philosophia: Muenchen 312pp
⋄ A05 B20 ⋄ REV Zbl 471 # 03005 • ID 55200

THOM, R. [1980] *Predication et grammaire universelle (English summary)* (S 3677) Fund Scientiae 1*23-34
⋄ B05 ⋄ REV MR 83d:00026a • ID 39590

THOMAS, I. [1952] *A new decision procedure for Aristotle's syllogistic* (J 0094) Mind 61*564-566
⋄ A10 B20 B25 ⋄ REV JSL 21.315 • ID 13541

THOMAS, I. [1957] *Eulerian syllogistic* (J 0036) J Symb Logic 22*15-16
⋄ A10 B20 ⋄ REV MR 19.626 Zbl 81.11 JSL 22.381
• ID 13543

THOMAS, I. [1960] *Functional completeness of Henkin's propositional fragments* (J 0047) Notre Dame J Formal Log 1*107-110
⋄ B20 ⋄ REV MR 26 # 10 Zbl 113.242 JSL 27.111
• ID 13549

THOMAS, I. [1960] *Independence of faris-rejection-axioms* (J 0047) Notre Dame J Formal Log 1*48-51
⋄ A10 B20 ⋄ REV Zbl 115.8 JSL 27.113 • ID 13547

THOMAS, I. [1960] *Independence of Tarski's law in Henkin's propositional fragments* (J 0047) Notre Dame J Formal Log 1*74-78
⋄ B20 ⋄ REV MR 26 # 9 Zbl 143.9 JSL 27.111 • ID 13548

THOMAS, I. [1962] *On the infinity of positive logic* (J 0047) Notre Dame J Formal Log 3*108
⋄ A05 B20 ⋄ REV MR 28 # 2039 Zbl 114.244 JSL 33.306
• ID 13550

THOMAS, I. [1962] *The rule of excision in positive implication* (J 0047) Notre Dame J Formal Log 3*64
⋄ A05 B20 ⋄ REV MR 33 # 3904 JSL 40.603 • ID 13551

THOMAS, I. [1964] *Universal variable non-Tarskian functors* (J 0047) Notre Dame J Formal Log 5*221-222
⋄ A05 B20 ⋄ REV MR 33 # 7238 • ID 13560

THOMAS, I. [1970] *Final word on a shortest implicational axiom* (J 0047) Notre Dame J Formal Log 11*16
⋄ B20 ⋄ REV MR 43 # 3087 Zbl 187.263 JSL 36.690
• ID 13568

THOMAS, I. [1974] *On Meredith's sole positive axiom* (J 0047) Notre Dame J Formal Log 15*477
⋄ B20 ⋄ REV MR 50 # 1833 Zbl 272 # 02041 • ID 13574

THOMAS, I. [1975] *Nice implicational axioms* (J 0047) Notre Dame J Formal Log 16*507-508
⋄ B20 ⋄ REV MR 54 # 9974 Zbl 283 # 02010 Zbl 311 # 02017 • ID 25600

THOMAS, I. [1975] *Shorter development of an axiom* (J 0047) Notre Dame J Formal Log 16*378
⋄ B20 ⋄ REV MR 52 # 7842 Zbl 283 # 02011 • ID 13576

THOMAS, I. [1975] *Simple implicational development* (J 0047) Notre Dame J Formal Log 16*268
⋄ B20 ⋄ REV MR 52#2811 Zbl 301#02011 • ID 13575

THOMAS, I. [1976] *One dimension in PS and PSI* (J 0047) Notre Dame J Formal Log 17*421-423
⋄ B20 ⋄ REV MR 55#12461 Zbl 331#02003 • ID 18416

THOMAS, I. see Vol. II, III, V for further entries

THOMAS, N.L. [1966] *Modern logic. An introduction* (X 1000) Barnes & Noble: Totowa xvii+236pp
⋄ B98 ⋄ REV JSL 36.544 • ID 13577

THOMAS, WILLIAM J. [1976] *Consistency of n-order logics* (J 0047) Notre Dame J Formal Log 17*257-262
⋄ B15 ⋄ REV MR 58#21424 Zbl 258#02019 • ID 18418

THOMAS, WILLIAM J. see Vol. IV, VI for further entries

THOMAS, WOLFGANG [1978] see EBBINGHAUS, H.-D.

THOMAS, WOLFGANG [1980] *On the bounded monadic theory of well-ordered structures* (J 0036) J Symb Logic 45*334-338
⋄ B15 B25 C85 E07 ⋄ REV MR 81h:03080 Zbl 437#03005 • ID 55873

THOMAS, WOLFGANG [1984] see BOERGER, E.

THOMAS, WOLFGANG see Vol. III, IV, V, VI for further entries

THOMASON, R.H. [1963] see BELNAP JR., N.D.

THOMASON, R.H. [1967] see LEBLANC, H.

THOMASON, R.H. [1968] see LEBLANC, H.

THOMASON, R.H. [1969] see JOHNSON, D.R.

THOMASON, R.H. [1970] *Symbolic logic. An introduction* (X 0843) Macmillan : New York & London 367pp
⋄ B98 ⋄ REV MR 43#4627 Zbl 214.6 JSL 36.678 • ID 13581

THOMASON, R.H. see Vol. II, III, VI for further entries

THOMBLINSON, W.S. [1977] see ARCHIE, L.C.

THOMPSON, C. [1985] *An introduction to nonstandard analysis* (J 4645) Irish Math Soc Newslett 15*31-41
⋄ H05 ⋄ ID 49155

THOMPSON, D. [1971] see MCKENZIE, R.

THOMPSON, P.B. [1981] *Bolzano's deducibility and Tarski's logical consequence* (J 2028) Hist & Phil Log 2*11-20
⋄ A10 B22 ⋄ REV MR 83g:03002 • ID 34837

THOMPSON, S.M. [1942] *Syllogistic logic in linear notation* (J 0153) Phil of Sci (East Lansing) 9*362-366
⋄ A10 B03 ⋄ REV MR 4.125 Zbl 60.19 • ID 13600

THOMSON, J. [1967] *Proof of the law of infinite conjunction using the perfect disjunctive normal form* (J 0036) J Symb Logic 32*196-197
⋄ B05 ⋄ REV MR 35#5289 Zbl 144.244 • ID 13602

THOMSON, J. see Vol. V for further entries

THURBER, J.K. [1974] see KATZ, JOSE

TICHY, P. [1971] *An approach to intensional analysis* (J 0097) Nous, Quart J Phil 5*273-297
⋄ B15 ⋄ ID 32497

TICHY, P. [1971] *On the vicious circle in definitions (Polish and Russian summaries)* (J 0063) Studia Logica 28*19-40
⋄ A05 B10 ⋄ REV MR 50#12642 Zbl 255#02054 • ID 13606

TICHY, P. [1971] *Synthetic components of infinite classes of postulates* (J 0009) Arch Math Logik Grundlagenforsch 14*167-178
⋄ A05 B15 ⋄ REV MR 49#2242 Zbl 248#02057 • ID 13607

TICHY, P. [1982] *Foundations of partial type theory* (J 0302) Rep Math Logic, Krakow & Katowice 14*59-72 • ERR/ADD ibid 16*55-56
⋄ B15 ⋄ REV MR 84e:03017 Zbl 488#03007 • ID 34355

TICHY, P. see Vol. II, IV for further entries

TIKEKAR, V. [1979] see CHAKRAPANI, N.

TIKEKAR, V. [1980] see CHAKRAPANI, N.

TIMOFEEVA, L.M. [1965] *A machine for proving geometrical functions (Russian)* (C 1413) Tekh Kibernetika 308-317
⋄ B35 ⋄ REV JSL 35.348 • ID 28693

TIRNOVEANU, M. [1964] *Elements of mathematical logic. Vol.1. Logic of bivalent propositions* (X 1569) Didact Ped: Bucharest 519pp
⋄ B05 ⋄ REV MR 35#1438 JSL 39.325 • ID 33700

TIRNOVEANU, M. [1965] *Sur quelques proprietes des propositions generales* (J 0147) An Univ Bucuresti, Acta Logica 7-8*151-178
⋄ B05 B10 ⋄ REV MR 32#5496 Zbl 231#02011 • ID 27426

TIRNOVEANU, M. see Vol. II, VI for further entries

TISON, P.I. [1967] *Generalization of consensus theory and application to the minimization of boolean functions* (J 4305) IEEE Trans Electr Comp EC-16*446-456
⋄ B05 ⋄ REV Zbl 158.161 JSL 33.468 • ID 13613

TITANI, S. [1973] *A proof of the cut-elimination theorem in simple type theory* (J 0036) J Symb Logic 38*215-226
⋄ B15 F05 F35 ⋄ REV MR 48#10777 Zbl 275#02017 • ID 13615

TITANI, S. see Vol. III, V, VI for further entries

TLAS-YAKO, EH. [1979] *Minimization of boolean functions using a new form of canonical decomposition (Russian)* (S 2850) Tr Ehnerg Inst Moskva 438*74-80,169
⋄ B35 ⋄ REV MR 82f:03057 • ID 79370

TODOROV, T.D. [1973] see KHRISTOV, KH.YA.

TODOROV, T.D. [1976] see KHRISTOV, KH.YA.

TODOROV, T.D. [1980] *Asymptotic numbers. I: Algebraic properties. II Order relations, infinitesimals and interval topology (Russian summaries)* (J 3072) Bulgar J Phys 7*450-468,547-562
⋄ H05 H20 ⋄ REV MR 82e:12036 MR 83b:46057 Zbl 478#03029 Zbl 478#03030 • ID 55644

TODOROV, T.D. [1981] *Asymptotic functions and the problem of multiplication of distributions* (J 3072) Bulgar J Phys 8*109-120
⋄ H05 ⋄ REV MR 82m:46041 Zbl 478#46035 • ID 82981

TODOROV, T.D. [1981] *Extended asymptotic functions - some examples (Russian summary)* (J 3072) Bulgar J Phys 8*216-226
⋄ H05 ⋄ REV MR 82m:46042 Zbl 478#46036 • ID 82980

TODT, G. [1983] see GLUBRECHT, J.-M.

TODT, G. see Vol. II, V for further entries

TOERNEBOHM, H. [1958] *Outlines of a boolean tensor algebra with applications to the lower functional calculus* (J 0105) Theoria (Lund) 24*39-47
⋄ B10 G05 G25 ⋄ REV JSL 25.367 • ID 13624

TOERNEBOHM, H. see Vol. II for further entries

TOGASHI, A. [1984] see NOGUCHI, S.

TOKARZ, M. & WOJCICKI, R. [1971] *The problem of reconstructability of propositional calculi (Polish and Russian summaries)* (J 0063) Studia Logica 28*119-129
⋄ B22 F50 ⋄ REV MR 45 # 8497 Zbl 257 # 02012 • ID 13627

TOKARZ, M. [1972] *Connections between some notions of completeness of structural propositional calculi* (J 0387) Bull Sect Logic, Pol Acad Sci 1/2*17-19
⋄ B22 ⋄ REV MR 50 # 12701 Zbl 345 # 02037 • ID 13629

TOKARZ, M. [1973] *Connections between some notions of completeness of structural propositional calculi (Polish and Russian summaries)* (J 0063) Studia Logica 32*77-91
⋄ B22 ⋄ REV MR 49 # 8850 Zbl 345 # 02037 • ID 13632

TOKARZ, M. [1976] *A strongly finite logic with infinite degree of maximality* (J 0063) Studia Logica 35*447-451
⋄ B22 ⋄ REV MR 55 # 10236 Zbl 375 # 02011 • ID 51588

TOKARZ, M. [1984] see DZIK, W.

TOKARZ, M. see Vol. II, VI for further entries

TOLEDO TOLEDO, R. [1946] *Mathematische Grundlagen einer strukturellen Logik (Spanisch)* (J 4341) Euclides Madrid 6*554-560,614-620
⋄ B20 ⋄ REV MR 8.430 Zbl 60.20 • ID 47937

TOMAS, F. [1975] *A test for consistency and its application to recursive arithmetic (Spanish)* (J 1507) An Inst Mat, Univ Nac Aut Mexico 15*133-190
⋄ B28 F30 ⋄ REV MR 54 # 7238 Zbl 345 # 02031 • ID 25028

TOMAS, F. [1979] *The absolute consistency of the geometry of the ruler and transporter of segments (Spanish)* (J 1507) An Inst Mat, Univ Nac Aut Mexico 19*41-107
⋄ B30 ⋄ REV MR 81i:03089 Zbl 531 # 03033 • ID 79413

TOMAS, F. see Vol. VI for further entries

TOMASZEWICZ, A. [1973] *Generators of propositional functions (Polish and Russian summaries)* (J 0063) Studia Logica 31*145-151
⋄ B05 ⋄ REV MR 48 # 3696 Zbl 288 # 02009 • ID 13635

TOMESCU, I. [1969] *Un algorithme pour la synthese des fonctions booleennes symetriques (Romanian) (French summary)* (J 0197) Stud Cercet Mat Acad Romana 21*675-681
⋄ B05 ⋄ REV Zbl 193.146 • ID 16241

TOMESCU, I. see Vol. IV for further entries

TOMS, R.M. [1966] *Systems of boolean equations* (J 0005) Amer Math Mon 73*29-35
⋄ B05 ⋄ REV Zbl 133.244 JSL 32.132 • ID 13640

TONCHEV, V.D. [1979] see DENEV, J.D.

TONOYAN, G.P. [1979] *The successive splitting of vertices of an n-dimensional unit cube into chains and decoding problems of monotonic Boolean functions (Russian)* (J 0199) Zh Vychisl Mat i Mat Fiz 19*1532-1542,1630
• TRANSL [1979] (J 1049) USSR Comput Math & Math Phys 19/6*179-191
⋄ B05 ⋄ REV MR 81b:94074 Zbl 426 # 68057 • ID 69961

TORALDO DI FRANCIA, G.G. [1976] see DALLA CHIARA SCABIA, M.L.

TORALDO DI FRANCIA, G.G. [1979] see DALLA CHIARA SCABIA, M.L.

TORBASOVA, V.P. [1981] *Decomposable machines over algebraic systems (Russian)* (S 3909) Mat Met Opt & Upravleniya Slozh Sist (Kalinin) 1981*3-12
⋄ B35 C05 ⋄ REV MR 85e:68082 Zbl 518 # 68030 • ID 38384

TORBASOVA, V.P. see Vol. IV for further entries

TOROSYAN, B.E. [1979] see BOZOYAN, SH.E.

TORRENS TORRELL, A. [1982] *The general deduction principle (Catalan) (English summary)* (P 3870) Congr Catala de Log Mat (1);1982 Barcelona 121-122
⋄ B22 ⋄ REV MR 84i:03003 Zbl 512 # 03030 • ID 36516

TORTORA, R. [1980] see GUCCIONE, S.

TORTORA, R. see Vol. II, VI for further entries

TORTORICI, M. [1967] *Calcolo inferenziale automatico* (J 0104) Atti Accad Sci Lett Arti Palermo, Ser 4/I 28*449-473
⋄ B35 ⋄ REV MR 41 # 8204 Zbl 235 # 02011 • ID 65762

TORTORICI, M. [1970] *Il calcolo inferenziale automatico nella trattazione delle equazioni logiche* (J 0104) Atti Accad Sci Lett Arti Palermo, Ser 4/I 29*263-284
⋄ B35 C05 ⋄ REV MR 45 # 6574 Zbl 235 # 02012 • ID 27847

TORTORICI, M. [1977] see MACALUSO, A.T.

TOSIC, R. [1972] *S-bases of propositional algebra* (J 0400) Publ Inst Math, NS (Belgrade) 14(28)*139-148
⋄ B05 ⋄ REV MR 53 # 7644 Zbl 268 # 94038 • ID 22969

TOSIC, R. [1978] *Some properties of monotone boolean functions over finite boolean algebras* (J 2855) Zbor Rad, Prir-Mat Fak, Ser Mat (Novi Sad) 8*63-68
⋄ B05 ⋄ REV MR 81a:06017 Zbl 441 # 06014 • ID 69963

TOSIC, R. [1979] *An optimal identification algorithm for some subclasses of monotone boolean functions* (J 2855) Zbor Rad, Prir-Mat Fak, Ser Mat (Novi Sad) 9*115-121
⋄ B05 ⋄ REV MR 82m:94062 Zbl 449 # 94028 • ID 69962

TOSIC, R. see Vol. II for further entries

TOTH, I. [1976] *Un probleme de logique et de linguistique concernant le rapport entre geometrie euclidienne (GE) et geometrie non-euclidienne (GNE)* (P 1748) Lang & Pensee Math;1976 Luxembourg 93-142
⋄ B30 ⋄ REV Zbl 373 # 01006 • ID 51464

TOTI RIGATELLI, L. [1968] *Sullo spazio dei modelli di un calcolo generale (French summary)* (J 0088) Ann Univ Ferrara, NS, Sez 7 13*55-65
⋄ B22 ⋄ REV MR 39 # 5451 Zbl 181.8 • ID 13651

TOTI RIGATELLI, L. see Vol. V for further entries

TOWNLEY, J.A. [1980] *A pragmatic approach to resolution-based theorem proving* (J 0435) Int J Comput & Inf Sci 9∗93-116
 ⋄ B35 ⋄ REV MR 81d:68129 Zbl 436#68059 • ID 82996

TRAKHTENBROT, B.A. [1953] *On recursively separability (Russian)* (J 0023) Dokl Akad Nauk SSSR 88∗953-956
 ⋄ B10 C13 D35 ⋄ REV MR 16.436 Zbl 50.8 JSL 19.60 • ID 19742

TRAKHTENBROT, B.A. [1956] *Algorithms and mechanical solution of problems I,II (Russian)* (J 2780) Mat v Shkole 1956/4∗3-10,1956/5∗5
 ⋄ B35 ⋄ ID 45719

TRAKHTENBROT, B.A. [1956] *The definition of a finite set and the deductive incompleteness of the theory of sets (Russian)* (J 0216) Izv Akad Nauk SSSR, Ser Mat 20∗569-582
 • TRANSL [1964] (J 0225) Amer Math Soc, Transl, Ser 2 39∗177-192
 ⋄ B28 D35 E30 ⋄ REV MR 18.269 Zbl 71.246 JSL 27.236 • ID 19739

TRAKHTENBROT, B.A. [1960] *Algorithms and machine solutions of problems (Russian)* (X 3709) Izdat Fiz-Mat Lit: Moskva 119pp
 • TRANSL [1977] (X 0885) Mir: Moskva 109pp (Spanish)
 ⋄ B25 B35 D20 ⋄ REV MR 22#10906 Zbl 80.114 JSL 47.702 • ID 42812

TRAKHTENBROT, B.A. [1961] *Certain constructions in the logic of one-place predicates (Russian)* (J 0023) Dokl Akad Nauk SSSR 138∗320-321
 • TRANSL [1961] (J 0062) Sov Math, Dokl 2∗623-625
 ⋄ B20 F35 ⋄ REV MR 25#1996 JSL 29.100 • ID 19735

TRAKHTENBROT, B.A. [1961] *Finite automata and the logic of one-place predicates (Russian)* (J 0023) Dokl Akad Nauk SSSR 140∗326-329
 • TRANSL [1961] (J 0470) Sov Phys, Dokl 6∗753-755
 ⋄ B20 D05 F35 ⋄ REV MR 26#1246 Zbl 115.7 JSL 29.100 • ID 19734

TRAKHTENBROT, B.A. [1962] *Finite automata and the logic of one-place predicates (Russian)* (J 0092) Sib Mat Zh 3∗103-131
 ⋄ B20 D05 F35 ⋄ REV MR 26#4908 Zbl 115.7 JSL 29.98 • ID 19733

TRAKHTENBROT, B.A. see Vol. II, III, IV, VI for further entries

TRAVIS, L.E. [1964] *Experiments with a theorem-utilizing program* (P 1640) AFIPS Spring Jt Computer Conf (25);1964 Washington 339-358
 ⋄ B35 ⋄ REV Zbl 156.180 • ID 29443

TRESKOV, S.A. [1965] see FRIDMAN, G.SH.

TREW, A. [1970] *Nonstandard theories of quantifications and identity* (J 0036) J Symb Logic 35∗267-293
 ⋄ B20 B60 ⋄ REV MR 46#3264 Zbl 208.9 • ID 13673

TREW, A. see Vol. II for further entries

TRILLAS, E. [1983] see PAWLOWSKY, V.

TRILLAS, E. see Vol. II, V for further entries

TROELSTRA, A.S. & DALEN VAN, D. (EDS.) [1982] *The L.E.J. Brouwer centenary symposium* (S 3303) Stud Logic Found Math 110∗ix+523pp
 ⋄ A05 A10 B97 F50 F55 F97 ⋄ REV MR 84i:03010 Zbl 505#00008 • ID 34494

TROELSTRA, A.S. see Vol. II, III, IV, V, VI for further entries

TROESCH, A. & URLACHER, E. [1978] *Perturbations singulieres et analyse non standard (English summary)* (J 2313) C R Acad Sci, Paris, Ser A-B 287∗A937-A939
 ⋄ H05 ⋄ REV MR 80c:34047 Zbl 399#34054 • ID 83012

TROESCH, A. [1982] *Etude macroscopique de l'equation de van der Pol* (P 4018) Th & Appl Sing Perturbations;1981 Oberwolfach 136-144
 ⋄ H05 ⋄ REV MR 83j:34003 Zbl 489#34034 • ID 46092

TROESCH, A. [1984] *Etude macroscopique de systemes differentiels* (J 3240) Proc London Math Soc, Ser 3 48∗121-160
 ⋄ H05 ⋄ REV MR 85f:58105 Zbl 544#34054 • ID 40031

TROYANOVSKIJ, S.V. [1964] *On the equivalence of formulae with respect to a given function of the algebra of logic (Russian)* (J 0468) Uch Zap Ped Inst (Alma Ata) 124∗106-111
 ⋄ B05 ⋄ REV MR 34#2447 • ID 13689

TRYBULEC, A. [1985] see BLAIR, H.A.

TSEJTIN, G.S. [1972] see CHUBARYAN, A.A.

TSEJTIN, G.S. [1975] see CHUBARYAN, A.A.

TSEJTIN, G.S. see Vol. IV, VI for further entries

TSEJTLIN, G.E. [1979] see KIRSANOV, G.M.

TSEJTLIN, G.E. see Vol. IV for further entries

TSELISHCHEV, V.V. (ED.) [1982] *Problems of logic and methodology of science (Russian)* (X 2642) Nauka: Novosibirsk 336pp
 ⋄ B98 ⋄ REV MR 83h:03005 Zbl 505#00009 • ID 38203

TSELISHCHEV, V.V. see Vol. II for further entries

TSINMAN, L.L. [1967] *On the complete induction axiom (Russian)* (J 0023) Dokl Akad Nauk SSSR 173∗273-274
 • TRANSL [1967] (J 0062) Sov Math, Dokl 8∗381-383
 ⋄ B28 F30 ⋄ REV MR 35#1468 Zbl 159.10 • ID 02176

TSINMAN, L.L. [1968] *The role of the principle of induction in a formal arithmetical system (Russian)* (J 0142) Mat Sb, Akad Nauk SSSR, NS 77(119)∗71-104
 • TRANSL [1968] (J 0349) Math of USSR, Sbor 6∗65-95
 ⋄ B28 F30 ⋄ REV MR 39#5312 Zbl 165.18 • ID 02177

TSINMAN, L.L. see Vol. III, IV, VI for further entries

TSITKIN, A.I. [1979] *On the question of an error in a famous work due to Wajsberg (Russian)* (C 2581) Issl Neklass Log & Teor Mnozh 240-256
 ⋄ A10 B20 F50 ⋄ REV MR 82d:03018 Zbl 422#03035 • ID 53496

TSITKIN, A.I. see Vol. II, VI for further entries

TSUKADA, H. [1981] *Category theory not based upon set theory* (J 0539) Sci Pap Coll Gen Educ, Univ Tokyo 31∗1-24
 ⋄ B30 G30 ⋄ REV MR 83d:18001 Zbl 477#18001 • ID 55603

TUCKER, J.V. [1982] see BERGSTRA, J.A.

TUCKER, J.V. see Vol. III, IV, VI for further entries

TUERKSEN, I.B. [1974] *Extensions on solutions of boolean equations via its conjunctions* (J 0193) Discr Math 8∗187-200
 ⋄ B05 ⋄ REV MR 49#2483 Zbl 278#90051 • ID 13733

TUERKSEN, I.B. see Vol. II, V for further entries

TUGUE, T. [1967] *A Lemma for negationless propositional logics and its applications* (J 0111) Nagoya Math J 30*303-308
⋄ B20 ⋄ REV MR 35#6521 Zbl 183.7 • ID 13712

TUGUE, T. [1981] see MUELLER, GERT H.

TUGUE, T. see Vol. IV, V for further entries

TUL'CHINSKIJ, G.L. [1982] *Some problems of the logical explication of "language games" (Russian)* (C 3743) Probl Log & Metodol Nauk 105-115
⋄ B10 ⋄ REV MR 84k:03009b • ID 34949

TULIPANI, S. [1974] see MARCJA, A.

TULIPANI, S. [1975] *Questioni di teoria dei modelli per linguaggi universali positivi II: Metodi di "back and forth" (English summary)* (J 0149) Atti Accad Naz Lincei Fis Mat Nat, Ser 8 59*328-335
⋄ B20 C07 C40 C50 ⋄ REV MR 56#5273 Zbl 363#02057 • REM Part I 1974 by Marcja,A. & Tulipani,S. • ID 50883

TULIPANI, S. [1978] *On solutions of algebraic equations whose coefficients are germs of continuous functions* (J 3495) Boll Unione Mat Ital, V Ser, B 15*931-938
⋄ H05 ⋄ REV MR 81b:39006 Zbl 449#39006 • ID 56749

TULIPANI, S. [1985] *An algorithm to determine for any prime p, a polynomial-sized Horn sentence which expresses "the cardinality is not p"* (J 0036) J Symb Logic 50*1062-1064
⋄ B35 D15 ⋄ ID 49361

TULIPANI, S. see Vol. III, IV, V for further entries

TURASHVILI, T.V. [1973] see GUREVICH, Y.

TURASHVILI, T.V. [1975] *The decidability problem of first order predicate logic (Russian)* (J 1477) Tr Vychisl Tsentr, Akad Nauk Gruz SSR 15*146-157
⋄ B20 D35 ⋄ REV MR 54#2415 • ID 24005

TURASHVILI, T.V. [1977] *On the undecidable minimal classes of first order predicate logic (Russian) (English summary)* (J 0233) Soobshch Akad Nauk Gruz SSR 86*589-591
⋄ B20 D35 ⋄ REV MR 56#15361 Zbl 362#02009 • ID 50731

TURASHVILI, T.V. see Vol. IV for further entries

TUREK, C. [1975] *Lehmann on the rules of the invalid syllogisms* (J 0047) Notre Dame J Formal Log 16*603-604
⋄ B20 ⋄ REV Zbl 311#02025 • ID 29602

TURI, L. [1982] see ECSEDI-TOTH, P.

TURI, L. see Vol. III for further entries

TURING, A.M. [1942] see NEWMAN, M.H.A.

TURING, A.M. [1942] *The use of dots as brackets in Church's system* (J 0036) J Symb Logic 7*146-156
⋄ B03 ⋄ REV MR 4.183 Zbl 60.20 JSL 8.85 • ID 13728

TURING, A.M. [1948] *Practical forms of type theory* (J 0036) J Symb Logic 13*80-94
⋄ B15 ⋄ REV MR 10.1 Zbl 54.6 JSL 14.182 • ID 13730

TURING, A.M. see Vol. III, IV, VI for further entries

TUROWICZ, A.B. [1960] *Sur une methode algebrique de verification des theoremes de la logique des enonces (Polish and Russian summaries)* (J 0063) Studia Logica 9*27-36
⋄ B05 ⋄ REV MR 24#A1819 Zbl 121.253 • ID 13734

TURQUETTE, A.R. [1962] *A general theory of k-place stroke functions in 2-valued logic* (J 0053) Proc Amer Math Soc 13*822-824
⋄ B05 ⋄ REV MR 25#3813 Zbl 109.6 JSL 29.143 • ID 13743

TURQUETTE, A.R. [1964] *Peirce's icons for deductive logic* (C 0707) Stud Phil of C.S. Pierce 95-108
⋄ A05 A10 B05 ⋄ REV JSL 39.354 • ID 13745

TURQUETTE, A.R. [1966] *A method for constructing implication logics* (J 0068) Z Math Logik Grundlagen Math 12*267-278
⋄ B20 ⋄ REV MR 35#1453 Zbl 147.249 JSL 33.308 • ID 13747

TURQUETTE, A.R. see Vol. II, IV for further entries

TURSMAN, R. [1968] *The shortest axioms of the implicational calculus* (J 0047) Notre Dame J Formal Log 9*351-358
⋄ B20 ⋄ REV MR 39#3964 Zbl 165.10 JSL 36.690 • ID 13752

TURZIK, D. [1982] see POLJAK, S.

TUSCHIK, H.-P. [1980] see BAUDISCH, A.

TUSCHIK, H.-P. see Vol. III, IV, V for further entries

TVERDOKHLEBOVA, N.N. [1973] *A number-theoretic method of comparing Boolean functions (Russian)* (J 0040) Kibernetika, Akad Nauk Ukr SSR 1973/3*1-5
• TRANSL [1973] (J 0021) Cybernetics 9*371-376
⋄ B05 ⋄ REV MR 50#9449 Zbl 267#94043 • ID 65819

TVERDOKHLEBOVA, N.N. see Vol. II for further entries

TYMIENIECKA, A.-T. (ED.) [1965] *Contributions to logic and methodology, in honor of I.M.Bochenski* (X 0809) North Holland: Amsterdam xviii+326pp
⋄ A05 B97 ⋄ REV MR 47#6424 Zbl 135.242 • ID 25257

TYTUS, F.J. [1967] *An elementary construction of the natural numbers* (J 0047) Notre Dame J Formal Log 8*297-300
⋄ B28 ⋄ REV MR 39#2629 Zbl 189.287 • ID 13757

TYTUS, F.J. see Vol. V for further entries

TYUGU, E.KH. [1982] see MINTS, G.E.

TZOUVARAS, A.D. [1980] *A non-standard characterization of the norm of free ultrafilters* (J 0465) Bull Greek Math Soc (NS) 21*81-86
⋄ C55 E05 H05 ⋄ REV MR 85i:03199 Zbl 548#03040 • ID 43212

TZOUVARAS, A.D. see Vol. III, V for further entries

UDALOV, V.I. [1975] *A method of determining whether boolean functions are monotypic (Russian)* (J 0474) Avtom Vychis Tekh, Akad Nauk Latv SSR 1975/6*9-10
• TRANSL [1975] (J 2666) Autom Control Comput Sci 9/6*9-11
⋄ B05 ⋄ REV MR 57#15771 Zbl 327#94047 • ID 65824

UDALOV, V.I. [1979] *The diagnosis of a programmed logical matrix with the use of algorithms for the minimization of boolean functions (Russian)* (S 0764) Teor Konech Avtom & Prilozh (Riga) 10*91-97
⋄ B35 ⋄ REV Zbl 443#94030 • ID 69970

UEDA, T. [1978] *The fixing groups for the 2-asummable boolean functions* (J 0193) Discr Math 20*77-82
⋄ B05 ⋄ REV MR 57#12325 Zbl 398#94031 • ID 52797

UEDA, T. [1979] *On the fixing group for a totally pre-ordered boolean function* (J 0193) Discr Math 26∗293-295
⋄ B05 ⋄ REV MR 80k:94060 Zbl 408 # 94021 • ID 69971

UEMOV, A.I. [1959] *Leere Klassen und aristotelische Logik (Russisch)* (C 1155) Log Issl (Moskva) 178-188
⋄ B20 E20 ⋄ REV Zbl 103.244 • ID 47995

UEMOV, A.I. [1964] *The problem of equivalence of logical structures (Russian)* (C 0570) Form Log & Metodol Nauk 52-64
⋄ B22 ⋄ ID 14572

UEMOV, A.I. [1970] *A construction of propositional logic without the affirmation principle (Russian)* (C 0668) Neklass Log 297-331
⋄ B20 ⋄ REV MR 49 # 2279 • ID 13759

UEMOV, A.I. [1971] *Logical foundations of the method of models (Russian)* (X 1470) Mysl': Moskva 311pp
⋄ A05 B30 ⋄ REV MR 50 # 4205 • ID 83023

UESU, T. [1966] *On Zermelo's set-theory and the simple type-theory with the axiom of infinity* (J 0407) Comm Math Univ St Pauli (Tokyo) 15∗49-59
⋄ B15 E30 E35 ⋄ REV MR 34 # 4116 Zbl 145.244 JSL 33.292 • ID 13760

UESU, T. [1971] *Simple type theory with constructive infinitely long expressions* (J 0407) Comm Math Univ St Pauli (Tokyo) 19∗131-163
⋄ B15 C75 F05 F35 ⋄ REV MR 45 # 8509 Zbl 255 # 02023 • ID 13764

UESU, T. see Vol. III, IV, V, VI for further entries

UGOL'NIKOV, A.B. [1976] *The realization of monotone functions by schemes of functional elements (Russian)* (J 0052) Probl Kibern 31∗167-185,238
⋄ B05 ⋄ REV MR 54 # 9885 Zbl 414 # 94046 • ID 69972

UGOL'NIKOV, A.B. [1979] *The synthesis of schemes and formulas in incomplete bases (Russian)* (J 0023) Dokl Akad Nauk SSSR 249∗60-63
• TRANSL [1979] (J 0062) Sov Math, Dokl 20∗1224-1227
⋄ B20 ⋄ REV MR 82d:94084 Zbl 442 # 94018 • ID 69973

UGOL'NIKOV, A.B. see Vol. II for further entries

UHLIG, D. [1972] *Ueber eine Kompliziertheitscharakteristik boolescher Funktionen* (J 0115) Wiss Z Humboldt-Univ Berlin, Math-Nat Reihe 21∗563-566
⋄ B05 ⋄ REV MR 48 # 8128 Zbl 257 # 94025 • ID 65829

UHLIG, D. [1973] *On relations between the complexity of realisations of boolean functions by circuits and the number of their subfunctions (Russian)* (J 0052) Probl Kibern 26∗183-201
⋄ B05 • ID 40964

UHLIG, D. [1974] *About a family of classes of easily realizable Boolean functions (Russian)* (J 0052) Probl Kibern 28∗25-42
⋄ B05 • ID 42311

UHLIG, D. [1978] *Boolean functions with linear combinational complexity* (S 2829) Rostocker Math Kolloq 10∗103-107
⋄ B05 ⋄ REV MR 81d:94045 Zbl 412 # 94028 • ID 69974

UHLIG, D. [1979] *A function of the algebra of logic with many subfunctions and a small complexity of realization (Russian)* (J 0052) Probl Kibern 35∗133-139,208
⋄ B05 ⋄ REV MR 80i:03052 Zbl 476 # 94023 • ID 55595

UHLIG, D. [1979] *Relations between the number of subfunctions and combinational complexity of boolean functions and vector functions* (P 2935) FCT'79 Fund of Comput Th;1979 Berlin/Wendisch-Rietz 476-482
⋄ B05 ⋄ REV Zbl 425 # 94022 • ID 69975

UHLIG, D. [1981] *On two classes of Boolean functions* (P 2552) Conf Finite Algeb & Multi-Val Log;1979 Szeged 739-752
⋄ B05 ⋄ REV MR 83g:94043 Zbl 519 # 94016 • ID 39166

UHLIG, D. [1982] see FISCHER, R.

ULAM, S.M. [1968] see KAC, M.

ULAM, S.M. see Vol. V for further entries

ULRICH, D. [1969] *Solution to a problem posed by Kalicki* (J 0053) Proc Amer Math Soc 22∗728-729
⋄ B22 ⋄ REV MR 39 # 5316 Zbl 207.5 • ID 13778

ULRICH, D. [1970] *Decidability results for some classes of propositional calculi* (J 0036) J Symb Logic 35∗353-354
⋄ B22 ⋄ ID 41535

ULRICH, D. [1972] *Some results concerning finite models for sentential calculi* (J 0047) Notre Dame J Formal Log 13∗363-368 • ERR/ADD ibid 15∗648
⋄ B22 ⋄ REV MR 46 # 8808 MR 50 # 1867 Zbl 212.16 • ID 16354

ULRICH, D. [1976] *On a property of matrices for subsystems of IC^+* (J 0068) Z Math Logik Grundlagen Math 22∗193-194
⋄ B20 F50 ⋄ REV MR 58 # 16173 Zbl 347 # 02018 • ID 18459

ULRICH, D. [1982] *Answer to a question raised by Harrop* (J 0387) Bull Sect Logic, Pol Acad Sci 11∗140-141
⋄ B22 ⋄ REV MR 84k:03092 • ID 36109

ULRICH, D. [1983] *The finite model property and recursive bounds on the size of countermodels* (J 0122) J Philos Logic 12∗477-480
⋄ B22 B25 C13 ⋄ REV MR 85m:03016 Zbl 575 # 03025 • ID 42640

ULRICH, D. see Vol. II for further entries

UMEZAWA, T. [1967] *A method of two-level simplification of boolean functions* (J 0111) Nagoya Math J 29∗201-210
⋄ B05 ⋄ REV MR 34 # 8889 Zbl 153.13 • ID 31272

UMEZAWA, T. [1967] *An absolute simplification of boolean functions* (J 1005) Rep Fac Sci, Shizuoka Univ 2∗1-15
⋄ B05 ⋄ REV MR 39 # 6680 Zbl 352 # 94024 • ID 31273

UMEZAWA, T. [1979] *A method for cut elimination in intuitionistic predicate logic and classical predicate logic* (J 1005) Rep Fac Sci, Shizuoka Univ 13∗1-13
⋄ B10 F05 F50 ⋄ REV MR 80k:03063 Zbl 415 # 03047 • ID 53149

UMEZAWA, T. see Vol. II, III, IV, V, VI for further entries

UNVERICHT, E. [1978] see SZABO, P.

URBANIK, K. [1978] *An axiomatic definition of information* (P 2944) Th de l'Inform CNRS;1977 Cachan 99-112
⋄ B30 G10 ⋄ REV MR 82a:94033 Zbl 482 # 94012 • ID 83024

URBANO, R.H. [1963] *Boolean matrices and the stability of neural nets* (P 0572) Inform Processing (2);1962 Muenchen 755-757
⋄ B05 D05 ⋄ REV Zbl 121.344 JSL 35.348 • ID 13791

URLACHER, E. [1978] see TROESCH, A.

URQUHART, A.I.F. [1971] see RESCHER, N.

URQUHART, A.I.F. [1974] *Proofs, snakes and ladders* (J 0488) Dialogue (Ottawa) 13*723-731
⋄ B22 F07 ⋄ REV MR 58 # 162 • ID 79550

URQUHART, A.I.F. [1977] *A finite matrix whose consequence relation is not finitely axiomatizable* (J 0302) Rep Math Logic, Krakow & Katowice 9*71-73
⋄ B22 ⋄ REV MR 58 # 27312 Zbl 391 # 03016 • ID 52343

URQUHART, A.I.F. [1981] *Decidability and the finite model property* (J 0122) J Philos Logic 10*367-370
⋄ B22 B25 D35 ⋄ REV MR 83i:03038 Zbl 465 # 03005 • ID 54908

URQUHART, A.I.F. see Vol. II, III, IV, VI for further entries

URSINI, A. [1980] *Dai numeri razionali ai numeri iperreali* (J 3495) Boll Unione Mat Ital, V Ser, B 17*340-351
⋄ H05 ⋄ REV MR 81h:03125 Zbl 428 # 03058 • ID 53817

URSINI, A. see Vol. II, III, IV, VI for further entries

USHENKO, A.P. [1936] *The theory of logic* (X 0837) Harper & Row: New York xii+197pp
⋄ B98 ⋄ REV JSL 1.113 FdM 62.1046 • ID 22603

USHENKO, A.P. [1939] *The calculus of propositions and self-contradiction* (J 0101) Phil Rev 48*322-325
⋄ B05 ⋄ REV FdM 65.1103 • ID 43784

USHENKO, A.P. [1941] *The problems of logic* (X 0857) Princeton Univ Pr: Princeton 225pp
⋄ A05 B98 ⋄ REV JSL 6.166 • ID 13802

USPENSKIJ, V.A. [1981] *Nonstandard analysis (Bulgarian)* (J 0477) Spis Bulgar Akad Nauk 23(56)*219-231
⋄ H05 H98 ⋄ REV MR 83i:03106 • REM Transl. from Russian of the preface of Davis,Martin D.:Applied Nonstandard analysis, 1980 • ID 34859

USPENSKIJ, V.A. see Vol. IV, V, VI for further entries

USTINOV, N.A. [1980] *The number of solutions of a system of logical equations (Russian)* (S 3911) Komb-Algeb Met Prikl Mat (Gor'kij) 1980*208-212,216
⋄ B05 ⋄ REV Zbl 516 # 03035 • ID 46423

UTKIN, A.A. & ZAKREVSKIJ, A.D. [1975] *Ueber die Loesung logischer Differenzengleichungen (Russian)* (J 0414) Dokl Akad Nauk Belor SSR 19*34-37
⋄ B05 ⋄ REV MR 52 # 13015 Zbl 317 # 94042 • ID 66203

VACALIUC, I. [1966] *Demonstration de quelques theoremes concernant l'addition et la multiplication dans l'arithmetique a logique elementaire* (J 0147) An Univ Bucuresti, Acta Logica 9*217-241
⋄ B28 ⋄ REV MR 34 # 5666 Zbl 257 # 02020 • ID 28994

VACCARINO, G. [1952] *La sillogistica. I,II* (J 1515) Archimede 4*58-60,156-159
⋄ B20 ⋄ REV Zbl 46.5 Zbl 47.7 • ID 48143

VACCARINO, G. see Vol. II, V for further entries

VAEAENAENEN, J. [1977] *Remarks on generalized quantifiers and second order logics* (P 1639) Set Th & Hierarch Th (1);1974 Karpacz 117-123
⋄ B15 C40 C80 C85 C95 ⋄ REV MR 58 # 5019 Zbl 382 # 03010 • ID 51934

VAEAENAENEN, J. see Vol. III, IV, V, VI for further entries

VAIDYANATHASWAMY, R. [1935] *On the arithmetico-logical symmetric functions of n attributes* (J 0545) Proc Indian Acad Sci, Sect A 2*54-62
⋄ B20 ⋄ REV Zbl 12.113 FdM 61.51 • ID 40777

VAIDYANATHASWAMY, R. see Vol. II, VI for further entries

VAIRAVAN, K. [1979] see BREJTBART, YU.YA.

VAJL', V.E. [1974] *Gentzen systems of postulates for set theory (Russian)* (C 2577) Issl Formaliz Yazyk & Neklass Log 109-130
⋄ B15 E30 E70 ⋄ REV MR 56 # 5236 • ID 79577

VAJL', V.E. see Vol. V for further entries

VAKARELOV, D. [1972] *Extensional logics (Russian)* (J 0137) C R Acad Bulgar Sci 25*1609-1612
⋄ B05 B25 ⋄ REV MR 48 # 55 Zbl 253 # 02024 Zbl 332 # 02012 • ID 00678

VAKARELOV, D. see Vol. II, VI for further entries

VALPOLA, V. [1947] *The position of negation in a language which expresses knowledge (Finnish)* (J 0963) Ajatus (Helsinki) 14*325-381
⋄ A05 B28 F30 F50 ⋄ REV JSL 13.125 • ID 42606

VALPOLA, V. [1955] *Ein System der negationslosen Logik mit ausschliesslich realisierbaren Praedikaten* (J 0096) Acta Philos Fenn 9*247pp
⋄ B20 F50 ⋄ REV MR 17.699 Zbl 66.8 JSL 22.355 • ID 16346

VALPOLA, V. [1955] *Eine Eigenschaft gewoehnlicher negationsloser Kalkuele der Propositionen- und Praedikatenlogik* (J 0132) Math Scand 3*107-114
⋄ B20 F50 ⋄ REV MR 17.447 Zbl 65.1 JSL 22.380 • ID 13817

VANDERVEKEN, D.R. [1976] *The Lesniewski-Curry theory of syntactical categories and the categorially open functors* (J 0063) Studia Logica 35*191-201
⋄ B15 D05 G30 ⋄ REV MR 56 # 5233 Zbl 356 # 02013 • ID 50285

VANDERVEKEN, D.R. see Vol. II, VI for further entries

VARGA, A. [1977] see ECSEDI-TOTH, P.

VARGA, A. [1980] see ECSEDI-TOTH, P.

VARGAS, T. [1970] *Mathematische Logik fuer Anfaenger. Aussagenlogik* (X 1036) Volk & Wissen: Berlin 172pp
⋄ B05 ⋄ REV MR 51 # 41 • ID 15176

VARINEAU, V.J. [1962] see HENKIN, L.

VARSAVSKY, O. [1974] *Quantifiers and equivalence relations* (S 0889) Notas Logica Mat 3*27-51
⋄ B10 G15 ⋄ REV Zbl 322 # 02054 JSL 24.252 • ID 65888

VASHCHENKO, V.P. [1967] *Mengentheoretischer Ansatz zur funktionalen Trennbarkeit (Russisch)* (J 0071) Met Diskr Analiz (Novosibirsk) 10*9-22
⋄ B05 ⋄ REV MR 36 # 4995 Zbl 236 # 02012 • ID 27852

VASHCHENKO, V.P. [1977] *The general case of simple functional decomposition (Russian)* (J 0023) Dokl Akad Nauk SSSR 234*509-512
• TRANSL [1977] (J 0062) Sov Math, Dokl 18*696-699
⋄ B05 ⋄ REV MR 58 # 26589 Zbl 386 # 94020 • ID 52216

VASHCHENKO, V.P. [1978] *A minimization of boolean functions by the synthesis of module schemes (Russian)* (J 3396) Mat Model Teor Ehlektr Tsepej 16*88-101
  ⋄ B35 ⋄ REV Zbl 439 # 94028 • ID 69986

VASHCHENKO, V.P. [1978] *Multiple separation of a function using a fixed adjoint function (Russian)* (J 0023) Dokl Akad Nauk SSSR 239*18-21
  • TRANSL [1978] (J 0062) Sov Math, Dokl 19*246-249
  ⋄ B05 ⋄ REV MR 58 # 27493 Zbl 395 # 03042 • ID 52593

VASHCHENKO, V.P. [1978] *Representation of functions of the algebra of logic by compositions of partial functions (Russian)* (S 2850) Tr Ehnerg Inst Moskva 357*26-30
  ⋄ B05 ⋄ REV MR 81j:03096 Zbl 449 # 94026 • ID 79669

VASHCHENKO, V.P. [1979] *On the computation of all nontrivial simple decompositions of a function of the algebra of logic (Russian)* (J 0023) Dokl Akad Nauk SSSR 247*15-18
  • TRANSL [1979] (J 0062) Sov Math, Dokl 20*629-632
  ⋄ B05 ⋄ REV MR 80j:03086 Zbl 437 # 94008 • ID 79670

VASHCHENKO, V.P. [1980] *Investigation of simple decomposition of a Boolean function (Russian)* (S 2850) Tr Ehnerg Inst Moskva 485*3-9,95
  ⋄ B05 ⋄ REV MR 84h:94028 • ID 39747

VASHCHENKO, V.P. see Vol. II, V for further entries

VASILACHE, S. [1977] *Ensembles, structures, categories, faisceaux* (X 2776) Pr Univ Laval: Ste-Foy xii+315pp
  • LAST ED [1977] (X 1752) Masson: Paris xii+315pp
  ⋄ B98 E98 G30 ⋄ REV MR 58 # 5445 Zbl 375 # 18001 • ID 51636

VASILACHE, S. see Vol. V for further entries

VASIL'EV, N.A. [1913] *Logic and metalogic (Russian)* (J 1198) Logos (Tuebingen) 2-3*53-81
  ⋄ A05 B05 ⋄ ID 42894

VASIL'EV, N.A. see Vol. II for further entries

VASIL'EV, S.N. [1979] see KARATUEV, V.G.

VASIL'EV, S.N. [1981] see MATROSOV, V.M.

VASIL'EV, YU.L. [1962] *Irreducible disjunctive normal forms for certain classes of truth functions (Russian)* (J 0023) Dokl Akad Nauk SSSR 142*263-265
  • TRANSL [1962] (J 0062) Sov Math, Dokl 3*43-46
  ⋄ B05 ⋄ REV MR 33 # 1226 Zbl 124.246 • ID 13834

VASIL'EV, YU.L. [1963] *Comparison of the complexity of terminal and minimal disjunctive normal forms (Russian)* (J 0052) Probl Kibern 10*5-61
  ⋄ B05 ⋄ REV MR 29 # 5717 • ID 13835

VASIL'EV, YU.L. [1964] *On the "superposition" of reduced disjunctive normal forms (Russian)* (J 0052) Probl Kibern 12*239-242
  ⋄ B05 ⋄ REV MR 30 # 4675 Zbl 261 # 94045 • ID 65891

VASIL'EV, YU.L. [1964] *On the number of terminal and minimal disjunctive normal forms (Russian)* (J 0071) Met Diskr Analiz (Novosibirsk) 2*3-9
  ⋄ B05 ⋄ REV MR 29 # 2173 • ID 13836

VASIL'EV, YU.L. [1973] *Ein Zusammenhang zwischen Abschaetzungen in der Theorie der disjunktiven Normalformen und in der kombinatorischen Analysis* (J 1670) Mitt Math Ges DDR 1973/2-3*152-160
  ⋄ B05 ⋄ REV Zbl 272 # 94023 • ID 65892

VASIL'EV, YU.L. [1978] *Massive classes of dense boolean functions. I (Russian)* (J 0071) Met Diskr Analiz (Novosibirsk) 32*21-33,95
  ⋄ B05 G05 ⋄ REV MR 81e:94046 Zbl 451 # 94011 • ID 83060

VASILIOU, P. [1937] *Ueber den gegenwaertigen Stand der axiomatischen Methode* (J 1699) Bull Soc Math Grece 17*119-136
  ⋄ B30 ⋄ REV FdM 63.820 • ID 41042

VASWANI, P.K.T. [1959] see PORTER, A.

VAUGHT, R.L. [1954] *On sentences holding in direct products of relational systems* (P 0575) Int Congr Math (II, 7);1954 Amsterdam 2*409-410
  ⋄ B20 B25 C30 ⋄ ID 41195

VAUGHT, R.L. [1958] see CRAIG, W.

VAUGHT, R.L. [1974] see ADDISON, J.W.

VAUGHT, R.L. see Vol. III, IV, V for further entries

VAVASIS, S.A. [1982] see LIEBERHERR, K.J.

VAZHENIN, YU.M. [1983] see ROZENBLAT, B.V.

VAZHENIN, YU.M. see Vol. III, IV, V for further entries

VEBLEN, O. [1904] *A system of axioms for geometry* (J 0064) Trans Amer Math Soc 5*343-384
  ⋄ B30 ⋄ REV FdM 35.495 • ID 13916

VEBLEN, O. see Vol. V, VI for further entries

VEENKER, G. [1967] *Beweisalgorithmen fuer die Praedikatenlogik* (J 0373) Comp Arch Inform & Numerik 2*263-283
  ⋄ B35 ⋄ REV MR 37 # 6059 Zbl 253 # 68022 • ID 65903

VEENKER, G. [1969] *Ein Entscheidungsverfahren fuer den Aussagenkalkuel und seine Realisierung in einem Rechenautomaten* (J 2068) Grundlstud Kybern Geistwiss 4*127-136
  ⋄ B35 ⋄ ID 46540

VEENKER, G. [1969] see NIETHAMMER, W.

VEENKER, G. [1971] *Maschinelles Beweisen* (J 1633) Angew Inf 13*276-282
  ⋄ B35 ⋄ REV Zbl 216.241 • ID 46541

VEENKER, G. [1971] see HOFFMANN, GEERD-RUEDIGER

VEITCH, E.W. [1952] *A chart method for simplifying truth functions* (P 0700) ACM Proc Conf;1952 Pittsburgh 127-133
  ⋄ B05 ⋄ REV JSL 19.56 • ID 42244

VEITCH, E.W. [1960] *Logical equation minimization involving higher order solutions* (P 0696) Inform Processing (1);1959 Paris 423-424
  ⋄ B15 ⋄ ID 27692

VELDMAN, W. [1975] see LOPEZ-ESCOBAR, E.G.K.

VELDMAN, W. see Vol. V, VI for further entries

VELTMAN, F. [1984] see LANDMAN, F.

VENINI, E. [1979] see CIPPO, C.P.

VENKATESH, G. [1985] *A decision method for temporal logic based on resolution* (P 4672) Found of Softw Tech & Th Comput Sci (5);1985 New Delhi 272-289
  ⋄ B35 B45 ⋄ ID 49739

VENN, J. [1881] *Symbolic logic* (X 1253) Crowell Collier & Macmillan: New York xxxix+446pp
- LAST ED [1971] (X 0848) Chelsea: New York xxxviii+540pp
  ⋄ A10 B20 B98 ⋄ REV MR 52#13290 Zbl 263#01030 JSL 37.614 • ID 21755

VENNE, M. [1967] *Ultraproduits de structures d'ordres superieurs* (J 2313) C R Acad Sci, Paris, Ser A-B 265*A305-A308
  ⋄ B15 C20 ⋄ REV MR 37#47 Zbl 204.4 • ID 13929

VENNE, M. see Vol. VI for further entries

VENNERI, B.M. [1975] *Semantic implications of Herbrand's theory of fields* (J 0286) Int Logic Rev 12*204-214,215-225
  ⋄ A05 B10 C07 ⋄ REV Zbl 332#02007 • ID 65905

VENNERI, B.M. [1984] see RONCHI DELLA ROCCA, S.

VENNERI, B.M. see Vol. VI for further entries

VENTURINI ZILLI, M. [1974] *Su alcune strategie della risoluzione* (J 3436) Quad, Ist Appl Calcolo, Ser 3 8*137-162
  ⋄ B35 ⋄ REV Zbl 427#68070 • ID 69989

VENTURINI ZILLI, M. [1976] *Complexity of the unification algorithm for first-order expressions* (J 0089) Calcolo 12*361-371
  ⋄ B35 ⋄ REV MR 54#6585 Zbl 332#68063 • ID 65906

VENTURINI ZILLI, M. see Vol. IV, VI for further entries

VERDU I SOLANS, V. [1979] *Distributive and boolean logics (Catalan) (English summary)* (J 2840) Stochastica, Univ Politec Barcelona 3/2*97-108
  ⋄ B22 G10 ⋄ REV MR 81d:03058 Zbl 419#03041 • ID 53384

VERDU I SOLANS, V. [1982] *Algebras characterized by means of inference rules (Catalan)* (P 3870) Congr Catala de Log Mat (1);1982 Barcelona 129-130
  ⋄ B22 C05 G05 G10 ⋄ REV MR 84i:03003 Zbl 518#03027 • ID 37524

VERDU I SOLANS, V. [1985] *Some algebraic structures determined by closure operators* (J 0068) Z Math Logik Grundlagen Math 31*275-278
  ⋄ B30 C05 G05 G10 ⋄ REV Zbl 553#03041 • ID 43358

VERDU I SOLANS, V. see Vol. II, III, V, VI for further entries

VERKHOZIN, O.M. [1972] see OTEPANOV, V.I.

VEROFF, R. [1983] see HENSCHEN, L.J.

VEROFF, R. [1984] see MCCUNE, W.

VERSHININ, K.P. [1972] see GLUSHKOV, V.M.

VERSHININ, K.P. [1973] *The connection of a formal language for writing down mathematical theories with the axiomatic systems of set theory (Russian) (English summary)* (J 0040) Kibernetika, Akad Nauk Ukr SSR 1973/4*68-73
  ⋄ B15 E30 E70 ⋄ REV MR 48#8234 Zbl 295#02036 • ID 13934

VERSHININ, K.P. [1978] see GERGELY, T.

VERSHININ, K.P. [1979] *About using of auxiliary statements during search of proof (Russian)* (S 2582) Semiotika & Inf, Akad Nauk SSSR 12*5-8
  ⋄ B35 ⋄ ID 33378

VERSHININ, K.P. [1981] see GERGELY, T.

VERSHININ, K.P. [1983] see MOROKHOVETS, M.K.

VERSHININ, K.P. see Vol. III, IV for further entries

VESELY, A. [1981] *Logically orientated Cluster analysis* (J 0156) Kybernetika (Prague) 17*82-92
  ⋄ B80 ⋄ REV MR 83e:62083 Zbl 456#62052 • ID 54320

VESLEY, R.E. [1981] *An intuitionistic infinitesimal calculus* (P 3146) Constr Math;1980 Las Cruces 208-212
  ⋄ H05 ⋄ REV MR 83a:03063 Zbl 463#03034 • ID 54574

VESLEY, R.E. see Vol. VI for further entries

VIKTOROVA, I.I. [1978] see LEONTOVICH, A.M.

VIKULIN, A.P. [1974] *An estimate of the number of conjunctions in abbreviated disjunctive normal forms (Russian)* (J 0052) Probl Kibern 29*151-166,245
  ⋄ B05 ⋄ REV MR 51#10026 Zbl 309#94050 • ID 17562

VILLA, G. [1972] *Su di una definizione dei numeri naturali* (J 0012) Boll Unione Mat Ital, IV Ser 6*71-79
  ⋄ B28 ⋄ REV MR 48#3672 Zbl 261#02001 • ID 30452

VILLA, G. see Vol. II for further entries

VILLARS, R. [1967] *Eine semantische Charakterisierung der durch die Implikation allein darstellbaren Wahrheitsfunktionen* (J 0009) Arch Math Logik Grundlagenforsch 10*34-36
  ⋄ B20 ⋄ REV MR 35#4087 Zbl 258#02011 • ID 13944

VINCENTI, R. [1979] see COLETTI, G.

VINNER, S. [1976] *Implicit axioms, ω-rule and the axiom of induction in high school mathematics* (J 0005) Amer Math Mon 83*561-566
  ⋄ B28 C75 ⋄ REV MR 54#4891 • ID 24084

VINNER, S. see Vol. III, IV, V for further entries

VINOGRADOV, A.P. [1979] *A theorem on the local rank of one-place predicates (Russian)* (J 0199) Zh Vychisl Mat i Mat Fiz 19*787-790,800
- TRANSL [1979] (J 1049) USSR Comput Math & Math Phys 19/3*248-251
  ⋄ B10 ⋄ REV MR 82h:03008 Zbl 417#68054 • ID 53284

VINOGRADOV, A.P. see Vol. IV for further entries

VINOGRADSKAYA, T.M. [1980] see RUBCHINSKIJ, A.A.

VISHKIN, U. [1984] see CHANDRA, A.K.

VISHKIN, U. see Vol. IV for further entries

VITYAEV, E.E. [1978] *Regularities in the language of empirical systems (Russian)* (S 0507) Vychisl Sist (Akad Nauk SSSR Novosibirsk) 76*3-14,154
  ⋄ B03 ⋄ REV MR 80k:68083 Zbl 436#62013 • ID 83078

VITYAEV, E.E. see Vol. II for further entries

VOBACH, A.R. [1977] *The weak topology on logical calculi* (J 0047) Notre Dame J Formal Log 18*436-440
  ⋄ B10 ⋄ REV MR 57#15948 Zbl 306#02013 • ID 24244

VOGHERA, N. [1960] see PRAWITZ, D.

VOJSHVILLO, E.K. [1958] *Method of simplification of the forms of truth functional expressions (Russian)* (J 4170) Nauch Dokl Vys Shk, Fiz Mat (Moskva) 1958/2*
  ⋄ B05 ⋄ ID 45783

VOJSHVILLO, E.K. see Vol. II, III for further entries

VOJTISHEK, V.V. [1966] *On an approach to the classification of Boolean functions (Russian)* (**J** 0071) Met Diskr Analiz (Novosibirsk) 8*35-41
 ⋄ B05 ⋄ REV MR 34 # 4061 Zbl 154.11 JSL 35.593
 • ID 43438

VOLKOV, V.A. [1978] *Elements of set theory and the development of the concept of number (Russian)* (**X** 0938) Leningrad Univ: Leningrad 83pp
 ⋄ A05 B28 E98 ⋄ REV MR 81h:04001 • ID 79749

VOLLMANN, F. [1983] see BENDA, W.

VOLLRATH, H.J. [1969] *Einige neuere Beweismethoden in der Logik* (**J** 2345) Ueberblicke Math 3*97-111
 ⋄ B05 ⋄ REV MR 42 # 4378 Zbl 236 # 02009 • ID 20960

VOROS, A. [1973] *Introduction to non-standard analysis* (**J** 0209) J Math Phys 14*292-296
 ⋄ H05 ⋄ REV MR 48 # 3378 Zbl 255 # 02060 • ID 13888

VOSAHLO, J. [1978] see HAVRANEK, T.

VRANESIC, Z.G. [1974] see SHEPPARD, D.A.

VRANESIC, Z.G. see Vol. II for further entries

VUCKOVIC, V. [1960] *Rekursive Modelle einiger nichtklassischer Aussagenkalkuele (Serbo-Croatian)* (**J** 1179) Filoz Jugos Chas (Belgrade) 4*69-84
 ⋄ B22 F30 ⋄ REV JSL 28.291 • ID 22133

VUCKOVIC, V. see Vol. II, III, IV, V, VI for further entries

VUJOSEVIC, S.T. [1979] *On the limits of the families of Lindenbaum algebras* (**J** 0400) Publ Inst Math, NS (Belgrade) 26(40)*293-296
 ⋄ B22 G05 ⋄ REV MR 81f:03073 Zbl 441 # 03025
 • ID 56078

VUJOSEVIC, S.T. see Vol. II for further entries

VUYSJE, D. [1954] *L'importance du point de depart psycho-linguistique pour l'application de la logistique aux sciences non mathematiques* (**P** 0646) Appl Sci de Log Math;1952 Paris 155-156
 ⋄ B30 ⋄ REV Zbl 56.249 • ID 27975

WAACK, S. [1985] see KRIEGEL, K.

WAACK, S. see Vol. IV for further entries

WADA, T. [1955] see IZUMI, Y.

WADLEIGH, H.J. [1970] *Expressibility in type theory* (**J** 0047) Notre Dame J Formal Log 11*257-288
 ⋄ B15 ⋄ REV MR 43 # 6034 Zbl 169.302 • ID 13961

WADLEIGH, H.J. [1974] *Translation of the simple theory of types into a first order language* (**J** 0047) Notre Dame J Formal Log 15*432-442
 ⋄ B15 ⋄ REV MR 50 # 6828 Zbl 225 # 02011 • ID 13962

WAESCHE, H. [1926] *Grundzuege zu einer Logik der Arithmetik* (**X** 1281) Heymanns: Koeln 43pp
 ⋄ B28 ⋄ REV FdM 52.44 • ID 25619

WAESCHE, H. [1974] *Ueber den indirekten Beweis* (**J** 0160) Math-Phys Sem-ber, NS 21*5-11
 ⋄ A05 B05 ⋄ REV MR 56 # 5189 • ID 79825

WAGNER, K. [1972] *Zur Axiomatisierung eines sequentiellen Aussagenkalkuels (Russian, English and French summaries)* (**J** 0115) Wiss Z Humboldt-Univ Berlin, Math-Nat Reihe 21*471-472
 ⋄ B22 D05 ⋄ REV MR 48 # 8212 Zbl 251 # 94032
 • ID 13969

WAGNER, K. see Vol. II, IV, V for further entries

WAHLSTER, W. (ED.) [1982] *GWAI-82* (**X** 0811) Springer: Heidelberg & New York vi+246pp
 ⋄ B97 ⋄ REV MR 85d:68009 Zbl 497 # 00016 • ID 38967

WAID, C.C. [1973] see KOMKOV, V.

WAID, C.C. see Vol. III for further entries

WAISMANN, F. [1936] *Einfuehrung in das mathematische Denken. Die Begriffsbildung der modernen Mathematik* (**X** 1007) Gerold: Wien viii+188pp
 • TRANSL [1939] (**X** 1262) Einaudi: Torino 326pp
 ⋄ A05 B98 ⋄ REV Zbl 29.242 JSL 13.117 FdM 62.1046 FdM 65.20 • REM Foreword by K.Menger. 2nd ed. 1947, vii+168pp • ID 40876

WAISMANN, F. [1951] *Introduction to mathematical thinking. The formation of concepts in modern mathematics* (**X** 1221) Ungar: New York & London xi+260pp
 ⋄ B98 ⋄ REV MR 13.899 Zbl 45.148 JSL 17.208
 • ID 25728

WAJSBERG, M. [1932] *Ein neues Axiom des Aussagenkalkuels in der Symbolik von Sheffer* (**J** 0124) Monatsh Math-Phys 39*259
 • TRANSL [1977] (**C** 4055) Wajsberg: Logical Works 37-39
 ⋄ B05 ⋄ REV Zbl 5.338 JSL 48.873 FdM 58.64 • ID 16536

WAJSBERG, M. [1933] *Beitrag zur Metamathematik* (**J** 0043) Math Ann 109*200-229
 • TRANSL [1977] (**C** 4055) Wajsberg: Logical Works 62-88 [1935] (**J** 4710) Pol Tow Mat, Wiad Mat 39*43-84
 ⋄ A05 B10 B30 B96 C35 F99 ⋄ REV Zbl 8.97 FdM 59.53 FdM 61.972 • ID 40808

WAJSBERG, M. [1933] *Ein erweiterter Klassenkalkuel* (**J** 0124) Monatsh Math-Phys 40*113-126
 • TRANSL [1977] (**C** 4055) Wajsberg: Logical Works 50-61
 ⋄ B20 E20 ⋄ REV Zbl 7.98 JSL 48.873 FdM 59.54
 • ID 16533

WAJSBERG, M. [1933] *Untersuchungen ueber den Funktionenkalkuel fuer endliche Individuenbereiche* (**J** 0043) Math Ann 108*218-228
 • TRANSL [1977] (**C** 4055) Wajsberg: Logical Works 40-49
 ⋄ B10 C13 ⋄ REV Zbl 6.242 JSL 48.873 FdM 59.864
 • ID 13979

WAJSBERG, M. [1937] *Metalogische Beitraege I* (**J** 4710) Pol Tow Mat, Wiad Mat 43*131-168
 • TRANSL [1977] (**C** 4055) Wajsberg: Logical Works 172-200 [1967] (**C** 0615) Polish Logic 1920-39 285-318
 ⋄ A05 B20 ⋄ REV Zbl 16.98 JSL 2.93 JSL 35.442 JSL 48.873 FdM 63.826 • REM Part II 1939 • ID 41049

WAJSBERG, M. [1939] *Metalogische Beitraege II* (**J** 4710) Pol Tow Mat, Wiad Mat 47*119-139
 • TRANSL [1977] (**C** 4055) Wajsberg: Logical Works 201-214 [1967] (**C** 0615) Polish Logic 1920-39 319-334
 ⋄ A05 B20 ⋄ REV Zbl 20.337 JSL 35.442 JSL 48.873 JSL 5.31 FdM 65.1105 • REM Part I 1937 • ID 13983

WAJSBERG, M. see Vol. II, III, VI for further entries

WAKITA, H. [1984] *Mathematical framework of quantum electrodynamics* (J 0352) Math Jap 29*199-217
⋄ H10 ⋄ REV MR 86b:81145a Zbl 559 # 46035 • ID 45292

WAKITA, H. [1984] *Solutions of the renormalized Tomonaga-Schwinger equation* (J 0352) Math Jap 29*321-328
⋄ H10 ⋄ REV MR 86b:81145b Zbl 547 # 46054 • ID 45291

WALDINGER, R. [1981] see MANNA, Z.

WALDINGER, R. [1985] see MANNA, Z.

WALIGORA, G. [1975] see KOPIEKI, R.

WALIGORSKI, S. [1961] *Calculation of the Quine's table for truth functions* (S 4556) Prace Zaklad Aparat Mat, Pol Akad Nauk, Ser A 15*5pp
⋄ B05 ⋄ REV MR 25 # 1100 Zbl 127.11 • ID 42861

WALIGORSKI, S. [1962] *On normal equivalents of truth functions* (J 0485) Algorytmy, Pol Akad Nauk 1*73-96
⋄ B05 G05 ⋄ REV MR 33 # 2528 Zbl 118.330 • ID 13991

WALIGORSKI, S. see Vol. II, IV for further entries

WALKOE JR., W.J. [1970] *Finite partially-ordered quantification* (J 0036) J Symb Logic 35*535-555
⋄ B15 C80 ⋄ REV MR 43 # 4646 Zbl 219 # 02008 JSL 40.239 • ID 13998

WALKOE JR., W.J. [1973] see KEISLER, H.J.

WALKOE JR., W.J. [1976] *A small step backwards* (J 0005) Amer Math Mon 83*338-344
⋄ B10 C80 ⋄ REV MR 58 # 16130 Zbl 339 # 02010 • ID 27272

WALLET, G. [1981] *Holonomie et cycle evanouissant* (J 0240) Ann Inst Fourier 31*181-186
⋄ H05 ⋄ REV MR 83c:57013 Zbl 469 # 57020 • ID 55184

WALSH, M.J. [1962] see HENKIN, L.

WALSH, T.R.S. [1983] see DEWDNEY, M.

WALSH, T.R.S. see Vol. IV for further entries

WALTER, M. [1973] see BAGINSKI, M.

WALTERS, A. [1962] see SUMMERSBEE, S.

WALTERS, A. [1963] see SUMMERSBEE, S.

WALTERS, R.F.C. [1973] see STREET, R.

WALTERS, R.F.C. [1978] see STREET, R.

WALTERS, R.F.C. see Vol. V for further entries

WALTHER, C. [1984] *Ein mehrsortiger Resolutionskalkuel mit Paramodulation* (S 3126) Ber, Fak Inf, Univ Karlsruhe 23/84*125pp
⋄ B35 B75 ⋄ REV Zbl 566 # 03006 • ID 48732

WALTHER, C. [1985] *A mechanical solution of Schubert's steamroller by many-sorted resolution* (J 0503) Artif Intell 26*217-224
⋄ B35 ⋄ ID 47250

WAN, ZHEXIAN [1960] see LU, YANGCI

WAN, ZHEXIAN [1980] see DAI, ZONGDUO

WANG, HAO [1947] *A note on Quine's principles of quantification* (J 0036) J Symb Logic 12*130-132
⋄ B10 ⋄ REV MR 9.403 Zbl 29.100 JSL 13.115 • ID 14004

WANG, HAO [1948] *A new theory of element and number* (J 0036) J Symb Logic 13*129-137
⋄ B28 E70 ⋄ REV MR 10.229 Zbl 32.99 JSL 14.69 • ID 14005

WANG, HAO [1949] *A theory of constructive types* (J 0175) Methodos 1*374-384
⋄ B15 F35 F65 ⋄ REV MR 12.384 Zbl 36.165 JSL 19.288 • ID 14006

WANG, HAO [1950] *A proof of independence* (J 0005) Amer Math Mon 57*99-100
⋄ B05 ⋄ REV MR 11.411 Zbl 35.8 JSL 15.138 • ID 14014

WANG, HAO [1950] see ROSSER, J.B.

WANG, HAO [1950] *Remarks on the comparison of axiom systems* (J 0054) Proc Nat Acad Sci USA 36*448-453
⋄ B15 F25 F30 ⋄ REV MR 12.578 Zbl 37.296 JSL 16.142 • ID 14011

WANG, HAO [1950] *Set-theoretical basis for real numbers* (J 0036) J Symb Logic 15*241-247
⋄ B28 E75 ⋄ REV MR 12.664 Zbl 40.293 JSL 16.216 • ID 14010

WANG, HAO [1952] *Logic of many-sorted theories* (J 0036) J Symb Logic 17*105-116
⋄ B10 ⋄ REV MR 14.3 Zbl 49.148 JSL 18.77 • ID 14017

WANG, HAO [1952] *Negative types* (J 0094) Mind 61*366-368
⋄ A05 B15 ⋄ REV MR 17.447 Zbl 49.148 JSL 19.288 • ID 14018

WANG, HAO [1953] *Between number theory and set theory* (J 0043) Math Ann 126*385-409
• REPR [1963] (C 1009) Wang: Survey Math Logic 478-506
⋄ B28 E30 F30 ⋄ REV MR 15.670 Zbl 51.246 JSL 22.82 JSL 30.100 • ID 14023

WANG, HAO [1953] *Quelques notions d'axiomatique* (J 0252) Rev Philos Louvain 51*409-443
⋄ B25 B30 C07 C35 E30 F30 ⋄ REV JSL 20.289 • ID 14024

WANG, HAO [1954] *The formalization of mathematics* (J 0036) J Symb Logic 19*241-266
• REPR [1963] (C 1009) Wang: Survey Math Logic 559-584
⋄ A05 B30 E70 F35 F65 ⋄ REV MR 16.661 Zbl 56.245 JSL 22.290 JSL 30.100 • ID 14025

WANG, HAO [1955] *On denumerable bases of formal systems* (P 1589) Math Interpr of Formal Systs;1954 Amsterdam 57-84
⋄ B10 E25 E30 E35 E70 ⋄ REV MR 17.700 Zbl 68.245 JSL 22.292 • ID 27722

WANG, HAO [1955] *On formalization* (J 0094) Mind 64*226-238
• LAST ED [1967] (C 0721) Contemp Readings in Log Th 29-40 • REPR [1963] (C 1009) Wang: Survey Math Logic 57-67
⋄ A05 B98 ⋄ REV JSL 22.292 • ID 38706

WANG, HAO [1955] *Undecidable sentences generated by semantic paradoxes* (J 0036) J Symb Logic 20*31-43
• REPR [1963] (C 1009) Wang: Survey Math Logic 68-81
⋄ A05 B28 F30 ⋄ REV MR 16.988 Zbl 64.245 JSL 30.100 • ID 14026

WANG, HAO [1957] *The axiomatization of arithmetic* (J 0036) J Symb Logic 22*145-158
- ◇ A05 B28 F30 ◇ REV MR 20 # 6963 Zbl 78.5 JSL 27.77
- • ID 33305

WANG, HAO [1958] *Eighty years of foundational studies (German and French summaries)* (J 0076) Dialectica 12*466-497
- ◇ A05 A10 B98 ◇ REV MR 21 # 9 Zbl 90.8 JSL 28.173
- • ID 14029

WANG, HAO [1959] *Circuit synthesis by solving sequential boolean equations* (J 0068) Z Math Logik Grundlagen Math 5*291-322
- • REPR [1963] (C 1009) Wang: Survey Math Logic 269-305
- ◇ B05 B70 ◇ REV MR 23 # A3090 Zbl 89.247 JSL 25.373
- • ID 21257

WANG, HAO [1960] *Proving theorems by pattern recognition I* (J 0212) ACM Commun 3*220-234
- • REPR [1983] (C 4659) Autom of Reasoning 1*229-243
- ◇ B35 ◇ REV Zbl 101.105 JSL 32.119 • REM Part II 1961
- • ID 14034

WANG, HAO [1960] *Toward mechanical mathematics* (J 0284) IBM J Res Dev 4*2-22
- • REPR [1983] (C 4659) Autom of Reasoning 1*244-266
- ◇ A05 B35 ◇ REV MR 22 # 4129 Zbl 97.4 JSL 30.249
- • ID 14035

WANG, HAO [1961] *Proving theorems by pattern recognition II* (J 0432) Bell Syst Tech J 40*1-41
- ◇ B35 ◇ REV JSL 32.119 • REM Part I 1960 • ID 14037

WANG, HAO [1962] *A survey of mathematical logic* (X 1876) Kexue Chubanshe: Beijing x+652pp
- • LAST ED [1971] (X 0848) Chelsea: New York x+651pp
- ◇ B96 E96 ◇ REV MR 27 # 2394 MR 42 # 4375 Zbl 106.236 Zbl 212.311 • REM Co-publ.: X0809; title of last edition: 'Logic, computers and sets' • ID 32531

WANG, HAO [1962] see DREBEN, B.

WANG, HAO [1962] see KAHR, A.S.

WANG, HAO [1963] *Dominoes and the AEA case of the decision problem* (P 0674) Symp Math Th of Automata;1962 New York 23-55
- ◇ B20 B25 D05 D35 ◇ REV MR 29 # 4688 Zbl 137.10
- • ID 14040

WANG, HAO [1963] *Many-sorted predicate calculi* (C 1009) Wang: Survey Math Logic 322-333
- ◇ B10 ◇ REV JSL 28.250 • ID 14052

WANG, HAO [1963] *Mechanical mathematics and inferential analysis* (C 0659) Computer Progr & Formal Syst 1-20
- ◇ B35 ◇ REV MR 30 # 1939 Zbl 112.9 JSL 32.120
- • ID 14042

WANG, HAO [1963] *Relative strength and reducibility* (C 1009) Wang: Survey Math Logic 432-442
- ◇ B10 E30 E70 F25 F30 ◇ REV JSL 29.208 • ID 14050

WANG, HAO [1963] *Some formal details on predicative set theories* (C 1009) Wang: Survey Math Logic 585-623
- ◇ B15 E70 F65 ◇ REV JSL 30.250 • ID 14044

WANG, HAO [1963] *Some partial systems* (C 1009) Wang: Survey Math Logic 507-534
- ◇ B28 E70 ◇ REV JSL 30.100 • ID 43035

WANG, HAO [1963] *The mechanization of mathematical arguments* (P 0719) Exper Arith, High Speed Comput & Math;1962 Chicago 31-40
- ◇ B35 ◇ REV MR 30 # 736 JSL 32.120 • ID 21987

WANG, HAO [1963] *The predicate calculus* (C 1009) Wang: Survey Math Logic 307-321
- ◇ B98 ◇ REV JSL 28.250 • ID 14053

WANG, HAO [1963] *Toward mechanical mathematics* (C 1128) Modelling of Mind: Computers & Intelligence 91-120
- ◇ B35 ◇ REV Zbl 106.8 JSL 30.249 • ID 22113

WANG, HAO [1964] see DREBEN, B.

WANG, HAO [1965] *Formalization and automatic theorem-proving* (P 0573) Inform Processing (3);1965 New York 1*51-58
- ◇ B35 ◇ REV Zbl 209.33 JSL 39.350 • ID 14055

WANG, HAO [1965] *Logic and computers* (J 0005) Amer Math Mon 72*135-140
- ◇ A05 B35 D03 D10 ◇ REV MR 30 # 4639 Zbl 123.335 JSL 31.264 • ID 14058

WANG, HAO [1965] *Note on rules of inference* (J 0068) Z Math Logik Grundlagen Math 11*193-196
- ◇ B05 B22 ◇ REV MR 31 # 3305 Zbl 143.249 JSL 40.604
- • ID 14056

WANG, HAO [1965] *Remarks on machines, sets, and the decision problem* (P 0688) Logic Colloq;1963 Oxford 304-320
- ◇ B20 B25 D03 D05 D10 D35 E30 ◇ REV MR 39 # 6729 Zbl 133.254 • ID 14057

WANG, HAO [1970] *On the long-range prospects of automatic theorem-proving* (P 0625) Symp Autom Demonst;1968 Versailles 101-111
- ◇ B35 ◇ REV Zbl 216.240 • ID 31303

WANG, HAO [1976] see DUNHAM, B.

WANG, HAO [1981] *Popular lectures on mathematical logic (Chinese)* (X 1876) Kexue Chubanshe: Beijing vii+257pp
- • TRANSL [1981] (X 0864) Van Nostrand: New York ix+273pp
- ◇ A05 B98 C98 E98 F30 F98 ◇ REV MR 82e:03001 MR 84g:03002 JSL 47.908 • ID 34116

WANG, HAO [1984] *Computer theorem proving and artificial intelligence* (P 3084) Autom Theor Prov After 25 Yea;1983 Denver 49-70
- ◇ B35 ◇ REV MR 85d:68005 Zbl 566 # 68074 • ID 45273

WANG, HAO see Vol. II, III, IV, V, VI for further entries

WANG, JUENTIN [1973] *On the representation of generative grammars as first-order theories* (P 0580) Int Congr Log, Meth & Phil of Sci (4,Sel Pap);1971 Bucharest 302-316
- ◇ B10 D05 ◇ REV MR 57 # 8200 Zbl 296 # 68079
- • ID 65982

WANG, SHIQIANG [1953] *An axiom system for the proposition calculus (Chinese)* (J 0418) Shuxue Xuebao 2*267-274
- ◇ B05 ◇ REV Zbl 51.5 JSL 37.185 • ID 43897

WANG, SHIQIANG see Vol. II, III, V for further entries

WANG, SHITIE [1981] *The HOI resolution and HOL resolution in Horn sets (Chinese) (English summary)* (J 3923) Xiamen Daxue Xuebao, Ziran Kexue 20*290-296
- ◇ B35 ◇ REV Zbl 484 # 03004 • ID 36858

WANG, SHUTANG [1979] *A generalized number system and its application I (Chinese) (English summary)* (J 1024) Zhongguo Kexue 22*1-11
⋄ H05 ⋄ REV MR 84j:03141 • ID 34734

WANG, SHUTANG [1985] *Generalized number system and its applications I* (J 2606) Tsukuba J Math 9*203-215
⋄ B28 ⋄ ID 49744

WANG, SHUTANG see Vol. V for further entries

WANG, XIANGHAO [1943] *A system of completely independent axioms for the sequence of natural numbers* (J 0036) J Symb Logic 8*41-44
⋄ B28 ⋄ REV MR 5.85 JSL 8.84 • ID 14067

WANG, XIANGHAO [1965] *Sequential Boolean equations (Chinese)* (P 4564) Math Logic;1963 Xi-An 58-64
⋄ B05 ⋄ ID 49311

WANG, XIANGHAO [1982] see LIU, XUHUA

WANG, XIANGHAO [1983] see LIU, XUHUA

WANG, XIANJUN [1956] *The truth functions in mathematical logic is the logical abstraction of composite propositions (Chinese)* (J 2521) Beijing Shifan Daxue Xuebao, Ziran Kexue 1956/2*47-57
⋄ B05 ⋄ ID 48539

WANG, XIANJUN [1982] *Introduction to mathematical logic (Chinese)* (X 3123) Beijing Shifan Daxue: Beijing 372pp
⋄ B98 ⋄ ID 48542

WARNER, M.W. [1983] *Lattices and lattice-valued relations in biology* (J 3073) Bull Math Biol 45*193-207
⋄ B80 G10 ⋄ REV MR 84k:92033 Zbl 518 # 92001
• ID 39286

WARNER, M.W. see Vol. IV, V for further entries

WARREN, D.W. [1954] see BURKS, A.W.

WARREN, D.W. see Vol. IV for further entries

WARTOFSKY, M.W. [1983] see COHEN, R.S.

WASILEWSKA, A. [1978] *Machines, logics and decidability* (J 2095) Fund Inform, Ann Soc Math Pol, Ser 4 1*291-303
⋄ B25 B35 ⋄ REV MR 58 # 9932 Zbl 386 # 68083
• ID 69996

WASILEWSKA, A. [1985] *Monadic second-order definability as a common characterization of finite automata, certain classes of programs and logics* (J 2095) Fund Inform, Ann Soc Math Pol, Ser 4 8/3-4*309-320
⋄ B15 B75 C40 D05 ⋄ ID 49467

WASILEWSKA, A. [1985] *Some remarks on theorem proving systems and Mazurkiewics algorithms associated with them* (J 0068) Z Math Logik Grundlagen Math 31*289-294
⋄ B35 ⋄ ID 47554

WASILEWSKA, A. see Vol. II, III, IV, VI for further entries

WASSERMAN, H.C. [1974] *Admissible rules, derivable rules, and extendible logistic systems* (J 0047) Notre Dame J Formal Log 15*265-278
⋄ B22 ⋄ REV MR 50 # 1868 Zbl 226 # 02007 • ID 14076

WASSERMAN, H.C. [1976] *A note of evaluation mappings* (J 0047) Notre Dame J Formal Log 17*613-614
⋄ B05 B45 ⋄ REV MR 55 # 5440 Zbl 258 # 02009
• ID 65990

WASSERMAN, H.C. [1978] *A second order axiomatic theory of strings* (J 0047) Notre Dame J Formal Log 19*629-633
⋄ B05 ⋄ REV MR 80d:03038 Zbl 316 # 02060 • ID 52117

WASSERMAN, H.C. see Vol. II, III for further entries

WATSON, F.R. [1978] see BAXANDALL, P.R.

WATTENBERG, F. [1969] see BERNSTEIN, A.R.

WATTENBERG, F. [1971] *Nonstandard topology and extensions of monad systems to infinite points* (J 0036) J Symb Logic 36*463-476
⋄ H05 ⋄ REV MR 45 # 2642 Zbl 239 # 54034 • ID 14093

WATTENBERG, F. [1973] *Monads of infinite points and finite product spaces* (J 0064) Trans Amer Math Soc 176*351-368
⋄ H05 ⋄ REV MR 47 # 1020 Zbl 255 # 54036 • ID 14094

WATTENBERG, F. [1974] see BERNSTEIN, A.R.

WATTENBERG, F. [1974] *Two topologies with the same monads* (P 1083) Victoria Symp Nonstand Anal;1972 Victoria 303-312
⋄ H05 ⋄ REV MR 58 # 31040 Zbl 272 # 54038 • ID 27652

WATTENBERG, F. [1977] *Nonstandard measure theory-Hausdorff measure* (J 0053) Proc Amer Math Soc 65*326-331
⋄ H05 ⋄ REV MR 56 # 2817 Zbl 365 # 28016 • ID 51068

WATTENBERG, F. [1977] *Topologies on the set of closed subsets* (J 0048) Pac J Math 68*537-551
⋄ H05 ⋄ REV MR 58 # 18400 Zbl 327 # 54005 • ID 83111

WATTENBERG, F. [1978] *Nonstandard analysis and the theory of Shape* (J 0027) Fund Math 98*41-60
⋄ H05 ⋄ REV MR 80c:54057 Zbl 375 # 55011 • ID 29208

WATTENBERG, F. [1979] *Nonstandard measure theory - avoiding pathological sets* (J 0064) Trans Amer Math Soc 250*357-368
⋄ H05 ⋄ REV MR 80h:03095 Zbl 377 # 02033 • ID 51750

WATTENBERG, F. [1981] see HENSON, C.W.

WATTS, D.E. [1980] see COHEN, JACK K.

WEAVER, G.E. [1973] see LEBLANC, H.

WEAVER, G.E. [1974] *Finite partitions and their generators* (J 0068) Z Math Logik Grundlagen Math 20*255-260
⋄ B20 B25 C07 C40 ⋄ REV MR 51 # 45 Zbl 309 # 02056
• ID 14096

WEAVER, G.E. [1975] *Uniform compactness and interpolation theorems in sentential logic* (J 0302) Rep Math Logic, Krakow & Katowice 5*93-95 • ERR/ADD ibid 12*67
⋄ B05 C40 ⋄ REV MR 57 # 9471 MR 82e:03018
Zbl 352 # 02013 • ID 21907

WEAVER, G.E. [1977] see LEBLANC, H.

WEAVER, G.E. [1980] *A note on the compactness theorem in first order logic* (J 0068) Z Math Logik Grundlagen Math 26*111-113
⋄ B10 C07 ⋄ REV MR 81i:03050 Zbl 441 # 03011
• ID 56064

WEAVER, G.E. see Vol. II, III, V, VI for further entries

WEBB, P. [1970] *A pair of primitive rules for the sentential calculus* (J 0068) Z Math Logik Grundlagen Math 16*439-446
⋄ B05 ⋄ REV MR 44 # 3831 Zbl 179.10 • ID 14102

WEBB, S.M. [1973] see CHAPIN JR., E.W.

WEBB, S.M. [1975] *Non-standard probability* (J 0047) Notre Dame J Formal Log 16*397-401
⋄ H05 ⋄ REV MR 51 # 11602 Zbl 272 # 02080 • ID 14103

WEBER, H. [1906] *Elementare Mengenlehre* (J 0157) Jbuchber Dtsch Math-Ver 15*173-184
⋄ B28 E10 ⋄ REV FdM 37.68 • ID 37891

WEBER, K. [1975] *Ueber verschiedene Kompliziertheitsmasse bei alternativen Normalformen* (J 0129) Elektr Informationsverarbeitung & Kybern 11*606-607
⋄ B05 ⋄ REV MR 58 # 32068 Zbl 316 # 02015 • ID 66007

WEBER, K. [1977] *Beziehungen zwischen verschiedenen Kompliziertheitsmassen bei alternativen Normalformen* (S 2829) Rostocker Math Kolloq 3*5-35
⋄ B05 ⋄ REV MR 58 # 32068 Zbl 397 # 03039 • ID 52705

WEBER, K. [1982] *The length of random Boolean functions (German and Russian summaries)* (J 0129) Elektr Informationsverarbeitung & Kybern 18*659-668
⋄ B05 ⋄ REV MR 85d:06014 Zbl 513 # 94024 • ID 39077

WEBER, K. [1983] *Irredundant disjunctive normal forms of random Boolean functions (German and Russian summaries)* (J 0129) Elektr Informationsverarbeitung & Kybern 19*529-534
⋄ B05 ⋄ REV MR 85f:03064 Zbl 542 # 94037 • ID 40809

WECHSLER, H. [1975] *Applications of fuzzy logic to medical diagnosis* (P 1805) Int Symp Multi-Val Log (5,Proc);1975 Bloomington 162-174
⋄ B80 ⋄ ID 35821

WECHSUNG, G. (ED.) [1984] *Frege conference 1984. Proceedings of the International Conference held at Schwerin, Sept.10-14, 1984* (X 0911) Akademie Verlag: Berlin 408pp
⋄ A05 A10 B97 C97 ⋄ REV MR 85m:03006 Zbl 544 # 00005 • ID 40976

WECHSUNG, G. see Vol. IV, V for further entries

WEESE, M. [1980] see BAUDISCH, A.

WEESE, M. see Vol. III, IV, V, VI for further entries

WEGENER, I. [1980] *A new lower bound on the monotone network complexity of boolean sums* (J 1431) Acta Inf 13*109-114
⋄ B05 B70 ⋄ REV MR 81g:68075 Zbl 427 # 94033 • ID 66301

WEGENER, I. [1981] *An improved complexity hierarchy on the depth of boolean functions* (J 1431) Acta Inf 15*147-152
⋄ B05 B70 D15 ⋄ REV MR 82i:94036 Zbl 431 # 94058 • ID 69392

WEGENER, I. [1981] *Boolean functions whose monotone complexity is of size $n^2 \log n$* (P 3475) Theor Comput Sci (5);1981 Karlsruhe 104*22-31
⋄ B05 D15 ⋄ REV Zbl 456 # 94023 • ID 66300

WEGENER, I. [1982] *Boolean functions whose monotone complexity is of size $n^2/\log n$* (J 1426) Theor Comput Sci 21*213-224
⋄ B05 ⋄ REV MR 83k:68040 Zbl 488 # 94035 • ID 40332

WEGENER, I. [1984] *Optimal decision trees and one-time-only branching programs for symmetric Boolean functions* (J 0194) Inform & Control 62*129-143
⋄ B05 ⋄ ID 46390

WEGENER, I. [1984] *Proving lower bounds of the monotone complexity of Boolean functions* (P 2342) Symp Rek Kombin;1983 Muenster 446-456
⋄ B05 ⋄ REV Zbl 513 # 03005 • ID 47033

WEGENER, I. [1985] *On the complexity of slice functions* (J 1426) Theor Comput Sci 38*55-68
⋄ B05 ⋄ ID 48605

WEGENER, I. see Vol. II for further entries

WEGLORZ, B. [1968] see PACHOLSKI, L.

WEGLORZ, B. see Vol. II, III, V for further entries

WEHLAU, D.L. [1983] see DEWDNEY, M.

WEHRFRITZ, H. [1974] *Techniques for the transformation of logic equations* (J 0187) IEEE Trans Comp C-23*477-480
⋄ B35 ⋄ REV MR 55 # 5319 Zbl 288 # 94020 • ID 66032

WEIDIG, I. [1971] *Einfuehrung der ganzen Zahlen durch eine Ordnungsstruktur* (J 0160) Math-Phys Sem-ber, NS 18*213-224
⋄ B28 E07 ⋄ REV MR 51 # 7875 Zbl 223 # 02001 • ID 17592

WEINBERG, G.M. [1967] *Computing machines* (C 0601) Encycl of Philos 2*168-173
⋄ B35 D10 ⋄ REV JSL 35.298 • ID 14125

WEINER, P. [1969] see MCKELLAR, A.C.

WEINER, P. see Vol. IV for further entries

WEINFELD, T. [1979] see SUSZKO, R.

WEINGARTNER, P. [1985] see DORN, G.

WEINGARTNER, P. see Vol. II, III, V for further entries

WEISS, M. [1967] *Axiomatische Untersuchungen zur elementaren Theorie der freien Halbgruppen mit Substitution als undefiniertem Grundbegriff* (J 0068) Z Math Logik Grundlagen Math 13*265-280
⋄ B03 D05 ⋄ REV MR 35 # 6540 Zbl 229 # 20056 • ID 14130

WEISS, M. [1977] *Bemerkungen zu Bewertungsmoeglichkeiten von Erzeugnissen, deren Funktionsweise mit Hilfe boolescher Funktionen beschrieben werden kann* (J 3464) Wiss Z, Hochsch Verkehrswesen Dresden 24*709-712
⋄ B80 ⋄ REV Zbl 381 # 90061 • ID 51924

WEISS, M. see Vol. V for further entries

WEISS, P. [1928] *The theory of types* (J 0094) Mind 37*338-348
⋄ B15 ⋄ ID 38711

WELDING, S.O. [1976] *Logic as based on truth-value relations* (J 2076) Rev Int Philos 30*151-166
⋄ A05 B10 ⋄ REV MR 58 # 27285 • ID 79906

WELHAM, B. [1981] see BUNDY, A.

WELLS, M.B. [1963] *Application of a finite set covering theorem to the simplification of boolean function expressions* (P 0572) Inform Processing (2);1962 Muenchen 731-735
⋄ B05 ⋄ REV Zbl 139.326 • ID 27613

WELSH JR., P.J. [1962] see CLARK, R.

WELSH JR., P.J. see Vol. II, V for further entries

WENDELIN, H. [1948] *Ein Kriterium fuer die Erweiterbarkeit einer Implikation zu einer Aequivalenz* (J 0931) Anz Oesterr Akad Wiss, Math-Nat Kl 85*66-69
⋄ B05 ⋄ REV MR 11.1 Zbl 31.195 JSL 23.215 • ID 28759

WENDELIN, H. [1971] *Kurzer Weg zur Bestimmung der durch eine Aussagenverbindung dargestellte Wahrheitsfunktionen* (J 0127) J Reine Angew Math 250*117-123 • ERR/ADD ibid 259*220
⋄ B05 ⋄ REV MR 46 # 3262 Zbl 229 # 02012 • ID 14133

WERNICK, W. [1939] *An enumeration of logical functions* (J 0015) Bull Amer Math Soc 45*885-887
⋄ B05 ⋄ REV MR 1.131 Zbl 22.291 JSL 5.31 FdM 65.30 • ID 14144

WERNICK, W. [1940] *Functional dependence in the calculus of propositions* (J 0005) Amer Math Mon 47*602-605
⋄ B05 ⋄ REV MR 2.65 Zbl 25.4 JSL 6.37 FdM 66.1193 • ID 14145

WERNICK, W. [1942] *Complete sets of logical functions* (J 0064) Trans Amer Math Soc 51*117-132
⋄ B05 ⋄ REV MR 3.131 Zbl 27.385 JSL 7.99 • ID 14146

WERNICK, W. see Vol. V for further entries

WERTZ, S.K. [1983] see LANGLEY, P.

WESLEY, E. [1971] *An application of non-standard analysis to game theory* (J 0036) J Symb Logic 36*385-394
⋄ H10 ⋄ REV MR 51 # 5041 Zbl 242 # 02063 • ID 26366

WESLEY, E. see Vol. V for further entries

WESSEL, H. [1972] *Eine dialogische Begruendung logischer Gesetze* (C 1533) Quantoren, Modal, Paradox 256-278
⋄ A05 B10 ⋄ REV Zbl 244 # 02001 • ID 27385

WESSEL, H. & ZINOV'EV, A.A. [1975] *Logische Sprachregeln. Eine Einfuehrung in die Logik* (X 0806) Dt Verlag Wiss: Berlin 592pp
⋄ A05 B98 ⋄ REV Zbl 307 # 02002 • REM Translation of the book of A. Sinowjew from Russian • ID 65327

WESSEL, H. see Vol. II, VI for further entries

WESTON, K. [1965] *On predicate letter formulas which have no substitution instances provable in a first order language* (J 0047) Notre Dame J Formal Log 6*296-300
⋄ B10 ⋄ REV MR 35 # 1472 Zbl 158.248 • ID 14152

WESTRHENEN VAN, S.C. [1968] *A computer programme for the first order predicate calculus without identity* (C 0684) Automation in Lang Transl & Theorem Prov 69-83
⋄ B35 ⋄ REV MR 39 # 5310 Zbl 193.337 • ID 14154

WESTRHENEN VAN, S.C. [1968] *A random generator for sentential calculus* (C 0684) Automation in Lang Transl & Theorem Prov 85-91
⋄ B35 ⋄ REV MR 38 # 5609 Zbl 193.338 • ID 14158

WESTRHENEN VAN, S.C. [1968] *A simple application of discrete Markov chains to mathematical logic* (C 0684) Automation in Lang Transl & Theorem Prov 7-18
⋄ B35 ⋄ REV MR 38 # 5606 Zbl 193.299 • ID 14155

WESTRHENEN VAN, S.C. [1968] *A simple stochastic model of the reduction of a semantic tableau* (C 0684) Automation in Lang Transl & Theorem Prov 3-6
⋄ B35 ⋄ REV MR 38 # 5605 Zbl 193.299 • ID 14156

WESTRHENEN VAN, S.C. [1968] *Some remarks on the statistical estimation of probability in first-order predicate calculus* (C 0684) Automation in Lang Transl & Theorem Prov 19-25
⋄ B35 ⋄ REV MR 38 # 5607 Zbl 193.312 • ID 14157

WESTRHENEN VAN, S.C. [1968] *Statistical estimation of definability* (C 0684) Automation in Lang Transl & Theorem Prov 27-39
⋄ B35 C40 ⋄ REV MR 38 # 5608 Zbl 193.299 • ID 14159

WESTRHENEN VAN, S.C. [1968] *Two programmes for the calculus of propositional logic* (C 0684) Automation in Lang Transl & Theorem Prov 49-67
⋄ B35 ⋄ REV MR 39 # 5309 Zbl 193.337 • ID 14153

WESTRHENEN VAN, S.C. [1969] *A probabilistic machine for the estimation of provability in the first order predicate calculus* (J 0068) Z Math Logik Grundlagen Math 15*291-297
⋄ B35 ⋄ REV MR 41 # 8236 Zbl 216.282 • ID 14160

WESTRHENEN VAN, S.C. [1972] *Statistical studies of theoremhood in classical propositional and first order predicate calculus* (J 0037) ACM J 19*347-365
⋄ B35 ⋄ REV MR 45 # 6578 Zbl 246 # 68017 • ID 14161

WESTRHENEN VAN, S.C. [1981] *Note on probabilistic proof procedure for the first order predicate calculus* (J 3150) Delft Prog Rep, Ser F 6*170-176
⋄ B35 ⋄ REV Zbl 477 # 68098 • ID 55610

WETTE, E. [1970] *Vom Unendlichen zum Endlichen* (J 0076) Dialectica 24*303-323
⋄ A05 B28 E30 F30 ⋄ REV Zbl 263 # 02008 • ID 14165

WETTE, E. [1974] *Contradiction within pure number theory because of a system-internal "consistency"-deduction* (J 0286) Int Logic Rev 9*51-62
⋄ B10 F30 ⋄ REV Zbl 335 # 02033 • ID 14955

WETTE, E. see Vol. II, III, IV, V, VI for further entries

WEYHRAUCH, R.W. [1980] *Prolegomena to a theory of mechanized formal reasoning* (J 0503) Artif Intell 13*133-170
⋄ B35 ⋄ REV MR 81j:68114 Zbl 435 # 68070 • ID 83143

WEYHRAUCH, R.W. [1982] *An example of FOL using metatheory: formalizing reasoning systems and introducting derived inference rules* (P 3840) Autom Deduct (6);1982 New York 151-158
⋄ B35 ⋄ REV MR 86e:68093 Zbl 488 # 68057 • ID 36747

WEYHRAUCH, R.W. [1984] see KETONEN, J.

WEYL, H. [1910] *Ueber die Definition der mathematischen Grundbegriffe* (J 3978) Math-Nat Blaetter 7*93-95,109-113
⋄ A05 B30 F99 ⋄ REV FdM 41.89 • ID 42999

WEYL, H. [1918] *Das Kontinuum* (X 2636) Veit: Leipzig iv+83pp
⋄ A05 B28 E30 F65 ⋄ REV MR 22 # 10886 JSL 25.282 FdM 46.56 • REM Also reprinted in: Weyl,H. 1966 "Das Kontinuum und andere Monographien" Chelsea: New York • ID 19791

WEYL, H. [1919] *Der "Circulus vitiosus" in der heutigen Begruendung der Analysis* (J 0157) Jbuchber Dtsch Math-Ver 28*85-92
⋄ A05 B28 F65 ⋄ REV FdM 47.895 • ID 14166

WEYL, H. [1924] *Randbemerkungen zu Hauptproblemen der Mathematik* (J 0044) Math Z 20*131-150
⋄ B30 F55 ⋄ REV FdM 50.37 • ID 38712

WEYL, H. [1946] *Mathematics and logic: a brief survey serving as a review of "The philosophy of Bertrand Russell"* (J 0005) Amer Math Mon 53*2-13
• REPR [1968] (C 1161) Weyl: Ges Abhandlungen 4*599-605
⋄ A05 B30 ⋄ REV MR 7.355 Zbl 60.19 • ID 22073

WEYL, H. [1985] *Axiomatic versus constructive procedures in mathematics* (J 2789) Math Intell 7/4*10-17,38
⋄ A05 B30 F99 ⋄ REV Zbl 569#01011 • ID 49136

WEYL, H. see Vol. II, V, VI for further entries

WHALEY, T.P. & WILLIFORD, J. [1980] *Differentiability and permutations of quantifiers* (J 0005) Amer Math Mon 87*745-788
⋄ B10 ⋄ REV MR 83g:26010 Zbl 475#26004 • ID 55501

WHALEY, T.P. see Vol. III, V for further entries

WHEELER, R.F. [1961] *Complete propositional connectives* (J 0068) Z Math Logik Grundlagen Math 7*185-198
⋄ B05 ⋄ REV MR 26#7 Zbl 116.5 JSL 33.127 • ID 14176

WHEELER, R.F. [1962] *An asymptotic formula for the number of complete propositional connectives* (J 0068) Z Math Logik Grundlagen Math 8*1-4
⋄ B05 ⋄ REV MR 26#3586 JSL 33.127 • ID 14177

WHEELER, R.F. see Vol. II for further entries

WHEELER, W.H. [1979] *The first order theory of N-colorable graphs* (J 0064) Trans Amer Math Soc 250*289-310
⋄ B30 C10 C25 C35 C65 ⋄ REV MR 80g:03035 Zbl 428#03028 • ID 53787

WHEELER, W.H. see Vol. III, IV for further entries

WHINSTON, A.B. [1980] see BULLERS, W.I.

WHITEHEAD, A.N. [1907] *The axioms of descriptive geometry* (X 0805) Cambridge Univ Pr: Cambridge, GB viii+74pp
⋄ B30 ⋄ REV FdM 38.502 • ID 22605

WHITEHEAD, A.N. & RUSSELL, B. [1910] *Principia mathematica. Vol.I* (X 0805) Cambridge Univ Pr: Cambridge, GB xv+666pp
• TRANSL [1981] (X 3560) Paraninfo: Madrid 471pp (Spanish)
⋄ B05 B10 B15 B28 E30 E98 ⋄ REV MR 24#A30 MR 82m:01082 Zbl 101.249 FdM 41.83 • REM 2nd ed. 1925, repr. 1957; xlvi+674pp. Part II 1912. The transl. is from a 1962 reprint which contains only the first 56 sections • ID 22607

WHITEHEAD, A.N. [1911] *Introduction to mathematics* (X 1236) Benn: Tunbridge 256pp
⋄ B98 ⋄ REV FdM 42.76 • ID 22606

WHITEHEAD, A.N. & RUSSELL, B. [1912] *Principia mathematica. Vol.II* (X 0805) Cambridge Univ Pr: Cambridge, GB xxiv+772pp
⋄ B15 B28 E10 E30 E98 ⋄ REV FdM 43.93 • REM 2nd ed. 1927, repr. 1957. Part I 1910. Part III 1913 • ID 22611

WHITEHEAD, A.N. & RUSSELL, B. [1913] *Principia mathematica. Vol.III* (X 0805) Cambridge Univ Pr: Cambridge, GB x+491pp
⋄ B15 B28 E10 E30 E98 ⋄ REV FdM 44.68 • REM 2nd ed. 1927, repr. 1957. Part II 1912 • ID 22612

WHITEHEAD, A.N. & RUSSELL, B. [1932] *Einfuehrung in die mathematische Logik* (X 1225) Drei Masken: Muenchen viii+168pp
⋄ A05 B98 ⋄ REV Zbl 4.1 FdM 58.997 • REM German translation of the introduction of the "Principia Mathematica" • ID 28556

WHITEHEAD, A.N. [1934] *Logical definitions of extension, class, and number* (J 0005) Amer Math Mon 41*129-131
⋄ A05 B28 E30 ⋄ ID 14183

WHITEHEAD, A.N. [1967] see RUSSELL, B.

WHITEHEAD, A.N. see Vol. V for further entries

WHITELEY, W. [1973] *Logic and invariant theory I. Invariant theory of projective properties* (J 0064) Trans Amer Math Soc 177*121-139
⋄ B30 C60 ⋄ REV MR 56#5279 Zbl 238#50002 • REM Part II 1978. Part III 1977 • ID 79943

WHITELEY, W. [1977] *Logic and invariant theory III. Axiom systems and basic syzygies* (J 3172) J London Math Soc, Ser 2 15*1-15
⋄ B30 C60 ⋄ REV MR 57#16057b Zbl 347#50001 • REM Part II 1978. Part IV 1979 • ID 79941

WHITELEY, W. [1978] *Logic and invariant theory II: Homogeneous coordinates, the introduction of higher quantities, and structural geometry* (J 0032) J Algeb 50*380-394
⋄ B30 C60 ⋄ REV MR 57#16057a Zbl 348#50003 • REM Part I 1973. Part III 1977 • ID 50914

WHITELEY, W. see Vol. III for further entries

WHOLEY, J.S. [1963] *Persistence and Herbrand expansions* (J 0036) J Symb Logic 28*280-282
⋄ B10 ⋄ REV MR 30#3837 Zbl 222#02062 JSL 31.127 • ID 14189

WICKS, M.J. [1983] see CHONG, C.T.

WICKSTEAD, A.W. [1977] see CHADWICK, J.J.M.

WIEDMER, E. [1974] *Ein neuer negationsloser Beweis eines Satzes von G.F.C.Griss* (J 0028) Indag Math 36*89-93 • ERR/ADD ibid 77(A)*412)
⋄ B20 F55 ⋄ REV MR 49#26a MR 49#26b Zbl 276#02017 • ID 14190

WIEDMER, E. see Vol. IV for further entries

WIENER, N. [1912] *A simplification of the logic of relations* (J 0171) Proc Cambridge Phil Soc Math Phys 17*387-390
• REPR [1967] (C 0675) From Frege to Goedel 224-227
⋄ B20 E07 • ID 14193

WIENER, N. see Vol. V for further entries

WIERZBICKI, E. & WOZNIAK, C. [1982] *On the formation of implicit constraints and free-boundary problems for elastodynamics by the nonstandard analysis technique (Russian summary)* (J 2677) Bull Acad Pol Sci, Ser Sci Tech 29*205-211
⋄ H10 ⋄ REV MR 83m:03078a Zbl 502#73003 • ID 35474

WIERZBICKI, E. [1984] *On the formation of internal constraints by the technique of nonstandard analysis* (J 2677) Bull Acad Pol Sci, Ser Sci Tech 32*389-393
⋄ H10 ⋄ REV Zbl 555#73028 • ID 44933

WIERZBICKI, E. [1984] see NOBIS, K.

WIETLISBACH, M.N. [1981] *Zur Komplexitaet von Entscheidungsalgorithmen, die auf dem Herbrand'schen Satz und regulaerer Resolution beruhen* (X 2710) Eidgen Techn Hochsch: Zuerich 112pp
⋄ B35 D15 ⋄ REV Zbl 516 # 03037 • ID 37266

WILCOX, W.C. [1968] see CARNES, R.D.

WILCOX, W.C. [1971] *A mistake in Copi's discussion of completeness* (J 0047) Notre Dame J Formal Log 12*459-460
⋄ B10 ⋄ REV MR 45 # 4939 Zbl 224 # 02039 • ID 14197

WILCOX, W.C. see Vol. II for further entries

WILDE, A.C. [1974] *A substitution property* (J 0047) Notre Dame J Formal Log 15*639-640
⋄ B05 ⋄ REV MR 50 # 6839 Zbl 254 # 02012 • ID 14198

WILDE, A.C. [1974] *Generalizations of the distributive and associative laws* (J 0047) Notre Dame J Formal Log 15*491-493
⋄ B05 ⋄ REV MR 55 # 2700 Zbl 245 # 02016 • ID 14199

WILDER, R.L. [1952] *Introduction to the foundations of mathematics* (X 0827) Wiley & Sons: New York xiv+305pp • LAST ED [1980] (X 2828) Krieger Publ Co: Melbourne xvi+327pp
⋄ A05 B98 ⋄ REV MR 14.441 MR 32 # 35 MR 81j:03002 Zbl 442 # 03001 Zbl 49.9 JSL 19.225 • ID 56358

WILHELMY, A. [1951] see ANGSTL, H.

WILHELMY, A. [1963] *Minimisierung mit nichtkombinatorischen Methoden* (P 1612) Schaltkreis & -werk Th (2);1961 Saarbruecken 9-48
⋄ B35 ⋄ REV Zbl 115.119 • ID 27835

WILHELMY, A. see Vol. II for further entries

WILKIE, A.J. [1985] see PARIS, J.B.

WILKIE, A.J. see Vol. III, IV, V, VI for further entries

WILKOSZ, W. [1937] *Remarque sur la notion de la definition conditionelle de Peano* (J 0283) Ann Soc Pol Math 16*176-178
⋄ A05 B10 ⋄ REV Zbl 18.337 JSL 3.163 FdM 64.925 • ID 14206

WILLIAMS, C.J.F. [1980] *Misinterpretations of quantifiers* (J 0094) Mind 89*420-422
⋄ A05 B10 ⋄ REV MR 81k:03012 • ID 79974

WILLIAMS, N.H. [1972] see ASH, C.J.

WILLIAMS, N.H. see Vol. V for further entries

WILLIAMS, P.M. [1974] *Certain classes of models for empirical systems* (J 0063) Studia Logica 33*73-90,121
⋄ A05 B30 ⋄ REV MR 54 # 4916 Zbl 288 # 02005 • ID 24092

WILLIAMS, P.M. [1976] *Indeterminate probabilities* (P 1804) Form Meth in Methodol of Emp Sci;1974 Warsaw 229-246
⋄ B30 ⋄ REV MR 58 # 27313 • ID 79980

WILLIAMSON, C. [1972] *Squares of opposition: comparisons between syllogistic and propositional logic* (J 0047) Notre Dame J Formal Log 13*497-500
⋄ B05 B20 ⋄ REV Zbl 242 # 02005 • ID 26336

WILLIFORD, J. [1980] see WHALEY, T.P.

WINKER, S.K. [1976] *Complete demodulation for automatic theorem proving* (J 2687) Comp Math Appl 2*161-179
⋄ B35 ⋄ REV Zbl 366 # 68061 • ID 51148

WINKER, S.K. [1981] see LUSK, E.L.

WINKER, S.K. [1982] *Generation and verification of finite models and counterexamples using an automated theorem prover answering two open questions* (J 0037) ACM J 29*273-284
⋄ B35 ⋄ REV MR 84g:03021 Zbl 488 # 68056 • ID 34135

WINKER, S.K. & Wos, L. [1982] *Procedure implementation through demodulation and related tricks* (P 3840) Autom Deduct (6);1982 New York 109-131
⋄ B35 ⋄ REV MR 85e:68002 • ID 45194

WINKER, S.K. [1983] see HENSCHEN, L.J.

WINKER, S.K. [1984] see SMITH, BRIAN

WINKER, S.K. & Wos, L. [1984] *Open questions solved with the assistance of AURA* (P 3084) Autom Theor Prov After 25 Yea;1983 Denver 73-88
⋄ B35 ⋄ REV MR 85d:68005 Zbl 566 # 68076 • ID 45274

WINKER, S.K. see Vol. II for further entries

WINKLER, R. [1976] *Ueber moegliche Erweiterungen der Analysis und ihre praktische Bedeutung* (J 2878) Wiss Z Tech Hochsch Karl-Marx-Stadt 18*446-454
⋄ H05 ⋄ REV MR 55 # 5439 Zbl 417 # 26010 • ID 53277

WINKLER, R. [1978] *Ein Existenzbeweis fur gewoehnliche Differentialgleichungen mit Methoden der non-standard-Analysis* (J 2878) Wiss Z Tech Hochsch Karl-Marx-Stadt 20*527-529
⋄ H05 ⋄ REV MR 80c:34006 Zbl 443 # 34006 • ID 83155

WINKOWSKI, J. [1978] see MAGGIOLO-SCHETTINI, A.

WINKOWSKI, J. see Vol. IV for further entries

WINNIE, J.A. [1970] *The completeness of Copi's system of natural deduction* (J 0047) Notre Dame J Formal Log 11*379-382
⋄ B10 B45 ⋄ REV MR 44 # 3833 Zbl 197.277 • ID 14236

WINTER, D.J. [1982] *Axiomatic game thoery* (P 3817) Algeb Homage:Ring Th & Rel Top (Jacobson);1981 New Haven 403-409
⋄ B30 E30 ⋄ REV MR 85d:03022 Zbl 528 # 90110 • ID 36766

WINTERSTEIN, G. [1977] *Unification in second order logic* (J 0129) Elektr Informationsverarbeitung & Kybern 13*399-411
⋄ B35 ⋄ REV MR 57 # 4684 Zbl 371 # 02009 • ID 51370

WIRSING, M. [1977] *Das Entscheidungsproblem der Klasse von Formeln, die hoechstens zwei Primformeln enthalten* (J 0504) Manuscr Math 22*13-25
⋄ B20 B25 D35 ⋄ REV MR 57 # 12199 Zbl 365 # 02035 • ID 51036

WIRSING, M. [1978] *Kleine unentscheidbare Klassen der Praedikatenlogik mit Identitaet und Funktionszeichen* (J 0009) Arch Math Logik Grundlagenforsch 19*97-109
⋄ B20 D35 ⋄ REV MR 80a:03054a Zbl 398 # 03005 • ID 29168

WIRSING, M. see Vol. II, IV for further entries

WISDOM, W.A. [1972] see LEBLANC, H.

WISDOM, W.A. see Vol. II for further entries

WISNIEWSKI, A. [1985] *Propositional logic and erotetic inferences* (J 0387) Bull Sect Logic, Pol Acad Sci 14*72-78
⋄ B05 ⋄ REV Zbl 575#03003 • ID 48818

WITTGENSTEIN, L. [1956] *Remarks on the foundations of mathematics* (X 0843) Macmillan : New York & London xix + xix + 204pp
• LAST ED [1983] (X 0865) MIT Pr: Cambridge, MA 444pp
⋄ A05 B30 ⋄ REV MR 19.1 MR 80f:00009 MR 84f:00024 Zbl 444#03001 Zbl 75.4 JSL 34.108 • REM Selections reprinted in C1105. Remarks by Ziff,P. in J0154 56*351-361, 1983 & by Bernays,P. in C1105 510-528, 1964
• ID 14250

WITTGENSTEIN, L. [1976] *Wittgenstein's lectures on the foundations of mathematics, Cambridge, 1939* (X 0992) Cornell Univ Pr: Ithaca 300pp
• REPR [1976] (X 3719) Harvester Pr: Sussex 300pp
⋄ A05 B30 B96 ⋄ REV MR 55#7728 • REM Ed. by Diamond,C. • ID 70145

WOJCICKI, R. [1969] *Logical matrices strongly adequate for structural sentential calculi (Russian summary)* (J 0014) Bull Acad Pol Sci, Ser Math Astron Phys 17*333-335
⋄ B22 ⋄ REV MR 41#3248 Zbl 198.19 • ID 14253

WOJCICKI, R. [1969] see PRZELECKI, M.

WOJCICKI, R. [1970] *Some remarks on the consequence operation in sentential logics* (J 0027) Fund Math 68*269-279
⋄ B22 ⋄ REV MR 42#1650 Zbl 206.274 • ID 14256

WOJCICKI, R. [1971] see TOKARZ, M.

WOJCICKI, R. [1972] *Some properties of strongly finite propositional calculi* (J 0387) Bull Sect Logic, Pol Acad Sci 1/1*69-71
⋄ B22 ⋄ REV MR 51#7824 • ID 17305

WOJCICKI, R. [1973] *Basic concepts of formal methodology of empirical sciences* (J 0963) Ajatus (Helsinki) 35*168-196
• REPR [1977] (C 3174) Twenty-five Years Log Meth Poland 681-708
⋄ A05 B30 ⋄ REV Zbl 315#02008 Zbl 372#02005
• ID 51413

WOJCICKI, R. [1973] *Dual counterparts of consequence operations* (J 0387) Bull Sect Logic, Pol Acad Sci 2*54-57
⋄ B22 ⋄ REV MR 52#13312 • ID 21775

WOJCICKI, R. [1973] *Matrix approach in methodology of sentential calculi (Polish and Russian summaries)* (J 0063) Studia Logica 32*7-39
⋄ B22 ⋄ REV MR 52#2837 Zbl 336#02012 • ID 14259

WOJCICKI, R. [1974] *Degree of maximality versus degree of completeness* (J 0387) Bull Sect Logic, Pol Acad Sci 3/2*41-43
⋄ B22 ⋄ REV MR 53#7760 • ID 23003

WOJCICKI, R. [1974] *Formal methods and epistemological problems (Russian)* (C 2578) Filos & Logika 102-146
⋄ A05 B30 ⋄ REV MR 58#27466 • ID 79778

WOJCICKI, R. [1974] *Set theoretic representations of empirical phenomena* (J 0122) J Philos Logic 3*337-343
• REPR [1977] (P 2116) Probl in Log & Ontology;1973 Salzburg 243-249
⋄ A05 B30 E75 ⋄ REV MR 55#76 MR 81d:03009 Zbl 281#02009 • ID 29567

WOJCICKI, R. [1975] *A theorem on strongly finite propositional calculi* (J 0387) Bull Sect Logic, Pol Acad Sci 4*2-8
⋄ B22 ⋄ REV MR 52#13326 • ID 21787

WOJCICKI, R. [1976] see MALINOWSKI, G.

WOJCICKI, R. & ZYGMUNT, J. [1977] *Results and open problems in the theory of strongly finite sentential calculi* (1111) Preprints, Manuscr., Techn. Reports etc. 25pp
⋄ B22 ⋄ REM Inst. Phil. and Soc., Pol. Acad. Sci., Technical Report No.1/sf, Wroclaw • ID 31573

WOJCICKI, R. [1977] *Strongly finite sentential calculi* (C 3026) Lukasiewicz Sel Pap on Sent Calc 53-77
⋄ B22 ⋄ REV MR 58#21454 Zbl 381#03017 • ID 80025

WOJCICKI, R. [1979] *A theorem on strictly finite propositional calculi (Russian)* (P 2554) All-Union Symp Th Log Infer;1974 Moskva 122-127
⋄ B22 ⋄ REV MR 85d:03065 • ID 45073

WOJCICKI, R. [1982] *Referential matrix semantics for propositional calculi* (P 3622) Int Congr Log, Meth & Phil of Sci (6,Proc);1979 Hannover 325-334
⋄ B22 B50 ⋄ REV MR 84b:03050 Zbl 505#03007
• ID 35669

WOJCICKI, R. [1984] *Lectures on propositional calculi* (X 2212) Ossolineum: Wroclaw 292pp
⋄ B98 ⋄ ID 45741

WOJCICKI, R. see Vol. II, III, VI for further entries

WOJCIECHOWSKA, A. [1979] *Elements of logic and set theory* (X 1034) PWN: Warsaw 200pp
⋄ B98 E98 ⋄ REV MR 81b:04002 • ID 80045

WOJCIECHOWSKA, A. see Vol. III, V for further entries

WOJCIECHOWSKI, W. & WOJCIK, A.S. [1979] *Multiple-valued logic design by theorem proving* (P 3003) Int Symp Multi-Val Log (9);1979 Bath 196-199
⋄ B35 ⋄ ID 35952

WOJCIK, A.S. [1972] *The minimization of higher-order Boolean functions* (P 2008) Symp Th & Appl of Multi-Val Log Design;1972 Buffalo 181-192
⋄ B35 ⋄ ID 42047

WOJCIK, A.S. [1979] see WOJCIECHOWSKI, W.

WOJCIK, A.S. see Vol. II for further entries

WOJTASIEWICZ, O. [1978] *The predicate calculus with extra-logical constants as an instrument of semantic description* (J 0063) Studia Logica 37*103-114
⋄ A05 B10 ⋄ REV MR 80a:03035 Zbl 381#03005
• ID 51873

WOJTYLAK, P. [1979] *Matrix representations for structural strengthenings of a propositional logic* (J 0063) Studia Logica 38*263-266
⋄ B22 G25 ⋄ REV MR 82b:03066 Zbl 422#03009
• ID 53470

WOJTYLAK, P. [1979] *Strongly finite logics: Finite axiomatizability and the problem of supremum* (J 0387) Bull Sect Logic, Pol Acad Sci 8*99-111
⋄ B22 ⋄ REV MR 80g:03027 Zbl 409#03017 • ID 56320

WOJTYLAK, P. [1981] *Mutual interpretability of sentential logics. I,II* (J 0302) Rep Math Logic, Krakow & Katowice 11*69-89,12*51-66
⋄ B22 F25 ⋄ REV MR 82k:03033 Zbl 468#03006 Zbl 468#03007 • ID 55071

WOJTYLAK, P. [1982] see POGORZELSKI, W.A.

WOJTYLAK, P. [1983] *Corrections to the paper: "Structural completeness of Lewis's system S5" by T. Prucnal* (J 0302) Rep Math Logic, Krakow & Katowice 15*67-70
⋄ B22 ⋄ REV MR 84g:03028 Zbl 515 # 03006 • REM The article was published in J0014 20(1972)* • ID 34142

WOJTYLAK, P. [1984] *An example of a finite though finitely non-axiomatizable matrix* (J 0302) Rep Math Logic, Krakow & Katowice 17*39-46
⋄ B22 ⋄ REV MR 86g:03051 Zbl 563 # 03011 • ID 44099

WOJTYLAK, P. see Vol. II, III, VI for further entries

WOLFF, M. [1972] *Nonstandard-Komplettierung von Cauchy-Algebren* (P 0732) Contrib to Non-Standard Anal;1970 Oberwolfach 178-213
⋄ C65 H05 ⋄ REV MR 58 # 2787 Zbl 262 # 54048
• ID 14270

WOLFF, M. [1984] *Spectral theory of group representations and their nonstandard hull* (J 0029) Israel J Math 48*205-224
⋄ C65 H10 ⋄ REV MR 86e:46046 Zbl 568 # 22005
• ID 44491

WOLNIEWICZ, B. [1980] *On the verifiers of disjunction* (J 0387) Bull Sect Logic, Pol Acad Sci 9*57-59
⋄ A05 B05 ⋄ REV MR 81h:03065 Zbl 438 # 03012
• ID 55924

WOLNIEWICZ, B. see Vol. II for further entries

WOLTER, H. [1968] *Eine Erweiterung der klassischen Analysis* (J 0068) Z Math Logik Grundlagen Math 14*167-184
⋄ C20 H05 ⋄ REV MR 37 # 5089 Zbl 159.18 • ID 14276

WOLTER, H. [1969] see LIPCZYNSKA, M.

WOLTER, H. [1972] *Ueber Mengen von Ausdruecken der Logik hoeherer Stufe, fuer die der Endlichkeitssatz und der Satz von Loewenheim-Skolem gelten* (J 0068) Z Math Logik Grundlagen Math 18*13-18
⋄ B15 C20 C85 ⋄ REV MR 45 # 3177 Zbl 238 # 02013
• ID 14277

WOLTER, H. [1972] *Untersuchung ueber Algebren und formalisierte Sprachen hoeherer Stufen (Russian, English and French summaries)* (J 0115) Wiss Z Humboldt-Univ Berlin, Math-Nat Reihe 21*487-495
⋄ B15 C20 C30 C40 C85 ⋄ REV MR 49 # 2354
Zbl 256 # 02027 • ID 14278

WOLTER, H. [1973] *Eine Erweiterung der elementaren Praedikatenlogik: Anwendungen in der Arithmetik und anderen mathematischen Theorien* (J 0068) Z Math Logik Grundlagen Math 19*181-190
⋄ B28 C10 C80 F30 ⋄ REV MR 51 # 7815
Zbl 302 # 02001 • ID 14280

WOLTER, H. [1980] see HERRMANN, E.

WOLTER, H. see Vol. III, IV, V, VI for further entries

WOLTER, U. [1985] see DASSOW, J.

WON, CHANGSA [1980] *On the computerization of minimizing the logic circuit by the MASK and the cost table method* (1111) Preprints, Manuscr., Techn. Reports etc. 11*195-207
⋄ B35 ⋄ REV Zbl 443 # 94033 • REM Institute of Technology. Ulsan, Rep. Korea • ID 69004

WONG, KAM [1975] see LIGHTSTONE, A.H.

WOODGER, J.-H. [1935] see FLOYD, W.F.

WOODGER, J.-H. [1937] *The axiomatic method in biology* (X 0805) Cambridge Univ Pr: Cambridge, GB x+174pp
⋄ B30 B80 ⋄ REV JSL 3.42 FdM 63.827 • ID 22620

WOODGER, J.-H. [1952] *From biology to mathematics* (J 0013) Brit J Phil Sci 3*1-21
⋄ B30 B80 ⋄ REV JSL 39.353 • ID 14291

WOODGER, J.-H. [1954] *Problems arising from the application of mathematical logic to biology* (P 0646) Appl Sci de Log Math;1952 Paris 133-140
⋄ B80 ⋄ REV Zbl 58.3 • ID 27971

WOODGER, J.-H. [1959] *Studies in the foundations of genetics* (P 0651) Axiomatic Method;1957 Berkeley 408-428
⋄ B30 B80 ⋄ REV MR 22 # 638 Zbl 104.383 • ID 27715

WOODGER, J.-H. see Vol. II for further entries

WORKS, C. & YOURGRAU, W. [1968] *Note on duality in propositional calculus* (J 0047) Notre Dame J Formal Log 9*284-288
⋄ B05 ⋄ REV MR 39 # 2602 Zbl 175.259 • ID 18074

WOS, L. [1964] see CARSON, D.F.

WOS, L. [1965] see CARSON, D.F.

WOS, L. [1966] see CARSON, D.F.

WOS, L. [1967] see CARSON, D.F.

WOS, L. [1969] see ROBINSON, G.A.

WOS, L. [1970] see ROBINSON, G.A.

WOS, L. [1973] see ROBINSON, G.A.

WOS, L. [1981] see LUSK, E.L.

WOS, L. [1982] see WINKER, S.K.

WOS, L. [1982] *Solving open questions with an automated theorem-proving program* (P 3840) Autom Deduct (6);1982 New York 1-31
• TRANSL [1984] (J 3079) Kiber Sb Perevodov, NS 21*235-263
⋄ B35 ⋄ REV MR 85k:68092 Zbl 481 # 68090 • ID 38463

WOS, L. [1983] see HENSCHEN, L.J.

WOS, L. [1984] see SMITH, BRIAN

WOS, L. [1984] see WINKER, S.K.

WOS, L. [1984] see MCCUNE, W.

WOS, L. [1985] *Automated reasoning* (J 0005) Amer Math Mon 92*85-92
⋄ B35 ⋄ ID 44462

WOS, L. see Vol. II for further entries

WOZNIAK, C. [1979] *Non-standard analysis and its application to mechanics* (P 3058) Trends Appl of Pure Math to Mech;1977 Kozubnik 327-341
⋄ H10 ⋄ REV MR 81e:03065 Zbl 423 # 70005 • ID 53583

WOZNIAK, C. [1980] *Nonstandard analysis and material systems in mechanics. I,II* (J 2677) Bull Acad Pol Sci, Ser Sci Tech 28*33-36,37-40
⋄ H10 ⋄ REV MR 82m:73002 Zbl 481 # 70022 • ID 83165

WOZNIAK, C. [1982] see NOBIS, K.

WOZNIAK, C. [1982] see WIERZBICKI, E.

WOZNIAK, C. [1982] *On the nonstandard interrelation between mass-point mechanics and continuum mechanics (Russian summary)* (J 2677) Bull Acad Pol Sci, Ser Sci Tech 29*219-224
 ◊ H10 ◊ REV MR 83m:03078c Zbl 536#73005 • ID 35476

WOZNIAK, C. [1982] *On the nonstandard model of the theory of elasticity (Russian summary)* (J 2677) Bull Acad Pol Sci, Ser Sci Tech 29*225-230
 ◊ H10 ◊ REV MR 83m:03078d Zbl 502#73006 • ID 35477

WOZNIAK, C. [1983] *Nonstandard analysis in Newtonian mechanics of a particle (Polish) (English and Russian summaries)* (J 3776) Mech Teor Stosow 19*355-374
 ◊ H10 ◊ REV MR 84e:03083 Zbl 523#70006 • ID 34415

WOZNIAK, C. [1983] *On the nonstandard analysis and the interrelation between mechanics of mass-point systems and continuum mechanics (Russian and Polish summaries)* (J 3776) Mech Teor Stosow 19*511-525
 ◊ H10 ◊ REV MR 84h:03143 Zbl 536#73005 • ID 34329

WOZNIAK, C. [1984] see NOBIS, K.

WOZNIAKOWSKA, B. [1978] *The representation theorem for the algebras determined by the fragments of infinite-valued logic of Lukasiewicz* (J 0387) Bull Sect Logic, Pol Acad Sci 7*176-178
 ◊ B20 B50 G20 G25 ◊ REV MR 80b:03108 Zbl 407#03029 • ID 31565

WOZNIAKOWSKA, B. see Vol. II for further entries

WRIGHT, C. [1983] *Frege's conception of numbers as objects* (X 2518) Aberdeen Univ Pr: Aberdeen 193pp
 ◊ A05 B28 ◊ REV MR 85g:00035 Zbl 524#03005 JSL 50.252 • ID 37589

WRIGHT, C. see Vol. VI for further entries

WRIGHT, J.B. [1954] see BURKS, A.W.

WRIGHT, J.B. [1954] *Quasi-projective geometry of two dimensions* (J 0133) Michigan Math J 2*115-122
 ◊ B30 ◊ REV MR 16.161 Zbl 58.141 • ID 28566

WRIGHT, J.B. [1962] see BURKS, A.W.

WRIGHT, J.B. [1968] see THATCHER, J.W.

WRIGHT, J.B. see Vol. III, IV, V for further entries

WRIGHT VON, G.H. [1948] *On the idea of logical truth I* (J 0990) Soc Sci Fennicae Comment Phys-Math 14/4*20pp
 ◊ A05 B05 B20 B25 ◊ REV MR 10.668 Zbl 32.385 JSL 15.58 • REM Part II 1950 • ID 14303

WRIGHT VON, G.H. [1950] *On the idea of logical truth II* (J 0990) Soc Sci Fennicae Comment Phys-Math 15/10*45pp
 ◊ A05 B05 B20 B25 ◊ REV MR 13.521 Zbl 38.5 JSL 16.147 • REM Part I 1948 • ID 14304

WRIGHT VON, G.H. [1952] see GEACH, P.T.

WRIGHT VON, G.H. [1952] *On double quantification* (J 0990) Soc Sci Fennicae Comment Phys-Math 16/3*14pp
 • REPR [1957] (C 1017) Logical Studies 44-57
 ◊ A05 B20 B25 ◊ REV MR 15.846 Zbl 47.8 Zbl 81.10 JSL 17.201 JSL 35.461 • ID 14309

WRIGHT VON, G.H. see Vol. II, III for further entries

WRIGHTSON, G. [1980] see SIEKMANN, J.

WRIGHTSON, G. [1982] see JANSOHN, H.-S.

WRIGHTSON, G. [1983] see SIEKMANN, J.

WRIGHTSON, G. [1984] see OHLBACH, H.-J.

WRIGHTSON, G. see Vol. II for further entries

WRONSKI, A. [1974] see PRUCNAL, T.

WRONSKI, A. [1974] *The degree of completeness of some fragments of the intuitionistic propositional logic* (J 0302) Rep Math Logic, Krakow & Katowice 2*55-61
 ◊ B20 F50 ◊ REV MR 52#5401 Zbl 312#02023 • ID 18432

WRONSKI, A. [1976] *On finitely based consequence operations* (J 0063) Studia Logica 35*453-458
 ◊ B22 ◊ REV MR 55#10237 Zbl 375#02012 • ID 51589

WRONSKI, A. [1977] *A method of axiomatizing an intersection of propositional logics* (J 0387) Bull Sect Logic, Pol Acad Sci 6*177-181
 ◊ B22 ◊ REV MR 58#21563 Zbl 404#03011 • ID 54798

WRONSKI, A. [1977] *On the depth of a consequence operation* (J 0387) Bull Sect Logic, Pol Acad Sci 6*96-101
 ◊ B22 ◊ REV MR 58#16253 Zbl 403#03019 • ID 54733

WRONSKI, A. [1979] *A three element matrix whose consequence operation is not finitely based* (J 0387) Bull Sect Logic, Pol Acad Sci 8*68-71
 ◊ B22 ◊ REV MR 80h:03047 Zbl 419#03017 • ID 53360

WRONSKI, A. see Vol. II, III, VI for further entries

WU, WENJUN [1978] *Mechanical theorem proving in elementary differential geometry (Chinese)* (J 2771) Kexue Tongbao 23*523-524
 ◊ B35 ◊ REV MR 80e:12028 Zbl 421#68084 • ID 66309

WU, WENJUN [1978] *On the decision problem and the mechanization of theorem-proving in elementary geometry (Chinese)* (J 1024) Zhongguo Kexue 21*159-172
 ◊ B35 ◊ REV MR 58#197 • ID 80129

WU, WENJUN [1979] *On the mechanization of theorem-proving in elementary differential geometry (Chinese) (English summary)* (J 1024) Zhongguo Kexue 22*94-102
 ◊ B35 ◊ REV MR 84h:03031 • ID 34243

WU, WENJUN [1982] *Mechanical theorem proving in elementary geometry and differential geometry* (P 3912) Diff Geom & Diff Equations;1980 Beijing 2*1073-1092
 ◊ B35 ◊ REV MR 85e:53001 Zbl 536#68076 • ID 36779

WU, WENJUN [1982] *Toward mechanization of geometry - some comments on Hilbert's "Grundlagen der Geometrie"* (J 4614) Acta Math (Engl Ed) 2/2*125-138
 ◊ B35 ◊ REV MR 84h:51007 Zbl 519#51001 • ID 39515

WU, WENJUN [1983] *Some remarks on mechanical theoremproving in elementary geometry* (J 4614) Acta Math (Engl Ed) 3/4*357-360
 ◊ B35 ◊ REV Zbl 529#51012 • ID 49469

WU, WENJUN [1984] *On the decision problem and the mechanization of theorem-proving in elementary geometry* (P 3084) Autom Theor Prov After 25 Yea;1983 Denver 213-234
 ◊ B35 ◊ REV MR 85d:68005 • ID 45260

WU, WENJUN [1984] *Some recent advances in mechanical theorem-proving of geometries* (P 3084) Autom Theor Prov After 25 Yea;1983 Denver 235-241
 ◊ B35 ◊ REV MR 85d:68005 • ID 45275

WU, WENJUN see Vol. VI for further entries

WU, YUNZENG [1979] *Two problems of modern mathematical logic and its philosophic significance (Chinese)* (J 4452) Zhexue Yanjiu 2*51-55
⋄ A05   B98   ⋄ ID 48545

WU, YUNZENG see Vol. IV, V for further entries

WYBRANIEC-SKARDOWSKA, U. [1970] see BRYLL, G.

WYBRANIEC-SKARDOWSKA, U. [1971] see BRYLL, G.

WYBRANIEC-SKARDOWSKA, U. [1972] see BRYLL, G.

WYBRANIEC-SKARDOWSKA, U. [1975] *Mutual definability of the notions of entailment and inconsistency (Polish) (English summary)* (S 1454) Zesz Nauk Wyz Szk Ped Mat, Opole 15*77-86
⋄ B20   ⋄ REV   MR 54#7198   Zbl 305#02064   • ID 24989

WYBRANIEC-SKARDOWSKA, U. see Vol. V for further entries

WYMAN JR., R.H. [1965] see LOOMIS JR., H.H.

XU, LIZHI & YUAN, XIANGWAN & ZHENG, YUXIN & ZHU, WUJIA [1982] *Antinomies and the foundational problem of mathematics I (Chinese) (English summary)* (J 3742) Shuxue Yanjiu yu Pinglun 2/3*99-108
⋄ A05   H05   ⋄ REV   MR 84g:03016a   • REM Part II 1982
• ID 34130

XU, LIZHI & YUAN, XIANGWAN & ZHENG, YUXIN & ZHU, WUJIA [1982] *Antinomies and the foundational problem of mathematics II (Chinese) (English summary)* (J 3742) Shuxue Yanjiu yu Pinglun 2/4*121-134
⋄ A05   B98   ⋄ REV   MR 84g:03016b   Zbl 505#03005   • REM Part I 1982. Part III 1983.   • ID 34131

XU, LIZHI [1983] *Generalized Moebius inversion theory associated with non-standard analysis (Chinese)* (J 3742) Shuxue Yanjiu yu Pinglun 3/1*51-53
⋄ H10   ⋄ REV   Zbl 518#03031   • ID 37528

XU, LIZHI [1983] *Generalized Moebius-Rota inversion theory associated with nonstandard analysis (Chinese summary)* (J 3594) Kexue Tansuo 3/1*1-8
⋄ H05   ⋄ REV   MR 86b:05010   • ID 48794

XU, LIZHI see Vol. VI for further entries

XU, YONGSHEN [1984] see MO, SHAOKUI

YABLONSKIJ, S.V. [1952] *On complete systems of functions of the algebra of logic (Russian)* (J 0067) Usp Mat Nauk 7/5(51)*197
⋄ B05   ⋄ ID 40894

YABLONSKIJ, S.V. [1952] *On the superpositions of functions of the algebra of logic (Russian)* (J 0142) Mat Sb, Akad Nauk SSSR, NS 30(72)*329-348
⋄ B05   ⋄ REV   MR 14.937   Zbl 46.6   JSL 20.175   • ID 18001

YABLONSKIJ, S.V. [1954] *Realization of a linear function in the class of Π-schemes (Russian)* (J 0023) Dokl Akad Nauk SSSR 94*805-806
⋄ B05   ⋄ REV   MR 16.203   Zbl 56.13   JSL 21.102   • ID 40895

YABLONSKIJ, S.V. [1957] *On classes of functions of the algebra of logic admitting a simple schematic realization (Russian)* (J 0067) Usp Mat Nauk 12/6(78)*189-196
⋄ B05   B70   ⋄ REV   MR 19.938   Zbl 79.9   JSL 24.75
• ID 40904

YABLONSKIJ, S.V. [1966] see GAVRILOV, G.P.

YABLONSKIJ, S.V. see Vol. II, IV, V for further entries

YADYKIN, S.A. [1979] see KARATUEV, V.G.

YAGLOM, I.M. [1978] *An unusual algebra (Russian)* (X 0885) Mir: Moskva 129pp
⋄ B05   E20   G05   ⋄ REV   Zbl 453#06001   • REM Transl. into English and Spanish also by X0885   • ID 54203

YAJIMA, K. [1965] *On the decision method for rules and regulations by propositional logic (Japanese)* (J 0425) Managmt Sci (Tokyo) 9*198-203
⋄ B05   ⋄ REV   MR 35#6522   • ID 16895

YAKUBAJTIS, EH.A. [1974] *Subclasses and classes of boolean functions (Russian)* (J 0474) Avtom Vychis Tekh, Akad Nauk Latv SSR 1974/1*1-8
• TRANSL [1974] (J 2666) Autom Control Comput Sci 8/1*1-6
⋄ B05   ⋄ REV   MR 50#4130   Zbl 278#94021   • ID 62725

YAKUBAJTIS, EH.A. see Vol. IV for further entries

YAKUBOVICH, A.M. [1981] *On the consistency of the theory of types with the axiom of choice relative to type theory (Russian)* (J 0023) Dokl Akad Nauk SSSR 261*825-828
• TRANSL [1981] (J 0062) Sov Math, Dokl 24*621-624
⋄ B15   E25   E35   F35   ⋄ REV   MR 83e:03092   Zbl 495#03008   • ID 46491

YAKUBOVICH, A.M. [1981] *Variants of the axiom of choice in the simple theory of types (Russian)* (J 0087) Mat Zametki (Akad Nauk SSSR) 30*269-276,315-316
• TRANSL [1981] (J 1044) Math Notes, Acad Sci USSR 30*622-626
⋄ B15   E25   E35   ⋄ REV   MR 82j:03007   Zbl 508#03024
• ID 74454

YAMASAKI, S. [1980] see DOSHITA, S.

YAMASAKI, S. [1983] see DOSHITA, S.

YAMASAKI, S. [1984] see DOSHITA, S.

YANG, ANZHOU [1979] *Problem of cardinal number and other problems of non-standard model *R (Chinese)* (J 4666) J Beijing Polytech Univ 5/1*3-5
⋄ E10   H05   ⋄ ID 49402

YANG, ANZHOU see Vol. V for further entries

YANOV, YU.I. [1959] see MUCHNIK, A.A.

YANOV, YU.I. see Vol. III, IV, VI for further entries

YANOVSKAYA, S.A. [1965] *Problems of introduction and elimination of abstractions of higher (than first) order (Russian)* (P 0563) Found of Statements & Decisions;1961 Warsaw 171-177
⋄ A05   B15   ⋄ ID 14849

YASUHARA, A. [1971] *Recursive function theory and logic* (X 0801) Academic Pr: New York xv+338pp
⋄ B98   D10   D20   D98   ⋄ REV   MR 47#1582   Zbl 254#02002   JSL 40.619   • ID 18028

YASUHARA, A. see Vol. IV for further entries

YASUMOTO, M. [1983] *Nonstandard arithmetic of function fields over H-convex subfields of *Q (Japanese)* (P 4113) Found of Math;1982 Kyoto 230-236
⋄ H05   H15   ⋄ REV   MR 85j:11028   Zbl 505#14019
• ID 47688

YASUMOTO, M. [1983] *Nonstandard arithmetic of function fields over H-convex subfields of *Q* (J 0127) J Reine Angew Math 342*1-11
⋄ H05 H15 ⋄ REV MR 85g:11028 Zbl 505 # 14019
• ID 38208

YASUMOTO, M. see Vol. III, V for further entries

YEH, CHEHCHIH [1977] see MURAKAMI, H.

YEH, CHEHCHIH [1983] see MURAKAMI, H.

YELICK, K. [1985] *Combining unification algorithms for confined regular equational theories* (P 4244) Rewriting Techn & Appl (1);1985 Dijon 365-380
⋄ B35 ⋄ ID 49791

YNTEMA, M.K. [1964] *A detailed argument for the Post-Linial theorems* (J 0047) Notre Dame J Formal Log 5*37-50
⋄ B20 D35 ⋄ REV MR 30 # 1031 Zbl 168.256 JSL 31.117
• ID 18050

YNTEMA, M.K. see Vol. IV for further entries

YOELI, M. [1964] see RINON, S.

YOELI, M. see Vol. IV for further entries

YOHE, J. [1979] *Implementing nonstandard arithmetics* (J 0163) SIAM Review 21*34-56
⋄ H05 H20 ⋄ REV Zbl 395 # 68040 • ID 52599

YOSHIDA, M. [1984] see DOSHITA, S.

YOUNG, L. [1972] *Functional analysis - a non-standard treatment with semifields* (P 0732) Contrib to Non-Standard Anal;1970 Oberwolfach 123-170
⋄ C65 H05 ⋄ REV MR 58 # 2161 Zbl 261 # 46002
• ID 18057

YOURGRAU, W. [1960] see LYNGHOLM, C.

YOURGRAU, W. [1968] see WORKS, C.

YUAN, XIANGWAN & ZHENG, YUXIN & ZHU, WUJIA [1980] *Concerning sets and point-set spaces (Chinese) (English summary)* (J 2804) Nanjing Daxue Xuebao, Ziran Kexue 1980/3*129-135,136
⋄ A05 B28 ⋄ REV MR 82b:03025 Zbl 448 # 04001
• ID 80261

YUAN, XIANGWAN [1982] see XU, LIZHI

YUAN, XIANGWAN see Vol. II, V, VI for further entries

YUGAI, S.A. [1982] see BAYACHOROVA, B.D.

YUKAMI, T. [1977] *Transfinite type theory and provability of second order formulas* (J 0090) J Math Soc Japan 29*634-654
⋄ B15 ⋄ REV MR 57 # 60 Zbl 359 # 02007 • ID 80180

YUKAMI, T. [1983] *A theorem on lengths of proof of Presburger formulas* (J 2606) Tsukuba J Math 7*169-184
⋄ B35 F20 ⋄ REV MR 85b:03103 Zbl 524 # 03047
• ID 37612

YUKAMI, T. see Vol. IV, VI for further entries

YULE, D. [1926] *Zur Grundlegung des Klassenkalkuels* (J 0043) Math Ann 95*446-452
⋄ B20 E30 G05 ⋄ ID 18075

YUNUSOV, D. [1975] see ABDULLAEV, D.A.

YUNUSOV, D. [1978] see ABDULLAEV, D.A.

YUSHCHENKO, E.L. [1979] see KIRSANOV, G.M.

YUSHCHENKO, E.L. see Vol. IV for further entries

ZABECH, F. [1974] see JACOBSON, A.

ZABSKI, E. [1983] *A formalization of the Weber-Fechner theory (Polish) (English summary)* (J 0481) Acta Univ Wroclaw 517(29)(Logika 8)*83-95
⋄ B80 ⋄ REV Zbl 533 # 03011 • ID 36541

ZABSKI, E. [1983] *Some elementary theories of empirical precedence (Polish) (English summary)* (J 0481) Acta Univ Wroclaw 517(29)(Logika 8)*97-108
⋄ B80 ⋄ REV Zbl 533 # 03012 • ID 36536

ZAHN, P. [1967] *Eine Einfuehrung der reellen Zahlen in der operativen Mathematik ohne die Unterscheidung von Sprachschichten* (S 0405) Mitt Math Sem Giessen 72*ii+75pp
⋄ B28 F35 F65 ⋄ REV MR 36 # 2469 Zbl 201.332
• ID 24851

ZAHN, P. [1982] *Ein argumentativer Weg zur Logik* (X 0890) Wiss Buchges: Darmstadt 212pp
⋄ B98 ⋄ REV MR 84f:03003 Zbl 502 # 03001 • ID 34423

ZAHN, P. see Vol. VI for further entries

ZAIONTZ, C. [1983] *Axiomatization of the monadic theory of ordinals $<\omega_2$* (J 0068) Z Math Logik Grundlagen Math 29*337-356
⋄ B25 B28 C65 C85 ⋄ REV MR 85j:03009
Zbl 553 # 03006 • ID 42214

ZAIONTZ, C. [1983] see BUECHI, J.R.

ZAKHAR'YASHCHEV, M.V. [1978] see POPOV, S.V.

ZAKHAR'YASHCHEV, M.V. see Vol. II, III, VI for further entries

ZAKON, E. [1969] see ROBINSON, A.

ZAKON, E. [1969] *Remarks on the nonstandard real axis* (P 0649) Appl Model Th to Algeb, Anal & Probab;1967 Pasadena 195-227
⋄ C65 H05 ⋄ REV MR 39 # 183 Zbl 195.14 • ID 16299

ZAKON, E. [1974] *A new variant of non-standard analysis* (P 1083) Victoria Symp Nonstand Anal;1972 Victoria 313-339
⋄ C20 H05 ⋄ REV MR 57 # 16060 Zbl 289 # 02041
• ID 30005

ZAKON, E. see Vol. III, V for further entries

ZAKREVSKIJ, A.D. [1965] *Algorithms of minimalization of weakly defined Boolean functions (Russian) (English summary)* (J 0040) Kibernetika, Akad Nauk Ukr SSR 1965/2*53-60
⋄ B35 D20 ⋄ REV MR 34 # 5588 Zbl 192.86 • ID 16244

ZAKREVSKIJ, A.D. [1969] *New algorithm for minimizing weakly defined boolean functions* (J 0040) Kibernetika, Akad Nauk Ukr SSR 1969/5*21-28
• TRANSL [1969] (J 0021) Cybernetics 5*553-560
⋄ B35 ⋄ REV Zbl 191.313 • ID 18079

ZAKREVSKIJ, A.D. [1974] *On the theory of polysyllogisms (Russian)* (J 0414) Dokl Akad Nauk Belor SSR 18*114-117,187
⋄ B20 ⋄ REV MR 52 # 13314 Zbl 293 # 02012 • ID 21777

ZAKREVSKIJ, A.D. [1975] see UTKIN, A.A.

ZAKREVSKIJ, A.D. [1978] *Determination of minimal disjunctive base of a boolean matrix (Russian)* (J 0414) Dokl Akad Nauk Belor SSR 22*39-41
⋄ B05 ⋄ REV Zbl 374 # 94017 • ID 51576

ZAKREVSKIJ, A.D. [1979] *On a formalization of polysyllogistics (Russian)* (P 2554) All-Union Symp Th Log Infer;1974 Moskva 300-309
⋄ B20 ⋄ REV MR 84k:03093 • ID 36110

ZAKREVSKIJ, A.D. [1980] *Some combinatorial problems of artificial intelligence (Russian)* (S 2582) Semiotika & Inf, Akad Nauk SSSR 15*3-16
⋄ B35 ⋄ REV MR 82a:68007 Zbl 453 # 68058 • ID 54208

ZALC, A. [1975] see PERRIN, M.J.

ZALC, A. see Vol. III, V, VI for further entries

ZALDOKAS, R. [1985] *On the question of the construction of complete term rewriting systems (Russian) (English and Lithuanian summaries)* (J 3939) Mat Logika Primen (Akad Nauk Litov SSR) 1985/4*113-121,143
⋄ B35 ⋄ ID 49042

ZALTA, E.N. [1982] *Meinongian type theory and its applications* (J 0063) Studia Logica 41*297-307
⋄ A05 B15 ⋄ REV MR 85i:03031 Zbl 531 # 03003 • ID 37671

ZALTA, E.N. see Vol. II for further entries

ZAMOV, N.K. [1968] see SHARONOV, V.I.

ZAMOV, N.K. [1969] see SHARONOV, V.I.

ZAMOV, N.K. [1970] see SHARONOV, V.I.

ZAMOV, N.K. [1972] *On a bound for the complexity of terms in the resolution method (Russian)* (S 0066) Tr Mat Inst Steklov 121*5-13,165
• TRANSL [1972] (S 0055) Proc Steklov Inst Math 121*1-10
⋄ B35 ⋄ REV MR 49 # 4335 Zbl 305 # 68064 • ID 66207

ZAMOV, N.K. [1973] see SHARONOV, V.I.

ZAMOV, N.K. [1974] *Beschraenkung der Kompliziertheit der Terme in Ableitungen und aufloesbare Fragmente der Praedikatenkalkuele (Russisch)* (S 2738) Issl Prikl Mat (Univ Kazan') 2*56-62
⋄ B35 F20 ⋄ REV MR 57 # 9503 Zbl 357 # 02011 • ID 50359

ZANDAROWSKA, W. [1966] *On certain connections between consequence, inconsistency and completeness (Polish) (Russian and English summaries)* (J 0063) Studia Logica 18*165-178
⋄ B22 ⋄ REV MR 33 # 5451 Zbl 294 # 02024 • ID 66208

ZANDAROWSKA, W. see Vol. II for further entries

ZANON, G. [1982] see MINKER, J.

ZARISKI, O. [1925] *Gli sviluppi piu recenti della teoria degli insiemi e il principio di Zermelo* (J 0313) Period Mat (Bologna) 5*57-80
⋄ A05 B10 E25 F99 ⋄ REV FdM 51.164 • ID 41579

ZARNECKA-BIALY, E. [1965] *Certain non-standard concept of consequence (Polish) (English and Russian summaries)* (S 0458) Zesz Nauk, Prace Log, Uniw Krakow 1*103-132
⋄ B22 ⋄ REV MR 37 # 1232 • ID 14359

ZARNECKA-BIALY, E. [1966] *On the equivalence of logical systems (Polish) (English summary)* (S 0458) Zesz Nauk, Prace Log, Uniw Krakow 2*55-62
⋄ B22 ⋄ ID 16510

ZARNECKA-BIALY, E. [1972] *The deduction theorems for propositional calculi when implication and falsum is present (Polish summary)* (S 0458) Zesz Nauk, Prace Log, Uniw Krakow 7*93-100
⋄ B20 F50 ⋄ REV MR 47 # 4757 • ID 16532

ZARNECKA-BIALY, E. [1973] *Negation in Ch.S.Peirce's propositional calculus* (J 0302) Rep Math Logic, Krakow & Katowice 1*99-101
⋄ A10 B20 F50 ⋄ REV MR 50 # 6782 Zbl 291 # 02002 • ID 14363

ZARNECKA-BIALY, E. [1973] *Wajsberg's algorithm for axiomatization of the classical propositional calculus* (J 0387) Bull Sect Logic, Pol Acad Sci 2*97-102
⋄ B05 ⋄ REV MR 55 # 7760 • ID 80224

ZARNECKA-BIALY, E. see Vol. II, VI for further entries

ZAWADZKI, W. [1974] *Condition for the existence of the decomposition of a product function* (J 0485) Algorytmy, Pol Akad Nauk 11*59-71
⋄ B05 ⋄ REV MR 51 # 12438 Zbl 307 # 94035 • ID 66222

ZAWISZA, B. [1984] see KRUPA, A.

ZBIERSKI, P. [1978] *Axiomatizability of second order arithmetic with ω-rule* (J 0027) Fund Math 100*51-57
⋄ B15 C62 E45 E70 F35 ⋄ REV MR 58 # 5161 Zbl 387 # 03021 • ID 29200

ZBIERSKI, P. see Vol. III, V, VI for further entries

ZBORAY, F. [1984] see BIUNDO, S.

ZELEZNIKAR, A. [1960] *Solvability problems of propositional equations (Serbo-Croatian summary)* (J 0371) Glas Mat-Fiz Astron, Ser 2 (Zagreb) 15*237-244
⋄ B05 ⋄ REV MR 24 # A34 Zbl 107.6 • ID 14380

ZELEZNIKAR, A. [1962] *Behandlung logistischer Probleme mit Ziffernrechner (Serbo-Croatian summary)* (J 0371) Glas Mat-Fiz Astron, Ser 2 (Zagreb) 17*171-179
⋄ B35 ⋄ REV MR 29 # 2188 Zbl 118.255 • ID 14382

ZELEZNIKAR, A. see Vol. IV for further entries

ZELICHENKO, A.I. [1979] *The correspondence of monotone boolean functions to systems of linear inequalities (Russian)* (J 0199) Zh Vychisl Mat i Mat Fiz 19*1543-1554,1630
• TRANSL [1979] (J 1049) USSR Comput Math & Math Phys 19/6*192-204
⋄ B05 ⋄ REV MR 81b:94075 Zbl 449 # 94029 • ID 83220

ZELINKA, B. [1966] *Independence of Birkhoff's postulate system for distributive lattices with a unit element (Czech) (Russian and English summaries)* (J 0086) Cas Pestovani Mat, Ceskoslov Akad Ved 91*472-477
⋄ B30 G10 ⋄ REV MR 34 # 7420 Zbl 161.14 • ID 22004

ZELINKA, B. see Vol. IV, V for further entries

ZERMELO, E. [1909] *Ueber die Grundlagen der Arithmetik* (P 0689) Int Congr Math (4);1908 Roma 2*8-11
⋄ B28 E20 ⋄ REV FdM 40.98 • ID 37941

ZERMELO, E. [1932] *Anmerkung* (C 1160) Cantor: Ges Abhandlungen 441-442
⋄ A05 A10 B28 ⋄ ID 22065

ZERMELO, E. [1935] *Grundlagen einer allgemeinen Theorie der mathematischen Satzsysteme (erste Mitteilung)* (J 0027) Fund Math 25*136-146
⋄ B05 C75 E20 ⋄ REV Zbl 12.241 FdM 61.972 • ID 14413

ZERMELO, E. see Vol. III, IV, V for further entries

ZHALDOKAS, R. [1983] *An approach to the construction of complete term rewriting systems (Russian) (English and Lithuanian summaries)* (J 3939) Mat Logika Primen (Akad Nauk Litov SSR) 3*9-22
⋄ B25 B35 B75 C05 ⋄ REV MR 85h:68043 Zbl 558 # 68029 • ID 43443

ZHANG, XIANG [1985] *An initial study of the use of hypernets to construct nonstandard models (Chinese)* (J 2771) Kexue Tongbao 30*415-417
⋄ H05 ⋄ ID 48568

ZHEGALKIN, I.I. [1927] *On the technique of proposition calculus in symbolic logic (Russian) (French summary)* (J 1404) Mat Sb, Akad Nauk SSSR 34*9-28
⋄ B05 ⋄ REV FdM 53.43 • ID 45785

ZHEGALKIN, I.I. [1928] *Arithmetization of symbolic logic (Russian) (French summary)* (J 1404) Mat Sb, Akad Nauk SSSR 35*311-378
⋄ B05 ⋄ REV FdM 54.58 • REM Part II 1929 • ID 45786

ZHEGALKIN, I.I. [1929] *Arithmetization of symbolic logic II (Russian) (French summary)* (J 1404) Mat Sb, Akad Nauk SSSR 36*205-238
⋄ B05 ⋄ REV FdM 55.628 • REM Part I 1928 • ID 45787

ZHEGALKIN, I.I. [1939] *Sur l'Entscheidungsproblem (Russian) (French summary)* (J 0142) Mat Sb, Akad Nauk SSSR, NS 6(48)*185-198
⋄ B20 B25 D35 ⋄ REV MR 1.322 Zbl 22.193 JSL 5.69 FdM 65.27 • ID 17903

ZHEGALKIN, I.I. [1946] *Sur le probleme de la resolubilite pour les classes finies (Russian) (French summary)* (S 1498) Uch Zap Univ, Moskva 100/1*155-211
⋄ B20 B25 C13 ⋄ REV MR 12.2 JSL 17.271 • ID 04825

ZHEGALKIN, I.I. see Vol. V, VI for further entries

ZHENG, XIANCHANG [1978] *Mechanical theorem proving and Dedekind's problem (Chinese)* (J 2879) Wuhan Daxue Xuebao, Ziran Kexue 1978/4*19-31
⋄ B35 ⋄ REV MR 83j:06014 • ID 39890

ZHENG, XIANCHANG [1982] *On the algorithm to realize theorem proving and solving problem (Chinese)* (J 3943) Shuxue Wuli Xuebao 2*459-464
⋄ B35 ⋄ ID 45214

ZHENG, YUXIN [1980] see YUAN, XIANGWAN

ZHENG, YUXIN [1982] see XU, LIZHI

ZHENG, YUXIN see Vol. II, VI for further entries

ZHEZHERUN, A.P. [1979] *Decidability of the unification problem for second-order languages with unary functional symbols (Russian) (English summary)* (J 0040) Kibernetika, Akad Nauk Ukr SSR 1979/5*120-125,155
• TRANSL [1979] (J 0021) Cybernetics 15*735-741
⋄ B15 B35 ⋄ REV MR 81h:03028 Zbl 443 # 03009
• ID 56415

ZHU, CHANGHONG [1982] *Subsemantic resolution (Chinese)* (J 3793) Jisuanjii Xuebao 5*67-69
⋄ B35 ⋄ REV MR 84m:03020 • ID 35725

ZHU, WUJIA [1980] see YUAN, XIANGWAN

ZHU, WUJIA [1982] see XU, LIZHI

ZHU, WUJIA see Vol. II, V, VI for further entries

ZHU, YANJUN [1934] *A summary on mathematical logic (Chinese)* (J 2879) Wuhan Daxue Xuebao, Ziran Kexue 5/2*
⋄ B98 ⋄ ID 48556

ZHU, YANJUN [1936] *Introduction to mathematical logic* (J 0409) Chinese J Math (Taipei) 1/1*
⋄ B98 ⋄ ID 48558

ZHURAVLEV, YU.I. [1957] *On the separability of subsets of the vertices of an n-dimensional unit cube (Russian)* (J 0023) Dokl Akad Nauk SSSR 113*264-267
⋄ B35 ⋄ REV MR 20 # 3774 Zbl 79.9 JSL 24.188
• ID 28552

ZHURAVLEV, YU.I. [1959] *Construction of minimal disjunctive normal forms for functions of the algebra of logic (Russian)* (J 0023) Dokl Akad Nauk SSSR 126*263-266
⋄ B35 ⋄ REV MR 23 # A1514 Zbl 100.9 JSL 31.140
• ID 14447

ZHURAVLEV, YU.I. [1960] *On simplification algorithms for disjunctive normal forms (Russian)* (J 0023) Dokl Akad Nauk SSSR 132*260-263
• TRANSL [1960] (J 0062) Sov Math, Dokl 1*526-529
⋄ B35 ⋄ REV MR 23 # A1515 Zbl 117.9 JSL 31.141
• ID 19779

ZHURAVLEV, YU.I. [1960] *On the impossibility of constructing minimal disjunctive normal forms for functions of the algebra of logic in a single class of algorithms (Russian)* (J 0023) Dokl Akad Nauk SSSR 132*504-506
• TRANSL [1960] (J 0062) Sov Math, Dokl 1*581-583
⋄ B35 ⋄ REV MR 22 # 9440 Zbl 117.10 JSL 30.379
• ID 21028

ZHURAVLEV, YU.I. [1960] *Various notions of minimality of disjunctive normal forms (Russian)* (J 0092) Sib Mat Zh 1*609-610
⋄ B35 ⋄ REV MR 23 # A1516 Zbl 111.8 JSL 33.630
• ID 14448

ZHURAVLEV, YU.I. [1961] *Finite index algorithms for the simplification of disjunctive normal forms (Russian)* (J 0023) Dokl Akad Nauk SSSR 139*1329-1331
• TRANSL [1961] (J 0470) Sov Phys, Dokl 6*686-687
⋄ B35 ⋄ REV MR 28 # 25 Zbl 124.245 • ID 14449

ZHURAVLEV, YU.I. [1962] *Set theoretical methods in the algebra of logic (Russian)* (J 0052) Probl Kibern 8*5-44
⋄ B35 ⋄ REV MR 29 # 4672 JSL 35.162 • ID 19777

ZHURAVLEV, YU.I. [1963] *Algorithms with finite memory on disjunctive forms (Russian)* (J 0071) Met Diskr Analiz (Novosibirsk) 1*5-12
⋄ B35 ⋄ REV MR 29 # 2169 JSL 33.630 • ID 19776

ZHURAVLEV, YU.I. [1963] *On a class of algorithms over finite sets (Russian)* (J 0023) Dokl Akad Nauk SSSR 151*1025-1028
• TRANSL [1963] (J 0062) Sov Math, Dokl 4*1135-1138
⋄ B35 ⋄ REV MR 28 # 26 Zbl 171.273 • ID 14450

ZHURAVLEV, YU.I. [1963] *On inessential variables of functions of a boolean algebra which are not defined everywhere (Russian)* (J 0071) Met Diskr Analiz (Novosibirsk) 1*28-31
⋄ B05 ⋄ REV MR 29 # 2172 • ID 14451

ZHURAVLEV, YU.I. [1964] *Appraisals of the complexity of algorithms for the construction of minimal disjunctive normal forms for the functions of the algebra of logic (Russian)* (J 0071) Met Diskr Analiz (Novosibirsk) 3*41-77
⋄ B35 ⋄ REV MR 32 # 3989 • ID 14453

ZHURAVLEV, YU.I. [1964] *Selection algorithms for sets of real variables of functions not everywhere defined in an algebra of logic (Russian)* (J 0052) Probl Kibern 11*271-275
- ⋄ B05 ⋄ REV MR 30#1046 • ID 14455

ZHURAVLEV, YU.I. [1964] *Set-theoretic methods in symbolic logic* (J 0052) Probl Kibern 11*1-72
- ⋄ B05 E75 ⋄ ID 28782

ZHURAVLEV, YU.I. [1974] see GUREVICH, I.B.

ZHURAVLEV, YU.I. [1979] *Local algorithms over disjunctive normal forms (Russian)* (J 0023) Dokl Akad Nauk SSSR 245*289-292
- • TRANSL [1979] (J 0062) Sov Math, Dokl 20*286-289
- ⋄ B35 G05 ⋄ REV MR 80h:03091 Zbl 449#68008
- • ID 80264

ZHURAVLEV, YU.I. [1985] see KOGAN, A.YU.

ZHURAVLEV, YU.I. see Vol. IV for further entries

ZHURKIN, V.A. [1972] see BOBROV, A.E.

ZHURKIN, V.A. see Vol. II for further entries

ZICH, O.V. [1948] *Sur la notion du nombre entier* (J 1040) Bull Math Nat Acad Tcheque Sci 49*139-149
- ⋄ A05 B28 ⋄ REV MR 12.792 Zbl 54.007 • ID 21115

ZICH, O.V. see Vol. II, V for further entries

ZIEMBINSKI, Z. [1970] *Conditions preliminaires de l'application de la logique deontique dans les raisonnements juridique* (J 0079) Logique & Anal, NS 13*107-124,125-154
- ⋄ B80 ⋄ REV Zbl 198.15 • ID 16207

ZIEMBINSKI, Z. [1976] *Practical logic (Polish)* (X 1034) PWN: Warsaw xv+437pp
- • TRANSL [1976] (X 0835) Reidel: Dordrecht xv+437pp
- ⋄ A05 B45 B98 ⋄ REV Zbl 372#02001 JSL 23.73 JSL 47.231 • REM With an appendix on "Deontic logic" by Ziemba,Z. • ID 51409

ZIEMBINSKI, Z. see Vol. II for further entries

ZIEWACZ, S. [1979] see CORCORAN, J.

ZIL'BER, B.I. [1981] *Solution of the problem of finite axiomatizability for theories that are categorical in all infinite powers (Russian)* (C 3806) Issl Teor Progr 69-75
- ⋄ B30 C35 C45 C52 ⋄ REV MR 84k:03095 Zbl 522#03019 • ID 36112

ZIL'BER, B.I. see Vol. III for further entries

ZINNES, J.L. [1963] see SUPPES, P.

ZINOV'EV, A.A. [1962] *Propositional logic and theory of deduction (Russian)* (X 0899) Akad Nauk SSSR : Moskva 152pp
- ⋄ B98 ⋄ REV Zbl 105.4 JSL 30.373 • ID 19782

ZINOV'EV, A.A. [1963] *Eine Verallgemeinerung der Syllogistik (Russian)* (C 1654) Probl Logiki 38-63
- ⋄ B20 ⋄ REV Zbl 285#02014 • ID 29944

ZINOV'EV, A.A. [1970] *Classical and nonclassical propositional relations (Russian)* (C 0668) Neklass Log 48-51
- ⋄ B05 ⋄ REV MR 48#1874 • ID 14424

ZINOV'EV, A.A. [1975] see WESSEL, H.

ZINOV'EV, A.A. see Vol. II, III, V, VI for further entries

ZIVALJEVIC, B. [1984] see MILLER, H.I.

ZIVALJEVIC, R. [1982] *A Loeb measure approach to the Riesz representation theorem* (J 0400) Publ Inst Math, NS (Belgrade) 32(46)*175-177
- ⋄ H10 ⋄ REV MR 84h:28020 Zbl 521#03053 • ID 37086

ZIVALJEVIC, R. [1983] *The notions of w-net and Y-compact space viewed under infinitesimal microscope* (J 0400) Publ Inst Math, NS (Belgrade) 34(48)*243-246
- ⋄ H05 ⋄ REV MR 86a:54059 Zbl 555#54030 • ID 44987

ZIVALJEVIC, R. [1985] *Loeb completion of internal vector-valued measures* (J 0132) Math Scand 56*276-286
- ⋄ H05 ⋄ ID 49589

ZIVALJEVIC, R. see Vol. V for further entries

ZLATIN, D.R. [1984] see CONSTABLE, R.L.

ZOLL, E.J. [1968] *Logic: A programmed text for 2-valued and 3-valued logics* (X 1330) Pitman Publ: Belmont & London 137pp
- ⋄ B50 B98 ⋄ ID 22544

ZUBIETA RUSSI, G. [1950] *On the substitution of functional variables in the functional calculus of the first order* (J 0250) Bol Soc Mat Mexicana 7*1-21
- ⋄ B10 ⋄ REV MR 12.790 JSL 16.291 • ID 14430

ZUBIETA RUSSI, G. [1951] *Sobre el calculo funcional de primer orden* (J 0250) Bol Soc Mat Mexicana 8*15-21
- ⋄ B10 ⋄ REV MR 13.309 JSL 15.200 • ID 14431

ZUBIETA RUSSI, G. [1951] *Some theorems in the theory of elementary quantification* (J 0250) Bol Soc Mat Mexicana 8*33-46
- ⋄ B10 ⋄ REV MR 16.555 JSL 29.56 • ID 24927

ZUBIETA RUSSI, G. [1957] *Clases aritmeticas definidas sin igualdad* (J 3127) Bol Soc Mat Mexicana, Ser 2 2*45-53
- ⋄ B10 B20 C07 C52 ⋄ REV MR 20#2272 Zbl 88.12 JSL 29.55 • ID 11688

ZUBIETA RUSSI, G. see Vol. III for further entries

ZUCKER, J.I. [1975] *Formalization of classical mathematics in automath* (1111) Preprints, Manuscr., Techn. Reports etc.
- ⋄ B35 ⋄ REM Automath-Report (Aut 40), Math. Dept., Technological University Eindhoven, Netherlands • ID 30596

ZUCKER, J.I. [1978] *The adequacy problem for classical logic* (J 0122) J Philos Logic 7*517-535
- ⋄ A05 B10 ⋄ REV MR 80b:03012 Zbl 408#03021 JSL 47.689 • ID 56260

ZUCKER, J.I. see Vol. II, IV, VI for further entries

ZUCKERMANN, M.M. [1973] *Formation sequences for propositional formulas* (J 0047) Notre Dame J Formal Log 14*134-138
- ⋄ B05 ⋄ REV MR 48#1875 Zbl 197.272 • ID 14442

ZUCKERMANN, M.M. see Vol. V for further entries

ZUEV, YU.A. [1978] *Approximation of a partial Boolean function by a monotonic boolean function (Russian)* (J 0199) Zh Vychisl Mat i Mat Fiz 18*1571-1578
- • TRANSL [1978] (J 1049) USSR Comput Math & Math Phys 18/6*212-218
- ⋄ B05 ⋄ REV MR 80a:94065 Zbl 401#94039 • REM TRl-year 1979 • ID 66320

ZVONKIN, A.K. [1984] see SHUBIN, M.A.

ZVONKIN, A.K. see Vol. IV for further entries

ZYGMUNT, J. [1972] *Direct products of consequences operations* (J 0387) Bull Sect Logic, Pol Acad Sci 1/4*61-64
⋄ A10 B22 ⋄ REV MR 58 # 27314 • ID 31568

ZYGMUNT, J. [1977] see WOJCICKI, R.

ZYGMUNT, J. [1980] see HAWRANEK, J.

ZYGMUNT, J. [1981] see HAWRANEK, J.

ZYGMUNT, J. [1981] *Notes on decidability and finite approximability of sentential logics* (J 0387) Bull Sect Logic, Pol Acad Sci 10*38-41
⋄ A10 B22 B25 ⋄ REV MR 82f:03024 Zbl 457 # 03008 • ID 54333

ZYGMUNT, J. [1983] *An application of the Lindenbaum method in the domain of strongly finite sentential calculi* (J 0481) Acta Univ Wroclaw 517(29)(Logika 8)*59-68
⋄ B22 ⋄ REV Zbl 537 # 03014 • ID 43733

ZYGMUNT, J. [1983] *On decidability and finite approximability of sentential logics* (J 0481) Acta Univ Wroclaw 517(29)(Logika 8)*69-81
⋄ B22 B25 ⋄ REV Zbl 547 # 03005 • ID 41843

ZYGMUNT, J. [1983] see HAWRANEK, J.

ZYGMUNT, J. [1983] *Some remarks on matrix consequence operations* (J 0481) Acta Univ Wroclaw 517(29)(Logika 8)*47-58
⋄ B22 ⋄ REV Zbl 548 # 03010 • ID 43193

ZYGMUNT, J. [1984] *An essay in matrix semantics for consequence relations* (J 0481) Acta Univ Wroclaw 741(43)(Logika 12)*76pp
⋄ B22 ⋄ REV Zbl 559 # 03012 • ID 46876

ZYGMUNT, J. [1984] see HAWRANEK, J.

ZYGMUNT, J. see Vol. II, III, VI for further entries

ZYKOV, A.A. [1953] *The spectrum problem in the extended predicate calculus (Russian)* (J 0216) Izv Akad Nauk SSSR, Ser Mat 17*63-76
• TRANSL [1956] (J 0225) Amer Math Soc, Transl, Ser 2 3*1-14
⋄ B15 C13 ⋄ REV MR 14.936 Zbl 50.7 JSL 22.360
• ID 19773

ZYKOV, A.A. [1959] *Remarks in connection with the reduction theorem for logical calculi (Russian)* (P 0607) All-Union Math Conf (3);1956 Moskva 4*85-86
⋄ B20 D35 ⋄ ID 30599

ZYKOV, A.A. see Vol. IV for further entries

ZYLINSKI, E. [1925] *Some remarks concerning the theory of deduction* (J 0027) Fund Math 7*203-209
⋄ B05 ⋄ REV FdM 51.47 • ID 43000

Source Index

Journals

J 0001 Acta Math Acad Sci Hung • H
Acta Mathematica Academiae Scientiarum Hungaricae
[1950-1982] ISSN 0001-5954
- CONT AS (J 4729) Acta Math Hung

J 0002 Acta Sci Math (Szeged) • H
Acta Scientiarum Mathematicarum [1947ff] ISSN 0001-6969
- CONT OF (J 0460) Acta Univ Szeged, Sect Mat

J 0003 Algebra i Logika • SU
Algebra i Logika (Algebra and Logic) [1962ff] ISSN 0373-9252
- TRANSL IN (J 0069) Algeb and Log

J 0004 Algeb Universalis • CDN
Algebra Universalis [1970ff] ISSN 0002-5240

J 0005 Amer Math Mon • USA
American Mathematical Monthly [1894ff] ISSN 0002-9890

J 0007 Ann Math Logic • NL
Annals of Mathematical Logic [1970-1982] ISSN 0003-4843
- CONT AS (J 0073) Ann Pure Appl Logic

J 0008 Arch Math (Basel) • CH
*Archiv der Mathematik * Archives of Mathematics * Archives Mathematiques* [1948ff] ISSN 0003-889X

J 0009 Arch Math Logik Grundlagenforsch • D
Archiv fuer Mathematische Logik und Grundlagenforschung [1950ff] ISSN 0003-9268

J 0010 Autom & Remote Control • USA
Automation and Remote Control [1958ff] ISSN 0005-1179
- TRANSL OF (J 0011) Avtom Telemekh

J 0011 Avtom Telemekh • SU
Avtomatika i Telemekhanika (Automation and Telemechanics) [1934ff] ISSN 0005-2310
- TRANSL IN (J 0010) Autom & Remote Control

J 0012 Boll Unione Mat Ital, IV Ser • I
Bolletino della Unione Matematica Italiana. Serie IV [1968-1975] ISSN 0041-7084
- CONT OF (J 4408) Boll Unione Mat Ital, III Ser • CONT AS (J 3285) Boll Unione Mat Ital, V Ser, A & (J 3495) Boll Unione Mat Ital, V Ser, B

J 0013 Brit J Phil Sci • GB
British Journal for the Philosophy of Science [1950ff] ISSN 0007-0882

J 0014 Bull Acad Pol Sci, Ser Math Astron Phys • PL
Bulletin de l'Academie Polonaise des Sciences. Serie des Sciences Mathematiques, Astronomiques et Physiques [1953-1978] ISSN 0001-4117
- CONT AS (J 3293) Bull Acad Pol Sci, Ser Math

J 0015 Bull Amer Math Soc • USA
Bulletin of the American Mathematical Society [1894-1978] ISSN 0002-9904
- CONT AS (J 0589) Bull Amer Math Soc (NS)

J 0016 Bull Austral Math Soc • AUS
Bulletin of the Australian Mathematical Society [1969ff] ISSN 0004-9727

J 0017 Canad J Math • CDN
*Canadian Journal of Mathematics * Journal Canadien de Mathematiques* [1949ff] ISSN 0008-414X

J 0018 Canad Math Bull • CDN
*Canadian Mathematical Bulletin * Bulletin Canadien de Mathematiques* [1958ff] ISSN 0008-4395

J 0020 Compos Math • NL
Compositio Mathematica [1933ff] ISSN 0010-437X

J 0021 Cybernetics • USA
Cybernetics [1965ff] ISSN 0011-4235
- TRANSL OF (J 0040) Kibernetika, Akad Nauk Ukr SSR

J 0022 Cheskoslov Mat Zh • CS
*Cheskoslovatskij Matematicheskij Zhurnal * Czechoslovak Mathematical Journal* [1951ff] ISSN 0011-4642
- REM From Vol. 19 (1969) on the title is only: Czechoslovak Mathematical Journal

J 0023 Dokl Akad Nauk SSSR • SU
Doklady Akademii Nauk SSSR (Reports of the Academy of Sciences of the USSR) [1933ff] ISSN 0002-3264
- TRANSL IN (J 0062) Sov Math, Dokl & (J 0470) Sov Phys, Dokl

J 0024 Dokl Akad Nauk Uzb SSR • SU
*Doklady Akademii Nauk Uzb SSR (DAN Uzb SSR) * UzSSR Fanlar Akademijasining Dokladlari (Reports of the Academy of Sciences of the Uzb SSR)* [1944ff] ISSN 0134-4307

J 0025 Duke Math J • USA
Duke Mathematical Journal [1935ff] ISSN 0012-7094

J 0026 Elem Math • CH
*Elemente der Mathematik * Revue de Mathematiques Elementaires * Rivista di Matematica Elementare* [1946ff] ISSN 0013-6018

J 0027 Fund Math • PL
Fundamenta Mathematicae [1920ff] ISSN 0016-2736

J 0028 Indag Math • NL
*Indagationes Mathematicae * Nederlandse Akademie van Wetenschappen. Proceedings* [1939ff] ISSN 0019-3577, ISSN 0023-3358
- REM Until 1950 part of Koninklijke Nederlandsche Akademie van Wetenschappen, Proceedings of the Section of Sciences; vol n+41 with separate pagination. Since 1951 same as Koninklijke Nederlandse Akademie van Wetenschappen, Proceedings of the Section of Sciences, Series A; vol n+41. Before 1951 page numbers in Proceedings and Indagationes different. Since 1951 the same page numbers as Proceedings Series A.

J 0029 Israel J Math • IL
Israel Journal of Mathematics [1963ff] ISSN 0021-2172
• CONT OF (J 0493) Bull Res Counc Israel Sect F

J 0030 Izv Akad Nauk Uzb SSR, Ser Fiz-Mat • SU
*Izvestiya Akademii Nauk UzSSR. Seriya Fiziko-Matematicheskikh Nauk (Fanlar Akademijasining Achboroti. Fizika-Matematika Fanlaris Serijas * Proceedings of the Academy of Sciences of the Uzb SSR. Series: Physics and Mathematics)* [1957ff] ISSN 0568-7144

J 0031 Izv Vyssh Ucheb Zaved, Mat (Kazan) • SU
Izvestiya Vysshikh Uchebnykh Zavedenij. Matematika (Proceedings of the University. Mathematics) [1957ff] ISSN 0021-3446
• TRANSL IN (J 3449) Sov Math

J 0032 J Algeb • USA
Journal of Algebra [1964ff] ISSN 0021-8693

J 0033 J Comb Th, Ser B • USA
Journal of Combinatorial Theory. Series B [1971ff] ISSN 0095-8956
• CONT OF (J 1669) J Comb Th

J 0034 J Math Anal & Appl • USA
Journal of Mathematical Analysis and Applications [1960ff] ISSN 0022-247X

J 0035 J Math Psychol • USA
Journal of Mathematical Psychology [1964ff] ISSN 0022-2496

J 0036 J Symb Logic • USA
The Journal of Symbolic Logic [1936ff] ISSN 0022-4812

J 0037 ACM J • USA
Journal of the ACM (= Association for Computing Machinery) [1954ff] ISSN 0004-5411

J 0038 J Austral Math Soc • AUS
Journal of the Australian Mathematical Society [1959-1975] ISSN 0004-9735
• CONT AS (J 3194) J Austral Math Soc, Ser A

J 0039 J London Math Soc • GB
The Journal of the London Mathematical Society [1926-1968]
• CONT AS (J 3172) J London Math Soc, Ser 2

J 0040 Kibernetika, Akad Nauk Ukr SSR • SU
Kibernetika. Akademiya Nauk Ukrainskoj SSR (Cybernetics. Academy of Sciences of the Ukrainian SSR) [1965ff] ISSN 0023-1274
• TRANSL IN (J 0021) Cybernetics

J 0041 Math Syst Theory • D
Mathematical Systems Theory. An International Journal [1967ff] ISSN 0025-5661

J 0042 Mat Vesn, Drust Mat Fiz Astron Serb • YU
Matematichki Vesnik. Drushtvo Matematichara, Fizichara i Astromoma SR Serbije, SFR Jugoslavija (Mathematical Publications. Society Serbe of Mathematicians, Physicists and Astronomers) [1964ff] ISSN 0025-5165
• CONT OF (J 4277) Vesn Drusht Mat Fiz Serbije

J 0043 Math Ann • D
Mathematische Annalen [1868ff] ISSN 0025-5831

J 0044 Math Z • D
Mathematische Zeitschrift [1918ff] ISSN 0025-5874

J 0045 Monatsh Math • A
Monatshefte fuer Mathematik [1943ff] ISSN 0026-9255
• CONT OF (J 0124) Monatsh Math-Phys

J 0046 Nieuw Arch Wisk • NL
Nieuw Archief voor Wiskunde [1875-1893]
• CONT AS (J 1793) Nieuw Arch Wisk, Ser 2

J 0047 Notre Dame J Formal Log • USA
Notre Dame Journal of Formal Logic [1960ff] ISSN 0029-4527

J 0048 Pac J Math • USA
Pacific Journal of Mathematics [1951ff] ISSN 0030-8730

J 0049 Period Math Hung • H
Periodica Mathematica Hungarica [1971ff] ISSN 0031-5303

J 0050 Port Math • P
Portugaliae Mathematica [1937ff] ISSN 0032-5155

J 0051 Commentat Math, Ann Soc Math Pol, Ser 1 • PL
*Annales Societatis Mathematicae Polonae. Series I. Commentationes Mathematicae * Roczniki Polskiego Towarzystwa Matematycznego. Seria I. Prace Matematyczne.* [1955ff] ISSN 0079-368X
• CONT OF (J 0611) Pol Tow Mat, Prace Mat-Fiz

J 0052 Probl Kibern • SU
Problemy Kibernetiki. Glavnaya Redaktsiya Fiziko-Matematicheskoj Literatury [1958ff] ISSN 0555-277X
• TRANSL IN (J 0471) Syst Th Res & (J 0449) Probl Kybern & (J 1195) Probl Cybernet

J 0053 Proc Amer Math Soc • USA
Proceedings of the American Mathematical Society [1950ff] ISSN 0002-9939

J 0054 Proc Nat Acad Sci USA • USA
Proceedings of the National Academy of Sciences of the United States of America [1915ff] ISSN 0027-8424

J 0056 Publ Dep Math, Lyon • F
Publications du Departement de Mathematiques. Faculte des Sciences de Lyon. [1964-1981] ISSN 0076-1656
• CONT AS (J 2107) Publ Dep Math, Lyon, NS

J 0057 Publ Math (Univ Debrecen) • H
Publicationes Mathematicae [1949ff] ISSN 0033-3883

J 0058 Rend Circ Mat Palermo • I
Rendiconti del Circolo Matematico di Palermo [1887-1940]
• CONT AS (J 3522) Rend Circ Mat Palermo, Ser 2

J 0060 Rev Roumaine Math Pures Appl • RO
Revue Roumaine de Mathematiques Pures et Appliquees. Academia Republicii Socialiste Romania [1956ff] ISSN 0035-3965

J 0061 Simon Stevin • B
Simon Stevin: A Quarterly Journal of Pure and Applied Mathematics [1921ff] ISSN 0037-5454
• CONT OF (J 1928) Mathematica B & (J 0291) Wis-natuur Tijdsch & (J 1379) Christiaan Huygens Internat Math Tijdschr
• REM The journal has incorporated J1928, J1379, J0291

J 0062 Sov Math, Dokl • USA
Soviet Mathematics. Doklady. [1960ff] ISSN 0038-5573
• TRANSL OF (J 0023) Dokl Akad Nauk SSSR

J 0063 Studia Logica • PL
Studia Logica [1953ff] ISSN 0039-3215

J 0064 Trans Amer Math Soc • USA
Transactions of the American Mathematical Society [1900ff]
ISSN 0002-9947

J 0065 Tr Moskva Mat Obshch • SU
Trudy Moskovskogo Matematicheskogo Obshchestva (Publications of the Moscow Mathematical Society) [1952ff]
ISSN 0134-8663
• TRANSL IN (**J 3279**) Trans Moscow Math Soc

J 0067 Usp Mat Nauk • SU
Uspekhi Matematicheskikh Nauk (Advances in Mathematical Sciences) [1936ff] ISSN 0042-1316
• TRANSL IN (**J 1399**) Russ Math Surv

J 0068 Z Math Logik Grundlagen Math • DDR
Zeitschrift fuer Mathematische Logik und Grundlagen der Mathematik [1955ff] ISSN 0044-3050

J 0069 Algeb and Log • USA
Algebra and Logic [1968ff] ISSN 0002-5232
• TRANSL OF (**J 0003**) Algebra i Logika

J 0070 Bull Soc Sci Math Roumanie, NS • RO
Bulletin Mathematique de la Societe des Sciences Mathematiques de la Republique Socialiste de Roumanie. Nouvelle Serie. [1957ff] ISSN 0007-4691
• CONT OF (**J 0494**) Bull Math Soc Sci Roumanie

J 0071 Met Diskr Analiz (Novosibirsk) • SU
Metody Diskretnogo Analiza. Sbornik Trudov (Methods of Discrete Analysis. Collected Papers) [1963ff] ISSN 0419-4160, ISSN 0136-1228

J 0072 IRE Trans Electr Comp • USA
Transactions on Electronic Computers. IRE (= Institute of Radio Engineers) [1952-1962]
• CONT AS (**J 4305**) IEEE Trans Electr Comp

J 0073 Ann Pure Appl Logic • NL
Annals of Pure and Applied Logic [1983ff] ISSN 0168-0072
• CONT OF (**J 0007**) Ann Math Logic

J 0074 Math Algor • USA
Mathematical Algorithms [1966-1968] ISSN 0025-5548

J 0075 Phil Phenom Research • USA
Philosophy and Phenomenological Research [1940ff] ISSN 0031-8205

J 0076 Dialectica • CH
Dialectica. International Review of Philosophy of Knowledge [1947ff] ISSN 0012-2017

J 0077 Proc London Math Soc • GB
Proceedings of the London Mathematical Society [1865-1903]
• CONT AS (**J 1910**) Proc London Math Soc, Ser 2

J 0079 Logique & Anal, NS • B
Logique et Analyse. Nouvelle Serie. Publication Trimestrielle du Centre National Belge de Recherche de Logique [1958ff] ISSN 0024-5836

J 0080 Izv Akad Nauk Ehston SSR, Fiz, Mat • SU
*Izvestiya Akademiya Nauk Ehstonskoj SSR. Fizika. Matematika * Eesti NSV Teaduste Akadeemia Toimetised. Fueuesika-Matemaatika (Proceedings of the Academy of Sciences of the Estonian SSR. Physics. Mathematics)* [1956ff] ISSN 0002-3140

J 0081 Proc Japan Acad • J
Proceedings of the Japan Academy [1925-1977] ISSN 0021-4280
• CONT AS (**J 3239**) Proc Japan Acad, Ser A

J 0082 Bull Soc Math Belg • B
Bulletin de la Societe Mathematique de Belgique [1948-1976] ISSN 0037-9476
• CONT AS (**J 3133**) Bull Soc Math Belg, Ser B & (**J 3824**) Bull Soc Math Belg, Ser A

J 0084 Gaz Mat (Lisboa) • P
Gazeta de Matematica. Jornal dos Concorrentes ao Exame de Aptidao e dos Estudantes de Matematica das Escolas Superiores [1940ff]

J 0085 Vest Ser Mat Mekh Astron, Univ Leningrad • SU
Vestnik Leningradskogo Universiteta. Seriya Matematika, Mekhanika, Astronomiya (Publications of the Leningrad University. Series: Mathematics, Mechanics, Astronomy) [1946ff] ISSN 0024-0850
• TRANSL IN (**J 3945**) Vest Math Univ Leningrad

J 0086 Cas Pestovani Mat, Ceskoslov Akad Ved • CS
Casopis pro Pestovani Matematiky. Ceskoslovenska Akademie Ved (Journal for the Cultivation of Mathematics. Czechoslovak Academy of Sciences) [1872ff]

J 0087 Mat Zametki (Akad Nauk SSSR) • SU
Matematicheskie Zametki (Mathematical Notes) [1967ff] ISSN 0025-567X
• TRANSL IN (**J 1044**) Math Notes, Acad Sci USSR

J 0088 Ann Univ Ferrara, NS, Sez 7 • I
Annali dell'Universita di Ferrara. Nuova Serie. Sezione 7. Scienze Matematiche [1966ff]

J 0089 Calcolo • I
Calcolo [1964ff] ISSN 0008-0624

J 0090 J Math Soc Japan • J
Journal of the Mathematical Society of Japan [1885ff] ISSN 0025-5645

J 0091 Sugaku • J
Sugaku (Mathematics) [1947ff] ISSN 0039-470X

J 0092 Sib Mat Zh • SU
Sibirskij Matematicheskij Zhurnal. Akademiya Nauk SSSR. Sibirskoe Otdelenie (Siberian Mathematical Journal. Academy of Sciences of the USSR. Siberian Section) [1960ff] ISSN 0037-4474
• TRANSL IN (**J 0475**) Sib Math J

J 0093 J Comp Syst • USA
The Journal of Computing Systems [1952-1954]

J 0094 Mind • GB
Mind. A Quarterly Review of Philosophy [1876ff] ISSN 0026-4423

J 0095 Philos Stud • NL
Philosophical Studies. An International Journal for Philosophy in the Analytic Tradition [1950ff] ISSN 0031-8116

J 0096 Acta Philos Fenn • SF
Acta Philosophica Fennica [1948ff] ISSN 0355-1792

J 0097 Nous, Quart J Phil • USA
Nous. A Quarterly Journal of Philosophy [1967ff] ISSN 0029-4624

J 0099 Ricerca, Riv Mat Pure & Appl • I
La Ricerca. Rivista di Matematiche Pure ed Applicate. [1950ff] ISSN 0048-8283

J 0100 Amer J Math • USA
American Journal of Mathematics [1878ff] ISSN 0002-9327

J 0101 Phil Rev • USA
The Philosophical Review [1896ff] ISSN 0031-8108

J 0103 Analysis (Oxford) • GB
Analysis [1933ff] ISSN 0003-2638

J 0104 Atti Accad Sci Lett Arti Palermo, Ser 4/I • I
Atti della Accademia di Scienze Lettere e Arti di Palermo. Serie Quarta. Parte I: Scienze [1940-1983]
• CONT AS (J 4186) Atti Accad Sci Lett Arti Palermo, Ser 5/I

J 0105 Theoria (Lund) • S
Theoria. A Swedish Journal of Philosophy [1934ff] ISSN 0040-5825

J 0106 Mem Fac Sci, Kyushu Univ, Ser A • J
*Memoirs of the Faculty of Science. Kyushu University. Series A. Mathematics * Kyushu Daigaku Rigakubu Kiyo. A. Sugaku* [1940ff] ISSN 0373-6385

J 0107 Abh Math Sem Univ Hamburg • D
Abhandlungen aus dem Mathematischen Seminar der Universitaet Hamburg [1922ff] ISSN 0025-5858

J 0109 C R Acad Sci, Paris • F
Academie des Sciences de Paris. Comptes Rendus Hebdomadaires des Seances [1835-1965]
• CONT AS (J 2313) C R Acad Sci, Paris, Ser A-B

J 0110 Anais Acad Bras Cienc • BR
Anais da Academia Brasileira de Ciencias. [1929ff] ISSN 0001-3765

J 0111 Nagoya Math J • J
Nagoya Sugaku Zashi (Nagoya Mathematical Journal) [1950ff] ISSN 0027-7630

J 0114 Math Nachr • DDR
Mathematische Nachrichten [1948ff] ISSN 0025-584X

J 0115 Wiss Z Humboldt-Univ Berlin, Math-Nat Reihe • DDR
Wissenschaftliche Zeitschrift der Humboldt-Universitaet Berlin. Mathematisch-Naturwissenschaftliche Reihe [1951ff] ISSN 0043-6852

J 0116 Electr & Comm Japan • USA
*Electronics and Communications in Japan * Scripta Electronica Japonica* [1963ff] ISSN 0036-9683
• TRANSL OF (J 0979) Denshi Tsushin Gakkai Ronbunshi, Sect A-D

J 0117 Arch Math Naturvid • N
Archiv for Matematikk og Naturvidenskab (Archives for Mathematics and Natural Sciences)

J 0119 J Comp Syst Sci • USA
Journal of Computer and System Sciences [1967ff] ISSN 0022-0000

J 0120 Ann of Math, Ser 2 • USA
Annals of Mathematics. 2nd Series [1899ff] ISSN 0003-486X

J 0121 Kon Norske Vidensk Selsk Forh • N
Kongelige Norske Videnskabers Selskabs. Forhandlinger. (Proceedings of the Royal Scandinavian Society of Sciences) [1926ff] ISSN 0368-6302

J 0122 J Philos Logic • NL
Journal of Philosophical Logic [1972ff] ISSN 0022-3611

J 0124 Monatsh Math-Phys • A
Monatshefte fuer Mathematik und Physik [1900-1942]
• CONT AS (J 0045) Monatsh Math

J 0127 J Reine Angew Math • D
Journal fuer die Reine und Angewandte Mathematik [1826ff] ISSN 0075-4102

J 0128 Acta Math Univ Comenianae (Bratislava) • CS
Universitas Comeniana. Acta Facultatis Rerum Naturalium. Mathematica [1956ff]

J 0129 Elektr Informationsverarbeitung & Kybern • DDR
Elektronische Informationsverarbeitung und Kybernetik [1965ff] ISSN 0013-5712

J 0130 BIT • DK
BIT. Nordisk Tidskrift for Informationsbehandling (BIT. Scandinavian Journal for Informatics) [1961ff] ISSN 0006-3835

J 0132 Math Scand • DK
Mathematica Scandinavica [1953ff] ISSN 0025-5521

J 0133 Michigan Math J • USA
The Michigan Mathematical Journal [1952ff] ISSN 0026-2285

J 0135 Izv Akad Nauk Azerb SSR, Ser Fiz-Tekh Mat • SU
Izvestiya Akademii Nauk Azerbajdzhanskoj SSR. Seriya Fiziko-Tekhnicheskikh i Matematicheskikh Nauk (Proceedings of the Academy of Sciences of the Azerbaijan SSR. Series: Physical-Technical and Mathematical Sciences) [1958ff] ISSN 0002-3108

J 0137 C R Acad Bulgar Sci • BG
Doklady Bolgarskoi Akademii Nauk (Comptes Rendus de l'Academie Bulgare des Sciences) [1948ff] ISSN 0001-3978

J 0140 Comm Math Univ Carolinae (Prague) • CS
Commentationes Mathematicae Universitatis Carolinae [1960ff] ISSN 0010-2628

J 0141 Arch Autom & Telemech • PL
Archiwum Automatyki i Telemechaniki (Archives of Automation and Telemechanics) [1956ff] ISSN 0004-072X

J 0142 Mat Sb, Akad Nauk SSSR, NS • SU
Matematicheskij Sbornik. Novaya Seriya. Akademiya Nauk SSSR i Moskovskoe Matematicheskoe Obshchestvo (Mathematical Collected Articles. New Series. Academy of Sciences of the USSR and Moskovian Mathematical Society) [1936ff] ISSN 0025-5157
• CONT OF (J 1404) Mat Sb, Akad Nauk SSSR • TRANSL IN (J 0349) Math of USSR, Sbor

J 0143 Mat Chasopis (Slov Akad Ved) • CS
Matematicky Chasopis (Journal of Mathematics) [1967-1975] ISSN 0025-5173
• CONT OF (J 4713) Mat Fyz Chasopis (Slov Akad Ved)
• CONT AS (J 1522) Math Slovaca

J 0144 Rend Sem Mat Univ Padova • I
Rendiconti del Seminario Matematico dell'Universita di Padova [1930ff] ISSN 0041-8994

J 0145 Rev Metaph Morale • F
Revue de Metaphysique et de Morale [1893ff] ISSN 0035-1571

J 0147 An Univ Bucuresti, Acta Logica • RO
Universitatea Bucuresti. Analele. Acta Logica (Universitaet Bukarest. Annalen. Acta Logica) [1958ff] ISSN 0524-823X
• REM From 1962-1967 Analele Universitatii Bucuresti. Seria Acta Logica. From 1958-1961 Analele Universitatii C.I. Parhon. Seria Acta Logica

J 0148 Math Gaz • GB
The Mathematical Gazette [1894ff] ISSN 0025-5572

J 0149 Atti Accad Naz Lincei Fis Mat Nat, Ser 8 • I
Atti della Accademia Nazionale dei Lincei. Rendiconti. Classe di Scienze Fisiche, Matematiche e Naturali. Serie VIII [1946ff] ISSN 0001-4435

J 0151 Arch Soc Belg Philos • B
Archives de la Societe Belge de Philosophie. [1928-1937]
• CONT AS (J 2076) Rev Int Philos

J 0152 Enseign Math • CH
L'Enseignement Mathematique: Revue Internationale [1899-1954]
• CONT AS (J 3370) Enseign Math, Ser 2

J 0153 Phil of Sci (East Lansing) • USA
Philosophy of Science [1934ff] ISSN 0031-8248

J 0154 Synthese • NL
Synthese. An International Journal for Epistemology, Methodology and Philosophy of Science [1936ff] ISSN 0039-7857

J 0155 Commun Pure Appl Math • USA
Communications on Pure and Applied Mathematics [1939ff] ISSN 0010-3640

J 0156 Kybernetika (Prague) • CS
Kybernetika (Cybernetics) [1965ff] ISSN 0023-5954
• REL PUBL (J 3524) Kybernetika Suppl (Prague)

J 0157 Jbuchber Dtsch Math-Ver • D
Jahresbericht der Deutschen Mathematiker-Vereinigung [1890ff] ISSN 0012-0456

J 0158 Erkenntnis (Dordrecht) • NL
Erkenntnis. The Journal of Unified Science: An International Journal of Analytic Philosophy [1975ff] ISSN 0165-0106
• CONT OF (J 3597) J Unif Sci

J 0160 Math-Phys Sem-ber, NS • D
Mathematisch-Physikalische Semesterberichte: Zur Pflege des Zusammenhangs von Schule und Universitaet. Neue Folge [1950-1979] ISSN 0340-4897
• CONT AS (J 2790) Math Sem-ber

J 0161 Bull London Math Soc • GB
The Bulletin of the London Mathematical Society [1926ff] ISSN 0024-6093

J 0162 Teorema (Valencia) • E
Teorema [1971ff]

J 0163 SIAM Review • USA
Review. SIAM (= Society for Industrial and Applied Mathematics) [1959ff] ISSN 0036-1445

J 0164 J Comb Th, Ser A • USA
Journal of Combinatorial Theory. Series A [1971ff] ISSN 0097-3165
• CONT OF (J 1669) J Comb Th

J 0168 Blaetter Deutsch Philos • D
Blaetter fuer Deutsche Philosophie [1927-1944]
• CONT OF (J 1095) Beitr Phil Deutsch Idealismus

J 0170 Ratio (Frankfurt) • D
Ratio [1957ff] ISSN 0342-1848
• REL PUBL (J 1022) Ratio (Oxford) • REM German Edition

J 0171 Proc Cambridge Phil Soc Math Phys • GB
Proceedings of the Cambridge Philosophical Society. Mathematical and Physical Sciences [1843-1974] ISSN 0008-1981
• CONT AS (J 0332) Math Proc Cambridge Phil Soc

J 0173 Riveon Lematemat • IL
Riveon Lematematika. A Quarterly Journal Intended to Promote Mathematical Research Among Students of Mathematics [1946ff]

J 0174 Bull Sect Sci Acad Roumaine • RO
Academie Roumaine. Bulletin de la Section Scientifique [1918-1948]
• CONT AS (J 0518) Bul Sti Mat-Fiz, Acad Romina

J 0175 Methodos • I
Methodos [1949ff]
• REM Does not seem to appear anymore

J 0178 Stud Gen • D
Studium Generale: Zeitschrift fuer die Einheit der Wissenschaften. Zusammenhang ihrer Begriffsbildungen und Forschungsmethoden. [1947-1971] ISSN 0039-4149

J 0179 Ann Fac Sci Clermont • F
Universite de Clermont. Faculte des Sciences. Annales [1952-1972]
• CONT AS (J 1934) Ann Sci Univ Clermont Math

J 0180 Rev Phil France & Etranger • F
Revue Philosophique de la France et de l'Etranger [1876ff] ISSN 0035-3833

J 0182 Kon Nederl Akad Wetensch Afd Let Med N S • NL
Koninklijke Nederlandse Akademie van Wetenschappen. Afdeeling Letterkunde: Mededelingen. N.S. [1921ff]
• REL PUBL (J 0028) Indag Math Mededelingen

J 0185 Unterrichtsbl Math Nat • D
Unterrichtsblaetter fuer Mathematik und Naturwissenschaften [1895-1947]
• CONT AS (J 0933) Math Nat Unterr

J 0186 Rev Franc Trait Info Chiffres • F
Revue Francaise de Traitement de l'Information, Chiffres [1963-1966]

J 0187 IEEE Trans Comp • USA
Transactions on Computers. IEEE (= Institute of Electrical and Electronics Engineers) [1968ff] ISSN 0018-9340
• CONT OF (J 4305) IEEE Trans Electr Comp

J 0188 Rev Union Mat Argentina • RA
Revista de la Union Matematica Argentina [1936ff] ISSN 0041-6932

J 0191 Inform Sci • USA
Information Sciences. An International Journal [1957ff] ISSN 0020-0255

J 0193 Discr Math • NL
Discrete Mathematics [1971ff] ISSN 0012-365X

J 0194 Inform & Control • USA
Information and Control [1958ff] ISSN 0019-9958

J 0195 Bull Calcutta Math Soc • IND
Bulletin of the Calcutta Mathematical Society [1908ff] ISSN 0008-0659

J 0197 Stud Cercet Mat Acad Romana • RO
Studii si Cercetari Matematice. Academia Republicii Socialiste Romania. (Mathematische Studien und Untersuchungen. Akademie der Sozialistischen Republik Rumaenien) [1950ff] ISSN 0567-6401
• CONT OF (J 0524) Disq Math Phys

J 0198 Bul Inst Politeh Iasi NS • RO
Buletinul Institutului Politehnic din Iasi. Serie Noua ∗ *Bulletin de l'Ecole Polytechnique de Jassy. Nouveau Serie* [1946-1976]
• CONT AS (J 3070) Bul Inst Politeh Iasi, Sect 1

J 0199 Zh Vychisl Mat i Mat Fiz • SU
Zhurnal Vychislitel'noj Matematiki i Matematicheskoj Fiziki (Journal of Computational Mathematical and Mathematical Physics) [1961ff] ISSN 0044-4669
• TRANSL IN (J 1049) USSR Comput Math & Math Phys

J 0202 Diss Math (Warsaw) • PL
Dissertationes Mathematicae. Polska Akademia Nauk, Instytut Matematyczny ∗ *Rozprawy Matematyczne* [1952ff] ISSN 0012-3862

J 0203 Rev Ser A Mat Fis, Univ Tucuman • RA
Universidad Nacional de Tucuman. Revista. Serie A. Matematica y Fisica Teorica. Facultad de Ciencias Exactas y Tecnologia

J 0205 Rev Franc Autom, Inf & Rech Operat • F
Revue Francaise d'Automatique, Informatique et Recherche Operationelle (RAIRO). Series: Bleue, Jaune, Rouge, Verte [1972-1976] ISSN 0399-0559
• CONT OF (J 3954) Rev Franc Inf & Rech Operat • CONT AS (J 4698) Rev Franc Autom, Inf & Rech Operat, Ser Rouge Inf Th & (J 2831) RAIRO Autom & (J 2832) RAIRO Inform
• REM In 1975 the Serie Rouge split into: Serie Rouge Analyse Numerique & J4698

J 0207 Ist Lombardo Accad Sci Rend, A (Milano) • I
Istituto Lombardo. Accademia di Science e Lettere. Rendiconti. A. Scienze Matematiche, Fisiche, Chimichze e Geologiche [1937ff] ISSN 0021-2504
• CONT OF (J 3986) Ist Lombardo Rend, Ser 2 (Milano)

J 0209 J Math Phys • USA
Journal of Mathematical Physics [1960ff] ISSN 0022-2488

J 0212 ACM Commun • USA
Communications of ACM (= Association for Computing Machinery) [1958ff] ISSN 0001-0782

J 0214 Math of Comp • USA
Mathematics of Computation [1960ff] ISSN 0025-5718
• CONT OF (J 0235) Math Tables Other Aids Comp

J 0215 Proc Irish Acad, Sect A • IRL
Proceedings of the Royal Irish Academy. Section A. Mathematical and Physical Sciences [1899ff] ISSN 0557-4056

J 0216 Izv Akad Nauk SSSR, Ser Mat • SU
Izvestiya Akademii Nauk SSSR. Seriya Matematicheskaya (Proceedings of the Academy of Sciences of the USSR. Mathematical Series) [1937ff] ISSN 0373-2436
• CONT OF (J 4717) Izv Akad Nauk SSSR • TRANSL IN (J 0448) Math of USSR, Izv

J 0218 Bul Sti Tehn Inst Politeh Timisoara, Ser Mat-Fiz-Mec • RO
Buletinul Stiintific si Tehnic al Institutului Politehnic "Traian Vuia" Timisoara. Seria Matematica-Fizica-Mecanica Teoretica si Aplicata (Wissenschaftliches und Technisches Bulletin des Polytechnischen Instituts "Traian Vuia" Timisoara. Serie Mathematik-Physik-Theoretische und Angewandte Mechanik) [1970-1977]
• CONT OF (J 4714) Bul Sti Tehn Inst Politeh Timisoara, NS
• CONT AS (J 4715) Bul Sti Tehn Inst Politeh Timisoara, Ser Mat-Fiz

J 0220 Atti Accad Sci Torino, Fis Mat Nat • I
Atti della Accademia delle Scienze di Torino. Classi di Scienze Fisiche, Matematiche e Naturali. Parte 1. ∗ *Acta Academiae Scientiarum Taurinensis* [1940ff] ISSN 0373-3033
• CONT OF (J 1742) Atti Accad Sci Torino, Fis Mat Nat

J 0224 God Vissh Tekh Ucheb Zaved Mat, Sofiya • BG
Godishnik na Visshite Tekhnicheski Uchebni Zavedeniya Matematika (Annuaire des Ecoles Techniques Superieures:Mathematiques) [1964-1973] ISSN 0436-1083
• CONT AS (J 3171) God Vissh Ucheb Zaved, Prilozhna Mat, Sofiya

J 0225 Amer Math Soc, Transl, Ser 2 • USA
American Mathematical Society. Translations. Series 2 [1955ff] ISSN 0065-9290

J 0226 Uch Zap Ped Inst, Ivanovo • SU
Ivanovskij Gosudarstvennyj Pedagogicheskij Institut imeni D.A.Furmanova. Uchenye Zapiski (Furmanov-Institute of Education in Ivanovo. Scientific Notes) [1941-1973] ISSN 0444-9681
• CONT AS (J 3536) Uch Zap Univ, Ivanovo

J 0230 An Univ Iasi, NS, Sect Ia • RO
Analele Stiintifice ale Universitatii Al.I. Cuza din Iasi. (Serie Noua) Sectiunea 1a: Matematica (Wissenschaftliche Annalen der Al.I. Cuza Universitaet Iasi. (Neue Serie) Sektion 1a: Mathematik) [1955ff] ISSN 0041-9109

J 0232 Inform Process Lett • NL
Information Processing Letters. Devoted to the Rapid Publication of Short Contributions to Information Processing [1971ff] ISSN 0020-0190

J 0233 Soobshch Akad Nauk Gruz SSR • SU
Soobshcheniya Akademii Nauk Gruzinskoj SSR ∗ *Sakaharth SSR Mecnierebatha Akademia Moambe (Communications of the Academy of Sciences of the Georgian SSR)* [1940ff] ISSN 0002-3167

J 0234 Rev Acad Cienc Exact Fis Nat Madrid • E
Revista de la Real Academia de Ciencias Exactas, Fisicas y Naturales de Madrid [1904ff] ISSN 0034-0596

J 0235 Math Tables Other Aids Comp • USA
Mathematical Tables and other Aids to Computation [1945-1959]
• CONT AS (J 0214) Math of Comp

J 0236 Rev Mat Hisp-Amer, Ser 4 • E
Revista Matematica Hispano-Americana. 4a Serie. Real Sociedad Matematica Espanola. [1941ff] ISSN 0373-0999
• CONT OF (J 3993) Rev Mat Hisp-Amer, Ser 2

J 0237 Stud Univ Cluj, Ser Math Phys Chem • RO
Studia Universitatis Babes-Bolyai. Series Mathematica-Physica-Chemia [1956-1963]
• CONT AS (J 0355) Studia Univ Babes-Bolyai, Math-Phys (Cluj) Kozlemenyei. Termeszettuclomanyi Sorozat

J 0238 Sitzb Oesterr Akad Wiss, Math-Nat Kl, Abt 2 • A
Oesterreichische Akademie der Wissenschaften. Mathematisch-Naturwissenschaftliche Klasse. Sitzungsberichte. Abteilung II. Mathematische, Physikalische und Technische Wissenschaft [1846ff] ISSN 0029-8816

J 0240 Ann Inst Fourier • F
Annales de l'Institut Fourier [1949ff] ISSN 0373-0956

J 0242 Language (Baltimore) • USA
Language [1924ff] ISSN 0097-8507

J 0243 Pi Mu Epsilon J • USA
Pi Mu Epsilon Journal. The Official Publication of the Honorary Mathematical Fraternity [1949ff] ISSN 0031-952X

J 0244 Trans Amer Inst Elect Engin • USA
American Institute of Electrical Engineers. Transactions [1846-1962]

J 0246 J Nat Sci Math • PAK
The Journal of Natural Sciences and Mathematics [1961ff] ISSN 0022-2941

J 0247 Bull Sci Math, Ser 2 • F
Bulletin des Sciences Mathematiques, Serie 2 [1870ff] ISSN 0007-4497

J 0248 Math Student • IND
The Mathematics Student [1933ff] ISSN 0025-5742

J 0249 J Math Pures Appl • F
Journal de Mathematiques Pures et Appliquees [1836-1921]
• CONT AS (**J 3941**) J Math Pures Appl, Ser 9

J 0250 Bol Soc Mat Mexicana • MEX
Boletin de la Sociedad Matematica Mexicana [1944-1955]
• CONT AS (**J 3127**) Bol Soc Mat Mexicana, Ser 2

J 0251 Proc Amer Acad Arts Sci • USA
American Academy of Arts and Sciences. Proceedings [1846-1958]

J 0252 Rev Philos Louvain • B
Revue Philosophique de Louvain [1946ff] ISSN 0035-3841
• CONT OF (**J 1720**) Rev Neoscolast Philos, Ser 2

J 0254 Gen Topology Appl • NL
General Topology and its Applications. A Journal Devoted to Set Theoretic, Axiomatic and Geometric Topology [1971-1979] ISSN 0016-660X
• CONT AS (**J 2635**) Topology Appl

J 0255 God Fak Mat & Mekh, Univ Sofiya • BG
Godishnik na Sofijskiya Universitet. Fakultet po Matematika i Mekhanika (Annuaire de l'Universite de Sofia. Faculte de Mathematiques.)

J 0259 Nyt Tidsskr Mat • DK
Nyt Tidsskrift for Matematik (New Journal for Mathematics) [1890-1918]
• CONT AS (**J 4510**) Norsk Mat Tidsskr

J 0260 Ann Jap Ass Phil Sci • J
Annals of the Japan Association for Philosophy of Science [1956ff]

J 0261 Tohoku Math J • J
Tohoku Mathematical Journal (Tohoku Sugaku Zashi) [1911ff] ISSN 0040-8735

J 0264 Collect Math (Barcelona) • E
Collectanea Mathematica [1948ff] ISSN 0010-0757

J 0267 Arch Hist Exact Sci • D
Archives for History of Exact Sciences [1960ff] ISSN 0003-9519

J 0269 Ann Sci Univ Jassy Sec 1 • RO
Universite de Jassy. Annales Scientifiques. Premiere Section. Mathematiques, Physique, Chimie [1937-1941]
• CONT OF (**J 4693**) Ann Sci Univ Jassy

J 0270 Dokl Akad Nauk Ukr SSR, Ser A • SU
Doklady Akademii Nauk Ukrainskoj SSR. Seriya A. Fiziko-Matematicheskie i Tekhnicheskie Nauki ∗ Dopovidi Akademii Nauk Uk'rainskoj RSR. Seriya A. Fiziko-Matematichni Ta Tekhnichni Nauki (Reports of the Academy of Sciences of the Ukrainian SSR. Series A. Physical-Mathematical and Engineering Sciences) [1939ff] ISSN 0002-3531, ISBN 0201-8446

J 0273 Australasian J Phil • AUS
Australasian Journal of Philosophy [1947ff] ISSN 0004-8402
• CONT OF (**J 4731**) Australasian J Psych & Phil

J 0274 Indian J Phys • IND
Indian Journal of Physics [1926ff] ISSN 0019-5480

J 0275 Proc Indian Assoc Cultivation Sci • IND
Indian Association for the Cultivation of Science. Proceedings [1926ff] ISSN 0019-5480

J 0277 Sitzb Preuss Akad Wiss Phys Math Kl • DDR
Die Preussische Akademie der Wissenschaften. Sitzungsberichte. Physikalisch-Mathematische Klasse [1922-1949]

J 0282 Sprawozd Towarz Nauk, Lwow • PL
Sprawozdania Towarzystwa Nauka we Lwowie (Reports of the Society of Sciences. Lwow) [1922-1939]

J 0283 Ann Soc Pol Math • PL
Societe Polonaise de Mathematique. Annales. ∗ Rocznik i Polskiego Towarzystwa Matematycznego [1922-1952]
• CONT AS (**J 1405**) Ann Pol Math

J 0284 IBM J Res Dev • USA
IBM (= International Business Machines) Journal of Research and Development [1957ff] ISSN 0018-8646

J 0286 Int Logic Rev • I
International Logic Review. ∗ Rassegna Internazionale di Logica [1970ff] ISSN 0048-6779

J 0287 Scripta Math • USA
Scripta Mathematica. [1932ff] ISSN 0036-9713

J 0288 Vest Ser Mat Mekh, Univ Moskva • SU
Vestnik Moskovskogo Universiteta. Seriya I. Matematika, Mekhanika (Publications of the Moscow University. Series I. Mathematics. Mechanics) [1946ff] ISSN 0201-7385, ISSN 0579-9368
• TRANSL IN (**J 0510**) Moscow Univ Math Bull & Moscow University. Mechanics Bulletin

J 0290 Euclides • NL
Euclides: Maandblad voor de Didactiek van de Wiskunde [1925ff]

J 0291 Wis-natuur Tijdsch • B
Wis- en Natuurkundig Tijdschrift: Orgaan van her Vlaamsch Natuur-, Wis- en Geneeskund Congres [1921-1944]
• CONT AS (**J 0061**) Simon Stevin

J 0296 Acta Salamantica Cienc • E
Acta Salamanticensia Ciencias. Iussu Senatus Universitatis Edita Ciencias [1954ff]

J 0297 Rend Mat Appl, Ser 5 • I
Rendiconti di Matematica e delle sue Applicazioni. Seria 5. Universita di Roma. Istituto Nazionale di Alta Matematica (INDAM) [1942-1967]
• CONT AS (J 3741) Rend Mat Appl, Ser 7

J 0299 Gac Mat, Ser 1a (Madrid) • E
Gaceta Matematica. Consejo Superior de Investigaciones Cientificas. Instituto "Jorge Juan". 1a Serie [1970ff] ISSN 0016-3805
• CONT OF (J 4443) Gac Mat (Madrid)

J 0300 Rozpr Akad Krakow Histor-Filoz Ser 2 • PL
Akademia Umiejetnosci. Krakow. Rozprawy i Sprawozdania z Posiedzen. Wydzialu Historyczno-Filozoficzny. Seria 2 (Academy of Sciences. Cracow. Papers of the Academy of Sciences. Department of History and Philosophy. 2. Series) [1891-1952]

J 0301 J Phil • USA
The Journal of Philosophy [1904ff] ISSN 0022-362X

J 0302 Rep Math Logic, Krakow & Katowice • PL
Reports on Mathematical Logic. The Jagiellonian University of Cracow. The Silesian University of Katowice [1973ff] ISSN 0083-4432
• CONT OF (S 0458) Zesz Nauk, Prace Log, Uniw Krakow

J 0304 An Univ Timisoara, Sti Mat • RO
Analele Universitatii din Timisoara. Stiinte Matematice. Facultatea de Stiinte Matematice ale Naturii. (Annalen der Universitaet Timisoara. Mathematische Wissenschaften. Fakultaet fuer mathematische Wissenschaften der Natur.) [1963ff] ISSN 0563-5608
• CONT OF (J 1152) Inst Ped Timisoara Lucr Sti Mat Fiz

J 0305 Invent Math • D
Inventiones Mathematicae [1966ff] ISSN 0020-9910

J 0306 Cah Topol & Geom Differ • F
Cahiers de Topologie et Geometrie Differentielle [1959ff] ISSN 0008-0004

J 0307 Rev Colomb Mat • CO
Revista Colombiana de Matematicas [1967ff] ISSN 0034-7426
• CONT OF (J 0348) Rev Mat Elementales

J 0308 Rocky Mountain J Math • USA
The Rocky Mountain Journal of Mathematics [1971ff] ISSN 0035-7596

J 0309 Wiss Z Univ Greifswald, Math-Nat Reihe • DDR
Wissenschaftliche Zeitschrift der Ernst Moritz Arndt-Universitaet Greifswald. Mathematisch-Naturwissenschaftliche Reihe [1951ff] ISSN 0075-7512

J 0310 Inquiry (Oslo) • N
Inquiry: An Interdisciplinary Journal of Philosophy and the Social Sciences [1958ff] ISSN 0020-174X

J 0311 Nordisk Mat Tidskr • N
Nordisk Matematisk Tidskrift (Scandinavian Mathematical Journal) [1953-1978] ISSN 0029-1412
• CONT OF (J 4510) Norsk Mat Tidsskr • CONT AS (J 3075) Normat

J 0312 Izv Akad Nauk Armyan SSR, Ser Mat • SU
Izvestiya Akademii Nauk Armyanskoj SSR. Seriya Matematika (Proceedings of the Academy of Sciences of the Armenian SSR. Series: Mathematics) [1965ff] ISSN 0002-3043
• TRANSL IN (J 3265) Sov J Contemp Math Anal, Armen Acad Sci

J 0313 Period Mat (Bologna) • I
Periodico di Matematiche

J 0315 Ann Sc Norm Sup Pisa Fis Mat, Ser 3 • I
Annali della Scuola Normale Superiore di Pisa. Classe di Science. Fisiche e Matematiche. Seria III [1947-1973] ISSN 0036-9918
• CONT OF (J 1568) Ann Sc Norm Sup Pisa, Fis Mat, Ser 2
• CONT AS (J 4702) Ann Sc Norm Sup Pisa Fis Mat, Ser 4

J 0316 Illinois J Math • USA
Illinois Journal of Mathematics [1957ff] ISSN 0019-2082

J 0319 Matematiche (Sem Mat Catania) • I
Le Matematiche [1946ff]

J 0320 Monist • USA
The Monist: International Quarterly of General Philosophical Inquiry [1890ff] ISSN 0026-9662

J 0324 Izv Akad Nauk Latv SSR • SU
*Izvestiya Akademii Nauk Latviskoj SSR * Latvijas PSR Zinatnu. Akademijas Vestis (Proceedings of the Academy of Sciences of the Latvian SSR)* [1947ff] ISSN 0023-8929

J 0325 Amer Phil Quart • GB
American Philosophical Quarterly [1964ff] ISSN 0003-0481

J 0326 J Pure Appl Algebra • NL
Journal of Pure and Applied Algebra [1971ff] ISSN 0022-4049

J 0329 Math Chron (Auckland) • NZ
Mathematical Chronicle [1969ff] ISSN 0581-1155

J 0330 Atti Ist Veneto, Fis Mat Nat • I
Istituto Veneto di Scienze, Lettere ed Arti. Venezia. Atti. Classe de Scienze Matematiche e Naturali

J 0332 Math Proc Cambridge Phil Soc • GB
Mathematical Proceedings of the Cambridge Philosophical Society [1975ff] ISSN 0305-0041
• CONT OF (J 0171) Proc Cambridge Phil Soc Math Phys

J 0335 Atti Accad Ligure Sci Lett (Genova) • I
Atti della Accademia Ligure di Scienze e Lettere [1890ff] ISSN 0392-2219

J 0336 Giorn Mat Battaglini • I
Giornale di Matematiche di Battaglini [1863ff] ISSN 0017-033X

J 0337 Mat Ezheg, Akad Nauk Latv SSR • SU
Latvijskij Matematicheskij Ezhegodnik. Latvijskij Ordena Trudovogo Krasnogo Znameni Gosudarstvennyj Universitet imeni P.Stuchki. Akademiya Nauk Latvijskoj SSR (Latvian Mathematical Yearbook) [1965ff] ISSN 0458-8223
• TRANSL IN Latvian Mathematical Yearbook

J 0338 Nauch-Tekh Inf, Ser 2, Akad Nauk SSSR • SU
Nauchno-Tekhnicheskaya Informatsiya. Seriya 2. Gosudarstvennyj Komitet SSSR po Nauke i Tekhnike. Akademiya Nauk SSSR. Vsesoyuznyj Institut Nauchnoj i Tekhnicheskoj Informatsii. Informatsionnye Protsesy i Sistemy (Scientific Technical Information. Series 2) [1967ff]
• TRANSL IN (J 2667) Autom Doc Math Linguist

J 0340 Mat Zap (Univ Sverdlovsk) • SU
Matematicheskie Zapiski (Mathematical Notes) ISSN 0076-5368

J 0343 Studia Math, Pol Akad Nauk • PL
Studia Mathematica. Polska Akademia Nauk, Instytut Matematyczny [1929ff] ISSN 0039-3223

J 0344 Izv Bulgar Akad Nauk Fiz Inst • BG
B"lgarska Akademiya na Naukite. Fizicheskij Institut. Izvestiya (Bulgarian Academy of Sciences. Institute of Physics. Reports) [?-1973]
• CONT AS (J 3072) Bulgar J Phys

J 0345 Adv Math • USA
Advances in Mathematics [1964ff] ISSN 0001-8708
• REL PUBL (S 3105) Adv Math, Suppl Stud

J 0346 Dokl Akad Nauk Armyan SSR • SU
Doklady Akademii Nauk Armyanskoj SSR (Reports of the Academy of Sciences of the Armenian SSR) [1944ff] ISSN 0321-1339

J 0347 Atti Accad Sci Bologna Fis, Ser 11 • I
Atti della Accademia delle Scienze dell'Istituto di Bologna. Classe di Scienze Fisiche. Rendiconti. Serie XI [?-1962]
• CONT AS (J 3069) Atti Accad Sci Bologna Fis Ser 12

J 0348 Rev Mat Elementales • CO
Revista de Matematicas Elementales [1952-1966]
• CONT AS (J 0307) Rev Colomb Mat

J 0349 Math of USSR, Sbor • USA
Mathematics of the USSR, Sbornik [1967ff] ISSN 0025-5734
• TRANSL OF (J 0142) Mat Sb, Akad Nauk SSSR, NS

J 0350 Sci Rep Tokyo Kyoiku Daigaku Sect A • J
Tokyo Kyoiku Daigaku (Tokyo University of Education. Science Reports. Section A.)

J 0351 Osaka J Math • J
Osaka Journal of Mathematics [1964ff] ISSN 0030-6126
• CONT OF (J 1770) Osaka Math J

J 0352 Math Jap • J
Mathematica Japonica [1948ff] ISSN 0025-5513

J 0354 Phil Trans Roy Soc London, Ser A • GB
Philosophical Transactions of the Royal Society of London. Series A. Mathematical and Physical Sciences. ISSN 0080-4614

J 0355 Studia Univ Babes-Bolyai, Math-Phys (Cluj) • RO
Studia Universitatis Babes-Bolyai. Mathematica - Physica [1964-1970]
• CONT OF (J 0237) Stud Univ Cluj, Ser Math Phys Chem
• CONT AS (J 3451) Stud Univ Cluj, Ser Math-Mech

J 0358 Versl Gewone Vergad Afd Natuurkd • NL
Koninklijke Nederlandse Akademie van Wetenschappen. Verslagee van de Gewone Vergaderingen der Afdling Natuurkunde (Royal Dutch Academy of Sciences. Reports of the Meetings of the Physics Section) [1892ff] ISSN 0023-3382

J 0359 Gregorianum (Vatican) • SCV
Gregorianum [1920ff] ISSN 0017-4114

J 0360 Ber Koenigl Ges Wiss Leipzig Math Kl • DDR
Koeniglich-Saechsische Gesellschaft der Wissenschaften zu Leipzig. Berichte ueber die Verhandlung. Mathematische Klasse

J 0361 Arch Systemat Phil • D
Archiv fuer Systematische Philosophie [1895-1916]

J 0362 Z Phil & Philos Kritik • D
Zeitschrift fuer Philosophie und Philosophische Kritik [1847-1918]

J 0363 Sitzb Jena Ges Med Natwiss • DDR
Jenaische Gesellschaft fuer Medizin und Naturwissenschaft. Sitzungsberichte.
• REM Supplement zur Zeitschrift fuer Naturwissenschaft (Jena)

J 0364 Publ Sci Univ Alger Ser A Math • DZ
Universite d'Alger. Publications Scientifiques. Serie A. Mathematiques [1954ff] ISSN 0002-5321

J 0366 Sitzber Berlin Math Ges • DDR
Die Berliner Mathematische Gesellschaft. Sitzungsberichte

J 0367 Scientia (Valparaiso) • RCH
Scientia ISSN 0036-8679

J 0370 Rev Sci (Paris) • F
La Revue Scientifique [1863-1954]

J 0371 Glas Mat-Fiz Astron, Ser 2 (Zagreb) • YU
Glasnik Matematichko-Fizichki i Astronomichki. Ser II (Publications of Mathematics, Physics, Astronomy. Ser II) [1946-1965]
• CONT AS (J 3519) Glas Mat, Ser 3 (Zagreb)

J 0372 Rad Jugosl Akad Znan Umjet, Mat Fiz Teh Znan • YU
Radovi Jugoslavenske Akademije Znanosti i Umjetnosti. Razred Za Matematichke, Fizichke i Tehnichke Znanosti. (Papers. Yugoslavian Academy of Sciences and Arts. Department of Mathematics, Physics and Engineering) [1940-1981]
• CONT OF (J 4735) Rad Jugosl Akad Znan Umjet • CONT AS (J 3944) Rad Jugosl Akad Znan Umjet, Mat Znan

J 0373 Comp Arch Inform & Numerik • A
Computing: Archiv fuer Informatik und Numerik ∗ Computing: Archives for Informatics and Numerical Computation [1966ff] ISSN 0010-485X

J 0374 SIAM J Appl Math • USA
Journal on Applied Mathematics. SIAM (= Society for Industrial and Applied Mathematics) [1966ff] ISSN 0036-1399
• CONT OF (J 0514) SIAM Journ

J 0377 Bol Mat (Bogota) • CO
Boletin de Matematicas [1967ff]

J 0378 Gaz Mat (Bucharest) Ser A • RO
Gazeta Matematica. Seria A (Mathematische Zeitschrift. Serie A) [1964-1974] ISSN 0016-5433
• CONT AS (J 3377) Gaz Mat (Bucharest)
• REM Combined with: Gazeta Matematica.Seria B to form Gazeta Mathematica

J 0379 Int J Electron • GB
International Journal of Electronics: Theoretical and Experimental [1965ff] ISSN 0020-7217

J 0380 Acta Cybern (Szeged) • H
Acta Cybernetica. Forum Centrale Publicationum Cyberneticarum Hungaricum [1969ff] ISSN 0324-721X

J 0381 J Tsuda College (Tokyo) • J
Journal of Tsuda College [1969ff]

J 0382 Int J Comput Math • GB
International Journal of Computer Mathematics. Section A: Programming Languages; Theory and Methods. Section B: Computational Methods [1964ff] ISSN 0020-7160

J 0384 Rend Mat, Ser 5 • I
Rendiconti di Matematica. Serie 5 [1940-1967]
• CONT AS (J 2311) Rend Mat, Ser 6

J 0387 Bull Sect Logic, Pol Acad Sci • PL
Bulletin of the Section of Logic. Polish Academy of Sciences. Institute of Philosophy and Sociology. [1972ff]
• REM Papers Published in the Bulletin are Generally: 1. Abstracts or preprints of papers submitted to other journals e.g. Studia Logica. 2. Abstracts of papers read at seminars or local conferences

J 0390 Publ Res Inst Math Sci (Kyoto) • J
Publications of the Research Institute for Mathematical Sciences [1965ff] ISSN 0034-5318
• REM Vols 1-4 Issued as: Kyoto Univ. Research Institute for Mathematical Sciences. Publications. Series A.

J 0391 Riv Mat Univ Parma • I
Rivista di Matematica della Universita di Parma [1891-1915]
• CONT AS (J 1526) Riv Mat Univ Parma, Ser 2

J 0392 Math Sci Hum • F
Mathematiques et Sciences Humaines [1962ff] ISSN 0025-5815

J 0396 Mat Lapok • H
Matematikai Lapok (Mathematical Papers) [1949ff] ISSN 0025-519X
• CONT OF (J 0461) Mat Fiz Lapok

J 0397 Proc Math Phys Soc Egypt • ET
Proceedings of the Mathematical and Physical Society of Egypt [1937ff] ISSN 0076-5317

J 0400 Publ Inst Math, NS (Belgrade) • YU
Institut Mathematique. Publications de l'Institut Mathematique. Nouvelle Serie [1961ff] ISSN 0522-828X
• CONT OF (J 4706) Publ Inst Math (Belgrade)

J 0401 J Number Th • USA
Journal of Number Theory [1968ff] ISSN 0022-314X

J 0402 Uch Zap Ped Inst, Kemerovo • SU
Kemerovskij Gosudarstvennyj Pedagogicheskij Institut. Uchenye Zapiski. (State Institute of Education in Kemerovo. Scientific Notes) [1956ff] ISSN 0453-4875

J 0403 Izv Akad Nauk Kazak SSR, Ser Fiz-Mat • SU
Izvestiya Akademii Nauk Kazakhskoj SSR. Seriya Fiziko-Matematicheskaya (Proceedings of the Academy of Sciences of the Kazakh SSR. Series: Physics & Mathematics) [1963ff] ISSN 0002-3191

J 0406 Bull Inst Math, Acad Sin (Taipei) • RC
Chung Yang Yen Chui y Uan Shu Hs Ueh Yen Chiu So (Bulletin of the Institute of Mathematics. Academia Sinica.) [1973ff]

J 0407 Comm Math Univ St Pauli (Tokyo) • J
Commentarii Mathematici Universitatis Sancti Pauli [1952ff] ISSN 0010-258X

J 0409 Chinese J Math (Taipei) • RC
Chinese Journal of Mathematics [1936ff]

J 0411 Studia Sci Math Hung • H
Studia Scientiarum Mathematicarum Hungarica. Auxilio Consilii Instituti Mathematici. Academiae Scientiarum Hungaricae [1966ff] ISSN 0081-6906

J 0413 Izv Akad Nauk Belor SSR, Ser Fiz-Mat • SU
*Vestsi Akademii Navuk BeSSR. Seriya Fizika-Matematychnykh Navuk * Izvestiya Akademii Nauk BSSR. Seriya Fiziko-Matematicheskikh Nauk (Proceedings of the Academy of Sciences of the Byelorussian SSR. Series: Physics, Mathematics)* [1964ff] ISSN 0002-3574

J 0414 Dokl Akad Nauk Belor SSR • SU
Doklady Akademii Nauk BSSR (Reports of the Academy of Sciences of the BSSR) [1957ff] ISSN 0002-354X

J 0415 J Differ Equations • USA
Journal of Differential Equations [1965ff] ISSN 0022-0396

J 0417 Sitzb Akad Wiss, Heidelberg, Math-Nat Kl • D
Sitzungsberichte der Heidelberger Akademie der Wissenschaften. Mathematisch-Naturwissenschaftliche Klasse [1909ff] ISSN 0073-1625

J 0418 Shuxue Xuebao • TJ
Shuxue Xuebao (Acta Mathematica Sinica) [1951ff]
• TRANSL IN (J 0419) Chinese Math Acta
• REM In 1951 published as: Journal of the Chinese Mathematical Society (N.S.)

J 0419 Chinese Math Acta • USA
Chinese Mathematics. Acta [1962-1967]
• TRANSL OF (J 0418) Shuxue Xuebao

J 0420 Shuxue Jinzhan • TJ
Shuxue Jinzhan (Advances in Mathematics) [1955ff] ISSN 0559-9326

J 0423 Pensee NS • F
La Pensee. Revue de Bon Rationalisme Moderne. Nouvelle Serie [1944ff]

J 0424 Wiss Z Univ Halle-Wittenberg, Math-Nat Reihe • DDR
Wissenschaftliche Zeitschrift der Martin-Luther-Universitaet Halle-Wittenberg. Mathematisch-Naturwissenschaftliche Reihe [1951ff]

J 0425 Managmt Sci (Tokyo) • J
Keiei-Kagaku (Management Science) [1956ff] ISSN 0451-5978

J 0426 Dt Math • DDR
Deutsche Mathematik [1936-1942]

J 0428 Proc Phys-Math Soc Japan • J
Physico-Mathematical Society of Japan. Proceedings [1919ff]

J 0432 Bell Syst Tech J • USA
The Bell System Technical Journal. [1922ff] ISSN 0005-8580

J 0434 J Fac Sci Univ Tokyo, Sect 1 • J
Journal of the Faculty of Science. University of Tokyo. Section 1 Mathematics, Astronomy, Physics, Chemistry [1925-1970] ISSN 0040-8980
• CONT AS (J 2332) J Fac Sci, Univ Tokyo, Sect 1 A

J 0435 Int J Comput & Inf Sci • USA
International Journal of Computer and Information Sciences [1972ff] ISSN 0091-7036

J 0437 Prog Math (Allahabad) • IND
Progress of Mathematics [1967ff] ISSN 0555-4330

J 0438 J Hokkaido Univ Educ Ser 1 • J
Journal of Hokkaido University of Education. Faculty of Science. Series 1. Mathematics [1930ff] ISSN 0018-3482

J 0441 J Indian Math Soc, NS • IND
The Journal of the Indian Mathematical Society. New Series [1934ff] ISSN 0019-5839

J 0446 Ann Acad Sci Fennicae, Ser A I, Diss • SF
Annales Academiae Scientiarum Fennicae. Series AI. Mathematica. Dissertationes [1975ff] ISSN 0066-1953
• CONT OF (**J 3994**) Ann Acad Sci Fennicae Ser A I

J 0448 Math of USSR, Izv • USA
Mathematics of the USSR, Izvestiya [1967ff] ISSN 0025-5726
• TRANSL OF (**J 0216**) Izv Akad Nauk SSSR, Ser Mat

J 0449 Probl Kybern • DDR
Probleme der Kybernetik [1958-1965]
• CONT AS (**J 0471**) Syst Th Res • TRANSL OF (**J 0052**) Probl Kibern

J 0451 Studia Soc Sci Torunensis Sect A • PL
Studia Societatis Scientiarum Torunensis, Sectio A

J 0452 Indiana Univ Math J • USA
Indiana University Mathematics Journal [1971ff] ISSN 0022-2518
• CONT OF (**J 4732**) J Math Mech (Indiana Univ)

J 0455 Phil Naturalis • D
Philosophia Naturalis: Archiv fuer Naturphilosphie und die Philosophischen Grenzgebiete der Exakten Wissenschaften und Wissenschaftsgeschichte [1950ff] ISSN 0031-8027

J 0459 C R Soc Sci Lett Varsovie Cl 3 • PL
*Societe des Sciences et des Lettres de Varsovie. Comptes Rendus des Seances. Classe III: Sciences Mathematiques et Physiques * Towarzystwo Naukowe Warszawskie. Sprawozdania z Posiedze. Wydzialu III: Nauk Matematyczno-Fizycznych (Warschauer Sitzungsberichte)* [1908-1950]

J 0460 Acta Univ Szeged, Sect Mat • H
Acta Litterarum ac Scientiarum Regiae Universitatis Hungaricae Francisco-Josephinae, Sectio Scientiarum Mathematicarum [1922-1946]
• CONT AS (**J 0002**) Acta Sci Math (Szeged)

J 0461 Mat Fiz Lapok • H
Matematikai es Fizikai Lapok (Mathematical and Physical Papers) [1892-1948]
• CONT AS (**J 0396**) Mat Lapok

J 0462 Mat Fiz Oszt Koezlem, Acad Sci Hung • H
Magyar Tudomanyos Akademia. Matematikai es Fizikai Tudomanyok Osztalyanak Koezlemenyek. (Hungarian Academy of Sciences. Bulletin of the Mathematical and Physical Sciences) [1952-1974]
• CONT AS (**J 1458**) Alkalmaz Mat Lapok

J 0464 Syst-Comp-Controls • USA
Systems - Computers - Controls [1970ff] ISSN 0096-8765
• TRANSL OF (**J 0979**) Denshi Tsushin Gakkai Ronbunshi, Sect A-D

J 0465 Bull Greek Math Soc (NS) • GR
Bulletin of the Greek Mathematical Society. New Series (Hellenike Mathematike Hetaireia. Deltion. Nea Seira.) [1960ff] ISSN 0072-7466
• CONT OF (**J 1699**) Bull Soc Math Grece

J 0466 Trans Illinois State Acad Sci • USA
Transactions of the Illinois State Academy of Science [1908ff] ISSN 0019-2252

J 0468 Uch Zap Ped Inst (Alma Ata) • SU
Kazakhskij Gosudarstvennyj Pedagogicheskij Institut im. Abaya. Uchenye Zapiski (Abaya-Institute of Education of Kazakhstan. Scientific Notes)

J 0470 Sov Phys, Dokl • USA
Soviet Physics. Doklady. [1956ff] ISSN 0038-5689
• TRANSL OF (**J 0023**) Dokl Akad Nauk SSSR

J 0471 Syst Th Res • USA
Systems Theory Research [1966ff] ISSN 0082-1255
• CONT OF (**J 1195**) Probl Cybernet & (**J 0449**) Probl Kybern
• TRANSL OF (**J 0052**) Probl Kibern

J 0472 Theory Decis • NL
Theory and Decision. An International Journal for Philosophy and Methodology of the Social Sciences [1970ff] ISSN 0040-5833

J 0474 Avtom Vychis Tekh, Akad Nauk Latv SSR • SU
Avtomatika i Vychislitel'naya Tekhnika. Akademiya Nauk Latvijskoj SSR (Automation and Computer Science. Academy of Sciences of the Latvian SSR) [1967ff] ISSN 0572-4538
• TRANSL IN (**J 2666**) Autom Control Comput Sci

J 0475 Sib Math J • USA
Siberian Mathematical Journal [1966ff] ISSN 0037-4466
• TRANSL OF (**J 0092**) Sib Mat Zh

J 0477 Spis Bulgar Akad Nauk • BG
B"lgarski Akademija na Naukite. Spisanie (Bulgarian Academy of Sciences. Journal) ISSN 0015-3265

J 0479 Kwart Filoz • PL
Kwartalnik Filozoficzny (Philosophical Quaterly)

J 0480 Roy Soc Bibl Mem Fellows • GB
The Royal Society. Bibliographical Memoirs of Fellows. [1955ff] ISSN 0080-4606

J 0481 Acta Univ Wroclaw • PL
Acta Universitatis Wratislaviensis

J 0482 Publ Math Univ Belgrade • YU
Universite de Belgrade. Publications Mathematiques [1932ff]

J 0483 Uch Zap, Univ Riga • SU
*Uchenye Zapiski Latvijskogo Gosudarstvennogo Universiteta imeni Petra Stutski. * Zinatniskie Raksti. Latvijas Valsts Universitate. (Scientific Notes of the Latvian State University.)*
• REM This journal consists of two series: Teoriya Algoritmov i Programm & Teoreticheskie Voprosy Avtomaticheskikh Sistem Upravleniya

J 0485 Algorytmy, Pol Akad Nauk • PL
Algorytmy (Algorithms) [1962ff] ISSN 0065-6240

J 0487 Math Unterricht • D
Der Mathematikunterricht: Beitraege zu seiner Wissenschaftlichen und Methodischen Gestaltung [1955ff] ISSN 0025-5807

J 0488 Dialogue (Ottawa) • CDN
*Dialogue: Canadian Philosophical Review * Dialogue: Revue Canadienne de Philosophie.* [1962ff] ISSN 0012-2173

J 0490 Collecteana Logica • PL
Collecteana Logica [1939ff]

J 0491 Publ Inst Mat Univ Nac Litoral (Rosario RA) • RA
Universidad Nacional del Litoral. Rosario. Instituto de Matematicas. Publicaciones [1939ff]

J 0493 Bull Res Counc Israel Sect F • IL
Research Council of Israel. Bulletin. Section F. Mathematics and Physics [1952-1962]
• CONT AS (**J 0029**) Israel J Math

J 0494 Bull Math Soc Sci Roumanie • RO
Societatea Romana de Stiinte, Sectia Mathematica. Bulletin Mathematiques de la Societe Roumaine des Sciences [1908-1956]
• CONT AS (J 0070) Bull Soc Sci Math Roumanie, NS

J 0497 Math Mag • USA
Mathematics Magazine [1947ff] ISSN 0025-570X
• CONT OF (J 1737) Nat Math Magazine (Louisiana)

J 0498 Ann Univ Turku, Ser A I • SF
Annales Universitatis Turkuensis. Series A.I: Astronomica, Chemica, Physica, Mathematica (Turun Yliopiston Julkaisuja. Sarja A.1.: Astronomica, Chemica, Physica, Mathematica) [1957ff] ISSN 0082-7002

J 0499 Uch Zap Univ, Gor'kij • SU
Uchenye Zapiski Gor'kovskogo Gosudarstvennogo Universiteta imeni N.I.Lobachevskogo (Scientific Notes of the Lobachevskij University of Gor'kij)

J 0501 Ann Univ Lublin, Sect Math • PL
Annales Universitatis Mariae Curie-Sklodowska. Sectio A. Mathematica [1946ff] ISSN 0365-1029

J 0503 Artif Intell • NL
Artificial Intelligence [1970ff] ISSN 0004-3702

J 0504 Manuscr Math • D
Manuscripta Mathematica [1969ff] ISSN 0025-2611

J 0505 Dominican Studies
Dominican Studies

J 0508 Machine Intelligence • GB
Machine Intelligence [1967ff] ISSN 0541-6418

J 0510 Moscow Univ Math Bull • USA
Moscow University Mathematics Bulletin [1969ff] ISSN 0027-1322
• TRANSL OF (J 0288) Vest Ser Mat Mekh, Univ Moskva

J 0512 Nature • GB
Nature. A Weekly Journal of Science [1869ff] ISSN 0028-0836

J 0513 Electronic Engin • GB
Electronic Engineering [1928ff] ISSN 0013-4902

J 0514 SIAM Journ • USA
Journal. SIAM (= Society for Industrial and Applied Mathematics) [1953-1965]
• CONT AS (J 0374) SIAM J Appl Math

J 0515 Bull Math Biophys • GB
Bulletin of Mathematical Biophysics [1939-1971] ISSN 0007-4985
• CONT AS (J 3073) Bull Math Biol

J 0517 Mathematica (Cluj) • RO
Mathematica. Revue d'Analyse Numerique et de Theorie de l'Approximation [1929ff] ISSN 0025-5505

J 0518 Bul Sti Mat-Fiz, Acad Romina • RO
*Academia Republicii Populare Romine. Buletinul Stiintific. Sectia de Stiinte Matematice si Fizice * Academie de la Republique Populaire Roumainie. Bulletin Scientifique. Section des Sciences Mathematiques et Physiques * Akademija Rumynskoi Respubliki. Nachnyi Vestnik. Otdelenie Matematicheskih i Fizicheskih Nauk* [1949-1965] ISSN 0515-1333
• CONT OF (J 0174) Bull Sect Sci Acad Roumaine

J 0519 Wiad Mat, Ann Soc Math Pol, Ser 2 • PL
*Annales Societatis Mathematicae Polonae. Seria 2. Wiadomosci Matematyczne * Roczniki Polskiego Towarzystwa Matematycznego. Seria 2. Wiadomosci Matematyczne* [1955ff] ISSN 0079-3698
• CONT OF (J 4710) Pol Tow Mat, Wiad Mat

J 0521 Semin Math, Inst Steklov • USA
Seminars in Mathematics. V.A.Steklov Mathematical Institute Leningrad [1967ff]
• TRANSL OF (S 0228) Zap Nauch Sem Leningrad Otd Mat Inst Steklov

J 0522 Engin Cybern • USA
Engineering Cybernetics. Soviet Journal of Computer and System Sciences. Essential Serials in Electronics and Systems Science [1963ff] ISSN 0013-788X
• TRANSL OF (J 0977) Izv Akad Nauk SSSR, Tekh Kibern

J 0523 Bull Nagoya Inst Tech • J
Bulletin of Nagoya Institute of Technology [1949ff]

J 0524 Disq Math Phys • RO
Disquisitiones Mathematicae et Physicae [1940-1949]
• CONT AS (J 0197) Stud Cercet Mat Acad Romana

J 0525 C R Acad Sci Roumanie • RO
Academie des Sciences de Roumanie. Comptes-Rendus des Seances [1937ff]
• REM Since 1939 called: Comptes-Rendus des Seances de l'Institut des Sciences de Roumanie

J 0534 Mem Fac Engin, Nagoya Univ • J
Nagoya Daigaku Kogakubu Kiyo (Nagoya University. Faculty of Engineering. Memoirs) [1948ff] ISSN 0027-7657

J 0537 Yokohama Math J • J
The Yokohama Mathematical Journal [1953ff] ISSN 0044-0523

J 0539 Sci Pap Coll Gen Educ, Univ Tokyo • J
Scientific Papers of the College of General Education. University of Tokyo. (Tokyo Daigaku Kyoyogakubu Shizenkagaku Kiyo) [1951ff] ISSN 0040-8964

J 0545 Proc Indian Acad Sci, Sect A • IND
Indian Academy of Sciences. Proceedings. Section A: Physical Sciences [1931ff] ISSN 0019-428X

J 0548 Stud Appl Math • USA
Studies in Applied Mathematics [1969ff] ISSN 0022-2526

J 0549 Riv Mat Univ Parma, Ser 4 • I
Rivista di Matematica della Universita di Parma. Serie 4 [1975ff] ISSN 0035-6298
• CONT OF (J 3254) Riv Mat Univ Parma, Ser 3

J 0589 Bull Amer Math Soc (NS) • USA
Bulletin of the American Mathematical Society. New Series [1979ff] ISSN 0273-0979
• CONT OF (J 0015) Bull Amer Math Soc

J 0611 Pol Tow Mat, Prace Mat-Fiz • PL
Polskie Towarzystwo Matematyczene. Prace Matematyczno-Fizyczne (Mathematische und Physikalische Abhandlungen) [1887-1954]
• CONT AS (J 0051) Commentat Math, Ann Soc Math Pol, Ser 1 & (J 2095) Fund Inform, Ann Soc Math Pol, Ser 4

J 0748 Erkenntnis (Leipzig) • DDR
Erkenntnis [1930-1939]
• CONT OF (J 1380) Ann Philos & Philos Kritik • CONT AS (J 3597) J Unif Sci

J 0752 Avh Norske Vid-Akad Oslo I • N
Avhandlinger. Norske Videnskaps-Akademi i Oslo. I: Matematisk-Naturvidenskapelig Klasse (Proceedings of the Scandinavian Academy of Sciences. Mathematical and Natural Sciences Class)

J 0931 Anz Oesterr Akad Wiss, Math-Nat Kl • A
Anzeiger der Oesterreichischen Akademie der Wissenschaften. Mathematisch-Naturwissenschaftliche Klasse [1864ff] ISSN 0065-535X

J 0933 Math Nat Unterr • D
Der Mathematische und Naturwissenschaftliche Unterricht [1948ff] ISSN 0025-5866
• CONT OF (J 0185) Unterrichtsbl Math Nat

J 0937 Bul Inst Politeh Brasov Ser A • RO
Institutul Politehnic din Brasov. Buletinul. Seria A. Mecanica (Das Polytechnische Institut Brasov. Bulletin. Serie A: Mechanik) [1966-1971] ISSN 0524-210X

J 0940 J Electron Control Ser 1 • GB
Journal of Electronics and Control Ser.1 [1957-1964]
• CONT AS (J 0379) Int J Electron & (J 1045) Int J Control

J 0945 Riv Filos • I
Rivista di Filosofia [1909ff] ISSN 0035-6239

J 0952 Rice Univ Studies • USA
Rice University Studies. Writings in All Scholarly Disciplines. [1915ff] ISSN 0035-4996

J 0954 Tr Inst Prikl Mat, Tbilisi • SU
*Tbilisskij Gosudarstvennyj Universitet. Institut Prikladnoj Matematiki. Trudy * Tbbilisis Sahelmcipho Universiteti Gamoqenebithi Matematikis Instituti Shromebi (State University of Tbilisi. Institute of Applied Mathematics. Publications)* [1969ff] ISSN 0082-2191

J 0956 Forsch Logik Grundl exakt Wiss • D
Forschungen zur Logik und zur Grundlegung der exakten Wissenschaften [1937-1943]

J 0963 Ajatus (Helsinki) • SF
Ajatus. Suomen Filosofisen Yhdislyksen Vuosikirja (Yearbook of the Philosopical Society of Finland) [1926ff]

J 0968 Uch Zap Univ, Kazan • SU
Uchenye Zapiski Kazanskogo (Ordena Trudovogo Krasnogo Znameni) Gosudarstvennogo Universiteta Imeni V.I.Ul'yanova-Lenina (Scientific Notes. State Universirty Kazan) [1843ff]

J 0974 Norsk Vid-Akad Oslo Mat-Natur Kl Skr • N
Norske Videnskaps - Akademi i Oslo. Matematisk-Naturvidenskapelig Klasse. Skrifter. (Monographs of the Scandinavian Academy of Sciences. Mathematical and Natural Sciences Class) [1929ff] ISSN 0029-2338
• CONT OF (J 1145) Vidensk Selsk Kristiana Skrifter Ser 1

J 0975 Norsk Mat Forenings Skr • N
Norsk Matematisk Forenings Skrifter

J 0977 Izv Akad Nauk SSSR, Tekh Kibern • SU
Izvestiya Akademii Nauk SSSR. Tekhnicheskaya Kibernetika. Otdelenie Mekhaniki i Protsessov Upravlenija (Proceedings of the Academy of Sciences of the USSR. Engineering Cybernetics. Department of Mechanics and Control Processes) [1963ff] ISSN 0002-3388
• TRANSL IN (J 0522) Engin Cybern

J 0979 Denshi Tsushin Gakkai Ronbunshi, Sect A-D • J
Denshi Tsushin Gakkai Ronbunshi. Sect. A-D (Reports of the University of Electro-Communications) [1949ff] ISSN 0493-4253
• TRANSL IN (J 0116) Electr & Comm Japan & (J 0464) Syst-Comp-Controls

J 0982 Z Wahrscheinltheor & Verw Geb • D
Zeitschrift fuer Wahrscheinlichkeitstheorie und Verwandte Gebiete [1962ff] ISSN 0044-3719

J 0983 Forsch Fortschritte • DDR
Forschungen und Fortschritte. Nachrichtenblatt der Deutschen Wissenschaft und Technik [1924-1967]

J 0985 Nova Acta Acad Caes Leopoldina (Halle) • DDR
Nova Acta Academiae Caesarea Leopoldino-Germanicae Naturae Curiosorum

J 0988 Rev Filos (Santiago) • RCH
Revista de Filosofia [1954ff] ISSN 0034-8236

J 0989 Z Allg Wissth • D
Zeitschrift fuer Allgemeine Wissenschaftstheorie (Journal for General Philosophy of Science) [1969ff] ISSN 0044-2216

J 0990 Soc Sci Fennicae Comment Phys-Math • SF
Societas Scientiarum Fennicae. Commentationes Physico-Mathematicae

J 0996 RAAG Res Notes • J
Research Notes and Memoranda of Applied Geometry for Prevenient Natural Philosophy. Post RAAG-Reports.

J 1005 Rep Fac Sci, Shizuoka Univ • J
Reports of the Faculty of Science. Shizuoka University. [1965ff] ISSN 0583-0923

J 1008 Demonstr Math (Warsaw) • PL
Demonstratio Mathematica [1969ff] ISSN 0420-1213

J 1019 Vest Ser Mekh Mat, Univ Khar'kov • SU
Vestnik Khar'kovskogo Gosudarstvennogo Universiteta. Seriya Mekhaniko-Matematicheskaya, Zapiski Mekhaniko-Matematicheskogo Fakul'teta i Khar'kovskogo Matematicheskogo Obshchestva (Publications of the Khar'kov State University. Series: Mechanics, Mathematics) ISSN 0453-8048, ISSN 0135-1850

J 1021 Itogi Nauki Ser Mat • SU
Itogi Nauki. Seriya Matematiki (Progress in Science. Mathematical Series) [1962-1971] ISSN 0579-1731
• CONT AS (J 1488) Itogi Nauki Tekh, Ser Probl Geom & (J 1501) Itogi Nauki Tekh, Ser Algeb, Topol, Geom & (J 1452) Itogi Nauki Tekh, Ser Sovrem Probl Mat & (J 3188) Itogi Nauki Tekh, Ser Teor Veroyat Mat Stat Teor Kibern & (J 4387) Itogi Nauki Tekh, Ser Tekh Kibern • TRANSL IN (J 1531) J Sov Math
• REM J1531 contains only selected translations

J 1022 Ratio (Oxford) • GB
Ratio [1957ff] ISSN 0034-0006
• REL PUBL (J 0170) Ratio (Frankfurt) • REM English Edition

J 1023 Perspectives of New Music • USA
Perspectives of New Music [1962ff] ISSN 0031-6016

J 1024 Zhongguo Kexue • TJ
Zhongguo Kexue (Scientia Sinica) [1950-1981]
• CONT AS (J 3766) Zhongguo Kexue, Xi A

J 1027 Proc & Addr Amer Phil Ass • USA
American Philosophical Association. Proceedings and Addresses [1927ff] ISSN 0065-972X

J 1030 Formalisation • F
La Formalisation: Cahier pour l'Analyse

J 1040 Bull Math Nat Acad Tcheque Sci • CS
Academie Tcheque des Sciences. Bulletin International. Classe des Sciences Mathematiques, Naturelles et de la Medicine

J 1044 Math Notes, Acad Sci USSR • USA
Mathematical Notes of the Academy of Sciences of the USSR [1967ff] ISSN 0001-4346
• TRANSL OF (**J** 0087) Mat Zametki (Akad Nauk SSSR)

J 1045 Int J Control • GB
International Journal of Control: The Theory of Process Control and Automation [1965ff] ISSN 0020-7179

J 1046 Z Angew Math Mech • DDR
Zeitschrift fuer Angewandte Mathematik und Mechanik: Ingeneurwissenschaftliche Forschungsarbeiten. Applied Mathematics and Mechanics [1921ff] ISSN 0044-2267

J 1048 Kiber Sb Perevodov • SU
Kibernetischeskij Sbornik: Sbornik Perevodov. (Collected Articles on Cybernetics: Collected Translations) [1960-1964]
• CONT AS (**J** 3079) Kiber Sb Perevodov, NS

J 1049 USSR Comput Math & Math Phys • GB
USSR Computational Mathematics and Mathematical Physics [1962ff] ISSN 0041-5553
• TRANSL OF (**J** 0199) Zh Vychisl Mat i Mat Fiz

J 1077 IEEE Proc • USA
Proceedings. IEEE (= Institute of Electrical and Electronics Engineers) [1963ff] ISSN 0018-9219
• CONT OF (**J** 4711) IRE Proc

J 1092 Ann Univ Paris • F
Universite de Paris. Annales ISSN 0041-9176

J 1093 Ruch Filoz • PL
Ruch Filozoficzny. Polskie Towarzystwo Filozoficzne (Philosophical Movement. Polish Philosophical Society) ISSN 0035-9599

J 1095 Beitr Phil Deutsch Idealismus • D
Beitraege zur Philosophie des Deutschen Idealismus [1918-1926]
• CONT AS (**J** 0168) Blaetter Deutsch Philos

J 1109 Nachr Akad Wiss Goettingen, Math-Phys Kl • D
Nachrichten der Akademie der Wissenschaften in Goettingen. II. Mathematisch-Physikalische Klasse [1893ff] ISSN 0065-5295

J 1113 Commun Math Phys • D
Communications in Mathematical Physics [1965ff] ISSN 0010-3616

J 1114 Arch Phil • D
Archiv fuer Philosophie [1911-1930, 1947-1964]

J 1124 Ergebn Math Kolloquium • A
Ergebnisse eines Mathematischen Kolloquiums [1929-1936]

J 1125 Przeglad Filoz • PL
Przeglad Filozoficzny (Revue Philosophique) [1897-1949]

J 1131 Schr Berlin Math Sem • D
Berliner Mathematisches Seminar. Schriften

J 1145 Vidensk Selsk Kristiana Skrifter Ser 1 • N
Videnskaps Selskapet i Kristiana. Skrifter Utgit. 1 Matematisk-Naturvidenskapelig Klasse [?-1928]
• CONT AS (**J** 0974) Norsk Vid-Akad Oslo Mat-Natur Kl Skr

J 1152 Inst Ped Timisoara Lucr Sti Mat Fiz • RO
Lucrarile Stiintifice ale Institutului Pedagogic Timisoara. Matematica-Fizica (Wissenschaftliche Arbeiten des Paedagogischen Instituts Timisoara. Mathematik-Physik) [1958-1962]
• CONT AS (**J** 0304) An Univ Timisoara, Sti Mat

J 1156 Izv Bulgar Akad Nauk Mat Inst • BG
B"lgarska Akademiya na Naukite. Otdelenie Za Fiziko-Matematicheski i Tekhnicheski Nauki. Izvestiya na Matematicheski Institut (Bulgarian Academy of Sciences. Department of Physics, Mathematics & Engineering. Reports of the Mathematical Institute) [1957-1974]
• CONT AS (**J** 2547) Serdica, Bulgar Math Publ

J 1179 Filoz Jugos Chas (Belgrade) • YU
Filozofija: Jugoslovenski Chasopis Za Filozofiju (Philosophy: Yugoslavian Journal of Philosophy) [1957ff] ISSN 0015-1866

J 1193 Comput J (London) • GB
The Computer Journal [1958ff] ISSN 0010-4620

J 1195 Probl Cybernet • GB
Problems of Cybernetics [1958-1965]
• CONT AS (**J** 0471) Syst Th Res • TRANSL OF (**J** 0052) Probl Kibern

J 1198 Logos (Tuebingen) • D
Logos [1910-1933]

J 1204 Boll Bibl Sci Mat • I
Bolletino di Bibliografia e Storia delle Scienze Matematiche

J 1208 Rend Ist Mat Univ Trieste • I
Rendiconti dell'Istituto di Matematica dell'Universita di Trieste [1969ff]

J 1209 Wiss Z Univ Leipzig, Math-Nat Reihe • DDR
Wissenschaftliche Zeitschrift der Karl-Marx-Universitaet Leipzig. Mathematisch-Naturwissenschaftliche Reihe [1951ff] ISSN 0043-6860

J 1377 Ann New York Acad Sci • USA
New York Academy of Sciences. Annals [1877ff] ISSN 0077-8923

J 1379 Christiaan Huygens Internat Math Tijdschr • NL
Christiaan Huygens. Internationaal Mathematisch Tijdschrift [1868-1942]
• CONT AS (**J** 0061) Simon Stevin

J 1380 Ann Philos & Philos Kritik • DDR
Annalen der Philosophie und Philosophischen Kritik [1925-1929]
• CONT AS (**J** 0748) Erkenntnis (Leipzig)

J 1399 Russ Math Surv • GB
Russian Mathematical Surveys [1946ff] ISSN 0036-0279
• TRANSL OF (**J** 0067) Usp Mat Nauk

J 1400 Econometrica • USA
Econometrica: Journal of the Econometric Society [1933ff] ISSN 0012-9682

J 1404 Mat Sb, Akad Nauk SSSR • SU
Matematicheskij Sbornik. Akademiya Nauk SSSR i Moskovskoe Matematicheskoe Obshchestvo (Mathematical Collected Articles. New Series. Academy of Sciences of the USSR and the Moscovian Mathematical Society) [1866-1935]
• CONT AS (**J** 0142) Mat Sb, Akad Nauk SSSR, NS

J 1405 Ann Pol Math • PL
Annales Polonici Mathematici. Polska Akademia Nauk, Instytut Matematyczny [1953ff] ISSN 0066-2216
• CONT OF (J 0283) Ann Soc Pol Math

J 1411 Math 20 Siecle • F
Mathematique du XXeme Siecle [1960]

J 1426 Theor Comput Sci • NL
Theoretical Computer Science [1975ff] ISSN 0304-3975

J 1428 SIAM J Comp • USA
Journal on Computing. SIAM (= Society for Industrial and Applied Mathematics) [1972ff] ISSN 0097-5397

J 1429 Kybernetes • GB
Kybernetes: An International Journal of Cybernetics and General Systems [1972ff] ISSN 0368-492X

J 1431 Acta Inf • D
Acta Informatica [1971ff] ISSN 0001-5903

J 1447 Houston J Math • USA
Houston Journal of Mathematics [1975ff] ISSN 0362-1588

J 1452 Itogi Nauki Tekh, Ser Sovrem Probl Mat • SU
Itogi Nauki i Tekhniki: Seriya Sovremennye Problemy Matematiki (Progress in Science and Technology: Series on Current Problems in Mathematics) [1972ff]
• CONT OF (J 1021) Itogi Nauki Ser Mat • TRANSL IN (J 1531) J Sov Math

J 1456 SIGACT News • USA
SIGACT (= ACM Special Interest Group on Automata and Computability Theory). News [1969ff]

J 1458 Alkalmaz Mat Lapok • H
Alkalmazott Matematikai Lapok (Papers in Applied Mathematics) [1951ff] ISSN 0133-3399
• CONT OF (J 0462) Mat Fiz Oszt Koezlem, Acad Sci Hung

J 1472 Sci Rep Saitama Univ, Ser A • J
The Science Reports of the Saitama University. Series A. Mathematics [1952-1982] ISSN 0558-2431
• CONT AS (J 3940) Saitama Math J Physics, and Chemistry

J 1477 Tr Vychisl Tsentr, Akad Nauk Gruz SSR • SU
Trudy Vychislitel'nogo Tsentra. Akademiya Nauk Gruzinskoj SSR Sakharthvelos SSR Mecnierebatha Akademis Gamothvlithi Centris Shromebi. (Publications of the Computational Centre. Academy of Sciences of the Georgian SSR) [1960ff] ISSN 0568-4900

J 1488 Itogi Nauki Tekh, Ser Probl Geom • SU
Itogi Nauki i Tekhnike. Seriya Problemy Geometrii (Progress in Science and Technology. Series Problems in Geometry) [1972ff] ISSN 0202-7461
• CONT OF (J 1021) Itogi Nauki Ser Mat • TRANSL IN (J 1531) J Sov Math

J 1501 Itogi Nauki Tekh, Ser Algeb, Topol, Geom • SU
Itogi Nauki i Tekhniki. Seriya Algebra, Topologiya, Geometriya. (Progress in Science and Technology. Series Algebra, Topology, Geometry) [1972ff] ISSN 0202-7445
• CONT OF (J 1021) Itogi Nauki Ser Mat • TRANSL IN (J 1531) J Sov Math & (C 4688) Prog in Math, Vol 12
• REM C4688 is Volume 1968

J 1507 An Inst Mat, Univ Nac Aut Mexico • MEX
Anales del Instituto de Matematicas. Universidad Nacional Autonoma de Mexico [1961ff] ISSN 0076-7441

J 1508 Math Sem Notes, Kobe Univ • J
Mathematics Seminar Notes. Kobe University [1973-1983] ISSN 0385-633X
• CONT AS (J 4390) Kobe J Math

J 1515 Archimede • I
Archimede. Rivista per gli Insegnanti e i Cultori di Matematiche Pure e Applicate [1949ff] ISSN 0003-8369

J 1516 Vest Ser Fiz Mat Mekh, Univ Minsk • SU
Vestnik Belorusskogo Gosudarstvennogo Universitet im. V.I.Lenina. Seriya I. Fizika, Matematika, Mekhanika (Publications of the Byelorussian State University. Series I: Physics, Mathematics, Mechanics) [1969ff] ISSN 0321-0367

J 1519 Kyungpook Math J (Taegu) • ROK
Kyungpook Mathematical Journal [1958ff] ISSN 0454-8124

J 1522 Math Slovaca • CS
Mathematica Slovaca [1976ff] ISSN 0025-5173
• CONT OF (J 0143) Mat Chasopis (Slov Akad Ved)

J 1526 Riv Mat Univ Parma, Ser 2 • I
Rivista di Matematica della Universita di Parma. Serie 2 [1960-1971] ISSN 0035-6298
• CONT OF (J 0391) Riv Mat Univ Parma • CONT AS (J 3254) Riv Mat Univ Parma, Ser 3

J 1527 Pokroky Mat Fyz Astron (Prague) • CS
Pokroky Matematiky, Fyziky a Astronomie (Progress in Mathematics, Physics and Astronomy) [1956ff] ISSN 0032-2423

J 1531 J Sov Math • USA
Journal of Soviet Mathematics [1973ff] ISSN 0090-4104
• TRANSL OF (J 1021) Itogi Nauki Ser Mat & (S 0228) Zap Nauch Sem Leningrad Otd Mat Inst Steklov & Problemy Matimaticheskogo Analiza & (J 1452) Itogi Nauki Tekh, Ser Sovrem Probl Mat & (J 1501) Itogi Nauki Tekh, Ser Algeb, Topol, Geom & (J 3188) Itogi Nauki Tekh, Ser Teor Veroyat Mat Stat Teor Kibern & (J 1488) Itogi Nauki Tekh, Ser Probl Geom
• REM This contains selected translations from each of the Russian Journals listed

J 1546 Rep Math Phys (Warsaw) • PL
Reports on Mathematical Physics [1970ff]

J 1550 Creation Math • IL
Creation in Mathematics [1970ff]

J 1568 Ann Sc Norm Sup Pisa, Fis Mat, Ser 2 • I
Annali della R. Scuola Normale Superiore di Pisa. Serie 2 Scienze Fisiche e Matematiche [1932-1946]
• CONT OF (J 1908) Ann Sc Norm Sup Pisa, Fis Mat • CONT AS (J 0315) Ann Sc Norm Sup Pisa Fis Mat, Ser 3

J 1572 Izv Vyssh Ucheb Zaved, Radiofizika (Moskva) • SU
Izvestiya Vysshikh Uchebnykh Zavednii. Radiofizika (Proceedings of the Universities. Radiophysics) [1958ff] ISSN 0021-3462
• TRANSL IN (J 4330) Radiophys & Quant Electr

J 1586 Rocz Pol Towarz Nauk Obczyznie • PL
Rocznik Polskiego Towarzystwa Naukowego na Obczyznie (Yearbook of the Polish Society of Arts and Sciences Abroad) [1951ff]

J 1615 Arch Geschichte Phil • D
Archiv fuer Geschichte der Philosophie [1888-1932, 1960ff] ISSN 0003-9101

J 1620 Asterisque • F
Asterisque [1973ff] ISSN 0303-1179

J 1633 Angew Inf • D
Angewandte Informatik ∗ Applied Informatics [1959ff] ISSN 0013-5704

J 1665 Appl Mathematicae, Pol Akad Nauk • PL
Zastosowania Matematiky. Polska Akademia Nauk. Instytut Matematyczny ∗ Applicationes Mathematicae (Applications of Mathematics. Polish Academy of Sciences. Mathematical Institute) [1960ff] ISSN 0044-1899

J 1669 J Comb Th • USA
Journal of Combinatorial Theory [1966-1970] ISSN 0021-9800
• CONT AS (J 0164) J Comb Th, Ser A & (J 0033) J Comb Th, Ser B

J 1670 Mitt Math Ges DDR • DDR
Mitteilungen der Mathematischen Gesellschaft der DDR

J 1680 Cienc Tecnol, Costa Rica • CR
Ciencia y Tecnologia, Revista de la Universidad de Costa Rica [1976ff] ISSN 0378-052X

J 1699 Bull Soc Math Grece • GR
(Bulletin de la Societe Mathematique Grece) [1921-1959]
• CONT AS (J 0465) Bull Greek Math Soc (NS)

J 1717 An Univ Chile • RCH
Universidad de Chile. Anales ISSN 0041-8358

J 1720 Rev Neoscolast Philos, Ser 2 • B
Revue Neo-Scolastique de Philosophie. Serie 2 [1910-1945]
• CONT AS (J 0252) Rev Philos Louvain

J 1725 Bull Daito-Bunka Univ • J
Bulletin of Daito-Bunka University (Liberal Arts), Japan [1968ff]

J 1735 Bull South East Asian Soc • SGP
Bulletin of the South East Asian Society [1977ff]

J 1737 Nat Math Magazine (Louisiana) • USA
National Mathematics Magazine [1926-1946]
• CONT AS (J 0497) Math Mag

J 1741 Int J Man-Mach Stud • USA
International Journal of Man-Machine Studies [1969ff] ISSN 0020-7373

J 1742 Atti Accad Sci Torino, Fis Mat Nat • I
Atti della Reale Accademia delle Scienze di Torino. Classe I: Scienze Fisiche, Matematiche e Naturali [1865-1939]
• CONT AS (J 0220) Atti Accad Sci Torino, Fis Mat Nat

J 1750 Atomes & Nucleus • F
Atomes & Nucleus ISSN 0029-5671

J 1770 Osaka Math J • J
Osaka Mathematical Journal [1949-1963]
• CONT AS (J 0351) Osaka J Math

J 1793 Nieuw Arch Wisk, Ser 2 • NL
Nieuw Archief voor Wiskunde. Reeks 2 [1894-1952]
• CONT OF (J 0046) Nieuw Arch Wisk • CONT AS (J 3077) Nieuw Arch Wisk, Ser 3

J 1813 Estudios • E
Estudios [1945ff]

J 1834 Arch Rechts- und Sozialphil • D
Archives de Philosophie du Droit et de Philosophie Sociale ∗ Archiv fuer Rechts- und Sozialphilosophie ∗ Archives for Philosophy of Law and Social Philosophy [1907ff] ISSN 0001-2343

J 1843 Amer Sci • USA
American Scientist. [1913ff] ISSN 0003-0996

J 1859 Naturwissenschaften • D
Die Naturwissenschaften [1913ff] ISSN 0028-1042

J 1885 J Theor Biol • USA
Journal of Theoretical Biology [1961ff] ISSN 0022-5193

J 1893 Relevance Logic Newslett • USA
The Relevance Logic Newsletter. [1976ff]

J 1908 Ann Sc Norm Sup Pisa, Fis Mat • I
Annali della Reale Scuola Normale Superiore Universitaria di Pisa. Scienze Fisiche e Matematiche [1913-1931]
• CONT AS (J 1568) Ann Sc Norm Sup Pisa, Fis Mat, Ser 2

J 1910 Proc London Math Soc, Ser 2 • GB
Proceedings of the London Mathematical Society. Serie 2 [1904-1951]
• CONT OF (J 0077) Proc London Math Soc • CONT AS (J 3240) Proc London Math Soc, Ser 3

J 1917 Rev Questions Sci • B
Revue des Questions Scientifiques [1924ff] ISSN 0035-2160

J 1927 Prace Inst Podstaw Inf, Pol Akad Nauk • PL
Prace Instytut Podstaw Informatyki Polskiej Akademii Nauk (Reports. Institute of Computer Science Polish Academy of Science)

J 1928 Mathematica B • NL
Mathematica : Tijdschrift voor Allen die de Hoogere Wiskunde Beoefenen. Afdeling B [1934-1948]
• CONT AS (J 0061) Simon Stevin
• REM Vols 1 - 3/1 appeared under the title 'Mathematica', which from 3/2 onwards was split into 'Mathematica A' and 'Mathematica B'

J 1929 Prace Centr Oblicz Pol Akad Nauk • PL
Polska Akademija Nauk. Centrum Obliczeniowe. Prace. (Polish Academy of Sciences. Computation Centre. Reports) ISSN 0079-3175

J 1934 Ann Sci Univ Clermont Math • F
Annales Scientifiques de l'Universite de Clermont-Ferrand II, Section Mathematiques (Clermont Ferrand) [1973ff] ISSN 0069-472X
• CONT OF (J 0179) Ann Fac Sci Clermont

J 1943 Recueil Travaux Inst Math (Beograd) • YU
Recueil des Travaux de l'Institut Mathematique (Beograd)

J 1944 Sitzb, Akad Wiss, Bayern, Math-Nat Kl • D
Bayerische Akademie der Wissenschaften. Sitzungsberichte [1931ff] ISSN 0340-7586

J 2022 Comm Math Helvetici • CH
Commentarii Mathematici Helvetici [1929ff] ISSN 0010-2571

J 2028 Hist & Phil Log • GB
History and Philosophy of Logic [1980ff] ISSN 0144-5340

J 2038 Rend Sem Mat, Torino • I
Rendiconti del Seminario Matematico (gia "Conferenze di Fisica e di Matematica"). Universita e Politecnico di Torino

J 2053 Math Medley • SGP
Mathematical Medley [1975ff]

J 2054 Math Scientist (Melbourne) • AUS
The Mathematical Scientist [1976ff] ISSN 0312-3685

J 2068 Grundlstud Kybern Geistwiss • D
Grundlagenstudien aus Kybernetik und Geisteswissenschaft
[1960ff] ISSN 0017-4939

J 2074 Sem-ber, Muenster • D
Semesterbericht Muenster [1931-1939]

J 2076 Rev Int Philos • B
Revue Internationale de Philosophie [1938ff] ISSN 0048-8143
• CONT OF (**J 0151**) Arch Soc Belg Philos

J 2093 Arbgem Forsch Nordrhein-Westfalen Nat, Ing, Ges-Wiss • D
Arbeitsgemeinschaft fuer Forschung des Landes Nordrhein-Westfalen. Natur-, Ingenieur- und Gesellschaftswissenschaften [1961ff] ISSN 0570-5657

J 2095 Fund Inform, Ann Soc Math Pol, Ser 4 • PL
Fundamenta Informaticae. Annales Societatis Mathematicae Polonae. Series 4 ∗ Roczniki Polskiego Towarzystwa Matematycznego. Seria 4 [1977ff] ISSN 0324-8429
• CONT OF (**J 0611**) Pol Tow Mat, Prace Mat-Fiz

J 2099 Boll Unione Mat Ital, VI Ser, A • I
Bolletino della Unione Matematica Italiana. Serie VI. A
[1982ff] ISSN 0041-7084
• CONT OF (**J 3285**) Boll Unione Mat Ital, V Ser, A

J 2100 Boll Unione Mat Ital, VI Ser, B • I
Bolletino della Unione Matematica Italiana. Serie VI. B
[1982ff] ISSN 0041-7084
• CONT OF (**J 3495**) Boll Unione Mat Ital, V Ser, B

J 2107 Publ Dep Math, Lyon, NS • F
Publications du Departement de Mathematiques. Nouvelle Serie. Faculte des Sciences de Lyon. [1982ff] ISSN 0076-1656
• CONT OF (**J 0056**) Publ Dep Math, Lyon

J 2128 C R Math Acad Sci, Soc Roy Canada • CDN
Comptes Rendus Mathematiques de l'Academie des Sciences. La Societe Royale du Canada ∗ Mathematical Reports of the Academy of Sciences [1979ff] ISSN 0706-1994

J 2146 Helv Phys Acta • CH
Helvetica Physica Acta [1928ff] ISSN 0018-0238

J 2157 Qinghua Daxue Xuebao • TJ
Qinghua Daxue Xuebao (Journal of Qinghua University)
[1955ff] ISSN 0577-9189

J 2293 Comp Linguist & Comp Lang • H
Computational Linguistics and Computer Languages

J 2307 Japan J Math • J
Japanese Journal of Mathematics [1924-1974]
• CONT AS (**J 2347**) Japan J Math, NS

J 2310 Obz Mat Fiz, Ljubljana • YU
Obzornik za Matematiko in Fiziko (Mathematical and Physical Reviews) [1951ff] ISSN 0473-7466

J 2311 Rend Mat, Ser 6 • I
Rendiconti di Matematica. Serie 6 [1968-1980] ISSN 0034-4427
• CONT OF (**J 0384**) Rend Mat, Ser 5

J 2313 C R Acad Sci, Paris, Ser A-B • F
Academie des Sciences de Paris. Comptes Rendus Hebdomadaires des Seances. Serie A: Sciences Mathematiques, Serie B: Sciences Physiques [1966-1980]
ISSN 0001-4036
• CONT OF (**J 0109**) C R Acad Sci, Paris • CONT AS (**J 3364**) C R Acad Sci, Paris, Ser 1 & (**J 2314**) C R Acad Sci, Paris, Ser 2

J 2314 C R Acad Sci, Paris, Ser 2 • F
Comptes Rendus des Seances de l'Academie des Sciences. Serie II. Mecanique-Physique, Chimie, Sciences de la Terre, Sciences de l'Univers [1981ff]
• CONT OF (**J 2313**) C R Acad Sci, Paris, Ser A-B

J 2320 Probl Peredachi Inf, Akad Nauk SSSR • SU
Problemy Peredachi Informatsii. Zhurnal Akademii Nauk SSSR (Probleme der Informationsuebertragung) [1965ff]
ISSN 0555-2923
• TRANSL IN (**J 3419**) Probl Inf Transmiss

J 2328 Ann Pont Ist Sup Sci Lett Napoli • I
Pontifico Istituto Superiore di Scienze et Lettere "S.Chiara" di Napoli Annali [1952ff]

J 2332 J Fac Sci, Univ Tokyo, Sect 1 A • J
Journal of the Faculty of Science. University of Tokyo. Section 1 A. Mathematics [1971ff] ISSN 0040-8980
• CONT OF (**J 0434**) J Fac Sci Univ Tokyo, Sect 1

J 2339 Vychisl Met & Progr, Univ Moskva • SU
Vychislitel'nye Metody i Programmirovanie. Sbornik Rabot Vychislitel'nogo Tsentra Moskovskogo Gosudarstvennogo Universiteta (Computational Methods and Programming)
[1962ff] ISSN 0507-5386

J 2345 Ueberblicke Math • D
Ueberblicke Math [1967-1974]
• CONT AS (**S 0780**) Jbuch Ueberblick Math

J 2347 Japan J Math, NS • J
Japanese Journal of Mathematics. New Series [1975ff] ISSN 0075-3432
• CONT OF (**J 2307**) Japan J Math

J 2349 Mat Struchno-Metod Chasopis • YU
Matematika, Struchno-Metod. Chasopis [1972ff]

J 2386 Epistemologia • I
Epistemologia [1978ff]

J 2521 Beijing Shifan Daxue Xuebao, Ziran Kexue • TJ
Beijing Shifan Daxue Xuebao. Ziran Kexue Ban (Journal of Natural Sciences of Beijing Normal University. Natural Science Edition)

J 2525 Chittagong Univ Stud Part II Sci • AUS
Chittagong University Studies. Part II: Science

J 2547 Serdica, Bulgar Math Publ • BG
Serdica. Bulgaricae Mathematicae Publicationes [1975ff]
ISSN 0204-4110
• CONT OF (**J 1156**) Izv Bulgar Akad Nauk Mat Inst

J 2551 J Log Progr • USA
Journal of Logic Programming [1984ff] ISSN 0743-1066

J 2562 Publ Sec Mat Univ Autonoma Barcelona • E
Publications de la Seccio de Matematiques. Universitat Autonoma de Barcelona (Veroeffentlichunge der Abteilung Mathematik)

J 2574 Litov Mat Sb (Vil'nyus) • SU
Litovskij Matematicheskij Sbornik ∗ *Lietuvos Matematikos Rinkinys (Lithuanian Mathematical Collected Articles)* [1961ff] ISSN 0132-2818
• TRANSL IN (**J 3283**) Lith Math J

J 2602 Su-Hak Kwa Mul-li, People's Rep Korea • PRK
Su-Hak Kwa Mul-li (Academy of Science of the People's Republic of Korea. Research Center for Physics and Mathematics)

J 2604 Progr Comput Software • USA
Programming and Computer Software [1975ff] ISSN 0361-7688
• TRANSL OF (**J 2605**) Programmirovanie

J 2605 Programmirovanie • SU
Programmirovanie (Programming) [1975ff] ISSN 0132-3474
• TRANSL IN (**J 2604**) Progr Comput Software

J 2606 Tsukuba J Math • J
Tsukuba Journal of Mathematics [1977ff] ISSN 0387-4982

J 2635 Topology Appl • NL
Topology and its Applications. A Journal Devoted to General, Geometric, Set-Theoretic and Algebraic Topology [1980ff] ISSN 0166-8641
• CONT OF (**J 0254**) Gen Topology Appl

J 2649 Acta Phys Austriaca • A
Acta Physica Austriaca [1947ff] ISSN 0001-6713

J 2650 Adv Appl Math • USA
Advances in Applied Mathematics. [1980ff] ISSN 0196-8858

J 2657 Ann of Sci • GB
Annals of Science. An International Review of the History of Science and Technology from the Thirteenth Century. [1963ff] ISSN 0003-3790

J 2661 Ann Probab • USA
The Annals of Probability. An Official Journal of the Institute of Mathematical Statistics. [1973ff] ISSN 0091-1798

J 2662 Appl Math Comp • USA
Applied Mathematics and Computation. [1975ff] ISSN 0096-3003

J 2666 Autom Control Comput Sci • USA
Automatic Control and Computer Sciences [1969ff] ISSN 0146-4116
• TRANSL OF (**J 0474**) Avtom Vychis Tekh, Akad Nauk Latv SSR

J 2667 Autom Doc Math Linguist • USA
Automatic Documentation and Mathematical Linguistics [1967ff] ISSN 0005-1055
• TRANSL OF (**J 0338**) Nauch-Tekh Inf, Ser 2, Akad Nauk SSSR

J 2668 Avtom Sist Upravl & Prib Avtom, Khar'kov • SU
Avtomatizirovannye Sistemy Upravleniya i Pribory Avtomatiki (Automatisierte Steuersysteme und Automatisierungsvorrichtungen) [1965ff] ISSN 0135-1710

J 2669 AIIE Trans • USA
Transactions. AIIE (= American Institute of Industrial Engineers) [1969-1982] ISSN 0569-5554

J 2677 Bull Acad Pol Sci, Ser Sci Tech • PL
Bulletin de l'Academie Polonaise des Sciences. Serie des Sciences Techniques [1953ff] ISSN 0001-4125

J 2678 Bull Aichi Univ Educ Nat Sci • J
Bulletin of Aichi University of Education. Natural Science

J 2679 Bull Inst Math Appl (Southend oS) • GB
Bulletin of the Institute of Mathematics and its Applications [1965ff]

J 2684 J Huazhong Inst Tech (Engl Ed) • TJ
Journal of Huazhong Institute of Technology. English Edition [1979-1981]
• CONT AS (**J 3218**) J Huazhong Univ Sci Tech (Engl Ed)
• TRANSL OF (**J 2754**) Huazhong Gongxueyuan Xuebao

J 2687 Comp Math Appl • USA
Computers and Mathematics with Applications. [1975ff] ISSN 0097-4943

J 2688 Conceptus (Wien) • A
Conceptus. Zeitschrift fuer Philosophie [1967ff] ISSN 0010-5155

J 2694 Cybern & Syst • USA
Cybernetics and Systems. An International Journal. [1980ff] ISSN 0196-9722

J 2695 Cybernetica • B
Cybernetica. Revue Trimestrielle de l'Association Internationale de Cybernetique [1958ff] ISSN 0011-4227

J 2699 Expo Math • D
Expositiones Mathematicae. International Journal for Pure and Applied Mathematics [1983ff]

J 2701 Digit Processes • CH
Digital Processes. An International Journal on the Theory and Design of Digital Systems [1975ff] ISSN 0301-4185

J 2702 Discr Appl Math • NL
Discrete Applied Mathematics [1979ff] ISSN 0166-218X

J 2703 Educ Stud Math • NL
Educational Studies in Mathematics. [1968ff] ISSN 0013-1954

J 2718 Fct Approximatio, Comment Math, Poznan • PL
Functiones et Approximatio. Commentarii Mathematici [1974ff]

J 2720 Fuzzy Sets Syst • NL
Fuzzy Sets and Systems [1978ff] ISSN 0165-0114

J 2734 Int J Math Educ Sci Technol • GB
International Journal of Mathematical Education in Science and Technology [1970ff] ISSN 0020-739X

J 2736 Int J Theor Phys • USA
International Journal of Theoretical Physics [1968ff] ISSN 0020-7748

J 2744 Izv Sev-Kavk Nauch Tsentra, Estestv (Rostov nD) • SU
Izvestiya Severo-Kavkazskogo Nauchnogo Tsentra Vysshej Shkoly. Estestvennye Nauki (Proceedings of the Scientific University Centre of the North-Caucasus. Natural Sciences)

J 2745 IEEE Trans Inf Theory • USA
Transactions on Information Theory. IEEE (= Institute of Electrical and Electronics Engineers) [1955ff] ISSN 0018-9448

J 2746 J Algor • USA
Journal of Algorithms [1980ff] ISSN 0196-6774

J 2748 J Comb Inf Syst Sci (New Dehli) • IND
Journal of Combinatorics, Information & System Sciences [1976ff] ISSN 0250-9628

J 2750 J Econ Th • USA
Journal of Economic Theory [1969ff] ISSN 0022-0531

J 2751 J Fac Lib Art Yamaguchi Univ • J
Journal of the Faculty of Liberal Arts. Yamaguchi University. Natural Science [1967ff]

J 2752 J Fct Anal • USA
Journal of Functional Analysis [1967ff] ISSN 0022-1236

J 2754 Huazhong Gongxueyuan Xuebao • TJ
*Huazhong Gongxueyuan Xuebao * Zhongguo Kexue Shuxue Zhuanji (Journal Huazhong (Central China) University of Science and Technology)*
 • TRANSL IN (**J 2684**) J Huazhong Inst Tech (Engl Ed) & (**J 3218**) J Huazhong Univ Sci Tech (Engl Ed)

J 2755 J Indian Inst Sci • IND
Journal of the Indian Institute of Science [1977ff] ISSN 0019-4964
 • REM Section A: Engineering & Technology. Section B: Physical & Chemical Sciences. Section C: Biological Sciences

J 2756 J Math Biol • A
Journal of Mathematical Biology [1974ff] ISSN 0303-6812

J 2760 J Phys A Math & Gen • GB
Journal of Physics. A. Mathematical and General [1968ff] ISSN 0305-4470

J 2761 J Recreational Math • USA
Journal of Recreational Mathematics [1968ff] ISSN 0022-412X

J 2770 Keio Math Sem Rep (Yokohama) • J
Keio Mathematical Seminar Reports [1976ff]

J 2771 Kexue Tongbao • TJ
Kexue Tongbao (Science Bulletin) [1950ff] ISSN 0023-074X
 • TRANSL IN (**J 3769**) Sci Bull, Foreign Lang Ed

J 2774 Koezlem MTA Szam & Autom: Kutat Intez • H
Koezlemenyek. Magyar Tudomanyos Akademia Szamitastechnikai es Automatizalasi Kutato Intezet Budapest (Bulletin of the Hungarian Academy of Sciences Budapest, Research Institut of Computer Science and Automatization) [1968ff]

J 2775 Nuovo Cimento B, Ser 2 • I
Il Nuovo Cimento. B. Serie II

J 2780 Mat v Shkole • SU
Matematika v Shkole. Nauchno-Metodicheskij Zhurnal Ministerstva Prosveshcheniya SSSR (Mathematics at School. Scientific Methodical Journal of the Ministry of Education of the USSR) [1934ff] ISSN 0130-9358

J 2789 Math Intell • D
The Mathematical Intelligencer [1978ff] ISSN 0343-6993

J 2790 Math Sem-ber • D
Mathematische Semesterberichte [1980ff] ISSN 0720-728X
 • CONT OF (**J 0160**) Math-Phys Sem-ber, NS

J 2794 Mem Fac Engin, Kyoto Univ • J
Memoirs of the Faculty of Engineering. Kyoto University [1914ff] ISSN 0023-6063

J 2803 Nachrichtentech Elektr • DDR
Nachrichtentechnik. Elektronik. Technisch-Wissenschaftliche Zeitschrift fuer die Gesamte Elektronische Nachrichtentechnik [1951ff] ISSN 0323-4657

J 2804 Nanjing Daxue Xuebao, Ziran Kexue • TJ
Nanjing Daxue Xuebao. Ziran Kexue Ban (Nanjing University Journal. Natural Sciences Edition) [1957ff]

J 2808 Organon • PL
Organon [1963ff] ISSN 0078-6500

J 2811 Phil Quart (St Andrews) • GB
The Philosophical Quarterly [1950ff] ISSN 0031-8094

J 2813 Phys Lett A • NL
Physics Letters. A [1962ff]

J 2815 Pol Tow Cybern, Kwart Nauk, Wroclaw • PL
Polskie Towarzystwo Cybernetyczne. Postepy Cybernetyki. Kwartalnik Naukowy (Polish Society of Cybernetics. Progress in Cybernetics. Scientific Quaterly) [1965ff] ISSN 0137-3595

J 2817 Prague Bull Math Linguist • CS
The Prague Bulletin of Mathematical Linguistics [1964ff] ISSN 0032-6585

J 2819 Probl Contr Inf Th, Akad Nauk SSSR & Acad Sci Hung • H
*Problems of Control and Information Theory. Academy of Sciences of the USSR & Hungarian Academy of Sciences * Problemy Upravlenija i Teorii Informacii* [1971ff] ISSN 0370-2529

J 2821 Pubbl Ist Mat App Univ Stud Roma • I
Pubblicazioni. Universita degli Studi di Roma. Facolta di Ingegneria. Istituto di Matematica Applicata.

J 2823 Quaest Math, S Africa • ZA
*Quaestiones Mathematicae. Journal of the South African Mathematical Society * Tydskrift van die Suid-Afrikaanse Wiskundevereniging* ISSN 0379-9468

J 2825 Rend Sem Fac Sci Univ Cagliari • I
Rendiconti del Seminario della Facolta di Scienze della Universita di Cagliari

J 2826 Rep Fac Sci, Kagoshima Univ, Math Phys Chem • J
Reports of the Faculty of Science of Kagoshima University. Mathematics, Physics, and Chemistry

J 2830 Rozpr, Mat Prirod, Cheskoslov Akad Ved • CS
Rozpravy Ceskoslovenske Akademie Ved Rada Matematickykh a Prirodnikh Ved. (Studies of the Czechoslovak Academy of Sciences. Series: Mathematics and Natural Sciences) [1953ff] ISSN 0069-228X
 • CONT OF (**J 3989**) Rozpr II, Tr Cesk Akad Ved

J 2831 RAIRO Autom • F
RAIRO Automatique. RAIRO (= Revue Francaise d'Automatique, d'Informatique et de Recherche Operationnelle). Series Automatique [1977ff] ISSN 0399-0524
 • CONT OF (**J 0205**) Rev Franc Autom, Inf & Rech Operat Ser Jaune

J 2832 RAIRO Inform • F
RAIRO Informatique. Revue Francaise d'Automatique, d'Informatique et de Recherche Operationnelle. Series Informatique [1977ff] ISSN 0399-0532
 • CONT OF (**J 0205**) Rev Franc Autom, Inf & Rech Operat Ser Bleue

J 2840 Stochastica, Univ Politec Barcelona • E
Stochastica. Revista de Matematica Pura y Aplicada del Departamento de Matematicas y Estadistica de la Escuela Tecnica Superior de Arquitectura [1975ff]

J 2845 Tanulmanyok • H
Tanulmanyok. Szamitastechnikai es Automatizalasi Kutato Intezete (Studies. Research Institut for Computer Science and Automatization)

J 2855 Zbor Rad, Prir-Mat Fak, Ser Mat (Novi Sad) • YU
Zbornik Radova Prirodno-Matematichkog Fakulteta. Serija za Matematiku (Collected Papers of the Faculty of Natural Sciences and Mathematics. Mathematical Series) [1971ff]

J 2867 Vest Politekh Inst Khar'kov • SU
Vestnik Khar'kovskogo Politekhnicheskogo Instituta (Publications of the Khar'kov Technical Institute) [1965ff] ISSN 0453-7998

J 2871 Vopr Fil, Moskva • SU
Voprosy Filosofii (Problems of Philosophy) [1947ff] ISSN 0042-8744

J 2878 Wiss Z Tech Hochsch Karl-Marx-Stadt • DDR
Wissenschaftliche Zeitschrift der Technischen Hochschule Karl-Marx-Stadt

J 2879 Wuhan Daxue Xuebao, Ziran Kexue • TJ
Wuhan Daxue Xuebao. Ziran Kexue Ban (Wuhan University Journal. Natural Sciences)

J 2887 Zbor Radova, NS • YU
Zbornik Radova. Nova Serija. (Collected Papers. New Series)

J 2889 Zesz Nauk, Mat, Politech Lodz • PL
Zeszyty Naukowe. Politechniki Lodzkiej. Matematyka (Scientific Papers. Technical University of Lodz. Mathematics) [1972ff]

J 2892 Molodoj Nauch Rabotnik, Erevan • SU
Molodoj Nauchnyj Rabotnik. Erevanskij Gosudarstvennyj Universitet (The Young Scientist. University of Erevan)

J 2985 Dongbei Gongxueyuan Xuebao • TJ
Dongbei Gongxueyuan Xuebao (Journal of Northeast Institute of Technology)

J 3068 Atti Accad Sci Bologna Fis Ser 13 • I
Atti della Accademia delle Scienze dell'Istituto di Bologna. Classe di Scienze Fisiche. Rendiconti. Serie XIII [1974ff]
• CONT OF (**J 3069**) Atti Accad Sci Bologna Fis Ser 12

J 3069 Atti Accad Sci Bologna Fis Ser 12 • I
Atti della Accademia delle Scienze dell'Istituto di Bologna. Classe di Scienze Fisiche. Rendiconti. Serie XII [1963-1973]
• CONT OF (**J 0347**) Atti Accad Sci Bologna Fis, Ser 11 • CONT AS (**J 3068**) Atti Accad Sci Bologna Fis Ser 13

J 3070 Bul Inst Politeh Iasi, Sect 1 • RO
Buletinul Institutului Politehnic din Iasi. Sectia 1. Matematica, Mecanica Teoretica, Fizica (Bulletin des Polytechnischen Instituts Jassy. Sektion 1. Mathematik, Theoretische Mechanik, Physik) [1977ff] ISSN 0304-5188
• CONT OF (**J 0198**) Bul Inst Politeh Iasi NS

J 3071 Bul Univ Brasov, Ser C • RO
Buletinul Universitatii din Brasov. Seria C. Matematica, Fizica, Chimie, Stiinte Naturale (Bulletin der Universitaet Brasov. Serie C. Mathematik, Physik, Chemie, Naturwissenschaften)

J 3072 Bulgar J Phys • BG
Bulgarian Journal of Physics [1974ff] ISSN 0323-9217
• CONT OF (**J 0344**) Izv Bulgar Akad Nauk Fiz Inst

J 3073 Bull Math Biol • GB
Bulletin of Mathematical Biology [1973ff] ISSN 0092-8240
• CONT OF (**J 0515**) Bull Math Biophys

J 3075 Normat • N
Normat. Nordisk Matematisk Tidskrift (Normat. Scandinavian Mathematical Journal) [1979ff]
• CONT OF (**J 0311**) Nordisk Mat Tidskr

J 3077 Nieuw Arch Wisk, Ser 3 • NL
Nieuw Archief voor Wiskunde. Derde Serie [1953-1982] ISSN 0028-9825
• CONT OF (**J 1793**) Nieuw Arch Wisk, Ser 2 • CONT AS (**J 3929**) Nieuw Arch Wisk, Ser 4

J 3079 Kiber Sb Perevodov, NS • SU
Kiberneticheskij Sbornik. Novaya Seriya. Sbornik Perevodov (Collected Articles on Cybernetics: New Series. Collected Translations) [1965ff] ISSN 0453-8382
• CONT OF (**J 1048**) Kiber Sb Perevodov

J 3101 Acta Tech Acad Sci Hung • H
Acta Technica Academiae Scientiarum Hungaricae [1950ff] ISSN 0001-7035

J 3109 Alxebra • E
Alxebra. Departamento de Algebra y Fundamentos,

J 3112 Annot Dokl, Inst Prikl Mat, Tbilisi • SU
Annotatsii Dokladov. Seminar Instituta Prikladnoj Matematiki Tbilisskogo Universiteta (Annotations of Papers. Seminar of the Institute of Applied Mathematics. University of Tbilisi) [1969ff] ISSN 0082-2191

J 3115 Gaz Austral Math Soc • AUS
The Australian Mathematical Society Gazette [1974ff]

J 3116 Austral Comp J • AUS
The Australian Computer Journal [1967ff] ISSN 0004-8917

J 3120 ACM Trans Program Lang & Syst • USA
ACM (= Association for Computing Machinery) Transactions on Programming Languages and Systems [1979ff] ISSN 0164-0925

J 3121 Izv Bulgar Akad Nauk Tekhn Kibern • BG
Bulgarska Akademiya na Naukite, Izvestiya Institut Tekhnicheskata Kibernetika (Bulgarian Academy of Sciences. Reports of the Institute of Engineering Cybernetics)

J 3127 Bol Soc Mat Mexicana, Ser 2 • MEX
Boletin de la Sociedad Matematica Mexicana. Segunda Serie [1956ff] ISSN 0037-8615
• CONT OF (**J 0250**) Bol Soc Mat Mexicana

J 3130 Bul Inst Politeh Bucuresti, Ser Electroteh • RO
Buletinul Institutului Politehnic "Gheorghe Gheorghiu-Dej" Bucuresti. Seria Electrotehnica (Bulletin des Polytechnischen Instituts "Gheorghe Gheorghiu-Dej" Bukarest. Serie Elektrotechnik)

J 3133 Bull Soc Math Belg, Ser B • B
Bulletin de la Societe Mathematique de Belgique. Serie B [1977ff] ISSN 0037-9476
• CONT OF (**J 0082**) Bull Soc Math Belg

J 3137 Shuxue Niankan • TJ
Shuxue Niankan (Chinese Annals of Mathematics) [1980-1982]
• CONT AS (**J 3187**) Shuxue Niankan, Xi A & (**J 4719**) Chinese Ann Math Ser B

J 3150 Delft Prog Rep, Ser F • NL
Delft Progress Reports. Series F: Mathematical Engineering, Mathematics and Information Engineering [1974ff] ISSN 0304-985X
• REM Preprint Series

J 3169 Gaz Math, Soc Math Fr • F
Gazette des Mathematiciens. Societe Mathematique de France
[1962ff] ISSN 0016-5549

J 3171 God Vissh Ucheb Zaved, Prilozhna Mat, Sofiya • BG
Godishnik na Visshite Uchebni Zavedeniya. Prilozhna Matematika (Annuaire de l'Universite, Mathematiques Appliquees) [1974ff]
• CONT OF (J 0224) God Vissh Tekh Ucheb Zaved Mat, Sofiya

J 3172 J London Math Soc, Ser 2 • GB
Journal of the London Mathematical Society. 2nd Series [1969ff] ISSN 0024-6107
• CONT OF (J 0039) J London Math Soc

J 3187 Shuxue Niankan, Xi A • TJ
Shuxue Niankan. Xi A (Chinese Annals of Mathematics. Series A) [1983ff]
• CONT OF (J 3137) Shuxue Niankan

J 3188 Itogi Nauki Tekh, Ser Teor Veroyat Mat Stat Teor Kibern • SU
Itogi Nauki i Tekhniki. Seriya Teoriya Veroyatnostej, Matematicheskaya Statistika, Teoreticheskaya Kibernetika (Progress in Science and Technology. Series Probability Theory, Mathematical Statistics, Theoretical Cybernetics) [1972ff] ISSN 0202-7488
• CONT OF (J 1021) Itogi Nauki Ser Mat • TRANSL IN (J 1531) J Sov Math

J 3191 IEEE Trans Pattern Anal & Mach Intell • USA
Transactions on Pattern Analysis and Machine Intelligence. IEEE (= Institute of Electrical and Electronics Engineers) [1979ff] ISSN 0162-8828

J 3192 IEEE Trans on Reliab • USA
Transactions on Reliability. IEEE (= Institute of Electrical and Electronics Engineers) [1952ff] ISSN 0018-9529

J 3194 J Austral Math Soc, Ser A • AUS
Journal of the Australian Mathematical Society. Series A [1976ff] ISSN 0263-6115
• CONT OF (J 0038) J Austral Math Soc

J 3200 Linear Algeb & Appl • USA
Linear Algebra and Its Applications [1968ff] ISSN 0024-3795

J 3217 Metroeconomica • I
*Metroeconomica. Rivista Internazionale di Economica * International Review of Economics * Revue Internationale d'Economique* [1949ff] ISSN 0026-1386

J 3218 J Huazhong Univ Sci Tech (Engl Ed) • TJ
Journal of Huazhong University of Science and Technology [1982ff]
• CONT OF (J 2684) J Huazhong Inst Tech (Engl Ed) • TRANSL OF (J 2754) Huazhong Gongxueyuan Xuebao

J 3239 Proc Japan Acad, Ser A • J
Proceedings of the Japan Academy. Series A. Mathematical Sciences [1978ff] ISSN 0386-2194

J 3240 Proc London Math Soc, Ser 3 • GB
Proceedings of the London Mathematical Society. 3rd Series [1951ff] ISSN 0024-6115
• CONT OF (J 1910) Proc London Math Soc, Ser 2

J 3254 Riv Mat Univ Parma, Ser 3 • I
Rivista di Matematica della Universita di Parma. Serie 3 [1972-1974] ISSN 0035-6298
• CONT OF (J 1526) Riv Mat Univ Parma, Ser 2 • CONT AS (J 0549) Riv Mat Univ Parma, Ser 4

J 3265 Sov J Contemp Math Anal, Armen Acad Sci • USA
Soviet Journal of Contemporary Mathematical Analysis. Armenian Academy of Sciences [1979ff] ISSN 0735-2719
• TRANSL OF (J 0312) Izv Akad Nauk Armyan SSR, Ser Mat

J 3274 Stud Leibnitiana • D
Studia Leibnitiana. Zeitschrift fuer Geschichte der Philosophie und der Wissenschaften. [1969ff] ISSN 0039-3185
• REL PUBL (S 3252) Stud Leibnitiana, Suppl

J 3279 Trans Moscow Math Soc • USA
Transactions of the Moscow Mathematical Society ISSN 0077-1554
• TRANSL OF (J 0065) Tr Moskva Mat Obshch

J 3283 Lith Math J • USA
Lithuanian Mathematical Journal. [1975ff] ISSN 0363-1672
• TRANSL OF (J 2574) Litov Mat Sb (Vil'nyus)
• REM Only selected translations of J2574

J 3285 Boll Unione Mat Ital, V Ser, A • I
Bollettino della Unione Matematica Italiana. Serie V. A [1976-1981] ISSN 0041-7084
• CONT OF (J 0012) Boll Unione Mat Ital, IV Ser • CONT AS (J 2099) Boll Unione Mat Ital, VI Ser, A

J 3287 Vopr Vychisl Prikl Mat (Tashkent) • SU
Voprosy Vychislitel'noj i Prikladnoj Matematiki (Problems of Computational and Applied Mathematics)

J 3289 Wiss Z Tech Hochsch Madgeburg • DDR
Wissenschaftliche Zeitschrift der Technischen Hochschule Otto von Guericke. Magdeburg [1957ff] ISSN 0541-8933

J 3293 Bull Acad Pol Sci, Ser Math • PL
Bulletin de l'Academie Polonaise des Sciences. Serie des Sciences Mathematiques [1979-1982] ISSN 0001-4117
• CONT OF (J 0014) Bull Acad Pol Sci, Ser Math Astron Phys
• CONT AS (J 3417) Bull Pol Acad Sci, Math

J 3298 Sci Comput Programming • NL
Science of Computer Programming [1981ff] ISSN 0167-6423

J 3350 Acta Polytech Prague CVUT, Ser 4 • CS
Acta Polytechnica Prace CVUT v Praze, Serie 4 [1977ff]

J 3359 Appl Comp Sci, Ber Prakt Inf • D
Applied Computer Science. Berichte zur praktischen Informatik

J 3360 Avtom Izchisl Tekh, Sofiya • BG
Avtomatika i Izchislitelna Tekhnika (Automation and Computational Technology)

J 3364 C R Acad Sci, Paris, Ser 1 • F
Comptes Rendus des Seances de l'Academie des Sciences. Serie I: Science Mathematique [1981ff] ISSN 0151-0509
• CONT OF (J 2313) C R Acad Sci, Paris, Ser A-B

J 3366 Comp & Electr Engin • USA
Computers & Electrical Engineering. An International Journal [1973ff] ISSN 0045-7906

J 3370 Enseign Math, Ser 2 • CH
L'Enseignement Mathematique. Revue Internationale. Serie 2 [1955ff] ISSN 0013-8584
• CONT OF (J 0152) Enseign Math

J 3377 Gaz Mat (Bucharest) • RO
Gazeta Matematica (Mathematische Zeitschrift) [1975ff]
• CONT OF (J 0378) Gaz Mat (Bucharest) Ser A

J 3396 Mat Model Teor Ehlektr Tsepej • SU
Matematicheskoe Modelirovanie i Teoriya Ehlektricheskikh Tsepej (Mathematical Modelling and Theory of Electric Circuits) [1962-1978]
• CONT AS (J 4526) Ehlektr Modelirovanie

J 3408 Nova Acta Akad Leopoldina, NF (Halle) • DDR
Nova Acta Leopoldina. Neue Folge. Abhandlungen der Deutschen Akademie der Naturforscher Leopoldina

J 3417 Bull Pol Acad Sci, Math • PL
Bulletin of the Polish Academy of Sciences. Mathematics [1983ff] ISSN 0001-4117
• CONT OF (J 3293) Bull Acad Pol Sci, Ser Math

J 3419 Probl Inf Transmiss • USA
Problems of Information Transmission [1965ff] ISSN 0032-9460
• TRANSL OF (J 2320) Probl Peredachi Inf, Akad Nauk SSSR

J 3434 Pubbl Ist Appl Calcolo, Ser 3 • I
Pubblicazioni. Serie III. Istituto per le Applicazioni del Calcolo "Mauro Picone" (IAC) Consiglio Nazionale delle Ricerche

J 3435 Pure Appl Math Sci • IND
Pure and Apppplied Mathematical Sciences [1974ff] ISSN 0379-3168

J 3436 Quad, Ist Appl Calcolo, Ser 3 • I
Quaderni Serie III. Istituto Applicazione Calcolo

J 3441 RAIRO Inform Theor • F
Revue Francaise d'Automatique, d'Informatique et de Recherche Operationnelle (RAIRO), Informatique Theorique [1977ff] ISSN 0399-0540
• CONT OF (J 4698) Rev Franc Autom, Inf & Rech Operat, Ser Rouge Inf Th

J 3449 Sov Math • USA
Soviet Mathematics [1974ff] ISSN 0197-7156
• TRANSL OF (J 0031) Izv Vyssh Ucheb Zaved, Mat (Kazan)

J 3450 Studia Univ Babes-Bolyai, Math (Cluj) • RO
Studia Universitatis Babes-Bolyai. Mathematica [1976ff]
• CONT OF (J 3451) Stud Univ Cluj, Ser Math-Mech

J 3451 Stud Univ Cluj, Ser Math-Mech • RO
Studia Universitatis Babes-Bolyai, Series Mathematica-Mechanica [1971-1975] ISSN 0370-8659
• CONT OF (J 0355) Studia Univ Babes-Bolyai, Math-Phys (Cluj) • CONT AS (J 3450) Studia Univ Babes-Bolyai, Math (Cluj)

J 3464 Wiss Z, Hochsch Verkehrswesen Dresden • DDR
Wissenschaftliche Zeitschrift. Hochschule fuer Verkehrswesen "Friedrich List", Dresden [1952ff] ISSN 0043-6844

J 3495 Boll Unione Mat Ital, V Ser, B • I
Bolletino della Unione Matematica Italiana. Serie V. B [1976-1981] ISSN 0041-7084
• CONT OF (J 0012) Boll Unione Mat Ital, IV Ser • CONT AS (J 2100) Boll Unione Mat Ital, VI Ser, B

J 3519 Glas Mat, Ser 3 (Zagreb) • YU
Glasnik Matematichki. Serija 3 (Publications of Mathematics. Series 3) [1966ff] ISSN 0017-095X
• CONT OF (J 0371) Glas Mat-Fiz Astron, Ser 2 (Zagreb)

J 3522 Rend Circ Mat Palermo, Ser 2 • I
Rendiconti del Circolo Matematico di Palermo. Serie II [1952ff] ISSN 0009-725X
• CONT OF (J 0058) Rend Circ Mat Palermo

J 3523 Period Mat, Ser 5 • I
Periodico di Matematiche. Serie V. (Roma)

J 3524 Kybernetika Suppl (Prague) • CS
Kybernetika. Supplement. (Prague) (Cybernetics. Supplement. (Prague))
• REL PUBL (J 0156) Kybernetika (Prague)

J 3536 Uch Zap Univ, Ivanovo • SU
Ivanovskij Gosudarstvennyj Universitet. Uchenye Zapiski (State University of Ivanovo. Scientific Notes) [1974ff]
• CONT OF (J 0226) Uch Zap Ped Inst, Ivanovo

J 3558 Uch Zap Ped Inst, Kaliningrad • SU
Kaliningradskij Gosudarstvennyj Pedagogicheskij Institut. Uchenye Zapiski (State Institute of Education in Kaliningrad. Scientific Notes) ISSN 0454-1359

J 3594 Kexue Tansuo • TJ
Kexue Tansuo (Science Exploration)

J 3596 Arch Rational Mech Anal • D
Archive for Rational Mechanics and Analysis [1957ff] ISSN 0003-9527

J 3597 J Unif Sci • USA
The Journal of Unified Science [1940-1974]
• CONT OF (J 0748) Erkenntnis (Leipzig) • CONT AS (J 0158) Erkenntnis (Dordrecht)

J 3665 Finsk Tidskrift • SF
Finsk Tidskrift. Kultur - Ekonomi - Politik (Finnische Zeitschrift. Kultur, Ekonomie, Politik) [1876ff] ISSN 0015-248X

J 3672 Philosophia Arhusiensis • DK
Philosophia Arhusiensis [1967-1976] ISSN 0556-0136

J 3685 Publ Inst Mat Univ Catolica Chile • RCH
Publicaciones del Instituto de Matematica de la Universidad Catolica de Chile

J 3741 Rend Mat Appl, Ser 7 • I
Rendiconti di Matematica e delle sue Applicazioni. Seria 7 [1981ff] ISSN 0034-4427
• CONT OF (J 0297) Rend Mat Appl, Ser 5

J 3742 Shuxue Yanjiu yu Pinglun • TJ
Shuxue Yanjiu yu Pinglun (Journal of Mathematical Research & Exposition) [1981ff]

J 3750 Jilin Daxue, Ziran Kexue Xuebao • TJ
*Jilin Daxue. Ziran Kexue Xuebao (Acta Scientiarum Naturalium Universitatis Jilinensis * Jilin University. Natural Sciences Journal)* [1975ff]
• CONT OF (J 4364) Dongbei Renmin Daxue Ziran Kexue Xuebao

J 3757 Libertas Math • USA
Libertas Mathematica [1981ff] ISSN 0278-5307

J 3766 Zhongguo Kexue, Xi A • TJ
*Zhongguo Kexue. Xi A. Mathematical, Physical, Astronomical & Technical Sciences * Scientia Sinica. Series A* [1982ff]
• CONT OF (J 1024) Zhongguo Kexue

J 3768 Boll Unione Mat Ital, VI Ser, D • I
Bolletino della Unione Matematica Italiana. Serie VI. D. Algebra e Geometria [1982ff] ISSN 0041-7084

J 3769 Sci Bull, Foreign Lang Ed • TJ
Kexue Tongbao. Foreign Language Edition (Science Bulletin) [1980ff]
• TRANSL OF (J 2771) Kexue Tongbao

J 3772 J Nigerian Math Soc • WAN
Journal of the Nigerian Mathematical Society [1982ff]

J 3776 Mech Teor Stosow • PL
Polskie Towarzystwo Mechaniki Teoretycznej i Stosowanej. Mechanika Teoretyczna i Stosowana (Polish Society of Theoretical and Applied Mechanics. Theoretical and Applied Mechanics) [1963ff] ISSN 0079-3701

J 3781 Topoi • NL
Topoi. An International Review of Philosophy [1982ff] ISSN 0167-7411

J 3789 Ann Hist of Comp • USA
Annals of the History of Computing [1979ff] ISSN 0164-1239

J 3793 Jisuanjii Xuebao • TJ
Jisuanji Xuebao (Chinese Journal of Computers) [1978ff]

J 3815 Acta Univ Lodz Folia Philos • PL
Acta Universitatis Lodziensis Folia Philosophica [1981ff] ISSN 0208-6107

J 3824 Bull Soc Math Belg, Ser A • B
Bulletin de la Societe Mathematique de Belgique. Serie A [1977ff]
• CONT OF (**J 0082**) Bull Soc Math Belg

J 3914 Math Soc Sci • NL
Mathematical Social Sciences [1980ff] ISSN 0165-4896

J 3923 Xiamen Daxue Xuebao, Ziran Kexue • TJ
*Xiamen Daxue Xuebao. Ziran Kexue Ban (Acta Scientarium Naturalium Universitatis Amoiensis * Journal of Xiamen University. Natural Science)*

J 3928 Stochastics • USA
Stochastics ISSN 0090-9491

J 3929 Nieuw Arch Wisk, Ser 4 • NL
Nieuw Archief voor Wiskunde. Vierde Serie. Uitgegeven door het Wiskunnid Genootschap te Amsterdam [1983ff] ISSN 0028-9825
• CONT OF (**J 3077**) Nieuw Arch Wisk, Ser 3

J 3930 Biometrics • USA
Biometrics. Journal of the Biometric Society. [1945ff] ISSN 0006-341X

J 3932 Comput Artif Intell (Bratislava) • CS
Pocitace a Umela Inteligencia (Computers and Artificial Intelligence) [1982ff]

J 3936 S Afrik Statist J • ZA
Suid-Afrikaanse Statistiese Tydskrif (South African Statistical Journal) [1967ff] ISSN 0038-271X

J 3937 Veroyat Met i Kibern (Kazan) • SU
Veroyatnostnye Metody i Kibernetika (Probabilistic Methods and Cybernetics)

J 3939 Mat Logika Primen (Akad Nauk Litov SSR) • SU
Matematicheskaya Logika i Ee Primeneniya (Mathematical Logic and its Applications) [1981ff]

J 3940 Saitama Math J • J
Saitama Mathematical Journal [1983ff]
• CONT OF (**J 1472**) Sci Rep Saitama Univ, Ser A

J 3941 J Math Pures Appl, Ser 9 • F
Journal de Mathematiques Pures et Appliquees. Neuvieme Serie [1922ff] ISSN 0021-7824
• CONT OF (**J 0249**) J Math Pures Appl

J 3942 Zhongshan Daxue Xuebao, Ziran Kexue • TJ
Zhongshan Daxue Xuebao. Ziran Kexue Ban (Acta Scientiarum Naturalium Universitatis Sunyatseni)

J 3943 Shuxue Wuli Xuebao • TJ
*Shuxue Wuli Xuebao * Acta Mathematica et Physica* [1981ff] ISSN 0252-9602
• TRANSL OF (**J 4614**) Acta Math (Engl Ed)

J 3944 Rad Jugosl Akad Znan Umjet, Mat Znan • YU
Radovi Jugoslavenske Akademije Znanosti i Umjetnosti Matematichke Znanosti (Papers. Academy of Sciences and Arts of Yugoslavia. Mathematics) [1982ff]
• CONT OF (**J 0372**) Rad Jugosl Akad Znan Umjet, Mat Fiz Teh Znan

J 3945 Vest Math Univ Leningrad • USA
Vestnik Leningrad University. Mathematics [1968ff] ISSN 0146-924X
• TRANSL OF (**J 0085**) Vest Ser Mat Mekh Astron, Univ Leningrad

J 3954 Rev Franc Inf & Rech Operat • F
Association Francaise pour la Cybernetique Economique et Technique. Revue Francaise d'Informatique et de Recherche Operationnelle [1967-1971]
• CONT AS (**J 0205**) Rev Franc Autom, Inf & Rech Operat

J 3955 Ann Univ Budapest, Sect Comp • H
Annales Universitatis Scientiarum Budapestinensis. Sectio Computatorica

J 3959 Chengdu Keji Daxue Xuebao • TJ
Chengdu Keji Daxue Xuebao (Journal of Chengdu University of Science and Technology)

J 3967 Boll Mathesis Soc Ital Mat • I
Bollettino della "Mathesis" Societa Italiana di Matematica

J 3975 Arch Math & Phys • D
Archiv der Mathematik und Physik mit Besonderer Ruecksicht auf die Beduerfnisse der Lehrer an Hoeheren Unterrichtsanstalten. 3.Reihe [1841ff]

J 3978 Math-Nat Blaetter • D
Mathematisch-Naturwissenschaftliche Blaetter. Organ des Verbandes Mathematischer und Naturwissenschaftlicher Vereine an Deutschen Hochchulen

J 3980 Karadeniz Univ Math J (Trabzon) • TR
Karadeniz University Mathematical Journal

J 3986 Ist Lombardo Rend, Ser 2 (Milano) • I
Reale Istituto Lombardo di Scienze e Lettre. Rendiconti. Series 2 [1868-1936]
• CONT AS (**J 0207**) Ist Lombardo Accad Sci Rend, A (Milano)

J 3989 Rozpr II, Tr Cesk Akad Ved • CS
Tridy Ceske Akademie Ved. Rozprawy II (Scientific Papers of the Royal Czech Academy of Sciences. Series: Mathematics & Natural Sciences.) [1891-1952]
• CONT AS (**J 2830**) Rozpr, Mat Prirod, Cheskoslov Akad Ved

J 3993 Rev Mat Hisp-Amer, Ser 2 • E
Revista Matematica Hispano-Americana. Serie 2 ISSN 0373-0999
• CONT AS (**J 0236**) Rev Mat Hisp-Amer, Ser 4

J 3994 Ann Acad Sci Fennicae Ser A I • SF
Annales Academiae Scientiarum Fennicae. Serie A I [1941-1974] ISSN 0066-1953
• CONT AS (**J 0446**) Ann Acad Sci Fennicae, Ser A I, Diss

J 3996 Boll Unione Mat Ital • I
Bollettino della Unione Matematica Italiana [1922-1945]
• CONT AS (**J** 4408) Boll Unione Mat Ital, III Ser

J 4077 New Scholast • USA
The New Scholasticism [1927ff] ISSN 0028-6621

J 4093 Byull Ob'ed Probl Mashin Perevodov • SU
Byulleten' Ob'edineniya po Problemam Mashinnogo Perevoda (Bulletin of the Association for Problems of Machine Translation)

J 4110 Zap Inst Nar Prosv, Dnepropetrovsk • SU
Dnepropetrovsk, Zapiski Instituta Narodnogo Prosveshcheniya (Notes of the Institute of Education. Dnepropetrovsk)

J 4122 Vest Komm Akad • SU
Vestnik Kommunisticheskoj Akademii (Publications of the Communist Academy)

J 4149 Publ Inst Rech Math Avancee • F
Publications de IRMA (= Institut de Recherche de Mathematiques Avancee)

J 4170 Nauch Dokl Vys Shk, Fiz Mat (Moskva) • SU
Nauchnye Doklady Vysshej Shkoly, Fiziko-Matematicheskie Nauki (Scientific Reports of the University. Physic-Mathematics)

J 4186 Atti Accad Sci Lett Arti Palermo, Ser 5/I • I
Atti della Accademia di Scienze. Lettere e Arti di Palermo. Serie Quinta. Parte I: Scienze [1984ff]
• CONT OF (**J** 0104) Atti Accad Sci Lett Arti Palermo, Ser 4/I

J 4277 Vesn Drusht Mat Fiz Serbije • YU
*Vesnik Drushtva Matematichara i Fizichara Narodne Republike Serbije * Vestnik Obshchestva Matematikov i Fizikov N.R. Serbie * Bulletin de la Societe des Mathematiciens et Physiciens de la R.P. Serbie (Publications of the Society of Mathematicians and Physicists of the P.R. Serbia)* [1949-1963]
• CONT AS (**J** 0042) Mat Vesn, Drust Mat Fiz Astron Serb

J 4279 Bull Soc Phys Math Kazan, Ser 3 • SU
(Bulletin de la Societe Physico-Mathematique de Kazan, Ser 3)

J 4305 IEEE Trans Electr Comp • USA
Transactions of Electronic Computers. IEEE (= Institute of Electrical and Electronics Engineers) [1963-1967]
• CONT OF (**J** 0072) IRE Trans Electr Comp • CONT AS (**J** 0187) IEEE Trans Comp

J 4330 Radiophys & Quant Electr • SU
Radiophysics and Quantum Electronics
• TRANSL OF (**J** 1572) Izv Vyssh Ucheb Zaved, Radiofizika (Moskva)

J 4341 Euclides Madrid • E
Euclides Madrid

J 4364 Dongbei Renmin Daxue Ziran Kexue Xuebao • TJ
*Dongbei Renmin Daxue Ziran Kexue Xuebao * Acta Scientiarum Naturalium (Natural Science Journal of Northeast Peoples's University)* [?-1974]
• CONT AS (**J** 3750) Jilin Daxue, Ziran Kexue Xuebao

J 4387 Itogi Nauki Tekh, Ser Tekh Kibern • SU
Itogi Nauki i Tekhniki. Seriya Tekhnicheskaya Kibernetika. Gosudarstvennyj Komitet SSSR po Nauke i Tekhnike. Akademiya Nauk SSSR. Vsesoyuznyj Institut Nauchnoj i Tekhnicheskoj Informatsii. (Progress in Science and Technology. Series: Engineering Cybernetics) [1972ff]
• CONT OF (**J** 1021) Itogi Nauki Ser Mat

J 4390 Kobe J Math • J
Kobe Journal of Mathematics [1984ff] ISSN 0289-9051
• CONT OF (**J** 1508) Math Sem Notes, Kobe Univ

J 4404 Vopr Mat Logiki & Pril • SU
Voprosy Matematicheskoj Logiki i Ee Prilozheniya (Problems of Mathematical Logic and its Applications)

J 4406 Zb Prats Obchis Mat Tekhn • SU
Zbirnik Prats' z Obchislyuv. Matematika i Tekhnika (Collected Periodic Papers. Mathematics and Technology)

J 4408 Boll Unione Mat Ital, III Ser • I
Bolletino della Unione Matematica Italiana, Ser III [1946-1967]
• CONT OF (**J** 3996) Boll Unione Mat Ital • CONT AS (**J** 0012) Boll Unione Mat Ital, IV Ser

J 4442 Kyushu Daigaku Kagaku Syuho • J
Kyushu Daigaku Kagaku Syuho (Technology Reports of the Kyushu University)

J 4443 Gac Mat (Madrid) • E
Gaceta Matematica [1949-1969]
• CONT AS (**J** 0299) Gac Mat, Ser 1a (Madrid)

J 4445 Moderni Logica
Moderni Logica

J 4452 Zhexue Yanjiu • TJ
Zhexue Yanjiu (Philosophical Research. Studies on Philosophy)

J 4467 Rocz Filoz • PL
Roczniki Filozoficzne (Philosophical Annals)

J 4510 Norsk Mat Tidsskr • N
Norsk Matematisk Tidsskrift (Scandinavian Mathematical Journal) [1919-1952]
• CONT OF (**J** 0259) Nyt Tidsskr Mat • CONT AS (**J** 0311) Nordisk Mat Tidskr

J 4526 Ehlektr Modelirovanie • SU
Ehlektronnoe Modelirovanie (Electronic Modelling) [1979ff]
• CONT OF (**J** 3396) Mat Model Teor Ehlekt Tsepej

J 4529 Rev A Tijdschrift • B
Revue A Tijdschrift. Revue Trimestrielle Editee par l'Institut Belge de Regulation et d'Automatisme

J 4535 Inst Politeh Timisoara Sem Mat Fiz • RO
Institutul Politehnic "Traian Vuia" Timisoara. Seminar Matematica - Fizica (Das Polytechnische Institut "Traian Vuia" Timisoara. Seminar Mathematik-Physik) [1982ff]
• CONT OF (**J** 0218) Bul Sti Tehn Inst Politeh Timisoara, Ser Mat-Fiz-Mec

J 4572 Shuxue Jikan, Sichuan Shang Shuxuehui • TJ
Shuxue Jikan, Sichuan Shang Shuxuehui (Collection Journal of Math., Mathematical Society of Sichuan Province)

J 4609 J Symb Comput • USA
Journal of Symbolic Computation [1985ff]

J 4614 Acta Math (Engl Ed) • TJ
Shuxue Wuli Xuebao. English Edition (Acta Mathematica)
• TRANSL IN (**J** 3943) Shuxue Wuli Xuebao

J 4645 Irish Math Soc Newslett • IRL
Irish Mathematical Society. Newsletter

J 4666 J Beijing Polytech Univ • TJ
Journal of the Beijing Polytechnic University [1975ff]

J 4693 Ann Sci Univ Jassy • RO
Annales Scientifiques de l'Universite de Jassy [?-1936]
• CONT AS (J 0269) Ann Sci Univ Jassy Sec 1

J 4698 Rev Franc Autom, Inf & Rech Operat, Ser Rouge Inf Th • F
Revue Francaise d'Automatique, d'Informatique et de Recherche Operationnelle (RAIRO). Serie Rouge Informatique Theorique [1975-1976] ISSN 0399-0540
• CONT OF (J 0205) Rev Franc Autom, Inf & Rech Operat Ser Rouge • CONT AS (J 3441) RAIRO Inform Theor

J 4701 Trudy Akad Nauk Ehston SSR, Fiz Astron (Tartu) • SU
*Trudy Instituta Fiziki i Astronomii Akademii Nauk Ehstonskoj SSR * Eesti NSV Teaduste Akadeemia Fueuesika ja Astronomia Instituudi Kurimused* [1951ff]

J 4702 Ann Sc Norm Sup Pisa Fis Mat, Ser 4 • I
Annali della Scuola Normale Superiore di Pisa. Classe di Science. Fisiche e Matematiche. Seria IV [1974ff]
• CONT OF (J 0315) Ann Sc Norm Sup Pisa Fis Mat, Ser 3

J 4706 Publ Inst Math (Belgrade) • YU
Academie Serbe des Sciences. Publications de l'Institut Mathematique [1948-1960]
• CONT AS (J 0400) Publ Inst Math, NS (Belgrade)

J 4710 Pol Tow Mat, Wiad Mat • PL
Polskie Towarzystwo Matematyczne. Wiadomosci Matematiczne [1899-1954]
• CONT AS (J 0519) Wiad Mat, Ann Soc Math Pol, Ser 2

J 4711 IRE Proc • USA
Proceedings. IRE (= Institute of Radio Engineers) [1911-1962]
• CONT AS (J 1077) IEEE Proc

J 4713 Mat Fyz Chasopis (Slov Akad Ved) • CS
Matematicky-Fyzikalny Chasopis (Journal of Mathematical Physics) [1951-1966]
• CONT AS (J 0143) Mat Chasopis (Slov Akad Ved)

J 4714 Bul Sti Tehn Inst Politeh Timisoara, NS • RO
Buletinul Stiintific si Tehnic al Institutului Politehnic Timisoara. Serie Noua (Wissenschaftliches und Technisches Bulletin des Polytechnischen Instituts Timisoara. Neue Serie) [1956-1969]
• CONT AS (J 0218) Bul Sti Tehn Inst Politeh Timisoara, Ser Mat-Fiz-Mec

J 4715 Bul Sti Tehn Inst Politeh Timisoara, Ser Mat-Fiz • RO
Buletinul Stiintific si Tehnic al Institutului Politehnic "Traian Vuia" Timisoara. Seria Matematica-Fizica (Wissenschaftliches und Technisches Bulletin des Polytechnischen Instituts "Traian Vuia" Timisoara. Serie Mathematik-Physik) [1978-1981]
• CONT OF (J 0218) Bul Sti Tehn Inst Politeh Timisoara, Ser Mat-Fiz-Mec • CONT AS (J 4535) Inst Politeh Timisoara Sem Mat Fiz

J 4716 Stud Philos, Leopolis (Poznan) • PL
Studia Philosophica. Commentarii Societatis Polonorum. Leopolis [1935ff]

J 4717 Izv Akad Nauk SSSR • SU
Izvestiya Akademii Nauk SSSR (Bulletin de l'Academie des Sciences Mathematiques et Naturelles. Leningrad) [?-1936]
• CONT AS (J 0216) Izv Akad Nauk SSSR, Ser Mat

J 4719 Chinese Ann Math Ser B • TJ
Shuxue Niankan. Xi B (Chinese Annals of Mathematics. Series B) [1983ff]
• CONT OF (J 3137) Shuxue Niankan

J 4729 Acta Math Hung • H
Acta Mathematica Hungarica [1983ff] ISSN 0236-5294
• CONT OF (J 0001) Acta Math Acad Sci Hung

J 4731 Australasian J Psych & Phil • AUS
The Australasian Journal of Psychology and Philosophy [1923-1946]
• CONT AS (J 0273) Australasian J Phil

J 4732 J Math Mech (Indiana Univ) • USA
Journal of Mathematics and Mechanics [1962-1970]
• CONT OF (J 0452) Indiana Univ Math J

J 4735 Rad Jugosl Akad Znan Umjet • YU
Radovi Jugoslavenske Akademije Znanosti i Umjetnosti. (Papers. Yugoslavian Academy of Sciences and Arts) [/??/-1939]
• CONT AS (J 0372) Rad Jugosl Akad Znan Umjet, Mat Fiz Teh Znan

Series

S 0019 Colloq Math (Warsaw) • PL
Colloquium Mathematicum [1947ff] • PUBL Academie Polonaise des Sciences, Institut Mathematique: Warsaw
• ISSN 0010-1354

S 0055 Proc Steklov Inst Math • USA
Proceedings of the Steklov Institute of Mathematics [1967ff]
• PUBL (X 0803) Amer Math Soc: Providence
• TRANSL OF (S 0066) Tr Mat Inst Steklov
• ISSN 0081-5438

S 0066 Tr Mat Inst Steklov • SU
Trudy Ordena Lenina Matematicheskogo Instituta imeni V.A.Steklova. Akademiya Nauk SSSR (Proceedings of the Mathematical Steklov-Institute of the Academy of Sciences SSSR) [1938ff] • PUBL (X 0899) Akad Nauk SSSR : Moskva
• CONT OF (S 1644) Tr Fiz-Mat Inst Steklov • TRANSL IN (S 0055) Proc Steklov Inst Math

S 0166 Mat Issl, Mold SSR • SU
Matematicheskie Issledovaniya. Akademiya Nauk Moldavskoj SSR. Ordena Trudovogo Krasnogo Znameni Institut Matematiki s Vychislitel'nym Tsentrom (Mathematical Studies) [1966ff] • PUBL (X 2741) Shtiintsa: Kishinev
• ISSN 0542-9994

S 0167 Mem Amer Math Soc • USA
Memoirs of the American Mathematical Society [1950-1974]
• PUBL (X 0803) Amer Math Soc: Providence
• CONT AS (S 2450) Mem Amer Math Soc, NS
• ISSN 0065-9266

S 0183 Publ Math Univ California • USA
University of California Publications in Mathematics PUBL (X 0926) Univ Calif Pr: Berkeley

S 0208 Uch Zap, Ped Inst, Moskva • SU
Uchenye Zapiski Moskovskogo Gosudarstvennogo Pedagogicheskogo Instituta imeni V.I.Lenina (Scientific Notes of the Moscow State Institute of Education) [1950ff] • PUBL (X 2802) Moskov Ped Inst: Moskva

S 0223 Nauch Trud NS, Politekh Inst Tashkent • SU
Tashkentskij Politekhnicheskij Institut. Nauchnye Trudy. Novaya Seriya (Polytechnic Institute of Tashkent. Scientific Papers. New Series) PUBL Polytekh Inst: Tashkent

S 0228 Zap Nauch Sem Leningrad Otd Mat Inst Steklov • SU
Zapiski Nauchnykh Seminarov Leningradskogo Otdeleniya Ordena Lenina Matematicheskogo Instituta imeni V.A.Steklova Akademii Nauk SSSR (LOMI) (Reports of the Scientific Seminars of the Leningrad Steklov Institute of Mathematics) PUBL (X 2641) Nauka: Leningrad
• TRANSL IN (J 1531) J Sov Math & (J 0521) Semin Math, Inst Steklov
• REM Transl. in J0521 up to vol. 19

S 0257 Monograf Mat • PL
Monografie Matematyczne (Mathematical Monography) [1932ff] • PUBL (X 1034) PWN: Warsaw • ALT PUBL (X 1758) Ars Polona: Warsaw
• ISSN 0077-0507

S 0281 Arch Towarz Nauk Lwow, Sect 3 • PL
Archivum Towarzystwa Naukowego we Lwowie. Dzial 3. Matematyczno-Przyrodniczy (Archive of the Scientific Society of Lwow. Section 3. Mathematics and Natural Sciences) [1920-1939] • PUBL Tow Nauk: Lwow

S 0393 Uch Zap Univ, Tartu • SU
*Uchenye Zapiski Tartuskogo Gosudarstvennogo Universiteta. * Tartu Riikliku Uelikooli Toimetised. * Acta et Commentationes Universitatis Tartuensis (Scientific Notes of the Tartu State University.)* [1961ff] • PUBL (X 2463) Tartusk Gos Univ: Tartu

S 0405 Mitt Math Sem Giessen • D
Mitteilungen aus dem Mathematischen Seminar Giessen PUBL (X 2464) Math Sem Selbstverl: Giessen

S 0410 Math Beitr Univ Halle-Wittenberg • DDR
Die Martin Luther-Universitaet Halle-Wittenberg. Mathematische Beitraege [1968ff] • PUBL (X 2446) Univ Halle-Wittenberg: Halle Wissenschaftliche Beitraege. ISSN 0440-1298
• ISSN 0441-6228

S 0422 Tr Vychisl Tsentra Akad Nauk Armyan SSR & Univ Erevan • SU
Trudy Vychislitel'nogo Tsentra Akademij Nauk Armyanskoj SSR i Erevanskogo Gosudarstvennogo Universiteta. Matematicheskie Voprosy Kibernetiki i Vychislitel'noj Tekhniki (Publications of the Computing Centre of the Academy of Sciences of the Armyan SSR and of the Erevan State University. Mathematical Problems of Cybernetics and Computational Techniques) [1963ff] • PUBL (X 2225) Akad Nauk Armyan SSR : Erevan

S 0445 Mem Muroran Inst Tech • J
Muroran Kogyo Daigaku.Kenkyu Hokoku (Memoirs of the Muroran Institute of Technology) [1950ff] • PUBL Muroran Institute of Technology: Muroran
• ISSN 0580-2393

S 0450 Istor-Mat Issl, Akad Nauk SSSR • SU
Istoriko-Matematicheskie Issledovaniya. Akademiya Nauk SSSR. Institut Istorii Estestvoznaniya i Tekhniki (Historico-Mathematical Investigations) [1947ff] • PUBL (X 2027) Nauka: Moskva

S 0458 Zesz Nauk, Prace Log, Uniw Krakow • PL
Zeszyty Naukowe Uniwersytetu Jagiellonskiego Prace z Logiki (Scientific Papers. Jagielleonian University. Reports on Logic) [1965-1972] • PUBL (X 1034) PWN: Warsaw
• CONT AS (J 0302) Rep Math Logic, Krakow & Katowice

S 0473 Contrib Cient, Ser Mat • RA
Contribuciones Cientificas. Serie Matematica. PUBL Facultad de Ciencias Exactas y Naturales, Univ. Buenos Aires

S 0478 Bonn Math Schr • D
Bonner Mathematische Schriften [1957ff] • PUBL (**X 0908**)
Univ Math Inst: Bonn
• ISSN 0524-045X

S 0502 Prace Wroclaw Tow Nauk, Ser B • PL
Prace Wroclawskiego Towarzystwa Naukowe. Series B. Nauki Scisle (Papers of the Scientific Society of Wroclaw. Series B. Exact Sciences) [1947ff] • PUBL (**X 1034**) PWN: Warsaw
• ISSN 0084-3024

S 0507 Vychisl Sist (Akad Nauk SSSR Novosibirsk) • SU
Vychislitel'nye Sistemy. Sbornik Trudov. Akademiya Nauk SSSR. Sibirskoe Otdelenie (Computer Systems. Collected Articles) [1962ff] • PUBL (**X 2652**) Akad Nauk Sibirsk Otd Inst Mat: Novosibirsk
• ISSN 0568-661X

S 0543 Zesz Nauk Wyz Ped Mat, Katowice • PL
Wyzsza Szkola Pedagogiczna W Katowicach. Zeszyty Naukowe. Sekcja Matematyki (Scientific Papers. University of Education in Katowice. Section: Mathematics) PUBL Szkola Ped: Katowice

S 0544 Prace Mat Uniw Katowice • PL
Prace Naukowe Uniwersytetu Slaskiego w Katowicach. Prace Matematyczne (Scientific Publications of the Silesian University in Katowice. Mathematical Papers) PUBL (**X 1425**) Univ Slaski: Katowice

S 0554 Issl Teor Algor & Mat Logik (Moskva) • SU
Issledovaniya po Teorii Algorifmov i Matematicheskoj Logike (Studies in the Theory of Algorithms and Mathematical Logic) [1973,1976,1979] • ED: MARKOV, A.A. & PETRI, N.V. (VOL 1). MARKOV, A.A. & KUSHNER, B.A. (VOL 2). MARKOV, A.A. & KHOMICH, V.I. (VOL 3) • PUBL (**X 2265**) Akad Nauk Vychis Tsentr: Moskva 287pp,160pp,133pp
• ISSN 0302-9085

S 0716 Vychisl Tekh Vopr Kibern (Univ Leningrad) • SU
Vychislitel'naya Tekhnika i Voprosy Kibernetiki (Computer Technology and Questions of Cybernetics) [1962ff] • ED: CHIRKOV, M.K. & MASLOV, S.P. & TSAR'KOVA, Z.I. (V 8). BRUSENTSOVA, N.P. & SHAUMAN, A.M. (V 7, V 15 - V 19)
• PUBL (**X 0938**) Leningrad Univ: Leningrad
• ISSN 0507-536X

S 0764 Teor Konech Avtom & Prilozh (Riga) • SU
Teoriya Konechnykh Avtomatov i ee Prilozheniya (Theory of Finite Automata and its Applications) [1972ff] • PUBL (**X 2230**) Zinatne: Riga
• LC-No 74-304069

S 0780 Jbuch Ueberblick Math • D
Jahrbuch Ueberblicke Mathematik [1975ff] • PUBL (**X 0876**) Bibl Inst: Mannheim
• CONT OF (**J 2345**) Ueberblicke Math

S 0889 Notas Logica Mat • RA
Notas de Logica Matematica. [1963ff] • PUBL Inst Mat, Univ Nacional del Sur: Bahia Blanca
• ISSN 0078-2017

S 1003 Publ Elektroteh, Ser Mat Fiz, Beograde • YU
Univerzitet u Beogradu. Publikacije Elektrotehnichkog Fakulteta. Serija Matematika i Fizika (University of Belgrade. Publications of the Faculty of Electrical Engineering. Series: Mathematics and Physics) PUBL Univ Beograd: Belgrade

S 1454 Zesz Nauk Wyz Szk Ped Mat, Opole • PL
Zeszyty Naukowe Wyzszej Szkoly Pedagogicznej W Opolu Matematyka (Scientific Papers. University of Education in Opole. Mathematics) [1961ff] • PUBL Wyzsza Szkola Pedagogiczna: Opole
• ISSN 0078-5431

S 1490 Stud Constructive Math • USA
Studies in Constructive Mathematics and Mathematical Logic [1969ff] • ED: SLISENKO, A.O. & MATIYASEVICH, YU. • SER (**J 0521**) Semin Math, Inst Steklov • PUBL (**X 2529**) Consult Bureau: New York
• TRANSL OF Issledovaniya po Konstruktvno Matematike i Matematichesko Logike
• REM Matiyasevich co-ed. starting with vol. 4

S 1498 Uch Zap Univ, Moskva • SU
Uchenye Zapiski Moskovskogo Gosudarstvennogo Universiteta (Scientific Notes of the Moscow State University) PUBL (**X 0898**) Moskov Gos Univ: Moskva

S 1562 Prace Inst Mat Fis, Politech Wroclaw, Ser Stud Mater • PL
Prace Naukowe. Instytut Matematyki i Fizyki Teoretycznej. Politechnika Wroclawska. Seria Studia i Materialy. (Scientific Publications. Institute of Mathematics and Theoretical Physics. Technical University of Wroclaw) [?-1969] • PUBL Politechnika Wroclawska: Wroclaw
• CONT AS (**S 4733**) Prace Inst Mat, Politech Wroclaw, Ser Stud Mater

S 1567 Semin Bourbaki • F
Seminaire Bourbaki [1948ff] • SER (**S 3301**) Lect Notes Math, (**J 1620**) Asterisque • PUBL (**X 0811**) Springer: Heidelberg & New York, (**X 2244**) Soc Math France: Paris, (**X 1623**) Univ Paris VI Inst Poincare: Paris Secretariat Mathematique: Paris

S 1605 Math Centr Tracts • NL
Mathematical Centre Tracts [1963-1983] • PUBL (**X 1121**) Math Centr: Amsterdam

S 1613 Probl Logic (Bucharest) • RO
Probleme de Logica [1956ff] • PUBL (**X 0871**) Acad Rep Soc Romania: Bucharest
• ISSN 0556-1655

S 1626 Oslo Preprint Ser • N
Oslo Preprint Series [1970ff] • PUBL (**X 2786**) Univ Oslo Mat Inst: Oslo

S 1642 Schr Inf Angew Math, Ber (Aachen) • D
Schriften zur Informatik und Angewandten Mathematik. Bericht PUBL (**X 3215**) TH Aachen Math Nat Fak: Aachen

S 1644 Tr Fiz-Mat Inst Steklov • SU
Trudy Fiziko-Matematicheskogo Instituta imeni V.A.Steklova Akademiya Nauk SSSR (Travaux de l'Institut Physico-Mathematique V.A.Stekloff de l'Academie des Sciences SSSR) [1930-1937] • PUBL (**X 0899**) Akad Nauk SSSR : Moskva
• CONT AS (**S 0066**) Tr Mat Inst Steklov

S 1655 Zesz Nauk, Mat, Politech Gdansk • PL
Zeszyty Naukowe Politechniki Gdanskiej. Matematyka (Scientific Papers. Technical University of Gdansk. Mathematics) [1963ff] • PUBL Politech Univ Gdanski: Gdansk
• ISSN 0072-0372

S 1747 Christiana Albertina (Kiel) • D
Christiana Albertina. Kieler Universitaetszeitschrift [1965ff]
• PUBL Christian Albrechts Universitaet, Presse und Informationsstelle: Kiel
• ISSN 0578-0160

S 1802 Colloq Int CNRS • F
Colloques Internationaux du Centre National de la Recherche Scientifique (CNRS) PUBL (X 0999) CNRS Inst B Pascal: Paris

S 1956 Tagungsbericht, Oberwolfach • D
Tagungsbericht. Mathematisches Forschungsinstitut Oberwolfach PUBL (X 0876) Bibl Inst: Mannheim

S 2043 Issl Mat Log & Teor Algor (Tbilisi) • SU
Issledovaniya po Matematicheskoj Logike i Teorii Algoritmov (Studies in Mathematical Logic and the Theory of Algorithms) [1975,1977,1978] • ED: PKHAKADZE, SH.S. & MAGNARADZE, L.G. • PUBL (X 1052) Tbilisi Univ: Tbilisi 152pp,39pp,54pp
• LC-No 75-404082

S 2073 Actualites Sci Indust • F
Actualites Scientifiques et Industrielles PUBL (X 0859) Hermann: Paris

S 2308 Symp Kyoto Univ Res Inst Math Sci (RIMS) • J
Surikaisekikenkyusho Kokyuroku (Kyoto University. Research Institute for Mathematical Sciences (RIMS). Proceedings of Symposia) PUBL (X 2441) Kyoto Univ Res Inst Math Sci: Kyoto

S 2348 Semin Th Nombres Bordeaux • F
Seminaire de Theorie des Nombres. 1972-1973 PUBL Univ Bourdeaux: Valence

S 2350 Zesz Nauk, Prace Mat, Uniw Krakow • PL
Zeszyty Naukowe Uniwersytetu Jagiellonskiego. Prace Matematyczne (Scientific Papers. Jagiellonian University. Reports on Mathematics) [1954-1984] • PUBL (X 2451) Uniw Jagiell Inst Mat: Krakow • ALT PUBL (X 1034) PWN: Warsaw
• ISSN 0083-4386

S 2450 Mem Amer Math Soc, NS • USA
Memoirs of the American Mathematical Society. New Series [1975ff] • PUBL (X 0803) Amer Math Soc: Providence
• CONT OF (S 0167) Mem Amer Math Soc
• ISSN 0065-9266

S 2579 Teor Mnozhestv & Topol (Izhevsk) • SU
Teoriya Mnozhestv i Topologiya (Set Theory and Topology) [1977,1979,1982] • ED: GRYZLOV, A.A. • PUBL (X 4562) Udmurtskij Gos Univ: Izhevsk 114pp,116pp,116pp
• REM Title of Vol.2: Sovremennaya Topologiya i Teoriya Mnozhestv, Vyp.2

S 2582 Semiotika & Inf, Akad Nauk SSSR • SU
Semiotika i Informatika. Gosudarstvennyj Komitet SSSR po Nauke i Tekhnike. Akademiya Nauk SSSR. Vsesoyuznyj Institut Nauchnoj i Tekhnicheskoj Informatsii (Semiotics and Information Science) ED: MIKHAJLOV, A.I. • PUBL (X 2235) VINITI: Moskva

S 2651 Prepr Inst Prikl Mat, Akad Nauk SSSR • SU
Preprint. Akademiya Nauk SSSR. Institut Prikladnoj Matematiki. (Preprint. Academy of Sciences of the USSR. Institute of Applied Mathematics) PUBL Akad Nauk SSSR, Inst Prikl Mat: Moskva

S 2653 Prepr Inst Kib, Akad Nauk Ukr SSR • SU
Preprint. Akademiya Nauk Ukrainskoj SSR. Institut Kibernetiki. (Preprint. Academy of Sciences of the Ukrainian SSR. Institute of Cybernetics) [1971ff] • PUBL (X 2522) Akad Nauk Inst Kibernet: Kiev

S 2738 Issl Prikl Mat (Univ Kazan') • SU
Issledovaniya po Prikladnoj Matematike (Studies in Applied Mathematics) [1973ff] • PUBL (X 3605) Kazan Gos Univ: Kazan'

S 2801 Sbor Nauch Trud (Ped Inst, Moskva) • SU
Sbornik Nauchnykh Trudov (Collected Scientific Papers) PUBL (X 2802) Moskov Ped Inst: Moskva

S 2824 Real Anal Exchange • USA
Real Analysis Exchange [1976ff] • PUBL Western Illinois Univ: Macomb
• ISSN 0147-1937

S 2829 Rostocker Math Kolloq • DDR
Rostocker Mathematisches Kolloquium. PUBL Wilhelm-Pieck-Univ Rostock, Sekt Math: Rostock

S 2848 Topology Proc • USA
Topology Proceedings [1976ff] • ED: KUPERBERG, W. & REED, G.M. & ZENOR, P. • PUBL Auburn University: Auburn, AL
• ISSN 0146-4124 • REM Contains Proceedings of Topology Conferences. Vol 1: 1976 Auburn, AL, USA. Vol 2: 1977 Baton Rouge, LA, USA. Vol 3: 1978 Norman, OK, USA. Vol 4: 1979 Athens, OH, USA. Vol 5: 1980 Birminghan, AL, USA. Vol 7: 1982 Annapolis, MD, USA. Vol 8: 1983 Houston, TX, USA. Vol 9: 1984 Auburn, AL, USA.

S 2850 Tr Ehnerg Inst Moskva • SU
Trudy Moskovskogo Ordena Lenina Ehnergeticheskogo Instituta Tematicheskij Sbornik (Proceedings of the Moscow Institute of Energetics) PUBL Moskovsk Ehnergeticheskij Instituta: Moskva

S 2874 Vopr Kibern, Akad Nauk SSSR • SU
Akademiya Nauk SSSR Nauchnyj Sovet po Kompleksnoj Probleme. "Kibernetika". Voprosy Kibernetiki (Problems of Cybernetics. Academy of Sciences. Scientific Council for Complexity Problems. Cybernetics) PUBL Akad Nauk SSSR Nauch Sovet Komplek Probl Kibern: Moskva

S 2877 Wiss Z Univ Leipzig, Ges-Sprachwiss Reihe • DDR
Wissenschaftliche Zeitschrift der Karl-Marx-Universitaet Leipzig. Gesellschafts- und Sprachwissenschaftliche Reihe [1951ff] • PUBL (X 2373) Karl-Marx-Univ: Leipzig
• ISSN 0043-6879

S 2890 Zesz Nauk, Mat Fiz, Politech Slask (Gliwice) • PL
Zeszyty Naukowe Politechniki Slaskiej. Matematyka, Fizyka (Scientific Papers. Silesian Technical University. Mathematics. Physics) [1961ff] • PUBL Politech Slask: Gliwice
• ISSN 0072-470X

S 2990 Rend Semin Mat Brescia • I
Rendiconti del Seminario Matematico di Brescia PUBL (X 1364) Vita e Pensiero: Milano

S 3105 Adv Math, Suppl Stud • USA
Advances in Mathematics. Supplementary Studies. [1965ff]
• PUBL (X 0801) Academic Pr: New York
• REL PUBL (J 0345) Adv Math

S 3126 Ber, Fak Inf, Univ Karlsruhe • D
Bericht. Institut fuer Informatik II PUBL (X 3159) TH Karlsruhe Fak Informatik: Karlsruhe

S 3180 Inform Ber, Inst Inf, Univ Bonn • D
Informatik Berichte PUBL (X 1512) Univ Bonn Inst Informatik: Bonn

S 3216 Mem Publ Soc Sci Arts Lett Hainaut • B
Memoires et Publications de la Societe des Sciences, des Arts et des Lettres du Hainaut. Recueil Scientifique Publie avec le Concours de la Province de Hainaut PUBL Maison Leon Losseau: Mons

S 3230 Prepr, Inst Math, Pol Acad Sci • PL
Preprint. Institute of Mathematics. Polish Academy of Sciences PUBL (X 2882) Pol Akad Nauk: Wroclaw

S 3252 Stud Leibnitiana, Suppl • D
Studia Leibnitiana. Supplementa [1968ff] • PUBL (X 2717) Steiner: Wiesbaden
• ISSN 0303-5980 • REL PUBL (J 3274) Stud Leibnitiana

S 3301 Lect Notes Math • D
Lecture Notes in Mathematics [1964ff] • PUBL (X 0811) Springer: Heidelberg & New York
• ISSN 0075-8434

S 3302 Lect Notes Comput Sci • D
Lecture Notes in Computer Science [1973ff] • PUBL (X 0811) Springer: Heidelberg & New York
• ISSN 0302-9743

S 3303 Stud Logic Found Math • NL
Studies in Logic and the Foundations of Mathematics [1954ff] • PUBL (X 0809) North Holland: Amsterdam
• ISSN 0049-237X

S 3304 Proc Symp Pure Math • USA
Proceedings of Symposia in Pure Mathematics PUBL (X 0803) Amer Math Soc: Providence

S 3305 Symposia Matematica • I
Symposia Matematica [1969ff] • PUBL (X 3604) INDAM: Roma
• ISSN 0082-0725

S 3307 Synth Libr • NL
Synthese Library. Studies in Epistemology, Logic, Methodology, and Philosophy of Science [1959ff] • PUBL (X 0835) Reidel: Dordrecht
• ISSN 0082-1128

S 3308 Univ Western Ontario Ser in Philos of Sci • NL
University of Western Ontario Series in Philosophy of Science [1972ff] • PUBL (X 0835) Reidel: Dordrecht

S 3309 AFIPS Conference Proc • USA
AFIPS Conference Proceedings PUBL (X 1354) Spartan Books: Sutton

S 3310 Lect Notes Pure Appl Math • USA
Lecture Notes in Pure and Applied Mathematics [1971ff] • PUBL (X 1684) Dekker: New York
• ISSN 0075-8469

S 3311 Boston St Philos Sci • NL
Boston Studies in the Philosophy of Science [1963ff] • PUBL (X 0835) Reidel: Dordrecht
• ISSN 0068-0346

S 3312 Coll Math Soc Janos Bolyai • H
Colloquia Mathematica Societatis Janos Bolyai PUBL (X 0809) North Holland: Amsterdam

S 3313 Contemp Math • USA
Contemporary Mathematics [1980ff] • PUBL (X 0803) Amer Math Soc: Providence
• ISSN 0271-4132

S 3314 Lect Notes Econ & Math Syst • D
Lecture Notes in Economics and Mathematical Systems [1968ff] • PUBL (X 0811) Springer: Heidelberg & New York
• ISSN 0075-8442

S 3358 Ann Discrete Math • NL
Annals of Discrete Mathematics [1977ff] • PUBL (X 0809) North Holland: Amsterdam

S 3382 Sem-ber, Humboldt-Univ Berlin, Sekt Math • DDR
Seminarberichte. Humboldt-Universitaet zu Berlin, Sektion Mathematik PUBL (X 2219) Humboldt-Univ Berlin: Berlin

S 3412 Prace Inst Mat, Politech Wroclaw, Ser Konf • PL
Prace Naukowe Instytutu Matematyki Politechniki Wroclaw. Serija: Konferencje (Scientific Papers of the Institute of Mathematics of Wroclaw Technical University. Series: Conferences) PUBL Politechnika Wroclawska: Wroclaw

S 3414 Prepr, Akad Wiss DDR, Inst Math • DDR
Preprint. Akademie der Wissenschaften der DDR. Institut fuer Mathematik PUBL (X 2655) Akad Wiss DDR Inst Math: Berlin

S 3416 Prepr Ser Dep Math Univ Ljubljana • YU
Preprint Series. University of Ljubljana, Department of Mathematics PUBL Univ Ljubljana, Dept Math: Ljubljana

S 3418 Pribory & Sist Avtomat (Khar'kov) • SU
Pribory i Sistemy Avtomatiki (Geraete und Systeme der Automation) [1962ff] • PUBL (X 2644) Vishcha Shkola: Khar'kov

S 3468 Tr Mat & Mekh (Tartu) • SU
*Matemaatika - ja Mekhaanika-Alaseid Toeid. * Trudy po Matematike i Mekhanike. (Works about Mathematics and Mechanics.)* SER (S 0393) Uch Zap Univ, Tartu • PUBL (X 2463) Tartusk Gos Univ: Tartu

S 3513 Ann Math Stud • USA
Annals of Mathematics Studies [1940ff] • PUBL (X 0857) Princeton Univ Pr: Princeton

S 3521 Mem Soc Math Fr • F
Memoire de la Mathematique de France. Supplement au Bulletin [1927ff] • PUBL (X 0834) Gauthier-Villars: Paris
• REM Since 1981: Nouvelle Serie

S 3677 Fund Scientiae • F
Fundamenta Scientiae. Cahiers du Seminaire sur les Fondements des Sciences PUBL Universite Louis Pasteur: Strasbourg

S 3726 Congressus Numerantium • CDN
Congressus Numerantium [1970ff] • PUBL (X 2420) Utilitas Mathematica Publ: Winnipeg

S 3909 Mat Met Opt & Upravleniya Slozh Sist (Kalinin) • SU
Matematicheskie Metody Optimizatsii i Upravleniya v Slozhnykh Sistemakh (Mathematical Methods of Optimization and Control in Complex Systems) [1981,1982] • ED: ABRAMOV, YU.A. • PUBL (X 1434) Kalinin Gos Univ: Kalinin 180pp,196pp

S 3911 Komb-Algeb Met Prikl Mat (Gor'kij) • SU
Kombinatorno-Algebraicheskie Metody v Prikladnoj Matematike (Combinatorical Algebraic Methods in Applied Mathematics) [1979ff] • ED: MARKOV, A.A. • PUBL Gor'kovskij Gos Univ: Gor'kij 124pp,219pp,165pp,184pp

S 3935 Cah Cent Log (Louvain) • B
Cahiers du Centre de Logique PUBL (X 0879) Inst Sup Philos: Louvain

S 4104 Sbor Rab Mat Razdela Komm Akad • SU
Sbornik Rabot Matematicheskogo Razdela. Kommunisticheskoj Akademiia (Collected Papers on Mathematics. Communist Academy) [1929-1935]

S 4293 Proc Dept Humanities • J
The Proceedings of the Department of Humanities [1951ff]
• PUBL (X 2434) Univ Tokyo Coll Gen Educ: Tokyo

S 4530 Stud Gnieseusia
Studia Gnieseusia

S 4550 Forsch Log & Grundl Exakt Wiss, NS • DDR
Forschungen zur Logik und zur Grundlegung der Exakten Wissenschaften. Neue Serie PUBL -?-: Leipzig

S 4556 Prace Zaklad Aparat Mat, Pol Akad Nauk, Ser A • PL
Prace Zakladu Aparat. Mat., Polsk. Akad. Nauk, Ser A (Publications of the Institute of Instrumentation)

S 4568 Hiroshima Univ Fac Engin, Mem • J
Memoirs of the Faculty of Engineering Hiroshima University

S 4733 Prace Inst Mat, Politech Wroclaw, Ser Stud Mater • PL
Politechnika Wroclawska Instytut Matematyki. Prace Naukowe. Studia i Materialy (Scientific Publications. Institute of Mathematics. Technical University of Wroclaw) [1970ff]
• PUBL Politechnika Wroclawska: Wroclaw
• CONT OF (S 1562) Prace Inst Mat Fis, Politech Wroclaw, Ser Stud Mater

Proceedings

P 0098 Skand Mat Kongr (11);1949 Trondheim • N
[1952] *den 11te Skandinaviske Matematikerkongress (Comptes-Rendus du 11eme Congres des Mathematiciens Scandinaves)* ED: LYCHE, R.T. • PUBL Johan Grund Tanums: Oslo xxxi+300pp
• DAT&PL 1949 Aug; Trondheim, N

P 0454 Math Founds of Comput Sci (4);1975 Marianske Lazne • CS
[1975] *Mathematical Foundations of Computer Science. Proceedings of the 4th Symposium* ED: BECVAR, J. • SER (S 3302) Lect Notes Comput Sci 32 • PUBL (X 0811) Springer: Heidelberg & New York x+476pp
• DAT&PL 1975 Sep; Marianske Lazne, CS • ISBN 3-540-07389-2, LC-No 75-22406

P 0559 Phil Probl in Logic;1968 Irvine • USA
[1970] *Philosophical Problems in Logic: Some Recent Developments. Colloquium on Free Logic, Modal Logic and Related Areas.* ED: LAMBERT, K. • SER (S 3307) Synth Libr • PUBL (X 0835) Reidel: Dordrecht viii+176pp • ALT PUBL (X 2696) Humanities Pr: Atlantic Highlands
• DAT&PL 1968 May; Irvine, CA, USA • LC-No 76-490025

P 0563 Found of Statements & Decisions;1961 Warsaw • PL
[1965] *The Foundation of Statements and Decision. International Colloquium on Methodology of Sciences* ED: AJDUKIEWICZ, K. • PUBL (X 1034) PWN: Warsaw xi+408pp
• DAT&PL 1961 Sep; Warsaw, PL • LC-No 65-76272

P 0572 Inform Processing (2);1962 Muenchen • D
[1963] *Information Processing '62. Proccedings of IFIP Congress* ED: POPPLEWELL, C.M. • PUBL (X 0809) North Holland: Amsterdam xiv+780pp
• DAT&PL 1962 Aug; Muenchen, D • LC-No 76-462349

P 0573 Inform Processing (3);1965 New York • USA
[1965-1966] *Information Processing '65. Proceedings of IFIP Congress* ED: KALENICH, W.A. • PUBL (X 0843) Macmillan: New York & London Vol 1: xv+1-304, Vol 2: viii+305-648 • ALT PUBL (X 1354) Spartan Books: Sutton
• DAT&PL 1965 May; New York, NY, USA • LC-No 65-24118

P 0575 Int Congr Math (II, 7);1954 Amsterdam • NL
[1954-1957] *Proceedings of the International Congress of Mathematicians 1954* ED: GERRETSEN, J.C.H. & GROOT DE, J. • PUBL (X 0809) North Holland: Amsterdam 3 Vols: 582pp,440pp,560pp • ALT PUBL (X 1317) Noordhoff: Groningen 1954-1957 & (X 3602) Kraus: Vaduz 1967
• DAT&PL 1954 Sep; Amsterdam, NL • LC-No 52-1808

P 0576 Raisonn en Math & Sci Exper;1955 Paris • F
[1958] *La Raisonnement en Mathematiques et en Sciences Experimentales.* SER (S 1802) Colloq Int CNRS 70 • PUBL (X 0999) CNRS Inst B Pascal: Paris 140pp
• DAT&PL 1955 Sep; Paris, F • LC-No 63-24106

P 0577 Automatenth & Formale Sprachen;1969 Oberwolfach • D
[1970] *Automatentheorie und Formale Sprachen.* ED: DOERR, J. & HOTZ, G. • SER (S 1956) Tagungsbericht, Oberwolfach 3 • PUBL (X 0876) Bibl Inst: Mannheim 505pp
• DAT&PL 1969 Oct; Oberwolfach, D • LC-No 76-857074

P 0578 Self Organizing Systs;1962 Chicago • USA
[1962] *Self Organizing Systems* ED: YOVITIS, M.C. & JACOBI, G.T. & GOLDSTEIN, G.D. • PUBL (X 1354) Spartan Books: Sutton ix+563pp
• DAT&PL 1962 -?-; Chicago, IL, USA • LC-No 62-20444

P 0580 Int Congr Log, Meth & Phil of Sci (4,Sel Pap);1971 Bucharest • RO
[1973] *Logic, Language and Probability.* ED: BOGDAN, R.J. & NIINILUOTO, I. • SER (S 3307) Synth Libr • PUBL (X 0835) Reidel: Dordrecht x+323pp
• DAT&PL 1971 Aug; Bucharest, RO • ISBN 90-277-0312-4, LC-No 72-95892 • REL PUBL (P 0793) Int Congr Log, Meth & Phil of Sci (4,Proc);1971 Bucharest
• REM A Selection of Papers Contributed to Sections IV, VI and XI of P0793

P 0593 Int Congr Math (II, 6);1950 Cambridge MA • USA
[1952] *Proceedings of the International Congress of Mathematicians* ED: GRAVES, L.M & HILLE, E. & SMITH, P.A. & ZARISKI, O. • PUBL (X 0803) Amer Math Soc: Providence 2 Vols: 769pp,461pp • ALT PUBL (X 3602) Kraus: Vaduz 1967 2 Vols
• DAT&PL 1950 Aug; Cambridge, MA, USA • LC-No 52-1808

P 0594 Inform Processing (4);1968 Edinburgh • GB
[1969] *Information Processing '68. Proceedings of IFIP Congress* ED: MORRELL, A.J.H. • PUBL (X 0809) North Holland: Amsterdam 2 Vols: xxvi+xi+1650pp • ALT PUBL (X 2696) Humanities Pr: Atlantic Highlands
• DAT&PL 1968 Aug; Edinburgh, GB • ISBN 0-7204-2032-6, LC-No 76-462349
• REM Vol 1: Mathematics, Software. Vol 2: Hardware, Applications

P 0603 Intuitionism & Proof Th;1968 Buffalo • USA
[1970] *Intuitionism and Proof Theory* ED: KINO, A. & MYHILL, J. & VESLEY, R.E. • PUBL (X 0809) North Holland: Amsterdam viii+516pp
• DAT&PL 1968 Aug,Buffalo, NY, USA • ISBN 0-7204-2257-4, LC-No 77-97196

P 0604 Scand Logic Symp (2);1970 Oslo • N
[1971] *Proceedings of the 2nd Scandinavian Logic Symposium* ED: FENSTAD, J.E. • SER (S 3303) Stud Logic Found Math 63 • PUBL (X 0809) North Holland: Amsterdam ii+405pp
• DAT&PL 1970 Jun; Oslo, N • ISBN 0-7204-2259-0, LC-No 71-153401

P 0606 Reunion Mat Espanoles (7);1966 Valladolid • E
[1967] *7 Reunion Anual de Matematicos Espanoles. Actas*
PUBL University of Valladolid: Valladolid 154pp
• DAT&PL 1966 Dec; Valladolid, E

P 0607 All-Union Math Conf (3);1956 Moskva • SU
[1959] *Trudy 3'go Vsesoyuznogo Matematicheskogo S'ezda (Proceedings of the 3rd All Union Mathematical Conference)* ED: NIKOL'SKIJ, S.M. & ABRAMOV, A.A. & BOLTYANSKIJ, V.G.
• PUBL (X 0899) Akad Nauk SSSR : Moskva 4 Vols
• DAT&PL 1956 Jun; Moskva, SU

P 0608 Logic Colloq;1966 Hannover • D
[1968] *Contributions to Mathematical Logic. Proceedings of the Logic Colloquium* ED: SCHMIDT, H.A. & SCHUETTE, K. & THIELE, H.-J. • SER (S 3303) Stud Logic Found Math • PUBL (X 0809) North Holland: Amsterdam ix+298pp
• DAT&PL 1966 Aug; Hannover, D • LC-No 68-24434

P 0610 Tarski Symp;1971 Berkeley • USA
[1974] *Proceedings of the Tarski Symposium. An International Symposium held to Honor Alfred Tarski on the Occasion of His 70th Birthday* ED: HENKIN, L. & ADDISON, J. & CHANG, C.C. & CRAIG, W. & SCOTT, D.S. & VAUGHT, R. • SER (S 3304) Proc Symp Pure Math 25 • PUBL (X 0803) Amer Math Soc: Providence ix+498pp
• DAT&PL 1971 Jun; Berkeley, CA, USA • ISBN 0-8218-1425-7, LC-No 74-8666
• REM Corrected Reprint 1979; xx+498pp

P 0612 Int Congr Log, Meth & Phil of Sci (1,Proc);1960 Stanford • USA
[1962] *Proceedings of the 1st International Congress for Logic, Methodology and Philosophy of Science* ED: NAGEL, E. & SUPPES, P. & TARSKI, A. • PUBL (X 1355) Stanford Univ Pr: Stanford ix+661pp
• DAT&PL 1960 Aug; Stanford, CA, USA • LC-No 62-9620
• TRANSL IN [1965] (P 2251) Mat Log & Primen;1960 Stanford

P 0613 Rec Fct Th;1961 New York • USA
[1962] *Recursive Function Theory* ED: DEKKER, J.C.E. • SER (S 3304) Proc Symp Pure Math 5 • PUBL (X 0803) Amer Math Soc: Providence vii+247pp
• DAT&PL 1961 Apr; New York, NY, USA • ISBN 0-8218-1405-2, LC-No 50-1183
• REM 2nd ed. 1970

P 0614 Th Models;1963 Berkeley • USA
[1965] *The Theory of Models.* ED: ADDISON, J.W. & HENKIN, L. & TARSKI, A. • SER (S 3303) Stud Logic Found Math
• PUBL (X 0809) North Holland: Amsterdam xv+494pp
• DAT&PL 1963 Jun; Berkeley, CA, USA • LC-No 66-7051

P 0616 Switch Circ Th & Log Design (3);1962 Chicago • USA
[1962] *Proceedings of the 3rd Annual Symposium on Switching Circuit Theory and Logical Design* ED: SEMON, W. • PUBL American Institute of Electrical Engineers (AIEE): New York;199pp
• DAT&PL 1962 Oct; Chicago, IL, USA

P 0623 Int Congr Log, Meth & Phil of Sci (2,Proc);1964 Jerusalem • IL
[1965] *Proceedings of the 2nd International Congress for Logic, Methodology and Philosophy of Science* ED: BAR-HILLEL, Y.
• SER (S 3303) Stud Logic Found Math • PUBL (X 0809) North Holland: Amsterdam viii+440pp
• DAT&PL 1964 Aug; Jerusalem, IL • ISBN 0-7204-2235-3, LC-No 66-7008
• REM 2nd ed. 1972

P 0624 Switch Circ Th & Log Design (1,2);1960 Chicago;1961 Detroit • USA
[1961] *Switching Circuit Theory and Logical Design. Proceedings of the 2nd Annual Symposium and Papers from the 1st Annual Symposium* ED: LEDLEY, R.S. • PUBL American Institute of Electrical Engineers: New York xi+341pp
• DAT&PL 1960 Oct; Chicago, IL, USA, 1961 Oct; Detroit, MI, USA

P 0625 Symp Autom Demonst;1968 Versailles • F
[1970] *Symposium on Automatic Demonstration.* ED: LAUDET, M. & LACOMBE, D. & NOLIN, L. & SCHUETZENBERGER, M.
• SER (S 3301) Lect Notes Math 125 • PUBL (X 0811) Springer: Heidelberg & New York v+310pp
• DAT&PL 1968 Dec; Versailles, F • ISBN 3-540-04914-2, LC-No 79-117526

P 0626 Ber Math-Tagung Berlin;1953 Berlin • DDR
[1953] *Bericht ueber die Mathematiker-Tagung* ED: GRELL, H. & SCHMID, H.L. • PUBL (X 0806) Dt Verlag Wiss: Berlin viii+302pp
• DAT&PL 1953 Jan; Berlin, D • LC-No 78-235228

P 0627 Int Congr Log, Meth & Phil of Sci (3,Proc);1967 Amsterdam • NL
[1968] *Proceedings of the 3rd International Congress for Logic, Methodology and Philosophy of Science* ED: ROOTSELAAR VAN, B. & STAAL, J.F. • SER (S 3303) Stud Logic Found Math
• PUBL (X 0809) North Holland: Amsterdam xii+553pp
• DAT&PL 1967 Aug; Amsterdam, NL • ISBN 0-444-85423-1, LC-No 68-29768

P 0632 Congr Int Phil des Sci;1935 Paris • F
[1936] *Actes du Congres International de Philosophie Scientifique* SER (S 2073) Actualites Sci Indust 388-395
• PUBL (X 0859) Hermann: Paris 8 Vol
• DAT&PL 1935 Sep; Paris, F
• REM Vol. II: Unite de la Science. Vol. III: Language et Pseudo-Problemes. Vol. IV: Induction et Probabilite. Vol. VI: Philosophie des Mathematiques. Vol. VII: Logique. Vol. VIII: Histoire de la Logique et de la Philosophie Scientifique.

P 0633 Infinitist Meth;1959 Warsaw • PL
[1961] *Infinitistic Methods. Proceedings of the Symposium on Foundations of Mathematics* PUBL (X 1034) PWN: Warsaw 362pp • ALT PUBL (X 0869) Pergamon Pr: Oxford 1961
• DAT&PL 1959 Sep; Warsaw, PL • LC-No 61-11351

P 0638 Logic Colloq;1969 Manchester • GB
[1971] *Logic Colloquium '69* ED: GANDY, R.O. & YATES, C.M.E. • SER (S 3303) Stud Logic Found Math 61 • PUBL (X 0809) North Holland: Amsterdam xiv+457pp
• DAT&PL 1969 Aug; Manchester, GB • ISBN 0-7204-2261-2, LC-No 71-146188

P 0639 Congr Nat des Sci (2);1935 Bruxelles • B
[1935] *2me Congres National des Sciences, Comptes Rendus*
• DAT&PL 1935 Jun; Bruxelles, B • LC-No 37-39347

P 0641 ACM Symp Th of Comput (2);1970 Northhampton • USA
[1970] *2nd Annual ACM Symposium on the Theory of Computation (Association for Computing Machinery)* PUBL (X 2205) ACM: New York
• DAT&PL 1970 May; Northhampton, MA, USA • LC-No 82-642181

P 0644 Meth Form en Axiom;1950 Paris • F
[1953] *Les Methodes Formelles en Axiomatiques* ED: DESTOUCHES, J.-L. & DESTOUCHES-FEVRIER, P. & JOFFRE, D. • SER (S 1802) Colloq Int CNRS 36 • PUBL (X 0999) CNRS Inst B Pascal: Paris 197pp
• DAT&PL 1950 Dec; Paris, F

P 0645 Int Congr Philos (11);1953 Bruxelles • B
[1953] *Actes du 11eme Congres International de Philosophie. * Proceedings of the 11th International Congress of Philosophy* PUBL (X 1313) Nauwelaerts: Louvain • ALT PUBL (X 0809) North Holland: Amsterdam 1953;14 Vols
• DAT&PL 1953 Aug; Bruxelles, B

P 0646 Appl Sci de Log Math;1952 Paris • F
[1954] *Applications Scientifiques de la Logique Mathematique. Actes du 2eme Colloque International de Logique Mathematique* ED: DESTOUCHES, J.-L. & DESTOUCHES-FEVRIER, P. • SER Collection de Logique Mathematique, Serie A 5 • PUBL (X 0834) Gauthier-Villars: Paris 176pp • ALT PUBL (X 1313) Nauwelaerts: Louvain 1954
• DAT&PL 1952 Aug; Paris, F

P 0649 Appl Model Th to Algeb, Anal & Probab;1967 Pasadena • USA
[1969] *Applications of Model Theory to Algebra, Analysis and Probability* ED: LUXEMBURG, W.A.J. • PUBL (X 0818) Holt Rinehart & Winston: New York vii+307pp
• DAT&PL 1967 May; Pasadena, CA, USA • LC-No 69-11203

P 0651 Axiomatic Method;1957 Berkeley • USA
[1959] *The Axiomatic Method. With Special Reference to Geometry and Physics* ED: HENKIN, L. & SUPPES, P. & TARSKI, A. • SER (S 3303) Stud Logic Found Math • PUBL (X 0809) North Holland: Amsterdam xi+488pp
• DAT&PL 1957 Dec; Berkeley, CA, USA • LC-No 58-63025

P 0652 Entretiens Zuerich Fond & Method Sci Math;1938 Zuerich • CH
[1941] *Les Entretiens de Zuerich sur les Fondements et la Methode des Sciences Mathematiques: Exposes et Discussions* ED: GONSETH, F. • PUBL (X 2220) Leemen: Zuerich 209pp
• DAT&PL 1938 Dec; Zuerich, CH • LC-No 42-650

P 0653 Int Congr Math (II, 4);1932 Zuerich • CH
[1932] *Verhandlungen des Internationalen Mathematiker-Kongresses Zuerich* ED: SAX, W. • PUBL (X 1268) Orell Fuessli: Zuerich 2 Vols: 335pp,365pp
• DAT&PL 1932 Sep; Zuerich, CH • LC-No 52-1808

P 0660 Int Congr Math (II, 8);1958 Edinburgh • GB
[1960] *Proceedings of the International Congress of Mathematicians* ED: TODD, J.A. • PUBL (X 0805) Cambridge Univ Pr: Cambridge, GB lxiv+573pp
• DAT&PL 1958 Aug; Edinburgh, GB • LC-No 52-1808

P 0669 Conv Teor Modelli & Geom;1969/70 Roma • I
[1971] *Convegni Teoria dei Modelli & Geometria* SER (S 3305) Symposia Matematica 5 • PUBL (X 3604) INDAM: Roma 475pp • ALT PUBL (X 0801) Academic Pr: New York
• DAT&PL 1969 Nov; Rome, I, 1970 Apr; Rome, I

P 0674 Symp Math Th of Automata;1962 New York • USA
[1963] *Proceedings of the Symposium on Mathematical Theory of Automata* ED: FOX, J. • SER Microwave Research Institute Symposia Series 12 • PUBL (X 2039) Poly Inst New York: Brooklyn xix+640pp • ALT PUBL (X 0827) Wiley & Sons: New York
• DAT&PL 1962 Apr; New York, NY, USA • LC-No 63-11286

P 0677 Int Congr Math (II, 9,Proc);1962 Djursholm • S
[1963] *Proceedings of the International Congress of Mathematicians* ED: STENSTROEM, V. • PUBL (X 1163) Almqvist & Wiksell: Stockholm 1+595pp
• DAT&PL 1962 Aug; Djursholm, S • 52-1808

P 0678 Word Probl: Decis & Burnside Probl in Group Th;1969 Irvine • USA
[1973] *Word Problems. Decision Problems and the Burnside Problem in Group Theory* ED: BOONE, W.W. & CANNONITO, F.B. & LYNDON, R.C. • SER (S 3303) Stud Logic Found Math 71 • PUBL (X 0809) North Holland: Amsterdam xii+646pp
• DAT&PL 1969 Sep; Irvine, CA, USA • ISBN 0-7204-2271-X, LC-No 70-146190

P 0682 Int Congr Philos (10);1948 Amsterdam • NL
[1949] *Library of the 10th International Congress of Philosophy* ED: BETH, E.W. & POS, H.J. & HOLLAK, H.J.A. • PUBL (X 0809) North Holland: Amsterdam Vol 1, L.J.Veen: Amsterdam Vol 2
• DAT&PL 1948 Aug; Amsterdam, NL • LC-No 50-35721

P 0683 Log & Founds of Sci (Beth);1964 Paris • F
[1967] *Logic and Foundations of Science. E.W.Beth Memorial Colloquium.* ED: DESTOUCHES, J.-L. • PUBL (X 0835) Reidel: Dordrecht viii+140pp
• DAT&PL 1964 May; Paris, F • LC-No 68-107520

P 0688 Logic Colloq;1963 Oxford • GB
[1965] *Formal Systems and Recursive Functions. Proceedings of the 8th Logic Colloquium* ED: CROSSLEY, J.N. & DUMMETT, M.A.E. • SER (S 3303) Stud Logic Found Math • PUBL (X 0809) North Holland: Amsterdam 320pp
• DAT&PL 1963 Jul; Oxford, GB • LC-No 66-2289

P 0689 Int Congr Math (4);1908 Roma • I
[1909] *Atti del 4 Congresso Internazionale dei Matematici* ED: CASTELNUOVO, G. • PUBL (X 2121) Accad Naz Linc: Roma 3 Vols: iv+216pp,319pp,588pp • ALT PUBL (X 3602) Kraus: Vaduz 1967
• DAT&PL 1908 Apr; Roma, I

P 0690 Comput Prob in Abstr Algeb;1967 Oxford • GB
[1970] *Computational Problems in Abstract Algebra.* ED: LEECH, J. • PUBL (X 0869) Pergamon Pr: Oxford x+402pp
• DAT&PL 1967 Aug; Oxford, GB • ISBN 0-08-012975-7, LC-No 75-84072

P 0691 Sets, Models & Recursion Th;1965 Leicester • GB
[1967] *Sets, Models and Recursion Theory. Proceedings of the Summer School in Mathematical Logic and 10th Logic Colloquium* ED: CROSSLEY, J.N. • SER (S 3303) Stud Logic Found Math • PUBL (X 0809) North Holland: Amsterdam v+331pp • ALT PUBL (X 0838) Amer Elsevier: New York 1967
• DAT&PL 1965 Aug; Leicester, GB • ISBN 0-7204-2242-6, ISBN 0-444-10696-0, LC-No 67-21973
• REM 2nd ed. 1974

P 0693 Axiomatic Set Th;1967 Los Angeles • USA
[1971-1974] *Axiomatic Set Theory* ED: SCOTT, D. (V 1) & JECH, T.J. (V 2) • SER (S 3304) Proc Symp Pure Math 13 • PUBL (X 0803) Amer Math Soc: Providence 2 Vols: vi+474pp,viii+222pp
• DAT&PL 1967 Jul; Los Angeles, CA, USA • ISBN 0-8218-0245-3 (V1), ISBN 0-8218-0246-1 (V2), LC-No 78-125172

P 0695 Rect Devel in Inform & Decis Processes;1961 West Lafayette • USA
[1962] *Recent Developments in Information and Decision Processes* ED: MACHOL, R. & GRAY, P. • PUBL (X 0843) Macmillan : New York & London x+197pp
• DAT&PL 1961 Apr; West Lafayette, IN, USA • LC-No 62-18380

P 0696 Inform Processing (1);1959 Paris • F
[1960] *Information Processing '59 (Unesco)* PUBL Unesco: Paris 520pp • ALT PUBL (X 0814) Oldenbourg: Muenchen 1960 & Butterworth: London 1960
• DAT&PL 1959 Jun; Paris, F • LC-No 76-462349

P 0698 Harvard Symp Digit Comput & Appl;1961 Cambridge MA • USA
[1962] *Proceedings of a Harvard Symposium on Digital Computers and Their Applications* SER Annals of the Computation Laboratory of Harvard University 31 • PUBL (X 0858) Harvard Univ Pr: Cambridge xiv+332pp • ALT PUBL (X 0894) Oxford Univ Pr: Oxford
• DAT&PL 1961 Apr; Cambridge, MA, USA • LC-No 62-19220

P 0700 ACM Proc Conf;1952 Pittsburgh • USA
[1952] *Proceedings of the Association for Computing Machinery* PUBL (X 2205) ACM: New York 305pp
• DAT&PL 1952 May; Pittsburgh, PA, USA • LC-No 53-3390

P 0710 Russell Mem Logic Conf;1971 Uldum • DK
[1973] *Proceedings of the Bertrand Russell Memorial Logic Conference* ED: BELL, J. & COLE, J. & PRIEST, G. & SLOMSON, A. • PUBL (X 2504) Russell Mem Conf: Leeds vi+404pp
• DAT&PL 1971 Aug; Uldum, DK • LC-No 74-193761
• REM Copies Available from A.Slomson, School of Math, Univ of Leeds, GB

P 0711 Concept & Role of Model in Math & Sci;1960 Utrecht • NL
[1961] *The Concept and the Role of the Model in Mathematics and Natural and Social Sciences* ED: FREUDENTHAL, H. • SER (S 3307) Synth Libr • PUBL (X 0835) Reidel: Dordrecht 194pp
• DAT&PL 1960 Jan; Utrecht, NL • LC-No 63-1436

P 0719 Exper Arith, High Speed Comput & Math;1962 Chicago • USA
[1963] *Experimental Arithmetic, High Speed Computing and Mathematics* ED: METROPOLIS, N. & TAUB, A.H. & TODD, J. & TOMPKINS, C.B. • SER (S 3304) Proc Symp Pure Math 15 • PUBL (X 0803) Amer Math Soc: Providence ix+396pp
• DAT&PL 1962 Apr; Chicago & Atlantic City, IL, USA
• LC-No 63-17582

P 0724 Skand Mat Kongr (15);1968 Oslo • N
[1970] *Proceedings of the 15th Scandinavian Congress* ED: AUBERT, K.E. & LJUNGGREN, W. • SER (S 3301) Lect Notes Math 118 • PUBL (X 0811) Springer: Heidelberg & New York iv+162pp
• DAT&PL 1968 Aug; Oslo, N • ISBN 3-540-04907-X, LC-No 70-112305

P 0725 Jorn Mat Luso-Espanol (1);1972 Lisbon • P
[1973] *Actas de las 1as Jornadas Matematicas Luso-Espanoles* PUBL (X 2091) Consejo Sup Invest Ci: Madrid 485pp • ALT PUBL (X 1407) Inst Mat J.Juan: Madrid
• DAT&PL 1972 Apr; Lisbon, P • ISBN 84-00-03948-3

P 0732 Contrib to Non-Standard Anal;1970 Oberwolfach • D
[1972] *Contributions to Non-Standard Analysis* ED: LUXEMBURG, W.A.J. & ROBINSON, A. • SER (S 3303) Stud Logic Found Math 69 • PUBL (X 0809) North Holland: Amsterdam vii+289pp
• DAT&PL 1970 Jul; Oberwolfach, D • LC-No 76-183275

P 0737 Math Aspects Comput Sci;1966 New York • USA
[1967] *Proceedings of Symposia in Applied Mathematics. Vol 19: Mathematical Aspects of Computer Science* ED: SCHWARTZ, J.T. • PUBL (X 0803) Amer Math Soc: Providence v+224pp
• DAT&PL 1966, Apr; New York, NY, USA • LC-No 67-16554

P 0741 Int Congr Math (II, 3);1928 Bologna • I
[1929-1932] *Atti del Congresso Internazionale dei Matematici* PUBL (X 1375) Zanichelli: Bologna 6 Vols: 338pp,365pp,472pp,429pp,494pp,554pp
• DAT&PL 1928 Sep; Bologna, I • LC-No 52-1808

P 0743 Int Congr Math (II,11,Proc);1970 Nice • F
[1971] *Actes du Congres International de Mathematiciens 1970* ED: BERGER, M. & DIEUDONNE, J. & LERAY, J. & LIONS, J.-L. & MALLIAVIN, M.P. & SERRE, J.-P. • PUBL (X 0834) Gauthier-Villars: Paris 3 Vols: xxxiii+532pp,959pp,iii+371pp
• DAT&PL 1970 Sep; Nice, F • REL PUBL (P 1158) Int Congr Math (II,11,Comm Ind);1970 Nice
• REM Vol 1: Documents.Medailles Fields.Conferences Generales.Logique.Algebre. Vol 2: Geometrie et Topologie. Analyse. Vol 3: Mathematiques Appliquees.Historie et Enseignement.

P 0756 Congr Int Phil (9);1937 Paris • F
[1937] *Travaux du IXe Congres International de Philosophie: Congres Descartes* ED: BAYER, R. • SER (S 2073) Actualites Sci Indust 530-541 • PUBL (X 0859) Hermann: Paris 12 Vols
• DAT&PL 1937 -?-; Paris, F

P 0762 Inform Th, Stat Decis Fcts & Random Proc (6);1971 Prague • CS
[1973] *Transactions of the 6th Prague Conference on Information Theory, Statistical Decision Functions & Random Processes* PUBL (X 1226) Academia: Prague 924pp • ALT PUBL (X 1553) Dokumentation Saur: Muenchen
• DAT&PL 1971 Sep; Prague, CS
• REM Dedicated to the Memory of A. Spacek (1911-1961)

P 0775 Logic Colloq;1973 Bristol • GB
[1975] *Logic Colloquium '73* ED: ROSE, H.E. & SHEPHERDSON, J.C. • SER (S 3303) Stud Logic Found Math 80 • PUBL (X 0809) North Holland: Amsterdam viii+513pp • ALT PUBL (X 0838) Amer Elsevier: New York
• DAT&PL 1973 Jul; Bristol, GB • ISBN 0-444-10642-1, LC-No 74-79302

P 0777 Grundl Geom & Algeb Meth;1973 Potsdam • DDR
[1974] *Grundlagen der Geometrie und Algebraischen Methoden. Internationales Kolloquium der Paedagogischen Hochschule "Karl Liebknecht"* ED: KLOTZEK, B. & QUAISSER, E. • SER Potsdam Forschungen, Reihe B 3 • PUBL Paed. Hochschule "Karl Liebknecht": Potsdam 208pp
• DAT&PL 1973 Aug; Potsdam, DDR • LC-No 75-405437

P 0783 Truth, Syntax & Modal;1970 Philadelphia • USA
[1973] *Truth, Syntax and Modality: Proceedings of the Temple University Conference on Alternative Semantics* ED: LEBLANC, H. • SER (S 3303) Stud Logic Found Math 68 • PUBL (X 0809) North Holland: Amsterdam vii+317pp
• DAT&PL 1970 Dec; Philadelphia, PA, USA • ISBN 0-7204-2269-8, LC-No 72-79730

P 0784 GI Jahrestag (5);1975 Dortmund • D
[1975] *GI - 5. Jahrestagung (Gesellschaft fuer Informatik)* ED: MUEHLBACHER, J. • SER (S 3302) Lect Notes Comput Sci 34
• PUBL (X 0811) Springer: Heidelberg & New York x+755pp
• DAT&PL 1975 Oct; Dortmund, D • ISBN 3-540-07410-4

P 0785 Scand Logic Symp (1);1968 Aabo • S
[1970] *Proceedings of the 1st Scandinavian Logic Symposium* SER Filosofiska Studier 8 • PUBL (X 0882) Univ Filos Foeren: Uppsala 171pp
• DAT&PL 1968 Sep; Aabo, S • LC-No 72-186670

P 0788 Skand Mat Kongr (12);1953 Lund • S
[1954] *12te Skandinaviska Matematikerkongressen (Comptes-Rendus du 12eme Congres des Mathematiciens Scandinaves)* PUBL Hakan Oh Issons Boktryckeri: Lund xvi+337pp
• DAT&PL 1953 Aug; Lund, S • LC-No 55-58514

P 0793 Int Congr Log, Meth & Phil of Sci (4,Proc);1971 Bucharest • RO
[1973] *Proceedings of the 4th International Congress for Logic, Methodology and Philosophy of Science* ED: SUPPES, P. & HENKIN, L. & MOISIL, G.C. & JOJA, A. • SER (S 3303) Stud Logic Found Math 74 • PUBL (X 0809) North Holland: Amsterdam x+981pp • ALT PUBL (X 0838) Amer Elsevier: New York & (X 1034) PWN: Warsaw
• DAT&PL 1971 Aug; Bucharest, RO • ISBN 0-444-10491-7, LC-No 72-88505 • REL PUBL (P 0580) Int Congr Log, Meth & Phil of Sci (4,Sel Pap);1971 Bucharest

P 0796 Congr Math Pays Slaves (1);1929 Warsaw • PL
[1930] *Sprawozdanie z 1 Kongresu Matematykow Krajow Slowianskich (Comptes Rendus du 1. Congres des Mathematiciens des Pays Slaves)* ED: LEJA, F. • PUBL (X 1764) Ksiaznica Atlas: Warsaw iii+395pp
• DAT&PL 1929 Sep; Warsaw, PL • LC-No 32-17553

P 0797 Fonds des Math, Machines Math & Appl;1962 Tihany • H
[1965] *Colloque sur les Fondements des Mathematiques, les Machines Mathematiques, et leurs Applications* ED: KALMAR, L. • SER Collection de Logique Mathematique, Serie A 19
• PUBL (X 0928) Akad Kiado: Budapest 320pp • ALT PUBL (X 0834) Gauthier-Villars: Paris & (X 1313) Nauwelaerts: Louvain
• DAT&PL 1962 Sep; Tihany, H

P 1072 Math Log & Founds of Set Th;1968 Jerusalem • IL
[1970] *Mathematical Logic and the Foundations of Set Theory* ED: BAR-HILLEL, Y. • SER (S 3303) Stud Logic Found Math
• PUBL (X 0809) North Holland: Amsterdam 145pp
• DAT&PL 1968 Nov; Jerusalem, IL • ISBN 0-7204-2255-8, LC-No 73-97195

P 1075 Logic Colloq;1976 Oxford • GB
[1977] *Logic Colloquium 76* ED: GANDY, R.O. & HYLAND, J.M.E. • SER (S 3303) Stud Logic Found Math 87 • PUBL (X 0809) North Holland: Amsterdam x+612pp • ALT PUBL (X 0838) Amer Elsevier: New York
• DAT&PL 1976 Jul; Oxford, GB • ISBN 0-7204-0691-9, LC-No 77-8943

P 1076 Latin Amer Symp Math Log (3);1976 Campinas • BR
[1977] *Non-Classical Logic, Model Theory and Computability. 3rd Latin American Symposium on Mathematical Logic* ED: ARRUDA, A.I. & COSTA DA, N.C.A & CHUAQUI, R. • SER (S 3303) Stud Logic Found Math 89 • PUBL (X 0809) North Holland: Amsterdam xviii+307pp • ALT PUBL (X 0838) Amer Elsevier: New York
• DAT&PL 1976 Jul; Campinas, BR • ISBN 0-7204-0752-4, LC-No 77-7366

P 1083 Victoria Symp Nonstand Anal;1972 Victoria • AUS
[1974] *Victoria Symposium on Non-Standard Analysis* ED: HURD, A. & LOEB, P. • SER (S 3301) Lect Notes Math 369
• PUBL (X 0811) Springer: Heidelberg & New York xviii+339pp
• DAT&PL 1972 May; Victoria, VIC, AUS • ISBN 3-540-06656-X, LC-No 73-22552

P 1086 Skand Mat Kongr (5);1922 Helsinki • SF
[1923] *Wissenschaftliche Vortraege gehalten auf dem 5ten Kongress der Skandinavischen Mathematiker* PUBL Akademische Buchhandlung: Helsinki 315pp
• DAT&PL 1922 Jul; Helsinki, SF

P 1091 Int Congr Math (3);1904 Heidelberg • D
[1905] *Verhandlungen des 3ten Internationalen Mathematikerkongresses* ED: KRAZER, A. • PUBL (X 1079) Teubner: Leipzig x+755pp
• DAT&PL 1904 Aug; Heidelberg, D

P 1129 Princeton Conf Inform Sci & Syst (3);1969 Princeton • USA
[1969] *Proceedings of the 3rd Annual Conference on Information Science and Systems* ED: THOMAS, J.B. & VALKENBURG VAN, M.E. & WEINER, P. • PUBL (X 2188) Princeton Univ Dept Elect Eng & Comp Sci: Princeton xiii+550pp
• DAT&PL 1969 Mar; Princeton, NJ, USA

P 1158 Int Congr Math (II,11,Comm Ind);1970 Nice • F
[1970] *Congres International des Mathematiciens 1970. Les 265 Communications Individuels* PUBL (X 0834) Gauthier-Villars: Paris vii+290pp
• DAT&PL 1970 Sep; Nice, F • LC-No 72-374601 • REL PUBL (P 0743) Int Congr Math (II,11,Proc);1970 Nice

P 1178 West Joint Comput Conf;1957 Los Angeles • USA
[1957] *Proceedings of the Western Joint Computer Conference*
• DAT&PL 1957 -?-; Los Angeles, CA, USA

P 1188 AFIPS Fall Jt Computer Conf (26);1964 San Francisco • USA
[1964] *AFIPS Conference Proceedings: 1964: Fall Joint Computer Conference* SER (S 3309) AFIPS Conference Proc 26 • PUBL (X 1354) Spartan Books : Sutton v+745pp • ALT PUBL (X 2508) Cleaver-Hume Pr: London
• DAT&PL 1964 Oct; San Francisco, CA, USA • LC-No 55-44701

P 1192 Symb Lang in Data Processing;1962 Roma • I
[1962] *Symbolic Languages in Data Processing* PUBL (X 0836) Gordon & Breach: New York xii+849pp
• DAT&PL 1962 Mar; Rome, I • LC-No 62-22085

P 1197 AFIPS Spring Jt Computer Conf (21);1962 San Francisco • USA
[1962] *AFIPS Conference Proceedings: 1962: Spring Joint Computer Conference* SER (S 3309) AFIPS Conference Proc 21 • PUBL (X 1354) Spartan Books : Sutton xv+392pp • ALT PUBL (X 0874) Nat Pr Books : Palo Alto
• DAT&PL 1962 May; San Francisco, CA, USA • LC-No 55-44701

P 1384 ACM Nat Conf (22);1967 • USA
[1967] *Proceedings of the 22nd National Conference ACM* PUBL (X 2205) ACM: New York
• DAT&PL 1967 -?-; -?- • LC-No 64-25615

P 1385 Int Symp Multi-Val Log (4);1974 Morgantown • USA
[1974] *Proceedings of the 1974 International Symposium on Multiple-Valued Logic.* PUBL (X 2179) IEEE: New York iv+551pp
• DAT&PL 1974 May; Morgantown, WV, USA • LC-No 79-641110

P 1390 Syst & Comput Sci;1965 London ON • CDN
[1967] *Systems and Computer Science* ED: HART, J.F. & TAKASU, S. • PUBL (X 1015) Univ Toronto Pr: Toronto x+249pp
• DAT&PL 1965 Sep; London, ON, CDN • Lc-No 68-114245

P 1401 Math Founds of Comput Sci (5);1976 Gdansk • PL
[1976] *Mathematical Foundations of Computer Science. Proceedings of the 5th Symposium* ED: MAZURKIEWICZ, A.
• SER (S 3302) Lect Notes Comput Sci 45 • PUBL (X 0811) Springer: Heidelberg & New York xi+606pp
• DAT&PL 1976 Sep; Gdansk, PL • ISBN 3-540-07854-1, LC-No 76-25494

P 1412 Union Mat Argentina Jorn Cientif (10);1957 Bahia Blanca • RA
[1957] *Actas de las X Jornadas: Union Matematicas Argentina* PUBL Instituto de Matematicas, Universidad Nacional del Sur: Bahia Blanca;73pp
• DAT&PL 1957 Oct; Bahia Blanca, RA

P 1440 ⊢ ISILC Proof Th Symp (Schuette);1974 Kiel • D
[1975] ⊢ *ISILC Proof Theory Symposium. Dedicated to Kurt Schuette on the Occasion of His 65th Birthday. Proceedings of the International Summer Institute and Logic Colloquium* ED: DILLER, J. & MUELLER, GERT H. • SER (S 3301) Lect Notes Math 500 • PUBL (X 0811) Springer: Heidelberg & New York viii+383pp
• DAT&PL 1974 Jul; Kiel, D • ISBN 3-540-07533-X, LC-No 75-40482 • REL PUBL (P 1442) ⊢ ISILC Logic Conf;1974 Kiel
• REM This Volume Contains Only the Proof Theory Part of the Conference.

P 1442 ⊢ ISILC Logic Conf;1974 Kiel • D
[1975] ⊢ *ISILC Logic Conference. Proceedings of the International Summer Institute and Logic Colloquium* ED: MUELLER, GERT H. & OBERSCHELP, A. & POTTHOFF, K. • SER (S 3301) Lect Notes Math 499 • PUBL (X 0811) Springer: Heidelberg & New York iv+651pp
• DAT&PL 1974 Jul; Kiel, D • ISBN 3-540-07534-8, LC-No 75-40431 • REL PUBL (P 1440) ⊢ ISILC Proof Th Symp (Schuette);1974 Kiel

P 1448 Math Founds of Comput Sci (2);1973 Strbske Pleso • CS
[1973] *Mathematical Foundations of Computer Science. Proceedings of the 2nd Symposium* ED: HAVEL, I.M. • PUBL (X 1773) Vydat Slov Akad: Bratislava 338pp
• DAT&PL 1973 Sep; Strbske Pleso, CS

P 1464 ACM Symp Th of Comput (6);1974 Seattle • USA
[1974] *Proceedings of 6th Annual ACM Symposium on Theory of Computing (Association for Computing Machinery)* PUBL (X 2205) ACM: New York iv+347pp
• DAT&PL 1974 Apr; Seattle, WA, USA • LC-No 82-642181

P 1476 Set Th & Hierarch Th (2) (Mostowski);1975 Bierutowice • PL
[1976] *Set Theory and Hierarchy Theory. A Memorial Tribute to Andrzej Mostowski. Proceedings of the 2nd Conference on Set Theory and Hierarchy Theory* ED: MAREK, W. & SREBRNY, M. & ZARACH, A. • SER (S 3301) Lect Notes Math 537
• PUBL (X 0811) Springer: Heidelberg & New York xiii+345pp
• DAT&PL 1975 Sep; Bierutowice, PL • ISBN 3-540-07856-8, LC-No 76-26536

P 1481 Probab Meth in Diff Equat;1974 Victoria • CDN
[1975] *Probabilistic Methods in Differential Equations* ED: PINSKY, M.A. • SER (S 3301) Lect Notes Math 451 • PUBL (X 0811) Springer: Heidelberg & New York vi+162pp
• DAT&PL 1974 Aug; Victoria, BC, CDN • ISBN 3-540-07153-9, LC-No 75-12982

P 1484 Int Congr Math (2);1900 Paris • F
[1902] *Comptes Rendus du 2eme Congres International des Mathematiciens. Proces Verbaux et Communications* ED: DUPORCQ, E. • PUBL (X 0834) Gauthier-Villars: Paris 455pp
• DAT&PL 1900 Aug; Paris, F

P 1513 IEEE Symp Founds of Comput Sci (16);1975 Berkeley • USA
[1975] *16th Annual IEEE Symposium on Foundations of Computer Science* PUBL (X 2179) IEEE: New York 193pp
• DAT&PL 1975 Oct; Berkeley, CA, USA • LC-No 80-646634

P 1556 Exact Philos: Probl, Tools & Goals;1971 Montreal • CDN
[1973] *Exact Philosophy: Problems, Tools, and Goals* ED: BUNGE, M. • SER (S 3307) Synth Libr, (J 0122) J Philos Logic 1/3-4 • PUBL (X 0835) Reidel: Dordrecht x+214pp
• DAT&PL 1971 Nov; Montreal, Que, CDN • ISBN 90-277-0251-9, LC-No 72-77872

P 1560 Tr Mezhdurn Semin Priklad Aspekt Teor Avtom;1971 Varna • BG
[1971] *Trudy Mezhdurnarodnogo Seminara po Prikladnym Aspektam Teorij Avtomatov (Proceedings of the International Seminar on Applied Aspects of the Automata Theory.)* PUBL (X 2237) Publ Bulg Acad Sci: Sofia 252pp
• DAT&PL 1971 May; Varna, BG

P 1589 Math Interpr of Formal Systs;1954 Amsterdam • NL
[1955] *Mathematical Interpretations of Formal Systems* SER (S 3303) Stud Logic Found Math 10 • PUBL (X 0809) North Holland: Amsterdam viii+113pp
• DAT&PL 1954 Sep; Amsterdam, NL • LC-No 56-3127
• REM 2nd ed. 1971

P 1601 Easter Conf on Model Th (1);1983 Diedrichshagen • DDR
[1983] *Proceedings of the 1st Easter Conference on Model Theory* ED: DAHN, B.I. • SER (S 3382) Sem-ber, Humboldt-Univ Berlin, Sekt Math 49 • PUBL (X 2219) Humboldt-Univ Berlin: Berlin 154pp
• DAT&PL 1983 Apr; Diedrichshagen, DDR

P 1607 Hungar Math Congr (2);1960 Budapest • H
[1961] *Magyar Matematikai Kongresszus (2nd Hungarian Mathematical Congress. Abstracts)* PUBL (X 0928) Akad Kiado: Budapest 7 Vols
• DAT&PL 1960 Aug; Budapest, H

P 1608 Int Congr Math (II, 5);1936 Oslo • N
[1937] *Comptes Rendus du Congres International des Mathematiciens* PUBL A.M. Broeggers Boktrykkeri a/S: Oslo 2 Vols: 316pp,vv+289pp • ALT PUBL (X 3602) Kraus: Vaduz 1967 2 Vols
• DAT&PL 1936 Jul; Oslo, N • LC-No 52-1808

P 1612 Schaltkreis & -werk Th (2);1961 Saarbruecken • D
[1963] *2. Colloquium ueber Schaltkreis- und Schaltwerk-Theorie* ED: DOERR, J. & PESCHL, E. & UNGER, H.
• SER International Series of Numerical Mathematics 4
• PUBL (X 0804) Birkhaeuser: Basel 152pp
• DAT&PL 1961 Oct; Saarbruecken, D • LC-No 63-48116

P 1618 ACM Symp Th of Comput (7);1975 Albuquerque • USA
[1975] *Proceedings of the 7th Annual ACM Symposium on the Theory of Computing (Association for Computing Machinery)* PUBL (X 2205) ACM: New York 265pp
• DAT&PL 1975 May; Albuquerque, NM, USA • LC-No 82-642181

P 1619 Coloq Log Simb;1975 Madrid • E
[1976] *Coloquio Sobre Logica Simbolica* PUBL Centro Calculo Univ. Complutense: Madrid 176pp
• DAT&PL 1975 Feb; Madrid, E • LC-No 77-555677

P 1639 Set Th & Hierarch Th (1);1974 Karpacz • PL
[1977] *Set Theory and Hierarchy Theory. Proceedings of the 1st Colloquium in Set Theory and Hierarchy Theory* SER (S 3412) Prace Inst Mat, Politech Wroclaw, Ser Konf 14/1 • PUBL Polytechnical Edition: Wroclaw 123pp
• DAT&PL 1974 Sep; Karpacz, PL

P 1640 AFIPS Spring Jt Computer Conf (25);1964 Washington • USA
[1964] *AFIPS Conference Proceedings : 1964: Spring Joint Computer Conference* SER (S 3309) AFIPS Conference Proc 25 • PUBL (X 1354) Spartan Books : Sutton 629pp • ALT PUBL (X 2508) Cleaver-Hume Pr: London
• DAT&PL 1964 -?-; Washington, DC, USA • LC-No 80-649584

P 1641 Kuenstl Intelligenzforschg BRD;1975 Bonn • D
[1975] *Kuenstliche Intelligenzforschung in der BRD. Bericht ueber ein 1tes Informelles Treffen* ED: VEENKER, G. • SER (S 3180) Inform Ber, Inst Inf, Univ Bonn 5 • PUBL (X 1512) Univ Bonn Inst Informatik: Bonn 58pp
• DAT&PL 1975 Feb; Bonn, D

P 1646 Int Congr Math (5);1912 Cambridge GB • GB
[1913] *Proceedings of the 5th International Congress of Mathematicians* ED: HOBSON, E.W. & LOVE, A.E.H. • PUBL (X 0805) Cambridge Univ Pr: Cambridge, GB 2 Vols: 500pp,657pp • ALT PUBL (X 3602) Kraus: Vaduz 1967
• DAT&PL 1912 Aug; Cambridge, GB • LC-No 52-1808

P 1657 Ancient Log & Modern Interpr;1972 Buffalo • USA
[1974] *Ancient Logic and Its Modern Interpretations* ED: CORCORAN, J. • SER Synthese Historical Library 9 • PUBL (X 0835) Reidel: Dordrecht x+211pp
• DAT&PL 1972 Apr; Buffalo, NY, USA • ISBN 90-277-0395-7, LC-No 73-88589

P 1675 Summer Inst Symb Log;1957 Ithaca • USA
[1957] *Summaries of Talks Presented at the Summer Institute for Symbolic Logic* PUBL Institute for Defense Analyses, Communications Research Division: Princeton; xvi+427pp
• DAT&PL 1957 Jul; Ithaca, NY, USA • LC-No 65-4418
• REM 2nd ed. 1960

P 1676 Int Joint Conf Artif Intell (5);1977 Cambridge MA • USA
[1977] *5th International Joint Conference on Artificial Intelligence*
• DAT&PL 1977 Aug; Cambridge, MA, USA

P 1681 Symp Inform Storage & Retrieval;1971 College Park • USA
[1971] *Proceedings of the Symposium on Information Storage and Retrieval* ED: MINKER, J. & ROSENFELD, S. • PUBL (X 2205) ACM: New York viii+285pp
• DAT&PL 1971 Apr; College Park, MD, USA • LC-No 72-178466

P 1691 Inform Processing (6);1974 Stockholm • S
[1974] *Information Processing '74. Proceedings of IFIP Congress* ED: ROSENFELD, J.L. • PUBL (X 0809) North Holland: Amsterdam xxi+1107pp • ALT PUBL (X 0838) Amer Elsevier: New York
• DAT&PL 1974 Aug; Stockholm, S • ISBN 0-444-10689-8, LC-No 74-76063

P 1694 Inform Processing (7);1977 Toronto • CDN
[1977] *Information Processing '77. Proceedings of IFIP Congress* ED: GILCHRIST, B. • SER IFIP Congress Series 7
• PUBL (X 0809) North Holland: Amsterdam xix+1004pp
• DAT&PL 1977 Aug; Toronto, ON, CDN • ISBN 0-7204-0755-9, LC-No 77-80624

P 1695 Set Th & Hierarch Th (3);1976 Bierutowice • PL
[1977] *Set Theory and Hierarchy Theory V. Proceedings of the 3rd Conference on Set Theory and Hierarchy Theory* ED: LACHLAN, A. & SREBRNY, M. & ZARACH, A. • SER (S 3301) Lect Notes Math 619 • PUBL (X 0811) Springer: Heidelberg & New York viii+358pp
• DAT&PL 1976 Sep; Bierutowice, PL • ISBN 3-540-08521-1, LC-No 78-309663

P 1704 Int Congr Log, Meth & Phil of Sci (5);1975 London ON • CDN
[1977] *Proceedings of 5th International Congress of Logic, Methodology and Philosophy of Science* ED: BUTTS, R.E. & HINTIKKA, J. • SER (S 3308) Univ Western Ontario Ser in Philos of Sci 9-12 • PUBL (X 0835) Reidel: Dordrecht 4 Vols: x+406pp, x+427pp, x+321pp, x+336pp
• DAT&PL 1975 Aug; London, ON, CDN • ISBN 90-277-0708-1 (V1), ISBN 90-277-0710-3 (V2), ISBN 90-277-0829-0 (V3), ISBN 90-277-0831-2 (V4), ISBN 90-277-0706-5 (Set of the 4 Vols), LC-No 77-22429 (V1), LC-No 77-22431 (V2), LC-No 77-22432 (V3), LC-No 77-22433 (V4)
• REM Vol 1: Logic, Foundations of Mathematics, and Computability Theory. Vol 2: Foundational problems in the Special Sciences. Vol 3: Basic Problems in Methodology and Linguistics. Vol 4: Historical and Philosophical Dimensions of Logic, Methodology and Philosophy of Science.

P 1705 Scand Logic Symp (4);1976 Jyvaeskylae • SF
[1979] *Essays on Mathematical and Philosophical Logic. Proceedings of the 4th Scandinavian Logic Symposium and of the 1st Soviet-Finnish Logic Conference* ED: HINTIKKA, J. & NIINILUOTO, I. & SAARINEN, E. • SER (S 3307) Synth Libr 122 • PUBL (X 0835) Reidel: Dordrecht viii+462pp
• DAT&PL 1976 Jun; Jyvaeskylae, SF • ISBN 90-277-0879-7, LC-No 78-14736

P 1740 ACM Symp Th of Comput (10);1978 San Diego • USA
[1978] *Conference Record of the 10th Annual ACM Symposium on Theory of Computing (Association for Computing Machinery)* PUBL (X 2205) ACM: New York 346pp
• DAT&PL 1978 May; San Diego, CA, USA • LC-No 79-101797

P 1748 Lang & Pensee Math;1976 Luxembourg • L
[1976] *Langage et Pensee Mathematiques. Actes du Colloque International* PUBL (X 3203) Cent Univ Luxembourg: Luxembourg 454pp
• DAT&PL 1976 Jun; Luxembourg, L • LC-No 79-380450

P 1766 Gen Topol & Rel to Mod Anal & Algeb (1);1961 Prague • CS
[1962] *General Topology and Its Relations to Modern Analysis and Algebra I. Proceedings of the 1st Prague Topological Symposium* ED: NOVAK, J. • PUBL (X 1034) PWN: Warsaw 363pp • ALT PUBL (X 0801) Academic Pr: New York
• DAT&PL 1961 Sep; Prague, CS • LC-No 63-751

P 1791 Proc Bienn Meet Phil of Sci Ass;1972 East Lansing • USA
[1974] *Proceedings of the 1972 Biennial Meeting: Philosophy of Science Association (PSA)* ED: SCHAFFNER, K.F.& COHEN, R.S. • SER (S 3307) Synth Libr 64, (S 3311) Boston St Philos Sci 20 • PUBL (X 0835) Reidel: Dordrecht ix+445pp
• DAT&PL 1972 Oct; East Lansing, MI, USA • ISBN 90-277-0408-2 (Cloth), ISBN 90-277-0409-0 (Pb.), LC-No 72-624169

P 1800 Brazil Conf Math Log (1);1977 Campinas • BR
[1978] *Proceedings of 1st Brazilian Conference on Mathematical Logic* ED: ARRUDA, A.I. & CHAQUI, R. & COSTA DA, N.C.A. • SER (S 3310) Lect Notes Pure Appl Math 39 • PUBL (X 1684) Dekker: New York xii+303pp
• DAT&PL 1977 Jul; Campinas, BR • LC-No 78-14488

P 1804 Form Meth in Methodol of Emp Sci;1974 Warsaw • PL
[1976] *Formal Methods in the Methodology of Empirical Sciences* ED: PRZELECKI, M. & SZANIAWSKI, K. & WOJCICKI, R. & MALINOWSKI, G. • SER (S 3307) Synth Libr 103 • PUBL (X 2212) Ossolineum: Wroclaw 457pp • ALT PUBL (X 0835) Reidel: Dordrecht
• DAT&PL 1974 Jun; Warsaw, PL • ISBN 90-277-0698-0, LC-No 76-4586

P 1805 Int Symp Multi-Val Log (5,Proc);1975 Bloomington • USA
[1975] *Proceedings of the 1975 International Symposium on Multiple-Valued Logic* PUBL (X 2179) IEEE: New York iv+475pp
• DAT&PL 1975 May; Bloomington, IN, USA • LC-No 76-370321 • REL PUBL (P 1894) Int Symp Multi-Val Log (5,Inv Pap);1975 Bloomington

P 1815 Congr Int Ling & Filol Roman (11);1965 Madrid • E
[1968] *Actas del XI Congreso Internacional de Linguistica y Filologia Romana.* ED: QUILIS, A. & CARRIL, R.B. & CANTANERO, M. • SER Revista de Filologia Espanola 86
• PUBL (X 2091) Consejo Sup Invest Ci: Madrid 4 Vols
• DAT&PL 1965 -?-; Madrid, E • LC-No 76-472491

P 1841 Fct Recurs & Appl;1967 Tihany • H
[1969] *Les Fonctions Recursives et leurs Applications* PUBL (X 0999) CNRS Inst B Pascal: Paris • ALT PUBL (X 3725) Bolyai Janos Mat Tars: Budapest
• DAT&PL 1967 Sep; Tihany, H

P 1864 Higher Set Th;1977 Oberwolfach • D
[1978] *Higher Set Theory* ED: MUELLER, GERT H. & SCOTT, D.S. • SER (S 3301) Lect Notes Math 669 • PUBL (X 0811) Springer: Heidelberg & New York xii+476pp
• DAT&PL 1977 Apr; Oberwolfach, D • ISBN 3-540-08926-8, LC-No 79-312135

P 1869 Automata, Lang & Progr (2);1974 Saarbruecken • D
[1974] *Automata, Languages and Programming: 2nd Colloquium* ED: LOECKX, J. • SER (S 3302) Lect Notes Comput Sci 14 • PUBL (X 0811) Springer: Heidelberg & New York viii+611pp
• DAT&PL 1974 Jul; Saarbruecken, D • ISBN 3-540-06841-4, LC-No 74-180345
• REM Also Abbreviated as ICALP 74

P 1870 Automata, Lang & Progr (3);1976 Edinburgh • GB
[1976] *Automata, Languages and Programming. 3rd Colloquium* ED: MICHAELSON, S. & MILNER, R. • PUBL (X 1261) Edinburgh Univ Pr: Edinburgh vi+559pp
• DAT&PL 1976 Jul; Edinburgh, GB • ISBN 0-85224-308-1, LC-No 77-359145
• REM Also Abbreviated as ICALP 76

P 1872 Automata, Lang & Progr (5);1978 Udine • I
[1978] *Automata, Languages and Programming. 5th Colloquium* ED: AUSIELLO, G. & BOEHM, C. • SER (S 3302) Lect Notes Comput Sci 62 • PUBL (X 0811) Springer: Heidelberg & New York viii+508pp
• DAT&PL 1978 Jul; Udine, I • ISBN 3-540-08860-1, LC-No 79-303999
• REM Also Abbreviated as ICALP 78

P 1873 Automata, Lang & Progr (6);1979 Graz • A
[1979] *Automata, Languages and Programming. 6th Colloquium* ED: MAURER, H.A. • SER (S 3302) Lect Notes Comput Sci 71 • PUBL (X 0811) Springer: Heidelberg & New York ix+682pp
• DAT&PL 1979 Jul; Graz, A • ISBN 3-540-09510-1, LC-No 79-15859
• REM Also Abbreviated as ICALP 79

P 1894 Int Symp Multi-Val Log (5,Inv Pap);1975 Bloomington • USA
[1977] *Modern Uses of Multiple-Valued Logic. Invited Papers from the 5th International Symposium on Multiple-Valued Logic* ED: DUNN, J.M. & EPSTEIN, G. • SER Epistime 2
• PUBL (X 0835) Reidel: Dordrecht x+338pp
• DAT&PL 1975 May; Bloomington, IN, USA • ISBN 90-277-0747-2, LC-No 77-23098 • REL PUBL (P 1805) Int Symp Multi-Val Log (5,Proc);1975 Bloomington
• REM With a Bibliography of Many-Valued Logic

P 1924 Polnisch Math Kongr;1953 Warsaw • PL
[1954] *Die Hauptreferate des 8. Polnischen Mathematikerkongresses* ED: GRELL, H. • PUBL (**X** 0806) Dt Verlag Wiss: Berlin 125pp
• DAT&PL 1953 Sep; Warsaw, PL
• REM Autorisierte Uebersetzung

P 1959 Int Congr Math (II,13);1978 Helsinki • SF
[1980] *Proceedings of the International Congress of Mathematicians* ED: LEHTO, O. • PUBL Academia Scientiarum Fennica: Helsinki 2 Vols: 1022pp
• DAT&PL 1978 Aug; Helsinki, SF • ISBN 951-41-0352-1

P 1984 Colloq on Diff Geom (4);1978 Santiago de Compostela • E
[1979] *Proceedings of the IV International Colloquium on Differential Geometry* PUBL (**X** 2856) Univ Santiago Sec Publ: Santiago de Compostela ix+294pp
• DAT&PL 1978 Sep; Santiago de Compostela, E • ISBN 84-7191-148-5

P 1986 Dt Kongr Philos (11);1975 Goettingen • D
[1977] *Logik, Ethik, Theorie der Geisteswissenschaften. XI. Deutscher Kongress fuer Philosophie* ED: PATZIG, G. & SCHEIBE, E. & WIELAND, W. • PUBL (**X** 1088) Meiner: Hamburg ix+544pp
• DAT&PL 1975 Oct; Goettingen, D • LC-No 78-338487

P 1998 Int Leibniz Kongr (2);1972 Hannover • D
[1973-1975] *Akten des 2. Internationalen Leibniz-Kongresses* ED: HEINEKAMP, A. & KALISCH, D. & WILUCKI VON, I. & ZACHER, H.J. • SER (S 3252) Stud Leibnitiana, Suppl 12-15
• PUBL (**X** 2717) Steiner: Wiesbaden 4 Vols: vi+164pp,325pp,415pp,v+303pp
• DAT&PL 1972 Jul; Hannover, D • ISBN 3-515-01848-4 (V2), ISBN 3-515-01924-3 (V3), ISBN 3-515-01925-1 (V4), LC-No 74-31763
• REM Vol 1: Begruessungsansprachen, Gesamtinterpretationen, Berichte, Geschichte-Recht-Gesellschaftstheorie, Historische Wirkung. Vol 2: Wissenschaftstheorie und Wissenschaftsgeschichte. Vol 3: Metaphysik-Ethik-Aesthetik-Monadenlehre. Vol 4: Logik, Erkenntnistheorie, Methodologie, Sprachphilosophie.

P 2008 Symp Th & Appl of Multi-Val Log Design;1972 Buffalo • USA
[1972] *Conference Records of the 1972 Symposium on the Theory and Applications of Multiple-Valued Logic Design* PUBL (**X** 2179) IEEE: New York 209pp
• DAT&PL 1972 May; Buffalo, NY, USA

P 2009 Int Symp Multi-Val Log (3);1973 Toronto • CDN
[1973] *Conference Records of the 1973 International Symposium on Multiple-Valued Logic* ED: ALLEN, C.M. & GIVONE, D.D. • PUBL (**X** 2179) IEEE: New York i+245pp
• DAT&PL 1973 May; Toronto, ON, CDN

P 2011 Int Symp Multi-Val Log (6);1976 Logan • USA
[1976] *Proceedings of the 6th International Symposium on Multiple-Valued Logic* PUBL (**X** 2205) ACM: New York vii+273pp • ALT PUBL (**X** 2179) IEEE: New York
• DAT&PL 1976 May; Logan, UT, USA • LC-No 79-641110
• REM IEEE Publication No. 76CH1111-4C

P 2013 Int Symp Multi-Val Log (7);1977 Charlotte • USA
[1977] *Proceedings of the 7th International Symposium on Multiple-Valued Logic* PUBL (**X** 2179) IEEE: New York iii+155pp • ALT PUBL (**X** 2205) ACM: New York
• DAT&PL 1977 May; Charlotte, NC, USA • LC-No 79-641110
• REM IEEE Publication No 77CH1222-9C

P 2014 Int Symp Multi-Val Log (8);1978 Rosemont • USA
[1978] *Proceedings of the 8th International Symposium on Multiple-Valued Logic* PUBL (**X** 2179) IEEE: New York 298pp
• DAT&PL 1978 May; Rosemont, IL, USA • LC-No 78-107956

P 2058 Kleene Symp;1978 Madison • USA
[1980] *The Kleene Symposium* ED: BARWISE, K.J. & KEISLER, H.J. & KUNEN, K. • SER (S 3303) Stud Logic Found Math 101 • PUBL (**X** 0809) North Holland: Amsterdam xx+425pp
• DAT&PL 1978 Jun; Madison, WI, USA • ISBN 0-444-85345-6, LC-No 79-20792

P 2059 Math Founds of Comput Sci (8);1979 Olomouc • CS
[1979] *Mathematical Foundations of Computer Science. Proceedings of the 8th Symposium* ED: BECVAR, J. • SER (S 3302) Lect Notes Comput Sci 74 • PUBL (**X** 0811) Springer: Heidelberg & New York ix+580pp
• DAT&PL 1979 Sep; Olomouc, CS • ISBN 3-540-09526-8, LC-No 79-17801

P 2080 Conf Math Log;1970 London • GB
[1972] *Conference on Mathematical Logic - London '70* ED: HODGES, W. • SER (S 3301) Lect Notes Math 255 • PUBL (**X** 0811) Springer: Heidelberg & New York viii+351pp
• DAT&PL 1970 Aug; London, GB • ISBN 3-540-05744-7, LC-No 70-189457

P 2116 Probl in Log & Ontology;1973 Salzburg • A
[1977] *Problems in Logic and Ontology. Internationales Forschungszentrum Salzburg. Forschungsgespraeche.* ED: MORSCHER, E. & CZERMAK, J. & WEINGARTNER, P. • PUBL (**X** 2596) Akad Druck-& Verlagsanstalt: Graz 310pp
• DAT&PL 1973 Sep; Salzburg, A • ISBN 3-201-01021-9, LC-No 80-487240
• REM Reprint of Vol. 3/3*171-343 of J0122

P 2153 Logic Colloq;1983 Aachen • D
[1984] *Proceedings of the Logic Colloquium, Part 1, 2* ED: MUELLER, GERT H. & RICHTER, M.M. (V1). BOERGER, E. & OBERSCHELP, W. & RICHTER, M.M. & SCHINZEL, B. & THOMAS, W. (V2) • SER (S 3301) Lect Notes Math 1103,1104
• PUBL (**X** 0811) Springer: Heidelberg & New York 2 Vols: viii+484pp,viii+475pp
• DAT&PL 1983 Jul; Aachen, D • ISBN 3-540-13900-1 (V1), ISBN 3-540-13901-X (V2), LC-No 84-26704
• REM Vol 1: Models and Sets. Vol 2: Computation and Proof Theory.

P 2160 Latin Amer Symp Math Log (6);1983 Caracas • YV
[1985] *Methods in Mathematical Logic. Proceedings of the 6th Latin American Symposium on Mathematical Logic* ED: PRISCO DI, C.A. • SER (S 3301) Lect Notes Math 1130
• PUBL (**X** 0811) Springer: Heidelberg & New York vii+407pp
• DAT&PL 1983 Aug; Caracas, YV • ISBN 3-540-15236-9, LC-No 85-14779

P 2167 Orders: Descr & Roles;1982 L'Arbresle • F
[1984] *Orders: Description and Roles in Set Theory, Lattices, Ordered Groups, Topology, Theory of Models and Relations, Combinatorics, Effectiveness, Social Sciences. Proceedings of a Conference on Ordered Sets and Their Applications* ED: POUZET, M. & RICHARD, D. • SER (S 3358) Ann Discrete Math 23, North Holland Mathematics Studies 99 • PUBL (X 0809) North Holland: Amsterdam xxviii + 548pp
• DAT&PL 1982 Jul; L'Arbresle, F • ISBN 0-444-87601-4, LC-No 84-13749

P 2180 Math Appl Categ Th;1983 Denver • USA
[1984] *Mathematical Applications of Category Theory. Proceedings of the Special Session of the 89th Annual Meeting of the American Mathematical Society (AMS)* ED: GRAY, J.W. • SER (S 3313) Contemp Math 30 • PUBL (X 0803) Amer Math Soc: Providence vii + 307pp
• DAT&PL 1983 Jan; Denver, CO, USA • ISBN 0-8218-5032-6, LC-No 84-9371

P 2251 Mat Log & Primen;1960 Stanford • USA
[1965] *Matematicheskaya Logika i Ee Primeneniya: Sbornik Statei (Mathematical Logic and Its Applications. Logic, Methodology and Philosophy of Science)* ED: MAL'TSEV, A.I. & NAGEL, E. & SUPPES, P. & TARSKI, A. • PUBL (X 0885) Mir: Moskva 341pp
• DAT&PL 1960 Aug; Stanford, CA, USA
• TRANSL OF [1962] (P 0612) Int Congr Log, Meth & Phil of Sci (1,Proc);1960 Stanford
• REM Only Parts of P0612 are translated.

P 2267 Obobsh Funk & Primen Mat Fiz;1980 Moskva • SU
[1981] *Obobshchennye Funktsii i Ikh Primeneniya b Matematicheskoj Fizike (Generalized Functions and Their Applications in Mathematical Physics)* ED: VLADIMIROV, V.S. • PUBL (X 2265) Akad Nauk Vychis Tsentr: Moskva 560pp
• DAT&PL 1980 Nov; Moskva, SU

P 2268 Int Colloq Philos of Sci;1965 London • GB
[1967-1970] *Proceedings of the International Colloquium in the Philosophy of Science. 4 Volumes* ED: LAKATOS, I. (V 1 - V 4) & MUSGRAVE, A. (V 3, V 4) • SER (S 3303) Stud Logic Found Math V 1 - V 3 • PUBL (X 0809) North Holland: Amsterdam V 1: xv + 241pp, V 2: viii + 420pp, V 3: ix + 448pp, (X 0805) Cambridge Univ Pr: Cambridge, GB V 4: viii + 282pp
• DAT&PL 1965, Jul; London, GB • ISBN 0-521-07826-1 (V4), LC-No 67-20007 (V1), LC-No 67-28648 (V2), LC-No 67-28649 (V3), LC-No 78-105496 (V4)
• REM Vol 1: Problems in the Philosophy of Mathematics. Vol 2: The Problem of Inductive Logic. Vol 3: Problems in the Philosophy of Science. Vol 4: Criticism and the Growth of Knowledge.

P 2323 Non-Numer Comput Adv Progr;1963 Oxford • GB
[1966] *Advance in Programming and Non-Numerical Computation. Proceedings of the Summer School Oxford University* ED: FOX, L. • PUBL (X 0869) Pergamon Pr: Oxford viii + 218pp
• DAT&PL 1963 -?-; Oxford, GB • LC-No 65-18420

P 2335 Canad Math Congr (25);1971 Thunder Bay • CDN
[1971] *Proceedings of the 25th Summer Meeting of the Canadian Mathematical Congress* ED: EAMES, W.R. & STANTON, R.G. & THOMAS, R.S.D. • PUBL Lakehead University: Thunder Bay viii + 648pp
• DAT&PL 1971 Jun; Thunder Bay, ON, CDN • LC-No 48-791

P 2342 Symp Rek Kombin;1983 Muenster • D
[1984] *Logic and Machines: Decision Problems and Complexity. Proceedings of the Symposium Rekursive Kombinatorik* ED: BOERGER, E. & HASENJAEGER, G. & ROEDDING, D. • SER (S 3302) Lect Notes Comput Sci 171 • PUBL (X 0811) Springer: Heidelberg & New York vi + 456pp
• DAT&PL 1983 May; Muenster, D • ISBN 3-540-13331-3, LC-No 84-55

P 2539 Frege Konferenz (1);1979 Jena • DDR
[1979] *"Begriffsschrift". Jenaer Frege-Konferenz* ED: BOLCK, F. • PUBL (X 2211) Schiller Univ: Jena iii + 548pp
• DAT&PL 1979 May; Jena, DDR

P 2552 Conf Finite Algeb & Multi-Val Log;1979 Szeged • H
[1981] *Proceedings of the Conference on Finite Algebra and Multiple-Valued Logic* ED: CSAKANY, B. & ROSENBERG, J.G. • SER (S 3312) Coll Math Soc Janos Bolyai 28 • PUBL (X 0809) North Holland: Amsterdam 880pp • ALT PUBL (X 3725) Bolyai Janos Mat Tars: Budapest
• DAT&PL 1979 Aug; Szeged, H • ISBN 0-444-85439-8, LC-No 81-214217

P 2554 All-Union Symp Th Log Infer;1974 Moskva • SU
[1979] *Logicheskij Vyvod. (Logical Inference. Proceedings of the All-Union Symposium on the Theory of Logical Inference.)* ED: SMIRNOV, V.A. • PUBL (X 2027) Nauka: Moskva 312pp
• DAT&PL 1974 Mar; Moskva, SU • LC-No 79-380844

P 2588 FCT'77 Fund of Comput Th;1977 Poznan • PL
[1977] *Fundamentals of Computation Theory. Proceedings of the International FCT '77 - Conference* ED: KARPINSKI, M. • SER (S 3302) Lect Notes Comput Sci 56 • PUBL (X 0811) Springer: Heidelberg & New York 542pp
• DAT&PL 1977 Sep; Poznan-Kornik, PL • ISBN 3-540-08442-8, LC-No 77-14022

P 2607 IBM Symp Math Log & Comput Sci (3); • USA
[1978] *Proceedings of the 3rd IBM Symposium on Mathematical Logic and Computer Science*

P 2611 Symp Founds of Math;
[1980] *Proceedings of the Symposium on the Foundations of Mathematics*

P 2614 Open Days in Model Th & Set Th;1981 Jadwisin • PL
[1981] *Proceedings of the International Conference "Open Days in Model Theory and Set Theory"* ED: GUZICKI, W. & MAREK, W. & PELC, A. & RAUSZER, C. • PUBL Selbstverlag 333pp
• DAT&PL 1981 Sep; Jadwisin, PL
• REM Available from John Derrick, School of Mathematics, University of Leeds, GB, or from Cecylia Rauszer, Institute of Mathematics, University of Warsaw, PL

P 2615 Scand Logic Symp (5);1979 Aalborg • DK
[1979] *Proceedings of the 5th Scandinavian Logic Symposium* ED: JENSEN, F.V. & MAYOH, B.H. & MOLLER, K.K. • PUBL (X 2646) Aalborg Univ Pr: Aalborg vii + 361pp
• DAT&PL 1979 Jan; Aalborg, DK • ISBN 87-7307-037-8, LC-No 80-464603

P 2633 Autom Deduct (7);1984 Napa • USA
[1984] *7th International Conference on Automated Deduction* ED: SHOSTAK, R.E. • SER (S 3302) Lect Notes Comput Sci 170 • PUBL (X 0811) Springer: Heidelberg & New York vi + 508pp
• DAT&PL 1984 May; Napa, CA, USA • ISBN 3-540-96022-8, LC-No 84-5441

P 2898 Algor Kompl, Lern-& Erkenn-Prozess;1976 Jena • DDR
[1976] *Algorithmische Kompliziertheit, Lern- und Erkennungsprozesse. 2tes Internationales Symposium* ED: BOLCK, F. • PUBL (X 2211) Schiller Univ: Jena 132pp
• DAT&PL 1976 Oct; Jena, DDR

P 2901 Appl Sheaves;1977 Durham • GB
[1979] *Applications of Sheaves. Proceedings of the Research Symposium on Applications of Sheaf Theory to Logic, Algebra and Analysis* ED: FOURMAN, M.P. & MULVEY, C.J. & SCOTT, D.S. • SER (S 3301) Lect Notes Math 753 • PUBL (X 0811) Springer: Heidelberg & New York xiv+779pp
• DAT&PL 1977 Jul; Durham, GB • ISBN 3-540-09564-0, LC-No 79-23219

P 2915 Compl Anal & Appl;1975 Trieste • I
[1976] *Complex Analysis and Its Applications. Lectures Presented at an International Seminar-Course* ED: LEWIS, M. • PUBL International Atomic Energy Agency: Wien 3 Vols:iii+343pp, ii+309pp,iii+320pp
• DAT&PL 1975 May; Trieste, I • ISBN 92-0-130376-9 (V1), ISBN 92-0-130476-5 (V2), ISBN 92-0-130576-1 (V3), LC-No 77-366644

P 2929 Form Descr of Progr Concepts (1);1977 St.Andrews • CDN
[1978] *Formal Description of Programming Concepts. Proceedings of the IFIP Working Conference* ED: NEUHOLD, E.J. • PUBL (X 0809) North Holland: Amsterdam xviii+648pp
• DAT&PL 1977 Aug; St. Andrews, NB, CDN • ISBN 0-444-85107-0, LC-No 81-455275

P 2935 FCT'79 Fund of Comput Th;1979 Berlin/Wendisch-Rietz • DDR
[1979] *Fundamentals of Computation Theory - FCT '79. Proceedings of the Conference on Algebraic, Arithmetic, and Categorical Methods in Computation Theory* ED: BUDACH, L. • SER Mathematical Research - Mathematische Forschung 2 • PUBL (X 0911) Akademie Verlag: Berlin 576pp
• DAT&PL 1979 Sep; Berlin/Wendisch-Rietz, DDR • LC-No 82-460828

P 2937 Gen Topol & Rel to Mod Anal & Algeb (4);1976 Prague • CS
[1977] *General Topology and Its Relations to Modern Analysis and Algebra IV. Proceedings of the 4th Prague Topological Symposium. Part A: Invited Papers* ED: NOVAK, J. • SER (S 3301) Lect Notes Math 609 • PUBL (X 0811) Springer: Heidelberg & New York xvi+225pp
• DAT&PL 1976 Aug; Prague, CS • ISBN 3-540-08437-1, LC-No 77-11622

P 2944 Th de l'Inform CNRS;1977 Cachan • F
[1978] *Theorie de l'Information. Developpements Recents et Applications* SER (S 1802) Colloq Int CNRS 276 • PUBL (X 0999) CNRS Inst B Pascal: Paris 506pp
• DAT&PL 1977 Jul; Cachan, F • ISBN 2-222-02258-4, LC-No 79-386165

P 2952 CAAP'79 Arbres en Algeb & Progr (4);1979 Lille • F
[1979] *Les Arbres en Algebre et en Programmation. 4eme Colloquium* PUBL Universite de Lille I: Lille iv+327pp
• DAT&PL 1979 Feb; Lille, F

P 2955 Int Summer Inst Theor Phys (8);1976 Bielefeld • D
[1978] *Many Degrees of Freedom in Field Theory. Proceedings of the 8th International Summer Institute of Theoretical Physics* ED: STREIT, L. • SER NATO Advanced Study Institute Series, Ser B 30 • PUBL (X 1332) Plenum Publ: New York vii+248pp
• DAT&PL 1976 Aug; Bielefeld, D • ISBN 0-306-35730-5, LC-No 77-29217

P 2958 Latin Amer Symp Math Log (4);1978 Santiago • RCH
[1980] *Mathematical Logic in Latin America. Proceedings of the 4th Latin American Symposium on Mathematical Logic* ED: ARRUDA, A.I. & CHUAQUI, R. & COSTA DA, N.C.A. • SER (S 3303) Stud Logic Found Math 99 • PUBL (X 0809) North Holland: Amsterdam xii+392pp
• DAT&PL 1978 Dec; Santiago, RCH • ISBN 0-444-85402-9, LC-No 79-20797

P 2965 Measure Th;1975 Oberwolfach • D
[1976] *Measure Theory* ED: BELLOW, A. & KOELZOW, D. • SER (S 3301) Lect Notes Math 541 • PUBL (X 0811) Springer: Heidelberg & New York xiv+430pp
• DAT&PL 1975 Jun; Oberwolfach, D • ISBN 3-540-07861-4, LC-No 76-40183

P 2975 Asilomar Conf Circ, Syst & Comput (9);1975 Pacific Grove • USA
[1976] *9th Asilomar Conference on Circuits, Systems, and Computers* ED: CHAN, SHU PARK • PUBL (X 1777) Western Periodicals: Hollywood xiii+638pp
• DAT&PL 1975 Nov; Pacific Grove, CA, USA • LC-No 86-643237

P 2980 Num Beh Variat & Steuerungsprobl;1974 Bonn • D
[1975] *Numerische Behandlung von Variations- und Steuerungsproblemen.* SER (S 0478) Bonn Math Schr 77 • PUBL (X 0908) Univ Math Inst: Bonn iv+126pp
• DAT&PL 1974 Feb; Bonn, D • LC-No 76-454743

P 2999 Proc Conf Databasis (Calzone);1985 Heidelberg • D
[1985] *Proceedings of a Conference on Databases* ED: GSCHNITZER, W. & SPRENGER, H. & STEIN, H. & THURN, K. • PUBL Apl & Wemding Co: Mosbach 294pp
• DAT&PL 1985 Dec; Heidelberg, D

P 3000 Conf Probab Th (4);1971 Brasov • RO
[1973] *Proceedings of the 4th Conference on Probability Theory.* ED: BEREANU, B. & IOSIFESCU, M. & POSTELNICU, T. & TAUTU, P. • PUBL (X 0871) Acad Rep Soc Romania: Bucharest 644pp
• DAT&PL 1971 Sep; Brasov, RO • LC-No 74-644019

P 3003 Int Symp Multi-Val Log (9);1979 Bath • GB
[1979] *Proceedings of the 9th International Symposium on Multiple-Valued Logic* PUBL (X 2179) IEEE: New York v+303pp
• DAT&PL 1979 May; Bath, GB • LC-No 79-641110

P 3006 Brazil Conf Math Log (3);1979 Recife • BR
[1980] *Proceedings of the 3rd Brazilian Conference on Mathematical Logic* ED: ARRUDA, A.I. & COSTA DA, N.C.A. & SETTE, A.M. • PUBL (X 2836) Soc Brasil Log: Sao Paulo vi+336pp
• DAT&PL 1979 Dec; Recife, BR

P 3013 Progr Symp;1974 Paris • F
[1974] *Programming Symposium. Proceedings. Colloque sur la Programmation* ED: ROBINET, B. • SER (S 3302) Lect Notes Comput Sci 19 • PUBL (X 0811) Springer: Heidelberg & New York v+425pp
• DAT&PL 1974 Apr; Paris, F • ISBN 3-540-06859-7, LC-No 75-19256

P 3042 Conv Algeb Commut & Conv Geom;1971/72 Roma • I
[1973] *Convegno di Algebra Commutativa, Convegno di Geometria. Istituto Nazionale di Alta Matematica (INDAM)* SER (S 3305) Symposia Matematica 11 • PUBL (X 3604) INDAM: Roma 456pp • ALT PUBL (X 0801) Academic Pr: New York
• DAT&PL 1971 Nov; Roma, I, 1972 Mar; Roma, I

P 3056 Inform Th, Stat Decis Fcts & Random Proc (7);1974 Prague • CS
[1978] *Transactions of the 7th Prague Conference on Information Theory, Statistical Decision Functions, Random Processes and of the 8th European Meeting of Statisticians* ED: KOZESNIK, J. • PUBL (X 1226) Academia: Prague 2 Vols:602pp,582pp • ALT PUBL (X 0835) Reidel: Dordrecht
• DAT&PL 1974 Aug; Prague, CS • ISBN 90-277-0852-5 (V 1)

P 3058 Trends Appl of Pure Math to Mech;1977 Kozubnik • PL
[1979] *Trends in Applications of Pure Mathematics to Mechanics. Vol. II* ED: ZORSKI, H. • SER Monographs and Studies in Mathematics 5 • PUBL (X 1330) Pitman Publ: Belmont & London viii+341pp
• DAT&PL 1977 Sep; Kozubnik, PL • ISBN 0-273-08421-6, LC-No 77-351685

P 3062 IEEE Symp Switch & Automata Th (14);1973 Iowa City • USA
[1973] *14th Annual IEEE Symposium on Switching and Automata Theory* PUBL (X 2179) IEEE: New York v+213pp
• DAT&PL 1973 Oct; Iowa City, IA, USA • LC-No 80-646635

P 3063 Autom Deduct (5);1980 Les Arcs • F
[1980] *5th Conference on Automated Deduction* ED: BIBEL, W. & KOWALSKI, R. • SER (S 3302) Lect Notes Comput Sci 87 • PUBL (X 0811) Springer: Heidelberg & New York vii+385pp
• DAT&PL 1980 Jul; Les Arcs, F • ISBN 3-540-10009-1, LC-No 80-18708

P 3074 Algor Lang;1981 Amsterdam • NL
[1981] *Algorithmic Languages* ED: BAKKER DE, J.W. & VLIET VAN, J.C. • PUBL (X 0809) North Holland: Amsterdam xxvi+431pp
• DAT&PL 1981 Oct; Amsterdam, NL • ISBN 0-444-86285-4, LC-No 81-14139

P 3084 Autom Theor Prov After 25 Yea;1983 Denver • USA
[1984] *Automated Theorem Proving After 25 Years. Proceedings of the Special Session on Automatic Theorem Proving, 89nd Annual Meeting of the American Mathematical Society (AMS)* ED: BLEDSOE, W.W. & LOVELAND, D.W. • SER (S 3313) Contemp Math 29 • PUBL (X 0803) Amer Math Soc: Providence ix+360pp
• DAT&PL 1983 Jan; Denver, CO, USA • ISBN 0-8218-5027-X, LC-No 84-9226

P 3087 Germ Worksh on Artif Intell (7);1983 Dassel • D
[1983] *GWAI-83. Proceedings of the 7th German Workshop on Artificial Intelligence* ED: NEUMANN, B. • SER Informatik-Fachberichte 76 • PUBL (X 0811) Springer: Heidelberg & New York vi+240pp
• DAT&PL 1983 Sep; Dassel/Solling, D • ISBN 3-540-12871-9, LC-No 84-130170

P 3088 Univer Alg & Link Log, Alg, Combin, Comp Sci;1983 Darmstadt • D
[1984] *Universal Algebra and Its Links with Logic, Algebra, Combinatorics and Computer Science. Proceedings of the '25. Arbeitstagung ueber Allgemeine Algebra'* ED: BURMEISTER, P. & GANTER, B. & HERRMAN, C. & KEIMEL, K. & POGUNTKE, W. WILLE, R. • SER Research and Exposition in Math 4 • PUBL Heldermann: Berlin vii+243pp
• DAT&PL 1983 -?-; Darmstadt, D • LC-No 85-129031

P 3092 Congr Naz Logica;1979 Montecatini Terme • I
[1981] *Atti del Congresso Nazionale di Logica* ED: BERNINI, S. • PUBL (X 1732) Bibliopolis: Napoli 735pp
• DAT&PL 1979 Oct; Montecatini Terme, I • LC-No 81-198713

P 3103 Adv Cybern Syst;1972 Oxford • GB
[1974] *Advances in Cybernetics and Systems* ED: ROSE, J. • PUBL (X 0836) Gordon & Breach: New York 3 Vols: xx+1730pp
• DAT&PL 1972 -?-; Oxford, GB • LC-No 75-316901

P 3108 AISB/GI Conf Artif Intel;1978 Hamburg • D
[1979] *Proceedings of the AISB/GI Conference on Artificial Intelligence* ED: SLEEMAN, D. & NAGEL, H.-H. • PUBL SSAISB and GI: Hamburg x+379pp
• DAT&PL 1978 Jul; Hamburg, D
• REM The Society for the Study of Artificial Intelligence and Simulation of Behaviour (SSAISB) and Gesellschaft fuer Informatik (GI)

P 3117 All-Union Conf Autom Contr-Eng Cybern (4);1968 Tbilisi • SU
[1972] *Avtomaty, Gibridnye Upravlyayushchie Mashiny. Trudy 4 Vsesoyuznogo Soveshchaniya po Avtomaticheskomu Upravleniyu (Teknicheskoj Kibernetike). Tom 3 (Automata, Hybrid and Control Machines. Proceedings of the 4th All-Union Conference on Automatic Control-Engineering Cybernetics. Vol.3)* ED: TRAPEZNIKOV, V.A. & GAVRILOV, M.A. & AJZERMAN, M.A. • PUBL (X 2027) Nauka: Moskva 268pp
• DAT&PL 1968 Sep; Tbilisi, SU • LC-No 72-338109

P 3143 Comput Orient Learn Process;1974 Bonas • F
[1976] *Computer Oriented Learning Processes* ED: SIMON, J.C. • SER NATO Advanced Study Institutes Series, Series E: Applied Science; 14 • PUBL (X 1317) Noordhoff: Groningen vi+595pp
• DAT&PL 1974 Aug; Bonas, F • LC-NO 76-378425

P 3146 Constr Math;1980 Las Cruces • USA
[1981] *Constructive Mathematics. Proceedings of the New Mexico State University Conference* ED: RICHMAN, F. • SER (S 3301) Lect Notes Math 873 • PUBL (X 0811) Springer: Heidelberg & New York vii+347pp
• DAT&PL 1980 Aug; Las Cruces, NM, USA • ISBN 3-540-10850-5, LC-No 81-9345

Proceedings

P 3165 FCT'81 Fund of Comput Th;1981 Szeged • H
[1981] *Fundamentals of Computation Theory. Proceedings of the 1981 International FCT-Conference* ED: GECSEG, F. • SER (S 3302) Lect Notes Comput Sci 117 • PUBL (X 0811) Springer: Heidelberg & New York xi+471pp
• DAT&PL 1981 Aug; Szeged, H • ISBN 3-540-10854-8, LC-No 81-13533

P 3185 Interpr & Found of Quantum Th;1979 Marburg • D
[1981] *Interpretations and Foundations of Quantum Theory* ED: NEUMANN, H. • SER Grundlagen der Exakten Naturwissenschaften 5 • PUBL (X 0876) Bibl Inst: Mannheim 144pp
• DAT&PL 1979 May; Marburg, D • ISBN 3-411-01601-9, LC-No 81-150650

P 3196 Kompl von Entscheid Probl;1973/74 Zuerich • CH
[1976] *Komplexitaet von Entscheidungsproblemen. Ein Seminar* ED: SPECKER, E. & STRASSEN, V. • SER (S 3302) Lect Notes Comput Sci 43 • PUBL (X 0811) Springer: Heidelberg & New York ii+217pp
• DAT&PL 1973 -?-; Zuerich, CH • ISBN 3-540-07805-3, LC-No 76-25088

P 3201 Logic Symposia;1979/80 Hakone • J
[1981] *Logic Symposia Hakone 1979,1980* ED: MUELLER, GERT H. & TAKEUTI, G. & TUGUE, T. • SER (S 3301) Lect Notes Math 891 • PUBL (X 0811) Springer: Heidelberg & New York xi+394pp
• DAT&PL 1979 Mar; Hakone, J, 1980 Feb; Hakone, J • ISBN 3-540-11161-1, LC-No 81-18424

P 3210 Math Founds of Comput Sci (9);1980 Rydzyna • PL
[1980] *Mathematical Foundations of Computer Science. Proceedings of the 9th Symposium* ED: DEMBINSKI, P. • SER (S 3302) Lect Notes Comput Sci 88 • PUBL (X 0811) Springer: Heidelberg & New York viii+723pp
• DAT&PL 1980 Sep; Rydzyna, PL • ISBN 3-540-10027-X, LC-No 80-20087

P 3234 Probl Founds of Physics;1977 Varenna • I
[1979] *Problems in the Foundations of Physics. Proceedings of the International School of Physics "Enrico Fermi". Course 72* ED: TORALDO DI FRANCIA, G. • PUBL (X 0809) North Holland: Amsterdam xii+497pp
• DAT&PL 1977 Jul; Varenna, I • ISBN 0-444-85285-9, LC-No 79-13069

P 3238 Conf Theoret Comput Sci;1977 Waterloo ON • CDN
[1977] *Proceedings of a Conference on Theoretical Computer Science* PUBL University of Waterloo, Computer Science Department: Waterloo iv+283pp
• DAT&PL 1977 Aug; Waterloo, ON, CDN

P 3242 Manitoba Conf Num Math (3);1973 Winnipeg • CDN
[1974] *Proceedings of the 3rd Manitoba Conference on Numerical Mathematics* ED: THOMAS, R.S.D. & WILLIAMS, H.C. • SER (S 3726) Congressus Numerantium 9 • PUBL (X 2420) Utilitas Mathematica Publ: Winnipeg viii+454pp
• DAT&PL 1973 Oct; Winnipeg, MB, CDN • LC-No 76-368374

P 3257 Semant of Concurr Comput;1979 Evian • F
[1979] *Semantics of Concurrent Computation* ED: KAHN, G. • SER (S 3302) Lect Notes Comput Sci 70 • PUBL (X 0811) Springer: Heidelberg & New York vi+368pp
• DAT&PL 1979 Jul; Evian, F • ISBN 3-540-09511-X, LC-No 79-15956

P 3269 Set Th Found Math (Kurepa);1977 Beograd • YU
[1977] *Set Theory. Foundations of Mathematics* SER (J 2887) Zbor Radova, NS 2(10) • PUBL (X 3727) Beograd Mat Inst: Belgrade 152pp
• DAT&PL 1977 Aug; Beograd, YU • LC-No 79-373865

P 3272 Strukt & Metodol Rek Soz Sist Avtom Ustr;1973 Budapest • H
[1973] *Struktura i Metodologicheskie Rekomendacii po Sozdaniju Sistem Avtomatizirovannogo Proektirovanija Diskretnyh Ustrojstv (Structure and Methodological Recommendations for the Creation of Systems of Automated Design of Dicrete Devices)* SER (J 2845) Tanulmanyok 21 • PUBL (X 0928) Akad Kiado: Budapest 118pp
• DAT&PL 1973 Feb; Budapest, H

P 3300 Asilomar Conf Circ, Syst & Comput (8);1974 Pacific Grove • USA
[1975] *8th Asilomar Conference on Circuits, Systems and Computers* ED: PARKER, S.R. • PUBL (X 1777) Western Periodicals: Hollywood xi+740pp
• DAT&PL 1974 Dec; Pacific Grove, CA, USA • LC-No 86-643237

P 3361 APL Congr '73;1973 Copenhagen • DK
[1973] *APL Congress 73* ED: GJERLOV, P. & HELMS, H.J. & NIELSEN, J. • PUBL (X 0809) North Holland: Amsterdam xv+506pp • ALT PUBL (X 0838) Amer Elsevier: New York
• DAT&PL 1973 Aug; Copenhagen, DK • LC-No 73-84650

P 3397 Mat & Mat Obrazov (7);1978 Sl"nchev Bryag • BG
[1978] *Matematika i Matematichesko Obrazovanie. Doklady na 7a Proletna Konferentsiya na S"juza na Matematicite v B"lgariya (Mathematics and Mathematical Education. Proceedings of the 7th Spring Conference of the Union of Bulgarian Mathematicians)* PUBL (X 2237) Publ Bulg Acad Sci: Sofia 551pp
• DAT&PL 1978 Apr; Sl"nchev Bryag (Sunny Beach), BG
• LC-No 80-643123

P 3411 Theor Comput Sci (3);1977 Darmstadt • D
[1977] *Theoretical Computer Science. 3rd GI Conference (Gesellschaft fuer Informatik)* ED: TZSCHACH, H. & WALDSCHMIDT, H. & WALTER, H.K.-G. • SER (S 3302) Lect Notes Comput Sci 48 • PUBL (X 0811) Springer: Heidelberg & New York vii+418pp
• DAT&PL 1977 Mar; Darmstadt, D • ISBN 3-540-08138-0, LC-No 77-3607

P 3425 SE Conf Combin, Graph Th & Comput (8);1977 Baton Rouge • USA
[1977] *Proceedings of the 8th Southeastern Conference on Combinatorics, Graph Theory and Computing* ED: HOFFMAN, F. & LESNIAK-FOSTER, L. & MCCARTHY, D. & MULLIN, R.C. & REID, K.B. & STANTON, R.G. • SER (S 3726) Congressus Numerantium 19 • PUBL (X 2420) Utilitas Mathematica Publ: Winnipeg vii+685pp
• DAT&PL 1977 Feb; Baton Rouge, LA, USA • ISBN 0-919628-19-2

P 3452 Symb & Algeb Comput;1979 Marseille • F
[1979] *EUROSAM '79. Symbolic and Algebraic Computation. An International Symposium on Symbolic and Algebraic Manipulation* ED: NG, E.W. • SER (S 3302) Lect Notes Comput Sci 72 • PUBL (X 0811) Springer: Heidelberg & New York xv+557pp
• DAT&PL 1979 Jun; Marseille, F • ISBN 3-540-09519-5, LC-No 79-16996

P 3475 Theor Comput Sci (5);1981 Karlsruhe • D
[1981] *Theoretical Computer Science. 5th GI-Conference (Gesellschaft fuer Informatik)* ED: DEUSSEN, P. • SER (S 3302) Lect Notes Comput Sci 104 • PUBL (X 0811) Springer: Heidelberg & New York vii + 261pp
• DAT&PL 1981 Mar; Karlsruhe, D • ISBN 3-540-10576-X, LC-No 81-4579

P 3479 Math Stud of Inform Process;1978 Kyoto • J
[1979] *Mathematical Studies of Information Processing* ED: BLUM, E.K. & PAUL, M. & TAKASU, S. • SER (S 3302) Lect Notes Comput Sci 75 • PUBL (X 0811) Springer: Heidelberg & New York viii + 629pp
• DAT&PL 1978 Aug; Kyoto, J • ISBN 3-540-09541-1, LC-No 80-494262

P 3482 Logic & Algor (Specker);1980 Zuerich • CH
[1982] *Logic and Algorithmic. An International Symposium Held in Honour of Ernst Specker* ED: ENGELER, E. & LAEUCHLI, H. & STRASSEN, V. • SER Monogr l'Enseign Math 30 • PUBL (X 3718) Enseign Math, Univ Geneve: Geneve 392pp
• DAT&PL 1980 Feb; Zuerich, CH • LC-No 82-184564

P 3512 Jorn Mat Luso-Espanol (7);1980 St Feliu de Guixois • E
[1980] *Actas de las Septimas Jornadas Luso-Espanolas de Matematica* SER (J 2562) Publ Sec Mat Univ Autonoma Barcelona 20,21,22 • PUBL Univ. Autonoma de Barcelona: Barcelona, E 276pp,263pp,329pp
• DAT&PL 1980 May; Sant Feliu de Guixois, E

P 3525 Int Comput Symp;1975 Antibes • F
[1975] *International Computing Symposium 1975* ED: GELENBE, E. & POTIER, D. • PUBL (X 0809) North Holland: Amsterdam vii + 266pp • ALT PUBL (X 0838) Amer Elsevier: New York
• DAT&PL 1975 Jun; Antibes, F • ISBN 0-7204-2839-4, LC-No 75-38988

P 3527 GI Jahrestag (4);1974 Berlin • D
[1975] *GI - 4. Jahrestagung (Gesellschaft fuer Informatik)* ED: SIEFKES, D. • SER (S 3302) Lect Notes Comput Sci 26 • PUBL (X 0811) Springer: Heidelberg & New York ix + 748pp
• DAT&PL 1974 Oct; Berlin, D • ISBN 3-540-07141-5

P 3535 IEEE Symp Founds of Comput Sci (20);1979 San Juan • PRI
[1979] *20th Annual IEEE Symposium on Foundations of Computer Science* PUBL (X 2179) IEEE: New York vii + 431pp
• DAT&PL 1979 Oct; San Juan, PRI • LC-No 80-646634
• REM IEEE Publication No. 79CH1471-26

P 3540 West-Coast Conf Combin, Graph Th & Comput;1979 Arcata • USA
[1980] *Proceedings of the West Coast Conference on Combinatorics, Graph Theory and Computing* ED: CHINN, P.Z. & MCCARTHY, D. • SER (S 3726) Congressus Numerantium 26 • PUBL (X 2420) Utilitas Mathematica Publ: Winnipeg vi + 322pp
• DAT&PL 1979 Sep; Arcata, CA, USA • ISBN 0-919628-26-5

P 3548 Generalized Fcts & Operat Calc;1975 Varna • BG
[1979] *(Generalized Functions and Operational Calculus)* ED: DIMOVSKI, I. • PUBL (X 2237) Publ Bulg Acad Sci: Sofia 216pp
• DAT&PL 1975 Oct; Varna, BG • LC-No 80-474830

P 3578 IEEE Symp Found of Comput Sci (19);1978 Ann Arbor • USA
[1978] *19th Annual IEEE Symposium on Foundations of Computer Science* PUBL (X 2179) IEEE: New York v + 290pp
• DAT&PL 1978 Oct; Ann Arbor, MI, USA • LC-No 80-646634

P 3580 Repr des Conn & Raison dans Sci Homme;1979 St Maximin • F
[1980] *Representation des Connaissances et Raisonnement dans les Sciences de l'Homme. Papers from the IRIA-LISH Colloquium* ED: BORILLO, M. • PUBL (X 2732) INRIA: Le Chesnay Cedex iii + 607pp
• DAT&PL 1979 Sep; Saint Maximin, F • ISBN 2-7261-0234-4

P 3589 Reconn des Formes & Intell Artif;1979 Toulouse • F
[1979] *Reconnaissance des Formes et Intelligence Artificielle. Tome I,II,III. Second AFCET-IRIA Congress* PUBL (X 2732) INRIA: Le Chesnay Cedex 3 Vols:423pp, 426pp, 368pp
• DAT&PL 1979 Sep; Toulouse, F

P 3621 Frege Konferenz (2);1984 Schwerin • DDR
[1984] *Frege Konferenz* ED: WECHSUNG, G. • SER Mathematical Research - Mathematische Forschungen 20 • PUBL (X 0911) Akademie Verlag: Berlin 408pp
• DAT&PL 1984 Sep; Schwerin, DDR • LC-No 84-248695

P 3622 Int Congr Log, Meth & Phil of Sci (6,Proc);1979 Hannover • D
[1982] *Proceedings of the 6th International Congress for Logic, Methodology and Philosophy of Science* ED: COHEN, L.J. & LOS, J. & PFEIFFER, H. & PODEWSKI, K.-P. • SER (S 3303) Stud Logic Found Math 104 • PUBL (X 0809) North Holland: Amsterdam xiv + 842pp
• DAT&PL 1979 Aug; Hannover, D • LC-No 80-12713

P 3634 Patras Logic Symp;1980 Patras • GR
[1982] *Patras Logic Symposium* ED: METAKIDES, G. • SER (S 3303) Stud Logic Found Math 109 • PUBL (X 0809) North Holland: Amsterdam ix + 391pp
• DAT&PL 1980 Aug; Patras, GR • ISBN 0-444-86476-8, LC-No 82-14107

P 3642 Colloq Math Log in Computer Sci;1978 Salgotarjan • H
[1981] *Proceedings of the Colloquium on Mathematical Logic in Computer Science* ED: DOEMALKI, B. & GERGELY, T. • SER (S 3312) Coll Math Soc Janos Bolyai 26 • PUBL (X 0809) North Holland: Amsterdam 758pp • ALT PUBL (X 3725) Bolyai Janos Mat Tars: Budapest
• DAT&PL 1978 Sep; Salgotarjan, H • ISBN 0-444-85440-1, LC-No 82-101557

P 3658 Math Founds of Comput Sci (11);1984 Prague • CS
[1984] *Mathematical Foundations of Computer Science 1984. Proceedings of the 11th Symposium* ED: CHYTIL, M.P. & KOUBEK, V. • SER (S 3302) Lect Notes Comput Sci 176 • PUBL (X 0811) Springer: Heidelberg & New York 581pp
• DAT&PL 1984 Sep; Prague, CS • ISBN 3-540-13372-0, LC-No 83-13980

P 3668 Log & Founds of Math;1983 Kyoto • J
[1984] *Logic and the Foundations of Mathematics* ED: TUGUE, T. • SER (S 2308) Symp Kyoto Univ Res Inst Math Sci (RIMS) 516 • PUBL (X 2441) Kyoto Univ Res Inst Math Sci: Kyoto ii + 195pp
• DAT&PL 1983 Oct; Kyoto, J

P 3669 SE Asian Conf on Log;1981 Singapore • SGP
[1983] *Southeast Asian Conference on Logic* ED: CHONG, CHI TAT & WICKS, M.J. • SER (S 3303) Stud Logic Found Math 111 • PUBL (X 0809) North Holland: Amsterdam xiv+210pp
• DAT&PL 1981 Nov; Singapore, SGP • ISBN 0-444-86706-6, LC-No 83-11458

P 3673 Int Symp Multi-Val Log (10);1980 Evanston • USA
[1980] *Proceeding of the 10th International Symposium on Multiple-Valued Logic* PUBL (X 2179) IEEE: New York iv+281pp
• DAT&PL 1980 May; Evanston, IL, USA • LC-No 79-641110

P 3696 Logic Colloq;1964 Bristol • GB
[1965] *Logic Colloquium 1964* SER (J 0096) Acta Philos Fenn 18 • PUBL (X 0809) North Holland: Amsterdam
• DAT&PL 1964 Aug; Bristol, GB

P 3705 Int Symp Multi-Val Log (11);1981 Oklahoma City & Norman • USA
[1981] *Proceedings of the 11th International Symposium on Multiple-Valued Logic.* PUBL (X 2179) IEEE: New York vi+298pp
• DAT&PL 1981 May; Oklahoma City, OK, USA & Norman, OK, USA • LC-No 79-641110

P 3708 Herbrand Symp Logic Colloq;1981 Marseille • F
[1982] *Proceedings of the Herbrand Symposium Logic. Colloquium '81* ED: STERN, J. • SER (S 3303) Stud Logic Found Math 107 • PUBL (X 0809) North Holland: Amsterdam xi+384pp
• DAT&PL 1981 Jul; Marseille, F • ISBN 0-444-86417-2, LC-No 82-6433

P 3710 Logic Colloq;1982 Firenze • I
[1984] *Proceedings of the Logic Colloquium '82* ED: LOLLI, G. & LONGO, G. & MARCJA, A. • SER (S 3303) Stud Logic Found Math 112 • PUBL (X 0809) North Holland: Amsterdam viii+358pp
• DAT&PL 1982 Aug; Firenze, I • ISBN 0-444-86876-3, LC-No 84-1630

P 3730 Actual Math;1981 Luxembourg • L
[1982] *Actualites Mathematiques. Actes du 6e Congres du Groupement des Mathematiciens d'Expression Latin* ED: PIER, J.P. • PUBL (X 0834) Gauthier-Villars: Paris vii+542pp
• DAT&PL 1981 Sep; Luxembourg, L • ISBN 2-04-015427-2, LC-No 83-150505

P 3738 Log of Progr;1981 Yorktown Heights • USA
[1982] *Logics of Programs. Papers Presented at the Workshop* ED: KOZEN, D. • SER (S 3302) Lect Notes Comput Sci 131 • PUBL (X 0811) Springer: Heidelberg & New York vi+429pp
• DAT&PL 1981 May; Yorktown Heights, NY, USA • ISBN 3-540-11212-X, LC-No 82-3219

P 3753 Penser Math;1981 Paris • F
[1982] *Penser les Mathematiques. Lectures from a Seminar on Philosophy and Mathematics* ED: GUENARD, F. & LELIEVRE, G. • SER Collection Points: Serie Sciences 29 • PUBL (X 1349) Seuil: Paris 277pp
• DAT&PL 1981; Paris, F • ISBN 2-02-006061-2, LC-No 82-144038

P 3754 Argumentation;1978 Groningen • NL
[1982] *Argumentation. Approaches to Theory Formation. Papers from the Groningen Conference on the Theory of Argumentation* ED: BARTH, E.M. & MARTENS, J.L. • SER Studies in Language Companion Series (SLCS) 8 • PUBL (X 2257) Benjamins: Amsterdam xvii+333pp
• DAT&PL 1978 Oct; Groningen, NL • ISBN 90-272-3007-2

P 3758 Algeb Conf (2);1981 Novi Sad • YU
[1982] *Algebraic Conference. Proceedings of the 2nd Conference* ED: GILEZAN, K. • PUBL (X 4030) Univ Novom Sadu, Inst Mat: Novi Sad vii+162pp
• DAT&PL 1982 May; Novi Sad, YU • LC-No 83-169879

P 3759 Phil of Econ;1981 Muenchen • D
[1982] *Philosophy of Economics* ED: STEGMUELLER, W. & BALZER, W. & SPOHN, W. • SER Studies in Contemporary Economics 2 • PUBL (X 0811) Springer: Heidelberg & New York viii+306pp
• DAT&PL 1981 Jul; Muenchen, D • ISBN 3-540-11927-2, LC-No 82-16947

P 3774 Th d'Ensembl de Quine;1981 Louvain-la-Neuve • B
[1982] *La Theorie des Ensembles de Quine* SER (S 3935) Cah Cent Log (Louvain) 4 • PUBL (X 3734) Cabay: Louvain-la-Neuve iii+76pp
• DAT&PL 1981 Oct; Louvain-la-Neuve, B • ISBN 2-87077-138-X

P 3808 Rect Trends in Math;1982 Reinhardsbrunn • D
[1982] *Recent Trends in Mathematics* ED: KURKE, H. & MECKE, J. & TRIEBEL, H. & THIELE, R. • SER Teubner-Texte zur Mathematik 50 • PUBL (X 0823) Teubner: Stuttgart 336pp
• DAT&PL 1982 Oct; Reinhardsbrunn, D • LC-No 83-168766

P 3817 Algeb Homage:Ring Th & Rel Top (Jacobson);1981 New Haven • USA
[1982] *Algebraists' Homage: Papers in Ring Theory and Related Topics. Proceedings of the Conference on Algebra in Honor of Nathan Jacobson* ED: AMITSUR, S.A. & SALTMAN, D.J. & SELIGMAN, G.B. • SER (S 3313) Contemp Math 13 • PUBL (X 0803) Amer Math Soc: Providence viii+409pp
• DAT&PL 1981 Jun; New Haven, CT, USA • ISBN 0-8218-5013-X, LC-No 82-18934

P 3829 Atti Incontri Log Mat (1);1982 Siena • I
[1983] *Atti Degli Incontri di Logica Matematica* ED: BERNARDI, C. • PUBL (X 3812) Univ Siena, Dip Mat: Siena 398pp
• DAT&PL 1982 Jan; Siena, I, 1982 Apr; Siena, I, 1982 Jun; Siena, I

P 3833 Symp Semigroups (6);1982 Kyoto • J
[1982] *Semigroup Theory and Its Related Fields. Proceedings of the 6th Symposium on Semigroups* ED: YOSHIDA, R. • PUBL Osaka College of Pharmacy ii+79pp • ALT PUBL Ritsumeikan Univ: Kyoto
• DAT&PL 1982 Oct; Kyoto, J

P 3836 Automata, Lang & Progr (9);1982 Aarhus • DK
[1982] *Automata, Languages and Programming. 9th Colloquium* ED: NIELSEN, M. & SCHMIDT, E.M. • SER (S 3302) Lect Notes Comput Sci 140 • PUBL (X 0811) Springer: Heidelberg & New York vii+614pp
• DAT&PL 1982 Jul; Aarhus, DK • ISBN 3-540-11576-5, LC-No 83-10430
• REM Also Abbreviated as ICALP '82

P 3840 Autom Deduct (6);1982 New York • USA
[1982] *6th Conference on Automated Deduction* ED: LOVELAND, D.W. • SER (S 3302) Lect Notes Comput Sci 138 • PUBL (X 0811) Springer: Heidelberg & New York vii+389pp
• DAT&PL 1982 Jun; New York, NY, USA • ISBN 3-540-11558-7, LC-No 82-5948

P 3842 AFCET Math de l'Inf;1982 Paris • F
[1982] *Les Mathematiques de l'Informatique. Association Francaise pour la Cybernetique Economique et Technique (AFCET)* ED: ROBINET, B. • PUBL AFCET: Paris vii+561pp
• DAT&PL 1982 Mar; Paris, F

P 3845 Conf Math Service of Man (2,Proc)(Feriet);1982 Las Palmas • E
[1982] *2nd World Conference on Mathematics at the Service of Man. Dedicated to the Remembrance of Joseph Kampe de Feriet (1893-1982)* ED: BALLESTER, A. & CARDUS, D. & TRILLAS, E. • PUBL Universidad Politecnica: Las Palmas xxx+696pp
• DAT&PL 1982 Jun; Las Palmas, E

P 3851 Automata, Lang & Progr (10);1983 Barcelona • E
[1983] *Automata, Languages and Programming. 10th Colloquium* ED: DIAZ, J. • SER (S 3302) Lect Notes Comput Sci 154 • PUBL (X 0811) Springer: Heidelberg & New York viii+734pp
• DAT&PL 1983 Jul; Barcelona, E • ISBN 3-540-12317-2, LC-No 83-10435
• REM Also Abbreviated as ICALP 83

P 3858 Adequate Modeling of Syst;1982 Bad Honnef • D
[1983] *Adequate Modeling of Systems. Proceedings of the International Working Conference on Model Realism* ED: WEDDE, H. • PUBL (X 0811) Springer: Heidelberg & New York xi+336pp
• DAT&PL 1982 Apr; Bad Honnef, D • ISBN 3-540-12567-1, LC-No 83-10335

P 3862 Theor Comput Sci (6);1983 Dortmund • D
[1982] *Theoretical Computer Science. 6th GI Conference. (Gesellschaft fuer Informatik)* ED: CREMERS, A.B. & KRIEGEL, H.P. • SER (S 3302) Lect Notes Comput Sci 145 • PUBL (X 0811) Springer: Heidelberg & New York x+365pp
• DAT&PL 1983 Jan; Dortmund, D • ISBN 3-540-11973-6, LC-No 82-19673

P 3867 Int Symp Progr (5);1982 Turin • I
[1982] *International Symposium on Programming, ISI'82* ED: DEZANI-CIANCAGLINI, M. & MONTANARI, U. • SER (S 3302) Lect Notes Comput Sci 137 • PUBL (X 0811) Springer: Heidelberg & New York vi+406pp
• DAT&PL 1982 Apr; Turin, I • ISBN 3-540-11494-7, LC-No 82-161894

P 3869 Stoch Diff Syts (2);1982 Bad Honnef • D
[1982] *Stochastic Differential Systems. Conference of the SFB of the DFG at the University at Bonn* ED: KOHLMANN, M. & CHRISTOPEIT, N. • SER (S 3302) Lect Notes Comput Sci 43 • PUBL (X 0811) Springer: Heidelberg & New York xii+377pp
• DAT&PL 1982 Jun; Bad Honnef, D • ISBN 3-540-07805-3, LC-No 83-137266

P 3870 Congr Catala de Log Mat (1);1982 Barcelona • E
[1982] *1er Congres Catala de Logica Matematica. Actes* PUBL Univ Politecnica & Univ Barcelona: Barcelona 130pp
• DAT&PL 1982 -?-; Barcelona, E

P 3877 Rab Sov Sist & Met Anal Vych EhVM & Prim Teor Fiz;1979 Dubna • SU
[1980] *Rabochee Soveshchanie po Sistemam i Metodam Analiticheskikh Vychislenij na EhVM i Ikh Primeneniyu v Teoreticheskoj Fizike (International Conference on Systems and Techniques of Analytical Computing and Their Applications in Theoretical Physics)* PUBL Ob"edinennyj Inst Yadernykh Issledovanij: Dubna 187pp
• DAT&PL 1979 Sep; Dubna, SU

P 3897 Inform Th (3);1980 Liblice • CS
[1980] *Proceedings of the 3rd Czechoslovak-Soviet-Hungarian Seminar on Information Theory* ED: DRIML, M. & VISEK, J.A. • PUBL (X 1226) Academia: Prague 224pp
• DAT&PL 1980 Jun; Liblice, CS

P 3906 Th Found of Progr Methodol;1981 Marktoberdorf • D
[1982] *Theoretical Foundations of Programming Methodology* ED: BROY, M. & SCHMIDT, G. • SER NATO Adv Study Inst Ser C 91 • PUBL (X 0835) Reidel: Dordrecht xiii+658pp
• DAT&PL 1981 -?-; Marktoberdorf, D • LC-No 82-12347

P 3908 Graph-Gram & Appl to Comput Sci (2);1982 Neunkirchen • D
[1983] *Graph-Grammars and Their Application to Computer Science.* ED: EHRIG, H. & NAGL, M. & ROZENBERG, G. • SER (S 3302) Lect Notes Comput Sci 153 • PUBL (X 0811) Springer: Heidelberg & New York vii+452pp
• DAT&PL 1982 Oct; Neunkirchen, D • ISBN 3-540-12310-5, LC-No 83-6677

P 3912 Diff Geom & Diff Equations;1980 Beijing • TJ
[1982] *(Proceedings of the 1980 Beijing Symposium of Differential Geometry and Differential Equations. Vol. 1,2,3)* ED: CHERN, S.S. & WU, WEN-TSUEN • PUBL (X 1876) Kexue Chubanshe: Beijing ix+1743pp • ALT PUBL (X 0836) Gordon & Breach: New York
• DAT&PL 1980 Aug; Beijing, TJ • ISBN 0-677-16420-3, LC-No 83-135012

P 4009 Europ Comput Algeb Conf;1983 London • GB
[1983] *Computer Algebra. EUROCAL '83. European Computer Algebra Conference* ED: HULZEN VAN, J.A. • SER (S 3302) Lect Notes Comput Sci 162 • PUBL (X 0811) Springer: Heidelberg & New York xiii+305pp
• DAT&PL 1983 Mar; London, GB • ISBN 3-540-12868-9, LC-No 83-20442

P 4015 SE Conf Combin, Graph Th & Comput (14);1983 Boca Raton • USA
[1983] *Proceedings of the 14th Southeastern Conference on Combinatorics, Graph Theory and Computing* SER (S 3726) Congressus Numerantium 39,40 • PUBL (X 2420) Utilitas Mathematica Publ: Winnipeg 2 Vols: 472pp, 449pp
• DAT&PL 1983 Apr; Boca Raton, FL, USA

P 4018 Th & Appl Sing Perturbations;1981 Oberwolfach • D
[1982] *Theory and Applications of Singular Perturbations* ED: ECKHAUS, W. & JAGER DE, E.M. • SER (S 3301) Lect Notes Math 942 • PUBL (X 0811) Springer: Heidelberg & New York v+363pp
• DAT&PL 1981 Aug; Oberwolfach, D • ISBN 3-540-11584-6, LC-No 82-10678

P 4037 Itog Nauch Konf Kazan Univ;1962 Kazan • SU
[1963] *Itogovaya Nauchnaya Konferentsiya Kazanskogo Gosudarstvennogo Universiteta imeni V.I. Ul'yanova-Lenina (Science Survey Conference. Kazan State University)* PUBL (X 3605) Kazan Gos Univ: Kazan' 194pp
• DAT&PL 1962 -?-; Kazan, SU

P 4041 Log, Inform, Law;1981 Firenze • I
[1982] *Selected Papers from the International Conference on Logic, Informatics, Law.* ED: CIAMPI, C. (VOL 1) & MARTINO, A.A. (VOL 2) • PUBL (X 0809) North Holland: Amsterdam 2 Vols: xii+476pp, xii+518pp
• DAT&PL 1981 Apr; Firenze, I • ISBN 0-444-86414-8 (V1), ISBN 0-444-86415-6 (V2), LC-No 82-6465
• REM Vol 1: Artificial Intelligence and Legal Information Systems. Vol 2: Deontic Logic, Computational Linguistics, and Legal Information Systems.

P 4043 Winter School on Abstract Anal (11);1983 Zelezna Ruda • CS
[1984] *Proceedings of the 11th Winter School on Abstract Analysis* SER (J 3522) Rend Circ Mat Palermo, Ser 2 Suppl 3 • PUBL (X 2528) Circolo Mat: Palermo 411pp
• DAT&PL 1983 Jan; Zelezna Ruda, CS

P 4048 Allerton Conf Commun, Control & Comput (16);1978 Monticello • USA
[1978] *16th Annual Allerton Conference on Communication, Control and Computing* PUBL (X 1285) Univ Ill Pr: Urbana xv+994pp
• DAT&PL 1978 Oct; Monticello, IL, USA

P 4051 Fuzzy Set & Possibility Th;1980 Acapulco • MEX
[1982] *Fuzzy Set and Possibility Theory. Recent Developments* ED: YAGER, R.R. • PUBL (X 0869) Pergamon Pr: Oxford xiv+633pp
• DAT&PL 1980 Dec; Acapulco, MEX • ISBN 0-08-026294-5, LC-No 81-10582

P 4072 Kuenstl Intell;1982 Teisendorf • D
[1982] *Kuenstliche Intelligenz* ED: BIBEL, W. & SIEKMANN, J.H. • SER Informatik-Fachberichte 59 • PUBL (X 0811) Springer: Heidelberg & New York xiii+383pp
• DAT&PL 1982 Mar: Teisendorf, D • ISBN 3-540-11974-4, LC-No 84-207604

P 4076 IFAC Symp Artif Intell;1983 Leningrad • SU
[1984] *Proceedings of the IFAC Symposium on Artificial Intelligence* SER IFAC Proc Series 9 • PUBL (X 0869) Pergamon Pr: Oxford vii+558pp
• DAT&PL 1983 Oct; Leningrad, SU • LC-No 84-9229

P 4081 Many-Val Log & Appl;1982 Kyoto • J
[1982] *Many-Valued Logic and Its Applications* ED: HASEGAWA, T. • SER (S 2308) Symp Kyoto Univ Res Inst Math Sci (RIMS) 455 • PUBL (X 2441) Kyoto Univ Res Inst Math Sci: Kyoto iii+280pp
• DAT&PL 1982 Jan; Kyoto, J

P 4109 B-Val Anal & Nonstand Anal;1978 Kyoto • J
[1978] *Boole Daisuchi no Kaisekigaku to Chojun Kaiseki (B-Valued Analysis and Nonstandard Analysis)* ED: NAMBA, K. • SER (S 2308) Symp Kyoto Univ Res Inst Math Sci (RIMS) 336 • PUBL (X 2441) Kyoto Univ Res Inst Math Sci: Kyoto ii+149pp
• DAT&PL 1978 May; Kyoto, J
• REM All Papers in Japanese

P 4113 Found of Math;1982 Kyoto • J
[1983] *Foundations of Mathematics* ED: SHINODA, J. • SER (S 2308) Symp Kyoto Univ Res Inst Math Sci (RIMS) 480 • PUBL (X 2441) Kyoto Univ Res Inst Math Sci: Kyoto iii+236pp
• DAT&PL 1982 Oct; Kyoto, J

P 4153 B-Val Anal & Nonstand Anal;1981 Kyoto • J
[1981] *Boolean-Algebra-Valued Analysis and Nonstandard Analysis* ED: NANBA, K. • SER (S 2308) Symp Kyoto Univ Res Inst Math Sci (RIMS) 441 • PUBL (X 2441) Kyoto Univ Res Inst Math Sci: Kyoto ii+158pp
• DAT&PL 1981 May; Kyoto, J

P 4180 Int Congr Log, Meth & Phil of Sci (7,Pap);1983 Salzburg • A
[1985] *Foundations of Logic and Linguistics. Problems and Their Solutions. Papers from the 7th International Congress of Logic, Methodology and Philosophy of Science* ED: DORN, G. & WEINGARTNER, P. • PUBL (X 1332) Plenum Publ: New York xi+715pp
• DAT&PL 1983 Jul; Salzburg, A • ISBN 0-306-41916-5, LC-No 84-26518

P 4200 Adv in Equilibrum Th;1984 Indianapolis • USA
[1985] *Advances in Equilibrum Theory. Proceedings of the Conference on General Equilibrium Theory* ED: ALIPRANTIS, C.P. & BURKINSHAW, O. & ROTHMAN, N.J. • SER (S 3314) Lect Notes Econ & Math Syst 244 • PUBL (X 0811) Springer: Heidelberg & New York v+235pp
• DAT&PL 1984 Feb; Indianapolis, --, USA • LC-No 85-8039, ISBN 3-540-15229-6

P 4201 AFIPS Proc;1971 • USA
[1971] *Proc. AFIPS*
• DAT&PL 1971 -?-; -?-

P 4202 AFIPS Spring Jt Computer Conf;1970 • USA
[1970] *AFIPS Spring Joint Computer Conference*
• DAT&PL 1970 -?-; -?- • LC-No 80-649584

P 4216 Artif Intell & Heuristic Progr;1970 Menaggio • I
[1971] *Artificial Intelligence and Heuristic Programming. Proceedings of the NATO Advance Study Institute* ED: FINDLER, N.V. & MELTZER, B. • PUBL (X 0838) Amer Elsevier: New York ii+327pp
• DAT&PL 1970 Aug; Menaggio, I • ISBN 0-444-19597-1, LC-No 71-164036

P 4223 Math Tool & Model:1981/82 Toulouse & Paris • F
[1983] *Mathematical Tools and Models for Control, System Analysis & Signal Process*
• DAT&PL 1981 -?-; Toulouse, F, 1982 -?-; Paris, F

P 4227 Modern Anal Probab;1982 New Haven • USA
[1984] *Conference on Modern Analysis Probability* ED: BEALS, R. • SER (S 3313) Contemp Math 26 • PUBL (X 0803) Amer Math Soc: Providence xi+432pp
• DAT&PL 1982 -?-; New Haven, • LC-No 84-484, ISBN 0-8218-5030-X

P 4230 Groupe Etude Anal Ultrametrique (9);1982 Marseille • F
[1983] *Groupe d'Etude d'Analyse Ultrametrique, 9e Annee, 1981/82, Fasc 3. Including Papers from the Conference on p-adic Analysis and Its Applications* ED: AMICE, Y. & CHRISTOL, G. & ROBBA, P. • PUBL (X 1623) Univ Paris VI Inst Poincare: Paris 125pp
• DAT&PL 1982 Sep; Marseille, F • ISBN 2-85926-281-4, LC-No 76-642267

P 4244 Rewriting Techn & Appl (1);1985 Dijon • F
[1985] *Rewriting Techniques and Applications. Papers from the 1st Conference* ED: JOUANNAUD, J.-P. • SER (S 3302) Lect Notes Comput Sci 202 • PUBL (X 0811) Springer: Heidelberg & New York vi+441pp
• DAT&PL 1985 May; Dijon, F • ISBN 3-540-15976-2, LC-No 85-22164

P 4248 Th Approaches Non-Num Probl Solving • USA
[1970] *Theoretical Approaches to Non-Numerical Problem Solving. Proceedings of the 4th Systems Symposium* ED: BANERIJ, R.B. & MESAROVIC, M.D. • SER Lect Notes in Operations Research & Math Systems 28 • PUBL (X 0811) Springer: Heidelberg & New York vi+466pp
• LC-No 79-121996

P 4250 Int Joint Conf Artif Intell (1);1969 Washington • USA
[1969] *Consequence - Finding. Proceedings of the 1st International Joint Conference Artificial Intelligence* ED: WALKER, D.E. & NORTON, L.M. • PUBL ?: Bedford ix+715pp
• DAT&PL 1969 May; Washington, DC, USA • LC-No 74-9045

P 4251 Int Joint Conf Artif Intell (2);1971 London • GB
[1971] *Proceedings of the 2nd International Joint Conference Artificial Intelligence*
• DAT&PL 1971 -?-; London, GB

P 4254 Joint Autom Control Conf;1980 San Francisco • USA
[1980] *Proceedings of the 1980 Joint Automatic Control Conference. Vol 1, 2. An ASME Century 2 Emerging Technology Conference* PUBL (X 2179) IEEE: New York 2 Vols
• DAT&PL 1980 Aug; San Francisco, CA, USA • LC-No 80-67739

P 4264 Math Probl in Th Phys;1981 Berlin • D
[1982] *Mathematical Problems in Theoretical Physics. Proceedings of the 6th International Conference on Mathematical Physics* ED: SCHRADER, R. & SEILER, R. & UHLENBROCK, P.A. • SER Lect Notes in Physics 153 • PUBL (X 0811) Springer: Heidelberg & New York xii+429pp
• DAT&PL 1981 Aug; Berlin, D • ISBN 3-540-11192-1, LC-No 82-817

P 4275 Trends in Math Psychol;1983 Bruxelles • B
[1984] *Trends in Mathematical Psychology. Papers Presented at the 14th European Mathematical Psychology Group Meeting* ED: DEGREEF, E. & BUGGENHAUT VAN, J. • SER Advances in Psychology 20 • PUBL (X 0809) North Holland: Amsterdam xiii+478pp
• DAT&PL 1983 Sep; Bruxelles, B • LC-No 84-6032

P 4300 Stoch Method & Comput Techn in Quant Dynam;1984 Graz • A
[1985] *Stochastic Methods and Computer Techniques in Quantum Dynamics* SER Acta Physica Austriaca Suppl 26
• PUBL (X 0902) Springer: Wien vi+452pp
• DAT&PL 1984 Apr; Graz, A • LC-No 84-14073

P 4306 Future Generat Comput Syst • NL
[1985] *Future Generation Computer Systems* PUBL (X 0809) North Holland: Amsterdam

P 4313 Int Congr Math (II,14);1983 Warsaw • PL
[1984] *Proceedings of the International Congress of Mathematicians, Vol 1,2* ED: CIESIELSKI, Z. & OLECH, C.
• PUBL (X 1034) PWN: Warsaw lxii+1730pp • ALT PUBL (X 0809) North Holland: Amsterdam
• DAT&PL 1983 Aug; Warsaw, PL • ISBN 83-01-05523-5, LC-No 84-13788

P 4446 Graph Th;1983 Novi Sad • YU
[1984] *Graph Theory* ED: CVETKOVIC, D. & GUTMAN, I. & PISANSKI, T. & TOSIC, R. • PUBL (X 4030) Univ Novom Sadu, Inst Mat: Novi Sad ii+319pp
• DAT&PL 1983 Apr; Novi Sad, YU

P 4474 Autom Demonstr;1968 Versailles • F
[1970] *Symposium on Automated Demonstration* ED: LACOMBE, L. & LAUDET, M. & NOLIN, L. & SCHUETZENBERGER, M. • SER (S 3301) Lect Notes Math 125
• PUBL (X 0811) Springer: Heidelberg & New York 310pp
• DAT&PL 1968 Dec; Versailles, F • LC-No 79-117526

P 4564 Math Logic;1963 Xi-An • TJ
[1965] *1963 Nian Cueanguo Shuli Luoji Zhuanye Xueshu Huiyi Lunwen Xuanji (Mathematical Logic. Proceedings of the National Symposium)* PUBL Defence Industry Press: Beijing
• DAT&PL 1963 Oct; Xi-An, TJ

P 4571 Log of Progr;1985 Brooklyn • USA
Logic of Programs ED: PARIKH, R. • SER (S 3302) Lect Notes Comput Sci 193 • PUBL (X 0811) Springer: Heidelberg & New York vi+424pp
• DAT&PL 1985 Jun; Brooklyn, NY, USA • ISBN 3-540-15648-8

P 4601 EUROCAL;1985 Linz • A
[1985] *Proceedings of European Conference on Computer Algebra (EUROCAL)* ED: BUCHBERGER, B. (V1) & CAVINESS, B.F. (V2) • SER (S 3302) Lect Notes Comput Sci 203,204
• PUBL (X 0811) Springer: Heidelberg & New York vi+233pp, xvi+650pp
• DAT&PL 1985 Apr; Linz, A • ISBN 3-540-15983-5 (V1), ISBN 3-540-15984-3 (V2)
• REM Vol 1: Invited Lectures. Vol 2: Research Contributions

P 4621 Log & Models of Concurrent Syst;1984 La Colle-sur-Loup • F
[1985] *Logics and Models of Concurrent Systems. Proceedings of the NATO Advanced Study Institute* ED: APT, K.R. • SER NATO Adv Sci Inst, Ser F 13 • PUBL (X 0811) Springer: Heidelberg & New York viii+498pp
• DAT&PL 1984 Oct; La Colle-sur-Loup, F • ISBN 3-540-15181-8

P 4628 Automata, Lang & Progr (12);1985 Nafplion • GR
[1985] *Automata, Languages and Programming.* ED: BRAUER, W. • SER (S 3302) Lect Notes Comput Sci 194 • PUBL (X 0811) Springer: Heidelberg & New York viii+520pp
• DAT&PL 1985 Jul; Nafplion, GR • ISBN 3-540-15650-X

P 4646 Atti Incontri Log Mat (2);1983/84 Siena • I
[1985] *Atti degli Incontri di Logica Matematica* ED: BERNARDI, C. & PAGLI, P. • PUBL (X 3812) Univ Siena, Dip Mat: Siena 648pp
• DAT&PL 1983 Jan; Siena, I, 1983 Apr; Siena, I, 1984 Jan; Siena, I, 1984 Apr; Siena, I

P 4647 FCT'85 Fund of Comput Th;1985 Cottbus • DDR
[1985] *Fundamentals of Computation Theory. FCT '85* ED: BUDACH, L. • SER (S 3302) Lect Notes Comput Sci 199
• PUBL (X 0811) Springer: Heidelberg & New York xii+542pp
• DAT&PL 1985 Sep; Cottbus, DDR • ISBN 3-540-15689-5

P 4661 Algeb & Log;1984 Zagreb • YU
[1985] *Proceedings of the Conference 'Algebra and Logic'* ED: STOJAKOVIC, Z. • PUBL (X 4030) Univ Novom Sadu, Inst Mat: Novi Sad vi+193pp
• DAT&PL 1984 Jun; Zagreb, YU

P 4672 Found of Softw Tech & Th Comput Sci (5);1985 New Delhi • IND
[1985] *Foundations of Software Technology and Theoretical Computer Science. Proceedings of the 5th Conference* ED: MAHESHWARI, S.N. • SER (S 3302) Lect Notes Comput Sci 206 • PUBL (X 0811) Springer: Heidelberg & New York x+522pp
• DAT&PL 1985 Dec; New Delhi, IND • ISBN 3-540-16042-6

Collection volumes

C 0552 Phil Contemp - Chroniques • I
[1968] *La Philosophie Contemporaine: Chroniques (Contemporary Philosophy: A Survey)* ED: KLIBANSKY, R.
• PUBL (**X** 1319) La Nuova Italia: Firenze 4 Vols
• LC-No 68-55649

C 0553 Morgan de: On Syllogism & other Log Writings • USA
[1966] *On the Syllogism and other Logical Writings by Augustus de Morgan* ED: HEATH, P. • PUBL (**X** 0875) Yale Univ Pr: New Haven xxxi+355pp • ALT PUBL (**X** 0866) Routledge & Kegan Paul: Henley on Thames
• LC-No 66-3469, LC-No 66-71119

C 0558 Contemp Phil Scand • USA
[1972] *Contemporary Philosophy in Scandinavia* ED: OLSON, R.E. & PAUL, A.M. • PUBL (**X** 1291) Johns Hopkins Univ Pr: Baltimore x+508pp
• ISBN 0-8018-1315-8, LC-No 70-148242

C 0567 Form & Strategy in Sci (Woodger) • NL
[1964] *Form and Strategy in Science: Studies Dedicated to Joseph Henry Woodger on the Occasion of His 70th Birthday* ED: GREGG, J.R. & HARRIS, F.T.C. • PUBL (**X** 0835) Reidel: Dordrecht vii+476pp
• LC-No 65-87332

C 0569 Phil of Math Oxford Readings • GB
[1969] *The Philosophy of Mathematics.* ED: HINTIKKA, K.J.J.
• SER Oxford Readings in Philosophy • PUBL (**X** 0894) Oxford Univ Pr: Oxford v+186pp
• ISBN 0-19-875011-0, LC-No 71-441791

C 0570 Form Log & Metodol Nauk • SU
[1964] *Formal'naya Logika i Metodologya Nauki (Formal Logic and Methodology of Science)* ED: TAVANETS, P.V.
• PUBL (**X** 2027) Nauka: Moskva 300pp
• LC-No 66-92737

C 0582 Log Semant & Modal Logika • SU
[1967] *Logicheskaya Semantika i Modal'naya Logika (Logical Semantics and Modal Logic)* ED: TAVANETS, P.V. • PUBL (**X** 2027) Nauka: Moskva 280pp
• LC-No 68-53796

C 0585 Struct, Meth & Meaning (Sheffer) • USA
[1951] *Structure, Method and Meaning. Essays in Honor of Henry M.Sheffer* ED: HENLE, P. & KALLEN, H.M. & LANGER, S.K. • PUBL New York: Liberal Arts Press xvi+306pp
• LC-No 51-2957

C 0587 Quine: Sel Logic Papers • USA
[1966] *Selected Logic Papers of Willard van Orman Quine* PUBL (**X** 0981) Random House: New York x+250pp
• LC-No 66-11147

C 0601 Encycl of Philos • USA
[1967] *Encyclopedia of Philosophy* ED: EDWARDS, P. • PUBL (**X** 0843) Macmillan : New York & London 8 Vols
• LC-No 67-10059

C 0609 Lukasiewicz: Log & Philos Papers • PL
[1961] *Logical and Philosophical Papers by Jan Lukasiewicz* ED: SLUPECKI, J. • PUBL (**X** 1034) PWN: Warsaw 249pp
• REM Translations and Reprints

C 0615 Polish Logic 1920-39 • GB
[1967] *Polish Logic 1920-39.* ED: MCCALL, S. • PUBL (**X** 0815) Clarendon Pr: Oxford viii+406pp
• LC-No 67-106639
• REM Translations of Several Articles

C 0621 Kontrolliertes Denken (Britzelmayr) • D
[1951] *Kontrolliertes Denken. Untersuchungen zum Logikkalkuel und zur Logik der Einzelwissenschaften. Festschrift fuer Wilhelm Britzelmayr* ED: MENNE, A. & WILHELMY, A. & ANGSTL, H. • PUBL (**X** 0826) Alber: Freiburg 122pp
• LC-No 56-30120

C 0622 Essays Found of Math (Fraenkel) • IL
[1961] *Essays on the Foundations of Mathematics: Dedicated to A.A.Fraenkel on His 70th Anniversary* ED: BAR-HILLEL, Y. & POZNANSKI, E.I.J. & RABIN, M.O. & ROBINSON, A. • PUBL (**X** 1299) Magnes Pr: Jerusalem x+351pp • ALT PUBL (**X** 0809) North Holland: Amsterdam & (**X** 0833) Acad Pr: Jerusalem & (**X** 0894) Oxford Univ Pr: Oxford
• LC-No 63-753

C 0640 Log, Autom, Inform • RO
[1971] *Logique, Automatique, Informatique* ED: MOISIL, G.C.
• PUBL (**X** 0871) Acad Rep Soc Romania: Bucharest 456pp
• LC-No 77-880254

C 0643 Philosophie 1946-48 • F
[1950] *Philosophie. Chronique des Annees d'apres la Guerre 1946-1948* ED: BAYER, R. • SER (S 2073) Actualites Sci Indust 1104 (V 12), 1110 (V 14) • PUBL (**X** 0859) Hermann: Paris at least 14 Vols
• REM V 12: Histoire de la Philosohie Metaphysique, Philosophie des Valeurs. V 13: Philosophie des Sciences. V 14: Psychologie, Phenomenologie et Existentialisme.

C 0647 Encycl Britannica • GB
[1950ff] *Encyclopaedia Britannica: A New Survey of Universal Knowledge* ED: YUST, W. • PUBL Encyclopaedia Britannica: Chicago-London-Toronto 24 Vols
• LC-No 50-4840

C 0648 Th of Sets and Topology (Hausdorff) • DDR
[1972] *Theory of Sets and Topology. In Honour of Felix Hausdorff (1868-1942)* ED: ASSER, G. & FLACHSMEYER, J. & RINOV, W. • PUBL (**X** 0806) Dt Verlag Wiss: Berlin 525pp
• LC-No 73-163576

C 0654 Stud in Model Th • USA
[1973] *Studies in Model Theory* ED: MORLEY, M.D. • SER Studies in Mathematics 8 • PUBL (**X** 1298) Math Ass Amer: Washington vii+197pp
• ISBN 0-88385-108-3, LC-No 73-86564

C 0656 Phil Mathematique • F
[1939] *Philosophie Mathematique.* ED: GONSETH, F. • SER (S 2073) Actualites Sci Indust 837 • PUBL (X 0859) Hermann: Paris 100pp

C 0659 Computer Progr & Formal Syst • NL
[1963] *Computer Programming and Formal Systems* ED: BRAFFORT, P. & HIRSCHBERG, D. • SER (S 3303) Stud Logic Found Math • PUBL (X 0809) North Holland: Amsterdam vi + 161pp • ALT PUBL (X 2696) Humanities Pr: Atlantic Highlands
• LC-No 63-3816

C 0668 Neklass Log • SU
[1970] *Neklassucheskaya Logika (Nonclassical Logic)* ED: TAVANETS, P.V. • PUBL (X 2027) Nauka: Moskva 384pp
• LC-No 70-592611

C 0675 From Frege to Goedel • USA
[1967] *From Frege to Goedel: A Source Book in Mathematical Logic 1879-1931* ED: HEIJENOORT VAN, J. • PUBL (X 0858) Harvard Univ Pr: Cambridge x + 660pp • ALT PUBL (X 0894) Oxford Univ Pr: Oxford
• LC-No 67-10905
• REM 2nd ed. 1971; xi + 660pp. Some Articles have been Reprinted 1970; iv + 117pp

C 0676 Handb of Math Psychology • USA
[1963-1965] *Handbook of Mathematical Psychology* ED: LUCE, R.D. & BUSH, R.R. & GALANTER, E. • PUBL (X 0827) Wiley & Sons: New York 3 Vols
• LC-No 63-9428

C 0684 Automation in Lang Transl & Theorem Prov • B
[1968] *Automation in Language Translation and Theorem Proving: Some Applications of Mathematical Logic* ED: BRAFFORT, P. & SCHEEPEN VAN, F. • PUBL (X 1714) Commiss Europ Comm : Bruxelles xv + 295pp
• LC-No 77-467790

C 0699 Logic & Ontology • USA
[1973] *Logic and Ontology* ED: MUNITZ, M.K. • PUBL (X 0924) New York Univ Pr: New York viii + 302pp
• ISBN 0-8147-5363-9, LC-No 72-96480

C 0705 Found of Math (Goedel) • D
[1969] *Foundations of Mathematics. Symposium Papers Commemorating the 60th Birthday of Kurt Goedel.* ED: BULLOFF, J.J. & HOLYOKE, T.C. & HAHN, S.W. • PUBL (X 0811) Springer: Heidelberg & New York xii + 195pp
• ISBN 3-540-04490-6, LC-No 68-28757

C 0707 Stud Phil of C.S. Pierce • USA
[1964] *Studies in the Philosophy of Charles Saunders Pierce. Second Series.* ED: MOORE, E.C. & ROBIN, R.S. • PUBL (X 1305) Univ Massachusetts Pr: Amerhurst xii + 525pp
• LC-No 65-3172

C 0712 Logik & Logikkalkuel • D
[1962] *Logik und Logikkalkuel* ED: KAESBAUER, M. & KUTSCHERA VON, F. • PUBL (X 0826) Alber: Freiburg 249pp

C 0718 The Undecidable • USA
[1965] *The Undecidable. Basic Papers on Undecidable Propositions, Unsolvable Problems and Computable Functions* ED: DAVIS, M. • PUBL (X 0887) Raven Pr: New York 440pp
• LC-No 65-3996

C 0721 Contemp Readings in Log Th • USA
[1967] *Contemporary Readings in Logical Theory* ED: COPI, I.M. & GOULD, J.A. • PUBL (X 0843) Macmillan : New York & London 342pp
• LC-No 67-15535

C 0727 Logic Found of Math (Heyting) • NL
[1968] *Logic and Foundations of Mathematics. Papers Dedicated to A.Heyting on the Occasion of His 70th Birthday* ED: DALEN VAN, D. & DYKMAN, J.G. & KLEENE, S.C. & TROELSTRA, A.S. • PUBL (X 0812) Wolters-Noordhoff : Groningen 249pp
• LC-No 70-408259
• REM Also published as J0020 vol. 20

C 0733 Izbr Vopr Algeb & Log (Mal'tsev) • SU
[1973] *Izbrannye Voprosy Algebry i Logiki: Sbornik Posvyashch Pamyati A.I.Mal'tsev (Selected Questions of Algebra and Logic: Volume Dedicated to the Memory of A.I.Mal'tsev)* ED: ERSHOV, YU.L. & KARGAPOLOV, M.I. & MERZLYAKOV, YU.I. & SMIRNOV, D.M. & SHIRSHOV, A.L. • PUBL (X 2642) Nauka: Novosibirsk 339pp
• LC-No 73-360316

C 0735 Logic & Value (Dahlquist) • S
[1970] *Logic and Value. Essays Dedicated to Thorild Dahlquist on His 50th Birthday* ED: PAULI, T. • SER Filosofiska Studier 9 • PUBL (X 0882) Univ Filos Foeren: Uppsala vi + 247pp
• LC-No 72-186683

C 0742 Phil Mid-Century • I
[1958-1959] *Philosophy in the Mid-Century. A Survey * La Philosophie au Milieu du Vingtieme Siecle. Chroniques* ED: KLIBANSKY, R. • PUBL (X 1319) La Nuova Italia: Firenze 4 Vols

C 0745 Herbrand: Log Writings • NL
[1971] *Logical Writings by J.Herbrand* ED: GOLDFARB, W.D. • PUBL (X 0835) Reidel: Dordrecht vii + 312pp • ALT PUBL (X 0858) Harvard Univ Pr: Cambridge vii + 312pp
• ISBN 90-277-0176-8, LC-No 74-146963
• TRANSL OF [1968] (C 2486) Herbrand: Ecrits Logiques

C 0749 Contrib Logic & Methodol (Bochenski) • NL
[1965] *Contributions to Logic and Methodology in Honour of J.M.Bochenski* ED: TYMIENIECKA, A.T. & PARSONS, C. • PUBL (X 0809) North Holland: Amsterdam xviii + 326pp
• LC-No 66-1566

C 0751 Th of Math Mach • GB
[1963] *Theory of Mathematical Machines.* ED: BAZILEVSKIJ, YU.YA. & JACKSON, J.M. • PUBL (X 0869) Pergamon Pr: Oxford xii + 264pp • ALT PUBL (X 0843) Macmillan : New York & London
• LC-No 65-2166, LC-No 60-10214
• TRANSL IN [1964] (C 3463) Vopr Teor Mat Mashin

C 0769 L'Antinom Ment nel Pensiero Contemp • I
[1961] *L'Antinomia del Mentitiore nel Pensiero Contemporaneo, Da Peirce a Tarski* ED: RIVETTI BARBO, F. • PUBL (X 1364) Vita e Pensiero: Milano xxxvii + 740pp
• LC-No 65-69081

C 0772 Model Th & Topoi • D
[1975] *Model Theory and Topoi* ED: LAWVERE, F.W. & MAURER, C. & WRAITH, G.C. • SER (S 3301) Lect Notes Math 445 • PUBL (X 0811) Springer: Heidelberg & New York 354pp
• ISBN 3-540-07164-4, LC-No 75-20007

C 1009 Wang: Survey Math Logic • TJ
[1963] *Wang,H.: Survey of Mathematical Logic* PUBL
(**X 1876**) Kexue Chubanshe: Beijing • ALT PUBL (**X 0809**) North Holland: Amsterdam 1963

C 1017 Logical Studies • USA
[1957] *Logical Studies* ED: WRIGHT VON, G.H. • SER International Library of Psychology, Philosophy and Scientific Method • PUBL (**X 2696**) Humanities Pr: Atlantic Highlands xi+195pp
• LC-No 58-732

C 1032 Philosopher of the Century (Russell) • GB
[1967] *Bertrand Russell: Philosopher of the Century. Essays in his Honour* ED: SCHOENEMANN, R. • PUBL (**X 0959**) Allen & Unwin: London 326pp • ALT PUBL (**X 0960**) Little, Brown & Co: Boston & Toronto
• LC-No 67-95441

C 1074 Tr Algeb Sem Riga 1969 • SU
[1969] *Trudy Rizhskogo Algebraicheskogo Seminara.* * *Rizhskij Algebraicheskij Seminar (Proceedings of the Riga Seminar on Algebra)* ED: PLOTKIN, B.I. ET AL. • PUBL (**X 0895**) Latv Valsts (Gos) Univ : Riga 310pp
• LC-No 74-549672

C 1098 Skolem: Select Works in Logic • N
[1970] *Skolem,T.A.: Selected Works in Logic* ED: FENSTAD, J.E. • SER Scandinavian University Books • PUBL (**X 1554**) Universitesforlaget: Oslo 732pp
• LC-No 74-485971

C 1105 Phil of Math. Sel Readings • USA
[1964] *Philosophy of Mathematics. Selected Readings.* ED: BENACERRAF, P. & PUTNAM, H. • SER Prentice Hall Philosophy Series • PUBL (**X 0819**) Prentice Hall: Englewood Cliffs vii+563pp • ALT PUBL (**X 1096**) Blackwell: Oxford
• LC-No 64-13252

C 1128 Modelling of Mind: Computers & Intelligence • USA
[1963] *The Modelling of Mind. Computers and Intelligence.* ED: SAYRE, K.M. & CROSSON, F.J. • PUBL (**X 0845**) Univ Notre Dame Pr: Notre Dame xi+275pp
• LC-No 63-19324

C 1134 Log Way of Doing Things • USA
[1969] *The Logical Way of Doing Things* ED: LAMBERT, K.
• PUBL (**X 0875**) Yale Univ Pr: New Haven xii+325pp
• LC-No 69-15450

C 1155 Log Issl (Moskva) • SU
[1959] *Logicheskie Issledovaniya (Logical Investigations)* PUBL (**X 0899**) Akad Nauk SSSR : Moskva
• LC-No 60-40373

C 1159 Tarski: Logic, Semantics, Metamathematics • GB
[1956] *Logic, Semantics, Metamathematics. Papers from 1923 to 1938 by Alfred Tarski* PUBL (**X 0815**) Clarendon Pr: Oxford 471pp
• LC-No 56-4171
• TRANSL IN [1964] (**C 1186**) Tarski: Logique, Semantique, Metamath
• REM Translations of Several Articles

C 1160 Cantor: Ges Abhandlungen • D
[1932] *Georg Cantors Gesammelte Abhandlungen Mathematischen und Philosophischen Inhalts mit Erlaeuternden Anmerkungen sowie mit Ergaenzungen aus dem Briefwechsel Cantor-Dedekind* ED: ZERMELO, E. • PUBL (**X 0811**) Springer: Heidelberg & New York 486pp • ALT PUBL (**X 0892**) Olms: Hildesheim 1966
• ISBN 3-540-09849-6
• REM 2nd ed. 1980

C 1161 Weyl: Ges Abhandlungen • D
[1968] *Hermann Weyls Gesammelte Abhandlungen* ED: CHANDRASEKHARAN, K. • PUBL (**X 0811**) Springer: Heidelberg & New York 4 Vols: xviii+2830pp
• ISBN 3-540-90330-5, LC-No 68-19815

C 1162 Hilbert: Ges Abhandlungen • D
[1935] *David Hilberts Gesammelte Abhandlungen* PUBL (**X 0811**) Springer: Heidelberg & New York 3 Vols: xvi+539pp, viii+453pp, vii+435pp
• ISBN 3-540-04877-4, LC-No 32-23172
• REM 2nd Edition 1970

C 1186 Tarski: Logique, Semantique, Metamath • F
[1964] *Tarski,A.: Logique, Semantique, Metamathematique. 1923-1944. Tome 1* PUBL (**X 0850**) Colin: Paris
• TRANSL OF [1956] (**C 1159**) Tarski: Logic, Semantics, Metamathematics
• REM Revised and extended French ed. 1972

C 1202 Rect Devel Switch Th • USA
[1971] *Recent Developments in Switching Theory* ED: MUKHAPODHYAY, A. • PUBL (**X 0801**) Academic Pr: New York xiii+436pp
• ISBN 0-12-509850-2, LC-No 70-137618

C 1217 Value Distr Th Compl Anal & Rel Topics Diff Geom • USA
[1974] *Value Distribution Theory Part A: Proc. Tulane Univ. Program on Value Distribution Theory in Complex Analysis and Related Topics in Differential Geometry 1972-1973.* ED: KUJALA, R.O. & VITTER III, A.L. • SER Pure and Applied Math 25 • PUBL (**X 1684**) Dekker: New York xi+269pp
• LC-No 73-89281

C 1389 Modality, Morality, Probl of Sense & Nonsense (Hallden) • S
[1973] *Modality, Morality and other Problems of Sense and Nonsense. Essays Dedicated to Soeren Hallden.* ED: HANSSON, B. ET AL. • PUBL (**X 1493**) Gleerup: Lund vi+221pp
• ISBN 91-40-02780-5, LC-No 73-180115

C 1409 Gentzen: Collected Papers • NL
[1969] *The Collected Papers of Gentzen* ED: SZABO, M.E.
• SER (**S 3303**) Stud Logic Found Math • PUBL (**X 0809**) North Holland: Amsterdam xii+338pp
• ISBN 0-7204-2254-X, LC-No 71-97201

C 1413 Tekh Kibernetika • SU
[1965] *Tekhnicheskaya Kibernetika (Technical Cybernetics)* ED: TSIPKIN, YA.Z. ET AL • PUBL (**X 2027**) Nauka: Moskva 429pp
• LC-No 67-115871

C 1443 Algebra, Vyp 2 (Irkutsk) • SU
[1973] *Algebra. Vyp 2* ED: FRIDMAN, E.I. • PUBL (**X 1006**) Irkutsk Gos Univ: Irkutsk 164pp
• LC-No 74-645580 • REL PUBL (**C 3549**) Algebra, Vyp 1 (Irkutsk)

Collection volumes

C 1467 Notas Inst Mat & Estatica Sao Paulo • BR
[1973] *Notas Do Instituto da Matematica e Estatica da Universidade de Sao Paulo* SER Serie Matematica 2 • PUBL Universidade de Sao Paulo: Sao Paulo 75pp
• LC-No 79-105015

C 1495 Encycl Universalis • F
[1968-1976] *Encyclopaedia Universalis* PUBL (X 2524) Encyclopaedia Universalis: Paris 20 Vols
• ISBN 2-85229-281-5, LC-No 75-516014

C 1523 Handb of Math Logic • NL
[1977 & 2nd ed. 1978] *Handbook of Mathematical Logic* ED: BARWISE, J. • SER (S 3303) Stud Logic Found Math 90
• PUBL (X 0809) North Holland: Amsterdam xi+1165pp
• ALT PUBL (X 0838) Amer Elsevier: New York
• ISBN 0-7204-2285-X, LC-No 76-26032
• TRANSL IN [1982] (C 1920) Spravochnaya Kniga po Mat Logike, Chast 1-4

C 1525 Semin Init Analyse (9-10) Paris 1969/71 • F
[1970ff] *Seminaire Choquet 9e-10e Annees: 1969-1971. Inititation a l'Analyse* ED: CHOQUET, G. • PUBL (X 1623) Univ Paris VI Inst Poincare: Paris

C 1530 Issl Log Sist (Yanovskaya) • SU
[1970] *Issledovaniya Logicheskikh Sistem (Investigations of Logical Systems)* ED: TAVANETS, P.V. • PUBL (X 2027) Nauka: Moskva 336pp
• LC-No 70-554105

C 1533 Quantoren, Modal, Paradox • DDR
[1972] *Quantoren, Modalitaeten, Paradoxien. Beitraege zur Logik* ED: WESSEL, H. • PUBL (X 0806) Dt Verlag Wiss: Berlin 524pp
• LC-No 72-3662278

C 1566 Semin IRIA Log & Automates 1971 • F
[1971] *Seminaires IRIA Logiques et Automates* PUBL (X 2732) INRIA: Le Chesnay Cedex 205pp
• LC-No 75-509899

C 1602 Lect on Modern Math • USA
[1963-1965] *Lectures on Modern Mathematics* ED: SAATY, T.L. • PUBL (X 0827) Wiley & Sons: New York 3 Vols
• LC-No 63-20369
• TRANSL IN [1965] (C 4652) Lect on Modern Math (Japanese)

C 1627 Etud Phil Sci (Gonseth) • CH
[1950] *Etudes de Philosophie des Sciences, en Hommage a F. Gonseth a l'Occasion de son 60eme Anniversaire* SER Bibliotheque Scientifique,Serie Dialectica 20 • PUBL (X 0272) Griffon: Neuchatel 176pp

C 1654 Probl Logiki • SU
[1963] *Problemy Logiki (Problems of Logic)* ED: YANOVSKAYA, S.A. ET AL. • PUBL (X 0899) Akad Nauk SSSR : Moskva 152pp

C 1698 Logic & Art (Goodman) • USA
[1972] *Logic and Art. Essays in Honor of Nelson Goodman* ED: RUDNER, R. & SCHEFFLER, J. • PUBL (X 1238) Bobbs-Merril: Indianapolis ix+330pp
• LC-No 76-140799

C 1706 Essays Honour J. Hintikka • NL
[1979] *Essays in Honour of Jaakko Hintikka: On the Occasion of His 50th Birthday on January 12, 1979* ED: SAARINEN, E. & HILPINEN, R. & NIINILUOTO, I. & PROVENCE, M. • SER (S 3307) Synth Libr 124 • PUBL (X 0835) Reidel: Dordrecht x+386pp
• ISBN 90-277-0916-5, LC-No 78-11364

C 1721 Sydney Category Th Sem, Sydney 1972/73 • AUS
[1974] *Proceedings Sydney Category Theory Seminar. 1972/73* ED: KELLY, G.M. • SER (S 3301) Lect Notes Math 420 • PUBL (X 0811) Springer: Heidelberg & New York vi+375pp
• ISBN 3-540-06966-6

C 1749 Kasusth, Klassifik, Semant Interpr • D
[1977] *Kasustheorie, Klassifikation, Semantische Interpretation. Beitraege zur Lexikologie und Semantik* ED: HEGER, K. & PETOETI, J.S. • SER Papiere zur Textlinguistik 11 • PUBL Buske Verlag: Hamburg vii+358pp
• ISBN 3-87118-297-4, LC-No 78-350284

C 1786 Lukasiewicz: Select Works • NL
[1970] *Selected Works by Jan Lukasiewicz* ED: BORKOWSKI, L. • SER (S 3303) Stud Logic Found Math • PUBL (X 0809) North Holland: Amsterdam xii+405pp • ALT PUBL (X 1034) PWN: Warsaw xii+405pp
• ISBN 0-7204-2252-3, LC-No 74-96734

C 1832 Katoba no Tetsugaku • J
[1972] *Kotoba no Tetsugaku (Philosophy of Language)* ED: SAKAMOTO, H. • SER Gendai Tetsugaku Sensho 4 • PUBL Gakubunsha: Tokyo 380pp
• LC-No 73-808077
• REM Translation of the Series-Title: Selected Modern Philosophical Works

C 1849 Grundzuege & Verfahren der Rechtslog • D
[1974-1977] *Grundzuege und Grundverfahren der Rechtslogik* ED: TAMMELO, T. & SCHREINER, H. • PUBL (X 1553) Dokumentation Saur: Muenchen 2 Vols
• ISBN 3-7940-2627-6 (V1), ISBN 3-7940-2659-4 (V2), LC-No 75-570416

C 1856 Log Enterprise • USA
[1975] *The Logical Enterprise* ED: MARCUS, R.B. & ANDERSON, A.R. & MARTIN, R.M. • PUBL (X 0875) Yale Univ Pr: New Haven x+261pp
• ISBN 0-300-01790-1, LC-No 74-20084

C 1891 Studien zu Frege • D
[1976] *Studien zu Frege* ED: SCHIRN, M. • SER Problemata 42-44 • PUBL (X 1267) Frommann: Stuttgart 3 Vols: 317pp, 303pp, 201pp
• ISBN 3-7728-0616-3 (V1), ISBN 3-7728-0618-X (V2), ISBN 3-7728-0620-1 (V3), LC-No 77-469242
• REM Vol 1: Logik und Philosophie der Mathematik. Vol 2: Logik und Sprachphilosophie. Vol 3: Logik und Semantik.

C 1920 Spravochnaya Kniga po Mat Logike, Chast 1-4 • SU
[1982] *Spravochnaya Kniga po Matematicheskoj Logike. Chast. 1-4 (Handbook of Mathematical Logic. Part 1-4)* ED: ERSHOV, YU.L. & PALYUTIN, E.A. & TAJMANOV, A.D. • PUBL (X 2027) Nauka: Moskva 4 Vols: 392pp, 375pp, 360pp, 391pp
• TRANSL OF [1977] (C 1523) Handb of Math Logic
• REM The Translation Contains a New Supplement by Palyutin,E.A. Part 1: Model Theory. Part 2: Set Theory. Part 3: Recursion Theory. Part 4: Proof Theory and Constructive Mathematics.

C 1936 Log Th & Semant Anal (Kanger) • NL
[1974] *Logical Theory and Semantic Analysis. Essays Dedicated to Stig Kanger on His 50th Birthday* ED: STENLUND, S. & HENSCHEN-DAHLQUIST, A.-M. & LINDAHL, L. & NORDENFELT, L. & ODELSTAD, J. • SER (S 3307) Synth Libr 63 • PUBL (X 0835) Reidel: Dordrecht v+217pp
• ISBN 90-277-0438-4, LC-No 73-94456

C 1990 Transl Phil Writings Frege • GB
[1952] *Translations from the Philosophical Writings of Gottlob Frege* ED: BLACK, M. & GEACH, P. • PUBL (X 1096) Blackwell: Oxford x+244pp • ALT PUBL (X 0920) Philos Lib: New York
• LC-No 52-11090

C 2012 Comput Sci & Multi-Val Logic • NL
[1977] *Computer Science and Multipled-Valued Logic. Theory and Applications* ED: RINE, D.C. • PUBL (X 0809) North Holland: Amsterdam xiv+548pp • ALT PUBL (X 0838) Amer Elsevier: New York
• ISBN 0-444-11052-6, LC-No 75-40064

C 2065 Teor Nereg Kriv Raz Geom Post • SU
[1979] *Teoriya Neregulyarnykh Krivykh v Razlichnykh Geometricheskykh Postranstvakh (Theorie der Nichtregulaeren Kurven in Verschiedenen Geometrischen Raeumen)* ED: TAJTSLIN, M.A. • PUBL (X 2769) Kazakh Gos Univ: Alma-Ata 136pp

C 2103 Develop of Log Probab (Lakatos) • NL
[1976] *The Development of Logical Probability. Essays in Honour of Imre Lakatos* ED: COHEN, R.S. & FEYERABEND, P.K. & WARTOFSKY, M.W. • SER (S 3311) Boston St Philos Sci 39, (S 3307) Synth Libr 99 • PUBL (X 0835) Reidel: Dordrecht xi+767pp
• LC-No 76-16770

C 2141 Filos Matematica • I
[1967] *La Filosofia della Matematica* ED: CELLUCCI, C.
• PUBL Laterza: Bari 320pp
• LC-No 68-120024

C 2318 Entwicklung Math in DDR • DDR
[1974] *Entwicklung der Mathematik in der DDR: Zum 25. Jahrestag der Gruendung der Deutschen Demokratischen Republik* ED: SACHS, H. & AHRENS, H. ET AL. • PUBL (X 0806) Dt Verlag Wiss: Berlin xx+756pp
• LC-No 75-569431

C 2319 Slozh Vychisl & Algor • SU
[1974] *Slozhnost' Vychislennij i Algorifmov. Sbornik Peredov (Die Kompliziertheit von Berechnungen und Algorithmen. Sammlung von Uebersetzungen)* ED: KOZMIDIADI, V.A. & MASLOV, A.N. & PETRI, N.V. • PUBL (X 0885) Mir: Moskva 392pp
• LC-No 75-568382

C 2351 Ustojchivost' Dvizheniya • SU
[1981] *Ustojchivost' Dvizheniya. Analiticheskaya Mekhanika. Upravlenie Dvizheniem (Stability of Motion. Analytic Mechanics. Control of Motion)* ED: MATROSOV, V.M. & DEMIN, V.G. • PUBL (X 2027) Nauka: Moskva 304pp
• LC-No 81-174927

C 2388 Log-Philos Studien • D
[1959] *Logisch-Philosophische Studien* ED: BOCHENSKI, I.M. & MENNE, A. • PUBL (X 0826) Alber: Freiburg vii+152pp
• TRANSL IN [1962] (C 2389) Logico-Philos Studies

C 2389 Logico-Philos Studies • NL
[1962] *Logico-Philosophical Studies* ED: BOCHENSKI, I.M. & MENNE, A. • PUBL (X 0835) Reidel: Dordrecht
• LC-No 63-3999
• TRANSL OF [1959] (C 2388) Log-Philos Studien

C 2486 Herbrand: Ecrits Logiques • F
[1968] *Herbrand,J.: Ecrits Logiques* ED: HEIJENOORT VAN, J.
• PUBL (X 0840) Pr Univ France: Paris
• LC-No 68-104359
• TRANSL IN [1971] (C 0745) Herbrand: Log Writings

C 2537 Spez Wissenschaftsth • D
[1979] *Spezielle Wissenschaftstheorie* ED: LORENZ, K. • PUBL (X 1174) Gruyter: Berlin 2 Vols
• REM Vol 1: Konstruktionen versus Positionen.

C 2542 Mat Teor Log Vyvoda • SU
[1967] *Matematicheskaya Teoriya Logicheskogo Vyvoda (Mathematical Theory of Logical Deduction)* ED: IDEL'SON, A.V. & MINTS, G.E. • PUBL (X 2027) Nauka: Moskva 351pp
• LC-No 68-41857

C 2546 Ehntsikl Kibern • SU
[1974] *Ehntsiklopediya Kibernetiki* ED: GLUSHKOV, V.M.
• PUBL Glavnaya Redaktsiya Ukrainskoj Sov. Ehntsiklopedii: Kiev
• LC-No 74-359262
• TRANSL IN Lexikon der Kybernetik

C 2577 Issl Formaliz Yazyk & Neklass Log • SU
[1974] *Issledovaniya po Formalizovannym Yazykam i Neklassicheskim Logikam (Investigations on Formalized Languages and Non-Classical Logics)* ED: BOCHVAR, D.A.
• PUBL (X 2027) Nauka: Moskva 275pp
• LC-No 75-554105

C 2578 Filos & Logika • SU
[1974] *Filosofiya i Logika. Filosofiya v Sovremennom Mire (Philosophy and Logic. Philosophy in the Modern World)* ED: TAVANETS, P.V. & SMIRNOV, V.A. • SER Filosofiya v Sovremennom Mire • PUBL (X 2027) Nauka: Moskva 479pp
• LC-No 74-358872

C 2581 Issl Neklass Log & Teor Mnozh • SU
[1979] *Issledovaniya po Neklassicheskim Logikam i Teorii Mnozhestv (Investigations on Non-Classical Logics and Set Theory)* ED: MIKHAJLOV, A.I. ET AL. • PUBL (X 2027) Nauka: Moskva 374pp
• LC-No 80-475529

C 2583 Aktual Probl Log & Metodol Nauki • SU
[1980] *Aktual'nye Problemy Logiki i Metodologii Nauki (Current Problems of Logic and the Methodology of Science. Collection of Scientific Works)* ED: POPOVICH, M.V. • PUBL (X 2199) Naukova Dumka: Kiev 336pp
• LC-No 80-506762

C 2617 Modern Log Survey • NL
[1981] *Modern Logic - A Survey. Historical, Philosophical and Mathematical Aspects of Modern Logic and Its Applications.* ED: AGAZZI, E. • SER (S 3307) Synth Libr 149 • PUBL (X 0835) Reidel: Dordrecht viii+475pp
• ISBN 90-277-1137-2, LC-No 80-22027

C 2621 Mal'tsev: Metamath of Algeb Syst • NL
[1971] *The Metamathematics of Algebraic Systems. Mal'tsev, A.I: Collected Papers, 1936-1967* ED: WELLS III, B.F. • SER (S 3303) Stud Logic Found Math 66 • PUBL (X 0809) North Holland: Amsterdam xviii+494pp
• LC-No 73-157020

C 2953 Log & Probab in Quant Mech • NL
[1976] *Logic and Probability in Quantum Mechanics* ED: SUPPES, P. • SER (S 3307) Synth Libr 78 • PUBL (X 0835) Reidel: Dordrecht xv+541pp

C 2962 Mat Voprosy Teor Intell Mashin • SU
[1975] *Matematicheskie Voprosy Teorii Intellektual'nykh Mashin. Sbornik Trudov (Mathematical Questions of the Theory of Machine Intelligence. Collection of Papers)* ED: KAPITONOVA, YU.V. & ASEL'DEROV, Z.M. • PUBL (X 2522) Akad Nauk Inst Kibernet: Kiev 84pp
• LC-No 77-507084

C 2967 Metodol Probl Mat • SU
[1979] *Metodologicheskie Problemy Matematiki (Methodological Problems of Mathematics)* ED: BORISOV, YU.F. • PUBL (X 2642) Nauka: Novosibirsk 303pp

C 2972 Natural Lang Commun with Computers • D
[1978] *Natural Language Communication with Computers* ED: BOLC, L. • SER (S 3302) Lect Notes Comput Sci 63 • PUBL (X 0811) Springer: Heidelberg & New York iii+292pp
• ISBN 3-540-08911-X, LC-No 78-15393

C 2987 Probab Anal & Rel Topics, Vol 2 • USA
[1979] *Probabilistic Analysis and Related Topics. Vol. 2* ED: BHARUCHA-REID, A.T. • PUBL (X 0801) Academic Pr: New York ix+207pp
• ISBN 0-12-095602-0, LC-No 78-106053

C 3018 Vopr Anal Vychisl Slozhnosti Algor • SU
[1979] *Voprosy Analiza Vychislitel'noj Slozhnosti Algoritmov (Questions of the Computational Complexity of Algorithms)* ED: ISHCHUK, V.A. • SER (S 2653) Prepr Inst Kib, Akad Nauk Ukr SSR 79/14 • PUBL (X 2522) Akad Nauk Inst Kibernet: Kiev 48pp

C 3026 Lukasiewicz Sel Pap on Sent Calc • PL
[1977] *Selected Papers on Lukasiewicz Sentential Calculi* ED: MALINOWSKI, G. & WOJCICKI, R. • PUBL (X 2885) Zakl Narod Wyd Pol Ak: Wroclaw 199pp
• LC-No 80-515606

C 3038 Log-Metodol Issl • SU
[1980] *Logiko-Metodologicheskie Issledovaniya (Studies in Logical Methodology)* ED: STARCHENKO, A.A. • PUBL (X 0898) Moskov Gos Univ: Moskva 375pp
• LC-No 81-482681

C 3046 Decid Theories II - Monad 2nd Ord Th Count Ordinals • D
[1973] *Decidable Theories II. The Monadic Second Order Theory of All Countable Ordinals* ED: MUELLER, GERT H. & SIEFKES, D. • SER (S 3301) Lect Notes Math 328 • PUBL (X 0811) Springer: Heidelberg & New York vi+217pp
• ISBN 3-540-06345-5, LC-No 73-82358

C 3050 Essays Combin Log, Lambda Calc & Formalism (Curry) • USA
[1980] *To H.B. Curry: Essays on Combinatory Logic, Lambda Calculus and Formalism* ED: SELDIN, J.P. & HINDLEY, J.R.
• PUBL (X 0801) Academic Pr: New York xxv+606pp
• ISBN 0-12-349050-2, LC-No 80-40139

C 3080 Asymptotic Anal, Vol 2 • D
[1983] *Asymptotic Analysis. Vol 2: From Theory to Application* ED: VERHULST, F. • SER (S 3301) Lect Notes Math 985 • PUBL (X 0811) Springer: Heidelberg & New York iii+497pp
• ISBN 3-540-12286-9, LC-No 83-6820

C 3094 Moisil: Essais Logiques Non Chrysippiennes • RO
[1972] *Moisil, G.C.: Essais sur les Logiques Non Chrysippiennes* PUBL (X 0871) Acad Rep Soc Romania: Bucharest 820pp
• LC-No 72-345318

C 3174 Twenty-five Years Log Meth Poland • NL
[1977] *Twenty-five Years of Logical Methodology in Poland* ED: PRZELECKI, M. & WOJCICKI, R. • SER (S 3307) Synth Libr 87 • PUBL (X 0835) Reidel: Dordrecht viii+735pp
• LC-No 76-7064
• REM Translated from the Polish

C 3202 Logik, Ethik & Sprache (Freundlich) • D
[1981] *Logik, Ethik und Sprache. Festschrift fuer Rudolf Freundlich* ED: WEINKE, K. • PUBL (X 0814) Oldenbourg: Muenchen viii+309pp
• LC-No 81-120890

C 3211 Mat Ling & Teor Algor • SU
[1978] *Matematicheskaya Lingvistika i Teoriya Algoritmov. Mezhvuzovskij Tematicheskij Sbornik (Mathematical Linguistics and Theory of Algorithms. Interuniversity Thematic Collection)* ED: GLADKIJ, A.V. • PUBL (X 1434) Kalinin Gos Univ: Kalinin 155pp
• LC-No 80-458139

C 3259 Semin Algeb Non Commutative Paris 1974/75 • F
[1975] *Seminaire d'Algebre Non Commutative. Annee 1974-1975. Exposes 1 a 13* SER Publications Mathematiques d'Orsay 154-7543 • PUBL (X 2854) Univ Paris XI UER Math: Paris 75pp

C 3263 Sint Diskr Avtom Upravl Ustr • SU
[1968] *Sintez Diskretnykh Avtomatov i Upravlyayushchikh Ustrojstv (Synthese Diskreter Automaten und Steuervorrichtungen)* PUBL (X 2027) Nauka: Moskva 92pp
• LC-No 70-379696

C 3271 Issl Teor Mnozh & Neklass Logik • SU
[1976] *Issledovaniya po Teorij Mnozhestv i Neklassicheskim Logikam. Sbornik Trudov (Studies in Set Theory and Nonclassical Logics. Collection of Papers)* ED: BOCHVAR, D.A. & GRISHIN, V.N. • PUBL (X 2027) Nauka: Moskva 328pp
• LC-No 77-501571

C 3354 Algeb & Teor Chisel ('77) • SU
[1977] *Algebra i Teoriya Chisel (Algebra and Number Theory)* ED: UNACHEV, KH.YA. • SER Subject Interuniversity Collection 2 • PUBL Kabardino-Balkarskij Gosudarstvennyj Universitet: Nal'chik 172pp

C 3409 Chisl Met Optim & Primen • SU
[1978] *Chislennye Metody Optimizatsij i Ikh Primenenie (Numerical Optimization Methods and Their Applications. Papers for the Seminar)* ED: ZHILINSKAS, A. • SER Optimal Decision Theory Issue 4 • PUBL Institute of Mathematics and Cybernetics of the Academy of Sciences of the Lithuanian SSR: Vilnius; 112pp
• REM Russian. English and Lithuanian Summaries

C 3463 Vopr Teor Mat Mashin • SU
[1964] *Voprosy Teorii Matematicheskikh Mashin (Questions in the Theory of Mathematical Machines)* ED: AKUSHKIJ, I.YA. & GLUZBERG, E.A. • PUBL (**X** 4513) Mashinostroenie: Moskva 247pp
• TRANSL OF [1963] (**C** 0751) Th of Math Mach

C 3494 Stud on Math Progr. Math Meth Oper Res, Vol 1 • H
[1980] *Studies on Mathematical Programming. Mathematical Methods of Operation Research. Vol 1* ED: PREKOPA, A.
• PUBL (**X** 0928) Akad Kiado: Budapest 200pp
• ISBN 963-05-1854-6, LC-No 80-496383

C 3515 Ital Studies in Phil of Sci • NL
[1981] *Italian Studies in the Philosophy of Science* ED: CHIARA SCABIA DALLA, M.L. • SER (**S** 3311) Boston St Philos Sci 47 • PUBL (**X** 0835) Reidel: Dordrecht xi+525pp
• ISBN 90-277-0735-9, LC-No 80-16665

C 3517 Found of Comput Sci, Vol 2, Part 2 • NL
[1976] *Foundations of Computer Science Vol 2, Part 2*. ED: APT, K.R. & BAKKER DE, J.W. • SER (**S** 1605) Math Centr Tracts 82 • PUBL (**X** 1121) Math Centr: Amsterdam 149pp
• ISBN 90-6196-141-6, LC-No 77-368695

C 3549 Algebra, Vyp 1 (Irkutsk) • SU
[1972] *Algebra. Vyp. 1* ED: KOKORIN, A.I. & PENZIN, YU.G.
• PUBL (**X** 1006) Irkutsk Gos Univ: Irkutsk 135pp
• LC-No 74-645580 • REL PUBL (**C** 1443) Algebra, Vyp 2 (Irkutsk)

C 3582 Adv Fuzzy Sets, Possibility Th & Appl • USA
[1983] *Advances in Fuzzy Sets, Possibility Theory, and Applications* ED: WANG, P.P. • PUBL (**X** 1332) Plenum Publ: New York xii+421pp
• ISBN 0-306-41390-6, LC-No 83-11077

C 3626 Rekursiv Mat Analiz • SU
[1970] *Rekursivnyj Matematicheskij Analiz (Recursive Mathematical Analysis)* ED: MINTS, G.E. • SER Prilozhenije 3 • PUBL (**X** 2027) Nauka: Moskva

C 3659 Intens Math • NL
[1985] *Intensional Mathematics* ED: SHAPIRO, S. • SER (**S** 3303) Stud Logic Found Math 113 • PUBL (**X** 0809) North Holland: Amsterdam
• ISBN 0-444-87632-4, LC-No 84-18856

C 3743 Probl Log & Metodol Nauk • SU
[1982] *Problemy Logiki i Metodologii Nauki (Problems of Logic and the Methodology of Science)* ED: TSELISHCHEV, V.V.
• PUBL (**X** 2027) Nauka: Moskva 336pp
• LC-No 82-197677

C 3786 Approx Reason in Decis Anal • NL
[1982] *Approximative Reasoning in Decision Analysis* ED: GUPTA, M.M. & SANCHEZ, E. • PUBL (**X** 0809) North Holland: Amsterdam xix+455pp
• ISBN 0-444-86492-X, LC-No 82-14308

C 3798 Mat Log, Mat Ling & Teor Algor • SU
[1983] *Matematicheskaya Logika, Matematicheskaya Lingvistike i Teoriya Algoritmov (Mathematical Logic, Mathematical Linguistics and Theory of Algorithms)* ED: GLADKIJ, A.V. • PUBL (**X** 1434) Kalinin Gos Univ: Kalinin 116pp
• LC-No 84-157035

C 3806 Issl Teor Progr • SU
[1981] *Issledovaniya po Teoreticheskomn Programmrovaniyu (Investigations in Theoretical Programming)* ED: TAJTSLIN, M.A. • PUBL (**X** 2769) Kazakh Gos Univ: Alma-Ata 104pp

C 3807 Issl Neklass Log & Formal Sist • SU
[1983] *Issledovaniya po Neklassicheskim Logikam i Formal'nym Sistemam (Studies in Nonclassical Logics and Formal Systems)* ED: MIKHAJLOV, A.I. • PUBL (**X** 2027) Nauka: Moskva 360pp
• LC-No 83-181942

C 3832 Avtom Issl Mat • SU
[1982] *Avtomatizatsiya Issledovanij v Matematike. Sbornik Nauchnykh Trudov (Automatization of Investigations in Mathematics. Collection of Scientific Works)* ED: KAPITONOVA, YU.V. • PUBL (**X** 2522) Akad Nauk Inst Kibernet: Kiev 98pp
• LC-No 83-165025

C 3838 Priklad Mat, Vyp 1 • SU
[1981] *Prikladnaya Matematika. Mezhvuzovskij Sbornik Nauchnykh Trudov Vyp. 1 (Applied Mathematics. Interuniversity Collection of Scientific Works No 1)* ED: TONOYAN, R.N. • PUBL (**X** 3559) Erevan Univ: Erevan 95pp

C 3849 Modal & Relevant Log, Vyp 1 • SU
[1982] *Modal'nye i Relevantnye Logiki. Trudy Nauchno-Issledovatel'skogo Seminara po Logike Institute Filosofii AN SSSR Vyp.1 (Modal and Relevant Logics. Vol.1)* ED: SMIRNOV, V.A. & KARPENKO, A.S. • PUBL (**X** 0899) Akad Nauk SSSR : Moskva 107pp

C 3865 Algeb Sistemy (Ivanovo) • SU
[1981] *Algebraicheskie Sistemy. Mezhvuzovskij Sbornik Nauchnykh Trudov (Algebraic Systems. Interuniversitary Collection of Scientific Works)* ED: MOLDAVANSKIJ, D.I.
• PUBL (**X** 2594) Ivanovo Gos Univ: Ivanovo 234pp

C 3881 Prog in Cybern & Syst Res, Vol 11 • USA
[1982] *Progress in Cybernetics and Systems Research Vol. 11. Data Base Design, International Information Systems, Semiotic Systems, Artificial Intelligence, Cybernetics and Philosophy, Special Aspects* ED: TRAPPL, R. & FINDLER, N.V. & HORN, W. • PUBL (**X** 2437) Hemisphere Publ: Washington xv+601pp • ALT PUBL (**X** 0822) McGraw-Hill: New York
• ISBN 0-89116-240-2, LC-No 75-6641

C 3884 Nonstandard Anal - Recent Develop • D
[1983] *Nonstandard Analysis - Recent Developments* ED: HURD, A.E. • SER (**S** 3301) Lect Notes Math 983 • PUBL (**X** 0811) Springer: Heidelberg & New York v+213pp
• ISBN 3-540-12279-6, LC-No 83-5837

C 4055 Wajsberg: Logical Works • PL
[1977] *Wajsberg,M.: Logical Works* ED: SURMA, S.J. • PUBL (**X** 2882) Pol Akad Nauk: Wroclaw 216pp
• LC-No 84-246718

C 4066 Sb Stat po Filos Mat • SU
[1936] *Sbornik Statej po Filosofii Matem.* PUBL -?-: Moskva

C 4082 Handb Wiss Begriffe • D
[1980] *Handbuch Wissenschaftstheoretischer Begriffe, 3 Baende* ED: SPECK, J. • PUBL (**X** 0903) Vandenhoeck & Ruprecht: Goettingen 3 Vols: 780pp
• ISBN 3-525-03313-3 (V 1), ISBN 3-525-03314-1 (V 2), ISBN 3-525-03316-8 (V 3)

C 4085 Handb Philos Log • NL
[1983 & 1984] *Handbook of Philosophical Logic* ED: GABBAY, D. & GUENTHNER, F. • SER (S 3307) Synth Libr 164,165
• PUBL (X 0835) Reidel: Dordrecht 2 Vols: xi+493pp,xi+776pp
• ISBN 90-277-1542-4 (V1), ISBN 90-277-1604-8 (V2), LC-No 83-4277

C 4103 Struktur Teor Relej Ustrojstv • SU
[1963] *Strukturnaya Teoria Relejnykh Ustrojstv* PUBL (X 0899) Akad Nauk SSSR : Moskva
• LC-No 64-44665

C 4106 Phil of Russell • USA
[1944] *The Philosophy of Bertrand Russell* ED: SCHILPP, P.A.
• PUBL (X 1318) Northwestern Univ Pr: Evanston 829pp
• LC-No 44-6786

C 4137 Kleene: Mat Logika • SU
[1974] *Kleene,S.C.: Matematicheskaya Logika (Mathematical Logic)* PUBL (X 0885) Mir: Moskva 480pp
• TRANSL OF Kleene: Mathematical Logic
• REM Contains 2 additional articles of Mints,G.E.

C 4140 Ifs • NL
[1981] *Ifs. Conditionals, Belief, Decision, Chance, and Time* ED: HARPER, W.L. & STALNAKER, R. & PEARCE, G. • SER (S 3308) Univ Western Ontario Ser in Philos of Sci 15
• PUBL (X 0835) Reidel: Dordrecht x+345pp
• ISBN 90-277-1184-4, ISBN 90-277-1220-4, LC-No 80-21638

C 4181 Math Log & Formal Syst (Costa da) • USA
[1985] *Mathematical Logic and Formal Systems. A Collection of Papers in Honor of Newton C.A. da Costa* ED: ALCANTARA DE, L.P. • SER (S 3310) Lect Notes Pure Appl Math 94 • PUBL (X 1684) Dekker: New York xv+297pp
• ISBN 0-8247-7330-6, LC-No 84-25984

C 4187 Math Struct, Comp Math, Math Modell • BG
[1984] *(Mathematical Structures - Computational Mathematics - Mathematical Modelling)* ED: SENDOV, B.
• PUBL (X 2237) Publ Bulg Acad Sci: Sofia 336pp

C 4203 Teor Avtom & Met Formal Sint Vychisl Mash & Sist Kiev 1968 • SU
[1968-1969] *Teoriya Avtomatov i Metody Formalizovannogo Sinteza Vychislitel'nykh Mashin i Sistem. Seminar, Kiev, 1968 g. Vyp. 3,4,5 (Automata Theory and Methods of Formalized Synthesis of Computers and Systems. Proceedings of a Seminar in Kiev 1968, No 3,4,5)* PUBL (X 2199) Naukova Dumka: Kiev 104pp,114pp,111pp

C 4213 Algeb, Combin & Log in Comput Sci • NL
[1985] *Algebra, Combinatorics, and Logic in Computer Science* PUBL (X 0809) North Holland: Amsterdam

C 4232 Altgeld Book • USA
[1976] *The Altgeld Book: Lect. Not. Funct. Anal.* ED: ROSENTHAL, H.P. & DOR, L. • PUBL (X 1285) Univ Ill Pr: Urbana

C 4283 Reduct, Time & Reality • GB
[1981] *Reduction, Time and Reality. Studies in the Philosophy of the Natural Science* ED: HEALEY, R. • PUBL (X 0805) Cambridge Univ Pr: Cambridge, GB xi+202pp
• ISBN 0-521-23708-4, LC-No 80-41535

C 4307 Comput & Thought • USA
[1963] *Computers and Thought* ED: FEIGENBAUM, E.A. & FELDMAN, J. • PUBL (X 0822) McGraw-Hill: New York
• LC-No 63-17596

C 4394 Kroneckers Werke • DDR
[1895-1931] *Kroneckers Werke, Vol I, II, III.1, III.2, IV, V* ED: HENSEL, K. • PUBL (X 1079) Teubner: Leipzig vi+474pp, viii+541pp, vii+474pp, i+216pp, x+509pp, x+528pp

C 4403 Logika, Pozn, Otrazh • SU
[1984] *Logika, Poznanie, Otrazhenie - Sverdlovsk (Logik, Erkenntnis, Reflektion - Sverdlovsk)*

C 4414 Math Syst in Econ • D
[1983] *(Mathematical Systems in Economics)* ED: BECKMANN, M.J. & EICHHORN, W. & KRELLE, W. • PUBL (X 2665) Athenaeum: Frankfurt
• ISBN 3-7610-8290-8

C 4422 Mathesis Univers • CH
[1961] *Mathesis Universalis. Abhandlungen zur Philosophy als Strenge Wissenschaft* ED: SCHOLZ, H. & HERMES, H. & KAMBARTEL, F. & RILLER, J. • PUBL (X 2088) Schwabe: Basel

C 4505 Royce's Log Ess • USA
[1951] *Royce's Logical Essays* PUBL Win.C. Brown Comp: Dubuque, IA, USA

C 4534 Kjoebenhavn Overs • DK
[1907] *Kjoebenhavn Oversettelser (Kjoebenhavn Translations)*

C 4594 Sel Pap Robinson • USA
[1979] *Selected Papers of Abraham Robinson* ED: KEISLER, H.J. & KOERNER, S. & LUXEMBURG, W.A.J. & YOUNG, A.D.
• PUBL (X 0875) Yale Univ Pr: New Haven xxxvii+694pp,xlv+582pp • ALT PUBL (X 0809) North Holland: Amsterdam
• REM Vol 1: Model Theory and Algebra. Vol 2: Nonstandard Analysis and Philosophy

C 4602 Metamath Meth in Geom • D
[1983] *Schwabhaeuser,W. & Szmielev,W. & Tarski,A.: Metamatematische Methoden in der Geometrie* PUBL (X 0811) Springer: Heidelberg & New York viii+482pp
• ISBN 3-540-12958-8

C 4649 Paradiso di Cantor • I
[1978] *Il Paradiso di Cantor* ED: CELLUCCI, C. • PUBL (X 1732) Bibliopolis: Napoli

C 4652 Lect on Modern Math (Japanese) • J
[1965] *(Lectures on Modern Mathematics)* ED: SAATY, T.L.
• PUBL (X 3552) Iwanami Shoten: Tokyo
• TRANSL OF [1963] (C 1602) Lect on Modern Math

C 4659 Autom of Reasoning • D
[1983] *Automation of Reasoning. Vol 1, 2* ED: SIEKMANN, J. & WRIGHTSON, G. • PUBL (X 0811) Springer: Heidelberg & New York xii+525pp,xii+637pp
• ISBN 3-540-12043-2 (V1), ISBN 3-540-12044-0 (V2)
• REM Vol 1: Classical Papers on Computational Logic 1957-1966. Vol 2: Classical Papers on Computational Logic 1967-1970

C 4675 Hintikka: Logiko-Epist Issled • SU
[1980] *Hintikka,K.J.J.: Logiko-Epistemologicheskie Issledovanniya (Hintikka,K.J.J.: Logical-Epistemological Investigations.)* ED: SADOVSKIJ, V.N. & SMIRNOV, V.A. • PUBL (X 2055) Progress: Moskva 448pp

C 4678 Enzykl Geisteswiss Arbeitsmethoden • D
[1968] *Enzyklopaedie der Geisteswissenschaftlichen Arbeitsmethoden. 3. Lieferung: Methoden der Logik und Mathematik. Statistische Methoden* PUBL (X 0814) Oldenbourg: Muenchen

C 4685 Gen Th Bound Value Probl • SU
[1983] *(General Theory of Boundary Value Problems)* ED: SKRYPNIK, A.V. • PUBL (X 2199) Naukova Dumka: Kiev 328pp

C 4688 Prog in Math, Vol 12 • USA
[1972] *Progress in Mathematics. Vol 12: Algebra and Geometry* ED: GAMKRELIDZE, R.V. • PUBL (X 1332) Plenum Publ: New York ix+254pp
• TRANSL OF [1972] (J 1501) Itogi Nauki Tekh, Ser Algeb, Topol, Geom 1968

C 4695 Game-Th Semantics • NL
[1979] *Game-Theoretical Semantics. Essays on Semantics by Hintikka, Carlson, Peacocke, Rantala, and Saarinen* ED: SAARINEN, E. • SER Synthese Language Library 5 • PUBL (X 0835) Reidel: Dordrecht xiv+392pp

Publishers

X 0272 *Editions du Griffon* (Neuchatel, CH) ISBN 2-88006

X 0740 *Shanghai Kexue Jishu Chubanshe (Scientific and Technical Press)* (Shanghai, TJ)

X 0801 *Academic Press* (New York, NY, USA & London, GB) ISBN 0-12

X 0802 *Allyn & Bacon* (London, GB & Boston, MA, USA & Spit Junction, NSW, AUS) ISBN 0-205, ISBN 0-695

X 0803 *American Mathematical Society* (Providence, RI, USA) ISBN 0-8218

X 0804 *Birkhaeuser Verlag* (Basel, CH & Stuttgart, D & Cambridge, MA, USA) ISBN 3-7643

X 0805 *The Cambridge University Press.* (Cambridge, GB & New York, NY, USA & Melbourne, Vic, AUS) ISBN 0-521

X 0806 *VEB Deutscher Verlag der Wissenschaften* (Berlin, DDR) ISBN 3-326

X 0807 *Duke University Press* (Durham, NC, USA) ISBN 0-8223

X 0808 *W. Kohlhammer* (Stuttgart, D & Koeln, D & Berlin, D & Mainz, D) ISBN 3-17

X 0809 *North-Holland Publishing Company.* (Amsterdam, NL & Oxford, GB) ISBN 0-7204 • REL PUBL (**X** 0838) Amer Elsevier: New York

X 0810 *W.B.Saunders Company* (Philadelphia, PA, USA & Eastbourne, GB & Toronto, ON, CDN) ISBN 0-7216 • REL PUBL (**X** 0818) Holt Rinehart & Winston: New York • REM In United Kingdom: Holt-Saunders: Eastbourne, GB

X 0811 *Springer-Verlag* (Heidelberg, D & Berlin, D & New York, NY, USA & Tokyo, J) ISBN 3-540, ISBN 0-387 • REL PUBL (**X** 1231) Barth: Leipzig & (**X** 0902) Springer: Wien

X 0812 *Wolters-Noordhoff* (Groningen, NL) ISBN 90-01 • REL PUBL (**X** 1317) Noordhoff: Groningen

X 0813 *Dover Publications* (New York, NY, USA) ISBN 0-486

X 0814 *R.Oldenbourg Verlag* (Muenchen, D & Wien, A) ISBN 3-486

X 0815 *The Clarendon Press* (Oxford, GB) ISBN 0-19 • REL PUBL (**X** 0894) Oxford Univ Pr: Oxford • REM This Imprint is Used for Academic Books Published by X0894.

X 0816 *Methuen & Company* (London, GB & New York, NY, USA & Agincourt, M, CDN & North Ryde, AUS) ISBN 0-416

X 0818 *Holt Rinehart & Winston* (New York, NY, USA & Toronto, ON, CDN & Artarmon, NSW, AUS & & Eastbourne,GB) ISBN 0-03 • REM In Australia & United Kingdom: Holt-Saunders: Eastbourne, GB & Artarmon, NSW, AUS

X 0819 *Prentice Hall* (Englewood Cliffs, NJ, USA & Brookvale, NSW, AUS & Scarborough, ON, CDN) ISBN 0-13 • REL PUBL (**X** 2040) Winthrop: Cambridge

X 0820 *Interscience Publishers* (New York, NY, USA & Chichester, GB) ISBN 0-470 • REL PUBL (**X** 0827) Wiley & Sons: New York

X 0821 *Wadsworth Publishing Co.* (Belmont, CA, USA & Crows Nest, NSW, AUS & Artamon, NSW, AUS) ISBN 0-534

X 0822 *McGraw-Hill Book Company* (New York, NY, USA & Roseville, NSW, AUS & Isando, ZA & Maidenhead, GB & Singapore, SGP & Scarborough, CDN & Sao Paulo, BR) ISBN 0-07 • REM Member Firms: 1) CRM/McGraw-Hill, Del Mar, CA, USA. 2) CTB/McGraw-Hill, Monterey, CA, USA. 3) Edutronics/McGraw-Hill, Los Angeles, CA, USA. 4) Instruto/McGraw-Hill, Paoli, PA, USA. 5) McGraw-Hill Continuing Education Center, Washington, DC, USA. 6) McGraw-Hill International Book Company, Singapore, SGP. 7) Sheperd's/McGraw-Hill, Colorado Springs, CO, USA. 8) McGraw-Hill do Brasil, Sao Paulo, BR.

X 0823 *B.G.Teubner* (Stuttgart, D) ISBN 3-519 • REM See also X1079

X 0824 *The Free Press* (New York, NY, USA) ISBN 0-02 • REL PUBL (**X** 0843) Macmillan : New York & London

X 0825 *Westkulturverlag Anton Hain* (Meisenheim am Glan, D & Koenigstein, D) ISBN 3-445

X 0826 *Verlag Karl Alber* (Freiburg, D & Muenchen, D) ISBN 3-495 • REL PUBL (**X** 1279) Herder: Freiburg

X 0827 *J.Wiley & Sons* (New York, NY, USA & Chichester, GB & Rexdale, ON, CDN & Auckland, NZ) ISBN 0-471 • REL PUBL (**X** 0942) Norton: New York & (**X** 0820) Intersci Publ: New York & (**X** 0880) Ronald Press: New York & (**X** 2737) Israel Progr Sci Transl: Jerusalem

X 0828 *Edwards Brothers* (Ann Arbor, MI, USA) ISBN 0-910546

X 0832 *Addison-Wesley Publishing Co.* (Reading, MA, USA & London, GB & Don Mills, ON, CDN & North Ryde, NSW, AUS) ISBN 0-201 • REL PUBL (**X** 0867) Benjamin: Reading

X 0833 *Jerusalem Academic Press* (Jerusalem, IL)

X 0834 *Gauthier-Villars Editeur* (Paris, F) ISBN 2-04

X 0835 *D.Reidel Publishing Company* (Dordrecht, NL & Hingham, MA, USA) ISBN 90-277

X 0836 *Gordon & Breach, Science Publishers* (New York, NY, USA & London, GB & Paris, F) ISBN 0-677

X 0837 *Harper & Row, Publishers* (New York, NY, USA & London, GB & Artamon, NSW, AUS & Petone, NZ) ISBN 0-06 • REM Member Firm: Lippincott, J.B.,Company: New York, NY, USA

X 0838 *American Elsevier Publishing Co.* (New York, NY, USA & Amsterdam, NL & London, GB) ISBN 0-444, ISBN 0-525, ISBN 0-87690 • REL PUBL (**X** 0809) North Holland: Amsterdam

X 0840 *Editions Presses Universitaires de France* (Paris, F) ISBN 2-13 • REL PUBL (**X** 2435) Pr Univ France Period: Paris

X 0841 *Blaisdell Publishing Company* (New York, NY, USA & London, GB & Toronto, CDN)

X 0842 *University of Wisconsin Press* (Madison, WI, USA & London, GB) ISBN 0-299 • REL PUBL (**X** 3828) Amer Univ Publ Group: London

X 0843 *Macmillan Publishing Company* (New York, NY, USA & Melbourne, Vic, AUS & London, GB & Toronto, Ont, CDN) ISBN 0-02 • REL PUBL (**X** 2375) Macmillan Journal: London & (**X** 0824) Free Press: New York

X 0844 *Giangiacomo Feltrinelli Editore* (Milano, I) ISBN 88-07

X 0845 *University of Notre Dame Press* (Notre Dame, IN, USA & London, GB) ISBN 0-268

X 0846 *Ferdinand Schoeningh* (Paderborn, D & Muenchen, D & Wien, A) ISBN 3-506

X 0847 *Houghton Mifflin Company* (Boston, MA, USA & Markham, CDN) ISBN 0-395, ISBN 0-89289

X 0848 *Chelsea Publishing Company* (New York, NY, USA) ISBN 0-8284

X 0849 *F.G.Kroonder, Uitgeverij* (Bussum, NL)

X 0850 *Armand Colin, Editeur* (Paris, F) ISBN 2-200

X 0851 *Edmund C. Berkeley and Associates* (New York, NY, USA)

X 0854 *Littlefield, Adams & Co.* (Totowa, NJ, USA) ISBN 0-8226

X 0855 *Hermann Schroedel Verlag* (Hannover, D) ISBN 3-507

X 0856 *Dunod, Editeur* (Paris, F) ISBN 2-04

X 0857 *Princeton University Press* (Princeton, NJ, USA & Guildford, GB) ISBN 0-691

X 0858 *Harvard University Press* (Cambridge, MA, USA & London, GB) ISBN 0-674

X 0859 *Hermann, Editeurs des Sciences et des Arts* (Paris, F) ISBN 2-7056

X 0860 *Cremonese Edizioni* (Firenze, I) ISBN 88-7083

X 0862 *University of Chicago Press* (Chicago, IL, USA & London, GB) ISBN 0-226

X 0863 *Harcourt Brace Jovanovich* (New York, NY, USA & London, GB & Boston, MA, USA & North Ryde, NSW, AUS) ISBN 0-15, ISBN 0-515 • REM Also: HBJ Press: Boston, MA, USA

X 0864 *Van Nostrand, Reinhold* (New York, NY, USA & Scarborough, ON, CDN & Mitcham, Vic, AUS & & Wokingham, GB & Florence, KY, USA) ISBN 0-442

X 0865 *The MIT Press* (Cambridge, MA, USA & London, GB) ISBN 0-262

X 0866 *Routledge & Kegan Paul* (Henley on Thames, GB & Boston, MA, USA) ISBN 0-7100, ISBN 0-7102

X 0867 *W.A. Benjamin* (Reading, MA, USA) ISBN 0-8053 • REL PUBL (**X** 0832) Addison-Wesley: Reading

X 0868 *Penguin Books* (Harmondsworth, GB & New York, NY, USA & Ringwood, Vic, AUS & & Auckland, NZ & Markham, CDN) ISBN 0-14

X 0869 *Pergamon Press* (Oxford, GB & Elmsford, NY, USA & Rushcutters Bay, NSW, AUS & & Willowdale, ON, CDN & Paris, F) ISBN 0-08 • REL PUBL (**X** 0900) Vieweg: Wiesbaden

X 0871 *Academiei Republicii Socialiste Romania Editura (RSR)* (Bucharest, RO)

X 0872 *Albert Blanchard* (Paris, F)

X 0873 *Mouton et Cie.* (Paris, F) ISBN 2-7193

X 0874 *National Press Books* (Palo Alto, CA, USA) ISBN 0-87484

X 0875 *Yale University Press* (New Haven, CT, USA & London, GB) ISBN 0-300

X 0876 *Bibliographisches Institut* (Mannheim, D & Wien, A & Zuerich, CH) ISBN 3-411

X 0877 *Max Niemeyer Verlag* (Tuebingen, D & Halle, DDR) ISBN 3-484

X 0878 *Doubleday & Co.* (London, GB & New York, NY, USA & Garden City, NY, USA & Auckland, NZ & Toronto, ON, CDN & Sydney, AUS) ISBN 0-385 • REM In Australia: Anchor Books: Sydney, AUS

X 0879 *Institut Superieur de Philosophie* (Louvain, B)

X 0880 *Ronald Press & Co.* (New York, NY, USA) ISBN 0-8260 • REL PUBL (**X** 0827) Wiley & Sons: New York

X 0881 *Societa Editrice Il Mulino S.R.I.* (Bologna, I)

X 0882 *Uppsala Universitet, Filosofiska Foereningen och Filosofiska Institutionen* (Uppsala, S)

X 0885 *Izdatel'stvo Mir* (Moskva, SU)

X 0886 *Leicester University Press* (Leicester, GB) ISBN 0-7185

X 0887 *Raven Press* (New York, NY, USA) ISBN 0-89004, ISBN 0-911216

X 0888 *New York University, Institute of Mathematical Sciences.* (New York, NY, USA)

X 0890 *Wissenschaftliche Buchgesellschaft* (Darmstadt, D) ISBN 3-534

X 0892 *Georg Olms Verlag* (Hildesheim, D & New York, NY, USA & Zuerich, CH) ISBN 3-487

X 0893 *Les Presses de l'Universite de Montreal* (Montreal, PQ, CDN) ISBN 2-7606

X 0894 *Oxford University Press* (Oxford, GB & London, GB & Melbourne, Vic, AUS & Don Mills, ON, CDN & Nairobi, EAK & Auckland, NZ & Petaling Jaya, MAL & New York, NY, USA & Karachi, PAK & Harare, ZW) ISBN 0-19 • REL PUBL (**X** 0815) Clarendon Pr: Oxford

X 0895 *Latvijas Valsts Universitate* (Riga, SU)

X 0898 *Izdatel'stvo Moskovskogo Gosudarstvennogo Universiteta* (Moskva, SU)

X 0899 *Izdatel'stvo Akademii Nauk SSSR* (Moskva, SU)

X 0900 *Vieweg, Friedrich & Sohn Verlagsgesellschaft* (Wiesbaden, D) ISBN 3-528 • REL PUBL (**X** 0869) Pergamon Pr: Oxford

X 0902 *Springer-Verlag* (Wien, A) ISBN 3-211 • REL PUBL (X 0811) Springer: Heidelberg & New York

X 0903 *Vandenhoeck & Ruprecht* (Goettingen, D) ISBN 3-525 • REM Member Firms: 1) E. Klotz Verlag, Goettingen, D. 2) Verlag der Medizinischen Psychologie Goettingen, Dr D.& Dr A. Ruprecht, Goettingen, D

X 0904 *Australian National University, Research School of Social Sciences, Department of Philosophy.* (Canberra, ACT, AUS) ISBN 0-909596

X 0905 *Boringhieri Editore* (Torino, I) ISBN 88-339

X 0907 *Westdeutscher Verlag* (Wiesbaden, D) ISBN 3-531

X 0908 *Universitaet Bonn, Mathematisches Institut* (Bonn, D)

X 0909 *Cedam* (Padova, I)

X 0910 *Aschendorffsche Verlagsbuchhandlung* (Muenster, D) ISBN 3-402

X 0911 *Akademie Verlag* (Berlin, DDR)

X 0913 *Novosibirskij Gosudarstvennyj Universitet* (Novosibirsk, SU)

X 0915 *Galois Institute of Mathematics and Art* (Brooklyn, NY, USA)

X 0917 *Nymphenburger Verlagshandlung* (Muenchen, D) ISBN 3-485

X 0918 *Ernst Klett Verlag* (Stuttgart, D) ISBN 3-12 • REL PUBL (X 1255) Deuticke: Wien

X 0920 *Philosophical Library* (New York, NY, USA) ISBN 0-8022

X 0924 *New York University Press* (New York, NY, USA) ISBN 0-8147

X 0926 *University of California Press* (Berkeley, CA, USA & London, GB) ISBN 0-520 • REL PUBL (X 1291) Johns Hopkins Univ Pr: Baltimore

X 0927 *University Tutorial Press* (Slough, GB) ISBN 0-7231

X 0928 *Akademiai Kiado, Publishing House of the Hungarian Academy of Sciences.* (Budapest, H) ISBN 963-05

X 0929 *Weizmann Science Press of Israel* (Jerusalem, IL) ISBN 965-270

X 0932 *Otto Salle Verlag* (Frankfurt, D) ISBN 3-7935

X 0935 *Southern Illinois University Press* (Carbondale, IL, USA) ISBN 0-8093

X 0938 *Izdatel'stvo Leningradskogo Universiteta* (Leningrad, SU)

X 0939 *Hutchinson Publishing Group* (London, GB & Willowdale, ON, CDN & Salem, NH, USA & Auckland, NZ & Richmond, Vic, AUS) ISBN 0-09

X 0942 *W.W.Norton & Co.* (New York, NY, USA & London, GB) ISBN 0-393 • REL PUBL (X 0827) Wiley & Sons: New York

X 0943 *Ginn and Company* (Boston, MA, USA & Aylesbury, GB & Scarborough, ON, CDN) ISBN 0-663

X 0944 *Harvard Cooperative Society* (Cambridge, MA, USA)

X 0949 *St. Martin's Press* (New York, NY, USA) ISBN 0-312, ISBN 0-320

X 0958 *Izdatel'stvo Saratovskogo Universiteta* (Saratov, SU)

X 0959 *George Allen & Unwin* (London, GB & North Sydney, NSW, AUS & Bombay, IND) ISBN 0-04

X 0960 *Little, Brown and Co.* (Boston, MA, USA & Toronto, ON, CDN) ISBN 0-316

X 0978 *Schulthess Polygraphischer Verlag* (Zuerich, CH) ISBN 3-7255

X 0981 *Random House* (New York, NY, USA & Mississauga, ON, CDN) ISBN 0-394

X 0992 *Cornell University Press* (Ithaca, NY, USA & London, GB) ISBN 0-8014

X 0994 *W.H.Freeman and Co.* (San Francisco, CA, USA & Oxford, GB) ISBN 0-7167

X 0995 *Beck'sche C.H. Verlagsbuchhandlung Oscar Beck* (Muenchen, D) ISBN 3-406

X 0997 *Queen's University* (Kingston, ON, CDN) ISBN 0-88911, ISBN 0-9690334

X 0999 *Centre National de la Recherche Scientifique (CNRS), Institut Blaise Pascal* (Paris, F)

X 1000 *Barnes & Noble Books* (Totowa, NJ, USA & New York, NY, USA) ISBN 0-389

X 1004 *D.C.Heath & Co.* (Lexington, MA, USA & Toronto, ON, CDN) ISBN 0-669

X 1006 *Irkutskij Gosudarstvennyj Universitet* (Irkutsk, SU)

X 1007 *Gerold* (Wien, A) ISBN 3-900190

X 1015 *University of Toronto Press* (Toronto, ON, CDN & Buffalo, NY, USA) ISBN 0-8020

X 1016 *Wydawnictwo Uniwersytetu im. Boleslawa Bieruta* (Wroclaw, PL)

X 1026 *Humboldt Verlag* (Frankfurt, D & Muenchen, D & Wien, A) ISBN 3-581

X 1034 *Panstwowe Wydawnictwo Naukowe (PWN)* (Warsaw, PL) ISBN 83-01

X 1035 *Thomas Nelson and Sons* (Walton on Thames, GB & Melbourne, Vic, AUS & Scarborough, ON, CDN & South Nashville, TN, USA) ISBN 0-17

X 1036 *Volk und Wissen, Volkseigener Verlag Berlin* (Berlin, DDR) ISBN 3-353

X 1039 *Moritz Diesterweg* (Frankfurt am Main & Berlin & Duesseldorf & Hamburg & Muenchen) ISBN 3-425

X 1042 *Espasa-Calpe* (Madrid, E & Buenos Aires, RA) ISBN 84-239

X 1052 *Izdatel'stvo Tbilisskogo Universiteta* (Tbilisi, SU)

X 1054 *Verlag Harri Deutsch* (Frankfurt, D & Thun, CH) ISBN 3-87144

X 1056 *Valention Bompiani* (Milano, I)

X 1071 *F. Alcan* (Paris, F)

X 1079 *B.G.Teubner Verlagsgesellschaft* (Leipzig, DDR) ISBN 3-519 • REM See also X0823

X 1088 *Felix Meiner Verlag* (Hamburg, D) ISBN 3-7873 • REL PUBL (X 4718) Meiner: Leipzig

X 1096 *Basil Blackwell* (Oxford, GB) ISBN 0-631, ISBN 0-632, ISBN 0-86286, ISBN 0-86793 • REM Also: Blackwell Scientific Publications: Oxford, GB & Carlton, Vic, AUS

X 1121 *Mathematisch Centrum* (Amsterdam, NL) ISBN 90-6196

X 1136 *Elanders Boktryckeri Aktiebolag* (Goteborg, S)

X 1163 *Almqvist & Wiksell Foerlag* (Stockholm, S & Bromma, S & Goeteborg, S & Malmoe, S) ISBN 91-20

X 1164 *Charles E. Merrill Publishing Co.* (Columbus, OH, USA & Wembley, GB & Pyrmont, NSW, AUS & Sydney, NSW, AUS) ISBN 0-675 • REM In Canada: Division by Bell & Howell

X 1165 *Hodder & Stroughton Educational* (London, GB & Sevenoaks, GB & Glenfield, NZ & Don Mills, ON, CDN & Lane Cove, NSW, AUS & Ashwood, Vic, AUS & Brisbane, Qld, AUS & Adelaide, SA, AUS & West Perth, WA, AUS) ISBN 0-340 • REM Formerly: The English University Press: London, GB

X 1166 *PZW (= Panstwowe Zaklady Wydawnictwo Szkolnych)* (Warsaw, PL)

X 1167 *Holden-Day* (San Francisco, CA, USA) ISBN 0-8162

X 1168 *Kyoritsu Shuppan Company* (Tokyo, J) ISBN 4-320

X 1172 *Les Editions de l'Office Central de Librairie S.A.R.L. (O.C.D.L.)* (Paris, F) ISBN 2-7043

X 1174 *Walter de Gruyter* (Berlin, D) ISBN 3-11 • REM Member Firms: 1) de Gruyter: Hawthorne, NY, USA. 2) Mouton Publishers: Berlin, D

X 1187 *Sankibo* (Tokyo, J)

X 1194 *University of Exeter* (Exeter, GB) ISBN 0-900771, ISBN 0-85989, ISBN 0-902414, ISBN 0-9501308

X 1201 *Dickenson Publishing Company* (Belmont, CA, USA) ISBN 0-8221

X 1205 *Junker und Duennhaupt Verlag* (Berlin, D)

X 1212 *Izdatel'stvo Belorusskogo Gosudarstvennogo Universiteta* (Minsk, SU)

X 1214 *New York University, Courant Institute of Mathematical Science.* (New York, NY, USA)

X 1221 *Frederick Ungar Publishing* (New York, NY, USA & London, GB) ISBN 0-8044

X 1225 *Drei Masken Verlag* (Muenchen, D & Berlin, D)

X 1226 *Academia* (Prague, CS) • REM Publishing House of the Czechoslovak Academy of Science

X 1228 *Appleton-Century-Crofts* (New York, NY, USA) ISBN 0-8385

X 1229 *Edizione della Libreria L'Ateneo* (Napoli, I)

X 1230 *Aufbau-Verlag* (Berlin, DDR & Weimar, DDR) ISBN 3-351

X 1231 *Johann Ambrosius Barth* (Leipzig, DDR) ISBN 3-335 • REL PUBL (**X 0811**) Springer: Heidelberg & New York

X 1233 *Bedminster Press* (Totowa, NJ, USA) ISBN 0-87087

X 1236 *Ernest Benn Ltd.-Benn Bros plc* (Tunbridge, GB) ISBN 0-510, ISBN 0-85314, ISBN 0-85459 • REM Also: Benn Business Information Services: Tunbridge Wells, GB

X 1237 *William Blackwood & Sons* (Edinburgh, GB) ISBN 0-85158

X 1238 *The Bobbs-Merril Company* (Indianapolis, IN, USA) ISBN 0-672 • REL PUBL (**X 1320**) Odyssey Pr: Indianapolis

X 1243 *William C.Brown Co., Publishers* (Dubuque, IA, USA) ISBN 0-697

X 1246 *California Institute of Technology (Caltech)* (Pasadena, CA, USA)

X 1248 *Chandler Publishing Company* (San Francisco, CA, USA & Clifton, NJ, USA & London, GB) ISBN 0-8102 • REL PUBL (**X 1288**) Intext Pr: New York

X 1253 *Crowell, Collier & Macmillan* (New York, NY, USA & West Drayton, GB) ISBN 0-02 • REM Also: 1) Macmillan Publishing: New York, NY, USA. 2) Collier-Macmillan: West Drayton, GB

X 1255 *Franz Deuticke, Verlagsgesellschaft* (Wien, A) ISBN 3-7005 • REL PUBL (**X 0918**) Klett: Stuttgart

X 1257 *Dryden Press* (Hinsdale, IL, USA & Darlinghurst, NSW, AUS) ISBN 0-8498

X 1258 *Duncker & Humblot* (Berlin, D) ISBN 3-428

X 1261 *Edinburgh University Press* (Edinburgh, GB) ISBN 0-85224

X 1262 *Giuliu Einaudi, Editore* (Torino, I) ISBN 88-06

X 1263 *N.G. Elwert Verlag* (Marburg, D) ISBN 3-7708

X 1265 *Gustav Fischer Verlag* (Stuttgart, D) ISBN 3-437

X 1267 *Friedrich Frommann Verlag Guenther Holzboog* (Stuttgart, D) ISBN 3-7728

X 1268 *Orell Fuessli Verlag* (Zuerich, CH) ISBN 3-280

X 1272 *Gutenberg-Gesellschaft* (Mainz, D) ISBN 3-7755

X 1274 *Hafner Press* (New York, NY, USA) ISBN 0-02 • REL PUBL (**X 1253**) Crowell Collier & Macmillan: New York & (**X 0843**) Macmillan : New York & London

X 1275 *Anton Hain K.G. Meisenheim Verlag* (Koenigstein, D & Meisenheim, D) ISBN 3-445 • REL PUBL (**X 1588**) Athenaeum/Hain/Hanstein: Koenigstein

X 1277 *Hayden Book Company* (Rochelle Park, NY, USA) ISBN 0-8104 • REL PUBL (**X 1354**) Spartan Books : Sutton

X 1279 *Herder & Co.* (Freiburg, D & Roma, I) ISBN 3-451

X 1281 *Carl Heymanns Verlag* (Koeln, D & Berlin, D & Muenchen, D & Bonn, D) ISBN 3-452

X 1282 *S.Hirzel Verlag* (Stuttgart, D) ISBN 3-7776

X 1283 *Ulrico Hoepli, Casa Editrice Libreria* (Milano, I) ISBN 88-203

X 1285 *University of Illinois Press* (Urbana, IL, USA & London, GB) ISBN 0-252 • REL PUBL (**X 3828**) Amer Univ Publ Group: London

X 1286 *Indiana University Press* (Bloomington, IN, USA & London, GB) ISBN 0-253 • REL PUBL (**X 3828**) Amer Univ Publ Group: London

X 1288 *Intext Press* (New York, NY, USA) ISBN 0-88444 • REL PUBL (**X 1248**) Chandler: San Francisco

X 1290 *Richard D.Irwin* (Homewood, IL, USA & London, GB) ISBN 0-256 • REL PUBL Pandemic: London, GB

X 1291 *Johns Hopkins University Press* (Baltimore, MD, USA) ISBN 0-8018

X 1292 *Alfred Kroener Verlag* (Stuttgart, D) ISBN 3-520

X 1296 *Longman Group* (Harlow, GB & New York, NY, USA & Melbourne, Vic, AUS & Surry Hills, NSW, AUS & Brisbane, Qld, AUS & Adelaide, SA, AUS & Perth, WA, AUS & Quarry Bay, HK & Auckland, NZ & Jurong Town, SGP & & Kampala, EAU & Dar es Salaam, EAT & Cape Town, ZA & Kingston, JA) ISBN 0-582 • REM Member Firm: Keesing's Reference Publications: Harlow, GB

X 1297 *J.B.Lippincott* (Philadelphia, PA, USA) ISBN 0-397, ISBN 0-06 • REL PUBL (**X** 0837) Harper & Row: New York & (**X** 1096) Blackwell: Oxford

X 1298 *Mathematical Association of America* (Washington, DC, USA) ISBN 0-88385

X 1299 *Magnes Press* (Jerusalem, IL) ISBN 965-223

X 1305 *University of Massachusetts Press* (Amerhurst, MA, USA) ISBN 0-87023

X 1306 *University of Miami Press* (Austin, TX, USA) ISBN 0-87024

X 1308 *E.S.Mittler & Sohn* (Herford, D & Bonn, D) ISBN 3-87547, ISBN 3-8132

X 1309 *J.C.B.Mohr (Paul Siebeck)* (Tuebingen, D) ISBN 3-16

X 1310 *Monarch Press* (New York, NY, USA) ISBN 0-671

X 1312 *Editions Aubier-Montaigne* (Paris, F) ISBN 2-7007

X 1313 *Nauwelaerts, Beatrice* (Louvain, B) ISBN 2-900014

X 1317 *P.Noordhoff International Publishing* (Groningen, NL & Leiden, NL) ISBN 90-01 • REL PUBL (**X** 0812) Wolters-Noordhoff : Groningen & (**X** 1352) Sijthoff: Leiden • REM Now: Sijthoff & Noordhoff International Publishers: Leiden, NL

X 1318 *Northwestern University Press* (Evanston, IL, USA) ISBN 0-8101

X 1319 *La Nuova Italia Editrice* (Firenze, I) ISBN 88-221

X 1320 *Odyssey Press* (Indianapolis, IN, USA) ISBN 0-8399 • REL PUBL (**X** 1238) Bobbs-Merril: Indianapolis

X 1321 *Oeffentliches Leben* (Frankfurt, D)

X 1323 *Oliver & Boyd* (Edinburgh, GB) ISBN 0-05

X 1324 *Open Court Publishing Co.* (LaSalle, IL, USA) ISBN 0-87548, ISBN 0-89688

X 1326 *Kegan Paul, Trench, Trubner & Co.* (London, GB) ISBN 0-7101

X 1328 *Pennsylvania State University, Department of Mathematics.* (University Park, PA, USA)

X 1330 *Pitman Publishing* (Belmont, CA, USA & London, GB & Toronto, ON, CDN & & Carlton, Vic, AUS & Johannesburg, ZA & Wellington, NZ & & Auckland, NZ) ISBN 0-273, ISBN 0-8224, ISBN 0-915092, ISBN 0-272

X 1331 *University of Pittsburgh Press* (Pittsburgh, PA, USA) ISBN 0-8229

X 1332 *Plenum Publishing Corporation* (New York, NY, USA & London, GB) ISBN 0-306 • REM Also Called Plenum Press

X 1334 *Praeger Publishers* (New York, NY, USA) ISBN 0-03, ISBN 0-275

X 1335 *Princeton University, Institute for Advanced Study (IAS)* (Princeton, NJ, USA)

X 1337 *Prindle, Weber & Schmidt* (Boston, MA, USA) ISBN 0-87150 • REM Also Called: PWS Publishers

X 1340 *Henry Regnery Co.* (Chicago, IL, USA) ISBN 0-8092

X 1342 *Royal Irish Academy* (Dublin, IRL) ISBN 0-901714

X 1343 *Russell & Russell Publishers* (New York, NY, USA) ISBN 0-8462

X 1347 *Charles Scribner's Sons Publishers* (New York, NY, USA) ISBN 0-684

X 1348 *La Scuola Editrice* (Brescia, I)

X 1349 *Editions du Seuil* (Paris, F) ISBN 2-02

X 1352 *A.W.Sijthoff International Publishing Co.* (Leiden, NL) ISBN 90-218, ISBN 90-286 • REL PUBL (**X** 1317) Noordhoff: Groningen

X 1354 *Spartan Books* (Sutton, GB & Rochelle Park, NJ, USA) ISBN 0-905532 • REL PUBL (**X** 1277) Hayden: Rochelle Park

X 1355 *Stanford University Press* (Stanford, CA, USA) ISBN 0-8047

X 1358 *University of Texas Press* (Austin, TX, USA & London, GB) ISBN 0-292 • REL PUBL (**X** 3828) Amer Univ Publ Group: London

X 1359 *Charles C. Thomas* (Springfield, IL, USA) ISBN 0-398

X 1364 *Vita e Pensiero, Publicazioni della Universita Cattolica* (Milano, I) ISBN 88-343

X 1367 *University of Washington Press* (Seattle, WA, USA & London, GB) ISBN 0-295 • REL PUBL (**X** 3828) Amer Univ Publ Group: London

X 1368 *C.A.Watts & Co.* (London, GB) ISBN 0-296

X 1371 *Wff'n Proof Publishers* (Ann Arbor, MI, USA) ISBN 0-911624

X 1372 *Carl Winter* (Heidelberg, D) ISBN 3-533

X 1375 *Nicola Zanichelli Editore* (Bologna, I) ISBN 88-08

X 1407 *Publicaciones del Instituto de Matematicas "Jorge Juan"* (Madrid, E) ISBN 84-00

X 1408 *Deerven F. Bohn* (Amsterdam, NL & Haarlem, Nl) ISBN 90-6051

X 1425 *Wydawnictwo Uniwersitetu Slaskiego* (Katowice, PL)

X 1434 *Kalininskij Gosudarstvennyj Universitet* (Kalinin, SU)

X 1453 *Moskovskij Gosudarstvennyj Istoriko-Arkh. Institut* (Moskva, SU)

X 1466 *Mueszaki Koenyvkiado (Institut of Technology)* (Budapest, H) ISBN 963-10

X 1470 *Izdatel'stvo Mysl'* (Moskva, SU)

X 1478 *Guadagnini* (Torino, I)

X 1493 *C.W.K.Gleerup Bokfoerlag a.B.* (Lund, S) ISBN 91-40

X 1494 *Munksgaard International Publishers* (Copenhagen, DK) ISBN 87-16

X 1512 *Rheinische Friedrich-Wilhelms-Universitaet Bonn, Institut fuer Informatik.* (Bonn, D)

X 1553 *Verlag Dokumentation Saur* (Muenchen, D & London, GB) ISBN 3-7940, ISBN 3-598 • REL PUBL (**X** 2797) Minerva Publ: Muenchen • REM Member Firm: Zell Publishers: Oxford, GB

X 1554 *Universitetsforlaget* (Oslo, N & Bergen, N & Tromsoe, N & New York, NY, USA) ISBN 82-00 • REM Also: UB-Forlaget: Oslo, N

X 1569 *Editura Didactica si Pedagogica* (Bucharest, RO)

X 1574 *Vyssheyshaya Shkola* (Minsk, SU)

X 1588 *Verlagsgruppe Athenaeum/ Hain/ Hanstein* (Koenigstein/Ts, D) • REL PUBL (**X** 1275) Hain: Koenigstein & (**X** 2665) Athenaeum: Frankfurt

X 1599 *Aarhus Universitet, Matematisk Institut* (Aarhus, DK) ISBN 87-87436

X 1623 *Universite de Paris VI, Institut Henri Poincare* (Paris, F) • REM The University Has Now Been Divided Into Smaller Units. The Mathematics Department Is Now: Universite de Paris VII-Pierre et Marie Curie, Secretariat Mathematique

X 1649 *Rusvozizdat* (Leningrad, SU)

X 1656 *Izdatel'stvo Inostr. Lit.* (Moskva, SU)

X 1684 *Marcel Dekker* (New York, NY, USA & Basel, CH) ISBN 0-8247 • REL PUBL (**X** 2442) Dekker Journal: New York

X 1714 *Commission of the European Communities.* (Bruxelles, B) ISBN 92-825, ISBN 92-826, ISBN 92-827, ISBN 92-828

X 1715 *Edunsa Editorial Universitaria* (Santiago, RCH)

X 1732 *Bibliopolis* (Napoli, I) ISBN 88-7088

X 1744 *Editorial Ariel* (Esplugas de Llobregat, E & Barcelona, E) ISBN 84-344

X 1752 *Masson, Editeur* (Paris, F) ISBN 2-225

X 1758 *Ars Polona* (Warsaw, PL)

X 1761 *Hekdosis Hellinikis Mathematikes Hetaireias-Greek Mathematical Society* (Athens, GR)

X 1763 *Royal Society* (London, GB) ISBN 0-85403

X 1764 *Ksiaznica Atlas T.N.S.W.* (Warsaw, PL & Lwow, PL)

X 1771 *American Philosophical Association* (College Park, MD, USA)

X 1772 *Mathematical Algorithms* (Cambridge, MA, USA)

X 1773 *VEDA - Vydavatelstvo Slovenskej Akademie Vied* (Bratislava, CS)

X 1776 *Les editions Sociales* (Paris, F) ISBN 2-209

X 1777 *Western Periodicals Company* (North Hollywood, CA, USA)

X 1781 *Tecnos* (Madrid, E) ISBN 84-309

X 1821 *Real Academia de Ciencias Exactas, Fisicas y Naturales* (Madrid, E)

X 1876 *Kexue Chubanshe (Science Press)* (Beijing, TJ)

X 1925 *Nauka i Izkustwo* (Sofia, BG)

X 1946 *University Press of America* (Lanham, MD, USA & Washington, DC, USA) ISBN 0-8191

X 1987 *A. Francke Verlag* (Muenchen, D) ISBN 3-7720

X 1994 *Koninklijke van Gorcum (Royal van Gorcum)* (Assen, NL & Amsterdam, NL) ISBN 90-232

X 2027 *Izdatel'stvo Nauka* (Moskva, SU & Alma-Ata, SU & Leningrad, SU & Novosibirsk, SU)

X 2036 *Franz Ehrenwirth Verlag* (Muenchen, D) ISBN 3-431

X 2039 *Polytechnic Institute of New York* (Brooklyn, NY, USA)

X 2040 *Winthrop* (Cambridge, MA, USA) ISBN 0-87626 • REL PUBL (**X** 0819) Prentice Hall: Englewood Cliffs

X 2045 *Perry Lane Press* (Stanford, CA, USA)

X 2055 *Progress* (Moskva, SU)

X 2075 *E.Flammarion & Cie, Librairie* (Paris, F) ISBN 2-08

X 2088 *Schwabe & Co.* (Basel, CH & Stuttgart, D) ISBN 3-7965

X 2091 *Consejo Superior de Investigaciones Cientificas.* (Madrid, E) ISBN 84-00

X 2109 *Angelicum* (Roma, I)

X 2110 *N.V. Dekker & van de Vegt* (Nijmegen, NL) ISBN 90-255

X 2111 *Staufenverlag* (Koeln, D)

X 2121 *Accademia Nazionale dei Lincei* (Roma, I) ISBN 88-218

X 2179 *IEEE (Institute of Electrical and Electronics Engineers* (New York, NY, USA & Long Beach, CA, USA & Piscataway, CA, USA) ISBN 0-87942 • REM Section: IEEE Computer Society: Long Beach, CA, USA. Section: IEEE United States Activities Commitee: Piscataway, CA, USA

X 2184 *C.S.I.R.O. (Commonwealth Scientific and Industrial Research Organisation).* (Melbourne, Vic, AUS) ISBN 0-643

X 2188 *Princeton University, Department of Electrical Engineering and Computer Science* (Princeton, NJ, USA)

X 2189 *Universitatea "Babes-Bolyai", Biblioteca Centrala Universitara* (Cluj-Napoca, RO)

X 2190 *Linguistic Society of America* (Arlington, VA, USA)

X 2191 *Pi Mu Epsilon* (Norman, SD, USA)

X 2192 *Government College, Research Council* (Lahore, PAK)

X 2193 *Indian Mathematical Society* (Bombay, IND)

X 2194 *Sociedad Matematica Mexicana* (Mexico City, MEX)

X 2195 *Peeters S.P.R.L.* (Louvain, B) ISBN 2-8017

X 2197 *Tohoku University, Mathematical Institute* (Sendai, J)

X 2198 *Universidad de Barcelona, Facultad de Ciencias, Seminario Matematico* (Barcelona, E)

X 2199 *Izdatel'stvo Naukova Dumka* (Kiev, SU)

X 2201 *Australasian Association of Philosophy.* (Bundoora, Vic, AUS) ISBN 0-9592545

X 2202 *Indian Association for the Cultivation of Science* (Calcutta, IND)

X 2203 *Philosophy of Science Association* (East Lansing, MI, USA) ISBN 0-917586

X 2204 *American Institute of Physics* (New York, NY, USA) ISBN 0-88318

X 2205 *(ACM) Association for Computing Machinery* (New York, NY, USA) ISBN 0-89791

X 2206 *Accademia delle Science* (Torino, I)

X 2207 *Universitatea "Al. I. Cuza" din Iasi* (Jassy (Iasi), RO)

Publishers

X 2208 *Taylor & Francis* (London, GB) ISBN 0-85066

X 2209 *Izdatel'stvo Akademii Nauk Gruzinskoj SSR* (Tbilisi, SU)

X 2211 *Friedrich Schiller Universitaet* (Jena, DDR)

X 2212 *Ossolineum, Publishing House of the Polish Academy of Sciences* (Wroclaw, PL & Warsaw, PL) • REL PUBL (X 2885) Zakl Narod Wyd Pol Ak: Wroclaw & (X 2882) Pol Akad Nauk: Wroclaw

X 2213 *State University of New York at Buffalo (SUNY)* (Buffalo, NY, USA)

X 2214 *Dialectica* (Bienne/Biel, CH)

X 2215 *Indiana University, Department of Philosophy* (Bloomington, IN, USA)

X 2216 *Cornell University, Sage School of Philosophy* (Ithaca, NY, USA)

X 2217 *Academia Brasileira de Ciencias.* (Rio de Janeiro, BR)

X 2219 *Humboldt-Universitaet zu Berlin* (Berlin, DDR)

X 2220 *Buchdruckerei und Verlag Leemen* (Zuerich, CH)

X 2224 *Matematisk Institut* (Bergen, N)

X 2225 *Izdatel'stvo Akademii Nauk Armyanskoj SSR* (Erevan, SU)

X 2227 *Scuola Normale Superiore di Pisa* (Pisa, I) ISBN 88-7642

X 2228 *Hegler Institute* (La Salle, IL, USA)

X 2230 *Isdevnieciba Zinatne* (Riga, SU)

X 2234 *B.Pellerano & S. del Gaudia Editrice Libraria* (Napoli, I)

X 2235 *Vsesoyuznyj Institut Nauchnoj i Tekhnicheskoj Informatsii (VINITI), Gosudarstvennyj Komitet SSSR po Nauke: Tekhnike (GKNT SSSR), Akademiya Nauk (AN) SSSR* (Moskva, SU)

X 2237 *Izdatelstvo na Bulgarskata Akademia na Naukite (Publishing House of the Bulgarian Academy of Sciences.)* (Sofia, BG)

X 2238 *Sociedad Colombiana de Matematicas* (Bogota, CO) ISBN 958-95081

X 2239 *Universidad de Salamanca Ediciones* (Salamanca, E) ISBN 84-7481

X 2241 *New York Academy of Sciences* (New York, NY, USA) ISBN 0-89072, ISBN 0-89766

X 2242 *Osaka University, Department of Mathematics* (Osaka, J)

X 2243 *University of Osaka Prefecture, Department of Mathematics* (Osaka, J)

X 2244 *Societe Mathematique de France* (Paris, F) ISBN 2-85629

X 2245 *Gregorian University Press* (Vatican City, SCV & Roma, I)

X 2246 *Universite d'Alger* (Alger, DZ)

X 2247 *Universidad Tecnica Federico Santa Maria.* (Valparaiso, RCH)

X 2248 *Drushtvo Matematichara i Fizichara Sr Hrvatske.* (Zagreb, YU)

X 2254 *Universidad Nacional Autonoma de Mexico, Instituto de Matematicas* (Mexico City, MEX)

X 2256 *Casa Editrice Felice le Monnier* (Firenze, I)

X 2257 *John Benjamins B.V.* (Amsterdam, NL & Philadelphia, PA, USA) ISBN 90-272

X 2258 *Periodika* (Tallinn, SU)

X 2261 *Ministerstvo Vysshego i Srednego Spetsial'nogo Obrazovaniya* (Moskva, SU)

X 2265 *Vychislitel'nyj Tsentr Akademii Nauk SSSR* (Moskva, SU)

X 2327 *Maki Shoten (Maki Publisher)* (Tokyo, J) ISBN 4-8375

X 2370 *British Computer Society* (London, GB) ISBN 0-901865

X 2373 *Karl-Marx-Universitaet, Direktorat fuer Forschung, Abteilung Wissenschaftliche Publikationen* (Leipzig, DDR)

X 2375 *MacMillan Journals* (London, GB) ISBN 0-333 • REL PUBL (X 0843) Macmillan : New York & London

X 2376 *The Econometric Society* (Evanston, IL, USA)

X 2378 *Society for Industrial and Applied Mathematics (SIAM)* (Philadelphia, PA, USA) ISBN 0-89871

X 2379 *University of Houston, Department of Mathematics* (Houston, TX, USA)

X 2382 *Institute of Mathematics and its Applications* (Southend-on-Sea, GB) ISBN 0-905091, ISBN 0-9501159

X 2392 *University of Electro-Communications* (Tokyo, J)

X 2393 *Universidad de Chile, Facultad de Filosofia y Letras.* (Santiago, RCH)

X 2394 *Polska Akademia Nauk, Institut Maszyn Matematycznych* (Warsaw, PL)

X 2395 *Turun Yliopsto* (Turku, SF) ISBN 951-641

X 2399 *Sigma Xi, Scientific Research Society of North America* (New Haven, CT, USA) ISBN 0-914446

X 2403 *AFIPS Press* (Montvale, NJ, USA) ISBN 0-88283

X 2409 *Suomalainen Tiedakatemia.* (Helsinki, SF) ISBN 951-41

X 2420 *Utilitas Mathematica Publications, University of Manitoba* (Winnipeg, MB, CDN) ISBN 0-919628

X 2421 *University of Tokyo, Faculty of Science* (Tokyo, J)

X 2422 *Canadian Philosophical Association* (Montreal, PQ, CDN) ISBN 0-9690153

X 2423 *Scripta Publishing Co.* (Silver Spring, MT, USA & Washington, DC, USA) ISBN 0-88380

X 2424 *Illinois Academy of Sciences* (Springfield, IL, USA)

X 2426 *Universa, P.V.B.A.* (Wetteren, B) ISBN 90-6281

X 2434 *University of Tokyo College of General Education* (Tokyo, J)

X 2435 *Presses Universitaires de France, Service Periodiques* (Paris, F) ISBN 2-13 • REL PUBL (X 0840) Pr Univ France: Paris

X 2436 *Indian Academy of Sciences, Publications Department* (Bangalore, IND)

X 2437 *Hemisphere Publishing* (Washington, DC, USA & New York, NY, USA & London, GB) ISBN 0-89116

X 2438 *Edizioni Scientifiche Inglesi Americane* (Roma, I)

X 2441 *Kyoto University, Research Institute for Mathematical Sciences* (Kyoto, J)

X 2442 *Marcel Dekker Journals.* (New York, NY, USA) ISBN 0-8247 • REL PUBL (**X 1684**) Dekker: New York

X 2443 *Izdatel'stvo Nauka, Otdelenie v Kazakhstane SSR* (Alma-Ata, SU)

X 2445 *Academia Sinica, Institute of Mathematics* (Taipei, RC)

X 2446 *Martin-Luther-Universitaet Halle-Wittenberg* (Halle, DDR)

X 2448 *Akademiya Navuk Belorusskaj SSR* (Minsk, SU)

X 2451 *Uniwersitet Jagiellonskie, Instytut Matematiyczny* (Krakow, PL)

X 2453 *Nagoya University, Faculty of Engineering* (Nagoya, J)

X 2455 *Yokohama City University, Department of Mathematics* (Yokohama, J)

X 2457 *Allerton Press* (New York, NY, USA) ISBN 0-89864

X 2461 *Jugoslovenska Akademija Zananosti i Umjetnosti Rad.* (Zagreb, YU) ISBN 86-407

X 2462 *Cairo University Press* (Cairo, ETH)

X 2463 *Tartuskij Gosudarstvennyj Universitet* (Tartu, SU)

X 2464 *Mathematisches Seminar, Selbstverlag* (Giessen, D)

X 2465 *Kinokuniya Company* (Tokyo, J) ISBN 4-314

X 2467 *Societatea de Stiinte Matematice din Republica Socialista Romania (RSR)* (Bucharest, RO)

X 2468 *National Research Council* (Washington, DC, USA) ISBN 0-309

X 2469 *Data, A/S af 2. april 1971* (Copenhagen, DK) ISBN 87-980512

X 2471 *IBM Corp.* (Armonk, NY, USA) ISBN 0-933186

X 2472 *Centro Superiore di Logica e Scienze Comporate, Editrice Franco Spisani* (Bologna, I)

X 2473 *Belfort Graduate School of Science; Yeshiva University* (New York, NY, USA)

X 2476 *Journal of Philosophy Inc.* (New York, NY, USA) ISBN 0-931206

X 2478 *Universitatea di Timisoara, Facultatea de Stiinte Ale Naturia* (Timisoara, RO)

X 2479 *Rocky Mountain Mathematics Consortium* (Tempe, AZ, USA)

X 2480 *Operations Research Society of Japan* (Tokyo, J)

X 2484 *American Telephone and Telegraph, Bell Laboratories* (Murray Hill, NY, USA) ISBN 0-88439

X 2492 *Japan Association for the Philosophy of Science* (Tokyo, J)

X 2504 *Bertrand Russell Memorial Logic Conference* (Leeds, GB) ISBN 0-9502983

X 2508 *Cleaver-Hume Press* (London, GB)

X 2510 *Elm (Izdatel'stvo Akademiya Nauk Azerbajdzhanskoj SSR)* (Baku, SU)

X 2511 *Universita Karlova, Matematicky Ustav* (Prague, CS)

X 2512 *Institut Matematiki Akademii Nauk SSSR* (Moskva, SU)

X 2513 *Mathematical Association* (Leicester, GB) ISBN 0-906588

X 2514 *Institut de Mathematiques* (Geneve, CH)

X 2515 *Universita Karlova, Fakulta Matematiky a Fyziky* (Prague, CS)

X 2516 *Union Matematica Argentina* (Buenos Aires, RA)

X 2517 *Calcutta Mathematical Society* (Calcutta, IND)

X 2518 *Aberdeen University Press* (Aberdeen, GB) ISBN 0-900015, ISBN 0-08

X 2519 *Izdatel'stvo Akademii Nauk Uzbekskoj SSR* (Tashkent, SU)

X 2520 *National Academy of Sciences, Transportation Research Board* (Washington, DC, USA) ISBN 0-309

X 2522 *Akademiya Nauk USSR, Nauchnyj Sovet po Kibernetike, Institut Kibernetiki.* (Kiev, SU)

X 2523 *Universidad Nacional de Tucuman* (Tucuman, RA) ISBN 950-554

X 2524 *Encyclopaedia Universalis* (Paris, F) ISBN 2-85229

X 2526 *Gazeta de Matematica* (Lisboa, P)

X 2528 *Circolo Matematico di Palermo* (Palermo, I)

X 2529 *Consultants Bureau* (New York, NY, USA) • REL PUBL (**X 1332**) Plenum Publ: New York

X 2530 *Societe Mathematique de Belgique* (Bruxelles, B)

X 2531 *Japan Academy* (Tokyo, J)

X 2593 *Nauchnyj Sovet po Kompleksnoj Problemy "Kibernetika" Akademij Nauk SSSR (Council for Problems of Complexity "Cybernetics", Academy of Science)* (Moskva, SU) LC-No 75-30834

X 2594 *Ivanovskij Gosudarstvennyj Universitet* (Ivanovo, SU)

X 2596 *Akademische Druck- und Verlagsanstalt Dr P. Struzl* (Graz, A) ISBN 3-201, ISBN 3-900144

X 2600 *DEB - Verlag. Das Europaeische Buch* (Berlin, DDR)

X 2636 *Verlag von Veit & Co.* (Leipzig, DDR)

X 2641 *Izdatel'stvo Nauka Leningradskoe Otdelenie* (Leningrad, SU)

X 2642 *Izdatel'stvo Nauka Sibirskoe Otdelenie* (Novosibirsk, SU)

X 2643 *Izdatel'stvo Sovetskoe Radio* (Moskva, SU)

X 2644 *Izdatel'skoe Ob"edineniya "Vishcha Shkola". Izdatel'stvo pri Khar'kovskom Gosudarskvennom Universitete* (Khar'kov, SU)

X 2645 *Izdatel'skoe Ob"edineniye "Vishcha Shkola". Izdatel'stvo pri Kievskom Gosudarstvennom Universitete* (Kiev, SU)

X 2646 *Aalborg University Press = Aalborg Universitetsforlag* (Aalborg, DK) ISBN 87-7307

X 2652 *Institut Matematiki Sibirskogo Otdeleniya Akademii Nauk SSSR (SOAN SSSR)* (Novosibirsk, SU)

X 2655 *Akademie der Wissenschaften der DDR, Institut fuer Mathematik* (Berlin, DDR)

X 2665 *Athenaeum Verlag* (Frankfurt, D) ISBN 3-7610 • REL PUBL (X 1588) Athenaeum/Hain/Hanstein: Koenigstein

X 2671 *Basic Books* (New York, NY, USA) ISBN 0-465

X 2673 *Biblioteka Akademiya Nauk SSSR* (Leningrad, SU)

X 2682 *Casa Editrice Tilgher* (Genova, I)

X 2686 *Compania Editorial Continental (CECSA)* (Mexico City, MEX)

X 2692 *University of Queensland Press* (St. Lucia, Qld, AUS) ISBN 0-7022

X 2696 *Humanities Press* (Atlantic Highlands, NY, USA) ISBN 0-391

X 2698 *Departamento de Logica y Filosofia de la Ciencia, Universidad de Valencia* (Valencia, E)

X 2710 *Eidgenoessische Technische Hochschule Zuerich* (Zuerich, CH)

X 2717 *Franz Steiner Verlag* (Wiesbaden, D & Stuttgart, D) ISBN 3-515

X 2719 *Fundacao Calouste Gulbenkian* (Lisbon, P) ISBN 972-15

X 2725 *Hackett Publishing Co.* (Indianapolis, IN, USA) ISBN 0-915144

X 2727 *Hochschul-Verlag* (Stuttgart, D)

X 2728 *Hoelder-Pichler-Tempsky* (Wien, A) ISBN 3-209

X 2732 *Institut National de Recherche en Informatique et en Automatique (INRIA)* (Le Chesnay Cedex, F) ISBN 2-7261 • REM Also Called: Institut de Recherche en Informatique et en Automatique (IRIA)

X 2733 *Polish Academy of Sciences, Institute of Philosophy and Sociology* (Wroclaw, PL)

X 2737 *Israel Program for Scientific Translations* (Jerusalem, IL) ISBN 0-7065 • REL PUBL (X 0827) Wiley & Sons: New York

X 2739 *Istituto Editoriale Internazionale* (Milano, I)

X 2741 *Shtiintsa* (Kishinev, SU)

X 2766 *Yaroslavskij Gosudarstvennyj Universitet* (Yaroslavl', SU)

X 2769 *Kazakhskij Gosudarstvennyj Universitet* (Alma-Ata, SU)

X 2776 *Les Presses de l'Universite Laval* (Ste-Foy, PQ, CDN) ISBN 0-7746, ISBN 2-7637

X 2786 *Universitet i Oslo, Matematisk Institut* (Oslo, N) ISBN 82-553

X 2797 *Minerva Publikation Saur* (Muenchen, D) ISBN 3-597 • REL PUBL (X 1553) Dokumentation Saur: Muenchen

X 2799 *Mladinska Knjiga* (Ljubljana, YU) ISBN 86-11

X 2802 *Gosudarstvennyj Pedagogicheskij Institut* (Moskva, SU)

X 2812 *The Philosophical Society of Finland* (Helsinki, SF) ISBN 951-95053

X 2828 *Robert E. Krieger Publishing Co.* (Melbourne, FL, USA) ISBN 0-88275, ISBN 0-89874

X 2836 *Sociedade Brasileira Logica* (Sao Paulo, BR)

X 2846 *Tioga Publishing Co.* (Palo Alto, CA, USA) ISBN 0-935382

X 2854 *U.E.R. (UER) Mathematique, Universite Paris XI* (Paris-Orsay, F)

X 2856 *Universidad de Santiago de Compostela, Secretariado de Publicationes* (Santiago de Compostela, E) ISBN 84-7191

X 2858 *Universitaet Fridericiana Karlsruhe. Technische Hochschule Karlsruhe* (Karlsruhe, D) • REL PUBL (X 3159) TH Karlsruhe Fak Informatik: Karlsruhe

X 2862 *Universitaetsverlag Fribourg* (Fribourg, CH) ISBN 3-7278, ISBN 2-8271

X 2865 *Verlag Peter Lang* (Frankfurt, D & Bern, CH, Las Vegas, USA) ISBN 3-8204, ISBN 3-261

X 2866 *Verlag Ullstein* (Berlin, D) ISBN 3-550

X 2870 *Vittorio Klostermann* (Frankfurt, D) ISBN 3-465

X 2880 *Wydawnictwa Naukowo-Techniczne* (Warsaw, PL)

X 2881 *Wydawnictwa Szkolne i Pedagogiczne* (Warsaw, PL) ISBN 83-02

X 2882 *Wydawnictwo Polskiej Akademii Nauk (Publisher of the Polish Academy of Science)* (Wroclaw, PL) • REL PUBL (X 2885) Zakl Narod Wyd Pol Ak: Wroclaw & (X 2212) Ossolineum: Wroclaw

X 2885 *Zaklad Narodowy imienia Ossolinskich, Wydawnictwo Polskiej Akademii Nauk* (Wroclaw, PL) ISBN 83-04 • REL PUBL (X 2882) Pol Akad Nauk: Wroclaw & (X 2212) Ossolineum: Wroclaw

X 3111 *Philosophia Verlag* (Muenchen, D & Wien, A) ISBN 3-88405

X 3119 *Intertext Books* (Aylesbury, GB)

X 3123 *Beijing Shifan Daxue (Beijing Normal University)* (Beijing, TJ)

X 3152 *Technische Hogeschool Eindhoven* (Eindhoven, NL)

X 3159 *Fakultaet fuer Informatik der Universitaet Karlsruhe (Technische Hochschule)* (Karlsruhe, D) • REL PUBL (X 2858) Univ Fridericiana : Karlsruhe

X 3175 *Faculte des Sciences de l'Universite Libanaise* (Hadath-Beyrouth, RL)

X 3176 *Fernuniversitaet Hagen* (Hagen, D)

X 3203 *Centre Universitaire de Luxembourg* (Luxembourg, L)

X 3215 *Mathematisch-Naturwissenschaftliche Fakultaet der Rheinisch-Westfaelischen Technischen Hochschule Aachen.* (Aachen, D)

X 3219 *Leningradskij Ordena Trudovogo Krasnogo Znameni Institut Tochnoj Mekhaniki i Optiki* (Leningrad, SU)

X 3223 *Carl Hanser Verlag* (Muenchen, D & Wien, A) ISBN 3-446

X 3249 *The Weizmann Institute of Science* (Rehovot, IL) ISBN 965-281

X 3290 *Skolska Knjiga* (Zagreb, YU) ISBN 86-03

X 3297 *Erkki Juhani Jussila* (Tampere, SF)

X 3333 *Unknown Publisher*

X 3357 *Institute of Education. University of Keele* (Keele, GB)

X 3407 *"Vysshaya Shkola"* (Moskva, SU)

X 3552 *Iwanami Shoten Publishers* (Tokyo, J) ISBN 4-00

X 3555 *Clarkson N. Potter, Publishers* (New York, NY, USA) ISBN 0-8257

X 3559 *Erevanskij Gosudarstvennyj Universitet* (Erevan, SU)

X 3560 *Paraninfo* (Madrid, E) ISBN 84-283

X 3561 *Rowman and Littlefield* (Totowa, NJ, USA) ISBN 0-87471, ISBN 0-8476

X 3581 *Computer Science Press* (Rockville, MD, USA) ISBN 0-914894

X 3602 *Kraus Reprint* (Vaduz, FL & Nendeln, FL) ISBN 3-262
• REM Parent Firm: Kraus Thomson Organisation: Vaduz, FL

X 3604 *Istituto Nazionale di Alta Matematica (INDAM)* (Roma, I)

X 3605 *Izdatel'stvo Kazanskogo Gosudarstvennogo Universiteta* (Kazan', SU)

X 3615 *Il Saggiatore* (Milano, I)

X 3636 *Tokyo Tosho* (Tokyo, J) ISBN 4-489

X 3709 *Gosudarstvennoye Izdatel'stvo Fiziko-Matematicheskoj Literatury* (Moskva, SU)

X 3711 *Koseisha Koseikaku* (Tokyo, J) ISBN 4-7699

X 3718 *L'Enseignement Mathematique, Universite de Geneve* (Geneve, CH)

X 3719 *The Harvester Press* (Sussex, GB) ISBN 0-85527, ISBN 0-901759, ISBN 0-7108

X 3725 *Bolyai Janos Matematikai Tarsulat (Janos Bolyai Mathematical Society)* (Budapest, H)

X 3727 *Beograd. Matematicki Institut* (Belgrade, YU)

X 3734 *Cabay Libraire-Editeur* (Louvain-la-Neuve, B) ISBN 2-87077

X 3775 *Radio i Svyaz* (Moskva, SU)

X 3777 *F. Angeli Editore* (Milano, I) ISBN 88-204

X 3782 *University of Maine at Orono* (Orono, MN, USA) ISBN 0-89101

X 3790 *Metsniereba* (Tbilisi, SU)

X 3805 *The Blindern Theoretic Research Team* (Oslo, N)

X 3812 *Universita di Siena, Dipartimento di Matematica, Scuola di Specializzazione in Logica Matematica* (Siena, I)

X 3828 *American University Publishers Group* (London, GB)
• REL PUBL (**X 0842**) Univ Wisconsin Pr: Madison & (**X 1285**) Univ Ill Pr: Urbana & (**X 1286**) Indiana Univ Pr : Bloomington & (**X 1358**) Univ Texas Pr: Austin & (**X 1367**) Univ Washington: Seattle

X 4030 *Univerzitet u Novom Sadu, Institut za Matematiku* (Novi Sad, YU)

X 4056 *Editorial Labor* (Barcelona, E) ISBN 84-335

X 4059 *Louis Nebert* (Halle, DDR)

X 4060 *Wilhelm Koebner* (Breslau, PL)

X 4061 *Hermann Pohle* (Jena, DDR)

X 4086 *Loescher Editore* (Torino, I) ISBN 88-201

X 4217 *Foris Publications* (Dordrecht, NL) ISBN 90-70176

X 4259 *Barron's Education Series* (Woodbury, NY, USA) ISBN 0-8120

X 4282 *The Rand Corporation* (Santa Monica, CA, USA) ISBN 0-8330

X 4322 *National Publishing House* (New Delhi, IND & Bihar, IND & Calcutta, IND & Ambala, IND)

X 4328 *Editora Pedagogica Universitaria* (Sao Paulo, BR) ISBN 85-202

X 4350 *Chr. Michelsens Institutt for Videnskap og Andsfrihet, Beretninger. A.S. John Griegs Boktrykkerei* (Bergen, N)

X 4380 *Posebna Izdanja Matematichok Instituta* (Belgrade, YU)

X 4435 *Macquarie University* (North Ryde, AUS) ISBN 0-85837

X 4457 *Presses Polytechniques* (Lausanne, CH)

X 4465 *Obshchestvennaya Pol'za i Provintsiya* (Leningrad, SU)

X 4499 *Ehlins Handboeker* (Stockholm, S)

X 4502 *University of Tehran* (Tehran, IR)

X 4512 *Alfred A. Knopf* (New York, NY, USA) ISBN 0-394

X 4513 *Mashinostroenie* (Moskva, SU)

X 4562 *Udmurtskij Gosudarstvennij Universiteta* (Izhevsk, SU)

X 4563 *Liberal Arts Press* (New York, NY, USA)

X 4569 *(Commercial Press)* (Beijing)

X 4579 *Gnodeng Jiaoyu Chubanshe (Higher Education Press)* (Beijing, TJ)

X 4589 *Shanghai Jiaoyu Chubanshe (Shanghai Education Press)* (Shanghai, TJ)

X 4643 *Universite de Paris VII, UER de Mathematiques* (Paris, F)

X 4718 *Felix Meiner Verlag* (Leipzig, DDR) • REL PUBL (**X 1088**) Meiner: Hamburg

X 4721 *Libreria Fratelli Bocca & C. Clausen* (Torino, I & Roma, I)

Miscellaneous Indexes

External classifications

This index complements the Subject Index at the beginning of this volume; it lists the items which, in addition to classifications in the present volume, have classifications *external to this volume*. These items are ordered by external classification code and within each code by author (the first alphabetically in the case of multi-author items), year and identification number (thus an item with, for example, two external classifications occurs twice in this listing). This index provides another way to search the bibliography. With it, the user can easily identify those items in this volume classified also in some area external to this volume.

B25

Aanderaa, S.O. [1973] 00010
Aanderaa, S.O. [1982] 33756
Aanderaa, S.O. [1984] 41791
Andrews, P.B. [1974] 03832
Ausiello, G. [1976] 51023
Ax, J. [1968] 00580
Baker, A. [1971] 42680
Baudisch, A. [1980] 56368
Behmann, H. [1922] 00942
Behmann, H. [1923] 00943
Behmann, H. [1950] 01559
Bel'tyukov, A.P. [1976] 29789
Bernays, P. [1928] 01076
Beth, E.W. [1950] 01156
Blanche, R. [1962] 22464
Boerger, E. [1974] 01368
Boerger, E. [1984] 40047
Boolos, G. [1979] 56312
Borkowski, L. [1958] 01482
Borkowski, L. [1960] 01483
Borkowski, L. [1961] 01484
Buechi, J.R. [1960] 01690
Buechi, J.R. [1965] 01694
Buechi, J.R. [1965] 01697
Buechi, J.R. [1969] 01700
Buechi, J.R. [1973] 01708
Buechi, J.R. [1973] 71377
Buechi, J.R. [1973] 71381
Buechi, J.R. [1983] 40515
Cegielski, P. [1984] 46773
Chapin Jr., E.W. [1971] 01987
Chen, Jiyuan [1984] 44191
Chlebus, B.S. [1980] 56482
Comer, S.D. [1985] 42290
Cooper, D.C. [1972] 61080
Denenberg, L. [1984] 43373
Doshita, S. [1983] 44209
Dreben, B. [1957] 03611
Dreben, B. [1962] 03127
Dreben, B. [1962] 03612
Dulac, M.-H. [1971] 03165
Elgot, C.C. [1966] 03291
Ferro, A. [1978] 56666
Fridman, Eh.I. [1972] 77209
Friedman, Joyce [1963] 04667
Friedman, Joyce [1966] 29439
Fuerer, M. [1981] 35169
Gabbay, D.M. [1974] 17555

Galvin, F. [1967] 04765
Geach, P.T. [1952] 16851
Gensler, H.J. [1973] 18156
Gentzen, G. [1935] 24221
Gladstone, M.D. [1979] 52622
Goedel, K. [1929] 22123
Goedel, K. [1933] 20885
Goldfarb, W.D. [1981] 55275
Goldfarb, W.D. [1984] 38765
Goodstein, R.L. [1955] 05177
Goodstein, R.L. [1963] 05189
Goodstein, R.L. [1966] 05190
Gurevich, R. [1984] 45002
Gurevich, R. [1985] 44633
Gurevich, Y. [1965] 05452
Gurevich, Y. [1966] 05455
Gurevich, Y. [1973] 32346
Gurevich, Y. [1976] 14777
Gurevich, Y. [1983] 33764
Gurevich, Y. [1983] 33766
Hamblin, C.L. [1973] 17398
Harrop, R. [1954] 05676
Harrop, R. [1958] 05678
Harrop, R. [1965] 05684
Harrop, R. [1976] 32206
Hartmanis, J. [1976] 24160
Hauschild, K. [1972] 05782
Hauschild, K. [1975] 29757
Henson, C.W. [1977] 30718
Herbrand, J. [1930] 20866
Herbrand, J. [1931] 16812
Ackermann, W. [1928] 00107
Joyner Jr., W.M. [1973] 81740
Joyner Jr., W.M. [1976] 23231
Kallick, B. [1969] 22190
Kalmar, L. [1928] 06835
Kalmar, L. [1932] 06837
Kalmar, L. [1933] 06838
Kanger, S. [1970] 23148
Kapetanovic, M. [1983] 40445
Kassler, M. [1976] 32363
Ketonen, J. [1984] 44002
Kopieki, R. [1975] 23964
Krejnovich, V.Ya. [1982] 35653
Krom, Melven R. [1964] 07570
Krom, Melven R. [1967] 07574
Krynicki, M. [1979] 52881
Krynicki, M. [1979] 53311
Ladner, R.E. [1977] 52252
Laeuchli, H. [1968] 07787

Langford, C.H. [1932] 22592
Lewis, H.R. [1978] 75527
Lewis, H.R. [1979] 53515
Lewis, H.R. [1980] 36546
Liogon'kij, M.I. [1970] 08182
Lisovik, L.P. [1984] 48409
Loewenheim, L. [1908] 08241
Longo, G. [1974] 53697
Loparic, A. [1977] 27299
Luckhardt, H. [1969] 08380
Makanin, G.S. [1964] 14845
Makanin, G.S. [1977] 51395
Makanin, G.S. [1977] 51863
Makanin, G.S. [1980] 82143
Manders, K.L. [1979] 53838
Marini, D. [1975] 22915
Maslov, S.Yu. [1966] 19334
Maslov, S.Yu. [1972] 08848
Maslov, S.Yu. [1977] 32672
McKinsey, J.C.C. [1943] 09021
McRobbie, M.A. [1980] 54132
Meredith, D. [1978] 31932
Mijajlovic, Z. [1974] 09242
Mizutani, C. [1977] 53069
Mostowski, A.Wlodzimierz [1979] 76601
Mostowski, A.Wlodzimierz [1981] 55152
Muchnik, A.A. [1985] 47256
Muller, D.E. [1981] 76648
Muller, D.E. [1985] 47646
Nakamura, A. [1981] 55427
Nelson, Greg [1984] 45270
Nieland, J.J.F. [1967] 09951
Norgela, S.A. [1976] 32671
Norgela, S.A. [1978] 32674
Norgela, S.A. [1980] 32676
Novikov, P.S. [1949] 19476
Oberschelp, A. [1958] 10042
Oppen, D.C. [1980] 55875
Orevkov, V.P. [1969] 10154
Paola di, R.A. [1973] 30684
Pepis, J. [1937] 19493
Pinus, A.G. [1979] 53695
Presburger, M. [1930] 10749
Quine, W.V.O. [1945] 10872
Rabin, M.O. [1968] 29625
Rabin, M.O. [1969] 10942
Rabin, M.O. [1970] 22233
Rabin, M.O. [1971] 14690
Rabin, M.O. [1972] 10943
Ramsey, F.P. [1930] 28781

Rasiowa, H. [1953] 11004
Reichbach, J. [1975] 32003
Robitashvili, N.G. [1965] 11330
Rousseau, G. [1966] 11593
Rousseau, G. [1968] 11598
Rozenblat, B.V. [1983] 34621
Schmidt, K. [1981] 35213
Schuette, K. [1933] 12997
Schuette, K. [1934] 12998
Schuette, K. [1950] 12999
Schuette, K. [1978] 56098
Schwabhaeuser, W. [1959] 13052
Seese, D.G. [1975] 50759
Seese, D.G. [1977] 27132
Seese, D.G. [1979] 53053
Shelah, S. [1977] 27202
Shostak, R.E. [1977] 69916
Shostak, R.E. [1979] 82799
Shostak, R.E. [1982] 38462
Siefkes, D. [1968] 21146
Siefkes, D. [1970] 12145
Siefkes, D. [1970] 27984
Siefkes, D. [1971] 12146
Siefkes, D. [1978] 29136
Siekmann, J. [1982] 36659
Skolem, T.A. [1919] 12383
Skolem, T.A. [1920] 12382
Skolem, T.A. [1928] 12385
Skolem, T.A. [1931] 12388
Skolem, T.A. [1933] 12391
Skolem, T.A. [1935] 12395
Skolem, T.A. [1936] 40864
Skolem, T.A. [1937] 27938
Slupecki, J. [1972] 15183
Stihi, T. [1973] 79027
Szczerba, L.W. [1965] 12831
Szczerba, L.W. [1979] 79177
Tajmanov, A.D. [1967] 13274
Tarski, A. [1931] 16924
Tarski, A. [1948] 19756
Tarski, A. [1967] 13463
Tauts, A. [1964] 13475
Tenney, R.L. [1975] 17647
Thatcher, J.W. [1968] 28596
Thomas, I. [1952] 13541
Thomas, Wolfgang [1980] 55873
Trakhtenbrot, B.A. [1960] 42812
Ulrich, D. [1983] 42640
Urquhart, A.I.F. [1981] 54908
Vakarelov, D. [1972] 00678
Vaught, R.L. [1954] 41195
Wang, Hao [1953] 14024
Wang, Hao [1963] 14040
Wang, Hao [1965] 14057
Wasilewska, A. [1978] 69996
Weaver, G.E. [1974] 14096
Wirsing, M. [1977] 51036
Wright von, G.H. [1948] 14303
Wright von, G.H. [1950] 14304
Wright von, G.H. [1952] 14309
Zaiontz, C. [1983] 42214
Zhaldokas, R. [1983] 43443
Zhegalkin, I.I. [1939] 17903
Zhegalkin, I.I. [1946] 04825
Zygmunt, J. [1981] 54333
Zygmunt, J. [1983] 41843

B40

Amer, M.A. [1976] 17946
Amer, M.A. [1977] 30619
Andrews, P.B. [1965] 00402
Barendregt, H.P. [1981] 55008
Baxter, L.D. [1978] 52222
Bellot, P. [1985] 42554
Bishop, P. [1974] 62501
Boehm, C. [1972] 60576
Bruijn de, N.G. [1978] 29191
Bruijn de, N.G. [1980] 72186
Bunder, M.W. [1980] 53699
Bunder, M.W. [1981] 55277
Bunder, M.W. [1983] 31631
Church, A. [1932] 02123
Church, A. [1933] 02124
Church, A. [1936] 23466
Cocchiarella, N.B. [1985] 47532
Courvoisier, M. [1976] 51687
Curry, H.B. [1957] 29324
Curry, H.B. [1958] 02627
Curry, H.B. [1967] 02637
Fitch, F.B. [1944] 04326
Fitch, F.B. [1944] 04328
Fitch, F.B. [1948] 04329
Fitch, F.B. [1949] 04333
Fitch, F.B. [1949] 04334
Fitch, F.B. [1950] 17113
Fitch, F.B. [1951] 17114
Fitch, F.B. [1952] 04335
Fitch, F.B. [1974] 24019
Giovannetti, E. [1983] 35624
Hindley, J.R. [1972] 23471
Kleene, S.C. [1935] 28412
Kuzichev, A.S. [1977] 52560
L'Abbe, M. [1953] 07730
Meredith, D. [1978] 31932
Meredith, D. [1980] 76287
Milner, R. [1979] 53582
Nederpelt, R.P. [1980] 55989
Newman, M.H.A. [1942] 09934
Plotkin, G.D. [1980] 77373
Quine, W.V.O. [1981] 55273
Robinson, John Alan [1969] 11291
Rosser, J.B. [1955] 23483
Salle, P. [1978] 52050
Salle, P. [1980] 56539
Schoenfinkel, M. [1924] 12955
Schuette, K. [1953] 13003
Scott, D.S. [1982] 38520
Scott, D.S. [1982] 38593
Staples, J. [1983] 37059

B45

Alchourron, C.E. [1971] 15012
Boolos, G. [1979] 56312
Bressan, A. [1974] 60704
Chellas, B.F. [1980] 53915
Copi, I.M. [1967] 04030
Cornides, T. [1974] 71948
Cresswell, M.J. [1968] 19031
Cresswell, M.J. [1972] 02551
Farinas del Cerro, L. [1982] 36159
Freudenthal, H. [1958] 04610
Hajek, P. [1981] 35379
Hajek, P. [1983] 34630
Hintikka, K.J.J. [1973] 44480
Hintikka, K.J.J. [1981] 42772
Lehmann, S.K. [1976] 14765
Lemmon, E.J. [1977] 52261
Lewis, C.I. [1918] 22590
Lewis, C.I. [1960] 25502
Makinson, D. [1974] 08573
Marsden, E.L. [1974] 63732
Mo, Shaokui [1950] 12028
Novikov, P.S. [1949] 10018
Pliuskeviciene, A. [1985] 49019
Pnueli, A. [1985] 49386
Pnueli, A. [1985] 49524
Prior, A.N. [1961] 10782
Prior, A.N. [1964] 10787
Quine, W.V.O. [1953] 10892
Rasiowa, H. [1953] 11004
Rautenberg, W. [1979] 77698
Rescher, N. [1969] 25168
Rescher, N. [1971] 11120
Roeper, P. [1980] 54022
Scedrov, A. [1985] 44771
Schwind, C.B. [1980] 78311
Smith, Henry Bradford [1936] 41315
Venkatesh, G. [1985] 49739
Wasserman, H.C. [1976] 65990
Winnie, J.A. [1970] 14236
Ziembinski, Z. [1976] 51409

B46

Anderson, A.R. [1975] 23056
Gabbay, D.M. [1973] 30381
Kalman, J.A. [1976] 62846
Lewis, C.I. [1918] 22590
Ohlbach, H.-J. [1984] 43199

B48

Adams, E.W. [1975] 60060
Carnap, R. [1971] 25461
Case, J. [1983] 37600
Cohen, L.J. [1970] 25404
Evans, H.P. [1939] 30933
Feys, R. [1937] 41048
Harper, W.L. [1981] 47401
Howson, C. [1976] 62614
Jeffrey, R.C. [1965] 25416
Leblanc, H. [1960] 07913
Los, J. [1963] 31415
Paulos, J.A. [1981] 56466
Reichenbach, H. [1932] 11080
Reichenbach, H. [1939] 43787
Skyrms, B. [1966] 21285

B50

Ackermann, R.J. [1967] 00099
Ackermann, R.J. [1967] 25260
Dreben, B. [1957] 29366
Foxley, E. [1962] 04474
Goddard, L. [1966] 15234
Goodstein, R.L. [1958] 05182
Khasin, L.S. [1969] 05733
Lindstroem, T.L. [1982] 36617
Los, J. [1949] 19296
Lukasiewicz, J. [1921] 28632
Muchnik, A.A. [1959] 00382
Murskij, V.L. [1965] 09654

Nieland, J.J.F. [1967] 09951
Post, E.L. [1921] 10681
Rasiowa, H. [1965] 29826
Rautenberg, W. [1979] 77698
Reichbach, J. [1962] 11065
Reichbach, J. [1963] 11067
Reichbach, J. [1964] 11068
Reichbach, J. [1965] 11070
Rose, A. [1951] 11403
Sanmartin Esplugues, J. [1973] 78125
Strazdins, I.E. [1981] 55620
Wojcicki, R. [1982] 35669
Wozniakowska, B. [1978] 31565
Zoll, E.J. [1968] 22544

B51

Dalla Chiara Scabia, M.L. [1979] 56529
Takeuti, G. [1981] 39565

B52

Gaines, B.R. [1981] 34823
Lee, R.C.T. [1972] 28860
Schmucker, K.J. [1984] 43453

B53

Arruda, A.I. [1980] 53627
Grant, J. [1975] 21774
Komori, Y. [1982] 37775

B55

Gabbay, D.M. [1974] 17555
Khomich, V.I. [1979] 53468
Maksimova, L.L. [1977] 52433
Renardel de Lavalette, G.R. [1981] 55207
Shekhtman, V.B. [1982] 38116

B60

Cirulis, J. [1982] 35570
Hermes, H. [1965] 06014
Lambert, K. [1967] 07823
Lieber, L.R. [1947] 22514
Los, J. [1963] 31415
Magari, R. [1966] 49174
Malinowski, G. [1985] 49715
Markwald, W. [1971] 08782
Markwald, W. [1974] 08784
Mueller, Gert H. [1957] 33928
Pliuskevicius, R. [1968] 10543
Quine, W.V.O. [1954] 10895
Silvestrini, D. [1981] 48725
Sobocinski, B. [1964] 21264
Stepien, T. [1985] 48971
Trew, A. [1970] 13673

B75

Andreka, H. [1982] 55505
Blass, A.R. [1984] 47716
Cartwright, Robert [1979] 56231
Degano, P. [1979] 56755
Leone, M. [1985] 49912
Melzi, G. [1976] 76279
Mints, G.E. [1982] 35718
Nazaryan, G.A. [1975] 53659

Paola di, R.A. [1971] 30683
Pnueli, A. [1985] 49386
Pnueli, A. [1985] 49524
Sain, I. [1983] 40674
Schnorr, C.-P. [1971] 23998
Schwartz, J.T. [1967] 23572
Sheng, C.L. [1969] 48529
Tajtslin, M.A. [1979] 82932
Wasilewska, A. [1985] 49467
Zhaldokas, R. [1983] 43443

B96

Dedekind, R. [1932] 38623
Frege, F.L.G. [1952] 25773
Frege, F.L.G. [1962] 22970
Gentzen, G. [1969] 29483
Herbrand, J. [1968] 24825
Lesniewski, S. [1967] 28535
Mostowski, Andrzej [1979] 82308
Neumann von, J. [1961] 25726
Novikov, P.S. [1979] 38498
Radu, E. [1978] 51580
Robinson, A. [1979] 53900
Skolem, T.A. [1970] 29109
Suppes, P. [1969] 12737
Tarski, A. [1956] 28568
Tarski, A. [1972] 30338
Wajsberg, M. [1933] 40808
Wang, Hao [1962] 32531

C05

Armbrust, M. [1972] 00468
Bacsich, P.D. [1975] 17658
Bloom, S.L. [1975] 21621
Czelakowski, J. [1979] 31660
Czelakowski, J. [1980] 56472
Czelakowski, J. [1980] 56473
Czelakowski, J. [1985] 47502
Felscher, W. [1969] 03737
Felscher, W. [1971] 03738
Fujiwara, T. [1971] 17104
Fujiwara, T. [1973] 04714
Givant, S. [1978] 29155
Givant, S. [1979] 55860
Goguen, J.A. [1985] 36903
Henkin, L. [1977] 27268
Herrera Miranda, J. [1978] 36172
Johnson, J.S. [1973] 06681
Kaiser, Klaus [1975] 06805
Kanger, S. [1970] 23148
Maksimova, L.L. [1977] 52433
Mal'tsev, A.I. [1959] 19260
McKenzie, R. [1975] 09009
McKinsey, J.C.C. [1943] 09021
Movsisyan, Yu.M. [1975] 16671
Nelson, Evelyn [1977] 26606
Richter, M.M. [1984] 43919
Rozenblat, B.V. [1979] 53949
Rozenblat, B.V. [1983] 34621
Siekmann, J. [1982] 36659
Suszko, R. [1972] 12781
Suszko, R. [1974] 12782
Tabata, H. [1969] 13247
Tabata, H. [1971] 13248
Zhaldokas, R. [1983] 43443

C07

Alcantara de, L.P. [1980] 54266
Andreka, H. [1973] 22879
Andreka, H. [1974] 03829
Andreka, H. [1974] 60135
Barwise, J. [1977] 24196
Bennett, D.W. [1973] 01038
Bennett, D.W. [1977] 21957
Benthem van, J.F.A.K. [1974] 03889
Beth, E.W. [1955] 01269
Beth, E.W. [1960] 01534
Blass, A.R. [1984] 40306
Bowen, K.A. [1972] 01505
Brady, R.T. [1977] 30639
Bullock, A.M. [1972] 01746
Bullock, A.M. [1973] 01748
Call, R.L. [1972] 71509
Canty, J.T. [1963] 01832
Casari, E. [1981] 56675
Craig, W. [1957] 02508
Craig, W. [1957] 02510
Crossley, J.N. [1966] 31222
Ehdel'man, G.S. [1974] 72569
Feferman, S. [1957] 03683
Felscher, W. [1969] 03737
Felscher, W. [1971] 03738
Fraisse, R. [1950] 04510
Fraisse, R. [1955] 04516
Fraisse, R. [1956] 04518
Fraisse, R. [1958] 04520
Fraisse, R. [1970] 04531
Fraisse, R. [1972] 04294
Fraisse, R. [1974] 04537
Gergely, T. [1978] 52253
Glebskij, Yu.V. [1969] 05032
Glubrecht, J.-M. [1982] 35619
Goedel, K. [1930] 20884
Goguen, J.A. [1985] 36903
Goodstein, R.L. [1972] 05195
Goodstein, R.L. [1974] 05197
Halkowska, K. [1975] 29801
Halkowska, K. [1976] 52421
Hanazawa, M. [1979] 54329
Hasenjaeger, G. [1952] 05722
Hasenjaeger, G. [1953] 05723
Hasenjaeger, G. [1955] 27719
Hauck, J. [1972] 62383
Henkin, L. [1949] 05892
Henkin, L. [1954] 05900
Henkin, L. [1957] 05910
Hennessy, M. [1980] 53954
Hintikka, K.J.J. [1953] 06111
Hintikka, K.J.J. [1964] 06122
Hintikka, K.J.J. [1965] 06123
Hintikka, K.J.J. [1973] 32405
Hook, J.L. [1985] 41804
Hugly, P. [1982] 54068
Jervell, H.R. [1973] 33265
Keisler, H.J. [1973] 07033
Kossowski, P. [1970] 07392
Kreisel, G. [1967] 07533
Kuehnrich, M. [1983] 36540
Ladner, R.E. [1977] 52252
Leblanc, H. [1969] 07936
Lindenbaum, A. [1935] 30921
Lis, Z. [1960] 08189
Los, J. [1955] 08332

Los, J. [1956] 08333
Mal'tsev, A.I. [1936] 08602
Mal'tsev, A.I. [1959] 19260
Manaster, A.B. [1975] 23120
Manin, Yu.I. [1977] 51984
Marcja, A. [1974] 21706
Markusz, Z. [1983] 44560
Mo, Shaokui [1964] 47141
Motohashi, N. [1982] 55534
Motohashi, N. [1984] 42487
Mundici, D. [1981] 90124
Novak, I.L. [1950] 09999
Paillet, J.L. [1973] 30495
Pogorzelski, W.A. [1961] 31969
Poizat, B. [1982] 35642
Rantala, V. [1975] 64743
Rasiowa, H. [1957] 11017
Rasiowa, H. [1960] 11024
Rautenberg, W. [1968] 11048
Reichbach, J. [1955] 33706
Reichbach, J. [1969] 11073
Ribeiro, H. [1969] 64815
Rieger, L. [1955] 19556
Rieger, L. [1964] 11193
Robinson, A. [1958] 25442
Robinson, A. [1959] 19551
Rodriguez Artalejo, M. [1981] 77840
Schroeter, K. [1958] 24841
Scott, D.S. [1979] 54069
Sikorski, R. [1961] 12306
Sikorski, R. [1962] 12308
Skolem, T.A. [1920] 12382
Skolem, T.A. [1923] 21221
Skolem, T.A. [1928] 12385
Skolem, T.A. [1929] 12386
Smirnov, V.A. [1972] 30359
Smolin, V.P. [1975] 51192
Smullyan, R.M. [1970] 12557
Specker, E. [1958] 12669
Surma, S.J. [1968] 12748
Surma, S.J. [1969] 12752
Surma, S.J. [1970] 12754
Surma, S.J. [1970] 79124
Suszko, R. [1972] 12781
Suszko, R. [1974] 12782
Szczerba, L.W. [1983] 39510
Tarski, A. [1932] 13406
Tarski, A. [1933] 28816
Tarski, A. [1935] 13409
Tarski, A. [1936] 13413
Tarski, A. [1936] 13415
Tarski, A. [1944] 13432
Tulipani, S. [1975] 50883
Venneri, B.M. [1975] 65905
Wang, Hao [1953] 14024
Weaver, G.E. [1974] 14096
Weaver, G.E. [1980] 56064
Zubieta Russi, G. [1957] 11688

C10

Baudisch, A. [1980] 56368
Beth, E.W. [1950] 01156
Borkowski, L. [1961] 01484
Buechi, J.R. [1977] 51165
Galda, K. [1974] 61899
Hodes, L. [1968] 24939
Langford, C.H. [1926] 07850
Langford, C.H. [1926] 07851
Langford, C.H. [1932] 22592
Lisovik, L.P. [1984] 48409
Marcja, A. [1974] 21706
McKee, T.A. [1980] 76216
McNaughton, R. [1965] 09073
Presburger, M. [1930] 10749
Schmerl, J.H. [1982] 35360
Skolem, T.A. [1919] 12383
Skolem, T.A. [1928] 12385
Skolem, T.A. [1931] 12388
Tarski, A. [1931] 16924
Tarski, A. [1948] 19756
Tarski, A. [1967] 13463
Wheeler, W.H. [1979] 53787
Wolter, H. [1973] 14280

C13

Aanderaa, S.O. [1982] 35210
Ax, J. [1968] 00580
Boolos, G. [1984] 40442
Bullock, A.M. [1972] 01746
Bullock, A.M. [1973] 01748
Christen, C. [1976] 52348
Deutsch, M. [1975] 02972
Gacs, P. [1977] 52753
Glebskij, Yu.V. [1969] 05032
Goedel, K. [1933] 20885
Gurevich, Y. [1966] 05456
Gurevich, Y. [1966] 05458
Gurevich, Y. [1973] 32346
Gurevich, Y. [1983] 33766
Hailperin, T. [1961] 05508
Hajek, P. [1973] 22921
Hajek, P. [1974] 24128
Hajek, P. [1978] 31131
Huntington, E.V. [1917] 41774
Huntington, E.V. [1917] 41776
Keisler, H.J. [1973] 07033
Kostyrko, V.F. [1971] 19122
Ladner, R.E. [1977] 52252
Liogon'kij, M.I. [1970] 08182
Lovasz, L. [1980] 53755
McKenzie, R. [1975] 09009
Poizat, B. [1982] 35642
Rautenberg, W. [1968] 11048
Roedding, D. [1972] 11343
Scarpellini, B. [1984] 39663
Scarpellini, B. [1984] 41786
Scarpellini, B. [1985] 47473
Scholz, H. [1952] 33304
Scott, D.S. [1958] 11906
Trakhtenbrot, B.A. [1953] 19742
Ulrich, D. [1983] 42640
Wajsberg, M. [1933] 13979
Zhegalkin, I.I. [1946] 04825
Zykov, A.A. [1953] 19773

C15

Fraisse, R. [1977] 52674
Fraisse, R. [1985] 48204
Hauschild, K. [1968] 05770
Herrmann, E. [1980] 56483
Tarski, A. [1948] 13436

C20

Abian, A. [1974] 03810
Adler, A. [1973] 00190
Armbrust, M. [1972] 00468
Ax, J. [1968] 00580
Barone, E. [1980] 80631
Bloom, S.L. [1975] 21621
Bunyatov, M.R. [1967] 02032
Button, R.W. [1979] 80839
Chadwick, J.J.M. [1977] 80867
Cherlin, G.L. [1972] 14684
Connes, A. [1970] 26249
Czelakowski, J. [1979] 31659
Czelakowski, J. [1980] 56473
Felscher, W. [1971] 03738
Flum, J. [1974] 15020
Frayne, T.E. [1962] 04557
Galvin, F. [1967] 04765
Galvin, F. [1970] 04766
Geiser, J.R. [1968] 04827
Giannone, A. [1982] 34858
Hauschild, K. [1971] 05779
Heinrich, S. [1980] 56660
Heinrich, S. [1980] 81516
Heinrich, S. [1981] 53749
Heinrich, S. [1982] 37826
Heinrich, S. [1982] 46460
Heinrich, S. [1983] 37053
Heinrich, S. [1983] 37054
Henson, C.W. [1976] 46653
Janssen, G. [1972] 06540
Kaiser, Klaus [1976] 06806
Kasahara, S. [1973] 30535
Keisler, H.J. [1965] 07008
Keisler, H.J. [1976] 33913
Krupa, A. [1984] 44797
Kuehnrich, M. [1983] 36540
Levy, M. [1984] 44350
Lin, Peikee [1985] 48745
Lolli, G. [1978] 55177
Luxemburg, W.A.J. [1962] 08426
Luxemburg, W.A.J. [1962] 21419
Luxemburg, W.A.J. [1969] 21147
Luxemburg, W.A.J. [1969] 24816
MacDowell, R. [1961] 19285
Magajna, B. [1983] 34592
Meisters, G.H. [1973] 09087
Osdol van, D.H. [1972] 22272
Potthoff, K. [1972] 10697
Richter, M.M. [1982] 34209
Robinson, A. [1964] 11258
Robinson, A. [1969] 11267
Roitman, J. [1982] 55446
Saito, M. [1976] 33414
Schwartz, Dietrich [1975] 13068
Skolem, T.A. [1933] 12392
Skolem, T.A. [1934] 12393
Skolem, T.A. [1955] 27718
Spivakov, Yu.L. [1974] 18401
Stone, A.L. [1965] 27645
Tajmanov, A.D. [1967] 13274
Takeuchi, Y. [1977] 79205
Venne, M. [1967] 13929
Wolter, H. [1968] 14276
Wolter, H. [1972] 14277
Wolter, H. [1972] 14278
Zakon, E. [1974] 30005

C25

Markovic, Z. [1983] 37386
Robinson, A. [1973] 26149
Wheeler, W.H. [1979] 53787

C30

Galvin, F. [1967] 04765
Galvin, F. [1970] 04766
Kaiser, Klaus [1974] 06804
Leblanc, H. [1984] 44672
Lyndon, R.C. [1959] 08439
Montali, T. [1971] 17210
Monteverdi, D. [1973] 21823
Oberschelp, A. [1958] 10042
Paillet, J.L. [1973] 30495
Richter, M.M. [1984] 43919
Rodriguez Artalejo, M. [1981] 77840
Shafaat, A. [1973] 18382
Vaught, R.L. [1954] 41195
Wolter, H. [1972] 14278

C35

Apelt, H. [1966] 00403
Buechi, J.R. [1983] 33901
Corcoran, J. [1980] 71934
Givant, S. [1978] 29155
Givant, S. [1979] 55860
Hauschild, K. [1968] 05770
Herring, J.M. [1976] 21916
Huntington, E.V. [1917] 41774
Huntington, E.V. [1917] 41776
Kochen, S. [1957] 07272
Langford, C.H. [1926] 07849
Langford, C.H. [1926] 07850
Langford, C.H. [1926] 07851
Langford, C.H. [1939] 07855
Lindenbaum, A. [1935] 30921
Palyutin, E.A. [1971] 64365
Presburger, M. [1930] 10749
Schwabhaeuser, W. [1956] 13051
Schwabhaeuser, W. [1959] 13052
Skolem, T.A. [1928] 12385
Surma, S.J. [1968] 12748
Szczerba, L.W. [1979] 79177
Tajtslin, M.A. [1979] 82932
Tarski, A. [1934] 30917
Tarski, A. [1936] 13415
Tarski, A. [1948] 19756
Tarski, A. [1967] 13463
Wajsberg, M. [1933] 40808
Wang, Hao [1953] 14024
Wheeler, W.H. [1979] 53787
Zil'ber, B.I. [1981] 36112

C40

Bacsich, P.D. [1975] 17658
Barwise, J. [1975] 60316
Beth, E.W. [1953] 01159
Borkowski, L. [1968] 01485
Bouvere de, K.L. [1965] 04261
Buechi, J.R. [1969] 01700
Chlebus, B.S. [1980] 56482
Craig, W. [1952] 02505
Craig, W. [1957] 02508
Craig, W. [1957] 02510
Craig, W. [1965] 02514
Czelakowski, J. [1982] 42743
Czelakowski, J. [1985] 47502
Dywan, Z. [1980] 54158
Ecsedi-Toth, P. [1982] 34907
Feferman, S. [1966] 03697
Ferro, R. [1978] 29281
Fujiwara, T. [1973] 04714
Fujiwara, T. [1975] 21821
Galvin, F. [1967] 04765
Galvin, F. [1970] 04766
Gao, Hengshan [1983] 45159
Garland, S.J. [1974] 24144
Gurevich, Y. [1983] 33768
Hajek, P. [1970] 05521
Hauschild, K. [1971] 05779
Hintikka, K.J.J. [1964] 06122
Hodes, L. [1968] 24939
Hodes, L. [1970] 06168
Kanovich, M.I. [1975] 18218
Kogalovskij, S.R. [1966] 07292
Kogalovskij, S.R. [1966] 24967
Kogalovskij, S.R. [1968] 07299
Kogalovskij, S.R. [1973] 74934
Krom, Melven R. [1963] 07571
Krom, Melven R. [1968] 07575
Krynicki, M. [1979] 53311
Lillo de, N.J. [1979] 56184
Lindenbaum, A. [1927] 30918
Lyndon, R.C. [1959] 08439
Maehara, S. [1960] 08520
Maehara, S. [1971] 08528
Maksimova, L.L. [1977] 52433
Mizutani, C. [1977] 53069
Mostowski, Andrzej [1962] 22070
Motohashi, N. [1972] 09599
Motohashi, N. [1973] 09600
Motohashi, N. [1973] 09602
Motohashi, N. [1978] 90092
Motohashi, N. [1982] 33418
Motohashi, N. [1982] 55534
Motohashi, N. [1984] 42488
Mundici, D. [1983] 37383
Nabebin, A.A. [1977] 52155
Nadiu, G.S. [1969] 09749
Oberschelp, A. [1958] 10042
Oikkonen, J. [1979] 52831
Pacholski, L. [1968] 10215
Padoa, A. [1902] 14911
Poizat, B. [1975] 17649
Rabin, M.O. [1970] 22233
Rabin, M.O. [1971] 14690
Reichbach, J. [1970] 27880
Renardel de Lavalette, G.R. [1981] 55207
Reznikoff, I. [1963] 11135
Reznikoff, I. [1965] 11136
Rickey, V.F. [1975] 11178
Rickey, V.F. [1975] 22975
Rickey, V.F. [1978] 51658
Robinson, A. [1969] 11269
Slagle, J.R. [1970] 12452
Smirnov, V.A. [1972] 30359
Stepien, T. [1981] 33327
Tarski, A. [1931] 16924
Tarski, A. [1932] 13406
Tarski, A. [1933] 28816
Tarski, A. [1934] 30917
Tarski, A. [1948] 13437
Tulipani, S. [1975] 50883
Vaeaenaenen, J. [1977] 51934
Wasilewska, A. [1985] 49467
Weaver, G.E. [1974] 14096
Weaver, G.E. [1975] 21907
Westrhenen van, S.C. [1968] 14159
Wolter, H. [1972] 14278

C45

Zil'ber, B.I. [1981] 36112

C50

Chikhachev, S.A. [1984] 42695
Hoover, D.N. [1984] 43234
Mijajlovic, Z. [1977] 50777
Pacholski, L. [1968] 10215
Poizat, B. [1982] 35642
Tulipani, S. [1975] 50883

C52

Armbrust, M. [1972] 00468
Bacsich, P.D. [1975] 17658
Chepurnov, B.A. [1976] 52949
Craig, W. [1958] 02511
Czelakowski, J. [1982] 42743
Czelakowski, J. [1985] 47502
Feferman, S. [1966] 03697
Fujiwara, T. [1971] 17104
Hanf, W.P. [1983] 35790
Hauschild, K. [1971] 05779
Herrera Miranda, J. [1978] 36172
Kaiser, Klaus [1974] 06804
Kaiser, Klaus [1975] 06805
Kaiser, Klaus [1976] 06806
Keisler, H.J. [1965] 07008
Krom, Melven R. [1963] 07571
Lightstone, A.H. [1957] 08149
Lyndon, R.C. [1959] 08439
Mal'tsev, A.I. [1959] 19260
Mundici, D. [1981] 90124
Nelson, Evelyn [1977] 26606
Oberschelp, A. [1958] 10042
Pouzet, M. [1979] 51038
Scott, D.S. [1958] 11906
Szczerba, L.W. [1965] 12831
Szczerba, L.W. [1979] 79177
Tabata, H. [1969] 13247
Tabata, H. [1971] 13248
Zil'ber, B.I. [1981] 36112
Zubieta Russi, G. [1957] 11688

C55

Asser, G. [1979] 70532
Barwise, J. [1972] 00839
Frayne, T.E. [1962] 04557
Garland, S.J. [1974] 24144
Hasenjaeger, G. [1967] 05731
Hort, C. [1984] 44206
Issel, W. [1969] 06459
Keisler, H.J. [1965] 07008
Krawczyk, A. [1977] 31404
Krynicki, M. [1979] 53311
Lindstroem, P. [1973] 08177

MacDowell, R. [1961] 19285
Malitz, J. [1980] 55717
Mizutani, C. [1977] 53069
Pinus, A.G. [1979] 53695
Seese, D.G. [1977] 27132
Shafaat, A. [1967] 12002
Shelah, S. [1973] 19627
Shelah, S. [1978] 31519
Tzouvaras, A.D. [1980] 43212

C57

Hasenjaeger, G. [1953] 05723
Kreisel, G. [1950] 07481
Kreisel, G. [1953] 20815
Manders, K.L. [1979] 53838
Rodriguez Artalejo, M. [1981] 77840

C60

Ax, J. [1968] 00580
Barwise, J. [1969] 00822
Baudisch, A. [1980] 56368
Bernays, P. [1955] 01090
Bernstein, A.R. [1973] 03898
Bunyatov, M.R. [1967] 02032
Charretton, C. [1973] 03982
Cozart, D. [1974] 02503
Dries van den, L. [1984] 41246
Ehdel'man, G.S. [1974] 72569
Frey, G. [1983] 37806
Grainger, A.D. [1975] 17507
Greniewski, H. [1950] 22000
Gurevich, Y. [1965] 05452
Hirschfeld, J. [1976] 26079
Janssen, G. [1972] 06540
Kasahara, S. [1973] 30535
Keisler, H.J. [1974] 30004
Kochen, S. [1957] 07272
Lightstone, A.H. [1957] 08149
Loeb, P.A. [1985] 44489
Lolli, G. [1978] 55177
Lorenzen, P. [1962] 08303
Luxemburg, W.A.J. [1970] 08429
Lyubetskij, V.A. [1985] 45424
Makowiecka, H. [1965] 08587
Makowiecka, H. [1975] 08588
Makowiecka, H. [1975] 08589
Makowiecka, H. [1975] 08590
Makowiecka, H. [1975] 18272
Makowiecka, H. [1975] 21816
Makowiecka, H. [1976] 24642
Mal'tsev, A.I. [1959] 19260
Manders, K.L. [1979] 53838
McKinsey, J.C.C. [1943] 09021
Meloni, G.C. [1973] 18293
Potthoff, K. [1983] 40113
Richter, M.M. [1982] 34209
Robinson, A. [1954] 11226
Robinson, A. [1963] 28804
Robinson, A. [1966] 22170
Robinson, A. [1969] 11269
Robinson, A. [1969] 21051
Robinson, A. [1973] 11276
Robinson, A. [1973] 11280
Robinson, A. [1973] 26149
Robinson, A. [1975] 17511
Shelah, S. [1973] 19627

Spivakov, Yu.L. [1974] 18401
Tarski, A. [1931] 16924
Tarski, A. [1938] 30925
Tarski, A. [1948] 19756
Tarski, A. [1967] 13463
Whiteley, W. [1973] 79943
Whiteley, W. [1977] 79941
Whiteley, W. [1978] 50914

C62

Belyakin, N.V. [1983] 41479
Boffa, M. [1977] 53983
Boffa, M. [1982] 34190
Boffa, M. [1983] 37638
Clote, P. [1983] 40361
Doepp, K. [1972] 04094
Girard, J.-Y. [1984] 48565
Grzegorczyk, A. [1971] 23119
Henson, C.W. [1984] 39860
Hinnion, R. [1979] 56023
Lindstroem, P. [1973] 08177
MacDowell, R. [1961] 19285
Maehara, S. [1957] 08516
Magidor, M. [1983] 35758
McNaughton, R. [1954] 09068
Mijoule, R. [1976] 25854
Montague, R. [1965] 24776
Motohashi, N. [1984] 42487
Nelson, George C. [1973] 17539
Pabion, J.F. [1981] 33728
Paris, J.B. [1978] 29247
Paris, J.B. [1985] 41796
Perrin, M.J. [1975] 23181
Potthoff, K. [1972] 10697
Pudlak, P. [1983] 36550
Rabin, M.O. [1961] 10934
Richard, D. [1984] 41779
Robinson, A. [1969] 11267
Rosser, J.B. [1950] 11558
Schmerl, J.H. [1982] 35360
Schwabhaeuser, W. [1979] 53471
Skolem, T.A. [1923] 21221
Skolem, T.A. [1933] 12392
Skolem, T.A. [1934] 12393
Skolem, T.A. [1955] 27718
Stojakovic, M. [1975] 79037
Tarski, A. [1933] 13407
Zbierski, P. [1978] 29200

C65

Becker, J.A. [1976] 29684
Bernays, P. [1955] 01090
Bernstein, A.R. [1973] 03898
Chadwick, J.J.M. [1977] 80867
Chlebus, B.S. [1980] 56482
Comer, S.D. [1985] 42290
Feferman, S. [1957] 03683
Fraisse, R. [1956] 04518
Grainger, A.D. [1975] 17507
Gurevich, R. [1984] 45002
Gurevich, R. [1985] 44633
Gurevich, Y. [1977] 50608
Gurevich, Y. [1979] 53793
Gurevich, Y. [1983] 33764
Gurevich, Y. [1983] 33765
Gurevich, Y. [1983] 33768

Gutierrez-Novoa, L. [1979] 54116
Hatcher, W.S. [1985] 48208
Hauschild, K. [1975] 29757
Heinrich, S. [1980] 56660
Heinrich, S. [1980] 81516
Heinrich, S. [1981] 53749
Heinrich, S. [1982] 37826
Heinrich, S. [1982] 46460
Heinrich, S. [1983] 37053
Heinrich, S. [1983] 37054
Henkin, L. [1959] 22387
Henson, C.W. [1972] 05952
Henson, C.W. [1972] 22286
Henson, C.W. [1973] 05956
Henson, C.W. [1974] 05959
Henson, C.W. [1974] 05960
Henson, C.W. [1974] 05962
Henson, C.W. [1974] 26632
Henson, C.W. [1974] 26633
Henson, C.W. [1975] 05963
Henson, C.W. [1975] 17439
Henson, C.W. [1976] 26078
Henson, C.W. [1976] 46653
Henson, C.W. [1977] 30718
Henson, C.W. [1983] 38543
Henson, C.W. [1984] 36539
Herrmann, R.A. [1978] 55117
Hoover, D.N. [1984] 43234
Kawai, T. [1979] 38297
Keisler, H.J. [1974] 30004
Krupa, A. [1984] 44797
Ladner, R.E. [1977] 52252
Laeuchli, H. [1968] 07787
Langford, C.H. [1926] 07849
Langford, C.H. [1926] 07850
Langford, C.H. [1926] 07851
Langford, C.H. [1939] 07855
Laugwitz, D. [1958] 48699
Levy, M. [1984] 44350
Lillo de, N.J. [1979] 56184
Lin, Peikee [1985] 48745
Lindenbaum, A. [1935] 30921
Loeb, P.A. [1985] 44489
Marek, W. [1973] 17662
Marek, W. [1973] 17663
Moore, S.M. [1982] 38856
Nelson, George C. [1973] 17539
Phillips, R.G. [1969] 32232
Rautenberg, W. [1962] 11041
Rautenberg, W. [1966] 11045
Rautenberg, W. [1968] 11048
Robinson, A. [1973] 11280
Scarpellini, B. [1963] 11875
Schwabhaeuser, W. [1956] 13051
Schwabhaeuser, W. [1959] 13052
Schwabhaeuser, W. [1979] 53471
Schwabhaeuser, W. [1983] 40225
Schwabhaeuser, W. [1983] 41875
Seese, D.G. [1977] 27132
Seese, D.G. [1979] 53053
Shelah, S. [1978] 31519
Skolem, T.A. [1920] 12382
Stone, A.L. [1965] 27645
Szczerba, L.W. [1965] 12831
Szczerba, L.W. [1979] 79177
Szmielew, W. [1962] 12850

Szmielew, W. [1974] 16578
Tarski, A. [1931] 16924
Tarski, A. [1967] 13463
Wheeler, W.H. [1979] 53787
Wolff, M. [1972] 14270
Wolff, M. [1984] 44491
Young, L. [1972] 18057
Zaiontz, C. [1983] 42214
Zakon, E. [1969] 16299

C70

Barwise, J. [1975] 60316
Oikkonen, J. [1979] 52831

C75

Barwise, J. [1969] 00822
Bochvar, D.A. [1940] 01348
Engeler, E. [1970] 03368
Ferro, R. [1978] 29281
Fraisse, R. [1974] 04537
Gao, Hengshan [1983] 45159
Geiser, J.R. [1974] 04831
Girard, J.-Y. [1984] 48565
Henson, C.W. [1977] 30718
Ito, Makoto [1934] 32185
Kaiser, Klaus [1976] 06806
Kreisel, G. [1975] 75092
Lopez-Escobar, E.G.K. [1967] 08262
Mostowski, Andrzej [1947] 09538
Motohashi, N. [1972] 09599
Motohashi, N. [1973] 09600
Motohashi, N. [1973] 09602
Motohashi, N. [1982] 55534
Motohashi, N. [1984] 42488
Novikov, P.S. [1939] 10016
Novikov, P.S. [1943] 19479
Oikkonen, J. [1979] 52831
Robinson, A. [1963] 11254
Shafaat, A. [1967] 12002
Skolem, T.A. [1920] 12382
Surma, S.J. [1970] 79124
Tauts, A. [1978] 54483
Uesu, T. [1971] 13764
Vinner, S. [1976] 24084
Zermelo, E. [1935] 14413

C80

Altham, J.E.J. [1971] 00285
Apelt, H. [1966] 00403
Asser, G. [1979] 70532
Baudisch, A. [1980] 56368
Blass, A.R. [1984] 47716
Chlebus, B.S. [1980] 56482
Cocchiarella, N.B. [1975] 21896
Corcoran, J. [1971] 27917
Corcoran, J. [1972] 02392
Daniels, C.B. [1978] 56537
Enderton, H.B. [1970] 03351
Flum, J. [1974] 15020
Hajek, P. [1973] 22921
Hajek, P. [1974] 24128
Hajek, P. [1978] 31131
Harel, D. [1979] 53965
Havranek, T. [1978] 54868
Hintikka, K.J.J. [1974] 50730
Issel, W. [1969] 06459

Johnson, D.R. [1969] 06648
Krynicki, M. [1979] 52881
Krynicki, M. [1979] 53311
Malitz, J. [1980] 55717
Mizutani, C. [1977] 53069
Mo, Shaokui [1964] 47141
Oikkonen, J. [1979] 52831
Pinus, A.G. [1979] 53695
Pudlak, P. [1977] 32425
Schmerl, J.H. [1982] 35360
Schock, R. [1964] 12942
Seese, D.G. [1977] 27132
Shelah, S. [1973] 12052
Skvortsov, D.P. [1983] 40156
Vaeaenaenen, J. [1977] 51934
Walkoe Jr., W.J. [1970] 13998
Walkoe Jr., W.J. [1976] 27272
Wolter, H. [1973] 14280

C85

Andreka, H. [1973] 22879
Andreka, H. [1974] 60135
Andreka, H. [1975] 55844
Andrews, P.B. [1972] 00366
Andrews, P.B. [1972] 00367
Armbrust, M. [1972] 00468
Asser, G. [1979] 70532
Barwise, J. [1969] 00822
Barwise, J. [1972] 00839
Boudreaux, J.C. [1979] 56096
Boudreaux, J.C. [1980] 53790
Buechi, J.R. [1960] 01690
Buechi, J.R. [1965] 01694
Buechi, J.R. [1965] 01697
Buechi, J.R. [1969] 01700
Buechi, J.R. [1973] 01708
Buechi, J.R. [1973] 71377
Buechi, J.R. [1973] 71381
Buechi, J.R. [1977] 51165
Buechi, J.R. [1983] 33901
Buechi, J.R. [1983] 40515
Chepurnov, B.A. [1976] 52949
Cocchiarella, N.B. [1980] 53911
Corcoran, J. [1980] 71934
Daniels, C.B. [1977] 30696
Doepp, K. [1972] 04094
Elgot, C.C. [1966] 03291
Enderton, H.B. [1970] 03351
Feferman, S. [1966] 03697
Ferro, R. [1978] 29281
Flum, J. [1971] 04388
Fraisse, R. [1958] 04520
Fraisse, R. [1974] 04537
Fraisse, R. [1977] 52674
Fraisse, R. [1985] 48204
Garland, S.J. [1974] 24144
Gurevich, Y. [1977] 50608
Gurevich, Y. [1979] 53793
Gurevich, Y. [1983] 33764
Gurevich, Y. [1983] 33765
Gurevich, Y. [1983] 33768
Harel, D. [1979] 53965
Hasenjaeger, G. [1967] 05731
Hauschild, K. [1975] 29757
Henkin, L. [1950] 05894
Herrmann, E. [1980] 56483
Hintikka, K.J.J. [1955] 06112

Hort, C. [1984] 44206
Kaiser, Klaus [1974] 06804
Kaiser, Klaus [1975] 06805
Kaiser, Klaus [1976] 06806
Kiselev, A.A. [1979] 32589
Kochen, S. [1957] 07272
Kogalovskij, S.R. [1966] 07292
Kogalovskij, S.R. [1966] 24967
Kogalovskij, S.R. [1968] 07299
Kogalovskij, S.R. [1973] 74934
Kogalovskij, S.R. [1974] 74929
Kogalovskij, S.R. [1974] 74930
Krawczyk, A. [1977] 31404
Krynicki, M. [1979] 52881
Krynicki, M. [1979] 53311
Ladner, R.E. [1977] 52252
Laeuchli, H. [1968] 07787
Langford, C.H. [1939] 07855
Lisovik, L.P. [1984] 48409
Lopez-Escobar, E.G.K. [1967] 08262
Mal'tsev, A.I. [1959] 19260
Malitz, J. [1980] 55717
Mostowski, Andrzej [1947] 09538
Mostowski, Andrzej [1961] 09570
Mostowski, Andrzej [1962] 22070
Motohashi, N. [1972] 09599
Motohashi, N. [1973] 09602
Muchnik, A.A. [1985] 47256
Oikkonen, J. [1979] 52831
Orey, S. [1959] 10165
Palyutin, E.A. [1971] 64365
Pinus, A.G. [1979] 53695
Rabin, M.O. [1968] 29625
Rabin, M.O. [1969] 10942
Rabin, M.O. [1970] 22233
Rabin, M.O. [1971] 14690
Rabin, M.O. [1972] 10943
Schreiber, P. [1965] 12976
Schwabhaeuser, W. [1967] 13058
Schwabhaeuser, W. [1979] 53471
Seese, D.G. [1977] 27132
Seese, D.G. [1979] 53053
Shelah, S. [1973] 12052
Shelah, S. [1973] 19627
Siefkes, D. [1968] 21146
Siefkes, D. [1970] 12145
Siefkes, D. [1970] 27984
Siefkes, D. [1971] 12146
Siefkes, D. [1978] 29136
Tajtslin, M.A. [1979] 82932
Tenney, R.L. [1975] 17647
Thomas, Wolfgang [1980] 55873
Vaeaenaenen, J. [1977] 51934
Wolter, H. [1972] 14277
Wolter, H. [1972] 14278
Zaiontz, C. [1983] 42214

C90

Amer, M.A. [1976] 17946
Amer, M.A. [1977] 30619
Andreka, H. [1982] 55505
Bernhardt, K. [1976] 60487
Bertossi, L. [1985] 41802
Czelakowski, J. [1982] 42743
Czelakowski, J. [1984] 41764
Daniels, C.B. [1977] 30696
Daniels, C.B. [1978] 56537

C95

Ellerman, D.P. [1978] 56695
Felscher, W. [1971] 03738
Fitting, M. [1969] 04349
Follesdal, D. [1968] 04443
Gabbay, D.M. [1984] 41831
Grayson, R.J. [1984] 41840
Hajek, P. [1973] 22921
Hajek, P. [1974] 24128
Hajek, P. [1977] 31130
Hajek, P. [1978] 31131
Josza, R. [1984] 45334
Leone, M. [1985] 49912
Lopez-Escobar, E.G.K. [1975] 24555
Lyubetskij, V.A. [1985] 45424
Maksimova, L.L. [1977] 52433
Markovic, Z. [1983] 37386
Mostowski, Andrzej [1949] 09545
Ono, H. [1985] 47510
Rantala, V. [1975] 64743
Rasiowa, H. [1953] 11004
Rasiowa, H. [1953] 11006
Rasiowa, H. [1955] 11010
Rautenberg, W. [1979] 77698
Reichbach, J. [1970] 27880
Reichbach, J. [1975] 32003
Rescher, N. [1969] 25168
Reznikoff, I. [1963] 11135
Sanmartin Esplugues, J. [1973] 78125
Shafaat, A. [1967] 12002
Takeuti, G. [1978] 52447

C95

Ebbinghaus, H.-D. [1978] 28201
Flum, J. [1971] 04388
Gol'dshtejn, B.G. [1983] 45227
Jankowski, A.W. [1985] 47505
Lindstroem, P. [1974] 29976
Tharp, L.H. [1973] 13527
Vaeaenaenen, J. [1977] 51934

C96

Mostowski, Andrzej [1979] 82308
Novikov, P.S. [1979] 38498
Robinson, A. [1979] 53900
Skolem, T.A. [1970] 29109
Tarski, A. [1956] 28568
Tarski, A. [1972] 30338

C97

Addison, J.W. [1974] 70206
Crossley, J.N. [1967] 31684
Crossley, J.N. [1975] 31725
Dejon, B. [1969] 23438
Delon, F. [1984] 43258
Gandy, R.O. [1977] 16612
Harrington, L.A. [1985] 49810
Hodges, W. [1972] 70226
Lerman, M. [1981] 54311
Lolli, G. [1984] 41493
Macintyre, A. [1978] 53596
Metakides, G. [1982] 35609
Meulen ter, A.G.B. [1983] 47464
Mueller, Gert H. [1984] 41750
Prisco di, C.A. [1985] 41792
Wechsung, G. [1984] 40976

C98

Barwise, J. [1975] 60316
Barwise, J. [1977] 24196
Barwise, J. [1977] 70117
Baudisch, A. [1980] 56368
Bucur, I. [1980] 46245
Dalen van, D. [1980] 55705
Delon, F. [1985] 47516
Ebbinghaus, H.-D. [1978] 28201
Enderton, H.B. [1972] 03355
Ershov, Yu.L. [1973] 32036
Fitting, M. [1969] 04349
Fraisse, R. [1967] 04529
Fraisse, R. [1972] 04535
Friedman, H.M. [1975] 04296
Gabbay, D.M. [1983] 41457
Gabbay, D.M. [1984] 41831
Goldblatt, R.I. [1979] 55754
Halkowska, K. [1976] 52421
Heinrich, S. [1980] 81516
Hodges, W. [1983] 39725
Kreisel, G. [1967] 22411
Langford, C.H. [1932] 22592
Lehmann, G. [1985] 42737
Lightstone, A.H. [1978] 51926
Malitz, J. [1979] 56167
Marek, W. [1972] 34216
Mostowski, Andrzej [1965] 09578
Rasiowa, H. [1963] 11028
Robinson, A. [1959] 19551
Robinson, A. [1963] 28804
Robinson, A. [1966] 22170
Robinson, A. [1973] 11276
Robinson, A. [1973] 26149
Rogers, R. [1971] 11365
Schwabhaeuser, W. [1983] 40225
Schwabhaeuser, W. [1983] 41875
Shoenfield, J.R. [1967] 22384
Smirnov, V.A. [1972] 30359
Stillwell, J.C. [1977] 52422
Tajtslin, M.A. [1970] 28675
Takahashi, S. [1974] 17488
Wang, Hao [1981] 34116

D03

Aanderaa, S.O. [1974] 03803
Boerger, E. [1978] 52807
Bukharaeva, Z.K. [1978] 52886
Cresswell, M.J. [1964] 02531
Gabrielian, A. [1981] 54258
Krom, Melven R. [1970] 07576
Kur'erov, Yu.N. [1981] 35760
Kuz'min, V.A. [1965] 29866
Markov, A.A. [1964] 08774
Markov, A.A. [1967] 08776
Maslov, S.Yu. [1972] 08850
Nagornyj, N.M. [1976] 76745
Nazaryan, G.A. [1975] 53659
Scognamiglio, G. [1963] 29833
Singletary, W.E. [1964] 12350
Skolem, T.A. [1954] 21133
Smullyan, R.M. [1961] 21025
Staples, J. [1983] 37059
Wang, Hao [1965] 14057
Wang, Hao [1965] 14058

D05

Aanderaa, S.O. [1974] 03803
Ajzerman, M.A. [1963] 00220
Anikeev, A.S. [1972] 00375
Braffort, P. [1963] 23565
Buechi, J.R. [1960] 01690
Buechi, J.R. [1965] 01694
Buechi, J.R. [1965] 01697
Buechi, J.R. [1969] 01700
Buechi, J.R. [1983] 33901
Buechi, J.R. [1983] 40515
Burks, A.W. [1954] 33359
Burks, A.W. [1962] 01780
Burks, A.W. [1962] 02718
Chomsky, N. [1972] 25314
Cooper, D.C. [1969] 28130
Dassow, J. [1985] 49482
Denning, P.J. [1978] 37736
Deutsch, M. [1977] 26483
Elgot, C.C. [1966] 03291
Fischer, R. [1982] 42341
Gabrielian, A. [1981] 54258
Godlevskij, A.B. [1974] 80465
Goessel, M. [1978] 69601
Gostev, Yu.G. [1981] 81380
Gurevich, Y. [1970] 22243
Hamblin, C.L. [1973] 62308
Havranek, T. [1974] 31143
Heringer, H.J. [1972] 21764
Hilton, A.M. [1963] 23532
Huzino, S. [1959] 06355
Kalmar, L. [1965] 48630
Kanovich, M.I. [1975] 18218
Klette, R. [1979] 53095
Kloetzer, G. [1972] 30031
Kuz'min, V.A. [1965] 29866
Ladner, R.E. [1977] 52252
Lewis, H.R. [1976] 14766
Lisovik, L.P. [1984] 48409
Melzi, G. [1976] 76279
Mostowski, A.Wlodzimierz [1981] 55152
Muchnik, A.A. [1985] 47256
Muller, D.E. [1981] 76648
Nabebin, A.A. [1977] 52155
Nakamura, A. [1981] 55427
Paltanea, R. [1980] 35233
Rabin, M.O. [1968] 29625
Rabin, M.O. [1969] 10942
Rabin, M.O. [1970] 22233
Rabin, M.O. [1971] 14690
Rabin, M.O. [1972] 10943
Ras, Z. [1971] 10995
Ras, Z. [1971] 27358
Ras, Z. [1972] 10996
Reynolds, J.C. [1969] 46610
Rohleder, H. [1961] 20926
Siefkes, D. [1970] 12145
Siefkes, D. [1970] 27984
Siefkes, D. [1978] 29136
Smullyan, R.M. [1961] 21025
Thatcher, J.W. [1968] 28596
Trakhtenbrot, B.A. [1961] 19734
Trakhtenbrot, B.A. [1962] 19733
Urbano, R.H. [1963] 13791
Vanderveken, D.R. [1976] 50285
Wagner, K. [1972] 13969
Wang, Hao [1963] 14040

Wang, Hao [1965] 14057
Wang, Juentin [1973] 65982
Wasilewska, A. [1985] 49467
Weiss, M. [1967] 14130

D10

Ajzerman, M.A. [1963] 00220
Boerger, E. [1978] 29169
Boerger, E. [1978] 52807
Buechi, J.R. [1962] 01691
Cannonito, F.B. [1962] 21294
Christen, C. [1976] 52348
Denning, P.J. [1978] 37736
Gabrielian, A. [1981] 54258
Haeussler, A.F. [1976] 52007
Kalmar, L. [1965] 48630
Kreisel, G. [1974] 63130
Kuz'min, V.A. [1965] 29866
Lewis, H.R. [1980] 36546
Marchenkov, S.S. [1982] 35236
Moshchenskij, V.A. [1973] 21271
Muller, D.E. [1985] 47646
Nurmeev, N.N. [1976] 52721
Roedding, D. [1972] 11343
Scarpellini, B. [1985] 47473
Schwartz, J.T. [1967] 23572
Simon, J. [1977] 16610
Solomonoff, R.J. [1978] 51980
Stillwell, J.C. [1977] 52422
Wang, Hao [1965] 14057
Wang, Hao [1965] 14058
Weinberg, G.M. [1967] 14125
Yasuhara, A. [1971] 18028

D15

Aanderaa, S.O. [1979] 53853
Aanderaa, S.O. [1981] 55613
Adleman, L.M. [1978] 69046
Apolloni, B. [1982] 38805
Aspvall, B. [1979] 52791
Aspvall, B. [1980] 69096
Ausiello, G. [1976] 51023
Ben-Ari, M. [1980] 55966
Bibel, W. [1979] 53410
Brown, Cynthia A. [1982] 39699
Bukharaeva, Z.K. [1978] 52886
Chen, Jiyuan [1984] 44191
Christen, C. [1976] 52348
Commentz-Walter, B. [1979] 53164
Cook, S.A. [1974] 25005
Cook, S.A. [1979] 56282
Daley, R.P. [1977] 31707
Dantsin, E.Ya. [1979] 81024
Dantsin, E.Ya. [1981] 55532
Denenberg, L. [1984] 43373
Doshita, S. [1980] 80151
Doshita, S. [1983] 44209
Dunham, B. [1976] 23648
Erni, W. [1981] 81184
Evangelist, M. [1982] 54618
Farat, V.M. [1979] 81930
Fischer, Michael J. [1982] 35363
Franco, J. [1983] 39281
Fuerer, M. [1981] 35169
Fuerer, M. [1984] 43160
Gacs, P. [1977] 52753

Galil, Z. [1977] 51297
Galil, Z. [1977] 52174
Haeussler, A.F. [1976] 52007
Haken, A. [1985] 49032
Harper, L.H. [1975] 62346
Hartmanis, J. [1976] 24160
Kanovich, M.I. [1971] 06900
Kanovich, M.I. [1974] 62867
Khomich, V.I. [1975] 17549
Klette, R. [1979] 53095
Kolmogorov, A.N. [1969] 63036
Kosovskij, N.K. [1978] 90060
Kramosil, I. [1979] 53285
Kramosil, I. [1980] 38493
Lewis, H.R. [1978] 75527
Lewis, H.R. [1978] 82072
Lewis, H.R. [1979] 53794
Lewis, H.R. [1980] 36546
Lieberherr, K.J. [1979] 38843
Lieberherr, K.J. [1981] 69677
Lieberherr, K.J. [1982] 46285
Longo, G. [1974] 53697
Lovasz, L. [1980] 53755
Luckhardt, H. [1984] 45393
Mostowski, A.Wlodzimierz [1979] 76601
Mostowski, A.Wlodzimierz [1981] 55152
Mundici, D. [1983] 37383
Mycielski, J. [1983] 37228
Nazaryan, G.A. [1976] 76769
Nazaryan, G.A. [1978] 52040
Nazaryan, G.A. [1979] 53426
Nazaryan, G.A. [1980] 76768
Nazaryan, G.A. [1982] 34748
Nurmeev, N.N. [1976] 52721
Oppen, D.C. [1980] 55875
Petri, N.V. [1969] 10436
Petri, N.V. [1969] 10438
Plaisted, D.A. [1980] 56667
Plaisted, D.A. [1984] 44042
Plotkin, J.M. [1982] 35398
Poljak, S. [1982] 39883
Pudlak, P. [1975] 21740
Pudlak, P. [1975] 29600
Pudlak, P. [1979] 54869
Pudlak, P. [1984] 43900
Roedding, D. [1972] 11343
Schaefer, T.J. [1978] 82718
Schnorr, C.-P. [1976] 69903
Schnorr, C.-P. [1977] 16638
Schuster, P. [1976] 69906
Simon, J. [1977] 16610
Springsteel, F. [1981] 82856
Stenger, H.-J. [1984] 43303
Stork, H.-G. [1979] 52752
Tarjan, R.E. [1978] 69946
Tulipani, S. [1985] 49361
Wegener, I. [1981] 66300
Wegener, I. [1981] 69392
Wietlisbach, M.N. [1981] 37266

D20

Ackermann, W. [1928] 00106
Boyer, R.S. [1984] 45265
Brady, J.M. [1977] 50984
Buck, R.C. [1963] 48047
Cartwright, Robert [1979] 56231
Case, J. [1978] 80855

Case, J. [1983] 37600
Church, A. [1932] 02123
Church, A. [1933] 02124
Church, A. [1936] 02132
Church, A. [1936] 23466
Crossley, J.N. [1965] 31683
Curry, H.B. [1941] 02612
Dedekind, R. [1888] 35601
Degano, P. [1979] 56755
Deutsch, M. [1977] 28193
Fisher, A. [1982] 36967
Friedman, H.M. [1978] 31762
Gandy, R.O. [1974] 17242
Godlevskij, A.B. [1974] 80465
Goedel, K. [1931] 15052
Gurevich, I.B. [1974] 62232
Hartmanis, J. [1976] 24160
Hermes, H. [1968] 14956
Hilbert, D. [1926] 45196
Kanovich, M.I. [1984] 45178
Khasin, L.S. [1969] 07080
Khomich, V.I. [1975] 17549
Kleene, S.C. [1967] 45895
Kolmogorov, A.N. [1969] 63036
Korshunov, A.D. [1969] 16309
Kosovskij, N.K. [1981] 55663
Kreisel, G. [1961] 07524
Lorenzen, P. [1962] 08303
Mayoh, B.H. [1970] 17641
Mints, G.E. [1971] 09286
Mints, G.E. [1982] 35718
Moore, J.S. [1979] 52930
Moshchenskij, V.A. [1973] 21271
Myhill, J.R. [1950] 09699
Nazaryan, G.A. [1978] 52040
Peter, R. [1969] 32551
Petri, N.V. [1969] 10438
Prida, J.F. [1982] 37105
Richter, M.M. [1985] 39589
Rosser, J.B. [1955] 23483
Rus, T. [1964] 16208
Semenov, A.L. [1977] 51323
Simon, J. [1977] 16610
Skolem, T.A. [1923] 38689
Slisenko, A.O. [1984] 48589
Smullyan, R.M. [1961] 21025
Stenger, H.-J. [1984] 43303
Trakhtenbrot, B.A. [1960] 42812
Yasuhara, A. [1971] 18028
Zakrevskij, A.D. [1965] 16244

D25

Bullock, A.M. [1973] 01748
Calude, C. [1981] 55399
Cannonito, F.B. [1962] 21294
Christen, C. [1976] 52348
Craig, W. [1960] 02513
Deutsch, M. [1975] 02972
Gladstone, M.D. [1965] 05010
Gladstone, M.D. [1968] 05013
Gurevich, Y. [1976] 23714
Hermes, H. [1951] 06000
Hughes, C.E. [1976] 14753
Kleene, S.C. [1967] 45895
Kripke, S.A. [1967] 31215
Kripke, S.A. [1967] 31216
Matiyasevich, Yu.V. [1970] 82198

Matiyasevich, Yu.V. [1977] 51718
Montague, R. [1957] 29361
Pour-El, M.B. [1968] 10705
Singletary, W.E. [1964] 12350
Singletary, W.E. [1974] 12356
Smullyan, R.M. [1961] 21025
Solon, B.Ya. [1981] 43887

D30

Gladstone, M.D. [1968] 05013
Hay, L. [1978] 30704
Jockusch Jr., C.G. [1976] 18212
Kreisel, G. [1953] 07487
Nerode, A. [1980] 54927
Prida, J.F. [1982] 37105
Singletary, W.E. [1964] 12350
Solon, B.Ya. [1981] 43887

D35

Aanderaa, S.O. [1971] 00087
Aanderaa, S.O. [1973] 00010
Aanderaa, S.O. [1974] 03803
Aanderaa, S.O. [1982] 33756
Aanderaa, S.O. [1982] 35210
Andrews, P.B. [1974] 03832
Applebee, R.C. [1970] 43922
Behmann, H. [1922] 00942
Behmann, H. [1923] 00943
Bernays, P. [1928] 01076
Bernays, P. [1958] 01541
Beth, E.W. [1950] 01156
Boerger, E. [1974] 01368
Boerger, E. [1974] 28158
Boerger, E. [1978] 29169
Boerger, E. [1978] 52807
Boerger, E. [1984] 40047
Bollman, D.A. [1967] 01419
Bollman, D.A. [1972] 01420
Buechi, J.R. [1962] 01691
Caviness, B.F. [1970] 03971
Chlebus, B.S. [1980] 56482
Christen, C. [1976] 52348
Church, A. [1936] 02132
Church, A. [1952] 02144
Cobham, A. [1956] 02278
Collins, G.E. [1970] 02332
Deutsch, M. [1975] 02972
Deutsch, M. [1977] 28193
Deutsch, M. [1981] 54907
Deutsch, M. [1984] 43185
Dreben, B. [1957] 03611
Dreben, B. [1962] 03127
Dulac, M.-H. [1971] 03165
Elgot, C.C. [1966] 03291
Fridman, Eh.I. [1972] 77209
Galvin, F. [1967] 04765
Galvin, F. [1970] 04766
Genenz, J. [1964] 21347
Gladstone, M.D. [1965] 05010
Gladstone, M.D. [1968] 05013
Goedel, K. [1931] 15052
Goedel, K. [1933] 20885
Goldfarb, W.D. [1973] 08111
Goldfarb, W.D. [1975] 05117
Goldfarb, W.D. [1981] 54331
Goldfarb, W.D. [1981] 55275

Goldfarb, W.D. [1984] 36555
Goldfarb, W.D. [1984] 42450
Gurevich, Y. [1965] 05452
Gurevich, Y. [1966] 05455
Gurevich, Y. [1966] 05456
Gurevich, Y. [1966] 05458
Gurevich, Y. [1970] 22243
Gurevich, Y. [1973] 25843
Gurevich, Y. [1976] 14777
Gurevich, Y. [1976] 23714
Gurevich, Y. [1982] 33757
Gurevich, Y. [1983] 33764
Gurevich, Y. [1983] 33765
Gurevich, Y. [1983] 33766
Haken, A. [1985] 49032
Harrop, R. [1964] 05682
Harrop, R. [1965] 05684
Henkin, L. [1959] 22387
Henson, C.W. [1977] 30718
Hermes, H. [1971] 06021
Ackermann, W. [1928] 00107
Hughes, C.E. [1976] 14752
Kahr, A.S. [1962] 19074
Kalicki, J. [1954] 06815
Kalmar, L. [1932] 06837
Kalmar, L. [1937] 06841
Kalmar, L. [1937] 06842
Kalmar, L. [1939] 06843
Kalmar, L. [1947] 06846
Kalmar, L. [1950] 06851
Kalmar, L. [1950] 06854
Kalmar, L. [1951] 42204
Kalmar, L. [1956] 06861
Kalmar, L. [1956] 16972
Kilmister, C.W. [1967] 22338
Kleine Buening, H. [1981] 55476
Kopieki, R. [1975] 23964
Kostyrko, V.F. [1964] 19124
Kostyrko, V.F. [1966] 19123
Kostyrko, V.F. [1971] 19122
Kripke, S.A. [1967] 31215
Kripke, S.A. [1967] 31216
Krom, Melven R. [1967] 07574
Krom, Melven R. [1970] 07576
Kuznetsov, A.V. [1963] 19144
Lewis, H.R. [1976] 14766
Lewis, H.R. [1979] 53515
Lifschitz, V. [1967] 08138
Loeb, M.H. [1976] 14746
Lorenzen, P. [1962] 08303
Manaster, A.B. [1975] 23120
Marchenkov, S.S. [1982] 35236
Maslov, S.Yu. [1965] 19335
Matiyasevich, Yu.V. [1970] 82198
Matiyasevich, Yu.V. [1977] 51718
Mayoh, B.H. [1970] 17641
McKenzie, R. [1975] 09009
Nagel, E. [1958] 09765
Nakamura, A. [1970] 09788
Nakamura, A. [1981] 55427
Nerode, A. [1980] 54927
Norgela, S.A. [1976] 32671
Norgela, S.A. [1977] 32680
Norgela, S.A. [1978] 32674
Novikov, P.S. [1949] 19476
Orevkov, V.P. [1968] 10153
Orevkov, V.P. [1971] 10157

Palyutin, E.A. [1971] 64365
Paola di, R.A. [1971] 30683
Pepis, J. [1936] 10360
Pepis, J. [1937] 19493
Pepis, J. [1938] 10361
Pepis, J. [1938] 10362
Pieczkowski, A. [1968] 10474
Pinus, A.G. [1973] 24124
Rautenberg, W. [1962] 11041
Rautenberg, W. [1968] 11048
Richard, D. [1984] 41779
Rogers Jr., H. [1956] 11355
Rosser, J.B. [1939] 11547
Rozenblat, B.V. [1979] 53949
Schuette, K. [1933] 12997
Schwabhaeuser, W. [1983] 40225
Seese, D.G. [1977] 27132
Sharonov, V.I. [1970] 14357
Shekhtman, V.B. [1982] 38116
Siefkes, D. [1971] 12146
Singletary, W.E. [1964] 12350
Skolem, T.A. [1919] 12383
Skolem, T.A. [1936] 40914
Skolem, T.A. [1937] 27938
Slomson, A. [1969] 12473
Smart, J.J.C. [1961] 31260
Stillwell, J.C. [1977] 52422
Suranyi, J. [1943] 19712
Suranyi, J. [1949] 20805
Suranyi, J. [1950] 12740
Suranyi, J. [1951] 12741
Suranyi, J. [1955] 12742
Suranyi, J. [1959] 12743
Suranyi, J. [1971] 12744
Szabo, P. [1978] 54642
Szczerba, L.W. [1965] 12831
Szczerba, L.W. [1979] 79177
Tajmanov, A.D. [1966] 19765
Trakhtenbrot, B.A. [1953] 19742
Trakhtenbrot, B.A. [1956] 19739
Turashvili, T.V. [1975] 24005
Turashvili, T.V. [1977] 50731
Urquhart, A.I.F. [1981] 54908
Wang, Hao [1963] 14040
Wang, Hao [1965] 14057
Wirsing, M. [1977] 51036
Wirsing, M. [1978] 29168
Yntema, M.K. [1964] 18050
Zhegalkin, I.I. [1939] 17903
Zykov, A.A. [1959] 30599

D40

Bergman, George M. [1978] 50464
Kreisel, G. [1975] 75092
Makanin, G.S. [1977] 51395
Makanin, G.S. [1977] 51863
Makanin, G.S. [1980] 82143
Rozenblat, B.V. [1979] 53949

D45

Calude, C. [1978] 52071
Kreisel, G. [1950] 07481
Kreisel, G. [1953] 20815
Mayoh, B.H. [1970] 17641

D50

Solon, B.Ya. [1981] 43887

D55

Belyakin, N.V. [1983] 41479
Boerger, E. [1974] 28158
Feferman, S. [1977] 27330
Garland, S.J. [1974] 24144
Hasenjaeger, G. [1953] 05723
Hasenjaeger, G. [1955] 27719
Jockusch Jr., C.G. [1976] 18212
Keisler, H.J. [1973] 07033
Kleene, S.C. [1967] 45895
Kondo, M. [1956] 07331
Kreisel, G. [1950] 07481
Kreisel, G. [1953] 07487
Kreisel, G. [1953] 20815
Krom, Melven R. [1963] 07571
Kuratowski, K. [1931] 07656
Markov, A.A. [1963] 08773
Martin-Loef, P. [1970] 20933
Nelson, George C. [1973] 17539
Tarski, A. [1931] 16924

D60

Barwise, J. [1975] 60316

D65

Feferman, S. [1977] 27330
Gandy, R.O. [1974] 17242
Normann, D. [1983] 37108

D70

Blass, A.R. [1984] 47716
Boyer, R.S. [1979] 56665
Karapetyan, B.K. [1977] 56756

D75

Bernays, P. [1969] 32552
Deutsch, M. [1977] 26483
Richter, M.M. [1985] 39589
Scott, D.S. [1982] 38593

D80

Aanderaa, S.O. [1974] 03803
Baxter, L.D. [1978] 52222
Degano, P. [1979] 56755
Fuerer, M. [1984] 43160
Hajek, P. [1978] 31131
Kolmogorov, A.N. [1969] 63036
Lehmann, G. [1985] 42737
Martin-Loef, P. [1970] 20933
Matiyasevich, Yu.V. [1977] 51718
Muller, D.E. [1981] 76648
Muller, D.E. [1985] 47646
Peters, F.E. [1974] 64464
Schnorr, C.-P. [1971] 23998
Schwartz, J.T. [1967] 23572

D96

Mostowski, Andrzej [1979] 82308

D97

Boerger, E. [1984] 40062
Boerger, E. [1984] 40072
Borga, M. [1980] 54440
Braffort, P. [1963] 23565
Crossley, J.N. [1965] 31683
Crossley, J.N. [1967] 31684
Dalen van, D. [1982] 36589
Delon, F. [1985] 47516
Gandy, R.O. [1977] 16612
Gladkij, A.V. [1983] 34944
Godlevskij, A.B. [1974] 80465
Harrington, L.A. [1985] 49810
Kalmar, L. [1965] 48630
Lolli, G. [1984] 41493
Magnaradze, D.G. [1975] 21507
Magnaradze, L.G. [1977] 70125
Mueller, Gert H. [1984] 41750
Schwartz, J.T. [1967] 23572
Stern, J. [1982] 36590

D98

Ajzerman, M.A. [1963] 00220
Barwise, J. [1975] 60316
Barwise, J. [1977] 70117
Boolos, G. [1974] 03933
Bucur, I. [1980] 46245
Cohen, L.J. [1982] 36588
Davis, Martin D. [1974] 21268
Denning, P.J. [1978] 37736
Friedman, H.M. [1975] 04296
Genenz, J. [1964] 21347
Kleene, S.C. [1952] 07173
Kleene, S.C. [1960] 07186
Kleene, S.C. [1967] 45895
Kloetzer, G. [1972] 30031
Korfhage, R. [1966] 07376
Kosovskij, N.K. [1981] 55663
Lavrov, I.A. [1970] 07889
Lavrov, I.A. [1975] 21765
Lewis, H.R. [1979] 53515
Lorenzen, P. [1962] 08303
Malitz, J. [1979] 56167
Manaster, A.B. [1975] 23120
Manin, Yu.I. [1977] 51984
Marek, W. [1972] 34216
Mostowski, Andrzej [1965] 09578
Peters, F.E. [1974] 64464
Porte, J. [1972] 17471
Shoenfield, J.R. [1967] 22384
Smullyan, R.M. [1961] 21025
Takahashi, S. [1974] 17488
Yasuhara, A. [1971] 18028

D99

Hofstadter, D.R. [1979] 74228
Lehmann, G. [1985] 42737

E05

Adler, A. [1973] 00189
Cherlin, G.L. [1972] 14684
Cowen, R.H. [1977] 21970
Eifrig, B. [1972] 17196
Luxemburg, W.A.J. [1962] 08426
Ramsey, F.P. [1930] 28781
Rudin, M.E. [1979] 53939
Skolem, T.A. [1933] 12391
Tzouvaras, A.D. [1980] 43212

E07

Asatryan, L.G. [1979] 54526
Balan, T. [1974] 70609
Chlebus, B.S. [1980] 56482
Cowen, R.H. [1977] 21970
Cuesta Dutari, N. [1953] 02588
Erne, M. [1981] 81183
Feys, R. [1949] 03761
Fraisse, R. [1950] 04510
Fraisse, R. [1955] 04516
Fraisse, R. [1958] 04520
Fraisse, R. [1977] 52674
Gonseth, F. [1933] 05128
Gurevich, Y. [1977] 50608
Gurevich, Y. [1979] 53793
Gurevich, Y. [1983] 33765
Harary, F. [1961] 05653
Hetper, W. [1938] 41420
Huntington, E.V. [1905] 37876
Huntington, E.V. [1917] 41774
Huntington, E.V. [1917] 41776
Khinchin, A.Ya. [1949] 06096
Korselt, A. [1914] 38011
Ladner, R.E. [1977] 52252
Langford, C.H. [1926] 07849
Sanchez, E. [1979] 52836
Schoenfeld, W. [1981] 37083
Tarski, A. [1941] 13430
Thomas, Wolfgang [1980] 55873
Weidig, I. [1971] 17592
Wiener, N. [1912] 14193

E10

Baer, Reinhold [1929] 00632
Buechi, J.R. [1965] 01694
Buechi, J.R. [1965] 01697
Buechi, J.R. [1973] 01708
Buechi, J.R. [1973] 71377
Buechi, J.R. [1973] 71381
Buechi, J.R. [1983] 33901
Buechi, J.R. [1983] 40515
Christian, C.C. [1976] 71770
Conway, J.H. [1976] 71909
Enriques, F. [1911] 37967
Feferman, S. [1957] 03683
Fraisse, R. [1950] 04510
Garland, S.J. [1974] 24144
Gurevich, Y. [1983] 33764
Harris, J.H. [1971] 05670
Hilbert, D. [1926] 45196
Huntington, E.V. [1905] 37876
Klaua, D. [1961] 21070
Koenig, J. [1914] 21392
Lillo de, N.J. [1979] 56184
Marek, W. [1973] 17662
Marek, W. [1973] 17663
Mueller, D.W. [1975] 21717
Plymen, R.J. [1961] 10567
Routley, R. [1965] 30839
Seckendorff von, V. [1937] 41030
Takeuti, G. [1957] 19761
Tarski, A. [1924] 13394

Tarski, A. [1948] 13436
Tauts, A. [1978] 54483
Weber, H. [1906] 37891
Russell, B. [1912] 22611
Russell, B. [1913] 22612
Yang, Anzhou [1979] 49402

E15

Kanovej, V.G. [1983] 44359
Kondo, M. [1956] 07331
Kondo, M. [1958] 37182
Kuratowski, K. [1931] 07656
Levin, A.M. [1975] 25810
Martin-Loef, P. [1970] 20933
Mostowski, Andrzej [1947] 09538
Tarski, A. [1931] 16924

E20

Anderson, D.E. [1965] 00336
Barros de, C.M. [1962] 00815
Bernstein, B.A. [1932] 01123
Both, N. [1974] 03941
Chimev, K.N. [1981] 36837
Cirulis, J. [1978] 56143
Cohen, Jack K. [1980] 54391
Cooley, J.E. [1975] 02360
Dywan, Z. [1984] 43341
Enriques, F. [1911] 37967
Finsler, P. [1933] 03795
Frege, F.L.G. [1904] 21337
Goodstein, R.L. [1972] 27766
Grzegorek, E. [1976] 32138
Hajek, O. [1960] 48087
Krom, Melven R. [1977] 50355
Kurepa, D. [1950] 28728
Laugwitz, D. [1958] 48699
Lines Escardo, E. [1962] 48076
Logan, G.J. [1976] 50056
Mollerup, J. [1907] 37914
Neumann, O. [1984] 42222
Ramsey, F.P. [1930] 28781
Robert, P. [1965] 26664
Rybak, Janet [1976] 64981
Salomaa, A. [1963] 11786
Schmidt, Garfield C. [1981] 54289
Schmidt, J. [1953] 12901
Schroeder, F.W.K.E. [1898] 12983
Schwabhaeuser, W. [1953] 27700
Schweizer, B. [1977] 56382
Seckendorff von, V. [1940] 43819
Tarski, A. [1924] 13394
Uemov, A.I. [1959] 47995
Wajsberg, M. [1933] 16533
Yaglom, I.M. [1978] 54203
Zermelo, E. [1909] 37941
Zermelo, E. [1935] 14413

E25

Abian, A. [1970] 00051
Andrews, P.B. [1972] 00367
Buechi, J.R. [1953] 01683
Buff, H.W. [1984] 43509
Chevallard, Y. [1971] 02082
Cirulis, J. [1978] 56143
Crabbe, M. [1983] 44943
Davis, Charles C. [1975] 02815
Dzik, W. [1981] 35320
Fitting, M. [1969] 04349
Germansky, B. [1961] 04911
Goedel, K. [1938] 05071
Hasenjaeger, G. [1960] 05727
Klimovsky, G. [1956] 07233
Kuroda, S. [1958] 07691
Leisenring, A.C. [1969] 07983
Levin, A.M. [1975] 25810
Luxemburg, W.A.J. [1969] 24816
Pincus, D. [1974] 21224
Rubchinskij, A.A. [1980] 55660
Schroeder, F.W.K.E. [1898] 12983
Surma, S.J. [1968] 12748
Tarski, A. [1924] 13394
Tarski, A. [1948] 13436
Wang, Hao [1955] 27722
Yakubovich, A.M. [1981] 46491
Yakubovich, A.M. [1981] 74454
Zariski, O. [1925] 41579

E30

Baer, Reinhold [1929] 00632
Bar-Hillel, Y. [1958] 00805
Barwise, J. [1972] 00839
Barwise, J. [1975] 60316
Belyakin, N.V. [1979] 70865
Bochvar, D.A. [1973] 03926
Bochvar, D.A. [1974] 71137
Boffa, M. [1977] 53983
Christian, C.C. [1972] 71771
Christian, C.C. [1976] 71770
Chwistek, L.B. [1926] 02159
Collins, G.E. [1970] 02332
Coppotelli, F. [1968] 02387
Coppotelli, F. [1977] 21956
Couture, J. [1983] 40942
Fraenkel, A.A. [1925] 04486
Frege, F.L.G. [1893] 21334
Frege, F.L.G. [1903] 21336
Frege, F.L.G. [1915] 04583
Frege, F.L.G. [1917] 04585
Frege, F.L.G. [1952] 25773
Frege, F.L.G. [1962] 22970
Friedman, H.M. [1973] 04650
Gandy, R.O. [1956] 21137
Glubrecht, J.-M. [1983] 36976
Gonseth, F. [1933] 05128
Haerlen, H. [1930] 05487
Harris, J.H. [1971] 05670
Hasenjaeger, G. [1977] 47066
Hatcher, W.S. [1968] 22644
Hauschild, K. [1968] 05770
Henson, C.W. [1984] 39860
Hilbert, D. [1926] 45196
Hinnion, R. [1979] 56023
Huntington, E.V. [1935] 06337
Keyser, C.J. [1902] 07078
Kreisel, G. [1967] 07533
Lake, J. [1975] 07802
Lebesgue, H. [1941] 37379
Lillo de, N.J. [1979] 56184
Loewenheim, L. [1940] 08247
McNaughton, R. [1954] 09068
Medvedev, F.A. [1978] 82222
Menger, K. [1928] 31950
Menger, K. [1928] 39109
Menger, K. [1928] 39110
Mollerup, J. [1907] 09387
Montague, R. [1961] 09424
Montague, R. [1965] 24776
Mueller, Gert H. [1981] 49862
Neumann von, J. [1928] 09902
Poincare, H. [1909] 37938
Pollock, J.L. [1970] 10627
Prior, A.N. [1955] 22325
Quine, W.V.O. [1937] 10863
Quine, W.V.O. [1955] 10896
Quine, W.V.O. [1963] 23295
Roedding, W. [1968] 11344
Rosser, J.B. [1950] 11558
Rosser, J.B. [1955] 23483
Rukhaya, Kh.M. [1977] 77999
Rukhaya, Kh.M. [1977] 78000
Russell, B. [1903] 25581
Schweizer, B. [1977] 56382
Scott, D.S. [1967] 22187
Shoenfield, J.R. [1954] 12087
Skolem, T.A. [1923] 21221
Skolem, T.A. [1929] 12386
Skvortsov, D.P. [1985] 47594
Suszko, R. [1950] 38697
Takeuti, G. [1957] 19761
Trakhtenbrot, B.A. [1956] 19739
Uesu, T. [1966] 13760
Vajl', V.E. [1974] 79577
Vershinin, K.P. [1973] 13934
Wang, Hao [1953] 14023
Wang, Hao [1953] 14024
Wang, Hao [1955] 27722
Wang, Hao [1963] 14050
Wang, Hao [1965] 14057
Wette, E. [1970] 14165
Weyl, H. [1918] 19791
Russell, B. [1910] 22607
Russell, B. [1912] 22611
Russell, B. [1913] 22612
Whitehead, A.N. [1934] 14183
Winter, D.J. [1982] 36766
Yule, D. [1926] 18075

E35

Ackermann, W. [1938] 00114
Ackermann, W. [1953] 00125
Andrews, P.B. [1972] 00366
Andrews, P.B. [1972] 00367
Baer, Reinhold [1929] 00632
Bochvar, D.A. [1973] 03926
Boffa, M. [1977] 26448
Boffa, M. [1977] 51446
Buff, H.W. [1984] 43509
Clay, R.E. [1968] 02255
Cocchiarella, N.B. [1976] 18458
Crabbe, M. [1975] 04051
Crabbe, M. [1978] 29204
Fitting, M. [1969] 04349
Friedman, H.M. [1973] 04650
Goedel, K. [1938] 05071
Gurevich, Y. [1983] 33764
Hasenjaeger, G. [1960] 05727
Hinnion, R. [1979] 56023
Hrbacek, K. [1978] 29206
Kawai, T. [1981] 47766
Levin, A.M. [1975] 25810

Manin, Yu.I. [1977] 51984
Manin, Yu.I. [1979] 54716
Marek, W. [1973] 17662
Marek, W. [1973] 17663
Scott, D.S. [1969] 24818
Shoenfield, J.R. [1954] 12087
Takeuti, G. [1957] 19761
Uesu, T. [1966] 13760
Wang, Hao [1955] 27722
Yakubovich, A.M. [1981] 46491
Yakubovich, A.M. [1981] 74454

E40

Amer, M.A. [1976] 17946
Amer, M.A. [1977] 30619
Gurevich, Y. [1983] 33768
Hajek, P. [1970] 05521
Jockusch Jr., C.G. [1976] 18212
Kakuda, Y. [1980] 43408
Scott, D.S. [1969] 24818
Takeuti, G. [1978] 52447
Takeuti, G. [1981] 39565

E45

Fitting, M. [1969] 04349
Fraisse, R. [1977] 52674
Fraisse, R. [1985] 48204
Goedel, K. [1938] 05071
Gurevich, Y. [1979] 53793
Hinnion, R. [1979] 56023
Jockusch Jr., C.G. [1976] 18212
Krawczyk, A. [1977] 31404
Magidor, M. [1983] 35758
Marek, W. [1973] 17662
Marek, W. [1973] 17663
Takeuti, G. [1957] 19761
Takeuti, G. [1968] 13334
Zbierski, P. [1978] 29200

E47

Deutsch, M. [1977] 28193
Gandy, R.O. [1974] 17242
Garland, S.J. [1974] 24144
Mostowski, Andrzej [1947] 09538
Tarski, A. [1931] 16924

E50

Bankston, P. [1983] 38418
Eifrig, B. [1972] 17196
Finsler, P. [1933] 03795
Fitting, M. [1969] 04349
Goedel, K. [1938] 05071
Goedel, K. [1947] 05073
Gurevich, Y. [1977] 50608
Gurevich, Y. [1979] 53793
Gurevich, Y. [1983] 33765
Hilbert, D. [1928] 06083
Kinokuniya, Y. [1973] 81856
Kreisel, G. [1967] 07533
Lourenco, M. [1979] 70030
Manin, Yu.I. [1977] 51984
Manin, Yu.I. [1979] 54716
Scott, D.S. [1969] 24818

E55

Facenda Aguirre, J.A. [1984] 44369
Gao, Hengshan [1983] 45159
Garland, S.J. [1974] 24144
Hort, C. [1984] 44206
Takeuti, G. [1968] 13334
Tauts, A. [1978] 54483

E60

Buechi, J.R. [1977] 51165
Cutland, N.J. [1984] 42232
Ladner, R.E. [1977] 52252

E65

Papic, P. [1971] 27414
Tauts, A. [1978] 54483

E70

Ackermann, W. [1938] 00114
Ackermann, W. [1941] 16704
Ackermann, W. [1950] 00122
Ackermann, W. [1952] 00124
Ackermann, W. [1953] 00125
Ackermann, W. [1958] 00131
Ackermann, W. [1961] 00192
Ackermann, W. [1965] 00135
Arruda, A.I. [1980] 53627
Asenjo, F.G. [1963] 00514
Bar-Hillel, Y. [1958] 00805
Berg van den, I. [1983] 43210
Bochvar, D.A. [1945] 01351
Bochvar, D.A. [1969] 01353
Boffa, M. [1977] 26448
Boffa, M. [1977] 51446
Boffa, M. [1981] 35424
Boffa, M. [1982] 34190
Boffa, M. [1983] 37638
Buff, H.W. [1984] 43509
Bunder, M.W. [1981] 55277
Bunder, M.W. [1982] 36080
Bunder, M.W. [1983] 31631
Burckhardt, J.J. [1938] 41071
Burgina, E.S. [1984] 45486
Chudacek, J. [1977] 52311
Chwistek, L.B. [1922] 41659
Chwistek, L.B. [1932] 02161
Chwistek, L.B. [1937] 02671
Clay, R.E. [1968] 02255
Cocchiarella, N.B. [1975] 32278
Cocchiarella, N.B. [1976] 18458
Cocchiarella, N.B. [1979] 56097
Conway, J.H. [1976] 71909
Coppotelli, F. [1968] 02387
Coppotelli, F. [1977] 21956
Crabbe, M. [1975] 04051
Crabbe, M. [1978] 29204
Crabbe, M. [1983] 44943
Cuda, K. [1980] 56506
Cuda, K. [1983] 37684
Ferraz de Aragon, D. [1979] 55098
Fitch, F.B. [1949] 04333
Fitch, F.B. [1950] 17113
Fitch, F.B. [1951] 17114
Freyd, P. [1972] 04618
Friedman, H.M. [1978] 31762

Grishin, V.N. [1976] 56149
Hatcher, W.S. [1968] 22644
Hauschild, K. [1968] 05770
Hetper, W. [1937] 41432
Hrbacek, K. [1978] 29206
Hrbacek, K. [1979] 74287
Jardine, C.J. [1971] 27501
Kawai, T. [1979] 56400
Kawai, T. [1981] 47766
Kawai, T. [1981] 55646
Krejnovich, V.Ya. [1982] 35653
Kuehnrich, M. [1983] 36566
Kuroda, S. [1958] 07681
Kuroda, S. [1958] 07683
Kuroda, S. [1958] 07691
Kuroda, S. [1959] 25455
Lake, J. [1975] 07802
Lambalgen van, M. [1983] 40731
Liu, Shichao [1980] 75591
Liu, Shichao [1984] 45423
Lopez-Escobar, E.G.K. [1972] 08267
Martin, R.M. [1950] 08817
Mlcek, J. [1979] 56719
Nelson, Edward [1977] 51504
Neumann von, J. [1928] 09902
Novak, I.L. [1950] 09999
Oberschelp, A. [1976] 31200
Orey, S. [1964] 10167
Pudlak, P. [1977] 32425
Quine, W.V.O. [1938] 10865
Quine, W.V.O. [1940] 10867
Rasiowa, H. [1964] 25051
Rasiowa, H. [1965] 29826
Resnik, M.D. [1969] 11127
Richter, M.M. [1982] 34209
Rosser, J.B. [1950] 11558
Rosser, J.B. [1953] 11565
Saito, M. [1978] 46816
Scedrov, A. [1985] 44771
Shoenfield, J.R. [1954] 12087
Smirnov, V.A. [1983] 42348
Sochor, A. [1976] 23811
Sochor, A. [1977] 31250
Sochor, A. [1979] 78823
Sols, I. [1975] 18397
Specker, E. [1958] 12669
Specker, E. [1962] 12670
Stahl, G. [1973] 13092
Takeuti, G. [1968] 13334
Takeuti, G. [1981] 39565
Vajl', V.E. [1974] 79577
Vershinin, K.P. [1973] 13934
Wang, Hao [1948] 14005
Wang, Hao [1954] 14025
Wang, Hao [1955] 27722
Wang, Hao [1963] 14044
Wang, Hao [1963] 14050
Wang, Hao [1963] 43035
Zbierski, P. [1978] 29200

E72

Cerruti, U. [1983] 44370
Sanchez, E. [1979] 52836

E75

Augenstein, B.W. [1984] 44872
Bankston, P. [1983] 38418
Bernstein, A.R. [1970] 01109
Bollman, D.A. [1973] 01422
Cerruti, U. [1983] 44370
Chuaqui, R.B. [1983] 38558
Chudacek, J. [1977] 52311
Conway, J.H. [1976] 71909
Cooper, W.S. [1974] 21633
Cresswell, M.J. [1975] 02553
Dedekind, R. [1888] 35601
Dedekind, R. [1932] 38623
Dedekind, R. [1948] 23344
Dedekind, R. [1967] 14914
Ebbinghaus, H.-D. [1974] 28198
Eifrig, B. [1972] 17196
Facenda Aguirre, J.A. [1984] 44369
Fenstad, J.E. [1967] 03742
Fenstad, J.E. [1971] 03744
Fletcher, P. [1972] 17123
Fraenkel, A.A. [1925] 04486
Frege, F.L.G. [1884] 21332
Frege, F.L.G. [1893] 21334
Frege, F.L.G. [1903] 21336
Frege, F.L.G. [1915] 04583
Gurevich, Y. [1977] 50608
Hamilton, N.T. [1961] 22640
Hausner, M. [1972] 05802
Huntington, E.V. [1905] 37876
Isaacs, G.L. [1968] 23369
Juhasz, I. [1972] 06764
Khinchin, A.Ya. [1923] 43534
Kinokuniya, Y. [1973] 81856
Kirk, R.B. [1975] 17233
Landau, E. [1930] 23388
Levin, A.M. [1975] 25810
Lindenbaum, A. [1927] 30918
Loeb, P.A. [1972] 19227
Luxemburg, W.A.J. [1962] 08426
Luxemburg, W.A.J. [1969] 24816
Ono, K. [1969] 10135
Peano, G. [1889] 19497
Pfanzagl, J. [1968] 10447
Pincus, D. [1974] 21224
Ptak, V. [1977] 51205
Pucciano, G. [1914] 38014
Quine, W.V.O. [1961] 10906
Ramsey, F.P. [1930] 28781
Richter, M.M. [1983] 44846
Robinson, A. [1969] 11267
Rudin, M.E. [1979] 53939
Seiler, E. [1973] 78359
Seiler, E. [1975] 78361
Sikorski, R. [1962] 42875
Sochor, A. [1976] 23811
Sochor, A. [1977] 31250
Stone, A.L. [1965] 27645
Takeuti, G. [1978] 52447
Wang, Hao [1950] 14010
Wojcicki, R. [1974] 29567
Zhuravlev, Yu.I. [1964] 28782

E96

Dedekind, R. [1932] 38623
Frege, F.L.G. [1952] 25773

Lesniewski, S. [1967] 28535
Mostowski, Andrzej [1979] 82308
Neumann von, J. [1961] 25726
Novikov, P.S. [1979] 38498
Wang, Hao [1962] 32531

E97

Bokshtejn, M.F. [1975] 80470
Gonseth, F. [1941] 48625
Harrington, L.A. [1985] 49810
Mueller, Gert H. [1981] 54903

E98

Artmann, B. [1983] 34479
Avenoso, F.J. [1970] 22631
Baginski, M. [1973] 75585
Bar-Hillel, Y. [1958] 00805
Barwise, J. [1975] 60316
Barwise, J. [1977] 70117
Bernays, P. [1958] 16969
Borkowski, L. [1963] 19509
Chenique, F. [1974] 23053
Christian, R.R. [1958] 22478
Combes, M. [1971] 02334
Detlovs, V.K. [1970] 17027
Dinkines, F. [1964] 22482
Dodge, C.W. [1969] 22633
Drake, F.R. [1985] 47704
Durnev, V.G. [1978] 72511
Edwards, R.E. [1979] 52995
Ehlers, F. [1968] 22634
Feferman, S. [1964] 03694
Felscher, W. [1978] 51762
Felscher, W. [1979] 72847
Fiorentini, M. [1970] 17901
Fraenkel, A.A. [1959] 28712
Freyd, P. [1972] 04618
Friedman, H.M. [1975] 04296
Gladkij, A.V. [1974] 21763
Goedel, K. [1947] 05073
Goldblatt, R.I. [1979] 55754
Grzegorczyk, A. [1955] 31447
Halkowska, K. [1976] 52421
Halkowska, K. [1978] 54789
Hamilton, N.T. [1961] 22640
Hanna, S.C. [1971] 22641
Hasse, M. [1966] 22643
Hatcher, W.S. [1968] 22644
Henkin, L. [1962] 05921
Karpov, V.G. [1977] 51362
Koenig, J. [1914] 21392
Krempa, J. [1977] 51983
Landau, E. [1930] 23388
Lavrov, I.A. [1970] 07888
Lavrov, I.A. [1975] 21765
Marek, W. [1972] 34216
Mostowski, Andrzej [1965] 09578
Oberschelp, A. [1970] 31198
Oberschelp, A. [1974] 31195
Oberschelp, A. [1978] 31196
Otepanov, V.I. [1972] 13931
Penzov, Yu.E. [1968] 16550
Quine, W.V.O. [1963] 23295
Rasiowa, H. [1971] 22289
Rasiowa, H. [1972] 31450
Rosser, J.B. [1953] 11565

Shoenfield, J.R. [1967] 22384
Stoll, R.R. [1961] 19683
Stoll, R.R. [1963] 13197
Vasilache, S. [1977] 51636
Volkov, V.A. [1978] 79749
Wang, Hao [1981] 34116
Russell, B. [1910] 22607
Russell, B. [1912] 22611
Russell, B. [1913] 22612
Wojciechowska, A. [1979] 80045

E99

Kreisel, G. [1973] 48845

F05

Anderson, A.R. [1975] 50675
Andrews, P.B. [1971] 00365
Andrews, P.B. [1974] 00368
Bernays, P. [1965] 16287
Bibel, W. [1969] 01198
Bibel, W. [1979] 53410
Bowen, K.A. [1972] 01505
Bowen, K.A. [1973] 01507
Bowen, K.A. [1974] 03943
Buchholz, W. [1975] 01681
Bunder, M.W. [1980] 53699
Craig, W. [1957] 02508
Craig, W. [1957] 02509
Craig, W. [1957] 02510
Craig, W. [1960] 02513
Craig, W. [1967] 02516
Dopp, J. [1962] 03085
Dragalin, A.G. [1980] 36552
Dreben, B. [1971] 17043
Gentzen, G. [1935] 24221
Gentzen, G. [1936] 04851
Gentzen, G. [1936] 22402
Gentzen, G. [1969] 29483
Glubrecht, J.-M. [1982] 35619
Hailperin, T. [1969] 05608
Hanazawa, M. [1979] 54329
Herbrand, J. [1930] 20866
Bernays, P. [1934] 01098
Bernays, P. [1939] 01082
Jervell, H.R. [1973] 33265
Kanger, S. [1957] 22373
Kodera, H. [1980] 74921
Kur'erov, Yu.N. [1979] 81977
Kuroda, S. [1959] 25622
Kuzichev, A.S. [1977] 52560
Leisenring, A.C. [1969] 07983
Los, J. [1956] 08333
Lyaletskij, A.V. [1981] 35594
Maehara, S. [1954] 08513
Maehara, S. [1960] 08520
Maehara, S. [1960] 08521
Maehara, S. [1962] 08526
Martin-Loef, P. [1973] 76088
Maslov, S.Yu. [1977] 32672
Mezhlumbekova, V.F. [1975] 63913
Mints, G.E. [1966] 09276
Mints, G.E. [1967] 37138
Mints, G.E. [1974] 37140
Mints, G.E. [1974] 37142
Motohashi, N. [1982] 33418
Orevkov, V.P. [1969] 10154

F07

Paeppinghaus, P. [1983] 40742
Prawitz, D. [1965] 22164
Prawitz, D. [1968] 10733
Prawitz, D. [1975] 27280
Raggio, A.R. [1974] 10972
Rasiowa, H. [1960] 11024
Richter, M.M. [1978] 51870
Sanchis, L.E. [1965] 11824
Sanchis, L.E. [1971] 11826
Sanchis, L.E. [1971] 11827
Scanlon, T.M. [1973] 11874
Schroeter, K. [1958] 24841
Schuette, K. [1950] 12999
Schuette, K. [1953] 13003
Schuette, K. [1958] 13008
Schuette, K. [1960] 13010
Schuette, K. [1960] 13011
Sikorski, R. [1958] 12303
Simauti, T. [1956] 12080
Smirnov, V.A. [1972] 30359
Smirnov, V.A. [1973] 21800
Smullyan, R.M. [1963] 12547
Smullyan, R.M. [1965] 12551
Smullyan, R.M. [1965] 16212
Smullyan, R.M. [1968] 12555
Statman, R. [1978] 52901
Tait, W.W. [1966] 13283
Takahashi, Moto-o [1968] 13295
Takeuti, G. [1955] 28446
Takeuti, G. [1956] 13307
Takeuti, G. [1956] 13308
Takeuti, G. [1956] 28447
Takeuti, G. [1957] 19761
Takeuti, G. [1958] 13312
Takeuti, G. [1958] 28448
Takeuti, G. [1960] 13316
Takeuti, G. [1961] 13321
Takeuti, G. [1978] 52447
Tang, Tonggao [1981] 34057
Titani, S. [1973] 13615
Uesu, T. [1971] 13764
Umezawa, T. [1979] 53149

F07

Ackermann, W. [1950] 00122
Anderson, J.M. [1962] 00341
Belnap Jr., N.D. [1963] 01010
Beneyto, R. [1971] 01031
Bennett, D.W. [1973] 01038
Bennett, D.W. [1977] 21957
Benthem van, J.F.A.K. [1974] 03889
Bernays, P. [1965] 16287
Beth, E.W. [1955] 01163
Beth, E.W. [1962] 04250
Bibel, W. [1975] 60499
Bibel, W. [1982] 40212
Bibel, W. [1982] 40242
Bing, K. [1968] 01210
Boolos, G. [1984] 40442
Call, R.L. [1984] 40582
Carnap, R. [1935] 03962
Chubaryan, A.A. [1972] 17157
Czermak, J. [1977] 24249
Dopp, J. [1962] 03085
Dreben, B. [1971] 17043
Funahashi, S. [1979] 77023
Gabbay, D.M. [1978] 32163

Gentzen, G. [1935] 24221
Gentzen, G. [1936] 22402
Gentzen, G. [1969] 29483
Gilbert, M.A. [1976] 62011
Gleason, G.G. [1974] 23612
Goad, C.A. [1980] 55988
Gregg, J.R. [1970] 05317
Gruenberg, T. [1983] 43736
Guillaume, M. [1958] 42813
Gupta, H.N. [1968] 14682
Hailperin, T. [1969] 05608
Hendry, H.E. [1975] 18186
Herbrand, J. [1928] 05972
Herbrand, J. [1930] 20866
Hintikka, K.J.J. [1973] 32405
Kalish, D. [1957] 24907
Kanger, S. [1957] 22373
Kanger, S. [1963] 06888
Kapur, D. [1984] 43561
Kreisel, G. [1977] 53079
Kuroda, S. [1958] 07681
Kuroda, S. [1958] 07683
Lambek, J. [1958] 07815
Lambert Jr., W.M. [1977] 31153
Leblanc, H. [1965] 75405
Leblanc, H. [1966] 07926
Leblanc, H. [1972] 07940
Leblanc, H. [1972] 19182
Lifschitz, V. [1980] 54190
London, F. [1925] 08256
Lyaletskij, A.V. [1975] 53444
Lyaletskij, A.V. [1975] 53445
Maslov, S.Yu. [1972] 08850
Matulis, V.A. [1963] 08901
Mayoh, B.H. [1974] 63802
Miller, Dale A. [1984] 43218
Mints, G.E. [1967] 19424
Mints, G.E. [1967] 37138
Mints, G.E. [1968] 19422
Mints, G.E. [1974] 37142
Nederpelt, R.P. [1977] 53110
Nieland, J.J.F. [1967] 09951
Norgela, S.A. [1978] 32674
Orevkov, V.P. [1969] 10154
Parks, R.Z. [1971] 11123
Parry, W.T. [1965] 10301
Popov, S.V. [1974] 15197
Prawitz, D. [1965] 22164
Prawitz, D. [1975] 27280
Prawitz, D. [1982] 34933
Prawitz, D. [1985] 45636
Price, Robert [1961] 10756
Raggio, A.R. [1964] 10969
Raggio, A.R. [1974] 10972
Raggio, A.R. [1977] 16607
Reichbach, J. [1967] 19564
Robitashvili, N.G. [1971] 11331
Rogava, M.G. [1972] 64879
Schoenfeld, W. [1983] 43253
Schroeder-Heister, P. [1984] 42500
Schroeter, K. [1955] 12989
Schroeter, K. [1958] 24841
Schuette, K. [1950] 12999
Schuette, K. [1956] 19693
Schuette, K. [1958] 13008
Shoesmith, D.J. [1978] 51869
Smirnov, V.A. [1972] 30359

Smullyan, R.M. [1963] 12547
Smullyan, R.M. [1963] 28562
Smullyan, R.M. [1965] 12551
Smullyan, R.M. [1965] 16212
Smullyan, R.M. [1968] 12555
Smullyan, R.M. [1970] 12557
Suvorov, P.Yu. [1979] 36649
Thiel, C. [1978] 30592
Urquhart, A.I.F. [1974] 79550

F10

Feferman, S. [1977] 27330
Kreisel, G. [1965] 27577
Takeuti, G. [1956] 13308

F15

Ackermann, W. [1928] 00106
Bibel, W. [1969] 01198
Feferman, S. [1977] 27330
Bernays, P. [1939] 01082
Schuette, K. [1960] 13010
Schuette, K. [1962] 22382
Schuette, K. [1963] 22383
Schuette, K. [1965] 13015
Takeuti, G. [1961] 13320

F20

Anikeev, A.S. [1972] 00375
Ben-Ari, M. [1980] 55966
Chubaryan, A.A. [1972] 17157
Chubaryan, A.A. [1975] 76827
Chubaryan, A.A. [1977] 52021
Chubaryan, A.A. [1982] 40752
Cook, S.A. [1974] 25005
Cook, S.A. [1975] 28181
Cook, S.A. [1979] 56282
Dantsin, E.Ya. [1981] 55532
Dardzhaniya, G.K. [1979] 53026
Evangelist, M. [1982] 54618
Goedel, K. [1934] 35971
Goldfarb, W.D. [1981] 54331
Hafner, I. [1979] 53656
Hartmanis, J. [1976] 24160
Justen, K. [1981] 55967
Kramosil, I. [1980] 56455
Longo, G. [1974] 53697
Norwood, F.H. [1982] 76885
Plaisted, D.A. [1980] 56667
Plotkin, J.M. [1982] 35398
Popov, S.V. [1975] 18344
Popov, S.V. [1976] 25598
Popov, S.V. [1977] 77440
Scarpellini, B. [1985] 47473
Schoenfeld, W. [1982] 34105
Schoenfeld, W. [1983] 43253
Statman, R. [1977] 16639
Statman, R. [1978] 52901
Yukami, T. [1983] 37612
Zamov, N.K. [1974] 50359

F25

Ackermann, W. [1950] 00122
Ackermann, W. [1952] 00124
Anderson, A.R. [1975] 50675
Andrews, P.B. [1974] 00368

Clay, R.E. [1968] 02255
Collins, G.E. [1970] 02332
Crabbe, M. [1978] 29204
Fitch, F.B. [1938] 04321
Furmanowski, T. [1982] 42298
Goedel, K. [1931] 15052
Gurevich, Y. [1965] 05452
Gurevich, Y. [1982] 33757
Gurevich, Y. [1983] 33765
Hajek, P. [1970] 05521
Hajek, P. [1981] 35379
Hanazawa, M. [1979] 54329
Hatcher, W.S. [1964] 32200
Herbrand, J. [1930] 20866
Hook, J.L. [1985] 41804
Ito, Makoto [1934] 32185
Ito, Makoto [1935] 28786
Kreisel, G. [1954] 22027
Kubinski, T. [1963] 07595
Lambert, K. [1967] 07823
Montague, R. [1957] 29356
Montague, R. [1961] 09424
Myhill, J.R. [1950] 09699
Pabion, J.F. [1980] 55872
Roessler, K. [1934] 19547
Rosser, J.B. [1950] 11558
Shoenfield, J.R. [1954] 12087
Szczerba, L.W. [1977] 51583
Szczerba, L.W. [1979] 56477
Szczerba, L.W. [1983] 39510
Wang, Hao [1950] 14011
Wang, Hao [1963] 14050
Wojtylak, P. [1981] 55071

F30

Ackermann, W. [1925] 00104
Ackermann, W. [1953] 00125
Ackermann, W. [1965] 00135
Apelt, H. [1966] 00403
Bel'tyukov, A.P. [1976] 29789
Bernays, P. [1970] 01284
Beth, E.W. [1962] 01176
Boffa, M. [1984] 42423
Bollman, D.A. [1973] 01422
Boolos, G. [1979] 56312
Boolos, G. [1980] 71208
Bosch, J.E. [1985] 48160
Both, N. [1967] 01491
Cannonito, F.B. [1962] 21294
Cegielski, P. [1984] 46773
Chauvin, A. [1949] 01996
Christian, C.C. [1972] 71771
Christian, C.C. [1976] 60952
Chubaryan, A.A. [1977] 52021
Chubaryan, A.A. [1982] 40752
Clote, P. [1983] 40361
Collins, G.E. [1970] 02332
Cooper, D.C. [1972] 61080
Crossley, J.N. [1975] 31689
Curry, H.B. [1941] 02612
Dapueto, C. [1982] 45889
Deutsch, M. [1975] 04086
Ding, Decheng [1984] 44972
Dragalin, A.G. [1972] 03610
Dreben, B. [1971] 17045
Dywan, Z. [1984] 45756
Elgot, C.C. [1966] 03291

Enderton, H.B. [1972] 03355
Fisher, A. [1982] 36967
Fogelis, E. [1950] 17125
Friedman, H.M. [1978] 31762
Friedman, H.M. [1980] 55240
Gentzen, G. [1935] 24221
Gentzen, G. [1936] 22402
Gentzen, G. [1954] 04853
Gentzen, G. [1969] 29483
Gentzen, G. [1974] 04855
Germano, G. [1970] 04906
Goedel, K. [1931] 15052
Goedel, K. [1934] 35971
Goodstein, R.L. [1945] 05166
Goodstein, R.L. [1951] 25669
Goodstein, R.L. [1955] 05177
Goodstein, R.L. [1957] 05180
Goodstein, R.L. [1958] 05182
Goodstein, R.L. [1963] 05189
Goodstein, R.L. [1966] 05190
Greniewski, H. [1951] 05333
Grzegorczyk, A. [1971] 23119
Hajek, P. [1981] 35379
Hauck, J. [1972] 62383
Hauschild, K. [1971] 05779
Henkin, L. [1954] 05900
Henkin, L. [1957] 05910
Henson, C.W. [1984] 36539
Henson, C.W. [1984] 39860
Herbrand, J. [1929] 05967
Herbrand, J. [1930] 20866
Herbrand, J. [1931] 16812
Hetper, W. [1934] 40820
Bernays, P. [1934] 01098
Bernays, P. [1939] 01082
Hischer, Horst [1976] 25590
Hotomski, P. [1982] 37063
Kalmar, L. [1928] 06835
Kleene, S.C. [1952] 07173
Kleene, S.C. [1967] 45895
Kosovskij, N.K. [1981] 55663
Kreisel, G. [1950] 07481
Kreisel, G. [1953] 20815
Kreisel, G. [1954] 22027
Kripke, S.A. [1967] 31215
Kripke, S.A. [1967] 31216
Kubinski, T. [1963] 07595
Kuroda, S. [1959] 15552
Kuroda, S. [1959] 25434
Kuroda, S. [1959] 25455
Ladriere, J. [1957] 07777
Lifschitz, V. [1985] 44843
Lourenco, M. [1979] 70030
Maehara, S. [1957] 08516
Maehara, S. [1962] 08526
Maliaukiene, L. [1985] 49123
Manin, Yu.I. [1977] 51984
Manin, Yu.I. [1979] 54716
Mayoh, B.H. [1970] 17641
McMinn, T.J. [1974] 09065
Mezhlumbekova, V.F. [1975] 63913
Mikolajewicz, B. [1983] 43117
Mints, G.E. [1971] 09286
Misercque, D. [1980] 76441
Montague, R. [1957] 29356
Montague, R. [1961] 09424
Myhill, J.R. [1950] 09699

Nagel, E. [1958] 09765
Nagel, E. [1961] 09768
Novikov, P.S. [1949] 19476
Pabion, J.F. [1974] 17648
Pabion, J.F. [1981] 33728
Paris, J.B. [1978] 29247
Paris, J.B. [1985] 41796
Peano, G. [1889] 19497
Prawitz, D. [1975] 27280
Presburger, M. [1930] 10749
Pudlak, P. [1983] 36550
Rabin, M.O. [1961] 10934
Richard, D. [1984] 39634
Richard, D. [1984] 41779
Richard, D. [1985] 45095
Richard, D. [1985] 46289
Robinson, R.M. [1957] 11315
Robinson, R.M. [1958] 48434
Roedding, W. [1968] 11344
Rosser, J.B. [1939] 11547
Rousseau, G. [1966] 11593
Rousseau, G. [1968] 11598
Sakai, H. [1974] 11772
Scanlon, T.M. [1973] 11874
Schuette, K. [1956] 19693
Schuette, K. [1960] 13010
Seckendorff von, V. [1937] 41030
Semenov, A.L. [1977] 51323
Shapiro, S. [1985] 44774
Shostak, R.E. [1977] 69916
Shostak, R.E. [1979] 82799
Singh, S. [1967] 12349
Skolem, T.A. [1923] 38689
Skolem, T.A. [1931] 12388
Skolem, T.A. [1933] 12392
Skolem, T.A. [1934] 12393
Skolem, T.A. [1936] 12396
Skolem, T.A. [1952] 21071
Skolem, T.A. [1953] 21072
Skolem, T.A. [1955] 27718
Skolem, T.A. [1956] 12416
Smart, J.J.C. [1961] 31260
Smullyan, R.M. [1957] 12538
Soudieux, C. [1960] 12654
Steen, S.W.P. [1972] 22386
Stepanov, V.A. [1985] 46515
Takeuti, G. [1978] 52447
Tarski, A. [1932] 13406
Tarski, A. [1933] 13407
Tarski, A. [1933] 28816
Tarski, A. [1939] 13428
Tomas, F. [1975] 25028
Tsinman, L.L. [1967] 02176
Tsinman, L.L. [1968] 02177
Valpola, V. [1947] 42606
Vuckovic, V. [1960] 22133
Wang, Hao [1950] 14011
Wang, Hao [1953] 14023
Wang, Hao [1953] 14024
Wang, Hao [1955] 14026
Wang, Hao [1957] 33305
Wang, Hao [1963] 14050
Wang, Hao [1981] 34116
Wette, E. [1970] 14165
Wette, E. [1974] 14955
Wolter, H. [1973] 14280

F35

Ackermann, W. [1953] 00125
Ackermann, W. [1965] 00135
Andrews, P.B. [1971] 00365
Andrews, P.B. [1972] 00366
Andrews, P.B. [1972] 00367
Andrews, P.B. [1974] 00368
Andrews, P.B. [1974] 03832
Belyakin, N.V. [1983] 41479
Beth, E.W. [1937] 41045
Bibel, W. [1969] 01198
Boffa, M. [1983] 37638
Boffa, M. [1984] 42423
Borkowski, L. [1958] 01479
Bowen, K.A. [1973] 01507
Bowen, K.A. [1974] 03943
Buchholz, W. [1975] 01681
Buechi, J.R. [1960] 01690
Buechi, J.R. [1969] 01700
Christian, C.C. [1976] 60952
Chwistek, L.B. [1912] 04231
Chwistek, L.B. [1938] 02162
Constable, R.L. [1985] 49190
Dragalin, A.G. [1980] 36552
Dreben, B. [1971] 17045
Elgot, C.C. [1966] 03291
Feferman, S. [1977] 27330
Fitch, F.B. [1948] 04329
Fitch, F.B. [1949] 04333
Friedman, H.M. [1980] 55240
Gastev, Yu.A. [1970] 17919
Gel'fond, M.G. [1972] 04841
Gentzen, G. [1936] 04851
Girard, J.-Y. [1984] 48565
Goedel, K. [1931] 15052
Goedel, K. [1934] 35971
Goetlind, E. [1952] 21134
Goguadze, D.F. [1970] 05093
Goldblatt, R.I. [1979] 55754
Grayson, R.J. [1984] 41840
Grzegorczyk, A. [1971] 23119
Hayashi, S. [1983] 37813
Henkin, L. [1950] 05894
Henson, C.W. [1984] 39860
Huet, G. [1973] 06291
Inoue, K. [1975] 48761
Jansohn, H.-S. [1982] 36620
Kogalovskij, S.R. [1968] 07299
Kogalovskij, S.R. [1970] 07300
Kreisel, G. [1965] 27577
Kreisel, G. [1969] 26648
Lambek, J. [1984] 44302
Levin, A.M. [1975] 25810
Levin, A.M. [1979] 53493
Loeb, M.H. [1968] 08229
Lopez-Escobar, E.G.K. [1967] 08262
Maehara, S. [1960] 08521
Maehara, S. [1962] 08526
Maehara, S. [1971] 08528
Martin-Loef, P. [1973] 76088
Mostowski, A.Wlodzimierz [1981] 55152
Myhill, J.R. [1951] 09702
Myhill, J.R. [1971] 28094
Myhill, J.R. [1974] 48354
Nabebin, A.A. [1977] 52155
Nagata, M. [1975] 21892
Nerode, A. [1980] 54927

Pabion, J.F. [1980] 55872
Paeppinghaus, P. [1983] 40742
Pietrzykowski, T. [1973] 10491
Prawitz, D. [1968] 10733
Robinson, John Alan [1969] 11291
Robinson, R.M. [1957] 11315
Robinson, R.M. [1958] 48434
Royse, James R. [1969] 11634
Sakai, H. [1974] 11772
Scarpellini, B. [1985] 47473
Scedrov, A. [1985] 44771
Schuette, K. [1960] 13010
Schuette, K. [1960] 13011
Schuette, K. [1965] 13015
Schwabhaeuser, W. [1979] 53471
Schwichtenberg, H. [1983] 40090
Seely, R.A.G. [1982] 34890
Shvarts, G.F. [1979] 56037
Siefkes, D. [1970] 12145
Siefkes, D. [1970] 27984
Siefkes, D. [1971] 12146
Simauti, T. [1956] 12080
Tait, W.W. [1966] 13283
Takahashi, Moto-o [1968] 13295
Takeuti, G. [1953] 37768
Takeuti, G. [1955] 28446
Takeuti, G. [1956] 13307
Takeuti, G. [1956] 13308
Takeuti, G. [1956] 28447
Takeuti, G. [1958] 13312
Takeuti, G. [1958] 28448
Takeuti, G. [1960] 13316
Takeuti, G. [1961] 13320
Takeuti, G. [1961] 13321
Takeuti, G. [1978] 52447
Tarski, A. [1931] 16924
Titani, S. [1973] 13615
Trakhtenbrot, B.A. [1961] 19734
Trakhtenbrot, B.A. [1961] 19735
Trakhtenbrot, B.A. [1962] 19733
Uesu, T. [1971] 13764
Wang, Hao [1949] 14006
Wang, Hao [1954] 14025
Yakubovich, A.M. [1981] 46491
Zahn, P. [1967] 24851
Zbierski, P. [1978] 29200

F40

Calude, C. [1978] 52071
Dapueto, C. [1982] 45889
Bernays, P. [1939] 01082
Nerode, A. [1980] 54927
Smullyan, R.M. [1957] 12538

F50

Anikeev, A.S. [1972] 00375
Barthelemy, J.-P. [1974] 04140
Belnap Jr., N.D. [1963] 01010
Bernays, P. [1970] 01284
Beth, E.W. [1960] 01171
Beth, E.W. [1962] 01176
Blanche, R. [1957] 01538
Bloom, S.L. [1977] 28146
Bowen, K.A. [1974] 03943
Brouwer, L.E.J. [1955] 02051
Buchholz, W. [1975] 01681

Church, A. [1928] 02122
Craig, W. [1967] 02516
Curry, H.B. [1950] 04222
Curry, H.B. [1952] 02620
Curry, H.B. [1952] 28059
Destouches-Fevrier, P. [1945] 02962
Dopp, J. [1962] 03085
Dragalin, A.G. [1972] 03610
Dragalin, A.G. [1980] 36552
Dzik, W. [1975] 21897
Fitting, M. [1969] 04349
Friedman, H.M. [1973] 04650
Friedman, H.M. [1978] 31762
Geiser, J.R. [1974] 04831
Gentzen, G. [1935] 24221
Gentzen, G. [1969] 29483
Gentzen, G. [1974] 04855
Goad, C.A. [1980] 55988
Goldblatt, R.I. [1979] 55754
Goodstein, R.L. [1957] 05180
Goodstein, R.L. [1958] 05182
Grayson, R.J. [1984] 41840
Grzegorczyk, A. [1972] 05418
Hayashi, S. [1983] 37813
Heyting, A. [1958] 06062
Hilbert, D. [1931] 21186
Hiz, H. [1946] 06149
Hosoi, T. [1966] 06242
Johansson, I. [1953] 06644
Josza, R. [1984] 45334
Kanovich, M.I. [1975] 18218
Khomich, V.I. [1970] 06201
Khomich, V.I. [1975] 17549
Kleene, S.C. [1945] 07169
Kleene, S.C. [1952] 07173
Kodera, H. [1980] 74921
Kreisel, G. [1965] 27577
Kreisel, G. [1967] 07533
Lambek, J. [1984] 44302
Leblanc, H. [1965] 75405
Leblanc, H. [1966] 07926
Leblanc, H. [1972] 07940
Lifschitz, V. [1985] 44843
Liu, Shichao [1984] 45423
Loeb, M.H. [1976] 14746
Lopez-Escobar, E.G.K. [1972] 08267
Lopez-Escobar, E.G.K. [1975] 24555
Lorenzen, P. [1955] 08290
Los, J. [1956] 08333
Luckhardt, H. [1969] 08380
Lukasiewicz, J. [1953] 25951
Marini, D. [1975] 22915
Markovic, Z. [1983] 37386
Martin-Loef, P. [1973] 76088
Maslov, S.Yu. [1965] 19335
Meredith, C.A. [1953] 09134
Meredith, C.A. [1968] 09143
Meredith, D. [1979] 31933
Meredith, D. [1980] 76287
Mezhlumbekova, V.F. [1975] 21799
Mezhlumbekova, V.F. [1975] 63913
Mints, G.E. [1967] 19424
Mints, G.E. [1968] 19422
Mints, G.E. [1982] 35718
Morgan, C.G. [1976] 27095
Mostowski, Andrzej [1953] 16376
Myhill, J.R. [1971] 28094

Myhill, J.R. [1974] 48354
Nelson, D. [1966] 09846
Ono, H. [1985] 47510
Orevkov, V.P. [1968] 10152
Orevkov, V.P. [1968] 19539
Plisko, V.E. [1978] 52084
Pogorzelski, W.A. [1960] 21001
Popper, K.R. [1947] 10648
Prawitz, D. [1965] 22164
Rasiowa, H. [1953] 11004
Rasiowa, H. [1955] 11010
Rasiowa, H. [1963] 11028
Rasiowa, H. [1974] 31451
Reichbach, J. [1975] 32003
Renardel de Lavalette, G.R. [1981] 55207
Renardel de Lavalette, G.R. [1984] 46669
Reznikoff, I. [1963] 11135
Rousseau, C. [1978] 51810
Rousseau, C. [1979] 82662
Scedrov, A. [1985] 44771
Schmidt, H.A. [1960] 12892
Schroeder-Heister, P. [1984] 42500
Schroeter, K. [1955] 12989
Schroeter, K. [1956] 12993
Schuette, K. [1950] 12999
Schwichtenberg, H. [1983] 40090
Seely, R.A.G. [1982] 34890
Shapiro, S. [1985] 44774
Shimada, K. [1974] 78527
Shvarts, G.F. [1979] 56037
Sikorski, R. [1962] 42875
Skolem, T.A. [1952] 21071
Skolem, T.A. [1953] 21072
Smirnov, V.A. [1979] 53921
Smyth, M.B. [1974] 12559
Takeuti, G. [1981] 39565
Tarski, A. [1938] 13420
Tokarz, M. [1971] 13627
Dalen van, D. [1982] 34494
Tsitkin, A.I. [1979] 53496
Ulrich, D. [1976] 18459
Umezawa, T. [1979] 53149
Valpola, V. [1947] 42606
Valpola, V. [1955] 13817
Valpola, V. [1955] 16346
Wronski, A. [1974] 18432
Zarnecka-Bialy, E. [1972] 16532
Zarnecka-Bialy, E. [1973] 14363

F55

Beth, E.W. [1956] 01166
Dienes, P. [1938] 03008
Heyting, A. [1958] 06062
Heyting, A. [1959] 27711
Heyting, A. [1961] 06065
Kronecker, L. [1887] 37377
Mitani, S. [1983] 41462
Rousseau, C. [1979] 82662
Skolem, T.A. [1954] 21133
Dalen van, D. [1982] 34494
Weyl, H. [1924] 38712
Wiedmer, E. [1974] 14190

F60

Demuth, O. [1967] 02915
Demuth, O. [1968] 16949
Demuth, O. [1968] 28592
Gel'fond, M.G. [1972] 04841
Goodstein, R.L. [1945] 05166
Goodstein, R.L. [1951] 25669
Goodstein, R.L. [1955] 05177
Goodstein, R.L. [1957] 05180
Kosovskij, N.K. [1981] 55663
Mayoh, B.H. [1970] 17641
Nederpelt, R.P. [1980] 55989
Scienza, G. [1979] 78312

F65

Ackermann, W. [1950] 00122
Ackermann, W. [1952] 00124
Ackermann, W. [1953] 00125
Beck, Jon M. [1979] 55969
Chwistek, L.B. [1924] 02670
Cocchiarella, N.B. [1974] 04010
Harms, S. [1977] 73840
Hintikka, K.J.J. [1956] 06114
Kondo, M. [1956] 07331
Kronecker, L. [1887] 37377
Lorenzen, P. [1955] 08290
Lorenzen, P. [1965] 08304
Maehara, S. [1962] 08526
Mycielski, J. [1981] 54194
Royse, James R. [1969] 11634
Sanchis, L.E. [1965] 11824
Schuette, K. [1953] 13003
Schuette, K. [1965] 13015
Takeuti, G. [1956] 13307
Wang, Hao [1949] 14006
Wang, Hao [1954] 14025
Wang, Hao [1963] 14044
Weyl, H. [1918] 19791
Weyl, H. [1919] 14166
Zahn, P. [1967] 24851

F96

Gentzen, G. [1969] 29483
Herbrand, J. [1968] 24825

F97

Boerger, E. [1984] 40062
Dejon, B. [1969] 23438
Gonseth, F. [1941] 48625
Harrington, L.A. [1985] 49810
Heijenoort van, J. [1967] 25903
Lacombe, D. [1970] 48620
Lolli, G. [1984] 41493
Lourenco, M. [1979] 70030
Matiyasevich, Yu.V. [1979] 70057
Orevkov, V.P. [1968] 37197
Dalen van, D. [1982] 34494

F98

Anderson, J.M. [1962] 00341
Bar-Hillel, Y. [1958] 00805
Barwise, J. [1977] 70117
Beth, E.W. [1962] 01176
Boolos, G. [1979] 56312
Curry, H.B. [1950] 04222
Dopp, J. [1962] 03085
Feferman, S. [1977] 27330
Fisher, A. [1982] 36967
Fitting, M. [1969] 04349
Friedman, H.M. [1975] 04296
Goldblatt, R.I. [1979] 55754
Goodstein, R.L. [1951] 25669
Goodstein, R.L. [1957] 05180
Grzegorczyk, A. [1971] 23119
Guillaume, M. [1958] 42813
Haas, G. [1984] 44527
Heyting, A. [1958] 06062
Bernays, P. [1934] 01098
Bernays, P. [1939] 01082
Hindley, J.R. [1972] 23471
Kanger, S. [1957] 22373
Kleene, S.C. [1952] 07173
Kleene, S.C. [1967] 45895
Kosovskij, N.K. [1981] 55663
Kreisel, G. [1965] 27577
Ladriere, J. [1957] 07777
Leisenring, A.C. [1969] 07983
Lorenzen, P. [1955] 08290
Lorenzen, P. [1962] 08303
Manin, Yu.I. [1979] 54716
Mostowski, Andrzej [1965] 09578
Nagel, E. [1958] 09765
Nicod, J. [1930] 09941
Prawitz, D. [1965] 22164
Rasiowa, H. [1963] 11028
Schmidt, H.A. [1960] 12892
Schuette, K. [1960] 13010
Shoenfield, J.R. [1967] 22384
Shoesmith, D.J. [1978] 51869
Siefkes, D. [1970] 12145
Smirnov, V.A. [1972] 30359
Smullyan, R.M. [1963] 28562
Soudieux, C. [1960] 12654
Steen, S.W.P. [1972] 22386
Wang, Hao [1981] 34116

F99

Barzin, M. [1935] 00914
Bernays, P. [1932] 01289
Bernays, P. [1950] 37320
Bernays, P. [1953] 47892
Chwistek, L.B. [1929] 02160
Fraenkel, A.A. [1951] 47908
Gruzintsev, G.A. [1927] 40886
Hadamard, J. [1954] 42157
Herbrand, J. [1929] 05968
Herbrand, J. [1931] 16812
Hilbert, D. [1899] 23454
Kreisel, G. [1970] 48003
Liu, Shichao [1980] 75591
Menger, K. [1928] 39109
Nepejvoda, N.N. [1979] 36869
Nicod, J. [1924] 25431
Smirnova, E.D. [1974] 78744
Smullyan, R.M. [1983] 41200
Wajsberg, M. [1933] 40808
Weyl, H. [1910] 42999
Weyl, H. [1985] 49136
Zariski, O. [1925] 41579

G30

Calude, C. [1978] 52071
Freyd, P. [1972] 04618
Goldblatt, R.I. [1979] 55754
Grayson, R.J. [1984] 41840
Hajek, P. [1970] 05521
Hatcher, W.S. [1968] 22644
Josza, R. [1984] 45334
Kock, A. [1974] 29047
Lambek, J. [1984] 44302
Mahe, L. [1973] 08550
Mijoule, R. [1976] 25854
Robert, P. [1965] 26664
Rousseau, C. [1978] 51810
Rousseau, C. [1979] 82662
Seely, R.A.G. [1982] 34890
Sols, I. [1975] 18397
Street, R. [1973] 32482
Street, R. [1974] 32478
Street, R. [1978] 32014
Street, R. [1978] 32484
Street, R. [1978] 32486
Takahashi, S. [1974] 17488
Tsukada, H. [1981] 55603
Vasilache, S. [1977] 51636

H15

Abian, A. [1974] 03810
Alling, N.L. [1985] 44331
Basarab, S.A. [1983] 38348
Belyakin, N.V. [1979] 70865
Connes, A. [1970] 02352
Conway, J.H. [1976] 71909
Cuda, K. [1980] 56506
Fenstad, J.E. [1970] 17153
Frey, G. [1983] 37806
Geiser, J.R. [1968] 04827
Geiser, J.R. [1974] 04831
Henson, C.W. [1984] 39860
Nelson, George C. [1973] 17539
Potthoff, K. [1972] 10697
Rabin, M.O. [1961] 10934
Robinson, A. [1961] 11251
Robinson, A. [1963] 11254
Robinson, A. [1963] 28804
Robinson, A. [1965] 11259
Robinson, A. [1966] 11263
Robinson, A. [1969] 21051
Robinson, A. [1973] 26149
Robinson, A. [1975] 17511
Rosser, J.B. [1950] 11558
Skolem, T.A. [1933] 12392
Skolem, T.A. [1934] 12393
Skolem, T.A. [1955] 27718
Yasumoto, M. [1983] 38208
Yasumoto, M. [1983] 47688

Alphabetization and alternative spellings of author names

The purpose of this index is to help the user find an author in whom he is interested. We begin by outlining both the general principles of alphabetization followed in the Author Index and the systems of transliteration used. The second half of this index addresses the problems which arise with author names for which there may be many variants in the literature. How do you find the primary form of a name used in the Bibliography? The ideal would be to have a table linking all the 'imaginable' versions of an author name to the unique primary form used here, but the obstacles to realizing this are obvious: one 'imaginable' form may correspond to two different authors and, worse, 'imaginable' itself depends on the linguistic background of the user. We have instead suggested some guidelines for identifying the primary form of a name from one of its variants. Finally, there is a list of alternative forms of names for those cases in which the difference between the alternative and the primary forms is particularly striking. For an author whose name has changed, each publication is listed under the name form used on that publication. Pointers to the other name form are given in the Author Index.

The Roman alphabet is as usual alphabetized in the following form:

A B C D E F G H I J K L M N O P Q R S T U V W X Y Z

Within this general framework, the ordering for hyphenated and double names is illustrated by the following example:

Ab,G. ; Ab-Aa,G. ; Ab Aa,G. ; Aba,G.

Apostrophes in a name are disregarded: Mal'tsev, for example, is treated as Maltsev for alphabetical purposes.

Titular prefixes such as von, du, de la, etc. come immediately after the surname (family name), and before the given name (or initials); so, e. g., J. von Neumann appears as Neumann von, J. Similarly J. Smith, Jr., and C.F. Miller III are given as Smith Jr, J. and Miller III, C.F., respectively.

In general, initials are used for given names. The full given name(s) are used only where necessary or helpful to distinguish between authors with the same surnames and initials.

As has been mentioned in the Preface, diacritical marks have, for practical reasons, mostly been disregarded. The following lists those diacritical marks of Scandinavian and German languages that have been transliterated:

- æ to ae,
- ø to oe,
- å to aa,
- ä to ae,
- ö to oe,
- ü to ue.

By the way one cannot infer that every ae, oe, or ue in German comes from ä, ö, or ü; e.g. Gloede is the correct spelling, not Glöde!

Note that the hacek in languages written in the Roman alphabet (e. g. Serbo-Croatian) has not been transliterated (so, for example, Šešelja appears here as Seselja).

The transliteration used for Cyrillic is explained in another index. (For a Russian author who has emigrated to the West the primary name is usually the form used by the author in Western publications. This form does not always agree with the transliteration of the Cyrillic name.) For Chinese names, the Pinyin system of transliteration has been used as far as possible, and commas have been added to separate the surname and given names (which are not abbreviated to initials) to accord with Western style. However, for Korean names no commas are used.

Over the last hundred years there have been in general use several different systems for transliterating Cyrillic into the Roman alphabet. This has given rise to many variants for author names originally written in Cyrillic. We list here our transliteration of those Cyrillic letters for which there have been several variants and give the most common alternative transliterations. If you are searching for an author name you suspect may be of Slavic origin this list will help you to find the most likely form used here: simply replace each (block of) letter(s) on the right occurring in your version by the appropriate letters given on the left.

Our transliteration	Possible alternatives			Our transliteration	Possible alternatives	
ya	ja	a		z	s	
yu	ju			j	i	y
eh	e			kh	h	
e	je	ye		v	w	ff
ts	c			"	y	
ch	c	tsh	tch tsch	ks	x	
sh	s	sch		u	ou	
zh	z					

The following is a selective listing of alternative forms of author names. It contains only those alternative forms from which the primary may not be guessed by using the guidelines above.

for	see
Abellanas Cebollero, P.	Abellanas, P.F.
Adams, M.M.	McCord Adams, M.
Albuquerque, J.	Ribeiro de Albuquerque, J.
Angulin, D.	Angluin, D.
Artalejo, R.M.	Rodriguez Artalejo, M.
Asjwiniekoemaar	Ashvinikumar
Avraham, U.	Abraham, U.
Barzdin', Ya.M.	Barzdins, J.
Benlahcen, D.	Benhalcen, D.
Bhaskara Rao, K.P.S.	Rao, K.P.S.Bhaskara
Bhaskara Rao, M.	Rao, M.Bhaskara
Bloch, A.S.	Blokh, A.Sh.
Blochina, G.N.	Blokhina, G.N.
Carroll, L.	Dodgson, C.L.
Chakan, B.	Csakany, B.
Char-Tung, R.	Lee, R.C.-T.
Chen, T.T.	Tang, Caozhen
Choodnovsky, D.V.	Chudnovsky, D.V.
Chu, W.J.	Zhu, Wujia
Cohen, E.L.	Longini Cohen, E.
Colburn, C.J.	Colbourn, C.J.
Colburn, M.J.	Colbourn, M.J.
Coppola, L.G.	Gonzalez Coppola, L.
Costa, A.A.	Almeida Costa, A.
Cresswell, M.M.	Meyerhoff Cresswell, Mary
Dao, D.H.	Dang Huu Dao
Decew, J.W.	Wagner Decew, J.
Dieu, P.D.	Phan Dinh Dieu
Duncan Luce, R.	Luce, R.D.
Dyson, V.H.	Huber-Dyson, V.
Fan Din' Zieu'	Phan Dinh Dieu
Foellesdal, D.	Follesdal, D.
Frejvald, R.V.	Freivalds, R.
Gegalkine, I.	Zhegalkin, I.I.
Gibbelato Valabrega, E.	Valabrega, E.G.
Greendlinger, M.	Grindlinger, M.
Hoo, T.-H.	Hu, Shihua
Hsu, L.C.	Xu, Lizhi
Hsueh, Yuang Cheh	Xueh, Yuangche
Jutting, L.S.B.	Benthem Jutting van, L.S.
Kao, H.	Gao, Hengshan
Kapinska, E.	Capinska, E.
Keldych, L.	Keldysh, L.V.
Khunyadvari, L.	Hunyadvari, L.
Kister, J.E.	Bridge, J.
Klein, F.	Klein-Barmen, F.
Kroonenberg, A.V.	Verbeek-Kroonenberg, A.
Kurepa, G.	Kurepa, D.
Kurkova-Pohlova, V.	Pohlova, V.
Kwei, M.S.	Mo, Shaokui
Lifshits, V.	Lifschitz, V.
Lo, Li Bo	Luo, Libo
Loewenthal, F.	Lowenthal, F.
Macdonald, S.O.	Oates MacDonald, S.
Malyaukene, L.K.	Maliaukiene, L.
Markus, S.	Marcus, S.
Moenting, J.S.	Schulte-Moenting, J.
Moh, S.-K.	Mo, Shaokui
Moura, J.E.A.	Almeida Moura de, J.E.
Nardzewski, C.R.	Ryll-Nardzewski, C.
Nash, W.C.S.J.A.	Nash-Williams, C.St.J.A.
Oates-Williams, S.	Oates MacDonald, S.
Plattner, A.	Pieczkowski, A.
Plyushkevichus, R.A.	Pliuskevicius, R.
Plyushkevichene, A.Yu.	Pliuskeviciene, A.
Poprougenko, G.	Popruzenko, J.
Puzio-Pol, E.	Pol, E.
R.-Salinas, B.	Rodriguez-Salinas, B.
Reymond, A.	Virieux-Reymond, A.
Riccioli, B.V.	Veit Riccioli, B.
Rucker, R.	Bitter-Rucker von, R.
Russi, G.Z.	Zubieta Russi, G.
Salinas, B.	Rodriguez-Salinas, B.
Schmir-Hay, L.	Hay, L.
Shain, B.M.	Schein, B.M.
Shaw, M.K.	Mo, Shaokui
Shih-Hua, H.	Hu, Shihua
Shlyakhovaya, N.I.	Slyakhova, N.I.
Solans, V.	Verdu i Solans, V.
Strazdin', I.Eh.	Strazdins, I.E.
Themaat, W.A.v.	Verloren van Themaat, W.A.
Toa van, T.	Tran van Toan
Toth, P.	Ecsedi-Toth, P.
Tsao-Chen, T.	Tang, Caozhen
Tseng, Y.X.	Zheng, Yuxin
Tsirulis, Ya.P.	Cirulis, J.
Tulcea, C.	Ionescu Tulcea, C.
Turksen, I.B.	Tuerksen, I.B.
Tzeng, O.C.	Tseng, O.C.
Vinter, H.	Winter, H.
Williams, C.St.J.A.N.	Nash-Williams, C.St.J.A.
Wou, Shou Zhi	Wu, Shouzhi
Wu, K.J.	Johnson Wu, K.
Yukna, S.P.	Jukna, S.
Yuting, S.	Shen, Y.-T.
Zhay, B.	Zhang, Bosheng
Zhen, Z.	Zhao, Zhen
Zilli, M.V.	Venturini Zilli, M.
Zou, Juan	Zhou, Juan

International vehicle codes

The following abbreviations are used as *codes for the country* in which a conference took place or in which a publishing company is located. (These abbreviations are those used internationally for vehicles.)

Code	Country
A	Austria
ADN	People's Dem. Rep. Yemen (South Yemen)
AFG	Afghanistan
AL	Albania
AND	Andorra
AUS	Australia
B	Belgium
BD	Bangladesh
BDS	Barbados
BG	Bulgaria
BH	Belize
BOL	Bolivia
BR	Brazil
BRN	Bahrain
BRU	Brunei
BS	Bahamas
BU	Burundi
BUR	Burma
C	Cuba
CDN	Canada
CH	Switzerland
CI	Ivory Coast
CL	Sri Lanka
CO	Columbia
CR	Costa Rica
CS	Czechoslovakia
CY	Cyprus
D	Fed. Rep. Germany (West Germany)
DDR	German Dem. Rep. (East Germany)
DK	Denmark
DOM	Dominican Republic
DZ	Algeria
E	Spain
EAK	Kenya
EAT	Tanzania
EAU	Uganda
EC	Ecuador
ES	El Salvador
ET	Egypt
ETH	Ethiopia
F	France
FJL	Fiji Islands
FL	Liechtenstein
FR	Faeroes
GB	Great Britain and Northern Ireland
GBA	Alderney
GBG	Guernsey
GBJ	Jersey
GBM	Isle of Man

Code	Country
GBZ	Gibraltar
GCA	Guatemala
GH	Ghana
GR	Greece
GUY	Guyana
H	Hungary
HK	Hong Kong
HV	Upper Volta
I	Italy
IL	Israel
IND	India
IR	Iran
IRL	Ireland (Eire)
IRQ	Iraq
IS	Iceland
J	Japan
JA	Jamaica
JOR	Jordan
K	Cambodia
KWT	Kuwait
L	Luxembourg
LAO	Laos
LAR	Libya
LB	Liberia
LS	Lesotho
M	Malta
MA	Morocco
MAL	Malaysia
MC	Monaco
MEX	Mexico
MS	Mauritius
MW	Malawi
N	Norway
NA	Netherlands Antilles
NIC	Nicaragua
NL	Netherlands
NZ	New Zealand
P	Portugal
PA	Panama
PAK	Pakistan
PE	Peru
PL	Poland
PNG	Papua-New Guinea
PRI	Puerto Rico
PRK	People's Rep. Korea (North Korea)
PY	Paraguay
Q	Qatar
RA	Argentina
RB	Botswana
RC	Taiwan
RCA	Central African Republic
RCB	Congo

Code	Country
RCH	Chile
RFC	Cameroon
RH	Haiti
RI	Indonesia
RIM	Mauritania
RL	Lebanon
RM	Madagascar
RMM	Mali
RN	Niger
RO	Romania
ROK	South Korea
ROU	Uruguay
RP	Philippines
RPB	Benin
RSM	San Marino
RWA	Ruanda
S	Sweden
SA	Saudi Arabia
SCV	Vatican
SD	Swaziland
SF	Finland
SGP	Singapore
SME	Surinam
SN	Senegal
SP	Somalia
STL	Windward Islands St. Lucia
SU	Soviet Union
SY	Seychelles
SYR	Syria
TG	Togo
THA	Thailand
TJ	People's Rep. China
TN	Tunisia
TR	Turkey
TT	Trinidad and Tobago
USA	United States of America
VN	Vietnam
WAG	Gambia
WAL	Sierra Leone
WAN	Nigeria
WD	Dominica
WG	Grenada
WS	Samoa
WV	Windward Islands St. Vincent
Y	Arabic Rep. Yemen (North Yemen)
YU	Yugoslavia
YV	Venezuela
Z	Zambia
ZA	South Africa
ZRE	Zaire
ZW	Zimbabwe

Transliteration scheme for Cyrillic

Author names and titles originally in *Cyrillic* have been transliterated into the Roman alphabet using the following scheme. (It is the same as the scheme curently used by Zbl and differs only slightly from that used by MR.)

Cyrillic		Roman
а	А	a
б	Б	b
в	В	v
г	Г	g
д	Д	d
е(ё)	Е(Ё)	e
ж	Ж	zh
з	З	z
и	И	i
й	Й	j
к	К	k

Cyrillic		Roman
л	Л	l
м	М	m
н	Н	n
о	О	o
п	П	p
р	Р	r
с	С	s
т	Т	t
у	У	u
ф	Ф	f
х	Х	kh

Cyrillic		Roman
ц	Ц	ts
ч	Ч	ch
ш	Ш	sh
щ	Щ	shch
ъ	Ъ	"
ы	Ы	y
ь	Ь	'
э	Э	eh
ю	Ю	yu
я	Я	ya

If you have any concerns about our products,
you can contact us on
ProductSafety@springernature.com

In case Publisher is established outside the EU,
the EU authorized representative is:
**Springer Nature Customer Service Center GmbH
Europaplatz 3, 69115 Heidelberg, Germany**

Printed by Libri Plureos GmbH
in Hamburg, Germany